Royal Military College of Science
Library

Shrivenham, Swindon, Wilts. SN6 8LA
Telephone: (01793) 785484 (Direct line)

Please return on or before the last date stamped below:

25 MAY 2000

WITHDRAWN

Progress in Nonlinear Differential Equations and Their Applications

Volume 35

Editor

Haim Brezis
Université Pierre et Marie Curie
Paris
and
Rutgers University
New Brunswick, N.J.

Editorial Board

Antonio Ambrosetti, Scuola Normale Superiore, Pisa
A. Bahri, Rutgers University, New Brunswick
Felix Browder, Rutgers University, New Brunswick
Luis Cafarelli, Institute for Advanced Study, Princeton
Lawrence C. Evans, University of California, Berkeley
Mariano Giaquinta, University of Pisa
David Kinderlehrer, Carnegie-Mellon University, Pittsburgh
Sergiu Klainerman, Princeton University
Robert Kohn, New York University
P.L. Lions, University of Paris IX
Jean Mahwin, Université Catholique de Louvain
Louis Nirenberg, New York University
Lambertus Peletier, University of Leiden
Paul Rabinowitz, University of Wisconsin, Madison
John Toland, University of Bath

Topics in Nonlinear Analysis

The Herbert Amann Anniversary Volume

Joachim Escher
Gieri Simonett
Editors

Birkhäuser
Basel · Boston · Berlin

Joachim Escher
Mathematik
Universität GH Kassel
Heinrich-Plett-Strasse 40
D-34132 Kassel
Germany

Gieri Simonett
Department of Mathematics
Vanderbilt University
1326 Stevenson Center
Nashville, TN 37240
USA

1991 Mathematics Subject Classification 35-06, 35-02, 46-06, 49-06

LC// Nonlinear analysis
Differential equations

A CIP catalogue record for this book is available from the Library of Congress, Washington D.C., USA

Deutsche Bibliothek Cataloging-in-Publication Data

Topics in nonlinear analysis : the Herbert Amann anniversary volume /
Joachim Escher ; Gieri Simonett ed. -
Basel ; Boston ; Berlin : Birkhäuser, 1999
(Progress in nonlinear differential equations and their applications ; Vol. 35)
ISBN 3-7643-6016-X (Basel...)
ISBN 0-8176-6016-X (Boston)

This work is subject to copyright. All rights are reserved, whether the whole or part of the material is concerned, specifically the rights of translation, reprinting, re-use of illustrations, broadcasting, reproduction on microfilms or in other ways, and storage in data banks. For any kind of use whatsoever, permission from the copyright owner must be obtained.

© 1999 Birkhäuser Verlag, P.O. Box 133, CH-4010 Basel, Switzerland
Printed on acid-free paper produced of chlorine-free pulp. TCF ∞
Printed in Germany
ISBN 3-7643-6016-X
ISBN 0-8176-6016-X

9 8 7 6 5 4 3 2 1

This collection of research papers is dedicated to

Herbert Amann

on the occasion of his sixtieth birthday

Contents

Preface

Herbert Amann's work is distinguished and marked by great lucidity and deep mathematical understanding. He has set high standards in research and teaching, and many colleagues and students has benefited from his erudition and extensive knowledge of mathematics.

His influential and lasting contributions to various fields of analysis comprise degree and fixed point theory, nonlinear elliptic boundary value problems, abstract evolution equations, quasi-linear parabolic systems, fluid dynamics, Fourier analysis, and the theory of function spaces.

Herbert Amann's personal qualities of commitment, modesty, and integrity leave lasting impression on his colleagues, students, and friends. This volume reflects their appreciation, gratitude, and friendship.

Kassel and Nashville, August 1998 J. Escher und G. Simonett

Acknowledgement
The editors would like to express their sincere gratitude to Dr. Thomas Hintermann and to Birkhäuser Verlag for their important support for this project.

Progress in Nonlinear Differential Equations
and Their Applications, Vol. 35
© 1999 Birkhäuser Verlag Basel/Switzerland

Remarks on Bifurcation from the Essential Spectrum

Antonio Ambrosetti and Marino Badiale

Dedicated to Herbert Amann for his 60th birthday.

Abstract. In this note we prove the existence of two families of solutions bifurcating from the essential spectrum for problem (1) below. We use a new perturbation method recently developed (see [1], [2], [3]).

1. Introduction

Bifurcation from the essential spectrum has been widely investigated, see for example [4, 5, 6] and references therein. More recently we have studied in [2] the bifurcation from the essential spectrum for some class of problems like

$$\begin{cases} \psi'' + \lambda\psi + h(x)|\psi|^{p-1}\psi &= 0, \quad x \in \mathbb{R}, \\ \lim_{|x|\to\infty} \psi(x) &= 0, \end{cases} \tag{1}$$

where $p > 1$, λ is the bifurcation parameter and there exists $L > 0$ such that either $h - L \in L^1(\mathbb{R})$ or $h - L \in L^1_{\text{loc}}(\mathbb{R})$ and $h \to L$ as $|x| \to \infty$. Under this kind of conditions, see [2] for precise statements and more details, it has been shown that (1) has a family of solutions bifurcating from the essential spectrum as $\lambda \to 0^-$. The assumptions on h made possible the use of an abstract perturbation method in critical point theory, developed in [1], see also [2], Sections 2 and 3. Actually, the change of variable $u(x) = \varepsilon^{2/(1-p)}\psi(x/\varepsilon)$, $\lambda = -\varepsilon^2$, transforms problem (1) into

$$\begin{cases} -u'' + u = L\,|u|^{p-1}u \;+\; (h(x/\varepsilon) - L)|u|^{p-1}u, \quad x \in \mathbb{R}, \\ \lim_{|x|\to\infty} u(x) = 0, \end{cases} \tag{2}$$

in which the term $(h(x/\varepsilon) - L)|u|^{p-1}u$ can be considered as a perturbation of the unperturbed equation $-u'' + u = L\,|u|^{p-1}u$, $u \in H^1(\mathbb{R})$.

The purpose of the present note is to show that the same abstract setting can be used to prove the bifurcation from the essential spectrum for (1) in the presence of a periodic coefficient h.

The paper is organized as follows: in Section 2 we recall the abstract functional frame and in Section 3 we state and prove the main bifurcation result.

Supported by M.U.R.S.T. National Project "Problemi non lineari … " and by E.E.C. contract n. ERBCHRXCT940494.

2. The Abstract Setting

We deal with a family of functionals f_ε, defined on a Hilbert space E, of the form

$$f_\varepsilon(u) = \frac{1}{2}\|u\|^2 - F(u) + G(\varepsilon, u),$$

where $\|\cdot\|$ denotes the norm in E, and with functions $F : E \to \mathbb{R}$ and $G : \mathbb{R} \times E \to \mathbb{R}$. We assume that

($\mathbf{F_0}$) $F \in C^2$;

($\mathbf{G_0}$) G is continuous in $(\varepsilon, u) \in \mathbb{R} \times E$ and $G(0, u) = 0$ for all $u \in E$; G is of class C^2 with respect to $u \in E$.

We will use the notation $F'(u)$, respectively $G'(\varepsilon, u)$, to denote the operators defined by setting

$$(F'(u)|v) = DF(u)[v], \quad \forall\, v \in E,$$

and, respectively,

$$(G'(\varepsilon, u)|v) = D_u G(\varepsilon, u)[v], \quad \forall\, v \in E,$$

where $(\cdot|\cdot)$ is the inner product corresponding to $\|\cdot\|$. Similarly, $F''(u)$, resp. $G''(\varepsilon, u)$, denote the maps in $L(E, E)$ defined by

$$\begin{aligned} (F''(u)v|w) &= D^2 F(u)[v, w] \\ (G''(\varepsilon, u)v|w) &= D^2_{uu} G(\varepsilon, u)[v, w]. \end{aligned}$$

Let us point out that, obviously, (G_0) implies that $G'(0, u) = 0$ and $G''(0, u) = 0$. The specific feature of our problem is that the unperturbed functional f_0 inherits a symmetry under the action of some non compact group. Precisely, we shall suppose that F satisfies

($\mathbf{F_1}$) there exists a d-dimensional C^2 manifold Z, $d \geq 1$, consisting of critical points of f_0, namely such that

$$z - F'(z) = 0, \quad \forall\, z \in Z. \tag{3}$$

Such a Z will be called a *critical manifold* of f_0. Let $T_z Z$ denote the tangent space to Z at z and I_E denote the identity map in E. We further suppose

($\mathbf{F_2}$) $F''(z)$ is compact $\forall\, z \in Z$;

($\mathbf{F_3}$) $T_z Z = \mathrm{Ker}[I_E - F''(z)]$, $\forall\, z \in Z$.

In addition to (G_0) we make the following further assumptions on G.

($\mathbf{G_1}$) The maps $(\varepsilon, u) \mapsto G'(\varepsilon, u)$, $(\varepsilon, u) \mapsto G''(\varepsilon, u)$ are continuous (as maps from $\mathbb{R} \times E$ to E, resp. to $L(E, E)$).

If the preceding assumptions hold true, we get the following main "abstract" theorem:

Theorem 2.1. *Assume* $(F_0$–$F_3)$ *and* $(G_0$–$G_1)$. *Let* $V \subset Z$ *be a relatively open set, contained in a relatively compact subset of* Z. *Then there exist an* $\varepsilon_0 > 0$ *and a* C^1 *function* $w : [-\varepsilon_0, \varepsilon_0] \times V \to E$ *such that*

i) *$w(\varepsilon, v)$ is orthogonal to Z in v.*
ii) *$f'_\varepsilon(v + w(\varepsilon, v))$ is tangent to Z in v.*

Furthermore $Z_\varepsilon = \{v + w(\varepsilon, v) \mid v \in V\}$ is a manifold diffeomorphic to V. Finally, if $u = v + w(\varepsilon, v)$ is a critical point for the constrained functional $f_\varepsilon|_{Z_\varepsilon}$, then it is also a free critical point for f_ε, i.e. $f'_\varepsilon(u) = 0$.

Theorem (2.1) allows us to look for critical points of the constrained functional, and in this way we have reduced an infinite dimensional variational problem to a finite dimensional one.

3. Bifurcation Results

Problem (1) has, for every $\lambda \in \mathbb{R}$, the trivial solution $\psi = 0$. The linearization of (1) around $\psi = 0$ is given by $-\phi'' + \lambda\phi = 0$, $\phi \in H^1(\mathbb{R})$, which has the positive real axis $[0, \infty)$ as spectrum. We will prove that, if h is periodic, $\lambda = 0$ (the infimum of the spectrum) is a bifurcation point. By this we mean that there are solutions (λ, ψ_λ) of problem (1), with $\psi_\lambda \neq 0$, such that $\|\psi_\lambda\|_{L^s} \to 0$, for some $s > 0$, as $\lambda \to 0^-$.

We assume that h satisfies

$$(H) \qquad h \in C(\mathbb{R}), \quad h \text{ is periodic with minimal period } T = 1$$
$$\text{and} \quad \int_0^1 h(s)ds = m > 0.$$

By the same change of variable $u(x) = \varepsilon^{2/(1-p)}\psi(x/\varepsilon)$, $\lambda = -\varepsilon^2$, and defining $g(x) = h(x) - m$, problem (1) becomes

$$\begin{cases} -u'' + u = m\,|u|^{p-1}u \;+\; g(x/\varepsilon)\,|u|^{p-1}u, & x \in \mathbb{R}, \\ \lim_{|x|\to\infty} u(x) \;=\; 0\,. \end{cases} \tag{4}$$

Note that g is a periodic continuous function with average zero.

The main result of this note is

Theorem 3.1. *Assume (H). Then for each $\varepsilon \neq 0$ small enough there are two distinct nontrivial solutions $u_{1,\varepsilon}, u_{2,\varepsilon}$ to (4). They give rise to two distinct families of nontrivial solutions $\psi_{1,\lambda}$, $\psi_{2,\lambda}$ of (1) bifurcating from $(0,0)$, in the sense that $\|\psi_\lambda\|_{L^s} \to 0$, for $s \in](p-1)/2, +\infty[\cap[2, +\infty[$, as $\lambda \to 0^-$.*

Theorem 3.1 will be proved in four steps.

Step 1: The functional setting. First of all let us show that our problem fits into the abstract frame discussed in the previous section. For this, we set $E = H^1(\mathbb{R})$

and

$$
\begin{aligned}
\|u\|^2 &= \int_{\mathbb{R}} (u'^2 + u^2)dx, \\
F(u) &= \frac{m}{p+1}\int_{\mathbb{R}} |u|^{p+1}dx, \\
G(\varepsilon, u) &= \begin{cases} \frac{-1}{p+1}\int_{\mathbb{R}} g(x/\varepsilon)|u|^{p+1}dx & \text{if} \quad \varepsilon \neq 0, \\ 0 & \text{if} \quad \varepsilon = 0 \end{cases}
\end{aligned}
$$

Obviously, (F_0) holds. It is also known that $(F_1 - F_3)$ are satisfied. More precisely, letting $z_0 \in E$ denote the positive, even solution of

$$
-u'' + u = mu^p,
$$

one has that $Z = \{z = z_0(x+\theta) \, : \, \theta \in \mathbb{R}\}$ and assumption (F_3) holds, see [3] and references therein. We will denote by z_θ the generic point of Z, i.e $z_\theta(x) = z_0(x+\theta)$.

Step 2: Proof of G_0, G_1. Here we shall prove that G_0, G_1 are satisfied, i.e., that G, G', G'' are continuous functions. We set

$$
\gamma(t) = \int_0^t g(s)ds.
$$

Let us remark that, since g is periodic with average zero, γ is periodic with the same period, and, in particular, it is bounded.

Proposition 3.2. *Assume that h statisfies (H). Then G, G', G'' are continuous functions.*

Proof. The claim is obvious if $(\epsilon, u) \to (\epsilon_0, u_0)$ and $\epsilon_0 \neq 0$. So we will only consider the case $(\epsilon, u) \to (0, u_0)$. Recalling the properties of γ, an integration by parts yields

$$
\begin{aligned}
G(\varepsilon, u) &= \frac{1}{p+1}\int_{\mathbb{R}} g\left(\frac{x}{\varepsilon}\right)|u|^{p+1}dx \\
&= \varepsilon\gamma\left(\frac{x}{\varepsilon}\right)|u|^{p+1}\Big|_{-\infty}^{+\infty} - \varepsilon\int_{\mathbb{R}} \gamma\left(\frac{x}{\varepsilon}\right)|u|^{p-1}uu'dx \\
&= -\varepsilon\int_{\mathbb{R}} \gamma\left(\frac{x}{\varepsilon}\right)|u|^{p-1}uu'dx.
\end{aligned}
$$

Recall that $H^1(\mathbb{R}) \subset L^q(\mathbb{R})$ for each $q \geq 2$, with continuous embeddings. Using Hölder's inequality, this implies that $u^p u' \to u_0 u_0'$ in L^1, as $u^p \to u_0$ in H^1. It is then easy to obtain

$$
G(\varepsilon, u) \to 0
$$

as $(\epsilon, u) \to (0, u_0)$.

Let us consider now G'. We recall that

$$
||G'(\varepsilon, u)|| = \sup_{||v||=1} |(G'(\varepsilon, u)|v)| = \sup_{||v||=1}\left|\int_{\mathbb{R}} g\left(\frac{x}{\varepsilon}\right)|u|^{p-1}uvdx\right|.
$$

With the same argument as before (integration by parts and Hölder's inequality) we obtain

$$\left|\int_{\mathbb{R}} g\left(\frac{x}{\varepsilon}\right)|u|^{p-1}uvdx\right| \leq \varepsilon C_1 \int_{\mathbb{R}} \left|p|u|^{p-1}u'v + |u|^{p-1}uv'\right| dx \leq$$

$$\varepsilon C_2 \left(\int_{\mathbb{R}} |u|^{2p}dx\right)^{\frac{p-1}{2p}} \cdot \left(\int_{\mathbb{R}} |u'|^{2p}dx\right)^{\frac{1}{2p}} \cdot \left(\int_{\mathbb{R}} |v|^2dx\right)^{\frac{1}{2}} +$$

$$\varepsilon C_2 \left(\int_{\mathbb{R}} |u|^{2p}dx\right)^{\frac{1}{2}} \cdot \left(\int_{\mathbb{R}} |v'|^2dx\right)^{\frac{1}{2}} \leq \varepsilon C_3$$

where $C_i > 0$ are independent of v. Hence

$$||G'(\varepsilon, u)|| \leq \varepsilon C$$

and the claim is proved also for G'.

The study of G'' is more complicated. Recall that

$$||G''(\varepsilon, u)|| = \sup_{||v||,||w||=1} |(G''(\varepsilon, u)w|v)| = \sup_{||v||,||w||=1} \left|p\int_{\mathbb{R}} g\left(\frac{x}{\varepsilon}\right)|u|^{p-1}vwdx\right|.$$

In fact, for $p \geq 2$ the previous arguments can be repeated with minor changes and the conclusion follows. If $1 < p < 2$ the function $|u|^{p-1}$ is not differentiable so we cannot integrate by parts. In order to bypass this problem we argue as follows.

Fix $\sigma > 0$ and take $M > 0$ such that

$$\left|\int_{|x|>M} g\left(\frac{x}{\varepsilon}\right)|u|^{p-1}vwdx\right| \leq \sigma. \tag{5}$$

Such M exists and is independent of ε, v, and w: indeed, by Hölder's inequality with exponents $q_1 = \frac{2}{p-1}, q_2 = q_3 = \frac{4}{3-p}$, noticing that $q_2, q_3 \geq 2$ and recalling that $||v||, ||w|| \leq 1$, we find

$$\left|\int_{|x|>M} g\left(\frac{x}{\varepsilon}\right)u^{p-1}vwdx\right| \leq C\left(\int_{|x|>M} u^2dx\right)^{\frac{p-1}{2}}.$$

Here C is independent of ε, v, and w. As u^2 is integrable, the result follows. Obviously, we can choose M to be an integer. Let us estimate the remainder term $\int_{-M}^{M} g\left(\frac{x}{\varepsilon}\right)|u|^{p-1}vwdx$. For $y \in \mathbb{R}$, we define $[y]$ to be its integer part. Then we have

$$\int_{-M}^{M} g\left(\frac{x}{\varepsilon}\right)|u|^{p-1}vwdx = \sum_{j=[-M/\varepsilon]+1}^{j=[M/\varepsilon]} \int_{\varepsilon(j-1)}^{\varepsilon j} g\left(\frac{x}{\varepsilon}\right)|u|^{p-1}vwdx$$
$$+ \int_{\varepsilon[M/\varepsilon]}^{M} g\left(\frac{x}{\varepsilon}\right)|u|^{p-1}vwdx - \int_{\varepsilon[-M/\varepsilon]}^{-M} g\left(\frac{x}{\varepsilon}\right)|u|^{p-1}vwdx. \tag{6}$$

Let us consider the different terms in (6). As for the latter two, since

$$M-\varepsilon\left[\frac{M}{\varepsilon}\right]\to 0,\quad -M-\varepsilon\left[-\frac{M}{\varepsilon}\right]\to 0,$$

as $\varepsilon\to 0$, then one has that

$$\int_{\varepsilon[M/\varepsilon]}^{M} g\left(\frac{x}{\varepsilon}\right)u^{p-1}vwdx\to 0,\quad \int_{\varepsilon[-M/\varepsilon]}^{M} g\left(\frac{x}{\varepsilon}\right)u^{p-1}vwdx\to 0 \tag{7}$$

as $\varepsilon\to 0$. Let us now consider the former term in (6). For each j we obtain, by an obvious change of variables

$$\int_{\varepsilon(j-1)}^{\varepsilon j} g\left(\frac{x}{\varepsilon}\right)|u|^{p-1}vwdx=\varepsilon\int_{(j-1)}^{j} g(y)|u|^{p-1}(\varepsilon y)v(\varepsilon y)w(\varepsilon y)dy. \tag{8}$$

Since $\int_{(j-1)}^{j} g(y)dy=0$ we obtain

$$\begin{aligned}&\varepsilon\int_{(j-1)}^{j} g(y)u^{p+1}(\varepsilon y)dy\\ &=\varepsilon\int_{(j-1)}^{j} g(y)\left(u^{p-1}(\varepsilon y)v(\varepsilon y)w(\varepsilon y)-u^{p-1}(\varepsilon(j-1))v(\varepsilon(j-1))w(\varepsilon(j-1))\right)dy.\end{aligned} \tag{9}$$

In the case $y\in[j-1,j]$ one has that $\big|\varepsilon y-\varepsilon(j-1)\big|\le\varepsilon$. On the other hand, thanks to the embedding of $H^1(\mathbb{R})$ in $C^{0,j}(\mathbb{R})$, we obtain that $|u|^{p-1},v,w$ are Hölder continuous, and $||v||_{C^{0,j}},||w||_{C^{0,j}}$ are bounded independently of v,w. This implies that for each $\sigma>0$ we can fix $\eta>0$, independent of v,w, such that $|x-y|\le\eta$ implies $|u(x)-u(y)|\le\sigma$. So we can fix ε_0 such that for $0<\varepsilon<\varepsilon_0$ we have

$$|u^{p-1}(\varepsilon y)-u^{p-1}(\varepsilon(j-1))|,\ |v(\varepsilon y)-v(\varepsilon(j-1))|,\ |w(\varepsilon y)-w(\varepsilon(j-1))|<\frac{\sigma}{M||g||_\infty}.$$

Let us now come back to (6). Using (8) and (9) we get

$$\Big|\sum_{j=[-M/\varepsilon]+1}^{j=[M/\varepsilon]}\int_{\varepsilon(j-1)}^{\varepsilon j} g\left(\frac{x}{\varepsilon}\right)|u|^{p-1}vwdx\Big|\le$$

$$\le\varepsilon\sum_{j=[-M/\varepsilon]+1}^{j=[M/\varepsilon]}\Big[\int_{j-1}^{j}|g|\,\big||u|^{p-1}(\varepsilon y)v(\varepsilon y)w(\varepsilon y)-|u|^{p-1}(\varepsilon(j-1))v(\varepsilon y)w(\varepsilon y)\big|\,dy+$$

$$\int_{j-1}^{j}|g|\,\big||u|^{p-1}(\varepsilon(j-1))v(\varepsilon y)w(\varepsilon y)-|u|^{p-1}(\varepsilon(j-1))v(\varepsilon(j-1))w(\varepsilon y)\big|\,dy+$$

$$\int_{j-1}^{j}|g|\big||u|^{p-1}(\varepsilon(j-1))v(\varepsilon(j-1))w(\varepsilon y)-|u|^{p-1}(\varepsilon(j-1))v(\varepsilon(j-1))w(\varepsilon(j-1))\big|dy\Big]$$

$$\leq \sum_{j=[-M/\varepsilon]+1}^{j=[M/\varepsilon]} C\frac{\varepsilon\sigma}{M||g||_\infty}||g||_\infty = 2C\frac{\varepsilon\sigma}{M}\left[\frac{M}{\varepsilon}\right] \leq 2C\sigma, \tag{10}$$

where C is independent of σ. Putting together (5), (7), (10), we get that for each $\sigma > 0$ there is an ε_0 such that, for $0 < \varepsilon < \varepsilon_0$, one has

$$|\int_{\mathbb{R}} g\left(\frac{x}{\varepsilon}\right)|u|^{p-1}(x)v(x)w(x)dx| \leq C\sigma,$$

where $C > 0$ is independent of σ, v, and w if $||v||, ||w|| \leq 1$. This implies

$$||G''(\varepsilon, u)|| \to 0$$

as $\varepsilon \to 0$. □

Step 3: Behaviour of $f_\varepsilon|_{Z_\varepsilon}$.
We have now to study the functional $\Phi_\varepsilon(\theta) := f_\varepsilon|_{Z_\varepsilon} = f_\varepsilon(z_\theta + w(\varepsilon\theta))$. Unlike the cases we dealt with in [2] we are not able here to write an explicit expansion of $f_\varepsilon|_{Z_\varepsilon}$ because, roughly speaking, $G(\varepsilon, z)$ is not infinitesimal of any finite order as $\varepsilon \to 0$. We will overcome this difficulty by proving that, for each fixed ε, $\Phi_\varepsilon(\theta)$ is periodic in θ. We first show

Lemma 3.3.

$$w(\varepsilon, \theta + \varepsilon)(x) = w(\varepsilon, \theta)(x + \varepsilon)$$

Proof. Recall that $w(\varepsilon, \theta)$ is defined to be the unique solution, in a neighbourhood of z_θ, of the equation

$$-(z_\theta + w)'' + (z_\theta + w) - |z_\theta + w|^{p-1}(z_\theta + w) = g\left(\frac{x}{\varepsilon}\right)|z_\theta + w|^{p-1}(z_\theta + w).$$

If we define $\widetilde{w}(x) = w(\varepsilon, \theta)(x + \varepsilon)$, we obtain that $\widetilde{w}$ satisfies

$$-(z(x+\theta+\varepsilon)+\widetilde{w})''+(z(x+\theta+\varepsilon)+\widetilde{w})-|z(x+\theta+\varepsilon)+\widetilde{w}|^{p-1}(z(x+\theta+\varepsilon)+\widetilde{w}) =$$

$$g\left(\frac{x+\varepsilon}{\varepsilon}\right)|z(x+\theta+\varepsilon)+\widetilde{w}|^{p-1}(z(x+\theta+\varepsilon)+\widetilde{w}) =$$

$$= g\left(\frac{x}{\varepsilon}\right)|z(x+\theta+\varepsilon)+\widetilde{w}|^{p-1}(z(x+\theta+\varepsilon)+\widetilde{w}).$$

This is the equation satisfied by $w(\varepsilon, \theta+\varepsilon)$ and hence by the uniqueness properties of w, the Lemma follows. □

Proposition 3.4. *Assume that h satisfies (H). Then $\Phi_\varepsilon(\theta)$ is periodic in θ, with period ε.*

Proof. Using Lemma 3.3, we conclude

$$\begin{aligned}
&\Phi_\varepsilon(\theta+\varepsilon)\\
&=\int_{\mathbb{R}}((z(x+\theta+\varepsilon)+w(\varepsilon,\theta+\varepsilon))'^2+(z(x+\theta+\varepsilon)+w(\varepsilon,\theta+\varepsilon)^2)dx\\
&\quad-\frac{1}{p+1}\int_{\mathbb{R}}|z(x+\theta+\varepsilon)+w(\varepsilon,\theta+\varepsilon|^{p+1}dx\\
&\quad-\frac{1}{p+1}\int_{\mathbb{R}}g(x/\varepsilon)|z(x+\theta+\varepsilon)+w(\varepsilon,\theta+\varepsilon)|^{p+1}dx\\
&=\int_{\mathbb{R}}((z(x+\theta+\varepsilon)+w(\varepsilon,\theta)(x+\varepsilon))'^2+(z(x+\theta+\varepsilon)+w(\varepsilon,\theta)(x+\varepsilon))^2)dx\\
&\quad-\frac{1}{p+1}\int_{\mathbb{R}}|z(x+\theta+\varepsilon)+w(\varepsilon,\theta)(x+\varepsilon)|^{p+1}dx\\
&\quad-\frac{1}{p+1}\int_{\mathbb{R}}g(x/\varepsilon)|z(x+\theta+\varepsilon)+w(\varepsilon,\theta)(x+\varepsilon)|^{p+1}dx\\
&=\int_{\mathbb{R}}((z_\theta+w(\varepsilon,\theta))'^2+(z_\theta+w(\varepsilon,\theta))^2)dx\\
&\quad-\frac{1}{p+1}\int_{\mathbb{R}}|z_\theta+w(\varepsilon,\theta)|^{p+1}dx\\
&\quad-\frac{1}{p+1}\int_{\mathbb{R}}g\left(\frac{x-\varepsilon}{\varepsilon}\right)|z_\theta+w(\varepsilon,\theta)|^{p+1}dx\\
&=\Phi_\varepsilon(\theta).
\end{aligned}$$

□

Step 4: Proof of Theorem 3.1 completed. According to the above discussion we are in a position to apply the abstract results collected in Section 2 before, in particular, Theorem 2.1. Thus we have only to find the critical points of $f_\varepsilon|_{Z_\varepsilon}$. Using Proposition 3.4 we can say that $f_\varepsilon|_{Z_\varepsilon}$ has a minimum and a maximum which give rise to the solutions $u_{1,\varepsilon}$, $u_{2,\varepsilon}$ of (4). Setting $\lambda=-\varepsilon^2$ and $\psi_{i,\lambda}=\varepsilon^{2/(p-1)}u_{i,\varepsilon}(\varepsilon x)$ it turns out that $(\lambda,\psi_{i,\lambda})$ solves (1). Moreover, one has $\|\psi_{i,\lambda}\|_{L^s}^s=\varepsilon^{2s/(p-1)-1}\|u_{1,\varepsilon}\|_{L^s}^s$. Since $u_{i,\varepsilon}\in L^s$ for all $s\geq 2$ and $\|u_{i,\lambda}\|_{L^s}\leq C$, it follows that $\|\psi_{i,\lambda}\|_{L^s}\to 0$ as $\lambda\to 0^-$, for all $s>(p-1)/2$ and all $s\geq 2$. This completes the proof of Theorem 3.1.

References

[1] A. Ambrosetti – M. Badiale, "Homoclinics: Poincaré-Melnikov type results via a variational approach", Annales I.H.P. – Analyse nonlin. **15** (1998), 233–252.

[2] A. Ambrosetti – M. Badiale, "Variational perturbative methods and bifurcation of bound states from the essential spectrum", Proc. Royal Soc. Edinburgh, to appear.

[3] A. Ambrosetti – M. Badiale – S. Cingolani, "Semiclassical states of nonlinear Schrödinger equations", Archive Rat. Mech. Anal. **140** (1997), 285–300.

[4] R.J. Magnus, "On perturbation of translationally invariant differential equations", Proc. Royal Soc. Edinburgh **110-A** (1988), 1–25.

[5] C. Stuart, "Bifurcation from the essential spectrum", *Topological Nonlinear Analysis, II.* M. Matzeu, A. Vignoli eds., PNDLE 27, Birkhäuser 1997, 397–444.

[6] C. Stuart, "Bifurcation of homoclinic orbits and bifurcation from the essential spectrum", SIAM J. Math. Anal. **20** (1989), 1145–1171.

Antonio Ambrosetti
Marino Badiale
Scuola Normale Superiore
I-56100 Pisa, Italy

Progress in Nonlinear Differential Equations
and Their Applications, Vol. 35
© 1999 Birkhäuser Verlag Basel/Switzerland

Constructions with Analytic Semigroups and Abstract Exponential Decay Results for Eigenfunctions

Sigurd Angenent

Presented on the occasion of H.Amann's 60th birthday

1. Introduction

In this note we present some constructions with generators of analytic semigroups which are an abstract version of the familiar method of "freezing the coefficients" to prove elliptic estimates for differential operators with continuous coefficients or Hölder-continuous coefficients. As a side result we obtain an abstract exponential decay result for, say, eigenfunctions corresponding to isolated eigenvalues.

After recalling some basic definitions in Section 1.1 we first introduce a *resolution* of an operator $A : E_1 \to E_0$ in a Banach couple (E_1, E_0) (Section 2) and then discuss *localized resolutions,* which allow us to formulate an abstract version of the method of "freezing the coefficients" (Section 3). In Section 4 we show how all this can be applied to show that systems of parabolic operators $A = \sum_{|j|\leq m} a_j(x)D^j$ generate analytic semigroups in L^p spaces. In this example the result is certainly not new, but it illustrates the theory and suggests generalizations. In Section 5 we show that the matrix elements of the resolvent of an $A \in Hol(E)$ with respect to a localized resolution are exponentially decaying, provided certain commutators are sufficiently small. (In the example of systems of differential operators on $\mathbb{R}^d$ this would imply that the integral kernel of the resolvent is exponentially decaying.) This result immediately implies exponential localization of generalized eigenvalues corresponding to isolated eigenvalues λ_0 of finite multiplicity, again assuming that certain commutators are small enough (see Section 6). For example it turns out one can always choose a localized resolution for which the relevant commutators are adequately small. This way we find that generalized eigenfunctions (corresponding to isolated eigenvalues with finite multiplicity) of parabolic operators $A = \sum_{|j|\leq m} a_j(x)D^j$ always decay exponentially at $x = \infty$. This fact has an implication for classical solutions of nonlinear elliptic systems, namely, if an entire solution $u(x)$ of an elliptic system of PDEs $F(u, Du, \dots, D^m u) = 0$ is such that the linearized equation at u has $\lambda = 0$ as isolated eigenvalue of finite multiplicity, then $\lim_{|x|\to\infty} u(x)$ exists (Section 8).

I have found the theory of analytic semigroups particularly useful to prove short time existence and regularity results for an ever increasing class of nonlinear diffusion equations. In the middle 80ies I found the works [1][2][3] particularly inspiring. The reader can find much information in [4], (see also [13][5][11][14][12]). There is some overlap between the theory presented here and the work in volumes I and II of [4]. I hope the reader will find the different perspective useful.

1.1. Some definitions and facts

A *Banach couple* $E = (E_1, E_0)$ is a pair of Banach spaces $E_1 \subset E_0$ with E_1 densely embedded in E_0. An operator $A \in \mathfrak{L}(E_1, E_0)$ generates an analytic semigroup if for some $\omega > 0$ all $\lambda - A : E_1 \to E_0$ with $\Re\lambda > \omega$ are invertible while $\sup_{\Re\lambda>\omega} \left\| (\lambda - A)^{-1} \right\|_{\mathfrak{L}(E_0, E_1)} < \infty$.

We denote the set of $A \in \mathfrak{L}(E_1, E_0)$ which generate an analytic semigroup by $Hol(E)$. This set is open in $\mathfrak{L}(E_1, E_0)$, meaning a small perturbation (in the $E_1 \to E_0$ operator norm) of a generator of an analytic semigroup also generates one. Other perturbations which do not destroy the property of generating an analytic semigroup are addition of a compact operator $K : E_1 \to E_0$, or an operator $L : E_1 \to E_0$ of zero relative bound (meaning for all $\varepsilon > 0$ there is an $C_\varepsilon < \infty$ such that $\|Lx\|_{E_0} \le \varepsilon \|x\|_{E_1} + C_\varepsilon \|x\|_{E_0}$ for any $x \in E_1$.)

2. Resolutions

Consider a Banach couple E. By definition a *resolution* $E \xrightarrow{\delta} F \xrightarrow{\varepsilon} E$ of E consists of another Banach couple F and two maps of Banach couples $\delta : E \to F$ and $\varepsilon : F \to E$ which satisfy

$$\varepsilon\delta = id_E.$$

A resolution of an operator $A \in \mathfrak{L}(E_1, E_0)$ is, by definition, an operator $A' \in \mathfrak{L}(F_1, F_0)$ where $E \xrightarrow{\delta} F \xrightarrow{\varepsilon} E$ is a resolution of the Banach couple E.

For any resolution A' of A we define the commutators

$$\begin{aligned} [A, \epsilon] &= A\epsilon - \epsilon A' \\ [A, \delta] &= A'\delta - \delta A. \end{aligned}$$

These are bounded operators: $[A, \epsilon] \in \mathfrak{L}(F_1, E_0)$ and $[A, \delta] \in \mathfrak{L}(E_1, F_0)$. In general a resolution will be quite useless, unless the corresponding commutators are in some sense small.

We will call the resolution A' *exact* if the commutators $[A, \epsilon]$ and $[A, \delta]$ vanish. If these commutators are compact operators (from F_1 to E_0, or E_1 to F_0) then we will call the resolution *compact.* If the commutators are operators with relative bound zero, then we'll say the resolution has *relative bound zero.*

Theorem 2.1. *Let $A \in \mathfrak{L}(E_1, E_0)$ have a resolution $A' \in \mathfrak{L}(F_1, F_0)$, and assume that this resolution is either compact, or of relative bound zero. Then $A' \in Hol(F)$ implies $A \in Hol(E)$.*

Proof. Let $A' \in \mathfrak{L}(F_1, F_0)$ be the given resolution and assume that the resolution is compact. It follows from $\epsilon\delta = id_E$ that

$$\pi_F = \delta\epsilon$$

is a bounded projection in F. This projection allows us to split F as

$$F = \text{range}(\delta) \oplus \text{kern}(\epsilon).$$

The matrix of the linear operator A' with respect to this splitting can be written as

$$A' = \begin{pmatrix} P & Q \\ R & S \end{pmatrix}$$

with $P \in \mathfrak{L}(range(\delta)_1, range(\delta)_0)$, $R \in \mathfrak{L}(range(\delta)_1, kern(\epsilon)_0)$, etc.

The off diagonal parts of A', R and Q, are given by

$$R = (1 - \pi_F)A'\pi_F = [A', \pi_F]\pi_F$$
$$Q = \pi_F A'(1 - \pi_F) = \pi_F[\pi_F, A']$$

and since

$$[A', \pi_F] = [A', \delta\epsilon] = [A, \delta]\epsilon + \delta[A, \epsilon]$$

the commutator $[A', \pi_F]$ is a compact operator. Therefore R and Q are also compact. Compact perturbations of an $A' \in Hol(F)$ still belong to $Hol(F)$, so that

$$\begin{pmatrix} P & 0 \\ 0 & S \end{pmatrix} \in Hol(F),$$

which implies that $P \in Hol(range(\delta))$. But δ is an isomorphism onto its range, with inverse ϵ, so

$$\epsilon A'\delta = \epsilon P\delta \in Hol(E).$$

Finally,

$$\epsilon A'\delta - A = \epsilon A'\delta - \epsilon\delta A = \epsilon[A, \delta]$$

is again a compact operator and we may conclude that $A \in Hol(E)$.

If the resolution is not compact but of vanishing relative bound instead, then the same proof will work. □

Given a resolution of $A \in \mathfrak{L}(E_1, E_0)$ one can try to express the resolvent of A in terms of the resolvent of A'. Let

$$R'(\lambda) = (\lambda - A')^{-1}$$

be the resolvent of A', for some $\lambda \in \mathbb{C}$.

As a first approximation to the resolvent of A we introduce

$$S(\lambda) = \epsilon R'(\lambda)\delta.$$

To measure by how much $S(\lambda)$ fails to be the resolvent of A, we consider

$$T(\lambda) = 1_E - (\lambda - A)S(\lambda) = [A, \epsilon]R'(\lambda)\delta.$$

Formally a right inverse for $\lambda - A$ is given by

$$\begin{aligned}
\text{right-}(\lambda - A)^{-1} &= S(\lambda)[(\lambda - A)S(\lambda)]^{-1} \\
&= S(\lambda)[1_E - T(\lambda)]^{-1} \\
&= S(\lambda)[1_E + T(\lambda) + T(\lambda)^2 + \cdots] \\
&= \sum_{k=1}^{\infty} \epsilon \circ R'(\lambda)\, \{\delta[A,\, \epsilon]R'(\lambda)\}^k\, \delta \\
&= \epsilon R'(\lambda) \left[\sum_{k=1}^{\infty} \{\delta[A,\, \epsilon]R'(\lambda)\}^k \right] \delta.
\end{aligned}$$

If the infinite series converges in the operator norm then the terms of the series decay at least exponentially, and the series represents an actual right inverse for $\lambda - A$. Thus the series converges if and only if for some $k \geq 1$ one has

$$\|\{\delta[A,\, \epsilon]R'(\lambda)\}^k\|_{\mathfrak{L}(F_0)} < 1,$$

or, what is equivalent, if the spectral radius of $\delta[A,\, \epsilon]R'(\lambda) \in \mathfrak{L}(E_0)$ is less than one.

To obtain a left inverse for $\lambda - A$ one considers

$$\hat{T}(\lambda) = 1_E - S(\lambda)(\lambda - A) = -\epsilon R'(\lambda)[A,\, \delta],$$

and one tries to sum the series

$$\begin{aligned}
&\text{left-}(\lambda - A)^{-1} = \\
&= [1_E + \hat{T}(\lambda) + \hat{T}(\lambda)^2 + \cdots]S(\lambda) \\
&= \epsilon \left[\sum_{k=1}^{\infty} \{-R'(\lambda)[A,\, \delta]\epsilon\}^k \right] R'(\lambda)\delta.
\end{aligned}$$

Again, a necessary and sufficient condition for summability of the series is that the spectrum of $R'(\lambda)[A,\, \delta]\epsilon$ as operator on F_0 is contained in the open unit disk. We can summarize this as follows.

Theorem 2.2. *If $A \in \mathfrak{L}(E_1,\, E_0)$ has a resolution A', and $\lambda \in \mathbb{C}$ lies in the resolvent set of A', then λ also belongs to the resolvent set of A if*

$$r_{spec}(R'(\lambda)[A,\, \delta]\epsilon) < 1 \quad \text{and} \quad r_{spec}(\delta[A,\, \epsilon]R'(\lambda)) < 1.$$

The resolvent of A at λ is then given by

$$(\lambda - A)^{-1} = \epsilon R'(\lambda) \left[\sum_{k=0}^{\infty} \{\delta[A,\, \epsilon]R'(\lambda)\}^k \right] \delta \tag{1}$$

$$= \epsilon \left[\sum_{k=0}^{\infty} \{-R'(\lambda)[A,\, \delta]\epsilon\}^k \right] R'(\lambda)\delta. \tag{2}$$

3. Localization

Let $E \xrightarrow{\delta} F \xrightarrow{\epsilon} E$ be a resolution of the Banach couple E. We will say that the resolution is *localized* if the Banach couple F can be written as a direct sum

$$F = \ell_p - \bigoplus_{\alpha \in \mathfrak{A}} F^{(\alpha)}$$

where $\mathfrak{A}$ is a finite or countable index set. More precisely, for each $\alpha \in \mathfrak{A}$ we have a Banach couple $F^{(\alpha)}$ for which $F_j^{(\alpha)}$ is a closed subspace of F_j $(j = 0, 1)$, and each $x \in F_j$ can be written in exactly one way as a norm convergent sum

$$x = \sum_{\alpha \in \mathfrak{A}} x^{(\alpha)}$$

and the F_j norm of x is equivalent to the expression

$$\left(\sum_{\alpha \in \mathfrak{A}} \|x^{(\alpha)}\|_{F_j}^p \right)^{1/p} .$$

Throughout this article all direct sums $\bigoplus_{\alpha \in \mathfrak{A}}$ will be understood to be ℓ_p sums in the sense just explained.

We will say that an $x \in F_i$ has *finite support* if $x \in F_i^{(\alpha_1)} \bigoplus \ldots \bigoplus F_i^{(\alpha_m)}$ for a finite number of $\alpha_1, \ldots, \alpha_m \in \mathfrak{A}$. We denote the space of x with finite support by F_i^{fin}. Since every $x \in F_i$ is a norm convergent sum of $x^{(\alpha)}$'s the space F_i^{fin} is dense in F_i.

Given such a direct sum, we define the canonical projections $p_\alpha : F \to F^{(\alpha)}$ and inclusions $i_\alpha : F^{(\alpha)} \to F$ by

$$\begin{cases} p_\alpha(x) = x^{(\alpha)} \text{ if } x = \sum_{\alpha \in \mathfrak{A}} x^{(\alpha)} \\ i_\alpha(x) = x \text{ for all } x \in F_j^{(\alpha)} \; (j = 0, 1)\,. \end{cases} \tag{3}$$

If $T \in \mathfrak{L}(F_i, F_j)$ $(i, j = 0, 1)$ is a bounded operator, then we can consider its associated matrix $\{T_{\alpha\beta}\}$, with $T_{\alpha\beta} \in \mathfrak{L}(F_i^{(\beta)}, F_j^{(\alpha)})$, defined by

$$T_{\alpha\beta} = p_\alpha T i_\beta. \tag{4}$$

For operators $T : E \to E$ we will also define the "matrix elements" $T_{\alpha\beta} \in \mathfrak{L}(F_i^{(\beta)}, F_j^{(\alpha)})$ by

$$T_{\alpha\beta} = p_\alpha \, \delta \, T \, \varepsilon \, i_\beta. \tag{5}$$

The graph of a localized resolution. We can define the matrix of the projection $\pi_F = \delta\epsilon$

$$\pi_{\alpha\beta} = p_\alpha \, \pi_F \, i_\beta,$$

and we can define a graph $\mathfrak{G}(E, F)$, whose vertices are the elements of $\mathfrak{A}$, and in which α and β are connected if either $\pi_{\alpha\beta} \neq 0$ or $\pi_{\beta\alpha} \neq 0$.

We will always assume that the graph $\mathfrak{G}(E, F)$ is *uniformly locally finite.* This means by definition that

$$n_{\mathfrak{G}} \stackrel{\text{def}}{=} \max_{\alpha\in\mathfrak{A}}\{\text{number of edges containing } \alpha\} \tag{6}$$

is finite.

Let $A \in \mathfrak{L}(E_1, E_0)$ be given. Then an operator $A' \in \mathfrak{L}(F_1, F_0)$ will be called a *localized resolution* of A, if

1. The decomposition $F = \bigoplus_{\alpha\in\mathfrak{A}} F^{(\alpha)}$ is A' invariant.
2. The operator $\delta A\epsilon$ is dominated by the projection π_F, in the sense that $(\delta A\epsilon)_{\alpha\beta} = 0$ whenever $\pi_{\alpha\beta} = 0$.

By A' invariance we mean:

$$A'(F_1^{(\alpha)}) \subset F_0^{(\alpha)} \quad (\forall\alpha \in \mathfrak{A}). \tag{7}$$

Lemma 3.1. *If $A' \in Hol(F)$, then the restriction $A^{(\alpha)}$ of A' to $F^{(\alpha)}$ belongs to $Hol(F^{(\alpha)})$.*

Proof. Let λ belong to the resolvent set of A'. Then (7) trivially implies

$$(\lambda - A')^{-1}(F_0^{(\alpha)}) \subset F_1^{(\alpha)} \quad (\forall\alpha \in \mathfrak{A}).$$

Indeed, if $f_\alpha \in F_0^{(\alpha)}$ then we can write $g = (\lambda - A')^{-1}f_\alpha$ as $g = \sum_{\beta\in\mathfrak{A}} g_\beta$, with $g_\beta \in F_1^{(\beta)}$, and it follows from

$$(\lambda - A')g = \sum_{\beta\in\mathfrak{A}}(\lambda - A')g_\beta = f_\alpha \in F_0^{(\alpha)}$$

and the uniqueness of the decomposition of g into the g_β's that $g_\beta = 0$ for all $\beta \neq \alpha$. Hence $g \in F_1^{(\alpha)}$.

Since the norm of the resolvent restricted to $F_0^{(\alpha)}$ can never be more than its norm on F_0, this shows that $A^{(\alpha)} \in Hol(F^{(\alpha)})$. □

As a converse to this lemma we have the following observation.

Lemma 3.2. *Assume that all the $A^{(\alpha)}$'s belong to $Hol(F^{(\alpha)})$, that an $\omega \in \mathbb{R}$ exists for which any $\lambda \in \mathbb{C}$ with $\Re(\lambda) \geq \omega$ belongs to the resolvent set of all the $A^{(\alpha)}$'s, and for which*

$$M = \sup_{\Re(\lambda)\geq\omega}\ \sup_{\alpha\in\mathfrak{A}} \|(\lambda - A^{(\alpha)})^{-1}\|_{\mathfrak{L}(F_0^{(\alpha)},\, F_1^{(\alpha)})} < \infty.$$

Then $A' \in Hol(F)$.

Indeed, if $\Re(\lambda) \geq \omega$ then λ is in the resolvent set of A'. The resolvent is given by

$$(\lambda - A')^{-1}\{f_\alpha\} = \{(\lambda - A^{(\alpha)})^{-1}f_\alpha\}$$

and its $\mathfrak{L}(F_0, F_1)$ norm does not exceed M.

4. An Example

Let $E = \big(W_p^m(\mathbb{R}^d;\,\mathbb{R}^n),\, L_p(\mathbb{R}^d;\,\mathbb{R}^n)\big)$ with $1 < p < \infty$, and consider the mth -order (matrix) differential operator

$$A = \sum_{|j|\le m} a_j(x) D^j \in \mathfrak{L}(E_1,\, E_0).$$

Here the coefficients $a_j(x)$ are continuous $d \times d$ matrix valued functions, and the subscript j is a multiindex. We will show that $A \in Hol(E)$, if the $a_j(x)$ are uniformly continuous, and if the principal symbol

$$A_m(x,\xi) \stackrel{\text{def}}{=} \sum_{|j|=m} a_j(x)(-i\xi)^j \tag{8}$$

satisfies

$$\left\|(\lambda I - A_m(x,\xi))^{-1}\right\| \le \frac{c}{|\lambda| + |\xi|^m} \tag{9}$$

for all $x \in \mathbb{R}^d$, all $\xi \in \mathbb{C}^d$, all $\lambda \in \quad C$ with $\Re\lambda \ge c$, and some constant $c > 0$. This is of course well known, but the proof we give below generalizes to many other situations and should also motivate the constructions which follow in this section.

Choose a bounded open symmetric neighbourhood $O \subset \mathbb{R}^d$ of the origin, whose translates $\{O + \alpha \,|\, \alpha \in \mathbb{Z}^d\}$ cover $\mathbb{R}^d$ (e.g. a sufficiently large sphere). We then define for any "scale" $r > 0$ and $\alpha \in \mathbb{Z}^d$ the set

$$O_{\alpha;\,r} \stackrel{\text{def}}{=} \{r(\alpha + x) \,|\, x \in O\}\ . \tag{10}$$

For any $r > 0$ the $O_{\alpha;\,r}$ cover $\mathbb{R}^d$. This covering is uniformly locally finite: since O is bounded it contains only a finite number N of lattice points, and hence any $O_{\alpha;\,r}$ intersects exactly N other $O_{\beta;\,r}$s.

Next we construct a suitable partition of unity. Choose $0 \le \psi \in C_c^\infty(O)$ so that the translates by $\mathbb{Z}^d$ of $\{x \,|\, \psi(x) > 0\}$ also cover $\mathbb{R}^d$. Define

$$\varphi_\alpha(x) \stackrel{\text{def}}{=} \frac{\psi(x-\alpha)}{\sqrt{\sum_{\beta\in\mathbb{Z}^d}\psi(x-\beta)^2}}$$

and

$$\varphi_{\alpha;\,r}(x) \stackrel{\text{def}}{=} \varphi_\alpha(\frac{x}{r}).$$

Then

$$\sum_{\alpha\in\mathbb{Z}^d} \varphi_{\alpha;\,r}(x-\alpha)^2 = 1 \qquad (\forall x \in \mathbb{R}^d). \tag{11}$$

We now describe the resolution. The index set will be $\mathfrak{A} = \mathbb{Z}^d$, and for each $\alpha \in \mathbb{Z}^d$ we put

$$F^{(\alpha)} = (W_p^m(\mathbb{R}^d;\,\mathbb{R}^n),\, L_p(\mathbb{R}^d;\,\mathbb{R}^n))$$

so that the Banach couple F is defined by

$$F = \ell_p - \bigoplus_{\alpha\in\mathfrak{A}} F^{(\alpha)}.$$

Thus F_j $(j = 0, 1)$ is the space of all sequences $\{f_\alpha\}_{\alpha\in\mathbb{Z}^d}$ of functions $f_\alpha \in F_j^{(\alpha)}$ for which the sum

$$\|\{f_\alpha\}_{\alpha\in\mathfrak{A}}\|_{F_j}^p = \sum_{\alpha\in\mathfrak{A}} \|f_\alpha\|_{F_j^{(\alpha)}}^p$$

is finite.

We construct the morphisms $E \xrightarrow{\delta} F \xrightarrow{\epsilon} E$ using the partition of unity $\{\varphi_\alpha\}$. For any $f \in E_j$ $(j = 0, 1)$ we define the αth component of δf to be

$$(\delta f)_\alpha = \varphi_{\alpha;\,r}(x) f(x), \tag{12}$$

and for any $\{f_\alpha\}_{\alpha\in\mathfrak{A}} \in F_j$ $(j = 0, 1)$ we put

$$(\epsilon\{f_\alpha\})(x) = \sum_{\beta\in\mathfrak{A}} \varphi_{\beta;\,r}(x) f_\beta(x). \tag{13}$$

Observe that this last sum is locally finite and that both $\epsilon : F_j^{(\alpha)} \to E_j$ and $\delta : E_j \to F_j^{(\alpha)}$ are bounded operators. As an example, let us verify that $\delta : F_1^{(\alpha)} \to E_1$ is bounded. Let $f \in W_p^m(\mathbb{R}^d;\, \mathbb{R}^n)$ be given. Then

$$\begin{aligned} \|\delta f\|_{F_1}^p &= \sum_{\alpha\in\mathfrak{A}} \|(\delta f)_\alpha\|_{F_1^{(\alpha)}}^p \\ &= \sum_{\alpha\in\mathfrak{A}} \|\varphi_\alpha f\|_{W_p^m(\ \ R^d;\mathbb{R}^n)}^p \\ &\le c \sum_{\alpha\in\mathfrak{A}} \|f\|_{W_p^m(O_{\alpha;\,r})}^p \\ &\le Nc\|f\|_{E_1} \end{aligned}$$

where, as above, N is the number of lattice points in O. So $\delta : E_1 \to F_1$ is indeed bounded.

For any $f \in F_j$ one has

$$(\epsilon\delta f)(x) = \sum_{\beta\in\mathfrak{A}} \varphi_\beta(x)^2 f(x) = f(x)$$

so that $\epsilon\delta = 1_E$, and therefore $E \xrightarrow{\delta} F \xrightarrow{\epsilon} E$ is indeed a localized resolution of E.

The projection $\pi_F = \delta\epsilon$ acts as follows:

$$(\pi_F\{f_\alpha\})_\alpha(x) = \sum_{\beta\in\mathfrak{A}} \varphi_\alpha(x)\varphi_\beta(x) f_\beta(x),$$

from which one can see what the graph $\mathfrak{G}(E, F)$ is: the vertices are the lattice points $\alpha \in \mathfrak{A}$, and $\alpha, \beta \in \mathfrak{A}$ are connected if and only if $\alpha - \beta \in O$. In particular this graph is uniformly locally finite, and we have $n_{\mathfrak{G}} = N$.

Interpretation of the matrix coefficients $T_{\alpha\beta} = p_\alpha \delta T \varepsilon i_\beta$ of an operator $T : E_1 \to E_0$. It is evident that for any $f \in E_1$ one has

$$T_{\alpha\beta}(f) = \varphi_{\alpha;r} T\left(\varphi_{\beta;r} f\right) \tag{14}$$

so that $T_{\alpha\beta}(f)$ more or less represents the component in $O_{\alpha;r}$ of the image under T of the component in $O_{\beta;r}$ of f.

Next, we introduce an operator $A' \in \mathfrak{L}(F_1, F_0)$ by saying how it acts on each $F_1^{(\alpha)}$:

$$\left(A^{(\alpha)} f_\alpha\right)(x) = \sum_{|j| \le m} a_j^{(\alpha)}(x) D^j f_\alpha(x) \qquad (\forall x \in \mathbb{R})$$

for any $f_\alpha \in F_1^{(\alpha)}$. The coefficients $a_j^{(\alpha)}(x)$ coincide with the coefficients $a_j(x)$ of the original operator A on the support of φ_α. Outside of these supports the $a_j^{(\alpha)}(x)$ are continuous extensions of the $a_j(x)$ for which

$$\operatorname*{osc}_{x \in \mathbb{R}^d} a_j^{(\alpha)}(x) = \operatorname*{osc}_{x \in \operatorname{spt} \varphi_\alpha} a_j^{(\alpha)}(x) \tag{15}$$

holds.[1]

The operator A' defined by $A'\{f_\alpha\} = \left\{A^{(\alpha)} f_\alpha\right\}$ is a resolution of A. Observe that the αth component of A', i.e. $A^{(\alpha)}$, coincides with A on the support of φ_α, but is chosen arbitrarily outside of this set.

In order to see how good the resolution A' is, we will compute the commutators $[A, \epsilon]$ and $[A, \delta]$. It turns out that these commutators do not depend on what the $A^{(\alpha)}$'s do outside of the support of φ_α. This gives us the freedom to choose $A^{(\alpha)}$ any way we like on the complement of the support of φ_α. The particular choice which we have made ensures that the coefficients $a_j^{(\alpha)}(x)$ of $A^{(\alpha)}$ do not oscillate more than the $a_j(x)$ do on $O_{\alpha;r}$.

In general the commutator of a differential operator with a multiplication operator is another differential operator of *lower* order. The point of the following computations is to show that $[A, \epsilon]$ and $[A, \delta]$ are made up of differential operators of order $m-1$.

---

[1]For vector valued functions f one should define $\operatorname{osc} f$ to be the smallest $r > 0$ such that the range of f is contained in a ball of radius $r/2$.

If $\{f_\alpha\} \in F_1$, then

$$\begin{aligned}
[A,\,\epsilon]\cdot\{f_\alpha\} &= A\left\{\sum_{\alpha\in\mathfrak{A}}\varphi_{\alpha;r}f_\alpha\right\} - \sum_{\alpha\in\mathfrak{A}}\varphi_{\alpha;r}A^{(\alpha)}\{f_\alpha\} \\
&= \sum_{|j|\le m}\sum_{\alpha\in\mathfrak{A}} a_j(x)[D^j,\,\varphi_{\alpha;\,r}]f_\alpha \\
&= \sum_{\alpha\in\mathfrak{A}} a_j(x)\sum_{|j|\le m}\sum_{0<k\le j}\gamma_{j,k}D^k\left(\varphi_{\alpha;\,r}\right)D^{j-k}\left(f_\alpha\right) \\
&= \sum_{|j|\le m}\sum_{0<k\le j}\sum_{\alpha\in\mathfrak{A}} b^{(\alpha)}_{j;k}(x)\frac{D^{j-k}\left(f_\alpha\right)}{r^{|k|}} \qquad (16)
\end{aligned}$$

where the $\gamma_{j,k}$ are fixed integers which appear as coefficients when one expands the commutator $[D^j,\,\varphi_{\alpha;\,r}]$, and where

$$b^{(\alpha)}_{j;k}(x) = a_j(x)\,\gamma_{j,k}\,\left(D^k\varphi\right)\left(\frac{x-\alpha}{r}\right). \qquad (17)$$

The coefficients $b^{(\alpha)}_{j;k}(x)$ are uniformly bounded, i.e. there is a constant $M<\infty$ such that

$$\left|b^{(\alpha)}_{j;k}(x)\right| < M$$

for all $x\in\mathbb{R}^d$ and all multiindices $0<k\le j$, and the support of $b^{(\alpha)}_{j;k}(x)$ is contained in $O_{\alpha;\,r}$. Consequently the operator $[A,\,\epsilon]$ is bounded from $\bigoplus_{\alpha\in\mathfrak{A}} W^{m-1}_p(\mathbb{R}^d;\mathbb{R}^n)$ to $L_p(\mathbb{R}^d;\mathbb{R}^n)$, and therefore $[A,\,\epsilon]$, by the interpolation inequality

$$\|u\|_{W^{m-1}_p} \le C\,\|u\|^{1-1/m}_{W^m_p}\,\|u\|^{1/m}_{L_p} \qquad (18)$$

is an operator of zero relative bound from F_1 to E_0.

As for $[A,\,\delta]$, we compute that for $f\in E_1$ one has

$$\begin{aligned}
([A,\,\delta]\cdot f)_\alpha &= A^{(\alpha)}(\varphi_\alpha(x)f(x)) - \varphi_\alpha(x)Af(x) \\
&= \sum_{|j|=m} a_j(x)\left[D^j,\,\varphi_{\alpha;\,r}\right]f \\
&= \sum_{|j|=m}\sum_{0<k\le j}\gamma_{j,k}D^k\left(\varphi_{\alpha;\,r}\right)D^{j-k}\left(f\right) \\
&= \sum_{|j|=m}\sum_{0<k\le j} b^{(\alpha)}_{j;k}(x)\frac{D^{j-k}\left(f\right)}{r^k}.
\end{aligned}$$

Again the coefficients $b^{(\alpha)}_{j;k}(x)$ are uniformly bounded, from which one deduces that $[A,\,\delta]$ is a bounded operator from $W^{m-1}_p(\mathbb{R}^d;\mathbb{R}^n)$ to $\bigoplus_{\alpha\in\mathfrak{A}} L_p(\mathbb{R}^d;\mathbb{R}^n)$. The interpolation inequality (18) implies that $[A,\,\delta] : E_1 \to F_0$ has zero relative bound.

So A' is a resolution of A with vanishing relative bound, and if we can show that $A'\in Hol(F)$, then we have also shown that $A\in Hol(E)$.

In order to verify that $A' \in Hol(F)$ we consider each $A^{(\alpha)}$ separately, with the intention of using lemma (3.2). By assumption the coefficients $a_j^{(\alpha)}(x)$ are almost constant if one chooses the scale $r > 0$ small enough. Thus, given any $\mu_0 > 0$ one can choose $r > 0$ so small that constant $d \times d$ matrices $\bar{a}_j^{(\alpha)}$ exist for which

$$\left| a_j^{(\alpha)}(x) - \bar{a}_j^{(\alpha)} \right| \leq \mu_0, \quad \forall x \in \mathbb{R}^d \tag{19}$$

holds for all $\alpha \in \mathfrak{A}$. So the $A^{(\alpha)}$'s are uniformly small perturbations of the constant coefficient operators

$$A_0^{(\alpha)} = \sum_{|j| \leq m} \bar{a}_j^{(\alpha)} D^j,$$

all of which belong to $Hol(F^{(\alpha)})$. The resolvent of $A_0^{(\alpha)}$ is given by convolution with

$$\mathfrak{F}^{-1}\left\{ \left(\lambda - \sum_{|j| \leq m} \bar{a}_j^{(\alpha)} (-i\xi)^j \right)^{-1} \right\}$$

where $\mathfrak{F}$ denotes the Fourier transform. An appropriate multiplier theorem then provides us with a constant $M < \infty$ such that

$$\|(\lambda - A_0^{(\alpha)})^{-1}\|_{\mathfrak{L}(F_0^{(\alpha)}, F_1^{(\alpha)})} \leq M$$

whenever $\Re(\lambda) \geq \omega$ for some fixed $\omega > 0$. Therefore if the size μ_0 of the perturbation $A^{(\alpha)} - A_0^{(\alpha)}$ is small enough, the $A^{(\alpha)}$ will satisfy the same estimate with $2M$ instead of M. By lemma (3.2) this implies that $A' \in Hol(F)$, and thus $A \in Hol(E)$.

5. Exponential decay of the resolvent

Consider a localized resolution $E \xrightarrow{\delta} F \xrightarrow{\varepsilon} E$, $A' \in Hol(F)$, of $A \in Hol(E)$ and assume that for some λ in the resolvent set of A' one has

$$\|\delta[A, \epsilon]R'(\lambda)\|_{L(F_0, F_1)} \leq \vartheta \quad \text{and} \quad \|R'(\lambda)[\delta, A]\epsilon\|_{L(F_0, F_1)} \leq \vartheta \tag{20}$$

for some $\vartheta < 1$. By Theorem 2.2 λ belongs to the resolvent set of A, and the resolvent is given by (1) or (2). One can write (2) as

$$(\lambda - A)^{-1} = \epsilon R'(\lambda) \left[1 + D + D^2 + \ldots + D^k + \ldots\right] \delta$$

where

$$D = \delta[A, \epsilon]R'(\lambda)$$

Using this expression we will estimate the matrix elements of $(\lambda - A)^{-1}$, as defined in (5), i.e.

$$\begin{aligned}\left\{(\lambda - A)^{-1}\right\}_{\alpha\beta} &= p_\alpha \delta\, (\lambda - A)^{-1}\, \varepsilon i_\beta \\ &= p_\alpha \delta \epsilon R'(\lambda)\left[1 + D + D^2 + \ldots + D^k + \ldots\right] \delta\varepsilon i_\beta \\ &= p_\alpha \pi_F R'(\lambda)\left[1 + D + D^2 + \ldots + D^k + \ldots\right] \pi_F i_\beta \quad (21)\end{aligned}$$

To begin we consider the operator $D : F_0 \to F_0$ and estimate the matrix elements of its powers D^k, $k = 1, 2, \ldots$ Recall that these matrix elements are defined as in (4), i.e. $D_{\alpha\beta} = p_\alpha\, D\, i_\beta$, etc. For the matrix elements of D^k we then have

$$\left(D^k\right)_{\alpha\beta} = \sum_{\alpha_1} \cdots \sum_{\alpha_k} D_{\alpha\alpha_1} D_{\alpha_1\alpha_2} \cdots D_{\alpha_k\beta} \tag{22}$$

Lemma 5.1. *The operator D is dominated by the projection π_F, i.e. $D_{\alpha\beta} \neq 0$ only if α and β are adjacent in the graph $\mathfrak{G}$.*

The only nonzero terms in the sum (22) are those for which the graph $\mathfrak{G}$ contains a path of length $k+1$ from α to β.

Proof. The resolvent $R'(\lambda)$ of the resolution A' is local, so we must show that $\delta[A, \epsilon]$ is dominated by the projection π_F. This follows from $\delta[A, \varepsilon] = \delta A\varepsilon - A'\delta\varepsilon = \delta A\varepsilon - A'\pi_F$. The first term in this expression is by assumption dominated by π_F, while the second is also dominated by π_F since A' is local.

If $D_{\alpha\alpha_1} D_{\alpha_1\alpha_2} \ldots D_{\alpha_k\beta} \neq 0$ then α is adjacent to α_1, α_1 is adjacent to α_2, etc, so $\{\alpha, \alpha_1, \ldots, \alpha_k, \beta\}$ is a path of length $k+1$ from α to β. □

Continuing the computation in (21) we find that the $\alpha\beta$ element of the resolvent $(\lambda - A)^{-1}$ is given by

$$\left\{(\lambda - A)^{-1}\right\}_{\alpha\beta} = \sum \pi_{\alpha\alpha_1} R'(\lambda) D_{\alpha_1\alpha_2} D_{\alpha_2\alpha_3} \cdots D_{\alpha_k\alpha_{k+1}}\, \pi_{\alpha_{k+1}\beta} \tag{23}$$

where the summation runs over all paths $\alpha \to \alpha_1 \to \ldots \to \alpha_{k+1} \to \beta$ of length $k+2$ at least 2.

Theorem 5.2. *If we assume that $\vartheta n_{\mathfrak{G}} < \frac{1}{2}$ then the operator norm from $F_0^{(\beta)}$ to $F_1^{(\alpha)}$ of $\left\{(\lambda - A)^{-1}\right\}_{\alpha\beta}$ is bounded by $C\,(\vartheta n_{\mathfrak{G}})^{-2}\, \|R'(\lambda)\|_{\mathfrak{L}(F_0,\, F_1)} \exp\left(-c\, d(\alpha, \beta)\right)$ where $c = -\ln \vartheta n_{\mathfrak{G}}$.*

Proof. Each term in (23) has norm at most

$$\|\pi\|^2_{\mathfrak{L}(F)}\, \|R'(\lambda)\|_{\mathfrak{L}(F_0,\, F_1)}\, \vartheta^k.$$

The number of such terms is the number of paths of length $k+2$ from α to β. Since each vertex in the graph $\mathfrak{G}$ is connected to at most $n_{\mathfrak{G}}$ other vertices (including itself) there at most $n_{\mathfrak{G}}^{k+2}$ paths emanating from α. On the other hand, by definition there are no paths from α to β of length less than $d(\alpha,\beta)$ so we can estimate the sum (23) by

$$\begin{aligned}\left\|\left\{(\lambda-A)^{-1}\right\}_{\alpha\beta}\right\|_{\mathfrak{L}(F_0^{(\beta)},\,F_1^{(\alpha)})} &\le \sum_{k+2\ge d(\alpha,\beta),\,k\ge 0} C\,\|R'(\lambda)\|_{\mathfrak{L}(F_0,\,F_1)}\,\vartheta^k n_{\mathfrak{G}}^{k+2}\\ &\le C'\,\|R'(\lambda)\|_{\mathfrak{L}(F_0,\,F_1)}\,\frac{(\vartheta n_{\mathfrak{G}})^{d(\alpha,\beta)-2}}{1-\vartheta n_{\mathfrak{G}}}\\ &\le \frac{2C'}{(\vartheta n_{\mathfrak{G}})^2}\,\|R'(\lambda)\|_{\mathfrak{L}(F_0,\,F_1)}\exp(-cd(\alpha,\beta))\\ &= C(\lambda,\,A',\vartheta n_{\mathfrak{G}})\exp\left(-c\,d(\alpha,\,\beta)\right)\end{aligned}$$

where $c=-\ln\vartheta n_{\mathfrak{G}}$, as promised. □

6. Exponential decay of eigenfunctions

Theorem 6.1. *Let λ_0 be an isolated eigenvalue with finite multiplicity of $A\in Hol(E)$. Assume there exists a localized resolution A' of A and a smooth closed curve $\Gamma\subset\mathbb{C}$ contained in the resolvent set of A', such that λ_0 is the only point in the spectrum of A lying inside of Γ. Keeping the same notation as in the previous section we assume that for some ϑ with $\vartheta n_{\mathfrak{G}}\le\frac{1}{2}$ the estimate (20) holds for all $\lambda\in\Gamma$.*

Then any generalized eigenfunction $v\in E_1$ of A corresponding to the eigenvalue λ_0 is exponentially localized, i.e. for some $\alpha_1\in\mathfrak{A}$ one has

$$\left\|(\delta v)_\beta\right\|_{F_0}\le C\exp(-cd(\beta,\alpha_1)).$$

Proof. The hypotheses imlpy that Γ also lies in the resolvent set of A.

Corresponding to any closed curve in the resolvent set of an operator A one can define the spectral projection P_Γ by a Dunford integral

$$P_\Gamma\overset{\text{def}}{=}\oint_\Gamma(\lambda-A)^{-1}\frac{d\lambda}{2\pi i}. \tag{24}$$

If Γ encloses one and only one eigenvalue λ_0 of A then P_Γ is the projection on the generalized eigenspace corresponding to λ_0. The matrix elements with respect to the localized resolution $E\xrightarrow{\delta}\bigoplus_{\alpha\in\mathfrak{A}}F^{(\alpha)}\xrightarrow{\varepsilon}E$ are then given by

$$(P_\Gamma)_{\alpha\beta}\overset{\text{def}}{=}\oint_\Gamma\left\{(\lambda-A)^{-1}\right\}_{\alpha\beta}\frac{d\lambda}{2\pi i} \tag{25}$$

By Theorem 5.2 we can estimate this operator by

$$\begin{aligned}\left\|(P_\Gamma)_{\alpha\beta}\right\| &\le C\,|\Gamma|\,(\vartheta n_{\mathfrak{G}})^{-2}\sup_{\lambda\in\Gamma}\|R'(\lambda)\|_{\mathfrak{L}(F_0,\,F_1)}\exp\left(-c\,d(\alpha,\,\beta)\right)\\ &= C(\Gamma,A',\vartheta n_{\mathfrak{G}})\exp\left(-c\,d(\alpha,\,\beta)\right)\end{aligned} \tag{26}$$

with $c=-\ln\vartheta n_{\mathfrak{G}}$.

Assume now that Γ encloses one and only one eigenvalue λ_0 of A, and that P_Γ has finite rank N (so λ_0 is an eigenvalue of finite multiplicity). Since the space F_1^{fin} of $x \in F_0$ with finite support is dense in F_0 there are $x_1, \dots, x_N \in F_0^{\text{fin}}$ such that $P_\Gamma x_1, \dots, P_\Gamma x_N$ span the range of P_Γ. If the x_k are contained in $F_0^{(\alpha_1)} \bigoplus \dots \bigoplus F_0^{(\alpha_m)}$ then (26) implies that the components of $v_k = P_\Gamma x_k$ decay exponentially:

$$\left\| (v_k)_\beta \right\|_{F_1} \le C \sum_{i=1}^{m} e^{-cd(\alpha_i,\beta)} \le C' e^{-cd(\alpha_1,\beta)} \tag{27}$$

(we have used the triangle inequality: $d(\alpha_i, \beta) \le d(\alpha_1, \beta) + d(\alpha_i, \alpha_1)$.)

The same estimate applies to any generalized eigenfunction $v = C_1 v_1 + \dots + C_m v_m$. □

7. Exponential decay in the Example

We return to the example in Section 4. In this section we choose a slightly different resolution $A' = \bigoplus A^{(\alpha)}$, namely we set all $A^{(\alpha)}$ equal to the same operator $A = \sum_{|j|\le m} a_j(x) D^j$. Thus the coefficients $a_j^{(\alpha)}(x)$ coincide with the coefficients $a_j(x)$ of A, and the computation (16) implies that the $\alpha\beta$-matrix element of $\delta[A, \epsilon]$ is given by

$$(\delta[A, \epsilon])_{\alpha\beta} = \sum_{|j|\le m} \sum_{0<k\le j} \varphi_{\alpha;r}(x) \frac{b_{j;k}^{(\beta)}(x)}{r^{|k|}} \circ D^{j-k}. \tag{28}$$

Since the $b_{j;k}^{(\beta)}(x)$ and $\varphi_{\alpha;r}(x)$ are uniformly bounded in $L_\infty\left(\mathbb{R}^d\right)$ (i.e. independently of $r > 0$) $(\delta[A, \epsilon])_{\alpha\beta}$ is a bounded operator from $W_p^{m-1}(\mathbb{R}^d; \mathbb{R}^n)$ to $L_p(\mathbb{R}^d; \mathbb{R}^n)$. Assuming that $r \gg 1$ the terms with the largest norm in (28) are those for which $|k| = 1$, so that we get the following estimate for the operator norm of $(\delta[A, \epsilon])_{\alpha\beta}$

$$\left\| (\delta[A, \epsilon])_{\alpha\beta} \right\|_{\mathfrak{L}\left(W_p^{m-1}(\mathbb{R}^d;\mathbb{R}^n);\, L_p(\mathbb{R}^d;\ R^n)\right)} \le \frac{C}{r} \quad \text{for } r \ge 1\,. \tag{29}$$

For small r the terms with $|k| = |j| = m$ dominate, and one gets

$$\left\| (\delta[A, \epsilon])_{\alpha\beta} \right\|_{\mathfrak{L}\left(W_p^{m-1}(\mathbb{R}^d;\mathbb{R}^n);\, L_p(\mathbb{R}^d;\ R^n)\right)} \le \frac{C}{r^m} \quad \text{for } 0 < r \le 1\,. \tag{30}$$

Composing $(\delta[A, \epsilon])_{\alpha\beta}$ from the left or right with $R'(\lambda)$ and keeping in mind that the action of $R'(\lambda)$ on $F^{(\alpha)}$ or $F^{(\beta)}$ is the same as that of $R(\lambda) = (\lambda - A)^{-1}$, we find that (20) is satisfied provided one takes $\vartheta = \frac{C}{r} \left\| R(\lambda) \right\|_{\mathfrak{L}(L_p(\mathbb{R}^d;\mathbb{R}^n); W_p^m(\mathbb{R}^d;\mathbb{R}^n))}$.

Theorem 7.1. *Let λ_0 be an isolated eigenvalue with finite multiplicity of the operator A of Section 4. Then any generalized eigenfunction $v(x) \in W_p^m(\mathbb{R}^d; \mathbb{R}^n)$ of A is exponentially decaying at ∞, i.e. for certain constants $c, C < \infty$ one has*

$$|v(x)| \le C e^{-c|x|}. \tag{31}$$

Proof. Choose a closed contour Γ around λ_0, e.g. one could let Γ be a sufficiently small circle centered at λ_0 . Then the resolvents $R(\lambda)$ are uniformly bounded from L_p to W_p^m for all $\lambda \in \Gamma$, and by (29) the condition (20) holds with $\vartheta = \frac{C}{r}$, where $\sup_{\lambda\in\Gamma} \|R(\lambda)\|$ is incorporated into the constant C. If one chooses r large enough, then Theorem 6.1 applies, and gives us exponential decay for $\|\varphi_{\alpha;r}(x)v(x)\|_{L_p}$. Since the $v(x)$ are generalized eigenfunctions of an elliptic operator with continuous coefficients one can easily translate the L_p estimates into pointwise estimates and thus obtain the pointwise exponential decay claimed in (31). □

So far we have not really exploited the fact that $(\delta[A,\,\epsilon])_{\alpha\beta}$ is a differential operator of order less than m. Choose an $\omega \in \mathbb{R}_+$ such that $\lambda - A$ is invertible with $\|R(\lambda)\|_{\mathfrak{L}(L_p;W_p^m)} \le C$ for all $\lambda \in \mathbb{C}$ with $\Re\lambda > \omega$. Then we also have $\|R(\lambda)\|_{\mathfrak{L}(L_p;L_p)} \le \frac{C}{|\lambda|}$ for all λ with $\Re\lambda \ge \omega$. Interpolation between W_p^m and L_p then implies that

$$\left\|D^{j-k}\circ R(\lambda)\right\|_{\mathfrak{L}(L_p,L_p)} \le C\,|\lambda|^{-1+\frac{|j|-|k|}{m}}\,. \tag{32}$$

Combine this with (28) and you get the following estimate for the $L_p \to L_p$ norm of the operator $(\delta[A,\,\epsilon]R(\lambda))_{\alpha\beta}$

$$\begin{aligned}
\left\|(\delta[A,\,\epsilon]R'(\lambda))_{\alpha\beta}\right\|_{\mathfrak{L}(L_p,L_p)} &\le \sum_{|j|\le m}\sum_{0<k\le j}\left\|\varphi_{\alpha;r}(x)\frac{b_{j;k}^{(\beta)}(x)}{r^{|k|}}\circ D^{j-k}\circ R(\lambda)\right\| \\
&\le C\sum_{|j|\le m}\sum_{0<k\le j} r^{-|k|}\,|\lambda|^{-1+\frac{|j|-|k|}{m}} \\
&\le C\sum_{|j|\le m}\sum_{0<k\le j}\left(r\,|\lambda|^{1/m}\right)^{-|k|} \\
&\le \frac{C}{r\,|\lambda|^{1/m}} \qquad (33)
\end{aligned}$$

assuming that $r\,|\lambda|^{1/m} \ge 1$. Since $\delta[A,\,\epsilon]R'(\lambda)$ is dominated by the projection π_F we also get

$$\|\delta[A,\,\epsilon]R'(\lambda)\|_{\mathfrak{L}(F_0)} \le \frac{n_{\mathfrak{G}}C}{r\,|\lambda|^{1/m}}. \tag{34}$$

Hence if $r\,|\lambda|^{1/m}$ is sufficiently large and if $\Re\lambda \ge \omega$, then the resolvent $R(\lambda)$ is given by (1) and its matrix elements are given by (23). With $r = \frac{2C}{n_{\mathfrak{G}}|\lambda|^{1/m}}$ we get $\vartheta n_{\mathfrak{G}} = \frac{1}{2}$ so that Theorem 5.2 implies

$$\begin{aligned}
\left\|\left((\lambda - A)^{-1}\right)_{\alpha\beta}\right\|_{\mathfrak{L}(E_0,E_1)} &\le C\,(\vartheta n_{\mathfrak{G}})^{-2}\,\|R'(\lambda)\|_{\mathfrak{L}(F_0,\,F_1)}\exp\left(-c\,d(\alpha,\,\beta)\right) \qquad (35)\\
&\le C'\exp\left(-cd(\alpha,\,\beta)\right). \qquad (36)
\end{aligned}$$

Refering to the interpretation of the matrix element given in (14) we see that the size of $(\lambda - A)^{-1} f$ in $O_{\alpha;r}$ where f is a function which is localized in $O_{\beta;r}$ is at most $C' e^{-cd(\alpha,\beta)} \|f\|$. The proper measurement of the exponential $e^{-cd(\alpha,\beta)}$ is in terms of the distance between the sets $O_{\alpha;r}$ and $O_{\beta;r}$, rather than between the indices $\alpha, \beta \in \mathfrak{A} = \mathbb{Z}^d$. The distance between $O_{\alpha;r}$ and $O_{\beta;r}$ is proportional to $r|\alpha - \beta|$, and the graph distance $d(\alpha, \beta)$ is also roughly proportional to $|\alpha - \beta|$. Hence we get the following estimate

$$\left\| \left((\lambda - A)^{-1} \right)_{\alpha\beta} \right\|_{\mathfrak{L}(E_0, E_1)} \leq C' \exp\left(-c |\lambda|^{1/m} |x_\alpha - x_\beta| \right) \tag{37}$$

for all λ with $\Re\lambda \geq \omega$. (We have written $x_\alpha = r\alpha$ for the centre of $O_{\alpha;r}$.) Using elliptic interior estimates for the equation $\lambda u - Au = f$ one can show that the resolvent $(\lambda - A)^{-1}$ has an integral kernel $K(\lambda; x, y)$ which is bounded by

$$|K(\lambda; x, y)| \leq C' \exp\left(-c |\lambda|^{1/m} |x - y| \right) \quad \text{for } |x - y| \geq C |\lambda|^{-1/m}. \tag{38}$$

8. Application to "Fredholm solutions" of Nonlinear Elliptic Systems

When studying solutions of systems of nonlinear elliptic PDE from the point of view of bifurcation theory (as in [7, 8]) it is of interest to consider the linearized equation. Theorems from bifurcation theory will in general only be applicable if the linearized operator is a Fredholm operator. The following theorem shows that a slightly stronger condition will not hold for most bounded entire solutions of elliptic PDEs.

Theorem 8.1. *Let $u \in BUC^2(\mathbb{R}^d; \mathbb{R}^n)$ be a solution of the nonlinear PDE*

$$F(u, Du, D^2 u, \ldots, D^m u) = 0 \tag{39}$$

where F is a C^1 function of its arguments, and assume that the linearized operator

$$A \stackrel{\text{def}}{=} \sum_{|j| \leq m} a_j(x) D^j, \qquad a_j(x) \stackrel{\text{def}}{=} \frac{\partial F}{\partial (D^j u)}(u(x), \ldots, D^m u(x)) \tag{40}$$

is uniformly elliptic. If $\lambda = 0$ does not belong to the essential spectrum of the operator $A : W_p^m(\mathbb{R}^d; \mathbb{R}^n) \to L_p(\mathbb{R}^d; \mathbb{R}^n)$ then $\lim_{|x|\to\infty} u(x)$ exists. (In case $d = 1$ the limits at $+\infty$ and $-\infty$ may be different.)

Proof. By differentiating the equation (39) with respect to x_k, one finds that $\partial_k u$ is an eigenfunction of A with eigenvalue 0. Hence by Theorem 7.1 all derivatives $\partial_k u$ decay exponentially as $|x| \to \infty$. By integrating the gradient along circles centered at the origin one sees that the oscillation of $u(x)$ on the sphere of radius r decays exponentially as $r \to \infty$ (at least if $d \geq 2$, otherwise such circles do not exist.) By integrating along rays one sees that $u(x)$ actually tends to a constant. □

References

[1] Amann, H., "Quasilinear evolution equations and parabolic systems", TAMS **293** (1986), 191–227.

[2] Amann, H., "On abstract parabolic fundamental solutions", J. Math. Soc. Japan, **39** (1987).

[3] Amann, H., "Quasilinear parabolic systems under nonlinear boundary conditions", Arch. Rat. Mech. Anal. **92** (1986), 153–192.

[4] Amann, H., *Linear and Quasilinear Parabolic Problems, Vol I*, Birkhäuser, 1995, *Vol II*, to appear.

[5] Amann, H., Hieber, M., Simonett, G.: "Bounded H^∞-calculus for elliptic operators", Differential and Integral Equations **7** (1994), 613–653.

[6] Angenent, S.B., "Nonlinear Analytic Semiflows", Proceedings of the Royal Society in Edinburgh **115A** (1990) pp. 91–107.

[7] Angenent, S.B., "The shadowing lemma for elliptic PDE", pp. 7–22 in "Dynamics of Infinite dimensional dynamical systems," NATO/ASI series F, vol.**37**, Springer Verlag.

[8] Angenent, S.B., "A variational interpretation of Melnikov's function and exponentially small separatrix splitting", pp. 5–35 in "Symplectic Geometry," London Mathematical Society Lecture Note Series vol. **192**, Cambridge University Press (1993)

[9] Bergh, J. and Löfström, J., *Interpolation Spaces, an Introduction*, Springer Verlag, Berlin, 1976.

[10] Butzer, P.L. and Berens, H., *Semigroups of Operators and Approximation*, Springer Verlag, Berlin, 1967.

[11] Clément, Ph., Heijmans, H.J.A.M., et. al., *One Parameter Semigroups*, CWI Monographs #5, North Holland, Amsterdam, 1987.

[12] DaPrato, G. and Grisvard, P., "Équations d'évolutions abstraites nonlinéaires de type parabolique", Annali di Mat. Pura ed Appl., **120** (1979), 329–396.

[13] Lunardi, A., "Quasilinear parabolic equations", Math. Annalen **267** (1984), 395–415.

[14] Pazy, A., *Semigroups of Linear Operators and Applications to Partial Differential Equations*, Springer Verlag, New York, 1983.

[15] Triebel, H., *Interpolation Theory, Function Spaces, Differential Operators*, North Holland, Amsterdam, 1978.

Progress in Nonlinear Differential Equations
and Their Applications, Vol. 35
© 1999 Birkhäuser Verlag Basel/Switzerland

Wiener Regularity and Heat Semigroups on Spaces of Continuous Functions

Wolfgang Arendt and Philippe Bénilan

Dedicated to Professor Herbert Amann on the occasion of his sixtieth birthday

Abstract. Let $\Omega \subset \mathbb{R}^N$ be open. It is shown that the Dirichlet Laplacian generates a (holomorphic) C_0-semigroup on $C_0(\Omega)$ if and only if Ω is regular in the sense of Wiener. The same result remains true for elliptic operators in divergence form.

0. Introduction

Let Ω be an open subset of $\mathbb{R}^N$. By $C_0(\Omega)$ we denote the space of all continuous scalar valued functions f on $\bar{\Omega}$ which are 0 on the boundary $\partial\Omega$ of Ω and 0 at infinity; i.e., $C_0(\Omega) = \{f : \Omega \to \mathbb{C}$ continuous: $\lim_{x \to z} f(x) = 0$ for all $z \in \partial\Omega$ and $\lim_{\substack{|x| \to \infty \\ x \in \Omega}} f(x) = 0\}$. Then $C_0(\Omega)$ is a Banach space for the supremum norm

$$\|f\|_\infty = \sup_{x \in \Omega} |f(x)| ,$$

and a Banach lattice for pointwise ordering. We consider the Laplacian Δ_0 on $C_0(\Omega)$ with maximal distributional domain; i.e.,

$$D(\Delta_0) = \{f \in C_0(\Omega) : \Delta f \in C_0(\Omega)\} \quad \Delta_0 f = \Delta f .$$

It is a closed, dissipative operator. Our aim is to characterize those open sets Ω for which Δ_0 is the generator of a C_0-semigroup on $C_0(\Omega)$.

We say that Ω is **regular (in the sense of Wiener)** if at each point $z \in \partial\Omega$ there exists a **barrier** (see Definition 3.1). If Ω is bounded then, by classical potential theory (see e.g. [DL], [GT]), Ω is regular if and only if the Dirichlet problem

$$D(\varphi) \left\{ \begin{array}{l} u \in C(\bar{\Omega}), \ u_{|\partial\Omega} = \varphi \\ \Delta u = 0 \quad \text{in} \quad \mathcal{D}(\Omega)' \end{array} \right.$$

has a solution for all $\varphi \in C(\partial\Omega)$. The purpose of this paper is to establish the following result:

AMS classification: 47D06, 35 K50

Theorem 0.1. *Let $\Omega \subset \mathbb{R}^N$ be open. The following assertions are equivalent:*

(i) *Ω is regular;*

(ii) *the resolvent set of Δ_0 is non-empty;*

(iii) *Δ_0 generates a holomorphic C_0-semigroup on $C_0(\Omega)$.*

The proof of Theorem 0.1 is based on classical potential theory and the methods are well known. Still it seems that the result is nowhere given explicitely in the literature. On the other hand, it is certainly of importance. For example, treating the semilinear heat equation and associated dynamical systems, the realization of the Laplacian on the space $C_0(\Omega)$ is precisely what is needed, (cf. Cazanave-Haraux [CH], where, even though merely the case of Lipschitz boundary is treated, only vague references and indications for the proof are given; see [CH], Lemma 2.6.5 and Remark 2.6, in particular). In this paper we give complete proofs and also describe in detail properties of the Laplacian we use.

The paper is organized as follows. Section 1 contains prerequisites concerning the heat semigroups on $L^p(\Omega)$. Then Theorem 0.1 is proved for bounded Ω in Section 3 by establishing the equivalence of *(ii)* and *(iii)* to well-posedness of the Dirichlet problem. In Section 4 we prove the result for arbitrary open sets establishing equivalence to regularity in the sense that there is a barrier at each point. These two sections are completely elementary and self-contained. In Section 4 we prove the corresponding result for elliptic operators. Here we use the fundamental result of De Giorgi and Nash on (Hölder-) continuity of weak solutions and further regularity results by Stampacchia. We also show several spectral and regularity results for the elliptic operators on $C_0(\Omega)$. In particular, we prove holomorphy of the semigroup on $C_0(\Omega)$ using that the adjoint semigroup is holomorphic on $L^1(\Omega)$ (see [AB], [AE], [Ou2]). For previous results exploiting this duality see Amann [Am] and Amann and Escher [AmE].

1. The Dirichlet Laplacian on $\mathrm{L}^p(\Omega)$

In this section we put together some known results on the Laplacian with Dirichlet boundary conditions. Let $\Omega \subset \mathbb{R}^N$ be an open set. By $\mathcal{D}(\Omega)$ we denote the space of all test functions and by $\mathcal{D}(\Omega)'$ the space of all distributions. The first Sobolev space is denoted by

$$H^1(\Omega) := \{u \in L^2(\Omega) : D_j u \in L^2(\Omega),\ j = 1 \dots N\}$$

where $D_j = \frac{\partial}{\partial x_j}$. Moreover, we let

$$H_0^1(\Omega) = \overline{\mathcal{D}(\Omega)}^{H^1(\Omega)}.$$

Recall that for real-valued $u \in H^1(\Omega)$ one has $u^+, u^-, (u-1)^+ \in H^1(\Omega)$ and

$$D_j u^+ = 1_{[u>0]} D_j u,\ \ D_j(u-1)^+ = 1_{[u>1]} D_j u\,. \tag{1.1}$$

Moreover, $u^+, (u-1)^+ \in H_0^1(\Omega)$ if $u \in H_0^1(\Omega)$. Note also that the mappings $u \mapsto u^+$ and $u \mapsto (u-1)^+$ at continuous from $H^1(\Omega)$ into $H^1(\Omega)$. See [GT], Chapter 7, for these simple facts. By Δ_2^Ω we denote the **Laplacian with Dirichlet boundary conditions** on $L^2(\Omega)$, also called the **Dirichlet Laplacian** for short; i.e.

$$D(\Delta_2^\Omega) = \{u \in H_0^1(\Omega) : \Delta u \in L^2(\Omega)\},\ \Delta_2^\Omega u = \Delta u\,.$$

This is the operator associated with the form

$$a(u,v) = \int_\Omega \nabla u \nabla u$$

on $H_0^1(\Omega)$; i.e., for $u \in H_0^1(\Omega)$, $\quad v \in L^2(\Omega)$ we have

$$u \in D(\Delta_2^\Omega), \Delta_2^\Omega u = v \;\Leftrightarrow\; -\int \nabla u \nabla \varphi = \int v\varphi \quad \forall\, \varphi \in H_0^1(\Omega)\,.$$

The operator Δ_2^Ω is self-adjoint and generates a positive contraction C_0-semigroup $(e^{t\Delta_2^\Omega})_{t\geq 0}$ on $L^2(\Omega)$. Moreover, $0 \in \varrho(\Delta_2^\Omega)$ (the resolvent set of Δ_2^Ω) if Ω is bounded. If $\Omega = \mathbb{R}^N$ we write $\Delta_2 = \Delta_2^{\mathbb{R}^N}$. Then $(e^{t\Delta_2})_{t\geq 0}$ is the *Gaussian semigroup* which we denote by G_2; i.e.,

$$(G_2(t)f)(x) = (4\pi t)^{-N/2} \int_{\mathbb{R}^N} f(y) e^{-(x-y)^2/4t} dy$$

for all $f \in L^2(\mathbb{R}^N)$.

We denote by $\varrho(A)$ the resolvent set of an operator A, and by $R(\lambda, A) = (\lambda - A)^{-1}$ its resolvent ($\lambda \in \varrho(A)$). The following domination property is well known (cf. [Da], Theorem 2.1.6).

For convenience of the reader we give a simple direct proof.

Proposition 1.1. *Let Ω_1, Ω_2 be open subsets of $\mathbb{R}^N$ such that $\Omega_1 \subset \Omega_2$. Then*

$$\begin{aligned} 0 &\leq e^{t\Delta_2^{\Omega_1}} f && \leq e^{t\Delta_2^{\Omega_2}} f\,, \\ 0 &\leq R(\lambda, \Delta_2^{\Omega_1}) f && \leq R(\lambda, \Delta_2^{\Omega_2}) f \end{aligned}$$

for $f \in L^2(\Omega)_+,\ \lambda > 0,\ t \geq 0$.

Here we identify $L^2(\Omega_1)$ with a subspace of $L^2(\Omega_2)$ extending functions by 0. We let $L^2(\Omega)_+ = \{f \in L^2(\Omega) : f(x) \geq 0 \quad \text{a.e.}\}, H^1(\Omega)_+ = L^2(\Omega)_+ \cap H^1(\Omega),\ H_0^1(\Omega) = H_0^1(\Omega) \cap L^2(\Omega)_+,\ \mathcal{D}(\Omega)_+ = \{\varphi \in \mathcal{D}(\Omega) : \varphi(x) \geq 0 \quad \text{for all} \quad x \in \Omega\}$. For $w_1, w_2 \in D(\Omega)'$ we define

$$w_1 \leq w_2 \;:\Leftrightarrow\; \langle w_1, \varphi\rangle \;\leq\; \langle w_2, \varphi\rangle \quad \text{for all} \quad \varphi \in \mathcal{D}(\Omega)_+\,.$$

Note that $\mathcal{D}(\Omega)_+$ is dense in $H_0^1(\Omega)_+$. Moreover, we recall: If $u \in H^1(\Omega)$ has compact support, then $u \in H_0^1(\Omega)$ (see [B], IX.5 p. 171).

The proof of Proposition 1.1 is based on the following.

Lemma 1.2. *Let $\Omega \subset \mathbb{R}^N$ be open, $\lambda > 0$. Let $u \in H_0^1(\Omega),\ v \in H^1(\Omega)$ such that $v \geq 0$ and*

$$\lambda u - \Delta u \leq \lambda v - \Delta v \quad in \quad \mathcal{D}(\Omega)'\,.$$

Then $u(x) \leq v(x)$ a.e.. Taking in particular $v = 0$, we obtain for $u \in H_0^1(\Omega)$

$$\lambda u - \Delta u \leq 0 \;\Rightarrow u \;\leq 0\,. \tag{1.2}$$

Proof. By hypothesis we have

$$\lambda \int (u-v)\varphi + \int \nabla(u-v)\nabla\varphi \leq 0 \tag{1.3}$$

for all $\varphi \in \mathcal{D}(\Omega)_+$. It follows by density that (1.3) remains true for all $\varphi \in H_0^1(\Omega)_+$. Note that $(u-v)^+ \in H_0^1(\Omega)$. In fact, let $u_n \in \mathcal{D}(\Omega)$, $u_n \to u$ in $H^1(\Omega)$. Then $(u_n - v)^+ \in H^1(\Omega)$ has compact support and thus $(u_n - v)^+ \in H_0^1(\Omega)$. Since $(u_n - v)^+ \to (u-v)^+ \quad (n \to \infty)$ in $H^1(\Omega)$, the claim follows. Now (1.3) for $\varphi = (u-v)^+$ yields

$$\lambda \int ((u-v)^+)^2 + \int |\nabla(u-v)^+|^2 \leq 0\,.$$

Hence $(u-v)^+ = 0$. □

Proof of Proposition 1.1. a) Let $\lambda > 0$, $f \in L^2(\Omega_1)_+$, $u = R(\lambda, \Delta_2^{\Omega_1})f$, $v = R(\lambda, \Delta_2^{\Omega_2})f$. We show that $0 \leq u \leq v$. In fact, $u \in H_0^1(\Omega_1)$, $\lambda u - \Delta u = f$ in $D(\Omega_1)'$, $v \in H_0^1(\Omega_2)$, $\lambda v - \Delta v = f$ in $\mathcal{D}(\Omega_2)'$. It follows from (1.2) that $v(x) \geq 0$ and $u(x) \geq 0$ a.e.. Since $\lambda u - \Delta u = \lambda v - \Delta v$ in $\mathcal{D}(\Omega_2)'$, it follows from Lemma 1.2 that $u \leq v$ a.e..
b) Let $0 \leq f \in L^2(\Omega_1)$. Then it follows from a) that $(I - t\Delta_2^{\Omega_1})^{-1} f \leq (I - t\Delta_2^{\Omega_2})^{-1} f$ for all $t > 0$. Hence $e^{t\Delta_2^{\Omega_1}} f = \lim_{n\to\infty} (I - \frac{t}{n}\Delta_2^{\Omega_1})^{-n} f \leq \lim_{n\to\infty} (I - \frac{t}{n}\Delta_2^{\Omega_2})^{-n} f = e^{t\Delta_2^{\Omega_2}} f$. □

It follows from Proposition 1.1 that

$$0 \leq e^{t\Delta_2^{\Omega}} f \leq G(t) f \quad (t \geq 0)\,, \tag{1.4}$$

$$0 \leq R(\lambda, \Delta_2^{\Omega}) f \leq R(\lambda, \Delta_2) f \quad (\lambda > 0) \tag{1.5}$$

for all $f \in L^2(\Omega)_+$.

Since the Gaussian semigroup is a contractive C_0-semigroup on $L^p(\mathbb{R}^N)$, it follows that there exist positive contraction C_0-semigroups $(e^{t\Delta_p^{\Omega}})_{t\geq 0}$ on $L^p(\Omega)$ such that

$$e^{t\Delta_p^{\Omega}} f = e^{t\Delta_p^{\Omega}} f \quad (t \geq 0)$$

for all $1 \leq p,\ q < \infty$, $f \in L^p(\Omega) \cap L^q(\Omega)$. Moreover,

$$0 \leq R(\lambda, \Delta_p^{\Omega}) f \leq R(\lambda, \Delta_p) f \tag{1.6}$$

for all $f \in L^p(\Omega)_+$, $\lambda > 0$, $p \in [1, \infty)$. Finally, defining Δ_∞^{Ω} as the adjoint of Δ_1^{Ω}, we have consistency in the sense that

$$R(\lambda, \Delta_p^{\Omega}) f = R(\lambda, \Delta_q^{\Omega}) f \quad (f \in L^p(\Omega) \cap L^q(\Omega))$$

for $\lambda > 0$, $1 \leq p, q \leq \infty$. Moreover, for each $p \in [1, \infty]$, one has $\Delta_p^{\Omega} f = \Delta f$ in $\mathcal{D}(\Omega)'$. It has been proved by Ouhabaz [Ou1] that Gaussian estimates (and in particular (1.4)) imply that the semigroup generated by Δ_1 is holomorphic of angle

$\frac{\pi}{2}$ (see also [AE]). Denoting by Δ_∞ the adjoint of Δ_1 we obtain in particular that $\sigma(\Delta_\infty) \subset (-\infty, 0]$ and a bound

$$\|\lambda R(\lambda, \Delta_\infty)\| \leq M_\theta \tag{1.7}$$

for all $\lambda \in \Sigma(\theta) = \{re^{i\alpha},\ r > 0,\ |\alpha| < \theta\}$ whenever $\theta \in [0, \pi)$. This estimate will be used later.

Next we establish some spectral properties. Let E, F be Banach spaces such that $F \hookrightarrow E$ (i.e. F is continuously injected into E). Let A be an operator on E. The **part** B of A in F is defined by

$$D(B) = \{x \in D(A) \cap F : Ax \in F\} \quad Bx = Ax\,.$$

This notation is motivated by the following observation. Let $\lambda \in \varrho(A)$. Then

$$\lambda \in \varrho(B) \ \Leftrightarrow\ R(\lambda, A)F \subset F\,, \tag{1.8}$$

and in this case, $R(\lambda, B) = R(\lambda, A)_{|F}$. A proof of the following easy result can be found in [ANS].

Proposition 1.3. *Assume that $\varrho(A) \neq \emptyset$. If there exists $k \in \mathbb{N}$ such that $D(A^k) \subset F$, then $\sigma(A) = \sigma(B)$.*

Now by (1.6) we have

$$0 \leq R(1, \Delta_p^\Omega)^k \leq R(1, \Delta_p)^k \tag{1.9}$$

for $1 \leq p < \infty$, $k \in \mathbb{N}$. Since $D(\Delta_2^k) = H^{2k}(\mathbb{R}^N) \subset L^\infty(\mathbb{R}^N)$ for $k > \frac{N}{4}$ we conclude that

$$D((\Delta_p^\Omega)^k) \subset L^\infty(\Omega) \quad \text{for} \quad k > \frac{N}{4},\ 2 \leq p \leq \infty\,. \tag{1.10}$$

If Ω has bounded Lebesgue measure, it follows from Proposition 1.3 that $\sigma(\Delta_p^\Omega) = \sigma(\Delta_\infty^\Omega)$ for $p \in [2, \infty]$. By duality and selfadjointness we conclude

Proposition 1.4. *Let $\Omega \subset \mathbb{R}^N$ be open and of finite measure. Then*

$$\sigma(\Delta_p^\Omega) = \sigma(\Delta_2^\Omega) \quad (1 \leq p \leq \infty)\,. \tag{1.11}$$

Moreover, $R(\lambda, \Delta_p^\Omega) = R(\lambda, \Delta_q^\Omega)_{|L^q}$ for all $\lambda \in \sigma(\Delta_p^\Omega)$, $\infty \geq p \geq q \geq 1$.

If Ω is arbitrary, then more elaborate arguments are needed, but (1.11) still remains true (see [Ar]). We also note that

$$R(1, \Delta_p)L^p(\mathbb{R}^N) \subset L^\infty(\mathbb{R}^N)$$

for $p > \frac{N}{2}$. Hence

$$D(\Delta_p^\Omega) \subset L^\infty(\Omega) \tag{1.12}$$

for $\frac{N}{2} < p \leq \infty$. Finally we recall the following result by de Pagter [dP].

Proposition 1.5. *Let A be the generator of a C_0-semigroup on a Banach space E. Assume that $R(\lambda, A)^k$ is compact for some $k \in \mathbb{N}$, $\lambda \in \varrho(A)$. Then A has compact resolvent.*

2. Bounded domains

Let $\Omega \subset \mathbb{R}^N$ be a bounded, open, nonempty set and $C_0(\Omega) = \{u \in C(\bar{\Omega}) : u_{|\partial\Omega} = 0\}$. By Δ_0^Ω we denote the Laplacian on $C_0(\Omega)$ with maximal domain; i.e.,

$$D(\Delta_0^\Omega) = \{u \in C_0(\Omega) : \Delta u \in C_0(\Omega)\} \quad \Delta_0^\Omega u = \Delta u\,.$$

Here, for $u \in L^1_{\text{loc}}(\Omega)$ we denote by $\Delta u \in \mathcal{D}(\Omega)'$ the distributional Laplacian of u. Note that $D(\Delta_0^\Omega) \not\subset C^2(\Omega)$ as is well known (cf. [DL], II. § 3 Remark 5). We recall the following easy local properties of the Laplacian (see e.g., [DL], II. § 3 Prop. 6 p. 336).

Lemma 2.1. *Let $u \in \mathcal{D}(\Omega)'$.*

a) *If $\Delta u \in L^p_{\text{loc}}(\Omega)$, $p > \frac{N}{2}$, then $u \in C(\Omega)$;*
b) *if $\Delta u \in L^p_{\text{loc}}(\Omega)$, $p > N$, then $u \in C^1(\Omega)$.*

In particular,

$$D(\Delta_0^\Omega) \subset C^1(\Omega)\,. \tag{2.1}$$

Next we show that Δ_0^Ω is the part of Δ_2^Ω in $C_0(\Omega)$.

Lemma 2.2. *a) Let $u \in C_0(\Omega)$ such that $\Delta u \in L^p(\Omega)$ where $p > N$. Then $u \in H_0^1(\Omega)$.*
b) In particular, the operator Δ_0^Ω is the part of Δ_2^Ω in $C_0(\Omega)$.

Proof. Let $u \in C_0(\Omega)$ such that $\Delta u = f \in L^p(\Omega)$ where $p > N$. We can assume that u is real-valued. Let $\varepsilon > 0$. Then $(u-\varepsilon)^+$ has compact support. Let $\omega \subset \Omega$ be open such that $\bar{\omega} \subset \Omega$ and $\operatorname{supp}(u-\varepsilon)^+ \subset \omega$. Since $u \in C^1(\Omega)$ (by (2.1)), we have $(u-\varepsilon)^+ \in H_0^1(\omega)$. By hypothesis, we have

$$\int \nabla u \nabla \varphi = \int f\varphi \quad \text{for all} \quad \varphi \in \mathcal{D}(\Omega)\,,$$

hence also for all $\varphi \in H_0^1(\omega)$ by density. Taking $\varphi = (u-\varepsilon)^+$ we obtain

$$\begin{aligned} \int (\nabla(u-\varepsilon)^+)^2 &= \int \nabla u \nabla (u-\varepsilon)^+ = \int f(u-\varepsilon)^+ \\ &\leq \|f\|_p \|u\|_q, \quad \text{where} \quad \frac{1}{p} + \frac{1}{q} = 1\,. \end{aligned}$$

Thus $\{(u-\varepsilon)^+ : 0 < \varepsilon \leq 1\}$ is bounded in $H_0^1(\Omega)$. Since $H_0^1(\Omega)$ is reflexive, we find a sequence $\varepsilon_n \downarrow 0 \quad (n \to \infty)$ such that $(u-\varepsilon_n)^+$ converges weakly in $H_0^1(\Omega)$, to $v \in H_0^1(\Omega)$, say. Then $(u-\varepsilon_n)^+$ converges weakly to v in $L^2(\Omega) \quad (n \to \infty)$. Since $(u-\varepsilon_n)^+ \to u^+$ in $L^2(\Omega)$, it follows that $u^+ = v \in H_0^1(\Omega)$. In the same way one sees that $u^- \in H_0^1(\Omega)$. □

We say that the **Dirichlet problem is well-posed** in Ω if for every $\varphi \in C(\partial\Omega)$ there exists a solution u of $(D(\varphi))$ (see Introduction). It is known that the solution is unique and in $C^\infty(\Omega)$. Moreover, the Dirichlet problem is well-posed whenever $D(\varphi)$ has a solution for all φ in a dense subspace of $C(\partial\Omega)$.

Theorem 2.3. *Let Ω be a bounded open set in $\mathbb{R}^N$ such the Dirichlet problem is well-posed. Assume that Ω is regular. Then Δ_0^Ω generates a holomorphic C_0-semigroup $T_0 = (T_0(t))_{t\geq 0}$ of angle $\frac{\pi}{2}$ on $C_0(\Omega)$.*

Moreover, $T_0(t)$ is compact for all $t > 0$ and $\sigma(\Delta_0^\Omega) = \sigma(\Delta_2^\Omega)$.

Proof. a) We show that $R(0, \Delta_2^\Omega)L^\infty(\Omega) \subset C_0(\Omega)$. Let $f \in L^\infty(\Omega)$. Let $v = E_N * f \in C(\mathbb{R}^N)$, where E_N is the Newtonian potential. Then $\Delta v = f$ in $\mathcal{D}(\Omega)'$. Let $\varphi = v_{|\partial\Omega}$. By hypothesis there exists $w \in C(\bar{\Omega})$ such that $w_{|\partial\Omega} = \varphi$ and $\Delta w = 0$ in $\mathcal{D}(\Omega)'$. Thus $u = w - v \in C_0(\Omega)$ and $-\Delta u = f$ in $\mathcal{D}(\Omega)'$. It follows from Lemma 2.2 that $u \in H_0^1(\Omega)$. Thus $R(0, \Delta_2^\Omega)f = u \in C_0(\Omega)$.
b) It follows from (1.8) that $0 \in \varrho(\Delta_0^\Omega)$. Moreover, $\varrho(\Delta_0^\Omega) = \varrho(\Delta_\infty^\Omega)$ by Proposition 1.3. By (1.8) again, we have $R(\lambda, \Delta_0^\Omega) = R(\lambda, \Delta_\infty^\Omega)_{|C_0(\Omega)}$ for all $\lambda \in \varrho(\Delta_\infty^\Omega)$. Now it follows from (1.7) that Δ_0^Ω generates a bounded holomorphic C_0-semigroup. Note that $D(\Delta_0^\Omega)$ is dense in $C_0(\Omega)$ since $\mathcal{D}(\Omega) \subset D(\Delta_0^\Omega)$.
c) We show that Δ_0 has compact resolvent. By (1.10) we have $R(0, \Delta_2)^k L^2(\Omega) \subset L^\infty(\Omega)$ where $k > \frac{N}{4}$. Thus by a) $R(0, \Delta_2)^{k+1} L^2(\Omega) \subset C_0(\Omega)$. Note that $R(0, \Delta_2) \in \mathcal{L}(L^2(\Omega))$ is compact. Thus we can write $R(0, \Delta_0)^{k+2} = R(0, \Delta_2)^{k+1} \circ R(0, \Delta_2) \circ j$ where $j : C_0(\Omega) \to L^2(\Omega)$ is the canonial embedding. It follows that $R(0, \Delta_0)^{k+2}$ is compact. Consequently, $R(0, \Delta_0)$ is compact by Proposition 1.5.
d) Finally, from Proposition 1.3 one sees that $\sigma(\Delta_0^\Omega) = \sigma(\Delta_2^\Omega)$. □

Let $C^b(\Omega)$ be the space of all bounded continuous scalar-valued functions on Ω with supremum norm. Denote by Δ_b the part of Δ_2 in $C^b(\Omega)$.
Since $R(0, \Delta_2)C^b(\Omega) \subset C^b(\Omega)$ is follows from Proposition 1.3 and 1.4 that

$$\sigma(\Delta_b) = \sigma(\Delta_2) = \sigma(\Delta_p) \tag{2.2}$$

for $1 \leq p \leq \infty$. Moreover, $R(\lambda, \Delta_b) = R(\lambda, \Delta_2)_{|C^b(\Omega)}$ for all $\lambda \in \varrho(\Delta_2)$.

We now consider necessary conditions for well-posedness of the Dirichlet problem.

Theorem 2.4. *Let $\Omega \subset \mathbb{R}^N$ be bounded and open. The following are equivalent:*

(i) *the Dirichlet problem is well-posed;*
(ii) *$\varrho(\Delta_0^\Omega) \neq \emptyset$;*
(iii) *$D(\Delta_p^\Omega) \subset C_0(\Omega)$ for all $p \in (\frac{N}{2}, \infty]$;*
(iv) *there exist $\frac{N}{2} < p \leq \infty$, $\lambda \in \varrho(\Delta_p^\Omega)$ and $f \in L^p(\Omega)$ such that $f(x) > 0$ a.e. and $R(\lambda, \Delta_p^\Omega)f \in C_0(\Omega)$.*

Proof. *(i)* ⇒ *(ii)* follows from Theorem 2.3.
(ii) ⇒ *(iii)* Since $\varrho(\Delta_b^\Omega)$ has non-empty interior, it follows from the hypothesis that $U := \varrho(\Delta_0^\Omega) \cap \varrho(\Delta_b^\Omega) \neq \emptyset$. The set U is open and relatively closed in $\varrho(\Delta_b^\Omega)$. In fact, let $\lambda \in \varrho(\Delta_b^\Omega)$ such that $\lambda = \lim_{n\to\infty} \lambda_n \in U$. Then $(R(\lambda_n, \Delta_b^\Omega))_{n\in\mathbb{N}}$ is bounded in $\mathcal{L}(C_b(\Omega))$, thus also $(R(\lambda_n, \Delta_0^\Omega))_{n\in\mathbb{N}}$ is bounded in $\mathcal{L}(C_0(\Omega))$. This implies that $\lambda \in \varrho(\Delta_0^\Omega)$. Since $\varrho(\Delta_b^\Omega) = \varrho(\Delta_2^\Omega)$ is connected, it follows that $\varrho(\Delta_b^\Omega) \subset \varrho(\Delta_0^\Omega)$. In particular, $0 \in \varrho(\Delta_0^\Omega)$. It follows from (1.8) that $R(0, \Delta_b^\Omega)C_0(\Omega) \subset C_0(\Omega)$. Let

$\frac{N}{2} < p \leq \infty$. Then $R(0, \Delta_p^\Omega)$ is a bounded operator from $L^p(\Omega)$ into $L^\infty(\Omega)$ (by (1.12)). Since $R(0, \Delta_p^\Omega) C_0(\Omega) \subset C_0(\Omega)$ and since $C_0(\Omega)$ is closed in $L^\infty(\Omega)$, it follows that $R(0, \Delta_p^\Omega) L^p(\Omega) \subset C_0(\Omega)$.
$(iii) \Rightarrow (i)$ We show that the Dirichlet problem has a solution for all $\varphi \in C(\partial\Omega)$. At first, assume that $\varphi = w_{|\partial\Omega}$ for some $w \in C^2(\bar{\Omega})$. Let $v = -R(0, \Delta_p^\Omega)\Delta w$. Then $v \in C_0(\Omega)$ and $\Delta v = \Delta w$ in $\mathcal{D}(\Omega)'$. Thus $u = w - v \in C(\bar{\Omega})$, $u_{|\partial\Omega} = \varphi$ and $\Delta u = 0$ in $\mathcal{D}(\Omega)'$; i.e. u is a solution of $D(\varphi)$. Since the set $\{w_{|\partial\Omega} : w \in C^2(\bar{\Omega})\}$ is dense in $C(\partial\Omega)$, it follows that $D(\varphi)$ has a (unique) solution for all $\varphi \in C(\partial\Omega)$.
$(iii) \Rightarrow (iv)$ is obvious.
$(iv) \Rightarrow (ii)$ We can assume that $N/2 < p < \infty$ (since $L^\infty(\Omega) \subset L^p(\Omega)$). Note that by (1.12) and Lemma 2.1, $R(\lambda, \Delta_p^\Omega)$ is a bounded operator from $L^p(\Omega)$ into $C^b(\Omega)$. Let $F = \{g \in L^p(\Omega) : \exists c \geq 0,\ |g| \leq cf\}$. Then the hypothesis implies that $R(\lambda, \Delta_p^\Omega) F \subset C_0(\Omega)$. Since F is dense in $L^p(\Omega)$, it follows that $R(\lambda, \Delta_p^\Omega) L^p(\Omega) \subset C_0(\Omega)$. In particular, $R(\lambda, \Delta_p^\Omega) C_0(\Omega) \subset C_0(\Omega)$. Now it follows from (1.8) that $\lambda \in \varrho(\Delta_0)$. Since Δ_0 is the part of Δ_p in $C_0(\Omega)$. □

Corollary 2.5. *Let Ω be a bounded, open, non-empty, connected subset of $\mathbb{R}^N$. Then Ω is regular if and only if the eigenfunction u_1 associated with the first eigenvalue of Δ_2^Ω is in $C_0(\Omega)$.*

We conclude this section by a remark concerning the realization of Dirichlet boundary conditions in $L^2(\Omega)$. There is another choice, namely to replace $H_0^1(\Omega)$ by

$$\tilde{H}_0^1(\Omega) := \{u \in L^2(\Omega),\ \tilde{u} \in H^1(\mathbb{R}^N)\} \quad \text{where} \quad \tilde{u}(x) = \begin{cases} u(x) & x \in \Omega \\ 0 & x \notin \Omega\,. \end{cases}$$

The operator associated with the form

$$a(u, v) = \int_\Omega \nabla u \nabla v$$

on $\tilde{H}_0^1(\Omega)$ is called the **pseudo-Dirichlet Laplacian** in [AB] and is denoted by $\tilde{\Delta}_\Omega$. Thus $D(\tilde{\Delta}_\Omega) = \{u \in \tilde{H}_0^1(\Omega) : \exists v \in L^2(\Omega)$ such that $\int_\Omega \nabla u \nabla \varphi = \int v\varphi \quad \forall \varphi \in \tilde{H}_0^1(\Omega)\}$ and $\tilde{\Delta}_\Omega u = v$. Equivalently, $D(\tilde{\Delta}_\Omega) = \{u \in \tilde{H}_0^1(\Omega) : \Delta u \in L^2(\Omega)\}$, $\tilde{\Delta}_\Omega u = \Delta u$. For example, if $N = 1$, $\Omega = (0, 1) \cup (1, 2)$, then $\tilde{\Delta}_\Omega \neq \Delta_2^\Omega$, but if Ω has Lipschitz boundary then $\tilde{\Delta}_\Omega = \Delta_2^\Omega$.

Now assume that the semigroup $(e^{t\tilde{\Delta}_\Omega})_{t\geq 0}$ generated by $\tilde{\Delta}_\Omega$ on $L^2(\Omega)$ leaves invariant $C_0(\Omega)$. Hence $R(1, \tilde{\Delta}_\Omega) C_0(\Omega) \subset C_0(\Omega)$. But then Lemma 2.2 implies that $R(1, \tilde{\Delta}_\Omega) C_0(\Omega) \subset H_0^1(\Omega)$. Thus $R(1, \tilde{\Delta}_\Omega) f = R(1, \Delta_2^\Omega) f$ for $f \in C_0(\Omega)$. It follows that $\tilde{\Delta}_\Omega = \Delta_2^\Omega$; i.e. $H_0^1(\Omega) = \tilde{H}_0^1(\Omega)$. Concluding we have the following result.

Proposition 2.6. *Let $\Omega \subset \mathbb{R}^N$ be a bounded open set. Assume that*

$$e^{t\tilde{\Delta}_\Omega} C_0(\Omega) \subset C_0(\Omega) \quad (t > 0)\,.$$

Then Ω is regular and $\tilde{\Delta}_\Omega = \Delta_2^\Omega$.

3. Unbounded open sets

Whereas in the preceding paragraph, for bounded open sets, we established an equivalence with well-posedness of the Dirichlet problem, in this section it will be more convenient to consider barriers. Let $\Omega \subset \mathbb{R}^N$ be open.

Definition 3.1. *a) Let $z \in \partial\Omega$. A* **barrier** *is a function $w \in C(\overline{\Omega \cap B})$ such that*

$$\Delta w \leq 0 \quad in \quad \mathcal{D}(\Omega \cap B)' \quad and \quad w(z) = 0,\ w(x) > 0 \quad for \quad x \in (\Omega \cap B) \setminus \{z\}$$

where $B = B(z, r)$ is a ball centered at z. We call w an $\mathbf{H^1}$**−barrier** *if in addition $w \in H^1(\Omega \cap B)$.*
b) We say that Ω is **regular**, *if at each boundary point $z \in \partial\Omega$ there exists a barrier.*

It is well known that a bounded open set Ω is regular if and only if the Dirichlet problem is well-posed (see [GT], § 2.8 or [DL], II). For unbounded Ω the situation is more complicated since the behaviour at infinity has to be taken into account (see [DL], II § 4). Still, the existence of a barrier at each boundary point is the right regularity property in order that Δ_0^Ω be a generator. This is the assertion of Theorem 0.1 which will be proved below. We first show that Δ_0^Ω is dispersive. This is not new (cf. [LP1], [LP2]). In the bounded case, it follows immediately from our arguments since Δ_0^Ω is the part of Δ_∞^Ω in $C_0(\Omega)$. We include a proof in the unbounded case for convenience of the reader. Recall that an operator A defined on the real space $C_0(\Omega)$ is called **dispersive** if for every $u \in D(A)$ such that $u^+ \neq 0$ there exists $x_0 \in \Omega$ such that

$$u^+(x_0) = \|u^+\|_\infty \quad \text{and} \quad (Au)(x_0) \leq 0\,. \tag{3.1}$$

Dispersiveness on $C_0(\Omega)$ implies dissipativity (as is easy to see). In particular, $(\lambda - A)$ is injective for $\lambda > 0$ whenever A is dispersive. This will be used in Theorem 3.7 below. More generally, a densely defined operator A generates a positive contractive C_0-semigroup if and only if A is dispersive and $I - A$ is surjective (see [N], C-II.Theorem 1.2).

Proposition 3.2. *The operator Δ_0 is closed and dispersive.*

Proof. It is obvious that Δ_0 is closed. Let $u \in D(\Delta_0)$, such that $u^+ \neq 0$. Let $(\varrho_n)_{n\in\mathbb{N}} \subset C^\infty(\mathbb{R}^N)$ be an approximate unit; i.e. $\varrho_n \geq 0$, $\int \varrho_n(x)dx = 1$, $\operatorname{supp} \varrho_n \subset B(0, \frac{1}{n})$. Let $u_n = \varrho_n * u$. Then $u_n \in C_0(\mathbb{R}^N)$ and $\lim\limits_{n\to\infty} u_n = u$ in $C_0(\Omega)$. Let $n_0 \in \mathbb{N}$ such that for all $n \geq n_0$, $\|u_n - u\|_\infty \leq \frac{\delta}{4}$, $\delta = \|u^+\|_\infty$. The set $K = \{x \in \Omega : u(x) \geq \frac{\delta}{2}\}$ is compact. Let $x_n \in \Omega$ such that $u_n(x_n) = \max\limits_{y\in\Omega} u_n(y)$. Then $u_n(x_n) \geq \frac{3}{4}\delta$ and so $u(x_n) \geq \frac{\delta}{2}$ for $n \geq n_0$. Hence $x_n \in K \quad (n \geq n_0)$. Since K is compact, we can assume that $x_0 = \lim\limits_{n\to\infty} x_n$ exists. Then $u(x_0) = \|u^+\|_\infty$. Since $u_n \in C^\infty(\Omega)$ we have $(\varrho_n * \Delta u)(x_n) = \Delta u_n(x_n) \leq 0$. Since $\varrho_n * \Delta u$ converges to u uniformly, it follows that $(\Delta u)(x_0) \leq 0$. □

Remark 3.3. *More generally the following* **maximum principle** *holds, which clearly implies dispersiveness: If* $u \in C(\Omega)$ *has a local maximum at* $x \in \Omega$ *and* $\Delta u \in C(\Omega)$, *then* $(\Delta u)(x) \leq 0$.

We deduce from Proposition 3.2 that the operator Δ_0^Ω is the only possible realization of the Laplacian in $C_0(\Omega)$ which might generate a semigroup.

Corollary 3.4. *Let* $A \subset \Delta_0^\Omega$ *be the generator of a* C_0*-semigroup on* $C_0(\Omega)$. *Then* $A = \Delta_0^\Omega$.

Proof. There exists $\lambda > 0$ such that $(\lambda - A)$ is surjective. Let $u \in D(\Delta_0^\Omega)$. There exists $v \in D(A)$ such that $(\lambda - A)v = (\lambda - \Delta_0^\Omega)u$. Since $Av = \Delta_0^\Omega v$ it follows that $u = v \in D(A)$. □

For the proof of Theorem 0.1 in the unbounded case we need some further preparation.

Lemma 3.5. *Let* Ω_1, Ω_2 *be regular, open subsets of* $\mathbb{R}^N$. *Then* $\Omega_1 \cap \Omega_2$ *is regular.*

Proof. Let $\Omega = \Omega_1 \cap \Omega_2$. Then $\partial\Omega \subset \partial\Omega_1 \cup \partial\Omega_2$. Let $z \in \partial\Omega$. Suppose that $z \in \partial\Omega_1$. Then there exists a barrier $w \in C(\overline{\Omega_1 \cap B})$ where $B = B(z,r)$. Then $w_{|\overline{\Omega \cap B}}$ is clearly a barrier on $\overline{\Omega \cap B}$ at z. □

We deduce from Lemma 3.5 the following

Lemma 3.6. *Let* Ω *be a regular open subset of* $\mathbb{R}^N$. *Then for each* $z \in \partial\Omega$ *there exists an* H^1*-barrier at* z.

Proof. It follows from Lemma 3.5 that $\Omega \cap B$ is regular, where $B = B(z,r)$, $z \in \partial\Omega$, $r > 0$. Thus the Dirichlet problem is well-posed on $\Omega \cap B$. Let $\varphi(x) = |x - z|^2$. Then there exists $v \in C(\overline{\Omega \cap B})$ such that $v(x) = \varphi(x)$ for all $x \in \partial(\Omega \cap B)$ and $\Delta v = 0$ in $\mathcal{D}(\Omega \cap B)'$. It follows from the maximum principle that $v(x) > 0$ for all $x \in B \cap \Omega$. We show that $v \in H^1(\Omega \cap B)$. Note that $\varphi \in C^2(\overline{\Omega \cap B})$ and $\Delta\varphi = \text{const}$. Let $u = v - \varphi$. Then $u \in C_0(\Omega \cap B)$ and $\Delta u = -\Delta\varphi$ in $\mathcal{D}(\Omega \cap B)'$. It follows from Lemma 2.2 that $u \in H_0^1(\Omega \cap B)$. Thus $v = u + \varphi \in H^1(\Omega \cap B)$. □

We recall the following simple fact which is easy to proof (cf. [B], IX).

Lemma 3.7. *Let* $u \in H_0^1(\Omega) \cap L^\infty(\Omega)$. *Then there exist* $u_n \in \mathcal{D}(\Omega)$ *such that* $\|u_n\|_\infty \leq \|u\|_\infty \quad (n \in \mathbb{N})$ *and* $\lim_{n\to\infty} u_n = u$ *in* $H_0^1(\Omega)$.

Moreover, we recall that for a bounded open set $\Omega_1 \subset \mathbb{R}^N$ one has

$$C_0(\Omega_1) \cap H^1(\Omega_1) \subset H_0^1(\Omega_1) \tag{3.2}$$

(see [B], Remark 20, p. 172 or [Da], Theorem 1.5.7).

Theorem 3.8. *Let* $\Omega \subset \mathbb{R}^N$ *be a regular, open set. Then* Δ_0^Ω *generates a holomorphic* C_0*-semigroup on* $C_0(\Omega)$. *Moreover,* Δ_0^Ω *is the part of* Δ_∞^Ω *in* $C_0(\Omega)$ *and* $\sigma(\Delta_p^\Omega) = \sigma(\Delta_0^\Omega)$ *for* $1 \leq p \leq \infty$.

Proof. a) Let $f \in C_0(\Omega) \cap L^2(\Omega)$, $u = R(1, \Delta_2^\Omega)f$, i.e. $u \in H_0^1(\Omega)$, $u - \Delta u = f$ in $\mathcal{D}(\Omega)'$. We show that $u \in C_0(\Omega)$. It follows from local regularity that u is continuous and bounded.

1. Let $z \in \partial\Omega$. We show that $\lim_{x \to z} u(x) = 0$. Let $\varepsilon > 0$. Choose $w \in C^1(\mathbb{R}^N)$ such that $\Delta w = u - f$ in $\mathcal{D}(\Omega)'$ and $w(z) = \varepsilon$ (one can take $w = E_N * (u - f) + \text{const}$). Let $v \in C(\overline{B \cap \Omega}) \cap H^1(\Omega \cap B)$ be an H^1-barrier where $B = B(z, r)$ is so small that $w \geq 0$ on $\bar{B}$. Multiplying v with a positive constant if necessary, we can assume that $v(x) > \|u\|_\infty$ on $\partial B \cap \partial(\Omega \cap B)$. Then $u - v - w \in H^1(\Omega \cap B)$ and $(u - v - w)^+ \in H_0^1(\Omega \cap B)$. To see the last point, choose $u_n \in \mathcal{D}(\Omega)$ such that $\|u_n\|_\infty \leq \|u\|_\infty$ and $u_n \to u$ in $H^1(\Omega)$ (by Lemma 3.7). Then $(u_n - v - w)^+ \to (u - v - w)^+$ in $H^1(\Omega \cap B)$. Note that $(u_n - v - w)^+ \in C(\overline{\Omega \cap B})$ vanishes on $\partial(\Omega \cap B) = (\partial(\Omega \cap B) \cap \partial\Omega) \cup (\partial(\Omega \cap B) \cap \partial B)$. Thus $(u_n - v - w)^+ \in H_0^1(\Omega \cap B)$ by (3.2). Thus $(u - v - w)^+ \in H_0^1(\Omega \cap B)$ and the claim is proved.

Now $\Delta(u - v - w) = -\Delta v \geq 0$ in $\mathcal{D}(\Omega \cap B)'$. Since $u - v - w \in H^1(\Omega)$, it follows that $\int \nabla(u - v - w)\nabla\varphi \leq 0$ for all $\varphi \in \mathcal{D}(\Omega \cap B)_+$ and hence for all $\varphi \in H_0^1(\Omega \cap B)_+$. Taking $\varphi = (u - v - w)^+$, we deduce that $\int |\nabla(u - v - w)^+|^2 \leq 0$. Hence $(u - v - w)^+ = 0$; i.e., $u \leq w + v$. Thus $\overline{\lim_{x \to z}} u(x) \leq \overline{\lim_{x \to z}} (v(x) + w(x)) = \varepsilon$. Since $\varepsilon > 0$ is arbitrary, it follows that $\overline{\lim_{x \to z}} u(x) \leq 0$. Replacing u by $-u$ we obtain $\lim_{\substack{x \to z \\ x \in \partial\Omega}} u(x) = 0$.

2. Since $|u| \leq R(1, \Delta_2)|f|$ (by (1.5)) and $R(1, \Delta_2)C_0(\mathbb{R}^N) \subset C_0(\mathbb{R}^N)$, it follows that $\lim_{|x| \to \infty} |u(x)| = 0$. Thus a) is proved.

b) Recall that $R(1, \Delta_2^\Omega)$ and $R(1, \Delta_\infty^\Omega)$ are consistent. It follows from a) and by density that $R(1, \Delta_\infty^\Omega)C_0(\Omega) \subset C_0(\Omega)$. We show that $1 \in \varrho(\Delta_0^\Omega)$ and $R(1, \Delta_0^\Omega) = R(1, \Delta_\infty^\Omega)_{|C_0(\Omega)}$. Let $f \in C_0(\Omega)$. Then $u = R(1, \Delta_\infty^\Omega)f \in C_0(\Omega)$ and $u - \Delta u = f$ in $\mathcal{D}(\Omega)'$. (In fact, for $\varphi \in \mathcal{D}(\Omega)$ we have $\langle u - \Delta u,\ \varphi\rangle = \langle u,\ \varphi - \Delta\varphi\rangle = \langle R(1, \Delta_\infty^\Omega)f,\ \varphi - \Delta\varphi\rangle = \langle f,\ R(1, \Delta_1^\Omega)(\varphi - \Delta\varphi)\rangle = \langle f, \varphi\rangle$). Thus $u \in D(\Delta_0)$ and $u - \Delta_0 u = f$. Conversely, let $u \in \mathcal{D}(\Delta_0^\Omega)$ and $f = u - \Delta u$. Let $v = R(1, \Delta_\infty^\Omega)f$. Then $w = u - v \in \mathcal{D}(\Delta_0)$ and $w - \Delta_0 w = 0$. Since Δ_0 is dissipative, it follows that $w = 0$. We have shown that $1 \in \varrho(\Delta_0^\Omega)$ and $R(1, \Delta_0^\Omega) = R(1, \Delta_\infty^\Omega)_{|C_0(\Omega)}$. Thus Δ_0^Ω is a generator and Δ_0^Ω is the part of Δ_∞^Ω in $C_0(\Omega)$.

c) It is clear from analyticity of the resolvent, that the set $U = \{\lambda \in \varrho(\Delta_\infty^\Omega) : R(\lambda, \Delta_\infty^\Omega)C_0(\Omega) \subset C_0(\Omega)\}$ is open and closed in $\varrho(\Delta_\infty^\Omega)$. Since $\varrho(\Delta_\infty^\Omega)$ is connected it follows that $U = \varrho(\Delta_\infty^\Omega)$; i.e. $\varrho(\Delta_\infty^\Omega) \subset \varrho(\Delta_0^\Omega)$. We have shown that $\sigma(\Delta_0^\Omega) \subset \sigma(\Delta_\infty^\Omega) = \sigma(\Delta_2^\Omega)$. In order to show the converse denote by B the adjoint of Δ_0^Ω on $C_0(\Omega)'$. Let $\lambda > 0$, $f \in L^1(\Omega) \cap L^2(\Omega)$. Then for $\varphi \in \mathcal{D}(\Omega)$, $\langle \varphi,\ R(\lambda, B)f\rangle = \langle R(\lambda, \Delta_0^\Omega)\varphi,\ f\rangle = \langle \varphi,\ R(\lambda, \Delta_2^\Omega)f\rangle = \langle \varphi,\ R(\lambda, \Delta_1^\Omega)f\rangle$. Hence $R(\lambda, B)L^1(\Omega) \subset L^1(\Omega)$ and $R(\lambda, B)_{|L^1(\Omega)} = R(\lambda, \Delta_1^\Omega)$. As before, it follows that $\varrho(\Delta_0^\Omega) = \varrho(B) \subset \varrho(\Delta_1^\Omega) = \varrho(\Delta_2^\Omega)$. Thus $\sigma(\Delta_2^\Omega) \subset \sigma(\Delta_0^\Omega)$. □

Of course, if Ω is regular, the semigroup $(e^{t\Delta_0^\Omega})_{t\geq 0}$ generated by Δ_0^Ω is consistent with $(e^{t\Delta_p^\Omega})_{t\geq 0}$ i.e.

$$e^{t\Delta_0^\Omega} f = e^{t\Delta_p^\Omega} f \quad (f \in C_0(\Omega)\cap L^p(\Omega),\ t\geq 0,\ 1\leq p\leq\infty). \tag{3.3}$$

This follows from consistency of the resolvents.

Next we show that Δ_0^Ω has compact resolvent whenever Ω is regular and has finite Lebesgue measure. This follows from the following more general result.

Proposition 3.9. *Let Ω be a regular open subset of $\mathbb{R}^N$ with finite Lebesgue measure. Then $e^{t\Delta_0^\Omega}$ is a compact operator on $C_0(\Omega)$ for all $t>0$.*

Proof. Let $t>0$. Then $e^{t\Delta_2^\Omega}$ is a bounded operator from $L^2(\Omega)$ into $L^\infty(\Omega)$. We have seen that $e^{t\Delta_2^\Omega}C_0(\Omega)\subset C_0(\Omega)$. Hence by density $e^{t\Delta_2^\Omega}L^2(\Omega)\subset C_0(\Omega)$. Denote by $j: C_0(\Omega)\to L^2(\Omega)$ the canonical injection. Then $e^{t\Delta_0^\Omega}$ can be written as

$$e^{\frac{t}{2}\Delta_2^\Omega}\circ e^{\frac{t}{2}\Delta_2^\Omega}\circ j : C_0(\Omega)\to L^2(\Omega)\to L^2(\Omega)\to C_0(\Omega).$$

Since $e^{\frac{t}{2}\Delta_2^\Omega}$ is a compact operator on $L^2(\Omega)$, the claim follows. □

In order to prove the converse of Theorem 3.1 we establish a spectral characterization of Δ_0^Ω being a generator.

Proposition 3.10. *Let $\Omega\subset\mathbb{R}^N$ be an open set. The following are equivalent:*

(i) *Δ_0^Ω is a generator;*
(ii) *$\varrho(\Delta_0^\Omega)\neq\emptyset$;*
(iii) *$R(\lambda,\Delta_2^\Omega)(L^2(\Omega)\cap C_0(\Omega))\subset C_0(\Omega)$.*

Proof. *(i) ⇒ (ii)* This is clear.
(ii) ⇒ (iii) If $\varrho(\Delta_0^\Omega)\neq\emptyset$, since $\sigma(\Delta_2^\Omega)\subset\mathbb{R}$, there exists $\lambda\in\varrho(\Delta_0^\Omega)\cap\varrho(\Delta_2^\Omega)$. Let B be the adjoint of Δ_0^Ω. We show that $R(\lambda,B)L^1(\Omega)\subset L^1(\Omega)$ and $R(\lambda,B)_{|L^1} = R(\lambda,\Delta_1^\Omega)$. Let $g\in L^1(\Omega)\cap L^2(\Omega)$, $u=R(\lambda,B)g$. Let $\varphi\in\mathcal{D}(\Omega)$. $v=R(\lambda,\Delta_0^\Omega)\varphi$. Then $v\in C_0(\Omega)$ and $\lambda v-\Delta v=\varphi$ in $\mathcal{D}(\Omega)'$. As in the proof of Lemma 2.2 one sees that $v\in H_0^1(\Omega)$. Thus $v=R(\lambda,\Delta_2^\Omega)\varphi$. Consequently, $\langle\varphi,R(\lambda,B)g\rangle = \langle R(\lambda,\Delta_0^\Omega)\varphi,g\rangle = \langle R(\lambda,\Delta_2^\Omega)\varphi,g\rangle = \langle\varphi,R(\lambda,\Delta_2^\Omega)g\rangle = \langle\varphi,R(\lambda,\Delta_1^\Omega)g\rangle$. We have shown that $R(\lambda,\Delta_1^\Omega)g = R(\lambda,B)g$ for all $g\in L^1(\Omega)\cap L^2(\Omega)$, hence for all $g\in L^1(\Omega)$. As in the proof of Theorem 3.8 we deduce that $\varrho(\Delta_1^\Omega)\subset\varrho(B)=\varrho(\Delta_0^\Omega)$ and $R(\lambda,B)_{|_{L^1(\Omega)}} = R(\lambda,\Delta_1^\Omega)$ for all $\lambda\in\varrho(\Delta_1^\Omega)$. Let $f\in C_0(\Omega)\cap L^2(\Omega)$. Then for $g\in L^1(\Omega)\cap L^2(\Omega)$, $\langle R(\lambda,\Delta_0^\Omega)f,g\rangle = \langle f,R(\lambda,B)g\rangle = \langle f,R(\lambda,\Delta_2^\Omega)g\rangle = \langle R(\lambda,\Delta_2^\Omega)f,g\rangle$. Hence $R(\lambda,\Delta_0^\Omega)f = R(\lambda,\Delta_2^\Omega)f$. Thus *(iii)* is proved.
(iii) ⇒ (i) Let $f\in L^2(\Omega)\cap C_0(\Omega)$. Then by hypothesis $u=R(\lambda,\Delta_2^\Omega)f\in C_0(\Omega)$. Thus $u\in D(\Delta_0^\Omega)$ and $u-\Delta_0 u=f$. We have shown that $\lambda-\Delta_0^\Omega$ has dense image. Since Δ_0^Ω is closed and dissipative, it follows that Δ_0^Ω is a generator. □

Corollary 3.11. *Let $\Omega_1,\Omega_2\subset\mathbb{R}^N$ be open. If $\Delta_0^{\Omega_1}$ and $\Delta_0^{\Omega_2}$ are generators, then $\Delta_0^{\Omega_1\cap\Omega_2}$ is a generator.*

Proof. Let $0 \le f \in C_0(\Omega) \cap L^2(\Omega)$, $\Omega = \Omega_1 \cap \Omega_2$. Since by Proposition 1.1 $R(\lambda, \Delta_2^\Omega)f \le R(\lambda, \Delta_2^{\Omega_1})f$ and $R(\lambda, \Delta_2^\Omega)f \le R(\lambda, \Delta_2^{\Omega_2})f$ and $\partial\Omega \subset \partial\Omega_1 \cup \partial\Omega_2$, it follows that $R(\lambda, \Delta_2^\Omega)f \in C_0(\Omega)$. □

Corollary 3.12. *Assume that $\Omega \subset \mathbb{R}^N$ is open. If Δ_0^Ω is a generator, then at each $z \in \partial\Omega$ there exists an H^1-barrier. In particular, Ω is regular.*

Proof. Let $z \in \partial\Omega$, $B = B(z,r)$, $r > 0$. By Corollary 3.11, the operator $\Delta_0^{\Omega\cap B}$ is a generator on $C_0(B \cap \Omega)$. By Theorem 2.4 this implies that the Dirichlet problem on $B \cap \Omega$ is well-posed. Taking $\varphi(x) = |x - z|^2$ on the boundary, the solution of the Dirichlet problem on $B \cap \Omega$ gives an H^1-barrier. □

Concluding, we mention that the proofs given in Section 2 and 3 are almost self-contained. In particular, for a bounded open set Ω in $\mathbb{R}^N$, we gave a complete proof of the equivalence of the following three assertions

(a) the Dirichlet problem on Ω is well-posed;
(b) at each $z \in \partial\Omega$ there exists an H^1-barrier;
(c) Δ_0^Ω is a generator.

For the fact that condition (b) can be replaces by the more general condition that Ω is regular we refer to classical potential theory. Concerning holomorphy of the semigroup generated by Δ_0 the following should be added: The case where Ω is bounded and of class C^∞ is due to Stewart [Stu]. Lumer and Paquet [LP1] (see also [LP2]) proved by a beautiful dissipativity argument that the semigroup $(e^{t\Delta_0})_{t>0}$ is holomorphic whenever Δ_0 is a generator. The duality argument we give here has the advantage to give the optimal angle. It was first used by Ouhabaz [Ou2].

4. Elliptic Operators on $\mathbf{C_0}(\Omega)$

Whereas Section 2 and 3 were self-contained, using only elementary results of potential theory, the following investigation of parabolic equations on $C_0(\Omega)$ depends on results by Stampacchia [Sta] among others who studied boundary behavior for solutions of elliptic equations. We consider merely bounded open sets generalizing the approach of Section 2.

Let $\Omega \subset \mathbb{R}^N$ be a bounded open set. We introduce elliptic operators using the notation of Gilbarg-Trudinger [GT], Chapter 8. Let $a_{ij} \in L^\infty(\Omega)$ $(i,j = 1,\dots,N)$ be real functions such that

$$\sum_{i,j=1}^N a_{ij}(x)\xi_i\xi_j \ge \alpha|\xi|^2$$

for all $\xi \in \mathbb{R}^N$, x-a.e., where $\alpha > 0$, and let $d, b_j, c_j \in L^\infty(\Omega)$ be real coefficients, $j = 1, \dots, N$. We consider the elliptic operator L, formally given by

$$Lu = \sum_{i=1}^N D_i(\sum_{j=1}^N a_{ij}D_ju + b_iu) + \sum_{i=1}^N c_iD_iu + du\,.$$

Defining the form

$$a(u,v) = \int_\Omega \{\sum_{i,j=1}^N a_{ij} D_j u D_i v + \sum_{i=1}^N (b_i u D_i v - c_i D_i u v) - duv\} dx \tag{4.1}$$

for $u \in H^1_{\text{loc}}(\Omega),\ v \in \mathcal{D}(\Omega)$, we can realize L as an operator. $L : H^1_{\text{loc}}(\Omega) \to \mathcal{D}(\Omega)'$ given by

$$\langle Lu,\ v\rangle = -a(u,v) \quad (u \in H^1_{\text{loc}}(\Omega),\ v \in \mathcal{D}(\Omega))\,. \tag{4.2}$$

We observe furthermore that

$$Lu \in (H^1_0(\Omega))' \quad \text{whenever} \quad u \in H^1(\Omega)\,. \tag{4.3}$$

We define the operator A_0 on $C_0(\Omega)$ as the part of L in $C_0(\Omega)$; i.e.

$$D(A_0) = \{u \in C_0(\Omega) \cap H^1_{\text{loc}}(\Omega) : Lu \in C_0(\Omega)\} \quad A_0 u = Lu\,.$$

In the following we will assume throughout that

$$\sum_{j=1}^N D_j b_j + d \le 0 \quad \text{in} \quad \mathcal{D}(\Omega)'\,. \tag{4.4}$$

Then the following is the main result of this section.

Theorem 4.1. *Assuming (4.4), the following are equivalent:*

(i) *Ω is regular;*
(ii) *$\varrho(A_0) \neq \emptyset$;*
(iii) *A_0 generates a positive, contractive C_0-semigroup on $C_0(\Omega)$.*

By A_2 we denote **the realization of L in $L^2(\Omega)$ with Dirichlet boundary conditions**; i.e.

$$D(A_2) = \{u \in H^1_0(\Omega) : Lu \in L^2(\Omega)\} \quad A_2 u = Lu\,.$$

Note that $-A_2$ is associated with the form a on the form domain $H^1_0(\Omega)$; i.e., for $u, v \in L^2(\Omega)$ one has

$$u \in D(A_2),\ A_2 u = v \ \Leftrightarrow\ u \in H^1_0(\Omega) \text{ and } a(u,\varphi) = -\int_\Omega v\varphi \text{ for all } \varphi \in H^1_0(\Omega)\,.$$

The form a is **elliptic**; i.e. for some $\beta > 0,\ w \ge 0$ one has

$$a(u,u) + w(u|u)_{L^2} \ge \beta \|u\|^2_{H^1} \tag{4.5}$$

for all $u \in H^1_0(\Omega)$. Thus A_2 generates a holomorphic C_0-semigroup T_2 on $L^2(\Omega)$. It follows from the first Beurling-Deny criterion that $T_2(t) \ge 0 \quad (t \ge 0)$ (see [Ou1]). In virtue of (4.4), by the second Beurling-Deny criterion one has for $f \in L^2(\Omega) \cap L^\infty(\Omega)$,

$$\|\lambda R(\lambda, A_2) f\|_\infty \le \|f\|_\infty \quad (\lambda > 0) \tag{4.6}$$

or equivalently,

$$\|T_2(t)f\|_\infty \le \|f\|_\infty \quad (t > 0) \tag{4.7}$$

(see [Ou1]). Actually (4.6) (and (4.7)) are equivalent to (4.4) (see [ABBO]). It follows from (4.6) that there are operators A_p on $L^p(\Omega)$, $2 \le p \le \infty$, such that $(0,\infty) \subset \varrho(A_p)$ and $R(\lambda, A_p) = R(\lambda, A_2)_{|L^p(\Omega)}$. Moreover, for $2 < p < \infty$, by (4.7) the restriction of T_2 to L^p is a C_0-semigroup T_p on $L^p(\Omega)$ whose generator is A_p.

Lemma 4.2. *Let $u \in H^1_{\text{loc}}(\Omega) \cap C_0(\Omega)$ such that $Lu \in L^2(\Omega)$. Then $u \in H^1_0(\Omega)$.*

Proof. Let $v = Lu$. Then

$$a(u,\varphi) = -\int v\varphi \quad \text{for all} \quad \varphi \in \mathcal{D}(\Omega)\,.$$

Let $\varepsilon > 0$. Then $(u-\varepsilon)^+ \in H^1_0(\Omega)$. Let ω be open such that $\bar\omega \subset \Omega$. Since $u \in H^1(\omega)$, $(u-\varepsilon)^+ \in H^1_0(\Omega)$, we have

$$a(u,(u-\varepsilon)^+) = -\int_\Omega v(u-\varepsilon)^+ dx \le \|v\|_{L^2}\|u\|_{L^2}\,.$$

Since $D_j(u-\varepsilon)^+ = 1_{\{u>\varepsilon\}}D_j u$ we have

$$\begin{aligned}
&a(u,(u-\varepsilon)^+) = \\
&a((u-\varepsilon)^+,\ (u-\varepsilon)^+) + \varepsilon \int_\Omega \sum_{i=1}^N b_i D_i(u-\varepsilon)^+ dx - \varepsilon\int_\Omega d(u-\varepsilon)^+ dx \\
&\ge a((u-\varepsilon)^+,\ (u-\varepsilon)^+) \qquad\qquad \text{(by (4.4))} \\
&\ge \beta\|(u-\varepsilon)^+\|^2_{H^1} - \omega\|(u-\varepsilon)^+\|^2_{L^2}\,.
\end{aligned}$$

Thus $\beta\|(u-\varepsilon)^+\|^2_{H^1} \le a(u,\ (u-\varepsilon)^+) + \omega\|u\|^2_{L^2} \le \|v\|_{L^2}\|u\|_{L^2} + \omega\|u\|^2_{L^2}$. As in the proof of Lemma 2.2 we deduce from this that $u^+ \in H^1_0(\Omega)$. Similarly, $u^- \in H^1_0(\Omega)$. □

Corollary 4.3. *The operator A_0 is the part of A_2 in $C_0(\Omega)$. In particular, A_0 is closed and dissipative.*

Proof. It follows from Lemma 4.2 that A_0 is the part of A_2 in $C_0(\Omega)$. Since A_2 is closed, A_0 is closed as well. In order to show dissipativity, let $\lambda > 0$, $u \in D(A_0)$, $\lambda u - A_0 u = f$. Since A_0 is the part of A_2 in $C_0(\Omega)$, it follows that $u = R(\lambda, A_2)f$. Now by (4.6), $\|\lambda u\|_\infty \le \|f\|_\infty$. □

Theorem 4.4. *Assume that Ω is regular, and that (4.4) is satisfied. Then A_0 generates a positive contractive C_0-semigroup T_0 on $C_0(\Omega)$. Moreover, $T_0(t) = T_2(t)_{|C_0(\Omega)}$ $(t \ge 0)$.*

Proof. 1. We extend the coefficients to $\mathbb{R}^N$ by setting

$$a_{ij}(x) = \begin{cases} 1 & \text{if} \quad i = j \\ 0 & \text{if} \quad i \neq j \end{cases}$$

for $x \in \mathbb{R}^N \setminus \Omega$ and $b_i(x) = c_i(x) = d(x) = 0$ for $x \in \mathbb{R}^N \setminus \Omega$. Denote by $\tilde{L} : H^1_{\text{loc}}(\mathbb{R}^N) \to \mathcal{D}(\mathbb{R}^N)'$ the corresponding elliptic operator and by $\tilde{A}_2$ its realization in $L^2(\mathbb{R}^N)$. Let $\lambda > \max\{0, \omega(\tilde{A}_2)\}$, where $\omega(\tilde{A}_2)$ is the type of $\tilde{A}_2$. We show that $\lambda - A_0$ is surjective, which proves our claim in view of Corollary 4.3. Let $g \in C_0(\Omega)$, $v = R(\lambda, \tilde{A}_2)g$. Then $v \in H^1(\mathbb{R}^N)$ and $v - Lv = g$ in $\mathcal{D}(\Omega)'$. By the famous result of De Giorgi and Nash [GT], Theorem 8.22, the function v is continuous on $\mathbb{R}^N$. Let $\varphi = v_{|\partial\Omega}$. By [GT], Theorem 8.31, (in the case where Ω is connected, see [Sta], Section 10 for the general case), there exists $w \in H^1_{\text{loc}}(\Omega) \cap C(\bar{\Omega})$ such that $w - Lw = 0$ in $\mathcal{D}(\Omega)'$ and $\omega_{|\partial\Omega} = \varphi$. Now let $u = v - w$. Then $u \in C_0(\Omega) \cap H^1_{\text{loc}}(\Omega)$ and $u - Lu = g$ in $\mathcal{D}(\Omega)'$. Thus $u \in D(A_0)$ and $u - A_0 u = g$. We have shown that A_0 generates a contractive C_0-semigroup. Since A_0 is the part of A_2, it follows that $R(\lambda, A_2)_{|C_0(\Omega)} = R(\lambda, A_0)$ $(\lambda > 0)$ which implies the last claim in the proposition.

We have shown that $(I - A_0)$ is surjective. Since A_0 is dissipative, it follows that $(0, \infty) \subset \varrho(A_0)$ and $\|\lambda R(\lambda, A_0)\| \leq 1$ $(\lambda > 0)$. Moreover, A_0 being the part of A_2 in $C_0(\Omega)$, we have $R(\lambda, A_0) = R(\lambda, A_2)_{|C_0(\Omega)}$ for $\lambda > 0$.

2. We show that $D(A_0)$ is dense in $C_0(\Omega)$. For that we have to show that $R(\lambda, A_0)' \in \mathcal{L}(C_0(\Omega)')$ is injective. Let $\mu \in C_0(\Omega)'$ such that $R(\lambda, A_0)'\mu = 0$. Take $\mu_n \in L^1(\Omega) \cap L^2(\Omega)$ such that $\sup_{n \in \mathbb{N}} \|\mu_n\| < \infty$ and $\lim_{n \to \infty} \mu_n = \mu$ for the topology $\sigma(C_0(\Omega)', C_0(\Omega))$. Let $v_n = R(\lambda, A_0)'\mu_n$. Let $1 < q < \frac{N}{N-1}$. By a result of Stampacchia [Sta], Théorème 4.4, $R(\lambda, A_0)'$ is a bounded operator from $L^1(\Omega)$ into $W_0^{1,q}(\Omega)$. By reflexivity, we can assume that $(v_n)_{n \in \mathbb{N}}$ converges weakly to v in $W_0^{1,q}(\Omega)$ (choosing a subsequence otherwise). Since $w^* - \lim_{n \to \infty} v_n = R(\lambda, A_0)'\mu = 0$, we have $v = 0$. Since $R(\lambda, A_0) = R(\lambda, A_2)_{|C_0(\Omega)}$ one has $v_n = R(\lambda, A_2)'\mu_n$.

Hence for $\varphi \in \mathcal{D}(\Omega)$,

$$\begin{aligned} &\textstyle\int \varphi d\mu = \lim_{n \to \infty} \int \mu_n \varphi = \\ &\lim_{n \to \infty} ((I - A_2)' v_n | \varphi)_{L^2} = \\ &\lim_{n \to \infty} \{\textstyle\int v_n \varphi + a(\varphi, v_n)\} = 0 \,. \end{aligned}$$

It follows that $\mu = 0$.

3. We have shown that A_0 generates a contractive C_0-semigroup T_0 on $C_0(\Omega)$. Since $R(\lambda, A_0) = R(\lambda, A_2)_{|C_0(\Omega)}$ $(\lambda > 0)$, it follows that

$$T_0(t)f = \lim_{n \to \infty} (I - \frac{t}{n} A_0)^{-n} f = \lim_{n \to \infty} (I - \frac{t}{n} A_2)^{-n} f = T_2(t) f \quad (f \in C_0(\Omega)) \,.$$

□

We introduce the following notation. Let $S \in \mathcal{L}(L^p(\Omega))$. Then

$$\|S\|_{\mathcal{L}(L^q, L^r)} = \sup\{\|Sf\|_r : f \in L^q \cap L^p, \ \|f\|_q \leq 1\}$$

where $1 \le p,\ q,\ r \le \infty$. We recall the following result on ultracontractivity (see [V], Théorème 2 or [C], Lemma 1).

Theorem 4.5. *Let T be a C_0-semigroup on $L^2(\Omega)$ such that*

$$\|T(t)f\|_\infty \le M_\infty \|f\|_\infty \quad (f \in L^2 \cap L^\infty,\ 0 < t \le 1) \tag{4.8}$$

and

$$\|T(t)f\|_q \le M_q t^{-\alpha} \|f\|_2 \quad (f \in L^2 \cap L^q,\ 0 < t \le 1) \tag{4.9}$$

where $2 < q < \infty,\ \alpha > 0,\ M_q \ge 0,\ M_\infty \ge 0$. Then there exists a constant $c > 0$ such that

$$\|T(t)f\|_\infty \le ct^{-\beta}\|f\|_2 \quad (0 < t \le 1) \tag{4.10}$$

where $\beta = \frac{\alpha}{1-\frac{2}{q}}$.

Corollary 4.6. *Assume that Ω is a bounded regular open set. Then the operators $T_0(t) \quad (t > 0)$ are compact. Moreover,*

$$\sigma(A_0) = \sigma(A_2)\,.$$

Proof. Since T_2 is holomorphic, one has $\sup\limits_{0<t\le 1} \|tA_2T_2(t)\|_{\mathcal{L}(L^2)} < \infty$. Since $H_0^1(\Omega) \hookrightarrow L^{2N/N-2}$, it follows that

$$\sup_{0<t\le 1} \|tT_2(t)\|_{\mathcal{L}(L^2, L^{2N/N-2})} < \infty\,.$$

Now it follows from Theorem 4.5 that

$$\|T_2(t)\|_{\mathcal{L}(L^2,L^\infty)} \le ct^{-N/2} \quad (0 < t \le 1)\,. \tag{4.11}$$

Since the injection $H_0^1(\Omega) \hookrightarrow L^2(\Omega)$ is compact, it follows as in the proof of Proposition 3.8 that $T_0(t)$ is compact $(t > 0)$. In particular, A_0 has compact resolvent. Now it follows from [Ar], Proposition 2.6 that $\sigma(A_0) = \sigma(A_2)$. □

Corollary 4.7. *Assume that $b_i, c_i \in W^{1,\infty}(\Omega)\ i = 1, \dots, N$. If Ω is a bounded, regular, open set, then A_0 generates a holomorphic C_0-semigroup on $C_0(\Omega)$.*

Proof. It follows from [Ou1] that there exists a C_0-semigroup T_1 on $L^1(\Omega)$ such that $T_1(t)_{|_{L^2(\Omega)}} = T_2(t) \quad (t \ge 0)$. Denote the generator of T_1 by A_1. It follows from [AE] that T_1 is holomorphic. By duality

$$\|\lambda R(\lambda, A_\infty)\| \le M \quad (\mathrm{Re}\ \lambda > \omega)$$

for some $\omega, M \ge 0$. This implies that $\|\lambda R(\lambda, A_0)\| \le M \quad (\mathrm{Re}\ \lambda > \omega)$, so A_0 generates a holomorphic C_0-semigroup. □

Remark 4.8. In Corollary 4.7 the hypothesis (4.4) is not needed a priori: it is satisfied if we replace L by $L-\omega$ for a suitable ω.

Lemma 4.9. *Let $\infty > p > N$. Then*

$$D(A_p) \subset L^\infty(\Omega)\,.$$

Moreover,

$$D(A_2^k) \subset L^\infty(\Omega)$$

for $k > \max\{\frac{N}{2},\ 1\}$.

Proof. By (4.11) we have

$$\|T_2(t)\|_{\mathcal{L}(L^2,L^\infty)} \le ct^{-N/2} \quad (0 < t \le 1)\,. \tag{4.12}$$

By interpolation we deduce from this that

$$\|T_p(t)\|_{\mathcal{L}(L^p,L^\infty)} \le c_p t^{-N/p} \quad (0 < t \le 1) \tag{4.13}$$

for some $c_p \ge 0$, $2 \le p < \infty$. Let $p > N$. Let $M \ge 0$, $\omega \ge 0$ such that $\|T_p(t)\|_{\mathcal{L}(L^p)} \le Me^{\omega t}$. Then for $\lambda > \omega$,

$$\begin{aligned}
\|R(\lambda, A_p)f\|_\infty &\le \textstyle\int_0^1 e^{-\lambda t}\|T(t)f\|_\infty\, dt + \int_0^\infty e^{-\lambda(t+1)}\|T(1)T(t)f\|_\infty\, dt\\
&\le c_p \textstyle\int_0^1 t^{-N/p}\, dt\; \|f\|_p + c_p \int_0^\infty e^{-\lambda(t+1)}\|T(t)f\|_p\, dt\\
&\le c_p\{\textstyle\int_0^1 t^{-N/p}\, dt + \int_0^\infty e^{-\lambda(t+1)}Me^{\omega t}\, dt\}\|f\|_p\\
&\le \text{const.} \cdot \|f\|_p\,.
\end{aligned}$$

Similarly, for λ sufficiently large, since

$$R(\lambda, A_2)^k = (k-1)! \int_0^\infty e^{-\lambda t} t^{k-1} T_2(t)dt$$

one obtains by (4.13) that $\|R(\lambda, A_2)^k\|_{\mathcal{L}(L^2,L^\infty)} < \infty$ if $k > \frac{N}{2}$. □

In order to prove the remaining implication $(ii) \Rightarrow (i)$ of Theorem 4.1 we consider the Dirichlet problem with respect to $L - I$. For this, fix $\tilde\Omega$ a large ball containing $\bar\Omega$. Let $h \in C(\partial\Omega)$ be the trace of a function $w \in H^1(\tilde\Omega) \cap C(\tilde\Omega)$; i.e. $h = w_{|\partial\Omega}$. Then there exists a unique function $u \in H^1(\Omega) \cap C^b(\Omega)$ such that

$$Lu - u = 0 \quad \text{in} \quad D(\Omega)' \quad \text{and} \quad u - w \in H_0^1(\Omega) \tag{4.14}$$

(see [Sta], Théorème 10.1 and [GT], 8.22). We interprete (4.14) as a weak form of $u_{|\partial\Omega} = h$. Then by the maximum principle (see [Sta], § 10 or [GT], 8.1),

$$\|u\|_{L^\infty(\Omega)} \le \|h\|_{C(\partial\Omega)}\,. \tag{4.15}$$

We set $u = Bh$. Since by the Stone-Weierstraß theorem traces of functions $w \in C(\tilde\Omega) \cap H^1(\tilde\Omega)$ are dense in $C(\partial\Omega)$, we find a unique linear extension

$$B : C(\partial\Omega) \to H^1(\Omega) \cap C^b(\Omega)$$

such that

$$\|Bh\|_{L^\infty(\Omega)} \le \|h\|_{C(\partial\Omega)}\,. \tag{4.16}$$

Recall that $L : H^1(\Omega) \to \mathcal{D}(\Omega)'$ is a continuous linear mapping. Let $h \in C(\partial\Omega)$, then $u = Bh$ is the solution of the Dirichlet problem

$$\begin{cases} u \in H^1(\Omega) \\ Lu - u = 0 \quad \text{in} \quad \mathcal{D}(\Omega)' \\ u_{|\partial\Omega} = h \end{cases} \tag{4.17}$$

where the last identity has to be understood in the sense of the construction.

It turns out that Ω is regular (which means by our definition regular with respect to the Laplacian) if and only if Ω is regular with respect to L. More precisely, the following remarkable result is due to Stampacchia [Sta], § 10 and Littmann, Stampacchia, Weinberger [LSW].

Theorem 4.10. *The following are equivalent:*

(i) Ω *is regular;*

(ii) $\lim\limits_{\substack{x \to z \\ x \in \Omega}} (Bh)(x) = h(z)$ *and all* $z \in \partial\Omega$ *for all* $h \in C(\partial\Omega)$.

Recall that here L is a fixed elliptic operator satisfying (4.4).

On the basis of this result we now deduce the following.

Proposition 4.11. *Assume (4.4). If* $\varrho(A_0) \neq \emptyset$, *then* Ω *is regular.*

Proof. 1. It follows from Lemma 4.8 and Proposition 1.3 that $\sigma(A_2) = \sigma(A_\infty)$. Since by hypothesis $\varrho(A_0) \neq \emptyset$ and since $\sigma(A_\infty) \subset \mathbb{R}$, it follows that $\varrho(A_0) \cap \varrho(A_\infty) \neq \emptyset$. As in the proof of Theorem 4.4 we conclude that $\lambda \in \varrho(A_0)$. Let $p > N$. Then by Lemma 4.8, $R(\lambda, A_p)L^p(\Omega) \subset L^\infty(\Omega)$. Since $R(\lambda, A_p)C_0(\Omega) = R(\lambda, A_0)C_0(\Omega) \subset C_0(\Omega)$, it follows by density that $R(\lambda, A_p)L^p(\Omega) \subset C_0(\Omega)$.
2. Let $\tilde{\Omega}$ be a ball containing $\bar{\Omega}$. Note that $\tilde{\Omega}$ is regular. Consider the elliptic operator $\tilde{L} : H^1_{\mathrm{loc}}(\tilde{\Omega}) \to \mathcal{D}(\tilde{\Omega})'$ one obtains by extending the coefficients as in the proof of Theorem 4.4 and let A_0 be its realization in $C_0(\tilde{\Omega})$. Then $1 \in \varrho(\tilde{A}_0)$ by Theorem 4.4. Since $\tilde{A}_0$ has dense domain in $C_0(\tilde{\Omega})$, and since each function in $C(\partial\Omega)$ has an extension to a function in $C_0(\tilde{\Omega})$, the space

$$F = \{h \in C(\partial\Omega) : \exists\, g \in C_0(\tilde{\Omega}),\ h = (R(\lambda, \tilde{A}_0)g)_{|\partial\Omega}\}$$

is dense in $C(\partial\Omega)$. Let $h \in F$, $h = w_{|\partial\Omega}$, $w = R(1, \tilde{A}_0)g$, $g \in C_0(\tilde{\Omega})$. Then $g_{|\Omega} \in L^p(\Omega)$. Let $v = R(\lambda, A_p)(g_{|\Omega})$. Then $v \in C_0(\Omega)$ by 1. Moreover, we have

$$\begin{aligned} v - Lv = g \quad &\text{in} \quad \mathcal{D}(\Omega)' \quad \text{and} \\ w - Lw = g \quad &\text{in} \quad \mathcal{D}(\Omega)'. \end{aligned}$$

Let $u = (w - v)_{|\Omega}$. Then $u \in H^1(\Omega) \cap C(\bar{\Omega})$ and $u - Lu = 0$ in $\mathcal{D}(\Omega)'$ and $u_{|\partial\Omega} = h$. Thus *(ii)* of Theorem 4.9 is satisfied for all $h \in F$ and hence for all $h \in C(\partial\Omega)$ by density. □

The proof of Theorem 4.1 is complete now.

Concluding Remark

For elliptic operators we restricted ourselves to bounded open sets. In order to carry over the strategy used for the Laplacian in Section 3 further arguments are needed. One can actually show that L_0 is dissipative under condition (4.4). However, one obtains barriers with respect to L merely in H^1_{loc} which presents an additional difficulty.

References

[AB] W. Arendt, C.J.K. Batty: "L'holomorphie du semi-groupe engendré par le Laplacien Dirichlet sur $L^1(\Omega)$". C.R. Acad. Sci. Paris, Sér I, **315** (1992), 31–35.

[ABBO] P. Auscher, L. Barthélemy, Ph. Bénilan, E.-M. Ouhabaz: *Absence de la L^∞-contractivité pour les semi-groupes associés aux opérateurs elliptiques complexes sous forme divergence.* Preprint.

[AE] W. Arendt, T. ter Elst: "Gaussian estimates for second order elliptic operators with boundary condition." J. Operator Theory **38** (1997), 87–130.

[Am] H. Amann: "Dual semigroups and second order linear elliptic boundary value problems." Israel J. Math. **45** (1983), 225–254.

[AmE] H. Amann, J. Escher: "Strongly continuous dual semigroups." Ann. Mat. Pura Appl. **161** (1996), 41–62.

[ANS] W. Arendt, F. Neubrander, U. Schlotterbeck: "Interpolation of semigroups and integrated semigroups." Semigroup Forum **45** (1992), 26–37.

[Ar] W. Arendt: "Gaussian estimates and interpolation of the spectrum in L^p." Diff. Int. Equ. **7** (1994), 1153–1168.

[B] H. Brézis: *Analyse Fonctionnelle.* Masson Paris 1983.

[C] T. Coulhon: "Dimension à l'infini d'un semi-groupe analytique". Preprint

[CH] T. Cazanave, A. Haraux: *Introduction aux Probèmes d'Évolution Semi-Linéaires.* Ellipses-Eidition, Paris 1990.

[Da] E.B. Davies: *Heat Kernels and Spectral Theory.* Cambridge University Press, Cambridge 1990.

[dP] B. de Pagter: "A characterization of sun-reflexivity." Math. Ann. **283** (1989), 511–518.

[DL] R. Dautray, J.-L. Lions: *Analyse Mathématique et Calcul Numérique.* Vol. 2, L'opérateur de Laplace. Masson, Paris 1987.

[DS] G. Da Prato, E. Sinestrari: "Differential operators with non-dense domain." Annali Scuola Normale Superiore Pisa **14** (1987), 285–344.

[GT] D. Gilbarg, N.S. Trudinger: *Elliptic Partial Differential Equations of Second Order.* Springer 1977.

[LP1] G. Lumer, L. Paquet: "Semi-groupes holomorphes, produit tensoriel de semi-groupes et équations d'évolution." Séminaire: Théorie du Potentiel Springer LN **563** (1976), 202–218.

[LP2] G. Lumer, L. Paquet: "Semi-groupes holomorphes et équation d'évolution." C.R. Acad. Sc. Paris **284** (1977) Série A, 237–240.

[LSW] W. Littmann, G. Stampacchia, H.F. Weinberger: "Regular points for elliptic operators with discontinuous coefficients." Ann. Scuola Normale Sup. Pisa **17** (1963), 43–77.

[N] R. Nagel (ed.): *One-parameter Semigroups of Positive Operators.* Springer LN **1184** (1986), Berlin.

[Ou1] E. Ouhabaz: "L^∞-contractivity of semigroups generated by sectorial forms." London Math. Soc. **46** (1992), 529–542.

[Ou2] E. Ouhabaz: "Gaussian estimates and holomorphy of semigroups." Proc. Amer. Math. Soc. **123** (1995), 1465–1474.

[Sta] G. Stampacchia: "Le Problème de Dirichlet pour les équations elliptiques du second ordre à coefficients discontinus." Ann. Inst. Fourier Grenoble **15** (1965) 189–258.

[Stu] H.B. Stewart: "Generation of analytic semigroups by strongly elliptic operators." Trans. Amer. Math. Soc. **199** (1974), 141–162.

[V] N. Varopoulos: "Hardy-Littlewood theory for semi-groups." J. Funct. Anal. **63** (1985), 240–260.

W. Arendt
Universität Ulm
Abteilung Mathematik V
D-89069 Ulm, Germany
arendt@mathematik.uni-ulm.de

Ph. Bénilan
Université de Franche-Comté
Equipe de Mathématiques
UMR CNRS 6623
F-25030 Besançon, France
benilan@math.univ-fcomte.fr

Progress in Nonlinear Differential Equations
and Their Applications, Vol. 35
© 1999 Birkhäuser Verlag Basel/Switzerland

Infinitely Many Solutions of Nonlinear Elliptic Systems

Thomas Bartsch* and Djairo G. de Figueiredo

1. Introduction

In this paper we study elliptic systems of the form

$$\begin{cases} -\Delta u &= H_v(x,u,v) \quad \text{in } \Omega \\ -\Delta v &= H_u(x,u,v) \quad \text{in } \Omega \end{cases} \tag{1.1}$$

where $\Omega \subset \mathbb{R}^N$, $N \geq 3$, is a smooth bounded domain and $H\colon \overline{\Omega} \times \mathbb{R} \times \mathbb{R} \to \mathbb{R}$ is a C^1-function. We shall also consider the case when $\Omega = \mathbb{R}^N$, and in this case the system takes the form

$$\begin{cases} -\Delta u + u &= H_v(x,u,v) \quad \text{in } \mathbb{R}^N \\ -\Delta v + v &= H_u(x,u,v) \quad \text{in } \mathbb{R}^N \end{cases} \tag{1.2}$$

In the bounded case, we look for solutions of (1.1) subject to Dirichlet boundary conditions $u = v = 0$ on $\partial\Omega$. In the case when $\Omega = \mathbb{R}^N$ we assume that some symmetry with respect to x holds; for instance, that the x-dependence of H is radial, or that H is invariant with respect to certain subgroups of $O(N)$ acting on $\mathbb{R}^N$. We shall obtain both radial and non-radial solutions in the radial symmetric case, thus observing a symmetry breaking effect.

In order to illustrate the kind of results obtained here, let us state two theorems. We first consider the case when Ω is bounded. In such a case, the following set of hypotheses is assumed. First, the regularity of the Hamiltonian:

(H_1) $H\colon \overline{\Omega} \times \mathbb{R} \times \mathbb{R} \to \mathbb{R}$ is C^1 and $H \geq 0$.

Next we assume conditions related to the growth of the right side of (1.1).

(H_2) There exist constants $p, q > 1$ and $c_1 > 0$ with

$$1 > \frac{1}{p} + \frac{1}{q} > 1 - \frac{2}{N} \tag{1.3}$$

such that

$$|H_u(x,u,v)| \leq c_1(|u|^{p-1} + |v|^{(p-1)q/p} + 1) \tag{1.4}$$

* Research supported by GMD–CNPQ. T.B. thanks the members of the Department of Mathematics at UNICAMP for their invitation and hospitality.

and

$$|H_v(x,u,v)| \le c_1(|v|^{q-1} + |u|^{(q-1)p/q} + 1) \tag{1.5}$$

for all $(x,u,v) \in \overline{\Omega} \times \mathbb{R} \times \mathbb{R}$.

The next condition is a "non-quadraticity" condition at infinity introduced by Costa-Magalhães [5]. It is related to the so-called Ambrosetti-Rabinowitz condition and it is devised to get some sort of Palais-Smale condition for the functionals involved.

(H$_3$) There exist $1 < \alpha < p$ and $1 < \beta < q$ with

$$\frac{1}{\alpha} + \frac{1}{\beta} = 1, \tag{1.6}$$

and such that

$$\frac{1}{\alpha} H_u(x,u,v)u + \frac{1}{\beta} H_v(x,u,v)v - H(x,u,v) \ge a(|u|^\mu + |v|^\nu - 1) \tag{1.7}$$

for all $(x,u,v) \in \overline{\Omega} \times \mathbb{R} \times \mathbb{R}$. Here a, μ, ν are positive constants satisfying

$$\mu > \frac{pN}{2} \max\left\{\frac{1}{2} - \frac{1}{p},\ 1 - \frac{1}{p} - \frac{1}{q}\right\}$$

and

$$\nu > \frac{qN}{2} \max\left\{\frac{1}{2} - \frac{1}{q},\ 1 - \frac{1}{p} - \frac{1}{q}\right\}$$

Remark 1.1. *Observe that* $1 - \frac{1}{p} - \frac{1}{q}$ *is always positive in view of* (H$_2$). *It follows from* (H$_3$) *that*

$$H(x,u,v) \ge c(|u|^\alpha + |v|^\beta - 1). \tag{1.8}$$

In fact, (1.8) *follows from a condition weaker than* (H$_3$), *namely*

$$\frac{1}{\alpha} H_u(x,u,v)u + \frac{1}{\beta} H_v(x,u,v)v \ge H(x,u,v)$$

for all $x \in \overline{\Omega}$ *and* $|(u,v)| \ge R$; *see Felmer* [10].

Remark 1.2. *Suppose* H *satisfies the following condition of Ambrosetti-Rabinowitz type: there is* $R > 0$ *and* $1 < \alpha' < p$ *and* $1 < \beta' < q$ *with* $\frac{1}{\alpha'} + \frac{1}{\beta'} < 1$ *and such that*

$$\frac{1}{\alpha'} H_u(x,u,v)u + \frac{1}{\beta'} H_v(x,u,v)v \ge H(x,u,v)$$

for $x \in \overline{\Omega}$ *and* $|(u,v)| \ge R$. *Then condition* (H$_3$) *holds. In this case, it follows that* H *is superquadratic, in the sense that*

$$H(x,u,v) \ge c_1(|u|^{\alpha'} + |v|^{\beta'}) - c_2.$$

The next condition provides the symmetry we assume here.

(H$_4$) $H(x,-u,-v) = H(x,u,v)$ for all $(x,u,v) \in \overline{\Omega} \times \mathbb{R} \times \mathbb{R}$.

Now we are prepared to state the result in the case of Ω bounded. For that matter we introduce a non-increasing sequence of constants δ_n, $n \in \mathbb{N}$, with $\delta_n \to 0$, which will be defined in Section 3, and which depend only on p, q, α and β.

Theorem 1.1. *Suppose that* (H_1)–(H_4) *hold. Then there is a* $k_0 \in \mathbb{N}$ *such that, if*

$$\liminf_{|(u,v)|\to\infty} \frac{2H(x,u,v)}{|u|^\alpha + |v|^\beta} > \frac{1}{\delta_K} \tag{1.9}$$

holds for $K \geq k_0$, *system* (1.1), *subject to Dirichlet boundary conditions, has* $K - k_0 + 1$ *pairs of nontrivial solutions.*
Moreover, if

$$\lim_{|(u,v)|\to\infty} \frac{H(x,u,v)}{|u|^\alpha + |v|^\beta} = +\infty,$$

(in particular, if H is superquadratic) then system (1.1), *subject to Dirichlet boundary conditions has infinitely many solutions.*

The solutions obtained in Theorem 1.1 are strong solutions in the sense that $u \in W^{2,p/(p-1)}(\Omega) \cap W_0^{1,p/(p-1)}(\Omega)$ and $v \in W^{2,q/(q-1)}(\Omega) \cap W_0^{1,q/(q-1)}(\Omega)$. The existence of at least one solution for the system (1.1), without symmetry assumptions, has been considered before. See the survey paper [6] for a list of references.

As a corollary of Theorem 1.1 we obtain solutions to two nonlinear eigenvalue problems. We consider first

$$\begin{cases} -\Delta u &= \delta u + \lambda |v|^{\beta-2} v + H_v(x,u,v), \\ -\Delta v &= \mu |u|^{\alpha-2} u + \delta v + H_u(x,u,v) \end{cases} \tag{1.10}$$

in a bounded domain $\Omega \subset \mathbb{R}^N$ subject to Dirichlet boundary conditions $u = v = 0$ on $\partial\Omega$. The constants α, β are those from (H_3).

Corollary 1.1. *If H satisfies* (H_1)–(H_4), *then for each* $k \in \mathbb{N}$, *there exists* $\Lambda_k > 0$, *such that* (1.10) *has k pairs of non-trivial solutions provided* λ, $\mu > \Lambda_k$.

Next we consider the eigenvalue problem

$$\begin{cases} -\Delta u &= \lambda H_v(x,u,v) \\ -\Delta v &= \lambda H_u(x,u,v) \end{cases} \tag{1.11}$$

in a bounded domain $\Omega \subset \mathbb{R}^N$ subject to Dirichlet boundary conditions $u = v = 0$ on $\partial\Omega$.

Corollary 1.2. *Suppose that H satisfies* $(\mathrm{H}_1$-$)$-(H_4), *and*

$$\liminf_{|(u,v)|\to\infty} \frac{H(x,u,v)}{|u|^\alpha + |v|^\beta} > 0.$$

Then for each $k \in \mathbb{N}$, *there exists* $\Lambda_k > 0$, *such that* (1.11) *has k pairs of non-trivial solutions provided* $\lambda > \Lambda_k$.

Let us now state a result for the case when system (1.2) is considered in the whole of $\mathbb{R}^N$. We need a distinct, but similar, set of hypotheses.

(H_1') $H : \mathbb{R}^N \times \mathbb{R} \times \mathbb{R} \to \mathbb{R}$ is C^1, $H \geq 0$, $H(x,u,v) > 0$ for $|(u,v)| > 0$ and H is radial in the variable x.

(H_2') There exist positive constants p, q, a, b and c_1 with

$$p, q > 2, \quad \frac{1}{p} + \frac{1}{q} > 1 - \frac{2}{N}, \quad 1 < a < p-1, \quad 1 < b < q-1, \tag{1.12}$$

such that

$$|H_u(x,u,v)| \leq c_1(|u|^{p-1} + |v|^{(p-1)q/p} + |u|^a) \tag{1.13}$$

and

$$|H_v(x,u,v)| \leq c_1(|v|^{q-1} + |u|^{(q-1)p/q} + |v|^b) \tag{1.14}$$

for all $(x,u,v) \in \mathbb{R}^N \times \mathbb{R} \times \mathbb{R}$.

(H_3') There exist $1 < \alpha < p$ and $1 < \beta < q$ with $\alpha^{-1} + \beta^{-1} < 1$ and such that

$$\frac{1}{\alpha} H_u(x,u,v)u + \frac{1}{\beta} H_v(x,u,v)v \geq H(x,u,v)$$

for all $(x,u,v) \in \mathbb{R}^N \times \mathbb{R} \times \mathbb{R}$.

(H_4') There are positive constants c and r such that

$$H(x,u,v) \geq c(|u|^p + |v|^q) \quad \text{for } x \in \mathbb{R}^N \text{ and } |(u,v)| \leq r.$$

(H_5') $H(x,u,v) = H(x,-u,-v)$ for all $(x,u,v) \in \mathbb{R}^N \times \mathbb{R} \times \mathbb{R}$.

Remark 1.3. *It follows from* (H_3') *that there are positive constants c and R such that*

$$H(x,u,v) \geq c(|u|^p + |v|^q) \quad \text{for } |(u,v)| \geq R. \tag{1.15}$$

Then (1.14) *and assumption* (H_4') *imply that*

$$H(x,u,v) \geq c(|u|^p + |v|^q) \quad \text{for all } (x,u,v) \in \mathbb{R}^N \times \mathbb{R} \times \mathbb{R}. \tag{1.16}$$

Theorem 1.2. *Assume that the Hamiltonian H satisfies the hypotheses* (H_1')–(H_5'). *Then system* (1.2) *has infinitely many radial solutions.*

The solutions obtained in Theorem 1.2 are strong solutions in the sense that they satisfy $u \in W^{2,p/(p-1)}_{\mathrm{loc}}(\mathbb{R}^N)$ and $v \in W^{2,q/(q-1)}_{\mathrm{loc}}(\mathbb{R}^N)$. They also satisfy $u \in H^s(\mathbb{R}^N) \subset L^\gamma(\mathbb{R}^N)$ and $v \in H^t(\mathbb{R}^N) \subset L^\delta(\mathbb{R}^N)$ for some $s, t > 0$ with $s+t = 2$ and $2 < \gamma, 2N/(N-2s)$, $2 < \delta < 2N/(N-2t)$. The existence of at least one solution has been obtained before for special cases of system (1.2) in [9] and recently in [14].

The next result exhibits the breaking of symmetry in certain dimensions. The result extends to the type of systems we have here a result that Bartsch-Willem [3] proved in the scalar case.

Theorem 1.3. *Suppose that* (H_1')–(H_5') *holds. If $N = 4$ or $N \geq 6$ then system* (1.2) *has infinitely many non-radial solutions.*

2. Some Abstract Critical Point Theory

We consider a Hilbert space E and a functional $\Phi \in C^1(E, \mathbb{R})$. Given a sequence $\mathcal{F} = (X_n)$ of finite dimensional subspaces $X_n \subset X_{n+1}$, with $\overline{\bigcup X_n} = E$, we say that Φ satisfies $(PS)_c^{\mathcal{F}}$, at level $c \in \mathbb{R}$, if every sequence z_j, $j \in \mathbb{N}$, with $z_j \in X_{n_j}$, $n_j \to \infty$, such that

$$\Phi(z_j) \to c \quad \text{and} \quad (1 + \|z_j\|)(\Phi|_{X_{n_j}})'(z_j) \to 0 \tag{2.1}$$

has a subsequence which converges to a critical point of Φ. In the case when $X_n = E$ for all $n \in \mathbb{N}$ this form of the Palais-Smale condition is due to Cerami [4]. It is closely related to the standard Palais-Smale condition and to the $(PS)^*$ condition of [1] and [11]. It also yields a deformation lemma. In the present form $(PS)_c^{\mathcal{F}}$ was introduced in Bartsch-Clapp [2].

Remark 2.1. *If Φ has the form*

$$\Phi(z) = \frac{1}{2}\langle Lz, z\rangle - \Psi(z)$$

with $L : E \to E$ a linear Fredholm operator of index zero and $\nabla\Psi : E \to E$ completely continuous, then a bounded $(PS)_c^{\mathcal{F}}$ sequence (z_j) has a convergent subsequence. By a $(PS)_c^{\mathcal{F}}$ sequence we mean a sequence as in (2.1). Let us prove the above statement. First select a subsequence, denoted again by (z_j) such that $z_j \rightharpoonup z$, weakly in E. Then $\nabla\Psi(z_j) \to \nabla\Psi(z)$, strongly in E. Let $P_n : E \to X_n$ denote the orthogonal projection onto X_n. We have that the sequence

$$P_{n_j}\nabla\Phi(z_j) = P_{n_j}Lz_j + P_{n_j}\nabla\Psi(z_j)$$

converges to zero in view of (2.1). So

$$P_{n_j}Lz_j \to -\nabla\Psi(z) = Lz.$$

Hence $Lz_j \to Lz$. And as a consequence, $z_j \to z$, because $\ker L$ *is finite dimensional.*

Now suppose that E splits as a direct sum $E = E^+ \oplus E^-$. Let $E_1^\pm \subset E_2^\pm \subset \dots$ be a strictly increasing sequence of finite dimensional subspaces of $E^\pm$ such that $\overline{\bigcup_{n=1}^\infty E_n^\pm} = E^\pm$. Setting $E_n = E_n^+ \oplus E_n^-$ we can formulate the hypotheses on Φ which are needed for our first abstract theorem.

(Φ_1) $\Phi \in C^1(E, \mathbb{R})$ and satisfies $(PS)_c^{\mathcal{F}}$ for $\mathcal{F} = (E_n)_{n\in\mathbb{N}}$ and $c > 0$.

(Φ_2) For some $k \geq 2$ and some $r > 0$ one has

$$b_k := \inf\{\Phi(z) : z \in E^+, z \perp E_{k-1}, \|z\| = r\} > 0. \tag{2.2}$$

(Φ_3) There exists an isomorphism $T : E \to E$ with $T(E_n) = E_n$, for all $n \in \mathbb{N}$, and there exist $K \geq k$ and $R > 0$ such that

$$\text{for } z = z^+ + z^- \in E_K^+ \oplus E^- \text{ with } \max\{\|z^+\|, \|z^-\|\} = R$$

one has

$$\|Tz\| > r \quad \text{and} \quad \Phi(Tz) \leq 0,$$

where k and r are the constants introduced in (Φ_2).

(Φ_4) $d_K := \sup\{\Phi \circ T(z^+ + z^-) : z^+ \in E_K^+, z^- \in E^-, \|z^+\|, \|z^-\| \le R\} < \infty$.
(Φ_5) Φ is even, i.e. $\Phi(-z) = \Phi(z)$.

A stronger condition that implies (Φ_4) and holds in our application is:

(Φ_6) Φ maps bounded sets to bounded sets.

Theorem 2.1. *Assume* (Φ_1)–(Φ_5). *Then, for every* $b < b_k$, Φ *has at least* $K - k + 1$ *pairs* $\pm z_i$ *of critical points with critical values in* $[b, d_K]$.

Proof. We need to recall the equivariant limit category defined in [2], specialized to our situation. We set $G = \mathbb{Z}/2$ which acts on E via the antipodal map. Given invariant subsets $Z \subset Y \subset X$ of E, we define the G–$\mathrm{cat}_X(Y, Z)$ to be the least integer m such that there exists a covering $Y \subset U_0 \cup \dots \cup U_m$ of Y with invariant open subsets $U_0, \dots, U_m$ of X with the properties:

(i) $Z \subset U_0$ and there exists a continuous family $h^t : U_0 \to X$, $0 \le t \le 1$, of odd maps satisfying $h^0(z) = z$ and $h^1(z) \in Z$ for every $z \in U_0$, and $h^t(z) = z$ for every $z \in Z$ and every $t \in [0, 1]$.

(ii) For $i = 1, \dots, m$ there exists a continuous family $h_i^t : U_i \to X$, $0 \le t \le 1$, of odd maps satisfying $h_i^0(z) = z$ for every $z \in U_i$ and such that $h_i^1(U_i) = \{\pm z_i\}$, for some $z_i \in X \setminus \{0\}$.

Now we define the equivariant limit category for G-invariant sets $Z \subset Y \subset E$ by

$$G\text{–}\mathrm{cat}_E^{\mathcal{F}}(Y, Z) := \limsup_{n \to \infty} G\text{–}\mathrm{cat}_{E_n}(Y \cap E_n, Z \cap E_n)\,.$$

Given $d > b > 0$ Theorem 2.8 of [2] says that Φ has at least G–$\mathrm{cat}_E^{\mathcal{F}}(\Phi^d, \Phi^b)$ pairs of critical points with critical values in $[b, d]$. Therefore it suffices to prove that G–$\mathrm{cat}_E^{\mathcal{F}}(\Phi^{d_K}, \Phi^b) \ge K - k + 1$ for $0 < b < b_k$. This follows from the next lemma. □

Lemma 2.1. *Fix* $0 \le b < b_k$ *and* $n \ge K$. *Then*

$$\gamma := G\text{–}\mathrm{cat}_{E_n}(\Phi^{d_K} \cap E_n, \Phi^b \cap E_n) \ge K - k + 1.$$

Proof. For simplicity we set $d := d_K$, and $B := B_R E_K^+ \times B_R E_n^-$ with $R > 0$ from (Φ_3). We also write $S_r E_n$ for the sphere of radius r in E_n. Let

$$\Phi_n^d := \Phi^d \cap E_n \subset U_0 \cup \dots \cup U_\gamma$$

be a covering as in the definition of G–$\mathrm{cat}_{E_n}(\Phi_n^d, \Phi_n^b)$. There are odd mappings $h^1 : U_0 \to \Phi_n^b$ and $h_i^1 : U_i \to \{\pm z_i\}$. Making U_0 smaller if necessary we may assume that h^1 extends continuously to $\overline{U_0}$. Then we can extend h^1 to an odd mapping $h^1 : E_n \to E_n$ by using Tietze's extension theorem. Now we set

$$\mathcal{O} := \{z \in B : \|h^1(Tz)\| < r\}.$$

For $z \in \partial B$ we have $\|Tz\| > r$ and $\Phi(Tz) \le 0$ by (Φ_3). Thus $Tz \in \Phi_n^0 \subset \Phi_n^b$ and $h^1(Tz) = Tz$, and hence $\|h^1(Tz)\| = \|Tz\| > r$. This implies that $\mathcal{O}$ is an open subset of B with $\overline{\mathcal{O}} \subset \operatorname{int} B$. Clearly $\mathcal{O}$ is an invariant neighborhood of 0 in $E_K^+ \oplus E_n^-$.

For $z \in T^{-1}(U_0)$ we have that $h^1(Tz) \in \Phi_n^b \subset E_n \backslash S_r(E_{k-1}^\perp \cap E_n^+)$, in virtue of (Φ_2). For $z \in \partial\mathcal{O}$, we have that $\|h^1(Tz)\| = r$. This implies that

$$h^1\big(\partial\mathcal{O} \cap T^{-1}(U_0)\big) \subset S_r E_n \setminus S_r\big(E_{k-1}^\perp \cap E_n^+\big).$$

The latter space has the sphere $S_r\big(E_{k-1} \oplus E_n^-\big)$ as a strong deformation retract. In particular, there exists an odd mapping

$$S_r E_n \setminus S_r\big(E_{k-1}^\perp \cap E_n^+\big) \longrightarrow S_r\big(E_{k-1} \oplus E_n^-\big).$$

Observe that $S_r\big(E_{k-1} \oplus E_n^-\big) \cong S^{k+n-2}$. Putting these mappings together we obtain an odd mapping

$$g_0 : \partial\mathcal{O} \cap T^{-1}(U_0) \longrightarrow S^{k+n-2} \subset \mathbb{R}^{k+n-1}.$$

The mappings $h_i^1 : U_i \to \{\pm z_i\}$ yield odd maps

$$g_i : \partial\mathcal{O} \cap T^{-1}(U_i) \longrightarrow S^0 = \{\pm 1\}.$$

By (Φ_4) we have that $T(B) \subset \Phi_n^d$. Therefore $T^{-1}(U_0), \dots, T^{-1}(U_\gamma)$ cover B. Setting $V_i := \partial\mathcal{O} \cap T^{-1}(U_i)$, we obtain an open invariant covering of $\partial\mathcal{O}$. Choose then a partition of the unity $\pi_i : \partial\mathcal{O} \to [0,1]$, $i = 0, \dots, \gamma$, subordinated to the covering $V_0, \dots, V_\gamma$ of $\partial\mathcal{O}$. Since the V_i's are invariant we may assume that the π_i's are even. Now we define the mapping

$$g : \partial\mathcal{O} \longrightarrow \mathbb{R}^{k+n-1} \times \mathbb{R}^\gamma, \quad g(z) := (\pi_0(z)g_0(z), \dots, \pi_\gamma(z)g_\gamma(z)).$$

First, observe that g is well defined. Namely, if $\pi_i(z) \neq 0$, then $z \in V_i$ and so $g_i(z)$ is defined. Obviously, g is odd, since the g_i are odd and the π_i are even. Also, g is continuous. In addition $g(z) \neq 0$ for every $z \in \partial\mathcal{O}$ because there exists $i \in \{0, \dots, \gamma\}$, with $\pi_i(z) \neq 0$, and hence $z \in V_i$ and $|g_i(z)| = 1$. Thus we have a continuous odd mapping $g : \partial\mathcal{O} \to \mathbb{R}^{k+n-1+\gamma} \setminus \{0\}$, where $\mathcal{O}$ is an invariant bounded open neighborhood of 0 in $E_K^+ \oplus E_n^-$. Now Borsuk's theorem implies that $k+n-1+\gamma \geq \dim E_K^+ \oplus E_n^- = K+n$. This shows that $\gamma \geq K-k+1$ as required. □

As an immediate corollary of Theorem 2.1, we obtain the Fountain Theorem, which we state below. First we introduce the following set of conditions.

(Φ_2') There exists a sequence $r_k > 0$, $k \in \mathbb{N}$, such that $b_k \to +\infty$ as $k \to \infty$. (Here b_k is defined as in (Φ_2) with r_k instead of r.)

(Φ_3') There exists a sequence of isomorphisms $T_k : E \to E$, $k \in \mathbb{N}$, with $T_k(E_n) = E_n$ for all k and n, and there exists a sequence $R_k > 0$, $k \in \mathbb{N}$, such that, for $z = z^+ + z^- \in E_k^+ \oplus E^-$ with $\max\{\|z^+\|, \|z^-\|\} = R_k$, one has

$$\|T_k z\| > r_k \quad \text{and} \quad \Phi(T_k z) < 0$$

where r_k is given in (Φ_2').

(Φ_4') $d_k := \sup\{\Phi\big(T_k(z^+ + z^-)\big) : z^+ \in E_k^+, z^- \in E^-, \|z^+\|, \|z^-\| \leq R_k\} < \infty$.

Theorem 2.2. (Fountain Theorem) *Suppose that (Φ_1), (Φ_2') – (Φ_4'), (Φ_5) hold. Then Φ has an unbounded sequence of critical values.*

Hypothesis (Φ_2') will be checked in the applications later on using the contents of the next remark.

Remark 2.2. *Let E be a Hilbert space and $E_1 \subset E_2 \subset \dots$ be finite dimensional subspaces such that $\overline{\bigcup_{n=1}^{\infty} E_n} = E$. Let $\Phi \in C^1(E,\mathbb{R})$ be of the form $\Phi = P - \Psi$ such that*

$$P(z) \geq \alpha \|z\|^p \quad \textit{for all } z \in E$$

and

$$|\Psi(z)| \leq \beta(1 + \|z\|_X^q) \quad \textit{for all } z \in E$$

Here X is a Banach space such that $E \subset X$ compactly, and $q > p$, α, β are positive constants.

First we prove that

$$\mu_k := \sup\{\|z\|_X : z \in E, z \perp E_{k-1}, \|z\| = 1\} \to 0 \quad \textit{as } k \to \infty.$$

Indeed, suppose by contradiction that this is not the case. So, there is $\varepsilon > 0$ and a sequence (z_j) in E with $z_j \perp E_{k_j-1}$, $\|z_j\| = 1$, $\|z_j\|_X \geq \varepsilon$ and $k_j \to \infty$ as $j \to \infty$. This implies that $z_j \rightharpoonup 0$, weakly in E hence $z_j \to 0$, strongly in X, which contradicting $\|z_j\|_X \geq \varepsilon$.

Next we prove that there are $r_k > 0$, $k \in \mathbb{N}$, so that

$$b_k := \inf\{\Phi(z) : z \in E, z \perp E_{k-1}, \|z\| = r_k\} \to \infty.$$

Indeed, for $z \in E$, $z \perp E_{k-1}$, we have

$$\begin{aligned} \Phi(z) &= P(z) - \Psi(z) \\ &\geq \alpha\|z\|^p - \beta(1 + \|z\|_X^q) \\ &\geq \alpha\|z\|^p - \beta - \beta\mu_k^q\|z\|^q. \end{aligned} \tag{2.3}$$

Taking $\|z\| = r_k$ with $r_k := (p\alpha/q\beta\mu_k^q)^{1/(q-p)}$, we obtain

$$\Phi(z) \geq c\mu_k^{p^2/(p-q)} \to +\infty,$$

where c depends only on p, q, α, β.

Although the Fountain Theorem is an immediate consequence of Theorem 2.1, we choose to give a direct proof of it which does not employ the equivariant limit category, since this is a result with many applications.

Proof of the Fountain Theorem. By (Φ_2') it suffices to show that Φ has a critical value in $[b_k, d_k]$, for every k with $b_k > 0$. Fix such a k and suppose that $[b_k, d_k]$ contains only regular values. By Proposition 2.6 in [2], for n large, there exists a continuous deformation $h_n^t : \Phi_n^{d_k} \to E_n$, $t \in [0,1]$, such that h_n^t is odd and $h_n^1(\Phi_n^{d_k}) \subset \Phi_n^{b_k-\varepsilon}$, for some $\varepsilon > 0$. Moreover $h_n^t(z) = z$ if $\Phi(z) \leq 0$. As above we set $B := B_{R_k}E_k^+ \times B_{R_k}E_n^-$, for $n \geq k$. Consider the set

$$\mathcal{O} = \{z \in B : \|h_1(T_k z)\| < r_k\}.$$

As in the proof of Theorem 2.1 one checks easily that $\mathcal{O}$ is an open invariant neighborhood of 0 in $E_k^+ \oplus E_n^-$, and that $\overline{\mathcal{O}} \subset \operatorname{int} B$. Now we set

$$g := P \circ h_1 \circ T_k : \partial\mathcal{O} \longrightarrow E_{k-1}^+ \oplus E_n^-,\ g(z) := P\big(h_1(T_k z)\big)$$

where $P : E_n \to E_{k-1}^+ \oplus E_n^-$ is the orthogonal projection. Since $\dim(E_k^+ \oplus E_n^-) > \dim(E_{k-1}^+ \oplus E_n^-)$, Borsuk's theorem tells us that g must have a zero. Now $z \in \partial\mathcal{O}$ implies that $\|h_1(Tz)\| = r_k$, and $g(z) = 0$ implies that $h_1(T_k z) \in E_n^+$, $h_1(T_k z) \perp E_{k-1}^+$. It follows from (Φ_2') that $\Phi(h_1(T_k z)) \geq b_k$. This contradicts the fact that $T_k z \in \Phi_n^{d_k}$ by (Φ_3') and $h_1(\Phi_n^{d_k}) \subset \Phi_n^{b_k-\varepsilon}$. □

3. The Variational Setting

3.1 The spaces in the case of a bounded domain in $\mathbb{R}^n$. Let φ_n, $n \in \mathbb{N}$, be an orthonormal basis of $L^2(\Omega)$ made up of eigenfunctions of the eigenvalue problem

$$-\Delta u = \lambda u \quad \text{in } \Omega, \qquad u = 0 \quad \text{on } \partial\Omega.$$

Let λ_n be the corresponding eigenvalues. For all real numbers $s > 0$ we define, for $u = \sum_{j=1}^\infty a_j\varphi_j$, $v = \sum_{j=1}^\infty b_j\varphi_j$:

$$H^s(\Omega) := \Big\{u \in L^2(\Omega) : \textstyle\sum_{j=1}^\infty \lambda_j^s |a_j|^2 < \infty\Big\}.$$

This is a Hilbert space with respect to the inner product $\langle u, v\rangle_s := \sum_{j=1}^\infty \lambda_j^s a_j b_j$. Clearly, the operator

$$A^s : H^s(\Omega) \longrightarrow L^2(\Omega), \quad u \longmapsto \sum_{j=1}^\infty \lambda_j^{s/2} a_j \varphi_j$$

is an isometric isomorphism. It is easy to see that

$$\int_\Omega A^s u\,\phi = \int_\Omega u\,A^s\varphi \qquad \text{for all } u, \phi \in H^s(\Omega)$$

which is used to prove the regularity of weak solutions. One has also the Sobolev embeddings

$$H^s(\Omega) \subset L^p(\Omega)$$

continuously if $1 \leq p \leq \frac{2N}{N-2s}$ and compactly if $1 \leq p < \frac{2N}{N-2s}$.

3.2 The spaces in the case $\Omega = \mathbb{R}^n$. In this case, the space $H^s(\mathbb{R}^n)$ is defined by

$$H^s(\mathbb{R}^N) := \big\{u \in L^2(\mathbb{R}^N) : \textstyle\int_{\mathbb{R}^N} (1 + |\xi|^s)^2 \cdot |\widehat{u}(\xi)|^2 d\xi < \infty\big\}$$

where $\widehat{u}$ denotes the Fourier transform of u. This is a Hilbert space with respect to

$$\langle u, v\rangle := \frac{1}{(2\pi)^n}\int_{\mathbb{R}^n} (1 + |\xi|^s)^2 \cdot \widehat{u}(\xi) \cdot \widehat{v}(\xi)\, d\xi.$$

The operator

$$A^s : H^s(\mathbb{R}^n) \longrightarrow L^2(\mathbb{R}^n), \quad u \longmapsto ((1 + |\xi|^s)\widehat{u})^\vee,$$

(where $\omega^\vee$ denotes the inverse Fourier transform of ω) is usually written as $A^s = (1 + |D|^s)$. It is readily seen that it is an isometric isomorphism. It is also easy to see that

$$\int_{\mathbb{R}^n} A^s u\, \phi = \int_{\mathbb{R}^n} u\, A^s \phi \qquad \text{for all } u, \phi \in H^s(\mathbb{R}^N).$$

We observe that $A^2 = u - \Delta u$ since

$$A^2 u = \big((1 + |\xi|^2)\widehat{u}\big)^\vee \quad \text{for } u \in H^2(\mathbb{R}^N).$$

This explains the form of the system (1.2) in the case of $\Omega = \mathbb{R}^n$.

If G is a subgroup of $O(N)$, then we set

$$L^2_G(\mathbb{R}^n) := \{u \in L^2(\mathbb{R}^n) : u(gx) = u(x) \text{ for } g \in G, \text{ and } x \in \mathbb{R}^n\, a.e.\}$$

and $H^s_G := H^s \cap L^2_G$. Clearly we have that

$$A^s(H^s_G(\mathbb{R}^n)) = L^2_G(\mathbb{R}^n)$$

In the case of $\mathbb{R}^n$ there is a loss of the compact embeddings. However, depending on the group G acting on $\mathbb{R}^n$, we can still recover them. We mention the following result due to P.-L. Lions [12]:

Proposition 3.1. *If*

$$G = O(N_1) \times \cdots \times O(N_k)$$

with $N_i \geq 2$ *and* $\sum_{i=1}^k N_i = N$, *then the embedding*

$$H^s_G(\mathbb{R}^n) \subset L^\gamma(\mathbb{R}^n) \quad \text{for } 2 < \gamma < \frac{2N}{N-2s}$$

is compact.

The case when $G = O(N)$ was first proved by Strauss [15].

3.3 The "quadratic" forms and the functionals. In the sequel we write E^s to denote both $H^s(\Omega)$ in the case of a bounded domain Ω, and $H^s(\mathbb{R}^N)$. Let us consider the Cartesian product $E := E^s \times E^t$ with $s, t \geq 0$, which is also a Hilbert space endowed with the inner product

$$\langle z, \eta \rangle := \langle u, \phi \rangle_s + \langle v, \psi \rangle_t, \quad \text{for } z = (u, v),\ \eta = (\phi, \psi) \in E.$$

We consider the bilinear form

$$B : E \times E \longrightarrow \mathbb{R},\ \ B[z, \eta] := \int (A^s u\, A^t \psi + A^s \phi\, A^t v),$$

where $\int$ denotes the integral in both cases, over Ω or over $\mathbb{R}^n$. Associated to B, we have the quadratic form

$$Q(z) := \frac{1}{2} B[z, z] = \int A^s u\, A^t v\,.$$

It is easy to see (cf. [7]) that the bounded self-adjoint operator $L : E \to E$ defined by $\langle Lz, \eta\rangle := B[z,\eta]$ has exactly two eigenvalues $+1$ and -1, and that the corresponding eigenspaces E^+ and E^- are given by

$$E^+ = \{(u, A^{-t}A^s u) : u \in E^s\} \quad \text{and} \quad E^- = \{(u, -A^{-t}A^s u) : u \in E^s\}$$

where we are using the notation $A^{-t} = (A^t)^{-1}$.

Now consider the Hamiltonian $H : \overline{\Omega} \times \mathbb{R} \times \mathbb{R} \to \mathbb{R}$ from Section 1. We consider the functional

$$\Phi(z) := Q(z) - \int H(x,u,v)\,dx \tag{3.1}$$

where $z = (u,v)$. By (H_2) or (H_2'), there exist $s, t > 0$, with $s + t = 2$ and

$$\frac{1}{p} > \frac{1}{2} - \frac{s}{n} \quad \text{and} \quad \frac{1}{q} > \frac{1}{2} - \frac{t}{n}. \tag{3.2}$$

This implies that we have continuous embeddings $E^s \subset L^p$ and $E^t \subset L^q$. We fix s and t with this property so that Φ is well defined in E by (H_2) or (H_2'). Moreover, (H_2) or (H_2') imply that $\Phi \in C^1(E, \mathbb{R})$ with

$$\langle \Phi'(z), \eta\rangle = B[z,\eta] - \int \Big(H_u(x,u,v)\phi + H_v(x,u,v)\psi \Big)\,dx$$

for $z, \eta \in E$. From this one deduces that a critical point $z = (u,v)$ of Φ corresponds to a weak solution of (1.1) or (1.2). Namely

$$\int A^s u\, A^t\psi = \int H_v(x,u,v)\,\psi \qquad \text{for all } \psi \in E^t$$

and

$$\int A^s \phi\, A^t v = \int H_u(x,u,v)\,\phi \qquad \text{for all } \phi \in E^s.$$

As shown in [7] for the case of Ω bounded these solutions are strong in the sense that

$$u \in W^{2,p/(p-1)}(\Omega) \cap W_0^{1,p/(p-1)}(\Omega) \;\text{ and }\; v \in W^{2,q/(q-1)}(\Omega) \cap W_0^{1,q/(q-1)}(\Omega),$$

and they satisfy (1.1). In the case of $\mathbb{R}^n$ we conclude that $u \in W_{\mathrm{loc}}^{2,p/(p-1)}(\mathbb{R}^N)$ and $v \in W_{\mathrm{loc}}^{2,q/(q-1)}(\mathbb{R}^N)$, and that they satisfy (1.2).

In order to apply Theorem 2.1 in the next sections, let us introduce the following notations. Let $(e_j)_{j\in\mathbb{N}}$ be an orthonormal basis of E^s. Clearly the $f_j := A^{-t}A^s e_j$, $j \in \mathbb{N}$, constitute an orthonormal basis of E^t. We set

$$E_n^s := \mathrm{span}\{e_j : j = 1,\dots,n\} \quad \text{and} \quad E_n^t := \mathrm{span}\{f_j : j = 1,\dots,n\}.$$

The following result can be readily seen.

Lemma 3.1. *With the above notations, we have*

$$E^\pm = \overline{\bigcup_{n=1}^\infty E_n^\pm}, \quad E = E^+ \oplus E^-, \quad E_n := E_n^+ \oplus E_n^- = E_n^s \times E_n^t$$

□

Next we check that the functional Φ defined in (3.1) satisfies the condition (Φ_2') in both cases, Ω bounded or $\mathbb{R}^N$.

Lemma 3.2. *Assume* (H_1)–(H_2) *or* (H_1')–(H_2'). *Then there exists a sequence of positive real numbers* r_k, $k \geq 1$, *such that*

$$b_k := \inf\{\Phi(z) : z \in E^+,\ z \perp E_{k-1}^+,\ \|z\| = r_k\} \to +\infty.$$

Proof. We use Remark 2.2 applied to the Hilbert space E^+ with $P(z) := Q(z)$, and

$$\Psi(z) := \int H(x,u,v)$$

where $z = (u,v)$. We know that $Q(z) = \frac{1}{2}\|z\|^2$ for $z \in E^+$, and from (H_2) we obtain

$$\begin{aligned} \left|\int H(x,u,v)\right| &\leq C\left(\int |u|^p + \int |v|^q + 1\right) \\ &\leq C'\left(\|z\|_X^r + 1\right) \end{aligned}$$

where $X = L^p(\Omega) \times L^q(\Omega)$ and $r = \max\{p,q\} > 2$. Also, assuming (H_2') we obtain:

$$\begin{aligned} \left|\int H(x,u,v)\right| &\leq C\left(\int |u|^p + \int |v|^q + \int |u|^{a+1} + \int |v|^{b+1}\right) \\ &\leq C'\|z\|_X^r \end{aligned}$$

for some $r > 2$. □

4. The case when Ω is bounded

In this section we prove Theorem 1.1. With the notation from Section 3, we want to apply Theorem 2.1 to the functional

$$\Phi : E \to \mathbb{R},\ \ \Phi(z) = Q(z) - \int H(x,u,v)\,dx \tag{4.1}$$

where $z = (u,v) \in E = E^s \times E^t$. First we show that Φ satisfies a Palais-Smale condition.

Lemma 4.1. *Assume* (H_1), (H_2) *and* (H_3). *Then* Φ *as defined in* (4.1) *satisfies* $(PS)_c^{\mathcal{F}}$ *for every* $c \in \mathbb{R}$ *and* $\mathcal{F} = (E_n)_{n\in\mathbb{N}}$.

Proof. In view of Remark 2.1, it suffices to prove that a $(PS)_c^{\mathcal{F}}$ sequence is bounded in E. This follows as in Section 3 of [8], up to the point where we are to get some bounds for $\|u_n\|_{E^s}$ and $\|v_n\|_{E^t}$. At this point we then use the fact that $\int A^s u_n\, A^t\psi = 0$, for all $\psi \in (E_n^t)^\perp$. □

Next we check the other assumptions of Theorem 2.1. It has already been proved that Φ satisfies condition (Φ_2'). In particular, condition (Φ_2) is satisfied.

For each $\lambda > 0$, let us define the isomorphism $T_\lambda : E \to E$ by

$$T_\lambda(u, v) = (\lambda^{\beta-1}u, \lambda^{\alpha-1}v).$$

Clearly $T_\lambda E_n = E_n$ for all $\in \mathbb{N}$. Observe however that $E_n^+ \cap T_\lambda E_n^+ = \{0\}$ and $E_n^- \cap T_\lambda E_n^- = \{0\}$ if $\alpha \neq \beta$.

For each $k \in \mathbb{N}$ consider the finite dimensional subspace E_k^s. Since all norms are equivalent in finite dimensional spaces, we have positive constants σ_k, σ_k', τ_k and τ_k' such that

$$\|u\|_{L^2} \geq \sigma_k \|u\|_{E^s} \quad \text{and} \quad \|u\|_{E^s} \geq \sigma_k' \|u\|_{L^\beta} \quad \text{for all } u \in E_k^s$$

and

$$\|v\|_{L^2} \geq \tau_k \|v\|_{E^t} \quad \text{and} \quad \|v\|_{E^t} \geq \tau_k' \|v\|_{L^\alpha} \quad \text{for all } v \in E_k^t.$$

These constants are going to compose the constants δ_k announced in the introduction.

Lemma 4.2. *Assume that there are constants c_2 and c_3, such that*

$$H(x, u, v) \geq \frac{1}{2} c_2(|u|^\alpha + |v|^\beta) - c_3 \tag{4.2}$$

for all $(x, u, v) \in \overline{\Omega} \times \mathbb{R} \times \mathbb{R}$. Then, for each $\lambda > 0$ and each $k \in \mathbb{N}$, one has

$$\Phi(T_\lambda z) \leq \lambda^{\alpha+\beta}(1 - c_2\delta_k) + c_3|\Omega|$$

for all $z = z^+ + z^- \in E_k^+ \oplus E^-$, with $\lambda := \|z^+\|$, where

$$\delta_k := \min\left\{ \left(\frac{\sigma_k^2 \sigma_k'}{2}\right)^\alpha, \left(\frac{\tau_k^2 \tau_k'}{2}\right)^\beta \right\}$$

Proof. For $z = z^+ + z^- \in E_k^+ \oplus E^-$ we write $z^- = z_1^- + z_2^-$ where $z_1^- \in E_k^-$, $z_2^- \perp E_k^-$. We also write $\overline{z} := z^+ + z_1^-$ and $z = (u, v)$, $z^\pm = (u^\pm, v^\pm), \ldots$ Using (4.2) we have

$$\int_\Omega H(x, T_\lambda z) \geq \frac{1}{2} c_2 \big(\lambda^{(\beta-1)\alpha} \|u\|_{L^\alpha}^\alpha + \lambda^{(\alpha-1)\beta} \|v\|_{L^\beta}^\beta\big) - c_3|\Omega|. \tag{4.3}$$

Since α and β are conjugate exponents and $u_2^- \perp \overline{u}$ in L^2, we obtain

$$\begin{aligned} \|u\|_{L^\alpha} &\geq |\langle u, \overline{u}\rangle_{L^2}| \cdot \|\overline{u}\|_{L^\beta}^{-1} \\ &= \|\overline{u}\|_{L^2}^2 \|\overline{u}\|_{L^\beta}^{-1} \\ &\geq \sigma_k^2 \sigma_k' \|\overline{u}\|_{E^s} \\ &=: \widetilde{\sigma}_k \|\overline{u}\|_{E^s}. \end{aligned} \tag{4.4}$$

Similarly

$$\|v\|_{L^\beta} \geq \tau_k^2\, \tau_k' \|\overline{v}\|_{E^t} =: \widetilde{\tau}_k \|\overline{v}\|_{E^t}. \tag{4.5}$$

Next observe that $\overline{u} = u^+ + u_1^-$ and $\overline{v} = A^{-t}\, A^s(u^+ - u_1^-)$. So

$$\|\overline{u}\|_{E^s} = \|u^+ + u_1^-\|_{E^s} \quad \text{and} \quad \|\overline{v}\|_{E^t} = \|u^+ - u_1^-\|_{E^s}. \tag{4.6}$$

Using (4.3)–(4.6) we obtain the following estimate on the Hamiltonian:

$$\begin{aligned}\int_\Omega & H(x, T_\lambda z)\\ &\geq \frac{1}{2} c_2 \Big(\lambda^{(\beta-1)\alpha} \widetilde{\sigma}_K^\alpha \|u^+ + u_1^-\|_{E^s}^\alpha + \lambda^{(\alpha-1)\beta} \widetilde{\tau}_K^\beta \|u^+ - u_1^-\|_{E^s}^\beta \Big) - c_3|\Omega|\end{aligned}$$

Since $\|u^+\|_{E^s} = \|z^+\|/2 = \lambda/2$, we have that either $\|u^+ + u_1^-\|_{E^s} \geq \lambda/2$ or $\|u^+ - u_1^-\|_{E^s} \geq \lambda/2$. In either case

$$\int_\Omega H(x, T_\lambda z) \geq \frac{1}{2} c_2\, \delta_k\, \lambda^{\alpha+\beta} - c_3|\Omega|. \tag{4.7}$$

On the other hand we have that

$$Q(T_\lambda z) = \lambda^{\alpha+\beta-2} Q(z) = \frac{1}{2} \lambda^{\alpha+\beta-2} (\|z^+\|^2 - \|z^-\|^2)$$

and so for $\|z^+\| = \lambda$ we obtain

$$Q(T_\lambda z) \leq \frac{1}{2} \lambda^{\alpha+\beta}. \tag{4.8}$$

Thus it follows from (4.7) and (4.8) that

$$\Phi(T_\lambda z) \leq \frac{1}{2}(1 - c_2\delta_k)\lambda^{\alpha+\beta} + c_3|\Omega| \quad \text{for all } \lambda = \|z^+\| \leq \|z^-\|. \tag{4.9}$$

□

An immediate consequence of Lemma 4.2 is the next result which establishes (Φ_3).

Lemma 4.3. *Suppose that* (4.2) *holds and there is* $K \in \mathbb{N}$ *such that*

$$1 - c_2\, \delta_K < 0 \tag{4.10}$$

Then, fixing $r > 0$ *there is a* $\lambda_0 > 0$ *such that, for all* $\lambda \geq \lambda_0$ *one has, for* $z = z^+ + z^- \in E_K^+ \oplus E^-$:

$$\Phi(T_\lambda z) < 0 \quad \text{if } \|z^+\| = \lambda, \quad \text{and} \quad \|T_\lambda z\| > r \quad \text{if } \max\{\|z^+\|, \|z^-\|\} = \lambda.$$

□

Proof of Theorem 1.1. As mentioned before we apply Theorem 2.1. First we observe that Lemma 4.1 gives condition (Φ_1). Lemma 3.2 implies that there exists a $k_0 \in \mathbb{N}$ such that $b_{k_0} > 0$, which then gives (Φ_2) and (Φ_2'). Now suppose that (1.9) holds for some $K \geq k_0$. Then we can apply Lemmas 4.2 and 4.3 and conclude that (Φ_3) holds. (Φ_4) is implied by the fact that Φ maps bounded sets of E into bounded sets of $\mathbb{R}$. Finally, condition (Φ_5) is a consequence of (H_4). □

5. The case $\Omega = \mathbb{R}^n$

With the notations introduced in Section 3 the weak solutions of (1.2) are the critical points of the functional

$$\Phi(z) = \int_{\mathbb{R}^N} A^s u\, A^t v \; - \; \int_{\mathbb{R}^N} H(x,u,v) \tag{5.1}$$

acting in $E = E^s \times E^t := H^s(\mathbb{R}^N) \times H^t(\mathbb{R}^N)$, where s and t satisfy (3.2). We shall consider the functional Φ restricted to certain subspaces of E where we have compact embeddings due to symmetry properties. Let us start with the group $G = O(N)$ acting in $\mathbb{R}^n$, and let us look for critical points of Φ in the subspace X of E given by $X = H^s_{O(N)}(\mathbb{R}^n) \times H^t_{O(N)}(\mathbb{R}^n)$. All subspaces introduced in Section 3 are now restricted to spherically symmetric functions. Observe that

$$X := \mathrm{Fix}(G) = \{(u,v) \in E : gu = u,\ gv = v,\ \text{ for all } g \in O(N)\}$$

where gu means $(gu)(x) = u(gx)$, for all $x \in \mathbb{R}^n$. We see also that Φ is invariant with respect to G, i.e. $\Phi(gu, gv) = \Phi(u,v)$. Hence, it follows from the Palais Principle of Symmetric Criticality, see [13] or [16], that the critical points of Φ restricted to X are critical points of Φ considered in the whole space E.

In order to prove Theorem 2.1, we have to check that $\Phi\big|_X$ satisfies the assumptions of the Fountain Theorem.

For each $\lambda > 0$, let us define the isomorphism $T_\lambda : E \to E$, by

$$T_\lambda(u,v) = (\lambda^\mu u,\ \lambda^\nu v),$$

where $\mu = \frac{m-p}{p}$ and $\nu = \frac{m-q}{q}$, and $m > \max(p,q)$. Clearly $T_\lambda E_n = E_n$.

Lemma 5.1. *Assume conditions* (H_1')-(H_4'). *Then, there is a sequence* $\lambda_k > 0$, $k \in \mathbb{N}$, *such that* (Φ_3') *holds with* $T_k := T_{\lambda_k}$ *and* $R_k := \lambda_k$.

Proof. Let us use the notation introduced in the proof of Lemma 4.2. It follows from (1.16) that for any $\lambda > 0$ we have

$$\int_{\mathbb{R}^N} H(x, T_\lambda z) \geq c \left(\lambda^{\mu p} \int_{\mathbb{R}^N} |u|^p + \lambda^{\nu q} \int_{\mathbb{R}^N} |v|^q \right). \tag{5.2}$$

Using Hölder's inequality, we obtain

$$\|u\|_{L^p} \, \|\overline{u}\|_{L^{p'}} \geq |\langle u, \overline{u} \rangle_{L^2}| = \|\overline{u}\|^2_{L^2} \quad \text{where } \frac{1}{p} + \frac{1}{p'} = 1,$$

Next, from the finite dimensionality of E_k^s we have that there is a positive constant γ_k such that

$$\|u\|_{L^p} \geq \gamma_k \|\overline{u}\|_{E^s} \quad \text{for } u \in E_k^s.$$

Similarly there is $\widetilde{\gamma} > 0$ with

$$\|v\|_{L^q} \geq \widetilde{\gamma}_k \|\overline{v}\|_{E^t} \quad \text{for } v \in E_k^t.$$

Thus it follows from (5.2) that

$$\int_{\mathbb{R}^N} H(x, T_\lambda z) \geq c\big(\lambda^{\mu p}\gamma_k^p \|\overline{u}\|_{E^s}^p + \lambda^{\nu q}\widetilde{\gamma}_k^q \|\overline{v}\|_{E^t}^q\big).$$

As in the proof of Lemma 4.2 we obtain

$$\int_{\mathbb{R}^N} H(x, T_\lambda z) \geq c \cdot \min\left\{\frac{1}{2^p}\lambda^{\mu p}\gamma_k^p\lambda^p,\ \frac{1}{2^q}\lambda^{\nu q}\widetilde{\gamma}_k^q\lambda^q\right\} \geq \sigma_k \lambda^m$$

provided $\|z^+\| = \lambda$.

On the other hand,

$$Q(T_\lambda z) = \lambda^{\mu+\nu}(\|z^+\|^2 - \|z^-\|^2) \leq \lambda^{\mu+\nu+2}$$

for $\|z^+\| = \lambda$. Consequently we have

$$\Phi(T_\lambda z) \leq \lambda^{\mu+\nu+2} - \sigma_k \lambda^m.$$

Since $m > \mu + \nu + 2$, it follows that there is a $\lambda_0(k) > 0$ such that $\Phi(T_{\lambda_k} z) < 0$ if $\lambda_k > \lambda_0(k)$. Also

$$\|T_\lambda z\| \geq \lambda^{\min\{\mu,\nu\}} \|z\|^2,$$

which implies that

$$\|T_{\lambda_k} z\| \geq \lambda_k^{\min\{\mu,\nu\}+2} \qquad \text{for } \max\{\|z^+\|, \|z^-\|\} = \lambda_k.$$

Therefore, we can choose λ_k such that

$$\Phi(T_{\lambda_k} z) < 0 \quad \text{and} \quad \|T_{\lambda_k} z\| \geq r_k$$

for any given r_k. □

Proof of Theorem 1.2. First we observe that hypotheses (H_1') and (H_2') imply that Φ is C^1 in E. And using (H_3') we prove easily that a $(PS)_c^{\mathcal{F}}$ sequence is bounded in X. So it follows from Remark 2.1 and Proposition 3.1 that $\Phi\big|_X$ satisfies the $(PS)_c^{\mathcal{F}}$ condition. Hence (Φ_1) holds. Condition (Φ_2') has already been checked in Lemma 3.2. Condition (Φ_3') is proved in Lemma 5.1. Condition (Φ_4') is trivially verified, and finally (Φ_5) is a consequence of (H_5). So we apply the Fountain Theorem and conclude. □

We omit the proof of Theorem 1.3 since it parallels a similar result of Bartsch-Willem [3] for the scalar case. The result in our case follows from an application of the Fountain Theorem, using Proposition 3.1.

References

[1] A. Bahri and H. Berestycki, "Existence of forced oscillations for some nonlinear differential equations", Comm. Pure Appl. Math. **37** (1984), 403–442.

[2] T. Bartsch and M. Clapp, "Critical point theory for indefinite functionals with symmetries", J. Fctl. Anal. **138** (1996), 107–136.

[3] T. Bartsch and M. Willem, "Infinitely many nonradial solutions of a Euclidean scalar field equation", J. Fctl. Anal. **117** (1993), 447–460.

[4] G. Cerami, "Un criterio di esistenza per i punti critici su varietà illimitate", Rec. Ist. Lomb. Sci. Lett. **112** (1978), 332–336.

[5] D.G. Costa and C.A. Magalhães, "A unified approach to a class of strongly indefinite functionals", J. Diff. Eq. **122** (1996), 521–547.

[6] D.G. de Figueiredo, "Semilinear elliptic systems", Proceedings of the Second School on Nonlinear Functional Analysis and Applications to Differential Equations, ICTP Trieste 1997, World Scientific Publishing Company (to appear).

[7] D.G. de Figueiredo and P. Felmer, "On superquadratic elliptic systems", Trans. AMS **343** (1994), 99–116.

[8] D.G. de Figueiredo and C.A. Magalhães, "On nonquadratic Hamiltonian elliptic systems", Advances in Diff. Eq. **1** (1996), 881–898.

[9] D.G. de Figueiredo and J. Yang, Decay, "Symmetry and existence of solutions of semilinear elliptic systems", to appear in Nonlinear Analysis, TMA.

[10] P. Felmer, "Periodic solutions of 'superquadratic' Hamiltonian systems", J. Diff. Eq. **102** (1993), 188–207.

[11] S. Li and J.Q. Liu, "Some existence theorems on multiple critical points and their applications", Kexue Tongbao **17** (1984), [in chinese].

[12] P.-L. Lions, "Symétrie et compacité dans les espaces de Sobolev", J. Funct. Anal. **49** (1982), 315–334.

[13] R.S. Palais, "The principle of symmetric criticality", Comm. Math. Phys. **69** (1979), 19–30.

[14] B. Sirakov, "On the existence of solutions of elliptic systems in $\mathbb{R}^N$", Preprint.

[15] W.A. Strauss, "Existence of solitary waves in higher dimensions", Comm. Math. Phys. (1977), 149–162.

[16] M. Willem, "Minimax Theorems", Progress in Nonlinear Differential Equations and their Applications, **24**, Birkhäuser 1996.

Thomas Bartsch
Mathematisches Institut
Universität Giessen
Arndtstr. 2
35392 Giessen, Germany

Djairo G. de Figueiredo
IMECC, UNICAMP
Caixa Postal 6065
13081-970 Campinas SP, Brazil

Progress in Nonlinear Differential Equations
and Their Applications, Vol. 35
© 1999 Birkhäuser Verlag Basel/Switzerland

Hölder Regularity for a Linear Fractional Evolution Equation

Ph. Clément, G. Gripenberg, and S-O. Londen

Dedicated to Professor Herbert Amann on the occasion of his 60th birthday

Abstract. The regularity of solutions of the equation

$$\left(D_t^\alpha(u-u_0)\right)(t,x)+c(t,x)u_x(t,x)=f(t,x),\quad t,\,x\geq 0,$$

where D_t^α denotes the fractional derivative, is studied. It is shown that if c and f are Hölder continuous in either t or x and c is strictly positive, then both u_x and $D_t^\alpha(u-u_0)$ are Hölder continuous as well (in either t or x).

1. Introduction

The purpose of this paper is to study the maximal regularity of the solutions of the partial differential equation

$$\left(D_t^\alpha(u-u_0)\right)(t,x)+c(t,x)u_x(t,x)=f(t,x),\quad t,\,x\geq 0, \tag{1}$$

with initial and boundary conditions

$$\begin{aligned} u(0,\underline{x}) &= u_0(\underline{x}),\\ u(\underline{t},0) &= u_1(\underline{t}). \end{aligned} \tag{2}$$

Here D_t^α denotes the fractional derivative of order $\alpha\in(0,1)$, see [8, p. 133], i.e.,

$$\begin{aligned} (D_t^\alpha v)(t) &\overset{\text{def}}{=} \frac{\mathrm{d}}{\mathrm{d}t}\int_0^t g_{1-\alpha}(t-s)v(s)\,\mathrm{d}s,\quad t>0,\\ (D_t^\alpha v)(0) &\overset{\text{def}}{=} \lim_{h\downarrow 0}\frac{1}{h}\int_0^h g_{1-\alpha}(h-s)v(s)\,\mathrm{d}s, \end{aligned}$$

where

$$g_\beta(t)\overset{\text{def}}{=}\frac{1}{\Gamma(\beta)}t^{\beta-1},\quad t>0,\quad \beta>0,$$

and where v is (at least) continuous and satisfies $v(0)=0$. If $w(\underline{t},\underline{x})$ is a function of two variables, then $(D_t^\alpha w)(\underline{t},\underline{x})$ is the function $(t,x)\mapsto(D_t^\alpha w(\cdot,x))(t)$.

The maximal regularity results that we prove state essentially that if c and f are Hölder continuous with respect to one variable, c is strictly positive and satisfies an additional continuity assumption, and the initial/boundary data satisfy certain compatibility conditions, then there is a unique solution of (1) such that u_x and $D_t^\alpha(u - u_0)$ are Hölder continuous with respect to the same variable. In addition, appropriate Schauder estimates are obtained. By integration and subsequent interpolation we obtain additional results on the regularity of u in both variables.

Our work on the smoothness of solutions of (1) is motivated by problems related to the nonlinear fractional conservation law

$$D_t^\alpha(u - u_0) + \sigma(u)_x = f. \tag{3}$$

Equations of this type can be employed to approximate solutions of nonlinear conservation laws, [3], and the properties of their solutions have been studied in e.g. [4] and [5]. A sufficiently smooth solution of (3) satisfies (1) with $c(\underline{t}, \underline{x}) = \sigma'(u(\underline{t}, \underline{x}))$. Thus the results we prove below can be used to show that if there is a solution of (3) that is Hölder continuous in one variable (continuously with respect to the other), then it follows, provided the functions f, u_0 and u_1 are sufficiently smooth, that it satisfies better Hölder conditions in both variables. Since this inference relies on quite strong assumptions on the smoothness of the solution it is, of course, not satisfactory. Unfortunately, there appear to be grave difficulties involved in trying to get the necessary bounds if one applies a direct bootstrapping technique. So it seems that one has to use a slightly different approach. We will return to this problem in future work.

Another motivation for studying equation (1) is that it is (when $c = 1$) the special case $\beta = 1$ of the equation

$$D_t^\alpha(u - u_0) + D_x^\beta(u - u_1) = f,$$

studied in [2] for the cases α, $\beta \in (0, 1)$.

Our proofs rely on an extension of [2, Thm. 6] to the nonconstant coefficient case and on an analogous extension of [7, Thms. 4.5, 5.5]. A key fact necessary for our analysis is that D_t^α is a positive operator with spectral angle $\alpha\pi/2 < \pi/2$. The condition $\alpha < 1$ is crucial for maximal regularity to hold. The Hölder spaces repeatedly appear in the analysis as interpolation spaces, see e.g. [1].

2. Statement of results

Let X be a (complex) Banach space and let I be an interval. The Hölder spaces $\mathcal{C}^{(\gamma)}(I; X)$, $\gamma \in [0, 1]$ are defined by

$$\mathcal{C}^{(\gamma)}(I; X) \stackrel{\text{def}}{=} \left\{ f : I \to X \,\middle|\, \sup_{\substack{s,t \in I \\ s \neq t}} \frac{\|f(t) - f(s)\|_X}{|t - s|^\gamma} < \infty \right\},$$

with norm

$$\|f\|_{\mathcal{C}^{(\gamma)}(I)} \stackrel{\text{def}}{=} \sup_{t \in I} \|f(t)\|_X + \sup_{\substack{s,t \in I \\ s \neq t}} \frac{\|f(t) - f(s)\|_X}{|t - s|^\gamma}.$$

If $\gamma \in (1,2]$, then $\mathcal{C}^{(\gamma)}(I;X) \stackrel{\text{def}}{=} \{ f \in \mathcal{C}^1(I;X) \mid f' \in \mathcal{C}^{(\gamma-1)}(I;X) \}$ with norm $\|f\|_{\mathcal{C}^{(\gamma)}(I)} \stackrel{\text{def}}{=} \sup_{t \in I} \|f(t)\|_X + \|f'\|_{\mathcal{C}^{(\gamma-1)}(I)}$. Observe that $\mathcal{C}^{(0)} \neq \mathcal{C}$ and $\mathcal{C}^{(1)} \neq \mathcal{C}^1$.

We consider a function of two variables to be a function of the first variable with values in a function space, that is, $f(\underline{t},\underline{x})$ is the function $t \mapsto (x \mapsto f(t,x))$. When we have to consider the other possibility, namely the function $x \mapsto (t \mapsto f(t,x))$ we use the notation $f_{\curvearrowleft}$ for the function, that is, we define

$$f_{\curvearrowleft}(\underline{x},\underline{t}) \stackrel{\text{def}}{=} f(\underline{t},\underline{x}).$$

In addition, in each particular case below, the domains of the variables indicate how the function is to be viewed. The letters τ, T, etc. refer to the first variable t and ξ, $\mathfrak{X}$, etc. to the second variable x.

Observe that the statement $f \in \mathcal{C}^{(\mu)}([0,\tau];\mathcal{C}([0,\xi];\mathbb{C}))$ is not equivalent to $f_{\curvearrowleft} \in \mathcal{C}([0,\xi];\mathcal{C}^{(\mu)}([0,\tau];\mathbb{C}))$. The second implies the first, but not conversely. We do, however, have $f \in \mathcal{C}^{(\mu)}([0,\tau];\mathcal{C}^{(\nu)}([0,\xi];\mathbb{C}))$ if and only if $f_{\curvearrowleft} \in \mathcal{C}^{(\nu)}([0,\xi];\mathcal{C}^{(\mu)}([0,\tau];\mathbb{C}))$

Our first result concerns Hölder continuity in t. We will use the notation $\mathcal{C}_{0 \mapsto 0}$ for functions vanishing at 0.

Theorem 1. *Assume that $\alpha \in (0,1)$, $\tau > 0$, $\xi > 0$, $\mu \in (0,\alpha)$, and that*

(i) $c_{\curvearrowleft} \in \mathcal{C}([0,\xi];\mathcal{C}^{(\mu)}([0,\tau];\mathbb{R}))$, *and*

$$c(t,x) > 0, \quad (t,x) \in [0,\tau] \times [0,\xi];$$

(ii) $f \in \mathcal{C}^{(\mu)}([0,\tau];\mathcal{C}([0,\xi];\mathbb{C}))$ *and* $f(0,0) = 0$;

(iii) $u_0 \in \mathcal{C}^1([0,\xi];\mathbb{C})$, $u_0(0) = u_0'(0) = 0$ *and*

$$c(0,\underline{x})u_0'(\underline{x}) - f(0,\underline{x}) \in \mathcal{C}^{(\mu/\alpha)}([0,\xi];\mathbb{C});$$

(iv) $u_1 \in \mathcal{C}_{0 \mapsto 0}([0,\tau];\mathbb{C})$ *and* $D_t^\alpha u_1 \in \mathcal{C}_{0 \mapsto 0}^{(\mu)}([0,\tau];\mathbb{C})$.

Then there exists a unique solution $u \in \mathcal{C}([0,\tau] \times [0,\xi];\mathbb{C})$ of (1) on $[0,\tau] \times [0,\xi]$ such that

(a) $u_x \in \mathcal{C}^{(\mu)}([0,\tau];\mathcal{C}([0,\xi];\mathbb{C}))$;

(b) $u \in \mathcal{C}^{(\alpha)}([0,\tau];\mathcal{C}([0,\xi];\mathbb{C}))$;

(c) $u \in \mathcal{C}^{(\theta\mu+(1-\theta)\alpha)}([0,\tau];\mathcal{C}^{(\theta)}([0,\xi];\mathbb{C}))$ *for every* $\theta \in (0,1)$.

Moreover, there is a constant M that depends on α, μ, τ, ξ, and c such that

$$\begin{aligned} &\|u_x\|_{\mathcal{C}^{(\mu)}([0,\tau];\mathcal{C}([0,\xi]))} + \|u\|_{\mathcal{C}^{(\alpha)}([0,\tau];\mathcal{C}([0,\xi]))} + \|u\|_{\mathcal{C}^{(\theta\mu+(1-\theta)\alpha)}([0,\tau];\mathcal{C}^{(\theta)}([0,\xi]))} \\ &\quad \leq M\Big(\|f\|_{\mathcal{C}^{(\mu)}([0,\tau];\mathcal{C}([0,\xi]))} + \|c(0,\underline{x})u_0'(\underline{x}) - f(0,\underline{x})\|_{\mathcal{C}^{\mu/\alpha}([0,\xi])} \\ &\qquad + \|D_t^\alpha u_1\|_{\mathcal{C}^{\mu}([0,\tau])}\Big). \end{aligned} \tag{4}$$

The claim (b) (and therefore also (c)) can be improved as follows: What we actuallly have is that

$$u(\underline{t},\underline{x}) - u_0(\underline{x}) + \frac{1}{\Gamma(\alpha+1)}\underline{t}^\alpha(c(0,\underline{x})u_0'(\underline{x}) - f(0,\underline{x})) \in \mathcal{C}^{(\alpha+\mu)}([0,\tau];\mathcal{C}([0,\xi];\mathbb{C})),$$

(except in the case where $\alpha+\mu=1$; then the space of Lipschitz functions should be replaced by the Zygmund class). In order to use interpolation to get an improved version of (c) one must make extra assumptions on the smoothness of the term $c(0,\underline{x})u_0'(\underline{x})-f(0,\underline{x})$ or not use the full force of (a).

Our second result concerns Hölder continuity in x.

Theorem 2. *Assume that $\alpha\in(0,1)$, $\tau>0$, $\xi>0$, $\nu\in(0,1)$, and that*

(i) $c\in\mathcal{C}([0,\tau],\mathcal{C}^{(\nu)}([0,\xi];\mathbb{R}))$ *and*

$$c(t,x)>0,\quad (t,x)\in[0,\tau]\times[0,\xi];$$

(ii) $f_{\curvearrowleft}\in\mathcal{C}^{(\nu)}([0,\xi];\mathcal{C}([0,\tau];\mathbb{C}))$ *and* $f(0,0)=0$;
(iii) $u_0\in\mathcal{C}^{(1+\nu)}([0,\xi];\mathbb{C})$, $u_0(0)=u_0'(0)=0$;
(iv) $u_1\in\mathcal{C}_{0\mapsto 0}([0,\tau];\mathbb{C})$ *and* $\frac{1}{c(\underline{t},0)}\left((D_t^\alpha u_1)(\underline{t})-f(\underline{t},0)\right)\in\mathcal{C}_{0\mapsto 0}^{(\nu\alpha)}([0,\tau];\mathbb{C})$.

Then there exists a unique solution $u\in\mathcal{C}([0,\tau]\times[0,\xi];\mathbb{C})$ of (1) such that

(a) $(u_x)_{\curvearrowleft}\in\mathcal{C}^{(\nu)}([0,\xi];\mathcal{C}([0,\tau];\mathbb{C}))$;
(b) $u\in\mathcal{C}^{(\alpha)}([0,\tau];\mathcal{C}^{(\nu)}([0,\xi];\mathbb{C}))$;
(c) $u\in\mathcal{C}^{((1-\theta)\alpha)}([0,\tau];\mathcal{C}^{(\theta+\nu)}([0,\xi];\mathbb{C}))$ *for each* $\theta\in(0,1)$.

Moreover, there is a constant M that depends on α, ν, τ, ξ and c such that

$$\begin{aligned}
&\|u_x\|_{\mathcal{C}^{(0)}([0,\tau];\mathcal{C}^{(\nu)}([0,\xi]))}+\|u\|_{\mathcal{C}^{(\alpha)}([0,\tau];\mathcal{C}^{(\nu)}([0,\xi]))}+\|u\|_{\mathcal{C}^{((1-\theta)\alpha)}([0,\tau];\mathcal{C}^{(\theta+\nu)}([0,\xi]))}\\
&\quad\le M\Bigg(\|f\|_{\mathcal{C}^{(0)}([0,\tau];\mathcal{C}^{(\nu)}([0,\xi]))}+\|u_0'\|_{\mathcal{C}^{\nu}([0,\xi])}\\
&\qquad+\left\|\frac{1}{c(\underline{t},0)}\Big((D_f^\alpha u_1)(\underline{t})-f(\underline{t},0)\Big)\right\|_{\mathcal{C}^{\nu\alpha}([0,\tau])}\Bigg).
\end{aligned}\tag{5}$$

We need the following simple lemma which is an extension of [6, Prop. 1.1.5, 2.2.12]. For the definition of the spaces $(X,Y)_{\theta,\infty}$ and the K-method, see, e.g., [1] or [6].

Lemma 3. *Let X and Y be Banach spaces that are continuously injected in a Hausdorff locally convex topological vector space. If I is an interval and $f\in\mathcal{C}^{(\alpha)}(I;X)\cap\mathcal{C}^{(\beta)}(I;Y)$ where α, $\beta\in(0,2]$, then $f\in\mathcal{C}^{((1-\theta)\alpha+\theta\beta)}(I;(X,Y)_{\theta,\infty})$ for each $\theta\in(0,1)$ and*

$$\begin{aligned}
\|f\|_{\mathcal{C}^{((1-\theta)\alpha+\theta\beta)}(I;(X,Y)_{\theta,\infty})}&\le 2\big(\|f\|_{\mathcal{C}^{(\alpha)}(I;X)}+\|f\|_{\mathcal{C}^{(\beta)}(I;Y)}\big)\\
&\quad+\begin{cases}2^\alpha|I|^{1-\alpha}\sup\limits_{t\in I}\|f'(t)\|_X, & \text{if } \alpha>1,\beta\le 1,\\ 2^\beta|I|^{1-\beta}\sup\limits_{t\in I}\|f'(t)\|_Y, & \text{if } \beta>1,\alpha\le 1.\end{cases}
\end{aligned}$$

Here $|I|$ denotes the length of the interval I.

3. Proofs

Proof of Lemma 3. First we recall that by [1, (2.2.2)] we have

$$\|v\|_{(X,Y)_{\theta,\infty}} \le \|v\|_X^{1-\theta}\|v\|_Y^{\theta}, \quad v \in (X,Y)_{\theta,\infty}. \tag{6}$$

Using this inequality we get

$$\sup_{t\in I}\|f(t)\|_{(X,Y)_{\theta,\infty}} \le \sup_{t\in I}\|f(t)\|_X + \sup_{t\in I}\|f(t)\|_Y.$$

Similarly, if α and $\beta > 1$ then

$$\sup_{t\in I}\|f'(t)\|_{(X,Y)_{\theta,\infty}} \le \sup_{t\in I}\|f'(t)\|_X + \sup_{t\in I}\|f'(t)\|_Y.$$

Furthermore, in the case where $0 < \alpha, \beta \le 1$, we get

$$\sup_{\substack{t,s\in I\\ s\ne t}} \frac{\|f(t)-f(s)\|_{(X,Y)_{\theta,\infty}}}{|t-s|^{(1-\theta)\alpha+\theta\beta}} \le \sup_{\substack{t,s\in I\\ s\ne t}} \frac{\|f(t)-f(s)\|_X}{|t-s|^{\alpha}} + \sup_{\substack{t,s\in I\\ s\ne t}} \frac{\|f(t)-f(s)\|_Y}{|t-s|^{\beta}},$$

and, in the case where $1 < \alpha, \beta \le 2$, we have

$$\sup_{\substack{t,s\in I\\ s\ne t}} \frac{\|f'(t)-f'(s)\|_{(X,Y)_{\theta,\infty}}}{|t-s|^{(1-\theta)\alpha+\theta\beta-1}} \le \sup_{\substack{t,s\in I\\ s\ne t}} \frac{\|f'(t)-f'(s)\|_X}{|t-s|^{\alpha-1}} + \sup_{\substack{t,s\in I\\ s\ne t}} \frac{\|f'(t)-f'(s)\|_Y}{|t-s|^{\beta-1}}.$$

Next we consider the case where $(\alpha-1)(\beta-1) < 0$ and we assume first that $\alpha > 1$ and $\beta < 1$. Furthermore, let us assume that $\theta = (\alpha-1)/(\alpha-\beta)$ so that $(1-\theta)\alpha+\theta\beta = 1$. This case is made more difficult by the fact that the estimates we obtain depend on the length of the interval. First we derive an estimate on f'.

Let $t \in I$ and let $h \ne 0$ be such that $t+h \in I$. Now we have

$$f'(t) = \int_0^1 \big(f'(t) - f'(t+\sigma h)\big)\,d\sigma + \frac{f(t+h)-f(t)}{h},$$

and we see that the first term belongs to X and the second to Y. If $\rho > 0$, then

$$K(\rho, f'(t), X, Y) \le |h|^{\alpha-1} \sup_{\substack{t,s\in I\\ s\ne t}} \frac{\|f'(t)-f'(s)\|_X}{|t-s|^{\alpha-1}}$$
$$+ \rho|h|^{\beta-1} \sup_{\substack{t,s\in I\\ s\ne t}} \frac{\|f(t)-f(s)\|_Y}{|t-s|^{\beta}}.$$

If I is unbounded we can for any $\rho > 0$ choose h such that $|h| = \rho^{1/(\alpha-\beta)}$. Then

$$\rho^{-\theta}K(\rho, f'(t), X, Y) \le \sup_{\substack{t,s\in I\\ s\ne t}} \frac{\|f'(t)-f'(s)\|_X}{|t-s|^{\alpha-1}} + \sup_{\substack{t,s\in I\\ s\ne t}} \frac{\|f(t)-f(s)\|_Y}{|t-s|^{\beta}}.$$

If I has length $|I|$ we can only be certain to be able to have $|h| < |I|/2$ and this means that we get the inequality above for $\rho < (|I|/2)^{\alpha-\beta}$. For larger values of ρ we have

$$\rho^{-\theta}K(\rho, f'(t), X, Y) \le \left(\frac{|I|}{2}\right)^{1-\alpha} \sup_{t\in I}\|f'(t)\|_X.$$

Thus we have shown that $f'(t) \in (X,Y)_{\theta,\infty}$ and, since $f(t) - f(s) = \int_s^t f'(\sigma)\,d\sigma$, and as we can treat $f'(\sigma)$ as above, we conclude that $f \in \mathcal{C}^{(1)}(I;(X,Y)_{\theta,\infty})$ and that we have the estimate

$$\|f\|_{\mathcal{C}^{(1)}(I;(X,Y)_{\theta,\infty})} \leq \sup_{\substack{t,s\in I\\ s\neq t}} \frac{\|f'(t)-f'(s)\|_X}{|t-s|^{\alpha-1}} + \sup_{\substack{t,s\in I\\ s\neq t}} \frac{\|f(t)-f(s)\|_Y}{|t-s|^{\beta}} + \left(\frac{|I|}{2}\right)^{1-\alpha} \sup_{t\in I}\|f'(t)\|_X.$$

If $\alpha < 1$ and $\beta > 1$, then we use the fact that $(X,Y)_{\theta,\infty} = (Y,X)_{1-\theta,\infty}$ and obtain a similar result.

Next we consider the case where $\alpha > 1$ and $\beta = 1$. Let $t \in I$. Since f is differentiable in X we see that $\{\frac{1}{h}(f(t+h)-f(t))\}$ is Cauchy sequence in X as $h \to 0$ (such that $t+h \in I$) and then it follows from (6) that it is a Cauchy sequence in $(X,Y)_{\theta,\infty}$ as well. Therefore it converges in this space and this shows that f is differentiable in $(X,Y)_{\theta,\infty}$. Applying inequality (6) to $\frac{1}{h}\big(f(t+h)-f(t)\big) - \frac{1}{k}\big(f(s+k)-f(s)\big)$ and then letting h and $k \to 0$ we get

$$\sup_{\substack{t,s\in I\\ s\neq t}} \frac{\|f'(t)-f'(s)\|_{(X,Y)_{\theta,\infty}}}{|t-s|^{(1-\theta)\alpha+\theta\beta}} \leq \sup_{\substack{t,s\in I\\ s\neq t}} \frac{\|f'(t)-f'(s)\|_X}{|t-s|^{\alpha}} + 2\sup_{\substack{t,s\in I\\ s\neq t}} \frac{\|f(t)-f(s)\|_Y}{|t-s|^{\beta}},$$

which is the desired result. If $\alpha = 1$ and $\beta > 1$ we again use the fact that $(X,Y)_{\theta,\infty} = (Y,X)_{1-\theta,\infty}$.

The remaining cases can be obtained by combining the results already obtained and using the reiteration theorem. □

Lemma 4. *Let*

$$\tau > 0,\ a \in \mathcal{C}([0,\tau];\mathbb{R}) \text{ with } a(\underline{t}) > 0, \text{ and let } Y = \mathcal{C}_{0\mapsto 0}([0,\tau];\mathcal{C}([0,\xi];\mathbb{C})).$$

Define the operator A_a in Y by $A_a u(\underline{t}) = \frac{1}{a(\underline{t})}(D_t^\alpha u)(\underline{t})$. Then there is for each $\eta > 0$ a constant N that depends on η, α, τ, $\inf_{t\in[0,\tau]} a(t)$, $\sup_{t\in[0,\tau]} a(t)$, and on the modulus of continuity of a, such that

$$\|(\lambda I + A_a)^{-1}\|_{\mathcal{L}(Y)} \leq \frac{N}{|\lambda|+1}, \quad \text{when } |\arg\lambda| \leq \pi(1-\frac{\alpha}{2}) - \eta.$$

Proof of Lemma 4. If we can prove this lemma with $\mathcal{C}([0,\xi];\mathbb{C})$ replaced by $\mathbb{C}$, then we get pointwise estimates for functions depending on a second variable, and these estimates also imply continuity, so we get the original claim. Therefore we shall consider this case only.

Let

$$a_0 \stackrel{\text{def}}{=} \min\{1, \inf_{t\in[0,\tau]} a(t)\} \quad \text{and} \quad a_1 \stackrel{\text{def}}{=} \sup_{t\in[0,\tau]} a(t).$$

For every $h \in Y$ we have to show that there is a solution g of the equation $\lambda g + A_a g = h$ such that $\|g\|_Y \leq \frac{N}{|\lambda|+1}\|h\|_Y$. We can rewrite this equation as

$$(D_t^\alpha g)(\underline{t}) + \lambda a(\underline{t}) g(\underline{t}) = a(\underline{t}) h(\underline{t}). \tag{7}$$

By [2, Lemma 11.(b)], there is a constant M_1 depending on η, α, and τ such that

$$\|(D_t^\alpha + \mu I)^{-1}\|_{\mathcal{L}(Y)} \leq \frac{M_1}{|\mu|+1}, \quad |\arg \mu| \leq \pi(1-\frac{\alpha}{2}) - \eta. \tag{8}$$

Let

$$\epsilon = \frac{a_0}{2M_1},$$

and let T_ϵ be a positive number such that

$$|a(t)-a(s)| \leq \epsilon, \quad t,s \in [0,\tau], \quad |t-s| \leq T_\epsilon.$$

Suppose that $T \in [0,\tau)$ and that there is a continuous solution g of (7) on $[0,T]$ such that for some constant N_T, depending on η, α, τ, and on the sup, inf, and modulus of continuity of a, one has

$$\sup_{t\in[0,T]} |g(t)| \leq \frac{N_T}{|\lambda|+1} \sup_{t\in[0,T]} |h(t)|, \tag{9}$$

for all $\lambda \in \mathbb{C}$ satisfying $|\arg \lambda| \leq \pi(1-\frac{\alpha}{2}) - \eta$. If $T=0$ we take $g=0$. Now we rewrite (7) in the form

$$(D_t^\alpha w)(\underline{t}) + \lambda a(T) w(\underline{t}) = a(\underline{t})h(\underline{t}) + \lambda\big(a(T)-a(\underline{t})\big)v(\underline{t}), \tag{10}$$

and we see that if $v=w$, then we have a solution of (7).

We define $\hat{T} = \min\{\tau, T+T_\epsilon\}$ and

$$U = \big\{\, v \in \mathcal{C}_{0\mapsto 0}([0,\hat{T}];\mathbb{C}) \,\big|\, v(t) = g(t), \quad t\in[0,T] \,\big\}.$$

Since $D_t^\alpha + \lambda a(T) I$ is invertible in $\mathcal{C}_{0\mapsto 0}([0,\hat{T}];\mathbb{C})$ there is a solution of (10) for each $v \in U$ and the uniqueness guarantees that we have $w \in U$. Denote $w = G(v)$. If v_1 and v_2 are two functions in U, then we have by (8) and our choices of T_ϵ and ϵ and by the fact that $v_1(t) - v_2(t) = 0$ when $t \in [0,T]$ that

$$\sup_{t\in[0,\hat{T}]} |G(v_1)(t) - G(v_2)(t)| \leq \frac{M_1}{|\lambda| a(T)+1} \sup_{t\in[0,\hat{T}]} |\lambda||a(T)-a(t)||v_1(t)-v_2(t)|$$

$$\leq \frac{M_1}{a_0} \sup_{t\in[T,\hat{T}]} |a(T)-a(t)| \sup_{t\in[0,\hat{T}]} |v_1(t)-v_2(t)| \leq \frac{1}{2} \sup_{t\in[0,\hat{T}]} |v_1(t)-v_2(t)|.$$

Thus G is a contraction on U and we get a unique fixed-point g that is the solution of (7) on $[0,\hat{T}]$. If we let $v_0 \in U$ be such that $v_0(t) = g(T)$ for $t \in [T,\hat{T}]$, then we get from (8) and (10) the estimate

$$\sup_{t\in[0,\hat{T}]} |G(v_0)(t)| \leq \frac{2M_1 a_1 |\lambda|}{|\lambda| a(T)+1} \sup_{t\in[0,T]} |g(t)| + \frac{M_1 a_1}{|\lambda| a(T)+1} \sup_{t\in[0,\hat{T}]} |h(t)|.$$

Because $\sup_{t\in[0,\hat{T}]}|v_0(t)| = \sup_{t\in[0,T]}|g(t)|$ and the contraction factor is $\frac{1}{2}$ we get the following estimate for the norm of the fixed-point,

$$\sup_{t\in[0,\hat{T}]} |g(t)| \leq \Big(3 + \frac{4M_1 a_1}{a_0}\Big) \sup_{t\in[0,T]} |g(t)| + \frac{2M_1 a_1}{a_0(|\lambda|+1)} \sup_{t\in[0,\hat{T}]} |h(t)|,$$

and we conclude that (9) holds with $[0,T]$ replaced by $[0,\hat{T}]$ and N_T replaced by $N_{\hat{T}}$, where

$$N_{\hat{T}} = N_T\Big(3+\frac{4M_1a_1}{a_0}\Big)+\frac{2M_1a_1}{a_0}.$$

Since the choice of T_ϵ only depends on M_1, the lower bound of a and on the modulus of continuity of a, we get the desired conclusion by induction. □

Proof of Theorem 1. By the linearity of the problem it can easily be split into two parts, where in the first part we have $u_1=0$ and $f(\underline{t},0)=0$ and in the second part we have $u_0=0$ and $f(0,\underline{x})=0$. We start by considering the case where $u_1=0$ and $f(\underline{t},0)=0$.

Let

$$c_0=\inf_{\substack{t\in[0,\tau]\\ x\in[0,\xi]}} c(t,x), \qquad c_1=\sup_{\substack{t\in[0,\tau]\\ x\in[0,\xi]}} c(t,x). \tag{11}$$

We begin by studying the following equation:

$$\big(D_t^\alpha(u-u_0)\big)(t,x)+b(x)u_x(t,x)=g(t,x), \quad t\in[0,\tau], \quad x\in[0,\xi], \tag{12}$$

with boundary condition $u(\underline{t},0)=0$ under the following assumption on the function b:

$$b\in\mathcal{C}(\mathbb{R}^+;\mathbb{R}) \text{ and } 0<c_0\le b(\underline{x})\le c_1<\infty. \tag{13}$$

We denote by B_b the linear operator in $\mathcal{C}_{0\mapsto 0}([0,\xi];\mathbb{C})$ with domain

$$\mathcal{D}_{B_b}=\big\{\, v\in\mathcal{C}^1([0,\xi];\mathbb{C}) \;\big|\; v(0)=v'(0)=0\,\big\}$$

and

$$(B_bv)(x)=b(x)v'(x), \quad x\in[0,\xi].$$

We denote by B the corresponding operator with $b(\underline{x})=1$ and ξ replaced by $\xi_0=\xi/c_0$.

Thus (12) can be written as

$$D_t^\alpha(u-u_0)+B_bu=g. \tag{14}$$

Next, perform a change of variable $y=\int_0^x \frac{1}{b(s)}\,\mathrm{d}s$, so that equation (14) is replaced by

$$D_t^\alpha(u^b-u_0^b)+Bu^b=g^b, \tag{15}$$

where

$$\begin{aligned} g^b(\underline{t},\underline{y})&=g(\underline{t},\rho(\underline{y})),\\ u_0^b(\underline{y})&=u_0(\rho(\underline{y})),\end{aligned} \quad y\in[0,\xi_b] \quad\text{and}\quad \begin{aligned} g^b(\underline{t},\underline{y})&=g(\underline{t},\xi),\\ u_0^b(\underline{y})&=u_0(\xi),\end{aligned} \quad y\in(\xi_b,\xi_0],$$

where $\xi_b=\int_0^\xi \frac{1}{b(s)}\,\mathrm{d}s$ and ρ is the inverse of the function $x\mapsto\int_0^x\frac{1}{b(s)}\,\mathrm{d}s$. By [2, Thm. 6.(a)] equation (15) has a unique solution u^b which satisfies the bound

$$\begin{aligned}&\|Bu^b(\underline{t})-g^b(0)\|_{\mathcal{C}^\mu([0,\tau];\mathcal{C}_{0\mapsto 0}([0,\xi_0]))}\\ &\qquad\le M_2\Bigg(\|Bu_0^b-g^b(0)\|_{\mathcal{C}^{\mu/\alpha}([0,\xi_0])}+\|g^b(\underline{t})-g^b(0)\|_{\mathcal{C}^\mu([0,\tau];\mathcal{C}_{0\mapsto 0}([0,\xi_0]))}\Bigg),\end{aligned}$$

where M_2 depends on α, μ, τ and ξ_0. Now we change variables back again, that is, we define

$$u(\underline{t}, x) = u^b\left(\underline{t}, \int_0^x \frac{1}{b(s)}\,\mathrm{d}s\right), \qquad \text{for } x \in [0, \xi].$$

We can therefore conclude that there is a unique solution u of (12) such that

$$\begin{aligned}\|u_x\|_{\mathcal{C}^{(\mu)}([0,\tau];\mathcal{C}([0,\xi];\mathbb{C}))} \le M_3\Big(&\|b(\underline{x})u_0'(\underline{x}) - g(0, \underline{x})\|_{\mathcal{C}^{\mu/\alpha}([0,\xi])} \\ &+ \|g\|_{\mathcal{C}^{(\mu)}([0,\tau];\mathcal{C}([0,\xi]))}\Big),\end{aligned} \tag{16}$$

where (with some crude estimates) $M_3 = \frac{1}{c_0}(M_2 \max\{1, c_1^{\mu/\alpha}\} + 2)$. Note that since M_2 depends on ξ_0, M_3 depends on c_0 as well.

Choose ϵ to be so small that

$$\frac{M_3 \epsilon}{c_0} \le \frac{1}{2}, \tag{17}$$

and choose $T_\epsilon > 0$ such that

$$\|c\|_{\mathcal{C}^{(\mu)}([0,\tau];\mathcal{C}([0,\xi]))} T_\epsilon^\mu \le \frac{\epsilon}{2}. \tag{18}$$

Suppose $T \in [0, \tau)$ and that we have found a solution of (1) on $[0, T] \times [0, \xi]$ that satisfies the inequality

$$\begin{aligned}&\|u_x\|_{\mathcal{C}^{(\mu)}([0,T];\mathcal{C}([0,\xi]))} \\ &\quad \le N_T\Big(\|f\|_{\mathcal{C}^{(\mu)}([0,\tau];\mathcal{C}([0,\xi]))} + \|c(0,\underline{x})u_0'(\underline{x}) - f(0,\underline{x})\|_{\mathcal{C}^{\mu/\alpha}([0,\xi])}\Big),\end{aligned} \tag{19}$$

for some constant N_T depending only on α, μ, τ, ξ, c_0, and c_1. If $T = 0$ we take this solution to be $u(0, \underline{x}) = u_0(\underline{x})$. We let $\hat{T} = \min\{\tau, T + T_\epsilon\}$ and let U denote the set

$$\begin{aligned}U \stackrel{\text{def}}{=} \{\, v \in \mathcal{C}^{(\mu)}([0, \hat{T}]; \mathcal{C}([0,\xi];\mathbb{C})) \mid\ & v(t, \underline{x}) = u_x(t, \underline{x}), \\ & 0 \le t \le T, \quad v(\underline{t}, 0) = 0\,\}.\end{aligned}$$

For each $v \in U$ we proceed to find a solution w of the equation

$$D_t^\alpha(w - u_0)(\underline{t}, \underline{x}) + c(T, \underline{x})w_x(\underline{t}, \underline{x}) = f(\underline{t}, \underline{x}) + \big(c(T, \underline{x}) - c(\underline{t}, \underline{x})\big)v(\underline{t}, \underline{x}), \tag{20}$$

with boundary condition $w(\underline{t}, 0) = 0$. Observe that the right-hand side of (20) evaluated at $t = 0$ is

$$f(0, x) + (c(T, x) - c(0, x))u_0'(x),$$

and therefore the element $b(x)u_0'(\underline{x}) - f(0, \underline{x})$ appearing in (16) when $b(\underline{x}) = c(T, \underline{x})$ is $c(0, x)u_0'(x) - f(0, x)$. Thus we conclude from (iii) and from the results above that we can find a solution w of (20) such that $w_x \in \mathcal{C}^{(\mu)}([0, \hat{T}]; \mathcal{C}([0, \xi]; \mathbb{C}))$ and the uniqueness guarantees that we have $w_x \in U$.

Let us denote the mapping $v \to w_x$ by $w_x = G(v)$. Using the linearity of equation (20), and (16) with $b(\underline{x}) = c(T, \underline{x})$ once more, we conclude that

$$\begin{aligned}&\|(G(v_1) - G(v_2))(\underline{t}, \underline{x})\|_{\mathcal{C}^{(\mu)}([0,\hat{T}];\mathcal{C}([0,\xi]))} \\ &\quad \leq M_3 \|(c(T, \underline{x}) - c(\underline{t}, \underline{x}))(v_1 - v_2)(\underline{t}, \underline{x})\|_{\mathcal{C}^{(\mu)}([0,\hat{T}];\mathcal{C}([0,\xi]))}.\end{aligned}$$

Let $v_\Delta = v_1 - v_2$ and $c_\Delta(\underline{t}, \underline{x}) = c(T, \underline{x}) - c(\underline{t}, \underline{x})$. Since v_1 and $v_2 \in U$ it follows that $v_\Delta(t, \underline{x}) = 0$ for $t \in [0, T]$ and therefore we can, when analyzing the term $(c(T, \underline{x}) - c(\underline{t}, \underline{x}))(v_1 - v_2)(\underline{t}, \underline{x})$ assume that $c(t, \underline{x}) = c(T, \underline{x})$ for $t \in [0, T]$. Thus we conclude from (18) that

$$\sup_{\substack{t\in[0,\hat{T}]\\ x\in[0,\xi]}} |c_\Delta(t, x) v_\Delta(t, x)| \leq \epsilon \sup_{\substack{t\in[0,\hat{T}]\\ x\in[0,\xi]}} |v_\Delta(t, x)|.$$

Furthermore, if we write $c_\Delta(t, x)v_\Delta(t, x) - c_\Delta(s, x)v_\Delta(s, x) = c_\Delta(t, x)(v_\Delta(t, x) - v_\Delta(s, x)) + (c_\Delta(t, x) - c_\Delta(s, x))(v_\Delta(s, x) - v_\Delta(T, x))$, using the fact that $v_\Delta(T, \underline{x}) = 0$, and use (18) once again, we conclude that

$$\begin{aligned}&\|(c(T, x) - c(t, x))(v_1 - v_2)(t, x)\|_{\mathcal{C}^{(\mu)}([0,\hat{T}];\mathcal{C}([0,\xi]))} \\ &\quad \leq \epsilon \|(v_1 - v_2)(t, x)\|_{\mathcal{C}^{(\mu)}([0,\hat{T}];\mathcal{C}([0,\xi]))}.\end{aligned}$$

Hence we have

$$\|(G(v_1) - G(v_2))(t, x)\|_{\mathcal{C}^{(\mu)}([0,\hat{T}];\mathcal{C}([0,\xi]))} \leq \frac{M_3\epsilon}{c_0} \|(v_1 - v_2)(t, x)\|_{\mathcal{C}^{(\mu)}([0,\hat{T}];\mathcal{C}([0,\xi]))},$$

and we see that the mapping G is a contraction and there is a unique fixed-point. Thus we get a solution on the interval $[0, \hat{T}]$.

If we take $v_0 \in U$ to be such that $v_0(t, \underline{x}) = u(T, \underline{x})$ for $t \in [T, \hat{T}]$ then

$$\|v_0\|_{\mathcal{C}^{(\mu)}([0,\hat{T}];\mathcal{C}([0,\xi]))} = \|u\|_{\mathcal{C}^{(\mu)}([0,\hat{T}];\mathcal{C}([0,\xi]))}$$

and using (16) to estimate

$$\|G(v_0)\|_{\mathcal{C}^{(\mu)}([0,\hat{T}];\mathcal{C}([0,\xi]))}$$

we conclude from (19) that (19) holds with T replaced by $\hat{T}$ and N_T replaced by

$$N_{\hat{T}} = \Big(3 + \frac{4M_2c_1}{c_0}\Big)N_T + 2M_2.$$

Since this procedure can be repeated with the same T_ϵ, we find a solution on $[0, \tau]$ that satisfies the desired bounds.

Next we consider the case where $u_0 = 0$ and $f(0, \underline{x}) = 0$. We proceed in the same manner as above but instead of using [2, Thm. 6] we use [7, Thm. 5.5] combined with Lemma 4. The conclusion we can draw is that for every $a \in \mathcal{C}^{(\mu)}([0, \tau]; \mathbb{C})$ with $a(\underline{t}) > 0$ there is a constant M_4 depending on α, μ, τ, and on $\inf_{t\in[0,\tau]} a(t)$ and $\|a\|_{\mathcal{C}^{(\mu)}([0,\tau])}$ such that if $g \in \mathcal{C}^{(\mu)}([0, \tau]; \mathcal{C}([0, \xi]; \mathbb{C}))$ with

$g(0,\underline{x}) = 0$ and $D_t^\alpha u_1 \in \mathcal{C}^{(\mu)}_{0\mapsto 0}([0,\tau];\mathbb{C})$, then there is a unique solution u of the equation

$$\begin{aligned} &\frac{1}{a(t)}(D_t^\alpha u)(t,x) + u_x(t,x) = g(t,x), \\ &u(t,0) = u_1(t), \quad t \in [0,\tau], \end{aligned} \tag{21}$$

such that

$$\|D_t^\alpha u\|_{\mathcal{C}^{(\mu)}([0,\tau];\mathcal{C}([0,\xi]))} \le M_4\Big(\|g\|_{\mathcal{C}^{(\mu)}([0,\tau];\mathcal{C}([0,\xi]))} + \|D_t^\alpha u_1\|_{\mathcal{C}^{(\mu)}([0,\tau])}\Big). \tag{22}$$

Choose ϵ to be so small that

$$\epsilon M_4 \left\| \frac{1}{c(\underline{t},\underline{x})} \right\|^2_{\mathcal{C}^{(\mu)}([0,\tau];\mathcal{C}([0,\xi]))} \le \frac{1}{2},$$

and choose $\mathfrak{X}_\epsilon > 0$ so that

$$\|c(\underline{t},x) - c(\underline{t},y)\|_{\mathcal{C}^{(\mu)}([0,\tau])} \le \epsilon, \quad |x-y| \le \mathfrak{X}_\epsilon, \quad x,y \in [0,\xi].$$

We assume that there is a solution of (1) on $[0,\tau]\times[0,\mathfrak{X}]$ for some $\mathfrak{X} \in [0,\xi)$ such that there is a constant $N_{\mathfrak{X}}$ satisfying

$$\|D_t^\alpha u\|_{\mathcal{C}^{(\mu)}([0,\tau];\mathcal{C}([0,\mathfrak{X}]))} \le N_{\mathfrak{X}}\Big(\|f\|_{\mathcal{C}^{(\mu)}([0,\tau];\mathcal{C}([0,\mathfrak{X}]))} + \|D_t^\alpha u_1\|_{\mathcal{C}^{(\mu)}([0,\tau])}\Big). \tag{23}$$

If $\mathfrak{X} = 0$ this is certainly the case. Let $\hat{\mathfrak{X}} = \min\{\xi, \xi + \mathfrak{X}_\epsilon\}$. In analogy with the previous case, we consider the equation

$$\begin{aligned} &\frac{1}{c(\underline{t},\mathfrak{X})}(D_t^\alpha w)(\underline{t},\underline{x}) + w_x(\underline{t},\underline{x}) = \frac{f(t,x)}{c(\underline{t},x)} + \left(\frac{1}{c(\underline{t},\mathfrak{X})} - \frac{1}{c(\underline{t},\underline{x})}\right) v(\underline{t},\underline{x}), \\ &w(\underline{t},0) = u_1(\underline{t}), \end{aligned} \tag{24}$$

and we see immediately that if $v = D_t^\alpha w$, then we have equation (1). We define the set U to be

$$\begin{aligned} U \stackrel{\text{def}}{=} \big\{ v \in \mathcal{C}^{(\mu)}([0,\tau];\mathcal{C}([0,\hat{\mathfrak{X}}];\mathbb{C})) \;\big|\; &v(\underline{t},x) = (D_t^\alpha u)(\underline{t},x), \\ &0 \le x \le \mathfrak{X}, \quad v(0,\underline{x}) = 0 \big\}. \end{aligned}$$

If v_1, $v_2 \in U$, then $v_\Delta = v_1 - v_2$ satisfies $v_\Delta(\underline{t},x) = 0$ when $x \in [0,\mathfrak{X}]$. Therefore we can use the fact that $\mathcal{C}^{(\mu)}$ is a Banach algebra with pointwise multiplication to see that

$$\begin{aligned} &\left\| \left(\frac{1}{c(\underline{t},\mathfrak{X})} - \frac{1}{c(\underline{t},\underline{x})}\right) v_\Delta(\underline{t},\underline{x}) \right\|_{\mathcal{C}^{(\mu)}([0,\tau];\mathcal{C}([0,\hat{\mathfrak{X}}]))} \\ &\quad \le \epsilon \left\| \frac{1}{c(\underline{t},\underline{x}))} \right\|^2_{\mathcal{C}^{(\mu)}([0,\tau];\mathcal{C}([0,\xi]))} \|v_\Delta\|_{\mathcal{C}^{(\mu)}([0,\tau];\mathcal{C}([0,\hat{\mathfrak{X}}]))}. \end{aligned}$$

By (22) this shows that the mapping $v \mapsto w_x$ is a contraction and therefore we get a fixed-point which is the solution u_x on $[0,\tau]\times[0,\hat{\mathfrak{X}}]$. We can take v_0 in U such that $v_0(\underline{t},x) = (D_t^\alpha u)(\underline{t},\mathfrak{X})$, for $x \in [\mathfrak{X},\hat{\mathfrak{X}}]$, and then we get the estimate (23) with $\mathfrak{X}$ replaced by $\hat{\mathfrak{X}}$ and $N_{\mathfrak{X}}$ replaced by $N_{\hat{\mathfrak{X}}} = N_{\mathfrak{X}}(3 + 4M_4\|1/c\|_{\mathcal{C}^{(\mu)}([0,\tau];\mathcal{C}([0,\xi]))}) +$

$2M_4 \max\{1, \|1/c\|_{\mathcal{C}^{(\mu)}([0,\tau];\mathcal{C}([0,\xi]))}\}$. Since we can choose $\mathfrak{X}_\epsilon$ independent of $\mathfrak{X}$ we get the desired result by iteration.

From the results established so far we know that (a) holds and that $D_t^\alpha(u - u_0) \in \mathcal{C}^{(\mu)}([0,\tau];\mathcal{C}([0,\xi];\mathbb{C}))$. The inverse of D_t^α is given by $(D_t^{-\alpha}h)(\underline{t}) = \int_0^t g_\alpha(\underline{t} - s)h(s)\,ds$ and so it follows from a straightforward calculation that $u - u_0 \in \mathcal{C}^{(\alpha)}([0,\tau];\mathcal{C}([0,\xi];\mathbb{C}))$. By assumption $u_0 \in \mathcal{C}^1([0,\xi];\mathbb{C})$, and therefore we have (b). Since (a) is equivalent to having $u \in \mathcal{C}^{(\mu)}([0,\tau];\mathcal{C}^1([0,\xi];\mathbb{C}))$, we get (c) from Lemma 3. The estimates in (4) follow immediately.

Finally observe that the arguments in the above paragraph can be used to validate the remark after Theorem 1. □

Proof of Theorem 2. First we consider the case where $u_1 = 0$ and $f(\underline{t}, 0) = 0$. We proceed in the same way as in the proof of Theorem 1 and by applying [2, Thm. 6.(c)] we conclude that there is a unique solution of (12) such that for some constant M_5 (depending on c_0 and c_1 and $\|b\|_{\mathcal{C}^\nu([0,\xi])}$ but not otherwise on b)

$$\|u_x\|_{\mathcal{C}([0,\tau];\mathcal{C}^{(\nu)}([0,\xi]))} \leq M_5\Big(\|u_0'\|_{\mathcal{C}^{(\nu)}([0,\xi])} + \|g\|_{\mathcal{C}([0,\tau];\mathcal{C}^{(\nu)}([0,\xi]))}\Big). \tag{25}$$

Choose ϵ to be so small that

$$M_5\epsilon \leq \frac{1}{2},$$

and choose T_ϵ so that

$$\|c(t,\underline{x}) - c(s,\underline{x})\|_{\mathcal{C}^{(\nu)}([0,\xi])} \leq \epsilon, \quad |t-s| \leq T_\epsilon, \quad t,s \in [0,\tau]. \tag{26}$$

Suppose that we have found a solution of (1) on $[0,T] \times [0,\xi]$ that satisfies the inequality

$$\|u_x\|_{\mathcal{C}([0,T];\mathcal{C}^{(\nu)}([0,\xi]))} \leq N_T\Big(\|f\|_{\mathcal{C}([0,\tau];\mathcal{C}^{(\nu)}([0,\xi]))} + \|u_0'(\underline{x})\|_{\mathcal{C}^{(\nu)}([0,\xi])}\Big), \tag{27}$$

If $T = 0$ we take this solution to be $u(0,\underline{x}) = u_0(\underline{x})$. Let $\hat{T} = \min\{\tau, T + T_\epsilon\}$ and let U denote the set

$$\begin{aligned} U \stackrel{\text{def}}{=} \{ & v_\frown \in \mathcal{C}^{(\nu)}([0,\xi];\mathcal{C}([0,\hat{T}];\mathbb{C})) \mid v(t,\underline{x}) = u_x(t,\underline{x}), \\ & 0 \leq t \leq T, \quad v(\underline{t},0) = 0\}. \end{aligned}$$

For each $v \in U$, we can by the results above find a solution w of (20) such that $(w_x)_\frown \in \mathcal{C}^{(\nu)}([0,\xi];\mathcal{C}([0,\hat{T}];\mathbb{C}))$ and we can apply inequality (25). The uniqueness of the solution guarantees that we have $w_x \in U$.

Let us denote the mapping $v \to w_x$ by $w_x = G(v)$. Using the linearity of the equation and (25) we conclude that

$$\begin{aligned} &\|(G(v_1) - G(v_2))(\underline{t},\underline{x})\|_{\mathcal{C}([0,\hat{T}];\mathcal{C}^{(\nu)}([0,\xi]))} \\ &\quad \leq M_5\|(c(T,\underline{x}) - c(\underline{t},\underline{x}))(v_1 - v_2)(\underline{t},\underline{x})\|_{\mathcal{C}([0,\hat{T}];\mathcal{C}^{(\nu)}([0,\xi]))}. \end{aligned}$$

Again we use the fact that since v_1 and $v_2 \in U$ it follows that $(v_1 - v_2)(t, \underline{x}) = 0$ for $t \in [0, T]$, together with the fact that $\mathcal{C}^{(\nu)}([0, \xi]; \mathcal{C}([0, \hat{T}]; \mathbb{C}))$ is a Banach algebra, to conclude that

$$\begin{aligned} &\|(c(T, \underline{x}) - c(\underline{t}, \underline{x}))(v_1 - v_2)(\underline{t}, \underline{x})\|_{\mathcal{C}([0,\hat{T}];\mathcal{C}^{(\nu)}([0,\xi]))} \\ &\quad \le \epsilon \|(v_1 - v_2)\|_{\mathcal{C}([0,\hat{T}];\mathcal{C}^{(\nu)}([0,\xi]))}. \end{aligned}$$

Hence we see that the mapping G is a contraction and there is a unique fixed-point. Therefore we get a solution on the interval $[0, \hat{T}]$ and since this procedure can be repeated with the same T_ϵ, we find a solution on $[0, \tau]$.

To see that the desired bounds hold, we observe that we can take $v_0 \in U$ such that $v_0(t, \underline{x}) = u_x(T, \underline{x})$ so that we have a bound on $\|v_0\|_{\mathcal{C}([0,\hat{T}];\mathcal{C}^{(\nu)}([0,\xi]))}$.

We proceed to consider the case where $u_0 = 0$ and $f(0, \underline{x}) = 0$. Now we apply [7, Thm 4.5] with A as in Lemma 4 and the x and t variables interchanged. The conclusion we can draw is that there is a constant M_6 depending on α, ν, τ, and on c such that if $g \in \mathcal{C}^{(\nu)}([0, \xi]; \mathcal{C}([0, \tau]; \mathbb{C}))$ with $g(0, \underline{x}) = 0$ and $g(\underline{t}, 0) - \frac{1}{a(\underline{t})}(D_t^\alpha u_1)(\underline{t}) \in \mathcal{C}_{0\mapsto 0}^{(\alpha\nu)}([0, \tau]); \mathbb{C})$ then there is a unique solution u of equation (21) such that

$$\begin{aligned} \|D_t^\alpha u\|_{\mathcal{C}([0,\tau];\mathcal{C}^{(\nu)}([0,\xi]))} \le M_6 \Big(&\|g\|_{\mathcal{C}([0,\tau];\mathcal{C}^{(\nu)}([0,\xi]))} \\ &+ \Big\| g(\underline{t}, 0) - \frac{1}{a(\underline{t})}(D_t^\alpha u_1)(\underline{t}) \Big\|_{\mathcal{C}^{(\alpha\nu)}([0,\tau])} \Big). \end{aligned}$$

Due to the uniqueness, the same inequality certainly holds with the same constant M_6 if τ and ξ are replaced by some smaller numbers.

Let ϵ be such that

$$M_6 \epsilon \le \frac{1}{2},$$

and let $\mathfrak{X}_\epsilon$ satisfy

$$\left\| \frac{1}{c(\underline{t}, \underline{x})} \right\|_{\mathcal{C}([0,\tau];\mathcal{C}^{(\nu)}([0,\xi]))} \mathfrak{X}_\epsilon^\nu \le \frac{\epsilon}{2}.$$

Again we suppose that we have found a solution on $[0, \tau] \times [0, \mathfrak{X}]$ and then consider (24). Observe that the term $g(\underline{t}, 0) - \frac{1}{a(\underline{t})}(D_t^\alpha u_1)(\underline{t})$ becomes $\frac{f(\underline{t},0)}{c(\underline{t},0)} - \frac{1}{c(\underline{t},0)} D_t^\alpha u_1(\underline{t})$.

Let $\hat{\mathfrak{X}} = \min\{\xi, \mathfrak{X} + \mathfrak{X}_\epsilon\}$. In this case we take the set U to be

$$\begin{aligned} U = \{\, v \in \mathcal{C}^{(\nu)}([0, \hat{\mathfrak{X}}]; \mathcal{C}([0, \tau]; \mathbb{C})) \mid v(\underline{t}, x) = (D_t^\alpha u)(\underline{t}, x), \\ x \in [0, \mathfrak{X}], \quad v(0, \underline{x}) = 0 \,\}. \end{aligned}$$

If v_1 and v_2 are two functions in U and $v_\Delta = v_1 - v_2$, then we get, using the fact that $v_\Delta(\underline{t}, x) = 0$ for $x \in [0, \mathfrak{X}]$ and the argument used in the first part of the proof of Theorem 1 that

$$\left\| \left(\frac{1}{c(\underline{t}, \mathfrak{X})} - \frac{1}{c(\underline{t}, \underline{x})} \right) v_\Delta(\underline{t}, \underline{x}) \right\|_{\mathcal{C}([0,\tau];\mathcal{C}^{(\nu)}([0,\hat{\mathfrak{X}}]))} \le \epsilon.$$

The desired conclusion follows.

From what we have shown so far and from the assumptions it follows that we have $(D_t^\alpha(u-u_0))_\curvearrowleft$ and $(u_x)_\curvearrowleft \in \mathcal{C}^{(\nu)}([0,\xi];\mathcal{C}([0,\tau];\mathbb{C}))$. The inverse of D_t^α is given by $(D_t^{-\alpha}h)(\underline{t}) = \int_0^{\underline{t}} g_\alpha(\underline{t}-s)h(s)\,\mathrm{d}s$; therefore a straightforward caluculation gives that $u-u_0 \in \mathcal{C}^{(\alpha)}([0,\tau];\mathcal{C}^{(\nu)}([0,\xi];\mathbb{C}))$. By assumption $u_0 \in \mathcal{C}^1([0,\xi];\mathbb{C})$ and we get (b). Since (a) is equivalent to having $u_\curvearrowleft \in \mathcal{C}^{(1+\nu)}([0,\xi];\mathcal{C}([0,\tau];\mathbb{C}))$ and (b) is equivalent to having $u_\curvearrowleft \in \mathcal{C}^{(\nu)}([0,\xi];\mathcal{C}^{(\alpha)}([0,\tau];\mathbb{C}))$ we obtain (c) from Lemma 3. The estimates in (5) follow immediately. □

References

[1] H. Amann, *Linear and Quasilinear Parabolic Problems*, Vol. I, Birkhäuser, Basel, 1995.

[2] Ph. Clément, G. Gripenberg, and S-O. Londen, "Schauder estimates for equations with fractional derivatives", Trans. Amer. Math. Soc., (to appear).

[3] B. Cockburn, G. Gripenberg, and S-O. Londen, "On convergence to entropy solutions of a single conservation law", J. Differential Equations **128** (1996), 206–251.

[4] E. Feireisl and H. Petzeltová, "Singular kernels and compactness in nonlinear conservation laws", (to appear).

[5] G. Gripenberg and S-O. Londen, "Fractional derivatives and smoothing in nonlinear conservation laws", Differential Integral Equations **8** (1995), 1961–1976.

[6] A. Lunardi, *Analytic Semigroups and Optimal Regularity in Parabolic Problems*, Birkhäuser, Basel, 1995.

[7] E. Sinestrari, "On the abstract Cauchy problem of parabolic type in spaces of continuous functions", J. Math. Anal. Appl. **107** (1985), 16–66.

[8] A. Zygmund, *Trigonometric Series II*, Cambridge University Press, Cambridge, 1959.

Ph. Clément
Faculty of Technical Mathematics,
and Informatics,
Delft University of Technology,
P.O. Box 5031,
2600 GA Delft,
The Netherlands
`clement@twi.tudelft.nl`

G. Gripenberg
Department of Mathematics,
P.O. Box 4,
FIN-00014 University of Helsinki,
Finland
www.helsinki.fi/~ggripenb
`gustaf.gripenberg@helsinki.fi`

S-O. Londen
Institute of Mathematics,
Helsinki University of Technology,
P.O. Box 1000,
FIN-02015 HUT, Finland
`stig-olof.londen@hut.fi`

Progress in Nonlinear Differential Equations
and Their Applications, Vol. 35
© 1999 Birkhäuser Verlag Basel/Switzerland

Non Gradient Perturbations of Ornstein-Uhlenbeck Semigroups

Giuseppe Da Prato

Dedicated to Herbert Amann for his 60th birthday

1. Introduction

We are concerned with the following linear operator N in a separable Hilbert space H,

$$Nu = \frac{1}{2} \text{ Tr } [CD^2 u] + \langle Ax + F(x), Du \rangle. \tag{1.1}$$

Here $A : D(A) \subset H \to H$ is the infinitesimal generator of a strongly continuous semigroup e^{tA} on H of negative type, $F : H \to H$ is bounded and $C : H \to H$ is symmetric and bounded together with C^{-1}.

This problem arises in the study of a deterministic dynamic system governed by a differential equation on H

$$\frac{dX(t)}{dt} = AX(t) + F(X(t)), \quad X(0) = x \in H, \tag{1.2}$$

perturbed by noise. Often one is in fact interested in a stochastic perturbation of the form

$$dX = (AX + F(X))dt + \sqrt{C}\, dW(t), \quad X(0) = x \in H. \tag{1.3}$$

Setting $u(t,x) = \mathbb{E}[\varphi(X(t,x))]$, one can show that, under suitable regularity assumptions and non-explosion, u is the unique solution to (1.1).

We are interested in studying problem (1.1) in a space $L^2(H,\nu)$, where ν is an invariant measure:

$$\int_H e^{tN} \varphi d\nu = \int_H \varphi d\nu, \; \forall\, t > 0, \; \varphi \in L^2(H,\nu), \tag{1.4}$$

or equivalently,

$$\int_H N\varphi d\nu = 0, \; \forall\; \varphi \in D(N). \tag{1.5}$$

When $F(x) = 0$, $X(\cdot, x)$ is the Ornstein-Uhlenbeck process and there is a unique invariant measure μ, that is the Borel-Gaussian measure in H $\mu = \mathcal{N}(0,Q)$

having mean 0 and covariance operator $Q = \int_0^{+\infty} e^{sA} C e^{sA^*} ds$. Then the operator

$$Lu = \frac{1}{2} \text{Tr}\,[CD^2 u] + \langle Ax, Du \rangle, \tag{1.6}$$

with a suitable domain $D(L)$, is m-dissipative on $L^2(H,\mu)$. In some cases $D(L)$ can be characterized as a closed subspace of the Sobolev space $W^{2,2}(H,\mu)$, see [5]; but the following continuous inclusion

$$D(L) \subset W^{1,2}(H,\mu), \tag{1.7}$$

always holds.

The important situation where A is self-adjoint, $C = I$, and F is a gradient, that is there exists a function $U : H \to \mathbb{R}$ such that $F(x) = -DU(x)$, has been extensively studied. In this case we have $Q = -\frac{1}{2}\, A^{-1}$, and a candidate for the invariant measure is

$$\nu(dx) = e^{-2U(x)}\; \mu(dx).$$

The goal is to show that N is essentially self-adjoint on $L^2(H,\mu)$.

Results of this kind have been obtained using Dirichlet Forms, see [1], [11], or some properties of dissipative operators, [6].

Here we develop this last method to cover more general situations where N is not self-adjoint.

Our main assumption is that F is bounded and regular (we think that this assumption can be weakened). In this case, due to the inclusion (1.7), we can show by a standard perturbation argument that N (with domain $D(N) = D(L)$) generates a strongly continuous semigroup on $L^2(H,\mu)$.

However, the natural space to study the asymptotic behaviour and ergodic properties of e^{tN} is $L^2(H,\nu)$.

Sections 2 and 3 are devoted to a review of several known results for the Ornstein-Uhlenbeck semigroup. These results, scattered in several papers, [3], [9], [10], [2], [5], [6], are collected here in a unified way, in some cases with simpler proofs. In §4 we finally study the operator N in $L^2(H,\nu)$, proving that it is dissipative and closable and that its closure is m-dissipative.

Let us end this section with some notations and preliminaries on Gaussian measures.

1.1. Preliminaries on Gaussian measures

We are given a trace class linear operator Q in H. We assume for simplicity that Ker $Q = \{0\}$. We denote by $\{e_k\}$ a complete orthonormal system in H and by $\{\lambda_k\}$ a sequence of positive numbers such that

$$Qe_k = \lambda_k e_k, \; k \in \mathbb{N}.$$

We set $x_k = \langle x, e_k \rangle$, $k \in \mathbb{N}$, and denote by $\mu_Q = \mu$ the Gaussian measure of mean 0 and covariance operator Q. We recall that μ coincides with the product of one-dimensional Gaussian measures,

$$\prod_{n=1}^{\infty} \mathcal{N}(0, \lambda_k),$$

where

$$\mathcal{N}(0, \lambda_k)(dx_k) = (2\pi\lambda_k)^{-1/2} e^{-\frac{x_k^2}{2\lambda_k}} \, dx_k, \; k \in \mathbb{N}.$$

We will need the space $\mathcal{E}(H)$ consisting of all exponential functions on H. $\mathcal{E}(H)$ is the the vector sub-space of $L^2(H, \mu)$ spanned by all functions

$$\varphi_h(x) = e^{i\langle h,x \rangle}, \; h \in H.$$

It is easy to see that $\mathcal{E}(H)$ is dense in $L^2(H, \mu)$. For any $h \in \mathbb{N}$ we denote by D_h the directional derivative in the direction e_h.

The following integration-by-parts formula is basic and can be easily proved by a direct verification,

$$\int_H D_h \varphi(x) d\mu = \frac{1}{\lambda_h} \int_H x_h \varphi(x) d\mu, \quad \varphi \in \mathcal{E}(H). \tag{1.8}$$

An immediate consequence of (1.8) is the following

$$\int_H [D_h\varphi(x) \, \psi(x) + \varphi(x) \, D_h \psi(x)] d\mu = \frac{1}{\lambda_h} \int_H x_h \varphi(x) \psi(x) d\mu, \quad \varphi, \psi \in \mathcal{E}(H). \tag{1.9}$$

From (1.9) it follows easily, see e.g. [2],[10], that for any $h \in \mathbb{N}$ the operator D_h is closable on $L^2(H, \mu)$. We shall still denote by D_h its closure.

We now recall the definition of Sobolev spaces $W^{1,2}(H, \mu)$ and $W^{2,2}(H, \mu)$. We set

$$W^{1,2}(H, \mu) = \Bigg\{ \varphi \in L^2(H, \mu) : \; D_h \varphi \in L^2(H, \mu), \; \forall \, h \in \mathbb{N}, \\ \sum_{h=1}^{\infty} \int_H |D_h \varphi|^2 d\mu < +\infty \Bigg\},$$

and

$$W^{2,2}(H, \mu) = \Bigg\{ \varphi \in W^{1,2}(H, \mu) : \; D_h D_k \varphi \in L^2(H, \mu), \; \forall \, h, k \in \mathbb{N}, \\ \sum_{h,k=1}^{\infty} \int_H |D_h D_k \varphi|^2 d\mu < +\infty \Bigg\}.$$

If $\varphi \in W^{1,2}(H,\mu)$ we define the gradient $D\varphi$ of φ by setting

$$D\varphi(x) = \sum_{k=1}^{\infty} D_k\varphi(x)e_k, \quad x \in H. \tag{1.10}$$

Notice that, since

$$\int_H |D\varphi(x)|^2 d\mu = \sum_{h=1}^{\infty} \int_H |D_h\varphi|^2 d\mu < +\infty,$$

the series in (1.10) is convergent for almost all $x \in H$. Therefore $D\varphi(x)$ is defined for almost all $x \in H$.

In a similar way if $\varphi \in W^{2,2}(H,\mu)$ we define the second derivative $D^2\varphi$ by setting

$$\langle D^2\varphi(x)y, z\rangle = \sum_{h,k=1}^{\infty} D_hD_k\varphi(x)\langle y, e_h\rangle\ \langle z, e_k\rangle, \ y, z \in H. \tag{1.11}$$

Since

$$\sum_{h,k=1}^{\infty} \int_H |D_hD_k\varphi|^2 d\mu < +\infty,$$

$D^2\varphi(x)$ is a well-defined linear bounded operator of Hilbert-Schmidt class in H for almost all $x \in H$, and we have

$$\int_H \text{Tr}\ [(D^2\varphi(x))^2] d\mu = \sum_{h,k=1}^{\infty} \int_H |D_hD_k\varphi|^2 d\mu < +\infty.$$

It is easy to check that $W^{1,2}(H,\mu)$ and $W^{2,2}(H,\mu)$ are Banach spaces with the norms

$$\|\varphi\|^2_{W^{1,2}(H,\mu)} = \int_H [|\varphi(x)|^2 + |D\varphi(x)|^2] d\mu,$$

$$\|\varphi\|^2_{W^{2,2}(H,\mu)} = \int_H (|\varphi(x)|^2 + |D\varphi(x)|^2 + \ \text{Tr}\ [(D^2\varphi(x))^2]) d\mu,$$

respectively.

We will need the following result; it is an easy consequence of the integration-by-parts formula and Hölder's inequality, see e.g. [10].

Proposition 1.1.

(i) *If $\varphi \in W^{1,2}(H,\mu)$ we have $|x|\varphi \in L^2(H,\mu)$. Moreover there exists $\kappa > 0$ such that*

$$\int_H |x|^2|\varphi(x)|^2 d\mu \le \kappa \|\varphi\|^2_{W^{1,2}(H,\mu)}. \tag{1.12}$$

(ii) *If $\varphi \in W^{2,2}(H,\mu)$ we have $|x|\varphi \in W^{1,2}(H,\mu)$. Moreover there exists $\kappa_1 > 0$ such that*

$$\int_H |x|^2 |D\varphi(x)|^2 d\mu \le \kappa_1 \|\varphi\|^2_{W^{2,2}(H,\mu)}. \tag{1.13}$$

(iii) *If $\varphi \in W^{2,2}(H,\mu)$ we have $|x|^2\varphi \in L^2(H,\mu)$. Moreover there exists $\kappa_2 > 0$ such that*

$$\int_H |x|^4 |\varphi(x)|^2 d\mu \le \kappa_2 \|\varphi\|^2_{W^{2,2}(H,\mu)}. \tag{1.14}$$

We finally recall that the embedding

$$W^{1,2}(H,\mu) \subset L^2(H,\mu).$$

is compact (see [7], [14]) whereas the embedding

$$W^{2,2}(H,\mu) \subset W^{1,2}(H,\mu),$$

is not, see [9].

2. The Ornstein-Uhlenbeck semigroup

We are given two linear operators A and C in a separable Hilbert space H, and we assume that

Hypothesis 2.1.

(i) *A is the infinitesimal generator of a strongly continuous semigroup e^{tA} on H. There exist $M > 0$ and $\omega > 0$ such that*

$$\|e^{tA}\| \le Me^{-\omega t},\ t \ge 0. \tag{2.1}$$

(ii) *C is a symmetric operator on H that is bounded together with its inverse.*

(iii) *The operator $Q^{a,c} := \int_0^{+\infty} e^{tA} C e^{tA^*} dt$ is of trace class.*

If Hypothesis 2.1 holds, we define the Ornstein-Uhlenbeck semigroup $R_t^{a,c}$, $t \ge 0$, on $L^2(H,\mu)$ by setting

$$R_t^{a,c}\varphi(x) = \int_H \varphi(e^{tA}x + y)\mu_t^{a,c}(dy),\ \varphi \in L^2(H,\mu). \tag{2.2}$$

where $\mu_t^{a,c}$ is the Gaussian measure $\mathcal{N}(0, Q_t^{a,c})$ with mean 0 and covariance operator $Q_t^{a,c}$ defined as follows

$$Q_t^{a,c} = \int_0^t e^{sA} C e^{sA^*} ds.$$

We set also $\mu^{a,c} = \mathcal{N}(0, Q^{a,c})$. Since we have

$$\lim_{t\to+\infty} R_t^{a,c}\varphi(x) = \int_H \varphi(y)\mu^{a,c}(dy),\ x \in H,$$

it follows that $\mu^{a,c}$ is an invariant measure for $R_t^{a,c}$. This means that

$$\int_H R_t^{a,c}\varphi(y)\mu^{a,c}(dy) = \int_H \varphi(y)\mu^{a,c}(dy), \tag{2.3}$$

for all $t > 0$ and $\varphi \in L^2(H,\mu)$.

Moreover it is easy to check that $R_t^{a,c}$, $t \geq 0$ is a strongly continuous semigroup on $L^2(H,\mu)$; we shall denote by $L^{a,c}$ its infinitesimal generator.

Let us introduce another class of exponential functions

$$\mathcal{E}_A(H) = \text{ span } \left\{ e^{i\langle h,\cdot\rangle},\ h \in D(A^*) \right\},$$

where A^* denotes the adjoint of A.

If $\varphi(x) = e^{i\langle h,x\rangle}$ we have

$$R_t^{a,c}\varphi(x) = e^{i\langle h,e^{tA}x\rangle}\ e^{-\frac{1}{2}\langle Q_t^{a,c}h,h\rangle},\ \ x \in H.$$

Therefore $\mathcal{E}_A(H)$ is invariant for $R_t^{a,c}$ and

$$L^{a,c}\varphi(x) = \frac{1}{2}\ \text{Tr}\ [CD^2\varphi] + \langle Ax, D\varphi\rangle,\ x \in D(A),\ \varphi \in \mathcal{E}_A(H).$$

Since $\mathcal{E}_A(H)$ is invariant for $R_t^{a,c}$ and dense in $L^2(H,\mu)$, it is a *core* for $L^{a,c}$.

Let us consider the *Lyapunov equation*

$$2\langle Xx, A^*x\rangle = -\langle Cx,x\rangle,\ \ x \in D(A^*), \tag{2.4}$$

that we will write also formally as

$$AX + XA^* = -C. \tag{2.5}$$

The following result is easy to prove.

Proposition 2.2. *$X = Q^{a,c}$ is the unique symmetric solution of Lyapunov equation (2.4).*

To study properties of $L^{a,c}$ the following proposition is basic, see [2], [10].

Proposition 2.3. *For any $\varphi, \psi \in \mathcal{E}_A(H)$ we have*

$$\int_H (L^{a,c}\varphi)(x)\ \psi(x)\mu^{a,c}(dx) = \int_H \langle AQ^{a,c}D\psi(x), D\varphi(x)\rangle\mu^{a,c}(dx), \tag{2.6}$$

and

$$\int_H (L^{a,c}\varphi)(x)\ \varphi(x)\mu^{a,c}(dx) = -\frac{1}{2}\int_H \langle CD\varphi(x), D\varphi(x)\rangle\mu^{a,c}(dx). \tag{2.7}$$

Proof. Let $h, k \in D(A^*)$ and set

$$\varphi(x) = e^{i\langle h,x\rangle},\ \psi(x) = e^{i\langle k,x\rangle},\ x \in H.$$

Then (2.6) follows by a direct verification. To prove (2.7) it is enough to notice that

$$\int_H \langle AQ^{a,c}D\psi, D\varphi\rangle \mu^{a,c}(dx) = \frac{1}{2}\int_H \langle [AQ^{a,c} + Q^{a,c}A^*]D\psi, D\varphi\rangle \mu^{a,c}(dx),$$

and then to use the Lyapunov equation. □

Proposition 2.4. *We have*

$$D(L^{a,c}) \subset W^{1,2}(H, \mu^{a,c}). \tag{2.8}$$

Proof. Let $\varphi \in D(L^{a,c})$. Since $\mathcal{E}_A(H)$ is a core there exists a sequence $\{\varphi_n\} \subset \mathcal{E}_A(H)$ such that

$$\varphi_n \to \varphi,\ L^{a,c}\varphi_n \to L^{a,c}\varphi, \text{ in } L^2_C(H, \mu^{a,c}).$$

Let $m, n \in \mathbb{N}$. Then by (2.7) it follows that

$$\int_H L^{a,c}(\varphi_n - \varphi_m)\,(\varphi_n - \varphi_m)d\mu^{a,c} = -\frac{1}{2}\int_H |C^{1/2}D(\varphi_n - \varphi_m)|^2 d\mu^{a,c}.$$

Since $C^{-1/2}$ is bounded, this implies that

$$\varphi_n \to \varphi, \text{ in } L^2_C(H, \mu^{a,c}).$$

□

Remark 2.5. By (2.7) it follows that, see [3], $L^{a,c}$ is symmetric if ad only if

$$AQ^{a,c} = Q^{a,c}A^*.$$

In this case (2.6) becomes

$$\int_H (L^{a,c}\varphi)\psi d\mu^{a,c} = -\frac{1}{2}\int_H \langle CD\varphi, D\psi\rangle d\mu^{a,c}. \tag{2.9}$$

If $L^{a,c}$ is not symmetric it is possible to find an operator A' such that $L^{a',c}$ is symmetric and $Q^{a',c} = Q^{a,c}$. We have in fact the result:

Proposition 2.6. *Assume that* $A' = \frac{1}{2}\,C(Q^{a,c})^{-1}$ *is the infinitesimal generator of a strongly continuous semigroup in* H. *Then* $L^{a',c}$ *is symmetric and* $Q^{a',c} = Q^{a,c}$. *In particular we have*

$$\int_H (L^{a',c}\varphi)(x)\psi(x)\mu^{a,c}(dx) = -\frac{1}{2}\int_H \langle CD\varphi(x), D\psi(x)\rangle \mu^{a,c}(dx). \tag{2.10}$$

Proof. Since

$$A'Q^{a,c} + Q^{a,c}(A')^* = -C$$

we have $Q^{a',c} = Q^{a,c}$ due to the uniqueness of solutions of Lyapunov equation (2.4). □

We now prove the following result, see [8].

Proposition 2.7. *For all $\varphi, \psi \in \mathcal{E}_A(H)$ we have*

$$\int_H \langle (L^{a',c} D\varphi)(x), D\psi(x)\rangle \mu^{a,c}(dx) = -\frac{1}{2}\int_H Tr\,[C(D^2\varphi(x))^2]\mu^{a,c}(dx). \quad (2.11)$$

Proof. We prove Proposition 2.7 under the additional assumption that there exist a complete orthonormal system $\{f_k\}$ in H and a sequence of positive numbers $\{c_k\}$ such that

$$Cf_k = c_k f_k,\ k \in \mathbb{N}.$$

The general case can be handled by approximating C by a sequence of linear operators having the above property. We have

$$\langle L^{a,c} D\varphi, D\psi\rangle = \sum_{k=1}^\infty c_k L^{a,c}(D_k\varphi)\ D_k\varphi.$$

Therefore by (2.7) it follows that

$$\int_H \langle L^{a,c} D\varphi, D\psi\rangle d\mu^{a,c} = -\frac{1}{2}\sum_{k=1}^\infty c_k \int_H \langle CDD_k\varphi, DD_k\varphi\rangle d\mu^{a,c}$$

$$= -\frac{1}{2}\sum_{k,j=1}^\infty c_k c_j \int_H (D_iD_j\varphi)^2 d\mu^{a,c} = -\frac{1}{2}\int_H \mathrm{Tr}\,[C(D^2\varphi)^2] d\mu^{a,c}.$$

□

We shall use the following formula for the commutator $[D, L^{a,c}] = DL^{a,c}\varphi - L^{a,c}D\varphi$, that can be easily checked,

$$[D, L^{a,c}]\varphi = A^* D\varphi,\ \varphi \in \mathcal{E}_A(H). \quad (2.12)$$

We conclude this section by giving a characterization of the adjoint $(L^{a,c})^*$ of $L^{a,c}$.

Proposition 2.8. *Assume that*

(i) *The linear operator A_1 defined as*

$$A_1 = Q^{a,c}A^*(Q^{a,c})^{-1} \quad (2.13)$$

is the infinitesimal generator of a strongly continuous semigroup in H.

(ii) *$Y := D(A^*) \cap D(A_1^*)$ is dense on H.*

Then we have

$$(L^{a,c})^* = L^{a_1,c}. \quad (2.14)$$

Proof. Denote by $\mathcal{F}$ the linear span of all functions

$$\varphi_h(x) = e^{i\langle h,x\rangle},\ h \in Y.$$

By the assumptions $\mathcal{F}$ is dense in $L^2(H, \mu^{a,c})$.

Note now that $Q^{a_1,c} = Q^{a,c}$, since $Q^{a,c}$ is a solution to the Lyapunov equation

$$A_1 Q + Q A_1^* = -C.$$

Then by (2.6) it follows, setting $Q^{a,c} = Q^{a_1,c}$ and $\mu = \mu^{a,c} = \mu^{a_1,c}$, that for any $\varphi, \psi \in \mathcal{F}$,

$$\int_H L^{a,c}\varphi(x)\psi(x)\mu(dx) = \int_H \langle AQD\psi(x), D\varphi(x)\rangle \mu(dx),$$

and

$$\int_H L^{A_1,C}\psi(x)\varphi(x)\mu(dx) = \int_H \langle A_1 QD\varphi(x), D\psi(x)\rangle \mu(dx)$$

$$\int_H \langle QA^* D\varphi(x), D\psi(x)\rangle \mu(dx),$$

and the conclusion follows. □

Remark 2.9. The assumptions of Proposition 2.8 are obviously fulfilled when H is finite-dimensional.

3. The equation $\lambda\varphi - L^{a,c}\varphi = f$.

Here we assume that Hypothesis 2.1 holds. If there is no possibility of confusion we write $L = L^{a,c}$, $Q = Q^{a,c}$, and $\mu = \mu^{a,c}$.

We fix $\lambda > 0$, $\varphi \in \mathcal{E}_A(H)$ and set $f = \lambda\varphi - L\varphi$.

The following result follows in a standard way from (2.6).

Proposition 3.1. *We have*

$$\int_H |\varphi(x)|^2 \mu(dx) \le \frac{1}{\lambda^2} \int_H |f(x)|^2 \mu(dx), \tag{3.1}$$

and

$$\int_H |C^{1/2} D\varphi(x)|^2 \mu(dx) \le \frac{2}{\lambda} \int_H |f(x)|^2 \mu(dx). \tag{3.2}$$

We want now to estimate second order derivatives of φ. For this we first prove two identities.

Proposition 3.2. *For any $\varphi \in \mathcal{E}_A(H)$ we have*

$$\begin{aligned}\lambda \int_H \langle CD\varphi, D\varphi\rangle d\mu - \frac{1}{2}\int_H \langle (AC + CA^*)D\varphi, D\varphi\rangle d\mu \\ +\frac{1}{2}\int_H Tr\,[C(D^2\varphi)^2] d\mu = \int_H \langle Df, CD\varphi\rangle d\mu.\end{aligned} \tag{3.3}$$

Proof. We have

$$\lambda D\varphi - [D, L]\varphi - LD\varphi = Df.$$

It follows by (2.12) that

$$\lambda D\varphi - A^* D\varphi - LD\varphi = Df.$$

Multiplying both sides by $D\varphi$, integrating in H, and taking into account Proposition 2.6, the conclusion follows. □

Proposition 3.3. *For any $\varphi \in \mathcal{E}_A(H)$ we have*

$$\begin{aligned} &\lambda \int_H \langle CD\varphi, D\varphi\rangle d\mu - \frac{1}{2} \int_H \langle (AC + CA^*)D\varphi, D\varphi\rangle d\mu \\ &+\frac{1}{2} \int_H Tr\,[C(D^2\varphi)^2] d\mu = 2 \int_H |f(x)|^2 d\mu - 2\lambda \int_H f(x)\varphi(x) d\mu \\ &+ \int_H f(x)\langle Ax + \frac{1}{2}\, CQ^{-1}x, D\varphi(x)\rangle d\mu \end{aligned} \tag{3.4}$$

Proof. Setting $A' = -\frac{1}{2}\, CQ^{-1}$, we have

$$\int_H \langle Df, CD\varphi\rangle d\mu = -2 \int_H L^{A',C}\varphi f d\mu.$$

Since

$$\begin{aligned} L^{A',C}\varphi &= \frac{1}{2}\, \mathrm{Tr}\,[CD^2\varphi] - \frac{1}{2}\, \langle CQ^{-1}x, D\varphi\rangle \\ &= L\varphi - \langle Ax + \frac{1}{2}\, CQ^{-1}x, D\varphi\rangle \\ &= \lambda\varphi - f - \langle Ax + \frac{1}{2}\, CQ^{-1}x, D\varphi\rangle, \end{aligned}$$

the conclusion follows. □

Using identity (3.4) we can characterize $D(L)$ under suitable assumptions. The following result was proved in [12], [5].

Theorem 3.4. *If H is finite-dimensional we have*

$$D(L) = W^{2,2}(H, \mu). \tag{3.5}$$

Proof. In fact the square of the last term in (3.4) can be estimated as

$$\left[\int_H |f(x)|\,|x|\,|D\varphi(x)|d\mu\right]^2 \le \varepsilon \int_H |x|^2|D\varphi(x)|^2 d\mu + C_\varepsilon \int_H |f(x)|^2 d\mu,$$

for any $\varepsilon > 0$ and for a suitable constant C_ε.

On the other hand by Proposition 1.1 we have

$$\int_H |x|^2|D\varphi(x)|^2 d\mu \le \kappa_1 \int_H [|\varphi(x)|^2 + \text{ Tr }[(CD^2\varphi)^2]d\mu.$$

Now the conclusion follows by a standard argument. □

Theorem 3.5. *If $AQ = QA^*$ we have:*

$$D(L) = \left\{\varphi \in W^{2,2}(H,\mu) : \int_H \langle Q^{-1}CD\varphi, CD\varphi\rangle d\mu < +\infty\right\}. \tag{3.6}$$

If in particular $A = A^*$ and $C = I$, we have

$$D(L) = \left\{\varphi \in W^{2,2}(H,\mu) : \int_H |(-A)^{1/2}D\varphi|^2 d\mu < +\infty\right\}. \tag{3.7}$$

Proof. It is enough to notice that, under the assumptions of the theorem the last term in in (3.4) vanishes. □

4. The nonlinear case

Here we are given, besides two linear operators A, C fulfilling Hypothesis 2.1 (we shall write again for simplicity $L = L^{a,c}$, $Q = Q^{a,c}$, and $\mu = \mu^{a,c}$), a mapping F from H into H such that

Hypothesis 4.1.

(i) *F belongs to $C_b^1(H,H)$, i.e. F is continuously differentiable and bounded together with its first derivative.*

(ii) *$DF(x)$ is of trace class for any $x \in H$. We set*

$$\textit{div}\, F(x) = \textit{Tr}\,[DF(x)],\ x \in H.$$

(iii) *There exists a continuous and bounded function $F_1 : H \to H$, such that*

$$F_1(x) = \langle Q^{-1}x, F(x)\rangle,\ x \in Q(H).$$

We set

$$N\varphi = L\varphi + \langle F(x), D\varphi\rangle,\ \varphi \in D(L). \tag{4.1}$$

Since $D(L) \subset W^{1,2}(H,\mu)$ the term $\langle F(x), D\varphi\rangle$ is a lower-order pertubation of L. Therefore it is easy to show that N generates a strongly continuous semigroup P_t, $t \ge 0$ on $L^2(H,\mu)$.

We now characterize the adjoint of N.

Proposition 4.2. *Assume that besides the assumptions of Proposition 2.8, Hypotheses* 2.1 *and* 4.1 *are fulfilled. Then the adjoint N^* of N is given by*

$$N^*\varphi = L^*\varphi - \langle F(x), D\varphi\rangle + (F_1(x) - \ div\, F(x))\,\varphi,\ \varphi \in D(L^*). \tag{4.2}$$

Proof. Let $\{e_k\}$ be a complete orthonormal system that diagonalizes Q and let $\{\lambda_k\}$ be a corresponding sequence of eigenvectors:

$$Qe_k = \lambda_k e_k.$$

Set

$$\mathcal{F}\varphi = \langle F(x), D\varphi\rangle,\ \varphi \in W^{1,2}(H,\mu).$$

We first show that the adjoint $\mathcal{F}^*$ of $\mathcal{F}$ is given by

$$\mathcal{F}^*\varphi = -\langle F(x), D\varphi\rangle + (F_1(x) - \ \text{div}\ F(x))\varphi,\ \varphi \in W^{1,2}(H,\mu). \tag{4.3}$$

In fact for any $\varphi, \psi \in W^{1,2}(H,\mu)$ we have, setting $F_k(x) = \langle F(x), e_k\rangle$ and using the integration-by-parts formula (1.9),

$$\int_H \langle F(x), D\varphi(x)\rangle \psi(x)\mu(dx) = \sum_{k=1}^\infty \int_H F_k(x) D_k\varphi(x)\ \psi(x)\mu(dx)$$

$$= -\sum_{k=1}^\infty \int_H D_k F_k(x)\varphi(x)\ \psi(x)\mu(dx) - \sum_{k=1}^\infty \int_H F_k(x) D_k\psi(x)\ \varphi(x)\mu(dx)$$

$$+\sum_{k=1}^\infty \frac{1}{\lambda_k} \int_H x_k F_k(x)\varphi(x)\ \psi(x)\mu(dx)$$

$$= -\int_H \text{div}\ F(x)\ \varphi(x)\ \psi(x)\ \mu(dx) - \int_H \langle F(x), D\psi(x)\rangle\varphi(x)\mu(dx)$$

$$+\int_H \langle Q^{-1}x, F(x)\rangle\varphi(x)\ \psi(x)\mu(dx),$$

that yields (4.3).

Let now $\varphi \in D(N^*)$. Since by Proposition 2.4

$$D(L^*) \subset W^{1,2}(H,\mu),$$

we obtain, taking into account (4.3), that $D(L^*) \subset D(N^*)$.

It remains to prove that the converse inclusion holds. Let $\psi \in D(N^*)$. Then the mapping

$$\Lambda(\varphi) = \int_H L\varphi(x)\psi(x)\ \mu(dx),\ \varphi \in D(L),$$

is continuous in $L^2(H,\mu)$. Moreover, since, again by Proposition 2.4

$$D(N) = D(L) \subset W^{1,2}(H,\mu) = D(\mathcal{F}^*),$$

it follows that the mapping

$$\varphi \to \int_H L\varphi\psi \, d\mu, \; \varphi \in D(L),$$

is continuous in $L^2(H,\mu)$. This implies $\psi \in D(L^*)$ as required. □

We will need the following result proved in [9].

Proposition 4.3. *Assume that Hypothesis 2.1 holds and that F is bounded and Borel. Then there exists a unique invariant measure ν for the semigroup P_t, $t \geq 0$. Moreover, ν is absolutely continuous with respect to μ and there exists a nonnegative function $\alpha \in D(N^*)$ such that $N^*\alpha = 0$ and $\nu(dx) = \alpha(x)\mu(dx)$.*

Remark 4.4. Setting $\psi = \log \alpha$, we find that ψ fulfills the following Hamilton-Jacobi equation

$$L^*\psi - \langle F(x), D\psi\rangle - \frac{1}{2}\,|C^{1/2}D\psi|^2 + V(x) = 0, \tag{4.4}$$

where

$$V(x) = \text{ div } F(x) - F_1(x), \; x \in H. \tag{4.5}$$

Example 4.5. Assume that $F = -DU$ where $U \in C_b^2(H)$. Then we can easily prove that $\alpha(x) = e^{-2U(x)}$ fulfills (4.4).

It is important to show that the semigroup P_t, $t \geq 0$ can be extended to a contraction semigroup on $L^2(H,\nu)$. In fact this is the appropriate space to study ergodic properties.

Proposition 4.6. *Assume that besides the assumptions of Proposition 2.8, Hypotheses 2.1 and 4.1 are fulfilled. Let $\nu(dx) = \alpha(x)\mu(x)$ be the invariant measure for e^{tN}, and suppose that $D\log\alpha \in C_b^1(H,H)$. Then for any $\varphi, \psi \in \mathcal{E}_A(H)$ we have*

$$\begin{aligned}\int_H N\varphi(x)\psi(x)\mu(dx) &= \int_H \langle AQD\psi(x), D\varphi(x)\rangle\nu(dx)\\ &+ \int_H \langle AQD\log\alpha(x) + F(x), D\varphi(x)\rangle\psi(x)\nu(dx),\end{aligned} \tag{4.6}$$

and

$$\int_H N\varphi(x)\varphi(x)\nu(dx) = -\frac{1}{2}\int_H \langle CD\varphi(x), D\varphi(x)\rangle\nu(dx). \tag{4.7}$$

Proof. We have, using (2.6),

$$\int_H L\varphi(x)\psi(x)\nu(dx) = \int_H L\varphi(x)(\alpha(x)\psi(x))\mu(dx)$$

$$= \int_H \langle AQD(\alpha(x)\psi(x)), D\varphi(x)\rangle\mu(dx)$$

$$= \int_H \langle AQD\psi(x), D\varphi(x)\rangle\nu(dx) + \int_H \langle AQD\log(x)\alpha(x), D\varphi(x)\rangle\nu(dx).$$

and (4.6) follows. To prove (4.7) we will show that for all φ the following identity holds:

$$\int_H \langle AQD\log(x)\alpha(x) + F(x), D\varphi(x)\rangle\varphi(x)\nu(x) = 0. \tag{4.8}$$

To prove (4.8) we set $G = AQD\log\alpha + F$ and we first show that

$$\int_H \langle G(x), D\varphi\rangle\varphi\nu(x) = \frac{1}{2}\int_H \varphi^2\left[\langle Q^{-1}x, G(x)\rangle - \operatorname{div} G - \langle G, D\log\alpha\rangle\right]\nu(x). \tag{4.9}$$

Setting

$$I = \int_H \langle G(x), D\varphi(x)\rangle\varphi(x)\nu(x) = \sum_{k=1}^{\infty}\int_H D_k\varphi(x)\, G_k(x)\varphi(x)\,\alpha(x)\mu(x),$$

we have

$$I = -\sum_{k=1}^{\infty}\int_H \varphi(x)(D_kG_k(x)\varphi(x)\,\alpha(x) + G_k(x)\alpha(x)D_k\varphi(x))\mu(dx)$$

$$+\sum_{k=1}^{\infty}\int_H\left(G_k(x)\varphi(x)D_k\alpha(x) + \frac{x_k}{\lambda_k}\,G_k(x)\varphi(x)^2\alpha(x)\right)\mu(dx)$$

$$= -\int_H \operatorname{div} G(x)\,\varphi(x)^2\nu(dx) - \int_H \langle G, D\varphi(x)\rangle\varphi(x)\nu(dx)$$

$$-\int_H \langle G(x), D\log\alpha(x)\rangle\varphi(x)\nu(dx) + \int_H \langle Q^{-1}x, G(x)\rangle\varphi(x)^2\nu(dx)$$

$$= \int_H \varphi^2\left[\langle Q^{-1}x, G(x)\rangle - \operatorname{div} G - \langle G, D\log\alpha\rangle\right]\nu(dx) - I,$$

and (4.9) follows.

Finally we show that

$$J := \langle Q^{-1}x, G(x)\rangle - \operatorname{div} G - \langle G, D\log\alpha\rangle = 0, \tag{4.10}$$

and that will imply (4.8). We have in fact

$$J = \langle Q^{-1}x, AQD\log\alpha(x)\rangle + \langle Q^{-1}x, F(x)\rangle$$

$$-\text{div}\,[AQD\log\alpha(x)] - \text{div}\,F(x)$$

$$-\langle AQD\log\alpha(x), D\log\alpha(x)\rangle - \langle F, D\log\alpha(x)\rangle$$

$$\langle A_1x - F(x)D\log\alpha(x)\rangle + \frac{1}{2}\,\text{Tr}\,[CD^2\log\alpha(x)]$$

$$+\langle Q^{-1}x, F(x)\rangle - \text{ div}\,F(x).$$

Setting $\psi = -\log\alpha$ and recalling that ψ fulfills the Hamilton-Jacobi equation (4.5), we find $J = 0$ as required. □

Remark 4.7. If

$$AQD\log\alpha + F = 0,$$

the identity (4.6) simplifies. This happens for instance when A, C commute and $F = -DU$.

Theorem 4.8. *Assume that the assumptions of Proposition 4.6 hold and that $\alpha \in L^\infty(H,\mu)$. Then N is closable and its closure is m-dissipative on $L^2(H,\nu)$.*

Proof. Notice first that since $\alpha \in L^\infty(H,\mu)$ we have

$$L^2(H,\mu) \subset L^2(H,\nu).$$

Now let $\lambda > 0$, $\varphi \in D(L)$, and $\lambda\varphi - N\varphi = f$. Recalling that $\mathcal{E}_A(H)$ is a core for L, we see that there exists a sequence $\{\varphi_n\} \subset \mathcal{E}_A(H)$ such that

$$\varphi_n \to \varphi, \;\; \lambda\varphi_n - L\varphi_n \to \lambda\varphi - L\varphi \text{ in } L^2(H,\mu).$$

Consequently

$$f_n := \lambda\varphi_n - N\varphi_n \to f \text{ in } L^2(H,\mu).$$

Moreover, by (3.4) it follows that

$$\|\varphi_n\|_{L^2(H,\nu)} \le \frac{1}{\lambda}\,\|f_n\|_{L^2(H,\nu)}.$$

Since convergence on $L^2(H,\mu)$ implies convergence on $L^2(H,\nu)$ it follows that

$$\|\varphi\|_{L^2(H,\nu)} \le \frac{1}{\lambda}\,\|f\|_{L^2(H,\nu)},$$

so that N is dissipative on $L^2(H,\nu)$.

Moreover, the image of $\lambda - N$ contains $L^2(H,\mu)$ which is dense on $L^2(H,\nu)$. Therefore the closure of N is m-dissipative. □

We end with some spectral properties of N.

Proposition 4.9. *Assume that the assumptions of Theorem 4.8 hold and that all functions $\varphi \in W^{1,2}(H,\nu)$ such that $D\varphi = 0$ are constant. Then $\{0\}$ is a simple eigenvalue of N so that the semigroup $P_t, t \geq 0$ is ergodic.*

Moreover, no number belonging to $i\mathbb{R}\backslash\{0\}$ is an eigenvalue of N.

Proof. The first statement follows immediately from (4.7). Let us prove the second one. Let $\lambda \neq 0$ and let $\varphi \in D(N)$ be such that $L\varphi = \lambda\varphi$. We have to show that $\varphi = 0$. Set $\varphi = \varphi_1 + i\varphi_2$ with φ_1, φ_2 real-valued. Then we have

$$L\varphi_1 = -\lambda\varphi_2, \quad L\varphi_2 = -\lambda\varphi_1.$$

By (4.7) it follows

$$\int_H L\varphi_1(x)\varphi_1(x)d\nu = -\frac{1}{2}\int_H |D\varphi_1(x)|^2\nu(x) = -\lambda\int_H \varphi_1(x)\varphi_2(x)\nu(x), \quad (4.11)$$

and

$$\int_H L\varphi_2(x)\varphi_2(x)\nu(x) = -\frac{1}{2}\int_H |D\varphi_2(x)|^2\nu(x) = \lambda\int_H \varphi_1(x)\varphi_2(x)\nu(x). \quad (4.12)$$

Summig up (4.11) and (4.12) we find

$$\int_H [|D\varphi_1(x)|^2 + |D\varphi_1(x)|^2]\nu(x) = 0,$$

which yields $\varphi = 0$. □

References

[1] Albeverio S., Bogachev V. I. & Röckner M., "Dirichlet operators via stochastic analysis", Journal of Functional Analysis, **128**, n.1 (1995), 102–138.

[2] Bogachev V. I. , Röckner M. & Schmuland B., "Generalized Mehler semigroups and applications", Probability and Related Fields, **114** (1996), 193–225.

[3] Chojnowska-Mihalik A. & Goldys B., "On Ornstein-Uhlenbeck generators", preprint S96-12 of the School of Mathematics, The University of New South Wales (1996).

[4] Da Prato G., *Applications Croissantes et Équations d'Évolutions dans les Espaces de Banach,* Academic Press (1976).

[5] Da Prato G., "Regularity results for Kolmogorov equations on $L^2(H,\mu)$ spaces and applications", Ukrainian Mathematical Journal, **49**, n. 3 (1997), 448–457.

[6] Da Prato G., "The Ornstein-Uhlenbeck generator perturbed by the gradient of a potential", Boll.U.M.I. (to appear).

[7] Da Prato G., Malliavin P. & Nualart D., "Compact families of Wiener functionals", C. R. Acad. Sci. Paris, **315** (1992), 1287–1291.

[8] Da Prato G. & Tubaro L., "Self-adjointness of generator of a gradient system", (in preparation).

[9] Da Prato G. & Zabczyk J., *Ergodicity for Infinite Dimensional Systems.* London Mathematical Society Lecture Notes, n.229, Cambridge University Press (1996).

[10] Fuhrman M., "Analyticity of transition semigroups and closability of bilinear forms in Hilbert spaces", Studia Mathematica, **115** (1995), 53–71.

[11] Liskevich V. & Röckner M., "Strong uniqueness for a class of infinite dimensional Dirichlet operators and application to stochastic quantization", preprint.

[12] Lunardi A., "On the Ornstein-Uhlenbeck operator in L^2 spaces with respect to invariant measures", Trans. Amer. Math. Soc. **349** (1995), 155–169.

[13] Ma Z. M. & Röckner M., *Introduction to the Theory of (Non Symmetric) Dirichlet Forms*, Springer Verlag (1992).

[14] Peszat S., "On a Sobolev space of function of infinite numbers of variables", Bull. Polish. Acad. Sci., **40** (1993), 55–60.

Giuseppe Da Prato
Scuola Normale Superiore di Pisa

Progress in Nonlinear Differential Equations
and Their Applications, Vol. 35
© 1999 Birkhäuser Verlag Basel/Switzerland

Remarks on Jumping Nonlinearities

E.N. Dancer

Dedicated to Herbert Amann on the occasion of his 60th birthday

In this paper, we discuss the jumping nonlinearity problem

$$-\Delta u = au^{+} + du^{-} - f \ \text{ in } \Omega\,, \qquad u = 0 \ \text{ on } \partial\Omega\,, \tag{1}$$

the special case where $f = 0$ and problems which behave asymptotically like (1).

We prove a number of results. Firstly, we obtain partial results on the critical groups of the solution 0 of (1) for $f = 0$. In particular, we completely determine the least critical group which could possibly be non-zero and the last critical group which could possibly be non-zero. In many but not all cases, we obtain complete information on the change of the critical groups when we increase a from d till we first cross the resonance set A_0 (which is the set of (a, d) for which (1) with $f = 0$ has a non-trivial solution). Our information is precise enough to get much more explicit results than in [9]. This shows that the result in [9] when (1) has at least 2 solutions for f close to a large positive multiple of the first eigenfunction of $-\Delta$ with Dirichlet boundary conditions is equivalent to the result obtained by combining [19] and [20]. Curiously the results in [9] and [19] and [20] on when there are 3 solutions are distinct. Our results also give extra information on the resonance set. We also use our techniques to understand better the structure of A_0 near (λ, λ) when λ is an eigenvalue of $-\Delta$. (We obtain complete local information in most cases.)

Lastly, we give two examples. Firstly, we show that the resonance set can be quite complicated locally (though this was implicit in [8]). Secondly, we show that the problem (1) with $f = 0$ can behave somewhat differently to the linear case by constructing an example where (a, d) is not in the resonance set and the map $u \longrightarrow u + (\Delta)^{-1}(au^{+} + du^{-})$ has index d at 0 where $|d| > 1$.

In Section 1, we make our main critical group calculations while in Section 2 we consider the case when a and d are both close to the same eigenvalue. Finally, in Section 3, we consider the examples.

1. Critical group computations

In this section we make some critical group computations. We consider the problem

$$-\Delta u = au^{+} + du^{-} \ \text{ in } \Omega\,, \qquad u = 0 \ \text{ on } \partial\Omega\,. \tag{2}$$

We assume that the problem has only the trivial solution, that is, $(a,d) \in \mathbb{R}^2 \backslash A_0$, where by definition $A_0 = \{(a,d) \in \mathbb{R}^2 : (2)$ has a non-trivial solution$\}$. Now we can easily write (2) as an equation $\nabla G = 0$ where $G : \mathring{W}^{1,2}(\Omega) \longrightarrow \mathbb{R}$ and $\nabla G = I + C$ where C is completely continuous. Hence the critical groups $C_k(G,0)$ as defined by Chang [5] and in [24] in terms of homotopy indices are equivalent. We will always take our field of coefficients to be $\mathbb{Z}$. Without loss of generality we can assume $d \geq a$ since it is easy to check that interchanging a and d does not affect the critical groups. We also assume $d > a$ since otherwise we are back in the linear case. Let $\lambda_1 < \lambda_2 < \lambda_3 < \cdots$ denote the *distinct* eigenvalues of $-\Delta$ under Dirichlet boundary conditions on Ω. Let ℓ denote the sum of the multiplicities of the eigenvalues less than or equal to a and u denote the sum of the multiplicities of the eigenvalues less than d. Then it is well known (cp. [9] or [24]) that $C_j(G,0) = 0$ if $j < \ell$ and $C_j(G,0) = 0$ if $j > u$. We will compute $C_j(G,0)$ for $j = \ell$ and $a \neq \lambda_i$ for all i (and for $j = u$ and $d \neq \lambda_i$ for all i).

Theorem 1. *(i) Assume that $a \neq \lambda_i$ for all i and $a > \lambda_1$ and let $t(a) = \inf \{d > a : (a,d) \in A_0\}$. Then $t(a)$ is defined, $C_\ell(G,0) = \mathbb{Z}$ (and $C_j(G,0) = 0$ if $j \neq \ell$) if $a < d < t(a)$ while $C_\ell(G,0) = 0$ if $d > t(a)$ and $(a,d) \notin A_0$.*

(ii) Assume that $d \neq \lambda_i$ for all i and $d > \lambda_1$ and let $s(d) = \sup\{a : a < d, (a,d) \in A_0\}$. Then $s(d)$ is defined and $C_u(G,0) = \mathbb{Z}$ if $s(d) < a < d$ while $C_u(G,0) = 0$ if $\lambda_1 < a < s(d)$ and $(a,d) \notin A_0$.

Remarks. Note that 0 is not an isolated critical point of G if $d = t(a)$ or if $a = s(d)$ and so the critical groups are not defined. Since A_0 is closed, $t > a$ and $s < d$. Note that ℓ is independent of d for $d > a$ and that u is independent of a for $a < d$.

Before starting the proof, we need to discuss a little the calculation of critical groups for these problems. Suppose W is a finite dimensional subspace of $\mathring{W}^{1,2}(\Omega)$ generated by eigenspaces of $-\Delta$ where W includes all the eigenspaces corresponding to eigenvalues in $[a,d]$. We write elements of $\mathring{W}^{1,2}(\Omega)$ as $w + n$ where $n \in W^\perp$ and let P be the natural projection onto W. It is a well-known simple consequence of the contraction mapping theorem (cp. [8]) that the problem $(I-P)A(w+n) = 0$ can be uniquely solved for n as a function of $w, n = S(w)$ and S is positive homogeneous. Here $A(u) = -\Delta u - au^+ - du^-$ (on $\mathring{W}^{1,2}(\Omega)$). Really we are writing our equation in the weak sense on $\mathring{W}^{1,2}(\Omega)$. Thus (2) is equivalent to the problem $PA(w + S(w)) = 0$ on W. It is also shown in [8] that the mapping $\widetilde{F}(w) = PA(w + S(w))$ is the gradient of $\widetilde{f}(w) = G(w + s(w))$ where $G(u) = \frac{1}{2}\int_\Omega |\nabla u|^2 - a(u^+)^2 - d(u^-)^2$. We write a and d as subscripts on mappings when the dependence is important.

Note that $\widetilde{F}$ is also positive homogeneous. Let t be the sum of the multiplicities of the eigenvalues of $-\Delta$ restricted to $W^\perp$ which are below a. Then as in [9], one easily sees that $C_i(G,0)$ is zero if $i < t$ and is $C_{i-t}(\widetilde{f},0)$ otherwise. Moreover, if 0 is a minimum of $\widetilde{f}$, then $C_i(\widetilde{f},0) = \delta_{i0}\mathbb{Z}$ while if zero is not a minimum, $C_i(\widetilde{f},0) = \widetilde{H}_{i-1}(Y,Z)$ where $Y = \{w \in W : \|w\| = 1, \widetilde{f}(w) < 0\}$ by p. 9 of [10]. It follows that if 0 is not a local minimum of $\widetilde{f}$, then $C_\ell(G,0) = 0$.

Proof of Theorem 1. We will only prove (i). The proof of (ii) is similar (using some of the computations in [10]). To prove the first statement of (i) note that since the critical groups are constant on components of A_0 and since we can decrease d to a with (a,d) staying in A_0 then it suffices to evaluate the critical groups when $d = a$ where we are in the linear case where the result is well known.

We choose W so as to exclude all the eigenvalues less than a (and hence t is ℓ). By our earlier comments, we see that our result follows if we prove that 0 is not a local minimum of $\widetilde{f}$ when $d > t(a)$ and $(a,d) \notin A_0$. We will prove a little later that S is a differentiable function at $\widetilde{w}$ if $\widetilde{w} + s(\widetilde{w})$ only vanishes on a set of measure zero, and hence $\widetilde{F}$ is a differentiable function of d (at such $\widetilde{w}$). We prove that $\frac{\partial \widetilde{f}}{\partial d}(\widetilde{w}) < 0$ whenever $\widetilde{F}(\widetilde{w}) = 0$. Now zero is a local minimum of $\widetilde{f}$ if $d < t(a)$ by our earlier comments. Thus, by continuity zero is a local (and hence global by homogeneity) minimum of $\widetilde{f}$ for $d = t(a)$. Since $(a, t(a)) \in A_0$, zero is a non-isolated solution of $A(u) = 0$ and hence of $\widetilde{F}(w) = 0$ (for $d = t(a)$). By homogeneity, any solution of $\widetilde{F}(w) = 0$ has $\widetilde{f}(w) = 0$ (where $\widetilde{f}(0) = 0$) and hence there is a non-zero $\hat{w}$ in W such that $\widetilde{F}(\hat{w}) = 0$ and $\widetilde{f}(\hat{w}) = 0$ (for $d = t(a)$). Now $\frac{\partial \widetilde{f}}{\partial d}(\hat{w}) < 0$ and hence if we increase d slightly, $\widetilde{f}(\hat{w}) < 0$. Since $\widetilde{f}(0) = 0$, this implies zero is not a local minimum of $\widetilde{f}$ if d is a little larger than $t(a)$. We can use a similar argument for d not close to $t(a)$. If there exists d larger than $t(a)$ for which zero is not a local minimum, let $d^* = \inf\{d > t(a) : \text{zero is a local minimum of } \widetilde{f}_d\}$. By continuity, 0 is a local minimum of $\widetilde{f}_{d^*}$. We prove that $\widetilde{f}_{d^*}$ has a non-zero critical point and we can then obtain a contradiction as before. If not, $\widetilde{f}_{d^*}(w) > 0$ on $\|w\| = \delta$ for some $\delta > 0$ and hence by homogeneity on $\|w\| = 1$. By compactness and continuity, it follows that $\widetilde{f}_d(w) > 0$ if $\|w\| = 1$ and d is near d^*. By homogeneity, it follows that $\widetilde{f}_d(w) > 0$ if $w \in W \backslash \{0\}$ and d is close to d^*. This contradicts the definition of d^*.

We still have to prove the differentiability properties. If $\hat{w}$ is a non-zero critical point of $\widetilde{f}, \hat{w} + S(\hat{w})$ is a non-zero solution of (1) and hence by Caffarelli and Friedman [3] the zero set of $w + S(\hat{w})$ has measure zero. Hence the map $u \longrightarrow au^+ + du^-$ is strictly differentiable at $\hat{w} + S(\hat{w})$ in the sense of Cartan [4] as a map of $\mathring{W}^{1,2}(\Omega) \longrightarrow L^2(\Omega)$ (cp. [12], p. 248). Hence A is strictly differentiable at $\hat{w} + S(\hat{w})$ considered as a map of $\mathring{W}^{1,2}(\Omega)$ into itself. Since the map $au^+ + du^-$ is smooth in d, it follows easily by a slight variant of the proof of the implicit function theorem that $S(w)$ is differentiable in d at $\hat{w}$ and

$$(I-P)(-\Delta(\frac{\partial S}{\partial d}) - h'(\hat{w} + S(\hat{w}))\frac{\partial S}{\partial d}) = -(I-P)(\hat{w} + S(\hat{w}))^- \tag{3}$$

where $h(u) = au^+ + du^-$. Now

$$\frac{\partial}{\partial d}\widetilde{f}_d(\hat{w}) = \frac{\partial}{\partial d}(G_d(\hat{w} + S_d(\hat{w}))$$

$$= -\frac{1}{2}\int_\Omega ((\hat{w} + S(\hat{w}^-)^2 + (-\Delta(\hat{w} + S(\hat{w})) - h(\hat{w} + S(\hat{w}))\,\frac{\partial S}{\partial d} \tag{4}$$
$$= -\frac{1}{2}\int_\Omega (\hat{w} + S(\hat{w}))^-)^2 \ \text{(since } \hat{w} + S(\hat{w}) \text{ is a solution of (2))} < 0\,,$$

as required. (If this integral were zero, $\hat{w} + S(\hat{w}) \geq 0$ in Ω which is only possible if $a = \lambda_1$.)

We still have to prove that $t(a)$ and $s(d)$ exist. That $s(d)$ exists follows from Theorem 1 in [8] or from [1] or [15]. To see that $t(a)$ exists is more complicated. If not, zero is a local minimum of $\widetilde{f}_d$ for any $d > a$, that is, $f_d(w + S_d(w)) \geq 0$ on W. We write $W^\perp = W_1 \oplus W_2$ where W_2 is spanned by eigenvectors corresponding to eigenvalues larger than d. Now by arguing as in the proof of Theorem 2 in [8] (cp. p. 38–39), we see that, if $f_d(w + S_d(w)) \geq 0$ on W, then $f_d(w + S_d(w) + w_2) \geq 0$ if $w \in W, w_2 \in W_2$, that is, $f_d(P_1 S_d(w) + w + w_2) \geq 0$ on $W \oplus W_2$ where P_1 is the orthogonal projection on W_1. Note that $W \oplus W_2$ and W_1 are independent of d. We will prove below that $P_1 S_d(w)$ are uniformly bounded as (in d) as a map into $\mathring{W}^{1,2}(\Omega)$ for fixed $\widetilde{w} = w + w_2$. Let $z_d = P_1 S_d(w)$. Then

$$\begin{aligned} 0 &\leq \int_\Omega (\nabla z_d)^2 + (\nabla \widetilde{w})^2 - a((\widetilde{w} + z_d)^+)^2 - d((\widetilde{w} + z_d)^-)^2 \\ &\leq \int_\Omega |\nabla z_d|^2 + \|\widetilde{w}\|_2^2 - a\|z_d + \widetilde{w}\|_2^2\,. \end{aligned} \tag{5}$$

Hence

$$0 \leq \int_\Omega |\nabla z_d|^2 - a z_d^2 + K\|\widetilde{w}\|_{1,2}^2 - (d-a)\|(\widetilde{w} + z_d)^-\|_2^2\,. \tag{6}$$

Now $z_d \in W_1$ which is spanned by eigenvectors corresponding to eigenvalues less than a and W_1 is finite dimensional, we easily see that $\int_\Omega |\nabla z_d|^2 - a z_d^2 \leq \mu\|z_d\|_{1,2}^2$ where $\mu < 0$. Hence (6) implies that $\|z_d\|_{1,2} \leq \widetilde{\mu}^{-\frac{1}{2}}\|\widetilde{w}\|_{1,2}$ where $\widetilde{\mu}$ is independent of d and $\widetilde{w}$ and

$$(d-a)\|(\widetilde{w} + z_d)^-\|_2^2 \leq \widetilde{K}\|\widetilde{w}\|_{1,2}^2\,.$$

Hence $(\widetilde{w} + z_d)^- \longrightarrow 0$ in $L^2(\Omega)$ as $d \longrightarrow \infty$. Now since W_1 is finite dimensional, we can choose a subsequence so that $z_{d_n} \longrightarrow \hat{z}$ in W_1 as $d_n \longrightarrow \infty$. Then $\hat{z} \in W_1$ and $\widetilde{w} + \hat{z} \geq 0$ on Ω. We show that this is impossible for suitable $\widetilde{w}$. Choose $v \in \mathring{W}_{1,2}(\Omega)$ so v is not essentially bounded above or below. Let $\widetilde{w} = v - \sum_{i=1}^k (v, \phi_i)\phi_i$ where ϕ_i are an orthogonal basis for W_1. Now the sum is in $L^\infty(\Omega)$ since $\phi_i \in L^\infty(\Omega)$ (by regularity results for $-\Delta$). Thus $\widetilde{w}$ is neither essentially bounded above or below. By the same argument, $\hat{z} \in W_1$ and hence in $L^\infty(\Omega)$ and thus $\widetilde{w} - \hat{z}$ is not essentially bounded below. This is a contradiction since $\hat{z} + \widetilde{w} \geq 0$ in Ω. Hence we have a contradiction and hence $t(a)$ exists. This argument is related to an argument in [15]. □

Remarks. 1. The proof clearly is valid for a much more general class of operators. With care, one does not even need that solutions only vanish on a set of measure zero. S is now possibly only Lipschitz in d but one can still show $\widetilde{f}_d(\widetilde{w})$ is differentiable in d by working with finite differences and showing that the term corresponding to the last term in (4) has zero limit as the difference tends to zero.

2. There is a similar result if a is an eigenvalue, $a = \lambda_i$ with $i > 1$. In this case, one can show that $t(a) \geq \lambda_{i+1}$ (by Theorem 4 in [7]), that $t(a)$ exists, that $C_j(f,0) = \delta_{j,\ell}\mathbb{Z}$ if $d \leq t(a)$ (once again starting from Theorem 4 in [7]), and that $C_j(f,0) = 0$ if $j \leq \ell, d > t(a)$ and $(a,d) \notin A_0$. The proof is similar to before because with care, for $d \geq \lambda_{i+1}$, one can show our reduction is still valid but now with W restricted to eigenvectors corresponding to eigenvalues in $[\lambda_{i+1}, d]$. One proves the reduction by a limiting argument (starting with a a little larger). There is a similar result if d is an eigenvalue. One has to be slightly more careful to prove that $s(d)$ exists if d is an eigenvalue. One uses that if d is close to λ_1 the critical groups are the same as for $a = d = \frac{1}{2}(\lambda_1 + \lambda_2)$ and a homotopy invariance argument.

The next result follows by combining Theorem 1 here with Theorem 1 in [9] and the remarks on p. 77 in [9]. (One can obtain a positive solution with $\frac{u}{\|u\|}$ near ϕ_1 and a changing sign solution by a homotopy index and approximation argument. The key point is that the homotopy index on the whole space is different to that of the solutions near $(a - \lambda_1)^{-1}t\phi_1$.)

Corollary 1. *Assume that $g : \mathbb{R} \longrightarrow \mathbb{R}$ is continuous $y^{-1}g(y) \longrightarrow a(d)$ as $y \longrightarrow \infty(-\infty)$ and $h \in L^\infty(\Omega)$. If $a > \lambda_1$ and a is not an eigenvalue of $-\Delta$ with Dirichlet boundary conditions if $d > t(a)$ and if $(a,d) \notin A_0$ then the equation $-\Delta u = g(u) - t\phi_1 - h$ (plus Dirichlet boundary conditions on $\partial\Omega$) has at least 2 solutions for large positive t. (Here ϕ_1 is the positive eigenfunction of $-\Delta$).*

Remarks. 1. A similar result holds if a is an eigenvalue and $a > \lambda_1$ (where λ_1 is the principal eigenvalue). The only difference is that the positive solutions for $a = \lambda_1$ are not isolated (they form a convex set) but one can use a deformation argument like that in Solimini [27] to calculate the homotopy index of the set of positive solutions for $g(y) = ay^+ + dy^-$.

2. Assume that the conditions of Corollary 1 hold, that g is continuously differentiable and $g'(y) \longrightarrow a(d)$ as $y \longrightarrow \infty(y \longrightarrow -\infty)$. Finally assume that the critical groups of G at 0 are not all trivial. (Remember that we are using reduced homology). Then the equation $-\Delta u = g(u) - t\phi_1 + h$ has at least 3 solutions for large positive t. This follows by combining Theorem 1 with the results in [9]. It makes one of the conditions in [9] much more explicit. It seems neither to contain or be contained in the results in [19] and [20] (even in the case where g satisfies the extra regularity conditions above).

3. Our argument that $t(a)$ exists when combined with the work of D'Aujourd'hui [2] gives an alternative proof of the result of de Figueredo-Gossez [15].

Our argument that $t(a)$ exists is easily modified to show it holds uniformly in a on $[\lambda_i, \lambda_{i+1} - \delta]$. (The only difficulty is to show that the bound for z_d holds uniformly in a. One can easily prove from (6) that $(\widetilde{w} + z_d)^- \longrightarrow 0$ and that the only component of z_d that can be large is the part in the direction of the eigenspace corresponding to λ_i and this would contradict that $(\widetilde{w}+z_d)^- \longrightarrow 0$.)

By combining this with D'Aujourd'hui [2], we obtain the following result which is an analogue of the de Figueredo-Gossez paper. (See also Magaelhaes [18].) It improves slightly work of Schechter [26] and [18]. A very similar result to the first part was also obtained in [20] but the proof is quite different.

Corollary 2. *If $i \geq 3$, there is a continuous decreasing function $\psi(a)$ on $[\lambda_{i-1}, \lambda_i]$ such that $\psi(\lambda_i) = \lambda_i, (a, \psi(a)) \in A_0$ and $(a, \lambda) \notin A_0$ if $a > \lambda_{i-1}$ and $a < d \leq \psi(a)$. Moreover the component of $A_0 \backslash \{(\lambda_i, \lambda_i)\}$ containing $\{(a, \psi(a)) : \lambda_{i-1} \leq a < \lambda_i\}$ is unbounded.*

The last part follows by a similar argument to that in the proof of Theorem 1 in [8] (but correcting an error in the statement of Lemma 3 in [8] by adding that $K_2 \cap A$ is empty). With a little care, one can also show $(\lambda_{i-1}, t) \notin A_0$ if $\lambda_{i-1} < t < \psi(\lambda_{i-1})$. Note that the last part of the statement of Corollary 2 improves Theorem 1 in [8]. An analogous result holds for the "upper curve" in A_0 bifurcating from (λ_i, λ_i) but the result seems less interesting in this case.

We now show that we can obtain much more precise information if we make an assumption on the behaviour at $t(a)$. Note that $\widetilde{F}$ is strictly differentiable at $\widetilde{w}$ if $\widetilde{w} \neq 0$ and $\widetilde{F}(\widetilde{w}) = 0$ by our earlier arguments (since if $\widetilde{F}(u) = 0, u + S(u)$ only vanishes on a set of measure zero and the map $u \longrightarrow au^+ + du^-$ is strictly differentiable at $u + S(u)$ as a map of $L^p(\Omega) \longrightarrow L^q(\Omega)$ if $p > q > 1$.) We say that $(a, t(a))$ is regular if whenever $u \neq 0$ and $\widetilde{F}(u) = 0$, $\widetilde{F}'_{t(a)}(u)$ has only a one-dimensional kernel. (By the positive homogeneity, u must be in the kernel.) To eliminate the degeneracy, we will usually consider $\widetilde{F}_d|\widetilde{W}$, where $\widetilde{W}$ is the unit sphere in W. We think of the regular case as the "generic" case. By using the strict differentiability of $\widetilde{F}|_{\widetilde{W}}$ at zeros of $\widetilde{F}$ we easily see that in the regular case, zeros of $\widetilde{F}_{t(a)}|_{\widetilde{W}}$ are isolated and hence, by compactness they are finite. Since $(a, t(a)) \in A_0$, there must be at least one such zero.

Theorem 2. *Assume that a is not an eigenvalue of $-\Delta, a > \lambda_1$ and $(a, t(a))$ is regular. Then for ϵ small and positive $(a, t(a) + \epsilon) \notin A_0$ and $C_i(G, 0)$ (where $d = t(a) + \epsilon$) is $\mathbb{Z} \oplus \mathbb{Z} \oplus \mathbb{Z}(m-1$ times) if $i = \ell + 1$ and is trivial otherwise. Here m is the number of zeros of $\widetilde{F}_{t(a)}|_{\widetilde{W}} = 0$.*

Proof. We first consider the structure of A_0 near $(a, t(a))$. It is convenient to change notation here and replace a by a_0 and consider a close to a_0. We do this for the rest of this proof.

If $\{u_i\}_{i=1}^m$ are the zeros of $\widetilde{F}_{t(a)}$ on $\widetilde{W}$, then $u_i + S(u_i)$ are the zeros of $A_{t(a_0)}$ and our assumption ensures that the zeros of (2) near $(a_0, t(a_0), u_i + S(u_i))$ in $\mathbb{R}^2 \times \mathring{W}^{1,2}(\Omega)$ are of the form of a continuous decreasing curve $d = \phi_i(a), u = \psi_i(a)$ (up to scalar multiples of u) where $t(a_0) = \phi_i(a_0), \psi_i(a_0) = u_i + S(u_i)$. Here we use Pope [21] or Pistoia [20] and the regularity. Since a simple compactness argument ensures that any solution of (2) with a near a_0, d near $t(a)$ (and $\|u\|_2 = 1$) must be close to one of the above solutions, it follows that near $(a_0, t(a_0)), A_0$ is locally $\cup_{i=1}^m \{(a, \phi_i(a)) : |a - a_0| \leq \delta\}$. Our first claim follows immediately.

Our second claim will follow if we show that if $d = t(a_0) + \epsilon$ with $\epsilon > 0$ small, then $T = \{w \in \widetilde{w} : \widetilde{f}_d(w) < 0\}$ is a union of disjoint "balls" one near each u_i (by an earlier formula for the critical groups in terms of T). Now if d increases from $t(a_0)$ an argument in the proof of Theorem 1 shows that $\widetilde{f}_d(u_i)$ decreases and hence becomes negative. Thus T contains u_i and points near u_i. By continuity we see that, if d is near $t(a_0)$, T is close to $\cup_{i=1}^m \{u_i\}$. Hence our result follows if we show that $T_i = \{u : u \text{ is near } u_i : \widetilde{f}_d(u_i) < 0\}$ is contractible if d is near $t(a_0)$. Now $\widetilde{f}_{t(a_0)}|_{\widetilde{W}}$ has a second derivative at u_i and the first derivative is strictly differentiable at u_i (by earlier arguments). Moreover $\widetilde{f}_{t(a_0)}|_{\widetilde{W}}$ is positive semi-definite (because $\widetilde{u}_i$ is a local minimum of $\widetilde{f}_{t(a_0)}|_{\widetilde{W}}$) and has no kernel by our regularity assumption. Hence there is a $K > 0$ such that $\widetilde{f}''_{t(a_0)}(u_i)(w, w) \geq K\|w\|^2$ for w tangent to $\widetilde{W}$ at u_i. By our comments above, it is easy to see that it suffices to prove that, if $d > t(a_0)$ and d is near $t(a_0)$ then $f_{t(a_0)}(u) = 0$ at most once on each half line $\{u_i + \alpha w : 0 \leq \alpha < \widetilde{a}\}$ where $w \in W, \|\hat{w}\| = 1, (w, u_i) = 0\}$. Here $\widetilde{a}$ is fixed but small. Note that, from our earlier comments, it must intersect at least once, and that the above result implies that T_i can be deformed down to $\{u_i\}$ and is hence contractible.) Our claim is obvious except for smoothness troubles (by considering derivatives on these lines). Now $\widetilde{F}_d(w) = Lw + Pg_d(w + S_d(w))$ where $g_d(u) = a_0 u^+ + du^-$.

The perturbation argument is easy but tedious if we show that

$$\begin{aligned}\widetilde{F}_d(w) - \widetilde{F}_{t(a_0)}(w) &= (d - t(a_0))P(w + S_d(w))^- \\ &\quad + P(g_{t(a_0)}(w + S_d w) - g_{t(a_0)}(w + S_{t(a_0)} w))\end{aligned}$$

satisfies a small Lipschitz condition in w for w near u_i and d near $t(a_0)$ since S_d is Lipschitz in w locally uniformly in d. This is trivial for the first term. For the second term which we denote by $\widetilde{Z}$ we need to be a little more careful. Now, since $Pg_{t(a_0)}$ is strictly differentiable at $u_i + S_{t(a_0)}(u_i), Pg_{t(a_0)}(v) = M(v - u_i - S_{t(a_0)}(u_i)) + h(v)$ where M is linear and h satisfies a small Lipschitz condition near $u_i + S_{t(a_0)} u_i$. It is easy to see that the terms in $\widetilde{Z}$ coming from h satisfy a small Lipschitz condition in w near u_i. Hence it suffices to consider the term

$$\begin{aligned}PM(w + S_d(w) - u_i - S_{t(a_0)}(u_i) &- PM(w - S_{t(a_0)}(w) - u_i - S_{t(a_0)}(u_i)) \\ &= PM(S_d(w) - S_{t(a_0)}(w)) .\end{aligned}$$

Hence we see that it suffices to prove that if d is near $t(a_0)$ the term $H(w) = S_d(w) - S_{t(a_0)}(w)$ satisfies a small Lipschitz condition in w. Now $(I - P)(LS_d(w) + g_d(w + S_d w)) = 0$. Therefore

$$\begin{aligned}V(w) \equiv (I - P)(LH(w) + g_{t(a_0)}(w + S_d(w)) - g_{t(a_0)}(w + S_{t(a_0)}(w))) \\ = -(t(a_0) - d)(w + S_d(w))^- .\end{aligned}$$

Now it is easy to see that the right-hand side satisfies a small Lipschitz condition in w for d near $t(a_0)$. Thus it suffices to prove that there is a $\mu > 0$ such that

$$\|V(w_1) - V(w_2)\|_{1,2} \geq k\|H(w_1) - H(w_2)\|_{1,2} .$$

This follows easily because the strict differentiability of $g_{t(a_a)}$ at $u_i + S_{t(a_0)}(u_i)$ easily implies that

$$V(w_1) - V(w_2) = (I - P)(L + g'_{t(a_0)}(u_i - S_{t(a_0)}(u_i)))(H(w_1) - H(w_2))$$

plus a term of $o(\|H(w_1) - H(w_2)\|_{1,2})$ and our claim follows since $(I - P)(L + g'_{t(a_0)}(u_i - S_{t(a_0)}(u_i))$ is invertible as a map of $\mathring{W}_{1,2}(\Omega) \cap W^{\perp}$ into itself. This completes the proof of the result. □

Remarks. 1. The last part of the argument is a little tedious because of our very limited smoothness. However, the idea is simple.

2. Normally we would only expect there to be m zeros of F for $d = t(a_0)$ with $m > 1$ if they are forced by a symmetry of Ω. We could also obtain an analogous result when there is a continuous group $\hat{G}$ acting on Ω and we replace the regularity assumption by the condition that $N(\widetilde{F}'_{t(a_0)}(u_i)|_{\widetilde{W}})$ is equal to the tangent space to the orbit at u_i of the action of $\hat{G}$ on W (where for simplicity we assume the group acts orthogonally on $L^2(\Omega)$). In this case, if the zero set of $\widetilde{F}'_{t(a_0)}|_{\widetilde{W}}$ is a single orbit $\widetilde{S}$, then for $d = t(a_0) + \epsilon$, $C_i(G, 0) = \widetilde{H}_{i-\ell-1}(\widetilde{S})$. (We need to show that T is some form of tubular neighbourhood of $\widetilde{S}$.)

3. Our methods can be used to obtain a number of other results. For example if the assumptions of Theorem 2 hold we can use our arguments and the results on A_0 at the start of the proof to obtain a formula for the critical groups for all $(a, d) \notin A_0$ near $(a, t(a))$. Our methods could also be used to obtain some information on the critical groups when d increases across a regular point (a_0, m) in A_0 with $m > t(a_0)$. The difficulty here is that there are more possibilities. (One case where one can easily obtain some information is when the critical groups are zero for $(a_0, m + \epsilon)$ or for $(a_0, m - \epsilon)$. For example the critical groups can not be trivial for both $(a_0, m + \epsilon)$ and $(a_0, m - \epsilon)$).

4. One can prove analogous results if we decrease a across $s(d)$ or if a is an eigenvalue.

5. Note that if $m = 1$, the critical groups are trivial for $d = t(a_0) + \epsilon$. We suspect that in this case $-\Delta u - a_0 u^+ - (t(a_0) + \epsilon)u^-$ has range not all of $L^2(\Omega)$ for small positive ϵ. Indeed, we suspect $u_1 + S_{t(a_0)+\epsilon}(u_1)$ is not in the range of $A_{t(a_0)+\epsilon}$. It is possible to prove that $u_1 + S_{t(a_0)}(u_1)$ is not the image of $A_{t(a_0)+\epsilon}$ of any point αw where $\alpha \geq 0$ and w is close to $u_1 + S_{t(a_0)}(u_1)$ in $\mathring{W}^{1,2}(\Omega)$.

Our methods can be used if we replace $-\Delta$ by certain higher order elliptic operators L. The only difficulty is to check that every non-trivial solution of the equation $Lu = b(x)u$ in Ω (with $b \in L^{\infty}(\Omega)$) only vanishes on a set of measure zero. This can be proved in a few cases where our problem can be written as a second order elliptic systems with principal part in diagonal form by combining the result of Chen [6] with the natural analogue of Arouszajn et al [1] for such systems. (This idea was pointed out to me by Chen.) Note that there are a few changes because it may be a higher eigenvalue which has a positive eigenfunction (or no eigenvalue may have a positive eigenfunction).

2. Behaviour near an eigenvalue

In this short section, we discuss the behaviour of A_0 and the critical groups for (a, d) near (λ_i, λ_i) where $i > 1$. We improve a little some results in [8] and, more importantly, we find a new formulation of the results in [8] which is much simpler to understand. Let $\widetilde{W}_i$ denote the unit sphere in the eigenspace corresponding to λ_i and let $N(u) = \int_\Omega (u^+)^2$ for $u \in \widetilde{W}_i$. It turns out that the behaviour of N on $\widetilde{W}_i$ almost completely determines the structure of A_0 locally and also the critical groups for (a, d) near (λ_i, λ_i). Since $u^+ \neq 0$ and $u^- \neq 0$ for $u \in \widetilde{W}_i, N(u) \in (0, 1)$ for $u \in \widetilde{W}_i$. Let S denote the compact set of critical values of N. Thus $S \subseteq (0, 1)$. Note that N is C^2 on $\widetilde{W}_i$ since every $u \in \widetilde{W}_i$ only vanishes on a set of measure zero. (This is similar to arguments in Section 1.)

Theorem 3. *(i) Assume that* $(a_n, d_n) \in A_0, (a_n, d_n) \neq (\lambda_i, \lambda_i)$ *for all* n *and* $(a_n, d_n) \longrightarrow (\lambda_i, \lambda_i)$ *as* $n \longrightarrow \infty$. *Then* $(a_n - \lambda_i)(d_n - \lambda_i) < 0$ *for large* n *and* $\frac{\lambda_i - d_n}{a_n - d_n} \longrightarrow S$ *as* $n \longrightarrow \infty$. *Moreover if* u_n *is a solution of (2), for* $a = a_n$ *and* $d = d_n$ *such that* $\|u_n\| = 1$ *and* $\frac{\lambda_i - d_n}{a_n - d_n} \longrightarrow \alpha$ *as* $n \longrightarrow \infty$, *then a subsequence of* u_n *converges to a critical point of* N *in* $N^{-1}(\alpha)$.

(ii) If z *is a non-degenerate critical point of* N *and* $N(z) = \gamma$, *then there exists a continuous curve* (u_t, a_t, d_t) *of solutions of (2) for small non-zero* t *such that* $\|u_t\|_1 = 1$, $\|u_t - z\| \longrightarrow 0$ *as* $t \longrightarrow 0$ *and* $(a_t, d_t) = (\lambda_i, \lambda_i) + t\frac{(\gamma-1, \gamma)}{\sqrt{(\gamma-1)^2 + \gamma^2}} + o(t)$ *as* $t \longrightarrow 0$. *Moreover this solution is locally unique for small non-zero* t *(provided we normalize* (a_t, d_t) *so that* $\frac{1}{t}(a_t - \lambda_i, d_t - \lambda_i)$ *has length 1).*

(iii) If α *is an isolated point of* S *and* $\{u \in \widetilde{W}_i : N(u) \leq \alpha + \epsilon\}$ *and* $\{u \in \widetilde{W}_i : N(u) \leq \alpha - \epsilon\}$ *have different homology, then for all small non-zero* t *there is a solution* (a_t, d_t, u_t) *of (2) such that* $\|u_t\|_2 = 1, (a_t, d_t) = (\lambda_i, \lambda_i) + t\frac{(\alpha-1, \alpha)}{\sqrt{(\alpha-1)^2 + \alpha^2}} + o(t)$ *and* u_t *is close to the critical points of* N *in* $N^{-1}(\alpha)$.

Remark. Theorem 3 gives a complete picture of the local structure of A_0 near (λ_i, λ_i) if N has only non-degenerate critical points and no two critical points have the same value of N. The last condition is not needed if there exist critical points with the same value of N but this is forced by a symmetry of Ω (and hence of $\widetilde{W}_i$). If all the critical points of N are non-degenerate, but different critical points of N have the same value of N, we can still obtain a good understanding of the structure of A_0 much as at the beginning of the proof of Theorem 2.

Theorem 4. *Assume that $B \in (0,1)\backslash S$. Then, for small non-zero t, $(\lambda_i, \lambda_i) + t(\gamma - 1, \gamma) \notin A_0$ and the critical groups $C_j(G,0) = \widetilde{H}_{j-k-1}(N^B, \mathbb{Z})$ if $t > 0$ and small where $N^B = \{u \in \widetilde{W}_i : N(u) < B\}$ while $C_j(A,0) = H_{j-k-1}(N_B, \mathbb{Z})$ if t is small and negative where $N_B = \{u \in \widetilde{W}_i : N(u) > B\}$. Here k is the sum of the multiplicities of the first $(i-1)$ eigenvalues of $-\Delta$.*

Remark. This gives an almost complete information on the critical groups near (λ_i, λ_i) and complete information if N has only non-degenerate critical points and no two critical points have the same value of N. Note that the cases where $B \notin (0,1)$ the critical groups follow from Theorem 1 since (a,d) is above the maximal curve through (λ_i, λ_i) in A_0 or below the minimal curve (in the sense of [1] or [18] or [26]). It is possible to refine our ideas to calculate the homotopy index at all points not in A_0 close to a direction determined by a γ where the critical points in $N^{-1}(\gamma)$ are all non-degenerate (but may contain more than one element). We can also handle cases where there is a continuous group of symmetries acting and an appropriate non-degeneracy condition is assumed (much as earlier).

Proof of Theorem 3. It is easy to prove (cp. p. 42 of [8]) that if $a = \lambda_i + t\mu, d = \lambda_i + t\nu$ (where $\mu^2 + \nu^2 = 1$) and if t is small and $W = W_i$, then $S_t(w)$ satisfies $\|S_t(w)\|_{1,2} \leq K|t|\|w\|_2$ for small t (uniformly in (μ,ν) and the corresponding Lipschitz estimate in w holds). Here we alter the notation for S slightly to emphasize the dependence upon t. Hence we see that (2) reduces to the problem

$$-tP(\mu(w + S_t w)^+ + \nu(w + S_t w)^-) = 0\,. \tag{7}$$

This equation (for $\|w\| = 1$) can be written as $P(\mu w^+ + \nu w^-) + o(t) = 0$ and hence we see that if $t_n(\mu_n, \nu_n) + (\lambda_i, \lambda_i) \in A_0$ where $t_n \longrightarrow 0$ and $(\mu_n, \nu_n) \longrightarrow (\mu, \nu)$ as $n \longrightarrow \infty$, then it is necessary that $P(\mu w^+ + \nu w^-) = 0$ has a non-zero solution $w \in W_i$; equivalently, $\widetilde{E}(w) = \frac{1}{2}\int_\Omega \mu(w^+)^2 + \nu(w^-)^2$ has a critical point on the sphere $\widetilde{W}_i$ in W_i with energy $\widetilde{E}(w) = 0$. (To see the necessity, we simply scalar multiply $P(\mu w^+ + \nu w^-) = 0$ by w.) However, if $w \in \widetilde{W}_i$,

$$\widetilde{E}(w) = \frac{1}{2}\mu\|w^+\|_2^2 + \frac{1}{2}\nu(1 - \|w^+\|_2^2) = \frac{1}{2}(\nu + (\mu - \nu)N(w))\,.$$

Hence we see that the necessary condition is that N has a critical point $\widetilde{w}$ on the sphere such that $\nu + (\mu - \nu)N(\widetilde{w}) = 0$ and this proves the first part of Theorem 3. (The result on the solution u_n follows easily from our estimate for $S_t(w)$ and a simple compactness argument.)

We now prove (ii). We need to look for solutions of (7) (or more strictly (7) divided by t) with zero energy (i.e. with $\int_\Omega \mu((w + S_t w)^+)^2 + \nu((w + S_t w)^-)^2 = 0$) with (μ,ν) near (μ_0, ν_0) on S^1 and w near w_0, $\|w\| = 1$. We can easily solve the equation that the energy is zero to obtain $(\mu,\nu) = h(w,t)$ where $h(w_0, 0) = (\mu_0, \nu_0)$ and $h(w,t) = h(w,0) + Z(w,t)$ where Z is Lipschitz in t with small Lipschitz constant, where $Z(w,0) = 0$ and where there is a simple formula for $h(w,0)$ (since $S_0 w = 0$). (Note that S_t also depends on (μ,ν) and we are assuming $\mu^2 + \nu^2 = 1$.)

Then our problem reduces to finding the zeros of the vector field on the sphere

$$P(h_1(w,t)(w+S_tw)^+ + h_2(w,t)(w+S_tw)^-) = 0\,.$$

(We are really splitting (7) into a radial and tangential term.) Note that h_1 and h_2 are the components of h.

This is a perturbation of the C^2 term $P_1(h_1(w,0)w^+ + h_2(w,0)w^-)$ which is seen by an easy but tedious calculation from our assumptions to have a non-degenerate critical point at w_0. Now we can use similar but easier arguments to that in the proof of Theorem 2 to prove that the perturbation term satisfies a small Lipschitz condition in w if t is small. Hence the result follows from the contraction mapping theorem.

(iii) Assuming Theorem 4 below, we see that the critical groups of zero for

$$(\lambda_i, \lambda_i) + t\frac{(\gamma_j - 1, \gamma_j)}{\sqrt{(\gamma_j - 1)^2 + \gamma_j^2}}$$

are different for small non-zero t where $\gamma_j = \gamma \pm \epsilon$ for $j = 1,2$. Since the critical groups change when we deform from $\gamma + \epsilon$ to $\gamma - \epsilon$, some point on this circular segment must belong to A_0 and our claim follows. □

Remark. We suspect that there is a generalization of (ii) where we drop the non-degeneracy but assume that the critical point has a non-trivial critical group on $\widetilde{W}_i$ (and do not expect uniqueness or continuity of u_t). The difficulty in the proof is to maintain the variational structure. Note that the homology of N^B always changes if B crosses a single non-degenerate critical point of N or indeed any isolated critical point with a non-trivial critical group but not two adjacent critical groups C_j and C_{j+1} both non-trivial (or if the critical point has non-zero degree).

Proof of Theorem 4. First assume $t > 0$. If $B \notin S$, the arguments in [8] (cp. the proof of Theorem 3) and the reduction type arguments of [9]) show that the result is true but with N^B replaced by $\{w \in \widetilde{W} : (B-1)\int_\Omega (u^+)^2 + B\int_\Omega (u^-)^2 > 0\}$. However, a simple calculation similar to the one above shows that this set is indeed N^B and our claim follows.

If $t < 0$, (7) is $|t|P(\mu w^+ + \nu w^-)$ + higher order terms. If we assume the same argument as before the only difference is we now have

$$\{(B-1)\int_\Omega (u^+)^2 + B\int_\Omega (u^-)^2 < 0\}$$

which is N_B. □

Remark. We could use duality theorems as in Dold [16] to express everything in terms of N^B (or N_B).

3. Examples

We first construct an example where $(a,d) \notin A_0$ and the map

$$u \longrightarrow u - (-\Delta)^{-1}(au^+ + du^-)$$

has degree γ where $|\gamma| > 1$. This shows how differently our problem behaves to the linear case. We also construct a simple example where there are more than 4 distinct curves in A_0 bifurcating from (λ_i, λ_i).

To construct the first example, we use an idea of D'Aujourd'hui [2]. We consider a domain D_ϵ in $\mathbb{R}^2$ which is invariant under rotations through $\frac{2\pi}{3}$ and is a connected approximation (as in [14]) to the union of three non-intersecting symmetrically placed balls B_1, B_2, B_3 (cp. Fig. 1).

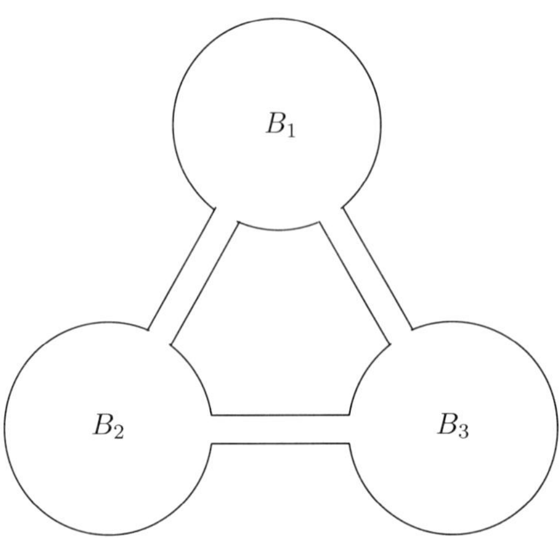

By the theory of domain variation (cp. Stummel [28] or Rauch and Taylor [23] or Dancer [14]), Ω will have 3 eigenvalues close to $\lambda_1(B)$ (counting multiplicity) and the corresponding eigenfunctions must be close in L^p to the subspace spanned by $\{\phi_i(x)\}_{i=1}^3$ where $\phi_i(x)$ is the first eigenfunction on B_i. Since our domain is connected, $\lambda_1^\epsilon \equiv \lambda_1(D_\epsilon)$ is simple. The corresponding eigenfunction must be invariant under rotations through $\frac{2\pi}{3}$ and hence must be L^2 close to $\frac{1}{\sqrt{3}}(\phi_1(x)+\phi_2(x)+\phi_3(x))$ where our eigenfunctions are normalized to have L^2 norm 1. (Note that ϕ_2 and ϕ_3 are simply rotations of ϕ_1 and ϕ_i is defined to be zero outside B_i. Now the eigenspace T_ϵ corresponding to $\lambda_2(D_\epsilon)$ must be invariant under rotations through $\frac{2\pi}{3}$ (because of the symmetry) and hence T_ϵ can not be one-dimensional. Moreover T_ϵ must be orthogonal to the first eigenfunction. Let $\Phi_1(x) \equiv \frac{1}{\sqrt{2}}(\phi_1(x) - \phi_3(x))$ and $\Phi_2(x) \equiv \frac{1}{\sqrt{5}}(\phi_1(x) - 2\phi_2(x) + \phi_3(x))$. Thus the unit sphere $\hat{T}_\epsilon$ in T_ϵ is L^2 close to $\{\sin\, s\ \Phi_1(x) + \cos\, s\Phi_2(x) : 0 \le 0 \le 2\pi\}$. Hence we see that for $\lambda_i = \lambda_2(D_\epsilon)$ the functional N_ϵ of Section 2 is close to $\int_{B_1 \cup B_2 \cup B_3} ((\sin\, s\Phi_1(x) + \cos\, s\Phi_2(x))^+)^2$ and a corresponding result holds for its derivative. Hence we can obtain information on $\int_{D_\epsilon}(u^+)^2$ on T_ϵ by considering it as a perturbation of the functional

$$H(s) = \int_{B_1\cup B_2\cup B_3} ((\sin\ s\Phi_1(x) + \cos\ s\Phi_2(x))^+)^2$$

defined for $0 \leq s \leq 2\pi$. The computations are easy here because the B_i are disjoint and the x dependence is the same on each B_i. Note that the symmetry is retained so that if s is a critical point so is $s+\frac{2\pi}{3}$. (In fact, the functional is unchanged if we translate s by $\frac{2\pi}{3}$.) As D'Aujourd'hui observed, it is easy to see that s is a critical point of H if and only if $(|\Phi(s)|, \Phi'(s)) = 0$ (since $(\Phi(s), \Phi'(s)) = 0$) where $\Phi(s) = \sin\ s\Phi_1(x) + \cos\ s\Phi_2(x)$ and where the scalar product is the L^2 scalar product on $B_1 \cup B_2 \cup B_3$. A simple computation shows that $(|\Phi_1(x)|, \Phi_2(x)) \neq 0$ and hence $(|\Phi(\frac{1}{2}\pi)|, \Phi'(\frac{1}{2}\pi)) \neq 0$ and thus $\Phi(\frac{1}{2}\pi)$ is not a critical point of H. Hence H is not constant on the sphere. Hence N_ϵ is not constant on $\hat{T}_\epsilon$ for small ϵ. Now N_ϵ is invariant under rotations through $\frac{2}{3}\pi$. Hence we see that, if $\min\ N_\epsilon|_{\hat{T}_\epsilon} < \alpha < \max\ N_\epsilon|_{\hat{T}_\epsilon}$, then $\{u \in \hat{T}_\epsilon : N_\epsilon(u) < \alpha\}$ must consist of k connected (and hence contractible) components where $k \neq 0$ and k is divisible by 3. Now N_ϵ is C^2 on $\hat{T}_\epsilon$ (by earlier arguments) and hence by Sard's theorem we can choose α to be a regular value of $N_\epsilon|_{\hat{T}_\epsilon}$ (for fixed α). By Theorem 4, if t is small and positive, $(\lambda_2(D_\epsilon), \lambda_2(D_\epsilon)) + t(\alpha - 1, \alpha) \notin A_0$ while the critical groups of A are zero except in dimension 2 where the critical group is $k\mathbb{Z}$. By the formula for the degree of A in terms of the homotopy index (cp [5] or [10]) the degree of A is $\pm k$ and hence our claim follows. This completes the construction of the example. This shows our problem is behaving quite differently to the asymptotically linear case. If (μ, ν) are as above, g is C^1 and $y^{-1}g(y) \longrightarrow \mu(\nu)$ as $y \longrightarrow \infty(-\infty)$, then the problem $-\Delta u = g(u) - f$ in D_ϵ plus the boundary condition has more than one solution for almost all f in $L^2(D_\epsilon)$.

We sketch on alternative way of constructing such an example. We consider the unit ball in $\mathbb{R}^2$. It is well known that the second eigenvalue λ_2 of $-\Delta$ is a double eigenvalue with non-radially symmetric eigenfunctions. Fix k a positive integer $k \geq 3$. Then, any domain $\widetilde{D}_\epsilon$ in $\mathbb{R}^2$ which is reasonably close to the ball and invariant under rotations of $\frac{2\pi}{k}$ will have its second eigenvalue $\lambda_2(\widetilde{D}_\epsilon)$ double (by using the symmetries). For this case the N of Section 3 (with $\lambda_i = \lambda_2(\widetilde{D}_\epsilon)$) will have similar properties to the last example but is invariant under rotations of $\frac{2\pi}{k}$. However, we do not know if N is constant on the unit sphere. It is almost certain that by transversality arguments (but we have not checked the details) we can make smooth small perturbation of $\widetilde{D}_\epsilon$ which retain the symmetry and for which N (for the perturbed domain) is non-constant on the sphere. This would give similar examples on a convex domain.

Lastly, we give simple examples where there are more than 4 curves of A_0 bifurcating from an eigenvalue. We consider domains $A \times B$ in $\mathbb{R}^{n+k}$ where A is a domain in $\mathbb{R}^n$ and B is a domain in $\mathbb{R}^k$. Note that by separation of variables the spectrum on $A \times B$ is $\{\lambda + \mu : \lambda \in \sigma(-\Delta_A), \mu \in \sigma(-\Delta_B)\}$ where Δ_A denoted the Laplacian on A. Assume $\lambda_1(A) < \lambda_k(A)$ are eigenvalues of $-\Delta_A$ and $\lambda_1(B) <$

$\lambda_r(B)$ are eigenvalues of $-\Delta_B$. By replacing A by αA, we can arrange that $\lambda_1(A)+\lambda_r(B)=\lambda_1(B)+\lambda_k(A)$. If $\Phi_{1,A}$ is the positive eigenfunction of $-\Delta_A$, then $(\lambda,\mu)\in A_0$ (on $A\times B$) if $(\lambda-\lambda_1(A),\mu-\lambda_1(A))\in A_0$ (on B). This follows because we can obtain solutions of $-\Delta u=\lambda u^+ + \mu u^-$ on $A\times B$ of the form $\Phi_{1,A}(x)\Psi(y)$ (for $x\in A, y\in B$) where

$$-\Delta_B\Psi=(\lambda-\lambda_1(A))\Psi^+ + (\mu-\lambda_1(A))\Psi^- \,. \tag{8}$$

Here it is crucial that $\Phi_{1,A}(x)>0$ on A (and we use that the map $z\longrightarrow \lambda z^+ + \mu z^-$ is positive homogeneous). Note that a number of people have used these separation of variables (for example, McKenna, D'Aujourd'hui and many others). Hence if $\lambda_r(B)$ is simple with normalized eigenfunction $\Phi_r^2(y)$, (and the simplicity is true for generic B), there will be 4 curves in A_0 bifurcating from $(\lambda_1(A)+\lambda_r(B),\lambda_1(A)+\lambda_r(B))$ bifurcating in the directions determined by $\pm(\alpha_2-1,\alpha_2)$, $\pm(\alpha_2,1-\alpha_2)$ where $\alpha_2=\|(\Phi_r^2)^+\|_2^2$. (Note that $\|(-\Phi_r^2)^+\|_2^2=1-\alpha_2$). To ensure that there are really 4 directions, we must assume $\alpha_2\neq\frac{1}{2}$, which we can ensure rather easily by choosing explicit examples or by using perturbation theory. Since $\lambda_1(A)+\lambda_r(B)=\lambda_1(B)+\lambda_k(A)$ we can reverse the roles of A and B and obtain 4 more curves emanating from $(\lambda_1(A)+\lambda_r(B),\lambda_1(A)+\lambda_r(B))$ which will be different if $\alpha_1\neq\alpha_2, 1-\alpha_2$ where $\alpha_1=\|(\Phi_k^1)^+\|_2^2$ (with the obvious notation). This gives the required example provided we can choose α_1,α_2 as above. One way we can achieve this is to choose A and B intervals in $\mathbb{R}$ and r,k both odd. (In this case, it is easy to compute α_1 explicitly.) However, it can be achieved in many other ways. This is the required example.

It might be objected that $A\times B$ has corners. We *sketch* how it is possible to obtain an example without corners by smoothing. Firstly if we can choose $r=k=2$, then $\lambda_1(A)+\lambda_2(B)$ will be a double eigenvalue. By perturbing A and B, it is possible with care to achieve this so that the α_i are distinct. Moreover, with care one can also arrange that on the two-dimensional eigenspace corresponding to $\lambda_1(A)+\lambda_2(B)$ N has 4 non degenerate critical points. This is not difficult but rather tedious and may require some extra smooth perturbations of A and B. Lastly one smooths $A\times B$ by a small perturbation. Most perturbations split the double eigenvalue but it is often possible to use a topological argument to show that some perturbations preserve the double eigenvalue (often made easier by any symmetries in the problem). In fact with care this can be done with a rectangle if we choose $r=3, k=7$ (and a good deal of computations). Preserving the double eigenvalue can be easily arranged in this case if we choose smooth approximating domains retaining the two reflection symmetries.

Lastly there is another way in which similar examples can almost certainly be constructed. We take the example above which showed that the degree is not always 0 or ±1 and perturb it smoothly breaking the symmetry so that the double eigenvalue is not split but the many critical points of N are perturbed to be at different energy levels. This is likely to be a rather technical transversality argument. (It can easily be checked that the critical points of N on $\hat{T}_\epsilon$ are non-degenerate.)

References

[1] N. Arouszajn, A. Krzywicki and I. Szarski, "A unique continuation theorem for exterior differential forms on Riemannian manifolds", Ark. Math. **4** (1962), 417–453.

[2] M. D'Aujourd'hui, "Sur l'ensemble de résonance d'un problème demi-linéaire", preprint, EPFL, Lausanne.

[3] L. Caffarelli and A. Friedman, "Partial regularity of the zero-set of solutions of linear and superlinear elliptic equations", J. Diff. Eqns. **60** (1985), 420–433.

[4] H. Cartan, *Calcul Différentiel*, Hermann, Paris, 1967.

[5] K.C. Chang, *Infinite Dimensional Morse Theory and Multiple Solution Problems*, Birkhäuser, Boston, 1993.

[6] X.Y. Chen, "Polynomial asymptotics near zero points of solutions of general elliptic equations", preprint.

[7] E.N. Dancer, "On the Dirichlet problem for weakly non-linear elliptic partial differential equations", Proc. Royal Soc. Edinburgh **76A** (1977), 283–300.

[8] E.N. Dancer, "On the existence of solutions of certain asymptotically homogeneous problems", Math. Z. **177** (1981), 33–48.

[9] E.N. Dancer, "Multiple solutions of asymptotically homogeneous problems", Annali di Matematica Pura ed Applicada **152** (1988), 63–78.

[10] E.N. Dancer, "Degenerate critical points, homotopy indices and Morse inequalities", J. Reine Ang. Math. **350** (1984), 1–22.

[11] E.N. Dancer, "Degenerate critical points, homotopy indices and Morse inequalities II", J. Reine Ang. Math **382** (1987), 145–164.

[12] E.N. Dancer, "On positive solutions of some pairs of differential equations II", J. Diff. Eqns. **60** (1985), 236–258.

[13] E.N. Dancer, "Generic domain dependence for non-smooth equations and the open set problem for jumping nonlinearities", Topological Methods in Nonlinear Analysis **1** (1993), 139–150.

[14] E.N. Dancer, "On the effect of domain shape on the number of positive solutions of certain nonlinear equations II", J. Diff. Eqns. **87** (1990), 316–339.

[15] D.G. de Figueredo and J. Gossez, "On the first curve of the Fucik spectrum of an elliptic operator", Diff. and Integral Eqns. **7** (1994), 1285–1302.

[16] A. Dold, *Algebraic Topology*, Springer-Verlag, Berlin, 1970.

[17] S. Fucik, *Solvability of Nonlinear Equations and Boundary Value Problems*, Reidel, Dordrecht, 1980.

[18] C.Magalhaes, "Semilinear elliptic problem with crossing of multiple eigenvalues", Comm. Partial Diff. Eqns. **15** (1990), 1265–1292.

[19] A Marino, A. Micheletti and A. Pistoia, "A non symmetric asymptotically linear problem", Topological methods in nonlinear analysis **4** (1994), 289–339.

[20] A Micheletti and A. Pistoia, "A note on the resonance set for a semilinear elliptic equation and an application to jumping nonlinearities", Topological methods in nonlinear analysis **6** (1995), 67–80.

[21] A. Pistoia, "A generic property of the resonance set of an elliptic operator with respect to the domain", to appear in Proc. Royal Soc. Edinburgh.

[22] P. Pope, *Solvability of Non Self-adjoint and Higher Order Differential Equations with Jumping Nonlinearities*, Ph.D. thesis, University of New England, 1984.

[23] J. Rauch and M. Taylor, "Potential and scattering theory in wildly perturbed domains", J. Funct. Anal. **18** (1975), 27–59.

[24] K.J. Rybakowski, *The Homotopy Index and Partial Differential Equations*, Springer-Verlag, Berlin, 1987.

[25] J. Saut and R. Temam, "Generic properties of nonlinear boundary value problems", Comm. Partial Diff. Eqns. **4** (1979), 293–319.

[26] M. Schechter, "The Fucik spectrum", Indiana University Math J. **43** (1994), 1139–1157.

[27] S. Solimini, "Some remarks on the number of solutions of some nonlinear elliptic problems", Analyse Nonlinéaire **2** (1985), 143–156.

[28] F. Stummel, *Perturbation of Domains in Elliptic Boundary Value Problems*, in Lecture Notes in Mathematics 503, 110–135, Springer-Verlag, Berlin, 1975.

E.N. Dancer
School of Mathematics and Statistics
University of Sydney
N.S.W. 2006, Australia

Progress in Nonlinear Differential Equations
and Their Applications, Vol. 35
© 1999 Birkhäuser Verlag Basel/Switzerland

Local Singular Variation of Domain for Semilinear Elliptic Equations

Daniel Daners

Dedicated to Herbert Amann on the occasion of his 60th birthday

Abstract. We prove domain perturbation results for semilinear elliptic equations subject to Neumann or Robin boundary conditions. We consider perturbations of a given domain localized near a set of capacity zero. In this case results can be proved which do not in general hold for Neumann boundary conditions.

1. Introduction

In this paper we study domain perturbation for elliptic equations subject to Robin or Neumann boundary conditions of the form

$$\begin{aligned} -\Delta u + \mu u &= \lambda f(u) && \text{in } \Omega \\ u &= 0 && \text{on } \Gamma_0 \\ \frac{\partial}{\partial \nu} u + b_0 u &= 0 && \text{on } \Gamma_1. \end{aligned} \tag{1.1}$$

Here, Ω is a bounded open set in $\mathbb{R}^N$ for some $N \geq 2$. Moreover, Γ_0, Γ_1 are disjoint, open and closed subsets of the boundary $\partial\Omega$ of Ω, and $\Gamma_0 \cup \Gamma_1 = \partial\Omega$. Further, ν is the outer unit normal to $\partial\Omega$, λ and μ real parameters, and f a given differentiable nonlinearity. Finally, we assume that $b_0 \in L_\infty(\partial\Omega)$ is nonnegative. By a solution of (1.1) we always mean a weak solution, and the boundary conditions are supposed to be satisfied in a weak sense.

We want to find out what happens to the solutions of (1.1) if Ω is slightly perturbed in some sense to be specified later. If $b_0 = 0$, or more generally if b_0 is not bounded away from zero by a positive constant, many domain-perturbation results which are valid for Dirichlet boundary conditions are no longer true in general. Results in the case of Dirichlet or Robin boundary conditions include those in [6, 7, 10, 12], for results in case of Neumann boundary conditions see e.g. [4, 5, 7, 8, 14, 15]. In this paper we want to consider some interesting cases where "positive" results can be proved. We shall see that under suitable conditions we are able to obtain results where the behaviour is much as in case of the Dirichlet problem.

One crucial point in proving domain perturbation results is to have domain-independent a priori estimates in L_p-norms for the solutions of the equation under consideration. Particularly complete results can be obtained if L_p-L_q estimates hold for solutions of the linear equation with constants independent of the domains Ω_n approaching Ω. Such estimates are always true for the Dirichlet problem, and the Robin problem with b_0 bounded from below by a positive constant (see [6, Lemma 1] and [11]). However, they are not in general true for the Neumann problem. They are a consequence of the Sobolev embedding theorem (or a generalized version thereof) for the Hilbert space appearing in the weak formulation of the elliptic problem under consideration. For the Neumann problem this is the Sobolev space $W_2^1(\Omega)$. In this case there is in general no control over the embedding constant (e.g. [21]) unless we put some geometric restrictions on Ω.

In Section 2 we discuss some conditions on quite singular perturbations of domains which allow to control the Sobolev embedding constant. The perturbations of Ω we consider are localized near a small set. By a small set we mean a set of capacity zero. In particular, these perturbations include cutting or drilling small holes of quite a general shape, or joining two or more touching balls.

The results on nonlinear problems we prove make use of convergence results for linear problems. In Section 3 we show that, as the domain converges, the resolvents of the linear equation converge in the uniform operator topology to the resolvent of the limiting problem. As a consequence any finite part of the spectrum converges to the spectrum of the limiting problem. Note that this is not true in general for the Neumann problem (see [16, 17]).

For the nonlinear equation we consider two cases. First, in Section 4, we consider nonlinear eigenvalue problems of the form (1.1) in a variational setting, and ask what happens to energy minimizing solutions and the corresponding eigenvalues as we perturb the domain. In order to prove these results we do not assume nondegeneracy of the solutions. These results considerably generalize and complement results in [23, 24] (see also the references therein). We do not get information on the rate of convergence of eigenvalues and eigenvectors of semilinear eigenvalue problems. However, we relax the conditions on the holes considerably, and can deal with all dimensions $N \geq 2$, not only $N = 3$. An example is given in the first part of Section 6.

Second, in Section 5, we consider (1.1) for $\lambda = 1$ fixed. Here the aim is to show that near any nondegenerate solution of (1.1) there exists a nondegenerate solution of the same equation on a slightly perturbed domain. The main observation here is that some results in [7, Section 4] and [8] hold under weaker assumptions on the nonlinearity and the solution of (1.1), provided we are willing to consider a more restrictive notion of domain perturbation. The results in Section 5 are also related to those in [10, Section 3 and 6], where the Robin problem with b_0 bounded from below by a positive constant is treated. In the present paper we make weaker assumptions on b_0, the nonlinearity and the solution of the original problem, but have a stronger notion of domain convergence. The perturbation results we prove in the present paper are good enough to be able to construct many examples and counterexamples concerning the multiplicity of solutions of elliptic boundary value problems with Robin or Neumann boundary conditions. An outline of the ideas is given in the second part of Section 6.

Basic Notation

We denote by $L_p(X)$ the Lebesgue spaces on a measurable set $X \subset \mathbb{R}^N$, and by $\|\cdot\|_{p,X}$ its norm. If no confusion seems likely we just write L_p and $\|\cdot\|_p$, respectively. Moreover, $W_p^k(\Omega)$ is the Sobolev space, which consists of all $u \in L_p$ such that all the distributional derivatives up to the order k lie in $L_p(\Omega)$. $\mathcal{D}(\Omega)$ is the space of smooth functions with compact support in Ω and $\mathring{W}_2^1(\Omega)$ its closure in $W_2^1(\Omega)$. Finally, $C(\Omega)$ and $C^k(\Omega)$ are the spaces of continuous and k-times continuously differentiable functions, respectively. The norm in the Banach space E we usually denote by $\|\cdot\|_E$. If E, F are Banach spaces, we write $E \hookrightarrow F$ if $E \subset F$, and the natural injection is continuous. Further, $\mathcal{L}(E,F)$ is the Banach space of all bounded linear operators from E to F equipped with the usual operator norm, and $\mathcal{L}(E) := \mathcal{L}(E,E)$. The norm in $\mathcal{L}(L_p, L_q)$ we denote by $\|\cdot\|_{p,q}$. For a given subset Ω of a topological space we denote by $\overline{\Omega}$, $\Omega^{\complement}$ and $|\Omega|$ the closure, complement, and Lebesgue measure of Ω, respectively. If Ω, Ω' are two subsets of a topological space we write $\Omega' \subset\subset \Omega$ if $\overline{\Omega}'$ is compact, and $\overline{\Omega}'$ is contained in the interior of Ω.

Acknowledgment

The main idea of the proof of Theorem 2.8 I owe to E.N. Dancer. I should like to thank him for permission to include the result here.

2. Sobolev Embedding Constants

Crucial for getting domain independent a priori estimates for solutions of linear elliptic equations with boundary conditions is the validity of the Sobolev embedding theorem for a given closed subspace of $W_2^1(\Omega)$ depending on the boundary conditions. For the boundary conditions under consideration this subspace is the closure of

$$\{u \in W_2^1(\Omega) \colon u = 0 \text{ in a neighbourhood of } \Gamma_0\}$$

in $W_2^1(\Omega)$, where Γ_0 is a subset of $\partial\Omega$. We denote this space by $W_{2,\Gamma_0}^1(\Omega)$. Clearly, we have that $W_{2,\partial\Omega}^1(\Omega) = \mathring{W}_2^1(\Omega)$ and $W_{2,\emptyset}^1(\Omega) = W_2^1(\Omega)$. The Sobolev embedding theorem asserts that

$$W_{2,\Gamma_0}^1(\Omega) \hookrightarrow L_{N^*}(\Omega) \tag{2.1}$$

for $N^* := 2N(N-2)^{-1}$ in case $N \geq 3$, and $N^* < \infty$ if $N = 2$. The norm of this embedding is usually called the Sobolev embedding constant and is given by

$$c(\Omega, \Gamma_0) := \sup\{\|u\|_{N^*} \colon u \in W_{2,\Gamma_0}^1(\Omega),\ \|u\|_{W_2^1(\Omega)} = 1\}. \tag{2.2}$$

If $\Gamma_0 = \emptyset$ we set $c(\Omega) := c(\Omega, \emptyset)$. If $\Gamma_0 = \partial\Omega$ then (2.1) always holds with $c(\Omega, \partial\Omega)$ only depending on the dimension N (e.g. [13, Theorem 7.10], so we concentrate on the other cases.

In many examples it is known that under certain geometric conditions on Ω the Sobolev embedding constant $c(\Omega)$ can be arbitrarily large (see e.g. [21]). One condition which guarantees control over the embedding constant is the interior cone condition.

2.1. Definition. An open set $\Omega \subset \mathbb{R}^N$ is said to satisfy an interior cone condition if each $x \in \Omega$ is the vertex of a cone whose closure is contained in Ω and which is congruent to a cone $\mathcal{C}$ represented by the inequalities $x_1^2 + \cdots + x_{N-1}^2 < bx_N^2$, $x_N \in (0, a)$ for some $a, b > 0$. The closed set H is said to satisfy an exterior cone condition if $H^{\complement}$ satisfies an interior cone condition.

From [2, Lemma 5.14 and Theorem 6.2] we have the following lemma.

2.2. Lemma. *Suppose that Ω satisfies an interior cone condition. Then the embedding*

$$W_2^1(\Omega) \hookrightarrow L_q(\Omega)$$

holds for $q \in [1, N^*]$, and $c(\Omega)$ is bounded by a constant depending only on N and the cone determining the cone condition (i.e. on a lower bound on its angle and length). If $N = 2$ it also depends on N^*. Finally, if Ω is a bounded domain, the above embedding is compact if $q \in [1, N^*)$.

We are particularly interested in controlling the Sobolev embedding constant on sequences of domains Ω_n which are obtained from a fixed domain by cutting small holes. In the following assumption we exactly define what we mean by that.

2.3. Assumption. Suppose that Ω is a domain in $\mathbb{R}^N$, that $K_0 \subset \Omega$ is a compact set of capacity zero, and that $H_n \subset \Omega$ are closed sets. (Here, capacity is $(1,2)$-capacity with respect to a bounded set.) We define

$$\Omega_n := \Omega \setminus H_n$$

for all $n \in \mathbb{N}$. Further, we assume that for all neighbourhoods U of K_0 the sets H_n are contained in U for n large enough.

Our aim is to give conditions on the holes H_n which guarantee that the Sobolev embedding constant $c(\Omega_n, \Gamma_0)$ remains bounded uniformly with respect to $n \in \mathbb{N}$. A first result of that kind is the following.

2.4. Proposition. *Suppose* (2.1) *holds, and that H_n satisfies an exterior cone condition with the same cone for all $n \in \mathbb{N}$. Then there exists $C > 0$ such that*

$$c(\Omega_n, \Gamma_0) < C \tag{2.3}$$

for all $n \in \mathbb{N}$.

Proof. Suppose that $\Omega' \subset \Omega$ is an open set satisfying an interior cone condition, and that $H_n \subset \Omega'$ for all $n \in \mathbb{N}$. Let $\varphi \in C^\infty(\overline{\Omega})$ be such that $0 \leq \varphi \leq 1$, and that $\varphi = 0$ on Ω'. Set $\Omega'_n := \Omega' \setminus H_n$. If $u \in W^1_{2,\Gamma_0}(\Omega_n)$ then, obviously $\varphi u \in W^1_{2,\Gamma_0}(\Omega)$. Therefore by assumption and a simple calculation using the product rule for Sobolev functions

$$\begin{aligned}\|u\|_{W^1_2(\Omega_n)} &\leq \|\varphi u\|_{W^1_2(\Omega)} + \|(1-\varphi)u\|_{W^1_2(\Omega'_n)}\\ &\leq c(\Omega,\Gamma_0)\|\varphi u\|_{N^*,\Omega} + \|(1-\varphi)u\|_{W^1_2(\Omega'_n)}\\ &\leq c(\Omega,\Gamma_0)\|u\|_{N^*,\Omega_n} + (1+\|\nabla\varphi\|_\infty)\|u\|_{W^1_2(\Omega'_n)}.\end{aligned}$$

Next observe that by our assumption and the choice of Ω' the sets Ω'_n satisfy an interior cone condition with the same cone for all $n \in \mathbb{N}$. (We possibly have to use a smaller cone than those for Ω' and H_n.) Hence, by Lemma 2.2 we have a bound on $c(\Omega'_n)$. As

$$\|u\|_{W^1_2(\Omega_n)} \leq \big(c(\Omega,\Gamma_0) + (1+\|\nabla\varphi\|_\infty)c(\Omega'_n)\big)\|u\|_{N^*,\Omega_n}$$

this implies that (2.3) holds. □

2.5. Remark. From the proof of the above lemma we see that, in order to prove a uniform bound on $c(\Omega_n,\Gamma_0)$, it is sufficient to choose an open neighbourhood Ω' of K_0 with smooth boundary, and then to prove a uniform bound on $c(\Omega'\setminus H_n)$ for n large enough. If K_0 has finitely many components K_{0i}, $i = 1,\ldots,m$ it is sufficient to prove a bound on $c(\Omega'_i \cap H_n^{\complement})$ for some smooth open neighbourhood Ω'_i of K_{0i} for all $i = 1,\ldots,m$.

We next want to allow holes which do not necessarily satisfy a uniform exterior cone condition. In this case we assume that H_n shrinks to a set of finitely many points in the sense of Assumption 2.3. By Remark 2.5 it is sufficient to consider the case where Ω is smooth, and H_n shrinks to one point. After translation and rescaling we may therefore assume without loss of generality that the following conditions hold.

2.6. Assumption. Suppose that Ω is smooth with $\bar{B}_2 \subset \Omega$, and that $H_n \subset B_1$, where B_r denotes the open ball of radius $r > 0$ in $\mathbb{R}^N$ centered at zero. Further, assume that $H_n \subset B_{\delta_n}$ with $\delta_n \in (0,1]$, and that δ_n converges to zero as n goes to infinity. Set $U_n := B_2 \setminus \delta_n^{-1}\tilde{H}_n$, and suppose that there exist constants C_1, $C_2 > 0$ such that for all $n \in \mathbb{N}$

$$\|u - \frac{1}{|U_n|}\int_{U_n} u\,dx\|_{2,U_n} \leq C_1\|\nabla u\|_{2,U_n} \tag{2.4}$$

for all $u \in W^1_2(U_n)$, and that

$$c(U_n) \leq C_2. \tag{2.5}$$

The first is a uniform Poincaré inequality, the second a uniform Sobolev inequality on U_n.

2.7. Example. The above assumption covers the case of a shrinking hole of fixed shape. More precisely, let $H \subset B_1$ be a closed set with smooth boundary such that $B_2 \setminus H$ is connected. We then set $H_n = \delta_n \overline{H}$ for some sequence $\delta_n \in (0,1]$ converging to zero. Then Assumption 2.6 is satisfied.

The main idea of the proof of the following theorem I owe to E.N. Dancer.

2.8. Theorem. *Under Assumption 2.6 the sequence of Sobolev embedding constants $c(\Omega_n)$ is bounded.*

Proof. Suppose that $c(\Omega_n)$ is not a bounded sequence. Then, for all $k \in \mathbb{N}$ there exists $n_k \in \mathbb{N}$ and $u_k \in W_2^1(\Omega_{n_k})$ such that such that

$$\sqrt{2}k\|u_{n_k}\|_{W_2^1(\Omega_{n_k})} \leq \|u_{n_k}\|_{N^*,\Omega_{n_k}}$$

for all $k \in \mathbb{N}$. Renumbering the subsequences we can assume that

$$\sqrt{2}k\|u_k\|_{W_2^1(\Omega_k)} \leq \|u_k\|_{N^*,\Omega_k} \tag{2.6}$$

for all $k \in \mathbb{N}$. By the reverse triangle inequality, (2.6), and the elementary inequality $(s^2+t^2)^{1/2} \geq 2^{-1/2}(s+t)$ we have that

$$\begin{aligned}\|u_k\|_{N^*,B_{2\delta_k}\setminus H_k} &\geq \|u_k\|_{N^*,\Omega_k} - \|u_k\|_{N^*,\Omega\setminus B_{2\delta_k}}\\ &\geq \sqrt{2}k\|u_k\|_{W_2^1(\Omega_k)} - c(\Omega \setminus B_{2\delta_k})\|u_k\|_{W_2^1(\Omega\setminus B_{2\delta_k})}\\ &\geq k\|u_k\|_{W_2^1(B_{2\delta_k}\setminus H_k)} + \big(k - c(\Omega\setminus B_{2\delta_k})\big)\|u_k\|_{W_2^1(\Omega\setminus B_{2\delta_k})}\end{aligned}$$

for all $k \in \mathbb{N}$. By Proposition 2.4 the sequence $c(\Omega \setminus B_{2\delta_k})$ is uniformly bounded and therefore

$$k\|u_k\|_{W_2^1(B_{2\delta_k}\setminus H_k)} \leq \|u_k\|_{N^*,B_{2\delta_k}\setminus H_k} \tag{2.7}$$

for all $k \in \mathbb{N}$ large enough. By (2.4) and (2.5) we have that for all $v \in W_2^1(U_n)$ with mean zero

$$\|v\|_{N^*,U_k} \leq C_2\big(\|v\|_{2,U_k} + \|\nabla v\|_{2,U_k}\big) \leq C_2(1+C_1)\|\nabla v\|_{2,U_k} = C_0\|\nabla v\|_{2,U_k}$$

where C_0 is independent of $k \in \mathbb{N}$. Rescaling this inequality we get

$$\|w\|_{N^*,B_{2\delta_k}\setminus H_k} \leq C_0\delta_k^{\frac{N}{N^*}-\frac{N}{2}+1}\|\nabla w\|_{2,B_{2\delta_k}\setminus H_k} \leq C_0\|\nabla w\|_{2,B_{2\delta_k}\setminus H_k} \tag{2.8}$$

for all $w \in W_2^1(B_{2\delta_k} \setminus H_k)$ with mean zero. Let

$$\bar{u}_k := \frac{1}{|B_{2\delta_k}\setminus H_k|}\int_{B_{2\delta_k}\setminus H_k} u_k(x)\,dx$$

be the mean of u_k. Then by (2.8) we have that

$$\|u_k - \bar{u}_k\|_{N^*,B_{2\delta_k}\setminus H_k} \leq C_0\|\nabla u_k\|_{2,B_{2\delta_k}\setminus H_k} \tag{2.9}$$

for all $k \in \mathbb{N}$. Now, (2.7) and (2.9) imply that there exist $\alpha_k \neq 0$ and $v_k \in W_2^1(B_{2\delta_k} \setminus H_k)$ with mean $\bar{v}_k = 0$ such that $u_k = \alpha_k(1+v_k)$. Replacing u_k by $\alpha_k^{-1}u_k$ we can assume without loss of generality that

$$u_k = 1 + v_k$$

with $\bar{v}_k = 0$. Then, from (2.8) and (2.7) we see that

$$\begin{aligned}\|v_k\|_{N^*,B_{2\delta_k}\setminus H_k} &\le C_0\|\nabla v_k\|_{2,B_{2\delta_k}\setminus H_k} \le C_0\|u_k\|_{W_2^1(B_{2\delta_k}\setminus H_k)}\\ &\le C_0k^{-1}\|u_k\|_{N^*,B_{2\delta_k}\setminus H_k}\\ &\le C_0k^{-1}|B_{2\delta_k}\setminus H_k|^{\frac{1}{N^*}} + C_0k^{-1}\|v_k\|_{N^*,B_{2\delta_k}\setminus H_k}.\end{aligned}$$

This implies that for k large enough

$$\|v_k\|_{N^*,B_{2\delta_k}\setminus H_k} \le \frac{C_0}{k-C_0}|B_2|^{\frac{1}{N^*}}\delta_k^{\frac{N}{N^*}}. \tag{2.10}$$

From this it follows that

$$\begin{aligned}\|u_k\|_{N^*,\Omega_k} &\le \|u_k\|_{N^*,\Omega\setminus B_{2\delta_k}} + \|1+v_k\|_{N^*,B_{2\delta_k}\setminus H_k}\\ &\le \|u_k\|_{N^*,\Omega\setminus B_{2\delta_k}} + |B_{2\delta_k}|^{\frac{1}{N^*}} + \|v_k\|_{N^*,B_{2\delta_k}\setminus H_k}\\ &\le \|u_k\|_{N^*,\Omega\setminus B_{2\delta_k}} + \Big(1+\frac{C_0}{k-C_0}\Big)|B_2|^{\frac{1}{N^*}}\delta_k^{\frac{N}{N^*}}\end{aligned} \tag{2.11}$$

for all k large enough. Moreover, by Proposition 2.4 and (2.6) we have that

$$\|u_k\|_{N^*,\Omega\setminus B_{\delta_k}} \le c(\Omega\setminus B_{\delta_k})\|u_k\|_{W_2^1(\Omega_k)} \le \frac{c_0}{k}\|u_k\|_{N^*,\Omega_k}$$

with c_0 independent of $k\in\mathbb{N}$. Taking into account (2.11) we have that

$$\|u_k\|_{N^*,\Omega\setminus B_{\delta_k}} \le \frac{c_0}{k}\Big(\|u_k\|_{N^*,\Omega\setminus B_{2\delta_k}} + \frac{k}{k-C_0}|B_2|^{\frac{1}{N^*}}\delta_k^{\frac{N}{N^*}}\Big) \tag{2.12}$$

for all k large enough. Next we prove a lower estimate on $\|u_k\|_{N^*,\Omega\setminus B_{\delta_k}}$. As the function $t\mapsto t^p$ is convex for $p\ge 1$ the elementary inequality $(s^p+s^p)^{1/p}\ge 2^{-(p-1)/p}(t+s)$ holds. Using this and (2.10) we see that

$$\begin{aligned}2^{1-\frac{1}{N^*}}\|u_k\|_{N^*,\Omega\setminus B_{\delta_k}} &\ge \|u_k\|_{N^*,\Omega\setminus B_{2\delta_k}} + \|1+v_k\|_{N^*,B_{2\delta_k}\setminus B_{\delta_k}}\\ &\ge \|u_k\|_{N^*,\Omega\setminus B_{2\delta_k}} + |B_{2\delta_k}\setminus B_{\delta_k}|^{\frac{1}{N^*}} - \|v_k\|_{N^*,B_{2\delta_k}\setminus H_k}\\ &\ge \|u_k\|_{N^*,\Omega\setminus B_{2\delta_k}} + \Big(\frac{1}{2}-\frac{C_0}{k-C_0}\Big)|B_2|^{\frac{1}{N^*}}\delta_k^{\frac{N}{N^*}}\end{aligned} \tag{2.13}$$

for k large enough. Here we used that

$$|B_{2\delta_k}\setminus B_{\delta_k}|^{\frac{1}{N^*}} = (1-2^{-N})^{\frac{1}{N^*}}|B_2|^{\frac{1}{N^*}} \ge \frac{1}{2}|B_2|^{\frac{1}{N^*}}.$$

Now, subtracting (2.12) and (2.13) we obtain

$$\begin{aligned}0 \le& \Big(\frac{c_0}{k} - 2^{\frac{1}{N^*}-1}\Big)\|u_k\|_{N^*,\Omega\setminus B_{2\delta_k}}\\ &+ \Big(\frac{c_0}{k-C_0} + 2^{\frac{1}{N^*}-1}\frac{C_0}{k-C_0} - 2^{\frac{1}{N^*}-2}\Big)|B_2|^{\frac{1}{N^*}}\delta_k^{\frac{N}{N^*}}\end{aligned}$$

which implies that for k sufficiently large

$$0 \le \frac{c_0}{k-C_0} + 2^{\frac{1}{N^*}-1}\frac{C_0}{k-C_0} - 2^{\frac{1}{N^*}-2} < 0$$

which is impossible. Therefore, we cannot find a sequence satisfying (2.6). This finishes the proof of the theorem. □

2.9. Remark. At least if $N \geq 3$ the validity of the Poincaré inequality (2.4) uniformly with respect to $n \in \mathbb{N}$ is a necessary condition in order for Theorem 2.8 to be true. To see this assume that there is no uniform bound on the constant in the Poincaré inequality. Then, for each $k \in \mathbb{N}$ there exists $u_{n_k} \in W_2^1(U_{n_k})$ with mean $\bar{u}_{n_k} = 0$ such that

$$k\|\nabla u_{n_k}\|_{2,U_{n_k}} \leq \|u_{n_k}\|_{2,U_{n_k}}$$

for all $k \in \mathbb{N}$. Renumbering the subsequences we can assume that

$$k\|\nabla u_k\|_{2,U_k} \leq \|u_k\|_{2,U_k} \leq |B_2|^{\frac{1}{2}-\frac{1}{N^*}}\|u_k\|_{N^*,U_k} \tag{2.14}$$

for all $k \in \mathbb{N}$. Clearly we can modify u_k such that it has trace zero on the boundary of B_2, its mean is zero, and (2.14) still holds. Now set

$$v_k(x) := u_k(\delta_k^{-1}x)$$

for all $x \in B_{2\delta_k} \setminus H_k$ and $k \in \mathbb{N}$. Then, v_k has zero trace on the boundary of $B_{2\delta_k}$. Hence we can extend v_k by zero outside $B_{2\delta_k}$ and consider it as an element of $W_2^1(\Omega_k)$ with support in $B_{2\delta_k}$. Rescaling (2.14) it follows that

$$k\|\nabla v_k\|_{2,\Omega_k} \leq |B_2|^{\frac{1}{2}-\frac{1}{N^*}}\delta_k^{\frac{N}{2}-\frac{N}{N^*}-1}\|u_k\|_{N^*,\Omega_k} = |B_2|^{\frac{1}{N}}\|u_k\|_{N^*,\Omega_k}$$

for all $k \in \mathbb{N}$. Here we used that for $N^* = 2N(N-2)^{-1}$ for $N \geq 3$. As the support of v_k lies in $B_{2\delta_k}$ this implies that

$$\|v_k\|_{W_2^1(\Omega_k)} \leq (k^{-1} + \delta_k)|B_2|^{\frac{1}{N}}$$

for all $k \in \mathbb{N}$. Therefore, $c(\Omega_k) \geq (k^{-1} + \delta_k)^{-1}|B_2|^{-\frac{1}{N}}$ for all $k \in \mathbb{N}$ and hence $c(\Omega_k)$ unbounded. This shows that (2.4) is necessary in Theorem 2.8.

2.10. Remark. Note that the Poincaré and Sobolev inequalities are independent of each other. It is possible that (2.4) fails but (2.5) is true. As an example look at U_n where part of U_n pinches off as n goes to infinity in such a way that U_n satisfies an interior cone condition uniformly with respect to $n \in \mathbb{N}$. By Lemma 2.2 condition (2.5) holds. On the other hand, (2.4) fails.

3. Linear Problems

In this section we consider linear problems of the form

$$\begin{aligned} -\Delta u + \mu u &= f && \text{in } \Omega \\ u &= 0 && \text{on } \Gamma_0 \\ \frac{\partial}{\partial \nu}u + b_0 u &= 0 && \text{on } \Gamma_1\,, \end{aligned} \tag{3.1}$$

where Γ_0, Γ_1 are open and closed disjoint subsets of $\partial\Omega$ with $\Gamma_0 \cup \Gamma_1 = \Omega$. We assume that $b_0 \in L_\infty(\partial\Omega)$ is nonnegative. We start by proving L_p a priori estimates of weak solutions of the above problem. In order to have good control of the constants over the domain we make the following assumptions.

3.1. Assumption. Ω is an open bounded set in $\mathbb{R}^N$ with the following properties.

(i) Ω satisfies an interior cone condition.
(ii) $C(\overline{\Omega}) \cap W_2^1(\Omega)$ is dense in $W_2^1(\Omega)$.

Condition (ii) is for instance satisfied for any extension domain, but holds under much weaker conditions. In two dimensions it is satisfied if $\partial\Omega$ is a Jordan curve (see [20]). It is also satisfied for the union of two touching balls. Further, we do not really need the cone condition for Ω. It is sufficient to assume that it holds in a neighbourhood of Γ_1. But for simplicity of the exposition we suppose it holds for Ω. We then have the following lemma.

3.2. Lemma. *Let Assumption 3.1 be satisfied. Then there exists a trace operator*

$$\gamma \in \mathcal{L}(W_2^1(\Omega), L_2(\Gamma_1))$$

with $\gamma(u) = u|_{\Gamma_1}$ *for all* $u \in W_2^1(\Omega) \cap C(\overline{\Omega})$. *Moreover,* γ *is compact. Here,* $L_2(\Gamma_1) := L_2(\Gamma_1, \mathcal{H}_{N-1})$ *and* $\mathcal{H}_{N-1}$ *is the* $(N-1)$*-dimensional Hausdorff measure restricted to* Γ_1.

Note that if the boundary of Ω is smooth then $\mathcal{H}_{N-1}$ coincides with the usual surface measure on $\partial\Omega$.

Proof. By a lemma due to Gagliardo (see [2, Theorem 4.8]) any bounded domain Ω satisfying an interior cone condition can be written as a union of finitely many Lipschitz domains Ω_i $(i = 1, \ldots, m)$. For Lipschitz domains we have trace inequalities $\|u\|_{L_2(\partial\Omega_i\cap\Gamma_1)} \leq c_i\|u\|_{W_2^1(\Omega_i)}$ for all $u \in W_2^1(\Omega) \cap C(\overline{\Omega})$. Moreover the trace operator is compact (e.g. [22, Théorème 6.6.2]). Setting $c := \max\{c_1, \ldots, c_m\}$ and observing that $\partial\Omega \subset \bigcup_{i=1}^m \partial\Omega_i$ this yields

$$\|u\|_{L_2(\Gamma_1)} \leq \sum_{i=1}^m \|u\|_{L_2(\partial\Omega_i\cap\Gamma_1)} \leq c\sum_{i=1}^m \|u\|_{W_2^1(\Omega_i)} \leq cm\|u\|^2_{W_2^1(\Omega)}$$

for all $u \in W_2^1(\Omega) \cap C(\overline{\Omega})$. By Assumption 3.1(ii) we can continue the trace operator to $W_2^1(\Omega)$. □

As usual we call u a weak solution of (3.1) if $u \in W^1_{2,\Gamma_0}(\Omega)$, and u satisfies the integral identity

$$a_\mu(u, v) := \int_\Omega \nabla u \nabla v + \mu uv\, dx + \int_{\Gamma_1} b_0 uv\, d\mathcal{H}_{N-1} = \int_\Omega fv\, dx =: \langle f, v\rangle \tag{3.2}$$

for all $v \in W^1_{2,\Gamma_0}(\Omega)$ (or a dense subset thereof). By $R(\mu)$ we denote the resolvent, i.e. $R(\mu)f = u$ is the solution of (3.1). Clearly, $a_\mu(\cdot\,,\cdot)$ is coercive for $\mu > 0$ and so any $\mu > 0$ is in the resolvent set. Moreover, by Lemma 2.2 and duality it turns out that $R(\mu) \in \mathcal{L}(L_{q'}(\Omega), L_q(\Omega))$ for all $q \in [2, 2N(N-2)^{-1}]$ if $N \geq 3$ and $q \in [2, \infty)$ for $N = 2$. Here, q' is the dual exponent to q, i.e. $1/q + 1/q' = 1$.

3.3. Theorem. *There exists a constant $C > 0$ depending only on N, p, and an upper bound for the embedding constant of* (2.1) *such that*

$$\|R(1)\|_{p,m(p)} \leq C, \tag{3.3}$$

where

$$m(p) := \begin{cases} Np(N-2p)^{-1} & \text{if } \ 2N(N+2)^{-1} < p < N/2 \\ \infty & \text{if } \ p > N/2 \end{cases}$$

If $N \geq 3$ the result also holds for $p = 2N(N+2)^{-1}$.

Proof. The proof of the theorem is essentially the same as that of [11, Theorem 4.1] but simpler as our present problem is self adjoint. In particular, the constant δ appearing in statement and proof of that result is zero. Due to Lemma 2.2 we can work with $d = N$ rather than $d = 2N$ (see also [11, Remark 4.4(c)]). □

Note that the bound (3.3) is not satisfied for bad shaped domains. For examples see [9, Proposition 3.2] or [10, Theorem 5.1]. The examples clearly can be modified to see that (3.3) fails if we cut bad shaped holes into a domain.

Let us now consider the problem

$$\begin{aligned} -\Delta u + \mu u &= f_n && \text{in } \Omega_n \\ u &= 0 && \text{on } \Gamma_{0n} \\ \frac{\partial}{\partial \nu} u + b_{0n} u &= 0 && \text{on } \Gamma_{1n}\,, \end{aligned} \tag{3.4}$$

on a sequence of domains Ω_n approaching Ω. We again assume that Γ_{0n}, Γ_{1n} are disjoint, open and closed subsets of $\partial\Omega$ with $\partial\Omega_n = \Gamma_{0n} \cup \Gamma_{1n}$. We then want to prove that the solutions of (3.4) converge to the solution of (3.1) if f_n and b_{0n} converge. We assume that the boundary of Ω is not perturbed too much in the following sense.

3.4. Assumption. Suppose $\Omega_n \subset \mathbb{R}^N$ are bounded domains satisfying Assumption 3.1, and that the Sobolev embedding constant $c(\Omega_n, \Gamma_0)$ is bounded by a constant independent of $n \in \mathbb{N}$. Moreover, we suppose that there exists a compact set $K \subset \overline{\Omega}$ of capacity zero such that for all neighbourhoods U of K

$$\Omega \cap U^{\mathsf{C}} = \Omega_n \cap U^{\mathsf{C}} \qquad \text{and} \qquad \Gamma_0 \cap U^{\mathsf{C}} = \Gamma_{0n} \cap U^{\mathsf{C}} \tag{3.5}$$

for n large enough.

One condition that the Sobolev embedding constant is bounded independently of $n \in \mathbb{N}$ is that the sets Ω_n satisfy an interior cone condition with the same cone for all $n \in \mathbb{N}$. This follows from Proposition 2.4. Another condition is Assumption 2.6, where a uniform bound is proved in Theorem 2.8. As the Sobolev embedding constant is determined by local properties we can consider various combinations of the two conditions.

We denote the resolvent of problem (3.1) by $R(\mu)$, and the resolvent of (3.4) by $R_n(\mu)$. By restricting $f \in L_p(\mathbb{R}^N)$ to Ω, applying $R(\mu)$, and then extending

$R(\mu)f$ trivially by zero outside Ω we may consider $R(\mu)$ as an operator on $L_p(\mathbb{R}^N)$. Similarly, proceed with $R_n(\mu)$. We have the following theorem.

3.5. Theorem. *Suppose $\mu \in \mathbb{R}$ is in the resolvent set of (3.1). Then, μ is in the resolvent set of (3.4) for n large enough. Moreover, suppose that f_n converges to f weakly in $L_p(\mathbb{R}^N)$ for some $p \in (2N(N+2)^{-1}, \infty)$, and that $b_{0n} \in L_\infty(\Gamma_{1n})$ converges weakly to b_0 in $L_2(\Gamma_{1n} \cap U^{\complement})$ for all neighbourhoods U of K. Then, the weak solutions $u_n = R_n(\mu)f_n$ of (3.4) converge strongly to the weak solution $u = R(\mu)f$ of (3.1) in $L_q(\mathbb{R}^N)$ for all $q \in [1, m(p))$, where $m(p)$ is the same as in Theorem 3.3.*

Proof. Using the same argument as in [12, Corollary 3.3] it follows from the assertion of the theorem that

$$\lim_{n\to\infty} R_n(\mu) = R(\mu) \tag{3.6}$$

in $\mathcal{L}(L_p(\mathbb{R}^N), L_q(\mathbb{R}^N))$, i.e. in the uniform operator topology. Due to [18, Remark IV.3.3.13] $R_n(\mu)$ exists for large n if $R(\mu)$ does, and (3.6) holds for all μ in the resolvent set of (3.1) if it holds for one particular μ. Hence, it is sufficient to prove the assertion of the theorem for $\mu = 1$ (see also [10, Remark 3.9]).

By Assumption 3.4 the Sobolev embedding constant $c(\Omega_n)$ is bounded independently of $n \in \mathbb{N}$. Hence, by Theorem 3.3 there exists a constant $C > 0$ independent of $n \in \mathbb{N}$ such that for $p > 2N(N+2)^{-1}$

$$\|R_n(1)\|_{p,m(p)} \le C \tag{3.7}$$

for all $n \in \mathbb{N}$. If we can prove convergence of $R_n(1)$ for $p > N/2$ an application of the Riesz-Thorin interpolation theorem then shows that convergence holds also for $2N(N+2)^{-1} < p < N/2$.

Suppose now that $\mu = 1$ and $p > N/2$. By (3.7) we have that

$$\|u_n\|_\infty \le C\|f_n\|_p \tag{3.8}$$

for all $n \in \mathbb{N}$, where u_n is the weak solution of (3.4). Fix a sequence U_n of open neighbourhoods of K with $\bigcap_{n\in\mathbb{N}} U_n = K$, $\Omega \cap U_n^{\complement} = \Omega_n \cap U_n^{\complement}$ and $\overline{U}_{n+1} \subset U_n$ for all $n \in \mathbb{N}$. This is possible by (3.5). Then, select a sequence of functions $\varphi_n \in C^\infty(\mathbb{R}^N)$ with $\varphi_n = 0$ on U_n and $\varphi_n = 1$ on $U_{n-1}^{\complement}$. As the capacity of K is zero we can choose them such that $\lim_{n\to\infty} \varphi_n = 1$ in $W_2^1(\Omega)$. We then set $w_n = \varphi_n u_n$. By (3.5) and (3.8) $w_n \in W_2^1(\Omega) \cap W_{2,\Gamma_{0n}}^1(\Omega_n)$. Now, an elementary calculation using the fact that u_n is the weak solution of (3.4) and (3.8) shows that

$$\begin{aligned}\|w_n\|_{W_2^1}^2 &\leq a_{1,n}(w_n, w_n) = a_{1,n}(u_n, \varphi_n^2 u_n) + \int_\Omega u_n^2 |\nabla\varphi_n|^2\,dx \\ &\leq \langle f_n, \varphi_n^2 u_n\rangle + \|u_n\|_\infty^2 \|\nabla\varphi_n\|_2^2 \\ &\leq |\Omega|^{1-1/p}\|f_n\|_p\|u_n\|_\infty + \|u_n\|_\infty^2\|\nabla\varphi_n\|_2^2 \\ &\leq C\|f_n\|_p^2\big(|\Omega|^{1-1/p} + C\|\nabla\varphi_n\|_2^2\big).\end{aligned}$$

Here, $a_{1,n}(\cdot\,,\cdot)$ is the form associated with (3.4) with $\mu = 1$. By assumption the norm $\|\nabla\varphi_n\|_2$ converges to zero as n goes to infinity. Further, as f_n converges weakly in $L_p(\mathbb{R}^N)$ the sequence is bounded in $L_p(\mathbb{R}^N)$. Therefore, w_n is a bounded sequence in $W_2^1(\Omega)$. Hence, it has a weakly convergent subsequence, which, by Lemma 2.2, converges strongly in $L_2(\Omega)$. Let us denote this subsequence again by w_n, and its limit by w. We show that w is a solution of (3.1) and therefore must coincide with u. To do that let $\varphi \in W_{2,\Gamma_0}^1(\Omega)$ be such that $\varphi = 0$ in a neighbourhood of K. Then $\lim_{n\to\infty} a_{1,n}(w_n, \varphi) = a(w, \varphi)$ and $\lim_{n\to\infty}\langle f_n, \varphi\rangle = \langle f, \varphi\rangle$. By definition of w_n we have $w_n = u_n$ on the support of φ if n is large enough, and thus $a_{1,n}(w_n, \varphi) = a_{1,n}(u_n, \varphi) = \langle f_n, \varphi\rangle$ for large enough n. Hence, passing to the limit this yields $a_1(w, \varphi) = \langle f, \varphi\rangle$. As the capacity of K is zero the test functions under consideration are dense in $W_{2,\Gamma_0}^1(\Omega)$ and thus $w = u$, and u is the solution of (3.1). Finally, it remains to show that u_n converges in $L_q(\mathbb{R}^N)$ for all $q < \infty$. By (3.8) and interpolation it is sufficient to show convergence in $L_2(\mathbb{R}^N)$. To do so note that

$$\begin{aligned}\|u_n - u\|_2 &\leq \|u_n - u\|_{2,\Omega\cap U_{n-1}^{\complement}} + \|u_n - u\|_{2,U_{n-1}} \\ &\leq \|w_n - u\|_{2,\Omega} + |U_{n-1}|^{1/2}\big(\|u_n\|_\infty + \|u\|_\infty\big).\end{aligned}$$

As w_n converges to u in $L_2(\Omega)$ the first term converges to zero. The second converges to zero by (3.8) and the fact that $|U_{n-1}|^{1/2}$ tends to zero as n goes to infinity. Finally observe that the whole sequence converges as the limit is unique. This finishes the proof of the theorem. □

The above theorem, as in case of the Dirichlet and Robin problem, implies convergence of the resolvent in the uniform operator topology, and continuity of any finite part of the spectrum. (Compare [12, Theorem 5.3] or [10, Corollary 3.7].)

3.6. Corollary. *Suppose the assumptions of the above theorem are satisfied. Then* (3.6) *holds in* $\mathcal{L}(L_p(\mathbb{R}^N), L_q(\mathbb{R}^N))$ *(i.e. in the uniform operator topology) for every* μ *in the resolvent set of* (3.1) *and* $q \in [1, m(p))$. *Moreover, the spectrum consists* *of eigenvalues of finite multiplicity, and if Σ is a finite system of eigenvalues of*

$$\begin{aligned} -\Delta u + \mu u &= \alpha u && \text{in } \Omega \\ u &= 0 && \text{on } \Gamma_0 \\ \frac{\partial}{\partial \nu} u + b_0 u &= 0 && \text{on } \Gamma_1, \end{aligned} \tag{3.9}$$

and U an open neighbourhood of Σ then R_n has the same number of eigenvalues in U for n large enough (counting multiplicity).

Proof. The first assertion we discussed at the beginning of the proof of Theorem 3.5, and the second is a consequence of this (see [18, Section IV.3.5]). □

4. Nonlinear Eigenvalue Problems

In this section we study domain perturbation for the nonlinear eigenvalue problem

$$\begin{aligned} -\Delta u + \mu u &= \lambda f(x, u) && \text{in } \Omega \\ u &= 0 && \text{on } \Gamma_0 \\ \frac{\partial}{\partial \nu} u + b_0 u &= 0 && \text{on } \Gamma_1, \end{aligned} \tag{4.1}$$

where $\mu \in \mathbb{R}$ is fixed. By a solution of this problem we mean a pair $(\lambda, u) \in \mathbb{R} \times W^1_{2,\Gamma_0}(\Omega)$ such that u is a weak solution of (4.1).

Throughout, we assume that the nonlinearity satisfies the following assumptions.

4.1. Assumption. We assume that $f \in C^1(\overline{\Omega} \times \mathbb{R})$, and that there exists $c_0 > 0$ and $\beta \in (1, (N+2)(N-2)^{-1})$ such that

$$\Big|\frac{\partial}{\partial \xi} f(x, \xi)\Big| \leq c_0(1 + |\xi|^{\beta - 1}) \tag{4.2}$$

for all $x \in \mathbb{R}^N$ and $\xi \in \mathbb{R}$. Next define

$$F(x, \xi) = \int_0^\xi f(x, \eta)\, d\eta$$

for all $\xi \in \mathbb{R}$, and assume that the set of functions $u \in W^1_{2,\Gamma_0}(\Omega)$ such that

$$G(u) := \int_\Omega F(x, u(x))\, dx \neq 0 \tag{4.3}$$

is nonempty. Finally, we assume that for all $u \in W^1_{2,\Gamma_0}(\Omega)$ with $G(u) \neq 0$

$$f(x, u(x)) \neq 0 \tag{4.4}$$

on a set of positive Lebesgue measure.

4.2. Remark. We really only need a growth condition on f, not its derivative. However, for our purposes we can always modify f for large values of ξ without changing the result. For more details see Remark 4.10.

Equation (4.1) can be put into variational form by considering the energy functional

$$J(u) := \frac{1}{2}\Big(\int_\Omega |\nabla u|^2 + \mu|u|^2\,dx + \int_{\Gamma_1} b_0 u^2\,d\mathcal{H}_{N-1}\Big) \tag{4.5}$$

on the set

$$M := \{u \in W^1_{2,\Gamma_0}(\Omega)\colon G(u) = 1\}.$$

Note that, due to (4.3), we can always replace f by a scalar multiple such that $M \neq \emptyset$. This just modifies the eigenvalue of (4.1) by a scalar multiple. The following theorem is known, but for completeness we include its proof.

4.3. Remark. If we are interested in positive solutions of (4.1) we replace M by

$$M_+ := \{u \in W^1_{2,\Gamma_0}(\Omega)\colon u > 0,\ G(u) = 1\}.$$

In this case we have to assume that $f(x,0) \geq 0$ for all $x \in \Omega$.

4.4. Theorem. *Suppose that f satisfies Assumption 4.1. If the first eigenvalue α_1 associated with* (3.9) *is positive then $J(\cdot)$ attains a positive minimum on M. Moreover, any minimizer u of $J(\cdot)$ is a weak solution of* (4.1), *and for the corresponding λ, the term $\lambda\langle f(\cdot,u(\cdot)),u\rangle$ is nonzero.*

A solution of (4.1) minimizing $J(\cdot)$ is said to be an energy minimizing solution.

Proof. By Lemma 2.2 any sequence converging weakly in $W^1_2(\Omega)$ converges strongly in $L_{\beta+1}(\Omega)$. Here, β is from (4.2). Moreover, due to (4.2) the substitution operator induced by F is continuous from $L_{\beta+1}(\Omega)$ to $L_1(\Omega)$ and thus G is continuous on $L_{\beta+1}(\Omega)$. This implies that the set M is weakly closed in $W^1_{2,\Gamma_0}(\Omega)$. By the Rayleigh-Ritz characterization of the first eigenvalue of (3.9) and the definition of $J(\cdot)$ we get

$$J(u) = \frac{1}{2}a_\mu(u,u) \geq \frac{1}{4}\inf\{1,\alpha_1\}\|u\|^2_{W^1_2(\Omega)} \tag{4.6}$$

for all $u \in M$. Hence $J(\cdot)$ is coercive. Further, by the weak lower semicontinuity of the norm in $W^1_2(\Omega)$ and Lemma 2.2 the energy functional $J(\cdot)$ is lower semicontinuous on M. Then, by [25, Theorem 1.2] it follows that $J(u)$ attains its minimum on M at a point u_0. Next note that $J(\cdot)$ is Fréchet differentiable on $W^1_{2,\Gamma_0}(\Omega)$, and $\langle DJ(u),v\rangle = a_\mu(u,v)$. By the growth condition (4.2), also the function $G\colon W^1_{2,\Gamma_0}(\Omega) \to \mathbb{R}$ defined by (4.3) is Fréchet differentiable (e.g. [3, Theorem 3.1 and 3.12]. Its derivative is given by $\langle DG(u),v\rangle = \langle f(\cdot,u(\cdot)),v\rangle$ for all $v \in W^1_{2,\Gamma_0}(\Omega)$. By assumption we may choose $v \in W^1_{2,\Gamma_0}(\Omega)$ such that $\langle DG(u),v\rangle \neq 0$ and therefore $DG(u)$ is surjective for all $u \in M$. Hence, 1 is a regular value of G which implies that $G^{-1}(1)$ is a C^1-manifold in $W^1_{2,\Gamma_0}(\Omega)$ (e.g. [1, Theorem 3.5.4]). By the Lagrange multiplier rule (e.g. [1, Proposition 3.5.25] there exists $\lambda \in \mathbb{R}$ such that

$$\langle DJ(u_0),v\rangle - \lambda\langle DG(u_0),v\rangle = a_\mu(u_0,v) - \lambda\langle f(\cdot,u_0(\cdot)),v\rangle = 0$$

for all $v \in W^1_{2,\Gamma_0}(\Omega)$. Hence, u_0 is a weak solution of (4.1). As $G(0) = 0$ and $G(u_0) = 1$ the function u_0 cannot vanish identically. Inserting $v = u_0$ in the above identity it follows from the coercivity of J that $J(u_0) > 0$. This also implies that λ and $\langle f(\cdot, u_0(\cdot)), u_0\rangle$ are nonzero. □

We next want to prove that the solutions of (4.1) are bounded. To do so we prove boundedness of weak solutions of (4.1) with $\lambda = 1$.

4.5. Proposition. *Suppose that u is a weak solution of* (4.1) *with $\lambda = 1$, and that* (4.2) *holds for some $\beta \in (1, \infty)$. If $u \in L_p(\Omega)$ for some*

$$p > \max\{\frac{N}{2}(\beta - 1), \frac{2N\beta}{N+2}\} \tag{4.7}$$

then $u \in L_\infty(\Omega)$, and

$$\|u\|_\infty \leq b(\|u\|_p), \tag{4.8}$$

where $b\colon \mathbb{R}_+ \to \mathbb{R}_+$ is an increasing function depending only on N, p and upper bounds for c_0, $\max\{0, 1 - \mu\}$, $\|f(\cdot\,, 0)\|_\infty$, $|\Omega|$ and the constant C in (3.3).

Proof. The proof is similar to that of [10, Proposition 3.3]. As we proved better a priori estimates in Theorem 3.3 we can replace N by $N/2$ in the present case. □

4.6. Remark. Note that if $\beta \in (1, (N+2)(N-2)^{-1})$ then (4.7) is automatically satisfied by (2.1).

After these preparations we are ready to prove the main results of this section.

4.7. Proposition. *Suppose that Assumption 3.1, 3.4 and 4.1 hold, and that $b_{0n} \in L_\infty(\Gamma_{1n})$ converges weakly to b_0 in $L_2(\Gamma_{1n} \cap U^{\complement})$ for all neighbourhoods U of K. If the first eigenvalue α_1 of* (3.9) *is positive then the problem*

$$\begin{aligned} -\Delta u &= \lambda_n f(x, u) && \text{in } \Omega_n \\ u &= 0 && \text{on } \Gamma_{0n} \\ \frac{\partial}{\partial \nu} u + b_{0n} u &= 0 && \text{on } \Gamma_{1n}. \end{aligned} \tag{4.9}$$

has an energy minimizing solution (λ_n, u_n) for large n.

Proof. By Corollary 3.6 the first eigenvalue α_{1n} of

$$\begin{aligned} -\Delta u + \mu u &= \alpha u && \text{in } \Omega_n \\ u &= 0 && \text{on } \Gamma_{0n} \\ \frac{\partial}{\partial \nu} u + b_{0n} u &= 0 && \text{on } \Gamma_{1n}\,, \end{aligned} \tag{4.10}$$

converges to the first eigenvalue α_1 of (3.9) and thus $\alpha_{1n} > 0$ for n large. Moreover, for n large Assumption 4.1 holds and therefore we conclude from Theorem 4.4 that (4.9) has an energy minimizing solution. In this case, the functional

$$J_n(v) = \frac{1}{2}\Big(\int_{\Omega_n} |\nabla v|^2 + \mu|v|^2\,dx + \int_{\Gamma_{1n}} b_{0n}v^2\,d\mathcal{H}_{N-1}\Big) \tag{4.11}$$

is minimized over

$$M_n := \{u \in W_2^1(\Omega_n) \colon G_n(u) = 1\} \tag{4.12}$$

where G_n is defined in the same way as G but with Ω replaced by Ω_n. □

4.8. Theorem. *Suppose (λ_n, u_n) and (λ, u) are energy minimizing solutions of (4.9) and (4.1), respectively. Further suppose that $b_{0n} \in L_\infty(\Gamma_{1n})$ converges to b_0 weakly in $L_2(\Gamma_{1n} \cap U^{\complement})$ for all neighbourhoods U of K. Then*

(i) *As n goes to infinity, $J_n(u_n)$ converges to $J(u)$.*
(ii) *(λ_n, u_n) has a subsequence (λ_{n_k}, u_{n_k}) converging to an energy minimizing solution of (4.1) in $\mathbb{R} \times L_p(\mathbb{R}^N)$ for all $p \in [1,\infty)$ as k tends to infinity.*
(iii) *If (λ, u) is an isolated minimizer, then there exist solutions $(\tilde\lambda_n, \tilde u_n)$ converging to (λ, u) in $\mathbb{R} \times L_p(\mathbb{R}^N)$ for all $p \in [1,\infty)$ as k tends to infinity.*

Proof. We prove (i) and (ii) simultaneously. As in the proof of Proposition 4.7 the first eigenvalue α_{1n} of (4.10) converges to α_1 as n goes to infinity. Hence, due to (4.6) we have that for all $n \in \mathbb{N}$

$$\|v\|^2_{W_2^1(\Omega_n)} \le 4\max\{1, \alpha_{1n}^{-1}\} J_n(v)$$

for all $v \in M_n$. If we pick $v \in \mathcal{D}(\Omega)$ with $G(v) = 1$ then $G_n(v) = 1$ for large n, and therefore

$$\|u_n\|^2_{W_2^1(\Omega_n)} \le 4\max\{1, \alpha_{1n}^{-1}\} J_n(u_n) \le CJ(v) \tag{4.13}$$

for large $n \in \mathbb{N}$ with a constant $C > 0$ independent of $n \in \mathbb{N}$. In particular, if U is any neighbourhood of K, this implies that $u_n|_{\Omega\cap U^{\complement}}$ is a bounded sequence in $W_2^1(\Omega \cap U^{\complement})$. By Lemma 2.2 it has a subsequence, which we denote the same way, converging in $L_2(\Omega\cap U^{\complement})$. We now consider u_n as a function on $\mathbb{R}^N$ by extending it by zero outside Ω_n. By a diagonalizing process we can find a further subsequence such that u_n converges to some function u_0 weakly in $W_2^1(\Omega \cap U^{\complement})$, and strongly in in $L_2(\mathbb{R}^N \cap U^{\complement})$ for any neighbourhoods U of K. Now, for any $q \ge 2$ we have

$$\begin{aligned}\|u_n - u_0\|_{2,\mathbb{R}^N} &\le \|u_n - u_0\|_{2,\Omega\cap U^{\complement}} + \|u_n - u_0\|_{2,U}\\ &\le \|u_n - u_0\|_{2,\Omega\cap U^{\complement}} + |U|^{\frac{1}{2}-\frac{1}{N^*}}\|u_n - u_0\|_{N^*}\end{aligned}$$

Taking into account (4.13) and the assumption that the Sobolev embedding constant is bounded independently of $n \in \mathbb{N}$ we conclude that

$$\|u_n\|_{N^*} \le c \quad\text{and}\quad \|u\|_{N^*} \le c \tag{4.14}$$

for some constant independent of $n \in \mathbb{N}$. As we can choose U to be of arbitrarily small measure it follows that $u_n \to u_0$ in $L_2(\mathbb{R}^N)$, and by interpolation in $L_p(\mathbb{R}^N)$ for all $p \in [1, N^*)$.

We next show that u_0 is a solution of (4.1) for some $\lambda_0 \neq 0$. To do so fix any $\varphi \in W^1_{2,\Gamma_0}(\Omega)$ with $\varphi = 0$ in a neighbourhood of K. Then $\varphi \in W^1_{2,\Gamma_{0n}}(\Omega_n)$ for n large enough and hence $a_{\mu,n}(u_n, \varphi) = \lambda_n \langle f(\cdot, u_n(\cdot)), \varphi\rangle$ for n large. By the weak convergence of u_n in $W^1_2(\Omega \cap U^{\complement})$ we get $a_{\mu,n}(u_n, \varphi) \to a_\mu(u_0, \varphi)$ as n goes to infinity. By (4.2) the substitution operator induced by f is continuous from $L_q(\mathbb{R}^N)$ to $L_{q'}(\mathbb{R}^N)$ if only q is close enough to $2N(N-2)^{-1}$. Here, $1/q + 1/q' = 1$. By Lemma 2.1 $\varphi \in L_q(\Omega)$ and therefore $\langle f(\cdot, u_n(\cdot)), \varphi\rangle \to \langle f(\cdot, u_0(\cdot)), \varphi\rangle$ as n goes to infinity. It also follows that $G_n(u_n) \to G(u_0) = 1$, i.e. $u_0 \in M$. By the assumptions on f we can choose φ such that $\langle f(\cdot, u_0(\cdot)), \varphi\rangle \neq 0$, and so

$$\lambda_n = \frac{a_{\mu,n}(u_n, \varphi)}{\langle f(\cdot, u_n(\cdot)), \varphi\rangle} \longrightarrow \lambda_0 := \frac{a_\mu(u_0, \varphi)}{\langle f(\cdot, u_0(\cdot)), \varphi\rangle}.$$

This implies that $a_\mu(u_0, \varphi) = \lambda_0 \langle f(\cdot, u_0(\cdot)), \varphi\rangle$ for all $\varphi \in W^1_{2,\Gamma_0}(\Omega)$ with $\varphi = 0$ in a neighbourhood of K. As this set of functions is dense in $W^1_{2,\Gamma_0}(\Omega)$ it follows that (λ_0, u_0) is a solution of (4.1). Also, as λ_n converges it is bounded, and therefore Proposition 4.5 and (4.14) show that

$$\|u_n\|_\infty \leq c \quad \text{and} \quad \|u\|_\infty \leq c \tag{4.15}$$

for some constant independent of $n \in \mathbb{N}$. By interpolation it therefore follows that u_n converges to u_0 in $L_p(\mathbb{R}^N)$ for all $p \in [1, \infty)$. This shows the convergence in part (ii) of the theorem, provided we can prove the remaining assertions.

We finally show that u_0 is a minimizer for $J(\cdot)$ on the manifold M. This follows if we can show that for any minimizer u of $J(\cdot)$ we can construct $v \in M_n$ such that $J_n(v)$ is close to $J(u)$ whenever n is large. To do so we suppose that u is a minimizer for $J(\cdot)$ on M, and define the function

$$H \colon W^1_2(\Omega) \times \mathbb{R} \to \mathbb{R}, \quad (\varphi, \delta) \mapsto G((1+\delta)u\varphi) - 1.$$

Then H is continuously differentiable, and

$$\frac{\partial}{\partial \delta} H(\mathbf{1}, 0) = \langle f(\cdot, u(\cdot)), u\rangle \neq 0$$

by Theorem 4.4. Here, $\mathbf{1}$ is the constant function with value 1. Moreover, $H(\mathbf{1}, 0) = 0$ and by the implicit function theorem there exists a neighbourhood W of $\mathbf{1}$ and $g \in C^1(W, \mathbb{R})$ such that $H(\varphi, g(\varphi)) = 0$ or equivalently $v_\varphi := (1 + g(\varphi))u\varphi \in M$ for all $\varphi \in W$. If $\varphi \in W^1_2(\mathbb{R}^N)$ is such that $\varphi = 0$ in a neighbourhood of K it follows that $G(v_\varphi) = G_n(v_\varphi) = 1$, i.e. $v_\varphi \in M_n$ for n large enough. Let $\varepsilon > 0$ be given. As K is of capacity zero we can choose such a $\varphi \in W^1_2(\mathbb{R}^N)$ with $0 \leq \varphi \leq 1$ so close to $\mathbf{1}$ in $W^1_2(\Omega)$ that

$$J(v_\varphi) \leq J(u) + \varepsilon. \tag{4.16}$$

Further, using the definition of J and J_n, we see that for large $n \in \mathbb{N}$

$$J_n(v_\varphi) \leq J(v_\varphi) + \frac{1}{2}\int_{\Gamma_1} v_\varphi(b_{0n} - b_0)\, d\mathcal{H}_{N-1} \tag{4.17}$$

The boundary integral converges to zero by the assumption on b_{0n}. (Note that $v_\varphi \in L_2(\Gamma_1)$ is zero in a neighbourhood of K.) Hence, for large $n \in \mathbb{N}$ the last term is arbitrarily small. Given $\varepsilon > 0$ we therefore have $J_n(v_\varphi) \leq J(v_\varphi) + \varepsilon$ for n large. Hence, combining (4.16) and (4.17) we see that

$$J_n(u_n) \leq J_n(v_\varphi) \leq J(u) + 2\varepsilon \tag{4.18}$$

for large $n \in \mathbb{N}$. Finally note that

$$J_n(u_n) = \frac{1}{2}\langle \lambda_n f(\cdot, u_n(\cdot)), u_n\rangle \longrightarrow \frac{1}{2}\langle \lambda f(\cdot, u_0(\cdot)), u_0\rangle = J(u_0)$$

as n goes to infinity. This shows that u_0 must be a minimizer for $J(\cdot)$ on M. This completes the proof of part (ii). Part (i) follows as for any converging subsequence of $J_n(u_n)$ the limit is the same.

It remains to prove part (iii). Suppose that $\varepsilon > 0$ is given. Then set

$$\tilde{M} = \{v \in M \colon \|u - v\|_2 \leq \varepsilon\} \quad \text{and} \quad \tilde{M}_n = \{v \in M_n \colon \|u - v\|_2 \leq \varepsilon\}$$

for all $n \in \mathbb{N}$. We choose ε so small that $\tilde{M}$ does not contain any other minimizer of J. Then, clearly $\tilde{M}_n \neq \emptyset$ if n is large. Also, $\tilde{M}_n$ and $\tilde{M}$ are weakly closed, and J_n and J are coercive and weakly lower semicontinuous. Hence, by the first part of the proof of Theorem 4.4 there exist minimizers $\tilde{u}_n$ of J_n in $\tilde{M}_n$. By choice of ε the only minimizer of J on $\tilde{M}$ is u. Note that (4.13) still holds, and by similar arguments as in the first part of the proof we find a subsequence of minimizers $\tilde{u}_n$ converging to some function $\tilde{u}_0$ weakly in $W^1_{2,\Gamma_0}(\Omega \cap U^\complement)$ and strongly in $L_2(\Omega \cap U^\complement)$ for all neighbourhoods U of K. Also, a bound similar to (4.14) holds. Hence, in a similar manner as above it follows that $\tilde{u}_n$ converges to u_0 in $L_2(\mathbb{R}^N)$, and by (4.14) and interpolation $\tilde{u}_n \to \tilde{u}_0$ in $L_p(\mathbb{R}^N)$ for all $p \in [1, N^*)$. The arguments leading to (4.18) also show that $\liminf_{n\to\infty} J_n(\tilde{u}_n) = J(u)$. It remains to show that $J(\tilde{u}_0) \leq J(u)$, and therefore $J(\tilde{u}_0) = J(u)$. As $\tilde{M}$ does only contain one minimizer we have $\tilde{u}_0 = u$. From this, it also follows that $\tilde{u}_n$ is in the interior of $\tilde{M}_n$ and thus it is a (possibly local) minimizer of J_n on M_n, and as such a solution of (4.9) for some $\lambda_n = \tilde{\lambda}_n$. Arguments similar to those in the previous paragraph show that $(\tilde{\lambda}_n, \tilde{u}_n) \to (\lambda, u)$ in $\mathbb{R} \times L_p(\mathbb{R}^N)$ for all $p \in [1, \infty)$. To show that $J(\tilde{u}_0) \leq J(u)$ we use that the Sobolev norm is weakly lower semicontinuous. Note also that due to Lemma 3.2 $\tilde{u}_n \to \tilde{u}_0$ in $L_2(\Gamma_1 \cap U^\complement)$ for any neighbourhood U of K. Hence, for any $k > 0$ and any neighbourhood U of K

$$\begin{aligned}\frac{1}{2}\Big(\int_{\Omega\cap U^{\mathsf{C}}} &|\nabla\tilde{u}_0|^2 + \mu|\tilde{u}_0|^2\,dx + \int_{\Gamma_1\cap U^{\mathsf{C}}} b_0 \min\{k,\tilde{u}_0\}^2\,d\mathcal{H}_{N-1}\Big)\\ &\leq \liminf_{n\to\infty} J_n(u_n) + \liminf_{n\to\infty}\int_{\Gamma_1\cap U^{\mathsf{C}}} (b_0 - b_{0n})\min\{k,\tilde{u}_n\}^2\,d\mathcal{H}_{N-1}\\ &\leq \liminf_{n\to\infty} J_n(u_n) + k^2 \liminf_{n\to\infty} \|b_0 - b_{0n}\|_{2,\Gamma_1\cap U^{\mathsf{C}}}\mathcal{H}_{N-1}(\Gamma_1)^{1/2}\\ &= \liminf_{n\to\infty} J_n(u_n)\end{aligned}$$

where we used the assumption on b_{0n} to get the last equality. As U and k were arbitrary this shows that $J(\tilde{u}_0) \leq \liminf_{n\to\infty} J_n(\tilde{u}_n) = J(u)$. Clearly, $J(u_0) \geq J(u)$ and thus $J(u_0) = J(u)$. This completes the proof of the theorem. □

4.9. Remark. The above theorem shows that if (4.1) has a unique energy minimizing solution then u_n converges to u in $L_p(\mathbb{R}^N)$ for all $p \in [1,\infty)$.

4.10. Remark. As mentioned in Remark 4.2 we only need a growth condition on f not its derivative in order for Theorem 4.8 to be true. More precisely, we only need to assume that

$$|f(x,\xi)| \leq c\big(1 + |\xi|^\beta\big) \tag{4.19}$$

for some $\beta \in (1,(N+2)(N-2)^{-1})$. To see this note first that Proposition 4.5 remains true under the growth condition (4.19). Taking into account Remark 4.6 Proposition 4.5 implies that any weak solution of (4.1) is in $L_\infty(\Omega)$. In particular, this is true for any minimizer u_0 of J. Suppose that $\eta > \|u_0\|_\infty$ is a fixed constant to be determined later. Then we truncate $f(x,\xi)$ for $|\xi| > \eta$ in such a way that (4.2) is satisfied. Clearly, we can make sure that the (4.19) still holds with the same constants c and β for the truncated f. Then Theorem 4.8 applies. Suppose that (λ_n, u_n) is a sequence of solutions of (4.9) converging to a solution of (4.1) such that $J_n(u_n)$ converges to $J(u)$. For any fixed $\varepsilon > 0$ we have

$$\|u_n\|_{W_2^1(\Omega_n)} \leq \frac{1}{4}\min\{1,\alpha_{1n}\}J_n(u_n) \leq \frac{1}{4}\min\{1,\alpha_1+\varepsilon\}\big(J(u)+\varepsilon\big)$$

for all n large enough. Here, α_{1n} and α_1 are the first eigenvalues of (4.10) and (3.9), respectively. We also used that α_{1n} converges to α_1 as n goes to infinity. Taking into account the assumption that the Sobolev embedding constant $c(\Omega_n)$ is bounded independently of $n \in \mathbb{N}$ the above inequality implies that

$$\|u_n\|_{N^*} \leq c(\Omega_n)\frac{1}{4}\min\{1,\alpha_1+\varepsilon\}\big(J(u)+\varepsilon\big) \leq \eta_0$$

for some η_0 independent of $n \in \mathbb{N}$ and the choice of η. By Proposition 4.5 this implies that

$$\|u_n\|_\infty \leq b(\eta_0)$$

As the truncated nonlinearity satisfies a similar growth condition as f the constant $b(\eta_0)$ does not depend on η. Hence, chosing $\eta := b(\eta_0)$ it follows that $\|u_n\|_\infty \leq \eta$

for n large enough and therefore u_n are solutions of the equation with the original nonlinearity.

5. Semilinear Problems

In this section we consider the boundary value problem

$$\begin{aligned} -\Delta u + \mu u &= f(x,u) && \text{in } \Omega \\ u &= 0 && \text{on } \Gamma_0 \\ \frac{\partial}{\partial \nu} u + b_0 u &= 0 && \text{on } \Gamma_1, \end{aligned} \tag{5.1}$$

and the corresponding perturbed problem

$$\begin{aligned} -\Delta u + \mu u &= \lambda f(x,u) && \text{in } \Omega_n \\ u &= 0 && \text{on } \Gamma_{0n} \\ \frac{\partial}{\partial \nu} u + b_0 u &= 0 && \text{on } \Gamma_{1n}. \end{aligned} \tag{5.2}$$

We throughout assume that Assumptions 3.1 and 3.4 hold. Our goal is to prove existence of solutions of (5.2) near a solution of (5.1). This can be done if u is a nondegenerate solution of (5.1). By a nondegenerate solution we mean a solution u that the linearized equation

$$\begin{aligned} -\Delta \varphi - f_u(x,u)\varphi &= 0 && \text{in } \Omega \\ \varphi &= 0 && \text{on } \Gamma_0 \\ \frac{\partial}{\partial \nu} \varphi + b_0 \varphi &= 0 && \text{on } \Gamma_1\,, \end{aligned}$$

has only the trivial solution. The following results are obtained by means of abstract arguments based on Corollary 3.6, Proposition 4.5 and the mapping properties of the substitution operator induced by the nonlinearity f. We don't want to repeat them here. For details we refer to the proof of [10, Theorem 6.1 and 6.2]. Note that, due to a better embedding theorem, in our case we have to replace N by $N/2$.

5.1. Theorem. *Suppose that u_0 is a nondegenerate solution of (5.1) and that f satisfies (4.2). If $u_0 \in L_p(\Omega)$ for some p satisfying (4.7) then, close to u_0 in $L_p(\mathbb{R}^N)$, (5.2) has a unique solution $u_n \in L_p(\Omega_n)$ if only n is large enough. In that case $u_n, u_0 \in L_\infty(\mathbb{R}^N)$ with $\|u_n\|_\infty$ being uniformly bounded, and u_n is close to u_0 in $L_q(\mathbb{R}^N)$ for all $q \in [1,\infty)$. Finally, u_n is nondegenerate if n is large enough.*

There is a modified version of this theorem in case the nonlinearity does not satisfy a growth condition.

5.2. Theorem. *Suppose that $f \in C^1(\mathbb{R}^N \times \mathbb{R})$ is arbitrary and $u_0 \in L_\infty(\Omega)$ is a nondegenerate solution of (5.1). Then, for n large enough there exists a solution u_n of (5.2) in $L_\infty(\Omega_n)$ close to u_0 in $L_q(\mathbb{R}^N)$ for all $q \in [1,\infty)$. Moreover, u_n is* *nondegenerate if n is large enough. Finally, this solution is unique among those solutions close to u_0 in $L_q(\mathbb{R}^N)$ satisfying $\|u_n\|_\infty \leq K$ for all $n \in \mathbb{N}$, where $K > \|u_0\|_\infty$ is fixed.*

Finally, it turns out that bounded sequences of solutions of the semilinear problem under consideration are close to solutions of the limiting problem.

5.3. Theorem. *Suppose that u_n are solutions of (5.2), and that $\|u_n\|_\infty$ is a bounded sequence. Then, for large n, u_n is close to some solution of (5.1) in $L_q(\mathbb{R}^N)$ for all $q \in [1, \infty)$.*

5.4. Remark. Nondegeneracy of the solution u_n is always true for the Dirichlet and Robin problem if b_0 is bounded from below by a positive constant (see [6, Theorem 1] and [10, Theorem 6.2]). For Neumann boundary conditions this is not always the case (e.g. [14])

5.5. Remark. One could also consider nonnegative solutions. Under suitable assumptions on the nonlinearity f it can be shown that the solution of (5.2) is positive if the nearby solution of (5.1) was positive. The arguments are the same as those used in [6, Theorem 2].

6. Some Examples

In this section we consider two examples, one illustrating the results from Section 4 and the other those from Section 5.

Nonlinear eigenvalue problems

First we study the equation

$$\begin{aligned} -\Delta u &= \lambda u|u|^{\beta-1} && \text{in } \Omega \\ u &= 0 && \text{on } \partial\Omega, \end{aligned} \tag{6.1}$$

where $\beta \in (1, (N+2)(N-2)^{-1})$, and Ω is a bounded domain. We consider the case where small holes are cut into Ω, generalizing some results of Ozawa and Roppongi [24]. Let $K_0 \subset \Omega$ be a compact set of capacity zero, and H_n sets shrinking to K_0 in the sense of Assumption 2.3. As in Section 2 we set $\Omega_n := \Omega \setminus H_n$. By Proposition 2.4 and Theorem 2.8 the Sobolev embedding constants $c(\Omega_n, \partial\Omega)$ are bounded independently of $n \in \mathbb{N}$ if either of the two following conditions are satisfied:

- H_n satisfies an exterior cone condition with a cone independent of $n \in \mathbb{N}$.
- $K_0 = \{x_1, \dots, x_m\}$ is a set of points, and for each of these points Assumption 2.6 holds after a suitable translation and rescaling.

Note that we could also consider a combination of the two cases. In both cases Assumptions 3.1 and 3.4 are satisfied. We then want to investigate the behaviour of solutions (λ, u) of

$$\begin{aligned} -\Delta u &= \lambda u|u|^{\beta-1} && \text{in } \Omega_n \\ u &= 0 && \text{on } \partial\Omega, \\ \frac{\partial}{\partial\nu}u + b_0 u &= 0 && \text{on } \partial H_n \end{aligned} \tag{6.2}$$

as n goes to infinity, where $b_0 \geq 0$ for simplicity is a constant. In [24] H_n is a ball shrinking to a point. For $N \geq 3$ we can for instance remove a cylinder with circular base shrinking to a line segment, that is, we can drill small holes through Ω. We can also remove holes of a fixed shape and shrink them to one point.

In the above situation the results from Section 4 apply. In this case we have

$$G(u) = \frac{1}{\beta+1}\|u\|_{\beta+1}^{\beta+1}$$

and the manifolds M_n the energy functional is minimized over is given by

$$M_n = \{u \in W_2^1(\Omega_n) \colon (\beta+1)^{-1}\|u\|_{\beta+1}^{\beta+1} = 1\}.$$

Finally, the energy functionals are given by

$$J_n(u) = \frac{1}{2}\Big(\int_{\Omega_n} |\nabla u|^2\, dx + \int_{\partial H_n} b_0|u|^2\, d\mathcal{H}_{N-1}\Big)$$

and

$$J(u) := \frac{1}{2}\int_\Omega |\nabla u|^2\, dx,$$

respectively. If we are only interested in positive solutions note that Remark 4.3 applies. Our first result considerably generalizes [24, Theorem 1]. It shows that there is a bound on all energy minimizing solutions of (6.2) uniformly with respect to $n \in \mathbb{N}$.

6.1. Theorem. *Let S_n be the set of energy minimizing solutions of* (6.2). *Then,*

$$\sup_{n\in\mathbb{N}}\sup_{u\in S_n}\|u\|_\infty < \infty$$

Proof. Let (λ_n, u_n) be an energy minimizing solution of (6.2). As u_n is a weak solution of (6.2) we have that

$$\lambda_n = \frac{\lambda_n}{\beta+1}\|u_n\|_{\beta+1}^{\beta+1} = \frac{\lambda_n}{\beta+1}\langle u|u|^\beta, u\rangle = \frac{2}{\beta+1}J_n(u_n).$$

Note that $J_n(u_n)$ is bounded independently of $n \in \mathbb{N}$ as shown in (4.13). On the other hand, (4.13) and the uniform bound on the Sobolev embedding constant make sure that $\|u_n\|_{N^*}$ is bounded independently of $n \in \mathbb{N}$. Combining both and invoking Proposition 4.5 we get a bound on $\|u_n\|_\infty$ not depending on n or the particular minimizer of $J_n(\cdot)$. This completes the proof of the theorem. □

According to Theorem 4.4 and 4.8 there always exists a sequence of energy minimizing solutions (λ_n, u_n) of (6.2) converging to an energy minimizing solution (λ_0, u_0) of (6.1). The following result complements those in [24]. We do not get the asymptotics, and u_n is shown to converge in a weaker sense, but on the other hand the holes under consideration are much more general, and we can deal with any dimension $N \geq 2$.

6.2. Theorem. *We have that (λ_n, u_n) converges to (λ_0, u_0) in $L_p(\mathbb{R}^N)$ for all $p \in [1, \infty)$. If Ω is smooth, and U is any neighbourhood of K_0 then u_n tends to u_0 in $C(\overline{\Omega \setminus U})$ as n goes to infinity.*

Proof. The first assertion directly follows from Theorem 4.8, we only need to show uniform convergence on $\overline{\Omega \setminus U}$ for any given neighbourhood U of K_0. By the above theorem we have uniform bounds on $\|u_n\|_\infty$. Using interior Hölder regularity of solutions of elliptic equations (e.g. [19, Section 3.14]) it follows that u_n is Hölder continuous with exponent $\gamma \in (0,1)$ independent of $n \in \mathbb{N}$, and u_n is bounded in the Hölder space $C^\gamma(\overline{\Omega \setminus U})$. As the embedding $C^\gamma(\overline{\Omega \setminus U}) \hookrightarrow C(\overline{\Omega \setminus U})$ is compact we get uniform convergence of u_n on $\overline{\Omega \setminus U}$. (We know that u_n converges in a weaker norm already.) □

Semilinear equations

Next we indicate how the results in Section 5 can be used to construct domains such that the boundary value problem

$$\begin{aligned} -\Delta u &= f(u) && \text{in } \Omega \\ \frac{\partial}{\partial \nu} u + b_0 u &= 0 && \text{on } \partial\Omega, \end{aligned} \tag{6.3}$$

has multiple solutions. Here as usual $b_0 \geq 0$ and Ω is a bounded open set, not necessarily connected, satisfying Assumption 3.1. The results we present are closely related to those in [7, Section 4] and [8]. In these papers it was assumed that $f(\xi)\xi < 0$ for sufficiently large values of ξ, leading to a L_∞-bound for all solutions of (6.3). We drop this condition, but in turn use a more restrictive notion of domain convergence. We suppose that Ω is the union of two touching balls. Clearly Assumption 3.1 is satisfied in this case. It is now clearly possible to join the two balls to get a sequence of domains Ω_n satisfying a cone condition with a cone independent of $n \in \mathbb{N}$, and that Ω_n converges to Ω in the sense of Assumption 3.4. We can even make sure that Ω_n is star shaped. Note that by Lemma 2.2 the Sobolev embedding constant $c(\Omega_n)$ is bounded independently of $n \in \mathbb{N}$. Hence the results from Section 5 apply. Suppose now that, for Ω a ball, (6.3) has two nondegenerate solutions. In many cases we have the trivial and a nontrivial solution. Then, by combining the different solutions we get four nondegenerate solutions of (6.3) if Ω is the union of two touching balls. Under the assumptions of Theorem 5.1 and 5.2 we get at least four nondegenerate solutions on Ω_n, one near each of the solutions on Ω. Hence there are more solutions than on the ball, even if Ω_n is star shaped.

References

[1] Ralph Abraham, Jerrold E. Marsden, and Tudor Ratiu, *Manifolds, Tensor Analysis and Applications*, Addison Wesley, Reading, Massachusetts, 1983.

[2] Robert A. Adams, *Sobolev Spaces*, Academic Press, New York, 1975.

[3] J. Appell and P.P. Zabrejko, *Nonlinear Superposition Operators*, Cambridge University Press, Cambridge, 1990.

[4] J.M. Arrieta, "Neumann eigenvalue problems on exterior perturbations of the domain", J. Differential Equations **118** (1995), 54–103.

[5] Dorin Bucur and Jean-Paul Zolésio, "Continuité par rapport au domaine dans le problème de Neumann", C. R. Acad. Sci. Paris Sér. I Math. **319** (1994), 57–60.

[6] E.N. Dancer, "The effect of domain shape on the number of positive solutions of certain nonlinear equations", J. Differential Equations **74** (1988), 120–156.

[7] ———, "The effect of domain shape on the number of positive solutions of certain nonlinear equations, II", J. Differential Equations **87** (1990), 316–339.

[8] ———, "Addendum to [7]", J. Differential Equations **97** (1992), 379–381.

[9] E.N. Dancer and D. Daners, *A priori bounds for solutions and spectra of Robin boundary value problems*, Tech. report, University of Sydney, 1994.

[10] ———, "Domain perturbation for elliptic equations subject to Robin boundary conditions", J. Differential Equations **138** (1997), 86–132.

[11] Daniel Daners, "Robin boundary value problems on arbitrary domains", Trans. Amer. Math. Soc., to appear.

[12] ———, "Domain Perturbation for Linear and Nonlinear Parabolic Equations", J. Differential Equations **129** (1996), 358–402.

[13] D. Gilbarg and N.S. Trudinger, *Elliptic Partial Differential Equations of Second Order*, 2nd ed., Springer, Berlin, 1983.

[14] Shuichi Jimbo, "Singular perturbation of domains and semilinear elliptic equations", J. Fac. Sci. Univ. Tokyo Sect. IA Math. **35** (1988), 27–76.

[15] ———, "Singular perturbation of domains and semilinear elliptic equations, II", J. Differential Equations **75** (1988), 264–289.

[16] ———, "The singularly perturbed domain and the characterization for the eigenfunctions with Neumann boundary condition", J. Differential Equations **77** (1989), 322–350.

[17] ———, "Erratum to [16]", J. Differential Equations **84** (1990), 204.

[18] T. Kato, *Perturbation Theory for Linear Operators*, Springer, Berlin, 1966.

[19] O.A. Ladyženskaja and N.N. Ural'ceva, *Linear and Quasilinear Elliptic Equations*, Academic Press, New York, 1968.

[20] John L. Lewis, "Approximation of Sobolev functions in Jordan domains", Ark. Mat. **25** (1987), 255–264.

[21] V.G. Maz'ja and S.V. Poborchi, "Extension of Functions in Sobolev Spaces on Parameter Dependent Domains", Math. Nachr. **178** (1996), 5–45.

[22] Jindřich Nečas, *Les Méthodes Directes en Théorie des Équations Elliptiques*, Academia, Prague, 1967.

[23] Shin Ozawa, "L^∞ boundedness of nonlinear eigenfunction under singular variation of domains", Osaka J. Math. **32** (1995), 363–371.

[24] Shin Ozawa and Susumu Roppongi, "Singular variation of domains and continuity property of eigenfunction for some semi-linear elliptic equations", Kodai Math. J. **18** (1995), 315–327.

[25] Michael Struwe, *Variational Methods*, Springer, Berlin, 1996.

Daniel Daners
Department of Mathematics
Brigham Young University
Provo, Utah 84602
daners@math.byu.edu

Progress in Nonlinear Differential Equations
and Their Applications, Vol. 35
© 1999 Birkhäuser Verlag Basel/Switzerland

Singular Semilinear Elliptic Equations in $L^1(\mathbb{R}^N)$

E. DiBenedetto* and D.J. Diller†

1. Introduction and Main Results

Consider the singular semilinear elliptic equation in $L^1(\mathbb{R}^N)$,

$$\lambda u - \Delta \ln u = f, \qquad \lambda > 0, \tag{1.1}$$

where $f \in L^1_{\text{loc}}(\mathbb{R}^N)$ is non-negative. We say that (u, f) is a solution pair to (1.1), if $u : \mathbb{R}^N \to \mathbb{R}$ is non-negative, u and $\ln u$ are in $L^1_{\text{loc}}(\mathbb{R}^N)$ and (1.1) holds in $\mathcal{D}'(\mathbb{R}^N)$. The main results of this note are:

Theorem 1.1 (Existence). *Let $N = 2$. Then (1.1) has a solution if and only if*

$$\|f\|_{1,\mathbb{R}^2} > 4\pi. \tag{1.2}$$

Theorem 1.2 (Lack of Uniqueness). *To every such non-negative f there corresponds a continuum of solutions u_s. Precisely, let $s_o > 1$ be defined by*

$$\|f\|_{1,\mathbb{R}^2} = 4\pi s_o.$$

Then for each $s \in [1, s_o)$ there exists a unique solution u_s to (1.1), satisfying

$$\lambda \|u_s\|_{1,\mathbb{R}^2} = \|f\|_{1,\mathbb{R}^2} - 4\pi s. \tag{1.3$_s$}$$

The family $\{u_s\}_{s\in[1,s_o)}$ is decreasing, i.e.,

$$u_s(x) \ge u_{s'}(x) \ \text{ for a.e. } x \in \mathbb{R}^2, \ \text{ whenever } s \le s'. \tag{1.4}$$

Theorem 1.3 (Behavior as $|x| \to \infty$). *The solution corresponding to $s = 1$ is maximal in the sense that it is the pointwise largest solution. Moreover, maximal solutions are characterized by the asymptotic behavior*

$$u(x) \ge \frac{C_o}{|x|^2 \ln^2 |x|}, \qquad \forall\, |x| \ge C_1, \tag{1.5}$$

for two positive constants C_o and $C_1 \ll 1$. That is, if $x \to u(x)$ is a maximal solution to (1.1), then there exist two positive constants C_o and C_1 that can be determined a priori only in terms of $\|f\|_{1,\mathbb{R}^2}$ such that (1.5) holds. Vice versa, if $x \to u(x)$ is a solution of (1.1) satisfying (1.5) for two positive constants C_o and C_1, then u is maximal.

* Partially supported by NSF Grant DMS–9706388.
† Partially supported by Sloan-DOE grant 97-2-5 CMB..

From Theorem 1.1 it follows that if $f = 0$, then (1.1) has no solution. The first non-existence results, for $f \leq 0$ are due to Osserman [20] and Oleinik [19]. Theorems 1.1–1.3 continue to hold if f is of variable sign, provided the positive part of f is sufficiently large in the sense

$$\int_{\mathbb{R}^2} f(x)dx > 4\pi. \tag{1.2$'$}$$

This can be interpreted, roughly speaking, as a measure of the size of the positive part of f for (1.1) to be solvable. To keep the presentation as streamlined as possible, we have chosen to present our results for non-negative f.

This investigation is part of the theory of elliptic singular, semilinear equations in $\mathbb{R}^N$ as originated in [2, 3, 4, 6, 10]. In particular, we give necessary and sufficient conditions of solvability, at least for $N = 2$, for the degenerate limiting cases left open in [2]. For (1.1) set in $\mathbb{R}^N$ for any $N \geq 3$, we establish a variety of a priori estimates. This study is also motivated by the corresponding parabolic problems and questions of thin film dynamics [5, 12, 13, 23]. We will discuss these issues in Sections 2 and 3.

It turns out, however, that Theorems 1.1–1.3 have also a geometric connection, which has developed in its own independent direction. In Appendix A we will discuss such a geometric aspect, the main available results and their relation to ours.

2. Some General Remarks about (1.1)

By setting $v = \ln u$, (1.1) takes the form

$$\beta(v) - \Delta v \ni f \quad \text{in } \mathbb{R}^N, \tag{1.1$'$}$$

where $\beta(v) = e^v$. Beside their intrinsic mathematical interest, solutions of (1.1)$'$ serve to generate, by semigroup methods, solutions of the corresponding parabolic Cauchy problem

$$u_t - \Delta \ln u = 0 \quad \text{in } \mathbb{R}^N \times \mathbb{R}^+, \quad u(\cdot, 0) = u_o \in L^1(\mathbb{R}^N). \tag{2.1}$$

This evolution problem arises in thin film dynamics and in the Ricci flow in complete $\mathbb{R}^2$ (see [13] and references therein). For $N = 2$, solutions to such a Cauchy problem have been constructed in [13], by means of the indicated generation technique. The construction is based on the solvability of (1.1) as established in Theorems 1.1–1.3. This motivates the investigation of (1.1).

More generally the term $\beta(\cdot)$ in (1.1)$'$ could be a maximal monotone graph in $\mathbb{R} \times \mathbb{R}$, and the differential inclusion means that $\{f - \Delta v\}$ is included in the set $\beta(v)$. In this setting (1.1)$'$ has been studied in [2,3,10] provided $0 \in \beta(0)$. It is shown in [2] that for $N \geq 3$, every $f \in L^1(\mathbb{R}^N)$ yields a unique solution to (1.1)$'$. If $N = 1, 2$ then every $f \in L^1(\mathbb{R}^N)$ generates a solution to (1.1)$'$ provided $\beta(\cdot)$ satisfies the coercivity condition

$$0 \in \{\text{interior of the set}\}\beta(0).$$

In addition, such solutions are unique up to a constant. In the case $\beta(v) = e^v$, the required condition fails. Moreover solutions of $(1.1)'$, for such a $\beta(\cdot)$ are not unique not even up to constants.

In the next section we will exhibit two explicit solutions of (1.1) with dramatically different behavior at infinity. In view of Theorem 1.3, it appears that the asymptotic behavior as $|x| \to \infty$ is crucial in identifying classes of uniqueness. This is in contrast with the case

$$\lambda u - \Delta u^m = f \text{ in } \mathbb{R}^N; \quad \beta(v) = v^{1/m},\ m > 0. \tag{$(1.1)_m$}$$

In [3] Brézis shows that, for such a $\beta(\cdot)$ the behavior at infinity is immaterial and $(1.1)_m$ has a *unique* solution for every non-negative $f \in L^1_{\text{loc}}(\mathbb{R}^N)$.[1] The strong role played by the asymptotic behavior in (1.1) and the absence of such a role in $(1.1)_m$ is somewhat surprising since (1.1) can be regarded, at least formally, as a limiting case of $(1.1)_m$ as $m \searrow 0$. Indeed the change of variables

$$\xi = \frac{x}{\sqrt{m}}, \qquad w(\xi) = u(\sqrt{m}\,\xi),$$

transforms $(1.1)_m$ into

$$\lambda w - \Delta_\xi \frac{w^m - 1}{m} = f \quad \longrightarrow \quad \lambda w - \Delta \ln w = f \quad \text{formally as } m \searrow 0.$$

To our knowledge the singular equation (1.1) has not been systematically investigated in the literature. We are only aware of an existence result in the case of one space dimension (see [10]) going back to the mid '70s. The result is based on some a priori estimates possible only for $N = 1$.

In Sections 5–8 we present a spectrum of a priori estimates valid in every dimension $N \geq 1$. Thus the difficulty of going from $N = 1$ to $N = 2$ is not one of compactness. It is rather due to the geometrical limitations indicated in the Appendix.

These remarks raise the question of a forcing term f of variable sign and that of the space dimension $N \geq 3$.

For $N = 2$ and f of variable sign Theorems 1.1–1.3 continue to hold, provided the solvability condition (1.2) is replaced by $(1.2)'$. Thus the negative part f^- is required to have relatively small mass when compared to f^+. We have chosen to present our main results for $N = 2$ and $f \geq 0$.

For $N \geq 3$, we have collected in Part IV, sharp non-existence results in terms of the behavior of f as $|x| \to \infty$, as well as some existence results. The theory however is far from as precise as for $N = 2$ and will be the object of further investigations.

[1] The result of Brézis includes general $f \in L^1(\mathbb{R}^N)$ of variable sign and corresponding solutions of variable sign.

3. Remarks on Rescaling and Uniqueness

If (u, f) is a solution to (1.1), so are the pairs $(u_\varepsilon, f_\varepsilon)$, where

$$u_\varepsilon(x) = \varepsilon u(\sqrt{\varepsilon}x), \qquad f_\varepsilon(x) = \varepsilon f(\sqrt{\varepsilon}x); \qquad \varepsilon > 0.$$

If $f \in L^1(\mathbb{R}^N)$, then $\|f_\varepsilon\|_{1,\mathbb{R}^N} = \varepsilon^{1-N/2}\|f\|_{1,\mathbb{R}^N}$. Therefore if $N \neq 2$ and (1.1) has a solution (u, f) with $f \in L^1(\mathbb{R}^N)$, then it also has a solution (v, g) with $\|g\|_{1,\mathbb{R}^N}$ arbitrarily small. No such phenomenon occurs for $N = 2$ and makes Theorem 1.1 possible.

3.1. Constructing Solutions and Lack of Uniqueness

The pair $u(x) = e^{-|x|^2}$ and $f(x) = 2N + \lambda e^{-|x|^2}$ is a solution to (1.1). However for the same f one can construct a solution v such that $v(x) \geq 2N/\lambda$, $\forall x \in \mathbb{R}^N$. For $n \in \mathbb{N}$, let v_n be the unique classical solution of

$$\lambda v_n - \Delta \ln v_n = 2N + \lambda e^{-|x|^2} \quad \text{in } \{|x| < n\}, \qquad \text{and } v_n \Big|_{|x|=n} = \frac{2N}{\lambda}.$$

By classical estimates one finds

$$v_n \geq \frac{2N}{\lambda} \quad \text{in} \quad \{|x| < n\} \quad \text{for all } n \in \mathbb{N}.$$

Thus a standard limiting process produces a solution v everywhere bounded below by $2N/\lambda$. More generally, for any non-negative $f \in L^1_{\text{loc}}(\mathbb{R}^N)$ and every $\varepsilon > 0$, one can find a solution pair to (1.1),

$$(v_\varepsilon f + \lambda\varepsilon) \quad \text{such that} \quad v_\varepsilon > \varepsilon \text{ in } \mathbb{R}^N.$$

This example exhibits the lack of uniqueness and raises the issue of a *maximal* solution to (1.1) associated to a given f. Such a solution is, roughly speaking, the largest possible solution associated to a given f. By their definition maximal solutions are unique. In Section 11 we give a criterion for a solution to be maximal.

The argument given above indicates that existence of solutions to (1.1) hinges on some uniform local lower bound on a suitable sequence of approximating solutions. We realize this by means of a Harnack-type estimate introduced in Section 7.

4. Some Explicit Solutions

4.1. Example I. On the Behavior of Solutions as $|x| \to \infty$

For $M > 0$ and parameters μ, $s > 0$, the following is a two-parameter family of solution pairs to (1.1),

$$w_{s,\mu}(x) = \frac{M}{(\mu+|x|^2)^s}; \qquad \phi_{s,\mu}(x) = w_{s,\mu} + \frac{4s\mu}{(\mu+|x|^2)^2}. \tag{4.1}$$

If $s > 1$ then both $w_{s,\mu}$ and $\phi_{s,\mu}$ are in $L^1(\mathbb{R}^2)$, and by direct computation one verifies that

$$\int_{\mathbb{R}^2} \{\phi_{s,\mu}(x) - \lambda w_{s,\mu}(x)\}\, dx = 4\pi s. \tag{4.2}$$

4.2. Example-II. On the Critical Behavior of Maximal Solutions

For a positive constant M and parameters $\mu > 1$ and $\alpha \in (1, 2]$, set

$$v_{\alpha,\mu}(x) = \frac{M}{(\mu+|x|^2)\,[\ln(\mu+|x|^2)]^\alpha}. \tag{4.3}$$

$$\begin{aligned}\varphi_{\alpha,\mu}(x) = \lambda v_{\alpha,\mu}(x) + \frac{4\mu}{(\mu+|x|^2)^2} + \frac{4\alpha\mu}{(\mu+|x|^2)^2\ln(\mu+|x|^2)} \\ - \frac{4\alpha|x|^2}{(\mu+|x|^2)^2\,[\ln(\mu+|x|^2)]^2}.\end{aligned} \tag{4.4}$$

By direct calculations one verifies that $(v_{\alpha,\mu}, \varphi_{\alpha,\mu})$ is a family of solution pairs, parametrized with $\mu > 1$ and $\alpha \in (1, 2]$. Moreover, for all $\alpha \in (1, 2]$, all $M > 0$ and all $\mu > 1$,

$$\int_{\mathbb{R}^2} \{\varphi_{\alpha,\mu}(x) - \lambda v_{\alpha,\mu}(x)\}\,dx = 4\pi. \tag{4.5}$$

Remark 4.1. If $\alpha > 1$, both $v_{\alpha,\mu}$ and $\varphi_{\alpha,\mu}$ are integrable in $\mathbb{R}^2$. If $\alpha \le 2$ and λM is sufficiently large then $\varphi_{\alpha,\mu} \ge 0$ in $\mathbb{R}^2$. If $\alpha > 2$, then for sufficiently large $|x|$, the forcing term $\varphi_{\alpha,\mu}$ is negative. This motivates the range $(1, 2]$ of the parameter α.

PART I: Sup-Estimates and Harnack-Type Inequalities

5. Some a Priori Estimates

In this section we assume that we have a classical solution of (1.1) and derive some a priori estimates for it. These in turn will enable us to construct weak solutions to (1.1). For $\rho > 0$ we let $B_\rho(x_o)$ denote the ball of radius ρ about x_o, and if x_o coincides with the origin we let $B_\rho(0) = B_\rho$.

Proposition 5.1. *Assume that $f \in L^r_{loc}(\mathbb{R}^N)$ for some $r > \max\{1; N/2\}$. There exists a constant γ depending only upon r, such that for every $x_o \in \mathbb{R}^N$ and every $\rho > 0$,*

$$\lambda\|u\|_{r,B_\rho(x_o)} \le \gamma\left(\|f\|_{r,B_{2\rho(x_o)}} + \rho^{\frac{N-2r}{r}}\right) \tag{5.1}$$

Proof. Without loss of generality we may assume that $x_o = 0$. Let ζ denote a non-negative smooth cutoff function from $B_{\rho(1+\sigma)}$ to B_ρ, for some $\sigma \in (0, 1)$, satisfying

$$\begin{aligned}\zeta \in (0,1) \text{ in } B_{\rho(1+\sigma)},\ \zeta = 1 \text{ in } B_\rho,\ \zeta\,\Big|_{\partial B_{\rho(1+\sigma)}} = 0; \\ |D\zeta|^2,\ |\Delta\zeta| \le \gamma/(\sigma\rho)^2 \text{ in } B_{\rho(1+\sigma)}\end{aligned} \tag{5.2}$$

for a positive constant γ. Multiply (1.1) by $u^{r-1}\zeta^2$, and integrate over $B_{\rho(1+\sigma)}$ to obtain, after standard calculations

$$\begin{aligned}\lambda\int_{B_{\rho(1+\sigma)}} u^r\zeta^2\,dx \,+\, (r-1)\int_{B_{\rho(1+\sigma)}} |Du|^2 u^{r-3}\zeta^2\,dx \\ = \frac{1}{r-1}\int_{B_{\rho(1+\sigma)}} Du^{r-1}\cdot D\zeta^2\,dx + \int_{B_{\rho(1+\sigma)}} fu^{r-1}\zeta^2\,dx \\ \le \frac{1}{r-1}\int_{B_{\rho(1+\sigma)}} u^{r-1}|\Delta\zeta^2|\,dx + \int_{B_{\rho(1+\sigma)}} fu^{r-1}\zeta^2\,dx \\ \le \left\{\frac{\gamma\,\rho^{\frac{N}{r}}}{(\sigma\rho)^2} + \|f\|_{r,B_{2\rho}}\right\}\|u\|^{r-1}_{r,B_{\rho(1+\sigma)}}\,. \qquad (5.3)\end{aligned}$$

For $n = 0, 1, 2, \dots$, define the positive numbers ρ_n, σ_n and M_n, by

$$\rho_n = \rho\sum_{i=0}^{n} 2^{-i}; \qquad \rho_{n+1} = \rho_n(1+\sigma_n); \qquad M_n = \int_{B_{\rho_n}} u^r\,dx.$$

Then from (5.3) and Young's inequality we derive, for all $\varepsilon > 0$,

$$\begin{aligned}M_n &\le \frac{\gamma 4^n}{\lambda}\left\{\|f\|_{r,B_{2\rho}} + \rho^{\frac{N-2r}{r}}\right\} M_{n+1}^{\frac{r-1}{r}} \\ &\le \varepsilon M_{n+1} + \frac{\gamma(\varepsilon,r)b^n}{\lambda^r}\left\{\|f\|_{r,B_{2\rho}} + \rho^{\frac{N-2r}{r}}\right\}^r, \qquad b = 4^r,\end{aligned}$$

for all $n = 0, 1, 2, \dots$. From this by iteration

$$M_n \le \varepsilon^n M_{n+1} + \gamma(\varepsilon,r)\lambda^{-r}\left\{\|f\|_{r,B_{2\rho}} + \rho^{\frac{N-2r}{r}}\right\}^r \sum_{i=0}^{n-1}(b\varepsilon)^i.$$

To prove (5.1) we choose $\varepsilon < 1/b$ and let $n \to \infty$. □

Corollary 5.1. *If $f \in L^r(\mathbb{R}^N)$ for some $r > N/2$, then $u \in L^r(\mathbb{R}^N)$, and*

$$\lambda\|u\|_{r,\mathbb{R}^N} \le \|f\|_{r,\mathbb{R}^N}\,. \qquad (5.4)$$

Proposition 5.2. *For every $\alpha, \delta \in (0,1)$, there exists a constant γ depending only upon N, α, and δ, such that for all $x_o \in \mathbb{R}^N$, and all $\rho, \eta > 0$,*

$$\lambda\int_{B_\rho(x_o)} u\chi[u>\eta]\,dx \le (1+\delta)\int_{B_{2\rho}(x_o)} f\chi[u>\eta]dx + \frac{\gamma\rho^{N-\frac{2}{1-\alpha}}}{(\lambda\eta)^{\frac{\alpha}{1-\alpha}}}. \qquad (5.5)$$

Proof. As before we may assume that $x_o = 0$. Let ζ be as in (5.2), and let $S_n(\cdot) : \mathbb{R} \to [0,1]$ be an approximation of the sign$(\cdot)$ function defined by

$$S_n(s) = \begin{cases} 1 & \text{for } s \ge 1/n; \\ sn & \text{for } 0 \le s \le 1/n; \\ 0 & \text{for } s \le 0. \end{cases}$$

Multiply (1.1) by $\zeta^2 S_n(\ln u - \ln\eta)$, integrate by parts over $B_{\rho(1+\sigma)}$, and let $n\to\infty$ to obtain

$$\lambda\int_{B_\rho} u\chi[u>\eta]\,dx \le \frac{\gamma}{(\sigma\rho)^2}\int_{B_{\rho(1+\sigma)}}(\ln u-\ln\eta)_+dx + \int_{B_{\rho(1+\sigma)}} f\chi[u>\eta]\,dx.$$

For all $\alpha\in(0,1)$, there exists a constant γ depending only upon α, such that

$$(\ln u-\ln\eta)_+ = \ln\left(1+\frac{(u-\eta)_+}{\eta}\right)\le\gamma\frac{(u-\eta)_+^\alpha}{\eta^\alpha}.$$

Combining this with the previous integral estimate gives

$$\lambda\int_{B_\rho}u\chi[u>\eta]dx \le \frac{\gamma\,\rho^{N(1-\alpha)}}{(\lambda\eta)^\alpha(\sigma\rho)^2}\left(\lambda\int_{B_{\rho(1+\sigma)}}u\chi[u>\eta]dx\right)^\alpha + \int_{B_{2\rho}}f\chi[u>\eta]dx. \tag{5.6}$$

Introduce the sequences ρ_n, σ_n as in the proof of Proposition 5.1, and set

$$M_n=\lambda\int_{B_{\rho_n}}u\chi[u>\eta]\,dx;\quad A=\frac{(4\varepsilon^\alpha\gamma)^{\frac{1}{1-\alpha}}\rho^{N-\frac{2}{1-\alpha}}}{(\lambda\eta)^{\frac{\alpha}{1-\alpha}}};\quad B=\int_{B_{2\rho}}f\chi[u>\eta]\,dx,$$

where $\varepsilon>0$ is to be chosen. Then (5.6) and Young's inequality imply the recursive inequalities

$$M_n\le\varepsilon M_{n+1}+Ab^n+B,\qquad n=0,1,2,\dots;\qquad b=4^{\frac{1}{1-\alpha}}.$$

From this by iteration we derive

$$\int_{B_\rho}u\chi[u>\eta] = M_o\le\frac{A}{\varepsilon}\sum_{i=0}^\infty(\varepsilon b)^i + B\sum_{i=0}^\infty\varepsilon^i.$$

The estimate follows by choosing ε sufficiently small. □

Corollary 5.2. *If $f\in L^1(\mathbb{R}^N)$, then also $u\in L^1(\mathbb{R}^N)$ and*

$$\lambda\|u\|_{1,\mathbb{R}^N}\le\|f\|_{1,\mathbb{R}^N}.$$

Proof. In (5.5) choose α so close to one that the exponent of ρ is negative. Then let $\rho\to\infty$, and make use of the arbitrariness of δ. □

6. Sup-Estimates

Theorem 6.1. *Assume that $f\in L^r_{\text{loc}}(\mathbb{R}^N)$ for some $r>\max\{1;N/2\}$. Then there exists a constant γ depending only upon r and N such that for all $x_o\in\mathbb{R}^N$,*

$$\sup_{B_\rho(x_o)}u\le\frac{\gamma}{\lambda}\left\{\|f\|_{r,B_{2\rho}(x_o)}+\rho^{\frac{N-2r}{r}}\right\}^{\frac{2r}{2r-N}},\qquad\textit{for all }\rho>0. \tag{6.1}$$

Furthermore if for some $r>\max\{1;N/2\}$,

$$|||f|||_r=\sup_{\rho>1}\fint_{B_\rho}\left[|x|^2f(x)\right]^r dx<\infty, \tag{6.2}$$

then, there exists a constant γ depending only upon N and r, such that

$$|x|^2 u(x) \le \gamma \left(1 + \|f\|_r^{\frac{2}{2r-N}}\right), \qquad \textit{for all } x \in \mathbb{R}^N. \tag{6.3}$$

Proof. If u is a solution of (1.1) then $w = \lambda u$ is a solution of (1.1) with $\lambda = 1$. Thus without loss of generality we may assume that $\lambda = 1$ and that $x_o = 0$. For $\sigma \in (0,1)$, $\rho > 0$ and $n = 0, 1, 2 \ldots$, define

$$\rho_n = \rho\left(1 + \sigma - 2^{-n}\sigma\right), \qquad B_n = B_{\rho_n}, \qquad k_n = k\left(1 - 2^{-n}\right) \tag{6.4}$$

where $k > 0$ is to be chosen. Multiply (1.1), with $\lambda = 1$, by $(u - k_{n+1})_+^{r-1} \zeta_n^2$, and integrate by parts over B_n. Here ζ_n is the cutoff function from B_n to B_{n+1} introduced in (5.2), with the proper redefinition of σ. Standard calculations give

$$\begin{aligned}
\int_{B_n} \left[(u - k_{n+1})_+^{\frac{r}{2}} \zeta_n\right]^2 dx + \frac{1}{\sup_{B_o} u} \int_{B_n} \left|D\left[(u - k_{n+1})_+^{\frac{r}{2}} \zeta_n\right]\right|^2 dx& \\
\le \frac{\gamma 4^n}{(\sigma\rho)^2 k} \int_{B_n} (u - k_n)_+^r \, dx + \int_{B_n} f (u - k_n)_+^{r-1} dx& \\
\le \frac{\gamma 4^n}{\sigma^2} \left\{ \frac{1}{\rho^2 k} \|u\|_{r,B_o} + \|f\|_{r,B_o} \right\} \left(\int_{B_n} (u - k_n)_+^r \, dx \right)^{\frac{r-1}{r}}&.
\end{aligned}$$

Next we take $k \ge \rho^{-2}$ and apply the estimate of Proposition 5.1. By setting

$$M = \sup_{B_o} u; \qquad \mathcal{A} = \left\{ \|f\|_{r,B_o} + \rho^{\frac{N-2r}{r}} \right\}; \qquad Y_n = \int_{B_n} (u - k_n)_+^r \, dx,$$

we arrive at the integral inequalities

$$\int_{B_n} \left[(u - k_{n+1})_+^{\frac{r}{2}} \zeta_n\right]^2 dx + \frac{1}{M} \int_{B_n} \left|D\left[(u - k_{n+1})_+^{\frac{r}{2}} \zeta_n\right]\right|^2 dx \le \frac{\gamma 4^n}{\sigma^2} \mathcal{A} Y_n^{\frac{r-1}{r}}. \tag{6.5}$$

In the remainder of the proof we assume first that $N \ge 3$. By Hölder's inequality and the Sobolev embedding Theorem,

$$\begin{aligned}
Y_{n+1} &\le \int_{B_n} \left[(u - k_{n+1})_+^{\frac{r}{2}} \zeta_n\right]^2 dx \\
&\le \left(\int_{B_n} \left[(u - k_{n+1})_+^{\frac{r}{2}} \zeta_n\right]^{\frac{2N}{N-2}} dx \right)^{\frac{N-2}{N}} |[u > k_{n+1}] \cap B_n|^{\frac{2}{N}} \\
&\le \gamma \left(\int_{B_n} \left|D\left[(u - k_{n+1})_+^{\frac{r}{2}} \zeta_n\right]\right|^2 dx \right) |[u > k_{n+1}] \cap B_n|^{\frac{2}{N}} \\
&\le \gamma 4^n \sigma^{-2} M \mathcal{A} Y_n^{1 - \frac{1}{r}} |[u > k_{n+1}] \cap B_n|^{\frac{2}{N}}.
\end{aligned}$$

Next we observe that

$$|[u > k_{n+1}] \cap B_n| \le \frac{\gamma 4^{rn}}{k^r} Y_n,$$

so that the previous inequality takes the form

$$Y_{n+1} \le \frac{\gamma b^n M}{\sigma^2 k^{2r/N}} \mathcal{A} Y_n^{1-\frac{1}{r}+\frac{2}{N}}; \qquad b = 4^{\frac{N+2}{N}}.$$

Since $(\frac{2}{N} - \frac{1}{r}) > 0$, it follows from these recursive inequalities that $Y_n \to 0$ as $n \to \infty$ if

$$k \ge \frac{\gamma M^{\frac{N}{2r}}}{\sigma^{\frac{N}{r}}} \mathcal{A}^{\frac{N}{2r}} \|u\|_{r,B_{\rho(1+\sigma)}}^{1-\frac{N}{2r}}.$$

This and Proposition 5.1 now imply that

$$\sup_{B_\rho} u \le \frac{\gamma M^{\frac{N}{2r}}}{\sigma^{\frac{N}{r}}} \left\{ \|f\|_{r,B_{4\rho}} + \rho^{\frac{N-2r}{r}} \right\} + \frac{1}{\rho^2},$$

where $M = \|u\|_{\infty,B_{\rho(1+\sigma)}}$. Since $r > N/2$, inequality (6.1) now follows from the usual interpolation technique. For the dimension $N = 1, 2$ the proof is completed by making use of the appropriate version of the Sobolev embedding Theorem. Inequality (6.3) is obvious if $|x| \le 1$. To prove it for all $x \in \mathbb{R}^N$, fix $|x| > 1$, and in (6.1) choose $\rho = |x|/8$. Then

$$\begin{aligned} |x|^2 u(x) &\le \gamma \left\{ 1 + \left(|x|^{2r-N} \int_{B_{4\rho}} f^r dy \right)^{\frac{2}{2r-N}} \right\} \\ &\le \gamma \left\{ 1 + \left(\fint_{B_{4\rho}} \left[|y|^2 f(y) \right]^r dy \right)^{\frac{2}{2r-N}} \right\}. \end{aligned}$$

□

Remark 6.1. If $N = 1$, a minor modification of these arguments shows that (6.1) and (6.3) continue to hold for $r = 1$.

7. Implicit Harnack Estimates

Proposition 7.1. *Let $u > 0$ be a classical solution to (1.1). There exists a constant γ depending only upon N and r, such that for all $x_o \in \mathbb{R}^N$ and all $\lambda\rho^2 \ge e$,*

$$-\inf_{B_\rho(x_o)} \ln u \le \gamma \fint_{B_{2\rho}(x_o)} (\ln u)_- \, dx + \ln \lambda\rho^2. \tag{7.1}$$

Proof. Setting $v = -\ln u$, it suffices to derive a local upper bound for $v_+ = \max\{0; v\}$. Such a function satisfies

$$-\Delta v_+ \le \lambda e^{-v} \qquad \text{in } \mathbb{R}^N. \tag{7.2}$$

For $\sigma \in (0,1)$ fixed and $n = 0,1,2,\dots$, let radii ρ_n, balls B_n and levels k_n be defined as in (6.4). We multiply (7.2) by $(v-k_{n+1})_+^{r-1}\zeta_n^2$, where ζ_n is the cutoff function from B_n to B_{n+1} introduced in (5.2). Standard calculations give

$$\left\| D\left[(v-k_{n+1})_+^{r/2}\zeta_n\right]\right\|_{2,B_n}^2 \le \frac{\gamma 4^n}{(\sigma\rho)^2}\mathcal{A}_\rho \|(v-k_n)_+\|_{r,B_n}^{r-1}, \tag{7.3}$$

where γ is an absolute constant depending only upon N and r, and

$$\mathcal{A}_\rho = \left(1 + e^{-k}\lambda\frac{\rho^2}{k}\right)\|v_+\|_{r,B_{(1+\sigma)\rho}}.$$

We stipulate to take $\lambda\rho^2 \ge 2$ and k so large that

$$e^{-k}\lambda\frac{\rho^2}{k} \le 1 \qquad \Longrightarrow \qquad k \ge \ln\lambda\rho^2. \tag{7.4}$$

Assume first $N \ge 3$. By the Sobolev embedding Theorem and (7.3),

$$\begin{aligned}
\|(v-k_{n+1})_+\|_{r,B_{n+1}}^r &\le \int_{B_n}\left[(v-k_{n+1})_+^{r/2}\zeta_n\right]^2 dx \\
&\le \left(\int_{B_n}\left[(v-k_{n+1})_+^{r/2}\zeta_n\right]^{\frac{2N}{N-2}}dx\right)^{\frac{N-2}{N}}\left(\int_{B_n}\chi\left[(v-k_{n+1})_+ > 0\right]dx\right)^{\frac{2}{N}} \\
&\le \gamma\int_{B_n}\left|D\left[(v-k_{n+1})_+\zeta_n\right]\right|^2 dx\left(\int_{B_n}\chi\left[(v-k_{n+1})_+ > 0\right]dx\right)^{\frac{2}{N}} \\
&\le \frac{\gamma 2^{(2+r)n}}{(\sigma\rho)^2 k^{2r/N}}\mathcal{A}_\rho\|(v-k_n)_+\|_{r,B_n}^{r-1+\frac{2r}{N}}.
\end{aligned}$$

Thus we obtain from these the recursive inequalities,

$$\|(v-k_{n+1})_+\|_{r,B_{n+1}} \le \frac{\gamma^{\frac{1}{r}}8^n}{(\sigma\rho)^{\frac{2}{r}}k^{\frac{2}{N}}}\mathcal{A}_\rho^{\frac{1}{r}}\|(v-k_n)_+\|_{r,B_n}^{1+\frac{2r-N}{Nr}}.$$

It follows from these, that if k is chosen from

$$\|v_+\|_{r,B_{(1+\sigma)\rho}} = (\gamma\mathcal{A}_\rho)^{-\frac{N}{2r-N}}8^{-\left(\frac{N}{2r-N}\right)^2}(\sigma\rho)^{\frac{2N}{2r-N}}k^{\frac{2r}{2r-N}}, \tag{7.5}$$

then, as $n \to \infty$,

$$\|(v-k_n)_+\|_{r,B_n} \to 0 \qquad \Longrightarrow \qquad \sup_{B_\rho} v_+ \le k.$$

Denoting with γ a generic positive constant that can be determined a priori only in terms of N and r, and taking into account (7.4) and the definition of $\mathcal{A}_\rho$, we compute from (7.5),

$$\sup_{B_\rho} v_+ \le \gamma(\sigma)^{\frac{N}{r}} \left(\fint_{B_{(1+\sigma)\rho}} v_+^r \, dx \right)^{\frac{1}{r}} + \ln \lambda\rho^2.$$

The estimate (7.1) now follows from a standard interpolation over the parameter σ. $\square$

Remark 7.1. The estimate (7.2) implies that if $\ln u \in L^1_{\mathrm{loc}}(\mathbb{R}^N)$ then u is bounded away from zero everywhere in $\mathbb{R}^N$.

7.1. The Intrinsic Harnack Estimate

As in [12] we introduce the Green's type functions

$$H_\rho(x) = \begin{cases} \dfrac{1}{\omega_N(N-2)} \left\{ |x|^{2-N} - \rho^{2-N} + \dfrac{N-2}{2} \dfrac{(|x|^2-\rho^2)}{\rho^N} \right\}, & \text{for } N \ge 3; \\ \dfrac{1}{2\pi} \left\{ \ln\rho - \ln|x| + \dfrac{1}{2} \dfrac{|x|^2-\rho^2}{\rho^2} \right\}, & \text{for } N = 2; \\ \dfrac{1}{2} \left\{ \rho - |x| + \dfrac{1}{2} \dfrac{|x|^2-\rho^2}{2\,\rho} \right\}, & \text{for } N = 1, \end{cases}$$

which satisfies

$$H_\rho(x) \,\big|_{\partial B_\rho} = 0; \qquad DH_\rho(x) \cdot \frac{x}{\rho} \,\big|_{\partial B_\rho} = 0; \qquad \Delta H_\rho(x) = -\delta_o + \frac{N}{\omega_N} \rho^{-N}.$$

Thus for any $w \in C^2(\overline{B}_\rho)$,

$$\fint_{B_\rho} w(x)\, dx = w(0) + \int_{B_\rho} H_\rho(x) \Delta w \, dx.$$

In particular for $w = \ln u$,

$$\fint_{B_\rho(x_o)} \ln u \, dx - \ln u(x_o) + \int_{B_\rho(x_o)} H_\rho(x - x_o) \{\lambda u \quad f\} \, dx.$$

From this and Proposition 7.1

$$\begin{aligned} -\inf_{B_{\rho/2}(x_o)} \ln u &\le \gamma \int_{B_\rho(x_o)} (\ln u)_- \, dx + \ln\lambda\rho^2 \le -\gamma \ln u(x_o) + \gamma \fint_{B_\rho(x_o)} (\ln u)_+ \, dx \\ &\quad + \gamma \int_{B_\rho(x_o)} H_\rho(x - x_o) \{f - \lambda u\} \, dx + \ln\lambda\rho^2. \end{aligned}$$

Thus for all $x \in B_{\rho/2}(x_o)$,

$$u(x) \ge \frac{\big(u(x_o)\big)^\gamma}{\lambda\rho^2} e^{-\gamma \mathcal{I}_\rho(x_o)}, \qquad \lambda\rho^2 \ge e \tag{7.6}$$

where

$$\mathcal{I}_\rho(x_o) = \fint_{B_\rho(x_o)} (\ln u)_+ \, dx + \int_{B_\rho(x_o)} H_\rho(x - x_o) \{f - \lambda u\} \, dx. \tag{7.7}$$

8. Some Consequences of the Sup and Harnack Estimates

Let (u, f) be a solution pair of (1.1). Since $u(\cdot)$ must be positive at some point, say for example $u(0) > 0$, by Remark 7.1 it is positive a.e. in the whole $\mathbb{R}^N$. If in addition f is bounded such a lower bound can be given a quantitative form.

Lemma 8.1. *Let (u, f) be a solution pair of (1.1) where*

$$f \in L^\infty(\mathbb{R}^N). \tag{8.1}$$

Then u is continuous in $\mathbb{R}^N$, and there exist two constants γ_i, $i = 0, 1$ that can be determined a priori only in terms of $\|f\|_{\infty,\mathbb{R}^N}$ and λ, such that

$$u(x) \geq \gamma_1 |x|^{-\gamma_o} \qquad \text{for all } |x| > 1. \tag{8.2}$$

Proof. We estimate above the quantity $\mathcal{I}_\rho(x_o)$ given in (7.7). Since f is bounded, by making use of (6.1) we have

$$\fint_{B_\rho(x_o)} (\ln u)_+ \, dx \leq \gamma\big(\|f\|_{\infty,\mathbb{R}^N}, N\big)\,(1 + \ln \rho), \qquad \rho \geq 1.$$

Next, since the Green's type function H_ρ is non-negative, we estimate

$$\begin{aligned}\int_{B_\rho(x_o)} H_\rho(x - x_o)\{f - \lambda u\}\, dx &\leq \int_{B_\rho(x_o)} H_\rho(x - x_o) f \, dx \\ &\leq \|f\|_{\infty,\mathbb{R}^N} \int_{B_\rho(x_o)} H_\rho(x - x_o) f \, dx.\end{aligned}$$

Therefore there exists a constant γ depending only upon N and $\|f\|_{\infty,\mathbb{R}^N}$, such that

$$\mathcal{I}_\rho(x_o) \leq \gamma\big(\|f\|_{\infty,\mathbb{R}^N}, N\big)(1 + \ln \rho), \qquad \rho \geq 1.$$

This in (7.7) proves (8.2). Since u is locally bounded above and below, it is also locally Hölder continuous in $\mathbb{R}^N$. □

8.1. Local Implicit Representations

Let (u, f) be a solution pair of (1.1) where f satisfies (8.1). Having fixed a ball B_ρ, by Lemma 8.1, the restriction of u to ∂B_ρ is well defined as a continuous function. Therefore by the Poisson representation formula of solutions of the Dirichlet problem in B_ρ,

$$\ln u(x) = \int_{\partial B_\rho} \ln u(y) \mathcal{P}_\rho(x; y) d\sigma(y) + \int_{B_\rho} G_\rho(x; y)\,\{f - \lambda u\}\, dy, \tag{8.4}$$

where $d\sigma(y)$ is the surface measure on ∂B_ρ, the function $\mathcal{P}_\rho(\cdot;\cdot)$ is the Poisson kernel

$$\mathcal{P}_\rho(x; y) = \frac{1}{\omega_N \rho} \frac{\rho^2 - |x|^2}{|x - y|^N}, \quad |y| = \rho, \quad \forall N \geq 2,$$

and $G_\rho(\cdot;\cdot)$ is the Green's function for the Laplacian in B_ρ. Such a function is given by

$$G_\rho(x;y) = \begin{cases} \dfrac{1}{\omega_N(N-2)}\left\{|x-y|^{2-N} - \Phi(x;y)\right\} & \text{for } N \geq 3; \\ \dfrac{-1}{2\pi}\{\ln|x-y| - \Phi(x;y)\} & \text{for } N = 2, \end{cases} \tag{8.5}$$

where, for all $x \in B_\rho$, the function $y \to \Phi(x;y)$ is the solution of

$$\begin{cases} \Delta_y \Phi(x;\cdot) = 0 \quad \text{in } B_\rho; \\ \Phi(x;y)\Big|_{y\in\partial B_\rho} = \begin{cases} |x-y|^{2-N} & \text{for } N \geq 3; \\ \ln|x-y| & \text{for } N = 2. \end{cases} \end{cases} \tag{8.6}$$

Lemma 8.2. *For all $|x| < \rho/2$, there holds*

$$\begin{cases} \ln\left(\dfrac{\rho}{2}\right) \leq \Phi(x;y) \leq \ln(2\rho), & \forall y \in B_\rho, \ \textit{for } N = 2; \\ \left(\dfrac{2}{3\rho}\right)^{N-2} \leq \Phi(x;y) \leq \left(\dfrac{2}{\rho}\right)^{N-2}, & \forall y \in B_\rho, \ \textit{for } N \geq 3. \end{cases} \tag{8.7}$$

Moreover for $N \geq 3$ there holds

$$0 \leq \Phi(x;y) \leq |x-y|^{2-N} \qquad \forall x,y \in B_\rho. \tag{8.8}$$

Proof. To prove the first of (8.7) we observe that for $|x| \leq \rho/2$ and $y \in \partial B_\rho$, we have

$$\ln\left(\frac{\rho}{2}\right) \leq \Phi(x;y) = \ln|x-y| \leq \ln 2\rho \qquad \forall y \in \partial B_\rho.$$

Since $y \to \Phi(x;y)$ is harmonic in B_ρ the same bounds continue to hold for all $y \in B_\rho$, by the maximum principle. The second of (8.7) as well as (8.8) are proved analogously. □

The local representation (8.4) holds for all $x \in B_\rho$. We compute it for $x = 0$ and subtract it from (8.4) to obtain,

$$\begin{aligned} \ln u(x) = \ln u(0) &+ \int_{\partial B_\rho} \{\mathcal{P}_\rho(x;y) - \mathcal{P}_\rho(0;y)\} \ln u(y)\, d\sigma(y) \\ &+ \int_{B_\rho} \{G_\rho(x;y) - G_\rho(0;y)\}\{f - \lambda u\}\, dy. \end{aligned} \tag{8.9}$$

8.2. Global Implicit Representations

Let (u,f) be a solution pair of (1.1) where f satisfies (8.1). We assume in addition that u and f are so that

$$(f - \lambda u) \in L^1(\mathbb{R}^N). \tag{8.10}$$

Then u can be implicitly represented by means of the kernel

$$\mathcal{K}_o(x;y) = \begin{cases} \dfrac{1}{2\pi}\{\ln|y| - \ln|x-y|\}, & \text{for } N = 2; \\ \dfrac{1}{\omega_N(N-2)}\left\{\dfrac{1}{|x-y|^{N-2}} - \dfrac{1}{|y|^{N-2}}\right\}, & \text{for } N \geq 3. \end{cases} \tag{8.11}$$

Proposition 8.1. *Let (u, f) be a solution pair of (1.1), and let (8.10) hold. Then*

$$\ln u(x) = \ln u(0) + \int_{\mathbb{R}^N} \mathcal{K}_o(x;y)\{f - \lambda u\}\, dy. \tag{8.12}$$

Proof. The representation is obtained from (8.9) by letting $\rho \nearrow \infty$. To compute such a limit we first observe that for all $x \in B_{\rho/2}$,

$$|\mathcal{P}_\rho(x;y) - \mathcal{P}_\rho(0;y)| \le \gamma(N)\, \frac{|x|}{\rho^N},$$

for a constant $\gamma(N)$ depending only upon the dimension N. Therefore taking also into account Lemma 8.1, we have

$$\lim_{\rho\nearrow\infty} \left| \int_{\partial B_\rho} \{\mathcal{P}_\rho(x;y) - \mathcal{P}_\rho(0;y)\} \ln u(y)\, d\sigma(y) \right| \le \gamma\, |x| \lim_{\rho\nearrow\infty} \frac{\ln \rho}{\rho} = 0,$$

for all $x \in \mathbb{R}^N$. Next, the difference $\{G_\rho(x;y) - G_\rho(0;y)\}$ is given by,

$$\mathcal{K}(x;y) = \begin{cases} \dfrac{1}{2\pi}\{\ln \rho - \Phi(x;y)\} & \text{for } N = 2; \\ \dfrac{1}{\omega_N(N-2)}\left\{\Phi(x;y) - \rho^{2-N}\right\} & \text{for } N \ge 3. \end{cases}$$

Consider first the case $N = 2$. Since $\{\Phi(x;\cdot) - \ln \rho\}$ is harmonic in B_ρ, by the Maximum Principle, for all $|x| < \rho/2$ we have

$$|\Phi(x;y) - \ln \rho| \le \big| \ln|x - y| - \ln|y| \big|_{y\in\partial B_\rho} \le 2\, \frac{|x|}{\rho},$$

for all $|x| < \rho/2$ and all $y \in B_\rho$. Therefore for any fixed $x \in \mathbb{R}^2$,

$$\lim_{\rho\nearrow\infty} \int_{B_\rho} (\Phi(x;y) - \ln \rho)\, \{f - \lambda u)\, dy = 0,$$

since $\{f - \lambda u\} \in L^1(\mathbb{R}^2)$. Analogously for $N \ge 3$ we estimate,

$$\left|\Phi(x;y) - \rho^{2-N}\right| \le \gamma(N)\, \frac{|x|}{\rho^{N-1}}, \qquad \forall |x| < \frac{\rho}{2}, \qquad \forall y \in B_\rho,$$

for a constant $\gamma(N)$ depending only upon N. Putting this in (8.9) and letting $\rho \nearrow \infty$ proves (8.12). □

PART II: Maximality and Stability

9. Maximal Solutions and the Maximum Principle

Let (u, f) be a solution pair to (1.1). We say that u is the *maximal* solution corresponding to f, if for any other solution pair (v, g) with $g \leq f$, then $v \leq u$. In this section we give a sufficient condition for u to be maximal.

Theorem 9.1 ($N = 2$). *Assume that u is continuous in $\mathbb{R}^2$ and satisfies,*

$$u(x) \geq \frac{C}{|x|^2 \ln |x|} \qquad \text{for all} \quad |x| > R, \tag{9.1}$$

for some C, $R > 0$. Then u is maximal.

Proof. Let (v, g) be a solution pair with $g \leq f$. Then

$$\lambda(v - u) - \Delta(\ln v - \ln u) = g - f \leq 0.$$

Multiply this by $S_n(\ln v - \ln u)\zeta^2$, where ζ is the cutoff function introduced in (5.2) and $S_n(\cdot)$ is the approximation to the sign$(\cdot)$ function introduced in Section 5. Proceeding as in that section we compute for $\alpha \in (0, 1]$

$$\lambda \int_{B_\rho} (v - u)_+ \, dx \leq \frac{\gamma_\alpha}{(\sigma\rho)^2} \int_{B_{\rho(1+\sigma)} \setminus B_\rho} \frac{(v - u)_+^\alpha}{u^\alpha} \, dx. \tag{9.2}$$

We write this for all $\rho > R$ and estimate

$$\begin{aligned} \lambda \int_{B_\rho} (v - u)_+ \, dx &\leq \frac{\gamma(\ln \rho)^\alpha}{C^\alpha \sigma^2 \rho^{2(1-\alpha)}} \int_{B_{\rho(1+\sigma)}} (v - u)_+^\alpha \, dx \\ &\leq \frac{\gamma(\ln \rho)^\alpha}{C^\alpha \sigma^2} \left(\int_{B_{\rho(1+\sigma)}} (v - u)_+ \, dx \right)^\alpha . \end{aligned}$$

Interpolating over the parameter σ gives

$$\int_{B_\rho} (v - u)_+ \, dx \leq \gamma_\alpha (\ln \rho)^{\frac{\alpha}{1-\alpha}}, \qquad \forall \rho > R. \tag{9.3}$$

Next from (9.2) with $\sigma, \alpha = 1$ and $\rho > R$, we derive

$$\int_{B_\rho} (v - u)_+ \, dx \leq \frac{\gamma \ln \rho}{C} \int_{B_{2\rho} \setminus B_\rho} (v - u)_+ \, dx. \tag{9.4}$$

Introduce the sequence of radii $\rho_n = 2^n \rho$, and set

$$M_n = \int_{B_{\rho_n}} (v - u)_+ \, dx.$$

Then from (9.4) we deduce for all $n = 0, 1, \ldots,$

$$M_n \leq \gamma n (M_{n+1} - M_n) \quad \Longrightarrow \quad M_n \leq \frac{\gamma n}{1 + \gamma n} M_{n+1},$$

where $\gamma = \gamma(C,\rho)$ is independent of α. From this by iteration,

$$M_o \le \prod_{i=1}^{n} \frac{\gamma\, i}{1+\gamma\, i} M_n. \tag{9.5}$$

We now observe that there are constants $\gamma_o \ge 1$ and $\eta \in (0,1)$ that can be determined a priori only in terms of $\gamma(C,\rho)$ such that

$$\prod_{i=1}^{n} \frac{\gamma\, i}{1+\gamma\, i} \le \gamma_o n^{-\eta}.$$

Then from (9.3) and (9.5),

$$M_o \le \gamma_o \gamma_a (\ln 2\rho)^{\frac{\alpha}{1-\alpha}} n^{\frac{\alpha}{1-\alpha}-\eta}.$$

To prove the Theorem, we choose α so small that the exponent of n is negative, and let $n \to \infty$. □

Theorem 9.2 ($N \ge 3$). *Assume that u is continuous in $\mathbb{R}^N$. There exists a constant C_o depending only upon N such that if*

$$u(x) \ge \frac{C_o}{|x|^2} \qquad \text{for all} \quad |x| > R, \tag{9.6}$$

for some $R > 0$, then u is maximal.

Proof. Let (v, g) be a solution pair with $g \le f$. Then from (9.3) which is valid in any number of dimensions we derive

$$\begin{aligned} \lambda \int_{B_\rho} (v-u)_+ \, dx &\le \frac{\gamma}{\sqrt{C_o}\sigma^2 \rho} \int_{B_{\rho(1+\sigma)}} (v-u)_+^{1/2} \, dx \\ &\le \frac{\gamma \rho^{\frac{N-2}{2}}}{\sqrt{C_o}\sigma^2} \left(\int_{B_{\rho(1+\sigma)}} (v-n)_+ dx \right)^{1/2}. \end{aligned}$$

Interpolating over the parameter σ gives

$$\int_{B_\rho} (v-u)_+ \, dx \le \frac{\gamma}{\tilde{\tilde{C}}} \rho^{N-2}, \qquad \forall \rho > R. \tag{9.7}$$

Introduce the sequence of radii $\rho_n = 2^n \rho$, and from (9.3) written for $\rho > R$ and $\alpha = 1$, deduce

$$\int_{B_{\rho_n}} (v-u)_+ \, dx \le \frac{\gamma}{C_o} \int_{B_{\rho_{n+1}}} (v-u)_+ \, dx.$$

Thus by iteration and (9.7),

$$\int_{B_\rho} (v-u)_+ \, dx \le \left(\frac{\gamma}{C_o}\right)^n \int_{B_{\rho_n}} (v-u)_+ \, dx \le \frac{\text{const}\, \rho^{N-2}}{C_o} \left(\frac{\gamma 2^{N-2}}{C_o}\right)^n.$$

Thus if C_o is chosen sufficiently large depending only upon N, the right hand side tends to zero as $n \to \infty$, thereby proving the Theorem. □

10. Stability and Maximality

A minor modification of the previous argument establishes the following stability estimate.

Proposition 10.1. *Let (u, f) and (v, g) be solution pairs of (1.1). Assume that $(f-g) \in L^1(\mathbb{R}^N)$, and that in addition u and v satisfy (9.1) if $N = 2$, and (9.6) if $N \geq 3$. Then $(u - v) \in L^1(\mathbb{R}^N)$ and*

$$\lambda \|u - v\|_{1,\mathbb{R}^N} \leq \|f - g\|_{1,\mathbb{R}^N}. \tag{10.1}$$

Corollary 10.1. *If (u, f) is a solution pair and $u \geq \varepsilon > 0$, then u is maximal.*

Corollary 10.2. *Let (u, f) be a solution pair to (1.1). Then there is a v such that (v, f) is a solution pair and v is maximal.*

Proof. Suppose (w, g) is a solution pair such that $g \leq f$. Consider the family of solution pairs $(v_\varepsilon, f+\lambda\varepsilon)$, where $\varepsilon > 0$ and v_ε is the unique solution satisfying $v_\varepsilon > \varepsilon$. The solution v_ε can be constructed by the approximation method indicated in Section 5.1, and it is unique since it is maximal, by Corollary 10.1. By maximality $v_\varepsilon \geq u$, and $v_\varepsilon \geq w$ for all $\varepsilon > 0$. Also $v_\varepsilon > v_{\varepsilon'}$ if $\varepsilon > \varepsilon'$. Thus $v_\varepsilon \searrow v$ as $\varepsilon \searrow 0$, and (v, f) is a solution pair. Since $v \geq w$ and (w, g) is an arbitrary solution pair with $g \leq f$, the solution v is maximal. □

Remark 10.1. A consequence of Corollary 10.2 is that all the maximal solutions to (1.1) are constructed as decreasing limits of $(v_\varepsilon, f + \lambda\varepsilon)$, starting from a given solution u. Since $v_\varepsilon \geq u$, the existence of such a u guarantees that the v_ε have a non trivial limit.

Corollary 10.3. *Let f, g be non-negative and in $L^1(\mathbb{R}^N)$. Then the maximal solutions (u, f) and (v, g), if any, satisfy the stability estimate (10.1).*

Proof. Indeed the maximal solutions $(u_\varepsilon, f + \lambda\varepsilon)$ and $(v_\varepsilon, g + \lambda\varepsilon)$ satisfy the assumptions of Proposition 10.1. Therefore they satisfy (10.1) uniformly in ε. □

10.1. An Application to the Case $N = 1$

Proposition 10.2. *Let $f \in L^1(\mathbb{R})$ be non-negative and not identically zero. Then (1.1) has a unique solution u satisfying $\lambda\|u\|_{1,\mathbb{R}} = \|f\|_{1,\mathbb{R}}$. Furthermore u is maximal.*

Proof. Without loss of generality we may assume that $\lambda = 1$ and for $k \gg 1$ define the family of solutions pairs (v_k, g_k), by

$$v_k(x) = 2(k + x^2)^{-1}, \qquad g_k(x) = 4k(k + x^2)^{-2}.$$

Construct now the solutions pairs $(u_k, f + g_k)$, where u_k is maximal. Such a maximal solution can be constructed by first finding the maximal solution pairs $(u_k^\varepsilon, f + g_k + \varepsilon)$ and then observing that such a u_k^ε is bounded below by the maximal solution $\widetilde{v}_k \geq v_k$ corresponding to g_k. Thus $u_k^\varepsilon \geq v_k$ uniformly in ε, and a standard limiting process establishes the maximality of the solution pair $(u_k, f + g_k)$. Moreover by Corollary 5.2, $u_k \in L^1(\mathbb{R})$. Multiplying the equation by the cutoff function ζ^2 introduced in (5.2), letting $\rho \to \infty$, and taking into account that $u_k \geq v_k$, yields $\|u_k\|_{1,\mathbb{R}} = \|f + g_k\|_{1,\mathbb{R}}$. Since g_k decreases to zero in $L^1(\mathbb{R})$ the sequence u_k decreases to some function $u \in L^1(\mathbb{R})$, which satisfies $\|u\|_{1,\mathbb{R}} = \|f\|_{1,\mathbb{R}} > 0$. Thus there exists a point $x_o \in \mathbb{R}$ and a positive constant δ, such that $u_k(x_o) > \delta$ for all $k \gg 1$. This suffices to complete the proof in view of the Harnack-type estimate of Section 7. In this respect we observe that the quantity $\mathcal{I}_\rho(x_o)$ defined in (7.8) is finite in view of Remark 6.1. □

This 1-dimensional result has been observed, by different techniques, in [10].

PART III
Existence and Non-Existence of Solutions for $N = 2$

11. Non-Existence of Solutions

Proposition 11.1. *Let $N = 2$, and assume that f satisfies,*

$$f \geq 0, \qquad f \in L^1(\mathbb{R}^2) \qquad \text{and} \qquad \|f\|_{1,\mathbb{R}^2} \leq 4\pi. \tag{11.1}$$

Then (1.1) has no solution.

Proof. First we establish the Proposition under the following more stringent assumptions:

$$\begin{cases} f \text{ satisfies (11.1) with strict inequality, i.e., } \|f\|_{1,\mathbb{R}^2} < 4\pi; \\ f \in L^\infty(\mathbb{R}^2) \text{ and satisfies the decay estimate (6.2).} \end{cases} \tag{11.2}$$

If (u, f) is a solution pair to (1.1), by the estimation (6.3),

$$u(x) \leq \frac{M}{1 + |x|^2}, \qquad \text{for some } M > 0, \quad \forall x \in \mathbb{R}^2.$$

Therefore for any $\rho > 0$ fixed, the pair (u, f) satisfies,

$$\Delta \ln u = \lambda u - f \quad \text{in } B_\rho, \qquad \text{and} \qquad u\Big|_{|x|=\rho} \leq \frac{M}{1 + \rho^2}.$$

From this and the local representation formula (8.4), we derive for all $|x| < \rho$,

$$\begin{aligned} \ln u(x) &\leq \ln \frac{M}{1+\rho^2} - \int_{B_\rho} G_\rho(x;y)\{\lambda u - f\}\, dy \\ &\leq \gamma(M) - 2\ln\rho + \int_{B_\rho} G_\rho(x;y) f(y)\, dy. \end{aligned} \tag{11.3}$$

In deriving (11.3) we have also taken into account the fact that the Poisson kernel has mass one on the unit circle and that $G_\rho(\cdot;\cdot) \geq 0$. To estimate the last integral on the right hand side of (11.3), we write

$$2\pi \int_{B_\rho} G_\rho(x;y) f(y) dy = \int_{B_\rho} \Phi(x;y) f(y) dy - \int_{B_\rho} \ln|x-y|\, f(y) dy = I_1 + I_2. \tag{11.4}$$

The first term is estimated by making use of the first of (8.7), i.e.,

$$I_1 \leq \ln\rho\, \|f\|_{1,\mathbb{R}^2} + \ln 2\, \|f\|_{1,\mathbb{R}^2}.$$

In estimating I_2 we recall that f is non-negative and is in $L^r_{\mathrm{loc}}(\mathbb{R}^2)$ for some $r > 1$. Then for $|x| \ll \rho$,

$$I_2 \leq -\int_{|x-y|<1} \ln|x-y|\, f(y)\, dy \leq \gamma\, \|f\|_{r,B_1(x)}.$$

Combining these estimates in (11.3) we conclude that there exists a constant γ depending only on M, such that,

$$\ln u(x) \leq \gamma(M)\left(\|f\|_{1,\mathbb{R}^2} + \|f\|_{r,B_1(x)}\right) - 2\ln\rho\left(1 - \frac{\|f\|_{1,\mathbb{R}^2}}{4\pi}\right). \tag{11.5}$$

Let $R > 0$ be arbitrary but fixed. We let $\rho \nearrow \infty$ in (11.5), by keeping $|x| < R$. Since $\|f\|_{1,\mathbb{R}^2} < 4\pi$ this implies that $u = 0$ in the ball B_R about the origin. Therefore $u = 0$ in the whole $\mathbb{R}^2$, since R is arbitrary. This establishes the Proposition under the assumptions (11.2).

11.1. An Auxiliary Lemma

Let f satisfy (11.2), and for $\varepsilon > 0$ consider the maximal solution pairs $(u_\varepsilon, f + \lambda\varepsilon)$. These can be constructed by the method indicated in Section 5.1 and are maximal by Corollary 10.1.

Lemma 11.2. *$\{u_\varepsilon\} \to 0$ in $L^1_{\mathrm{loc}}(\mathbb{R}^2)$.*

Proof. By construction $u_\varepsilon \geq \varepsilon$ and $\varepsilon \to \{u_\varepsilon\}$ is decreasing. Thus $u \to v$ in $L^1_{\mathrm{loc}}(\mathbb{R}^2)$ for some non-negative $v \in L^1_{\mathrm{loc}}(\mathbb{R}^2)$. If $v(x_o) > 0$ for some $x_o \in \mathbb{R}^2$ then by the Harnack estimate (7.7) the functions u_ε would be bounded below, uniformly in ε, on every compact subset of $\mathbb{R}^2$. Thus (v, f) would be a solution pair of (1.1), contrary to Proposition 11.1, established under the assumptions (11.2). □

11.2. Proof of Proposition 11.1 Concluded

Let f satisfy (11.1), and suppose that (1.1) has a solution pair (u, f). In such a case, by Corollary 10.2, the equation (1.1) admits a maximal solution pair (v, f) constructed as the decreasing limit of $(v_\varepsilon, f + \lambda\varepsilon)$, as indicated in Remark 10.1. Construct a sequence of functions $\{f_j\}_{j\in\mathbb{N}}$ such that

$$f_j \quad \text{satisfy (11.2) for all} \quad j \in \mathbb{N} \quad \text{and} \quad f_j \nearrow f.$$

In correspondence to each f_j consider also the maximal solutions $(u_{\varepsilon,j}, f_j + \lambda\varepsilon)$. By Corollary 10.3 $(v_\varepsilon - u_{\varepsilon,j}) \in L^1(\mathbb{R}^2)$, and the pairs $(v_\varepsilon, f + \lambda\varepsilon)$ and $(u_{\varepsilon,j}, f_j + \lambda\varepsilon)$, satisfy the stability estimate (10.1), i.e.,

$$\lambda\|v_\varepsilon - u_{\varepsilon,j}\|_{1,\mathbb{R}^2} \le \|f - f_j\|_{1,\mathbb{R}^2} \qquad \forall j \in \mathbb{N}.$$

Letting $\varepsilon \searrow 0$ for $j \in \mathbb{N}$ fixed, and taking into account Lemma 11.2, yields

$$\lambda\|v\|_{1,\mathbb{R}^2} \le \|f - f_j\|_{1,\mathbb{R}^2}, \quad \text{for all} \quad j \in \mathbb{N}.$$

□

11.3. Some Consequences of Proposition 11.1 and Its Proof

Corollary 11.1. *Let f be non-negative and satisfy*

$$f \in L^1(\mathbb{R}^2) \quad \text{and} \quad \|f\|_{1,\mathbb{R}^2} \le 4\pi s, \quad \text{for some } s \ge 1. \tag{11.6}$$

Then (1.1) has no solution u satisfying

$$u(x) \le \frac{M}{(1+|x|^2)^s}, \qquad \text{for some } M > 0 \text{ and for } |x| \gg 1. \tag{11.7}$$

Proof. For such a solution we may derive an analog of (11.3) in the form

$$\begin{aligned} \ln u(x) &\le \ln \frac{M}{(1+\rho^2)^s} - \int_{B_\rho} G_\rho(x;y)\,\{\lambda - f\}\,dx \\ &\le \gamma(M) - 2s\ln\rho + \int_{B_\rho} G_\rho(x;y) f(y)\,dy. \end{aligned}$$

Proceeding as before we arrive at the analog of (11.5) which now reads

$$\ln u(x) \le \gamma(M)\left(\|f\|_{1,\mathbb{R}^2} + \|f\|_{r,B_1(x)}\right) - 2s\ln\rho\left(1 - \|f\|_{1,\mathbb{R}^2}/4\pi s\right).$$

□

12. Existence of Solutions for $N = 2$

Theorem 12.1. *Let $s \geq 1$, and assume that $f \in L^1(\mathbb{R}^2)$ and satisfies*

$$\|f\|_{1,\mathbb{R}^2} > 4\pi s. \tag{12.1}$$

For each such s, there exists a solution $u_s \in L^1(\mathbb{R}^2)$, satisfying

$$\|u_s\|_{1,\mathbb{R}^2} = \|f\|_{1,\mathbb{R}^2} - 4\pi s. \tag*{(12.1)$_s$}$$

Furthermore the maximal solution corresponding to such a f is $u = u_1$.

Proof. Assume first that f is bounded and has compact support and that $s > 1$. For such a s let (w_s, ϕ_s) be the solution pair introduced in Section 6. Since f is bounded and compactly supported, we may choose M so large that $f \leq \phi_s$. For $n \in \mathbb{N}$ let u_n be the unique solution of

$$\begin{cases} \lambda u_n - \Delta \ln u_n = f & \text{in } B_n; \\ u_n(x) = \dfrac{M}{(1+n^2)^s} & \text{for } x \in \partial B_n. \end{cases} \tag{12.2}$$

Since $f \leq \phi_s$ and $u_n = w_s$ on ∂B_n, we also have $u_n \leq w_s$ in B_n. Therefore

$$\lambda \int_{B_\rho} (w_s - u_n)dx \leq \lambda \int_{B_n} (w_s - u_n)\, dx \leq \int_{\mathbb{R}^2} (\phi_s - f)\, dx, \tag{12.3}$$

for all $n > \rho > 0$. From this we compute

$$\begin{aligned} \lambda \int_{B_\rho} u_n\, dx &\geq \int_{\mathbb{R}^2} f dx - \int_{\mathbb{R}^2} \frac{4s}{(1+|x|^2)^2}\, dx - \int_{\mathbb{R}^2 \setminus B_\rho} \frac{\lambda M}{(1+|x|^2)^s}\, dx \\ &= \|f\|_{1,\mathbb{R}^2} - 4\pi s - \frac{\lambda M}{(s-1)(1+\rho^2)^{s-1}}. \end{aligned} \tag{12.4}$$

Thus by choosing ρ sufficiently large we can guarantee that

$$\int_{B_\rho} u_n\, dx \geq \eta > 0 \qquad \text{for some } \eta \text{ independent of } n.$$

Since $\{u_n\}_{n\in\mathbb{N}}$ is decreasing, there exists some $x_o \in B_\rho$ such that

$$u_n(x_o) \geq \frac{\eta}{\lambda |B_\rho|}, \qquad \text{for all } n \in \mathbb{N}.$$

In light of the Harnack estimate of Section 10, this is sufficient to guarantee the existence of a solution. From (12.4) it follows that

$$\lambda \|u\|_{1,\mathbb{R}^2} \geq \|f\|_{1,\mathbb{R}^2} - 4\pi s. \tag{12.5}$$

To prove (12.1)$_s$ we establish, by contradiction, that (12.5) cannot hold with strict inequality. Thus assume that there exists some $\varepsilon \in (0,1)$ such that

$$\lambda \|u\|_{1,\mathbb{R}^2} \geq \|f\|_{1,\mathbb{R}^2} - 4\pi s + \varepsilon. \tag*{(12.5)$_\varepsilon$}$$

Choose $\delta \in (0,1)$ satisfying

$$\delta \|f\|_{1,\mathbb{R}^2} = 4\pi s - \tfrac{1}{2}\varepsilon, \tag{12.6}$$

and construct a solution u_δ to (1.1) corresponding to the datum δf. Such a solution can be constructed by the same approximating procedure indicated in (12.2), with the role of the comparison function w_s now being taken by u. If we denote by $\{u_{\delta,n}\}_{n\in\mathbb{N}}$ the corresponding approximating sequence, the analog of (12.3) now reads

$$\lambda \int_{B_\rho} (u - u_{\delta,n})\, dx \le (1-\delta) \int_{\mathbb{R}^2} f\, dx. \tag{12.3$_\delta$}$$

From this and the previous choices of ε and δ,

$$\lambda \int_{B_\rho} u_{\delta,n}\, dx \ge \lambda \int_{B_\rho} u\, dx \;-\; (1-\delta)\, \|f\|_{1,\mathbb{R}^2} \ge \tfrac{1}{2}\varepsilon \;-\; \int_{\mathbb{R}^2\setminus B_\rho} u\, dx.$$

Now, since $u \in L^1(\mathbb{R}^2)$, we may choose ρ so large that

$$\int_{B_\rho} u_{\delta,n}\, dx \ge \tfrac{1}{4}\varepsilon \quad \text{uniformly in } n.$$

This supplies the bound below needed to establish the existence of u_δ. Notice however that by the indicated construction procedure, such a u_δ lies below u which in turn lies below w_s. Thus u_δ is a solution to (1.1) decaying as $|x| \to \infty$ as in (11.7), and corresponding to a datum δf satisfying (12.6). By Corollary 11.1 this yields a contradiction and establishes $(12.1)_s$.

For a general non-negative $f \in L^1(\mathbb{R}^2)$ satisfying (12.1), we may construct an increasing sequence $\{f_j\}_{j\in\mathbb{N}}$ such that $\|f_j\|_{1,\mathbb{R}^2} > 4\pi s$ and each f_j is bounded and has compact support. The corresponding solutions $\{u_j\}_{j\in\mathbb{N}}$ form an increasing sequence. Therefore, still for $s > 1$, the Theorem is established by a standard limiting argument. If $\{s_i\}_{i\in\mathbb{N}}$ is a decreasing sequence tending to 1, the corresponding solutions $\{u_i\}_{i\in\mathbb{N}}$, constructed by the previous method, form an increasing sequence. A standard limiting process now establishes the existence of $u = u_1 \in L^1(\mathbb{R}^2)$ solution of (1.1) and satisfying $(12.1)_1$. By Corollary 10.2, starting from (u_1, f) we may construct a maximal solution v, which would satisfy

$$v \ge u_1 \quad \Longrightarrow \quad \|v\| \ge \|f\|_{1,\mathbb{R}^2} \;-\; 4\pi.$$

An argument similar to that leading to $(12.1)_s$ proves that strict inequality cannot hold. Therefore $v = u_1$ and u_1 is maximal. □

PART IV
Asymptotic Behavior and Uniqueness for $N = 2$

13. Asymptotic Behavior of Maximal Solutions

Let $(v_{\alpha,\mu}, \varphi_{\alpha,\mu})$ be the solution pairs introduced in Section 7. For $\alpha = 2$ we obtain the one among these that decreases the fastest when $|x| \to \infty$, while the corresponding forcing term $\varphi_{2,\mu}$ remains positive. Also for the special choice $\lambda M = 8$ of the parameter M we obtain the maximal solution whose corresponding forcing term $\varphi_{2,\mu}$ decreases the fastest as $|x| \to \infty$. Indeed for these choices of the parameters α and M we have

$$\begin{aligned} \lambda v_{2,\mu} &= \frac{8}{(\mu+|x|^2)\,[\ln(\mu+|x|^2)]^2}; \\ \varphi_{2,\mu} &= \frac{4\mu}{(\mu+|y|^2)^2}\left\{1 + \frac{2}{\ln(\mu+|y|^2)} + \frac{2}{[\ln(\mu+|y|^2)]^2}\right\}. \end{aligned} \tag{13.1}$$

These solutions, parametrized with $\mu > 1$, determine a lower bound on the possible decay of maximal solutions as $|x| \to \infty$. That is maximal solutions to (1.1) corresponding to some non-negative $f \in L^1(\mathbb{R}^2)$ decay, as $|x| \to \infty$ no faster than (13.1). Conversely if (u, f) is a solution pair so that $x \to u(x)$ decays no faster than indicated in (13.1) then u is maximal.

Theorem 13.1. *Let $f \in L^1(\mathbb{R}^2)$ be non-negative, and let (u, f) be a solution pair to (1.1). Then u is maximal if and only if there exist constants C_o and C_1, such that*

$$u(x) \geq \frac{C_o}{|x|^2 \ln^2 |x|}, \qquad \forall\, |x| \geq C_1. \tag{13.2}$$

14. Proof of Theorem 13.1. Sufficient Condition

We assume that u is maximal and prove that it satisfies (13.2). We first establish such a lower estimate in a particular case.

Proposition 14.1. *Let (u, f) be a solution pair to (1.1), and assume that u is maximal. Assume moreover that*

$$f(x) \leq \varphi_{2,\mu}(x) \qquad \forall x \in \mathbb{R}^2, \qquad \textit{for some } \mu > 1. \tag{14.1}$$

Then, there exists a constant $C = C(\mu)$ such that for all $|x| \geq 4$,

$$u(x) \geq \frac{C(\mu)\, u(0)}{(\mu+|x|^2)\,[\ln(\mu+|x|^2)]^2}, \qquad C(\mu) = \left(\frac{\ln \mu}{\mu}\right)^2 e^{-\gamma(1+1/\ln^2 \mu)},$$

for an absolute constant γ.

Proof. In the proof we will work with the generic pair $(v_{\alpha,\mu}, \varphi_{\alpha,\mu})$, to single out the special role played by the pair in (13.1). Thus assume that $f \le \varphi_{\alpha,\mu}$ for some $\varphi_{\alpha,\mu}$. Then by the representation formula (8.12), with $\mathcal{K}_o(\cdot;\cdot)$ defined by (8.11),

$$\begin{aligned}\ln u(x) &= \ln u(0) + \int_{\mathbb{R}^2} \mathcal{K}_o(x;y)\,\{f - \lambda u\}\,dy \\ &= \ln u(0) - \frac{\ln|x|}{2\pi}\int_{\mathbb{R}^2}\{f-\lambda u\}\,dy + \int_{\mathbb{R}^2}\mathcal{K}(x;y)\,\{f-\lambda u\}\,dy,\end{aligned}$$

where

$$2\pi\mathcal{K}(x;y) = \{\ln|x| + \ln|y| - \ln|x-y|\}\,. \tag{14.2}$$

Since u is maximal we have

$$\frac{\ln|x|}{2\pi}\int_{\mathbb{R}^2}\{f-\lambda u\}\,dy = 2\ln|x|.$$

Therefore

$$\ln u(x) = \ln u(0) - 2\ln|x| + \int_{\mathbb{R}^2}\mathcal{K}(x;y)\,\{f-\lambda u\}\,dy.$$

In a similar manner we obtain,

$$\ln v_{\alpha,\mu}(x) = \ln v_{\alpha,\mu}(0) - 2\ln|x| + \int_{\mathbb{R}^2}\mathcal{K}(x;y)\,\{f-\lambda u\}\,dy,$$

for the same kernel $\mathcal{K}(\cdot;\cdot)$. From these by difference

$$\begin{aligned}\ln\frac{u(x)}{v_{\alpha,\mu}(x)} &= \ln\frac{u(0)}{v_{\alpha,\mu}(0)} + \lambda\int_{\mathbb{R}^2}\mathcal{K}(x;y)(v_{\alpha,\mu}-u)\,dy - \int_{\mathbb{R}^2}\mathcal{K}(x;y)(\varphi_{\alpha,\mu}-f)\,dy \\ &\ge \ln\frac{u(0)}{v_{\alpha,\mu}(0)} - \lambda\int_{\mathbb{R}^2}\mathcal{K}_-(x;y)v_{\alpha,\mu}\,dy - \int_{\mathbb{R}^2}\mathcal{K}_+(x;y)\varphi_{\alpha,\mu}\,dy.\end{aligned} \tag{14.3}$$

To estimate the last two integrals, we stipulate to take $|x| \ge 4$ and subdivide the domain of integration into the three portions

$$S_1 = \{|y| \le |x|-1\}\,;\quad S_2 = \{|x|-1 \le |y| \le |x|+1\}\,;\quad S_3 = \{|y| \ge |x|+1\}\,.$$

Within each of the S_i, $i = 1,2,3$ we have

$$\begin{aligned}\ln\left(\frac{|y|}{2}\right) &\le \mathcal{K}(x;y) \le \ln(1+|y|), && \text{in } S_1;\\ \ln\frac{|y||x|}{2(|x|+1)} &\le \mathcal{K}(x;y) \le 2\ln(1+|y|) + (\ln|x-y|)_-\,, && \text{in } S_2;\\ \ln\frac{|y||x|}{|x|+|y|} &\le \mathcal{K}(x;y) \le \ln(1+|y|), && \text{in } S_3.\end{aligned}$$

Therefore $\mathcal{K}(\cdot;\cdot)$ can be negative only in the disc $\{|y|<2\}$. On such a disc,

$$\int_{\{|y|<2\}}\mathcal{K}(x;y)v_{\alpha,\mu}\,dy \le \gamma v_{\alpha,\mu}(0),$$

for an absolute constant γ. Next

$$\int_{\mathbb{R}^2} \mathcal{K}_+(x;y)\varphi_{\alpha,\mu}\,dy \le \int_{\mathbb{R}^2} \ln(1+|y|)\varphi_{\alpha,\mu}\,dy - \int_{|x-y|<1} \ln|x-y|\varphi_{\alpha,\mu}(y)\,dy$$
$$\le \int_{\mathbb{R}^2} \ln(1+|y|)\varphi_{\alpha,\mu}(y)\,dy + \gamma\varphi_{\alpha,\mu}(0),$$

provided the last integral converges. Assuming it for the moment, we combine these remarks in (14.3) to deduce that there exists an absolute constant γ such that for all $|x| \ge 4$,

$$\ln\frac{u(x)}{v_{\alpha,\mu}(x)} \ge \ln\frac{u(0)}{v_{\alpha,\mu}(0)} - \gamma\left\{\varphi_{\alpha,\mu}(0) + v_{\alpha,\mu}(0)\right\} - \int_{\mathbb{R}^2} \ln(1+|y|)\varphi_{\alpha,\mu}(y)\,dy. \tag{14.4}$$

This estimate holds for every $\mu > 1$, every $M > 0$ and every $\alpha \in (1,2]$, provided $f \le \varphi_{\alpha,\mu}$. However the last integral on the right-hand side of (14.4) is convergent only for the choices $\alpha = 2$ and $\lambda M = 8$. In such a case

$$\int_{\mathbb{R}^2} \ln(1+|y|)\varphi_{\alpha,\mu}(y)\,dy \le \frac{\gamma}{\ln^2\mu},$$

for an absolute constant γ. Also for the indicated choices of the parameters M and α we have

$$\gamma\left\{\varphi_{\alpha,\mu}(0) + v_{\alpha,\mu}(0)\right\} \le \bar{\gamma}\left(1 + 1/\ln^2\mu\right),$$

for all $|x| \ge 4$. We conclude that there exists an absolute constant γ, such that for all $|x| \ge 4$,

$$\ln\frac{u(x)}{v_{2,\mu}(x)} \ge \ln\frac{u(0)}{v_{2,\mu}(0)} - \gamma\left(1 + \frac{1}{\ln^2\mu}\right).$$

□

Lemma 14.1. *Let (u, f) be a solution pair to (1.1) where u is maximal, and assume that*

$$f \in L^\infty(\mathbb{R}^2), \qquad \operatorname{supp}\{f\} \subset B_R \qquad \text{for some } R > 0.$$

There exists constants C_o and C_1 depending only upon R and $\|f\|_{\infty,B_R}$, such that

$$u(x) \ge \frac{C_o}{|x|^2\ln^2|x|}, \qquad \forall\,|x| \ge C_1.$$

Proof. Consider the pairs (u_h, f_h) defined for $h > 0$, by

$$u_h(x) = hu(\sqrt{h}x), \qquad f_h(x) = hf(\sqrt{h}x).$$

By the rescaling arguments of Section 3, these are still solution pairs. Moreover they are still maximal, since for $N = 2$ the L^1-norm is preserved under such a rescaling. We will show that we may select $h > 0$ and $\mu > 1$ depending upon R and $\|f\|_{\infty,B_R}$, such that $f_h \le \varphi_{2,\mu}$. The Lemma will then be a consequence of

Proposition 14.1. We begin by observing that the support of f_h is contained in the ball $B_{R/\sqrt{h}}$. Then we take $\sqrt{h} > R$, so that within the support of f_h,

$$\varphi_{2,\mu} \geq \frac{8}{9\left[\ln(\mu+|x|^2)\right]^2} \geq \frac{8}{9\left[\ln\left(\mu + R^2/h\right)\right]^2}, \qquad \forall\, |x| \leq \frac{R}{\sqrt{h}}.$$

Therefore it suffices to show that $h > R^2$ and $\mu > 1$ can be chosen so that

$$f_h(x) \leq h\|f\|_{\infty,B_R} \leq \frac{8}{9\left[\ln\left(\mu + R^2/h\right)\right]^2}, \qquad \forall\, |x| \leq \frac{R}{\sqrt{h}}.$$

We take $\mu = 1 + R^2/h$ and estimate

$$\frac{8}{9\left[\ln\left(\mu + R^2/h\right)\right]^2} = \frac{8}{9\left[\ln\left(1 + 2R^2/h\right)\right]^2} \geq \frac{8h^2}{36R^4} \geq h\|f\|_{\infty,B_R},$$

for the choice of $h \geq 9\|f\|_{\infty,B_R} R^4/2$. □

Proof of Theorem 13.1. Sufficient Condition. By Proposition 11.1 we must have $\|f\|_{1,\mathbb{R}^2} > 4\pi$. There exist $R > 0$ and $f_o \leq f$, supported in B_R such that

$$\|f_o\|_{1,B_R} > 4\pi \qquad \text{and} \qquad \|f_o\|_{\infty,B_R} < \infty.$$

Let u_o be the maximal solution corresponding to such a f_o. Then since u is maximal $u \geq u_o$, and u_o satisfies the lower bound of Lemma 14.1. □

15. Proof of Theorem 13.1. Necessary Condition

Let (u, f) be a solution pair to (1.1) where $f \in L^1(\mathbb{R}^2)$ and u satisfies the lower bound (13.2). We will show that such a u is maximal. First, by Proposition 11.1 we must have $\|f\|_{1,\mathbb{R}^2} > 4\pi$. Let $\{f_j\}_{j\in\mathbb{N}}$ be an increasing sequence of functions such that

$$0 \leq f_j \leq j, \qquad \operatorname{supp}\{f_j\} \subset \{|x| < j\}, \qquad \|f_j\|_{1,B_j} > 4\pi, \qquad f_j \nearrow f.$$

Let also $\{s_j\}_{j\in\mathbb{N}}$ be the increasing sequence of numbers defined by

$$\|f_j\|_{1,\mathbb{R}^2} = 4\pi\, s_j, \qquad j = 1, 2, \ldots.$$

For each $j \in \mathbb{N}$ and $s \in (1, s_j)$, let $u_{j,s}$ be the solution constructed in Section 16 corresponding to the datum f_j and satisfying

$$\|u_{j,s}\|_{1,\mathbb{R}^2} = \|f_j\|_{1,\mathbb{R}^2} - 4\pi s. \tag{15.1}$$

The functions $u_{j,s}$, for j and s fixed, can be constructed in turn as decreasing limits, as $n \to \infty$, of the solutions $u_{j,s}^{(n)}$ of the problems (12.2). In these, M is a constant depending upon j and chosen so that $f_j \leq \phi_s$, where (w_s, ϕ_s) are the solution pairs introduced in Section 4. Therefore u and $u_{j,s}^{(n)}$ satisfy

$$\begin{cases} \lambda\left(u - u_{j,s}^{(n)}\right) - \Delta\left(\ln u - \ln u_{j,s}^{(n)}\right) = f - f_j, & \text{in } B_n; \\ \left(u - u_{j,s}^{(n)}\right)\Big|_{\partial B_n} \geq \dfrac{C_o}{n^2 \ln^2 n} - \dfrac{M(j)}{(1+n^2)^s}. \end{cases}$$

It follows that for n sufficiently large $u \geq u_{j,s}^{(n)}$ in B_n. Letting $n \to \infty$ we also obtain

$$u(x) \geq u_{j,s}(x), \qquad \text{for all } j \in \mathbb{N} \quad \text{and for all } s > 1.$$

It follows from (15.1), upon letting $j \nearrow \infty$ and $s \searrow 1$, that

$$\|u\|_{1,\mathbb{R}^2} \geq \|f\|_{1,\mathbb{R}^2} - 4\pi. \qquad \square$$

16. Uniqueness of the Solutions u_s

The estimations leading to (14.4) permit us to establish that the solutions u_s constructed in Theorem 12.1, are unique.

Theorem 16.1. *Let $f \in L^1(\mathbb{R}^2)$ be non-negative. If u is a solution to (1.1) corresponding to such a f and satisfying*

$$\|u\|_{1,\mathbb{R}^2} = \|f\|_{1,\mathbb{R}^2} - 4\pi\, s, \qquad \text{for some } 1 \leq s < \|f\|_{1,\mathbb{R}^2}/4\pi, \tag{16.1}$$

then $u = u_s$ as constructed in Theorem 12.1.

Proof. We first prove the Theorem under the additional assumption

$$0 \leq f(x) \leq \frac{M}{(1+|x|^2)^{s+1}} \qquad \text{for some } M \gg 1. \tag{16.2}$$

Consider the solution pair $(v_{2,\mu}, \varphi_{2,\mu})$ introduced in Section 4. If (16.2) holds, we may choose constants M and μ so that $f \leq \varphi_{2,\mu}$. Since by (7.3) the solutions $v_{2,\mu}$ are maximal, we also have $u \leq v_{2,\mu}$. By the representation formula (8.12) with the kernel $\mathcal{K}(\cdot;\cdot)$ given in (14.2), we have for all $|x| > 4$,

$$\begin{aligned} \ln u(x) &= \ln u(0) - \frac{\ln|x|}{2\pi}\int_{\mathbb{R}^2} \{f - \lambda u\}\, dy + \int_{\mathbb{R}^2} \mathcal{K}(x;y)\{f - \lambda u\}\, dy \\ &= \ln u(0) - 2s\ln|x| + \int_{\mathbb{R}^2} \mathcal{K}(x;y)\{f - \lambda u\}\, dy \\ &\geq \ln u(0) - 2s\ln|x| - \lambda\int_{\mathbb{R}^2} \mathcal{K}_+(x;y) v_{2,\mu}(y)\, dy - \int_{\mathbb{R}^2} \mathcal{K}_-(x;y) f(y)\, dy \\ &\geq \gamma\{u(0), s, M, \lambda, \mu\} - 2s\ln|x|. \end{aligned} \tag{16.3}$$

Thus there exists a positive constant C, such that

$$u(x) \geq \frac{C}{|x|^{2s}} \qquad \forall\, |x| > 4. \tag{16.4}$$

Let $\varepsilon > 0$ be defined by

$$\|f\|_{1,\mathbb{R}^2} = 4\pi(s+\varepsilon),$$

and for $s' \in (s, s+\varepsilon)$ consider the solution $u_{s'}$ constructed in Section 12. By the construction procedure of Theorem 12.1, such a solution is obtained as the increasing limit of solutions $\{u_{s',j}, f_j\}_{j\in\mathbb{N}}$, where f_j are bounded, have compact support and increase to f. Moreover the $u_{s',j}$ satisfy

$$u_{s',j}(x) \le \frac{C_j}{(1+|x|^2)^{s'}}, \qquad \forall x \in \mathbb{R}^2,$$

for some constant C_j. This and (16.4) imply that $u \ge u_{s',j}$ for all $j \in \mathbb{N}$ and therefore $u \ge u_{s'}$, for all $s \in (s, s+\varepsilon)$. Since $u_{s'}$ is increasing as $s' \searrow s$, we also have $u \ge u_s$. On the other hand it follows from (16.1) that $\|u\|_{1,\mathbb{R}^2} = \|u_s\|_{1,\mathbb{R}^2}$. Thus $u = u_s$. To remove the assumption (16.2), consider the increasing sequence of functions

$$f_n(x) = \min\left\{ f(x) ; \frac{n}{(1+|x|^2)^{s+1}} \right\},$$

and let n_o be so large that $\|f_n\|_{1,\mathbb{R}^2} > 4\pi s$ for all $n \ge n_o$. Then by the procedure of Theorem 12.1 we may construct a solution u_n corresponding to f_n, satisfying $u_n \le u$ and

$$\int_{\mathbb{R}^2} u_n\, dy \ge \int_{\mathbb{R}^2} f_n\, dy - \int_{\mathbb{R}^2} f\, dy + \int_{\mathbb{R}^2} u\, dy = \int_{\mathbb{R}^2} f_n\, dy - 4\pi s.$$

Therefore for all $|x| > 4$

$$-\frac{\ln|x|}{2\pi} \int_{\mathbb{R}^2} \{f_n - \lambda u_n\}\, dy \ge -2s\ln|x|.$$

By estimates similar to (16.3), this implies that u_n satisfies (16.4) for a constant C depending upon n. Thus if $u_{s,n}$ is the solution constructed by Theorem 12.1 corresponding to f_n, we must have $u \ge u_n \ge u_{s,n}$. Letting $n \nearrow \infty$ proves the assertion. $\square$

PART V
Existence and Non-Existence of Solutions for $N \geq 3$

17. Some Examples

The explicit solutions introduced in Section 4 for $N = 2$, have a counterpart for $N \geq 3$.

17.1. Explicit Solutions in $L^1(\mathbb{R}^N)$

For μ, $s > 0$, the following is a two-parameter solution pair to (1.1) in $\mathbb{R}^N$,

$$w_{s,\mu}(x) = \frac{M}{(\mu + |x|^2)^s}, \tag{17.1}$$

$$\phi_{s,\mu}(x) = \frac{\lambda M}{(\mu + |x|^2)^s} + \frac{4s\mu}{(\mu + |x|^2)^2} + \frac{2(N-2)s}{(\mu + |x|^2)}. \tag{17.2}$$

If $s > N/2$, then $w_{s,\mu} \in L^1(\mathbb{R}^N)$. However $\phi_{s,\mu} \notin L^1(\mathbb{R}^N)$ for any value of s. By direct calculation one verifies that

$$\lim_{\rho\to\infty} \frac{1}{\ln\rho} \int_{B_\rho} \frac{\phi_{s,\mu}(x)}{|x|^{N-2}}\, dx = \begin{cases} 2s(N-2)\omega_N, & \text{if } s > 1; \\ \{\lambda M + 2(N-2)\}\omega_N, & \text{if } s = 1; \\ \infty, & \text{if } s < 1. \end{cases} \tag{17.3}$$

Therefore, for any $s > 0$,

$$\lim_{\rho\to\infty} \frac{1}{\ln\rho} \int_{B_\rho} \frac{\phi_{s,\mu}(x)}{|x|^{N-2}}\, dx \geq 2(N-2)\omega_N. \tag{17.4}$$

17.2. Explicit Solutions not in $L^1(\mathbb{R}^N)$

For a positive constant M and parameters $\mu > 1$ and $\alpha \in \mathbb{R}$, set

$$v_{\alpha,\mu}(x) = \frac{M}{(\mu + |x|^2)\,[\ln(\mu + |x|^2)]^\alpha}; \tag{17.5}$$

$$\begin{aligned} \varphi_{\alpha,\mu}(x) &= \lambda v_{\alpha,\mu}(x) + \frac{4\mu}{(\mu + |x|^2)^2} + \frac{4\alpha\mu}{(\mu + |x|^2)^2 \ln(\mu + |x|^2)} \\ &+ \frac{2(N-2)}{(\mu + |x|^2)} - \frac{4\alpha|x|^2}{(\mu + |x|^2)^2\,[\ln(\mu + |x|^2)]^2} + \frac{2(N-2)\alpha}{(\mu + |x|^2)\ln(\mu + |x|^2)}. \end{aligned} \tag{17.6}$$

One verifies that $(v_{\alpha,\mu}, \varphi_{\alpha,\mu})$ is a family of solution pairs, parametrized with $\mu > 1$ and $\alpha \in \mathbb{R}$. For any value of α, neither $v_{\alpha,\mu}$ nor $\varphi_{\alpha,\mu}$ are in $L^1(\mathbb{R}^N)$. The functions $v_{\alpha,\mu}$ are always positive for any $\alpha \in \mathbb{R}$, and the forcing terms $\varphi_{\alpha,\mu}$ are always positive for any $\alpha \in \mathbb{R}$, provided the parameter M is taken sufficiently large, depending on α and μ. By direct calculation one verifies that

$$\lim_{\rho\to\infty} \frac{1}{\ln\rho} \int_{B_\rho} \frac{\varphi_{\alpha,\mu}(x)}{|x|^{N-2}}\, dx = \begin{cases} 2(N-2)\omega_N, & \text{if } \alpha > 0; \\ \{\lambda M + 2(N-2)\}\omega_N, & \text{if } \alpha = 0; \\ \infty, & \text{if } \alpha > 0. \end{cases} \tag{17.7}$$

In either case we have

$$\lim_{\rho\to\infty}\frac{1}{\ln\rho}\int_{B_\rho}\frac{\varphi_{\alpha,\mu}(x)}{|x|^{N-2}}\,dx \ge 2(N-2)\omega_N, \qquad \forall\alpha\in\mathbb{R}, \tag{17.8}$$

and in addition

$$\lim_{\rho\to\infty}\frac{1}{\ln\rho}\int_{B_\rho}\frac{\{\varphi_{\alpha,\mu}(x)-\lambda v_{\alpha,\mu}(x)\}}{|x|^{N-2}}\,dx = 2(N-2)\omega_N, \qquad \forall\alpha\in\mathbb{R}, \qquad (17.8)'$$

18. Non-Existence of Solutions for $N \geq 3$

The non-existence results of Proposition 11.1 have a N-dimensional version provided f satisfies the decay estimate (6.2).

Proposition 18.1. *Let $N \geq 3$, and assume that $f \in L^1_{\mathrm{loc}}(\mathbb{R}^N)$ satisfies (6.2) and*

$$\lim_{\rho\to\infty}\frac{1}{\ln\rho}\int_{B_\rho}\frac{f(x)}{|x|^{N-2}}\,dx < 2\omega_N(N-2). \tag{18.1}$$

Then (1.1) has no solution.

Proof. Since f satisfies (6.2), by the estimation (6.3) we have

$$u(x) \le \frac{M}{1+|x|^2}, \qquad \text{for some } M>0, \quad \forall x\in\mathbb{R}^N. \tag{18.2}$$

Proceeding as in the proof of Proposition 11.1, we arrive at the analog of (11.3), i.e.,

$$\ln u(x) \le \gamma(M) - 2\ln\rho + \int_{B_\rho} G_\rho(x;y)f(y)\,dy, \qquad \forall |x|<\rho. \tag{18.3}$$

In this formula $(x,y) \to G_\rho(x;y)$ is the Green's function for the Laplacian in $B_\rho \subset \mathbb{R}^N$ constructed in (8.5)–(8.6). By (8.8) the function $\Phi(\cdot;\cdot)$ is non-negative. We evaluate (18.3) at $x=0$ and discard the non positive term arising from $\Phi(0;y)$. This gives

$$\begin{aligned}\ln u(0) &\le \gamma(M) - 2\ln\rho + \frac{1}{(N-2)\omega_N}\int_{B_\rho}\frac{f(y)}{|y|^{N-2}}\,dy\\ &\le \gamma(M) - \frac{\ln\rho}{(N-2)\omega_N}\left\{2(N-2)\omega_N - \frac{1}{\ln\rho}\int_{B_\rho}\frac{f(y)}{|y|^{N-2}}\,dy\right\},\end{aligned}$$

for all $\rho>0$. If (18.1) holds, letting $\rho\to\infty$ in this inequality, yields $u(0)=0$. Thus $x \to u(x) = 0$ by the Harnack estimate (7.7). □

Similar non-existence results where the right hand side of (18.1) is a generic constant depending upon N, have been given in [11].

18.1. Remarks on the Limiting Case of (18.1)

The proposition is in agreement with the example in (17.1)–(17.2). Indeed, by (17.4), the forcing terms $\phi_{s,\mu}$ violate (18.1). It also agrees with the example in (17.5)–(17.6), as the forcing terms $\varphi_{\alpha,\mu}$ satisfy (17.8).

The non-existence condition of Proposition 18.1 is sharp in the sense that, unlike the case $N = 2$, equality in (18.1) is not allowed. Indeed, the first of (17.7) exhibits a family of forcing terms $\varphi_{\alpha,\mu}$ satisfying the limiting case of (18.1), to which correspond the explicit solutions $v_{\alpha,\mu}$.

On the other hand there exist non-negative functions $f \in L^1_{\text{loc}}(\mathbb{R}^N)$ satisfying (18.1) with equality, i.e.,

$$\lim_{\rho\to\infty} \frac{1}{\ln \rho} \int_{B_\rho} \frac{f(x)}{|x|^{N-2}}\, dx = 2(N-2)\omega_N, \tag{18.1$'$}$$

for which (1.1) has no solution. To construct such a function we take $N = 3$ and denote by $\overline{x} = (x_1, x_2)$ points in $\mathbb{R}^2$ and by $x = (\overline{x}, x_3)$ points in $\mathbb{R}^3$. Let f be the characteristic function of the disc of radius 2 centered at the origin, i.e.,

$$f(\overline{x}) = \begin{cases} 1 & \text{if } |\overline{x}| < 2; \\ 0 & \text{otherwise}; \end{cases} \qquad \text{which satisfies} \quad \int_{\mathbb{R}^2} f(\overline{x})\, d\overline{x} = 4\pi. \tag{18.4}$$

Regarding now f as a function of the three variables $(\overline{x}, x_3)$, one verifies that

$$\lim_{\rho\to\infty} \frac{1}{\ln \rho} \int_{\{|x|<\rho\}} \left(\frac{f(\overline{x})}{|x|} \right) dx8.5 \tag{1}$$

$$= 2 \int_{\{|\overline{x}|<2\}} \left(\lim_{\rho\to\infty} \frac{1}{\ln \rho} \int_0^{\sqrt{(\rho/|\overline{x}|)^2-1}} \left(1+t^2\right)^{-1/2} dt \right) d\overline{x} = 8\pi.$$

Thus f satisfies (18.1)$'$. Assume that (1.1), viewed in $\mathbb{R}^3$ and with f given by (18.4), has a solution $(\overline{x}, x_3) \to u(\overline{x}, x_3)$. By Corollary 10.2 we may assume that such a u is maximal. By Remark 10.1 the solution pair (u, f), can be constructed as the decreasing limit of the maximal solutions pairs $(u_\varepsilon, f + \lambda\varepsilon)$. On the other hand, regarding (1.1) as in $\mathbb{R}^2$, to each $f + \lambda\varepsilon$ there exists a unique maximal solution $\overline{x} \to \widetilde{u}_\varepsilon(\overline{x})$. Regarding now $\widetilde{u}$ as a solution in $\mathbb{R}^3$, it follows that by maximality $u(\overline{x}, x_3) = \widetilde{u}(\overline{x})$. Consequently, the maximal solution of (1.1) in $\mathbb{R}^3$ is independent of x_3 and therefore it can be regarded as a solution of (1.1) in $\mathbb{R}^2$. We conclude that if (1.1) with f given by (18.4) has a solution in $\mathbb{R}^3$, then it has a solution also in $\mathbb{R}^2$. This however contradicts Proposition 11.1, for a datum f satisfying the second of (18.4).

18.2. Remarks on Non-Existence of Solutions

Let $f \in L^1_{\text{loc}}(\mathbb{R}^N)$ be non-negative, assume that the limit in (18.1) exists, finite or infinite, and set $J_2 = \|f\|_{1,\mathbb{R}^2}$ and

$$J_N = \lim_{\rho\to\infty} \frac{1}{\ln \rho} \int_{\{|x|<\rho\}} \frac{f(x)}{|x|^{N-2}}\, dx, \qquad \text{for} \quad N \geq 3.$$

Regarding now f as a function of the $(N+1)$ variables (x,η), we set

$$J_{N+1} = \lim_{\rho\to\infty} \frac{1}{\ln\rho} \int_{\{|(x,\eta)|<\rho\}} \frac{f(x)}{|(x,\eta)|^{N-1}}\, dx d\eta.$$

The numbers J_N and J_{N+1} satisfy the recursive equalities,

$$J_{N+1} = J_N \frac{N-1}{N-2} \frac{\omega_{N+1}}{\omega_N}, \qquad N = 2,3,\dots. \tag{18.6}$$

Assume first that $N \geq 3$. Then from the definition of J_{N+1} we compute,

$$\begin{aligned} J_{N+1} &= 2 \lim_{\rho\to\infty} \frac{1}{\ln\rho} \int_{\{|x|<\rho\}} \frac{f(x)}{|x|^{N-2}} \left(\int_0^{\sqrt{(\rho/|x|)^2-1}} (1+t^2)^{-\frac{N-1}{2}} dt \right) dx \\ &\leq \lim_{\rho\to\infty} \frac{1}{\ln\rho} \int_{\{|x|<\rho\}} \frac{f(x)}{|x|^{N-2}} \left(\int_0^{\infty} (1+t^2)^{-\frac{N-1}{2}} dt \right) dx \\ &= 2\, J_N \int_0^\infty (1+t^2)^{-\frac{N-1}{2}} dt. \end{aligned}$$

On the other hand, for any $n>1$,

$$\begin{aligned} J_{N+1} &\geq 2 \lim_{\rho\to\infty} \frac{1}{\ln\rho} \int_{\{|x|<\rho/n\}} \frac{f(x)}{|x|^{N-2}} \left(\int_0^{\sqrt{(\rho/|x|)^2-1}} (1+t^2)^{-\frac{N-1}{2}} dt \right) dx \\ &\geq \lim_{\rho\to\infty} \frac{2}{\ln\rho} \int_{\{|x|<\rho/n\}} \frac{f(x)}{|x|^{N-2}} \left(\int_0^{\sqrt{n^2-1}} (1+t^2)^{-\frac{N-1}{2}} dt \right) dx \\ &= 2\, J_N \int_0^{\sqrt{n^2-1}} (1+t^2)^{-\frac{N-1}{2}} dt. \end{aligned}$$

Therefore, since $n>1$ is arbitrary,

$$\begin{aligned} J_{N+1} &= 2\, J_N \int_0^\infty (1+t^2)^{-\frac{N-1}{2}} dt = 2\, J_N \int_0^{\pi/2} \sin^{N-3}\tau d\tau \\ &= 2\, J_N \frac{N-1}{N-2} \int_0^{\pi/2} \sin^{N-1}\tau d\tau = J_N \frac{N-1}{N-2} \frac{\omega_{N+1}}{\omega_N}. \end{aligned}$$

For $N=2$ the recursive formula is proved by a calculation similar to (18.5).

Let $(x_1,x_2) \to f(x_1,x_2) \in L^1_{\mathrm{loc}}(\mathbb{R}^2)$ be non-negative and satisfying $\|f\|_{1,\mathbb{R}^2} > 4\pi$. Then, by Theorem 12.1, there is a solution u of (1.1) in $\mathbb{R}^2$. Such a u can also be regarded as a solution of (1.1) in $\mathbb{R}^N, \forall N \geq 2$. Thus for every function f of two variables, such that $J_2 > 2\omega_2$, there is a solution of (1.1) in $\mathbb{R}^N$. Regarding f as a function of N variables, by virtue of (18.6), we have $J_N > 2(N-2)\omega_N$, in agreement with (18.1).

19. Constructing Solutions from Given Ones

Suppose that for a given non-negative $\psi \in L^1_{\rm loc}(\mathbb{R}^N)$, a solution v to (1.1) is known. Then for any non-negative $f \in L^1_{\rm loc}(\mathbb{R}^N)$, a solution u to (1.1) can be constructed provided f is not too small with respect to ψ in the sense that

$$\lim_{\rho\nearrow\infty} \frac{1}{\omega_N(N-2)} \int_{B_\rho} \frac{(\psi-f)_+}{|y|^{N-2}}\,dy = C < \infty. \tag{19.1}$$

Proposition 19.1. *Let (v,ψ) be a known solution pair to (1.1). Then for every f satisfying (19.1) for some $C>0$, there exists a solution u to (1.1) corresponding to such a f.*

Proof. Having fixed a non-negative $f \in L^1_{\rm loc}(\mathbb{R}^N)$, set

$$\varphi = \max\{\psi;\, f\}. \tag{19.2}$$

A solution pair (w,φ) can be constructed by the method indicated in Section 3.1. Indeed the known solution v corresponding to ψ serves as a uniform lower bound on a suitable sequence of approximating solutions monotonically increasing to w. To construct a solution u corresponding to f, consider the sequence of problems

$$\lambda u_\rho - \Delta \ln u_\rho = f \quad \text{in } B_\rho; \qquad u_\rho\big|_{\partial B_\rho} = w\big|_{\partial B_\rho}. \tag{19.3}$$

Since w satisfies the the same problem with the same boundary data, we obtain, by difference

$$-\Delta \ln(u_\rho/w) = (f-\varphi) + \lambda(m-u_\rho).$$

By the Poisson representation formula

$$\begin{aligned}
\ln\frac{u_\rho(x)}{v(x)} &= \int_{B_\rho} G_\rho(x;y)(f-\varphi)\,dy + \lambda\int_{B_\rho} G_\rho(x;y)(w-u_\rho)\,dy\\
&\ge \frac{1}{\omega_N(N-2)}\int_{B_\rho}\left\{\frac{1}{|x-y|^{N-2}} - \Phi(x;y)\right\}(f-\varphi)\,dy\\
&\ge \frac{1}{\omega_N(N-2)}\int_{B_\rho}\frac{1}{|x-y|^{N-2}}(f-\varphi)\,dy.
\end{aligned} \tag{19.4}$$

In these estimations we have used the fact that $G_\rho(\cdot;\cdot)\ge 0$, that $\Phi(\cdot;\cdot)\ge 0$ and that $w \ge u_\rho$ by the maximum principle. The representation (19.4) holds for all $x \in B_\rho$. Computing it for $x=0$ yields

$$\ln\frac{u_\rho(0)}{v(0)} \ge \frac{1}{\omega_N(N-2)}\int_{B_\rho}\frac{(f-\varphi)}{|y|^{N-2}}\,dy. \tag{19.5}$$

From the definition (19.2) it follows that

$$(f-\varphi) = (f-\max\{f;\psi\}) = -(\psi-f)_+.$$

Now if (19.1) holds, from (19.5) we derive

$$u_\rho(0) \ge c_o\varphi(0) \qquad \forall\,\rho>1; \qquad c_o = e^{-C}.$$

Therefore all the approximating solutions u_ρ in the balls B_ρ, are uniformly bounded away from zero at the origin. By the Harnack estimate (7.7) they are all bounded away from zero over compact subsets of $\mathbb{R}^N$ for ρ sufficiently large. The existence of a solutions u corresponding to f now follows by a standard limiting argument. □

In particular, if f is a non-negative function in $L^1_{\rm loc}$, satisfying (19.1) for any of the functions $\phi_{s,\mu}$ or $\varphi_{\alpha,\mu}$ introduced in (17.2) and (17.6), then there is a solution u to (1.1), corresponding to such a f.

Appendix: Some Geometric Aspects of (1.1) for $N = 2$

A. Complete Metrics in $\mathbb{R}^2$

Let $\{\Sigma; ds\}$ be a 2-dimensional orientable, simply connected, non compact Riemaniann surface with metric

$$ds^2 = g_{ij}dx_i dx_j, \qquad i,j = 1,2,$$

where g_{ij} is the first fundamental form of Σ. By the *Theorema Egregium*, the Gauss curvature K can be computed only in terms of the g_{ij}, and in a system of rectangular coordinates ($g_{12} \equiv 0$) takes the form ([22] Pag. 139),

$$K = \frac{1}{2\sqrt{g_{11}g_{22}}}\left\{\frac{\partial}{\partial x_1}\frac{1}{\sqrt{g_{11}g_{22}}}\frac{\partial}{\partial x_1}g_{11} + \frac{\partial}{\partial x_2}\frac{1}{\sqrt{g_{11}g_{22}}}\frac{\partial}{\partial x_2}g_{22}\right\}. \tag{A.1}$$

If the g_{ij} are sufficiently regular, there exists a local system of rectangular coordinates and a positive, smooth function u, defined on a subset of $\mathbb{R}^2$, such that ([14] p. 15)

$$ds^2 = u(x)d|x|^2, \qquad d|x| = \sqrt{dx_1^2 + dx_2^2}, \tag{A.2}$$

From (A.1) with $g_{ii} = u$, $i = 1,2$ we compute $2K = -u^{-1}\Delta \ln u$. Moreover if $dA = u dx_1 dx_2$ denotes the element of area on Σ we have

$$K\,dA = -\tfrac{1}{2}(\Delta \ln u)\,dx_1 dx_2. \tag{A.3}$$

Therefore if u in (A.2) is defined in the whole $\mathbb{R}^2$, the total curvature of $\{\Sigma; ds\}$ is

$$\int_\Sigma K dA = -\frac{1}{2}\int_{\mathbb{R}^2} \Delta \ln u(x)\,dx, \tag{A.3$'$}$$

provided the integrals are convergent. The metric ds is complete if Σ, endowed with such a metric is a complete metric space. Equivalently ds is complete if every divergent path on Σ has infinite length ([22] p. 220). A sufficient condition on the function u in (1.5), for $\{\Sigma; ds\}$ to be complete is that ([14] p. 55) for every divergent locally rectifiable path $\sigma \in \Sigma$

$$\int_\sigma \sqrt{u(x)}\, ds = \infty. \tag{A.4}$$

We report here an example of A. Huber ([14] p. 61) i.e.,

$$ds^2 = \frac{|dx|^2}{[|x| \ln(1+|x|)]^2} \quad \text{is complete since} \quad \int_1^\infty \frac{d|x|}{|x|\ln(1+|x|)} = \infty. \qquad \text{(A.4)}'$$

Therefore completeness of $\{\Sigma; ds\}$ is related to the asymptotic behavior of $u(x)$ as $|x| \to \infty$. Completeness is also related to the total curvature by the Cohn-Vossen Theorem [9],

$$\{\Sigma; ds\} \text{ complete} \quad \Longrightarrow \quad \{\text{total curvature}\} \le 2\pi\chi, \tag{A.5}$$

where χ is the Euler-Poincaré characteristic of Σ. For a surface Σ homeomorphic to $\mathbb{R}^2$ we have $\chi(\mathbb{R}^2) = 1$.

We now take $\Sigma \equiv \mathbb{R}^2$ and combine the previous remarks by interpreting the results of our Theorems 1.1–1.3, as follows. Given any non-negative $f \in L^1(\mathbb{R}^2)$, satisfying (1.2) and any $s \in [1, s_o)$ there exists a unique non-negative function $u_s \in L^1(\mathbb{R}^2)$ such that

$$\int_{\mathbb{R}^2} 2K_s\, dA_s = -\int_{\mathbb{R}^2} \Delta \ln u_s(x)\, dx = \int_{\mathbb{R}^2} \{f - u_s\}(x)\, dx = 4\pi s. \tag{A.6}$$

Such a solution u_s generates a metric in $\mathbb{R}^2$ via the formula (A.2), with Gauss curvature K_s and element of surface area dA_s given by

$$2K_s = \frac{1}{u_s}\Delta \ln u_s, \qquad dA_s = u_s\, dx_1 dx_2.$$

If $s > 1$ such a metric cannot be complete in $\mathbb{R}^2$ since (A.6) would violate the Cohn-Vossen Theorem (A.5). The maximal solution, corresponding to $s = 1$ satisfies (A.5) with equality. However this is not sufficient to conclude that the corresponding metric is complete, since the Cohn-Vossen Theorem (A.5) only provides a necessary condition for completeness. The asymptotic behavior of Theorem 1.3, coupled with (A.4)–(A.4)$'$, permits to conclude that the metric generated by a maximal solution u is complete.

To summarize, any non-negative $f \in L^1(\mathbb{R}^2)$ with sufficiently large mass in the sense of (1.2), generates a continuous family of metrics in $\mathbb{R}^2$ and of these only the maximal is complete.

B. Surfaces Σ Conformally Equivalent to the Unit Circle

Proposition B.1. *Let $\{\Sigma; ds\}$ be a simply-connected surface with finite area, finite total curvature, and satisfying,*

$$\begin{cases} K \geq -\frac{1}{2}\lambda \text{ for some positive number } \lambda; \\ \{\text{total curvature of } \Sigma \text{ is}\} < 2\pi, \end{cases} \tag{B.1}$$

Then $\{\Sigma; ds\}$ is conformally equivalent to the unit disc.

Proof. The surface $\{\Sigma; ds\}$ can be mapped conformally either onto the unit disc or the whole plane. In the latter case ds is given by (A.2) with u defined in the whole $\mathbb{R}^2$, and with Gauss curvature given by (A.3). By the assumptions, both u and $\ln u$ are in $L^1(\mathbb{R}^2)$. By the first of (B.1), the function $f \equiv \{\lambda u - \Delta \ln u\}$ is non-negative and is in $L^1(\mathbb{R}^2)$. Moreover by the second of (B.1),

$$\{\text{total curvature}\} = \frac{1}{2}\int_{\mathbb{R}^2} \Delta \ln u(x)\, dx = \frac{1}{2}\int_{\mathbb{R}^2} \{f - \lambda u\}(x)\, dx < 2\pi.$$

Therefore (u, f) would be a solution pair of (1.1) contradicting Theorems 1.1 and 1.2. □

The non-existence part of Theorem 1.1 asserts in particular that (1.1) has no solution if $f \equiv 0$. This fact is contained in a result of Ossermann [20], i.e., the differential inequality

$$\Delta v \geq \varepsilon e^v \quad \text{has no solution for any } \varepsilon > 0.$$

Osserman then used this lack of existence to prove that a simply connected surface $\{\Sigma; ds\}$ with Gauss curvature K everywhere uniformly negative on Σ, is conformally equivalent to the unit disc ([20], Theorem 2, pp. 1646). The assertion of Proposition B.1 is in the same spirit.

If (A.5) holds with equality, the Cohn-Vossen Theorem can be regarded as some sort of the Gauss-Bonnet Theorem, when compactness of Σ is replaced with completeness. The assertion in (A.5), even with equality, is only a necessary condition of completeness. It is natural to ask for which surfaces $\{\Sigma; ds\}$, condition (A.5) holding with equality, would be a necessary and sufficient condition for completeness.

Proposition B.2. *Let $\{\Sigma; ds\}$ be a simply connected surface, with finite area, homeomorphic to $\mathbb{R}^2$, and satisfying*

$$\begin{cases} K \geq -\frac{1}{2}\lambda \text{ for some positive number } \lambda; \\ \{\text{total curvature of } \Sigma \text{ is}\} = 2\pi. \end{cases} \tag{B.1}$$

Then $\{\Sigma; ds\}$ is complete.

Proof. Let u be the positive function that determines the metric ds by (A.2). Both u and $\ln u$ are in $L^1(\mathbb{R}^2)$ and by the first of (B.1)$'$, the function $f = \{\lambda u - \Delta \ln u\}$ is non-negative and is in $L^1(\mathbb{R}^2)$. Thus, by the second of (B.1)$'$, u is a solution of (1.1) for such an f and satisfies $(1.3)_s$ for $s = 1$. By Theorem 1.3 and (A.4)–(A.4)$'$, the metric ds generated by such a u is complete. □

Remark B.1. For a geometric interpretation of the parabolic equation associated with (1.1) we refer to [13].

C. Prescribing the Gaussian Curvature in $\mathbb{R}^2$

Let $f \in C_o(\mathbb{R}^2)$ be non-negative and set,

$$h(x) = \frac{-1}{2\pi} \int_{\mathbb{R}^2} \ln|x-y|\, f(y)\, dy. \tag{C.1}$$

If u is a solution of (1.1) for such a f, setting $\ln u = 2v + h$, transforms (1.1) into,

$$-\Delta v + K\, e^{2v} = 0, \qquad \text{where} \qquad K = e^h. \tag{C.2}$$

or equivalently,

$$2\,K(x) = \frac{1}{w}\Delta \ln w \quad \text{in } \mathbb{R}^2, \qquad \text{where} \qquad w = e^{2v}. \tag{C.3}$$

From (A.7) it follows that K is the Gauss curvature of the metric generated by w. It is then natural to seek conditions for a prescribed function $K(\cdot)$ to be the Gaussian curvature of a conformal meetric in $\mathbb{R}^2$. In this form the question has been investigated in [1,6,7,8,16,17,18,21]. Perhaps the main points of the theory is the non-existence result of [21], and the existence results of [18, 7,8]. It is shown by Sattinger [21] that (C.3) has no solution if

$$K(x) \geq \frac{c}{|x|^2} \quad \text{for} \quad |x| \gg 1, \tag{C.4}$$

for some $c > 0$. From (C.1) and (C.2) one computes for $|x| \gg 1$,

$$h(x) = \alpha \ln\frac{1}{|x|} + O(1) \quad \text{i.e.} \quad K(x) = \frac{c}{|x|^\alpha} \quad \text{where } \alpha = \frac{\|f\|_{1,\mathbb{R}^2}}{2\,\pi}. \tag{C.5}$$

Therefore if $\|f\|_{1,\mathbb{R}^2} \leq 4\pi$, then (C.4) holds true and (C.2) has no solution. This corresponds to the non-existence part of Theorem 1.1. The existence of infinitely many solutions to (C.2), when (C.4) is violated, was established in [18], and the existence of a maximal solution was proved in [7,8].

We do not know of an analog of our Theorem 1.2, establishing a sharp relationship between the mass of the solutions u_s, $s \in [1, s_o)$, and the mass of f.

Thus in the cases when (1.1) can be cast in the form (C.2) we recover all the known results and we give, in addition, a precise description of the possible solutions.

These analogies however, hold whenever the integral defining $h(\cdot)$ on the right hand side of (C.1) is convergent and, in addition whenever one can establish the asymptotic behaviour (C.5). This is clearly not the case for a general $f \in L^1(\mathbb{R}^2)$.

Acknowledgement

We would like to thanks R. Schoen and Jing-Yi Chen, for enlightening conversations on the geometrical aspects of the problem.

References

[1] P. Aviles, "Conformal Complete Metrics with Prescribed non-negative Gaussian Curvature in $\mathbb{R}^2$", Inven. Math. **83** (1986), 519–544.

[2] P. Bénilan, H. Brézis and M. G. Crandall, "A Semilinear Equation in $L^1(\mathbb{R}^N)$", Ann. Scuola Norm. Sup. Pisa **2** (1975), 523–555.

[3] H. Brézis, "Semilinear Equations in $\mathbb{R}^N$ Without Conditions at Infinity", Appl. Math. Optim. **12** (1984), 271–282.

[4] H. Brézis and F. Merle, "Uniform Estimates and Blow-up for Solutions of $-\Delta u = v(x)e^u$ in Two Dimensions", Commun. in Part. Diff. Equ. **16** (8&9) (1991), 1223–1253.

[5] J. P. Burelbach, S. G. Bankoff and S. H. Davis, "Non Linear Stability of Evaporating/ Condensing Liquid Films", J. Fluid Mech. **195** (1988), 463–494.

[6] K-S. Cheng and T. C. Lin, "The Structure of Solutions of a Semilinear Elliptic Equation, Trans. AMS **332** (2) (1992), 535–554.

[7] K-S. Cheng and W.-M. Ni, "On the Structure of the Conformal Gaussian Curvature Equation in $\mathbb{R}^2$-I", Duke Math. J. **62** (1991), 721–737.

[8] K-S. Cheng and W.-M. Ni, "On the Structure of the Conformal Gaussian Curvature Equation in $\mathbb{R}^2$-II", Math. Ann. **290** (1991), 681–687.

[9] S. Cohn-Vossen, "Kürzeste Wege und Totalkrümmung auf Flächen", Compositio Math. **2** (1935), 69–133.

[10] M. G. Crandall and L. C. Evans, "A Singular Semilinear Equation in $L^1(\mathbb{R})$", Trans. AMS **225** (1977), 188–203.

[11] P. Daskalopuolos and M. A. Del Pino, "On Fast Diffusion Nonlinear Heat Equations and a Related Singular Elliptic Problem", Indiana Univ. Math. J. **43** (2) (1994), 703–728.

[12] S.H. Davis, E. DiBenedetto and D. J. Diller, "Some A Priori Estimates For A Singular Evolution Equation Arising In Thin Film Dynamics", SIAM J. on Math. Analysis **27** (3) (1996), 638–660.

[13] E. DiBenedetto and D. J. Diller, *About a Singular Parabolic Equation arising in Thin Film Dynamics and in the Ricci flow for Complete* $\mathbb{R}^2$. (Special volume for the 70th birthday of C. Pucci. Edited by G. Talenti, E. Vesentini and P. Marcellini; Marcel Dekker New York Jan. 1996).

[14] A. Huber, "On Subharmonic Functions and Differential Geometry in the Large", Comm. Math. Helvetici **32** (1957), 13–72.

[15] J.B. Keller, "Electrohydrodynamics I", J. Rat. Mech. Analysis **5** (1956), 715–724.

[16] R. McOwen, "On the Equation $\Delta u + K^{2u} = f$ and Prescribed Negative Curvature in $\mathbb{R}^2$", J. Math. Anal. Appl. **103** (1984), 365–370.

[17] R. McOwen, "Conformal Metrics in $\mathbb{R}^2$ with Prescribed Gaussian Curvature and Positive Total Curvature", Ind. Univ. Math. J. **34** (1985), 97–104.

[18] W.-M. Ni, "On the Elliptic Equation $\Delta u + Ke^{2u} = 0$ and Conformal Metrics with Prescribed Gaussian Curvatures", Invent. Math. **66** (1982), 343–352.

[19] O. A. Oleinik, "On the Equation $\Delta u + k(x)e^u$", Russian Math. Surveys **33** (2) (1978), 243–244.

[20] R. Osserman, "On the Inequality $\Delta u \geq f(u)$", Pacific J. Math. **7** (1957), 1641–1647.

[21] D.H. Sattinger, "Conformal Metrics in $\mathbb{R}^2$ with Prescribed Curvature", Indiana Univ. Math. J. **22** (1972), 1–4.

[22] J.J. Stoker, *Differential Geometry*, Wiley-Interscience, New York, 1969.

[23] M. B. Williams and S.H. Davis, "Non Linear Theory of Film Rupture", Jour. of Colloidal and Interface Sc. **90** (1) (1982), 220–228.

E. DiBenedetto
Department of Mathematics
Northwestern University
Evanston, Illinois
60208 USA

D.J. Diller
Department of Biological Structure
University of Washington
Seattle
WA 98195 USA

Progress in Nonlinear Differential Equations
and Their Applications, Vol. 35
© 1999 Birkhäuser Verlag Basel/Switzerland

Moving Surfaces and Abstract Parabolic Evolution Equations

Joachim Escher and Gieri Simonett

Dedicated to Herbert Amann on the occasion of his 60th birthday

1. Introduction

It is the purpose of this paper to give a survey over some recent developments in the theory of classical solutions to elliptic and parabolic problems involving moving surfaces. Problems of this type do not satisfy a superposition principle for solutions and, hence, carry an inherent nonlinear structure. In fact, it turns out that most of the equations describing the evolution of surfaces are of quasilinear or even of fully nonlinear type. Additionally, these equations are often of a nonlocal nature.

From a mathematical point of view it therefore seems tempting – as a first step in the analysis – to look for weak solutions to these nonlinear equations. In many applications, however, the corresponding mathematical model is obtained under the assumption of the existence of a sharp and smooth surface or moving boundary. Of course, one can try to follow a two-step procedure of first constructing weak solutions and then, in a second step, analyzing the actual regularity of a weak solution. Both steps are often closely tied to a comparison principle. Hence, for problems without the luxury of a comparison or maximum principle the above program is not at all obvious to realize. We mention the quasi-stationary Stefan problem with surface tension, the Mullins-Sekerka model, or the surface diffusion flow which do not satisfy a comparison principle. For these problems neither existence of (even weak) solutions nor uniqueness of (even classical) solutions was established until quite recently, see [34, 35, 36, 40, 20, 21, 22]. On the other hand, there are moving boundary problems for which one can guarantee existence of weak solutions (e.g. the Hele-Shaw flow without surface tension, the Stefan problem with Gibbs-Thomson corrections) but for which the actual regularity of weak solutions is still far from being understood.

There is a different approach to problems with moving surfaces in which one seeks a unique classical solution from the very beginning. In this approach one

---

The research of the second author has been partially supported by the Vanderbilt University Research Council and by NSF Grant DMS-9801337.

transforms a given moving surface problem by means of appropriate diffeomorphisms into a problem on a fixed reference domain and then evolves the diffeomorphisms. In this paper we present several problems with moving surfaces for which the transformed equations lead to nonlinear evolution equations. It turns out that these equations are of parabolic type (in the sense that the corresponding linearized equations can be treated by means of analytic semigroup theory) and hence are accessible to well-established techniques for nonlinear parabolic evolution equations, see [5, 6, 9, 24, 56]. This approach does not rely on a comparison or maximum principle and, in fact, will provide existence and uniqueness results of classical solutions to problems for which neither existence of weak solutions nor uniqueness of classical solutions was previously known, see the results in Sections 2 and 3.

A further advantage of this method is that we are able to specify 'large' function spaces of initial data on which the evolution equations under consideration generate local semiflows. This means in particular that there is no loss of regularity as long as one does not encounter singularities. Furthermore, we show that these semiflows possess a regularizing property in the sense that for positive time the solutions are smooth or even real analytic in space and time.

Any information on the dynamical behavior of these semiflows immediately translates into information about the corresponding moving surface. This approach allows, for instance, to determine stability properties of equilibria, leading to new global existence results [36, 38, 40], even for well-studied problems like the averaged mean curvature flow [39].

As already mentioned above, the analysis of the linearized problems is fundamental for this approach. In particular, we need sharp parameter-dependent a priori estimates for the corresponding linear operators in order to ensure that these operators generate analytic semigroups on appropriate function spaces. However, we cannot rely on well-established a priori estimates for elliptic differential operators, since the linear operators coming from nonlocal moving surface problems are usually not differential operators, but rather pseudo-differential operators. For example, the linearization of the one-sided Hele-Shaw operator without surface tension involves a first order hyperbolic differential operator and a singular Green operator, see [37]. A similar situation, with an additional first order elliptic pseudo-differential operator, is also described in Lemma 2.2 below. In order to obtain sharp resolvent estimates for operators of this type we strongly rely on results and techniques of H. Amann's theory of Fourier multipliers on Besov spaces [6, 7].

2. Flows through porous media

We consider the flow of a quasi-incompressible Newtonian fluid through a porous medium. More precisely, let $\mathbb{R}^3$ represent a homogeneous, isotropic, and deformable porous medium. Let furthermore $\Sigma := \mathbb{R}^2 \times \{0\}$ be a fixed impermeable layer in $\mathbb{R}^3$ and assume that some part of the region above the layer Σ is occupied by the fluid. Finally, we assume that there is a sharp interface Γ_t separating at every instant t the wet part Ω_t from the dry part $\mathbb{R}^3 \setminus cl(\Omega_t)$.

In order to describe a hydrological model for this flow, let ρ, v, p, and ε denote the density, the velocity, the pressure, and the strain tensor of the fluid phase, respectively. Under suitable simplifying hypotheses, in particular assuming that spatial changes of ρ and ε are much smaller than the corresponding temporal changes, the basic mass balance equation of the fluid phase is given by

$$\nabla \cdot v + n\beta\partial_t p + \partial_t \varepsilon = 0, \tag{2.1}$$

where β is the so-called fluid's coefficient of compressibility and n is the solid's porosity, cf. [12], p.300. We now assume that

$$\rho = \text{const.} \qquad \text{and} \qquad n = \text{const.}$$

This assumption particularly permits to introduce the so-called velocity potential or piezometric head

$$u(x,y) := \frac{p(x,y)}{g\rho} + y, \qquad (x,y) \in \mathbb{R}^2 \times \mathbb{R}, \tag{2.2}$$

where g is the gravity acceleration, see [12], p. 175, [23], p. 32, or [29], p. 24. Assuming that the motion is governed according to Darcy's law $v = -k\nabla u$, where $k > 0$ stands for the hydraulic conductivity, the mass balance equation (2.1) becomes

$$-k\Delta u + \rho g n\beta\partial_t u + \partial_t \varepsilon = 0. \tag{2.3}$$

Finally, we introduce the compressibility α of the porous medium (cf. [12], p. 308) and we assume that the dilatation-pressure relation $\varepsilon = \alpha p$ holds. This relation is fulfilled if, e.g., one assumes that displacements of the solid phase occur in vertical direction only, an assumption usually taken for granted in hydrology of groundwater, see [12], p. 310. Using the relation $\varepsilon = \alpha p$ in (2.3) we find

$$S_0\partial_t u - k\Delta u = 0 \quad \text{in} \quad \Omega_t, \tag{2.4}$$

where $S_0 := \rho g(\alpha + n\beta) \geq 0$ is called the specific storativity. Observe that $S_0 = 0$ for the flow of an incompressible Newtonian fluid through a rigid porous medium, since in this case $\alpha = \beta = 0$.

Equation (2.4) governs the motion of the fluid in the domain Ω_t. Let us now discuss the boundary conditions on $\partial\Omega_t = \Sigma \cup \Gamma_t$. First, recall that Σ is assumed to be impermeable for the fluid. Hence we have the no-flux condition

$$\partial_3 u = 0 \quad \text{on} \quad \Sigma.$$

Next, we assume that the interface separating the fluid from the air is given as a graph over $\mathbb{R}^2$ of a time-dependent function f, i.e.,

$$\Gamma_t := \Gamma_{f(t)} := \text{graph}(f(t,\cdot)).$$

Moreover, the free interface $\Gamma_{f(t)}$ is characterized by

$$(x,y) \in \Gamma_{f(t)} \qquad \Longleftrightarrow \qquad y = \sup\{z \in \mathbb{R}\,;\, p(t,x,z) > 0\}.$$

If we neglect surface tension effects the pressure p is continuous, see [29], p. 18. Hence, normalizing the atmospheric pressure to be equal to 0, we find

$$p(t,x,y)=0 \quad \text{for} \quad (x,y)\in\Gamma_{f(t)}. \tag{2.5}$$

Because of $y=f(t,x)$, we obtain from (2.2) and (2.5) the boundary condition

$$u(t,x,f(t,x))=f(t,x), \quad (x,t)\in\mathbb{R}^2\times[0,\infty). \tag{2.6}$$

Finally, setting $F(t,z):=y-f(t,x)$ for $z=(x,y)\in\mathbb{R}^2\times\mathbb{R}$ and $t\geq 0$, the interfaces $\{\Gamma_{f(t)}\,;\,t\geq 0\}$ are characterized by the conservation property that F is identically equal to 0 on $\{\Gamma_{f(t)}\,;\,t\geq 0\}$. Consequently, we get

$$\frac{d}{dt}F(t,z)=\partial_t F(t,z)+\nabla f(t,z)z'=0 \quad \text{on} \quad \Gamma_{f(t)}. \tag{2.7}$$

Again by Darcy's law the velocity z' is given by $-k\nabla u$. Since $\partial_t F=-\partial_t f$ and $\nabla F=(-\nabla_x f,1)$, equation (2.7) becomes

$$\partial_t f+k\sqrt{1+|\nabla f|^2}\partial_\nu u=0 \quad \text{on} \quad \Gamma_{f(t)},$$

where ∂_ν stands for differentiation with respect to the outer unit normal. Summarizing, we obtain the following set of equations:

$$\begin{cases} S_0\partial_t u-k\Delta u=0 & \text{in} \quad \Omega_t\\ \partial_3 u=0 & \text{on} \quad \Sigma\\ u=f & \text{on} \quad \Gamma_t\\ \partial_t f+k\sqrt{1+|\nabla f|^2}\partial_\nu u=0 & \text{on} \quad \Gamma_t \end{cases} \tag{2.8}$$

where the constants k, S_0 satisfy $k>0$ and $S_0\geq 0$. Of course, system (2.8) has to be completed by initial conditions and, as it turns out, by a normalization at ∞. We will specify appropriate initial conditions later depending on whether $S_0>0$ or $S_0=0$.

Let us first take (2.8) as a model for the following general situation: Let $n\in\mathbb{N}$ with $n\geq 1$ be fixed and set

$$A_0:=\{f\in BC^1(\mathbb{R}^n,\mathbb{R})\,;\,\inf_{x\in\mathbb{R}^n}f(x)>0\}.$$

Given $f\in A_0$, define

$$\Omega_f:=\{(x,y)\in\mathbb{R}^{n+1}\,;\,0<y<f(x)\}.$$

The boundary of this unbounded C^1-domain is given by $\Sigma\cup\Gamma_f$, where $\Sigma:=\mathbb{R}^n\times\{0\}$ and $\Gamma_f:=\text{graph}(f)$.

2.1. Incompressible fluids in rigid porous media

If we consider the flow of an incompressible fluid in a rigid porous medium, the specific storativity S_0 vanishes, cf. (2.4). In this case (2.8) is a model case for the following moving boundary problem

$$\begin{cases} \Delta u = 0 & \text{in } \Omega_f \\ \partial_\nu u = 0 & \text{on } \Sigma \\ u = f & \text{on } \Gamma_f \\ \lim_{|(x,y)|\to\infty} u(\cdot,(x,y)) = c & \text{on } [0,T] \\ \partial_t f + k\sqrt{1+|\nabla f|^2}\partial_\nu u = 0 & \text{on } \Gamma_f \\ f(0,\cdot) = f_0 & \text{on } \mathbb{R}^n, \end{cases} \tag{2.9}$$

where $f_0 \in A_0$ is a given initial data and c is a given positive constant. To formulate our results for system (2.9), let $h^{k+\alpha}(\mathbb{R}^n)$, $k \in \mathbb{N}$, $\alpha \in (0,1)$, be the closure of the Schwartz space $\mathcal{S}(\mathbb{R}^n)$ in the usual Hölder norm of $BUC^{k+\alpha}(\mathbb{R}^n)$. It is also convenient to write $h_c^{k+\alpha} := h^{k+\alpha} + c$. Let now

$$U := \{f \in A_0 \,;\, f \in h_c^{2+\alpha}(\mathbb{R}^n)\}$$

and consider for a given $f \in U$ the following elliptic boundary value problem on the unbounded domain Ω_f

$$\begin{cases} \Delta u = 0 & \text{in } \Omega_f \\ \partial_\nu u = 0 & \text{on } \Sigma \\ u = f & \text{on } \Gamma_f \\ \lim\limits_{|(x,y)|\to\infty} u(x,y) = c. \end{cases} \tag{2.10}$$

It can be shown that there exists a unique classical solution u_f of (2.10), see [33], Section 2. We are now prepared to introduce the following domain of parabolicity for problem (2.9):

$$V := \{f \in U \,;\, \partial_{n+1} u_f(x, f(x)) < (1+|\nabla f(x)|^2)^{-1}, \quad x \in \mathbb{R}^n\}. \tag{2.11}$$

Obviously, $f \equiv c$ belongs to V. Additionally, it can be shown that V is an open neighborhood of c in $BUC^{2+\alpha}(\mathbb{R}^n)$ and that its diameter in $BUC^{2+\alpha}(\mathbb{R}^n)$ is unbounded. Our main result for the moving boundary problem (2.9) reads as follows:

Theorem 2.1. *Given $f_0 \in V$, there exists a unique maximal classical solution to (2.9), i.e., there exists a maximal $t^+ = t^+(f_0) > 0$ and a unique pair (u, f) such that*

$$u(t,\cdot) \in h_c^{2+\alpha}(\Omega_{f(t)}),\ t \in [0,t^+),$$
$$f \in C([0,t^+), V) \cap C^1([0,t^+), h_c^{1+\alpha}(\mathbb{R}^n)).$$

In addition, we have the following smoothing property for the solution:

$$u(t,\cdot) \in C^\omega(\overline{\Omega}_{f(t)}), \quad t \in (0,t^+), \quad f \in C^\omega((0,t^+) \times \mathbb{R}^n),$$

where C^ω stands for the set of all real analytic functions.

Let us explain some steps of the proof of Theorem 2.1. The full details can be found in [32, 33]. First we reduce system (2.9) to a single evolution equation for the moving interface. For this let $V_c := V - c$, and observe that V_c is an open neighborhood of 0 in $h^{2+\alpha}(\mathbb{R}^n)$. Given $g \in V_c$, define

$$\varphi_g(x,y) := (x, (1-y)(c+g(x))) \quad \text{for} \quad (x,y) \in \Omega,$$

where $\Omega := \mathbb{R}^n \times (0,1)$ is now a fixed reference domain. It is not difficult to see that φ_g is a diffeomorphism of class $C^{2+\alpha}$, mapping Ω onto Ω_{c+g}. Let $\varphi_g^* := u \circ \varphi_g$ and $\varphi_*^g v := v \circ \varphi_g^{-1}$ denote the corresponding pull-back and push-forward operator, respectively, and let

$$A(g)v := -\varphi_g^* \Delta(\varphi_*^g v), \qquad B_i(g)v := k\varphi_g^*(\gamma_i \nabla(\varphi_*^g v)|n_i), \quad i = 0, 1, \tag{2.12}$$

for $v \in h^{2+\alpha}(\Omega)$ be the corresponding transformed operators. Here, γ_i is the trace operator with respect to Σ and Γ_{c+g}, respectively, and $n_1 := (-\nabla g, 1)$ and $n_0 := (0,-1)$ stand for the outer normal direction on Γ_{c+g} and on Σ, respectively. The solution operator for the transformed elliptic boundary value problem on Ω

$$\begin{cases} A(g)v &= w & \text{in } \Omega \\ v &= h & \text{on } \Gamma_0 \\ B_1(g)v &= 0 & \text{on } \Gamma_1 \\ \lim\limits_{|(x,y)|\to\infty} v(x,y) = 0, \end{cases} \tag{2.14}$$

is denoted by $R(g)$, where $\Gamma_i := \mathbb{R}^n \times \{i\}$, $i = 0, 1$. More precisely, given $(w,h) \in h^\alpha(\Omega) \times h^{2+\alpha}(\Gamma_0)$, it can be shown by classical methods (cf. Appendix C in [32]) that there exists a unique solution $R(g)(w,h)$ of (2.14) in $h^{2+\alpha}(\Omega)$. It is convenient to split the operator $R(g)$ into $R(g) = S(g) \oplus T(g)$, where $S(g) := R(g)|h^\alpha(\Omega)\times\{0\}$ and $T(g) := R(g)|\{0\} \times h^{2+\alpha}(\Gamma_0)$. Of course, using the notation from (2.10), we have $\varphi_g^* u_f - c = T(g)g$ with $g = f - c$. Hence it is not difficult to see that the moving boundary problem (2.9) is equivalent to the abstract evolution equation

$$\frac{d}{dt}g + \Phi(g) = 0, \qquad g(0) = g_0 \qquad \text{in} \quad h^{1+\alpha}(\mathbb{R}^n), \tag{2.15}$$

where $\Phi(g) := B_0(g)T(g)g$, $g_0 := f_0 - c$, and where we identify in the following Γ_0 with $\mathbb{R}^n$. More precisely, if (u, f) is a classical solution to (2.9) with initial data $f_0 \in V$, then $g := f - c$ is a solution to (2.15) with initial data $g_0 := f_0 - c$, and vice versa: if g is a solution to (2.15) with initial data $g_0 \in V_c$ belonging to $C([0,t^+),V_c) \cap C^1([0,t^+), h^{1+\alpha}(\mathbb{R}^n))$, then the pair $(\varphi_*^g T(g)g + c, g + c)$ is a classical solution to (2.9) with initial data $f_0 := g_0 + c$. Consequently, we will focus our attention on the abstract evolution equation (2.15).

Observe that the diffeomorphism φ_g depends analytically on g. Hence, expressing the coefficients of the transformed operators $A(g)$ and $B_0(g)$ in terms of φ_g and using the inverse function theorem, one can verify that Φ depends analytically on $g \in V_c$. More precisely, Proposition 3.1 [33] and Lemma 4.3 [32] yield:

Lemma 2.2. *We have $\Phi \in C^\omega(V_c, h^{1+\alpha}(\mathbb{R}^n))$ with*

$$\partial\Phi(g)h = B_0(g)T(g)h + \partial B_0(g)[h, T(g)g] - B_0(g)S(g)\partial A(g)[h, T(g)g],$$

for $g \in V_c$ and $h \in h^{2+\alpha}(\mathbb{R}^n)$. Here

$$\partial B_0(g)[h,v] := \frac{d}{ds}B_0(g+sh)v\Big|_{s=0}, \quad \partial A(g)[h,v] := \frac{d}{ds}A(g+sh)v\Big|_{s=0},$$

for $v \in h^{2+\alpha}(\Omega)$.

In order to further investigate the linearization of Φ, it is convenient to introduce the following notation. Given two Banach spaces E_1 and E_0 such that E_1 is continuously and densely injected into E_0, we denote by $\mathcal{H}(E_1, E_0)$ the set of all $A \in \mathcal{L}(E_1, E_0)$ such that $-A$, considered as an unbounded operator in E_0 with domain E_1, generates a strongly continuous analytic semigroup on E_0.

In the following we fix $g \in V_c$. The operator $B_0(g)T(g)$ is a pseudo-differential operator of first order and it is called the generalized Dirichlet-Neumann operator, [8, 30, 48]. It can be shown that $B_0(g)T(g) \in \mathcal{H}(h^{2+\alpha}(\mathbb{R}^n), h^{1+\alpha}(\mathbb{R}^n))$, see Corollary 6.3 in [32]. The operator $\partial B_0(g)[\cdot, T(g)g]$ is a first order hyperbolic differential operator, whereas $B_0(g)S(g)\partial A(g)[\cdot, T(g)g]$ is a first order pseudo-differential operator. Hence, the full linearization $\partial\Phi(g)$ is the sum of three first order pseudo-differential operators. However, the following result holds true:

Theorem 2.3. *Given $g \in V_c$, we have*

$$\partial\Phi(g) \in \mathcal{H}(h^{2+\alpha}(\mathbb{R}^n), h^{1+\alpha}(\mathbb{R}^n)).$$

The main ideas of the proof of Theorem 2.3 can be summarized as follows: First we associate to each of the operators

$$B_0(g)T(g), \quad \partial B_0(g)[\cdot, T(g)g], \quad \text{and} \quad - B_0(g)S(g)\partial A(g)[\cdot, T(g)g]$$

Fourier multiplier operators F_1, F_2, and F_3 by freezing the spatial variable $x \in \mathbb{R}^n$. Using then the fact that each $g \in V_c$ satisfies

$$\partial_{n+1}[T(g)g](x, 0) + \frac{c + g(x)}{1 + |\nabla g(x)|^2} > 0, \quad x \in \mathbb{R}^n,$$

cf. (2.14), it can be shown that $F := F_1 + F_2 + F_3$ is a regularly elliptic Fourier multiplier of first order in the sense of H. Amann, see [6, 7]. For details we refer to Section 3 in [33]. Subtle perturbation techniques then allow to carry over the generation property of F to $\partial\Phi(g)$, cf. Section 6 in [32].

Based on Theorem 2.3 we can now rely on maximal regularity results in the sense of Da Prato and Grisvard [24], see also [9, 56, 65], to find a unique solution

$$g \in C([0, t^+), V_c) \cap C^1([0, t^+), h^{1+\alpha}(\mathbb{R}^n))$$

of the nonlinear equation (2.15). The regularizing property of the solution, i.e., the fact that g is actually real analytic in $(0, t^+) \times \mathbb{R}^n$, follows from the equivariance of problem (2.9) with respect to spatial translations, and again from maximal regularity. We will explain this method, which relies on an idea of S.B. Angenent [9, 10], in detail for the flow through a deformable porous medium discussed below.

Remarks 2.4. a) The first analytic results for problem (2.9) are due to Kawarada and Koshigoe [51], see also [52]. These authors construct in the case $n = 1$ for suitable initial data in the Sobolev space $H^{18}(\mathbb{R})$ a solution in $C([0, T], H^{14}(\mathbb{R})) \cap C^1([0, T], H^1(\mathbb{R}))$ by a Nash-Moser iteration. This approach leads to a serious loss of regularity for solutions. In addition, there are no uniqueness results in [51, 52].

b) We treat (2.15) as a fully nonlinear equation, although the operator $\Phi(g) = B_0(g)T(g)g$ has a quasi-linear structure in the sense that, given $g \in V_c$, the mapping $[h \mapsto B_0(g)T(g)h]$ is a first order linear operator. This quasi-linear structure is not useful in our situation, since the nonlinear dependence in Φ is of first order as well.

c) Given $g \in V_c$, let $f = g + c$. Then we have

$$\partial_{n+1} u_f(x, f(x)) < (1 + |\nabla f(x)|^2)^{-1}, \qquad x \in \mathbb{R}^n, \tag{2.16}$$

where u_f is the solution of the elliptic problem (2.10). Relation (2.16) should be seen as a parabolicity condition for (2.9) in the sense that the corresponding linearized problem is well-posed and induces a strongly continuous analytic semigroup on $h^{1+\alpha}(\mathbb{R}^n)$, provided (2.16) holds true, whereas the same linearized problem is backwards parabolic if

$$\partial_{n+1} u_f(x, f(x)) > (1 + |\nabla f(x)|^2)^{-1}, \qquad x \in \mathbb{R}^n.$$

On the other hand, letting $p(x, y) := u_f(x, y) - y$ for $(x, y) \in \mathbb{R}^n \times [0, \infty)$, it follows from the strong maximum principle that $\partial_{n+1} u_f(x, f(x)) < 1$, since ∇p is orthogonal to Γ_f. We do not know whether or not problem (2.9) is still well-posed if condition (2.16) is replaced by the condition $\partial_{n+1} u_f(x, f(x)) < 1$.

d) It can be shown that

$$\partial B_0(g)[\cdot, T(g)g] - B_0(g)S(g)\partial A(g)[\cdot, T(g)g] \to 0$$

in $\mathcal{L}(h^{2+\alpha}(\mathbb{R}^n), h^{1+\alpha}(\mathbb{R}^n))$ as $\|g\|_{h^{2+\alpha}(\mathbb{R}^n)} \to 0$. Since $B_0(g)T(g)$ belongs to $\mathcal{H}(h^{2+\alpha}(\mathbb{R}^n), h^{1+\alpha}(\mathbb{R}^n))$ it follows therefore from Lemma 2.2 and well-known perturbation results for the class $\mathcal{H}(h^{2+\alpha}(\mathbb{R}^n), h^{1+\alpha}(\mathbb{R}^n))$ that there is an $\varepsilon_0 > 0$ such that

$$\partial\Phi(g) \in \mathcal{H}(h^{2+\alpha}(\mathbb{R}^n), h^{1+\alpha}(\mathbb{R}^n)),$$

provided $\|g\|_{h^{2+\alpha}(\mathbb{R}^n)} < \varepsilon_0$. Let us emphasize that there are arbitrarily large $g \in V_c$ and that Theorem 2.3 does not rely on a smallness condition.

2.2. Quasi-incompressible fluids in deformable porous media

We consider now the flow of a quasi-incompressible fluid in a deformable porous medium. In this case the specific storativity S_0 is positive, cf. (2.4), and (2.8) leads us to the problem

$$\begin{cases} \partial_t u - \Delta u = 0 & \text{in } \Omega_f \\ \partial_\nu u = 0 & \text{on } \Sigma \\ u = f & \text{on } \Gamma_f \\ \lim_{|(x,y)| \to \infty} u(\cdot, (x, y)) = c & \text{on } [0, T] \\ \partial_t f + k\sqrt{1 + |\nabla f|^2}\,\partial_\nu u = 0 & \text{on } \Gamma_f \\ f(0, \cdot) = f_0 & \text{on } \mathbb{R}^n \\ u(0, \cdot) = u_0 & \text{in } \Omega_{f_0}, \end{cases} \tag{2.17}$$

where now $f_0 \in A_0$ and $u_0 \in BUC^1(\Omega_{f_0})$ are given initial data. As before, c and k are positive constants. In contrast to problem (2.9) we treat system (2.17) in appropriate Sobolev spaces rather than in Hölder spaces. To formulate our results, we fix $p > n+2$ and denote by W_p^σ the Sobolev-Slobodeckii spaces of order $\sigma \geq 0$, i.e., W_p^σ is the usual Sobolev space if $\sigma \in \mathbb{N}$, whereas W_p^σ stands for the Besov space B_{pp}^σ if $\sigma > 0$ is not an integer. Furthermore, in the following we fix $s \in (0, 1/p)$, $\alpha \in [1/p, 1-((n+1)/p)]$, and $\gamma \in (0, c)$. Given $g \in W^{2+s-1/p}(\mathbb{R}^n)$, it follows from well-known trace theorems that there exists an extension $\tilde{g} \in W^{2+s}(\mathbb{R}^{n+1})$ of g such that $\operatorname{supp}(\tilde{g}) \subset \mathbb{R}^n \times (-3/4, 3/4)$ and such that $\tilde{g}|\mathbb{R}^n \times \{0\} = g$. For technical reasons we introduce some further notation. Given $f \in A_0$, let $g := f - c$ and set

$$\omega(f) := \sup_{(x,y)\in\Omega} [|\nabla_x \tilde{g}(x,y)|^2 + |(1-y)\partial_{n+1}\tilde{g}(x,y) - \tilde{g}(x,y)|^2]^{1/2}.$$

Later on, the quantity $\omega(f)$ will be useful to construct an appropriate diffeomorphism, mapping again the reference domain Ω onto Ω_{c+g}. Now let

$$V := \{f \in A_0 \; ; \; f \in W_{p,c}^{2+s-1/p}(\mathbb{R}^n), \; \omega(f) < \min(1, \gamma)\},$$

where, for simplicity, we write $W_{p,c}^\sigma := W_p^\sigma + c$, $\sigma \geq 0$.

Theorem 2.5. *Let $\rho > 0$ be given. Then there exists a $\delta > 0$ such that for any $f_0 \in V$ and $u_0 \in W_{p,c}^{2+s}(\Omega_{f_0})$ satisfying*

$$\|u_0 - c\|_{W^{2+s}(\Omega_{f_0})} < \rho \qquad \|\partial_{n+1}u_0\|_{BUC^\alpha(\Omega_{f_0})} < \delta, \\ \|\partial_{n+1}u_0|\Gamma_{f_0}\|_{W_p^{1+s-1/p}(\mathbb{R}^n)} < \delta$$

and the compatibility conditions

$$u_0|\Gamma_{f_0} = f_0, \qquad \partial_{n+1}u|\Sigma = 0,$$

there exists $t^+ := t^+(u_0, f_0)$ and a unique maximal classical solution (u, f) of problem (2.17) in the class

$$f \in C([0,t^+), V) \cap C^1([0,t^+), W_{p,c}^{1+s-1/p}(\mathbb{R}^n)) \cap C^\omega((0,t^+) \times \mathbb{R}^n) \\ u(\cdot,\cdot) \in C^\infty(\overline{\Omega}_{f,T}), \quad u(t,\cdot) \in BUC^{1+\alpha}(\Omega_{f(t)}), \; t \in [0,T),$$

where $\overline{\Omega}_{f,T} := \{(t,(x,y)) \in (0,T) \times \mathbb{R}^{n+1} \; ; \; (x,y) \in \overline{\Omega}_{f(t)}\}$.

Before we prove Theorem 2.5, let us add the following comments.

Remarks 2.6. a) Theorem 2.5 extends earlier results obtained in [31], since we are now able to prove the analyticity of the free interface Γ_f and the smoothness of the potential u.

b) There is a different approach in weighted Hölder spaces to problem (2.17) outlined in [14, 13]. Recently, a (different) L_p-theory for problems of type (2.17) was proposed in [66, 67].

c) Our regularity result for the moving boundary Γ_f also extends results in [53], where, under the assumption of the existence of classical C^1-solutions, the C^∞-smoothness of the moving boundary is proved. No existence results are presented in [53]. In contrast, Theorem 2.5 guarantees the existence of classical solutions and we further establish the actual regularity of the moving boundary.

d) Observe that the constant function $(u, f) \equiv (c, c)$ is a solution to (2.17). This trivial solution should be regarded as an equilibrium of system (2.17). Furthermore it is not difficult to verify that V is an open neighborhood of c in $W_{p,c}^{2+s-1/p}(\mathbb{R}^n)$.

Proof of Theorem 2.5 (i) In a first step we provide an appropriate extension for $g \in W_p^{k+s-1/p}(\mathbb{R}^n)$ to a function on the whole of $\mathbb{R}^{n+1}$. For this, let γ_0 be the trace operator with respect to $\Sigma \equiv \mathbb{R}^n$ and let

$$\gamma_0^c \in \mathcal{L}(W_p^{k+s-1/p}(\mathbb{R}^n), W_p^{k+s}(\mathbb{R}^{n+1})), \quad k \in \mathbb{N}^* \tag{2.18}$$

be the coretraction of γ_0 constructed in [69], Theorem 2.7.2. It follows from formula (2.7.2.42) in [69] that γ_0^c possesses the following translation equivariance

$$\tau_a \gamma_0^c g = \gamma_0^c \tau_a g, \quad a \in \mathbb{R}^n, \ g \in W_p^{k+s-1/p}(\mathbb{R}^n),$$

where

$$(\tau_a g)(x) := g(x+a) \quad \text{and} \quad (\tau_a v)(x,y) := v(x+a, y)$$

for $g \in W_p^{k+s-1/p}(\mathbb{R}^n)$, $v \in W_p^{k+s}(\mathbb{R}^{n+1})$, and $(x,y) \in \mathbb{R}^n \times \mathbb{R}$. Next we pick $\eta \in C^\infty(\mathbb{R})$ with $\operatorname{supp}(\eta) \subset (-3/4, 3/4)$ and $\eta(y) = 1$ for $y \in [-1/2, 1/2]$. Given $g \in W_p^{k+s-1/p}(\mathbb{R}^n)$, set

$$(\rho_0 g)(x,y) := \eta(y)(\gamma_0^c g)(x,y), \quad (x,y) \in \Omega. \tag{2.19}$$

Recall that $\Omega := \mathbb{R}^n \times (0,1)$. Obviously, we have

$$\tau_a \rho_0 g = \rho_0 \tau_a g, \quad a \in \mathbb{R}^n, \ g \in W_p^{k+s-1/p}(\mathbb{R}^n). \tag{2.20}$$

For simplicity we often write $\tilde{g} := \rho_0 g$.

(ii) Let $V_c := V - c$, pick $g \in V_c$, and set

$$\varphi_g(x,y) := (x, (1-y)(c + \tilde{g}(x,y))), \quad (x,y) \in \Omega.$$

As in [31] Lemma 2.2 one shows that V_c is an open neighborhood of 0 in $W_p^{2+s-1/p}(\mathbb{R}^n)$, and that, given $g \in V_c$, we have

$$\varphi_g \in \mathit{Diff}^{1+\alpha}(\Omega, \Omega_{c+g}), \quad \gamma_0 \varphi_g \in \mathit{Diff}^{1+\alpha}(\Sigma, \Gamma_{c+g}).$$

As in (2.12) we now define the transformed operators

$$A(g)v := -\varphi_g^*(\Delta \varphi_*^g v), \quad B_i(g)v := k\varphi_g^*(\gamma_i \nabla(\varphi_*^g) | n_i), \ i = 0, 1$$

for $v \in W_p^2(\Omega)$, where $n_0 := (-\nabla g, 1)$ and $n_1 := (0, -1)$ denote the outer normals on Γ_{c+g} and Γ_0, respectively. To express the coefficients of these operators in terms of $g \in V_c$, let $G(g) := c + \tilde{g} - \pi \partial_{n+1} \tilde{g}$, where $\pi(x,y) := 1 - y$ for $(x,y) \in \Omega$, and denote by $G_{jk}(g) := (\partial_j \varphi_g | \partial_k \varphi_g)$, $1 \le j,\ k \le n+1$, the components of the metric tensor induced by φ_g. It is not difficult to verify that $\det[G_{jk}(g)] = G(g)$ and that the inverse $[G^{jk}(g)]$ of $[G_{jk}(g)]$ is given by

$$[G^{jk}(g)] = \begin{bmatrix} 1 & 0 & \cdots & 0 & \frac{\pi\partial_1\tilde{g}}{G(g)} \\ 0 & 1 & \cdots & 0 & \frac{\pi\partial_2\tilde{g}}{G(g)} \\ \vdots & & \ddots & & \vdots \\ 0 & 0 & \cdots & 1 & \frac{\pi\partial_n\tilde{g}}{G(g)} \\ \frac{\pi\partial_1\tilde{g}}{G(g)} & \frac{\pi\partial_2\tilde{g}}{G(g)} & \cdots & \frac{\pi\partial_n\tilde{g}}{G(g)} & \frac{1+|\pi\nabla_x\tilde{g}|^2}{G^2(g)} \end{bmatrix} \tag{2.21}$$

Then we have

$$A(g)v = -\frac{1}{G(g)}\sum_{j,k=1}^{n+1} \partial_j(G(g)G^{jk}(g)\partial_k v), \quad v \in W_p^2(\Omega). \tag{2.22}$$

By the construction of the extension operator ρ_0, in particular by the choice of η, one easily verifies that the boundary operator B_0 is represented as

$$B_0(g)v = k\sum_{j=1}^{n} -\partial_j g\gamma_0(\partial_j v) - \frac{k}{\gamma_0 G(g)}(1+|\nabla g|^2)\gamma_0(\partial_{n+1}v), \tag{2.23}$$

whereas $B_1(g)v = (k/c)\gamma_1(\partial_{n+1}v)$. Finally, we need

$$F(v,g) := -\frac{\pi\partial_{n+1}v}{G(g)}\rho_0(B_0(g)v), \quad (v,g) \in W_p^2(\Omega) \times V_c.$$

Assume now that (u,f) is a solution to problem (2.17). Setting $g := f - c$ and $v := \varphi_g^*(u-c)$, one shows that (v,g) is formally a solution to

$$\begin{cases} \partial_t v + A(g)v + F(v,g) = 0 & \text{in} \quad (0,T] \times \Omega \\ v = g & \text{on} \quad [0,T] \times \Gamma_0 \\ \partial_{n+1}v = 0 & \text{on} \quad [0,T] \times \Gamma_1 \\ \lim_{|(x,y)|\to\infty} v(\cdot,(x,y)) = 0 & \text{on} \quad [0,T] \\ \partial_t g + B_0(g)v = 0 & \text{on} \quad (0,T] \times \Gamma_0 \\ v(0,\cdot) = v_0 & \text{in} \quad \Omega \\ g(0,\cdot) = g_0 & \text{on} \quad \Gamma_0, \end{cases} \tag{2.24}$$

(iii) We shall now focus our attention on the transformed system (2.24). In order to introduce an appropriate notion of solutions to (2.24), let

$$\begin{aligned} E_0 &:= W_p^s(\Omega) \times W_p^{1+s-1/p}(\Gamma_0) \\ E_1 &:= \{(v,g) \in W_p^{2+s}(\Omega) \times W_p^{2+s-1/p}(\Gamma_0)\,;\ \gamma_0 v = g,\ \gamma_1\partial_{n+1}v = 0\}. \end{aligned}$$

Each of these spaces is given the natural topology, i.e., the product topology for E_0 and the relative topology for E_1. It follows from the trace theorem that E_1 is a closed subspace of $W_p^{2+s}(\Omega) \times W_p^{2+s-1/p}(\Gamma_0)$ and it can be shown that E_1 is dense in E_0, see Lemma 3.1 in [31]. Observe further that the "stationary" boundary conditions of (2.24) are incorporated in the space E_1.

Next let $D_1 := \{(v,g) \in E_1\,;\ g \in V_c\}$ and define

$$\Pi(z) := (A(g)v + F(v,g), B_0(g)v) \quad \text{for} \quad z = (v,g) \in D_1.$$

It is not difficult to see that D_1 is an open subset of E_1. Moreover, using the representations (2.22) and (2.23) one verifies that

$$\Pi \in C^\omega(D_1, E_0). \tag{2.25}$$

Recalling the definition of E_1, system (2.24) is equivalent to the abstract evolution equation

$$\frac{d}{dt}z + \Pi(z) = 0, \quad z(0) = z_0 := (v_0, g_0) \quad \text{in} \quad E_0. \tag{2.26}$$

More precisely, given $z_0 = (v_0, g_0) \in D_1$, we call $z = (v, g)$ a classical W^s_p-solution to (2.24) if and only if

$$z \in C([0,T], D_1) \cap C^1([0,T], E_0) \tag{2.27}$$

and z satisfies the equations in (2.26) pointwise on $[0,T]$. Using the diffeomorphism φ_g one shows that each classical W^s_p–solution $z = (v, g)$ of (2.24) gives rise to a classical solution to (2.17) by setting $f := g + c$ and

$$u(t,(x,y)) := (\varphi_*^{g(t)} v(t))(x,y) + c, \quad t \in [0,T], \ (x,y) \in \Omega_{f(t)},$$

cf. Lemma 2.4 in [31].

(iv) We know from (2.25) that, given $z = (v, g) \in D_1$, we have $\partial\Pi(z) \in \mathcal{L}(E_1, E_0)$. Unfortunately, in order to guarantee that $\partial\Pi(z)$ belongs to the class $\mathcal{H}(E_1, E_0)$, we need a smallness assumption for v. More precisely, given $\rho > 0$ and $\delta > 0$, let

$$W_{\rho,\delta} := \{(v,g) \in D_1 \; ; \; \|v\|_{2+s,p} < \rho, \ \|\partial_{n+1} v\|_\alpha < \delta, \ \|\gamma_0 \partial_{n+1} v\|_{1+s,p} < \delta\},$$

where $\|\cdot\|_{\sigma,p}$ and $\|\cdot\|_\alpha$ stands for the norm in W^σ_p and BUC^α, respectively. The following crucial result for $\partial\Pi(z)$ was proved in [31], Corollary 3.6: Given $\rho > 0$, there is a $\delta > 0$ such that

$$\partial\Pi(z) \in \mathcal{H}(E_1, E_0) \quad \text{for all} \quad z \in W_{\rho,\delta}. \tag{2.28}$$

Based on (2.28) we can now apply results of A. Lunardi to find a unique solution of system (2.24). Indeed, given $z_0 = (v_0, g_0) \in W_{\rho,\delta}$, Theorem 2 in [55] and an extension argument (cf. Theorem 3.8 in [31]) show that there exists a unique maximal classical W^s_p-solution to (2.24) on $[0, t^+(v_0, g_0))$.

(v) Let us now verify that the interface constructed above depends analytically on the space and time variables. For this we fix $\rho > 0$ and choose $\delta > 0$ such that (2.28) holds. Let $W := W_{\rho,\delta}$. We fix $(v_0, g_0) \in W$ and let $z = (v, g) \in C(I, W) \cap C^1(I, E_0)$ denote the unique solution of (2.24) on $[0,T]$, where $T \in (0, t^+(v_0, g_0))$ is fixed and $I := [0,T]$. Additionally, let $I_0 := (0,T]$.

Given $\beta \in (0,1)$ and a Banach space E, let

$$C^\beta_\beta(I,E) := \{u \in BUC(I,E) \cap C^\beta(I_0,E) \; ; \; \lim_{\varepsilon \to 0} \varepsilon^\beta \sup_{\varepsilon \le s < t \le 2\varepsilon} \frac{\|u(s) - u(t)\|_E}{|s-t|^\beta} = 0\},$$

where $C^\beta(I_0, E)$ denotes the space of all locally β-Hölder continuous functions from I_0 to E. It is not difficult to verify that $C^\beta_\beta(I,E)$, equipped with the norm

$$\|u\|_{C^\beta_\beta(I,E)} := \|u\|_\infty + \sup_{2\varepsilon\in I_0} \varepsilon^\beta \sup_{\varepsilon\le s<t\le 2\varepsilon} \frac{\|u(s)-u(t)\|_E}{|s-t|^\beta}, \quad u \in C^\beta_\beta(I,E),$$

is a Banach space. The space $C^{1+\beta}_\beta(I,E)$ consists of those $u \in C^\beta_\beta(I,E)$ such that u' belongs to $C^\beta_\beta(I,E)$ too, given the obvious norm. Of course, $C^{1+\beta}_\beta(I,E)$ is also a Banach space. With this notation, it follows from (2.28) and Theorem III.2.5.6 in [6] that

$$(C^\beta_\beta(I,E_0), C^\beta_\beta(I,E_1)\cap C^{1+\beta}_\beta(I,E_0))$$

is a pair of maximal regularity for the family $\{\partial\Pi(w)\,;\, w\in W\}$, that is,

$$(\frac{d}{dt}+\partial\Pi(w), \mathrm{tr}) \in \mathrm{Isom}(C^\beta_\beta(I,E_1)\cap C^{1+\beta}_\beta(I,E_0), C^\beta_\beta(I,E_0)\times E_1), \tag{2.29}$$

for every $w\in W$, where $\mathrm{tr}u := u(0)$ for $u\in C^\beta_\beta(I,E_1)$. For later purposes we also need the following additional regularity of the solution z of (2.26)

$$z \in C^\beta_\beta(I,E_1)\cap C^{1+\beta}_\beta(I,E_0), \tag{2.30}$$

which is guaranteed by Theorem 8.1.1 in [56].

(vi) Next we collect some useful properties of the solution $z=(v,g)$ and Π concerning translations in space and dilation in time. Given $a\in\mathbb{R}^n$ and $w=(r,h)\in E_0$, let $\tau_a w := (\tau_{(a,0)}r, \tau_a h)$, cf. (2.20). It follows from (2.20)–(2.23) that

$$\tau_a(W)\subset W \quad\text{and}\quad \tau_a\Pi(w) = \Pi(\tau_a w),\ a\in\mathbb{R}^n,\ w\in W. \tag{2.31}$$

Let now $(\lambda,\mu)\in(1-\varepsilon,1+\varepsilon)\times\mathbb{R}^n$, with $\varepsilon>0$ sufficiently small, be given and set

$$z_{\lambda,\mu}(t) := \tau_{t\mu}z(\lambda t), \quad t\in I.$$

Since $\tau_a(W)\subset W$, we find that $z_{\lambda,\mu}(I)\subset W$. Additionally, we have

$$z_{\lambda,\mu}\in C(I,W)\cap C^1(I,E_0). \tag{2.32}$$

Examplarily, let us show that the first component $v_{\lambda,\mu} := \tau_{t\mu}v(\lambda t)$ of $z_{\lambda,\mu}$ belongs to $C(I,W^{2+s}_p(\Omega))$. Obviously, we have

$$\begin{aligned} v_{\lambda,\mu}(t+h) - v_{\lambda,\mu}(t) &= \tau_{(t+h)\mu}(v(\lambda(t+h)) - v(\lambda t)) \\ &\quad + (\tau_{(t+h)\mu} - \tau_{t\mu})v(\lambda t). \end{aligned}$$

Hence, using the fact that the set of all translations $\{\tau_a\,;\, a\in\mathbb{R}^n\}$ forms a strongly continuous group of contractions on $W^{2+s}_p(\Omega)$, and the fact that $v\in C(I,W^{2+s}_p(\Omega))$, we conclude that $v_{\lambda,\mu}\in C(I,W^{2+s}_p(\Omega))$. Let now

$$\Pi_{\lambda,\mu}(w) := \lambda\Pi(w) - D_\mu w, \quad w\in W,$$

where $D_\mu w := ((\mu|\nabla_x v), (\mu|\nabla g))$. Using the translation equivariance of Π, cf. (2.31), one shows that $z_{\lambda,\mu}$ solves the evolution equation

$$\frac{d}{dt}z_{\lambda,\mu} + \Pi_{\lambda,\mu}(z_{\lambda,\mu}) = 0, \quad t\in I, \quad z_{\lambda,\mu}(0) = z_0. \tag{2.33}$$

Moreover, since $(\mu|\nabla g)$ involves only tangential derivatives and since $[v \mapsto (\mu|\nabla_x v)]$ is a first order operator, it is not difficult to see that solutions to (2.33) are unique in the class (2.32), cf. the proof of Theorem 3.8 in [31].

(vii) In the following, let Λ be an open neighborhood of (1,0) in $(1-\varepsilon, 1+\varepsilon)\times\mathbb{R}^n$. Moreover, we set $D := C(I,W)\cap C^\beta_\beta(I,E_1)\cap C^{1+\beta}_\beta(I,E_0)$. Given $(\lambda,\mu)\in\Lambda$ and $w\in D$, we define

$$F(w,(\lambda,\mu)) := (\frac{d}{dt}w + \lambda\Pi(w) - D_\mu w, w(0) - z_0).$$

Recall that I is compact and that W is open in E_1. Moreover, one has the embedding $C^\beta_\beta(I,E_1)\hookrightarrow C(I,E_1)$, cf. Proposition III.2.1.1 in [6]. Hence we find that D is open in $C^\beta_\beta(I,E_1)\cap C^{1+\beta}_\beta(I,E_0)$. In summary, we see that $\mathrm{dom}(F) = D\times\Lambda$ is an open subset of $(C^\beta_\beta(I,E_1)\cap C^{1+\beta}_\beta(I,E_0))\times\mathbb{R}^{n+1}$. In addition, it follows from (2.25) that

$$F\in C^\omega(D\times\Lambda, C^\beta_\beta(I,E_0)\times E_1).$$

The derivative $\partial_1 F$ of F with respect to $w\in D$ is given by

$$\partial_1 F(w,(1,0))h = (\frac{d}{dt}h + \partial\Pi(w)h, h(0)). \tag{2.34}$$

Observe that $F(w,(\lambda,\mu)) = 0$ holds true if and only if w is a solution to the evolution equation (2.33). Observe also that we have $(z,(1,0))\in D\times\Lambda$, cf. (2.30), with $F(z,(1,0)) = 0$. In addition, it follows from (2.34), (2.29) and Theorem 2.6.1 in [6] that

$$\partial_1 F(z,(1,0))\in \mathrm{Isom}(C^\beta_\beta(I,E_1)\cap C^{1+\beta}_\beta(I,E_0), C^\beta_\beta(I,E_0)\times E_1)).$$

Consequently, the implicit function theorem guarantees the existence of an open neighborhood Λ_0 of Λ and a unique mapping

$$[(\lambda,\mu)\mapsto w_{\lambda,\mu}]\in C^\omega(\Lambda_0, C(I,W)\cap C^\beta_\beta(I,E_1)\cap C^{1+\beta}_\beta(I,E_0))$$

such that $F(w_{\lambda,\mu},(\lambda,\mu)) = 0$. Since $C^{1+\beta}_\beta(I,E_0)\subset C^1(I,E_0)$, we find by the unique solvability of (2.33) in the class (2.32) that

$$[(\lambda,\mu)\mapsto z_{\lambda,\mu}]\in C^\omega(\Lambda_0, C(I,W)\cap C^\beta_\beta(I,E_1)\cap C^{1+\beta}_\beta(I,E_0)). \tag{2.35}$$

(viii) We now show that the interface depends analytically on the spatial and temporal variables. For this let $g_{\lambda,\mu} := \tau_{t\mu}g(\lambda t)$ be the second component of $z_{\lambda,\mu}$. By (2.35) we particularly have

$$[(\lambda,\mu)\mapsto g_{\lambda,\mu}]\in C^\omega(\Lambda_0, C^1(I, W^{1+s-1/p}_p(\Gamma_0)). \tag{2.36}$$

Moreover, given $(k,\gamma)\in\mathbb{N}\times\mathbb{N}^n$, an induction argument shows that

$$t^{k+|\gamma|}\partial^k_t\partial^\gamma_x g(t) = \partial^k_\lambda\partial^\gamma_\mu g_{\lambda,\mu}|_{(\lambda,\mu)=(1,0)}(t), \quad t\in I,$$

cf. the proof of Theorem 4.4 in [33]. In particular, we have $g\in C^\infty((0,T)\times\mathbb{R}^n)$. Also the fact that

$$g\in C^\omega((0,T)\times\mathbb{R}^n) \tag{2.37}$$

follows from (2.36). Indeed, representing $g_{\lambda,\mu}$ by its Taylor series, and using Sobolev's embedding theorem, one shows that, given $(t_0, x_0) \in (0,T) \times \mathbb{R}^n$, there are positive constants r and M such that

$$|\partial_t^k \partial_x^\gamma g(t,x)| r^{k+|\gamma|} \le M k! \gamma!$$

for all $(k,\gamma) \in \mathbb{N} \times \mathbb{N}^n$ and all $(t,x) \in \mathbb{B}((t_0,x_0),r)$.

To verify the smoothness of v, consider the semilinear parabolic equation

$$\begin{cases} \partial_t v + A(g)v + F(v,g) = 0 & \text{in} \quad (0,T] \times \Omega \\ v = g & \text{on} \quad [0,T] \times \Gamma_0 \\ \partial_{n+1} v = 0 & \text{on} \quad [0,T] \times \Gamma_1 \\ v(0,\cdot) = v_0 & \text{in} \quad \Omega \end{cases}$$

for the function v. Combining (2.37) with (2.18), (2.19), (2.22), and (2.23) we find that all coefficients of $A(g)$ and $F(\cdot,g)$ are smooth. Hence, it follows from well-known regularity results for semilinear parabolic initial boundary value problems, cf. Corollary 9.4 in [4], that v belongs to $C^\infty((0,T)\times\overline{\Omega})$. This completes the proof. □

Remark 2.7. Although we prove that the function g appearing in (2.24) (and yielding the interface $f = g + c$ in (2.17)) is real analytic, we only get C^∞-regularity of the corresponding potential v. Since we use a cut-off function in our construction of the extension operator ρ_0, the regularity of v cannot be improved to real analyticity in the framework presented here. We leave it as an open problem to find an extension operator sharing all properties of ρ_0 but implying in addition the analyticity of the coefficients $A(g)$ and $F(\cdot,g)$. If such an extension operator exists the analyticity of v follows from [41] Theorem 3.3.1.

3. Volume-preserving mean curvature flows

Let $\Gamma = \{\Gamma(t)\,;\, t \ge 0\}$ be a family of closed compact hypersurfaces in $\mathbb{R}^{n+1}$, $n \ge 1$. In this section we consider the evolution of Γ under the assumption that the normal velocity V of Γ is given as a function of the mean curvature $H(t)$ of each individual surface $\Gamma(t)$, i.e., $V(t) = F(H(t))$. Of course, both quantities V and H have to be given an appropriate orientation. More precisely, we adopt the sign convention that $V(t)$ and $H(t)$ are positive for a locally expanding family of hypersurfaces and for a locally convex hypersurface, respectively. Let $A(t) = \int_{\Gamma(t)} d\sigma(t)$ denote the surface area of $\Gamma(t)$ and assume that each surface $\Gamma(t)$ is smooth. Then it follows from the first variation of the area functional, cf. Theorem 4 in [54], that

$$\frac{d}{dt}A(t) = n \int_{\Gamma(t)} H(t)V(t)d\sigma(t) = n \int_{\Gamma(t)} H(t)F(H(t))d\sigma(t), \;\; t \ge 0. \tag{3.1}$$

In particular, we observe that $A(t)$ is monotone as a function of t if $\int H(t)V(t)$ has a sign. The most prominent and most simple example of this type is certainly the mean curvature flow, where the normal velocity is given by $V(t) = -H(t)$. We shall present below some further examples for which $\int V(t)H(t)$ has a sign.

Assume now furthermore that each surface $\Gamma(t)$ encloses a well-defined domain $\Omega(t)$ in $\mathbb{R}^{n+1}$ and let $vol(t) := \int_{\Omega(t)} dx$ denote the volume of $\Omega(t)$. Then one has

$$\frac{d}{dt} vol(t) = \int_{\Gamma(t)} V(t) d\sigma(t) = \int_{\Gamma(t)} F(H(t)) d\sigma(t), \ t \geq 0, \tag{3.2}$$

cf. Theorem 2E in [47]. Therefore the evolution equation $V = F(H)$ is volume-preserving, provided $\int_{\Gamma(t)} F(H) d\sigma(t) = 0$. This holds true for instance if $F = div(X)$ for some vector field X, since $\Gamma(t)$ is closed.

3.1. The averaged mean curvature flow

We consider the evolution equation

$$V(t) = \overline{H}(t) - H(t), \quad \Gamma(0) = \Gamma_0, \tag{3.3}$$

where $\overline{H} := |\Gamma(t)|^{-1} \int_{\Gamma(t)} H \, d\sigma(t)$ is the average of the mean curvature. Obviously, this flow is volume-preserving. Moreover, (3.1) shows that

$$\frac{1}{n} \frac{d}{dt} A(t) = \int_{\Gamma(t)} H(\overline{H} - H) d\sigma(t) = -\int_{\Gamma(t)} (H - \overline{H})^2 d\sigma(t) \leq 0,$$

since $\overline{H} \int (H - \overline{H}) d\sigma = 0$. Thus the flow (3.3) decreases the area $A(t)$. The averaged mean curvature flow has been identified as the singular limit of a nonlocal Ginzburg-Landau equation [15].

3.2. The surface diffusion flow

We consider now the evolution equation

$$V(t) = \Delta_{\Gamma(t)} H(t), \quad \Gamma(0) = \Gamma_0, \tag{3.4}$$

where $\Delta_{\Gamma(t)}$ stands for the Laplace-Beltrami operator on $\Gamma(t)$. Again this flow is volume-preserving, i.e.,

$$\frac{d}{dt} vol(t) = \int_{\Gamma(t)} V \, d\sigma(t) = \int_{\Gamma(t)} div_{\Gamma(t)} \, grad_{\Gamma(t)} \, H(t) \, d\sigma = 0,$$

and area decreasing

$$\frac{1}{n} \frac{d}{dt} A(t) = \int_{\Gamma(t)} HV \, d\sigma(t) = \int_{\Gamma(t)} [\Delta_{\Gamma(t)} H] H \, d\sigma(t) = -\int_{\Gamma(t)} |grad_{\Gamma(t)} \, H|^2 d\sigma(t) \leq 0.$$

The surface diffusion flow (3.4) was first introduced by Mullins [60] to model surface dynamics for phase interfaces when the evolution is governed only by mass diffusion in the interface. It has also been examined in a more general mathematical and physical context by Davi and Gurtin [26], and by Cahn and Taylor [19].

3.3. The intermediate surface diffusion flow

The surface diffusion flow (3.4) and the averaged mean curvature flow (3.3) are formally connected by the so-called intermediate surface diffusion flow given by

$$V(t) = \Delta_{\Gamma(t)}(\alpha - \beta\Delta_{\Gamma(t)})^{-1}(H(t) - \overline{H}(t)),\ t > 0, \quad \Gamma(0) = \Gamma_0, \tag{3.5}$$

where α and β are positive constants. Indeed, fix $t \geq 0$, set $\Sigma := \Gamma(t)$ and let Δ denote the $L_2(\Sigma)$-realization of Δ_Σ. Furthermore, let $H := 1^\perp$ and write $\Delta_\perp := \Delta|H$.

We first consider the case $\beta = 1$ and let $\alpha > 0$ tend to 0. Since $0 \notin \sigma(\Delta_\perp)$ we have

$$\lim_{\alpha\to 0}(\alpha - \Delta_\perp)^{-1} = -\Delta_\perp^{-1} \quad \text{in} \quad \mathcal{L}(H),$$

suggesting that solutions to (3.5) should converge to solutions of the averaged mean curvature flow (3.3) as $\alpha \to 0$ and $\beta = 1$.

To consider the case $\alpha = 1$ and $\beta \to 0$, recall that $-\Delta_\perp$ is m-accretive on H. Hence the Lumer-Phillips theorem implies that

$$(1 - \beta\Delta_\perp)^{-1} = \frac{1}{\beta}(\frac{1}{\beta} - \Delta_\perp)^{-1} \to 1_H \quad \text{in} \quad \mathcal{L}_s(H)$$

as $\beta \to 0$, cf. Lemma 1.3.2 in [61]. Here $\mathcal{L}_s(H)$ denotes the space $Hom(H)$ given the strong operator topology. In this case one therefore expects that the solution to (3.5) should converge to solutions of the surface diffusion flow. We mention that this formal connection between (3.5) and the flows induced by (3.3) and (3.6), respectively, was first formulated in [19].

A rigorous study of these singular limits will be the topic of a separate paper. Here we shall establish, as a first step in the analysis, well-posedness of (3.5) and some global existence results. Hence, for the sake of simplicity, set $\alpha = \beta = 1$ and consider the evolution equation

$$V(t) = \Delta_{\Gamma(t)}(1 - \Delta_{\Gamma(t)})^{-1}H(t),\ t > 0, \quad \Gamma(0) = \Gamma_0. \tag{3.6}$$

Observe that

$$\Delta_{\Gamma(t)}(1 - \Delta_{\Gamma(t)})^{-1}H(t) = \operatorname{div}_{\Gamma(t)}[\operatorname{grad}_{\Gamma(t)}(1 - \Delta_{\Gamma(t)})^{-1}H(t)],$$

so that the flow (3.6) preserves the volume, cf. the remark following (3.2).

Observe also that $\Delta_{\Gamma(t)}(1 - \Delta_{\Gamma(t)})^{-1}$ is a non-positive self-adjoint operator in $L_2(\Gamma(t))$. Hence the flow (3.6) is area shrinking too, see (3.1).

Problem (3.5) has recently been derived by Cahn and Taylor [19] to model growth laws for morphological change for a class of problems where surface diffusion is the transport mechanism and the only driving force is the reduction of total surface free energy.

3.4. The Mullins-Sekerka flow

The Mullins-Sekerka model is a nonlocal evolution law in which V is given by the jump across the interface of the normal derivative of a function being harmonic on either side and which equals the mean curvature of the moving interface. More precisely, let Ω be a bounded connected domain in $\mathbb{R}^{n+1}$ with a smooth boundary $\partial\Omega$. Suppose Γ_0 is a compact hypersurface being the boundary of an open set Ω_0^- which is compactly contained in Ω. Additionally, we set $\Omega_0^+ := \Omega \setminus cl(\Omega_0^-)$. We are looking for a family $\Gamma := \{\Gamma(t);\, t \geq 0\}$ of hypersurfaces, separating at each instant the domain Ω into the domain $\Omega^-(t)$ enclosed by $\Gamma(t)$ and $\Omega^+(t) := \Omega \setminus cl(\Omega^-(t))$, and a function $u : \Omega \to \mathbb{R}$ such that

$$\left\{\begin{aligned} \Delta u &= 0 && \text{in} \quad \Omega \setminus \Gamma(t) \\ u &= H && \text{on} \quad \Gamma(t) \\ \partial_\nu u &= 0 && \text{on} \quad \partial\Omega \\ V &= [\partial_\nu u] && \text{on} \quad \Gamma(t) \\ \Gamma(0) &= \Gamma_0, \end{aligned}\right. \tag{3.7}$$

where $\partial_\nu u$ denotes the normal derivative of u on $\partial\Omega$ and

$$[\partial_\nu u] := \partial_\nu u^+ - \partial_\nu u^-$$

stands for the jump of the normal derivative of u across $\Gamma(t)$, with $u^\pm := u^\pm(\cdot, t)$ being the restriction of u to $\Omega^\pm(t)$.

Let us now reduce system (3.7) to a single evolution equation of the form $V(t) = F(H(t))$. For this, fix t and let $h \in h^{1+\alpha}(\Gamma(t))$ be given. It can be shown (cf. Lemma 2.2 in [35]) that there exists a unique solution $u_h \in h^{1+\alpha}(\Omega)$ satisfying

$$\left\{\begin{aligned} \Delta u &= 0 && \text{in} \quad \Omega \setminus \Gamma(t) \\ u &= h && \text{on} \quad \Gamma(t) \\ \partial_\nu u &= 0 && \text{on} \quad \partial\Omega. \end{aligned}\right.$$

Hence, letting $F(h) := [\partial_\nu u_h]$, system (3.7) reduces to

$$V(t) = F(H(t)), \; t > 0, \qquad \Gamma(0) = \Gamma_0. \tag{3.8}$$

It is easy to see that F is an operator of first order in the sense that the inclusion $F(h^{1+\alpha}(\Gamma(t))) \subset h^\alpha(\Gamma(t))$ holds true. In addition, we mention that F is a nonlocal operator.

Assume now that there is a smooth solution Γ to (3.8). Then we have

$$\frac{1}{n}\frac{d}{dt}A(t) = \int_{\Gamma(t)} HV\, d\sigma(t) = \int_{\Gamma(t)} u_H[\partial_\nu u_H]\, d\sigma(t) = -\int_\Omega |\nabla u_H|^2 dx \leq 0,$$

and

$$\frac{d}{dt}vol(t) = \int_{\Gamma(t)} V\, d\sigma(t) = \int_{\Gamma(t)} [\partial_\nu u_H]\, d\sigma(t) = -\int_\Omega \Delta u_H dx = 0,$$

showing that (3.8) is volume-preserving and area decreasing.

System (3.7) was introduced by Mullins and Sekerka [59] to study solidification and liquiditation of materials of zero specific heat. This model is also closely related to the quasi-stationary two-phase Stefan problem with surface tension, cf. [36].

In the calculations above we always assumed existence of a smooth solution. In order to justify these arguments, let us first establish the following existence and uniqueness result for the flows induced by (3.3), (3.4), (3.6), and (3.8).

Theorem 3.1. *Assume that $0 < \beta < 1$ and let Γ_0 be a compact, closed, connected, embedded hypersurface in $\mathbb{R}^{n+1}$ of class $h^{1+\beta}$. Assume additionally in the cases of the surface diffusion flow (3.4), the intermediate surface diffusion flow (3.6), and the Mullins-Sekerka model (3.8) that Γ_0 is of class $h^{2+\beta}$.*

a) Each of the flows induced by (3.3), (3.4), (3.6), and (3.8) has a unique classical solution $\Gamma = \{\Gamma(t)\,;\, t \in [0,T)\}$, where $T := T(\Gamma_0)$ is the maximal existence time. Moreover, the mapping $[t \mapsto \Gamma(t)]$ is smooth on $(0,T)$ with respect to the C^∞-topology and continuous on $[0,T)$ with respect to the $h^{1+\beta}$-topology in the case (3.3) and with respect to the $h^{2+\beta}$-topology in the cases (3.4), (3.6), and (3.8), respectively.

b) Let Σ be a smooth hypersurface and suppose that the initial data Γ_0 is a $h^{1+\beta}$-graph over Σ in the case (3.3), and a $h^{2+\beta}$-graph over Σ in the cases (3.4), (3.6), and (3.8), respectively. Then the mapping $\varphi := [(t,\Gamma_0) \mapsto \Gamma(t)]$ defines a smooth local semiflow on an open subset of $h^{1+\beta}(\Sigma)$ in the case (3.3), and on an open subset of $h^{2+\beta}(\Sigma)$ in the cases (3.4), (3.6), and (3.8), respectively.

Let us explain the proof of Theorem 3.1 in the case of the intermediate surface diffusion flow (3.6). A proof of the assertions concerning the flows induced by (3.3), (3.4), and (3.8) can be found in [34, 35, 39, 40].

(i) We first parameterize an appropriate neighborhood of Γ_0. More precisely, given $a > 0$, there exists a smooth hypersurface Σ, with outer unit normal μ, and $\rho_0 \in h^{2+\beta}(\Sigma)$ with $\|\rho_0\|_{C^1(\Sigma)} < a/2$ such that $id_\Gamma + \rho_0\mu$ is a diffeomorphism of class $C^{2+\beta}$, mapping Σ onto Γ_0. For $a > 0$ small enough the mapping

$$X : \Sigma \times (-a,a) \to \mathbb{R}^{n+1}, \quad X(s,r) := s + r\mu(s)$$

is a smooth diffeomorphism onto its image $\mathcal{R} := im(X)$. We split the inverse of X into $X^{-1} = (S,\Lambda)$, where

$$S \in C^\infty(\mathcal{R},\Sigma) \quad \text{and} \quad \Lambda \in C^\infty(\mathcal{R},(-a,a))$$

is the metric projection of $\mathcal{R}$ onto Σ and the signed distance function with respect to Σ, respectively.

Let now $0 < \beta_0 < \beta < \alpha < 1$ be fixed and set

$$U := \{\rho \in h^{2+\beta_0}(\Sigma)\,;\, \|\rho\|_{C^1(\Sigma)} < a\}.$$

For $T > 0$ and $\rho \in C^1((0,T], U \cap C^\infty(\Sigma))$, define

$$\Phi_\rho : \mathcal{R} \times (0,T] \to \mathbb{R}, \quad (x,t) \mapsto \Lambda(x) - \rho(S(x),t).$$

At any instant $t \in (0,T]$, the zero-level set $\Gamma_{\rho(t)} := \Phi_\rho^{-1}(\cdot,t)(0)$ is a smooth compact connected hypersurface. The normal velocity of $\{\Gamma_{\rho(t)}\,;\, t\in(0,T]\}$ is then given by

$$V(s,t) = \frac{\partial_t \rho(s,t)}{|\nabla_x \Phi_\rho(x,t)|}\Big|_{x=X(s,\rho(s,t))}, \qquad (s,t)\in\Sigma\times(0,T]. \tag{3.9}$$

Observe that

$$\Gamma_{\rho(t)} = \{x\in\mathbb{R}^{n+1}\,;\, x = X(s,\rho(s,t)),\ s\in\Sigma\}, \quad t\in(0,T]. \tag{3.10}$$

Hence, letting $\theta_{\rho(t)}(s) := X(s,\rho(s,t))$ for $s\in\Sigma$, we see that $\theta_{\rho(t)}$ is a diffeomorphism, mapping Σ onto $\Gamma_{\rho(t)}$. We need some further notation. First let

$$L_\rho(s,t) := |\nabla_x\Phi_\rho(x,t)|\big|_{x=X(s,\rho(s,t))}, \quad (s,t)\in\Sigma\times(0,T].$$

Moreover, let $\theta_\rho^*\eta$ be the pull-back metric on Σ, where η is the usual Euclidean metric. We denote by Δ_ρ and H_ρ the Laplace-Beltrami operator and the mean curvature of $(\Sigma,\theta_\rho^*\eta)$, respectively. Then we have

$$H_\rho = \theta_\rho^* H_{\Gamma_\rho} \qquad \text{and} \qquad \Delta_\rho\theta_\rho^* = \theta_\rho^*\Delta_{\Gamma_\rho}, \quad \rho\in U, \tag{3.11}$$

where Δ_{Γ_ρ} and H_{Γ_ρ} stands for the mean curvature and the Laplace-Beltrami of (Γ_ρ,η), respectively. Finally, we set

$$G(\rho) := -L_\rho\theta_\rho^*(\Delta_{\Gamma_\rho}(1-\Delta_{\Gamma_\rho})^{-1}H_{\Gamma_\rho}), \quad \rho\in U$$

and we consider the evolution equation

$$\frac{d}{dt}\rho + G(\rho) = 0, \qquad \rho(0)=\rho_0, \tag{3.12}$$

where ρ_0 is determined by Γ_0. We set $W := h^{2+\beta}(\Sigma)\cap U$. A function $\rho : [0,T)\to W$ is called a classical solution to (3.12) if

$$\rho\in C([0,T),W)\cap C^\infty((0,T)\times C^\infty(\Sigma))$$

and if ρ satisfies (3.12) pointwise. By construction, the intermediate surface diffusion flow and (3.12) are equivalent in $\mathcal{R}$: If ρ is a classical solution to (3.12) then $\Gamma := \{\Gamma_{\rho(t)}\,;\, t\in[0,T)\}$ is a classical solution to (3.6) such that $\Gamma_{\rho(t)}\subset\mathcal{R}$, $t\in[0,T)$. Conversely, if $\Gamma := \{\Gamma(t)\,;\, t\in[0,T)\}$ is a classical solution to (3.6) with $\Gamma(t)\subset\mathcal{R}$, then the above construction yields a classical solution to (3.12).

(ii) From (3.11) we easily deduce that

$$(1-\Delta_\rho)^{-1}\theta_\rho^* = \theta_\rho^*(1-\Delta_{\Gamma_\rho})^{-1}, \quad \rho\in U.$$

Hence we find $G(\rho) = -L_\rho\Delta_\rho(1-\Delta_\rho)^{-1}H_\rho$, implying that

$$G(\rho) = L_\rho H_\rho - L_\rho(1-\Delta_\rho)^{-1}H_\rho, \quad \rho\in U.$$

It is known, see Lemma 3.1 in [38], that the mean curvature operator carries a quasi-linear structure in the sense that there exist functions

$$P\in C^\infty(U,\mathcal{L}(h^{2+\alpha}(\Sigma),h^\alpha(\Sigma))) \quad \text{and} \quad Q\in C^\infty(U,h^{1+\beta_0}(\Sigma)) \tag{3.14}$$

such that

$$H_\rho = P(\rho)\rho + Q(\rho) \quad \text{for} \quad \rho \in h^{2+\alpha}(\Sigma) \cap U.$$

Additionally, given $\rho \in U$, the linear operator $[h \mapsto P(\rho)h]$ is a uniformly elliptic operator of second order. Since L_ρ belongs to $h^{1+\beta_0}(\Sigma)$ and is strictly positive, it follows that $L_\rho P(\rho) \in \mathcal{H}(h^{2+\alpha}(\Sigma), h^\alpha(\Sigma))$. Finally, let

$$F(\rho) := L_\rho[(1 - \Delta_\rho)^{-1} H_\rho - Q(\rho)], \quad \rho \in U.$$

Then it is not difficult to verify that $F \in C^\infty(U, h^{1+\beta_0}(\Sigma))$, cf. the proof of Lemma 2.1 in [40]. We are now prepared to apply H. Amann's general theory of abstract quasilinear parabolic evolution equations. In order to verify the hypotheses of [5], let $E_0 := h^\alpha(\Sigma)$, $E_1 := h^{2+\alpha}(\Sigma)$, and let $E_\theta := (E_0, E_1)^0_{\theta,\infty}$, $\theta \in (0,1)$, be the continuous interpolation spaces between E_0 and E_1. Since the little Hölder spaces are stable under continuous interpolation, we find $0 < \theta_0 < \theta_1 < \theta_2 < 1$ such that

$$E_{\theta_0} = h^{1+\beta_0}(\Sigma), \quad E_{\theta_1} = h^{2+\beta_0}(\Sigma), \quad E_{\theta_2} = h^{2+\beta}(\Sigma).$$

The above considerations show that

$$[\rho \mapsto (L_\rho P(\rho), F(\rho))] \in C^\infty(U, \mathcal{H}(h^{2+\alpha}(\Sigma), h^\alpha(\Sigma)) \times h^{1+\beta_0}(\Sigma)).$$

Hence Theorem 12.1 in [5] guarantees the existence of a unique solution $\rho \in C([0,T), W) \cap C^1((0,T), h^\alpha(\Sigma))$ to (3.12). An additional bootstrapping argument as in the proof of Theorem 1 in [35] shows that $\rho \in C^\infty((0,T), C^\infty(\Sigma))$. This completes the proof. □

Using Alexandrov's characterization of embedded surfaces of constant mean curvature, cf. [1], it is not difficult to verify that Euclidean spheres are the only embedded equilibria for each of the flows induced by (3.3), (3.4), (3.6), (3.8). Since these flows also preserve the volume and decrease the surface area, the isoperimetric inequality suggests that global smooth solutions should converge to spheres. In fact, in certain situations it is possible to rigorously justify this heuristic argument. The main idea here is to use the framework of Theorem 3.1 and some techniques from maximal regularity to construct a finite dimensional stable center manifold to each of the above flows. In a second step we then shall identify this stable manifold as the set of all equilibria. Therefore we conclude that if a solution starts close to this manifold, i.e., close to some sphere, it exists globally and converges to the manifold, i.e, to some sphere. The precise statement reads as follows:

Theorem 3.2. *Let $\mathcal{S}$ be a fixed Euclidean sphere and let $\mathcal{M}$ denote the set of all spheres which are sufficiently $C^{1+\beta}$-close to $\mathcal{S}$ in the case (3.3) and sufficiently $C^{2+\beta}$-close to $\mathcal{S}$ in the cases (3.4), (3.6), (3.8). Then $\mathcal{M}$ attracts at an exponential rate all solutions which are $C^{1+\beta}$-close in the case (3.3) and sufficiently $C^{2+\beta}$-close to $\mathcal{M}$ in the cases (3.4), (3.6), and (3.8). In particular, all solutions starting from such a neighborhood exist globally and converge exponentially fast to some sphere as $t \to \infty$. The convergence is in the C^k-topology for any fixed $k \in \mathbb{N}$.*

Proof. We provide a proof for the intermediate surface diffusion flow (3.6). The details for the cases (3.3.), (3.4) and (3.8) can be found in [38, 39, 40].

(i) For simplicity we assume that $\Sigma = \mathcal{S}$ is the Euclidean sphere of radius 1 with center at the origin. We first calculate the Fréchet derivative of G at 0. For this we note that the mapping $[\rho \mapsto G(\rho)] : h^{2+\alpha}(S) \cap U \to h^\alpha(S))$ is smooth. We obtain

$$\partial G(0)h = -\partial(L_\rho \Delta_\rho (1-\Delta_\rho)^{-1})|_{\rho=0}[h, H_0] - L_0 \Delta_0 (1-\Delta_0)^{-1} \partial H_\rho|_{\rho=0} h,$$

for $h \in h^{2+\alpha}(\Sigma)$. But $L_0 \equiv 1$ and $H_0 \equiv 1$, so that

$$L_\rho \Delta_\rho (1-\Delta_\rho)^{-1} H_0 = L_\rho (1-\Delta_\rho)^{-1} \Delta_\rho H_0 = 0, \quad \rho \in h^{2+\alpha}(S) \cap U.$$

Consequently,

$$\partial(L_\rho \Delta_\rho (1-\Delta_\rho)^{-1})|_{\rho=0}[h, H_0] = \frac{d}{d\varepsilon}(L_{\varepsilon h} \Delta_{\varepsilon h}(1-\Delta_{\varepsilon h})^{-1}) H_0\Big|_{\varepsilon=0} = 0.$$

The derivative of the mean curvature operator is given by

$$\partial H_\rho|_{\rho=0} = -\frac{1}{n}(n + \Delta_0),$$

see [38], Lemma 3.1. Hence we find

$$\partial G(0) = \frac{1}{n} \Delta_0 (1-\Delta_0)^{-1}(n+\Delta_0).$$

ii) Next we locate the spectrum of $A := -\partial G(0)$. For this let $\{Y_k\,;\, 1 \le k \le n+1\}$ be the spherical harmonics of degree 1 and set $Y_0 \equiv 1$. Then it is known that

$$N := \operatorname{span}\{Y_0, \ldots, Y_{n+1}\} = ker(\Delta_0(1-\Delta)^{-1}(n+\Delta_0)).$$

We conclude that 0 is an eigenvalue of A of multiplicity $n+2$. Assume now that $\lambda \in \mathbb{C}^*$ and $h \in h^{2+\alpha}(\mathcal{S})$ satisfy $(\lambda + A)g = 0$. It follows that $h \in N^\perp$, where the orthogonal complement is taken with respect to the $L_2(\mathcal{S})$ inner product. Next, observe that there are positive constants c_1 and c_2 such that

$$(\Delta_0^{-1} g|g) \le -c_1(g|g), \qquad ((n+\Delta_0)g|g) \le -c_2(g|g) \tag{3.15}$$

for all $g \in h^{2+\alpha}(\mathcal{S}) \cap N^\perp$, where $(\cdot|\cdot)$ denotes the inner product in $L_2(\mathcal{S})$. Writing $w := (1-\Delta_0)^{-1/2} h$ we find

$$0 = ((\lambda + A)h|\Delta_0^{-1}h) = \lambda(h|\Delta_0^{-1}h) + \frac{1}{n}((n+\Delta_0)w|w).$$

(3.15) implies that $\lambda < 0$. Since $h^{2+\alpha}(\mathcal{S})$ is compactly embedded in $h^\alpha(\mathcal{S})$ we have that the spectrum of A consists of a sequence $\{\mu_k\,;\, k \in \mathbb{N}\}$ of eigenvalues with $\mu_k < \mu_{k-1} < \cdots < \mu_1 < \mu_0$, where $\mu_0 = 0$ has multiplicity $(n+2)$.

(iii) In a next step we briefly sketch the construction of a locally invariant center manifold $\mathcal{M}^c$ over N. Let $Y_0 := |\mathcal{S}|^{-1}\mathbf{1}$ and let $Pg := \sum_{k=0}^{n+1}(g|Y_k)Y_k$ for $g \in h^r(\mathcal{S})$. Then P is a continuous projection of $h^r(\mathcal{S})$ onto N parallel to $ker(P)$, and it is easy to verify that P commutes with A, that is, $PAg = APg = 0$ for every $g \in h^{2+\alpha}(\mathcal{S})$. Therefore, $N = im(P)$ and $ker(P)$ are complimentary subspaces of $h^{2+\alpha}(\mathcal{S})$ that reduce A. To simplify the notation we write $\pi^c = P$ and $\pi^s = (1-P)$, and we define $h_s^{2+\alpha}(\mathcal{S}) := \pi^s(h^{2+\alpha}(\mathcal{S}))$. It follows that $\sigma(\pi^c A) = \{0\}$ and $\sigma(\pi^s A) \subset (-\infty, 0)$. For this reason, N and $h_s^{2+\alpha}(\mathcal{S})$ are called the *center subspace* and the *stable subspace* of A, respectively. We can now apply Theorem 4.1 in [65], see also [56] Theorem

9.2.2. These results imply that, given $m \in \mathbb{N}^*$, there exists an open neighborhood U_0 of 0 in N and a mapping

$$\gamma \in C^m(U_0, h_s^{2+\alpha}(\mathcal{S})) \quad \text{with} \quad \gamma(0) = 0, \ \partial\gamma(0) = 0$$

such that $\mathcal{M}^c := \text{graph}(\gamma)$ is a locally invariant manifold for the semiflow generated by the quasilinear evolution equation (3.12). $\mathcal{M}^c$ is an $(n+2)$-dimensional submanifold of $h^{2+\alpha}(\mathcal{S})$. Moreover, $\mathcal{M}^c$ attracts solutions of (3.12) that start in a sufficiently small $h^{2+\beta}(\mathcal{S})$-neighborhood $W_0 \subset W$ of 0 at an exponential rate, and $\mathcal{M}^c$ contains all small equilibria of (3.12), see [65] Theorems 4.1 and 5.8.

iv) Step (iii) shows that $\mathcal{M}^c$ contains all small equilibria of (3.12). We show that $\mathcal{M}^c$ in fact coincides with $\mathcal{M}$ near 0. Suppose that $\mathcal{S}'$ is a sphere which is sufficiently close to $\mathcal{S}$. Let $(z_1, \ldots, z_{n+1})$ be the coordinates of its center and let r be its radius. Recall that $\mathcal{S} \subset \mathbb{R}^{n+1}$ is the unit sphere centered at the origin and let $z_0 := 1 - r$. If ρ measures the distance of $\mathcal{S}$ to $\mathcal{S}'$ in normal direction with respect to $\mathcal{S}$, then it can be verified that

$$(1 + z_0)^2 = \sum_{k=1}^{n+1} \big((1+\rho)Y_k - z_k\big)^2.$$

Here we used that the spherical harmonics Y_k, $k = 1, \cdots, n+1$, are given as the restrictions of the harmonic coordinate functions $[x \mapsto x_k]$. Let $Y_0 := \mathbf{1}$. Solving the above identify for ρ, we obtain that $\mathcal{S}'$ can be parameterized over $\mathcal{S}$ by the distance function

$$\rho(z) = \sum_{k=1}^{n+1} z_k Y_k - 1 + \sqrt{\Big(\sum_{k=1}^{n+1} z_k Y_k\Big)^2 + (1+z_0)^2 - \sum_{k=1}^{n+1} z_k^2}, \tag{3.16}$$

where $z := (z_0, \ldots, z_{n+1}) \in \mathbb{R}^{n+2}$. If O is a sufficiently small neighborhood of 0 in $\mathbb{R}^{n+2}$, then it is clear that any sphere $\mathcal{S}'$ which is close to $\mathcal{S}$ can be characterized by (3.16) with $z \in O$. The mapping $[z \mapsto \rho(z)] : O \to h^{2+\alpha}(\mathcal{S})$ is well-defined and smooth. Let $\mathcal{M} := \{\rho(z)\,;\, z \in U_0\}$. We conclude that $\mathcal{M} \subset \mathcal{M}^c$, since $\mathcal{M}$ consists of spheres, which are the equilibria of the intermediate surface diffusion flow. We intend to show that $\mathcal{M} = \mathcal{M}^c$. This follows, for instance, if we can verify that $\pi^c(\mathcal{M})$ is an open neighborhood of 0 in N. In order to show this we investigate the mapping

$$F : O \to N, \qquad F(z) := \pi^c \rho(z).$$

It is easy to see that the partial derivatives of F with respect to z_j at $0 \in O$ are given by $\partial_{z_0} F(0) = \mathbf{1}$ and $\partial_{z_k} F(0) = Y_k$ for $1 \le k \le n+1$. We conclude that the Fréchet derivative $\partial F(0)$ of F at 0 is given by

$$\partial F(0)h = \sum_{k=0}^{n+1} h_k Y_k, \qquad h \in \mathbb{R}^{n+2}. \tag{3.17}$$

Since the set $\{Y_k\}$ is a basis of N, we conclude that $\partial F(0) \in L(\mathbb{R}^{n+2}, N)$ is an isomorphism. Consequently, the Inverse Function Theorem implies that F is a smooth diffeomorphism from O onto its image $V := \text{im}(F)$, provided O is small enough. Therefore, $\pi^c(\mathcal{M})$ is an open neighborhood of 0 in N which can be assumed to coincide with the open neighborhood U_0 constructed in step (iii).

v) It follows from step (iv) that the reduced flow of (3.12) on $\mathcal{M}^c$ consists exactly of equilibria. Therefore, 0 is a stable equilibrium for the reduced flow and we conclude that 0 is also stable for the evolution equation (3.12), see Theorem 3.3 in [64]. In particular, there exists a neighborhood W_0 of 0 in $h^{2+\beta}(\mathcal{S})$ such that solutions of (3.12) exist globally and converge to $\mathcal{M}^c$ exponentially fast for every initial value $\rho_0 \in W_0$.

(vi) As in [38] Theorem 6.5 and Proposition 6.6, one shows the following result. Given $k \in \mathbb{N}$ and $\omega \in (0, -\mu_1)$ there exists a neighborhood $W_0 = W_0(k, \omega)$ of 0 in $h^{2+\beta}(\mathcal{S})$ with the following property: Given $\rho_0 \in W_0$, the solution $\rho(\cdot, \rho_0)$ of (3.12) exists globally and there exist $c = c(k, \omega) > 0$, $T = T(k, \omega) > 0$, and a unique $z_0 = z_0(\rho_0) \in U_0$ such that

$$\|\big(\pi^c \rho(t, \rho_0), \pi^s \rho(t, \rho_0)\big) - \big(z_0, \gamma(z_0)\big)\|_{C^k} \leq c e^{-\omega t} \|\pi^s \rho_0 - \gamma(\pi^c \rho_0)\|_{h^{2+\beta}}$$

for $t > T$. According to step (iv), $(z_0, \gamma(z_0))$ is a sphere and the proof is now complete. □

Remarks 3.3. a) By Theorem 3.1 the averaged mean curvature flow (3.3) generates a smooth semiflow on an open subset of $h^{1+\beta}(\Sigma)$. Moreover, Theorem 3.2 ensures that solutions starting sufficiently $C^{1+\beta}$-close to a sphere exist globally. Since in every $C^{1+\beta}$-neighborhood we also find non-convex surfaces, we get global solutions to (3.3) emerging from non-convex initial data. Of course, these global in time solutions are obtained as small perturbations of equilibria. However, it seems that this result is not reachable by the techniques in [42, 43, 49, 50].

b) It was shown in [18] by formal asymptotics that the surface diffusion flow is the singular limit of the zero level set of solutions to the Cahn-Hilliard equation with a concentration dependent mobility. Analytically, the surface diffusion flow was investigated for curves in two space dimensions by [11, 28, 44].

c) The results for the intermediate surface diffusion flow (3.6) obtained in Theorem 3.1 and Theorem 3.2 are new. The only other analytic results we are aware of are contained in [28]. As for the surface diffusion flow, the methods in [28] seem restricted to curves in $\mathbb{R}^2$.

d) In contrast to the Mullins-Sekerka model (3.7), equations (3.3), (3.4), and (3.6) make perfectly sense for immersed hypersurfaces. In fact, our methods are general and flexible enough to extend Theorem 3.1 to the case of compact closed immersed orientable hypersurfaces, see e.g. [40]. Observe that there are immersed surfaces of constant mean curvature which are not Euclidean spheres, cf. [70]. Consequently, the flows induced by (3.3), (3.4), and (3.6) admit equilibria which are not spheres. The stability properties of these non-embedded equilibria are not known. We also refer to the numerical simulations in [40] which show, for instance, that a four-leaf rose in $\mathbb{R}^2$ evolves in such a way as to approach a limiting configuration which is a triply covered immersed circle.

e) It is well known that the maximum principle prevents embedded hypersurfaces from developing self-intersections under the mean curvature flow. This is no longer true for the averaged mean curvature flow and the surface diffusion flow. It is shown in [58] that both flows can drive embedded hypersurfaces to a self-intersection in finite time. This behavior was conjectured for the averaged mean curvature flow in [42]. Moreover, it was suggested in [28] and later proved [44] that the surface diffusion flow can drive a dumbbell curve of appropriate shape to a self-intersection. The methods in [44] seem restricted to curves in $\mathbb{R}^2$.

f) It is known that the averaged mean curvature flow preserves strict convexity, cf. Theorem 1.3 in [50] and Theorem 4.1 in [42]. In contrast, the surface diffusion flow and the Mullins-Sekerka flow do not share this property, see [57, 45].

g) The Mullins-Sekerka model arises as a singular limit of the Cahn-Hilliard equation. This was formally derived in [62] and rigorously proved in [2] under the assumption that there exist classical solutions to (3.7). The first existence and uniqueness results for classical solutions to the Mullins-Sekerka model were obtained in [34, 35] and independently for initial data in $C^{3+\beta}$ in [21]. It should be mentioned that even existence of weak solutions to (3.7) in higher space dimensions was not established previously. In two dimensions, existence of global weak solutions for initial curves that are small perturbations of circles was shown in [20]. In [22] the authors prove existence of classical solutions starting from small analytic perturbation of a circle. However, the methods of these papers seem restricted to the plane setting. The Mullins-Sekerka model was also analyzed in [27] for strip-like domains in $\mathbb{R}^2$. Observe that Theorem 3.1 guarantees local existence for arbitrarily large initial data. Finally, the Mullins-Sekerka model can also be obtained as an asymptotic limit of some phase field models [16, 17, 68].

h) The Mullins-Sekerka model (3.7) describes solidification and liquiditation phenomena of two materials which are separated by a connected interface. Of course, in applications the situation is considerably more involved. In particular, one usually has to deal with multi-component processes. The Mullins-Sekerka model has also been proposed to account for aging or Ostwald ripening in phase transitions. In general, the kinetics of a first order phase transition is characterized by a first stage where small droplets of a new phase are created out of the old phase, e.g., solid formation in an undercooled liquid. The first stage, called nucleation, yields a large number of small particles. During the next stage the nuclei grow rapidly at the expense of the old phase. When the phase regions are formed, the mass of the new phase is close to equilibrium and the amount of undercooling is small, but large surface area is present. At the next stage, the configuration of phase regions is coarsened, and the geometric shape of the phase regions become simpler and simpler, eventually tending to regions of minimum surface area with given volume. The driving force of this process comes from the need to decrease the interfacial energy. There have been considerable efforts in finding a theory which describes Ostwald ripening, and the Mullins-Sekerka model is a prominent candidate.

Since the mechanism of the Mullins-Sekerka flow shows a distinct nonlocal feature, the corresponding mathematical formulation leads to a strongly coupled system of nonlinear evolution equations of third order. This multi-component system will be the topic of the forthcoming paper [3].

i) The construction of center manifolds for finite dimensional dynamical system is well known. Its extension to quasilinear and fully nonlinear infinite-dimensional semiflows (e.g. for φ) is considerably more involved. To overcome the technical difficulties involved with this situation we strongly rely on the theory of maximal regularity, see [25, 56, 65].

j) Observe that the averaged mean curvature flow induces a nonlinear and nonlocal operator of second order, meaning that the principal part of its linearization is a second order elliptic operator. The same is true for the intermediate surface diffusion flow. In this sense the surface diffusion flow is of forth order and the Mullins-Sekerka model is of third order, where in the latter case the principle part of the linearization is represented by an elliptic pseudo-differential operator of third order. It is worthwhile to note that there are also first order evolution equations driven by mean curvature, which conserve the volume and decrease the area. Let us mention the Stokes problem with surface tension

$$\begin{cases} -\Delta v + \nabla p = 0 & \text{in } \Omega(t) \\ div\, v = 0 & \text{in } \Omega(t) \\ T(v,p)\nu(t) = -H(t)\nu(t) & \text{on } \Gamma(t) \\ V = \partial_\nu u & \text{on } \Gamma(t) \\ \Gamma(0) = \Gamma_0, & \end{cases}$$

where v, p, and T stand for the velocity field, the pressure field, and the stress tensor of a liquid drop $\Omega(t)$ with boundary $\Gamma(t)$ moving freely under the influence of an exterior field and of surface tension. For a detailed study of this problem we refer to [46, 63].

Acknowledgement

The authors are grateful to G. Prokert for carefully reading the manuscript to this paper.

References

[1] A.D. Alexandrov, "Uniqueness theorems for surfaces in the large I", Vestnik Leningrad Univ. **11** (1956), 5–7, (Math. Rev. **19**, 167).

[2] N. Alikakos, P.W. Bates, X. Chen, "Convergence of the Cahn-Hilliard Equation to the Hele-Shaw model", Arch. Rational Mech. Anal. **128** (1994), 164–205.

[3] N. Alikakos, J. Escher, G. Fusco, G. Simonett, "On the Mullins-Sekerka model", *in preparation.*

[4] H. Amann, "Dynamic theory of quasilinear parabolic equations II". Reaction-Diffusion systems, Diff. Int. Equs. **3** (1990), 13–75.

[5] H. Amann, "Nonhomogeneous linear and quasilinear elliptic and parabolic boundary value problems", in H. J. Schmeisser, H. Triebel, editors, *Function Spaces, Differential Operators and Nonlinear Analysis*, Teubner, Stuttgart, Leipzig, 1993, 9–126.

[6] H. Amann, *Linear and Quasilinear Parabolic Problems, Vol. I*, Birkhäuser, Basel, 1995. *Vol. II, III*, in preparation.

[7] H. Amann, "Operator-valued Fourier multipliers, vector-valued Besov spaces, and applications", Math. Nachr. **186** (1997), 5–56.

[8] H. Amann, J. Escher, "Strongly continuous dual semigroups", Ann. Mat. Pura Appl. **161** (1996), 41–62.

[9] S.B. Angenent, "Nonlinear analytic semiflows", Proc. Roy. Soc. Edinburgh **115A** (1990), 91–107.

[10] S.B. Angenent, "Parabolic equations for curves on surfaces, Part I. Curves with $p-$integrable curvature". Ann. of Math. **132** (1990), 451–483.

[11] P. Baras, J. Duchon, R. Robert, "Evolution d'une interface par diffusion de surface", Comm. Partial Differential Equations **9** (1984), 313–335.

[12] J. Bear, Y. Bachmat, *Introduction to Modeling of Transport Phenomena in Porous Media*, Kluwer Academic Publisher, Boston, 1990.

[13] G.I. Bizhanova, V.A. Solonnikov, "On the initial boundary value problems for second order parabolic equations with time derivative in the boundary and conjugate conditions in weighted Hölder spaces of functions", *Abstracts ICM 94*, 164 (1994).

[14] G.I. Bizhanova, "Solution in a weighted Hölder function space of multidimensional two-phase Stefan and Florin problems for a second order parabolic equation in bounded domains", Algebra Analiz. **7** (1995), 46–76.

[15] L. Bronsard and B. Stoth, "Volume-preserving mean curvature flow as a limit of a nonlocal Ginzburg-Landau equation", SIAM J. Math. Anal. **28** (1997), 769–807.

[16] G. Caginalp, "Stefan and Hele-Shaw type models as asymptotic limits of the phase-field equations", Phys. Rev. A (3) **39** (1989), 5887–5896.

[17] G. Caginalp, "The dynamics of a conserved phase field system: Stefan-like, Hele-Shaw and Cahn-Hilliard models as asymptotic limits", IMA J. Appl. Math. **44** (1990), 77–94.

[18] J.W. Cahn, C.M. Elliott, A. Novick-Cohn, "The Cahn-Hilliard equation with a concentration dependent mobility: motion by minus the Laplacian of the mean curvature", European J. Appl. Math. **7** (1996), 287–301.

[19] J.W. Cahn, J.E. Taylor, "Surface motion by surface diffusion", Acta Metallurgica **42** (1994), 1045–1063.

[20] X. Chen, "The Hele-Shaw problem and area-preserving curve shortening motion", Arch. Rational Mech. Anal. **123** (1993), 117–151.

[21] X. Chen, J. Hong, F. Yi, "Existence, uniqueness, and regularity of classical solutions of the Mullins-Sekerka problem", Comm. Partial Differential Equations **21** (1996), 1705–1727.

[22] P. Constantin, M. Pugh, "Global solutions for small data to the Hele-Shaw problem", Nonlinearity **6** (1993), 393–415.

[23] J. Crank, *Free and Moving Boundary Problems*, Clarendon Press, Oxford, 1984.

[24] G. Da Prato, P. Grisvard, "Equations d'évolution abstraites nonlinéaires de type parabolique", Ann. Mat. Pura Appl. (4) **120** (1979), 329–396.

[25] G. Da Prato, A. Lunardi, "Stability, instability and center manifold theorem for fully nonlinear autonomous parabolic equations in Banach space", Arch. Rational Mech. Anal. **101** (1988), 115–144.

[26] F. Davì and M.E. Gurtin, "On the motion of a phase interface by surface diffusion", Z. Angew. Math. Phys. **41** (1990), 782–811.

[27] J. Duchon, R. Robert, "Evolution d'une interface par capillarité et diffusion de volume I. Existence locale en temps", Ann. Inst. H. Poincaré, Anal. Non Linéaire **1** (1984), 361–378.

[28] C.M. Elliott & H. Garcke, "Existence results for geometric interface models for surface diffusion", Adv. Math. Sci. Appl. **7** (1997), 467–490.

[29] C.M Elliott, J.R. Ockendon, *Weak and Variational Methods for Moving Boundary Problems*, Pitman, Boston, 1982.

[30] J. Escher, "The Dirichlet-Neumann operator on continuous functions", Ann. Scuola Norm. Sup. Pisa (4) **XXI** (1994), 235–266.

[31] J. Escher, "On moving boundaries in deformable media". Adv. Math. Sci. Appl. **7** (1997), 275–316.

[32] J. Escher, G. Simonett, "Maximal regularity for a free boundary problem", NoDEA **2** (1995), 463–510.

[33] J. Escher, G. Simonett, "Analyticity of the interface in a free boundary problem", Math. Ann. **305** (1996), 439–459.

[34] J. Escher, G. Simonett, "On Hele-Shaw models with surface tension", Math. Res. Lett. **3** (1996), 467–474.

[35] J. Escher, G. Simonett, "Classical solutions for Hele-Shaw models with surface tension", Adv. Differential Equations **2** (1997), 619–642.

[36] J. Escher, G. Simonett, "Classical solutions for the quasi-stationary Stefan Problem with surface tension", Proc. Differential Equations, Asymptotic Analysis, and Mathematical Physics, Mathematical Research **100** (1997), 98–104.

[37] J. Escher, G. Simonett, "Classical solutions of multidimensional Hele-Shaw models", SIAM, J. Math. Anal. **28** (1997), 1028–1047.

[38] J. Escher, G. Simonett, "A center manifold analysis for the Mullins-Sekerka model", J. Differential Equations **143** (1998), 267–292.

[39] J. Escher, G. Simonett, "The volume-preserving mean curvature flow near spheres", Proc. Amer. Math. Soc. **126** (1998), 2789–2796.

[40] J. Escher, G. Simonett, U.F. Mayer, "The surface diffusion flow for immersed hypersurfaces", to appear in *SIAM J. Math. Anal.*

[41] A. Friedman, *Partial Differential Equations*, Holt, Rinehart and Winston, New York, 1969.

[42] M. Gage, "On an area-preserving evolution equation for plane curves", *Nonlinear Problems in Geometry*, D.M. DeTurck, editor, Contemp. Math. 51, AMS, Providence, 51–62 (1986).

[43] M. Gage, R. Hamilton, "The shrinking of convex curves by the heat equation", J. Differential Geom. **23** (1986), 69–96.

[44] Y. Giga, K. Ito, "On pinching of curves moved by surface diffusion", Comm. Appl. Anal. **2** (1998), 393–405.

[45] Y. Giga, K. Ito, *Loss of convexity of simple closed curves moved by surface diffusion*, in this volume.

[46] M. Günther, G. Prokert, "Existence results for the quasistationary motion of a capillary liquid drop", Zeitschrift für Analysis und ihre Anwendungen **16** (1997), 311–348.

[47] M.E. Gurtin, *Thermomechanics of Evolving Phase Boundaries in the Plane*, Clarendon Press, Oxford, 1993.

[48] T. Hintermann, "Evolution equations with dynamic boundary conditions", Proc. Roy. Soc. Edinburgh, **113 A** (1989), 43–60.

[49] G. Huisken, "Flow by mean curvature of convex surfaces into spheres", J. Differential Geom. **20** (1984), 237–266.

[50] G. Huisken, "The volume-preserving mean curvature flow", J. Reine Angew. Math. **382** (1987), 35–48.

[51] H. Kawarada, H. Koshigoe, "Unsteady flow in porous media with a free surface", Japan J. Indust. Appl. Math. **8** (1991), 41–82.

[52] H. Kawarada, H. Koshigoe, A. Sasamoto, "The behaviour of free surface appearing in the flow through porous media", Adv. Math. Sci. Appl. **1** (1992), 157–174.

[53] H. Koch, "Classical solutions to phase transition problems are smooth", preprint (1997).

[54] H.B. Lawson, *Lectures on Minimal Surfaces*, Publish or Perish, Berkeley, 1980.

[55] A. Lunardi, "On the local dynamical system associated to a fully nonlinear abstract parabolic equation", in *Nonlin. Anal. & Appl.*, ed. V. Lakshmikantham, M. Dekker, New York, 316–326 (1987).

[56] A. Lunardi, *Analytic Semigroups and Optimal Regularity in Parabolic Problems*, Birkhäuser, Basel, 1995.

[57] U.F. Mayer, "Two-sided Mullins-Sekerka flow does not preserve convexity", *submitted.*

[58] U.F. Mayer, G. Simonett, "Self-intersections for the surface diffusion and the volume-preserving mean curvature flow", *submitted.*

[59] W.W. Mullins, R.F. Sekerka, "Morphological stability of a particle growing diffusion or heat flow", J. Appl. Math. **34** (1963), 323–329.

[60] W.W. Mullins, "Theory of thermal grooving", J. Appl. Phys. **28** (1957), 333–339.

[61] A. Pazy, *Semigroups of Linear Operators and Applications to Partial Differential Equations*, Springer Verlag, New York, 1983.

[62] R.L. Pego, "Front migration in the nonlinear Cahn-Hilliard equation", Proc. Roy. Soc. London Ser. A **422** (1989), 261–278.

[63] G. Prokert, *Parabolic Evolution Equations for Quasistationary Free boundary Problems in Capillary Fluid Mechanics*, PhD thesis, Tech. Univ. Eindhoven, The Netherlands, 1997.

[64] G. Simonett, "Invariant manifolds and bifurcation for quasilinear reaction-diffusion systems", Nonlinear Anal. **23** (1994), 515–544.

[65] G. Simonett, "Center manifolds for quasilinear reaction-diffusion systems", Differential Integral Equations **8** (1995), 753–796.

[66] V.A. Solonnikov, E.V. Frolova, "L_p-theory for the Stefan problem", to appear in Journal of Mathematical Sciences.

[67] V.A. Solonnikov, "Free boundary problems for second order parabolic equations", preprint (1997).

[68] H.M. Soner, "Convergence of the phase-field equations to the Mullins-Sekerka problem with kinetic undercooling", Arch. Rational Mech. Anal. **131** (1995), 139–197.

[69] H. Triebel, *Theory of Function Spaces*, Birkhäuser Verlag, Basel, 1983.

[70] H. Wente, "Counter-example to a conjecture of H. Hopf", Pacific J. Math. **121** (1986), 193–243.

Joachim Escher
Department of Mathematics
University of Kassel
D-34132 Kassel, Germany
escher@mathematik.uni-kassel.de

Gieri Simonett
Department of Mathematics
Vanderbilt University
Nashville, TN 37240, USA
simonett@math.vanderbilt.edu

Progress in Nonlinear Differential Equations
and Their Applications, Vol. 35
© 1999 Birkhäuser Verlag Basel/Switzerland

Evolution Equations with Dissipation of Memory Type

Eva Fašangová* and Jan Prüss

Dedicated to Professor H. Amann on the occasion of his 60th birthday

Abstract. Well-posedness of semilinear evolution equations with linear dissipation of fractional order or memory type is studied in a Hilbert space framework. Under the main assumptions that the energy is coercive, the set of stationary solutions is discrete, the underlying linear problem is stable, and if the nonlinearity is subject to a compactness condition, convergence of the mild solutions to a steady state is shown. The methods employed include stability of the linear problem, energy estimates, and harmonic analysis of vector-valued functions. The results apply to a variety of problems in mathematical physics, like semilinear beam and plate models, and semilinear wave equations.

1. Introduction

During the last decades much work has been devoted to the study of semilinear wave equations with instantaneous dissipation, i.e. on equations of the form

$$\ddot{u} + B\dot{u} + Au = F(u), \quad u(0) = u_0, \tag{1}$$

in a Hilbert space X, where A is self-adjoint and positive definite, B is accretive, and F nonlinear. Local and global existence and uniqueness, but in particular the asymptotic behaviour of solutions have been under consideration. We refer to Hale [7] as a general reference.

However, in many situations of physical relevance, instantaneous dissipation is not the right concept. In fact, dissipation may result from constitutive laws which are nonlocal in time, given by convolutions over the past history like a fractional derivative. We discuss now some examples which also may serve as test cases for the theory to be developed in this paper.

* Supported by the Deutscher Akademischer Austauschdienst Bonn, by Grant Nr. 201/96/0432 of the Grant Agency of the Czech Republic and by Grant Nr. 14/97 of the Grant Agency of Charles University Prague.

Example 1.1 Consider the semilinear initial-boundary value problem

$$\begin{aligned} \partial_t^2 w + e_0\partial_t\partial_x^4 w + e_1 * \partial_t\partial_x^4 w + e_\infty\partial_x^4 w = f(w), \quad t>0,\ x\in(0,\pi),\\ w(t,x) = \partial_x^2 w(t,x) = 0, \quad t>0,\ x=0,\pi, \\ w(0,x) = w_0(x),\ \partial_t w(0,x) = w_1(x), \quad x\in(0,\pi). \end{aligned} \tag{2}$$

Here ∂_t resp. ∂_x denote the partial derivative w.r.t. time t resp. space x, and the $*$ indicates convolution over the halfline, i.e.

$$[e_1 * u](t,x) = \int_0^t e_1(t-s)u(s,x)ds, \quad t>0.$$

This problem serves as a model for a simply supported viscoelastic Euler-Bernoulli beam with nonlinear load. Here $e_\infty > 0$ denotes the elasticity modulus of the beam, $e_0 \geq 0$ a Newtonian viscosity, and $e_1 \in L^1_{\text{loc}}(\mathbb{R}_+) \cap C_0(0,\infty)$ the viscoelastic stress relaxation modulus. According to the fading memory hypothesis, e_1 should be nonnegative, nonincreasing, and of positive type. The nonlinear load f will later on assumed to be of class $C^2(\mathbb{R})$ and such that $f(0)=0$, so we have the trivial solution $w\equiv 0$. The initial values w_0 and w_1 are given.

Example 1.2 If instead of the viscoelastic Euler-Bernoulli beam, the viscoelastic Timoshenko beam model is used, the following problem results.

$$\begin{aligned} \partial_t^2 w - da * \partial_t(\partial_x^2 w + \partial_x\phi) - a_\infty(\partial_x^2 w + \partial_x\phi) = \partial_w f(w,\phi), \\ \partial_t^2\phi - [de * \partial_t + e_\infty]\partial_x^2\phi + \gamma[da * \partial_t + a_\infty](\partial_x w + \phi) = \partial_\phi f(w,\phi). \end{aligned} \tag{3}$$

Boundary conditions are e.g. $w=\phi=0$ at a clamped end, $\partial_x\phi = \partial_x w + \phi = 0$ at a free end, and $w = \partial_x\phi = 0$ at a simply supported end. For the meaning of the variables w and ϕ as well as a derivation and discussion of viscoelastic beam models we refer to [10], Section 9.1. The numbers $\gamma, e_\infty, a_\infty$ must all be positive, the functions e, a are *creep functions*, i.e. nonnegative, nondecreasing and concave. The right-hand side can be interpreted as a nonlinear load which admits the potential $f(w,\phi)$.

Example 1.3 One may of course consider such problems also in space dimensions N greater than 1. An extension of Example 1.1 to $N>1$ is the following problem.

$$\begin{aligned} \partial_t w + de * \Delta^2\partial_t w + e_\infty\Delta^2 w = f(w),\ x\in\Omega \\ w(t,x) = \partial_n w(t,x) = 0, \quad x\in\partial\Omega,\ t>0, \end{aligned} \tag{4}$$

where $\Omega\subset\mathbb{R}^N$ is an open bounded domain with smooth boundary $\partial\Omega$, de as above. For $N=2$ this problem is related to models for viscoelastic plates.

Example 1.4 As a last example consider a semilinear wave equation with boundary dissipation of memory type.

$$\begin{aligned} \partial_t w - \Delta w = f(w),\ x\in\Omega \\ \partial_n w + dk * \partial_t w = g(w), \quad \text{on } \partial\Omega,\ t>0, \end{aligned} \tag{5}$$

where as before $\Omega \subset \mathbb{R}^N$ is an open, bounded, smooth domain, and k a creep function.

A general framework for the above problems is the following. Let H and V be Hilbert spaces such that V is continuously and densely embedded into H, let α be a bounded, bilinear, symmetric form on V which is positive definite, $\beta : \mathbb{R}_+ \times V \times V \to \mathbb{R}$ uniformly of bounded variation in t, bilinear, bounded, symmetric for each fixed t, and let $\Phi : V \to \mathbb{R}$ be of class C^1. Consider the problem

$$(\ddot{u}(t), v)_{V^*,V} + \int_0^t d\beta(s, \dot{u}(t-s), v) + \alpha(u(t), v) = (\Phi'(u(t), v)_{V^*,V} \quad (6)$$
$$u(0) = u_0 \in V, \quad \dot{u}(0) = u_1 \in H, \quad t \geq 0, \text{ for all } v \in V.$$

Representing the forms α and $\beta(t)$ by operators $A, B(t) \in \mathcal{B}(V, V^*)$, the problem may be also stated as a semilinear evolution equation in V^*.

$$\ddot{u}(t) + \int_0^t dB(s)\dot{u}(t-s) + Au(t) = \Phi'(u(t)),\ t \geq 0, \quad (7)$$
$$u(0) = u_0,\ \dot{u}(0) = u_1.$$

A particularly nice special case is that of a problem of *scalar type*, where

$$dB = dbA^\eta, \qquad \text{for some } \eta \in [0, 1],$$

and b a creep function. It will turn out that in this situation much more can be said than in general.

It is the purpose of this contribution to establish well-posedness of (6), and in particular the convergence of the mild solutions of (6) to an equilibrium solution, i.e. to a solution of the stationary problem

$$Aw = \Phi'(w), \quad (8)$$

provided the energy

$$E(u(t)) = \frac{1}{2}[|\dot{u}(t)|_H^2 + \alpha(u(t), u(t))] - \Phi(u(t))$$

of the problem is coercive, the set $\mathcal{E}$ of stationary solutions is discrete, the linear problem is "stable" (in a sense to be specified), $\Phi' : V \to H$ is locally Lipschitz and compact, and the dissipation is "strong enough"(in a sense to be specified).

For the case of *instantaneous damping*, when no memory is present, i.e. $dB(t) = \delta(t)B_0$, ($\delta(t)$ denotes the Dirac measure at zero) such a result can be obtained by well-known methods. In fact, in this case the problem gives rise to a semiflow in the natural phase space, the energy of the system serves as a Ljapunov function, and by means of a variant of LaSalle's invariance principle, convergence of the solutions in V can be shown, as soon as the set of stationary solutions $\mathcal{E}$ is discrete. As a general reference for this approach the monograph Hale [7] may serve.

However, this method does not work for (6) for two reasons. Because of memory dependence, equation (6) does not generate a semiflow in the natural phase space, and the energy is in general not decreasing, and so the theory of Ljapunov functions and in particular LaSalle's principle cannot be applied. Therefore some other tools have to be developed which serve as a substitute of the dynamical systems approach.

Combining properties of the linear problem, in particular stability properties, methods from nonlinear Volterra equations in finite dimensions, cf. Gripenberg, Londen and Staffans [6], and results on harmonic analysis of vector-valued functions, it was recently shown by the authors [4] that in Example 1.1 the solutions converge to a steady state, as soon as $e_0 > 0$ or $e_1 \not\equiv 0$ is subject to (Hb); cf. Section 2. The presence of any whatsoever small viscous or viscoelastic damping forces convergence of the solution to a stationary solution.

Here we prove such a result in the general setting introduced above. This allows for a much broader class of applications, and shows the strength of our approach. In particular, our method applies to the case of instantaneous damping, and it is possible to recover many results known for this case. It will turn out that a good understanding of the underlying linear problem, in particular of its stability, is crucial for the proof.

Replacing e.g. in Example 1.1 the damping term $de * \partial_t \partial_x^4$, where $e(t) = e_0 + \int_0^t e_1(s)ds$, by $-de * \partial_t \partial_x^2 w$ or by $de * \partial_t w$, we obtain two more interesting examples. Of special interest are the fractional kernels $e_1(t) = t^{-\gamma}/\Gamma(1-\gamma)$ for $\gamma \in (0,1)$. Then the time-operator $e_1 * \partial_t$ can be interpreted as ∂_t^γ, a derivative of fractional order. In a similar way we may replace ∂_x^4 in Example 1.1 by the general fractional order operator $(i\partial_x)^{4\eta}$. Then an interesting question arises, namely, for which orders γ, η the linear damping operator $\partial_t^\gamma \partial_x^\eta$, or in the general setting of scalar type, $\partial_t^\gamma A^\eta$, is strong enough to enforce stability of the linear problem, and/or convergence of the solutions of the nonlinear problem. We show below that these properties are valid for $\gamma + 2\eta \geq 1$. Thus there is a nice interplay between the spatial and time orders of the damping operator.

The plan for this paper is as follows. In Section 2 we introduce some basic notation and hypotheses which are needed throughout. The well-posedness of the linear problem is proved in Section 3, while Section 4 contains the corresponding result for the nonlinear problem. The main result is stated and proved in Section 5. The theory developed in Sections 3–5 is fairly general and applies to many concrete problems. However, the limiting item seems to be linear stability. In fact, in contrary to the semigroup case, there is no complete characterization for the stability of the linear problem, only partial answers are available at the moment. Therefore, we discuss in Section 6 several results in this direction, some which are already known but also some which are new. Finally, for the convenience of the reader, in an appendix, Section 7, a number of tools are collected which are used in the main body of the paper.

2. Preliminaries

We begin with some

2.1. Assumptions and Notations

We will denote $\mathbb{R}_+ = [0, \infty)$, $\mathbb{C}_+ = \{\Re\lambda > 0\}$, $\mathcal{B}(X, Y)$ the space of bounded linear operators from X to Y, where X and Y are Banach spaces, $\mathcal{C}_0(\mathbb{R}_+, X) := \{u : \mathbb{R}_+ \to X \text{ continuous }, u(t) \to 0 \text{ as } t \to \infty\}$. Subscript "s" will mean strong topology, for example $U \in \mathcal{C}_s(\mathbb{R}_+, \mathcal{B}(X, Y))$ if and only if for any $x \in X$ the function $U(.)x \in \mathcal{C}(\mathbb{R}_+, Y)$. The spaces $\mathcal{BUC}(\mathbb{R}_+, X)$ and $L^p(\mathbb{R}_+, X)$ are defined as usual.

δ means the Dirac measure at zero, $\widehat{}$ indicates Laplace transform with respect to the variable t, so we define

$$\widehat{u}(\lambda) := \int_0^\infty e^{-\lambda t} u(t)dt, \quad \Re\lambda > 0,$$

for functions $u \in L^1_{\text{loc}}(\mathbb{R}_+, X)$. c_i will denote various positive constants, without referring to relations between them. $*$ denotes convolution on the halfline with respect to t.

The following hypotheses will be employed below.

(HS) V, H are (real) Hilbert spaces with continuous and dense embedding $V \hookrightarrow H$. Identifying H with its dual H^* via the Riesz representation theorem, we also have the continuous and dense embedding $H \hookrightarrow V^*$ by the relation $(u, v)_{H,H} = (u, v)_{V^*,V}$, for all $u \in H$, $v \in V$.

(Hα) $\alpha : V \times V \to \mathbb{R}$ is a bilinear form, symmetric (i.e. $\alpha(u, v) = \alpha(v, u)$), bounded (i.e. $|\alpha(u, v)| \le c|u|_V|u|_V$), and positive definite (i.e. $|\alpha(u, u)| \ge c|u|^2_V$). Then $\alpha(u, u)$ defines an equivalent inner product on V and in the sequel we will work with this inner product.

(Hβ) $\beta : \mathbb{R}_+ \times V \times V \to \mathbb{R}$ satisfies: $\beta(t, \cdot, \cdot)$ is a bilinear form which is symmetric and bounded, for any $t \ge 0$; there exists a function $\beta_0 : \mathbb{R}_+ \to \mathbb{R}_+$ which is nondecreasing, left-continuous, $\int_0^\infty e^{-\lambda t} d\beta_0(t)dt < \infty$, for all $\lambda > 0$, $\beta_0(0) = 0$, and

$$|\beta(t, u, v) - \beta(s, u, v)| \le |\beta_0(s) - \beta_0(t)||u|_V|v|_V, \quad t, s \ge 0, \ u, v \in V.$$

Hypotheses (Hα) and (Hβ) imply the existence of bounded linear operators A and $B(t)$ from V to V^* satisfying (HA) and (HB) below.

(HA) The operator $A : V \to V^*$ defined by $\alpha(u, v) = (Au, v)_{V^*,V}$, $u, v \in V$, is linear, bounded, self-adjoint, positive definite, hence it is an isomorphism between V and V^*. Denoting $D(A_H) = \{u \in V; \ Au \in H\}$, the part A_H of A in H is an unbounded operator on H which is linear, self-adjoint, positive definite, hence also boundedly invertible.

(HB) Let $B : \mathbb{R}_+ \to \mathcal{B}(V, V^*)$ be defined by $\beta(t, u, v) = (B(t)u, v)_{V^*,V},\ u, v \in V$. Then for each $t \geq 0$, $B(t)$ is a linear, self-adjoint, bounded operator from V to V^*, $B \in \mathcal{BV}_{\mathrm{loc}}(\mathbb{R}_+, \mathcal{B}(V, V^*))$ has subexponential growth, is left continuous, $B(0) = 0$, and

$$|B(t) - B(s)|_{\mathcal{B}(V,V^*)} \leq \beta_0(t) - \beta_0(s), \qquad \text{for all } 0 \leq s \leq t.$$

In the main results we will also assume $\lim_{t\to\infty} \beta_0(t)/t = 0$. This means that the measure $d\beta_0$ tends to zero in the sense of Cesaro, hence dB tends to zero in $\mathcal{B}(V, V^*)$ in the sense of Cesaro as well. A physically plausible explanation for this assumption is the absence of damping forces in steady states, damping is a dynamical process. In particular, the steady states of (7) are given by the equation $Au = \Phi'(u)$.
(Hb) $b \not\equiv 0$ is a creep function, i.e. nonnegative, nondecreasing, and concave. Then there is a unique decomposition

$$b(t) = b_0 + b_\infty t + \int_0^t b_1(s)ds, \quad t > 0,$$

with $b_0, b_\infty \geq 0$, $b_1 \in L^1_{\mathrm{loc}}(\mathbb{R}_+)$ nonnegative, nonincreasing, $\lim_{t\to\infty} b_1(t) = 0$. In addition we assume $b_\infty = 0$, b_1 convex (hence $b_1 \in C_0(0, \infty)$) and either that $\log b_1$ or $-\dot{b}_1$ is convex.

When using the vector valued Laplace transform with respect to the variable t, we have to consider complexifications of spaces and forms. We define $H_{\mathbb{C}} = H + iH$, $V_{\mathbb{C}} = V + iV$, denote by superscript $*$ the antidual space resp. adjoint operator, and let $\alpha_{\mathbb{C}}(.,.)$, $\beta_{\mathbb{C}}(t,.,.)$ be the sesquilinear extensions of α, β, and $A_{\mathbb{C}}$, $B_{\mathbb{C}}(t)$ the linear extensions of A, $B(t)$. Then (HS), (Hα), (Hβ), (HA), (HB) are equivalent to their complex form, where "bilinear" is replaced by "sesquilinear". In the sequel we will not distinguish in notation between the real and the complex setting.

2.2. Dissipativity Concepts

We next introduce two notions of dissipativity of operator-valued kernels which are basic for a priori estimates for solutions of problem (6).

Definition 2.1. *Let (HS) and (Hβ) be satisfied. The form $d\beta$ is said to be*

- of positive type *if*

$$2\Re \int_0^T \left[\int_0^t d\beta(\tau, u(t-\tau), u(t))\right] dt \geq 0,$$

for each $u \in \mathcal{C}(\mathbb{R}_+, V)$ and $T > 0$;
- dissipative *if there exists either a function $a_1 \in L^1(\mathbb{R}_+)$ with $\widehat{a_1}(i\varrho) \neq 0$ for $\varrho \in \mathbb{R}$, or $a_1 = a_0\delta$ with $a_0 > 0$ such that for any $u \in \mathcal{C}(\mathbb{R}_+, V)$, $T > 0$ we have*

$$2\Re \int_0^T \left[\int_0^t d\beta(\tau, u(t-\tau), u(t))\right] dt \geq \int_0^T |a_1 * u(t)|_H^2 dt;$$

Remark 2.2. It is well known that forms of positive type can be characterized by frequency domain conditions. More precisely,

$$d\beta \text{ is of positive type} \Leftrightarrow 2\Re\widehat{d\beta}(\lambda, v, v) \geq 0, \quad v \in V,\ \Re\lambda > 0;$$

cf. [10] Proposition 6.7. The notion of a dissipative form is a generalization of strictly b-positive kernels from [4] to the operator-valued case. It is not difficult to show that the condition

$$2\Re\widehat{d\beta}(\lambda, v, v) \geq |\widehat{a}_1(\lambda)|^2 |v|_H^2, \quad \Re\lambda > 0,\ v \in V,$$

implies dissipativity of $d\beta$, cf. [4].

We want to mention especially the so-called *scalar case* where

$$\beta(t, u, v) = b(t)\beta_0(u, v), \quad t \geq 0,\ u, v \in V,$$

with a bounded symmetric form β_0 on V and $b \in BV_{\text{loc}}(\mathbb{R}_+)$ with $b(0) = 0$. Then $d\beta$ is of positive type iff the measure db is of positive type and β_0 is nonnegative. As a general reference for measures of positive type see [6]. $d\beta$ is dissipative if $\beta_0(v, v) \geq c|v|_H^2$ for all $v \in V$, and b satisfies (Hb); cf. Lemma 7.3 when b_1 is log-convex, and [6], Proposition 16.4.3 in case $b_1, -\dot{b}_1$ are convex.

2.3. Energy Inequality

Proposition 2.3. *Let (HS), (Hα), (Hβ) be satisfied, let A and $B(t)$ be defined by (HA) and (HB), and suppose $g \in L^1_{\text{loc}}(\mathbb{R}_+, V^*)$, $u_0 \in V$. Then for a solution $u \in C(\mathbb{R}_+, V) \cap C^1(\mathbb{R}_+, V^*)$ of*

$$\dot{u} + dB * u + 1 * Au = g \text{ in } V^* \text{ for a.a. } t \in \mathbb{R}_+,\ u(0) = u_0, \tag{9}$$

the following assertions are true.

(i) *if dB is of positive type, then for any $T \geq 0$*

$$|u(T)|_H^2 + |1 * u(T)|_V^2 \leq |u_0|_H^2 + 2\Re \int_0^T (g(t), u(t))_{V^*,V}\, dt;$$

(ii) *if dB is dissipative then for any $T \geq 0$*

$$|u(T)|_H^2 + |1 * u(T)|_V^2 + \int_0^T |a_1 * u(t)|_H^2 dt \leq |u_0|_H^2 + 2\Re \int_0^T (g(t), u(t))_{V^*,V}\, dt.$$

Proof. Taking the duality $(.,.)_{V^*,V}$ between the equation (9) and u, integrating over t, and taking real parts we obtain

$$\Re \int_0^T (\dot{u}(t), u(t))_{V^*,V}\, dt + \Re \int_0^T \int_0^t d\beta(s, u(t-s), u(t))dt + \int_0^T (A \int_0^t u(s)ds, u(t))_{V^*,V}\, dt = \Re \int_0^T (g(t), u(t))_{V^*,V}\, dt.$$

By means of

$$\begin{aligned}\frac{d}{dt}|u(t)|_H^2 &= \frac{d}{dt}(u(t),u(t))_{H,H}\\ &= \frac{d}{dt}(u(t),u(t))_{V^*,V} = 2\Re(\dot u(t),u(t))_{V^*,V},\end{aligned}$$

and

$$2\int_0^T (A\int_0^t u(s)ds, u(t))_{V^*,V}dt = \int_0^T \frac{d}{dt}|\int_0^t u(s)ds|_V^2 ds = |\int_0^T u(s)ds|_V^2,$$

as well as by Definition 2.1, the assertions follow. □

3. Well-posedness of the Linear Problem

The main goal of this section is to prove existence, uniqueness and continuous dependence on initial data of solutions of the homogeneous linear equation

$$\ddot u + Au + dB * \dot u = 0,\ t>0, \quad u(0)=u_0, \quad \dot u(0) = u_1, \tag{10}$$

as well as a priori estimates. Here A and $B(t)$ are the representations of the bilinear forms α and $\beta(t)$ as in (HA) and (HB). Throughout we assume that $d\beta$ is of positive type.

3.1. Existence of the Solution Families

Denoting (formally) $R(t)u_1 = u(t)$, $S(t)u_1 = \dot u(t)$ when u is a solution of (10) with $u_0 = 0$; and $C(t)u_0 = u(t)$, $\dot C(t)u_0 = \dot u(t)$ when u is a solution of (10) with $u_1 = 0$; the main result is the following theorem, which is taken from Propst and Prüss [8].

Theorem 3.1. *Suppose (HS), (Hα), (Hβ) from Section 1 hold, and assume $d\beta$ is of positive type. Let A, B be defined via the representations (HA) and (HB), and let*

$$K(\lambda) = (\lambda^2 I + \lambda\widehat{dB}(\lambda) + A)^{-1}, \quad \Re\lambda > 0,$$

where the hat indicates Laplace transform.

Then there exist operator families S, R, T, C such that:

(i) $S \in \mathcal{C}_s(\mathbb{R}_+, \mathcal{B}(H,H))$, $|S(t)|_{\mathcal{B}(H,H)} \le 1$, $\widehat S(\lambda) = \lambda K(\lambda)$, $\lambda > 0$, *and for $u_1 \in H$ the function $v(t) = S(t)u_1$ satisfies*

$$v + d(tA+B) * 1 * v = u_1 \ \text{in}\ V^*;$$

(ii) $R = 1 * S \in \mathcal{C}_s(\mathbb{R}_+, \mathcal{B}(H,V))$, $|R(t)|_{\mathcal{B}(H,V)} \le 1$, $\widehat R(\lambda) = K(\lambda)$, $\lambda > 0$, *and for $u_1 \in H$ the function $u(t) = R(t)u_1$ satisfies*

$$\dot u + d(tA+B) * u = u_1 \ \text{in}\ V^*, \quad u(0) = 0;$$

(iii) $R \in \mathcal{C}_s(\mathbb{R}_+, \mathcal{B}(V^*,H))$, $|R(t)|_{\mathcal{B}(V^*,H)} \le 1$, $\widehat R(\lambda) = K(\lambda)$, $\lambda > 0$, *and for $u_1 \in V^*$ the function $u(t) = R(t)u_1$ satisfies*

$$u + d(tA+B) * 1 * u = tu_1 \ \text{in}\ V^*;$$

(iv) $T = 1 * R \in \mathcal{C}_s(\mathbb{R}_+, \mathcal{B}(V^*, V))$, $|T(t)|_{\mathcal{B}(V^*,V)} \le 2$, $\widehat{T}(\lambda) = \frac{1}{\lambda}K(\lambda)$, $\lambda > 0$, *and for* $u_1 \in V^*$ *the function* $w(t) = T(t)u_1$ *satisfies*

$$\dot{w} + d(tA + B) * w = tu_1 \text{ in } V^*, \quad w(0) = 0;$$

(v) $C = I - TA \in \mathcal{C}_s(\mathbb{R}_+, \mathcal{B}(V, V))$, $|C(t)|_{\mathcal{B}(V,V)} \le 1$, $\widehat{C}(\lambda) = \frac{1}{\lambda}I - \frac{1}{\lambda}K(\lambda)A$, *and for* $u_0 \in V$ *the function* $u(t) = C(t)u_0$ *satisfies*

$$\dot{u} + d(tA + B) * (u - u_0) = -tAu_0 \text{ in } V^*, \quad u(0) = u_0.$$

(vi) $\dot{C} = -RA \in \mathcal{C}_s(\mathbb{R}_+, \mathcal{B}(V, H))$, $|\dot{C}(t)|_{\mathcal{B}(V,H)} \le 1$, $\widehat{\dot{C}}(\lambda) = -K(\lambda)A$, $\lambda > 0$, *and for* $u_0 \in V$ *the function* $v(t) = \dot{C}(t)u_0$ *satisfies*

$$v + d(tA + B) * 1 * v = -tAu_0 \text{ in } V^*;$$

We will give here some more details of the proof of this theorem which have not been carried out in [8].

The idea of the proof is the following. We approximate (10) by the parabolic problem

$$\ddot{u} + \varepsilon A\dot{u} + Au + dB * \dot{u} = 0, \ t > 0, \quad u(0) = u_0, \quad \dot{u}(0) = u_1, \tag{11}$$

with $\varepsilon > 0$. Using results of Prüss [9] on the existence of solution to (11) the Laplace transform of the function $S_\varepsilon(.)u_1$ is well defined. Passing to the limit in the frequency domain $\widehat{S_\varepsilon}(.)u_1$ we obtain a holomorphic function $H(.)u_1$ satisfying some estimates which follow from estimates for the solutions of the approximating equations (11) in the time domain. Then the vector-valued version of Widder's inversion theorem for the Laplace transform due to Arendt [1] is employed to find an operator family $S(t)$ such that $\widehat{S} = H$ strongly in the right half-plane. In a similar way one obtains R and T.

3.1.1. The Approximating Problem. Formally, (11) is equivalent to

$$\ddot{u} + d(\varepsilon A + tA + B) * \dot{u} = -Au_0, \ t > 0, \quad u(0) = u_0, \quad \dot{u}(0) = u_1, \tag{12}$$

and also to its integrated form, which reads

$$\dot{u} + d(\varepsilon A + tA + B) * u = u_1, \ t > 0, \quad u(0) = 0, \tag{13}$$

provided $u_0 = 0$ (we will use only this case). These two equations are Volterra equations of parabolic type studied in [9]:

$$\dot{v} + \varepsilon Av + d(tA + B) * v = f, \ t > 0, \quad v(0) = v_0.$$

Proposition 3.2. *Suppose (HS), (Hα), (Hβ) hold and let* $\varepsilon > 0$. *Then*

(i) *for any* $u_0, u_1 \in V$ *there exists a unique function* $u \in \mathcal{C}^1(\mathbb{R}_+, V) \cap \mathcal{C}^2(\mathbb{R}_+, V^*)$ *such that* u *is a solution of (12) in* V^*; *if* $u_0 = 0$ *then* u *solves (13) too.*

(ii) *If $d\beta$ is of positive type then the function u from (i) satisfies*

$$|\dot{u}(t)|_H^2 + |u(t)|_V^2 \le |u_1|_H^2 + |u_0|_V^2, \quad t \ge 0;$$

and if $u_0 = 0$ then also

$$|u(t)|_H^2 + |1 * u(t)|_V^2 \le 2\Re(u_1, 1 * u(t))_{V^*,V}, \quad t \ge 0.$$

(iii) *If $d\beta$ is of positive type then:*
(a) *for any $u_0 \in V$, $u_1 \in H$ there exists a unique function $u \in \mathcal{C}(\mathbb{R}_+, V) \cap \mathcal{C}^1(\mathbb{R}_+, H)$ depending continuously on $(u_0, u_1) \in V \times H$ in the norm of $\mathcal{C}([0,T], V) \cap \mathcal{C}^1([0,T], H)$ for any $T > 0$, such that $u(0) = u_0$, $\dot{u}(0) = u_1$, and u coincides with the solution in (i) provided $u_1 \in V$;*
(b) *for any $u_1 \in V^*$ ($u_0 = 0$) there exists a unique function u such that $(u, 1 * u) \in \mathcal{C}(\mathbb{R}_+, H) \times \mathcal{C}(\mathbb{R}_+, V)$, depends continuously on $u_1 \in V^*$ locally uniformly in t, $u(0) = 0$, $\dot{u}(0) = u_1$, and coincides with the solution in (i) provided $u_1 \in V$.*

Proof. **(i)** By (HA) the operator $-A$ generates a bounded analytic semigroup on V^* which is even of negative type. By Theorem 1 of [9] there exists a resolvent for (9) defining the solution u of (12) resp. the solution u of (13). By Proposition 3 of [9], $\dot{u}$ resp. u belongs to $\mathcal{C}(\mathbb{R}_+, V) \cap \mathcal{C}^1(\mathbb{R}_+, V^*)$.
(ii) $d\beta$ of positive type implies that $d(\varepsilon A + tA + B)$ is of positive type as well. We arrange the terms in the energy inequality Proposition 2.3 applied to $\dot{u}$

$$|\dot{u}(T)|_H^2 + |1 * \dot{u}(T)|_V^2 \le |\dot{u}(0)|_H^2 + 2\Re \int_0^T (-Au(0), \dot{u}(t))_{V^*,V} dt$$

as follows

$$\begin{aligned} |1 * \dot{u}(T)|_V^2 &= |u(T)|_V^2 - 2\Re(Au(0), u(T))_{V^*,V} + |u(0)|_V^2, \\ 2\Re \int_0^T (-Au(0), \dot{u}(t))_{V^*,V} dt &= -2\Re(Au(0), u(T))_{V^*,V} + 2|u(0)|_V^2; \end{aligned}$$

if $u_0 = 0$ then with Proposition 2.3 we have from (13):

$$|u(T)|_H^2 + |1 * u(T)|_V^2 \le 2\Re \int_0^T (u_1, u(t))_{V^*,V} dt.$$

(iii) Follows from (ii). □

Proposition 3.3. *Suppose (HS), (Hα), (Hβ) hold, $d\beta$ is of positive type and let $\varepsilon > 0$. Then there exist operator families S_ε, R_ε, T_ε, C_ε for (11) having the properties as in Theorem 3.1 with the exception that $K(\lambda) = (\lambda^2 + A + \lambda\widehat{dB})^{-1}$ is replaced by*

$$K_\varepsilon(\lambda) = (\lambda^2 + \varepsilon\lambda A + A + \lambda\widehat{dB})^{-1}, \quad \lambda > 0.$$

Moreover, for $n = 0, 1, \dots$ and $\Re\lambda > 0$ we have

$$|\frac{d^n}{d\lambda^n}\widehat{S_\varepsilon}(\lambda)|_{\mathcal{B}(H,H)},\ |\frac{d^n}{d\lambda^n}\widehat{R_\varepsilon}(\lambda)|_{\mathcal{B}(H,V)},\ |\frac{d^n}{d\lambda^n}\widehat{R_\varepsilon}(\lambda)|_{\mathcal{B}(V^*,H)},$$
$$\frac{1}{2}|\frac{d^n}{d\lambda^n}\widehat{T_\varepsilon}(\lambda)|_{\mathcal{B}(V^*,V)},\ |\frac{d^n}{d\lambda^n}\widehat{\dot{C}_\varepsilon}(\lambda)|_{\mathcal{B}(V,H)},\ |\frac{d^n}{d\lambda^n}\widehat{C_\varepsilon}(\lambda)|_{\mathcal{B}(V,V)} \quad \le \quad \frac{n!}{(\Re\lambda)^{n+1}}.$$

Proof. Strong continuity and boundedness in time domain follow from Theorem 3.2; observe that

$$2\Re(u_1, 1 * u(t))_{V^*,V} \le |u_1|^2_{V^*} + |1 * u(t)|^2_V,$$
$$2\Re(u_1, 1 * u(t))_{V^*,V} \le 2|u_1|^2_{V^*} + \frac{1}{2}|1 * u(t)|^2_V.$$

It follows that the functions $S_\varepsilon(.)u_1$, etc. are Laplace transformable and that the asserted estimates in frequency domain hold. In fact, we have e.g. for S_ε

$$|\frac{d^n}{d\lambda^n}\widehat{S_\varepsilon}(\lambda)|_{\mathcal{B}(H,H)} = |\int_0^\infty e^{-\lambda t} t^n S_\varepsilon(t)dt|_{\mathcal{B}(H,H)} \le \widehat{t^n}(\Re\lambda) = n!/(\Re\lambda)^{n+1}.$$

(a) For $u_1 \in H$ let $v(t) = S_\varepsilon(t)u_1$, then

$$\dot{v} + d(\varepsilon A + tA + B) * v = 0 \text{ in } V^*, \quad v(0) = u_1,$$

and taking the Laplace transform we obtain

$$(\lambda I + [d(\varepsilon A + tA + B)]\,\widehat{}\,(\lambda))\,\widehat{v} = u_1,$$

i.e. the operator $\lambda I + [d(\varepsilon A + tA + B)]\,\widehat{}\,(\lambda) : H \to H$ is bijective and $\widehat{S_\varepsilon}(\lambda)$ is its inverse.

(b) For $u_1 \in H$ let $u(t) = R_\varepsilon(t)u_1$, then

$$\ddot{u} + d(\varepsilon A + tA + B) * \dot{u} = 0 \text{ in } V^*, \quad u(0) = u_1, \quad \dot{u}(0) = u_1,$$

and taking the Laplace transform we obtain $(\lambda I + [d(\varepsilon A + tA + B)]\,\widehat{}\,)\widehat{u} = \frac{1}{\lambda}u_1$, i.e. the operator $\lambda(\lambda I + [d(\varepsilon A + tA + B)]\,\widehat{}\,) : V \to H$ is bijective and $\widehat{R_\varepsilon}(\lambda)$ is its inverse.

(c) For $u_1 \in V^*$ let $u(t) = R_\varepsilon(t)u_1$, then

$$\dot{u} + d(\varepsilon A + tA + B) * u = u_1 \text{ in } V^*, \quad u(0) = 0,$$

and taking the Laplace transform we obtain $(\lambda I + [d(\varepsilon A + tA + B)]\,\widehat{}\,)\widehat{u} = \frac{1}{\lambda}u_1$, i.e. the operator $\lambda(\lambda I + [d(\varepsilon A + tA + B)]\,\widehat{}\,) : H \to V^*$ is bijective and $\widehat{R_\varepsilon}(\lambda)$ is its inverse.

(d) For $u_1 \in V^*$ let $w(t) = T_\varepsilon(t)u_1$, then

$$\dot{w} + d(\varepsilon A + tA + B) * w = tu_1 \text{ in } V^*, \quad w(0) = 0,$$

and taking the Laplace transform we obtain $(\lambda I + [d(\varepsilon A + tA + B)]\,\widehat{}\,)\widehat{w} = \frac{1}{\lambda^2}u_1$, i.e. the operator $\lambda^2(\lambda I + [d(\varepsilon A + tA + B)]\,\widehat{}\,) : V \to V^*$ is bijective and $\widehat{T_\varepsilon}(\lambda)$ is its inverse.

(e) For $u_0 \in V$ let $v(t) = \dot{C}_\varepsilon(t)u_0$, then

$$\dot{v} + d(\varepsilon A + tA + B) * v = -Au_0 \text{ in } V^*, \quad v(0) = 0,$$

and taking the Laplace transform we obtain $(\lambda I+[d(\varepsilon A+tA+B)]\,\widehat{}\,)\widehat{v}=-\frac{1}{\lambda}Au_0$, i.e. the operator $-A^{-1}\lambda(\lambda I+[d(\varepsilon A+tA+B)]\,\widehat{}\,):V\to H$ is bijective and $\widehat{\dot{C}_\varepsilon}(\lambda)$ is its inverse.

(f) For $u_0\in V$ let $u(t)=C_\varepsilon(t)u_1$, then

$$\ddot{u}+d(\varepsilon A+tA+B)*\dot{u}=-Au_0 \text{ in } V^*,\quad u(0)=u_0,\quad \dot{u}(0)=0,$$

and taking the Laplace transform we obtain

$$\left(\lambda^2 I+\lambda\,[d(\varepsilon A+tA+B)]\,\widehat{}\,\right)\left(\widehat{u}-\frac{1}{\lambda}u_0\right)=-\frac{1}{\lambda}Au_0,$$

i.e. the operator $-\lambda A^{-1}(\lambda^2 I+\lambda \widehat{d(\varepsilon A+tA}+B)):V\to V$ is bijective and $\widehat{C_\varepsilon}(\lambda)-\frac{1}{\lambda}I$ is its inverse. □

3.1.2. The Limiting Problem

Proposition 3.4. *Suppose (HS), (Hα), (Hβ), dβ is of positive type. Let S_ε, R_ε, T_ε be given by Proposition 3.3. Then there exists a holomorphic function $K(\lambda)$ on $\mathbb{C}_+$ such that we have as $\varepsilon\to 0$:*

$$\widehat{S_\varepsilon}(\lambda)\to\lambda K(\lambda)\quad in\quad \mathcal{B}(H,H),\quad \widehat{T_\varepsilon}(\lambda)\to\frac{1}{\lambda}K(\lambda)\quad in\quad \mathcal{B}(V^*,V),$$
$$\widehat{R_\varepsilon}(\lambda)\to K(\lambda)\quad in\quad \mathcal{B}(H,V)\cap\mathcal{B}(V^*,H),$$

locally uniformly in $\lambda\in\mathbb{C}_+$. In addition, $K(\lambda)=(\lambda^2+A+\lambda\widehat{dB}(\lambda))^{-1}$.

Proof. We verify the Cauchy-Bolzano condition, e.g. for S.

$$\begin{aligned}
|\widehat{S_{\varepsilon_1}}(\lambda)-\widehat{S_{\varepsilon_2}}(\lambda)|_{\mathcal{B}(H,H)} &= |\widehat{S_{\varepsilon_1}}(\lambda)(\varepsilon_2-\varepsilon_1)A\widehat{S_{\varepsilon_2}}(\lambda)|_{\mathcal{B}(H,H)}\\
&\le |\varepsilon_2-\varepsilon_1||\lambda^2||\frac{\widehat{S_{\varepsilon_1}}(\lambda)}{\lambda}|_{\mathcal{B}(V^*,H)}|\frac{\widehat{S_{\varepsilon_2}}(\lambda)}{\lambda}|_{\mathcal{B}(H,V)}\\
&\le |\varepsilon_2-\varepsilon_1|\frac{|\lambda^2|}{(\Re\lambda)^2},
\end{aligned}$$

This shows that the limit of $\widehat{S}_\varepsilon(\lambda)$ for $\varepsilon\to 0$ exists in $\mathcal{B}(H)$. In a similar way, convergence of the other operator families is obtained. To identify the limit, let $x\in V, y\in V^*$.

$$\begin{aligned}
&\lim_{\varepsilon\to 0+}(\lambda^2+\varepsilon\lambda A+A+\lambda\widehat{dB})^{-1}(\lambda^2+A+\lambda\widehat{dB})x\\
&=\lim_{\varepsilon\to 0+}x-\varepsilon\lambda(\lambda^2+\varepsilon\lambda A+A+\lambda\widehat{dB})^{-1}Ax=x \text{ in } V,\\
&\lim_{\varepsilon\to 0+}(\lambda^2+A+\lambda\widehat{dB})(\lambda^2+\varepsilon\lambda A+A+\lambda\widehat{dB})^{-1}y\\
&=\lim_{\varepsilon\to 0+}y-\varepsilon\lambda A(\lambda^2+\varepsilon\lambda A+A+\lambda\widehat{dB})^{-1}y=y \text{ in } V^*.
\end{aligned}$$

So the both sided inverse operator $(\lambda^2+A+\lambda\widehat{dB})^{-1}$ exists in $\mathcal{B}(V^*,V)$ and is equal to $K(\lambda)$. □

Proof of Theorem 3.1. By the vector valued Widder inverse Laplace transform theorem due to Arendt [1] (see also [10], Theorem 0.2), and proceeding as in the proof of Theorem 4.1 of [8], we obtain the existence of strongly continuous bounded operator families S, R, T, C with Laplace transforms as asserted.

For $u_1 \in V^*$ let $u(t) = R(t)u_1$. Since

$$(\lambda^2 + A + \lambda \widehat{dB}(\lambda))\widehat{u}(\lambda) = u_1,$$

we have by the inverse Laplace transform

$$u + d(tA + B) * 1 * u = tu_1 \text{ in } V^*.$$

If $u_1 \in H$ then the latter equation can be differentiated (since $u \in \mathcal{C}(\mathbb{R}_+, V)$ then):

$$\dot{u} + d(tA + B) * u = u_1 \text{ in } V^*, \quad u(0) = 0.$$

The assertions for $C, \dot{C}$ can be obtained from the relations $\dot{C} = -RA$, $C = I - TA$. □

Corollary 3.5. *The operator families from Proposition 3.1 satisfy in addition: $S \in \mathcal{B}(H, H)$ is a self-adjoint operator, and the adjoint of $R \in \mathcal{B}(H, V)$ is $R \in \mathcal{B}(V^*, H)$.*

Proof. Follows from uniqueness of Laplace transform; take S for example:

$$\begin{aligned}\widehat{S^*}(\lambda) &= \left[\widehat{S}(\overline{\lambda})\right]^* = \left[\left(\overline{\lambda} + \frac{1}{\overline{\lambda}}A + \widehat{dB}(\overline{\lambda})\right)^{-1}\right]^* \\ &= \left(\lambda + \frac{1}{\lambda}A^* + \left[\widehat{dB}(\overline{\lambda})\right]^*\right)^{-1} = \left(\lambda + \frac{1}{\lambda}A + \widehat{dB^*}(\lambda)\right)^{-1} \\ &= \left(\lambda + \frac{1}{\lambda}A + \widehat{dB}(\lambda)\right)^{-1} = \widehat{S}(\lambda).\end{aligned}$$

□

Corollary 3.6. *If S, R, T, C, $\dot{C}$ resp. S_ε, R_ε, T_ε, C_ε, $\dot{C}_\varepsilon$ are the operator families from Propositions 3.1 resp. 3.3, then the operator families for the approximating equation (11) converge to the operator families for the limiting equation (10) in the strong topologies of the corresponding spaces, locally uniformly in $t \in \mathbb{R}_+$.*

Proof. For the proof we employ the following Trotter Kato type result.

Lemma 3.7. *Let X be a Banach space, $\{g_n\}_{n \in \mathbb{N}_0} \subset \mathcal{C}(\mathbb{R}_+, X)$ a sequence of continuous functions which is uniformly bounded, such that*

$$\widehat{g}_n(\lambda) \to \widehat{g}_0(\lambda) \quad \text{as} \quad n \to \infty, \quad \text{for each} \quad \lambda > 0.$$

Then for any $\phi \in L^1(\mathbb{R}_+)$,

$$\int_0^\infty \phi(t) g_n(t)dt \to \int_0^\infty \phi(t) g_0(t)dt \quad \text{in } X.$$

If, in addition, $\{g_n\}_{n \in \mathbb{N}_0}$ is equi-continuous, then

$$g_n(t) \to g_0(t) \text{ in } X, \quad \text{uniformly for bounded } t.$$

Proof. For $n \in \mathbb{N}_0$, define operators $T_n : L^1(\mathbb{R}_+) \to X$ by means of

$$T_n\phi = \int_0^\infty \phi(t)g_n(t)dt, \quad \phi \in L^1(\mathbb{R}_+).$$

Since the functions g_n are continuous and uniformly bounded, the T_n are well defined and bounded, uniformly in n. Now, with $\phi_\lambda(t) = e^{-\lambda t}$ we have

$$T_n\phi_\lambda = \widehat{g}_n(\lambda) \to \widehat{g}_0(\lambda) = T_0\phi_\lambda$$

by assumption, for each $\lambda > 0$. The Banach-Steinhaus theorem then implies $T_n \to T_0$ strongly, by density of $\mathrm{span}\{\phi_\lambda, \lambda > 0\}$ in $L^1(\mathbb{R}_+)$.

To prove the second statement, choose $\phi \in \mathcal{C}^1(\mathbb{R})$, $\phi \geq 0$, $\mathrm{supp}(\phi) \subset (0,1)$, such that $\int_0^\infty \phi(t)dt = 1$, and let $\phi_{\varepsilon,\tau}(t) = \varepsilon\phi((t-\tau)/\varepsilon)$. Then $\{\phi_{\varepsilon,\tau} : \tau \in [0,T]\}$ is a relatively compact subset of $L^1(\mathbb{R}_+)$, hence $T_n\phi_{\varepsilon,\tau} \to T_0\phi_{\varepsilon,\tau}$ as $n \to \infty$, uniformly for $\tau \in [0,T]$, for each fixed $\varepsilon > 0$. On the other hand $T_n\phi_{\varepsilon,\tau} \to g_n(\tau)$ as $\varepsilon \to 0$, uniformly in $\tau \in [0,T]$ and $n \in \mathbb{N}$, by equicontinuity of the g_n. From these properties the second assertion follows. □

To complete the proof of Corollary 3.6, take an arbitrary sequence $\varepsilon_n \to 0$ and set $g_n(t) = T_{\varepsilon_n}(t)x$ for a fixed $x \in H$. Then $\widehat{g}_n(\lambda) \to \widehat{g}_0(\lambda)$, where $g_0(t) = T(t)x$, the functions g_n are uniformly bounded in V and uniformly Lipschitz since $\dot{g}_n(t) = R_{\varepsilon_n}(t)x$ is bounded in V. Therefore Lemma 3.7 yields $T_\varepsilon(t)x \to T(t)x$ in V, uniformly for bounded t, for each $x \in H$. By density of H in V^* and uniform boundedness of T_ε in $\mathcal{B}(V^*, V)$, the Banach-Steinhaus theorem yields $T_\varepsilon \to T$ in the strong topology of $\mathcal{B}(V^*, V)$.

In a similar manner one obtains $R_\varepsilon \to R$ strongly in $\mathcal{B}(V^*, H)$, hence the identities $C_\varepsilon(t) = I - T_\varepsilon(t)A$ and $\dot{C}_\varepsilon(t) = -R_\varepsilon(t)A$ imply the corresponding convergence of C_ε and $\dot{C}_\varepsilon$. To derive the remaining convergence properties one can use the identities $R = tI - 1 * TA - T * dB$ and $S = 1 - TA - R * dB$. □

Definition 3.8. *Let R, C be the operator families from Proposition 3.1. Then for $u_0 \in V$, $u_1 \in H$, $f \in L^1_{\mathrm{loc}}(\mathbb{R}_+, V^*)$ the function*

$$u(t) := R(t)u_1 + C(t)u_0 + R * f(t)$$

will be called the mild solution *of*

$$\ddot{u} + Au + dB * \dot{u} = f, \ t > 0, \quad u(0) = u_0, \quad \dot{u}(0) = u_1. \tag{14}$$

Proposition 3.9. *Suppose (HS), (Hα), (Hβ), $d\beta$ is of positive type. Let $u_0 \in V$, $u_1 \in H$. The mild solution enjoys the following properties:*

(i) *if $f \in L^1_{\mathrm{loc}}(\mathbb{R}_+, H)$ then $u \in \mathcal{C}(\mathbb{R}_+, V) \cap \mathcal{C}^1(\mathbb{R}_+, H)$, $u(0) = u_0$, $\dot{u}(0) = u_1$, and we have in V^* :*

$$\dot{u} + A\ 1 * [u - u_0] + dB * [u - u_0] = u_1 - tAu_0 + 1 * f, \ t \geq 0.$$

(ii) *if $f \in L^1_{\mathrm{loc}}(\mathbb{R}_+, V^*)$ then $u \in \mathcal{C}(\mathbb{R}_+, H)$ and we have in V^* :*

$$u + A\ t * [u - u_0] + dB * 1 * [u - u_0] = tu_1 - \frac{t^2}{2}Au_0 + u_0 + t * f, \ t \geq 0;$$

Proof. $\dot{u}(t) = \dot{R}(t)u_1 + \dot{C}(t)u_0 + \dot{R} * f(t), \dot{R} = S$. The equations in (ii), (iii) are formally integrated versions of (14). The assertions follow from properties of R, C, cf. Proposition 3.1: R and C satisfy the equations (in the strong topology of $\mathcal{B}(V, V^*)$):

$$\begin{aligned} \dot{R} + d(tA + B) * R &= I, \quad R(0) = 0, \\ \dot{C} + d(tA + B) * (C - I) &= -tA, \quad C(0) = I, \\ R + d(tA + B) * 1 * R &= tI, \\ C - I + d(tA + B) * 1 * (C - I) &= -(t^2/2)A. \end{aligned}$$

We convolve the equations for R with f and sum up the resulting three equations for $R, C - I, R * f$. □

3.2. Energy Inequality

Proposition 3.10. *Suppose (HS), (Hα), (Hβ) are valid, and assume that $d\beta$ is dissipative. Let $u_0 \in V$, $u_1 \in H$, $f \in \mathcal{C}(\mathbb{R}_+, H)$. Then the mild solution satisfies*

$$\begin{aligned} &|\dot{u}(T)|_H^2 + |u(T)|_V^2 + 2\Re \int_0^T |a_1 * \dot{u}(t)|_H^2 dt \\ &\leq |u_1|_H^2 + |u_0|_V^2 + 2\Re \int_0^T (f(t), \dot{u}(t))_{H,H} dt \quad , \end{aligned}$$

for any $T \geq 0$, with a_1 from Definition 2.1.

Proof. The proof is done via approximation. Let $u_{1,n} \in V$, $u_{1,n} \to u_1$ in H, $\varepsilon > 0$,

$$u_{\varepsilon,n} := C_\varepsilon(t)u_0 + R_\varepsilon(t)u_{1,n} + R_\varepsilon * f(t),$$

where C_ε, R_ε are operator families from Proposition 3.3. Then $u_{\varepsilon,n} \in \mathcal{C}^1(\mathbb{R}_+, V) \cap \mathcal{C}^2(\mathbb{R}_+, V^*)$. Since

$$\begin{aligned} \frac{d^2}{dt^2}(R_\varepsilon * f)(t) &= \dot{S}_\varepsilon * f(t) + f(t), \\ d(\varepsilon A + tA + B) * \frac{d}{dt}(R_\varepsilon * f) &= d(\varepsilon A + tA + B) * S_\varepsilon * f = -\dot{S}_\varepsilon * f \end{aligned}$$

in V^*, $u_{\varepsilon,n}$ satisfies

$$\begin{aligned} \ddot{u}_{\varepsilon,n} + d(\varepsilon A + tA + B) * \dot{u}_{\varepsilon,n} &= f - Au_0 \quad \text{in} \quad V^*, \\ u_{\varepsilon,n}(0) = u_0, \quad \dot{u}_{\varepsilon,n}(0) &= u_{1,n}. \end{aligned}$$

Since $d(\varepsilon A + B)$ is dissipative as well, we obtain from Proposition 2.3

$$\begin{aligned} &|\dot{u}_{\varepsilon,n}(T)|_H^2 + |1 * \dot{u}_{\varepsilon,n}(T)|_V^2 + 2\Re \int_0^T |a_1 * \dot{u}_{\varepsilon,n}(t)|_H^2 dt \\ &\leq |\dot{u}_{\varepsilon,n}(0)|_H^2 + 2\Re \int_0^T (-Au_0 + f(t), \dot{u}_{\varepsilon,n}(t))_{V^*,V} dt, \end{aligned}$$

or rearranging

$$\begin{aligned} &|\dot{u}_{\varepsilon,n}(T)|_H^2 + |u_{\varepsilon,n}(T)|_V^2 + 2\Re \int_0^T |a_1 * \dot{u}_{\varepsilon,n}(t)|_H^2 dt \\ &\leq |\dot{u}_{1,n}|_H^2 + |u_0|_V^2 + 2\Re \int_0^T (f(t), \dot{u}_{\varepsilon,n}(t))_{H,H} dt. \end{aligned}$$

Passing to the limit first as $n \to \infty$ then as $\varepsilon \to 0$ we obtain the required inequality. □

4. Well-posedness of the Nonlinear Problem

The main goal of this section is to prove existence, uniqueness and continuous dependence on initial data of solutions to the nonlinear equation with potential right-hand side

$$\ddot{u} + dB * \dot{u} + Au = \Phi'(u), \; t > 0, \quad u(0) = u_0, \quad \dot{u}(0) = u_1, \tag{15}$$

and also a priori estimates, under the assumption that the linear problem is well posed (i.e. if $d\beta$ is of positive type).

4.1. Local Existence

Consider the nonlinear evolution equation

$$\begin{aligned} \ddot{u}(t) + Au(t) + dB * \dot{u}(t) &= f(t, u(t), \dot{u}(t)), \quad t > 0, & (16)\\ u(0) = x_0, \quad \dot{u}(0) &= x_1. & (17)\end{aligned}$$

Theorem 4.1. *Assume (HS), (Hα), (Hβ) and let $d\beta$ be of positive type. Let R, S, C be the operator families given by Proposition 3.1, representing the solution of the linear homogeneous equation (10), corresponding to (16), (17). Suppose that the function $f : \mathbb{R}_+ \times V \times H \to H$ is locally Lipschitz continuous in the following sense: for any $r, t >$ there exists a constant $L_f(t, r) \geq 1$ such that for $s \in [0, t]$, $|u_i|_V \leq r$, $|v_i|_H \leq r$ we have*

$$|f(s, u_1, v_1) - f(s, u_2, v_2)|_H \leq L_f(t, r)\left(|u_1 - u_2|_V + |v_1 - v_2|_H\right).$$

Then for any $x_0 \in V$, $x_1 \in H$ there exists a maximal time $T_m \in (0, \infty]$ and a unique function u such that

$$\begin{aligned} u &\in \mathcal{C}([0, T_m), V) \cap \mathcal{C}^1([0, T_m), H), & (18)\\ u(t) &= C(t)x_0 + R(t)x_1 + R * f(., u(.), \dot{u}(.))(t) \text{ in } V, \quad t \in [0, T_m), & (19)\\ \dot{u}(t) &= \dot{C}(t)x_0 + S(t)x_1 + S * f(., u(.), \dot{u}(.))(t) \text{ in } H, \quad t \in [0, T_m). & (20)\end{aligned}$$

If $f = f(u, \dot{u})$ then $T_m < \infty$ implies $\limsup_{t \to T_m -} [|u(t)|_V + |\dot{u}(t)|_H] = \infty$, and if $f = f(u)$ then $T_m < \infty$ implies $\limsup_{t \to T_m -} |u(t)|_V = \infty$. Moreover, the solution depends continuously on $(x_0, x_1, f) \in V \times H \times \mathcal{C}([0, T] \times V \times H, H)$ in the norm of $\mathcal{C}([0, T], V) \cap \mathcal{C}^1([0, T], H)$, for any $T < T_m$.

Proof. The proof is based on the contraction mapping principle. Let

$$\begin{aligned} Y &:= \mathcal{C}([t_1, t_2], V) \cap \mathcal{C}^1([t_1, t_2], H),\\ \|u\|_Y &:= \sup_{t \in [t_1, t_2]} e^{-\omega(t - t_1)}\left(|u(t)|_V + |\dot{u}(t)|_H\right),\\ M &:= \{u \in Y : \sup_{t \in [t_1, t_2]} \max[|u(t)|_V, |\dot{u}(t)|_H] \leq r\},\end{aligned}$$

where $0 \le t_1 < t_2 \le \infty,\ r > 0,\ \omega > 0$ will be chosen. For $x_0 \in V,\ x_1 \in H, w \in \mathcal{C}([0,t_1],V) \cap \mathcal{C}^1([0,t_1],H)$ given and $u \in Y$, we set

$$\begin{aligned} K(u)(t) := C(t)x_0 + R(t)x_1 \quad &+ \quad \int_0^{t_1} R(t-s)f\left(s, w(s), \dot{w}(s)\right) ds \\ &+ \quad \int_{t_1}^{t} R(t-s)f(s,u(s),\dot{u}(s))ds, \quad t \in [t_1,t_2]. \end{aligned}$$

Step I M is a closed subset of the Banach space Y, and given $x_0,\ x_1$ and w, K maps M into M, provided r is large enough and $t_2 - t_1$ is small enough.

In fact, choose r large such that

$$\begin{aligned} &|C(t)x_0 + R(t)x_1 + \textstyle\int_0^{t_1} R(t-s)f(s,w(s),\dot{w}(s)ds|_V \\ &\quad \le |x_0|_V + |x_1|_H + \textstyle\int_0^{t_1} |f(s,w(s),\dot{w}(s))|_H ds \le \frac{r}{2}, \\ &|\dot{C}(t)x_0 + \dot{R}(t)x_1 + \textstyle\int_0^{t_1} \dot{R}(t-s)f(s,w(s),\dot{w}(s))ds|_H \\ &\quad \le |x_0|_V + |x_1|_H + \textstyle\int_0^{t_1} |f(s,w(s),\dot{w}(s))|_H ds \le \frac{r}{2}. \end{aligned}$$

On the other hand, let $u \in Y$. Then for $t \in [t_1,t_2]$ we have

$$\begin{aligned} &|\int_{t_1}^{t} R(t-s)f(s,u(s),\dot{u}(s))ds|_V \le \int_{t_1}^{t} |f(s,u(s),\dot{u}(s)) \pm f(s,0,0)|_H ds \\ &\le L_f(t_2,r) \int_{t_1}^{t} [|u(s)|_V + |\dot{u}(s)|_H + |f(s,0,0)|_H] ds \\ &\le L_f(t_2,r)(t_2-t_1)\left(2r + \sup_{\tau \in [t_1,t_2]} |f(\tau,0,0)|_H\right) \le \frac{r}{2}, \\ &|\int_{t_1}^{t} \dot{R}(t-s)f(s,u(s),\dot{u}(s))ds|_H \le \int_{t_1}^{t} |f(s,u(s),\dot{u}(s)) \pm f(s,0,0)|_H ds \le \frac{r}{2}, \end{aligned}$$

provided $t_2 - t_1$ is small.

Step II For given $x_0,\ x_1$ and w, K is a contraction on Y, provided ω is large enough.

In fact,

$$\begin{aligned} &|K(u) - K(v)|_Y \\ &\le \sup_{t \in [t_1,t_2]} e^{-\omega(t-t_1)} 2 \int_{t_1}^{t} |f(s,u(s),\dot{u}(s)) - f(s,v(s),\dot{v}(s))|_H ds \\ &\le 2 \sup_{t \in [t_1,t_2]} e^{-\omega(t-t_1)} L_f(t_2,r) \int_{t_1}^{t} |u(s)-v(s)|_V + |\dot{u}(s)-\dot{v}(s)|_H ds \\ &\le 2L_f(t_2,r)|u-v|_Y \sup_{t \in [t_1,t_2]} e^{-\omega(t-t_1)} \int_{t_1}^{t} e^{\omega(s-t_1)} ds \\ &= 2L_f(t_2,r)|u-v|_Y \sup_{t \in [t_1,t_2]} e^{-\omega(t-t_1)} \frac{e^{\omega(t-t_1)}-1}{\omega} \le \frac{2L_f(t_2,r)}{\omega}|u-v|_Y. \end{aligned}$$

Step III If $K_{x,f}$ corresponds to data x_0, x_1, f then

$$\begin{aligned}
&|K_{x,f}(u) - K_{y,g}(v)|_Y \leq |K_{x,f}(u) - K_{x,g}(u)|_Y \\
&+|K_{x,g}(u) - K_{y,g}(u)|_Y + |K_{y,g}(u) - K_{y,g}(v)|_Y \\
&\leq 2 \sup_{t\in[t_1,t_2]} e^{-\omega(t-t_1)} \int_0^{t_1} |f(s,w(s),\dot{w}(s)) - g(s,w(s),\dot{w}(s))|_H ds \\
&+2 \sup_{t\in[t_1,t_2]} e^{-\omega(t-t_1)} \int_{t_1}^{t} |f(s,u(s),\dot{u}(s)) - g(s,u(s),\dot{u}(s))|_H ds \\
&+|C(.)x_0 + R(.)x_1 - C(.)y_0 - R(.)y_1|_Y \\
&+2 \sup_{t\in[t_1,t_2]} e^{-\omega(t-t_1)} \int_{t_1}^{t} |g(s,u(s),\dot{u}(s)) - g(s,v(s),\dot{v}(s))|_H ds \\
&\leq 2t_1 \sup_{s\in[0,t_1],|w_1|_V\leq r_1,|w_2|_H\leq r_2} |f(s,w_1,w_2) - g(s,w_1,w_2)|_H \\
&+2|t_2 - t_1| \sup_{s\in[t_1,t_2],|w_1|_V\leq r,|w_2|_H\leq r} |f(s,w_1,w_2) - g(s,w_1,w_2)|_H \\
&+2|x_0 - y_0|_V + 2|x_1 - y_1|_H + \frac{2L_g(t_2,r)}{\omega}|u - v|_Y.
\end{aligned}$$

From steps I, II and III we obtain the following conclusions:
(a) By the contraction mapping principle, for given $x_0,\ x_1$ there exists a $\tau > 0$ such that there exists a unique solution u of (18)-(20) on $[0,\tau]$. Moreover, u depends continuously on $(x_0, x_1, f) \in V \times H \times \mathcal{C}([0,\tau] \times V \times H, H)$, since for the solutions u resp. v, corresponding to data $x_0,\ x_1,\ f$ resp. $y_0,\ y_1,\ g$, we have

$$\begin{aligned}
|u - v|_Y = |K_{x,f}(u) - K_{y,g}(v)|_Y \\
\leq 2\tau \sup |f - g|_H + 2|x_0 - y_0|_V + 2|x_1 - y_1|_H + 2\tfrac{L_g(\tau,r)}{\omega}|u - v|_Y,
\end{aligned}$$

where we choose $\omega > 2L_g(\tau, r)$.
(b) Let $x_0,\ x_1,\ t_0 > 0$ be given and w be a solution of (18)-(20) on $[0, t_0]$. Then there exists a $\tau > 0$ such that there exists a function $u \in \mathcal{C}([t_0 - \tau, t_0 + \tau], V) \cap \mathcal{C}^1([t_0 - \tau, t_0 + \tau], H)$ solving

$$\begin{aligned}
u(t) = C(t)x_0 + R(t)x_1 \quad &+ \int_0^{t_0-\tau} R(t-s)f(s,w(s),\dot{w}(s))ds \\
&+ \int_{t_0-\tau}^{t} R(t-s)f(s,u(s),\dot{u}(s))ds,
\end{aligned}$$

for $t \in [t_0 - \tau, t_0 + \tau]$. In addition, u depends continuously on $x_0,\ x_1,\ f$.
(c) Let $x_0,\ x_1$ and $t_0 \in [0,\infty)$ be given, and let w be a solution of (18)-(20) on $[0, t_0]$. Then for any $\tau > t_0$ there exists at most one function $u \in \mathcal{C}([t_0,\tau], V) \cap$

$\mathcal{C}^1([t_0,\tau],H)$ which solves

$$u(t)=C(t)x_0+R(t)x_1 + \int_0^{t_0} R(t-s)f(s,w(s),\dot w(s))ds + \int_{t_0}^{t} R(t-s)f(s,u(s),\dot u(s))ds, \quad t\in[t_0,\tau].$$

By a continuation argument we obtain from conclusions (a), (b), (c) a unique solution to (18)–(20) defined on some maximal time interval and depending continuously on data. Finally, suppose that the solution is bounded. Then we can fix an r for steps I, II and III and obtain a fixed τ, any time we use them in proving the conclusions (a), (b), (c). In this way we obtain a globally defined solution. □

Definition 4.2. *The function u from Theorem 4.1 is called the* mild solution *of (16), (17) .*

Remark 4.3. Further regularity properties of mild solutions of (16), (17) could be formulated using Proposition 3.9 by considering $g(t)=f(t,u(t),\dot u(t))$ and posing assumptions on f. We will not go into details here.

4.2. The Nonlinear Energy Inequality

Proposition 4.4. *Suppose (HS), (Hα), (Hβ) are valid, assume $d\beta$ is dissipative, and let $\Phi: V\to\mathbb{R}$ be such that $\Phi': V\to H$ is Lipschitz continuous on bounded subsets of V.*

Then the mild solution u of (15) satisfies:

$$E_u(t)+\int_0^t |a_1 * \dot u(s)|_H^2 ds \le E_u(0), \quad t\in[0,T_m),$$

where

$$E_u(t)=\frac{1}{2}|\dot u(t)|_H^2+\frac{1}{2}|u(t)|_V^2-\Phi(u(t))$$

is called the energy *of u, and a_1 is given by Definition 2.1.*

Proof. Theorem 4.1 is applicable with $f(u)=\Phi'(u)$. We know that $\Phi'(u(.))\in\mathcal{C}([0,T_m),H)$ as a composite function, so by Proposition 3.10 we have

$$|\dot u(t)|_H^2+|u(t)|_V^2+2\int_0^t |a_1*\dot u(s)|_H^2 ds \le |u_1|_H^2+|u_0|_V^2+2\int_0^t (\Phi'(u(s)),u(s))_{H,H}ds$$

(we skipped the real part $\Re$ since the functions under consideration are real valued), where

$$\int_0^t (\Phi'(u(s)),\dot u(s))_{H,H}ds=\int_0^t \frac{d}{ds}\Phi(u(s))ds=\Phi(u(t))-\Phi(u(0)). \qquad \square$$

Corollary 4.5. *Suppose (HS), (Hα), (Hβ) are valid and that $d\beta$ is of positive type. Let Φ be as in Proposition 4.4, and assume moreover the growth condition*

(HG) *there exist constants* $\varepsilon \in (0, \frac{1}{2})$, $c \in \mathbb{R}$ *such that*

$$\Phi(v) \leq \left(\frac{1}{2} - \varepsilon\right) |v|_V^2 + c \quad \textit{for any } v \in V.$$

Then the mild solution of (15) is defined on $[0, \infty)$ *and is bounded, i.e. there exist a constant* $C = C(u_0, u_1)$ *such that*

$$|u(t)|_V + |\dot{u}(t)|_H \leq C \quad \textit{for all } t \geq 0.$$

Proof. From the energy inequality (Proposition 4.4 for kernels of positive type) and (HG) we obtain

$$E_u(0) \geq E_u(t) \geq \varepsilon \left(\frac{1}{2} |\dot{u}(t)|_H^2 + \frac{1}{2} |u(t)|_V^2 \right) - c,$$

hence $|\dot{u}(t)|_H$ and $|u(t)|_V$ are uniformly bounded for $t \geq 0$. □

Remark 4.6. The factor 1/2 in the growth condition results from the particular choice of the norm $|v|_V = \sqrt{\alpha(v, v)}$ in V.

5. Asymptotic Behaviour

In this section we study the convergence of mild solutions of (15) to a stationary solution, as $t \to \infty$, for dissipative forms $d\beta$. We prove that the nonlinear problem (15) is stable provided its linear part (10) is stable in an approriate sense. This is the main result of this paper.

5.1. The Main Result

Theorem 5.1. *Let (HS), (H*α*), (H*β*) be valid,* $d\beta$ *be dissipative, and moreover, let* $\lim_{t\to\infty} \beta_0(t)/t = 0$, *with* β_0 *from (H*β*). Suppose further that:*

(i) *The linear problem (10) is stable in the following sense:* S *and* R *from Proposition 3.1 satisfy*
  (a) $R \in \mathcal{C}_{0,s}(\mathbb{R}_+, \mathcal{B}(H, V))$, $\quad S \in \mathcal{C}_{0,s}(\mathbb{R}_+, \mathcal{B}(H, H))$;
  (b) $R \in L^1_s(\mathbb{R}_+; \mathcal{B}(H, V) \cap \mathcal{B}(V^*, H))$, $\quad S \in L^1_s(\mathbb{R}_+; \mathcal{B}(H, H))$.

(ii) *The function* $\Phi : V \to \mathbb{R}$ *is of class* C^1 *and such that*
  (a) $\Phi' : V \to H$ *is Lipschitz continuous on bounded sets of* V,
  (b) $\Phi' : V \to H$ *is compact,*
  (c) Φ *satisfies the growth condition (HG) in Corollary 4.5.*

(iii) *The set of stationary solutions* $\mathcal{E} := \{w \in V,\ Aw = \Phi'(w)\}$ *is a discrete subset of* V.

Then for all initial values $u_0 \in V$, $u_1 \in H$ *there exists a unique stationary solution* $w \in \mathcal{E}$ *such that the mild solution* u *of (15) converges to* w *in* V, *and* $\dot{u}(t) \to 0$ *in* H, *as* $t \to \infty$.

Proof. $R \in L^1_s(\mathbb{R}_+; \mathcal{B}(H,V))$ implies $T(t)v = \int_0^t R(s)vds \to \widehat{R}(0)v$ in V, for all $v \in H$. We evaluate $\widehat{R}(0)$ as follows. Using the strong integrability of R and Lebesgue's theorem, the Laplace transform of R given by $\widehat{R}(\lambda) = (\lambda^2 + \lambda \widehat{dB}(\lambda) + A)^{-1}$, and by assumption

$$\lim_{\lambda \to 0} |\lambda \widehat{dB}(\lambda)|_{\mathcal{B}(V,V^*)} \leq \lim_{\lambda \to 0} \lambda \widehat{d\beta_0}(\lambda) = \lim_{t \to \infty} \beta_0(t)/t = 0,$$

yields

$$\begin{aligned} \int_0^\infty R(s)xds &= \lim_{\lambda \to 0+} \int_0^\infty e^{-\lambda s} R(s)xds = \lim_{\lambda \to 0+} \widehat{R(.)x}(\lambda) \\ &= \lim_{\lambda \to 0+} (\lambda^2 + \lambda \widehat{dB}(\lambda) + A)^{-1}x = A^{-1}x \text{ in } H, \end{aligned}$$

for any $x \in H$.

Hence $T(t) \to A^{-1}$ strongly in $\mathcal{B}(V^*, V)$ by boundedness of $T(t)$; this yields also $C(t) = I - T(t)A \to 0$ strongly in $\mathcal{B}(V,V)$. Similarly, since by assumption $S \in L^1_s(\mathbb{R}_+; \mathcal{B}(H,H))$, we obtain $R(t)v = \int_0^t S(s)vds \to \widehat{S}(0)v = 0$ in H for each $v \in H$, hence $R \in \mathcal{C}_{0,s}(\mathbb{R}_+, \mathcal{B}(V^*, H))$; this gives also $\dot{C}(t) = -R(t)A \to 0$ strongly in $\mathcal{B}(V,H)$.

By Corollary 4.5 we know that there exists an r so that $|\dot{u}(t)|_H$, $|u(t)|_V \leq r$ for $t \geq 0$. From the compactness assumption (ii.b) we have that the set $\{\Phi'(u(t));\ t \geq 0\}$ is relatively compact in H. The variation of parameters formulae

$$\begin{aligned} u(t) &= R(t)u_1 + C(t)u_0 + R * \Phi'(u(\cdot))(t), \\ \dot{u}(t) &= S(t)u_1 + \dot{C}(t)u_0 + S * \Phi'(u(\cdot))(t), \end{aligned}$$

and Lemma 7.13 imply, using the assumptions in (i),

$$u \in \mathcal{BUC}(\mathbb{R}_+, V), \quad \dot{u} \in \mathcal{BUC}(\mathbb{R}_+, H),$$

and the sets $\{u(t);\ t \geq 0\} \subset V$, $\{\dot{u}(t);\ t \geq 0\} \subset H$ are relatively compact. From the energy inequality Proposition 4.4 we deduce that $a_1 * \dot{u} \in L^2(\mathbb{R}_+, H)$. So by Corollary 7.10 we have $\dot{u} \in \mathcal{C}_0(\mathbb{R}_+, H)$.

Now we turn to the proof of convergence of u. Referring to the theory of dynamical systems (see Appendix D) we will show that the omega limit set (with respect to the shift dynamical system) of $u \in \mathcal{BUC}(\mathbb{R}_+, V)$ consists of a unique constant (in time) function, which is a stationary solution. To do so, let $w(.) \in \mathcal{BUC}(\mathbb{R}_+, V)$ be such that there exists a sequence $t_n \to \infty$ with $u(t_n + .) \to w(.)$ in V, uniformly on bounded time intervals. Since $\dot{u}(t) \to 0$ in H as $t \to \infty$ we conclude that w is time independent. Take a fixed $t \geq 0$, $y \in V^*$, and consider the duality $(.,.)_{V,V^*}$ in the variation of parameters formula at time $t_n + t$ with y .

$$\begin{aligned} (u(t_n + t), y)_{V,V^*} &= (C(t_n + t)u_0, y)_{V,V^*} + (R(t_n + t)u_1, y)_{V,V^*} \\ &+ (R * \Phi'(u)(t_n + t), y)_{V,V^*}. \end{aligned}$$

To pass to the limit as $n \to \infty$, we have to investigate the last term in this identity, namely we will show

$$(R * \Phi'(u)(t_n + t), y)_{V,V^*} \to \left(\int_0^\infty R(s)\Phi'(w)ds, y \right)_{V,V^*}.$$

To this end we compute

$$\begin{aligned}
&(R * \Phi'(u)(t_n + t), y)_{V,V^*} - \left(\int_0^\infty R(s)\Phi'(w)ds, y \right)_{V,V^*} \\
&= \int_0^{t+t_n} (R(s)\Phi'(u(t + t_n - s)), y)_{V,V^*}\, ds - \int_0^\infty (R(s)\Phi'(w), y)_{V,V^*}\, ds \\
&= \int_0^T (\Phi'(u(t + t_n - s)) - \Phi'(w), R(s)^* y)_{H,H} ds \\
&+ \int_T^{t+t_n} (\Phi'(u(t + t_n - s)), R(s)^* y)_{H,H} ds - \int_T^\infty (\Phi'(w), R(s)^* y)_{H,H} ds.
\end{aligned}$$

The last two terms of the latter expression can be estimated by

$$\sup_{|v|_V \le r} |\Phi'(v)|_H \int_T^\infty |R(s)^* y|_H ds,$$

so, by the hypothesis (i.b), they can be made arbitrarily small, provided T is large enough. In the first term we can use the Lebesgue dominated convergence theorem (for a fixed T), since $\Phi'(u(t + t_n - s))$ converges to $\Phi'(w)$ pointwise and the integrand can be estimated by $2 \sup_{|v|_V \le r} |\Phi'(v)|_H \; |y|_{V^*}$. Using the hypothesis (i.a), we obtain by passing to the limit

$$w = \int_0^\infty R(s)\Phi'(w)ds = A^{-1}\Phi'(w) \text{ in } V.$$

This implies $w \in \mathcal{E}$, hence $\omega(u) \subset \{c_w;\ w \in \mathcal{E}\}$ (cf. Appendix D). Since by Lemma 7.15 and Proposition 7.17, $\omega(u)$ is a connected subset of $\mathcal{BUC}\,(\mathbb{R}_+, V)$, Lemma 7.16 implies $\omega(u) = \{c_w\}$ for some $w \in \mathcal{E}$. This yields $\lim_{t\to\infty} u(t) = w$ in V, by Lemma 7.14. □

Remark 5.2. Note that the a growth condition for Φ' similar to (HG) yields boundedness of $\mathcal{E}$ as a subset of V. Namely suppose

$$(HG') \quad (\Phi'(u), u)_{V^*,V} \le (1 - \varepsilon)|u|_V^2 + c|u|_V, \quad u \in V, \tag{21}$$

is valid. Then $\mathcal{E} \subset \{u \in V : \ |u|_V \le c/\varepsilon\}$. If moreover $\Phi' : V \to H$ is Lipschitz on bounded sets, then $\phi'(\mathcal{E}) \subset H$ is bounded, hence even $\mathcal{E} \subset D(A_H)$ is bounded. And if $\Phi' : V \to H$ is compact then $\Phi'(\mathcal{E}) \subset H$ is relatively compact, hence $\mathcal{E} \subset D(A_H)$ is also compact. Note that $\mathcal{E}$ is finite if it is compact and discrete in V.

Observe that (HG') is stronger than (HG). In fact, the identity (w.o.l.g. $\Phi(0) = 0$)

$$\Phi(u) = \int_0^1 \frac{d}{dt}\Phi(tu)dt = \int_0^1 (\Phi'(tu), u)_{V^*,V},$$

hence

$$\Phi(u) \leq \int_0^1 (1-\varepsilon)t|u|_V^2 + c|u|_V dt \leq \frac{1-\varepsilon}{2}|u|_V^2 + c|u|_V,$$

which implies (HG).

5.2. Examples and Applications

Let us first consider the case $dB = dbA$ with b subject to (Hb). In this situation the stability properties (i) in Theorem 5.1 are fulfilled, and dB is dissipative, by Lemma 7.3 and [6], Proposition 16.4.3. As a consequence we obtain the following extension of the main result of [4].

Corollary 5.3. *Let A be a linear, self-adjoint, positive definite operator in a Hilbert space H, let b be subject to (Hb), suppose Φ satisfies assumption (ii) of Theorem 5.1 with $V = D(A^{\frac{1}{2}})$, and let the set $\mathcal{E} = \{v \in D(A) : Av = \Phi'(v)\}$ be discrete.*

Then every mild solution u of the problem

$$\ddot{u} + db * A\dot{u} + Au = \Phi'(u),\ t \geq 0,\ u(0) = u_0 \in V,\ \dot{u}(0) = u_1 \in H, \tag{22}$$

converges in V to a stationary solution $u_\infty \in \mathcal{E}$.

This result applies to Example 1.1, as shown in [4]. It can be used in Example 1.3 as well, with $H = L^2(\omega)$, $V = H_0^2(\Omega)$, $\alpha(u,v) = \int_\Omega \Delta u(x)\overline{\Delta v(x)}dx$, and $\Phi(u) = \int_\Omega f(u(x))dx$, where $F' = f$. Obviously α is a bounded sesquilinear form on V which is positive, and Φ is of class C^1 with $\Phi'(u)(x) = f(u(x))$, hence $\Phi' : V \to H$ is well defined and Lipschitz on bounded sets, provided $f \in C(\mathbb{R})$ is locally Lipschitz, and in space dimensions $N \geq 4$ the additional growth condition

$$|f(r)| \leq C(1+|r|)^p, \quad r \in \mathbb{R},$$

is satisfied, where $p \geq 1$ is arbitrary for $N = 4$, and $p < N/(N-4)$ for $N > 4$. In space dimensions $N \leq 3$ no growth condition is required because then V is compactly embedded into $C(\overline{\Omega})$. Assume the condition

$$F(r) \leq \frac{\eta}{2}r^2 + C, \quad r \in \mathbb{R},$$

with $\eta < \lambda_1$, the first eigenvalue of Δ^2 with Dirichlet boundary conditions. Then in case b is subject to (Hb) and the solution set of the stationary problem is discrete, Corollary 5.3 yields the convergence of the mild solutions of Example 1.3 to steady states.

To apply the corollary to Example 1.2, we have to impose an additional assumption on the kernels a and e namely $e(t) = \kappa a(t)$ on $\mathbb{R}_+$ where $\kappa > 0$. In viscoelasticity this is the so-called case of *synchronous materials*; cf. [10], Section 5. Consider e.g. a clamped boundary condition on the left end $x = 0$ and a free boundary condition at the right end $x = 1$. Let $H := L^2(0,1)^2$ and

$$A := \begin{pmatrix} -a_\infty \partial_x^2 & -a_\infty \partial_x \\ \gamma a_\infty \partial_x & -\kappa a_\infty \partial_x^2 + \gamma a_\infty \end{pmatrix},$$

with domain

$$D(A) = \{(w,\phi) \in H^2(0,1)^2 : \; w(0) = \phi(0) = 0, \; \partial_x\phi(1) = \partial_x w(1) + \phi(1) = 0\}.$$

Then it is not difficult to verify that A is self-adjoint and positive definite, w.r.t. the inner product on H defined by

$$((w_1,\phi_1),(w_2,\phi_2)) := \gamma \int_0^1 w_1\overline{w}_2 dx + \int_0^1 \phi_1\overline{\phi}_2 dx.$$

The form $\alpha(u,v) = (Au,v)$ is then given by

$$\alpha((w_1,\phi_1),(w_2,\phi_2)) = a_\infty\kappa \int_0^1 \phi_1'\overline{\phi_2'}dx + a_\infty\gamma \int_0^1 [w_1'+\phi_1]\overline{[w_2'+\phi_2]}dx$$

If we set $b(t) = a(t)/a_\infty$ and let

$$\Phi(w,\phi) = \int_0^1 f(w(x),\phi(x))dx$$

then problem (3) can be written abstractly as (22). Since $H^1(0,1) \hookrightarrow C[0,1]$, the functional Φ is well-defined and it is C^2 whenever $f \in C^2$. The growth condition (HG) is implied by the condition

$$f(r,s) \le (\frac{\lambda_1}{2} - \varepsilon)[\gamma r^2 + s^2] + c, \quad r,s \in \mathbb{R},$$

where $\lambda_1 > 0$ denotes the first eigenvalue of A. So in this case Theorem 5.1 yields convergence of the solutions once the function a is subject to (Hb), and once we know that the set of stationary solutions is discrete in V. Concerning the latter, since the stationary problem for the Timoshenko beam is a nonlinear boundary value problem for a system of ode's, it is possible to show that $\mathcal{E}$ is discrete in case f e.g. is analytic.

If the assumption $e(t) = \kappa a(t)$ does not hold, it is still possible to obtain convergence of solutions if $a(0+) < \infty$ and $e(0+) < \infty$ hold. In fact, in Prüss [10], Section 13, it has been shown that the assumptions of Theorem 6.5 and Propositions 6.1 and 6.3 are satisfied, hence by means of these results the linear stability assumptions in the main result hold. The details will be carried out elsewhere.

We continue with the special case $dB = dbA^\eta$ where $\eta \in [0,1)$ for which we characterize the stability properties (i) of Theorem 5.1 in the next section.

Corollary 5.4. *Let A be a linear, self-adjoint, positive definite operator in a Hilbert space H, let $b_\gamma(t) = \Gamma(1-\gamma)^{-1}t^{-\gamma}e^{-\varepsilon t}$, $t > 0$, for some $\varepsilon \ge 0$ and $\gamma \in (0,1]$, (with the interpretation $b_1(t) = \delta(t)$.) Suppose Φ satisfies assumption (ii) of Theorem 5.1 with $V = D(A^{\frac{1}{2}})$, and let the set $\mathcal{E} = \{v \in D(A) : Av = \Phi'(v)\}$ be discrete. Let $\eta \in [0,1)$ be such that $2\eta + \gamma \ge 1$.*

Then every mild solution u of the problem

$$\ddot{u} + b_\gamma * A^\eta\dot{u} + Au = \Phi'(u), \; t \ge 0, \; u(0) = u_0 \in V, \; \dot{u}(0) = u_1 \in H, \tag{23}$$

converges in V to a stationary solution $u_\infty \in \mathcal{E}$.

Note that this includes the case of instantaneous damping for $\gamma = 1$; the condition then becomes $\eta \geq 1/2$.

This result can be applied to the simply supported Euler-Bernoulli beam with damping term $-b * \partial_x^2$ since for these special boundary values we have $-\partial_x^2 = A^{\frac{1}{2}}$, where $A = \partial_x^4$. Since $\eta = 1/2$ one obtains convergence for all $\gamma \in (0, 1]$.

Finally, a brief comment on the last example, Example 1.4, is in order. Here our main result does not apply for several reasons. The underlying linear problem can be shown to satisfy the stability assumptions (i) of Theorem 5.1, provided the kernel k is subject to (Hb) and $k_0 > 0$. However, in this example the form $d\beta$ defined by

$$\beta(t, u, v) = k(t)\beta_1(u, v) = k(t) \int_{\partial\Omega} u(x)\overline{v(x)} d\sigma(x)$$

is not dissipative because $\beta_1(u, u) = 0$ for all $u \in H_0^1(\Omega)$, i.e. β_1 is nonnegative but its kernel is nontrivial. Another obstruction in case $g \not\equiv 0$ is the fact that Φ' does not map $V = H^1(\Omega)$ into $H = L^2(\Omega)$ but only into V^*.

6. Linear Stability

The most important ingredients in the proof of Theorem 5.1 are the stability properties of the underlying linear problem. So far a general characterization is not available, only partial results are known. In this section we recall some known results and add some new ones, in this respect.

6.1. Convergence to Zero

We begin with a fairly general result about uniform continuity of the operator families associated with the linear problem

$$\ddot{u} + dB * \dot{u} + Au = f, \; t \geq 0, \quad u(0) = u_0, \; \dot{u}(0) = u_1, \tag{24}$$

which together with a spectral hypothesis will imply the C_0-properties assumed in Theorem 5.1.

Proposition 6.1. *Let (HS), (Hα), and (Hβ) be valid, suppose $d\beta$ is of positive type, and let S, R, T, C denote the operator families associated to (24) from Proposition 3.1. Assume in addition*

$$dB(t) = dB_1(t) + B_2(t)dt, \quad t \geq 0,$$

with $B_1 \in \mathcal{BV}(\mathbb{R}_+, \mathcal{B}(V, V^*))$, $B_2 \in \mathcal{BV}_s(\mathbb{R}_+, \mathcal{B}(V, V^*))$. Then the operator families are bounded uniformly continuous in the strong topologies:

$$\begin{aligned} S &\in \mathcal{BUC}_s(\mathbb{R}_+, \mathcal{B}(H, H)), & R &\in \mathcal{BUC}_s(\mathbb{R}_+, \mathcal{B}(V^*, H)), \\ R &\in \mathcal{BUC}_s(\mathbb{R}_+, \mathcal{B}(H, V)), & T &\in \mathcal{BUC}_s(\mathbb{R}_+, \mathcal{B}(V^*, V)), \\ C &\in \mathcal{BUC}_s(\mathbb{R}_+, \mathcal{B}(V, V)), & \dot{C} &\in \mathcal{BUC}_s(\mathbb{R}_+, \mathcal{B}(V, H)). \end{aligned}$$

Proof. We are left to prove the uniform continuity, cf. Proposition 3.1.
(a) Since $|\dot{T}(t)|_{\mathcal{B}(H,V)} \leq 1$, we have $T \in \mathcal{UC}_s(\mathbb{R}_+, \mathcal{B}(H,V))$, and by boundedness of T and density $H \hookrightarrow V^*$ we have $T \in \mathcal{UC}_s(\mathbb{R}_+, \mathcal{B}(V^*,V))$.
(b) Since $|\dot{R}(t)|_{\mathcal{B}(H,H)} \leq 1$, we have $R \in \mathcal{UC}_s(\mathbb{R}_+, \mathcal{B}(H,H))$, and by boundedness of R and density $H \hookrightarrow V^*$ we have $R \in \mathcal{UC}_s(\mathbb{R}_+, \mathcal{B}(V^*,H))$.
(c) The assertions for C and $\dot{C}$ follow from the relations

$$C = I - TA, \quad \dot{C} = -RA.$$

(d) Since $\widehat{R}(\lambda) = (\lambda^2 + A + \lambda \widehat{dB}(\lambda))^{-1}$, uniqueness of the Laplace transform yields in time domain

$$Sx + R * dBx + TAx = x \text{ for all } \ x \in V.$$

Hence

$$\begin{aligned} Sx &= x - TAx - R * dB_1 x - R * B_2 x \\ &= x - TAx - R * dB_1 x - T * dB_2 x, \ x \in V. \end{aligned}$$

By steps (a) and (b), the assumptions on B_j, each term on the right-hand side of the latter identity is in $\mathcal{UC}(\mathbb{R}_+, H)$ since V, H are Hilbert spaces which have the Radon-Nikodym property. Therefore density $V \hookrightarrow H$ and boundedness of S, imply the assertion for S.
(e) We proceed in a similar way as in (d) to prove that $R \in \mathcal{UC}_s\ (\mathbb{R}_+, \mathcal{B}(H,V))$. The identity

$$\begin{aligned} Rx &= tx - 1 * T * Ax - T * dB_1 - T * B_2 x \\ &= tx - 1 * TAx - T * dB_1 x - 1 * T * dB_2 x, \ x \in V, \end{aligned}$$

implies $Rx \in \mathcal{UC}(\mathbb{R}_+, V)$ since $T * dB_1 x$ has this property and the function $tx - 1 * T(Ax + B_2(0)x) - 1 * T * \dot{B}_2 x$ has a bounded derivative in V, hence is even Lipschitz. By density $V \hookrightarrow H$ and boundedness of R the last assertion follows. □

Corollary 6.2. *If b satisfies (Hb), $b_0 \geq 0$, and $\beta(t,u,v) = b(t)\beta_0(u,v)$ with a bounded, symmetric, nonnegative, bilinear form β_0 on V then the assertions of Lemma 6.1 hold.*

Proof. $d\beta$ is of positive type and

$$b(t) = (b(t) - b(1))\chi_{[0,1)}(t) + [b(t)\chi_{[1,\infty)}(t) + b(1)\chi_{[0,1)}(t)]$$

gives the decomposition of $d\beta$ assumed in Lemma 6.1. □

Uniform continuity of the operator families associated to (24) together with Lemma 7.9 will lead to the C_0-properties, once we know that the spectra of these families are empty. But this will be ensured provided that $K(\lambda)$ admits an extension in $\mathcal{C}_s(\overline{\mathbb{C}}_+; \mathcal{B}(V^*,V))$.

Proposition 6.3. *Let (HS), (Hα), and (Hβ) be valid, suppose $d\beta$ is of positive type, $\lim_{t\to\infty} \beta_0(t)/t = 0$, and let S, R, T, C, $K(\lambda)$ be given by Proposition 3.1. Assume*

$$dB(t) = dB_1(t) + B_2(t)dt, \ t \geq 0,$$

with kernels $B_1 \in \mathcal{BV}(\mathbb{R}_+, \mathcal{B}(V, V^*))$, *and* $B_2 \in \mathcal{BV}_s(\mathbb{R}_+; \mathcal{B}(V, V^*))$. *In addition, suppose that* $K(\lambda)$ *admits an extension in* $\mathcal{C}_s(\overline{\mathbb{C}}_+; \mathcal{B}(V^*, V))$ *to the closed right half-plane* $\overline{\mathbb{C}}_+$, *and*

$$\int_{-1}^{1} |K(i\rho)v - K(0)v|_{V^*} d\rho/|\rho| < \infty, \qquad \textit{for each } v \in V. \tag{25}$$

Then

$$\begin{array}{ll} S \in \mathcal{C}_{0,s}(\mathbb{R}_+, \mathcal{B}(H, H)), & R \in \mathcal{C}_{0,s}(\mathbb{R}_+, \mathcal{B}(V^*, H)), \\ R \in \mathcal{C}_{0,s}(\mathbb{R}_+, \mathcal{B}(H, V)), & T - A^{-1} \in \mathcal{C}_{0,s}(\mathbb{R}_+, \mathcal{B}(V^*, V)), \\ C \in \mathcal{C}_{0,s}(\mathbb{R}_+, \mathcal{B}(V, V)), & \dot{C} \in \mathcal{C}_{0,s}(\mathbb{R}_+, \mathcal{B}(V, H)). \end{array}$$

Proof. Since $\lambda K(\lambda)$ and $K(\lambda)$, are the Laplace transforms of S and R, we may conclude by assumption that the spectra (cf. Definition 7.7) of S and R in the appropriate spaces are empty, so by Lemma 7.9 and Lemma 6.1 they converge to 0 strongly as $t \to \infty$. Moreover, by the assumptions on B we have

$$\lim_{\lambda \to 0} |\lambda \widehat{dB}(\lambda)|_{\mathcal{B}(V,V^*)} = \lim_{\lambda \to 0} (\lambda \widehat{d\beta}_0(\lambda) = \lim_{t \to \infty} \beta_0(t)/t = 0,$$

hence $K(0) = A^{-1}$, and so the identity

$$\widehat{C}(\lambda) = [K(\lambda) - K(0)]/\lambda$$

shows by assumption that the spectrum of $C(\cdot)v$ in V is empty as well, hence Lemma 6.1 and Lemma 7.9 implies $C(t)v \to 0$ in V, for each $v \in V$. The assertions for T and $\dot{C}$ follow from the relations $\dot{C} = -RA$ and $C = I - TA$. $\square$

Remark 6.4. Observe that (25) is implied by the moment condition

$$\int_0^\infty t|dB_2(t)v|_{V^*} < \infty, \qquad \text{for each } v \in V.$$

6.2. Integrability

A general characterization of the integrability properties of R amd S which have been essential to the results in Section 5 is not known. However, there is a useful characterization of strong integrability of these operator families in terms of bounds on $K(\lambda)$. In fact, suppose $R \in L^1_s(\mathbb{R}_+; \mathcal{B}(H, V) \cap \mathcal{B}(V^*, H))$, $S \in L^1_s(\mathbb{R}_+; \mathcal{B}(H, H))$ and in addition assume $e^{-\eta t} * R \in L^1_s(\mathbb{R}_+; \mathcal{B}(V^*, V))$, for some $\eta > 0$. Then the Laplace transform $K(\lambda)$ of R admits a strongly continuous extension in $\mathcal{B}(V^*, V)$ to the closed right half-plane, which is uniformly bounded in $\mathcal{B}(V^*, H)$ and in $\mathcal{B}(H, V)$; in addition $\lambda K(\lambda)$ is uniformly bounded in $\mathcal{B}(H, H)$ and $(\lambda + \eta)^{-1} K(\lambda)$ has this property in $\mathcal{B}(V^*, V)$. Under an additional hypothesis on the kernel $dB(t)$ the converse of this statement holds true.

Theorem 6.5. *Suppose (HS), (Hα), and (Hβ) are valid, and let* S, R, T, K *be as in Proposition 3.1. In addition, assume that*

$$dB(t) = db(t)B_1 + B_2(t)dt, \quad t \geq 0,$$

with $B_1 \in \mathcal{B}(V,V^*)$, b *subject to (Hb),* $B_2 \in \mathcal{BV}(\mathbb{R}_+,\mathcal{B}(V,V^*))$, *and that there exists a number* $N > 0$ *such that*

$$\sup_{|\lambda|\geq N,\ \Re\lambda>0} |\lambda \widehat{dB_2}'(\lambda)|_{\mathcal{B}(V,V^*)} < \infty. \tag{26}$$

Then the following assertions are equivalent.

(A) $K(.) \in \mathcal{C}(\overline{\mathbb{C}_+},\mathcal{B}(V^*,V))$, *and there exists a constant* $M < \infty$ *such that for any* $\lambda \in \overline{\mathbb{C}}_+$ *we have*

$$\begin{array}{ll} |\lambda K(\lambda)|_{\mathcal{B}(H,H)} \leq M, & |K(\lambda)|_{\mathcal{B}(H,V)} \leq M, \\ |K(\lambda)|_{\mathcal{B}(V^*,H)} \leq M, & |\frac{1}{1+\lambda}K(\lambda)|_{\mathcal{B}(V^*,V)} \leq M \end{array}$$

(B) *The operator families* $S, R, T_\eta := e^{-\eta t} * R$ *are strongly integrable, i.e.* $S \in L^1_s(\mathbb{R}_+,\mathcal{B}(H,H))$, $R \in L^1_s(\mathbb{R}_+,\mathcal{B}(H,V))$, $R \in L^1_s(\mathbb{R}_+,\mathcal{B}(V^*,H))$, $e^{-\eta t} * R \in L^1_s(\mathbb{R}_+,\mathcal{B}(V^*,V))$.

The structure of the proof follows ideas in [10], Sections 11 and 12, in particular the proofs of Lemma 11.1 and Theorem 12.3. Recall that the functions $\lambda K, K, \frac{K}{\eta+\lambda}$ are the Laplace transforms of the operator families $S, R, T_\eta := e^{-\eta t} * R$ respectively. Let φ be a smooth function with compact support which is identically 1 on an interval $[-N,N]$. By means of this function we decompose $K(\lambda)$ as

$$K(i\rho) = \phi(\rho)K(i\rho) + (1-\phi(\rho))K(i\rho).$$

In the first part of the proof we employ the vector valued Paley-Wiener theorem to obtain an $R^\phi \in W^{\infty,1}(\mathbb{R},\mathcal{B}(V^*,V))$ such that $\widetilde{R^\phi}(i\rho) = \phi(\rho)K(i\rho)$ holds. Here the moment condition (26) obviously plays no role. In the second part we show by the vector-valued Plancherel theorem the asserted integrability properties for the families associated to $(1-\phi(\rho))K(i\rho)$. By uniqueness of the Fourier transform this then shows the integrability properties asserted in (B) of the Theorem 6.5.

Proof of Theorem 6.5. **(i)** For $m \in \mathbb{N}$, let $B_m(t) = B(t)$ for $t \in [0,m]$, $B_m(t) = B(m)$ for $t > m$. Obviously $B_m \in \mathcal{BV}(\mathbb{R}_+,\mathcal{B}(V,V^*))$ and $\widehat{dB_m}$ is an entire function. Then $K_m(\lambda) = (\lambda^2 + \lambda\widehat{dB}_m(\lambda) + A)^{-1}$ is well defined provided $m \in \mathbb{N}$ is large enough. In fact, we have

$$\begin{aligned} \lambda[\widehat{dB}(\lambda) - \widehat{dB_m}(\lambda)] &= \lambda\int_m^\infty e^{-\lambda t}dB(t) \\ &= \lambda\int_m^\infty e^{-\lambda t}db(t)B_1 + \int_m^\infty e^{-\lambda t}dB_2(t), \quad \Re\lambda \geq 0, \end{aligned}$$

which tends to zero as $m \to \infty$ since $B_2 \in \mathcal{BV}(\mathbb{R}_+,\mathcal{B}(V,V^*))$ and b satisfies (Hb) by assumption. Next we set $K^\phi(\rho) = \phi(\rho)K(i\rho)$ and $K^\phi_m(\rho) = \phi(\rho)K_m(i\rho)$. These functions are continuous with values in $\mathcal{B}(V^*,V)$ and moreover, K^ϕ_m belongs to

C^2 and has compact support, the derivatives are given by

$$K_m^{\phi\prime}(\rho) = K_m^\phi(\rho)(\rho - (i\rho\widehat{dB}_m(i\rho))')K_m^\phi(\rho),$$
$$K_m^{\phi\prime\prime}(\rho) = K_m^\phi(\rho)(\rho - (i\rho\widehat{dB}_m(i\rho))')K_m^\phi(\rho)(\rho - (i\rho\widehat{dB}_m(i\rho))')K_m^\phi(\rho)$$
$$+K_m(\rho)((1 - (i\rho\widehat{dB}_m(i\rho))'')K_m^\phi(\rho).$$

Define

$$R_m^\phi(t) = \frac{1}{2\pi}\int_{-\infty}^{\infty} e^{i\rho t}K_m^\phi(\rho)d\rho, \quad t \in \mathbb{R},$$

i.e. R_m^ϕ denotes the inverse Fourier transform of K_m^ϕ. It is well defined and even belongs to $W^{\infty,\infty}(\mathbb{R}, \mathcal{B}(V^*, V))$, since ϕ and therefore K_m^ϕ have compact supports. Integrating by parts twice in his formula we obtain

$$R_m^\phi(t) = \frac{-1}{2\pi t^2}\int_{-\infty}^{\infty} e^{i\rho t}K_m^{\phi\prime\prime}(\rho)d\rho, \quad t \in \mathbb{R},$$

hence $t^2R_m^\phi$ belongs to $W^{\infty,\infty}(\mathbb{R}, \mathcal{B}(V^*, V))$ as well. This yields by interpolation $R_m^\phi \in W^{\infty,1}(\mathbb{R}, \mathcal{B}(V^*, V))$, for $m \in \mathbb{N}$ sufficiently large.

Next let ψ be another C^∞-function with compact support such that $\psi \equiv 1$ on the support of ϕ. Multiply the identity

$$K(\lambda) = K_m(\lambda) + K_m(\lambda)[\lambda(\widehat{dB}(\lambda) - \widehat{dB}_m(\lambda))]K_1(\lambda)$$

for $\lambda = i\rho$ with $\phi(\rho)\psi(\rho)$ to obtain

$$K^\phi(\rho) = K_m^\phi(\rho) + K_m^\psi(\rho)[i\rho(\widehat{dB}(i\rho) - \widehat{dB}_m(i\rho))]K^\phi(\rho).$$

In time domain this corresponds to the Volterra convolution equation

$$R^\phi(t) = R_m^\phi(t) + \int_{-\infty}^{\infty} G_m(t-s)R^\phi(s)ds, \quad t \in \mathbb{R},$$

where G_m is defined by

$$\widetilde{G}_m(\rho) = K_m^\psi(\rho)[i\rho(\widehat{dB}(i\rho) - \widehat{db}_mB_1(i\rho))]$$
$$= i\rho K_m^\psi(\rho)[\widehat{dB}_{1,m}(i\rho)] + K_m^\psi(\rho)[\widehat{dB}_{2,m}(i\rho)],$$

with $dB_{j,m} = \chi_{[m,\infty)}dB_j$. By the first part of this proof, $R_m^\phi, R_m^\psi, R_m^{\psi\prime}$ belong to $L^1(\mathbb{R}, \mathcal{B}(V^*, V))$, and since $B_j \in BV(\mathbb{R}_+, \mathcal{B}(V, V^*))$ by assumption, we have also $G_m \in L^1(\mathbb{R}, \mathcal{B}(V, V))$. Now we are in position to apply the vector-valued Paley-Wiener lemma; cp. Prüss [10], Theorem 0.6. For $m \in \mathbb{N}$ sufficiently large, $I - \widetilde{G}_m(\rho)$ is easily seen to be invertible for each $\rho \in \mathbb{R}$, hence by the theorem just cited there is a unique solution $R^\phi \in L^1(\mathbb{R}, \mathcal{B}(V^*, V))$ of the Volterra equation, i.e. we have $\widetilde{R}^\phi(\rho) = \phi(\rho)K(i\rho)$ for all $\rho \in \mathbb{R}$. To obtain the integrability of the derivatives of R^ϕ one has only to replace $\phi(\rho)$ by $(i\rho)^n\phi(\rho)$ and to use uniqueness of the vector-valued Fourier transform. This ends the first part of the proof.

(ii) Fix $h \in H$, $v^* \in V^*$ arbitrarily, let $\psi(\rho) = 1 - \phi(\rho)$, and set $K_2(\rho) = (1 - \phi(\rho))K(i\rho)$. We are going to show first $\psi K_2h \in L^2(\mathbb{R}, V)$, $\psi K_2v^* \in L^2(\mathbb{R}, H)$, $i\rho\psi K_2h \in L^2(\mathbb{R}, H)$, and $(i\rho)^{-1}\psi K_2v^* \in L^2(\mathbb{R}; V)$. For this purpose fix an $\omega >$

0 and define $S_\omega(t) = e^{-\omega t}S(t)$, $R_\omega(t) = e^{-\omega t}R(t)$. Then by the boundedness properties of S, R, T obtained in Section 3 we obtain $S_\omega h \in L^2(\mathbb{R}_+, H)$, $R_\omega h \in L^2(\mathbb{R}_+, V)$, $R_\omega v^* \in L^2(\mathbb{R}_+, H)$, and $e^{-\omega\cdot}Rv^* \in L^2(\mathbb{R}_+, V)$. Plancherels theorem in the vector-valued case then implies that their Fourier transforms belong to the corresponding L^2-space, which gives us with $K_\omega(\lambda) = K(\omega+\lambda)$

$$(\omega + i\cdot)K_\omega(i\cdot)h \in L^2(\mathbb{R}, H), \qquad K_\omega(i\cdot)h \in L^2(\mathbb{R}, V),$$
$$K_\omega(i\cdot)v^* \in L^2(\mathbb{R}, H), \qquad (\omega + i\cdot)^{-1}K_\omega(i\cdot)v^* \in L^2(\mathbb{R}, V).$$

Now write

$$\begin{aligned}
&K(i\rho) = K_\omega(i\rho) + K(i\rho)[\omega^2 + 2i\omega\rho]K_\omega(i\rho)\\
&+K(i\rho)[\widehat{dB}_2(\omega+i\rho) - \widehat{dB}_2(i\rho)]K_\omega(i\rho)\\
&+\frac{(\omega+i\rho)\widehat{db}(\omega+i\rho) - i\rho\widehat{db}(i\rho)}{(\omega+i\rho)\widehat{db}(\omega+i\rho))}[I - AK_\omega(i\rho) - \widehat{dB}_2(\omega+i\rho)K_\omega(i\rho)],
\end{aligned}$$

and use this identity together with the assumed bounds from (A) to obtain the desired L^2-properties. For this purpose we employ the moment condition (26) in the following way.

$$\begin{aligned}
&|\widehat{dB}_2(\omega+i\rho) - \widehat{dB_2}(i\rho)|_{\mathcal{B}(V,V^*)} \le \int_0^\omega |[\widehat{dB}_2(r+i\rho)]'|_{\mathcal{B}(V,V^*)}dr\\
&\le M\int_0^\omega |r+i\rho|^{-1}dr \le C/|\rho|,
\end{aligned}$$

for $|\rho| \ge N$. This estimate, Lemma 7.4 and the asssumed bounds (A) for $K(\lambda)$ yield

$$|K(i\rho)v^*|_H \le M|K_\omega(i\rho)h|_H + M|v^*/\rho|_{V^*}, \quad |\rho| > N,$$

as well as

$$|K(i\rho)h|_V \le M|K_\omega(i\rho)h|_V + M|h/\rho|_H, \quad |\rho| > N.$$

With $\psi = 1-\phi$ this yields by Plancherel's theorem existence of functions $g \in L^2(\mathbb{R}, H)$ and $v \in L^2(\mathbb{R}, V)$ such that $\widetilde{g} = \psi K(i\rho)v^*$ and $\widetilde{v} = \psi K(i\rho)h$. Similarly, dividing the identity for K by $1+i\rho$, resp. multiplying by $i\rho$ we obtain $u \in L^2(\mathbb{R}, H)$, resp. $w \in L^2(\mathbb{R}, V)$ such that $\widetilde{u}(\rho) = \psi(\rho)i\rho K(i\rho)h$ and $\widetilde{w}(\rho) = \psi(\rho)(1+i\rho)^{-1}K(i\rho)v^*$, $\rho \in \mathbb{R}$. Differentiating $\psi K(i\rho)$ we obtain

$$\begin{aligned}
&(\psi(\rho)K(i\rho))' = \psi'(\rho)K(i\rho) - \psi(\rho)K(i\rho)[-2\rho + \widehat{dB}_2'(i\rho)]K(i\rho)\\
&+\frac{(i\rho\widehat{db}'(i\rho))}{\widehat{db}(i\rho)}K(i\rho)(i\rho)^{-1}[I - AK(i\rho) - \widehat{dB}_2(i\rho)K(i\rho)].
\end{aligned}$$

Using once more estimates (A), Lemma 7.4, and uniqueness of the Fourier transform we may deduce $tg, tu \in L^2(\mathbb{R}, H)$ and $tw, tv \in L^2(\mathbb{R}, V)$. By interpolation this in turn implies

$$\int_{-\infty}^{\infty} |u(t)|_H dt = \int_{-1}^{1} |u(t)|_H^2 dt + \int_{|t|>1} |u(t)|_H dt \le |u|_{L^2(\mathbb{R},H)} + |tu|_{L^2(\mathbb{R},H)},$$

i.e. $u \in L^1(\mathbb{R}, H)$, and similarly $v, w \in L^1(\mathbb{R}, V)$ and $g \in L^1(\mathbb{R}, H)$. Together with first part of this proof and uniqueness of the Fourier transform we may conclude the integrability properties (B) as asserted in the theorem. □

6.3. Fractional Orders and Derivatives

We will now consider kernels of the special type

$$B(t) = b(t)A^\eta dt, \; t \geq 0 \quad \eta \in [0,1],$$

in which case we are able to verify the necessary a priori estimates for the operator family $K(\lambda) = (\lambda^2 + \lambda\widehat{db}(\lambda)A^\eta + A)^{-1}$, by means of the functional calculus for the self-adjoint positive definite operator A.

Proposition 6.6. *Let (HS) and (Hα) be valid, and let $B(t) = 1 * b(t)A^\eta$ for some $\eta \in [0,1]$, where b is subject to (Hb). Then the following conditions (A) and (F) are equivalent:*

(A) *There exists a constant M such that for $\lambda \in \overline{C}_+$ we have*
(i) $|\lambda K(\lambda)|_{\mathcal{B}(H,H)} \leq M$; *(ii)* $|K(\lambda)|_{\mathcal{B}(H,V)} \leq M$;
(iii) $|K(\lambda)|_{\mathcal{B}(V^*,H)} \leq M$; *(iv)* $|\frac{K(\lambda)}{1+\lambda}|_{\mathcal{B}(V^*,V)} \leq M$.

(F) *(α)* $\liminf_{\varrho\to\infty} \varrho^{2\eta}\Re\widehat{db}(i\varrho) > 0$, *(β)* $\limsup_{\varrho\to\infty} \frac{-\Im\widehat{db}(i\varrho)}{\varrho\Re\widehat{db}(i\varrho)} < \infty$.

If $b \in BV(\mathbb{R}_+)$ then $|[K(\lambda) - A^{-1}]/\lambda|_{\mathcal{B}(V^*,V)} \leq M$ as well.

Proof. We will prove the following implications: (ii) ⇔ (iii), (i)& (iv) ⇒ (ii), (F)⇒ (i) ⇒ (iv), (i) ⇒ (F).

Since K is continuous at 0, the expressions in (A) are bounded in a neighbourhood of 0.

The equivalence of (ii) and (iii) follows immediately from the definitions of the spaces and operators (A, B commute). The remaining implications will be proved using functional calculus for the self-adjoint operator A. Define the function

$$g_\varrho(x) = (i\varrho + x^\eta\widehat{db}(i\varrho) + \frac{x}{i\varrho})^{-1}, \; g_0(x) \equiv 0.$$

Then $K(i\varrho) = \frac{1}{i\varrho}g_\varrho(A)$, and by the definition of V we have

$$|i\varrho K(i\varrho)|_{\mathcal{B}(H,H)} = \sup_{\mu\in\sigma(A)} |g_\varrho(\mu)|, \quad |K(i\varrho)|_{\mathcal{B}(H,V)} = \sup_{\mu\in\sigma(A)} |\frac{\sqrt{\mu}}{i\varrho}g_\varrho(\mu)|,$$
$$|\frac{1}{i\varrho}K(i\varrho)|_{\mathcal{B}(V^*,V)} = \sup_{\mu\in\sigma(A)} |\frac{\mu}{(i\varrho)^2}g_\varrho(\mu)|.$$

It is straightforward to see that (i) and (iv) imply (ii). In fact, one has just to observe that in case f, g are nonnegative real valued functions such that f and f^2g are bounded, then fg is bounded as well: $\sup fg \leq \sup_{f>1} fg + \sup_{f\leq 1} fg \leq$

$\sup f^2 g + \sup g$. We compute in advance

$$|g_\varrho(x)| = \frac{1}{\sqrt{|x^\eta \Re\widehat{db}|^2 + |\varrho - \frac{x}{\varrho} + x^\eta \Im\widehat{db}|^2}} \le c\frac{1}{|x^\eta \Re\widehat{db}| + |\varrho - \frac{x}{\varrho} + x^\eta \Im\widehat{db}|},$$

and observe $\Re\widehat{db} > 0$, $\Im\widehat{db} \le 0$, in virtue of (Hb).

(i)$\Rightarrow$ (iv) The expression in the infimum

$$\inf_{x\ge\mu_1,\ |\varrho|\ge\delta} \frac{\varrho^2}{x}\left(|x^\eta \Re\widehat{db}| + |\varrho - \frac{x}{\varrho} + x^\eta \Im\widehat{db}|\right)$$

is bounded away from zero by (i) when $\frac{\varrho^2}{x} \ge \frac{1}{2}$, and in the opposite case we can estimate it from below by

$$\frac{\varrho^2}{x}\left(-\varrho + \frac{x}{\varrho} + x^\eta(-\Im\widehat{db})\right) \ge \frac{\varrho^2}{x}(-\varrho + \frac{x}{\varrho}) > \frac{\varrho}{2} \ge \frac{\delta}{2},$$

provided $\varrho \ge \delta > 0$. The expression in the infimum is even with respect to ϱ (since $\Im\widehat{db}(-i\varrho) = -\Im\widehat{db}(i\varrho)$,) hence (iv) is implied by (i).

(F)$\Rightarrow$ (i) We have

$$\inf_{x\ge\mu_1,\ |\varrho|\ge\delta}[|x^\eta \Re\widehat{db}| + |\varrho - \tfrac{x}{\varrho} + x^\eta \Im\widehat{db}|] \ge$$
$$\ge \min\left\{\varepsilon_0, \inf_{x\ge\mu_1,\ |\varrho|\ge\delta,\ |x^\eta\Re\widehat{db}|\le\varepsilon_0} |\varrho - \tfrac{x}{\varrho} + x^\eta\Im\widehat{db}|\right\},$$

and in case $|x^\eta\Re\widehat{db}| \le \varepsilon_0$ we obtain

$$|\varrho - \frac{x}{\varrho} + x^\eta \Im\widehat{db}| \ge |\varrho|\left(1 - \frac{x}{\varrho^2} - \frac{x^\eta}{\varrho}(-\Im\widehat{db})\right) \ge$$
$$\ge |\varrho|\left(1 - \left(\frac{\varepsilon_0}{\Re\widehat{db}\varrho^{2\eta}}\right)^{\frac{1}{\eta}} - \frac{-\Im\widehat{db}\varepsilon_0}{\varrho\Re\widehat{db}}\right) \ge \delta(1 - \varepsilon_0^{\frac{1}{\eta}}\frac{1}{\varepsilon} - \frac{\varepsilon_0}{\varepsilon}) > 0,$$

by (α) and (β), provided $\varepsilon_0 > 0$ is chosen small enough, depending on $\varepsilon > 0$ from (F).

(i) $\Rightarrow$ (F) Suppose

$$\inf_{x\ge\inf\sigma(A),\ \varrho\in\mathbb{R}}[|x^\eta\Re\widehat{db}| + |\varrho - \frac{x}{\varrho} + x^\eta\Im\widehat{db}|] > 0. \tag{27}$$

We will show that this implies (F). To this end we are going to exploit the zeros of $\Im g_\varrho(x) = \varrho((1 - \frac{x}{\varrho^2} - (\frac{x}{\varrho^2})^\eta(\frac{-\Im\widehat{db}}{\varrho^{1-2\eta}}))$. Let $x(\varrho)$ denote the (unique) solution of

$$1 - \frac{x}{\varrho^2} - \left(\frac{x}{\varrho^2}\right)^\eta\left(\frac{-\Im\widehat{db}}{\varrho^{1-2\eta}}\right) = 0. \tag{28}$$

Then (27) yields the necessary condition

$$x(\varrho)^\eta\Re\widehat{db}(i\varrho) \ge \delta > 0 \text{ as } \varrho \to \infty. \tag{29}$$

There are two possibilities for the behaviour of $\frac{x(\varrho)}{\varrho^2}$ as $\varrho \to \infty$ (to be precise for a sequence ϱ_n):

Case I $0 < \varepsilon \leq \frac{x(\varrho)}{\varrho^2} \leq c < \infty, \quad \varrho \to \infty.$
In this case we deduce from (28) that

$$\infty > c_1 \geq \left(\frac{x(\varrho)}{\varrho^2}\right)^\eta \left(\frac{-\Im\widehat{db}}{\varrho^{1-2\eta}}\right) = x(\varrho)^\eta \Re\widehat{db}\left(\frac{-\Im\widehat{db}}{\varrho\Re\widehat{db}}\right)$$

which, together with (29), implies (β). On the other hand, (29) itself implies (α).
Case II $\frac{x(\varrho)}{\varrho^2} \to 0$ as $\varrho \to \infty$.
In this case we obtain from (28):

$$x(\varrho)^\eta \Re\widehat{db}\left(\frac{-\Im\widehat{db}}{\varrho\Re\widehat{db}}\right) = \left(\frac{x(\varrho)}{\varrho^2}\right)^\eta \left(\frac{-\Im\widehat{db}}{\varrho^{1-2\eta}}\right) \to 1 \text{ as } \varrho \to \infty.$$

So (29) implies (β). On the other hand, (29) also gives

$$0 < \delta \leq x(\varrho)^\eta \Re\widehat{db}(i\varrho) = \left(\frac{x(\varrho)}{\varrho^2}\right)^\eta \varrho^{2\eta}\Re\widehat{db}(i\varrho),$$

hence (α). □

An interesting special case of Proposition 6.6 arises in case $db(t)$ is a fractional derivative kernel.

Corollary 6.7. *Suppose (HS), (Hα), (Hb), let* $B(t) = (b_0 + 1 * b_1(t))A^\eta$, $\eta \in [0,1]$, *where either*

(i) $b_1(t) = \frac{1}{\Gamma(1-\gamma)} t^{-\gamma} e^{-\varepsilon t}$, $\gamma \in (0,1)$, $\varepsilon \geq 0$, *with* $2\eta + \gamma \geq 1$, *or*
(ii) $b_0 > 0$.

Then $d\beta$ is dissipative and assumptions (i) and (ii) of Theorem 5.1 are satisfied.

Proof. (i) For $\varepsilon > 0$ we have

$$\begin{aligned}\widehat{b}_1(i\varrho) &= \frac{1}{(i\varrho+\varepsilon)^{1-\gamma}} = \\ &= \left(\varrho^2+\varepsilon^2\right)^{\frac{\gamma-1}{2}}\left[\cos(1-\gamma)\arctan\frac{\varrho}{\varepsilon} - i\sin(1-\gamma)\arctan\frac{\varrho}{\varepsilon}\right],\end{aligned}$$

while for $\varepsilon = 0$

$$\widehat{b}_1(i\varrho) = \frac{1}{(i\varrho)^{1-\gamma}} = \varrho^{\gamma-1}\left[\cos(1-\gamma)\frac{\pi}{2} - i\sin(1-\gamma)\frac{\pi}{2}\right].$$

These representations and $\gamma \in (0,1)$ easily imply (F). (ii) $\widehat{db}(i\varrho) = b_0 + \Re\widehat{b}_1(i\varrho) + i\Im\widehat{b}_1(i\varrho) \to b_0$ as $|\varrho \to \infty|$ by the Riemann-Lebesgue lemma and Proposition 7.2. Therefore, by Proposition 6.6, the assumptions of Theorem 6.5 and Proposition 6.3 satisfied, we may use Corollary 6.2. □

7. Appendix

In this section we collect some results in various directions which are used in the main body of this paper.

7.1. 2-monotone Kernels

Definition 7.1 ([10], Definition 3.4). *A function $b \in L^1_{\mathrm{loc}}(\mathbb{R}_+)$ is called* 2-monotone, *if $b \in \mathcal{C}((0,\infty))$ is nonnegative, nonincreasing, and convex.*

Proposition 7.2. *If $b \not\equiv 0$ is 2-monotone and $b(\infty) = 0$ then*

(i) $\Re\widehat{b}(\lambda) \geq 0$ *for any* $\Re\lambda > 0$;
(ii) $-\varrho\Im\widehat{b}(i\varrho) > 0$ *for any* $\varrho \in \mathbb{R}$, $\varrho \neq 0$;
(iii) *If in addition b is not piecewise linear, then* $\Re\widehat{b}(\lambda) > 0$ *for any* $\Re\lambda \geq 0$.

Proof. For (i), (ii) see Propositions 3.10 and 3.3. from [10]. For (iii) see Proposition 16.4.3 and Theorem 16.4.5 from [6]. □

Lemma 7.3 ([4], Lemma 2). *Suppose $b \in L^1_{\mathrm{loc}}(\mathbb{R}_+)$ is nonnegative, nonincreasing, and log-convex, $b \not\equiv 0$. Then there exists a function $a_1 \in L^1(\mathbb{R}_+)$ with $\widehat{a_1}(i\rho) \neq 0$ for all $\rho \in \mathbb{R}$ such that*

$$2\Re\int_0^T b * u(t)\overline{u(t)}dt \geq \int_0^T |a_1 * u(t)|^2 dt \text{ for any } u \in \mathcal{C}(\mathbb{R}_+).$$

Lemma 7.4. *Suppose (Hb). Then for $\omega > 0$ and $\varepsilon > 0$ the functions*

$$h_1(\lambda) = \lambda\left(1 - \frac{\lambda\widehat{b}(\lambda)}{(\lambda+\omega)\widehat{b}(\lambda+\omega)}\right), \quad h_2(\lambda) = \lambda\frac{(\lambda\widehat{b})'}{\lambda\widehat{b}}$$

are bounded on $i\mathbb{R} \setminus [-i\varepsilon, i\varepsilon]$.

Proof. By (Hb), b_1 2-monotone, so 1-regular (see [10], Proposition 3.3, Definition 3.3). Then the function

$$h_2 = \lambda\frac{(\lambda\widehat{b})'}{\lambda\widehat{b}} = \frac{b_0 + \widehat{b_1} + \lambda\widehat{b_1}'}{b_0 + \widehat{b_1}} = 1 + \frac{\lambda\widehat{b_1}'}{\widehat{b_1}}\frac{\widehat{b_1}}{b_0 + \widehat{b_1}}$$

is bounded by Lemma 8.1 of [10]. The function h_1 can be shown to be bounded in the same way as it is done in the proof of Theorem 10.5 in [10]. □

Remark 7.5. Any nontrivial ($\not\equiv 0$) completely monotonic (cf. Definition 4.1 in [10]) function b on $(0,\infty)$ is log-convex (i.e. log b is convex); see Theorem 5.2.8 in [6]. Any log-convex function is convex.

Remark 7.6. The function $b_1(t) = \frac{1}{\Gamma(1-\gamma)}t^{-\gamma}e^{-\varepsilon t}$, $\gamma \in (0,1)$, $\varepsilon \geq 0$ satisfies the assumption in (Hb) and every statement in this appendix.

7.2. Spectrum of a Function

Definition 7.7. *Let $u \in \mathcal{BUC}(\mathbb{R}_+, X)$, X a Banach space. We define the* spectrum *of u, $\mathrm{Sp}(u) \subset \mathbb{R}$ as follows. We say that $\varrho \notin \mathrm{Sp}\,(u)$ if there exists a neighbourhood U of ϱ and a function $g \in L^1(U, X)$ such that*

$$\widehat{u}(\eta + i.) \to g(i.) \quad in \quad \mathcal{D}'(U) \quad as \quad \eta \to 0_+.$$

Remark 7.8. Let $\tilde{u}$ denote the Fourier transform of u in the sense of distributions. If $\tilde{u} \in L^1(\varrho - \varepsilon, \varrho + \varepsilon)$ then $\varrho \notin \mathrm{Sp}(u)$.

The following simple form of a Tauberian theorem has been presented in Chill [3]; see also Arendt and Batty [2].

Lemma 7.9 ([3]). *Let $u \in \mathcal{BUC}(\mathbb{R}_+, X)$ be such that $\mathrm{Sp}(u) = \emptyset$. Then $u \in \mathcal{C}_0(\mathbb{R}_+, X)$.*

Proof. First we show that for any $\varphi \in \mathcal{D}(\mathbb{R})$ we have $u * \tilde{\varphi} \in \mathcal{C}_0(\mathbb{R}_+, X)$. In fact, $\tilde{\varphi} \in L^1(\mathbb{R})$ and we can compute, using the Lebesgue dominated convergence theorem and Fubini's theorem:

$$\begin{aligned}
u * \tilde{\varphi}(t) &= \lim_{\eta \to 0+} \int_0^\infty e^{-\eta s} u(s) \tilde{\varphi}(t-s) ds = \\
&= \lim_{\eta \to 0+} \int_0^\infty e^{-\eta s} u(s) \int_{supp(\varphi)} e^{-i(t-s)z} \varphi(z) dz ds = \\
&= \lim_{\eta \to 0+} \int_{supp(\varphi)} \varphi(z) e^{-itz} \int_0^\infty u(s) e^{-s(\eta - iz)} ds dz = \\
&= \lim_{\eta \to 0_+} \int_{supp(\varphi)} \varphi(z) e^{-itz} \widehat{u}(\eta - iz) dz = \\
&= \int_{supp(\varphi)} g(-z) \varphi(z) e^{-itz} dz = [\widetilde{g(-.)\varphi(.)}](t),
\end{aligned}$$

where the last term converges to zero as $t \to \infty$ by the Riemann-Lebesgue lemma.

In the second step we show that $u * f \in \mathcal{C}_0(\mathbb{R}_+, X)$ for any $f \in L^1(\mathbb{R})$. $\mathcal{S}(\mathbb{R})$ is dense in $L^1(\mathbb{R})$, so it is enough to show this for $f \subset \mathcal{S}(\mathbb{R})$. Since the Fourier transform is an isomorphism on $\mathcal{S}(\mathbb{R})$ and $\mathcal{D}(\mathbb{R})$ is dense in $\mathcal{S}(\mathbb{R})$, any $f \in \mathcal{S}(\mathbb{R})$ can be written as $\tilde{g}$, $g \in \mathcal{S}(\mathbb{R})$, and approximated by $\widetilde{\varphi_n}$ in the norm of $L^1(\mathbb{R})$, where $\varphi_n \in \mathcal{D}(\mathbb{R})$. Finally, $u * \widetilde{\varphi_n}$ converges to $u * f$ uniformly.

In the third step we conclude that $u \in \mathcal{C}_0(\mathbb{R}_+, X)$. We define $f_n(t) := n$ for $t \in (0, \frac{1}{n})$, and $f_n(t) := 0$ otherwise. Then for $t \geq 1$ we have

$$u(t) = n \int_0^{\frac{1}{n}} [u(t) - u(t-s)] ds + u * f_n(t),$$

where the first term can be made arbitrarily small by choosing n large (since u is uniformly continuous), and $u * f_n \in C_0(\mathbb{R}_+, X)$ then imply $u(t) \to 0$ as $t \to \infty$. □

Corollary 7.10. *Let X be a Hilbert space. Suppose $u \in \mathcal{BUC}(\mathbb{R}_+, X)$ is such that $a * u \in L^2(\mathbb{R}_+, X)$ for some $a \in L^1(\mathbb{R}_+)$ with $\widehat{a}(i\rho) \neq 0$ for all $\rho \in \mathbb{R}$, or $a = a_0\delta,\ a_0 > 0$. Then $u \in \mathcal{C}_0(\mathbb{R}_+, X)$.*

Proof. Denote $h := a*u \in L^2(\mathbb{R}_+, X)$. Then $\tilde{u} = \frac{\tilde{h}}{\tilde{a}} \in L^1_{\rm loc}(\mathbb{R}, X)$, since the Fourier transform is an isomorphism on $L^2(\mathbb{R}, X)$. □

7.3. Convolutions with Operator-Valued Kernels

Let X, Y be Banach spaces.

Lemma 7.11. *If $U \in L^1_s(\mathbb{R}_+;\mathcal{B}(X,Y)) \cap \mathcal{C}_s(\mathbb{R}_+;\mathcal{B}(X,Y))$ then $K_U : X \to L^1(\mathbb{R}_+;Y)$ defined by $x \mapsto U(.)x$ is a bounded linear operator, and the convolution operator $f \mapsto U * f$ is bounded from $L^1(\mathbb{R}_+; X)$ to $L^1(\mathbb{R}_+; Y)$.*

Proof. We use the closed graph theorem. Assume $x_n \to x$ in X, $Kx_n \to h(.)$ in $L^1(\mathbb{R}_+, Y)$. Then for any $t \in \mathbb{R}_+$ we have $U(t)x_n \to U(t)x$ in Y, and passing to a subsequence if necessary, $U(t)x_{n_k} \to h(t)$ in Y, for a.e. $t \in \mathbb{R}_+$. Hence $h(t) = U(t)x$ a.e. on $\mathbb{R}_+$.

Let $f = \sum_{i=1}^n f_i\chi_i$ be an simple function $\in L^1(\mathbb{R}_+; X)$, where $f_i \in X$, and χ_i denote the characteristic function of mutually disjoint measurable subsets $A_i \subset \mathbb{R}_+$. Then $U * f = \sum U f_i * \chi_i$ belongs to $L^1(\mathbb{R}_+; Y)$ and

$$\begin{aligned} |U * f|_1 = \int_0^\infty |U * f(t)|dt &\leq \sum_{i=1}^n |Uf_i * \chi_i|_1 \leq \sum_{i=1}^n |Uf_i|_1 \mathrm{mes}(A_i) \\ &\leq |K_U| \sum_{i=1}^n |f_i| \mathrm{mes}(A_i) = |K_U||f|_1. \end{aligned}$$

Since simple functions are dense in $L^1(\mathbb{R}_+; X)$, the second assertion follows by continuous extension. □

Lemma 7.12. *Let X, Y be Banach spaces, $U \in L^1_s \cap \mathcal{C}_s(\mathbb{R}_+, \mathcal{B}(X,Y))$ and $U^* \in L^1_s \cap \mathcal{C}_s(\mathbb{R}_+, \mathcal{B}(Y^*, X^*))$. Then the mapping $f \mapsto U * f$ is bounded from $L^\infty(\mathbb{R}_+; X)$ to $L^\infty(\mathbb{R}_+, Y))$.*

Proof. Let $f \in L^\infty(\mathbb{R}_+, X)$, $t > 0$, $y^* \in Y^*$. We have

$$\begin{aligned} |(U * f(t)|_Y &= \sup_{|y^*|\leq 1} |\int_0^t (U(t-s)f(s), y^*)_{Y,Y^*} ds| \\ &= \sup_{|y^*|\leq 1} |\int_0^t (f(s), U^*(t-s)y^*)_{X,X^*} ds| \\ &\leq \sup_{|y^*|\leq 1} |f|_\infty |\Phi_{U^*(.)}|_{\mathcal{B}(Y^*, L^1(\mathbb{R}_+, X^*))} |y^*|_{Y^*} \\ &\leq |f|_\infty |\Phi_{U^*(.)}|_{\mathcal{B}(Y^*, L^1(\mathbb{R}_+, X^*))}. \end{aligned}$$

Therefore $U * f \in L^\infty(\mathbb{R}_+; Y)$. □

Lemma 7.13. *Let X, Y be Banach spaces,*

$$U \in L^1_s \cap \mathcal{C}_s(\mathbb{R}_+, \mathcal{B}(X,Y)), \quad U^* \in L^1_s \cap \mathcal{C}_s(\mathbb{R}_+, \mathcal{B}(Y^*, X^*)),$$

and $g \in \mathcal{C}(\mathbb{R}_+, X)$ such that $\gamma(g) := \{g(t);\ t \geq 0\}$ is relatively compact in X. Then

(i) *$U * g \in \mathcal{BUC}(\mathbb{R}_+, Y)$;*
(ii) *$\gamma(U * g)$ is relatively compact in Y.*

Proof. **(i)** Since $g(\mathbb{R}_+)$ is relatively compact in X, for each $\varepsilon > 0$ there are finitely many disjoint Borel sets $I_i \subset \mathbb{R}_+$ and $x_i \in X$ such that with $g_\varepsilon(t) := \sum_{i=1}^n x_i \chi_{I_i}(t)$ we have $\sup_{t\geq 0} |g(t) - g_\varepsilon(t)|_X \leq \varepsilon$. Then $U * g_\varepsilon = \sum_{i=1}^n U(.)x_i * \chi_{I_i}$ is a finite sum of convolutions of L^1-functions with scalar L^∞-functions, hence it is in $\mathcal{BUC}(\mathbb{R}_+, Y)$. It follows from Lemma 7.12 that $\sup_{t\geq 0} |U * g_\varepsilon(t) - U * g(t)| \leq c\varepsilon$, so passing to the limit we obtain $U * g \in \mathcal{BUC}(\mathbb{R}_+, Y)$.
(ii) If $x \in X$ and $\phi \in L^\infty(\mathbb{R})$ then the decomposition

$$Ux * \phi(t) = \int_0^T U(s)x\phi(t-s)ds + \int_T^\infty U(s)x\phi(t-s)ds$$

shows that $\gamma(Ux * \phi)$ is relatively compact; in fact, the first integral belongs to the set $T|\phi|_\infty \overline{\mathrm{conv}}\{U(s)x : 0 \leq s \leq T\}$ which is compact by strong continuity of U, and the second term is smaller than $|\phi|_\infty \int_T^\infty |U(s)x|ds$ which tends to zero as $T \to \infty$. Employing the same approximation as in (i), we obtain $\gamma(U * g_\varepsilon)$ relatively compact and with $\varepsilon \to 0$ the assertion follows. □

7.4. The Shift System

For a Banach space X let $Y = \mathcal{BUC}(\mathbb{R}_+, X)$ be equipped with the topology of local uniform convergence, and let $S : \mathbb{R}_+ \times Y \to Y$ be the shift dynamical system on Y defined by $(S(t)\varphi)(s) = \varphi(s+t)$, $t, s \geq 0$, $\varphi \in Y$. We denote by $\gamma(\varphi) = \{S(t)\varphi;\ t \geq 0\} \subset Y$ the trajectory, and $\omega(\varphi) = \{\psi \in Y;\ S(t_n)\varphi \to \psi$ in Y for some sequence $t_n \to \infty\}$ the ω-limit set of an element $\varphi \in Y$. For $x \in X$ let $c_x \in Y$ denote the constant function $t \mapsto x$. In [6] finite dimensional spaces X are considered, however, the results below are true for any Banach space X.

Lemma 7.14 ([6], Theorem 15.3.1). *For $\varphi \in Y$, $x \in X$ we have: $\omega(\varphi) = \{c_x\}$ if and only if $\lim_{t\to\infty} \varphi(t) = x$.*

Lemma 7.15 ([6], Theorem 15.2.3). *If $\{\varphi(t);\ t \geq 0\}$ is relatively compact in X then $\gamma(\varphi)$ is relatively compact in Y.*

Lemma 7.16. *If $\mathcal{E}$ is a discrete subset of X then $\{c_x;\ x \in \mathcal{E}\}$ is a discrete subset of Y.*

Proof. If $U_\delta(x)$ is a neighbourhood of $x \in \mathcal{E}$ then $V = \{\varphi \in Y;\ |\varphi(t) - x| \leq \delta$ for some $t \geq 0\}$ is a neighbourhood of c_x in Y. □

Proposition 7.17 ([6] Theorem 15.2.3). *If $\gamma(\varphi)$ is relatively compact in Y, then $\omega(\varphi)$ is a nonempty, compact, connected subset of Y.*

References

[1] W. Arendt, "Vector-valued Laplace transforms and Cauchy problems", Israel Journal of Mathematics **59** (1987), 327–352.

[2] W. Arendt and C.J.K. Batty, "Almost periodic solutions of first and second order Cauchy problems", J. Diff. Eqns., **137** (1997), 363–383.

[3] R. Chill, "Tauberian theorems for vector-valued Fourier and Laplace transforms". Preprint, 1997.

[4] E. Fašangová, J. Prüss, "Asymptotic behaviour of a semilinear viscoelastic beam model". Preprint, 1998.

[5] I. M. Gel'fand, D. A. Raikov and G. E. Shilov, *Commutative Normed Rings*. Chelsea, New York, 1964 (Russian edition: Moscow, 1960).

[6] G. Gripenberg, S.O. Londen, and O.J. Staffans, *Volterra Integral and Functional Equations*. Cambridge Univ. Press, Cambridge, 1990.

[7] J. Hale, *Asymptotic Behaviour of Dissipative Systems*, vol. 25 of *Mathematical Surveys and Monographs*. Amer. Math. Soc., Providence, Rhode Island, 1988.

[8] G. Propst, J. Prüss, "On wave equations with boundary dissipation of memory type", Journal of Integral Equations and Applications **8** (1996), 99–123.

[9] J. Prüss, "On linear Volterra equations of parabolic type in Banach spaces", Transactions of the AMS **301** (1987), 691–721.

[10] J. Prüss, *Evolutionary Integral Equations and Applications*. Birkhäuser Verlag, Basel, 1993.

Eva Fašangová
Department of Mathematical Analysis
Charles University
Sokolovská 83
Cz-18600 Praha 8
Czech Republic

Jan Prüss
Fachbereich Mathematik und Informatik
Martin-Luther-Universität
Halle-Wittenberg
Theodor-Lieser Str. 5
D-60120 Halle, Germany

Progress in Nonlinear Differential Equations
and Their Applications, Vol. 35
© 1999 Birkhäuser Verlag Basel/Switzerland

Large Time Behavior of Solutions of a Semilinear Parabolic Equation with a Nonlinear Dynamical Boundary Condition

Marek Fila and Pavol Quittner

1. Introduction

We consider the problem

$$u_t = \Delta u + \lambda |u|^{p-1} u, \qquad x \in \Omega,\ t > 0, \tag{1.1}$$

$$u_t + u_\nu = \mu |u|^{q-1} u, \qquad x \in \partial\Omega,\ t > 0, \tag{1.2}$$

$$u(x,0) = u_o(x), \qquad x \in \Omega, \tag{1.3}$$

where Ω is a bounded domain in $\mathbb{R}^N$, ν is the outer normal on $\partial\Omega$, $p, q > 1$, $\lambda, \mu \in \{-1, 0, 1\}$, $\max\{\lambda, \mu\} = 1$ and $\lambda p + \mu q > 0$.

Local existence and uniqueness of solutions of more general problems were established in [E2], [E3], [E4]. Under our assumptions on p, q, λ and μ blow-up in finite time occurs for some initial data. More precisely, there exist u_o and $T = T(u_o) > 0$ such that

$$\limsup_{t \to T-} \|u(\cdot, t; u_o)\|_{L^\infty(\Omega)} = \infty.$$

We are interested in the following three questions:

(i) Are global solutions uniformly bounded?
(ii) Can one derive an *a priori* bound for global solutions?
(iii) If $\lambda = 0$ and $\mu = 1$, what can be said about the blow-up rate and the blow-up set?

If the dynamical term u_t is not present in the boundary condition (1.2) then similar questions have been satisfactorily answered before. Our main aim here is to study the impact of the term u_t in (1.2) on the dynamics.

We shall show that if

$$(N-2)p < N+2 \qquad \text{and} \qquad (N-2)q < N \tag{1.4}$$

then all global solutions are uniformly bounded. In other words, only two types of behavior of solutions are possible. A solution either blows up in finite time or it exists globally, stays uniformly bounded and its ω-limit set consists of stationary solutions.

To obtain an *a priori* bound for global solutions in terms of suitable norms of u_o we need a stronger condition on p and q. Namely, if

$$\Big(N-\frac{4}{3}\Big)p < N+\frac{8}{3} \qquad \text{and} \qquad \Big(N-\frac{5}{3}\Big)q < N+\frac{1}{3}, \tag{1.5}$$

then

$$\sup_{t>0} \|u(\cdot,t;u_o)\|_{L^\infty(\Omega)} \le C\big(\|u_o\|_{L^\infty(\Omega)}, \|u_o\|_{H^1_2(\Omega)}\big)$$

provided u is a global solution. As an application of this *a priori* bound we shall prove the existence of sign changing solutions of the problem

$$\Delta u - |u|^{p-1}u = 0, \qquad x\in\Omega, \tag{1.6}$$

$$u_\nu = |u|^{q-1}u, \qquad x\in\partial\Omega, \tag{1.7}$$

if $1<p<q$, $(N-\frac{5}{3})q < N+\frac{1}{3}$. The existence of positive solutions of Problem (1.6)–(1.7) was studied before in [CFQ] and [Q1].

Before we discuss Question (iii), let us introduce two related problems:

$$u_t = \Delta u, \qquad x\in\Omega,\ t>0, \tag{1.8}$$

$$u_\nu = u^q, \qquad x\in\partial\Omega,\ t>0, \tag{1.9}$$

$$u(x,0) = u_o(x)\ge 0, \qquad x\in\Omega, \tag{1.10}$$

and

$$\Delta u = 0, \qquad x\in\Omega,\ t>0, \tag{1.11}$$

$$u_t + u_\nu = u^q, \qquad x\in\partial\Omega,\ t>0, \tag{1.12}$$

$$u(x,0) = u_o(x)\ge 0, \qquad x\in\partial\Omega, \tag{1.13}$$

where Ω is a bounded domain in $\mathbb{R}^N$ and $q>1$. Problem (1.8)–(1.10) was studied for example in [LP],[F1],[FQ1] and [HY], see [FF] for more references. It is known that for Problem (1.8)–(1.10) blow-up may occur only on the boundary and under suitable assumptions on q and u_o the blow-up rate is $(T-t)^{-\frac{1}{2(q-1)}}$, this means

$$c_1 \le (T-t)^{\frac{1}{2(q-1)}} \max_{x\in\bar\Omega} u(x,t) \le C_1 \tag{1.14}$$

for some constants c_1 and C_1, see [HY] and [H2]. For Problem (1.11)–(1.13), the blow-up behavior is completely different. Notice that

$$u^*(x,t) = \big((q-1)(T-t)\big)^{-\frac{1}{q-1}}$$

is an explicit solution of Problem (1.11)–(1.13). It blows up everywhere and the blow-up rate is $(T-t)^{-\frac{1}{q-1}}$. This blow-up rate can be established under suitable assumptions on u_o and q also for spatially nonhomogeneous solutions (see Remark 5.2(ii) and Proposition 5.3), that is

$$c_2 \le (T-t)^{\frac{1}{q-1}} \max_{x\in\bar\Omega} u(x,t) \le C_2 \tag{1.15}$$

for some constants c_2 and C_2. It is also easy to see that for $N=1$ positive solutions of (1.11)–(1.13) blow up also in Ω (see Proposition 5.3).

A natural question that arises in connection with the problem

$$u_t = \Delta u, \qquad x \in \Omega,\ t > 0, \tag{1.16}$$

$$u_t + u_\nu = u^q, \qquad x \in \partial\Omega,\ t > 0, \tag{1.17}$$

$$u(x,0) = u_o(x) \geq 0, \qquad x \in \Omega, \tag{1.18}$$

is: Where and with which rate do solutions blow-up? We show that under some suitable assumptions on u_o and q, solutions blow up only on the boundary (as for (1.8)–(1.10)) but the blow-up rate is $(T-t)^{-\frac{1}{q-1}}$ (as for (1.11)–(1.13)).

The paper is organized as follows. Section 2 contains preliminary results which are concerned mainly with the regularity of solutions. In Section 3 we prove that global solutions are bounded if (1.4) holds and we derive an *a priori* bound for global solutions if (1.5) is satisfied. In Section 4 we study the existence of sign changing solutions of (1.6)–(1.7). Finally, results on the blow-up rates and blow-up sets for Problems (1.11)–(1.13) and (1.16)–(1.18) are established in Section 5.

Acknowledgement

The authors were partially supported by VEGA Grant 1/4190/97 and by the Swiss National Science Foundation under project 7IP 051638.

2. Local existence and regularity

In this section we derive some existence and regularity results for solutions of the problem (1.1)–(1.3). These results are based on the functional analytic approach of H. Amann and J. Escher (see [A1–2], [AE], [E2–4]) and they can be extended to much more general problems in a straightforward way. A different approach to the study of regularity of solutions of problems with dynamical boundary conditions can be found in [BS].

Throughout this section we consider numbers $r > 1$, $\omega \geq 0$, differential operators

$$\mathcal{A}u = -\Delta u + \omega u,$$

$$\mathcal{B}v = v_\nu + \omega v,$$

and Nemytskii operators

$$G_p(u) = \omega u + \lambda |u|^{p-1} u,$$

$$G_q(v) = \omega v + \mu |v|^{q-1} v,$$

acting on functions u or v defined on Ω or $\partial\Omega$, respectively.

We assume that Ω is a smoothly bounded domain in $\mathbb{R}^N$ and we denote by Γ its boundary $\partial\Omega$ and by γ the trace operator. For $s \in \mathbb{R}$ we consider the Bessel potential spaces $H_r^s(\Omega)$ with norm $|\cdot|_{s,r}$ and Besov spaces $B_r^s(\Gamma) := B_{rr}^{s-1/r}(\Gamma)$ with norm $\|\cdot\|_{s,r}$. The norm in the Lebesgue space $H_r^0(\Omega) = L_r(\Omega)$ will be denoted simply by $|\cdot|_r$.

By $D(A)$ we denote the domain of definition of a linear operator A. If X and Y are Banach spaces then we denote by $\mathcal{L}(X,Y)$ the space of continuous linear operators $A : X \to Y$ and we set $Isom(X,Y) := \{A \in \mathcal{L}(X,Y) : A \text{ is an isomorphism}\}$. We write $X \doteq Y$ if X, Y are isomorphic and $X \hookrightarrow Y$ if X is continuously embedded into Y.

It is known (see [T]) that if $s > 1/r$ then $\gamma \in \mathcal{L}\big((H^s_r(\Omega), B^s_r(\Gamma)\big)$ and there exists a continuous right inverse $\mathcal{T} \in \mathcal{L}\big(B^s_r(\Gamma), H^s_r(\Omega)\big)$, i.e. $\gamma(\mathcal{T}u) = u$ for $u \in B^s_r(\Gamma)$. Following [E4], we identify the dual space $\big(L_r(\Omega)\big)'$ of $L_r(\Omega)$ with $L_{r'}(\Omega)$, $r' := r/(r-1)$ and the dual space of $B^s_r(\Gamma)$ with $B^{1-s}_{r'}(\Gamma)$. If $Z := (u,v) \in L_r(\Omega) \times L_{\tilde{r}}(\Gamma)$ and $\Upsilon := (\varphi, \psi) \in L_{r'}(\Omega) \times L_{\tilde{r}'}(\Gamma)$, we set

$$\langle Z, \Upsilon \rangle = \langle u, \varphi \rangle + \langle v, \psi \rangle = \int_\Omega u\varphi \, dx + \int_{\partial\Omega} v\psi \, dS.$$

We set also

$$\begin{aligned} E_1 &:= \{(u,v) \in H^2_r(\Omega) \times B^2_r(\Gamma) : \gamma u = v\}, \\ F_1 &:= L_r(\Omega) \times B^1_r(\Gamma), \\ A_1 &: E_1 \to F_1 : (u,v) \mapsto (\mathcal{A}u, \mathcal{B}u). \end{aligned}$$

It is shown in [E2] that choosing ω sufficiently large, the triplet $(A,E,F) := (A_1, E_1, F_1)$ has the following properties:

$$\left.\begin{aligned} &(\lambda + A) \in Isom(E,F) \text{ for } \operatorname{Re}\lambda \geq 0, \\ &(-A) \text{ generates a strongly continuous analytic semigroup } e^{-tA} \text{ in } F. \end{aligned}\right\} \tag{2.1}$$

Moreover, A_1 admits a unique continuous extension to $A_0 \in Isom(E_0, F_0)$, where

$$E_0 = L_r(\Omega) \times B^0_r(\Gamma),$$

F_0 is the completion of E_0 with respect to an appropriate norm and the triplet $(A,E,F) := (A_0, E_0, F_0)$ satisfies again (2.1). In fact, the space F_0 can be characterized as the dual of the space

$$E_1^{\#} := \{(\varphi, \psi) \in H^2_{r'}(\Omega) \times B^2_{r'}(\Gamma) : \gamma\varphi = \psi\}$$

with respect to the duality pairing induced by $\langle \cdot, \cdot \rangle$ (in the notation from [E2], one has

$$(A_1, E_1, F_1, A_0, E_0, F_0) = \big(\overline{A}_\varepsilon(\sigma), \overline{E}_{1,\varepsilon}(\sigma), \overline{E}_0, \tilde{A}_{-1,\varepsilon}(\sigma), \tilde{E}_0, \tilde{E}_{-1,\varepsilon}(\sigma)\big)$$

with r replaced by p everywhere).

Let $[\cdot,\cdot]_\theta$, $0 < \theta < 1$, be the complex interpolation functor. Set $E_\theta = [E_0, E_1]_\theta$, $F_\theta = [F_0, F_1]_\theta$, and let A_θ be the F_θ-realization of A_0, i.e. $A_\theta Z = A_0 Z$ for $Z \in D(A_\theta) := \{W \in D(A_0) : A_0 W \in F_\theta\}$. Then it is shown in [E2] that choosing $\theta \in I_r := (\frac{1}{2r}, \frac{1}{2} + \frac{1}{2r})$,

$$\left.\begin{aligned} &\text{the triplet } (A,E,F) := (A_\theta, E_\theta, F_\theta) \text{ fulfils (2.1)}, \\ &E_\theta \doteq \{(u,v) \in H^{2\theta}_r(\Omega) \times B^{2\theta}_r(\Gamma) : \gamma u = v\}, \end{aligned}\right\} \tag{2.2}$$

and

$$F_\theta \doteq \left[\{(\varphi,\psi) \in H^{1-2\theta}_{r'}(\Omega) \times B^{1-2\theta}_{r'}(\Gamma) \,:\, \gamma\varphi = \psi\}\right]'. \tag{2.3}$$

The reduction of admissible θ to the interval I_r in [E2] was required by the fact that the author of [E2] considered a more general situation including a classical (non-dynamical) nonlinear boundary condition on a part Γ_2 of $\partial\Omega$. In our situation, one can repeat the considerations from [E2, Section 3] in order to verify (2.2) also for $\theta > \frac{1}{2} + \frac{1}{2r}$. Moreover, one obtains

$$\left[H^{1-2\theta}_{r'}(\Omega) \times B^{1-2\theta}_{r'}(\Gamma)\right]' \hookrightarrow F_\theta \quad \text{if } \theta > \frac{1}{2} + \frac{1}{2r} \tag{2.4}$$

and

$$\left.\begin{aligned} E_\theta &\doteq H^{2\theta}_r(\Omega) \times B^{2\theta}_r(\Gamma) \\ \mathbb{E}_\theta &\doteq H^{2\theta}_r(\Omega) \times B^{1+\theta}_r(\Gamma) \end{aligned}\right\} \quad \text{if } 2\theta < \frac{1}{r}, \tag{2.5}$$

where $\mathbb{E}_\theta := [F_1, E_1]_\theta$.

Since $F_1 \hookrightarrow E_0$, an obvious modification of the results in the Appendix of [A2] yields

$$\|e^{-tA_\eta}\|_{\mathcal{L}(F_\eta,E_\theta)} \leq Ct^{\eta-\theta-1} \quad \text{for } 1 \geq \eta > \theta > \frac{1}{2r} \tag{2.6}$$

(cf. also [E4, Lemma 4.1]).

Now set

$$G(Z) := G(u,v) := \big(G_p(u), G_q(v)\big)$$

and let $PZ = P(u,v) := u$. If $r = 2$, $\theta = \frac{1}{2}$ and p,q satisfy (1.4) or if $r > N$ and $\theta \in I_r$ then the proof of [E4, Lemma 5.1] implies that

$$G : E_\theta \to F_\eta \quad \text{is uniformly Lipschitz continuous on bounded sets} \tag{2.7}$$

for some $\eta > \theta$, $\eta \in I_r$. The property (2.7) enabled the author of [E4] to show that if $u_o \in H^{2\theta}_r(\Omega)$ then (a generalization of) (1.1)–(1.3) possesses a unique maximal weak solution $u = PZ$, where $Z \in C(J, E_\theta)$ satisfies the variation of constants formula

$$Z(t) = e^{-A_\theta t}Z_0 + \int_0^t e^{-A_\eta(t-\tau)}G\big(Z(\tau)\big)\,d\tau, \tag{2.8}$$

$Z_o = (u_o, \gamma u_o)$ and the maximal interval of existence $J = J(Z_o)$ is open in $[0,\infty)$. Moreover, the solution Z exists globally provided Z is bounded in E_θ.

Our aim is to show (2.7) for bigger values of θ and η. This and (2.6) enable us to use a bootstrap argument in order to show that (2.8) is valid in E_1 for $t \in J$, $t > 0$.

Lemma 2.1.

(i) *Let $r \geq 2$, let p,q satisfy (1.4), $2\theta \geq 1$ and $2\eta < \min\{1 + \frac{1}{r}, 2(\theta+\varepsilon)\}$, where*

$$2\varepsilon := \min\Big\{2 - (p-1)\frac{N-2}{2}, 1 - (q-1)\frac{N-2}{2}\Big\} > 0. \tag{2.9}$$

Then (2.7) is satisfied.

(ii) *Let* $r > N\max\{\frac{1}{q-1},1\}$, $2\theta \in [1, 1+\frac{N}{r}]$ *and* $\eta < \theta+\varepsilon$, *where*

$$2\varepsilon := \hat{q} - 1 - \frac{N}{r} > 0, \qquad \hat{q} := \min\{q,2\}. \tag{2.10}$$

Then (2.7) *is satisfied.*

(iii) *Let* $r > N\max\{\frac{1}{q-1},1\}$, $2\theta > 1+\frac{N}{r}$, $0 < \delta < \min\{\frac{1}{2r}, 2\theta-1-\frac{N}{r}, q-1\}$. *Then*

$$G : E_\theta \to \mathbb{E}_\delta \quad \textit{is continuous and bounded on bounded sets.} \tag{2.11}$$

Proof. (i) The proof of (i) is a modification of the proof of [E5, Lemma 4.1] (see also [A3], [Q3]).

The operator G_p can be written in the form

$$H_r^{2\theta}(\Omega) \hookrightarrow L_z(\Omega) \xrightarrow{G_p} L_{z/p}(\Omega) \hookrightarrow \big[H_{r'}^{2(1-\eta)}(\Omega)\big]',$$

provided $z > p$, $2\theta - \frac{N}{r} \geq -\frac{N}{z}$ and $2(1-\eta) - \frac{N}{r'} \geq (\frac{z}{p})'$, where the last embedding follows from $H_{r'}^{2(1-\eta)}(\Omega) \hookrightarrow L_{(z/p)'}(\Omega)$. It can be easily seen that the existence of a suitable z is equivalent to the condition

$$2(\eta-\theta) < 2 - (p-1)\Big(\frac{N}{r} - 2\theta\Big). \tag{2.12}$$

Similarly, the operator G_q can be written in the form

$$B_r^{2\theta}(\Gamma) \hookrightarrow L_t(\Gamma) \xrightarrow{G_q} L_{t/q}(\Gamma) \hookrightarrow \big[B_{r'}^{2(1-\eta)}(\Gamma)\big]',$$

provided $t > q$, $2\theta - \frac{N-1}{r} \geq -\frac{N-1}{t}$ and $2(1-\eta) - \frac{N-1}{r'} \geq (\frac{t}{q})'$. The existence of a suitable t is now equivalent to the condition

$$2(\eta-\theta) < 1 - (q-1)\Big(\frac{N}{r} - 2\theta\Big). \tag{2.13}$$

Since $(\frac{N}{r} - 2\theta) \leq \frac{N-2}{2}$ (due to $2\theta \geq 1$ and $r \geq 2$), the conditions (2.12) and (2.13) are guaranteed by (2.9) and the assertion follows from the Lipschitz continuity of the Nemytskii mappings G_p, G_q in the corresponding Lebesgue spaces.

(ii) The operators G_p, G_q can be written in the form

$$H_r^{2\theta}(\Omega) \hookrightarrow C(\bar{\Omega}) \xrightarrow{G_p} C(\bar{\Omega}) \hookrightarrow \big[H_{r'}^{2(1-\eta)}(\Omega)\big]',$$
$$B_r^{2\theta}(\Gamma) \hookrightarrow C(\Gamma) \xrightarrow{G_q} C(\Gamma) \hookrightarrow \big[B_{r'}^{2(1-\eta)}(\Gamma)\big]',$$

where the last embedding holds provided $2\eta < 1+\frac{1}{r}$ (cf. also [E4, Lemma 5.1]). Hence, we have to verify only the Lipschitz continuity of G_q in the case $2\eta > 1+\frac{1}{r}$. Notice that (2.10) implies $\eta < 1$.

Without loss of generality we may assume $\omega = 0$. We shall briefly write u^q instead of $|u|^{q-1}u$ (if no confusion seems likely) and we define $\tilde{G}_q(u) = u^q$ for $u : \Omega \to \mathbb{R}$. The operator G_q can be written in the form

$$B_r^{2\theta}(\Gamma) \xrightarrow{\mathcal{T}} H_r^{2\theta}(\Omega) \xrightarrow{\tilde{G}_q} H_r^{2\eta-1}(\Omega) \xrightarrow{\gamma} B_r^{2\eta-1}(\Gamma),$$

so that it suffices to show the Lipschitz continuity of $\tilde{G}_q : H_r^{2\theta}(\Omega) \to H_r^{2\eta-1}(\Omega)$.

The norm $|\cdot|_{2\eta-1,r}$ is equivalent to $|\cdot|_r + |||\cdot|||_{2\eta-1,r}$, where $|||\cdot|||$ is the Slobodeckii seminorm,

$$|||u|||_{\sigma,r} = \Big(\int_\Omega \int_\Omega \frac{|u(x)-u(y)|^r}{|x-y|^{N+\sigma r}}\,dx\,dy\Big)^{1/r}.$$

Since $H_r^{2\theta}(\Omega) \hookrightarrow C(\bar\Omega)$, $\tilde G_q : C(\bar\Omega) \to C(\bar\Omega)$ is uniformly Lipschitz continuous on bounded sets and $C(\bar\Omega) \hookrightarrow L_r(\Omega)$, it suffices to show

$$|||\tilde G_q(u_2) - \tilde G_q(u_1)|||_{2\eta-1,r} \le C|u_2-u_1|_{2\theta,r}, \tag{2.14}$$

where the constant C may depend on $|u_1|_{2\theta,r}$ and $|u_2|_{2\theta,r}$.

Let $u_1, u_2 \in H_r^{2\theta}(\Omega)$, $u_\xi := u_1 + \xi(u_2-u_1)$ for $\xi \in (0,1)$, and let $x, y \in \Omega$, $x \neq y$. In what follows we shall denote by C various constants which may vary from step to step and which may depend on $|u_1|_{2\theta,r}$ and $|u_2|_{2\theta,r}$. The Mean Value Theorem and the embedding $H_r^{2\theta}(\Omega) \hookrightarrow C^{0,\alpha}(\bar\Omega)$ for $\alpha < 2\theta - \frac{N}{r}$ imply

$$\begin{aligned}
&\big|u_2^q(x) - u_1^q(x) - \big(u_2^q(y) - u_1^q(y)\big)\big| \\
&\quad = \Big|\int_0^1 q\big(|u_\xi(x)|^{q-1}[u_2(x)-u_1(x)] - |u_\xi(y)|^{q-1}[u_2(y)-u_1(y)]\big)\,d\xi\Big| \\
&\quad \le \int_0^1 q\big(|u_\xi(x)|^{q-1}|(u_2-u_1)(x)-(u_2-u_1)(y)| \\
&\qquad\quad + \big||u_\xi(x)|^{q-1} - |u_\xi(y)|^{q-1}\big|\cdot|(u_2-u_1)(y)|\big)\,d\xi \\
&\quad \le C\big(\|u_2-u_1\|_{C^{0,\alpha}(\bar\Omega)}\cdot|x-y|^\alpha + |u_\xi(x)-u_\xi(y)|^{\hat q-1}\cdot\|u_2-u_1\|_{C(\bar\Omega)}\big) \\
&\quad \le C|u_2-u_1|_{2\theta,r}|x-y|^{(\hat q-1)\alpha}.
\end{aligned}$$

Hence we obtain

$$\begin{aligned}
|||u_2^q - u_1^q|||_{2\eta-1,r} &\le C|u_2-u_1|_{2\theta,r}\Big(\int_\Omega\int_\Omega |x-y|^{-N-r[2\eta-1-(\hat q-1)\alpha]}\,dx\,dy\Big) \\
&\le C|u_2-u_1|_{2\theta,r},
\end{aligned}$$

since (2.10) implies $2\eta - 1 < (\hat q-1)\alpha$ if α is close to $2\theta - \frac{N}{r}$. This shows (2.14) and concludes the proof of (ii).

(iii) Since $2\theta > 1 + \frac{N}{r}$, we have $H_r^{2\theta}(\Omega) \hookrightarrow C^{1,\alpha}(\bar\Omega)$ for $\alpha < 2\theta - 1 - \frac{N}{r}$. Moreover, $G_p : C^1(\bar\Omega) \to C^1(\bar\Omega)$ is continuous and bounded on bounded sets and $C^1(\bar\Omega) \hookrightarrow H_r^{2\delta}(\Omega)$, thus it suffices to show the continuity and boundedness of G_q.

As in the proof of (ii), we shall consider the operator $\tilde G_q : H_r^{2\theta}(\Omega) \to H_r^{1+\delta}(\Omega)$ instead of $G_q : B_r^{2\theta}(\Gamma) \to B_r^{1+\delta}(\Gamma)$. The norm $|u|_{1+\delta,r}$ is equivalent to $|u|_{1,r} + \sum_{i=1}^N |||\partial_i u|||_{\delta,r}$. The operator $\tilde G_q : C^1(\bar\Omega) \to C^1(\bar\Omega)$ is continuous (and bounded on bounded sets) and $C^1(\bar\Omega) \hookrightarrow H_r^1(\Omega)$, so that we need only to check

$$|||\partial_i u^q|||_{\delta,r} \le C|u|_{2\theta,r} \tag{2.15}$$

for any $i = 1, 2, \dots, N$ and

$$|||\partial_i(u_n^q - u^q)|||_{\delta,r} \to 0 \quad \text{if } u_n \to u \text{ in } H_r^{2\theta}(\Omega). \tag{2.16}$$

We have

$$\begin{aligned}
&|\partial_i u^q(x) - \partial_i u^q(y)| \\
&\quad = q\big|\partial_i u(x)[|u(x)|^{q-1} - |u(y)|^{q-1}] + \big(\partial_i u(x) - \partial_i u(y)\big)|u(y)|^{q-1}\big| \\
&\quad \le C\big(|u(x) - u(y)|^{\hat q-1} + |\partial_i u(x) - \partial_i u(y)|\big) \\
&\quad \le C\big(|x-y|^{\hat q-1} + |x-y|^{\alpha}\big),
\end{aligned}$$

which implies (2.15) provided $\delta < \min\{\alpha, \hat q - 1\}$. The continuity assertion (2.16) follows from the Lebesgue theorem. □

In what follows we denote $\dot J := J \setminus \{0\}$ and we write $E_\theta = E_\theta(r)$ in order to emphasize the dependence of E_θ on r. We set also $\hat q = \min\{q, 2\}$.

Theorem 2.2. *Let either $r = 2$ and p, q satisfy (1.4) or $r > N/(\hat q - 1)$.*

(i) *Let $u_o \in H^1_r(\Omega)$ and let $Z \in C(J, E_{1/2}(r))$ be the unique maximal solution corresponding to (1.1)–(1.3) and satisfying (2.8). Then $Z \in C(\dot J, E_1(R)) \cap C^1(\dot J, F_1(R))$ for any $R > 1$ and (2.8) holds in $E_1(R)$ for $t \in \dot J$. If, in addition, $u_o \in H^2_R(\Omega)$ then $Z \in C(J, E_1(R))$.*

(ii) *If $u_o \in H^{2+\varepsilon}_r(\Omega) \cap H^2_R(\Omega)$ for some $\varepsilon > 0$ and $R > N/(\hat q - 1)$ then $Z \in C^1(J, F_1(r)) \cap C^1(J, \mathbb{E}_\delta(r))$ for some $\delta > 0$. If, in addition, $\Delta u_o \ge 0$ in Ω and $|u_o|^{q-1}u_o - (u_o)_\nu \ge 0$ on $\partial\Omega$ then $u_t(t, u_o) \ge 0$ for any $t \in J$.*

(iii) *If $u_o, \tilde u_o \in H^1_r(\Omega)$ and $u_o \ge \tilde u_o$, $u_o \not\equiv \tilde u_o$, then $u(t, u_o) > u(t, \tilde u_o)$ for any $t \in \dot J(u_o) \cap \dot J(\tilde u_o)$.*

Proof. (i) If $u_o \in H^1_r(\Omega)$ then Lemma 2.1(i), (2.6), (2.8) and an obvious bootstrap argument imply $Z \in C(\dot J, E_\theta(r))$ for any $\theta < \frac12 + \frac{1}{2r}$. We have $E_\theta(r) \hookrightarrow E_{1/2}(\tilde r)$ provided $\tilde r < \frac{N}{N-1} r$ and θ is close to $\frac12 + \frac{1}{2r}$, so that we can repeat the above arguments with r replaced by $\tilde r$ (and u_o by $u(t, u_o)$ with $t > 0$). After finitely many steps we obtain $Z \in C(\dot J, E_\theta(R))$ for any $R \ge r$ and $\theta < \frac12 + \frac{1}{2R}$. Now Lemma 2.1(ii), (2.6) and (2.8) imply $Z \in C(\dot J, E_\theta(R))$ for $2\theta < \hat q$. Finally, Lemma 2.1(iii), (2.6) (2.8) and [A1, Theorem II.1.2.2] conclude the proof.

(ii) The C^1-continuity of Z is a consequence of [A1 Theorem II.1.2.2] and Lemma 2.1(iii). Since the first component w of $W := Z_t \in C(J, L_r(\Omega) \times L_r(\Gamma))$ is the unique weak solution of

$$\left.\begin{aligned}
w_t &= \Delta w + \lambda p|u|^{p-1}w, && x \in \Omega,\ t > 0, \\
w_t + w_\nu &= \mu q|u|^{q-1}w, && x \in \Gamma,\ t > 0, \\
w(x,0) &= \Delta u_o(x), && x \in \Omega, \\
w(x,0) &= |u_o|^{q-1}u_o(x) - (u_o)_\nu(x), && x \in \Gamma,
\end{aligned}\right\} \tag{2.17}$$

(i.e. W is the solution of the corresponding variation of constants formula in $L_r(\Omega) \times L_r(\Gamma)$), the assertion follows from [AE, Theorem 3.2].

(iii) The inequality $u(t, u_o) \ge u(t, \tilde u_o)$ follows from [AE, Theorem 3.2]. The corresponding strict inequality is a consequence of the Hopf maximum principle. (Recall that $u \in C^1(J, C^{1,\alpha}(\bar\Omega))$ and u is of class C^2 inside Ω.) □

Now assume that p, q satisfy (1.4), $u \in H_2^1(\Omega)$ and set

$$\mathcal{E}(u) = \frac{1}{2}\int_\Omega |\nabla u|^2\,dx - \frac{\lambda}{p+1}\int_\Omega |u|^{p+1}\,dx - \frac{\mu}{q+1}\int_{\partial\Omega} |u|^{q+1}\,dS. \tag{2.18}$$

If $u_o \in H_2^1(\Omega)$ then the corresponding solution Z fulfils $Z \in C^1(\dot{J}, F_1) \cap C(\dot{J}, E_1)$ due to Theorem 2.2(i). Let $t_1, t_2 \in \dot{J}$, $t_2 > t_1$. If $\tilde{Z} = \tilde{Z}_n$ is a piecewise linear (in t) approximation of Z on $[t_1, t_2]$ (i.e. $\tilde{Z}_n(\tau_k) = Z(\tau_k)$ for $\tau_k = t_1 + k\frac{t_2-t_1}{n}$, $k = 0, 1, \ldots, n$ and $\tilde{Z}_n$ is linear on $[\tau_{k-1}, \tau_k]$, $k = 1, 2, \ldots, n$) then the Mean Value Theorem and integration by parts imply

$$\begin{aligned}\mathcal{E}\big(\tilde{u}(t_2)\big) - \mathcal{E}\big(\tilde{u}(t_1)\big) = \int_{t_1}^{t_2} \Big[\int_\Omega (-\Delta\tilde{u})\tilde{u}_t\,dx + \int_{\partial\Omega} \tilde{u}_\nu \tilde{u}_t\,dS \\ - \lambda\int_\Omega |\tilde{u}|^{p-1}\tilde{u}\tilde{u}_t\,dx - \mu\int_{\partial\Omega} |\tilde{u}|^{q-1}\tilde{u}\tilde{u}_t\,dS\Big]\,dt.\end{aligned} \tag{2.19}$$

Since $\tilde{Z}_n \to Z$ in $C([t_1, t_2], E_1)$, $(\tilde{Z}_n)_t(t) \to Z_t(t)$ for a.e. $t \in [t_1, t_2]$ and $(\tilde{Z}_n)_t(t)$ is uniformly bounded in F_1, we see that (2.19) holds also for Z. Now using (1.1)–(1.3) we obtain the energy identity

$$\mathcal{E}\big(u(t_2)\big) - \mathcal{E}\big(u(t_1)\big) = -\int_{t_1}^{t_2} \Big[\int_\Omega u_t^2\,dx + \int_{\partial\Omega} u_t^2\,dS\Big]\,dt. \tag{2.20}$$

Letting $t_1 \to 0+$ we see that (2.20) holds for all $t_1, t_2 \in J$, $t_2 \ge t_1$.

3. Boundedness of global solutions

Throughout this section we assume $u_o \in H_2^1(\Omega)$. We call a solution global if it is defined for all $t \ge 0$. By the ω-limit set of a global solution we mean the set

$$\begin{aligned}\omega(u_o) = \{v \in H_2^1(\Omega) :\ &\text{there is } \{t_n\}_{n=1}^\infty,\ t_n \to \infty, \\ &\text{such that } u(\cdot, t_n; u_o) \to v \text{ in } H_2^1(\Omega)\}.\end{aligned}$$

Lemma 3.1. *If u is a global solution then there is a constant $K = K\big(\mathcal{E}(u_o)\big) > 0$ such that*

(i)

$$\int_\Omega u^2\,dx + \int_{\partial\Omega} u^2\,dS \le K,$$

(ii)

$$\liminf_{t\to\infty} \int_\Omega |\nabla u|^2\,dx \le K.$$

Proof. Multiplying (1.1) by u and integrating by parts we obtain

$$\begin{aligned}&\frac{1}{2}\frac{d}{dt}\Big(\int_\Omega u^2\,dx + \int_{\partial\Omega} u^2\,dS\Big) \\ &\qquad = -\int_\Omega |\nabla u|^2\,dx + \lambda\int_\Omega |u|^{p+1}\,dx + \mu\int_{\partial\Omega} |u|^{q+1}\,dS.\end{aligned} \tag{3.1}$$

Using the inequality $\mathcal{E}\big(u(\cdot,t)\big) \le \mathcal{E}(u_o)$ we have for every $\delta > 0$ that

$$\frac{1}{2}\frac{d}{dt}\Big(\int_\Omega u^2\,dx + \int_{\partial\Omega} u^2\,dS\Big) \ge -(2+\delta)\mathcal{E}(u_o) + \frac{\delta}{2}\int_\Omega |\nabla u|^2\,dx$$
$$+\Big(1-\frac{2+\delta}{p+1}\Big)\lambda\int_\Omega |u|^{p+1}\,dx + \Big(1-\frac{2+\delta}{q+1}\Big)\mu\int_{\partial\Omega}|u|^{q+1}\,dS.$$

If we now choose

$$\delta = \begin{cases} q-1 & \text{if } (\lambda,\mu)=(1,-1),\\ \frac{p-1}{2} & \text{if } (\lambda,\mu)=(1,0),\\ p-1 & \text{if } (\lambda,\mu)=(-1,1),\\ \frac{q-1}{2} & \text{if } (\lambda,\mu)=(0,1),\\ \frac{1}{2}(\min\{p,q\}-1) & \text{if } (\lambda,\mu)=(1,1),\end{cases}$$

then it is easy to see that there are constants $c_1, c_2 > 0$ such that

$$\frac{d}{dt}\Big(\int_\Omega u^2\,dx + \int_{\partial\Omega} u^2\,dS\Big) \ge c_1\Big(\int_\Omega u^2\,dx + \int_{\partial\Omega} u^2\,dS - c_2\Big).$$

Hence,

$$h(t) := \int_\Omega u^2\,dx + \int_{\partial\Omega} u^2\,dS - c_2$$

satisfies

$$h' \ge c_1 h$$

which means that $h \to \infty$ exponentially as $t \to \infty$, provided there is $t_o \ge 0$ such that $h(t_o) > 0$. Our next aim is to prove by contradiction that such t_o does not exist. To do that we shall use the classical concavity method introduced in [L] and used later by many authors, see [CL],[F1],[F2], for example.

Denote

$$M(t) := \int_0^t\int_\Omega u^2\,dx\,d\tau + \int_0^t\int_{\partial\Omega} u^2\,dS\,d\tau.$$

Using the energy identity (2.20) we obtain from (3.1) that

$$\frac{1}{2}M'' \ge (2+\delta)\Big(\int_0^t\int_\Omega u_t^2\,dx\,d\tau + \int_0^t\int_{\partial\Omega} u_t^2\,dS\,d\tau - \mathcal{E}(u_o) + \varepsilon M' - c_3\Big),$$

where δ is as before and ε, c_3 are positive constants. If $M' = h + c_2 \to \infty$ as $t \to \infty$ then there is $t_1 \ge 0$ such that

$$M'' \ge 2(2+\delta)\Big(\int_0^t\int_\Omega u_t^2\,dx\,d\tau + \int_0^t\int_{\partial\Omega} u_t^2\,dS\,d\tau\Big) \quad \text{for } t \ge t_1.$$

Multiplying the last inequality by M and using the Cauchy-Schwarz inequality we obtain

$$M(t)M''(t) \ge \Big(1+\frac{\delta}{2}\Big)\big(M'(t)-M'(0)\big)^2 \ge \Big(1+\frac{\delta}{4}\Big)\big(M'(t)\big)^2 \quad \text{for } t \ge t_2,$$

where $t_2 \ge t_1$ is sufficiently large. This means that the function $M^{-\delta/4}$ is concave for $t \ge t_2$. Since $M' \to \infty$ as $t \to \infty$, we have that $M^{-\delta/4} \to 0$ as $t \to \infty$.

Obviously, $M^{-\delta/4} > 0$, and we arrive at a contradiction. Hence, $h(t) \leq 0$ for $t \geq 0$ and (i) holds.

To prove (ii) we use a similar contradiction argument as before. With the same choice of M and δ as before we obtain from (3.1) that

$$\begin{aligned}\frac{1}{2}M'' \geq (2+\delta)\Big(\int_0^t\int_\Omega u_t^2\,dx\,d\tau + \int_0^t\int_{\partial\Omega} u_t^2\,dS\,d\tau - \mathcal{E}(u_o)\Big)\\ + \frac{\delta}{2}\int_\Omega |\nabla u|^2\,dx.\end{aligned} \tag{3.2}$$

If

$$\liminf_{t\to\infty}\int_\Omega |\nabla u|^2\,dx = k > \Big(2+\frac{4}{\delta}\Big)\mathcal{E}(u_o),$$

then there is $t_3 \geq 0$ such that

$$M'' \geq \delta k - 2(2+\delta)\mathcal{E}(u_o) > 0 \quad \text{for } t \geq t_3.$$

This means that $M'(t) \to \infty$ as $t \to \infty$ and the concavity of $M^{-\delta/4}$ yields a contradiction as before. □

Proposition 3.2. *If $\mathcal{E}(u_o) < 0$ then u blows up in finite time.*

Proof. If we choose M and δ as in the proof of the previous lemma then we obtain from (3.2) that

$$M'' > -2(2+\delta)\mathcal{E}(u_o)$$

and the concavity argument gives again a contradiction with global existence. □

Remark 3.3. It is easy to see that under our assumptions on p, q, λ and μ there are initial data u_o for which $\mathcal{E}(u_o) < 0$. If $\min\{\lambda,\mu\} \geq 0$ then one can take arbitrary non-zero constants and if $\min\{\lambda,\mu\} = -1$ then one can take constants that are sufficiently large.

Lemma 3.4. *Assume (1.4). If u is a global solution and*

$$\limsup_{t\to\infty}|u(t;u_o)|_{1,2} = \infty,$$

$$\liminf_{t\to\infty}|u(t;u_o)|_{1,2} < \infty,$$

then for every constant B large enough there is a stationary solution $w \in \omega(u_o)$ such that $|w|_{1,2} = B$.

Proof. The proof is almost the same as those of Lemma 2 in [F1] or Lemma 2.2 in [F2]. □

Theorem 3.5. *Assume* (1.4). *If u is a global solution then*

(i)
$$\sup_{t>0} |u(t;u_o)|_{1,2} < \infty,$$

(ii)
$$\sup_{t>t_o} |u(t;u_o)|_\infty < \infty.$$

for any $t_o > 0$.

Proof. Suppose
$$\limsup_{t\to\infty} |u(t,u_o)|_{1,2} = \infty.$$
Lemma 3.1(ii) and Lemma 3.4 imply that for every B large enough there is a stationary solution $w \in \omega(u_o)$ such that $|w|_{1,2} = B$. Choosing δ as in the proof of Lemma 3.1 and using the fact that
$$\int_\Omega |\nabla w|^2\,dx = \lambda \int_\Omega |w|^{p+1}\,dx + \mu \int_{\partial\Omega} |w|^{q+1}\,dS,$$
we obtain
$$\begin{aligned}(2+\delta)\mathcal{E}(u_o) &\ge (2+\delta)\mathcal{E}(w)\\ &= \frac{\delta}{2}\int_\Omega |\nabla w|^2\,dx + \Big(1-\frac{2+\delta}{p+1}\Big)\lambda\int_\Omega |u|^{p+1}\,dx\\ &\quad + \Big(1-\frac{2+\delta}{q+1}\Big)\mu\int_{\partial\Omega}|u|^{q+1}\,dS\\ &\ge \frac{\delta}{2}\int_\Omega |\nabla w|^2\,dx.\end{aligned}$$
This and Lemma 3.1(i) yield a contradiction with $|w|_{1,2} = B$.

To see that (i) implies (ii) one can use the variation of constants formula and standard bootstrap arguments. □

In the next theorem we give an *a priori* bound for global solutions under a stronger assumption on p and q.

Theorem 3.6. *Assume* (1.5). *If u is a global solution then there is a constant $K = K(|u_o|_{1,2}) > 0$ such that*
$$\sup_{t>0} |u(t,u_o)|_{1,2} \le K.$$

Proof. The proof is a straightforward modification of the proof of Theorem 4.1 in [FQ2]. In fact, the estimates following (3.1) and Lemma 3.1(i) imply
$$\begin{aligned}\int_\Omega |\nabla u|^2\,dx &\le K_1\Big(1+\int_\Omega uu_t\,dx + \int_{\partial\Omega} uu_t\,dS\Big)\\ &\le K_2\Big(1+\Big(\int_\Omega u_t^2\,dx\Big)^{1/2} + \Big(\int_{\partial\Omega} u_t^2\,dS\Big)^{1/2}\Big).\end{aligned}$$

From (2.20) it now follows

$$\int_t^{t+1}\Big(\int_\Omega |\nabla u|^2\,dx\Big)^2\,dt \le K_3,$$

which implies uniform bounds for $u \in L_4(t,t+1;L_{2^*}(\Omega))$ and $\gamma u \in L_4(t,t+1;L_{2^{**}}(\Gamma))$ (where $2^* = \frac{2N}{N-2}$ and $2^{**} = \frac{2(N-1)}{N-2}$ if $N > 2$). Since we have also $u_t \in L_2(0,\infty;L_2(\Omega))$ and $\gamma u_t \in L_2(0,\infty;L_2(\Gamma))$, the interpolation estimates in [CL, Appendix] yield uniform bounds for $u \in L_\infty(0,\infty;L_z(\Omega))$ and $\gamma u \in L_\infty(0,\infty;L_t(\Gamma))$ provided $z < \frac{2N}{N-4/3}$ and $t < \frac{2(N-1)}{N-5/3}$. Repeating the arguments from the proofs of [Q3, Theorem and Remark 2] we obtain uniform bounds in $H_2^1(\Omega)$. □

4. Stationary solutions

In this section we shall use dynamical methods in order to show the existence of sign-changing solutions of (1.6)–(1.7) if $1 < p < q$ and $(N-\frac{5}{3})q < N+\frac{1}{3}$. Moreover, using ODE methods we show that the problem (1.6)–(1.7) admits exactly two sign-changing solutions $u_1 = -u_2$ provided $N = 1$ and $1 < p < 2q-1$.

Proposition 4.1. *Let $1 < p < q$ and $(N-\frac{5}{3})q < N+\frac{1}{3}$. Then (1.6)–(1.7) admits a sign-changing solution.*

Proof. The proof is a modification of the proof of Theorem 5.1 in [FQ2] (see also [Q2]). Denote by W^+ and W^- the set of all nonnegative and nonpositive nontrivial solutions of (1.6)–(1.7), respectively. The maximum principle implies that these solutions are positive and negative, respectively.

First we show that the zero solution of (1.1)–(1.3) is asymptotically stable in $H_2^1(\Omega)$. We have $\mathcal{E}(u) \ge C_1|u|_{1,2}^{p+1} - C_2|u|_{1,2}^{q+1} \ge \varepsilon(\rho) > 0$ for $|u|_{1,2} = \rho$ small and $\mathcal{E}(u) < \varepsilon(\rho)$ for $|u|_{1,2} < \delta(\rho)$, where $\delta(\rho)$ is small enough. Consequently, if $|u_o| < \delta(\rho)$ then the corresponding solution stays in the set $B_\rho(0) := \{u : |u|_{1,2} < \rho\}$ and its ω-limit set $\omega(u_o)$ consists of equilibria contained in $B_\rho(0)$. Hence $u(t,u_o) \to 0$ provided we show that 0 is an isolated equilibrium. Assume the contrary, i.e. there exist $u_n \in H_2^1(\Omega)$, $0 \not\equiv u_n \to 0$. Then $u_n = c_n + \tilde{u}_n$ where c_n are constant functions and $|\tilde{u}_n|_{1,2} = o(c_n)$ due to $0 = \mathcal{E}'(u_n)u_n = \int_\Omega |\nabla u_n|^2\,dx + o(|u_n|_{1,2})$. Hence

$$0 = \mathcal{E}'(u_n)u_n \ge \int_\Omega |u|^{p+1}\,dx - \int_{\partial\Omega} |u|^{q+1}\,dS = |c_n|^{p+1} - |c_n|^{q+1} + o(|c_n|^{p+1}),$$

which gives a contradiction.

Consequently, the domain of attraction of the zero solution,

$$D_A := \{u_o \in H_2^1(\Omega)\,;\; u(t,u_o) \text{ is global and } u(t,u_o) \to 0 \text{ as } t \to +\infty\}$$

is an open subset of $H_2^1(\Omega)$ and $\mathcal{E}(u) \ge 0$ if $u \in \overline{D_A}$. Moreover, the set $D_A \cap X$ is bounded for any finite dimensional linear subspace X of $H_2^1(\Omega)$ since $\mathcal{E}(u) < 0$ for $u \in X$ large enough. Choose any two-dimensional subspace $Y \subset C^2(\bar{\Omega})$ containing a positive function. Denote by D^+ or D^- the set of all initial values u_o in $\partial D_A \cap X$ such that their ω-limit set is contained in W^+ or W^-, respectively. The maximum principle and the continuous dependence of solutions of (1.1)–(1.3) in $L_\infty(\Omega)$ imply that the sets D^+,D^- are nonempty disjoint relatively open subsets of the compact set $\partial D_A \cap X$. If we assume that there are no sign-changing solutions for (1.6), (1.7) then we obtain a decomposition $\partial D_A = D^+ \cup D^-$ and an easy topological degree argument (see [Q2, Appendix]) leads to a contradiction. □

Proposition 4.2. *If $1 < p < 2q - 1$ then the problem*

$$\left.\begin{aligned} u'' &= |u|^{p-1}u, \qquad && |x| < L \\ u' &= \pm|u|^{q-1}u, \qquad && x = \pm L \end{aligned}\right\} \tag{4.1}$$

has for any $L > 0$ exactly two solutions $u_1 = -u_2$ that change sign.

Proof. Consider the initial value problem

$$\begin{aligned} u'' &= |u|^{p-1}u, \\ u(0) &= 0, \\ u'(0) &= m > 0. \end{aligned}$$

Obviously, the solution is positive for $x > 0$ as long as it exists. Let $L(m)$ denote any positive number for which

$$u'\big(L(m)\big) = u^q\big(L(m)\big).$$

We shall show that $L(m)$ is uniquely defined for all $m > 0$, it is a decreasing function, and

$$\lim_{m\to 0+} L(m) = \infty, \qquad \lim_{m\to\infty} L(m) = 0. \tag{4.2}$$

Let $R(m)$ denote the unique positive root of the equation

$$x^{2q} - \frac{2}{p+1}x^{p+1} = m^2.$$

Then $L(m)$ is given explicitly by the formula

$$L(m) = \int_0^{R(m)} \frac{ds}{\sqrt{\frac{2}{p+1}s^{p+1} + m^2}},$$

see [CV]. From the proof of Lemma 2.4 in [CV] it follows that

$$\Big(1 - \frac{2}{p+1}\Big)L(m) + mL'(m) = \Big(1 - \frac{2q}{p+1}\Big)\big(qR^{q-1}(m) - R^{p-q}(m)\big)^{-1}.$$

Since

$$qR^{q-1} - R^{p-q} > \frac{p+1}{2}R^{q-1} - R^{p-q} = \frac{(p+1)m^2}{2R^{q+1}} > 0,$$

we obtain that $L'(m) < 0$ for $m > 0$.

It is clear that

$$R(m) > q^{\frac{1}{p+1-2q}} =: Q,$$

therefore

$$L(m) > \int_0^Q \frac{ds}{\sqrt{\frac{2}{p+1}s^{p+1} + m^2}} =: I(m),$$

and $I(m) \to \infty$ as $m \to 0+$. On the other hand,

$$L(m) \le \int_0^\infty \frac{ds}{\sqrt{\frac{2}{p+1}s^{p+1} + m^2}} = m^{\frac{2}{p+1}-1} \int_0^\infty \frac{dz}{\sqrt{\frac{2}{p+1}z^{p+1} + 1}},$$

hence $L(m) \to 0$ as $m \to \infty$.

Since $L(m)$ is a decreasing function of m, every solution of (4.1) is odd. From (4.2) we obtain that the problem

$$\left.\begin{aligned} u'' &= u^p, \qquad x \in (0, L) \\ u(0) &= 0, \\ u'(L) &= u^q(L) \end{aligned}\right\} \tag{4.3}$$

has a positive solution for every L. The uniqueness of this solution follows again from the fact that $L(m)$ is decreasing. We remark that the existence of a unique positive solution of (4.3) for $1 < p < 2q - 1$ was shown in [CV] using a different argument. □

5. Blow-up rate and blow-up set

In this section we study blow-up of solutions of Problems (1.11)–(1.13) and (1.16)–(1.18).

Theorem 5.1. *Let u be the solution of (1.16)–(1.18), where $q > 1$, $u_o \in H_2^1(\Omega)$, $u_o \ge 0$, $u_o \not\equiv 0$. Then:*

(i) *u blows up in a finite time T.*

(ii) *If*

$$u_o \in C^{2+\varepsilon}(\bar\Omega), \quad \Delta u_o \ge 0 \ \text{ in } \Omega \quad \text{and} \quad u_o^q \ge (u_o)_\nu \ \text{ on } \partial\Omega, \tag{5.1}$$

then there is a constant $c > 0$ such that

$$\max_{x \in \bar\Omega} u(x,t) \ge c(T-t)^{-\frac{1}{q-1}}. \tag{5.2}$$

(iii) *If $(N-2)q \le N$ and*

$$\left.\begin{aligned} &u_o \in C^{4+\varepsilon}(\bar\Omega), \quad \Delta u_o = u_o^q - (u_o)_\nu \ \text{ on } \partial\Omega, \\ &\Delta u_o, \Delta^2 u_o \ge 0 \ \text{ in } \Omega \quad \text{and} \quad (\Delta u_o)^q \ge (\Delta u_o)_\nu \ \text{ on } \partial\Omega, \end{aligned}\right\} \tag{5.3}$$

then there is a constant $C > 0$ such that

$$\max_{x \in \bar\Omega} u(x,t) \le C(T-t)^{-\frac{1}{q-1}}. \tag{5.4}$$

Moreover, u blows up only on $\partial\Omega$.

Proof. (i) If $(N-2)q < N$ then this is a consequence of Theorem 3.5. Namely, if u was global then it would be bounded and its ω-limit set would consist of nonnegative equilibria. Since the only nonnegative equilibrium is zero and this is unstable from above, we obtain a contradiction. If $(N-2)q \geq N$, then a simple comparison argument yields the result. More precisely, we modify u^q for large values of u so that the growth at infinity becomes subcritical (but superlinear) and we observe that the proof of Theorem 3.5 works for the problem:

$$\begin{aligned} u_t &= \Delta u, && x \in \Omega,\ t > 0, \\ u_t + u_\nu &= f(u), && x \in \partial\Omega,\ t > 0, \\ u(x,0) &= u_o(x), && x \in \Omega, \end{aligned}$$

if f is sufficiently smooth and

$$\begin{aligned} &|f(u)| \leq k(|u|^q + 1) && \text{for some } k > 0 \text{ and } (N-2)q < N, \\ &uf(u) \geq (2+\varepsilon) \int_0^u f(v)\, dv && \text{for } u \text{ large enough and some } \varepsilon > 0. \end{aligned}$$

(ii) The assumption (5.1) and Theorem 2.2(ii) guarantee that $u_t = \Delta u \geq 0$. Let

$$U(t) := \max_{0 \leq \tau \leq t} \max_{x \in \bar{\Omega}} u(x,\tau). \tag{5.5}$$

Since $u(\cdot,t)$ attains its maximum on $\partial\Omega$, we have

$$U(t) = \max_{x \in \partial\Omega} u(x,t).$$

Let $t_2 > t_1$ and let $x_2, x_1 \in \partial\Omega$ be points for which $U(t_i) = u(x_i,t_i)$, $i = 1,2$. As in [FML, Proof of Theorem 4.5] one can see that

$$\frac{U(t_2) - U(t_1)}{t_2 - t_1} \leq u_t(x_2,t_2) + o(1) \qquad \text{if } t_1 \to t_2.$$

Since $u_\nu(x_2,t_2) \geq 0$, we obtain

$$u_t(x_2,t_2) \leq u^q(x_2,t_2) = U^q(t_2),$$

and

$$U'(t) \leq U^q(t)$$

at any point of differentiability. The result follows by integration.

(iii) The proof is similar to the proof of Theorem 2.1 in [H2]. Anyhow, we have to use a different scaling. In the first part of the proof, we will not use the assumption (5.3) (cf. Remark 5.2(i)).

As in the proof of (ii), we define $U(t)$ by (5.5). Since u attains its maximum on the parabolic boundary and $U(t) \to \infty$ as $t \to T$, we have that $U(t^*) = u(\hat{x},\hat{t})$ for some $(\hat{x},\hat{t}) \in \partial\Omega \times (0,t^*]$ provided $t^* \in (\frac{T}{2},T)$ is such that $U(t^*) > \max_{x\in\bar{\Omega}} u_o(x)$.

Define (cf. [GS], [H2] and [HY])

$$\varphi^\lambda(y,s) = \frac{1}{U(t^*)} u(\lambda R y + \hat{x}, \lambda s + \hat{t})$$

for $y \in \overline{\Omega_\lambda}$, $-\hat{t}/\lambda \le s < (T-\hat{t})/\lambda$, where $\Omega_\lambda = \{y : \lambda R y + \hat{x} \in \Omega\}$, R is a rotation operator such that $(-1, 0, \dots, 0)$ is the outer normal vector of $\partial\Omega_\lambda$ at 0 and

$$\lambda = \big(U(t^*)\big)^{1-q}.$$

Then φ^λ satisfies

$$\left.\begin{aligned}
\lambda\varphi_s^\lambda &= \Delta\varphi^\lambda, && y \in \Omega_\lambda,\ -\frac{\hat{t}}{\lambda} < s < \frac{T-\hat{t}}{\lambda},\\
\varphi_s^\lambda + \varphi_\nu^\lambda &= (\varphi^\lambda)^q, && y \in \partial\Omega_\lambda,\ -\frac{\hat{t}}{\lambda} < s < \frac{T-\hat{t}}{\lambda},\\
\varphi^\lambda(0,0) &= 1, && \\
0 \le \varphi^\lambda &\le 1, && y \in \Omega_\lambda,\ -\frac{\hat{t}}{\lambda} \le s \le 0.
\end{aligned}\right\} \tag{5.6}$$

Since φ^λ blows up when $s = (T-\hat{t})/\lambda$, there exists $s_\lambda > 0$ such that

$$\left.\begin{aligned}
&\max_{y\in\overline{\Omega_\lambda}} \varphi^\lambda(y,s) < 2, \qquad -\infty < s < s_\lambda,\\
&\max_{y\in\overline{\Omega_\lambda}} \varphi^\lambda(y,s_\lambda) = 2.
\end{aligned}\right\} \tag{5.7}$$

We show that there is a constant K such that

$$s_\lambda \le K \qquad \text{for } 0 < \lambda \ll 1. \tag{5.8}$$

Suppose the contrary, that is $s_{\lambda_j} \to \infty$ for some sequence $\{\lambda_j\}$, $\lambda_j \to 0$. Due to Theorem 2.2(ii) and the corresponding analogue for the linear problem (2.17), the assumption (5.3) guarantees that $u_t \ge 0$ and $u_{tt} \ge 0$, hence $\varphi_s^\lambda \ge 0$ and $\varphi_{ss}^\lambda \ge 0$.

We shall look at the problem (5.6) for $s = 0$ as an elliptic problem with $\lambda\varphi_s^\lambda$ and φ_s^λ considered as data. Since

$$\varphi^\lambda(y,s) = \varphi^\lambda(y,0) + \int_0^s \varphi_s^\lambda(y,\sigma)\,d\sigma \ge \varphi^\lambda(y,0) + s\varphi_s^\lambda(y,0),$$

the boundedness of φ^λ for $s \le s_\lambda$ and $s_{\lambda_j} \to \infty$ imply that $\varphi_s^{\lambda_j}(y,0) \to 0$ uniformly in y. Consequently, after straightening $\partial\Omega$ in a neighbourhood of $\hat{x} = \hat{x}(\lambda)$ and using appropriate cut-off functions as in [AQ, the case $x_\infty \in \Gamma_1$ in the proof of Lemma 13.2 and Theorem 3.1] one obtains local *a priori* estimates for $\varphi^{\lambda_j}(\cdot,0)$ in $C^{0,\alpha}$. These estimates enable us to pass to the limit with the sequence $\varphi^{\lambda_j}(\cdot,0)$. In the limit, we obtain a $C^{0,\alpha}$-function φ^∞ defined on $\mathbb{R}_+^N := \{y \in \mathbb{R}^N : y_1 \ge 0\}$ which is a solution of the problem

$$\left.\begin{aligned}
\Delta\varphi &= 0, && y_1 > 0,\\
\varphi_\nu &= \varphi^q, && y_1 = 0,\\
\varphi(0) &= 1, && \\
0 \le \varphi &\le 1, && y_1 \ge 0.
\end{aligned}\right\} \tag{5.9}$$

Anyhow, this a contradiction to the nonexistence results in [H1] provided $(N-2)q < N$. If $q(N-2) = N$ then there exists a unique solution of (5.9) and it is given explicitly by the formula

$$\varphi(y) = \Big(\frac{N-2}{|y-a_N|}\Big)^{N-2}, \qquad a_N = (2-N, 0, \dots, 0),$$

see [CSF] or [LZ]. Since $\varphi^{\lambda_j}(0,0) = 1$ and $\varphi^{\lambda_j}(0, -\hat{t}/\lambda_j) \to 0$ as $j \to \infty$, there exist $s_j < 0$ such that $\varphi^{\lambda_j}(0, s_j) = \frac{1}{2}$. Now passing to the limit with the sequence of functions $\varphi^{\lambda_j}(\cdot, s_j)$ we obtain (similarly as in the case of functions $\varphi^{\lambda_j}(\cdot, 0)$) a limit function $\tilde{\varphi}$ which satisfies (5.9) with the condition $\varphi(0) = 1$ replaced by $\tilde{\varphi}(0) = \frac{1}{2}$. Moreover, $\tilde{\varphi} \le \varphi$, and this is a contradiction to [H2, Lemma 3.2]. Consequently, (5.8) holds.

For $t_1 = \lambda s_\lambda + \hat{t} < T$ we obtain from (5.7) and (5.8) that

$$U(t_1) = 2U(t^*),$$
$$U^{q-1}(t^*)(t_1 - t^*) \le U^{q-1}(t^*)(t_1 - \hat{t}) = s_\lambda \le K.$$

Using t_1 instead of t^* we obtain $t_2 < T$ such that

$$U(t_2) = 2U(t_1),$$
$$U^{q-1}(t_1)(t_2 - t_1) \le K,$$

and by induction we construct a sequence $\{t_n\}$, $t_n \to T$, that satisfies the assumptions of Lemma 3.1 in [H2]. Hence

$$U(t^*) \le C(T - t^*)^{-\frac{1}{q-1}}.$$

The fact that u blows up only on the boundary $\partial\Omega$ is an immediate consequence of (5.4) and Theorem 4.1 in [HY]. □

Remark 5.2. (i) The assumption (5.3) in Theorem 5.1(iii) was used only in passing to the limit in the problem (5.6) for $s = 0$. If this assumption fails to hold then one can still expect that (assuming the contrary to (5.8)) the solutions φ^λ of (5.6) converge to the solution of

$$\begin{aligned} \Delta\varphi &= 0, && y_1 > 0,\ s \in \mathbb{R}, \\ \varphi_s + \varphi_\nu &= \varphi^q, && y_1 = 0,\ s \in \mathbb{R}, \\ \varphi(0,0) &= 1, && \\ 0 \le \varphi &\le 2, && y_1 \ge 0,\ s \in \mathbb{R}. \end{aligned}$$

Anyhow, this contradicts the nonexistence result in [AF] provided $(N-1)q \le N$.

(ii) A similar approach as in the proof of Theorem 5.1 can be used also for the problem (1.11)–(1.13) in order to show that the blow-up rate (1.15) holds also for (some) spatially nonhomogeneous solutions if $N > 1$, $(N-2)q \le N$. However, one cannot expect that blow-up will take place only on the boundary; see the next proposition.

Proposition 5.3. *Let u be the solution of (1.11)–(1.13), where $q > 1$, $N = 1$, $\Omega = (0,1)$, $u_o(1) \geq u_o(0) \geq 0$, $u_o(1) > 0$. Then u blows up in a finite time T, (1.15) holds and u blows up everywhere in $(0,1]$.*

If $q \leq 2$ then u blows up also for $x = 0$. If $q > 2$ and $u_o(1) \gg 1 \gg u_o(0)$ then $u(0,t)$ stays bounded as $t \to T$.

Proof. We deal with the problem

$$\begin{aligned} u_{xx} &= 0, && x \in (0,1),\ t > 0,\\ u_t + u_x &= u^q, && x = 1,\ t > 0,\\ u_t - u_x &= u^q, && x = 0,\ t > 0,\\ u(0,1) &= u_o(1), &&\\ u(0,0) &= u_o(0). && \end{aligned}$$

First assume

$$u_o(0) < u_o(1).$$

The solution is of the form

$$u(x,t) = \varphi(t) + \psi(t)x, \tag{5.10}$$

where

$$\begin{aligned} \varphi' &= \varphi^q + \psi, && (5.11)\\ \psi' &= -2\psi + (\varphi + \psi)^q - \varphi^q, && (5.12)\\ \varphi(0) &= u_o(0), &&\\ \psi(0) &= u_o(1) - u_o(0) > 0. && \end{aligned}$$

Since the set $\{(\varphi,\psi) \in \mathbb{R}^2 : \varphi \geq 0,\ \psi > 0\}$ is positively invariant for system (5.11)–(5.12), we have that $\psi(t)$ is positive as long as it exists. It follows from (5.11) that u blows up in a finite time T and (5.10) implies that the blow-up set contains the interval $(0,1]$. Setting $\eta(t) = \varphi(t) + \psi(t) = \max_{x\in[0,1]} u(x,t)$, (5.11) and (5.12) imply $\eta^q > \eta' \geq \eta^q - \eta \geq \frac{1}{2}\eta^q$ if t is close to T, hence (1.15) holds.

If $q \leq 2$ and φ was bounded for $t \to T$, then (1.15) would imply $\psi(t) \geq c(T-t)^{-\frac{1}{q-1}}$ for some $c > 0$ and (5.11) would yield

$$\varphi(T) - \varphi(t) \geq \int_t^T \psi(\tau)\,d\tau = \infty,$$

which is a contradiction. Hence $\varphi(t) = \min_{x\in[0,1]} u(x,t)$ blows up as $t \to T$.

Let $q > 2$ and $u_o(1) \gg 1 \gg u_o(0)$. From (5.12) it follows

$$\psi' \geq \psi^q - 2\psi \geq \frac{1}{2}\psi^q, \tag{5.13}$$

due to $\psi(0) = u_o(1) - u_o(0) \gg 1$. Integrating (5.13) we obtain

$$T \le \frac{2}{q-1}\psi^{1-q}(0) \ll 1, \tag{5.14}$$

$$\psi(t) \le c_q(T-t)^{-\frac{1}{q-1}}, \tag{5.15}$$

where $c_q^{q-1} = \frac{2}{q-1}$. Set $t_1 := \sup\{t \in [0,T) : \varphi(t) \le 1\}$. Then (5.11) and (5.15) imply

$$\begin{aligned}\varphi(t_1) &= \varphi(0) + \int_0^{t_1} (\varphi^q + \psi)\, d\tau \\ &\le u_o(0) + \int_0^{t_1} \left(1 + c_q(T-\tau)^{-\frac{1}{q-1}}\right) d\tau \\ &\le u_o(0) + T + c_q \int_0^T (T-\tau)^{-\frac{1}{q-1}}, d\tau \\ &= u_o(0) + T + \frac{q-1}{q-2} c_q T^{\frac{q-2}{q-1}} \ll 1,\end{aligned}$$

due to $u_o(0) \ll 1$ and (5.14). Consequently, $t_1 = T$ and $\varphi(t) \le 1$ for any $t \in [0,T)$.

If

$$u_o(0) = u_o(1) > 0,$$

then $u(x,t) = \varphi(t)$ and this case is trivial. □

References

[A1] H. Amann, *Linear and Quasilinear Parabolic Problems*, Volume I: Abstract Linear Theory, Birkhäuser, Basel, 1995.

[A2] H. Amann, "Semigroups and nonlinear evolution equations", Linear Algebra Appl. **84** (1986), 3–32.

[A3] H. Amann, "Parabolic evolution equations and nonlinear boundary conditions", J. Differ. Equations **72** (1988), 201–269.

[AE] H. Amann and J. Escher, "Strongly continuous dual semigroups", Ann. Mat. Pura Appl. **CLXXI** (1996), 41–62.

[AF] H. Amann and M. Fila, "A Fujita-type theorem for the Laplace equation with a dynamical boundary condition", Acta Math. Univ. Comenianae **66** (1997), 321–328.

[AQ] H. Amann and P. Quittner, "Elliptic boundary value problems involving measures: Existence, regularity, and multiplicity", Preprint 1998.

[BS] G. I. Bizhanova and V. A. Solonnikov, "*О разрешимости начально-краевой задачи для параболического уравнения второго порядка с производной по времени в граничном условии в весовом гёльдеровском пространстве функций*", Алгебра и анализ **5** (1993), 109–142.

[CL] T. Cazenave and P.-L. Lions, "Solutions globales d'équations de la chaleur semi linéaires", Comm. Partial Differential Equations **9** (1984), 955–978.

[CFQ] M. Chipot, M. Fila and P. Quittner, "Stationary solutions, blow up and convergence to stationary solutions for semilinear parabolic equations with nonlinear boundary conditions", Acta Math. Univ. Comenianae **60** (1991), 35–103.

[CSF] M. Chipot, I. Shafrir and M. Fila, "On the solutions to some elliptic equations with nonlinear Neumann boundary conditions", Adv. Differ. Equations **1** (1996), 91–110.

[CV] M. Chipot and F. Voirol, "On a class of nonlinear elliptic problems with Neumann boundary conditions growing like a power", Z. Anal. Anwendungen **14** (1995), 853–868.

[DFL] K. Deng, M. Fila and H. A. Levine, "On critical exponents for a system of heat equations coupled in the boundary conditions", Acta Math. Univ. Comenianae **63** (1994), 169–192.

[E1] J. Escher, "Nonlinear elliptic systems with dynamic boundary conditions", Math. Z. **210** (1992), 413–439.

[E2] J. Escher, "Quasilinear parabolic systems with dynamical boundary conditions", Comm. Partial Differential Equations **18** (1993), 1309–1364.

[E3] J. Escher, "On quasilinear fully parabolic boundary value problems", Diff. Integral Equations **7** (1994), 1325–1343.

[E4] J. Escher, "On the qualitative behaviour of some semilinear parabolic problems", Diff. Integral Equations **8** (1995), 247–267.

[E5] J. Escher, "Global existence and nonexistence for semilinear parabolic systems with nonlinear boundary conditions", Math. Ann. **284** (1989), 285–305.

[F1] M. Fila, "Boundedness of global solutions for the heat equation with nonlinear boundary conditions", Comm. Math. Univ. Carolinae **30** (1989), 479–484.

[F2] M. Fila, "Boundedness of global solutions of nonlinear diffusion equations", J. Differ. Equations **98** (1992), 226–240.

[FF] M. Fila and J. Filo, "Blow-up on the boundary: a survey", Banach Center Publ. **33** (1996), 67–78.

[FQ1] M. Fila and P. Quittner, "The blow-up rate for the heat equation with a nonlinear boundary condition", Math. Meth. Appl. Sci. **14** (1991), 197–205.

[FQ2] M. Fila and P. Quittner, "Global solutions of the Laplace equation with a nonlinear dynamical boundary condition", Math. Meth. Appl. Sci. **20** (1997), 1325–1333.

[FML] A. Friedman and B. McLeod, "Blow up of positive solutions of semilinear heat equations", Indiana Univ. Math. J. **34** (1985), 425–447.

[GS] B. Gidas, J. Spruck, "A priori bounds for positive solutions of nonlinear elliptic equations", Comm. Partial Differential Equations **6** (1981), 883–901.

[H1] B. Hu, "Nonexistence of a positive solution of the Laplace equation with a nonlinear boundary condition", Diff. Integral Equations **7** (1994), 301–313.

[H2] B. Hu, "Remarks on the blowup estimate for solutions of the heat equation with a nonlinear boundary condition", Diff. Integral Equations **9** (1996), 891–901.

[HY] B. Hu and H.-M. Yin, "The profile near blowup time for solution of the heat equation with a nonlinear boundary condition", Trans. Am. Math. Soc. **346** (1994), 117–135.

[L] H. A. Levine, "Some nonexistence and instability theorems for solutions of formally parabolic equations of the form $Pu_t = -Au + F(u)$", Arch. Rat. Mech. Anal. **51** (1973), 371–386.

[LP] H. A. Levine and L. E. Payne, "Nonexistence theorems for the heat equation with nonlinear boundary conditions and for the porous medium equation backward in time", J. Diff. Equations **16** (1974), 319–334.

[LZ] Y. Li and M. Zhu, "Uniqueness theorems through the method of moving planes", Duke Math. J. **80** (1995), 383–417.

[Q1] P. Quittner, "On global existence and stationary solutions for two classes of semilinear parabolic problems", Comment. Math. Univ. Carolinae **34** (1993), 105–124.

[Q2] P. Quittner, "Signed solutions for a semilinear elliptic problem", Differ. Integral Equations, to appear.

[Q3] P. Quittner, "Global existence of solutions of parabolic problems with nonlinear boundary conditions", Banach Center Publ. **33** (1996), 309–314.

[T] H. Triebel, *Interpolation Theory, Function Spaces, Differential Operators*, North-Holland, Amsterdam, 1978.

Marek Fila and Pavol Quittner
Institute of Applied Mathematics
Comenius University
Mlynská dolina
84215 Bratislava
Slovakia

Progress in Nonlinear Differential Equations
and Their Applications, Vol. 35
© 1999 Birkhäuser Verlag Basel/Switzerland

Functional Properties of the Navier-Stokes Operator and Bifurcation of Stationary Solutions: Planar Exterior Domains

Giovanni P. Galdi and Patrick J. Rabier

1. Introduction

The well-posedness of the stationary Navier-Stokes problem has been thoroughly investigated in the work of Leray [17], [18], Finn [8], Fujita [11], Ladyzhenskaya [16], Babenko [1] and many others. However, it is only fairly recently, with the input of Farwig [7], Galdi [12], Kobayashi and Shibata [14] and Shibata [25] that a functional setting has begun to emerge that seems suitable for the formulation of the steady-state Navier-Stokes problem on exterior domains. What we call here a suitable functional setting is a pair of Banach spaces X and Z and a mapping $F: X \to Z$ such that every solution of the equation $F(x) = z, x \in X, z \in Z$, yields a solution of the Navier-Stokes problem of interest. In this form, the problem may be given many unsubtle answers, but it becomes much trickier if, in addition, one insists on getting a mapping F having as many good properties as possible. Since mappings are primarily used to solve equations, a good property of a mapping F is anyone that opens up the possibility of taking advantage of known results to clarify the questions of existence, uniqueness (or nonuniqueness) or behavior of the solutions of the equation $F(x) = z$.

In the case of the Navier-Stokes system on bounded domains, the same issue as above is resolved in a completely straightforward way: The spaces X and Z can be chosen among the standard Sobolev spaces with F being a (nonlinear) real-analytic and compact perturbation of a linear isomorphism. Things are hardly as simple when the domain is unbounded, for then standard Sobolev spaces are inadequate to handle even the Oseen linearization. Fortunately, as shown in Galdi [12], the introduction of "graded" (or "nonuniform") Sobolev spaces delivers the fundamental isomorphism property of the Oseen linearization that fails to hold when ordinary Sobolev spaces are used. The grading consists in requiring different summability properties of the various derivatives. This leads to spaces $X^q(\Omega)$ (for the velocities) and $Y^q(\Omega)$ (for the pressure) indexed by the real number $q \geq 1$, where $\Omega \subset \mathbb{R}^n$ is the exterior domain of interest. The precise definition of these spaces (when $n = 2$) is postponed until later (see (2.3), (2.4) and (5.1)) For the purpose of this introduction, it suffices to say that the Oseen linearization is a linear isomorphism from $X^q(\Omega) \times Y^q(\Omega)$ into $(L^q(\Omega))^n$ for q in some interval depending upon n.

The above suggests that the functional setting which is adequate for the Oseen linearization might also be suitable for the full, nonlinear, Navier-Stokes system. However, the unboundedness of the domain Ω makes it hopeless to expect the nonlinear problem to appear as a simple compact perturbation of the Oseen linearization. This is due to the fact that all the embeddings among Sobolev or Sobolev-like spaces loose their compactness properties when large unbounded domains are involved.

Differentiable mappings between Banach spaces which are compact perturbations of linear isomorphisms have two fundamental properties which, by themselves, account for most of the reason why such mappings are so important regarding existence and bifurcation issues: They are proper on closed bounded subsets and their derivative at each point is a linear compact perturbation of an isomorphism. The first property refers to the fact that the intersection of the inverse image of any compact subset with every closed bounded subset is compact. The second one has the usual implications regarding the finite dimensionality of the null-space and the closedness and finite codimensionality of the range. These features contribute a lot to making compact perturbations of linear isomorphisms look like mappings between finite dimensional spaces. On the other hand, even jointly, they can still be found in substantially more general mappings.

Recall that a linear operator $T \in \mathcal{L}(X, Z)$ is said to be Fredholm if $\dim \ker T < \infty$ and codim rge $T < \infty$. It is well known that this implies that rge T is a closed subspace of Z (for this as well as other results mentioned below, see e.g. Lindenstrauss and Tzafriri [19] or Deimling [5]). The index of a Fredholm operator is defined as the integer dim ker $T-$codim rge $T \in \mathbb{Z}$. Fredholm operators of a given index k form an open subset of $\mathcal{L}(X, Z)$ – hence the index is locally constant – and invariant under arbitrary linear compact perturbations. A nonlinear mapping $F : X \to Z$ is called a Fredholm mapping if it is differentiable and $DF(x)$ is Fredholm for every $x \in X$. The local constancy of the index and the connectedness of X imply that a C^1 Fredholm mapping has an index, defined as the index of $DF(x)$ for any $x \in X$.

The main two results of this paper are that the Navier-Stokes system for a planar exterior domain and constant *nonzero* velocity at infinity can be viewed as a functional equation $F(u,p) = f$ where $F : X^q(\Omega) \times Y^q(\Omega) \to (L^q(\Omega))^2$ is Fredholm of index 0 if $1 < q \leq 6/5$ and also proper on the closed bounded subsets of $X^q(\Omega) \times Y^q(\Omega)$ if $1 < q < 6/5$. From our earlier remarks, this means that, in those spaces, F is almost as good as a compact perturbation of a linear isomorphism and that the unboundedness of the domain does not necessarily rule out the validity of basic properties.

Among other things, these results make available new degree theories for Fredholm mappings of index 0 generalizing the Leray-Schauder degree. (A degree for such maps has been available since the 1970 work of Elworthy and Tromba [6], which however was not quite complete and, especially, not very user-friendly.)

This falls short of providing the notoriously missing existence theory, because existence via degree arguments would require properness, not merely properness on closed bounded sets. The properness question remains open at this time, but it is not needed for the study of local or global bifurcation of stationary solutions, which we investigate as an application of our theoretical results. For the case of bounded domains, similar results have been known for a long time and are already reported, at least for their most part, in the 1973 review article by Kielhofer and Kirchgässner [13]. Temam [27] gives a treatment based on the Leray-Schauder degree. For a more recent account, see the book by Chossat and Iooss [4, pp. 20–24], which also contains other relevant references.

Since the local aspects of bifurcation are the same for Fredholm mappings of index 0 and for compact perturbations of linear isomorphisms (the latter assumption does not provide anything more as regards any local property), the mathematical tools needed for the analysis of the local phenomena are exactly the same as those required in problems over bounded domains, or in fact in any bifurcation problem involving a parametrized family of Fredholm mappings of index 0.

In this respect, it should be emphasized that the trouble associated with 0 being in the essential spectrum of the linearized operator, as reported by Babenko [2] in the three-dimensional case, and making the bifurcation analysis difficult even locally, is eliminated here. Far from being a miracle, this merely reflects the fact that the functional framework chosen to formulate a bifurcation (or other) problem has considerable bearing upon the properties inherited by the linear or nonlinear operators involved. The difficulty noticed by Babenko is real when the bifurcation problem is formulated in too broad a class of generalized solutions, but it disappears in our setting because we confine attention to a subclass in which the operators involved have much better functional properties. The same remark is true for the three–dimensional problem, which will be investigated in detail elsewhere.

Remark 1.1. It may be informative to point out that the spaces $X^q(\Omega)$ used here are *not* canonically embedded in $(L^q(\Omega))^2$. As a result, the very concept of spectrum of the linearized operator makes no sense whatsoever. Of course, since bifurcation of stationary solutions *never* has anything to do with the spectrum of the linearized operator (contrary to a frequent misconception arising from unfortunate coincidences in the most elementary, but also most notorious, examples of bifurcations; see Remark 6.1 for more along this line) this is in no way an obstacle to a successful mathematical investigation of the bifurcation phenomena. The study of Hopf-like bifurcation is a whole different story. □

All the background material is introduced in the next section, which also contains a few related basic results needed at various stages later on in the paper. In Section 3, we give the proof of the Fredholm property. The properness on closed bounded subsets is significantly harder to establish. We proceed in two steps: Section 4 deals with the nonphysical but mathematically crucial case $\Omega = \mathbb{R}^2$. There, we follow a procedure from Rabier and Stuart [23] to obtain the desired result. This procedure is based in part upon translation invariance arguments and can only be used in the whole plane. Next, in Section 5, we use a cut-off argument to reduce the general problem over an exterior domain to that of the whole plane resolved in Section 4. The applications to local and global bifurcation are described in Section 6.

2. Notation, background, and preliminaries

In what follows, $\Omega \subset \mathbb{R}^2$ is an exterior domain with compact, possibly empty, complement and Lipschitz-continuous boundary $\partial\Omega$. Given $1 < q < 3/2$, we introduce

$$s_1 = s_1(q) := \frac{3q}{3-q}, \quad s_2 = s_2(q) := \frac{3q}{3-2q}, \quad q^* := \frac{2q}{2-q}. \tag{2.1}$$

Note that

$$q < s_1 < q^* < s_2. \tag{2.2}$$

Since s_1, s_2 and q^* depend only upon q, we may define the Sobolev-like spaces $X^{1,q}(\Omega)$, $X^{2,q}(\Omega)$ and $Y^q(\Omega)$ by

$$X^{1,q}(\Omega) := \{u = (u_1, u_2) : \nabla \cdot u = 0, \quad u \in (L^{s_2}(\Omega))^2, \nabla u \in (L^{s_1}(\Omega))^4, \nabla u_2 \in (L^q(\Omega))^2\} \tag{2.3a}$$

$$X^{2,q}(\Omega) := \{u = (u_1, u_2) \in X^{1,q}(\Omega) : \partial^2_{jk} u_i \in L^q(\Omega), \ 1 \le i, j, k \le 2\}, \tag{2.3b}$$

$$Y^q(\Omega) := \{p \in L^{q^*}(\Omega) : \partial_i p \in L^q(\Omega), i = 1, 2\} \tag{2.4}$$

The spaces $X^{1,q}(\Omega), X^{2,q}(\Omega)$ and $Y^q(\Omega)$ have "natural" norms, denoted by $||\cdot||_{X,1,q,\Omega}$, $||\cdot||_{X,2,q,\Omega}$ and $||\cdot||_{Y,q,\Omega}$, respectively, and defined in the obvious way by

$$||u||_{X,1,q,\Omega} := |u|_{0,s_2,\Omega} + |u|_{1,s_1,\Omega} + |u_2|_{1,q,\Omega}, \tag{2.5a}$$

$$||u||_{X,2,q,\Omega} := ||u||_{X,1,q,\Omega} + |u|_{2,q,\Omega} \tag{2.5b}$$

$$||p||_{Y,q,\Omega} := |p|_{0,q^*,\Omega} + |\nabla p|_{0,q,\Omega}. \tag{2.6}$$

The basic properties of the spaces $X^{1,q}(\Omega), X^{2,q}(\Omega)$ and $Y^q(\Omega)$ follow easily from the treatment given in Galdi [12, Vol. I]: They are Banach spaces, and $X^{m,q}(\Omega) \cap (\mathcal{D}(\bar{\Omega}))^2$ is dense in $X^{m,q}(\Omega), m = 1, 2$. (The latter result is somewhat complicated by the fact that $X^{m,q}(\Omega)$ incorporates a divergence-free condition.) For future use, we also note that $X^{1,q}(\Omega)$ and $X^{2,q}(\Omega)$ are reflexive and separable since they are canonically isomorphic to closed subspaces of reflexive separable Banach spaces. Likewise, $\mathcal{D}(\bar{\Omega})$ is dense in $Y^q(\Omega)$ and $Y^q(\Omega)$ is reflexive and separable.

From (2.3b) and Sobolev's theorem, we have that $u = (u_1, u_2) \in X^{2,q}(\Omega)$ implies $\partial_j u_i \in L^{q^*}(\Omega)$, $1 \le i, j \le 2$, whence by (2.2) and (2.3a), that

$$u \in X^{2,q}(\Omega) \Rightarrow \begin{cases} \partial_1 u_1, \partial_1 u_2, \partial_2 u_2 \in L^r(\Omega), & q \le r \le q^*, \\ \partial_2 u_1 \in L^s(\Omega), & s_1 \le s \le q^*. \end{cases} \tag{2.7}$$

Furthermore, for every pair (r, s) as in (2.7), there is a constant $C_{r,s} > 0$ independent of $u \in X^{2,q}(\Omega)$ such that

$$|\nabla u_2|_{0,r,\Omega} + |\partial_1 u_1|_{0,r,\Omega} + |\partial_2 u_1|_{0,s,\Omega} \le C_{r,s}||u||_{X,2,q,\Omega}. \tag{2.8}$$

Since $q^* > 2$ and $\nabla u \in (L^{q^*}(\Omega))^2, u \in (L^{s_2}(\Omega))^2$, it follows from [12, Vol. I, Remark 7.2, p 80] that every element $u \in X^{2,q}(\Omega)$ is in $(C^0(\bar{\Omega}) \cap L^\infty(\Omega))^2$ and tends uniformly to 0 at infinity. As a result,

$$u \in X^{2,q}(\Omega) \Rightarrow \begin{cases} u_1 \in L^r(\Omega), & s_2 \le r \le \infty, \\ u_2 \in L^s(\Omega), & q^* \le s \le \infty, \end{cases} \tag{2.9}$$

and there is a constant $C_{r,s} > 0$ independent of $u \in X^{2,q}(\Omega)$ such that

$$|u_1|_{0,r,\Omega} + |u_2|_{0,s,\Omega} \le C_{r,s}||u||_{X,2,q,\Omega}. \tag{2.10}$$

Remark 2.1. It should also be noted that for $u \in X^{1,q}(\Omega)$, we have $\partial_1 u \in (L^q(\Omega))^2$ (because $\nabla u_2 \in (L^q(\Omega))^2$ and $\nabla \cdot u = 0$) and $u_2 \in L^{q*}(\Omega)$, whence $u_2 \in L^s(\Omega), q^* \le s \le s_2$. □

From (2.1), it follows at once that if $\omega \subset \Omega$ is a bounded open subset, then

$$X^{1,q}(\Omega) \hookrightarrow (W^{1,q}(\omega))^2, \; X^{2,q}(\Omega) \hookrightarrow (W^{2,q}(\omega))^2 \tag{2.11}$$

with continuous embeddings. In turn, this implies that for $m = 1, 2$, every element of $X^{m,q}(\Omega)$ has a trace $u_{|\partial\Omega} \in (W^{m-\frac{1}{q},q}(\partial\Omega))^2$ and that the mapping

$$X^{m,q}(\Omega) \longmapsto u_{|\partial\Omega} \in (W^{m-\frac{1}{q},q}(\partial\Omega))^2 \tag{2.12}$$

is continuous. Consistent with these remarks, we may now introduce the space

$$X_0^{1,q}(\Omega) := \{u \in X^{1,q}(\Omega) : u = 0 \text{ on } \partial\Omega\}. \tag{2.13}$$

From [12, Vol. I], we have that

$$\{u \in (\mathcal{D}(\Omega))^2 : \nabla \cdot u = 0\} \text{ is dense in } X_0^{1,q}(\Omega). \tag{2.14}$$

On the other hand, we have

$$\{u \in W^{2,q}(\Omega) : \nabla \cdot u = 0\} \hookrightarrow X^{2,q}(\Omega) \tag{2.15}$$

(continuous embedding), as is obvious from (2.2) and the standard Sobolev embedding theorems.

The importance of the above spaces is due to their close relationship with the Oseen operator

$$L_{\mathcal{R}}(u, p) := -\Delta u + \mathcal{R}\partial_1 u + \nabla p, \tag{2.16}$$

where $\mathcal{R} > 0$ is given. Indeed, it is clear that $L_{\mathcal{R}}$ maps $X^{2,q}(\Omega) \times Y^q(\Omega)$, and hence $[X^{2,q}(\Omega) \cap X_0^{1,q}(\Omega)] \times Y^q(\Omega)$, continuously into $(L^q(\Omega))^2$. The following much stronger result [12, Vol. I, Theorem 7.1, p. 410] is true:

Theorem 2.1. *The Oseen operator $L_{\mathcal{R}}$ in (2.16) is a linear isomorphism from $[X^{2,q}(\Omega) \cap X_0^{1,q}(\Omega)] \times Y^q(\Omega)$ into $(L^q(\Omega))^2$.*

In essence, Theorem 2.1 ascertains that the problem

$$\begin{cases} \Delta u = \mathcal{R}\partial_1 u + \nabla p + f \text{ in } \Omega, \\ \nabla \cdot u = 0 \text{ in } \Omega, \\ u = 0 \text{ on } \partial\Omega, \end{cases} \tag{2.17}$$

has a unique solution in $[X^{2,q}(\Omega) \cap X_0^{1,q}(\Omega)] \times Y^q(\Omega)$ for every $f \in (L^q(\Omega))^2$. Since the elements of $X^{2,q}(\Omega)$ tend uniformly to 0 at infinity, every solution $(u,p) \in [X^{2,q}(\Omega) \cap X_0^{1,q}(\Omega)] \times Y^q(\Omega)$ of (2.17) satisfies the implicit condition $\lim_{|x|\to\infty} u(x) = 0$. More generally, Theorem 7.1 in [12, Vol. I] also yields:

Theorem 2.2. *Let $u_\infty \in \mathbb{R}^2$ and $u_* \in (W^{2-\frac{1}{q}}(\partial\Omega))^2$ be given. The problem*

$$\begin{cases} \Delta u = \mathcal{R}\partial_1 u + \nabla p + f \textit{ in } \Omega, \\ \nabla \cdot u = 0 \textit{ in } \Omega, \\ u = u_* \textit{ on } \partial\Omega, \\ \lim_{|x|\to\infty} u(x) = u_\infty, \end{cases} \tag{2.18}$$

has a unique solution (u,p) such that $(u - u_\infty, p) \in X^{2,q}(\Omega) \times Y^q(\Omega)$ for every $f \in (L^q(\Omega))^2$.

Theorem 7.1 in [12, Vol. I] actually provides estimates for $||u - u_\infty||_{X,2,q,\Omega} + ||p||_{Y,q,\Omega}$ generalizing the isomorphism property of Theorem 2.1, and more precise information about the constants involved is given.

Remark 2.2. Theorems 2.1 and 2.2 *break down* for $\mathcal{R} = 0$. □Part (i)

of the following lemma will be used in various places later on. Part (ii) is needed only for the proof of Lemma 2.2 next.

Lemma 2.1. *Suppose that $1 \le q \le 6/5$. Then*
(i) The mapping $(u,v) \in X^{2,q}(\Omega) \times X^{2,q}(\Omega) \longmapsto u \cdot \nabla v \in (L^q(\Omega))^2$ is well defined, continuous and weakly sequentially continuous (i.e. $u_n \rightharpoonup u$ in $X^{2,q}(\Omega)$ and $v_n \rightharpoonup v$ in $X^{2,q}(\Omega)$ imply $u_n \cdot \nabla v_n \rightharpoonup u \cdot \nabla v$ in $(L^q(\Omega))^2$).
(ii) For $0 \le \epsilon \le (6-5q)/(6-3q)$, the mapping $u = (u_1, u_2) \in X^{2,q}(\Omega) \longmapsto |u_2|^{1-\epsilon}|\partial_2 u_1| \in L^q(\Omega)$ is well defined and maps bounded subsets onto bounded subsets.

Proof. (i) That the map is well defined is shown in Galdi [12, Vol. II]; here we include the proof for completeness. First, we show that with $u = (u_1, u_2), v = (v_1, v_2)$, the bilinear mappings $(u,v) \in X^{2,q}(\Omega) \times X^{2,q}(\Omega) \longmapsto u_i \partial_i v_j \in L^q(\Omega)$, $1 \le i,j \le 2$, are well defined and continuous. This is obvious if $i = 1$, for $\partial_1 v_j \in L^q(\Omega)$ by definition of $X^{2,q}(\Omega)$, while $u_i \in C^0(\bar{\Omega}) \cap L^\infty(\Omega)$, as noticed earlier. The continuity thus follows from (2.10) with $r = s = \infty$. Since $\nabla \cdot v = 0$ for every $v \in X^{2,q}(\Omega)$, we have $\partial_2 v_2 \in L^q(\Omega)$ and $|\partial_2 v_2|_{0,q,\Omega} \le ||v||_{X,2,q,\Omega}$. Hence, the same argument as above handles the case $i = j = 2$.

It remains to consider the case $i = 2, j = 1$. From (2.7), we have $\partial_2 v_1 \in L^s(\Omega)$ for $s_1 \leq s \leq q^*$ and (see (2.8)) $|\partial_2 v_1|_{0,s,\Omega} \leq C_s||v||_{X,2,q,\Omega}$ where $C_2 > 0$ is a constant independent of $v \in X^{2,q}(\Omega)$. Next, from (2.9), $u_2 \in L^t(\Omega)$ for every $q^* \leq t \leq \infty$ and (see (2.10)) there is a constant $C_t > 0$ such that $|u_2|_{0,t,\Omega} \leq C_t||u||_{X,q,\Omega}$. Since the multiplication is a continuous operation from $L^s(\Omega) \times L^t(\Omega)$ into $L^\sigma(\Omega)$, where $\sigma^{-1} = s^{-1} + t^{-1} \leq 1$, it suffices to show that $q^{-1} = s^{-1} + t^{-1}$ for some pair (s,t) with $s_1 \leq s \leq q^*$ and $q^* \leq t \leq \infty$. This follows from the inequalities $(q^*)^{-1} \leq q^{-1} \leq s_1^{-1} + (q^*)^{-1}$, i.e., $(2-q)/2q \leq 1/q \leq (12-5q)/6q$ (recall $q \leq 6/5$).

The above shows that the mapping $(u,v) \longmapsto u \cdot \nabla v$ is well defined and continuous from $X^{2,q}(\Omega) \times X^{2,q}(\Omega)$ into $(L^q(\Omega))^2$. To prove its weak sequential continuity, consider (bounded) sequences (u_n) and (v_n) from $X^{2,q}(\Omega)$ such that $u_n \rightharpoonup u$ and $v_n \rightharpoonup v$ in $X^{2,q}(\Omega)$. Then, from part (i) of the lemma, the sequence $(u_n \cdot \nabla v_n)$ is bounded in $(L^q(\Omega))^2$, and hence it contains a weakly convergent subsequence: There is $w \in (L^q(\Omega))^2$ such that $u_{n_k} \cdot \nabla u_{n_k} \rightharpoonup w$ in $(L^q(\Omega))^2$. We claim that $w = u \cdot \nabla v$. To see this, it suffices to show that $w_{|\omega} = (u \cdot \nabla v)_{|\omega}$ for every bounded open subset $\omega \subset \mathbb{R}^2$ of the form $\omega = \Omega \cap B_R$ where $B_R \subset \mathbb{R}^2$ is an open ball with center 0 and sufficiently large radius $R > 0$.

For $R > 0$ large enough, we have $\partial\omega = \partial\Omega \cup \partial B_R$ and $\partial\Omega \cap \partial B_R = \emptyset$, so that $\partial\omega$ is Lipschitz-continuous. Therefore, the embeddings $W^{2,q}(\omega) \hookrightarrow W^{1,q}(\omega)$ and $W^{2,q}(\omega) \hookrightarrow C^0(\bar{\omega})$ (since $2q > 2$) are compact. If thus follows from (2.11) that $u_{n_{|\omega}} \to u$ in $(C^0(\bar{\omega}))^2$ and that $\nabla(v_{n|_\omega} \to \nabla(v_{n_{|\omega}})$ in $(L^q(\omega))^4$. Hence, $u_{n_{|\omega}} \cdot \nabla(v_{n_{|\omega}}) \to u_{|\omega} \cdot \nabla(v_{n_{|\omega}}) = (u \cdot \nabla v)_{|\omega}$ in $(L^q(\omega))^2$, and, a fortiori, $u_{n_k|\omega} \cdot \nabla(v_{n_{k|\omega}}) \to (u \cdot \nabla v)_{|\omega}$ in $(L^q(\omega))^2$. Thus, as claimed, $w_{|\omega} = (u \cdot \nabla v)_{|\omega}$ and $w = u \cdot \nabla v$.

The above shows that $u \cdot \nabla v$ is the only weak cluster point of the sequence $(u_n \cdot \nabla v_n)$ in $(L^q(\Omega))^2$. Since the sequence $(u_n \cdot \nabla v_n)$ is bounded in $(L^q(\Omega))^2$ (as observed earlier) and every closed ball in $(L^q(\Omega))^2$ is a compact metrizable topological space for the weak topology of $(L^q(\Omega))^2$ (since $L^q(\Omega)$ is reflexive and separable), this property suffices to ascertain that $u_n \cdot \nabla v_n \rightharpoonup u \cdot \nabla v$ in $(L^q(\Omega))^2$.

(ii) This follows from arguments similar to those used in the first half of the proof of part (i) above. Specifically, we have $|\partial_2 u_1| \in L^s(\Omega)$ for $s_1 \leq s \leq q^*$ and $|\, |\partial_2 u_1|\, |_{0,s,\Omega} \leq C_s||u||_{X,2,q,\Omega}$ by (2.7) and (2.8), where $C_s > 0$ is a constant independent of $u \in X^{2,q}(\Omega)$. Also, $|u_2| \in L^t(\Omega)$ for $q^* \leq t \leq \infty$ and $||u_2||_{0,t,\Omega} \leq C_t||u||_{X,2,q,\Omega}$ by (2.9) and (2.10), where $C_t > 0$ is a constant independent of $u \in X^{2,q}(\Omega)$. This implies that $|u_2|^{1-\epsilon} \in L^\tau(\Omega)$ with $q^*/(1-\epsilon) \leq \tau \leq \infty$ and, since $|\, |u_2|\, |_{0,\tau,\Omega}^{1-\epsilon} = |\, |u_2|\, |_{(1-\epsilon)\tau,\Omega}^{1-\epsilon}$, we have $||u_2|^{1-\epsilon}|_{0,\tau,\Omega} \leq C_{(1-\epsilon)\tau}^{1-\epsilon}||u||_{X,2,q,\Omega}^{1-\epsilon}$. Thus, the product $|u_2|^{1-\epsilon}|\partial_2 u_1|$ is in $L^q(\Omega)$ if $q^{-1} = s^{-1} + \tau^{-1}$ for some pair (s,τ) with $s_1 \leq s \leq q^*$ and $q^*/(1-\epsilon) \leq \tau \leq \infty$. Such a pair exists if $(q^*)^{-1} \leq q^{-1} \leq s_1^{-1} + (1-\epsilon)(q^*)^{-1}$, which in turn is ensured by the hypotheses $q \leq 6/5$ and $\epsilon \leq (6-5q)/(6-3q)$. Furthermore, for such a pair (s,τ), we have $|\, |u_2|^{1-\epsilon}|\partial_2 u_1|\, |_{0,q,\Omega} \leq ||u_2|^{1-\epsilon}|_{0,\tau,\Omega}|\partial_2 u_1|_{0,s,\Omega} \leq C_{(1-\epsilon)\tau}^{1-\epsilon} C_s ||u||_{X,2,q,\Omega}^{1-\epsilon}$, which implies the desired boundedness property. □

Our next lemma will be crucial to the proof of Theorem 4.1, and hence to the whole properness issue. Before stating it, we introduce a useful notation.

$$C_0^0(\bar{\Omega}) := \{v \in C^0(\bar{\Omega}) : \lim_{|x|\to\infty} |v(x)| = 0\}. \tag{2.19}$$

We emphasize that the subscript "0" does not refer to "compact support" but to the more general property that members of $C_0^0(\bar{\Omega})$ tend to 0 uniformly at infinity. It is trivial to check that $C_0^0(\bar{\Omega})$ is a Banach space for the (uniform) norm induced by $L^\infty(\Omega)$. From earlier remarks, we have for $1 < q < 3/2$,

$$X^{2,q}(\Omega) \hookrightarrow (C_0^0(\bar{\Omega}))^2, \tag{2.20}$$

(continuous embedding).

Lemma 2.2. *Suppose $1 < q < 6/5$ and let (u_n) be a bounded sequence from $X^{2,q}(\Omega)$. Suppose that $u_n \to 0$ in $C_0^0(\bar{\Omega})$ (i.e. uniformly). Then, $u_n \cdot \nabla u_n \to 0$ in $(L^q(\Omega))^2$.*

Proof. From Lemma 2.1 (i) we already know that $u_n \cdot \nabla u_n \in (L^q(\Omega))^2$. To prove the desired result, it suffices to show, with $u_n = (u_{n1}, u_{n2})$, that $u_{ni}\partial_i u_{nj} \to 0$ in $L^q(\Omega), 1 \le i, j \le 2$. Once again, this is obvious for $i = 1$ since $\partial_1 u_{nj}$ is bounded in $L^q(\Omega)$ by definition of $X^{2,q}(\Omega)$ while $u_{n1} \to 0$ uniformly in $\bar{\Omega}$ by hypothesis. Since $\nabla \cdot u_n = 0$, the same argument shows that $u_{n2}\partial_2 u_{n2} \to 0$ in $L^q(\Omega)$. To show that $u_{n2}\partial_2 u_{n1} \to 0$ in $L^q(\Omega)$, we use Lemma 2.1 (ii): Since $1 < q < 6/5$, choose $0 < \epsilon < (6-5q)/(6-3q)(< 1)$, and write $|u_{n2}\partial_2 u_{n1}| = |u_{n2}|^\epsilon[|u_{n2}|^{1-\epsilon}|\partial_2 u_{n1}]$. By Lemma 2.1 (ii), the bracketed term is bounded in $L^q(\Omega)$ (since (u_n) is bounded in $X^{2,q}(\Omega)$). On the other hand, $|u_{n12}|^\epsilon \to 0$ uniformly in $\bar{\Omega}$ since $\epsilon > 0$ and $u_{n1} \to 0$ uniformly in $\bar{\Omega}$. Thus, the product $|u_{n2}|^\epsilon[|u_{n2}|^{1-\epsilon}|\partial_2 u_{n1}|]$, i.e. $|u_{n2}\partial_2 u_{n1}|$, tends to 0 in $L^q(\Omega)$. This, of course, is the same as saying that $u_{n2}\partial_2 u_{n1}$ tends to 0 in $L^q(\Omega)$. □

3. The index of nonlinear Oseen operators

In the remainder of this paper, we shall primarily be concerned with the problem of solving the Navier-Stokes system

$$\begin{cases} -\Delta v + \mathcal{R} v \cdot \nabla v + \nabla p = f \text{ in } \Omega, \\ \nabla \cdot v = 0 \text{ in } \Omega, \\ v = v_* \text{ on } \partial\Omega, \\ \lim_{|x|\to\infty} v(x) = v_\infty, \end{cases} \tag{3.1}$$

for given Reynolds number $\mathcal{R} > 0, f \in (L^q(\Omega))^2$, $v_* \in W^{2-\frac{1}{q},q}(\partial\Omega)$ and $v_\infty \in \mathbb{R}\backslash\{0\}$. Our first task will be to transform (3.1) into a suitable functional equation. To begin with, observe that since $v_\infty \neq 0$, it is not restrictive to assume, after a possible change of coordinates in $\mathbb{R}^2$ and a rescaling of v, to assume that $v_\infty = e_1 = (1, 0)$ (the rescaling of v amounts to changing $\mathcal{R}$ into $\mathcal{R}|v_\infty|$). Next, we may eliminate both the boundary condition $v = v_*$ on $\partial\Omega$ and the condition $\lim_{|x|\to\infty} v(x) = v_\infty (= e_1)$ by introducing a change of variable

$$v = V + u, \tag{3.2}$$

where V satisfies the conditions $V = v_*$ on $\partial\Omega$ and $\lim_{|x|\to\infty} V(x) = v_\infty = e_1$. In this respect, note that when $1 < q < 3/2$, Theorem 2.2 guarantees the existence of such a field V: Just solve the linear problem (2.18) with (say) $f = 0, u_* = v_*$ and $u_\infty = e_1$. This produces a field V satisfying, in addition, the (important) condition

$$V - e_1 \in X^{2,q}(\Omega) \tag{3.3}$$

where the space $X^{2,q}(\Omega)$ is given by (2.3).

From now on, we assume that $1 < q < 3/2$ and that V satisfies condition (3.3) along with

$$V = v_* \text{ on } \partial\Omega.$$

Note that due to the embedding $X^{2,q}(\Omega) \hookrightarrow C_0^0(\bar{\Omega})$ (see (2.20) and the definition of $C_0^0(\bar{\Omega})$ in (2.19)), condition (3.3) embodies the requirement $\lim_{|x|\to\infty} V(x) = e_1$ (which therefore holds uniformly).

Through the change of variable (3.2), we obtain an equivalent formulation of the problem (3.1) (with $v_\infty = e_1$) as

$$\begin{cases} -\Delta u + \mathcal{R}V \cdot \nabla u + \mathcal{R}u \cdot \nabla V + \mathcal{R}u \cdot \nabla u + \nabla p = f + \Delta V - \mathcal{R}V \cdot \nabla V \text{ in } \Omega, \\ \nabla \cdot u = 0 \text{ in } \Omega, \\ u = 0 \text{ on } \partial\Omega, \\ \lim_{|x|\to\infty} u(x) = 0. \end{cases} \tag{3.4}$$

It should be observed that, in (3.4), the condition $\nabla \cdot u = 0$ arises from (3.2), (3.3) and the condition $\nabla \cdot v = 0$ in (3.1) since, by (3.3) we have $\nabla \cdot V = \nabla \cdot (V - e_1) = 0$. On the other hand, the assumption $1 < q < 3/2$ does *not* suffice to ensure that the right-hand side $f + \Delta V - \mathcal{R}V \cdot \nabla V$ is in $(L^q(\Omega))^2$ when $f \in (L^q(\Omega))^2$. However, by Lemma 2.1 (i), it is so under the more restrictive condition $1 < q \leq 6/5$. In that case, Lemma 2.1 (i) also ensures that the terms $V \cdot \nabla u$, $u \cdot \nabla V$ and $u \cdot \nabla u$ are in $(L^q(\Omega))^2$ as soon as $u \in X^{2,q}(\Omega)$: This is plain for $u \cdot \nabla u$, and also for $u \cdot \nabla V$ since $u \cdot \nabla V = u \cdot \nabla (V - e_1)$ and both u and $V - e_1$ are in $X^{2,q}(\Omega)$ by (3.3). To see that $V \cdot \nabla u \in (L^q(\Omega))^2$, write $V \cdot \nabla u = (V - e_1) \cdot u + e_1 \cdot \nabla u = (V - e_1) \cdot \nabla u + \partial_1 u$, and recall that $\partial_1 u \in (L^q(\Omega))^2$ by definition of $X^{2,q}(\Omega)$.

From the above and the fact that $u = 0$ on $\partial\Omega$ amounts to the relation $u \in X_0^{1,q}(\Omega)$ (see (2.13)) while both $\nabla \cdot u = 0$ and $\lim_{|x|\to\infty} u(x) = 0$ are accounted for by the relation $u \in X^{2,q}(\Omega)$, it follows that when $1 < q \leq 6/5$, the system (3.4), and hence the system (3.1) as well, take the simple form

$$F_{\mathcal{R},V}(u,p) = f + \Delta V - \mathcal{R}V \cdot \nabla V \in (L^q(\Omega))^2, \tag{3.5}$$

where $F_{\mathcal{R},V} : [X^{2,q}(\Omega) \cap X_0^{1,q}(\Omega)] \times Y^q(\Omega) \to (L^q(\Omega))^2$ (see (2.4) for the definition of $Y^q(\Omega)$) is the operator defined by

$$F_{\mathcal{R},V}(u,p) := -\Delta u + \mathcal{R}V \cdot \nabla u + \mathcal{R}u \cdot \nabla V + \mathcal{R}u \cdot \nabla u + \nabla p. \qquad (3.6)$$

For breviety, the operator $F_{\mathcal{R},V}$ will be referred to as a *nonlinear Oseen operator.* (In this definition, it is always understood that V satisfies the condition (3.3) (and that $1 < q \leq 6/5$).)

From Lemma 2.1, it follows that the operator $F_{\mathcal{R},V}$ is continuous from the space $[X^{2,q}(\Omega) \cap X_0^{1,q}(\Omega)] \times Y^q(\Omega)$ into $(L^q(\Omega))^2$. Since also $F_{\mathcal{R},V}(u,p)$ is a (quadratic) polynomial in (u,p), $F_{\mathcal{R},V}$ is analytic and hence of class C^∞. In particular, its (total) derivative is given by

$$DF_{\mathcal{R},V}(u,p)(v,\pi) = -\Delta v \; + \; \mathcal{R}(V+u) \cdot \nabla v \; + \mathcal{R}v \cdot \nabla(V+u) \; + \; \nabla \pi. \qquad (3.7)$$

The main result of this section is that when $1 < q \leq 6/5, F_{\mathcal{R},V}$ is Fredholm of index 0, i.e., $DF_{\mathcal{R},V}(u,p) \in \mathcal{L}([X^{2,q}(\Omega) \cap X_0^{1,q}(\Omega)] \times Y^q(\Omega), (L^q(\Omega))^2)$ is Fredholm of index 0 for every $(u,p) \in [X^{2,q}(\Omega) \cap X_0^{1,q}(\Omega)] \times Y^q(\Omega)$. This will follow easily from:

Lemma 3.1. *Suppose $1 < q \leq 6/5$ and let $w \in X^{2,q}(\Omega)$. Then, both the operators $v \in X^{2,q}(\Omega) \longmapsto w \cdot \nabla v \in (L^q(\Omega))^2$ and $v \in X^{2,q}(\Omega) \longmapsto v \cdot \nabla w \in (L^q(\Omega))^2$ are compact.*

Proof. Set $Tv := w \cdot \nabla v$ and $Sv := v \cdot \nabla w$ for every $v \in X^{2,q}(\Omega)$. From the denseness of $X^{2,q}(\Omega) \cap (\mathcal{D}(\bar{\Omega}))^2$ in $X^{2,q}(\Omega)$ (Section 2), there is a sequence (w_n) from $X^{2,q}(\Omega)$ with Supp w_n bounded in Ω such that $w_n \to w$ in $X^{2,q}(\Omega)$. If the operators T_n and S_n are defined by $T_n v := w_n \cdot \nabla v$ and $S_n v := v \cdot \nabla w_n$ for every $v \in X^{2,q}(\Omega)$, respectively, we have by Lemma 2.1 (i) and bilinearity that $|(T_n - T)v|_{0,q,\Omega} \leq ||w_n - w||_{X,2,q,\Omega} ||v||_{X,2,q,\Omega}$ and $|(S_n - S)v|_{0,q,\Omega} \leq ||w_n - w||_{X,2,q,\Omega} ||v||_{X,2,q,\Omega}$. Hence, $T_n \to T$ and $S_n \to S$ in $\mathcal{L}(X^{2,q}(\Omega), (L^q(\Omega))^2)$.

Since limits of compact operators are compact, it suffices to show that T_n and S_n are compact for every $n \in \mathbb{N}$. In turn, this follows from T and S being compact when Supp w is bounded in Ω. To see this, observe that, in that case, $Tv = w \cdot \nabla(v_{|\omega})$ and $Sv = (v_{n_{|\omega}}) \cdot \nabla w$ where $\omega \subset \Omega$ is any bounded open subset such that Supp $w \subset \bar{\omega}$. Since $\partial\Omega$ is Lipschitz-continuous, we may assume that $\partial\omega$ is Lipschitz-continuous with no loss of generality (e.g. by letting ω be the intersection of Ω with an open ball with sufficiently large radius). The desired compactness property thus follows from the embedding $X^{2,q}(\Omega) \hookrightarrow (W^{2,q}(\omega))^2$ (see (2.11)) and the compactness of the embedding $W^{2,q}(\omega) \hookrightarrow W^{m,q}(\omega), m = 0, 1$. □

Theorem 3.1. *Suppose $\mathcal{R} > 0$ and $1 < q \leq 6/5$ and let $V : \Omega \to \mathbb{R}^2$ be a vector field such that $V - e_1 \in X^{2,q}(\Omega), e_1 = (1,0)$. Then, the nonlinear Oseen operator $F_{\mathcal{R},V}$ in (3.6) is Fredholm of index 0 from $[X^{2,q}(\Omega) \cap X_0^{1,q}(\Omega)] \times Y^q(\Omega)$ into $(L^q(\Omega))^2$ (*[1]*).*

---

[1]The same result is true, but nonphysical, if $\mathcal{R} < 0$ but *not* true for $\mathcal{R} = 0$.

In the expression (3.7) for $DF_{\mathcal{R},V}(u,p)(v,\pi)$, write $V = e_1 + (V - e_1)$. This yields

$$DF_{\mathcal{R},V}(u,p)(v,\pi) = L_{\mathcal{R}}(v,\pi) \ + \mathcal{R}(V - e_1 + u)\cdot\nabla v \ + \mathcal{R}v\cdot\nabla(V - e_1 + u),$$

where $L_{\mathcal{R}}$ is the operator (2.16). By Theorem 2.1, $L_{\mathcal{R}}$ is an isomorphism of $[X^{2,q}(\Omega)\cap X_0^{1,q}(\Omega)]\times Y^q(\Omega)$ onto $(L^q(\Omega))^2$, and by (3.3) and Lemma 3.1, the operators $v \longmapsto (V - e_1 + u)\cdot\nabla v$ and $v \longmapsto v\cdot\nabla(V - e_1 + u)$ are compact. Therefore, $DF_{\mathcal{R},V}(u,p)$ is a compact perturbation of an isomorphism and hence is Fredholm of index 0 for every $(u,p)\in[X^{2,q}(\Omega)\cap X_0^{1,q}(\Omega)]\times Y^q(\Omega)$. □

Remark 3.1. Naturally, Theorem 3.1 has a generalization when the vector $e_1 = (1,0)$ is replaced by an arbitrary *nonzero* vector $v_\infty\in\mathbb{R}^2$. In this case, the definition of the spaces $X^{m,q}(\Omega)$ must be modified accordingly, i.e. the partial derivatives $\partial/\partial x_1$ and $\partial/\partial x_2$ must be replaced by the directional derivatives along v_∞ and $v_\infty^\perp$, respectively, where $v_\infty^\perp$ is any nonzero vector orthogonal to v_∞. With this modification, the condition $V - e_1 \in X^{2,q}(\Omega)$ becomes $V - v_\infty \in X^{2,q}(\Omega)$. We emphasize that the condition $v_\infty \neq 0$ is essential, because this condition is necessary to reduce the problem to the case $v_\infty = e_1$ by a change of variable in $\mathbb{R}^2$ and rescaling of u. When $v_\infty = 0$, the derivative $DF_{\mathcal{R},V}(u,p)$ is still given by (3.7), but since $V\in X^{2,q}(\Omega)$ it is now a compact perturbation of the *Stokes* operator, for which no isomorphism result such as Theorem 2.1 is available. □

4. Properness of nonlinear Oseen operators: The case $\Omega = \mathbb{R}^2, V = e_1$

Although the case $\Omega = \mathbb{R}^2$ is mostly deprived of any physical interest in the original problem (3.1), its careful investigation is crucial to establishing the properness properties of nonlinear Oseen operators in exterior domains $\Omega \neq \mathbb{R}^2$. The general result, Theorem 5.1 in the next section, is that such operators are proper on the closed bounded subsets of $[X^{2,q}(\Omega)\cap X_0^{1,q}(\Omega)]\times Y^q(\Omega)$ when $1 < q < 6/5$. This will be obtained as a by-product of the same property in the special case $\Omega = \mathbb{R}^2, V = e_1$, proved here in Theorem 4.1. The noncompactness of the quadratic map $u\in X^{2,q}(\mathbb{R}^2)\longmapsto u\cdot\nabla u\in(L^q(\mathbb{R}^2))^2$, easily demonstrated by the usual translation argument, shows that the properness issue is not a trivial one. It is noteworthy that it is the very same translation invariance property of the spaces $X^{2,q}(\mathbb{R}^2)$ and $Y^q(\mathbb{R}^2)$, which disappears as soon as $\Omega\neq\mathbb{R}^2$, that will eventually allow us to prove the desired properness result.

The procedure for the proof of properness developed in this section is neither short (it has several stages) nor standard, and it is derived from a recent work by Rabier and Stuart [23] devoted to scalar quasilinear elliptic equations on $\mathbb{R}^N$. Accordingly, it seems appropriate to highlight the main ideas of our approach before embarking on the technicalities.

As pointed out in Section 2, the spaces $X^{2,q}(\Omega)$ and $Y^q(\Omega)$ are reflexive and separable when $1 < q < 3/2$. As a result, to say that a nonlinear Oseen operator

$F_{\mathcal{R},V} : [X^{2,q}(\Omega) \cap X_0^{1,q}(\Omega)] \times Y^q(\Omega) \to (L^q(\Omega))^2$ is proper on the closed bounded subsets of $[X^{2,q}(\Omega) \cap X_0^{1,q}(\Omega)] \times Y^q(\Omega)$ amounts to saying that every bounded sequence (u_n, p_n) from $[X^{2,q}(\Omega) \cap X_0^{1,q}(\Omega)] \times Y^q(\Omega)$ such that $F_{\mathcal{R}}(u_n, p_n)$ converges (strongly) in $(L^q(\Omega))^2$ contains a subsequence (u_{n_k}, p_{n_k}) converging (strongly) in $[X^{2,q}(\Omega) \cap X_0^{1,q}(\Omega)] \times Y^q(\Omega)$. When $\Omega = \mathbb{R}^2, V = e_1$ and $1 < q < 6/5$, the existence of (u_{n_k}, p_{n_k}) will be obtained in two main steps. First, we shall prove that under the condition $1 < q \leq 6/5$ (and $\Omega = \mathbb{R}^2$), we can always find a subsequence (u_{n_k}) converging *uniformly* to some $u \in X^{2,q}(\mathbb{R}^2) \cap X_0^{1,q}(\mathbb{R}^2) = X^{2,q}(\mathbb{R}^2)$. Due to the embedding $X^{2,q}(\mathbb{R}^2) \hookrightarrow (C_0^0(\mathbb{R}^2))^2$ (see Section 2), this result is a priori *weaker* than the desired one (that is, $u_{n_k} \longrightarrow u$ in $X^{2,q}(\mathbb{R}^2)$), and it doesn't say anything about the convergence of the corresponding sequence (p_{n_k}).

The second main step will be to prove that the sequence (u_{n_k}, p_{n_k}) is actually convergent in $X^{2,q}(\mathbb{R}^2) \times Y^q(\mathbb{R}^2)$ provided that $1 < q < 6/5$. The trick here is to notice that it is not restrictive to assume that (u_{n_k}, p_{n_k}) converges weakly to $(u,p) \in X^{2,q}(\mathbb{R}^2) \times Y^q(\mathbb{R}^2)$, and that, necessarily, $DF_{\mathcal{R},e_1}(u,p)(u_{n_k} - u, p_{n_k} - p)$ tends (strongly) to 0 in $(L^q(\Omega))^2$. This uses both the weak sequential continuity result in Lemma 2.1 (i) and Lemma 2.2, and this is where the restriction $q < 6/5$ appears. The conclusion, i.e. $u_{n_k} \to u$ in $X^{2,q}(\mathbb{R}^2)$ and $p_{n_k} \to p$ in $Y^q(\mathbb{R}^2)$ then follows from the (well-known) result that *linear* Fredholm operators are proper on closed bounded subsets and from Theorem 3.1 ensuring that $DF_{\mathcal{R},e_1}(u,p)$ is Fredholm.

Part of the material needed in the first main step referred to above has to do with the relationship between boundedness in the space $X^{2,q}(\mathbb{R}^2)$ and convergence in the space $(C_0^0(\mathbb{R}^2))^2$ (see (2.18) for the definition of $C_0^0(\mathbb{R}^2)$; recall that the subscript "0" does *not* refer to "compact support"). Of course, the embedding $X^{2,q}(\mathbb{R}^2) \hookrightarrow (C_0^0(\mathbb{R}^2))^2$ is not compact, but the very different nature of the norms of $X^{2,q}(\mathbb{R}^2)$ (of "mean" type) and of $C_0^0(\mathbb{R}^2)$ (of "uniform" type) still provides useful properties, summarized in Lemmas 4.1 and 4.2 below. We shall need the following simple concept.

Definition 4.1. *The sequence (u_n) from $(C_0^0(\mathbb{R}^2))^2$ will be said to vanish uniformly at infinity if for every $\epsilon > 0$, there are $R > 0$ and $n_0 \in \mathbb{N}$ such that $\sup_{|x| \geq R} |u_n(x)| \leq \epsilon$ for every $n \geq n_0$.*

Lemma 4.1. *Suppose $1 < q < 3/2$, and let (u_n) be a bounded sequence from $X^{2,q}(\mathbb{R}^2)$. Given $u \in X^{2,q}(\mathbb{R}^2)$, the following conditions are equivalent:*
(i) $u_n \to u$ in $(C_0^0(\mathbb{R}^2))^2$ (i.e. uniformly).
(ii) $u_n \rightharpoonup u$ in $X^{2,q}(\mathbb{R}^2)$ and the sequence (u_n) vanishes uniformly at infinity.

Proof. (i) $\Rightarrow$ (ii): That (i) implies $u_n \rightharpoonup u$ in $X^{2,q}(\mathbb{R}^2)$ is clear from the embedding $X^{2,q}(\mathbb{R}^2) \hookrightarrow (C_0^0(\mathbb{R}^2))^2$ (because (i) implies that u is the only weak cluster point of the sequence (u_n) in $X^{2,q}(\mathbb{R}^2)$, and $X^{2,q}(\mathbb{R}^2)$ is reflexive and separable). To prove that (u_n) vanishes uniformly at infinity, let $\epsilon > 0$ be given. Choose $R > 0$

such that $\sup_{|x|\geq R} |u(x)| \leq \epsilon/2$. Since (u_n) tends to u uniformly on $\mathbb{R}^2$, this implies $\sup_{|x|\geq R} |u_n(x)| \leq \epsilon$ for $n \in \mathbb{N}$ large enough.

(ii) $\Rightarrow$ (i). Let $\epsilon > 0$, and let $R > 0$ and $n_0 \in \mathbb{N}$ be such that $\sup_{|x|\geq R} |u_n(x)| \leq \epsilon/2$ for $n \geq n_0$. Since $u \in X^{2,q}(\mathbb{R}^2) \hookrightarrow C_0^0(\mathbb{R}^2)$, we may also assume that $\sup_{|x|\geq R} |u(x)| \leq \epsilon/2$ after increasing R if necessary. Hence,

$$\sup_{|x|\geq R} |u_n(x) - u(x)| \leq \epsilon, \quad \forall n \geq n_0. \tag{4.1}$$

Call B_R the open ball with center 0 and radius R in $\mathbb{R}^2$. Since $u_n \rightharpoonup u$ in $X^{2,q}(\mathbb{R}^2)$ and $X^{2,q}(\mathbb{R}^2) \hookrightarrow (W^{2,q}(B_R))^2$ by (2.11), and since the embedding $W^{2,q}(B_R) \hookrightarrow C^0(\overline{B}_R)$ is compact, it follows that $u_n \to u$ in $(C^0(\overline{B}_R))^2$. As a result, there is $n_1 \in \mathbb{N}$ such that

$$\sup_{|x|\leq R} |u_n(x) - u(x)| \leq \epsilon, \ \ \forall n \geq n_1. \tag{4.2}$$

From (4.1) and (4.2), we have $\sup_{x\in\mathbb{R}^2} |u_n(x) - u(x)| \leq \epsilon$ for $n \geq \max(n_0, n_1)$ and hence $u_n \to u$ in $(C_0^0(\mathbb{R}^2))^2$ since $\epsilon > 0$ is arbitrary. □

The implication (ii) $\Rightarrow$ (i) in Lemma 4.1 shows that the proof that a weakly convergent sequence (u_n) from $X^{2,q}(\mathbb{R}^2)$ converges uniformly to its weak limit reduces to showing that the sequence (u_n) vanishes uniformly at infinity. This is important, because the uniform vanishing at infinity happens to be one among two options for *every* bounded sequence (u_n) from $X^{2,q}(\mathbb{R}^2)$. Specifically,

Lemma 4.2. *Suppose $1 < q < 3/2$, and let (u_n) be a bounded sequence from $X^{2,q}(\mathbb{R}^2)$. Then, either*
(i) (u_n) vanishes uniformly at infinity,
or
(ii) there is a sequence (z_k) from $\mathbb{R}^2$ with $\lim_{k\to\infty} |z_k| = \infty$ and a subsequence (u_{n_k}) such that the "shifted" sequence $\tilde{u}_{n_k}(x) := u_{n_k}(x + z_k)$ has a nonzero weak limit $\tilde{u}$ in $X^{2,q}(\mathbb{R})$.$^{(2)}$[2]

Proof. Suppose that (i) does not hold, so that there is $\epsilon > 0$, and for every $k \in \mathbb{N}$ there are $n_k \in \mathbb{N}$, $n_k > n_{k-1}$, and $z_k \in \mathbb{R}^2$ with $|z_k| > k$ such that $|u_{n_k}(z_k)| > \epsilon$. Set $\tilde{u}_{n_k}(x) := u_{n_k}(x + z_k)$. By the translation invariance of the Lebesgue measure (and of the divergence-free condition), we have not only $\tilde{u}_{n_k} \in X^{2,q}(\mathbb{R}^2)$ but also $||\tilde{u}_{n_k}||_{X,2,q,\mathbb{R}^2} = ||u_{n_k}||_{X,2,q,\mathbb{R}^2}$. Hence, the sequence $(\tilde{u}_{n_k})$ is bounded in $X^{2,q}(\mathbb{R}^2)$ and by passing to a subsequence we may assume that $\tilde{u}_{n_k} \rightharpoonup \tilde{u}$ in $X^{2,q}(\mathbb{R}^2)$.

We claim that $\tilde{u} \neq 0$. Indeed, the continuity of the embedding $X^{2,q}(\mathbb{R}^2) \hookrightarrow (C_0^0(\mathbb{R}^2))^2$ shows that the evaluation operator $v \in X^{2,q}(\mathbb{R}^2) \longmapsto v(0) \in \mathbb{R}^2$ is linear and continuous, and hence weakly continuous. But then the finite dimensionality

[2] The whole point here is of course that $\tilde{u} \neq 0$.

of $\mathbb{R}^2$ along with the condition $\tilde{u}_{n_k} \rightharpoonup \tilde{u}$ in $X^{2,q}(\mathbb{R}^2)$ shows that $\tilde{u}_{n_k}(0) \to \tilde{u}(0)$ in $\mathbb{R}^2$. Since $|\tilde{u}_{n_k}(0)| = |u_{n_k}(z_k)| \geq \epsilon > 0$ for every k, it follows that $\tilde{u}(0) \neq 0$, whence $\tilde{u} \neq 0$ since $\tilde{u}$ is continuous. □

Before completing the first main step referrred to at the beginning of this section, we need two more lemmas. The first one is folklore; a quick proof is included for completeness:

Lemma 4.3. *Let $1 < q < \infty$ and let $f \in (L^q(\mathbb{R}^2))^2$. Given a sequence (z_k) from $\mathbb{R}^2$ with $\lim_{k\to\infty} |z_k| = \infty$, set $\tilde{f}_k(x) := f(x + z_k)$. Then, $\tilde{f}_k \rightharpoonup 0$ in $(L^q(\mathbb{R}^2))^2$.*

Proof. If f has compact support and $\omega \subset \mathbb{R}^2$ is a bounded subset, then $\tilde{f}_{k|_\omega} = 0$ for k large enough, whence $\tilde{f}_k \to 0$ in $(L^q(\omega))^2$ for every bounded open subset ω. It is easy to see that this property implies $\tilde{f}_k \rightharpoonup 0$ in $(L^q(\mathbb{R}^2))^2$. The general case follows from this and the denseness of $\mathcal{D}(\mathbb{R}^2)$ in $L^q(\mathbb{R}^2)$. □

Lemma 4.4. *Suppose $\mathcal{R} > 0$ and $1 < q \leq 6/5$, and let $F_{\mathcal{R},e_1} : X^{2,q}(\mathbb{R}^2) \times Y^q(\mathbb{R}^2) \to (L^q(\mathbb{R}^2))^2$ be the nonlinear Oseen operator corresponding to the choice $V = e_1$ in (3.6). Then, we have $F_{\mathcal{R},e_1}(u,p) = 0$ if and only if $u = 0$ and $p = 0$.*

Proof. The equation $F_{\mathcal{R},e_1}(u,p) = 0$ is simply

$$-\Delta u + \mathcal{R}\partial_1 u + \mathcal{R} u \cdot \nabla u + \nabla p = 0, \tag{4.3}$$

where both sides are in $L^q(\mathbb{R}^2)$. Let $q' = q/(q-1)$ denote the Hölder conjugate of q. A simple verification reveals that for $1 < q \leq 6/5$, we have $q' \geq s_2 \geq q^*$, where the notation (2.1) was used. It then follows from (2.9) that $X^{2,q}(\mathbb{R}^2) \hookrightarrow (L^{q'}(\mathbb{R}^2))^2$ (continuous embedding). This means that it is legitimate to multiply both sides of (4.3) by u and integrate over $\mathbb{R}^2$ to get

$$-\int_{\mathbb{R}^2} \Delta u \cdot u + \mathcal{R} \int_{\mathbb{R}^2} \partial_1 u \cdot u + \mathcal{R} \int_{\mathbb{R}^2} (u \cdot \nabla u) \cdot u + \int_{\mathbb{R}^2} (\nabla p) \cdot u = 0.$$

Each of the last three terms in the left-hand side vanishes when u has compact support, hence in general, due to the denseness of $X^{2,q}(\mathbb{R}^2) \cap (\mathcal{D}(\mathbb{R}^2))^2$ in $X^{2,q}(\mathbb{R}^2)$. We are thus left with the relation $-\int_{\mathbb{R}^2} \Delta u \cdot u = 0$. When u has compact support, we have $-\int_{\mathbb{R}^2} \Delta u \cdot u = \int_{\mathbb{R}^2} |\nabla u|^2$. Actually, this relation continues to hold for every $u \in X^{2,q}(\mathbb{R}^2)$ by the denseness of $X^{2,q}(\mathbb{R}^2) \cap (\mathcal{D}(\mathbb{R}^2))^2$ in $X^{2,q}(\mathbb{R}^2)$ and the remark that $\partial_j u_i \in L^2(\mathbb{R}^2)$ for every $u \in X^{2,q}(\mathbb{R}^2)$, $1 \leq i,j \leq 2$, with $\sum_{i,j=1}^{2} |\partial_j u_i|^2_{0,2,\mathbb{R}^2} = \int_{\mathbb{R}^2} |\nabla u|^2 \leq C||u||^2_{X,2,q,\mathbb{R}^2}$, where $C > 0$ is a constant independent of u (whence $u \in X^{2,q}(\mathbb{R}^2) \longmapsto \int_{\mathbb{R}^2} |\nabla u|^2 \in \mathbb{R}$ is continuous). This follows at once from (2.7), (2.8) and $q < s_1 \leq 2 < q^*$ due to the hypothesis $1 < q \leq 6/5$.

From the above, we infer that $\nabla u = 0$, whence u is constant, i.e. $u = 0$ since $u \in X^{2,q}(\mathbb{R}^2)$. In turn, from (4.3), we have $\nabla p = 0$ in $(L^q(\mathbb{R}^2))^2$, so that p is constant. Since $p \in L^{q^*}(\Omega)$, we conclude $p = 0$. □

The result of the main first step is given in our next lemma.

Lemma 4.5. *Suppose $\mathcal{R} > 0$ and $1 < q \leq 6/5$, and let $F_{\mathcal{R},e_1} : X^{2,q}(\mathbb{R}^2) \times Y^q(\mathbb{R}^2) \to (L^q(\mathbb{R}^2))^2$ be the nonlinear Oseen operator corresponding to the choice $V = e_1$ in (3.6). Suppose that (u_n, p_n) is a bounded sequence from $X^{2,q}(\mathbb{R}^2) \times Y^q(\mathbb{R}^2)$ such that $F_{\mathcal{R},e_1}(u_n, p_n) \to f$ in $(L^q(\mathbb{R}^2))^2$ for some $f \in (L^q(\mathbb{R}^2))^2$. Then, there is $u \in X^{2,q}(\mathbb{R}^2)$ and a subsequence (u_{n_k}) such that $u_{n_k} \to u$ in $(C_0^0(\mathbb{R}^2))^2$.*

Proof. With no loss of generality, assume that $(u_n, p_n) \rightharpoonup (u, p)$ in $X^{2,q}(\mathbb{R}) \times Y^q(\mathbb{R})$. We shall prove that $u_n \to u$ in $(C_0^0(\mathbb{R}^2))^2$ by using the implication (ii) $\Rightarrow$ (i) of Lemma 4.1, i.e. by showing that (u_n) vanishes uniformly at infinity. To do this, it suffices to show that case (ii) of Lemma 4.2 cannot occur. Our strategy is to argue by contradiction and indeed obtain a nonzero solution of the equation $F_{\mathcal{R},e_1}(u, p) = 0$, thereby contradicting Lemma 4.4.

Suppose then that case (ii) of Lemma 4.2 does occur, so that there is a sequence (z_k) from $\mathbb{R}^2$ with $\lim_{k\to\infty} |z_k| = \infty$ and a subsequence (u_{n_k}) such that $\tilde{u}_{n_k} \rightharpoonup \tilde{u} \neq 0$ in $X^{2,q}(\mathbb{R}^2)$, where $\tilde{u}_{n_k}(x) := u_{n_k}(x + z_k)$. The hypothesis $F_{\mathcal{R},e_1}(u_n, p_n) \to f$ in $(L^q(\mathbb{R}^2))^2$ also reads

$$-\Delta u_n + \mathcal{R}\partial_1 u_n + \mathcal{R} u_n \cdot \nabla u_n + \nabla p_n - f \to 0 \text{ in } (L^q(\mathbb{R}^2))^2. \tag{4.4}$$

By the translation invariance of the Lebesgue measure, the L^q-norm of the left-hand side of (4.4) is not affected by arbitrary shifts of the variable x. Furthermore, since the "coefficients" of the left-hand side of (4.4) are constant and since partial differentiations and translations commute (i.e. $\partial_j v(x+h) = \partial_j v_h(x)$ where $v_h(x) := v(x + h)$), any such shift amounts to shifting the functions u_n, p_n and f *before* differentiations are performed[(3)].[3] In particular, for $n = n_k$ in (4.4), and with $\tilde{u}_{n_k}$ as above along with $\tilde{p}_{n_k}(x) := p_{n_k}(x + z_k)$, $\tilde{f}_k(x) := f(x + z_k)$, we get

$$-\Delta \tilde{u}_{n_k} + \mathcal{R}\partial_1 \tilde{u}_{n_k} + \mathcal{R}\tilde{u}_{n_k} \cdot \nabla \tilde{u}_{n_k} + \nabla \tilde{p}_{n_k} - \tilde{f}_k \to 0 \text{ in } (L^q(\mathbb{R}^2))^2.$$

By Lemma 4.3, we have $\tilde{f}_k \rightharpoonup 0$ in $(L^q(\mathbb{R}^2))^2$, whence

$$-\Delta \tilde{u}_{n_k} + \mathcal{R}\partial_1 \tilde{u}_{n_k} + \mathcal{R}\tilde{u}_{n_k} \cdot \nabla \tilde{u}_{n_k} + \nabla \tilde{p}_{n_k} \rightharpoonup 0 \text{ in } (L^q(\mathbb{R}^2))^2. \tag{4.5}$$

Since $||\tilde{p}_{n_k}||_{Y,q,\mathbb{R}^2} = ||p_{n_k}||_{Y,q,\mathbb{R}^2}$ and the sequence (p_n) is bounded in $Y^q(\mathbb{R}^2)$, we may assume with no loss of generality that $\tilde{p}_{n_k} \rightharpoonup \tilde{p} \in Y^q(\mathbb{R}^2)$, so that $\nabla \tilde{p}_{n_k} \rightharpoonup \nabla \tilde{p}$ in $(L^q(\mathbb{R}^2))^2$. Evidently, we also have $\Delta \tilde{u}_{n_k} \rightharpoonup \Delta \tilde{u}$ and $\partial_1 \tilde{u}_{n_k} \rightharpoonup \partial_1 \tilde{u}$ in $(L^q(\mathbb{R}^2))^2$. Lastly, the weak sequential continuity result proved in Lemma 2.1 (i) yields $\tilde{u}_{n_k} \cdot \nabla \tilde{u}_{n_k} \rightharpoonup \tilde{u} \cdot \nabla \tilde{u}$ in $(L^q(\mathbb{R}^2))^2$. Thus, by taking the weak limit in (4.5), we obtain

$$-\Delta \tilde{u} + \mathcal{R}\partial_1 \tilde{u} + \mathcal{R}\tilde{u} \cdot \nabla \tilde{u} + \nabla \tilde{p} = 0,$$

i.e. $F_{\mathcal{R},e_1}(\tilde{u}, \tilde{p}) = 0$. This contradicts Lemma 4.4 since $\tilde{u} \neq 0$. □

Rather fortunately, the second main step can be handled in one stroke and requires fewer digressions than the first one:

---

[3]This is *not* true when the coefficients are variable, and this is the reason why Lemma 4.5 is not given for a general operator $F_{\mathcal{R},V}$.

Theorem 4.1. *Suppose $\mathcal{R} > 0$ and $1 < q < 6/5$. The nonlinear Oseen operator $F_{\mathcal{R},e_1} : X^{2,q}(\mathbb{R}^2) \times Y^q(\mathbb{R}^2) \to (L^q(\mathbb{R}^2))^2$ corresponding to the choice $V = e_1$ in (3.6) is proper on the closed bounded subsets of $X^{2,q}(\mathbb{R}^2) \times Y^q(\mathbb{R}^2)$.*

Proof. We must show that every bounded sequence (u_n, p_n) from $X^{2,q}(\mathbb{R}^2) \times Y^q(\mathbb{R}^2)$ such that $F_{\mathcal{R},e_1}(u_n, p_n) \to f$ in $(L^q(\mathbb{R}^2))^2$ for some $f \in (L^q(\mathbb{R}^2))^2$ contains a subsequence that converges (strongly) in $X^{2,q}(\mathbb{R}^2) \times Y^q(\mathbb{R}^2)$. From Lemma 4.5, it is not restrictive to assume that there is $u \in X^{2,q}(\mathbb{R}^2)$ such that $u_n \to u$ in $(C_0^0(\mathbb{R}^2))^2$. Of course, since (p_n) is bounded in $Y^q(\mathbb{R}^2)$, it is not restrictive either to assume that $p_n \rightharpoonup p$ in $Y^q(\mathbb{R}^2)$. In this case, by (i) $\Rightarrow$ (ii) of Lemma 4.1, we have $(u_n, p_n) \rightharpoonup (u, p)$ in $X^{2,q}(\mathbb{R}^2) \times Y^q(\mathbb{R}^2)$.

Everything from this point on hinges upon the remark that

$$F_{\mathcal{R},e_1}(u_n, p_n) - f - DF_{\mathcal{R},e_1}(u,p)(u_n - u, p_n - p) \to 0 \text{ in } (L^q(\mathbb{R}^2))^2. \tag{4.6}$$

Before proving (4.6), we show how it yields the result claimed in the lemma: Together with the assumption $F_{\mathcal{R},e_1}(u_n, p_n) \to f$ in $(L^q(\mathbb{R}^2))^2$, (4.6) implies that $DF_{\mathcal{R},e_1}(u,p)(u_n - u, p_n - p) \to 0$ in $(L^q(\mathbb{R}^2))^2$. Now, the sequence $(u_n - u, p_n - p)$ is bounded in $X^{2,q}(\mathbb{R}^2) \times Y^q(\mathbb{R}^2)$, and the operator $DF_{\mathcal{R},e_1}(u,p)$ is Fredholm of index 0 by Theorem 3.1 (since $X^{2,q}(\mathbb{R}^2) \cap X_0^{1,q}(\mathbb{R}^2) = X^{2,q}(\mathbb{R}^2)$). It is a well-known property of *linear* Fredholm operators of any index that they are proper on closed bounded subsets (Yood's criterion; see e.g. Deimling [5]). Thus, the sequence $(u_n - u, p_n - p)$, and hence the sequence (u_n, p_n), contains a subsequence converging (strongly) in $X^{2,q}(\mathbb{R}^2) \times Y^q(\mathbb{R}^2)$.

To complete the proof, we establish the validity of (4.6). First, note that by the weak sequential continuity result of Lemma 2.1 (i), we have $F_{\mathcal{R},e_1}(u,p) = f$, whence (4.6) also reads

$$F_{\mathcal{R},e_1}(u_n, p_n) - F_{\mathcal{R},e_1}(u,p) - DF_{\mathcal{R},e_1}(u,p)(u_n - u, p_n - p) \to 0 \text{ in } (L^q(\mathbb{R}^2))^2. \tag{4.7}$$

But there is no difficulty in checking that the left-hand side of (4.7) is simply

$$F_{\mathcal{R},e_1}(u_n, p_n) - F_{\mathcal{R},e_1}(u,p) - DF_{\mathcal{R},e_1}(u,p)(u_n - u, p_n - p) = \mathcal{R}(u_n - u) \cdot \nabla(u_n - u)$$

(to see this, it is simpler to notice right away that the left-hand side of (4.7) "kills" all the linear terms of $F_{\mathcal{R},e_1}$, whence only the quadratic term has to be dealt with). Now, since the sequence $(u_n - u)$ is bounded in $X^{2,q}(\mathbb{R}^2)$ and $u_n \to u$ in $(C_0^0(\mathbb{R}^2))^2$, hence $u_n - u \to 0$ in $(C_0^0(\mathbb{R}^2))^2$, the conclusion that $(u_n - u) \cdot \nabla(u_n - u) \to 0$ in $(L^q(\mathbb{R}^2))^2$ follows from the assumption $1 < q < 6/5$ via Lemma 2.2. □

5. Properness of nonlinear Oseen operators: The general case

We now return to the case of an exterior domain $\Omega \subset \mathbb{R}^2$. The space $X^{2,q}(\Omega) \cap X_0^{1,q}(\Omega)$ will be involved repeatedly throughout this section, and it will be convenient to set

$$X^q(\Omega) := X^{2,q}(\Omega) \cap X_0^{1,q}(\Omega), \tag{5.1}$$

although we shall refrain from using this abbreviation in the statement of the main results.

The operator $F_{\mathcal{R},e_1}$ obtained by choosing $V = e_1$ in (3.6) continues to be well defined from $X^q(\Omega) \times Y^q(\Omega)$ into $(L^q(\Omega))^2$. In Theorem 4.1, we saw that this operator is proper on the closed bounded subsets of $X^q(\Omega) \times Y^q(\Omega)$ when $\Omega = \mathbb{R}^2$ and $1 < q < 6/5$. We shall prove here that the same result holds for a general exterior domain Ω and, next, that it also holds for "arbitrary" operators $F_{\mathcal{R},V}$.

The properness of $F_{\mathcal{R},e_1}$ on the closed bounded subsets of $X^q(\Omega) \times Y^q(\Omega)$ is equivalent to the statement that every bounded sequence (u_n, p_n) from $X^q(\Omega) \times Y^q(\Omega)$ such that $F_{\mathcal{R},e_1}(u_n, p_n) \to f$ (strongly) in $(L^q(\Omega))^2$ for some $f \in (L^q(\Omega))^2$ contains a subsequence (u_{n_k}, p_{n_k}) converging (strongly) in $X^q(\Omega) \times Y^q(\Omega)$. Once again, the unboundedness of Ω rules out the compactness of the term $u \cdot \nabla u$ and hence the result is not trivial (see Remark 5.2).

The approach we shall take in this section is based upon the remark that the existence of (u_{n_k}, p_{n_k}) follows at once from Theorem 4.1 if, in addition, u_n and p_n vanish identically on some (fixed) neighborhood of $\partial\Omega$ in $\bar{\Omega}$. In that case, u_n, p_n and f can be extended by 0 outside $\bar{\Omega}$, and such an extension permits to view u_n, p_n and f as members of $X^q(\mathbb{R}^2), Y^q(\mathbb{R}^2)$ and $(L^q(\mathbb{R}^2))^2$, respectively, for which, in addition, it can be said that $F_{\mathcal{R},e_1}(u_n, p_n) \to f$ in $(L^q(\mathbb{R}^2))^2$. Then, Theorem 4.1 yields a convergent (extended) sequence (u_{n_k}, p_{n_k}) which, by restriction to Ω produces the desired subsequence.

In general, we reduce the problem to the above case, which of course involves a few technicalities, mostly about cut-off functions. This is made a little more delicate due to the divergence-free condition.

Lemma 5.1. *Let $\varphi \in \mathcal{D}(\mathbb{R}^2)$ be such that $\varphi = 1$ on some open neighborhood U of $\partial\Omega$ in $\mathbb{R}^2$, and let $B \subset \mathbb{R}^2$ be an open ball with Supp $\varphi \subset B$ (so that $\bar{U} \subset B$). Given any open neighborhood U' of $\partial\Omega$ in $\mathbb{R}^2$ such that $\bar{U}' \subset U$ and $1 < q < 3/2$, there is a constant $C_q > 0$ such that for every $u \in X^q(\Omega)$, there is $v \in X^q(\Omega)$ satisfying the following conditions:*
(i) Supp $v \subset B \cap \bar{\Omega}$,
(ii) $v - u = 0$ in $U' \cap \bar{\Omega}$,
(iii) $v - \varphi u \in W^{3,q}(\Omega)$ and $||v - \varphi u||_{3,q,\Omega} \leq C_q ||u||_{X,2,q,\Omega}$.

Proof. Let B' be another ball satisfying the same conditions as B with $\bar{B}' \subset B$. Since Supp $\nabla\varphi \subset B' \setminus \bar{U}'$, we have $(\nabla\varphi) \cdot u \in W_0^{2,q}(B' \setminus \bar{U}')$. From [ref], there is $w \in (W_0^{3,q}(B' \setminus \bar{U}'))^2$ such that $\nabla \cdot w = -(\nabla\varphi) \cdot u$ and $||w||_{3,q,B'\setminus\bar{U}'} \leq C_q ||u||_{2,q,B'\setminus\bar{U}'}$ where $C_q > 0$ is a constant independent of u. By extending w by 0 outside $B' \setminus \bar{U}'$ we obtain $w \in W^{3,q}(\Omega)$ with Supp $w \subset B \setminus U', \nabla \cdot w = -(\nabla\varphi) \cdot u$ in Ω and, by (2.15), $||w||_{3,q,\Omega} \leq C_q ||u||_{X,2,q,\Omega}$ after modifying C_q if necessary. Thus, (iii) holds with $v := w + \varphi u$. This also yields (i) since both w and φ vanish outside $B' \subset B$. Next, $\nabla \cdot v = \nabla \cdot w + (\nabla\varphi) \cdot u = 0$ (recall $\nabla \cdot u = 0$), whence $v \in X^{2,q}(\Omega)$ by (iii) and (2.15). On U', we have $\varphi = 1$ and $w = 0$, so that $v = u$, as claimed in (ii). In particular, $v = u$ on $\partial\Omega$, and since $u = 0$ on $\partial\Omega$, it follows that $v = 0$ on $\partial\Omega$, i.e. $v \in X_0^{1,q}(\Omega)$. This completes the proof. □

We also have the following result, as a consequence of Sobolev embedding theorem.

Lemma 5.2. *Let φ, U, U', B and q be as in Lemma 5.1. Given a bounded sequence (p_n) from $Y^q(\Omega)$, the sequence $(p_n\nabla\varphi)$ is a bounded sequence of $(W^{1,q}(\Omega))^2$. In particular, since Supp $(p_n\nabla\varphi)$ is bounded and contained in $B \cap \Omega$ for every n, the sequence $(p_n\nabla\varphi)$ contains a subsequence $(p_{n_k}\nabla\varphi)$ converging (strongly) in $(L^q(\Omega))^2$.*

Lemma 5.3. *Let φ, U, U', B and q be as in Lemmas 5.1 and 5.2, with the additional restriction that $1 < q \leq 6/5$, so that the nonlinear Oseen operator $F_{\mathcal{R},e_1}(\mathcal{R} > 0)$ is well defined from $X^q(\Omega) \times Y^q(\Omega)$ into $(L^q(\Omega))^2$. Let (u_n, p_n) be a bounded sequence from $X^q(\Omega) \times Y^q(\Omega)$ such that $F_{\mathcal{R},e_1}(u_n, p_n) \to f$ in $(L^q(\Omega))^2$ for some $f \in (L^q(\Omega))^2$. For every $n \in \mathbb{N}$, let $v_n \in X^q(\Omega)$ be given by Lemma 5.1 with $u = u_n$. Then, the sequence $(v_n, \varphi p_n)$ contains a subsequence converging (strongly) in $X^q(\Omega) \times Y^q(\Omega)$.*

Proof. We begin with a short discursion: If (u_n, p_n) is a bounded sequence from $X^q(\Omega) \times Y^q(\Omega)$ such that $F_{\mathcal{R},e_1}(u_n, p_n) \to f$ in $(L^q(\Omega))^2$ *and* if the quadratic term $u_n \cdot \nabla u_n$ contains a subsequence converging in $(L^q(\Omega))^2$ (this makes sense because of Lemma 2.1 (i) and $1 < q \leq 6/5$), then the sequence (u_n, p_n) contains a convergent subsequence in $X^q(\Omega) \times Y^q(\Omega)$. It is so because if $u_{n_k} \cdot \nabla u_{n_k} \to g \in (L^q(\Omega))^2$, the condition $F_{\mathcal{R},e_1}(u_n, p_n) \to f$ in $(L^q(\Omega))^2$ implies $-\Delta u_{n_k} + \mathcal{R}\partial_1 u_{n_k} + \nabla p_{n_k} \to f - \mathcal{R}g$ in $(L^q(\Omega))^2$, i.e. $L_{\mathcal{R}}(u_{n_k}, p_{n_k}) \to f - \mathcal{R}g$ in $(L^q(\Omega))^2$ if the notation (2.16) is used, whence (u_{n_k}, p_{n_k}) converges in $X^q(\Omega) \times Y^q(\Omega)$ since $L_{\mathcal{R}}$ is an isomorphism of $X^q(\Omega) \times Y^q(\Omega)$ onto $(L^q(\Omega))^2$ by Theorem 2.1 (and recall the notation (5.1)).

Now, the sequence $(u_n \cdot \nabla u_n)$ certainly contains a convergent subsequence in $(L^q(\Omega))^2$ when u_n vanishes outside a fixed bounded open subset $\omega \subset \Omega$, due to (2.11) and the compactness of the embeddings $W^{2,q}(\omega) \hookrightarrow W^{1,q}(\omega)$ and $W^{2,q}(\omega) \hookrightarrow C^0(\bar{\omega})$. The condition that v_n vanishes outside a fixed bounded open subset $\omega \subset \Omega$ holds in light of part (i) of Lemma 5.1. Thus, from the above introductory remark, the lemma will be proved if we show that the sequence $(v_n, \varphi p_n)$ is bounded in $X^q(\Omega) \times Y^q(\Omega)$ and that $F_{\mathcal{R},e_1}(v_n, \varphi p_n)$ contains a subsequence converging in $(L^q(\Omega))^2$. Indeed, it suffices to replace (u_n, p_n) by the subsequence $(v_{n_k}, \varphi p_{n_k})$ such that $F_{\mathcal{R},e_1}(v_{n_k}, \varphi p_{n_k})$ converges in the above arguments.

We begin with the boundedness of the sequence (v_n) in $X^q(\Omega)$: By Lemma 5.1 (iii), we have $v_n - \varphi u_n \in W^{2,q}(\Omega)$ and $||v_n - \varphi u_n||_{2,q,\Omega} \leq C||u_n||_{X,2,q,\Omega}$, where $C > 0$ is a constant independent of n, whence $||v_n||_{2,q,\Omega} \leq ||\varphi u_n||_{2,q,\Omega} + C||u_n||_{X,2,q,\Omega}$. Since it is clear that the compactness of Supp φ yields a constant $C_\varphi > 0$ such that $||\varphi u||_{2,q,\Omega} \leq C_\varphi ||u||_{X,2,q,\Omega}$ for every $u \in X^{2,q}(\Omega)$, the boundedness of (v_n) thus follows from the boundedness of (u_n) and (2.15).

Now, $||\varphi p_n||_{Y,q,\Omega} = |\nabla(\varphi p_n)|_{0,q,\Omega} + |\varphi p_n|_{0,q^*,\Omega} \leq |\varphi \nabla p_n|_{0,q,\Omega} + |p_n \nabla \varphi|_{0,q,\Omega}$ $+|\varphi p_n|_{0,q^*,\Omega} \leq (|\varphi|_{0,\infty,\Omega} + |\varphi|_{1,2,\Omega})||p_n||_{Y,q,\Omega}$. The boundedness of the sequence (p_n) in $Y^q(\Omega)$ yields the boundedness of the sequence (φp_n) in $Y^q(\Omega)$.

It remains to show that the sequence $(F_{\mathcal{R},e_1}(v_n, \varphi p_n))$ contains a subsequence converging in $(L^q(\Omega))^2$. This is not difficult, but rather cumbersome. To make things a little simpler, observe that it is not restrictive (by Lemma 5.2) to assume that the sequence $(p_n \nabla \varphi)$ converges in $(L^q(\Omega))^2$. It will also be convenient to use the notation $F_{\mathcal{R},e_1}(u,p) = L_{\mathcal{R}}(u,p) + \mathcal{R}u \cdot \nabla u$, where $L_{\mathcal{R}}$ is the (linear) operator (2.16). Thus, what we must show is that the sequence $L_{\mathcal{R}}(v_n, \varphi p_n) + \mathcal{R}v_n \cdot \nabla v_n$ contains a convergent subsequence in $(L^q(\Omega))^2$. Because v_n vanishes outside a fixed bounded subset ω of Ω and $v_n \in X^q(\Omega)$, the sequence $(v_n \cdot \nabla v_n)$ contains a subsequence converging in $(L^q(\Omega))^2$ (by (2.11) and the compactness of the embeddings $W^{2,q}(\omega) \hookrightarrow W^{1,q}(\omega)$ and $W^{2,q}(\omega) \hookrightarrow C^0(\bar{\omega})$). Thus, it suffices to show that the sequence $(L_{\mathcal{R}}(v_n, \varphi p_n))$ contains a subsequence converging in $(L^q(\Omega))^2$.

To see this, we write $v_n = \varphi u_n + w_n$ so that, by the linearity of $L_{\mathcal{R}}$, we have $L_{\mathcal{R}}(v_n, \varphi p_n) = L_{\mathcal{R}}(\varphi u_n, \varphi p_n) + L_{\mathcal{R}}(w_n, 0)$. From Lemma 5.1 (iii) and the boundedness of (u_n) in $X^q(\Omega)$, we have that (w_n) is bounded in $(W^{3,q}(\Omega))^2$. Furthermore, w_n vanishes outside a fixed bounded open subset $\omega \subset \Omega$ (because it is so of v_n and φu_n). It thus follows from the compactness of the embedding $W^{3,q}(\omega) \hookrightarrow W^{2,q}(\omega)$ that (w_n) contains a subsequence converging in $(W^{2,q}(\omega))^2$, hence in $(W^{2,q}(\Omega))^2$. Since the operator $L_{\mathcal{R}}$ in (2.16) is continuous from $(W^{2,q}(\Omega) \times Y^q(\Omega))^2$ into $(L^q(\Omega))^2$, it follows that the sequence $(L_{\mathcal{R}}(w_n, 0))$ contains a subsequence converging in $(L^q(\Omega))^2$.

Thus, the problem reduces to showing that the sequence $(L_{\mathcal{R}}(\varphi u_n, \varphi p_n))$ contains a subsequence converging in $(L^q(\Omega))^2$. We have

$$
\begin{aligned}
&L_{\mathcal{R}}(\varphi u_n, \varphi p_n) = -\Delta(\varphi u_n) + \mathcal{R}\partial_1(\varphi u_n) + \nabla(\varphi p_n) \\
&= \varphi[-\Delta u_n + \mathcal{R}\partial_1 u_n + \nabla p_n] - \nabla\varphi \cdot \nabla u_n - (\Delta\varphi)u_n + \mathcal{R}(\partial_1\varphi)u_n + p_n \nabla\varphi.
\end{aligned}
$$

Above, the sequences $(\nabla\varphi \cdot \nabla u_n), ((\Delta\varphi)u_n)$ and $((\partial_1\varphi)u_n)$ all vanish outside a fixed bounded open subset $\omega \subset \Omega$, and since they involve only first or zeroth order terms (in u_n), they all contain subsequences converging in $(L^q(\Omega))^2$. Earlier in the proof, we reduced the problem to the case when $(p_n \nabla\varphi)$ converges in $(L^q(\Omega))^2$, so that this term raises no difficulty either.

To complete the proof, we only have to show that the sequence $(\varphi[-\Delta u_n + \mathcal{R}\partial_1 u_n + \nabla p_n])$ contains a convergent subsequence in $(L^q(\Omega))^2$. Since, by the now usual argument, the sequence $(\varphi u_n \cdot \nabla u_n)$ does contain such a subsequence, this is equivalent to showing that the sum sequence $(\varphi[-\Delta u_n + \mathcal{R}\partial_1 u_n + \nabla p_n] + \mathcal{R}\varphi u_n \cdot \nabla u_n)$ contains a convergent subsequence in $(L^q(\Omega))^2$. But this is just the sequence $(\varphi F_{\mathcal{R},e_1}(u_n, p_n))$, and this sequence converges to φf in $(L^q(\Omega))^2$ due to the hypothesis $F_{\mathcal{R},e_1}(u_n, p_n) \to f$ in $(L^q(\Omega))^2$. □

In our next lemma, we extend Theorem 4.1 to the setting of this section

Lemma 5.4. *Suppose $\mathcal{R} > 0$ and $1 < q < 6/5$. The nonlinear Oseen operator $F_{\mathcal{R},e_1} : [X^{2,q}(\Omega) \cap X_0^{1,q}(\Omega)] \times Y^q(\Omega) \to (L^q(\Omega))^2$ corresponding to the choice $V = e_1$ in (3.6) is proper on the closed bounded subsets of $[X^{2,q}(\Omega) \times X_0^{1,q}(\Omega)] \times Y^q(\Omega)$.*

Proof. We return to the notation (5.1): The problem is to show that every bounded sequence (u_n, p_n) from $X^q(\Omega) \times Y^q(\Omega)$ such that $F_{\mathcal{R},e_1}(u_n, p_n) \to f$ in $(L^q(\Omega))^2$ for some $f \in (L^q(\Omega))^2$ contains a convergent subsequence.

Choose v_n and p_n as in Lemma 5.3. Since by Lemma 5.3 we know that the sequence $(v_n, \varphi p_n)$ contains a subsequence converging in $(L^q(\Omega))^2$, it suffices to show that the sequence $(u_n, p_n) - (v_n, \varphi p_n) = (u_n - v_n, (1 - \varphi)p_n)$ has the same property. This sequence is bounded in $X^q(\Omega) \times Y^q(\Omega)$ because both (u_n) and (v_n) are bounded in $X^q(\Omega)$ (for the latter, see the proof of Lemma 5.3) and (p_n) is bounded in $Y^q(\Omega)$. We claim that the sequence $(F_{\mathcal{R},e_1}(u_n - v_n, (1-\varphi)p_n))$ contains a subsequence converging in $(L^q(\Omega))^2$. This yields the desired result, for by Lemma 5.1 (ii), $u_n - v_n$ vanishes in the fixed open neighborhood $U' \cap \bar{\Omega}$ of $\partial\Omega$ in $\bar{\Omega}$ and so does $(1 - \varphi)p_n$. Hence, the extension by 0 of $u_n - v_n$ and $(1 - \varphi)p_n$ outside Ω produces a bounded sequence $(u_n - v_n, (1 - \varphi)p_n)$ from $X^{2,q}(\mathbb{R}^2) \times Y^q(\mathbb{R}^2)$ such that $F_{\mathcal{R},e_1}(u_n - v_n, (1 - \varphi)p_n) \to f$ in $(L^q(\mathbb{R}^2))^2$, where f also denotes the extension of f by 0 outside Ω. Thus, the conclusion follows from Theorem 4.1 (and the trivial remark that the restriction to Ω of the limit $(u, p) \in X^{2,q}(\mathbb{R}^2) \times Y^q(\mathbb{R}^2)$ satisfies $u = 0$ on $\partial\Omega$, whence is in $X^q(\Omega) \times Y^q(\Omega)$).

To show that the sequence $(F_{\mathcal{R},e_1}(u_n - v_n, (1 - \varphi)p_n))$ contains a convergent subsequence in $(L^q(\Omega))^2$, we use once again $F_{\mathcal{R},e_1}(u, p) = L_{\mathcal{R}}(u, p) + \mathcal{R}u \cdot \nabla u$, with $L_{\mathcal{R}}$ from (2.16). This yields

$$
\begin{aligned}
F_{\mathcal{R},e_1}(u_n - v_n, (1-\varphi)p_n) &= L_{\mathcal{R}}(u_n - v_n, (1-\varphi)p_n) + \mathcal{R}(u_n - v_n) \cdot \nabla(u_n - v_n) \\
&= L_{\mathcal{R}}(u_n, p_n) - L_{\mathcal{R}}(v_n, \varphi p_n) + \mathcal{R}u_n \cdot \nabla u_n - \mathcal{R}v_n \cdot \nabla u_n - \mathcal{R}u_n \cdot \nabla v_n + \mathcal{R}v_n \cdot \nabla v_n \\
&= F_{\mathcal{R},e_1}(u_n, p_n) - L_{\mathcal{R}}(v_n, \varphi p_n) - \mathcal{R}v_n \cdot \nabla u_n - \mathcal{R}u_n \cdot \nabla v_n + \mathcal{R}v_n \cdot \nabla v_n.
\end{aligned}
$$

By hypothesis, $F_{\mathcal{R},e_1}(u_n, p_n) \to f$ in $(L^q(\Omega))^2$, and since the sequence $(v_n, \varphi p_n)$ contains a convergent subsequence in $X^q(\Omega) \times Y^q(\Omega)$ from Lemma 5.3, it follows from the continuity of $L_{\mathcal{R}}$ that $(L_{\mathcal{R}}(v_n, \varphi p_n))$ contains a convergent subsequence in $(L^q(\Omega))^2$. All three sequences $(v_n \cdot \nabla u_n)$, $(u_n \cdot \nabla v_n)$ and $(v_n \cdot \nabla v_n)$ vanish outside a fixed bounded open subset $\omega \subset \Omega$ and contain no term of order greater than one in u_n or v_n, hence they all contain convergent subsequences in $(L^q(\Omega))^2$ by the arguments already used in the proof of Lemma 5.3. This completes the proof of Lemma 5.4. □

It is very easy to extend the result of Lemma 5.4 to general nonlinear Oseen operators:

Theorem 5.1. *Suppose $\mathcal{R} > 0$ and $1 < q < 6/5$. Let $V : \Omega \to \mathbb{R}^2$ be a vector field such that $V - e_1 \in X^{2,q}(\Omega)$ where $e_1 = (1, 0)$. Then, the nonlinear Oseen operator $F_{\mathcal{R},V}$ in (3.6) is proper on the closed bounded subsets of $[X^{2,q}(\Omega) \cap X_0^{1,q}(\Omega)] \times Y^q(\Omega)$.*

Proof. By writing $V = e_1 + \mathcal{R}(V - e_1)$, we obtain $F_{\mathcal{R},V}(u,p) = F_{\mathcal{R},e_1}(u,p) + (V - e_1) \cdot \nabla u + \mathcal{R}u \cdot \nabla(V - e_1)$. Since (3.3) holds, Lemma 3.1 ensures that both operators $u \longmapsto (V - e_1) \cdot \nabla u$ and $u \longmapsto u \cdot \nabla(V - e_1)$ are compact. Hence, $F_{\mathcal{R},V}$ is a compact perturbation of $F_{\mathcal{R},e_1}$. The conclusion follows from Lemma 5.4 since, as is well-known (and trivial), compact perturbations do not affect the properness on closed bounded subsets. □

Remark 5.1. Just like Theorem 3.1, Theorem 5.1 has an obvious generalization when e_1 is replaced by any nonzero vector v_∞, provided that the spaces $X^{m,q}(\Omega)$, $m = 1, 2$, and $Y^q(\Omega)$ are defined accordingly. (See Remark 3.1). □

Remark 5.2. Although translations can no longer be used to prove that the term $u \cdot \nabla u$ is not compact (i.e. that the mapping $u \in X^q(\Omega) \to u \cdot \nabla u \in L^q(\Omega)$ is not compact), the procedure of using cut-off functions easily shows that if it were compact, then the same thing would be true when $\Omega = \mathbb{R}^2$. Since it is not so, it follows that, indeed, the term $u \cdot \nabla u$ is never compact when Ω is an exterior domain. □

6. Local and global bifurcation in Banach spaces: A brief review

Here, we discuss an abstract global bifurcation theorem for nonlinear Fredholm mappings that generalizes the well-known Rabinowitz' global bifurcation theorem for compact perturbations of the identity [24]. We purposely give a somewhat simplified and restricted version to avoid having to discuss the concept of parity of a curve of linear Fredholm operators of index 0. The main result (Theorem 6.1) is taken from Fitzpatrick, Pejsachowicz and Rabier [10, Theorem 7.2] and Pejsachowicz and Rabier [20, Theorem 6.1]. The application to the Navier-Stokes system described in the next section will follow at once from the Fredholm (Theorem 3.1) and properness (Theorem 5.1) properties established earlier in this paper.

Let X and Y be real Banach spaces and $G(= G(\lambda, x)) : \mathbb{R} \times X \to Y$ a C^1 Fredholm mapping of index 1. Equivalently, the partial mapping $G(\lambda, \cdot) : X \to Y$ is Fredholm of index 0 for every $\lambda \in \mathbb{R}$. We shall assume that

$$G(\lambda, 0) = 0, \ \forall \ \lambda \in \mathbb{R} \tag{6.1}$$

and refer to the set $\mathbb{R} \times \{0\} \subset \mathbb{R} \times X$ as the trivial branch of solutions of the equation $G(\lambda, x) = 0$.

Let $\lambda_0 \in \mathbb{R}$ be a point such that $D_x G(\lambda_0, 0)$ is not an isomorphism of X onto Y. Since $D_x G(\lambda_0, 0)$ is Fredholm of index 0, we have that

$$r_0 := \dim \ker D_x G(\lambda_0, 0) = \text{ codim rge } D_x G(\lambda_0, 0) \tag{6.2}$$

is finite and $r_0 \geq 1$.

Let Y_0 be some chosen complement of rge $D_x G(\lambda_0, 0)$, so that $\dim Y_0 = r_0$ and, by continuity, Y_0 remains a complement of rge $D_x G(\lambda, 0)$ for $\lambda \in [\lambda_-, \lambda_+]$, a small enough interval with $\lambda_- < \lambda_0 < \lambda_+$. A simple but important remark is that the spaces

$$X_\lambda := D_x G(\lambda, 0)^{-1}(Y_0), \quad \lambda \in [\lambda_-, \lambda_+], \tag{6.3}$$

are all isomorphic to Y_0 and depend continuously upon λ. (The collection $(X_\lambda)_{\lambda \in [\lambda_-, \lambda_+]}$ is a C^0 vector bundle over $[\lambda_-, \lambda_+]$ with fiber isomorphic to Y_0.) This implies that there are (nonunique) continuous functions $\varphi_i (= \varphi_i(\lambda)) : [\lambda_-, \lambda_+] \to X, 1 \leq i \leq r_0 = \dim Y_0$, such that $\{\varphi_1(\lambda), \cdots, \varphi_{r_0}(\lambda)\}$ is a basis of X_λ for every $\lambda \in [\lambda_-, \lambda_+]$.

If we now choose a basis $\{\psi_1, \cdots, \psi_{r_0}\}$ of Y_0, the mapping $D_x G(\lambda, \cdot)_{|X_\lambda}$ is linear from X_λ to Y_0 hence has a matrix $\Gamma(\lambda)$ in the bases $\{\varphi_1(\lambda), \cdots, \varphi_{r_0}(\lambda)\}$ and $\{\psi_1, \cdots, \psi_{r_0}\}$. For the calculation of $\Gamma(\lambda)$, we need the (unique) vectors $\psi_i^* \in Y^*$, the topological dual of Y, characterized by the condition that rge $D_x G(\lambda_0, 0) \subset \ker \psi_i^*$ and $\langle \psi_i^*, \psi_j \rangle = \delta_{ij}$ (Kronecker delta), $1 \leq i, j \leq r_0$. Then,

$$\Gamma(\lambda) = (g_{ij}(\lambda)), \tag{6.3a}$$

where

$$g_{ij}(\lambda) := \langle \psi_j^*, D_x G(\lambda, 0)\varphi_i(\lambda) \rangle, \quad 1 \leq i, j \leq r_0. \tag{6.3b}$$

The global bifurcation criterion will depend upon the number

$$\gamma(\lambda) := \det\ \Gamma(\lambda), \quad \lambda \in [\lambda_-, \lambda_+]. \tag{6.4}$$

Note that $\Gamma(\lambda_0) = 0$ (since $X_{\lambda_0} = \ker D_x G(\lambda_0, 0)$), whence $\gamma(\lambda_0) = 0$, and that $\gamma(\lambda)$ is a continuous function of λ.

We denote by $\mathcal{C}$ the connected component of the closure of $G^{-1}(0) \setminus \mathbb{R} \times \{0\}$ that contains $(\lambda_0, 0)$ (if it exists; otherwise, $\mathcal{C} = \emptyset$). The following theorem ensures not only that $\mathcal{C} \neq \emptyset$, i.e. that bifurcation occurs from $(\lambda_0, 0)$ in the equation $G(\lambda, x) = 0$, but also that $\mathcal{C}$ cannot be contained in arbitrarily small neighborhoods of $(\lambda_0, 0)$ in $\mathbb{R} \times X$.

Theorem 6.1. *Suppose that $\gamma(\lambda)$ changes sign as λ crosses λ_0. Then, either*
(i) $\mathcal{C}$ is noncompact,
or
(ii) $\mathcal{C}$ contains a point $(\lambda_*, 0)$ with $\lambda_* \neq \lambda_0$.
Furthermore, if G is proper on the closed bounded subsets of $\mathbb{R} \times X$, then (i) may be replaced by
(i′) $\mathcal{C}$ is unbounded.

Theorem 6.1 with G of class C^2 and proper on closed bounded sets can be found in [10]. The above refinement (most notably allowing for G to be only of class C^1) is from [20]. In both papers, the condition regarding the sign change of $\gamma(\lambda)$ is replaced by the more general requirement that the parity of the curve $(D_x G(\lambda, 0))_{\lambda \in [\lambda_-, \lambda_+]}$ is negative. In that case, it may only be ascertained that bifurcation occurs from some point of $(\lambda_-, \lambda_+) \times \{0\}$, not necessarily λ_0, in case (λ_-, λ_+) contains other points where $D_x G(\lambda, 0)$ is singular. We will try to explain this better later on, without going into the definition of the parity.

The remainder of this section includes various comments about Theorem 6.1, notably the discussion of some special cases when the sign change condition for $\gamma(\lambda)$ is easier to check. We shall also give a straightforward corollary useful when the mapping G is not defined over $\mathbb{R} \times X$ but only over $J \times X$ where J is an open interval (Corollary 6.1).

The number $\gamma(\lambda)$ in (6.4) depends upon the choice of some complement Y_0 of rge $D_xG(\lambda_0, 0)$ and upon the choice of bases in Y_0 and $X_\lambda, \lambda \in [\lambda_-, \lambda_+]$. However, whether or not it changes sign as λ crosses λ_0 is independent of such choices, provided that, of course, the basis of X_λ depends continuously upon λ.

The nonlocal nature of the conclusion (i′) of Theorem 6.1 is obvious. We now explain why (i) and (ii) are also nonlocal properties. As regards part (i), let us recall that Fredholm mappings are locally proper (Smale [26]). Hence, there is a closed neighborhood N of $(\lambda_0, 0)$ in $\mathbb{R} \times X$ such that $G_{|_N}$ is proper. When part (i) of the theorem holds, $\mathcal{C}$ is not contained in N, for otherwise it would be compact (since $\mathcal{C} \subset G^{-1}(0)$). This brings the remark that it is the exact same argument which yields (i′) instead of (i) when G is proper on the closed bounded subsets of $\mathbb{R} \times X$.

When part (ii) of Theorem 6.1 applies, the point $(\lambda_*, 0)$ is itsefl a point of bifurcation of the equation $G(\lambda, x) = 0$, and hence $D_xG(\lambda_*, 0)$ is not an isomorphism of X onto Y. On the other hand, the condition that $\gamma(\lambda)$ changes sign as λ crosses λ_0 incorporates the implicit assumption that $\gamma(\lambda)$ has constant (nonzero) sign on $(\lambda_0 - \epsilon, \lambda_0)$ and on $(\lambda_0, \lambda_0 + \epsilon)$ for some $\epsilon > 0$. This will be further commented upon below. Since $\gamma(\lambda) \neq 0$ if and only if $D_xG(\lambda, 0)$ is an isomorphism, the closest possible point λ_*, if any, is at strictly positive distance from λ_0 and depends only upon the linearizations $D_xG(\lambda, 0)$ rather than upon G itself. Thus, $\mathcal{C}$, which is connected and must contain both $(\lambda_0, 0)$ and $(\lambda_*, 0)$, cannot be confined to any neighborhood of $(\lambda_0, 0)$ leaving out the closest point $\lambda \neq \lambda_0$ where $D_xG(\lambda, 0)$ is singular. In fact, a stronger result is true (see Remark 6.2).

Remark 6.1. Contrary to some popular belief, the generalized null-space of $D_xG(\lambda_0, 0)$ has *usually* nothing to do with the bifurcation criterion that $\gamma(\lambda)$ changes sign. It is only when $X = Y$ *and* $D_xG(\lambda, 0)$ has the very special structure $D_xG(\lambda, 0) = I - \lambda K$ with $K \in \mathcal{L}(X)$ compact that the sign change of $\gamma(\lambda)$ is equivalent to λ_0^{-1} being an eigenvalue of K of odd algebraic multiplicity. See [22] for further details. Of course, when $X = Y$, the spectrum of $D_xG(\lambda_0, 0)$ is even *less* relevant to the bifurcation issue. Although the case $D_xG(\lambda, 0) = I - \lambda K$ was the first one studied in the early days of bifurcation theory (Krasnosell'skii [15], Rabinowitz [24]), such a simple structure is not always available in concrete and non academic problems, and certainly not in the Navier-Stokes system (see Section 7). □

Due to the fact that the calculation of $\gamma(\lambda)$ in (6.4) involves a continuously parametrized basis $\{\varphi_i(\lambda)\}_{\lambda \in [\lambda_-, \lambda_+]}$, it may be difficult to calculate. On the other hand, since the explicit knowledge of $\gamma(\lambda)$ is needed only to assess whether its sign changes as λ crosses λ_0, it may be replaced by approximations $\tilde{\gamma}(\lambda)$ as long as these approximations are known to change or not to change sign simultaneously with $\gamma(\lambda)$. When the curve $D_xG(\lambda,0)$ is analytic in λ, the basis $\{\varphi_i(\lambda)\}$ can be chosen analytic in λ, and the sign change condition can be checked after replacing $\varphi_i(\lambda)$ by its Taylor polynomial $\tilde{\varphi}_i(\lambda)$ of "sufficiently large" degree κ_i, thereby leading to $\tilde{\gamma}(\lambda) := \det(\tilde{g}_{ij}(\lambda))$, where $\tilde{g}_{ij}(\lambda) := \langle \psi_j^*, D_xG(\lambda,0)\tilde{\varphi}_i(\lambda)\rangle, 1 \leq i,j \leq r_0$. The suitable degree κ_i *can* be calculated from the derivatives $\partial^k G/\partial\lambda^k$ at $\lambda = \lambda_0$, by using the concept of generalized Jordan chain; see Rabier [22]. Rather unfortunately, this approach antedates the introduction of the parity (due to Fitzpatrick and Pejsachowicz [9]) and hence the connection with parity is not explored in [22].

Although constructive, the method of [22] to assess the sign change of $\gamma(\lambda)$ may be extremely laborious. However, under the frequently encountered condition that

$$\{h \in \ker D_xG(\lambda_0,0), h \neq 0\} \Rightarrow D^2_{\lambda x}G(\lambda_0,0)h \notin \text{ rge } D_xG(\lambda_0,0), \tag{6.5}$$

which may be viewed as a "nondegeneracy" condition, we have the important equivalence

$$\{\gamma(\lambda) \text{ changes sign as } \lambda \text{ crosses } \lambda_0\} \Leftrightarrow r_0 = \dim\ker D_xG(\lambda_0,0) \text{ is odd.} \tag{6.6}$$

This equivalence is the content of relation (5.1) in [22]; that (6.5) above implies *local* bifurcation from $(\lambda_0, 0)$ was observed much earlier by Westreich [28] and others. The even more particular case when $r_0 = 1$ is of special interest: Assume

$$\dim\ker D_xG(\lambda_0,0) = 1, \tag{6.7}$$

and let $\varphi_0 \in \ker D_xG(\lambda_0,0), \varphi_0 \neq 0$, be any null-vector. Given $\psi_0 \in Y\setminus$ rge $D_xG(\lambda_0,0)$, call $\psi_0^* \in Y^*$ the (unique) vector such that rge $D_xG(\lambda_0,0) = \ker\psi_0^*$ and $\langle\psi_0^*, \psi_0\rangle = 1$. Then, condition (6.5) becomes the simple

$$\langle\psi_0^*, D^2_{\lambda x}G(\lambda_0,0)\varphi_0\rangle \neq 0, \tag{6.8}$$

and, by (6.6) and Theorem 6.1, it is a sufficient condition for global bifurcation since $r_0 = 1$ is odd. Naturally, condition (6.8) is independent of the choices made for φ_0 and ψ_0^*.

The above shows that the sign change condition for $\gamma(\lambda)$ may reduce to a rather simple criterion in special, yet important, cases. At this stage, we would also like to point out the superiority of the concept of parity versus the simplified sign change condition for $\gamma(\lambda)$, which, sometimes, may be ambiguous: For instance, suppose that $\gamma(\lambda) = \lambda^2 \sin(\lambda^{-1})$ and $\lambda_0 = 0$. Does the sign of $\gamma(\lambda)$ change as λ crosses 0? The question does not make sense since the sign of $\gamma(\lambda)$ oscillates as λ approaches 0. However, choose $\lambda_- < 0$ and $\lambda_+ > 0$ such that $\gamma(\lambda_-)$ and $\gamma(\lambda_+)$ have opposite signs. This is certainly possible. In that case, as it turns out, the parity of $D_xG(\lambda,0)$ over $[\lambda_-,\lambda_+]$ will be negative, and (global) bifurcation will be ensured from *some* point $(\lambda,0)$ of $(\lambda_-,\lambda_+)\times\{0\}$ where $\gamma(\lambda) = 0$ (hence $D_xG(\lambda,0)$ is singular). This point may be $(0,0)$ or any other point $((k\pi)^{-1},0), k \in \mathbb{Z}$, lying in (λ_-,λ_+). By shrinking $[\lambda_-,\lambda_+]$ around 0, a (global) bifurcation point can be shown to exist arbitrarily close to $(0,0)$, but that point may not be $(0,0)$ (although of course bifurcation from $(0,0)$ does occur; but it need not be global in the sense of Theorem 6.1).

Remark 6.2. In the above situation, the bifurcation theorems of [10] and [20] also ensure that, in case (ii) of Theorem 6.1, the point λ_* lies *outside* the interval $[\lambda_-, \lambda_+]$. Thus, the nonlocal nature of the result is not only preserved but even further emphasized. □

Such examples as $\gamma(\lambda) = \lambda^2 \sin(\lambda^{-1})$ *never* occur when $D_x G(\lambda, 0)$ is analytic in λ, but it is possible, though rather atypical, that $\gamma(\lambda) \equiv 0$. If so, the bifurcation criterion of Theorem 6.1 fails.

It often happens that the mapping G is not defined over the entire space $\mathbb{R} \times X$ but, instead, over $J \times X$ where $J = (a, b)$ is an open interval containing λ_0 and (6.1) holds for $\lambda \in J$. This case can be reduced to the previous one when $J = \mathbb{R}$ by simply changing the variable λ through a diffeomorphism $\theta : \mathbb{R} \to J$, thereby changing λ into $\theta(\mu), \mu \in \mathbb{R}$. Then, Theorem 6.1 will apply with $G(\lambda, x)$ being replaced by $G(\theta(\mu), x)$. It is straightforward to check that, going back to the λ variable, cases (i) and (i′) of Theorem 6.1 now split into two cases: Either $\mathcal{C}$ remains noncompact (or unbounded), or $\mathcal{C}$ contains a point (a, x) or (b, x) for some $x \in X$. For the record, we state this result as a corollary:

Corollary 6.1. *Suppose that G is defiend only on $J \times X$ where $J = (a, b)$ is an open interval containing λ_0. Suppose also that $\gamma(\lambda)$ changes sign as λ crosses λ_0. Then, one of the following three conditions holds:*
(i) $\mathcal{C}$ contains a point (a, x) or (b, x) for some $x \in X$.
(ii) $\mathcal{C}$ is noncompact.
(iii) $\mathcal{C}$ contains a point $(\lambda_*, 0)$ with $\lambda_* \neq \lambda_0$.
Furthermore, if G is proper on the closed bounded subsets of $[a + \epsilon, b - \epsilon] \times X$ for every $\epsilon > 0$ small enough(4),[4] *then (ii) may be replaced by*
(ii′) $\mathcal{C}$ is unbounded.

Remark 6.3. If, as is often the case in practice, G is uniformly continuous on the bounded subsets of $[a + \epsilon, b - \epsilon] \times X$ (e.g. if the derivative of G is bounded on those bounded subsets), the properness condition for G in the "furthermore" part of Corollary 6.1 is equivalent to $G(\lambda, \cdot)$ being proper on the closed bounded subsets of X for every $\lambda \in J$. □

In (Theorem 6.1 and) Corollary 6.1, it is assumed that the equation $G(\lambda, x) = 0$ has the trivial branch $J \times \{0\}$ of solutions. This suffices to handle the more general and more realistic case when

$$G(\lambda, \xi(\lambda)) = 0, \quad \forall\, \lambda \in J, \tag{6.9}$$

where $\xi : J \to X$ is a C^1 curve. Indeed, it suffices to replace G in Corollary 6.1 by

$$\hat{G}(\lambda, x) := G(\lambda, x + \xi(\lambda)), \tag{6.10}$$

[4] In particular, proper on the closed bounded subsets of $J \times X$, but this is more restrictive.

which, indeed, satisfies

$$\hat{G}(\lambda,0) = G(\lambda,\xi(\lambda)) = 0, \quad \forall\, \lambda \in J. \tag{6.11}$$

Evidently, the condition that $D_x\hat{G}(\lambda_0,0)$ is singular amounts to the assumption that $D_xG(\lambda_0,\xi(\lambda_0))$ is singular, and it is obvious how the global bifurcation criterion of Corollary 6.1 for the mapping $\hat{G}$ can be completely rephrased in terms of the mapping G. For breviety, we translate here only the simplest form of such a criterion, namely conditions (6.7) and (6.8), that now become

$$\dim\ker D_xG(\lambda_0,\xi(\lambda_0)) = 1 \tag{6.12}$$

and

$$\langle \psi_0^*, D^2_{\lambda x}G(\lambda_0,\xi(\lambda_0))\varphi_0 + D^2_{xx}G(\lambda_0,\xi(\lambda_0))(\dot{\xi}(\lambda_0),\varphi_0)\rangle \neq 0, \tag{6.13}$$

respectively, where $\varphi_0 \in \ker D_xG(\lambda_0,\xi(\lambda_0))$ is any nonzero vector, $\psi_0^* \in Y^*$ is any (nonzero) vector such that $\ker\psi_0^* = \text{rge}\; D_xG(\lambda_0,\xi(\lambda_0))$ and $\dot{\xi}(\lambda_0) \in X$ is the derivative $(d\xi/d\lambda)(\lambda_0)$. Equivalently, condition (6.13) means that

$$D^2_{\lambda x}G(\lambda_0,\xi(\lambda_0))\varphi_0 + D^2_{xx}G(\lambda_0,\xi(\lambda_0))(\dot{\xi}(\lambda_0),\varphi_0) \notin \text{ rge } D_xG(\lambda_0,\xi(\lambda_0)). \tag{6.13$'$}$$

While the above shows how the case (6.9) reduces to the case when $G(\lambda,0) = 0$ in a straightforward way *if the curve* $\xi(\lambda)$ *is known*, a much more interesting and less trivial question is to find conditions ensuring the existence of the curve $\xi(\lambda)$ in (6.9) without ruling out the possibility of bifurcation from the point $(\lambda_0,\xi(\lambda_0))$. Let us explain this point: Suppose that $G(\lambda_0,\xi_0) = 0$ for some point $(\lambda_0,\xi_0) \in J \times X$. The existence of a C^1 curve $\xi(\lambda)$ for $\lambda \in J$ such that (6.9) holds after shrinking J around λ_0 if necessary follows at once from the implicit function theorem if it is assumed that $D_xG(\lambda_0,\xi_0)$ is an isomorphism. But at the same time, this assumption, consistent with the uniqueness part of the implicit function theorem, prevents bifurcation from occurring from the point $(\lambda_0,\xi_0) = (\lambda_0,\xi(\lambda_0))$.

From the above, the condition that $D_xG(\lambda_0,\xi_0)$ is singular is necessary for the existence of a C^1 curve $\xi(\lambda)$ such that (6.9) holds and from which bifurcation occurs from $(\lambda_0,\xi_0) = (\lambda_0,\xi(\lambda_0))$. Note also that by differentiation of (6.9), we find $D_\lambda G(\lambda_0,\xi_0) + D_xG(\lambda_0,\xi_0)\dot{\xi}(\lambda_0) = 0$, whence

$$D_\lambda G(\lambda_0,\xi_0) \in \text{ rge } D_xG(\lambda_0,\xi_0) \tag{6.14}$$

is also a necessary condition for the curve $\xi(\lambda)$ to exist. Under the simplifying assumption that

$$\dim\ker D_xG(\lambda_0,\xi_0) = 1, \tag{6.15}$$

and if G is of class C^2, the Morse lemma (used with the so-called Lyapunov-Schmidt reduction) yields a simple sufficient condition for the zero set of G near (λ_0,ξ_0) to consist of *two* transversal cruves (hence bifurcation does occur from (λ_0,ξ_0)). This condition is that a suitable Hessian is negative. Specifically, with φ_0 and ψ_0^* as above, let $\psi_0 \in Y$ be such that $\langle\psi_0^*,\psi_0\rangle = 1$ and call $P_0 \in \mathcal{L}(Y)$ the projection operator onto rge $D_xG(\lambda_0,\xi_0)$ associated with the splitting $Y = \mathbb{R}\psi_0 \oplus$ rge $D_xG(\lambda_0,\xi_0)$. Choose a closed complement of $\ker D_xG(\lambda_0,\xi_0)$ in X, so that $P_0 D_x G(\lambda_0, \xi_0)$ is an isomorphism of X_0 onto rge $D_x G(\lambda_0, \xi_0)$. Due to (6.14), $[P_0 D_x G(\lambda_0, \xi_0)]^{-1} D_\lambda G(\lambda_0, \xi_0) \in X_0$ is well defined. Set

$$\theta_0 := [P_0 D_x G(\lambda_0, \xi_0)]^{-1} D_\lambda G(\lambda_0, \xi_0) \in X_0.$$

The Hessian condition referred to above is that

$$\det H(\lambda_0, \xi_0) < 0, \tag{6.16a}$$

where $H(\lambda_0, \xi_0)$ is the 2×2 matrix associated with the quadratic form

$$(\mu, t) \in \mathbb{R}^2 \longmapsto \langle \psi_0^*, D^2 G(\lambda_0, \xi_0)(\mu, t\varphi_0 - \mu\theta_0)^2 \rangle \in \mathbb{R}. \tag{6.16b}$$

In (6.16b), $D^2 G$ refers to the total second derivative of G, i.e. with respect to (λ, x). The possible values for $\dot{\xi}(t_0)$ are the vectors $\dot{\xi}_0 \in X$ given by

$$\dot{\xi}_0 = \bar{t}\varphi_0 - \theta_0 \in X, \tag{6.17}$$

where $\bar{t}$ is a root of the polynomial in t obtained by setting $\mu = 1$ in (6.16b). By (6.16a), there are two real roots if $\langle \psi_0^*, D^2_{xx} G(\lambda_0, \xi_0)(\varphi_0)^2 \rangle \neq 0$, i.e. if $D^2_{xx} G(\lambda_0, \xi_0)(\varphi_0)^2 \notin$ rge $D_x G(\lambda_0, \xi_0)$, and only one otherwise. This reflects the fact that when $D^2_{xx} G(\lambda_0, \xi_0)(\varphi_0)^2 \in$ rge $D_x G(\lambda_0, \xi_0)$, only *one* of the two curves in the zero set of G near (λ_0, ξ_0) can be *smoothly* parametrized, if at all, by the variable λ.

The above is essentially folklore from "elementary" bifurcation theory, where everything involving condition (6.15) is by definition elementary, though the verification of (6.16a) may be somewhat tedious. Few expositions however go into the discussion of when exactly the curves can be parametrized by λ. Generalizations to the case when $\dim \ker D_x G(\lambda_0, \xi_0) > 1$ can be found in Rabier [21] and Buchner, Marsden and Schecter [3].

Let us summarize what the above means in the perspective of Corollary 6.1: Suppose that $(\lambda_0, \xi_0) \in J \times X$ is a point where $G(\lambda_0, \xi_0) = 0$ and the conditions (6.14) and (6.15) hold (both (6.14) and at least (6.15) being necessary for bifurcation). Suppose also that condition (6.16a) holds, and denote by $\xi(\lambda)$ any C^1 curve such that (6.9) holds with $\xi(\lambda_0) = \xi_0$. Recall that (6.16a) ensures that $G^{-1}(0)$ consists of exactly two curves in the vicinity of (λ_0, ξ_0), but that the second one cannot be (smoothly, at least) parametrized by λ if $D^2_{xx} G(\lambda_0, \xi_0)(\varphi_0)^2 \in$ rge $D_x G(\lambda_0, \xi_0)$. Thus, bifurcation does occur from the curve $(\lambda, \xi(\lambda))$ at the point (λ_0, ξ_0), but this result is only local. Denote by J any interval on which the curve $\xi(\lambda)$ is defined. Certainly, $\xi(\lambda)$ can be extended as long as no other singularity lies on $(\lambda, \xi(\lambda))$, i.e. as long as $D_x G(\lambda, \xi(\lambda))$ remains invertible, but it may also happen that $\xi(\lambda)$ may be extended beyond other bifurcation points, for instance beyond those at which all the conditions required at (λ_0, ξ_0) above are satisfied.

Suppose now that condition (6.13) also holds, where $\dot{\xi}(t_0) = \dot{\xi}_0$ is given by (6.17) (for the appropriate root $\bar{t}$). Then, Corollary 6.1 explains what happens to the *other* curve through (λ_0, ξ_0), which obviously is contained in $\mathcal{C}$: It is contained in a continuum (that is, $\mathcal{C}$) which either is noncompact (or even unbounded), or at least intersects $\{a\} \times X$ or $\{b\} \times X$ where a and b are the endpoints of J, or else returns to the curve $(\lambda, \xi(\lambda))$ at some other bifurcation point $(\lambda_*, \xi(\lambda_*))$.

Remark 6.4. If G is smoother than C^2, the curve $\xi(\lambda)$ above inherits added smoothness, up to and including analyticity. There is, in general, a loss of only one degree of smoothness between G and the curves obtained by the Morse lemma or its generalizations. □

Remark 6.5. When $D^2_{xx}G(\lambda_0,\xi_0)(\varphi_0)^2 \in \operatorname{rge} D_xG(\lambda_0,\xi_0)$ and hence only one curve through (λ_0,ξ_0) may be smoothly parametrized by λ, Corollary 6.1 applies to the other curve after a change of parametrization: Just use Corollary 6.1 with G replaced by $\hat{G}(s,x) := G(\lambda(s),\xi(s)+x)$ where $(\lambda(s),\xi(s))$ is a C^1 parametrization of the curve. Of course, the physical meaning of the parameter s may be lost in that process. □

7. Application to the Navier-Stokes system on exterior domains

In this section, we shall apply the general results from the previous section to the specific case of the problem (3.1). The only thing to do is to spell out the exact functional setting and to identify both the parameter λ of relevance and the mapping G in the problem at hand. The Fredholm and properness properties established in Theorems 3.1 and 5.1, respectively, along with the polynomial, hence analytic, nature of the Navier-Stokes equations convey the idea that the tools described in Section 6 are perfectly adequate for the study of bifurcations. On the other hand, the nominal amount of information currently available regarding existence and the lack of any explicitly known curve of solutions in any setting make it impossible to check the validity of even the simplest conditions. For the time being, one may only expect that, since all the basic ingredients are present, the general theory is indeed relevant, at least in most cases, to describe the bifurcation phenomena in the Navier-Stokes system on two dimensional exterior domains.

As explained in Section 3, the system (3.1) with fixed $v_\infty \in \mathbb{R}^2, v_\infty \neq 0$, reduces to the case $v_\infty = e_1$, and hence to the equation

$$F_{\mathcal{R},V}(u,p) = f + \Delta V - \mathcal{R}V\cdot\nabla V, \tag{7.1}$$

where $F_{\mathcal{R},V}$ is given by (3.6) and the field V satisfies the conditions $V = v_*$ on $\partial\Omega$ and $\lim_{|x|\to\infty} V(x) = e_1$. Every solution of (7.1) satisfying $u = 0$ on $\partial\Omega$ and $\lim_{|x|\to\infty} u(x) = 0$ yields a solution of (3.1) through $v := V + u$.

We now consider the equation (7.1) for $f \in L^q(\Omega)$ with $1 < q \leq 6/5$ chosen once and for all, but we do allow for f to depend upon $\mathcal{R}$ in a C^1 way. Provided that V satisfies the nonrestrictive condition $V - e_1 \in X^{2,q}(\Omega)$ (see (2.3) for the definitin of $X^{2,q}(\Omega)$), the problem (7.1) is well-posed in the space $X^q(\Omega)\times Y^q(\Omega)$ where $X^q(\Omega) := X^{2,q}(\Omega)\cap X^{1,q}_0(\Omega)$ and $Y^q(\Omega)$ is given by (2.4). In that setting, both requirements $u = 0$ on $\partial\Omega$ and $\lim_{|x|\to\infty} u(x) = 0$ hold without any further assumption.

Suppose we are given a C^1 curve $(u(\mathcal{R}),p(\mathcal{R})) \in X^q(\Omega)\cap Y^q(\Omega)$ of (7.1), defined for $\mathcal{R}$ in some open interval $J \subset \mathbb{R}_+$. With the field V being fixed once and for all, the equation (7.1) is equivalent to

$$G(\mathcal{R},(u,p)) = 0, \tag{7.2}$$

with $G : J \times X^q(\Omega) \times Y^q(\Omega) \to L^q(\Omega)$ defined by

$$G(\mathcal{R},(u,p)) := F_{\mathcal{R},V}(u,p) - f(\mathcal{R}) - \Delta V + \mathcal{R}V \cdot \nabla V, \tag{7.3a}$$

or, in more explicit form

$$G(\mathcal{R},(u,p)) := -\Delta u + \mathcal{R}V\cdot\nabla u + \mathcal{R}u\cdot\nabla V + \mathcal{R}u\cdot\nabla u + \nabla p - f(\mathcal{R}) - \Delta V + \mathcal{R}V\cdot\nabla V. \tag{7.3b}$$

We are in the situation described in Section 6 with $X = X^q(\Omega) \times Y^q(\Omega), Y = L^q(\Omega), \lambda = \mathcal{R}$ and $x = (u,p)$. The trivial branch of solutions is the known branch $(\mathcal{R}, u(\mathcal{R}), p(\mathcal{R})), \mathcal{R} \in J$.

By (7.3a) and Theorem 3.1, $G(\mathcal{R},\cdot)$ is Fredholm of index 0 for every $\mathcal{R} > 0$, and by Theorem 5.1, $G(\mathcal{R},\cdot)$ is also proper on the closed bounded subsets of $X^q(\Omega) \times Y^q(\Omega)$ for every $\mathcal{R} > 0$ when also $1 < q < 6/5$. Since the uniform continuity of G on the bounded subsets of $\mathbb{R} \times X^q(\Omega) \times Y^q(\Omega)$ is obvious, it follows from Remark 6.3 that the extra properness property needed for the validity of part (ii') of Corollary 6.1 is satisfied in this case.

From (7.3b), we have

$$D_{(u,p)}G(\mathcal{R},(u,p))(h,\pi) = -\Delta h + \mathcal{R}(V+u)\cdot\nabla h + \mathcal{R}h\cdot\nabla(V+u) + \nabla\pi, \tag{7.4}$$

for every $(h,\pi) \in X^q(\Omega) \times Y^q(\Omega)$, and

$$D_{\mathcal{R}}G(\mathcal{R},(u,p)) = (V+u)\cdot\nabla(V+u) - \dot{f}(\mathcal{R}) \tag{7.5}$$

This yields

$$D^2_{\mathcal{R},(u,p)}G(\mathcal{R},(u,p))(h,\pi) = h\cdot\nabla V + V\cdot\nabla h, \tag{7.6}$$

and

$$D^2_{(u,p),(u,p)}G(\mathcal{R},(u,p))((h,\pi),(h',\pi')) = \mathcal{R}h'\cdot\nabla h + \mathcal{R}h\cdot\nabla h', \tag{7.7}$$

for every $(h,\pi),(h',\pi) \in X^q(\Omega) \times Y^q(\Omega)$.

For instance, conditions (6.12) and (6.13$'$) ensuring global bifurcation from the point $(\mathcal{R}_0, u(\mathcal{R}_0), p(\mathcal{R}_0)) = (\mathcal{R}_0, u_0, p_0)$ for short mean that the operator

$$(h,\pi) \in X^q(\Omega)\times Y^q(\Omega) \longmapsto -\Delta h + \mathcal{R}(V+u_0)\cdot\nabla h + \mathcal{R}h\cdot\nabla(V+u_0) + \nabla\pi \in L^q(\Omega) \tag{7.8}$$

has a one-dimensional null-space, and that, given $(h_0,\pi_0) \neq (0,0)$ in this null-space, we have (by (7.6) and (7.7))

$$h_0\cdot\nabla(V+\mathcal{R}\dot{u}_0) + (V+R\dot{u}_0)\cdot\nabla h_0 \neq -\Delta h + \mathcal{R}(V+u_0)\cdot\nabla h + \mathcal{R}h\cdot\nabla(V+u_0) + \nabla\pi, \tag{7.9}$$

for every $(h,\pi) \in X^q(\Omega) \times Y^q(\Omega)$, $\dot{u}_0 := \dot{u}(\mathcal{R}_0) = (du/d\mathcal{R})(\mathcal{R}_0)$.

If no C^1 curve $(\mathcal{R},(u(\mathcal{R}),p(\mathcal{R})))$ of solutions is known, but merely a point $(\mathcal{R}_0,(u_0,p_0))$ where $G(\mathcal{R}_0,(u_0,p_0)) = 0$, then the conditions ensuring that two C^1 transversal curves of solutions pass through the point $(\mathcal{R}_0,(u_0,p_0))$, that is, (6.14), (6.15) and (6.16) can easily be formulated explicitly (this is cumbersome, but straightforward). Note that, from (7.5), $D^2_{\mathcal{R}\mathcal{R}}G = 0$, but that is the only simplification.

While the above discusses bifurcation in the problem (3.1) when the Reynolds number $\mathcal{R}$ is taken as a parameter, a different problem consists in studying bifurcation when the velocity v_∞ is varied. We consider only the case when $v_\infty = \lambda e_1, \lambda > 0$, and since $\mathcal{R}$ is now fixed, we assume $\mathcal{R} = 1$ for notational convenience. However, we now allow for f depending upon λ. The corresponding problem obtains by letting $\mathcal{R} = 1$ in (7.3) and replacing V by λV. Thus, the (new) operator $G : \mathbb{R}_+ \times X^q(\Omega) \times Y^q(\Omega) \to L^q(\Omega)$ is given by

$$G(\lambda, (u,p)) := -\Delta u + \lambda V\cdot\nabla u + \lambda u\cdot\nabla V + u\cdot\nabla u + \nabla p - f - \lambda\Delta V + \lambda^2 V\cdot\nabla V. \quad (7.10)$$

This yields

$$D_{(u,p)}G(\lambda,(u,p))(h,\pi) = -\Delta h + (\lambda V + u)\cdot\nabla h + h\cdot\nabla(\lambda V + u) + \nabla\pi, \quad (7.11)$$

$$D_\lambda G(\lambda,(u,p)) = V\cdot\nabla u + u\cdot\nabla V - \Delta V + 2\lambda V\cdot\nabla V - \dot{f}(\lambda), \quad (7.12)$$

$$D^2_{\lambda,(u,p)}G(\lambda,(u,p))(h,\pi) = V\cdot\nabla h + h\cdot\nabla V, \quad (7.13)$$

$$D^2_{(u,p),(u,p)}G(\lambda,(u,p))((h,\pi),(h',\pi')) = h'\cdot\nabla h + h\cdot\nabla h'. \quad (7.14)$$

We leave it to the reader to formulate the conditions (6.12) and (6.13′) (and others) based on the formulas (7.11) through (7.14).

Remark 7.1. Above, $\lambda = 0$ corresponds to $v_\infty = 0$ in (3.1), a value for which both Theorems 3.1 and 5.1 break down. Hence, Corollary 6.1 is not available when J is an open interval containing 0. □

References

[1] Babenko, K.I., "On stationary solutions of the problem of flow past a body of a viscous incompressible fluid", Math. SSSR Sbornik **20** (1973), 1–25.

[2] Babenko, K.I., "Spectrum of the linearized problem of flow of a viscous incompressible liquid round a body", Soviet Phys. Dokl. **27** (1982), 25–27.

[3] Buchner, M., Marsden, J. and Schecter, S., "Application of the blowing-up construction and algebraic geometry to bifurcation problems", J. Diff. Eq. **48** (1983), 404–433.

[4] Chossat, P. and Iooss, G., *The Couette-Taylor Problem*, Applied Math. Sci., **102**, Springer (1994).

[5] Deimling, K. *Nonlinear Functional Analysis*, Springer, Berlin (1985).

[6] Elworthy, K.D. and Tromba, A.J., "Degree theory on Banach manifolds", Proc. Symp. Pure Math. Part I **18** (1970), 86–94.

[7] Farwig, R., "The stationary exterior $3D$-problem of Oseen and Navier-Stokes equations in anysotropically weighted Sobolev spaces", Math. Z. **211** (1992), 409–447.

[8] Finn, R., "Stationary solutions of the Navier-Stokes equations", Proc. Symp. Appl. Math **17**; Amer. Math. Soc., Providence R.I. (1965), 121–153.

[9] Fitzpatrick, P.M. and Pejsachowicz, J., "Parity and generalized multiplicity", Trans. AMS **326** (1991), 281–305.

[10] Fitzpatrick, P.M., Pejsachowicz, J. and Rabier, P.J., "The degree of proper C^2 Fredholm mappings I.", J. reine angew. Math. **427** (1992), 1–33.

[11] Fujita, H., "On the existence and regularity of the steady-state solutions of the Navier-Stokes equations", J. Fac. Sci. Univ. Tokyo (1A) **9** (1961), 59–102.

[12] Galdi, G.P., *An Introduction to the Mathematical Theory of the Navier-Stokes Equations*, Vol. I and II, Springer Tracts in Natural Philosophy (1994).

[13] Kielhofer, H. and Kirchgässner, K., "Stability and bifurcation in fluid mechanics", Rocky Mountain J. Math. **3** (1973), 275–318.

[14] Kobayashi, T. and Shibata, Y., "On the Oseen equation in three dimensional exterior domains", Math. Ann., (to appear).

[15] Krasnosel'skii, M.A., *Topological methods in the theory of nonlinear integral equations*, Pergamon Press, Oxford (1964).

[16] Ladyzhenskaya, O., "Investigation of the Navier-Stokes equation for a stationary flow of an incompressible fluid", Uspehi Mat. Nauk, (3) **14** (1959), 75–97.

[17] Leray, J., "Etude de diverses equations intégrales non linéaires et de guelgues problémes que pose l'hydrodynamique", J. Math. Pures Appl. **12** (1933), 1–82.

[18] Leray, J., "Sur le mouvement d'un liquide visqueux emplissant l'espace", Acta. Math. **63** (1934), 193–248.

[19] Lindenstrauss, J. and Tzafriri, L., *Classical Banach spaces I and II*, Springer-Verlag, Berlin (1996).

[20] Pejsachowicz, J. and Rabier, P.J., "Degree theory for C^1 Fredholm mappings of index 0", (preprint).

[21] Rabier, P.J., "A generalization of the implicit function theorem for mappigns from $\mathbb{R}^{n+1}$ into $\mathbb{R}^n$ and its applications", J. Funct. Anal. **56** (1984), 145–170.

[22] Rabier, P.J., "Generalized Jordan chains and two bifurcation theorems of Krasnosell'skii", Nonlinear Analysis, TMA **13** (1989), 903–934.

[23] Rabier, P.J. and Stuart, C.A., "Fredholm and properness properties of quasilinear operators on $\mathbb{R}^N$", (preprint).

[24] Rabinowitz, P.H., "Some global results for nonlinear eigenvalue problems", J. Funct. Anal. **7** (1971), 487–513.

[25] Shibata, Y., "An exterior initial boundary value problem for Navier-Stokes equations", Quart. Appl. Math., (to appear).

[26] Smale, S., "An infinite dimensional version of Sard's theorem", Amer. J. Math. **87** (1965), 861–866.

[27] Temam, R., *Navier-Stokes equations*, North-Holland, Amsterdam (1977).

[28] Westreich, D., "Bifurcation at eigenvalues of odd multiplicity", Proc. Am. Math. Soc. **41** (1973), 609–614.

Giovanni P. Galdi
Department of Mathematics
University of Pittsburgh
Pittsburgh, PA 15260
galdimath.pitt.edu

Patrick J. Rabier
Department of Mathematics
University of Pittsburgh
Pittsburgh, PA 15260
rabiervms.cis.pitt.edu

Progress in Nonlinear Differential Equations
and Their Applications, Vol. 35
© 1999 Birkhäuser Verlag Basel/Switzerland

Loss of Convexity of Simple Closed Curves Moved by Surface Diffusion

Yoshikazu Giga* and Kazuo Ito†

Dedicated to Professor Herbert Amann on the Occasion of His 60th Birthday

Abstract. We rigorously prove that there exists a simple, strictly convex, smooth closed curve which loses convexity but stays simple without developing singularities when it moves by its surface diffusion for a short time.

1. Introduction

We consider the surface diffusion equation of the form

$$\begin{cases} V = -\kappa_{ss} & \text{on } \Gamma(t),\ t > 0, \\ \Gamma(0) = \Gamma_0. \end{cases} \tag{1}$$

Here t denotes the time variable and $\Gamma(t)$ denotes an unknown evolving closed curve embedded in $\mathbf{R}^2$; Γ_0 is a given initial closed curve. The quantities V, κ, and s denote the outward normal velocity, the outward curvature, and the arc-length parameter of $\Gamma(t)$, respectively. For consistency of notation we take s so that $\Gamma(t)$ is parametrized clock-wise by s. The subscript s in (1) denotes the partial derivative with respect to s. The main goal of this paper is to prove that there exists a simple, strictly convex, smooth closed curve Γ_0 such that the solution curve $\Gamma(t)$ of (1) loses its convexity for t belonging to some interval (t_0, T) with $t_0 > 0$ while $\Gamma(t)$ stays simple without developing singularities for $t \in (0, T)$. This answers a conjecture posed by J. Escher at the conference "Nonlinear Evolution Equations" held at the end of June 1997 in Oberwolfach. Loss of convexity was also suggested by numerical studies by B. D. Coleman, R. S. Falk and M. Moakher [7, 8]. We actually prove a stronger statement. Our initial data Γ_0 (so that $\Gamma(t)$ loses its convexity) is obtained by deforming any strict convex smooth curve C in the xy-plane symmetric with respect to the y-axis such that Γ_0 agrees with C outside a small neighborhood of one of the intersections of C and the y-axis. In particular,

* Partially supported by NISSAN SCIENCE FOUNDATION and The Japan Ministry of Education, Science, Sports and Culture through Grant No. 08874005.
† Partially supported by The Japan Ministry of Education, Science, Sports and Culture through Grant No. 08740082.

a curve Γ_0 obtained by a slight modification of a circle C near a point on C leads to a loss of convexity when moved by surface diffusion.

Equation (1) was first proposed by Mullins [17] to explain thermal grooving in material sciences. We refer to J. W. Cahn and J. E. Taylor [5] for a derivation of (1) as well as other related equations. Recently, J. W. Cahn, C. M. Elliott and A. Novick-Cohen [6] derived (1) from the Cahn-Hilliard equation with a concentration dependent mobility as a singular limit in formal basis. The equation (1) is a fourth order fully nonlinear parabolic equation. C. Elliott and H. Garcke [9] constructed, without uniqueness, a local-in-time classical solution $\Gamma(t)$ of (1) which is simple for arbitrary smooth, simple closed initial curves Γ_0. The unique existence of a local solution, allowing that $\Gamma(t)$ may develop self-intersection, is established by the authors [13]. Their proof is elementary in the sense that they only use the coerciveness property in L^2 Sobolev spaces. A slightly modified version is also presented in the present paper. Recently, J. Escher, U. F. Mayer and G. Simonett [10] established the unique existence of a local solution for any immersed initial data not only for (1) but also for a higher-dimensional version of (1). Their proof, however, uses a sophisticated semi-group theory in small Hölder spaces developed by H. Amann [2]. Besides the parabolicity, the equation (1) has two important structures: preservation of the area $A(t)$ enclosed by $\Gamma(t)$ and decrease of the total length $L(t)$ of $\Gamma(t)$, which is easily observed by

$$\begin{aligned} \frac{dA(t)}{dt} &= \int_{\Gamma(t)} V ds = -\int_{\Gamma(t)} \kappa_{ss} ds = 0, \\ \frac{dL(t)}{dt} &= -\int_{\Gamma(t)} \kappa V ds = -\int_{\Gamma(t)} \kappa_s^2 ds \leq 0. \end{aligned} \tag{2}$$

In fact, C. Elliott and H. Garcke [9] utilized property (2) to establish the global existence of a solution $\Gamma(t)$ of (1) if the initial data Γ_0 is close to a circle. They also proved that $\Gamma(t)$ converges to a circle with the enclosed area equal to that of Γ_0 as $t \to \infty$. These results have been extended to a higher-dimensional version by J. Escher, U. F. Mayer and G. Simonett [10].

The equation (1) is a nonlinear fourth order parabolic equation, so there are several phenomena which are different from those of a second order model such as the curve shortening equation

$$\begin{cases} V = \kappa \quad \text{on } \Gamma(t),\ t > 0, \\ \Gamma(0) = \Gamma_0. \end{cases} \tag{3}$$

For the curve shortening equation (3), if Γ_0 is a simple, closed, smooth curve, then $\Gamma(t)$ stays simple and smooth and becomes convex in finite time (Grayson [14]). Once $\Gamma(t)$ becomes convex, it stays convex until it shrinks to a point (M. Gage and R. Hamilton [12]). For the surface diffusion equation (1) it is conjectured by C. Elliott and H. Garcke [9] that $\Gamma(t)$ may cease to be embedded even if the initial data Γ_0 is simple, i.e., embedded. This conjecture has been proved by the authors [13]. In fact, it is shown in [13] that if the initial curve Γ_0 is a dumbbell-like shape, then $\Gamma(t)$ ceases to be embedded in finite time before it develops singularities.

Our method in [13] yields an explicit example that the order of solutions may not be preserved. In fact, consider a small circle contained in the neck of the dumbbell. Since the circle is a stationary solution of (1) and since our solution pinches its neck, the order of solutions is not preserved. A numerical evidence of such 'pinching' is presented in [10] for various closed curves. In the present paper we show that (1) does not preserve convexity. This shows a strong contrast with the order-preserving curvature flow equation

$$V = \kappa - \frac{1}{L}\int_{\Gamma(t)} \kappa ds,$$

where convexity is preserved [11], but embeddedness is not preserved. Loss of convexity has been proved for somewhat nonlocal models such as the Mullins-Sekerka problem by U.F. Mayer [15, 16]. We note that convexity may be lost also by the effect of a nonlocal lower order term for a class of spatially homogeneous surface evolution equations related to chemotaxis [4].

The loss of both embeddedness and convexity reflects the fact that a fourth order parabolic equation does not fulfill the maximum principle or the comparison principle which are main properties of a second order parabolic equation. We explain why the convexity may not be preserved for a fourth order problem by giving a simple linear example. Consider the initial boundary value problem

$$\begin{cases} u_t = -u_{xxxx}, & \text{in } (0,\infty)\times(-1,1), \\ u(t,\pm 1) = 0, \quad u_{xx}(t,\pm 1) = 0, & t > 0, \\ u(0,x) = f(x), & x \in (-1,1). \end{cases}$$

The second derivative $v = u_{xx}$ solves the same problem with initial data $v(0,x) = f_{xx}(x)$ since

$$v_{xx}(t,\pm 1) = u_{xxxx}(t,\pm 1) = -u_t(t,\pm 1) = 0.$$

So the nonpreserving of concavity is reduced to the nonpreserving of negativity of the solution v, which is easy to imagine, at least heuristically. For $\varepsilon > 0$ we deform the concave function f near zero so that $\partial_x^6 f^\varepsilon(0) = -1$ and $0 > \partial_x^2 f^\varepsilon(0) \geq -\varepsilon$ where f^ε denotes the deformed concave function. Let u^ε denote the solution with initial data f^ε and $v^\varepsilon = u^\varepsilon_{xx}$. The mean value theorem implies

$$v^\varepsilon(t,0) = v^\varepsilon(0,0) + v^\varepsilon_t(0,0)t + v^\varepsilon_{tt}(\zeta,0)t^2/2,$$

for some $\zeta \in (0,t)$. Since $v^\varepsilon_t = -v^\varepsilon_{xxxx}$, this implies

$$v^\varepsilon(t,0) \geq -\varepsilon + t - \sup_{0\leq\tau<t} |v^\varepsilon_{tt}(\tau,0)|t^2/2.$$

Thus $v^\varepsilon(t,0)$ becomes positive for some t if ε is sufficiently small, provided that the size of v^ε_{tt} is bounded independently of ε. This is a rough idea of proving the loss of concavity. However, one should take into account that v^ε_{tt} may depend on ε significantly unless we specify the way of deformation. (For this particular problem v_{tt} depends only on f_{xxxx} so it is easy to specify the way of deformation.) Also, if the problem is nonlinear, one should bear in mind that the maximal existence time T^ε may tend to zero as $\varepsilon \to 0$. We introduce a specific way of deformation so that the local existence results guarantee a positive lower bound for T^ε for (1). To deform f we are tempted to replace f_{xx} by $-\varepsilon - x^4/4!$ and integrate in x twice. However, this simple integration does not work since the deformation affects the first derivatives in a long range. One should localize the deformation also in the first derivative. Our way of deformation is given in Section 2. In contrast to the fourth order equation, if we consider the second order initial-boundary value problem

$$\begin{cases} u_t = u_{xx}, & \text{in } (0,\infty)\times(-1,1), \\ u(t,\pm 1) = 0, & t>0, \\ u(0,x) = f(x), & x\in(-1,1), \end{cases}$$

then $u(t,\cdot)$ is concave in $(-1,1)$ if f is concave. This is easy to prove by the maximum principle for $v = u_{xx}$ since v solves the same equation with the same boundary condition with initial data $v(0,x) = f(x)$.

This paper is organized as follows. In Section 2 we specify a way to deform a concave function and estimate several quantities of deformed functions. In Section 3, we use this deformation to deform a given simple convex closed curve Γ_0 to another simple convex closed curve Γ_0' whose local convexity is sufficiently weak compared with $-\kappa_{ssss}$. In Section 4 we present the fact that the solution curve $\Gamma'(t)$ of (1) starting from Γ_0' exists in a finite time interval $(0,T)$ depending on Γ_0 which, however, does not shrink by the above deformation of Γ_0. In Section 5 we prove that $\Gamma'(t)$ obtained as above loses its convexity in $(0,T)$ with a suitable choice of parameter of deformation. In the last section we give the proof of the unique existence results in Section 4.

After this work was completed we were informed of two recent interesting works related to ours. In [1] S. A. Alvarez and C. Liu proved the analytic dependence of solutions of (1) with respect to initial data as well as unique local existence of solutions. In [3] several interesting self-similar patterns of pinching-off are presented. However, the formation of pinching for (1) is not proved there.

2. Deformation of a concave function

We give a way to modify an even strictly concave function defined in the interval $I := (-1,1)$ near zero so that its concavity is weak near zero while the minus of its sixth derivative is not small at zero.

For a given concave function f on I and parameters ε, $\delta > 0$ we define a new function

$$(M_{\varepsilon,\delta}f)(x) := f(-\frac{1}{2}) + \int_{-1/2}^{x} v_{\varepsilon,\delta}(\xi)d\xi, \quad x\in I$$

with

$$v_{\varepsilon,\delta}(x) = \int_0^x w_{\varepsilon,\delta}(\xi)d\xi\varphi_{1/4}(x) + f'(x)(1-\varphi_{1/4}(x)),$$

$$w_{\varepsilon,\delta}(x) = (-\varepsilon - \frac{x^4}{4!})\varphi_\delta(x) + f''(x)(1-\varphi_\delta(x)),$$

where $f' = df/dx$, $f'' = d^2f/dx^2$. Here φ_δ is a cut-off function near zero defined by

$$\varphi_\delta(x) = \varphi(|x|/\delta),$$

where φ is given, for example, by

$$\varphi(x) = q(2-x), \quad q(x) = \frac{p(x)}{p(x)+p(1-x)}, \quad x \in \mathbf{R} \tag{4}$$

with $p(x) = e^{-1/x}$ for $x > 0$; $p(x) = 0$ for $x \leq 0$, so that $\varphi \in C_0^\infty[0,\infty)$ satisfies $0 \leq \varphi \leq 1$, $\varphi(x) = 1$ for $x \leq 1$ and $\varphi(x) = 0$ for $x \geq 2$. (We won't use the explicit formula of φ but we need the properties in the sequel.) The parameter ε measures the concavity of $M_{\varepsilon,\delta}f$ near zero while δ measures the size of the region where the sixth derivative of $M_{\varepsilon,\delta}f$ equals -1. We shall estimate $M_{\varepsilon,\delta}f$ in the L^2-Sobolev space $H^m(I)$ for f in $H^m(I)$; the norm of f in $H^m(I)$ is denoted by $\|f\|_m$. By the above deformation we should be afraid that $M_{\varepsilon,\delta}f$ may lose its concavity. Fortunately, it turns out that $M_{\varepsilon,\delta}f$ is strictly concave if δ is taken sufficiently small independently of $0 < \varepsilon \leq 1$. We shall state this in a precise way.

Lemma 1. *Let f be in $H^m(I)$ with an integer $m \geq 2$. Assume that $\sup_I f'' < 0$ and that f is even, i.e., $f(x) = f(-x)$ for $x \in I$. Assume that $0 < \varepsilon \leq 1$ and that $0 < \delta < 1/8$.*
(i) The function $f_{\varepsilon,\delta} = M_{\varepsilon,\delta}f$ is an even function.
(ii) $f_{\varepsilon,\delta} \in H^m(I)$ and there exists a positive constant $C_\delta^m = C(m,\delta,f,\varphi)$ such that

$$\|f_{\varepsilon,\delta}\|_m \leq C_\delta^m \quad \text{for all } \varepsilon \in (0,1].$$

(iii) $f_{\varepsilon,\delta}(x) = f(x)$ for $1/2 \leq |x| < 1$.
(iv) $f_{\varepsilon,\delta}^{(2)}(x) = -\varepsilon - x^4/4!$ for $|x| \leq \delta$. In particular,

$$f_{\varepsilon,\delta}^{(2)}(0) = -\varepsilon, \quad f_{\varepsilon,\delta}^{(4)}(0) = 0, \quad f_{\varepsilon,\delta}^{(6)}(0) = -1,$$

where $f^{(k)}$ denotes the k th derivative of f.
(v) There exists $\delta_0 = \delta_0(f,\varphi) > 0$ such that

$$f_{\varepsilon,\delta}^{(2)}(x) \leq -\min(\varepsilon, \frac{1}{2}\inf_I |f^{(2)}|) < 0$$

for all $x \in I$, $\delta \in (0,\delta_0)$ and $\varepsilon \in (0,1]$. In particular,

$$f_{\varepsilon,\delta}(x) \leq f(\frac{1}{2}) \quad \text{for } |x| \leq \tfrac{1}{2}$$

for all $\delta \in (0,\delta_0)$ and $\varepsilon \in (0,1]$.

Proof. (i) Since φ_δ and f are even, $w_{\varepsilon,\delta}$ is even, and therefore $v_{\varepsilon,\delta}$ is odd. Thus $f_{\varepsilon,\delta}$ is even.

(ii) By definition and the Schwarz inequality we see that

$$\begin{aligned}&\|f_{\varepsilon,\delta}-f(-\frac{1}{2})\|_0 \le \sqrt{2}\|f_{\varepsilon,\delta}-f(-\frac{1}{2})\|_{L^\infty(I)}\\ &\le \sqrt{2}\|f'_{\varepsilon,\delta}\|_{L^1(I)} \le 2\|f'_{\varepsilon,\delta}\|_0 = 2\|v_{\varepsilon,\delta}\|_0,\\ &\|v_{\varepsilon,\delta}\|_0 \le \sqrt{2}\|v_{\varepsilon,\delta}\|_1 \le \sqrt{2}(\|w_{\varepsilon,\delta}\|_{L^1(I)}+\|f^{(2)}\|_{L^1(I)}).\end{aligned}$$

Since

$$\|w_{\varepsilon,\delta}\|_0 \le \sqrt{2}(1+\frac{1}{4!})+\|f^{(2)}\|_0 \quad \text{for } 0<\varepsilon\le 1$$

and

$$\|f^{(2)}_{\varepsilon,\delta}\|_0 \le C_\varphi(\|w_{\varepsilon,\delta}\|_0+\|f'\|_0)+\|f^{(2)}\|_0$$

for a constant depending only on φ, the Schwarz inequality yields a bound for $\|f_{\varepsilon,\delta}\|_0$, $\|f_{\varepsilon,\delta}\|_1$, $\|f_{\varepsilon,\delta}\|_2$ independent of $\varepsilon\in(0,1]$, $\delta\in(0,1/8)$.

We next estimate $f'_{\varepsilon,\delta}$ in $H^{m-1}(I)$ for $m\ge 3$. Since $H^{m-1}(I)$ is a Banach algebra for $m\ge 2$,

$$\begin{aligned}&\|\int_0^x w_{\varepsilon,\delta}(\xi)d\xi\|_{m-1} \le \|\int_0^x w_{\varepsilon,\delta}(\xi)d\xi\|_0+\|w_{\varepsilon,\delta}\|_{m-2}\\ &\le \sqrt{2}\|w_{\varepsilon,\delta}\|_0 + C^m(\|\varepsilon+\frac{x^4}{4!}\|_{m-2}\|\varphi_\delta\|_{m-2}+\|f^{(2)}\|_{m-2}\|1-\varphi_\delta\|_{m-2})\\ &\le C^m_\delta(1+\|f^{(2)}\|_{m-2})\end{aligned}$$

with some constant C^m_δ independent of $\varepsilon\in(0,1]$ but dependent on δ through φ_δ, while C^m depends only on m. Since

$$\|f'_{\varepsilon,\delta}\|_{m-1} \le C^{m+1}(\|\int_0^x w_{\varepsilon,\delta}(\xi)d\xi\|_{m-1}\|\varphi_{1/4}\|_{m-1}+\|f'\|_{m-1}\|1-\varphi_{1/4}\|_{m-1}),$$

we obtain

$$\|f'_{\varepsilon,\delta}\|_{m-1} \le C^m_\delta(1+\|f'\|_{m-1})$$

with some constant C^m_δ independent of $\varepsilon\in(0,1]$. This completes the proof of (ii).

(iii) Since both $f_{\varepsilon,\delta}$ and f are even it suffices to prove that $f_{\varepsilon,\delta}(x)=f(x)$ for $x\in[1/2,1)$. Since $\varphi_{1/4}(\xi)=0$ for $|\xi|\ge 1/2$,

$$\begin{aligned}f_{\varepsilon,\delta}(x)-f(x) &= \int_{-1/2}^x (v_{\varepsilon,\delta}(\xi)-f'(\xi))d\xi\\ &= \int_{-1/2}^{1/2}(\int_0^\xi w_{\varepsilon,\delta}(\eta)d\eta - f'(\xi))\varphi_{1/4}(\xi)d\xi \quad \text{for } x\ge 1/2.\end{aligned}$$

The integrand is an odd function so that $f_{\varepsilon,\delta}(x)=f(x)$ for $x\in[1/2,1)$.

(iv) Since $\delta<1/8$, $2\delta<1/4$, and we see that $\varphi_\delta(x)\equiv\varphi_{1/4}(x)\equiv 1$ for $|x|\le\delta$. Thus

$$f^{(2)}_{\varepsilon,\delta}(x)=v'_{\varepsilon,\delta}(x)=w_{\varepsilon,\delta}(x)=-\varepsilon-\frac{x^4}{4!} \quad \text{for } |x|\le\delta.$$

(v) We differentiate $f_{\varepsilon,\delta}$ twice to get

$$f^{(2)}_{\varepsilon,\delta}(x)=w_{\varepsilon,\delta}(x)\varphi_{1/4}(x)+\int_0^x w_{\varepsilon,\delta}(\xi)d\xi\varphi'_{1/4}(x)+f^{(2)}(x)(1-\varphi_{1/4}(x))-f'(x)\varphi'_{1/4}(x).$$

For $|x|\le 1/4$ we see that

$$f^{(2)}_{\varepsilon,\delta}(x)=w_{\varepsilon,\delta}(x)\le -\min(\varepsilon,\inf_I|f^{(2)}|)<0,$$

since $\varphi_{1/4}(x)=1$. For $1/2\le|x|<1$ we see that

$$f^{(2)}_{\varepsilon,\delta}(x)=f^{(2)}(x)\le -\inf_I|f^{(2)}|<0,$$

since $\varphi_{1/4}(x)=0$. It remains to estimate $f^{(2)}_{\varepsilon,\delta}(x)$ for x satisfying $1/4<|x|<1/2$. We may assume that $1/4<x<1/2$ since $f_{\varepsilon,\delta}$ is even. For such an x, since $\delta<1/8$, we see that

$$\begin{aligned} w_{\varepsilon,\delta}(x) &= f^{(2)}(x),\\ f^{(2)}_{\varepsilon,\delta}(x) &= f^{(2)}(x)+(\int_0^x w_{\varepsilon,\delta}(\xi)d\xi - f'(x))\varphi'_{1/4}(x). \end{aligned}$$

Because of $f'(0)=0$, we see that

$$\begin{aligned} \int_0^x w_{\varepsilon,\delta}(\xi)d\xi &= \int_0^x\{(-\varepsilon-\frac{\xi^4}{4!})\varphi_\delta(\xi)+f^{(2)}(\xi)(1-\varphi_\delta(\xi))\}d\xi\\ &= -\int_0^x(\varepsilon+\frac{\xi^4}{4!}+f^{(2)}(\xi))\varphi_\delta(\xi)d\xi+f'(x). \end{aligned}$$

Combining these two identities we arrive at

$$\begin{aligned} f^{(2)}_{\varepsilon,\delta}(x) &= f^{(2)}(x)-\int_0^x(\varepsilon+\frac{\xi^4}{4!}+f^{(2)}(\xi))\varphi_\delta(\xi)d\xi\varphi'_{1/4}(x)\\ &\le -\inf_I|f^{(2)}|+\int_0^{2\delta}\varphi_\delta(\xi)d\xi\sup_I|\varphi'_{1/4}|M \quad \text{for } \varepsilon\in(0,1] \end{aligned}$$

with $M=\sup_I|f^{(2)}|+1+1/4!$. Since

$$\int_0^{2\delta}\varphi_\delta(\xi)d\xi=\delta\int_0^2\varphi(\xi)d\xi\le 2\delta,$$

we obtain

$$f^{(2)}_{\varepsilon,\delta}(x)\le -\frac{1}{2}\inf_I|f^{(2)}|$$

for all $\varepsilon\in(0,1]$ and $x\in(1/4,1/2)$ provided that δ is sufficiently small, say,

$$0<\delta<\delta_0:=\frac{\inf_I|f^{(2)}|}{4\sup_I|\varphi'_{1/4}|M}.$$

We have thus proved (v). $\square$

Remark 2. From the proof we see that the constant C_δ^m in (ii) may tend to infinity as δ tends to zero for $m \geq 3$ since we have to estimate $\|\varphi_\delta\|_{m-2}$. For $m = 0, 1, 2$ the constant C_δ^m stays bounded as δ tends to zero.

3. Deformation of a strictly convex closed curve

We introduce a parametrization of simple, closed, strictly convex curves in the plane $\mathbf{R}^2$ by using a reference curve. Let M^0 be a smooth closed convex curve embedded in $\mathbf{R}^2$. We assume that M^0 is symmetric with respect to the y-axis and that M^0 is contained in $\{(x, y); y \leq 0\}$. We further assume that M^0 contains a straight line segment on the x-axis. Let η be the arc-length parameter of M^0 so that it parametrizes M^0 clock-wise. By definition M^0 is of the form

$$M^0 = \{X^0(\eta) = (X_1^0(\eta), X_2^0(\eta)) \in \mathbf{R}^2; \eta \in \mathbf{T} := \mathbf{R}/2L\mathbf{Z}\},$$

where $2L$ is the length of M^0. By symmetry the arc-length parameter η is taken so that

$$\begin{aligned} &X_1^0(\eta) = -X_1^0(-\eta) \geq 0, \quad |\eta| \leq L, \\ &X_2^0(\eta) = X_2^0(-\eta) \leq 0, \quad |\eta| \leq L, \\ &X^0(\eta) = (\eta, 0) \quad \text{for } |\eta| \leq a \text{ with some } a \in (0, L/2). \end{aligned}$$

Since M^0 is convex, the outward curvature $\kappa^0(\eta)$ is always nonpositive for $\eta \in \mathbf{T}$. We fix M^0 and call M^0 a reference curve. For an integer $m \geq 2$ let $E^m(\mathbf{T})$ be

$$E^m(\mathbf{T}) = \{d_0 \in H^m(\mathbf{T}); d_0(\eta) = d_0(-\eta) > 0 \quad \text{for } \eta \text{ satisfying } |\eta| \leq L\}.$$

Given $d_0 \in E^m(\mathbf{T})$, we associate a (at least) C^1 curve

$$\Gamma[d_0] = \{X^0(\eta) + d_0(\eta) n^0(\eta); \eta \in \mathbf{T}\},$$

where $n^0(\eta)$ is the outward normal of M^0. By symmetry of M^0 and $d_0 \in E^m(\mathbf{T})$, $\Gamma[d_0]$ is simple closed since M^0 is convex and $d_0(\eta) > 0$. Let Σ^m be the set of all $\Gamma[d_0]$ with $d_0 \in E^m(\mathbf{T})$ such that the outward curvature κ_0 of $\Gamma[d_0]$ is negative everywhere. Of course, for each $\Gamma_0 \in \Sigma^m$ there is a unique $d_0 \in E^m(\mathbf{T})$ such that $\Gamma_0 = \Gamma[d_0]$.

We shall deform $\Gamma[d_0] \in \Sigma^m$. Since the curvature κ_0 of $\Gamma[d_0]$ is negative,

$$\sup_{|\eta| \leq a} d_{0\eta\eta} < 0.$$

Since the restriction $r_a d_0$ of d_0 on $(-a, a)$ is even, $r_a d_0 \in H^m(-a, a)$. We deform $r_a d_0$ by $M_{\varepsilon,\delta}$ and dilation. For a function $f \in H^m(-\alpha, \alpha)$ and $\beta > 0$ let $D_\beta f$ denote

$$(D_\beta f)(x) = f(\beta x), \quad |x| < \frac{\alpha}{\beta}.$$

Evidently, D_β gives an isomorphism from $H^m(-\alpha, \alpha)$ onto $H^m(-\alpha/\beta, \alpha/\beta)$. It also preserves uniform nonnegativity of the second derivative of a function. For even $f \in H^m(-a, a)$ with $\sup_{|x| \leq a} f^{(2)} < 0$ we set

$$M_{\varepsilon,\delta}^a f = (D_{1/a} \circ M_{\varepsilon,\delta} \circ D_a) f \in H^m(-a, a),$$

where $M_{\varepsilon,\delta}$ is the deformation operator given in Section 2 with parameters $\varepsilon \in (0,1]$, $\delta \in (0,1/8)$. We deform d_0, defining $d_0^{\varepsilon,\delta}$ by

$$d_0^{\varepsilon,\delta}(\eta) = \begin{cases} (M_{\varepsilon,\delta}^a(r_a d_0))(\eta), & |\eta| \le a, \\ d_0(\eta), & a < |\eta| \le L. \end{cases} \tag{5}$$

The next results immediately follow from Lemma 1 and Remark 2.

Theorem 3. *Let $d_0 \in E^m(\mathbf{T})$ such that $\Gamma[d_0] \in \Sigma^m(\mathbf{T})$ with an integer $m \ge 2$. Let $0 < \varepsilon \le 1$ and $0 < \delta < 1/8$ and let $d_0^{\varepsilon,\delta}$ be given as in (5).*

(i) $\Gamma[d_0^{\varepsilon,\delta}]$ is symmetric with respect to the y-axis.

(ii) $d_0^{\varepsilon,\delta} \in H^m(\mathbf{T})$ and there exists a positive constant $b_{a,\delta}^m$ independent of ε such that

$$\|d_0^{\varepsilon,\delta}\|_{H^m(\mathbf{T})} \le b_{a,\delta}^m \quad \textit{for all } \varepsilon \in (0,1]$$

and $b_{a,\delta}^m$ is bounded as $\delta \to 0$ for $m = 2$ but unbounded as $\delta \to 0$ for $m \ge 3$.

(iii) Outside the set $\{(x,y); |x| \le \delta a, y > 0\}$ the set $\Gamma[d_0]$ agrees with $\Gamma[d_0^{\varepsilon,\delta}]$.

(iv) $\dfrac{d^2}{d\eta^2} d_0^{\varepsilon,\delta}(0) = -\varepsilon/a^2$, $\dfrac{d^4}{d\eta^4} d_0^{\varepsilon,\delta}(0) = 0$, $\dfrac{d^6}{d\eta^6} d_0^{\varepsilon,\delta}(0) = -1/a^6$.

(v) There exists $\delta_0 > 0$ such that

$$\sup_{|\eta| \le a} \frac{d^2}{d\eta^2} d_0^{\varepsilon,\delta}(\eta) < 0,$$

$$\inf_{|\eta| \le a} d_0^{\varepsilon,\delta}(\eta) \ge d_0(a) > 0$$

for all $\varepsilon \in (0,1]$ and $\delta \in (0,\delta_0)$. In particular, $\Gamma[d_0^{\varepsilon,\delta}] \in \Sigma^m(\mathbf{T})$ with $\inf_{\mathbf{T}} d_0^{\varepsilon,\delta} \ge \inf_{\mathbf{T}} d_0 > 0$ for all $\varepsilon \in (0,1]$ and $\delta \in (0,\delta_0)$.

For $\Gamma_0 = \Gamma[d_0] \in \Sigma^m$ we set

$$S^{\varepsilon,\delta}(\Gamma_0) := \Gamma[d_0^{\varepsilon,\delta}],$$

where $d_0^{\varepsilon,\delta}$ is given as in (5). If δ is taken small ($\delta < \delta_0$) as in Theorem 3 (v), so that $\Gamma[d_0^{\varepsilon,\delta}]$ has negative outward curvature everywhere, $S^{\varepsilon,\delta}$ gives a mapping from Σ^m into itself for $\varepsilon \in (0,1]$ and $\delta \in (0,\delta_0)$. $S^{\varepsilon,\delta}(\Gamma_0)$ is a deformation of Γ_0 near the point where Γ_0 crosses the positive y-axis. The outward curvature $\kappa^{\varepsilon,\delta}$ of $S^{\varepsilon,\delta}(\Gamma_0)$ and its derivatives at the intersection point A of $S^{\varepsilon,\delta}(\Gamma_0)$ and the positive y-axis can be calculated by Theorem 3 (iv). In fact,

$$\kappa^{\varepsilon,\delta} = -\varepsilon/a^2, \quad \kappa_{\eta\eta}^{\varepsilon,\delta} = 0, \quad \kappa_{\eta\eta\eta\eta}^{\varepsilon,\delta} = -1/a^6 \quad \text{at point } A.$$

This shows that the convexity of $S^{\varepsilon,\delta}(\Gamma_0)$ can be weakened compared with the negative of the fourth order derivative of the curvature near A by taking ε small.

4. Existence of solutions for motion by surface diffusion

We establish unique local existence of solutions of (1) which is useful to prove the loss of convexity. As in [9, 13], to find a local solution $\Gamma(t)$ of (1) for $\Gamma_0 = \Gamma[d_0] \in \Sigma^m$, it is equivalent to find $d = d(t,\eta)$ solving

$$\begin{cases} \dfrac{1-\kappa^0 d}{J} d_t = -\dfrac{1}{J}(\dfrac{1}{J}\kappa_\eta)_\eta, & (t,\eta) \in (0,T) \times \mathbf{T}, \\ d(0,\eta) = d_0(\eta), & \eta \in \mathbf{T} \end{cases} \tag{6}$$

at least for a short time by setting $\Gamma(t) = \Gamma[d(t,\cdot)]$; here J and κ are the arc-length element and the outward curvature of $\Gamma(t)$, respectively. Their explicit forms are

$$J = (d_\eta^2 + (1-\kappa^0 d)^2)^{1/2},$$
$$\kappa = \frac{1}{J^3}\{(1-\kappa^0 d)d_{\eta\eta} + 2\kappa^0 d_\eta^2 + \kappa_\eta^0 d d_\eta + \kappa^0(1-\kappa^0 d)^2\},$$

where κ^0 denotes the outward curvature of the reference curve M^0. The problem (6) is now written as

$$\begin{cases} d_t + J^{-4} d_{\eta\eta\eta\eta} + P d_{\eta\eta\eta} + Q = 0, & (t,\eta) \in (0,T) \times \mathbf{T}, \\ d(0,\eta) = d_0(\eta), & \eta \in \mathbf{T} \end{cases} \tag{7}$$

where $P = P(\eta, d, d_\eta, d_{\eta\eta})$ and $Q = Q(\eta, d, d_\eta, d_{\eta\eta})$ are expressed as polynomials of $(1-\kappa^0 d)^{-1}$, J^{-1}, $d^i\kappa^0/d\eta^i$ $(i = 0,1,2,3)$, $\partial^j d/\partial\eta^j$ $(j = 0,1,2)$. For η with $\kappa^0(\eta) = 0$ the evolution equation (6) is of the form

$$d_t = -\left\{\frac{1}{(1+d_\eta^2)^{1/2}}\left(\frac{d_{\eta\eta}}{(1+d_\eta^2)^{3/2}}\right)_\eta\right\}_\eta. \tag{8}$$

Proposition 4. *Let $m \geq 4$ be an integer. Let K_i be a positive constant for $i = 1,2$. Then there exists $T > 0$ such that for any $d_0 \in E^m(\mathbf{T})$ with $\|d_0\|_{H^m(\mathbf{T})} \leq K_1$ and $\min_{\mathbf{T}} d_0 \geq 1/K_2$ there exists a unique solution $d = d(t,\eta)$ of (7) that satisfies*

$$d \in L^2(0,T; H^{m+2}(\mathbf{T})), \quad d_t \in L^2(0,T; H^{m-2}(\mathbf{T})),$$
$$\sup_{0<t\leq T} \|d(t,\cdot)\|_{H^m(\mathbf{T})} \leq \mathcal{A}(K_1),$$
$$\inf_{[0,T]\times\mathbf{T}} d \geq 1/(2K_2)$$

with some nondecreasing function $\mathcal{A} = \mathcal{A}(K_1)$. Moreover, $d(t,\eta) = d(t,-\eta)$, $|\eta| \leq L$ so that $d(t,\cdot) \in E^m(\mathbf{T})$ for $t \in [0,T]$.

Unfortunately, the first assertion on unique existence of solutions does not directly follow from that of [13, Theorem 3] since the convexity of M^0 guarantees that a solution always exists without imposing smallness of d_0. The second assertion of Proposition 4 is not difficult to prove. Indeed, $d(t,-\eta)$ solves (7) with initial data $d_0(-\eta) = d_0(\eta)$. Then by uniqueness of solutions $d(t,-\eta) = d(t,\eta)$ for $(t,\eta) \in [0,T] \times \mathbf{T}$. This shows $d(t,\cdot) \in E^m(\mathbf{T})$. We shall prove the first assertion in Section 6 for reader's convenience although the basic idea for the proof is similar to that of [13, Theorem 3]. Applying Proposition 4 for deformed initial data $S^{\varepsilon,\delta}(\Gamma_0) = \Gamma[d_0^{\varepsilon,\delta}]$, we have uniform local existence for (1) by virtue of Theorem 3. We state this fact explicitly.

Theorem 5. *Let $m \geq 4$ be an integer and $\Gamma_0 = \Gamma[d_0] \in \Sigma^m$. Let δ_0 be the positive constant (determined by d_0) in Theorem 3 (v). Then there exists a $T_0^\delta > 0$ (independent of ε) such that for any $\varepsilon \in (0,1]$, $\delta \in (0, \delta_0)$ there exists a unique solution $d^{\varepsilon,\delta}$ of (7) in $(0, T_0^\delta) \times \mathbf{T}$ with initial data $d_0^{\varepsilon,\delta}$ (defined in Theorem 3) that satisfies*

$$d^{\varepsilon,\delta} \in L^2(0, T_0^\delta; H^{m+2}(\mathbf{T})), \quad d_t^{\varepsilon,\delta} \in L^2(0, T_0^\delta; H^{m-2}(\mathbf{T})),$$

$$\sup_{0<t\leq T_0^\delta} \|d^{\varepsilon,\delta}(t,\cdot)\|_{H^m(\mathbf{T})} \leq \mathcal{A}(b_{a,\delta}^m), \qquad \inf_{[0,T_0^\delta]\times\mathbf{T}} d^{\varepsilon,\delta} \geq \frac{1}{2} \min_{\mathbf{T}} d_0,$$

where $b_{a,\delta}^m$ is given in Theorem 3 (ii) and $\mathcal{A}$ is given in Proposition 4. Moreover, $d^{\varepsilon,\delta}(t,\cdot) \in E^m(\mathbf{T})$.

We postpone the proof of Proposition 4 and Theorem 5 to Section 6.

5. Nonpreserving of convexity

For any strictly convex smooth curve Γ_0 symmetric with respect to y-axis, it is easy to take a reference curve M^0 and $d_0 \in E^m(\mathbf{T})$ so that $\Gamma_0 = \Gamma[d_0]$ up to translation in the direction of the y-axis. By this interpretation the next result shows the existence of an initial convex curve (even nearly strictly convex curve) which loses its convexity during evolution by (1).

Theorem 6. *(Loss of convexity). Let $m \geq 11$ be an integer and $\Gamma_0 = \Gamma[d_0] \in \Sigma^m$. Let δ_0 be the positive constant defined in Theorem 3 and let δ be in $(0, \delta_0)$. Let $d_0^{\varepsilon,\delta}$ be the deformed function of d_0 (defined in Theorem 3) for $\varepsilon \in (0,1]$. Let T_0^δ be the time (defined in Theorem 5) such that there is a unique solution $d^{\varepsilon,\delta}$ of (7) in $(0, T_0^\delta) \times \mathbf{T}$ with initial data $d_0^{\varepsilon,\delta}$. Then there is ε_0^δ such that for any $\varepsilon \in (0, \varepsilon_0^\delta)$ there are $t_0^{\varepsilon,\delta}$ and $t_1^{\varepsilon,\delta}$ with $t_0^{\varepsilon,\delta} < T_0^\delta$ and $t_0^{\varepsilon,\delta} < t_1^{\varepsilon,\delta}$ with the property that the solution $\Gamma[d^{\varepsilon,\delta}(t,\cdot)]$ of (1) starting from $S^{\varepsilon,\delta}(\Gamma_0) \subset \Sigma^m$ loses its convexity at least for $t \in (t_0^{\varepsilon,\delta}, \min(T_0^\delta, t_1^{\varepsilon,\delta}))$. Moreover, $t_0^{\varepsilon,\delta} \to 0$ as $\varepsilon \to 0$.*

Proof. First step. We shall prove

$$d_{\eta\eta t}^{\varepsilon,\delta}(0,0) = a^{-6} + 54\varepsilon^5 a^{-10} \geq a^{-6} \quad \text{for } \varepsilon \in (0,1].$$

(Smallness of $\delta < \delta_0$ is not invoked in the first two steps.) Note that equation (7) is of the form (8) for $|\eta| < a$. Differentiating (8) in η twice and using the property $d^{\varepsilon,\delta}(t,\eta) = d^{\varepsilon,\delta}(t,-\eta)$, $|\eta| < L$, we arrive at

$$d_{\eta\eta t}^{\varepsilon,\delta}(t,0) = -(\frac{\partial^6}{\partial\eta^6} d^{\varepsilon,\delta} - 33(d_{\eta\eta}^{\varepsilon,\delta})^2 \frac{\partial^4}{\partial\eta^4} d^{\varepsilon,\delta} + 54(d_{\eta\eta}^{\varepsilon,\delta})^5)(t,0). \tag{9}$$

We set $t = 0$ and use the values of the derivatives of $d_0^{\varepsilon,\delta}$ calculated in Theorem 3 to get the desired value of $d_{\eta\eta t}^{\varepsilon,\delta}(0,0)$.

Second step. There is a bound K^δ depending on δ such that

$$\sup_{\varepsilon\in(0,1]}\sup_{t\in[0,T_0^\delta]}|d_{\eta\eta tt}^{\varepsilon,\delta}(t,0)|\le K^\delta.$$

Indeed, differentiating (9) twice in t and using (8), we observe that

$$\sup_{t\in[0,T_0^\delta]}|d_{\eta\eta tt}^{\varepsilon,\delta}(t,0)|\le G(\sum_{k=0}^{10}\sup_{t\in[0,T_0^\delta]}|\frac{\partial^k}{\partial\eta^k}d^{\varepsilon,\delta}(t,0)|)$$

with a positive nondecreasing function $G = G(\lambda)$. By the Sobolev inequality and the estimate of $d^{\varepsilon,\delta}(t,\cdot)$ in $H^m(\mathbf{T})$ in Theorem 5, we now obtain

$$\sup_{t\in[0,T_0^\delta]}|d_{\eta\eta tt}^{\varepsilon,\delta}(t,0)|\le G(C'b_{a,\delta}^{11})$$

with a constant C' depending only on L. Setting $K^\delta = G(C'b_{a,\delta}^{11})$ yields the desired estimate.

Third step. We shall complete the proof of Theorem 6. We take $0 < \delta < \delta_0$ so that the outward curvature is negative everywhere on $\Gamma[d_0^{\varepsilon,\delta}]$. By Taylor's expansion and the second step, we see that

$$\begin{aligned} d_{\eta\eta}^{\varepsilon,\delta}(t,0) &= d_{\eta\eta}^{\varepsilon,\delta}(0,0)+d_{\eta\eta t}^{\varepsilon,\delta}(0,0)t+\int_0^t(\int_0^\sigma d_{\eta\eta\tau\tau}^{\varepsilon,\delta}(\tau,0)d\tau)d\sigma \\ &\ge d_{0\eta\eta}^{\varepsilon,\delta}(0)+d_{\eta\eta t}^{\varepsilon,\delta}(0,0)t-t^2K^\delta \end{aligned}$$

for $0\le t\le T_0^\delta$. Since $d_{0\eta\eta}^{\varepsilon,\delta}(0) = -\varepsilon a^{-2}$ by Theorem 3 and $d_{\eta\eta}^{\varepsilon,\delta}(t,0)\ge a^{-6}$ by the first step, we now arrive at

$$d_{\eta\eta}^{\varepsilon,\delta}(t,0)\ge -\varepsilon a^{-2}+a^{-6}t-K^\delta t^2 \quad \text{for } t\in[0,T_0^\delta]. \tag{10}$$

We can take ε so small that the quadratic polynomial on the right-hand side has the smallest positive zero less than T_0^δ. In fact, we can take $\varepsilon_0^\delta > 0$ so small that

$$0<\frac{a^{-6}-(a^{-12}-4K^\delta\varepsilon_0^\delta a^{-2})^{1/2}}{2K^\delta}<T_0^\delta.$$

Then for each $\varepsilon\in(0,\varepsilon_0^\delta)$ the polynomial $-\varepsilon a^{-2}+a^{-6}t-K^\delta t^2$ has two positive zeros $t_0^{\varepsilon,\delta}<t_1^{\varepsilon,\delta}$ such that $t_0^{\varepsilon,\delta}<T_0^\delta$. By (10) for $\varepsilon\in(0,\varepsilon_0^\delta)$

$$d^{\varepsilon,\delta}(t,0)\ge -(t-t_0^{\varepsilon,\delta})(t-t_1^{\varepsilon,\delta}) \quad \text{for } t\in[0,T_0^\delta].$$

This implies that

$$d^{\varepsilon,\delta}(t,0)>0 \quad \text{for } t_0^{\varepsilon,\delta}<t<\min(t_1^{\varepsilon,\delta},T_0^\delta).$$

This shows that $\Gamma[d^{\varepsilon,\delta}(t,\cdot)]$ loses its convexity at least for $t_0^{\varepsilon,\delta}<t<\min(t_1^{\varepsilon,\delta},T_0^\delta)$. The assertion $t_0^{\varepsilon,\delta}\to 0$ as $\varepsilon\to 0$ follows from its definition. □

6. Proof of Proposition 4 and Theorem 5

Proposition 4 is based on the convexity $\kappa^0 \leq 0$ of the reference curve M^0. It guarantees that $1 - \kappa^0 d$ does not take zero for $d \geq 0$. Thus J^{-4}, P and Q in (7) are always regular for $d \geq 0$. To prove Proposition 4 we begin with a general result guaranteeing the lower bound of solutions.

General framework. Let $\mathbf{T} = \mathbf{R}/\omega\mathbf{Z}$, where ω is a positive constant. We consider a general equation of the form:

$$\begin{cases} u_t + a(x,u,u_x)u_{xxxx} + b(x,u,u_x,u_{xx})u_{xxx} + c(x,u,u_x,u_{xx}) = 0, \\ u(0,x) = u_0(x) \end{cases} \tag{11}$$

for $t \geq 0$ and $x \in \mathbf{T}$. We assume the following:

(H1) There are positive constants Λ_*, Λ and a_1 with $\Lambda_* < \Lambda$ such that $a(x,\alpha_0,\alpha_1) \geq a_1$ for $x \in \mathbf{T}$, $\Lambda_* \leq \alpha_0 \leq \Lambda$ and $|\alpha_1| \leq \Lambda$.

(H2) The functions $a(x,\alpha_0,\alpha_1)$, $b(x,\alpha_0,\alpha_1,\alpha_2)$ and $c(x,\alpha_0,\alpha_1,\alpha_2)$ are regular in $\mathbf{T} \times \mathbf{R}^3$.

Then we have

Lemma 7. *Let $m \geq 4$ be an integer. Assume (H1) and (H2). Then, for any $u_0 \in H^m(\mathbf{T})$ with $\min_{x\in\mathbf{T}} u_0(x) \geq 2\Lambda_*$, there are a*

$$T_0 = T_0(\|u_0\|_{H^m(\mathbf{T})}, \min_{x\in\mathbf{T}} u_0(x)) > 0$$

and a unique solution $u(t,x)$ of (11) satisfying

$$\begin{gathered} u \in L^2(0,T_0;H^{m+2}(\mathbf{T})), \quad u_t \in L^2(0,T_0;H^{m-2}(\mathbf{T})), \\ \sup_{t\in[0,T_0]} \|u(t)\|_{H^m(\mathbf{T})} \leq \mathcal{A}(\|u_0\|_{H^m(\mathbf{T})}), \\ \min_{(t,x)\in[0,T_0]\times\mathbf{T}} u(t,x) \geq \frac{1}{2}\min_{x\in\mathbf{T}} u_0(x) \end{gathered} \tag{12}$$

where $\mathcal{A}$ is some nondecreasing function.

Remark 8. (i) $T_0(\mu_0,\mu_1)$ is nonincreasing in $\mu_0 \geq 0$ and nondecreasing in $\mu_1 \geq 0$.
(ii) Proposition 4 follows directly from Lemma 7.

Proof. Set

$$Z_T^m = \{u \in L^2(0,T;H^{m+2}(\mathbf{T})); u_t \in L^2(0,T;H^{m-2}(\mathbf{T})), u(0,x) = u_0(x)\}.$$

As in the proof of Theorem 1 of [13], we can show by virtue of (H1) and (H2) that for any $T > 0$ and given $v \in Z_T^m$ the linear inhomogeneous equation for $w(t,x)$

$$\begin{cases} w_t + a(x,u_0,u_{0x})w_{xxxx} = & \{a(x,u_0,u_{0x}) - a(x,v,v_x)\}v_{xxxx} \\ & -b(x,v,v_x,v_{xx})v_{xxx} - c(x,v,v_x,v_{xx}), \\ w(0,x) = u_0(x) & \end{cases}$$

admits a unique solution $w \in Z_T^m$. It can also be shown that there are positive constants T_1 nonincreasing in $\|u_0\|_{H^m(\mathbf{T})}$ and R nondecreasing in $\|u_0\|_{H^m(\mathbf{T})}$ such that the mapping

$$B \to B; \quad v \mapsto w$$

admits a unique fixed point u satisfying

$$\sup_{t\in[0,T_1]} \|u(t)\|_{H^m(\mathbf{T})} \le \mathcal{A}(\|u_0\|_{H^m(\mathbf{T})})$$

with some nondecreasing function $\mathcal{A}$, where

$$B = \{u \in Z^m_{T_1}; \|u\|_{Z^m_{T_1}} \le R\},$$

$$\|u\|_{Z^m_{T_1}} = \|u\|_{L^2(0,T_1;H^{m+2}(\mathbf{T}))} + \|u_t\|_{L^2(0,T_1;H^{m-2}(\mathbf{T}))}.$$

Furthermore, since

$$\|u(t) - u_0\|_{L^\infty(\mathbf{T})} \le C_L\|u(t) - u_0\|_{H^1(\mathbf{T})} \le C_L \int_0^t \|u_\tau(\tau)\|_{H^1(\mathbf{T})}d\tau$$

$$\le C_L \left(\int_0^{T_1} \|u_t(t)\|^2_{H^{m-2}(\mathbf{T})}dt\right)^{1/2} t^{1/2} \le C_L R t^{1/2}$$

for some constant $C_L > 0$, we have

$$u(t,x) \ge \min_{x\in\mathbf{T}} u_0(x) - C_L R t^{1/2}$$

for $t \in [0, T_1]$ and $x \in \mathbf{T}$. Thus, if we put

$$T_0 = \left\{\min\left(\frac{\min_{x\in\mathbf{T}} u_0(x)}{2C_L R}, T_1\right)\right\}^2,$$

then we obtain (12). □

Proof of Theorem 5. We put

$$\alpha_* := \frac{1}{2}\min_{\eta\in\mathbf{T}} d_0(\eta) > 0,$$

$$A(\eta,\alpha_0,\alpha_1) := q(\frac{2}{\alpha_*}\alpha_0)(\alpha_1^2 + (1-\kappa^0(\eta)\alpha_0)^2)^{-2},$$

$$B(\eta,\alpha_0,\alpha_1,\alpha_2) := q(\frac{2}{\alpha_*}\alpha_0)P(\eta,\alpha_0,\alpha_1,\alpha_2),$$

$$C(\eta,\alpha_0,\alpha_1,\alpha_2) := q(\frac{2}{\alpha_*}\alpha_0)Q(\eta,\alpha_0,\alpha_1,\alpha_2)$$

for $\eta \in \mathbf{T}$ and $(\alpha_0,\alpha_1,\alpha_2) \in \mathbf{R}^3$, where q is in (4). We note that

$$A(\eta,\alpha_0,\alpha_1) = (\alpha_1^2 + (1-\kappa^0(\eta)\alpha_0)^2)^{-2}, \tag{13}$$

$$B(\eta,\alpha_0,\alpha_1,\alpha_2) = P(\eta,\alpha_0,\alpha_1,\alpha_2), \tag{14}$$

$$C(\eta,\alpha_0,\alpha_1,\alpha_2) = Q(\eta,\alpha_0,\alpha_1,\alpha_2) \tag{15}$$

for $\eta \in \mathbf{T}$, $\alpha_0 \ge \alpha_*/2$ and $(\alpha_1,\alpha_2) \in \mathbf{R}^2$, since $q(2\alpha_0/\alpha_*) = 1$ for $\alpha_0 \ge \alpha_*/2$.

We consider the following equation:

$$\begin{cases} d_t + A(\eta,d,d_\eta)d_{\eta\eta\eta\eta} + B(\eta,d,d_\eta,d_{\eta\eta})d_{\eta\eta\eta} + C(\eta,d,d_\eta,d_{\eta\eta}) = 0, \\ d(0,\eta) = d_0^{\varepsilon,\delta}(\eta) \end{cases} \tag{16}$$

for $t \ge 0$ and $\eta \in \mathbf{T}$.

We check (H1) and (H2) for (16). $A(\eta,\alpha_0,\alpha_1)$ satisfies (H1) with $\Lambda_* = \alpha_*$, $\Lambda = 2C_L\|d_0^{\varepsilon,\delta}\|_{H^m(\mathbf{T})}$ (where C_L is the constant in the Sobolev inequality $\|f\|_{L^\infty} \le C_L\|f\|_{H^1(\mathbf{T})}$) and

$$a_1 = (\Lambda^2 + (1 + \|\kappa^0\|_{L^\infty(\mathbf{T})}\Lambda)^2)^{-2}.$$

Since $\kappa^0(\eta) \le 0$, $A(\eta,\alpha_0,\alpha_1)$, $B(\eta,\alpha_0,\alpha_1,\alpha_2)$ and $C(\eta,\alpha_0,\alpha_1,\alpha_2)$ also satisfy (H2).

Therefore, it follows from Lemma 7 that there are

$$T_0^{\varepsilon,\delta} := T_0(\|d_0^{\varepsilon,\delta}\|_{H^m(\mathbf{T})}, \min_{\eta\in\mathbf{T}} d_0^{\varepsilon,\delta}(\eta)) > 0$$

and a unique solution $d^{\varepsilon,\delta}(t,\eta)$ of (16) satisfying

$$d^{\varepsilon,\delta} \in L^2(0,T_0^{\varepsilon,\delta}; H^{m+2}(\mathbf{T})), \quad d_t^{\varepsilon,\delta} \in L^2(0,T_0^{\varepsilon,\delta}; H^{m-2}(\mathbf{T}))$$

$$\sup_{t\in[0,T_0^{\varepsilon,\delta}]} \|d^{\varepsilon,\delta}(t)\|_{H^m(\mathbf{T})} \le \mathcal{A}(\|d_0^{\varepsilon,\delta}\|_{H^m(\mathbf{T})}),$$

$$\min_{(t,\eta)\in[0,T_0^{\varepsilon,\delta}]\times\mathbf{T}} d^{\varepsilon,\delta}(t,\eta) \ge \frac{1}{2}\min_{\eta\in\mathbf{T}} d_0^{\varepsilon,\delta}(\eta).$$

Put

$$T_0^{\delta} := T_0(b_{a,\delta}^m, \min_{\eta\in\mathbf{T}} d_0(\eta)) > 0,$$

with $b_{a,\delta}^m$ as in Theorem 3 (ii). Then it follows from Remark 8 and Theorem 3 (ii) and (v) that

$$T_0^{\varepsilon,\delta} \ge T_0^{\delta},$$

$$\sup_{t\in[0,T_0^{\delta}]} \|d^{\varepsilon,\delta}(t)\|_{H^m(\mathbf{T})} \le \mathcal{A}(b_{a,\delta}^m),$$

$$\min_{(t,\eta)\in[0,T_0^{\delta}]\times\mathbf{T}} d^{\varepsilon,\delta}(t,\eta) \ge \frac{1}{2}\min_{\eta\in\mathbf{T}} d_0(\eta) = \alpha_*$$

for $\varepsilon \in (0,1]$. Furthermore, the last inequality and (13)–(15) yield

$$A(\eta, d^{\varepsilon,\delta}, d_\eta^{\varepsilon,\delta}) = ((d_\eta^{\varepsilon,\delta})^2 + (1 - \kappa^0 d^{\varepsilon,\delta})^2)^{-2},$$

$$B(\eta, d^{\varepsilon,\delta}, d_\eta^{\varepsilon,\delta}, d_{\eta\eta}^{\varepsilon,\delta}) = P(\eta, d^{\varepsilon,\delta}, d_\eta^{\varepsilon,\delta}, d_{\eta\eta}^{\varepsilon,\delta}),$$

$$C(\eta, d^{\varepsilon,\delta}, d_\eta^{\varepsilon,\delta}, d_{\eta\eta}^{\varepsilon,\delta}) = Q(\eta, d^{\varepsilon,\delta}, d_\eta^{\varepsilon,\delta}, d_{\eta\eta}^{\varepsilon,\delta}).$$

Thus, we conclude that $d^{\varepsilon,\delta}(t,\eta)$ for $t \in [0,T_0^{\delta}]$ and $\eta \in \mathbf{T}$ is the desired solution of (7). The proof of Theorem 5 is complete. □

References

[1] S. A. Alvarez and C. Liu, "Motion of a closed curve by minus the surface Laplacian of curvature", preprint.

[2] H. Amann, "Nonhomogeneous linear and quasilinear elliptic and parabolic boundary value problems", in H. J. Schmeisser, H. Triebel, editors, *Function Spaces, Differential Operators and Nonlinear Analysis*, Teubner, Stuttgart, Leipzig, 9–126 (1993).

[3] A. J. Bernoff, A. L. Bertozzi and T. P. Witelski, "Axisymmetric surface diffusion: Dynamics and stability of self-similar pinch-off", preprint.

[4] A. Bonami, D. Hilhorst, E. Logak and M. Mimura, "Singular limit of a chemotaxis-growth model", preprint, 1998.

[5] J. W. Cahn and J. E. Taylor, "Surface motion by surface diffusion", Acta Metallurgica **42** (1994), 1045–1063.

[6] J. W. Cahn, C. M. Elliott and A. Novick-Cohen, "The Cahn-Hilliard equation: Surface motion by the Laplacian of the mean curvature", Euro. J. Appl. Math. **7** (1996), 287–301.

[7] B. D. Coleman, R. S. Falk and M. Moakher, "Stability of cylindrical bodies in the theory of surface diffusion", Physica D **89** (1995), 123–135.

[8] B. D. Coleman, R. S. Falk and M. Moakher, "Space-time finite element methods for surface diffusion with applications to the theory of the stability of cylinders", SIAM J. Sci. Comput. **17** (1996), 1434–1448.

[9] C. M. Elliott and H. Garcke, "Existence results for diffusive surface motion laws", Adv. Math. Sci. Appl. **7** (1997), 467–490.

[10] J. Escher, U. F. Mayer and G. Simonett, "The surface diffusion flow for immersed hypersurfaces", SIAM J. Math. Anal., to appear.

[11] M. Gage, "On an area preserving evolution equation for plane curves", Contemporary Mathematics **51** (1986), 51–62.

[12] M. Gage and R. S. Hamilton, "The heat equation shrinking convex plane curves", J. Differential Geometry **23** (1986), 69–96.

[13] Y. Giga and K. Ito, "On pinching of curves moved by surface diffusion", Comm. Appl. Anal. **2** (1998), 393–405.

[14] M. Grayson, "The heat equation shrinks embedded plane curves to round points", J. Differential Geometry **26** (1987), 285–314.

[15] U. F. Mayer, "One-sided Mullins-Sekerka flow does not preserve convexity", Electronic J. Differential Equations **1993** (1993), 1–7.

[16] U. F. Mayer, "Two-sided Mullins-Sekerka flow does not preserve convexity", preprint.

[17] W. W. Mullins, "Theory of thermal grooving", J. Appl. Phys. **28** (1957), 333–339.

Yoshikazu Giga, Kazuo Ito
Department of Mathematics, Faculty of Science
Hokkaido University
Sapporo 060-0810, Japan

Progress in Nonlinear Differential Equations
and Their Applications, Vol. 35
© 1999 Birkhäuser Verlag Basel/Switzerland

On the Equations of Fiber Spinning in Nonisothermal Viscous Flow

Thomas Hagen and Michael Renardy

Abstract. We discuss the equations of nonisothermal fiber spinning for viscous fluids. Existence, uniqueness and regularity of solutions will be established. Our solution strategy employs weak* compactness arguments together with the Contraction Mapping Principle in suitable Banach spaces. Energy estimates of the solutions for certain boundary-initial value problems for the first-order hyperbolic equation form the basis of our results.

1. Introduction

Fiber spinning is a widely used industrial procedure to manufacture long thin polymeric filaments. The exact spinning method is usually dictated by the chemical and physical properties of the polymer. For polymers with thermally stable melts melt-spinning is a particularly simple and effective technique (cf. [8], [12]). In this nonisothermal process fibers are formed by extruding the molten polymer from a pressurized reservoir through a small circular orifice (spinneret), stretching and cooling the liquid jet, and winding the solidified filament on a take-up device (spool). Solidification is solely caused by heat transfer. Dry-spinning and wet-spinning are similar spinning procedures that, however, rely on chemical rather than thermal properties of the polymer.

In this exposition we shall discuss the equations governing nonisothermal fiber spinning for Newtonian fluids. We assume that the flow is axisymmetric and vertically downward. Hence a cylindrical coordinate system, centered at the spinneret exit, is reduced to a radial (r) and an axial (z) component. The flow is completely described by the fluid velocity $\mathbf{v}$, the fluid radius R and the fluid temperature T. Although the process is intended to be steady-state (in the Eulerian sense), we allow time-dependence (t) of all quantities. Following [2, p. 2545], [7, pp. 427–429] and [8, pp. 58–63], we suppose a thin filament approximation: velocity $\mathbf{v}$, radius R and temperature T depend on the axial coordinate z and time t only. In addition, $\mathbf{v}$ has a nonvanishing axial component v only, i.e. $\mathbf{v} = (0, 0, v)$. This

This research was supported by the Department of Mathematics at VPI&SU and by the National Science Foundation under Grant DMS-9622735.

assumption is reasonable if the fluid radius is small compared to the length l of the draw-down region.

The fluid exits the spinneret at $z = 0$ with radius R_E, velocity v_E and temperature T_E. The velocity v_S of the take-up spool is larger than the exit velocity, so that the thread is actually strechted. In the flow region between extrusion and take-up the liquid jet is cooled by the ambient gas until the polymer reaches its solidification temperature T_S $(< T_E)$ and freezes. We normalize the temperature of the environment to 0, so that T_S is positive. As soon as the polymer solidifies, it is no longer extended and moves with the wind-up velocity v_S on to the spool.

Finally, following [2, pp. 2542–2545] and [8, p. 56], we shall assume that the viscous forces are the dominating forces. Then we obtain the following equations of change for the uniaxial extension of the fluid before it solidifies:

Balance of Mass

$$\frac{\partial}{\partial t}\left(R^2(t,z)\right) + \frac{\partial}{\partial z}\left(v(t,z)\, R^2(t,z)\right) = 0 \tag{1.1}$$

Balance of Momentum

$$\frac{\partial}{\partial z}\left(R^2(t,z)\,\frac{\partial}{\partial z}v(t,z)\right) = 0 \tag{1.2}$$

Balance of Energy

$$\frac{\partial}{\partial t}T(t,z) + v(t,z)\,\frac{\partial}{\partial z}T(t,z) + \beta\,\frac{T(t,z)}{R(t,z)} = 0 \tag{1.3}$$

The β in Eq. (1.3) is a constant, positive heat transfer coefficient. The tensile viscosity is assumed as constant and hence factors out. Eqs. (1.1)–(1.3) are complemented by the boundary and initial conditions

$$\text{at } z = 0: \qquad v = v_E, \quad R = R_E, \quad T = T_E, \tag{1.4}$$

$$\text{at } T = T_S: \qquad v = v_S, \tag{1.5}$$

$$\text{at } t = 0: \qquad R = R^0, \quad T = T^0. \tag{1.6}$$

We remark that the boundary condition (1.5) is given implicitly, and that initial and boundary conditions have to satisfy certain compatibility conditions.

In the following sections we will discuss the (local in time) existence, uniqueness and regularity of solutions to the boundary-initial value problem (1.1)–(1.6). We first derive an equivalent set of equations in Section 2. These transformed equations will have the advantage that all boundary conditions are stated explicitly. Our existence proof uses the fixed point principle for contractions. We begin with the discussion of the linear first-order hyperbolic equation

$$u_t = p\,u_x + f, \quad t \geq 0,\ x \in [a,b], \tag{1.7}$$

subject to the conditions

$$u(0,x) = u^0(x) \quad \text{and} \quad u(t,b) = u^b(t). \tag{1.8}$$

This problem will be treated in Section 3. Our approach employs an approximation scheme for Eq. (1.7) through smooth solutions, and weak* compactness arguments, based on certain energy estimates. Similar strategies can be found elsewhere (cf. [4], [5], [9]). In the final section we shall obtain the sought-for existence, uniqueness and regularity results for the reformulated equations of fiber spinning.

2. The model equations

2.1. Reformulated Equations

Eqs. (1.1)–(1.3) are to be satisfied as long as the temperature T does not reach the freezing point T_S. This condition like the boundary condition (1.5) is stated implicitly and complicates any further analysis. Hence it is useful to replace the free variable z by T. To this end we formally assume $\frac{\partial}{\partial z}T \neq 0$. Thus for $T_S \leq \tilde{T} \leq T_E$, the equation

$$T(t,z) = \tilde{T} \tag{2.1}$$

has a unique solution $z = \tilde{z}(t,\tilde{T})$. Now we redefine:

$$\begin{aligned} \tilde{R}(t,\tilde{T}) &= R(t,\tilde{z}(t,\tilde{T})), & (2.2)\\ \tilde{v}(t,\tilde{T}) &= v(t,\tilde{z}(t,\tilde{T})). & (2.3) \end{aligned}$$

In general, if we let $\tilde{y}(t,\tilde{T}) = y(t,\tilde{z}(t,\tilde{T}))$, then we obtain:

$$\begin{aligned} \tilde{y}_t(t,\tilde{T}) &= y_t(t,z) + y_z(t,z)\,\tilde{z}_t(t,\tilde{T}), & (2.4)\\ \tilde{y}_{\tilde{T}}(t,\tilde{T}) &= y_z(t,z)\,\tilde{z}_{\tilde{T}}(t,\tilde{T}), & (2.5)\\ \tilde{y}_{\tilde{T}\tilde{T}}(t,\tilde{T}) &= y_z(t,z)\,\tilde{z}_{\tilde{T}\tilde{T}}(t,\tilde{T}) + y_{zz}(t,z)\,\tilde{z}^2_{\tilde{T}}(t,\tilde{T}). & (2.6) \end{aligned}$$

We also remark that

$$T_z(t,z)\,\tilde{z}_{\tilde{T}}(t,\tilde{T}) = 1, \tag{2.7}$$
$$T_t(t,z) + T_z(t,z)\,\tilde{z}_t(t,\tilde{T}) = 0. \tag{2.8}$$

Replacing y, $\tilde{y}$ by R^2, $\tilde{R}^2$ and v, $\tilde{v}$, resp., in Eqs. (2.4)–(2.6), inserting the resultant equations together with Eqs. (2.7)–(2.8) into Eqs. (1.1)–(1.3), dropping all tildes, and finally setting $A \stackrel{\text{def}}{=} \pi R^2$, $\alpha \stackrel{\text{def}}{=} \sqrt{\pi}\,\beta$, we recover the equations

$$A_t(t,T) = \frac{\alpha T}{\sqrt{A(t,T)}}\,A_T(t,T) - \frac{v_T(t,T)}{z_T(t,T)}\,A(t,T), \tag{2.9}$$

$$\frac{\partial}{\partial T}\left(\frac{v_T(t,T)}{z_T(t,T)}\,A(t,T)\right) = 0, \tag{2.10}$$

$$z_t(t,T) = \frac{\alpha T}{\sqrt{A(t,T)}}\,z_T(t,T) + v(t,T). \tag{2.11}$$

Eqs. (2.9)–(2.11) are accompanied by the corresponding boundary conditions

$$A(t,T_E) = A_E(t), \tag{2.12}$$

$$z(t,T_E) = 0, \tag{2.13}$$

$$v(t,T_E) = v_E(t), \tag{2.14}$$

$$v(t,T_S) = v_S(t), \tag{2.15}$$

and the initial conditions

$$A(0,T) = A^0(T), \tag{2.16}$$

$$z(0,T) = z^0(T). \tag{2.17}$$

Here we set $A_E \stackrel{\text{def}}{=} \pi R_E^2$. In lieu of discussing Eqs. (1.1)–(1.6), we shall analyze the boundary-initial value problem posed in Eqs. (2.9)–(2.17). If a solution z of the transformed problem is continuously differentiable in both arguments, the formal transformation (2.2)–(2.3) is indeed justified as long as z_T does not vanish. To this end, we should obviously assume $z_T^0 < 0$.

2.2. Validity of the Model

Eqs. (2.9)–(2.17) form an idealized model for the dynamics of fiber spinning. Nonetheless, this model provides a reasonable qualitative description of the flow behavior from the standpoint of continuum mechanics and is applicable to both viscous fluids (e.g. glasses) and weakly viscoelastic fluids (e.g. polycondensates) with Newtonian-like response in elongation. Our assumption that the flow is uniaxial and extensional is not true close to the orifice, since the fluid, before exiting the spinneret, is exposed to a strong shear stress. As soon as the fluid leaves the orifice, the velocity profile undergoes a transition from shear flow to extensional flow. This velocity rearrangement is accompanied by a change of the fluid jet diameter (die swell or contraction). However, for Newtonian fluids this effect is negligible if the draw-down region is long (cf. [11, p. 43]). To accomodate die swell, we could consider the flow farther downstream where a fully extensional behavior has developed, and impose appropriate upstream conditions on the flow. Then we would have to replace the boundary condition (2.13) by

$$z(t,T_E) = z_E(t) \tag{2.18}$$

with some prescribed function z_E. Our existence results will actually hold for this boundary condition.

In the momentum balance, Eq. (1.2), we have left out inertia, gravity, aerodynamical friction and surface tension. Surface tension is small for melt-spinning, thus being dropped (cf. [8, p. 95]). For fluids that remain viscous, even at large temperatures, it is reasonable to neglect inertia as well. However, we can expect an exponentially decreasing tensile viscosity η^* with increasing temperatures (Arrhenius equation), so that inertial forces may dominate at least throughout the high-temperature region of the flow. On the other hand, fluids of low viscosity may not be spinnable at all. In Eq. (2.10) the tensile viscosity does not appear explicitly since we have assumed it independent of temperature (and hence independent of axial displacement). If we included a temperature-dependent viscosity η^*, the momentum balance, Eq. (2.10), would change to

$$\frac{\partial}{\partial T}\left(\frac{v_T(t,T)}{z_T(t,T)}\,\eta^*(T)\,A(t,T)\right)=0. \tag{2.19}$$

Since Eq. (2.19) is qualitatively not different from Eq. (2.10) for sufficiently smooth η^* and since we prefer to avoid yet more technicalities in this already detail-laden exposition, we shall assume the tensile visosity η^* (and also the heat-transfer coefficient α) as constant. However, we point out that straightforward, but tedious calculations would show that our results remain correct if this assumption is dropped.

For more details on the validity of the thin filament approximation and related topics we refer the reader to [2], [7], [8], [11] and [12].

3. The first order hyperbolic equation

Eqs. (2.9) and (2.11) suggest to study boundary-initial value problems of the form

$$u_t(t,x)=p(t,x)\,u_x(t,x)+f(t,x),\quad t\in[0,t_0],x\in[a,b], \tag{3.1}$$
$$u(0,x)=u^0(x),\quad x\in[a,b], \tag{3.2}$$
$$u(t,b)=u^b(t),\quad t\in[0,t_0]. \tag{3.3}$$

The interval $[a,b]$ shall be chosen once and for all. Thus any dependence on the quantities a and b will be suppressed. Also if the situation is clear, we will frequently omit the arguments of functions.

In this section we shall discuss existence, uniqueness and regularity of solutions for (3.1)–(3.3). Our main result is Theorem 3.5 .

We shall interpret a function r of the form $r=r(t,x)$ as a function of t, that takes values in a function space over the spatial variable x. Therefore we may write $r(t)$ in lieu of $r(t,\cdot)$ at our convenience. All function spaces will be real

We shall also use several abbreviations: Let $r_1<r_2$, $s_1<s_2$, $t_1>0$ and m, n, $k\in\mathbb{N}_0$. Then we write

1. $||\cdot||_p$ for the norm on the Lebesgue space $L^p(r_1,r_2)$, $1\le p\le\infty$,
2. $||\cdot||_{H^k}$ for the norm on the Sobolev space $H^k(r_1,r_2)$,
3. $||\cdot||_{m,n}$ for the norm on the Sobolev space $W^{m,\infty}([r_1,r_2];H^n(s_1,s_2))$,
4. $||\cdot||_{H^{m,n}}$ for the norm on the Sobolev space $H^m([r_1,r_2];H^n(s_1,s_2))$.

By convention, the preceding norms will always be understood with respect to (w.r.t.) the entire domain of the particular functions (unless noted otherwise), so that the numbers r_1, r_2, s_1 and s_2 are clear from the context. For the definition of Sobolev spaces see [1]. We shall also use the notation

1. $||\cdot||_{p,[t]}$ for the seminorms on the space $L^p(0,t_1)$, $1 \le p \le \infty$, defined for $0 \le t \le t_1$ by

$$||f||_{p,[t]} \stackrel{\text{def}}{=} ||f|_{[0,t]}||_p, \tag{3.4}$$

2. $||\cdot||_{H^k[t]}$ for the seminorms on the space $H^k(0,t_1)$, defined for $0 \le t \le t_1$ by

$$||f||_{H^k[t]} \stackrel{\text{def}}{=} ||f|_{[0,t]}||_{H^k}, \tag{3.5}$$

3. $||\cdot||_{m,n,[t]}$ for the seminorms on the space $W^{m,\infty}([0,t_1];H^n(s_1,s_2))$, defined for $0 \le t \le t_1$ by

$$||f||_{m,n,[t]} \stackrel{\text{def}}{=} ||f|_{[0,t]}||_{m,n}, \tag{3.6}$$

4. $||\cdot||_{H^{m,n}[t]}$ for the seminorms on the space $H^m([0,t_1];H^n(s_1,s_2))$, defined for $0 \le t \le t_1$ by

$$||f||_{H^{m,n}[t]} \stackrel{\text{def}}{=} ||f|_{[0,t]}||_{H^{m,n}}. \tag{3.7}$$

3.1. Statement of the Main Result

Definition 3.1. *We shall call a function u a solution of* (3.1)–(3.3) *if and only if*

$$u \in W^{1,\infty}([0,t_0];L^2(a,b)) \cap L^\infty([0,t_0];H^1(a,b)), \tag{3.8}$$
$$u \textit{ satisfies Eq. } (3.1), \tag{3.9}$$
$$u \textit{ satisfies Eqs. } (3.2) \textit{ and } (3.3) \textit{ pointwise.} \tag{3.10}$$

Remark 3.2 Condition (3.10) for a function u, satisfying (3.8), requires that $u(t,b)$ exist for each $t \in [0,t_0]$. This requirement is indeed satisfied according to the following lemma and the Sobolev embedding theorem.

Lemma 3.3. *Given the Hilbert spaces V and H such that V is continuously and densely embedded in H. Let V^* be the dual space of V, paired through the inner product $(\cdot,\cdot)$ on H. Suppose that $h \in L^\infty([0,t_0];V) \cap C([0,t_0];H)$. Then $h(t) \in V$ for every $t \in [0,t_0]$, and the map $t \mapsto (h(t),\phi)$ is continuous on $[0,t_0]$ for every $\phi \in V^*$.*

Proof. See [10, p. 389]. □

Definition 3.4. *We shall call a function h on $[0,t_0] \times [a,b]$ boundary-regular if and only if*

$$h \in W^{1,\infty}([0,t_0];H^1(a,b)) \cap L^\infty([0,t_0];H^2(a,b)), \tag{3.11}$$
$$h_x(\cdot,a), h_x(\cdot,b) \in H^1(0,t_0). \tag{3.12}$$

Theorem 3.5. *Given functions f and p on $[0,t_0]\times[a,b]$, u^0 on $[a,b]$ and u^b on $[0,t_0]$ such that*

$$p \text{ and } f \text{ are boundary-regular,} \tag{3.13}$$
$$p>0 \text{ on } [0,t_0]\times[a,b], \tag{3.14}$$
$$u^0\in H^2(a,b), \tag{3.15}$$
$$u^b\in H^2(0,t_0), \tag{3.16}$$
$$u^0(b)=u^b(0), \tag{3.17}$$
$$u_t^b(0)=p(0,b)\,u_x^0(b)+f(0,b). \tag{3.18}$$

Then the boundary-initial value problem (3.1)–(3.3) has a boundary-regular solution u such that

$$u\in C^1([0,t_0];H^1(a,b))\cap C([0,t_0];H^2(a,b)), \tag{3.19}$$
$$u \text{ is unique in } W^{1,\infty}([0,t_0];L^2(a,b))\cap L^\infty([0,t_0];H^1(a,b)). \tag{3.20}$$

Remark 3.6

(a) Theorem 3.5 yields the existence and uniqueness of classical solutions for problem (3.1)–(3.3).

(b) If u and w are any two solutions of (3.1)–(3.3) in $W^{1,\infty}([0,t_0];L^2(a,b))\cap L^\infty([0,t_0];H^1(a,b))$, then they satisfy

$$\int_a^b (u(t,x)-w(t,x))^2\,dx \le C\,||p||_{0,2}\int_0^t\int_a^b (u(s,x)-w(s,x))^2\,dx\,ds \tag{3.21}$$

with some constant C. By Gronwall's inequality, $u\equiv w$. Hence the uniqueness claim (3.20) is obvious.

(c) Conditions (3.17)–(3.18) are obviously necessary for the existence of classical solutions. Similar compatibility conditions will be studied later on.

(d) The condition of boundary-regularity (3.13) is the analogue to the condition

$$p,f\in W^{1,\infty}([0,t_0];H^1(\mathbb{R}))\cap L^\infty([0,t_0];H^2(\mathbb{R})) \tag{3.22}$$

for the pure initial value problem, stated on the whole horizon.

The proof of Theorem 3.5 will be split up in several parts. In Sections 3.2–3.3 we shall obtain smooth solutions for C^∞-functions f and p. Section 3.4 discusses the approximation of the coefficients f and p, satisfying (3.11)–(3.12), through C^∞-coefficients. In Section 3.5 we shall derive energy estimates for the solutions of Section 3.3 . These estimates allow the application of weak* compactness arguments, which yield solutions to (3.1)–(3.3) (Section 3.6). Finally, in Section 3.7 we shall discuss the regularity of solutions.

3.2. A Preliminary Result

We consider the following modified problem:

$$u_t(t,x) = p(t,x)\, u_x(t,x) + q(t,x)\, u(t,x) + f(t,x), \;\; t \in [0,t_0], x \in [a,b], \tag{3.23}$$

$$u(0,x) = u^0(x), \quad x \in [a,b], \tag{3.24}$$

$$u(t,b) = u^b(t), \quad t \in [0,t_0]. \tag{3.25}$$

Definition 3.7. *We shall call a function u a continuous solution of* (3.23)–(3.25) *if and only if*

$$u \in C^1([0,t_0]; L^2(a,b)) \cap C([0,t_0]; H^1(a,b)), \tag{3.26}$$

$$u \text{ satisfies Eq. (3.23)}, \tag{3.27}$$

$$u \text{ satisfies Eqs. (3.24) and (3.25) pointwise.} \tag{3.28}$$

Remark 3.8 A continuous solution of (3.23)–(3.25) is obviously continuous on $[0,t_0] \times [a,b]$. Moreover, every continuous solution of (3.23)–(3.25) with $q \equiv 0$ is a solution of (3.1)–(3.3). Hence, by (3.21), a unique continuous solution of (3.1)–(3.3) is also unique as a solution of (3.1)–(3.3).

Theorem 3.9. *Given functions f, p and q on $[0,t_0] \times [a,b]$, u^0 on $[a,b]$, and u^b on $[0,t_0]$ such that*

$$p \in C^1([0,t_0]; L^2(a,b)) \cap C([0,t_0]; C^1([a,b])), \tag{3.29}$$

$$p > 0 \text{ on } [0,t_0] \times [a,b], \tag{3.30}$$

$$q \in C^1([0,t_0]; L^2(a,b)) \cap C([0,t_0] \times [a,b]), \tag{3.31}$$

$$f \in C^1([0,t_0]; L^2(a,b)), \tag{3.32}$$

$$u^0 \in H^1(a,b), \tag{3.33}$$

$$u^b \in C^2([0,t_0]), \tag{3.34}$$

$$u^0(b) = u^b(0). \tag{3.35}$$

Then (3.23)–(3.25) *has a unique continuous solution u.*

Proof. It suffices to assume $u^b \equiv 0$. Let

$$D = \left\{ v \in H^1(a,b) \;\middle|\; v(b) = 0 \right\}. \tag{3.36}$$

Then D is dense in $L^2(a,b)$. Define a family $\{A(t)\}_{0 \le t \le t_0}$ of operators on D by

$$[A(t)\, v]\,(x) = p(t,x)\, v_x(x) + q(t,x)\, v(x). \tag{3.37}$$

Let $(\cdot,\cdot)$ be the inner product on $L^2(a,b)$. Since

$$(v, A(t)\, v) \le \int_a^b \left(q(t,x) - \frac{1}{2}\, p_x(t,x) \right) v^2(x)\, dx, \tag{3.38}$$

each operator $A(t)$ is quasidissipative in $L^2(a,b)$. Moreover, the theory of ordinary differential equations yields immediately that the spectrum of $A(t)$, $\sigma(A(t))$, is empty. Hence, by the Lumer-Philips theorem, each operator $A(t)$ generates a quasicontraction semigroup, $\{T_t(s)\}_{s\geq 0}$, on $L^2(a,b)$. Since p_x and q are bounded on $[0,t_0]\times[a,b]$, there exists $\omega\in\mathbb{R}$ such that

$$||T_t(s)||\leq e^{\omega s}\quad \text{for } s\geq 0 \text{ and } 0\leq t\leq t_0. \tag{3.39}$$

Hence the theory of evolution systems (cf. [6, pp. 139–149]) yields the desired result. □

3.3. The Compatibility Conditions

Definition 3.10. *Given functions f and p on $[0,t_0]\times[a,b]$, u^0 on $[a,b]$ and u^b on $[0,t_0]$ such that*

$$f,p\in C^\infty([0,t_0]\times[a,b]), \tag{3.40}$$
$$p>0, \tag{3.41}$$
$$u^0\in C^\infty([a,b]), \tag{3.42}$$
$$u^b\in C^\infty([0,t_0]). \tag{3.43}$$

Then we shall say that, for $N\in\mathbb{N}_0$, the functions u^0 and u^b are compatible of order N w.r.t. f and p if and only if the sequences (V^n) and (W^n), defined for $n\geq 0$ by

$$V^n(x)\stackrel{\text{def}}{=}[D_x^n u^0](x) \tag{3.44}$$

and

$$W^0(t)\stackrel{\text{def}}{=}u^b(t), \tag{3.45}$$
$$W^{n+1}(t)\stackrel{\text{def}}{=}(p(t,b))^{-1}\left(W_t^n(t)-[D_x^n f](t,b)\right) - (p(t,b))^{-1}\left(\sum_{k=0}^{n-1}\binom{n}{k}W^{k+1}(t)\,[D_x^{n-k}p](t,b)\right), \tag{3.46}$$

satisfy the compatibility conditions

$$V^n(b)=W^n(0)\quad \textit{for } 0\leq n\leq N. \tag{3.47}$$

If the compatibility conditions (3.47) *hold for every $N\in\mathbb{N}_0$, we shall say that u^0 and u^b are infinitely compatible w.r.t. f and p.*

Lemma 3.11. *Given functions f, p, u^0 and u^b, satisfying* (3.40)–(3.43), *such that u^0 and u^b are compatible of order N w.r.t. f and p for some $N\in\mathbb{N}_0$. Then the boundary-initial value problem* (3.1)–(3.3) *has a unique solution u such that*

$$u\in C^1([0,t_0];H^N(a,b))\cap C([0,t_0];H^{N+1}(a,b)). \tag{3.48}$$

Proof. We shall proceed by induction over the order of compatibility N. To this end, we consider the continuous solutions $u^{[n]}$, $0\leq n\leq N$, of the problems

$$u_t^{[n]}(t,x)=p(t,x)\,u_x^{[n]}(t,x)+n\,p_x(t,x)\,u^{[n]}(t,x)+f^n(t,x), \tag{3.49}$$
$$u^{[n]}(0,x)=V^n(x), \tag{3.50}$$
$$u^{[n]}(t,b)=W^n(t), \tag{3.51}$$

where V^n and W^n are defined by (3.44)–(3.46), and f^n is given by

$$f^n(t,x) \stackrel{\text{def}}{=} [D_x^n f](t,x) + \sum_{k=0}^{n-2} \binom{n}{k} [D_x^{n-k} p](t,x)\, u^{[k+1]}(t,x). \tag{3.52}$$

We shall prove the following claim:
For each $n \in \mathbb{N}_0$, $0 \le n \le N$, problem (3.49)–(3.51) has a unique solution $u^{[n]}$ in $C^1([0,t_0]; H^{N-n}(a,b)) \cap C([0,t_0]; H^{N+1-n}(a,b))$ such that

$$u^{[n]} \equiv D_x^n u^{[0]}. \tag{3.53}$$

The case $N = 0$ was proven in Theorem 3.9. Assume the claim holds for $N \ge 0$. Then (3.52) well-defines f^{N+1} as an element of $C^1([0,t_0]; L^2(a,b))$. Hence, by Theorem 3.9, problem (3.49)–(3.51) with $n = N+1$ has a unique continuous solution $u^{[N+1]}$ in $C^1([0,t_0]; L^2(a,b)) \cap C([0,t_0]; H^1(a,b))$. Let

$$U(t,x) \stackrel{\text{def}}{=} W^N(t) + \int_b^x u^{[N+1]}(t,y)\, dy. \tag{3.54}$$

Then we have

$$\begin{aligned} &U(0,x) = V^N(x),\\ &U(t,b) = W^N(t),\\ &U \in C^1([0,t_0]; H^1(a,b)) \cap C([0,t_0]; H^2(a,b)). \end{aligned} \tag{3.55}$$

Moreover, (3.46), (3.49) and (3.52) yield

$$\begin{aligned} U_t(t,x) \;=\;& p(t,x)\, U_x(t,x) + N\, p_x(t,x)\, U(t,x) \\ &+ f^N(t,x) + \int_b^x p_{xx}(t,y) \left(u^{[N]}(t,y) - U(t,y)\right) dy. \end{aligned} \tag{3.56}$$

Set

$$W \stackrel{\text{def}}{=} U - u^{[N]}. \tag{3.57}$$

Then $W(0,x) = W(t,b) = 0$, and

$$\begin{aligned} W_t(t,x) \;=\;& p(t,x)\, W_x(t,x) + N\, p_x(t,x)\, W(t,x) \\ &- \int_b^x p_{xx}(t,y)\, W(t,y)\, dy. \end{aligned} \tag{3.58}$$

We multiply (3.58) by W and integrate over $[a,b]$. If $M>0$ is a constant that is sufficiently large, we obtain

$$\begin{aligned}\frac{1}{2}\frac{d}{dt}\int_a^b W^2\,dy &= \frac{p}{2}W^2\Big|_a^b - \int_a^b \frac{p_x}{2}W^2\,dy + \int_a^b N\,p_x\,W^2\,dy \\ &\quad + \int_a^b W\int_b^z p_{xx}\,W\,dy\,dz \\ &\le M\int_a^b W^2\,dy + M\left(\int_a^b |W|\,dy\right)^2 \\ &\le M\,(b-a+1)\int_a^b W^2\,dy.\end{aligned} \tag{3.59}$$

For this estimate we have used (3.41) and the Cauchy-Schwarz inequality. By Gronwall's inequality, it follows easily from (3.59) that $W\equiv 0$. Hence $u^{[N+1]}\equiv D_x u^{[N]}$. The induction hypothesis (3.53) yields

$$u^{[n]}\in C^1([0,t_0];H^{N+1-n}(a,b))\cap C([0,t_0];H^{N+2-n}(a,b)) \quad \text{for } 0\le n\le N+1. \tag{3.60}$$

Hence the claim is proven. □

Theorem 3.12. *Suppose that the functions f, p, u^0 and u^b satisfy hypotheses* (3.40)–(3.43). *Then, for $N\in\mathbb{N}_0$, the boundary-initial value problem* (3.1)–(3.3) *has a unique solution u, satisfying*

$$u\in\bigcap_{k=0}^{N+1} C^k([0,t_0];H^{N+1-k}(a,b)), \tag{3.61}$$

if and only if u^0 and u^b are compatible of order N w.r.t. f and p.

Corollary 3.13. *Suppose that the functions f, p, u^0 and u^b satisfy hypotheses* (3.40)–(3.43). *Then:*

(i) *For $N\in\mathbb{N}_0$, the boundary-initial value problem* (3.1)–(3.3) *has a unique solution u in $C^N([0,t_0]\times[a,b])$ if and only if u^0 and u^b are compatible of order N w.r.t. f and p.*

(ii) *The boundary-initial value problem* (3.1)–(3.3) *has a unique solution u in $C^\infty([0,t_0]\times[a,b])$ if and only if u^0 and u^b are infinitely compatible w.r.t. f and p.*

Proof of Theorem 3.12. Let V^n and W^n be given by (3.44)–(3.46). Firstly assume that u is a solution, satisfying (3.61) for some $N\in\mathbb{N}_0$. If $N=0$, then necessarily $u^0(b)=V^0(b)=W^0(0)=u^b(0)$. In case $N>0$, we define

$$\omega^n \stackrel{\text{def}}{=} D_x^n u \quad \text{for } 0\le n\le N. \tag{3.62}$$

Then $\omega^n(0,x) = V^n(x)$ and $\omega^0(t,b) = W^0(t)$. Apply D_x^n, $0 \le n \le N-1$, to (3.1) to get

$$\omega_t^n(t,x) = p(t,x)\,\omega^{n+1}(t,x) + [D_x^n f](t,x) + \sum_{k=0}^{n-1} \binom{n}{k}\,\omega^{k+1}(t,x)\,[D_x^{n-k} p](t,x). \tag{3.63}$$

Hence, comparing (3.63) with (3.46), we conclude that

$$\omega^n(t,b) = W^n(t) \quad \text{for } 0 \le n \le N. \tag{3.64}$$

Therefore, by regularity of u,

$$V^n(b) = \omega^n(0,b) = W^n(0) \quad \text{for } 0 \le n \le N. \tag{3.65}$$

Secondly assume that u^0 and u^b are compatible of order N w.r.t. f and p. By Lemma 3.11, problem (3.1)–(3.3) has a unique solution u which belongs to $C^1([0,t_0];H^N(a,b)) \cap C([0,t_0];H^{N+1}(a,b))$. Hence (3.61) is obvious for $N=0$. Assume $N>0$. By Eq. (3.1) and regularity of u,

$$D_t D_x^n u(t,x) = D_x^n\,(p(t,x)\,u_x(t,x) + f(t,x)) \quad \text{for } 0 \le n \le N. \tag{3.66}$$

Thus, if $D = D_t^m D_x^n$ with $0 \le m+n \le N+1$, Du assumes the form

$$[Du](t,x) = \sum_{k=0}^{m+n} \beta_k(t,x)\,[D_x^k u](t,x) \tag{3.67}$$

with some functions $\beta_k \in C^\infty([0,t_0]\times[a,b])$, depending on D. The operator D with $m=n=0$ is understood as the identity. Since the right-hand side of Eq. (3.67) is a function in $C([0,t_0];L^2(a,b))$ for every choice of D, (3.61) is proven. □

3.4. Approximating Sequences

Lemma 3.14. *Given a boundary-regular function h on $[0,t_0]\times[a,b]$. Let M be a constant such that*

$$M \ge ||h||_{1,1} + ||h||_{0,2}. \tag{3.68}$$

Then there exists a sequence $(h_n)_{n\in\mathbb{N}}$ of functions on $[0,t_0]\times[a,b]$ such that, for $n\in\mathbb{N}$,

$$h_n \in C^\infty([0,t_0]\times[a,b]), \tag{3.69}$$

$$h_n \longrightarrow h \text{ in } C([0,t_0];H^1(a,b)) \text{ as } n\to\infty, \tag{3.70}$$

$$||h_n||_{H^{1,1}} \le \sqrt{2\,t_0}\,(M+1), \tag{3.71}$$

$$||h_n||_{0,2} \le M+1. \tag{3.72}$$

Moreover, for each $t\in(0,t_0]$, there exists $N\in\mathbb{N}$ such that

$$||h_n||_{H^{1,1}[t]} \le \sqrt{2\,t}\,(M+1) \quad \text{for } n \ge N. \tag{3.73}$$

Proof. Define $\hat{h}$ by

$$\hat{h}(t) \stackrel{\text{def}}{=} \begin{cases} h(t_0) & \text{for } t > t_0, \\ h(t) & \text{for } 0 \le t \le t_0, \\ h(0) & \text{for } t < 0. \end{cases} \tag{3.74}$$

Then it is readily seen that $\hat{h} \in W^{1,\infty}(\mathbb{R}; H^1(a,b)) \cap L^\infty(\mathbb{R}; H^2(a,b))$. Now let $h_a \stackrel{\text{def}}{=} \hat{h}_x(\cdot, a)$, $h_b \stackrel{\text{def}}{=} \hat{h}_x(\cdot, b)$, and define H by

$$H(t,x) \stackrel{\text{def}}{=} \begin{cases} h_b(t)\,(x-b) + \hat{h}(t,b) & \text{for } t \in \mathbb{R},\ x > b, \\ \hat{h}(t,x) & \text{for } t \in \mathbb{R},\ a \le x \le b, \\ h_a(t)\,(x-a) + \hat{h}(t,a) & \text{for } t \in \mathbb{R},\ x < a. \end{cases} \tag{3.75}$$

It follows immediately that $H \in L^\infty(\mathbb{R}; H^2_{loc}(\mathbb{R}))$. Moreover, by assumption (3.12), $H \in H^1_{loc}(\mathbb{R}; H^1_{loc}(\mathbb{R}))$. Let J be a nonnegative function in $C^\infty(\mathbb{R})$, having compact support in $[-1,1]$ and unit integral. We define the sequence $(J_n)_{n\in\mathbb{N}}$ of Friedrichs mollifiers by

$$J_n(x) \stackrel{\text{def}}{=} n\,J(n\,x) \quad \text{for } x \in \mathbb{R},\ n \in \mathbb{N}. \tag{3.76}$$

Using the sequence (J_n), define the sequence $(H_n)_{n\in\mathbb{N}}$ by

$$H_n(t,x) \stackrel{\text{def}}{=} \int_{\mathbb{R}}\int_{\mathbb{R}} J_n(t-s)\,J_n(x-y)\,H(s,y)\,dy\,ds \quad \text{for } t,\ x \in \mathbb{R}. \tag{3.77}$$

Then $H_n \in C^\infty(\mathbb{R}^2)$. For s, $y \in \mathbb{R}$, let L_s, R_y be the shift operators, defined for functions f on $\mathbb{R}^2$ by

$$L_s f(t,y) = f(t-s,x), \tag{3.78}$$

$$R_y f(t,x) = f(t,x-y). \tag{3.79}$$

Given an arbitrary interval $[t_1,t_2]$. Let $\epsilon > 0$. Since $H \in C(\mathbb{R}; H^1_{loc}(\mathbb{R}))$ and $[t_1,t_2]$ is compact, there exists $m \in \mathbb{N}$ such that $|y| \le m^{-1}$ implies

$$\int_a^b |R_y H(t,x) - H(t,x)|^2\,dx < \epsilon \quad \text{for } t \in [t_1,t_2], \tag{3.80}$$

$$\int_a^b |R_y H_x(t,x) - H_x(t,x)|^2\,dx < \epsilon \quad \text{for } t \in [t_1,t_2]. \tag{3.81}$$

Also, there exists $k \in \mathbb{N}$ such that $|s| \le k^{-1}$ implies

$$\int_a^b |L_s H(t,x) - H(t,x)|^2\,dx < \epsilon \quad \text{for } t \in [t_1,t_2], \tag{3.82}$$

$$\int_a^b |L_s H_x(t,x) - H_x(t,x)|^2\,dx < \epsilon \quad \text{for } t \in [t_1,t_2]. \tag{3.83}$$

Denote $||\cdot||_{k,(a,b)}$, $k \in \mathbb{N}_0$, the norm in $H^k(a,b)$. Since

$$\begin{aligned} ||H_n(t) - H(t)||_{1,(a,b)} &\le \int_{\mathbb{R}}\int_{\mathbb{R}} J_n(s)\, J_n(y)\, ||L_s\, R_y H(t) - H(t)||_{1,(a,b)}\, dy\, ds \\ &\le \int_{\mathbb{R}}\int_{\mathbb{R}} J_n(s)\, J_n(y)\, ||L_s H(t) - H(t)||_{1,(a,b)}\, dy\, ds \\ &\quad + \int_{\mathbb{R}}\int_{\mathbb{R}} J_n(s)\, J_n(y)\, ||L_s\, R_y H(t) - L_s H(t)||_{1,(a,b)}\, dy\, ds, \end{aligned} \tag{3.84}$$

estimates (3.80)–(3.83) prove that H_n (restricted to $[a,b]$) converges to H (restricted to $[a,b]$) in $C([0,t_0];H^1(a,b))$ as $n \to \infty$. The estimates

$$\begin{aligned} \int_a^b |R_y H(t,x)|^2\, dx &\le \int_{a-|y|}^a \Big(h_a(t)\,(x-a) + \hat{h}(t,a)\Big)^2\, dx \\ &\quad + \int_b^{b+|y|} \Big(h_b(t)\,(x-b) + \hat{h}(t,b)\Big)^2\, dx \\ &\quad + \int_a^b \hat{h}^2(t,x)\, dx, \end{aligned} \tag{3.85}$$

$$\begin{aligned} \int_a^b |R_y H_x(t,x)|^2\, dx &\le \int_{a-|y|}^a h_a^2(t)\, dx + \int_b^{b+|y|} h_b^2(t)\, dx \\ &\quad + \int_a^b \hat{h}_x^2(t,x)\, dx, \end{aligned} \tag{3.86}$$

$$\int_a^b |R_y H_{xx}(t,x)|^2\, dx \le \int_a^b \hat{h}_{xx}^2(t,x)\, dx \tag{3.87}$$

imply that

$$||R_y H(t)||_{2,(a,b)} \le M+1 \quad \text{for all } t \in \mathbb{R} \text{ if } |y| \text{ is sufficiently small.} \tag{3.88}$$

Since

$$||H_n(t)||_{2,(a,b)} \le \int_{\mathbb{R}}\int_{\mathbb{R}} J_n(t-s)\, J_n(y)\, ||R_y H(s)||_{2,(a,b)}\, dy\, ds, \tag{3.89}$$

it follows that

$$||H_n(t)||_{2,(a,b)} \le M+1 \quad \text{for all } t \in \mathbb{R} \text{ if } n \text{ is large enough.} \tag{3.90}$$

Next we deduce from (3.90) for large n that

$$\int_0^{t_0} ||H_n(t)||_{1,(a,b)}^2\, dt \le t_0\,(M+1)^2. \tag{3.91}$$

Moreover,

$$\int_0^{t_0}\int_a^b |L_s\,R_y H_t(t,x)|^2\,dx\,dt \;\le\; \int_0^{t_0}\int_a^b \hat{h}_t^2(t-s,x)\,dx\,dt + \int_0^{t_0}\int_{a-|y|}^a \Big(\dot{h}_a(t-s)\,(x-a)+\hat{h}_t(t-s,a)\Big)^2\,dx\,dt + \int_0^{t_0}\int_b^{b+|y|} \Big(\dot{h}_b(t-s)\,(x-b)+\hat{h}_t(t-s,b)\Big)^2\,dx\,dt, \tag{3.92}$$

$$\int_0^{t_0}\int_a^b |L_s\,R_y H_{xt}(t,x)|^2\,dx\,dt \;\le\; \int_0^{t_0}\int_a^b \hat{h}_{xt}^2(t-s,x)\,dx\,dt + \int_0^{t_0}\int_{a-|y|}^a \dot{h}_a^2(t-s)\,dx\,dt + \int_0^{t_0}\int_b^{b+|y|} \dot{h}_b^2(t-s)\,dx\,dt. \tag{3.93}$$

Since

$$||H_n||_{H^1([0,t_0];H^1(a,b))} \le \int_{\mathbb{R}}\int_{\mathbb{R}} J_n(s)\,J_n(y)\,||L_sR_yH||_{H^1([0,t_0];H^1(a,b))}\,dy\,ds, \tag{3.94}$$

estimates (3.91)–(3.93) show that

$$||H_n||_{H^1([0,t_0];H^1(a,b))} \le \sqrt{2\,t_0}\,(M+1) \quad \text{if } n \text{ is large enough.} \tag{3.95}$$

The same calculation also holds with $t' \in (0,t_0]$ replacing t_0. Hence, for sufficiently large $K \in \mathbb{N}$, the sequence $(H_{n+K})_{n\in\mathbb{N}}$, restricted to $[0,t_0]\times[a,b]$, has the required properties (3.69)–(3.73). □

3.5. Energy Estimates

Lemma 3.15. *Given $\tilde{u}^b \in C^2([0,t^*])$ for some $t^* > 0$. Let $t_0 \in (0,t^*]$, and define $u^b \stackrel{\text{def}}{=} \tilde{u}^b|_{[0,t_0]}$. Suppose the boundary-initial value problem (3.1)–(3.3) has a solution u in $C^3([0,t_0]\times[a,b])$ with $f,\,p \in C^2([0,t_0]\times[a,b])$ and $u^0 \in C^2([a,b])$. Let $p^a \stackrel{\text{def}}{=} p(\cdot,a)$ and $p^b \stackrel{\text{def}}{=} p(\cdot,b)$. Then there exists a positive constant C, depending only on t^*, such that, for $0 \le t \le t_0$, the solution u obeys the estimates:*

$$||u(t)||_{H^1}^2 \le \Big(||u^0||_{H^1}^2 + C\,||p||_{0,1}\Big(||u^b||_{2,[t]}^2 + ||u_x(\cdot,b)||_{2,[t]}^2\Big) + t\,||f||_{0,1}^2\Big)\cdot\exp\{(C\,||p||_{0,2}+1)\,t\}, \tag{3.96}$$

$$||u(t)||_{H^2}^2 \le \Big(||u^0||_{H^2}^2 + C\,||p||_{0,1}\Big(||u^b||_{2,[t]}^2 + ||u_x(\cdot,b)||_{2,[t]}^2 + ||u_{xx}(\cdot,b)||_{2,[t]}^2\Big) + t\,||f||_{0,2}^2\Big)\exp\{(C\,||p||_{0,2}+1)\,t\}, \tag{3.97}$$

$$||u||_{1,1,[t]}^2 \le \Big(||u^0||_{H^1}^2 + ||p(0)\,u_x^0 + f(0)||_{H^1}^2 + C\,||u||_{0,2}\,||p||_{H^{1,1}[t]}^2 + ||f||_{H^{1,1}[t]}^2 + C\,||p||_{0,1}\Big(||u^b||_{H^1[t]}^2 + ||u_x(\cdot,b)||_{H^1[t]}^2\Big)\Big)\cdot\exp\{(C\,||p||_{0,2} + C\,||u||_{0,2}+1)\,t\}, \tag{3.98}$$

where

$$\begin{aligned}
\|u_x(\cdot,b)\|_{2,[t]} &\le \|(p^b)^{-1}\|_\infty \, (\|u^b\|_{H^1[t]} + C\sqrt{t}\,\|f\|_{0,1}), && (3.99)\\
\|u_x(\cdot,b)\|_{H^1[t]} &\le C\,\|(p^b)^{-1}\|_\infty \, (\|u^b\|_{H^2[t]} + \|f\|_{H^{1,1}[t]} \\
&\quad + \|(p^b)^{-1}\|_\infty \, (\|\tilde{u}^b\|_{H^2} + \|f\|_{0,1})\,\|p\|_{H^{1,1}[t]}), && (3.100)\\
\|u_{xx}(\cdot,b)\|_{2,[t]} &\le \|(p^b)^{-1}\|_\infty \, (\|u_x(\cdot,b)\|_{H^1[t]} \\
&\quad + C\,\|p\|_{0,2}\,\|u_x(\cdot,b)\|_{2,[t]} + C\sqrt{t}\,\|f\|_{0,2}). && (3.101)
\end{aligned}$$

Moreover

$$\begin{aligned}
\|u_x(\cdot,a)\|_{H^1}^2 &\le \big(\|u^0\|_{H^1}^2 + \|p(0)\,u_x^0 + f(0)\|_{H^1}^2 + C\,\|u\|_{0,2}\,\|p\|_{H^{1,1}}^2 \\
&\quad + \|f\|_{H^{1,1}}^2 + C\,\|p\|_{0,1}\,\big(\|u^b\|_{H^1}^2 + \|u_x(\cdot,b)\|_{H^1}^2\big)\big) && (3.102)\\
&\quad \cdot \|(p^a)^{-1}\|_\infty \exp\{(C\,\|p\|_{0,2} + C\,\|u\|_{0,2} + 1)\,t_0\}.
\end{aligned}$$

Proof. The constant C of the theorem will be a generic constant that absorbs the embedding constants of various Sobolev estimates of functions over $[a,b]$ and $[0,t^*]$. Let t_0 be in $(0,t^*]$. Since

$$\begin{aligned}
u_x(t,b) &= \left(p^b(t)\right)^{-1}\left(u_t^b(t) - f(t,b)\right), && (3.103)\\
u_{xt}(t,b) &= \left(p^b(t)\right)^{-1}\left(u_{tt}^b(t) - p_t^b(t)\,u_x(t,b) - f_t(t,b)\right), && (3.104)\\
u_{xx}(t,b) &= \left(p^b(t)\right)^{-1}\left(u_{xt}(t,b) - p_x(t,b)\,u_x(t,b) - f_x(t,b)\right), && (3.105)
\end{aligned}$$

estimates (3.99)–(3.101) follow immediately if we use in (3.100) that

$$\|u_x(\cdot,b)\|_\infty \le C\,\|(p^b)^{-1}\|_\infty \left(\|\tilde{u}^b\|_{H^2} + \|f\|_{0,1}\right). \qquad (3.106)$$

Differentiation of (3.1) yields

$$\begin{aligned}
u_{tx} &= p\,u_{xx} + p_x\,u_x + f_x, && (3.107)\\
u_{txx} &= p\,u_{xxx} + 2\,p_x\,u_{xx} + p_{xx}\,u_x + f_{xx}, && (3.108)\\
u_{tt} &= p\,u_{tx} + p_t\,u_x + f_t, && (3.109)\\
u_{ttx} &= p\,u_{txx} + p_x\,u_{tx} + p_t\,u_{xx} + p_{tx}\,u_x + f_{tx}. && (3.110)
\end{aligned}$$

We multiply (3.1) by u and (3.107) by u_x, integrate over $[a,b]$ and add the two equations to obtain

$$\begin{aligned}
\frac{1}{2}\frac{d}{dt}\int_a^b u^2 + u_x^2\,dx &\le \frac{p^b(t)}{2}\left(\left(u^b(t)\right)^2 + u_x^2(t,b)\right) + \frac{1}{2}\,\|f(t)\|_{H^1}^2 \\
&\quad + \left(\frac{C}{2}\,\|p\|_{0,2} + \frac{1}{2}\right)\int_a^b u^2 + u_x^2\,dx. && (3.111)
\end{aligned}$$

Integration in t yields

$$\begin{aligned}
\|u(t)\|_{H^1}^2 &\le \|u^0\|_{H^1}^2 + \|p^b\|_\infty\left(\|u^b\|_{2,[t]}^2 + \|u_x(\cdot,b)\|_{2,[t]}^2\right) \\
&\quad + t\,\|f\|_{0,1}^2 + (C\,\|p\|_{0,2} + 1)\int_0^t \|u(s)\|_{H^1}^2\,ds. && (3.112)
\end{aligned}$$

We apply Gronwall's inequality to (3.112) to deduce estimate (3.96). Next we multiply (3.108) by u_{xx} and integrate again. Proceeding as before leads to the estimate

$$\begin{aligned} ||u(t)||_{H^2}^2 &\leq ||u^0||_{H^2}^2 + t\,||f||_{0,2}^2 + (C\,||p||_{0,2} + 1) \int_0^t ||u(s)||_{H^2}^2\,ds \\ &+ ||p^b||_\infty \left(||u^b||_{2,[t]}^2 + ||u_x(\cdot,b)||_{2,[t]}^2 + ||u_{xx}(\cdot,b)||_{2,[t]}^2 \right). \end{aligned} \tag{3.113}$$

By Gronwall's inequality, estimate (3.97) follows. For (3.113) we have used that $H^1(a,b)$ is a Banach algebra. To derive estimate (3.98), we observe that

$$\begin{aligned} \frac{1}{2}\frac{d}{dt}\int_a^b u^2 + u_x^2 + u_t^2 + u_{tx}^2\,dx &\leq \frac{1}{2}||f(t)||_{H^1}^2 + \frac{1}{2}||f_t(t)||_{H^1}^2 \\ &+ \frac{p^b(t)}{2}\left(\left(u^b(t)\right)^2 + u_x^2(t,b) + u_t^2(t,b) + u_{xt}^2(t,b) \right) \\ &+ \int_a^b p_t\,u_x\,u_t\,dx + \int_a^b u_{tx}\,\frac{\partial}{\partial x}(p_t\,u_x)\,dx \\ &+ \left(\frac{C}{2}||p||_{0,2} + \frac{1}{2}\right)\int_a^b u^2 + u_x^2 + u_t^2 + u_{tx}^2\,dx \\ &- \frac{p^a(t)}{2}\left(u_x^2(t,a) + u_{tx}^2(t,a)\right). \end{aligned} \tag{3.114}$$

In this estimate we have intentionally included the boundary terms at a. We integrate (3.114) in t, noting that

$$\int_a^b p_t\,u_x\,u_t\,dx \leq C\,||u||_{0,2}\left(||u_t(t)||_2^2 + ||p_t(t)||_2^2\right), \tag{3.115}$$

$$\int_a^b u_{tx}\,\frac{\partial}{\partial x}(p_t\,u_x)\,dx \leq C\,||u||_{0,2}\left(||u_{tx}(t)||_2^2 + ||p_t(t)||_{H^1}^2\right). \tag{3.116}$$

Gronwall's inequality yields now estimate (3.98) when the boundary terms at a in (3.114) are neglected. Having shown (3.98), we obtain the bound at a (3.102) from estimate (3.114). □

3.6. Proof of Theorem 3.5 – Existence and Uniqueness of Solutions

Theorem 3.16. *Suppose that the functions f, p, u^0 and u^b satisfy hypotheses* (3.13)–(3.18). *Then the boundary-initial value problem* (3.1)–(3.3) *has a unique boundary-regular continuous solution u such that*

$$u \in W^{1,\infty}([0,t_0]; H^1(a,b)) \cap L^\infty([0,t_0]; H^2(a,b)). \tag{3.117}$$

Proof. Let $M \stackrel{\text{def}}{=} ||p||_{1,1} + ||p||_{0,2} + ||f||_{1,1} + ||f||_{0,2}$. By Lemma 3.14, for p and f there are C^∞-sequences (p^k) and (f^k), resp., with properties (3.69)–(3.72). By density, there exist sequences $\left(u_k^b\right)$ and $\left(u_k^0\right)$ such that, for $k \in \mathbb{N}$,

$$u_k^b \in C^\infty([0,t_0]), \tag{3.118}$$
$$u_k^0 \in C^\infty([a,b]), \tag{3.119}$$
$$u_k^0 \text{ and } u_k^b \text{ are compatible of order 3 w.r.t. } f^k,\ p^k, \tag{3.120}$$
$$u_k^b \longrightarrow u^b \text{ in } H^2(0,t_0) \text{ as } k \to \infty, \tag{3.121}$$
$$u_k^0 \longrightarrow u^0 \text{ in } H^2(a,b) \text{ as } k \to \infty. \tag{3.122}$$

The problems

$$\begin{aligned} u_t^{[k]}(t,x) &= p^k(t,x)\,u_x^{[k]}(t,x) + f^k(t,x),\\ u^{[k]}(0,x) &= u_k^0(x),\\ u^{[k]}(t,b) &= u_k^b(t), \end{aligned} \tag{3.123}$$

have solutions $u^{[k]}$ in $C^3([0,t_0]\times[a,b])$ by Corollary 3.13. For n, $m \in \mathbb{N}$ define

$$w^{n,m} \overset{\text{def}}{=} u^{[n]} - u^{[m]}. \tag{3.124}$$

Then $w^{n,m}$ solves

$$\begin{aligned} w_t^{n,m}(t,x) &= p^n(t,x)\,w_x^{n,m}(t,x) + (p^n(t,x) - p^m(t,x))\,u_x^{[m]}(t,x)\\ &\quad + f^n(t,x) - f^m(t,x),\\ w^{n,m}(0,x) &= u_n^0(x) - u_m^0(x),\\ w^{n,m}(t,b) &= u_n^b(t) - u_m^b(t). \end{aligned} \tag{3.125}$$

By Lemma 3.15, $w^{n,m}$ obeys the estimate

$$\begin{aligned} ||w^{n,m}(t)||_{H^1}^2 &\le (||u_n^0 - u_m^0||_{H^1}^2 + C\,(1 + ||p^n||_{0,1}\\ &\quad + ||p^n||_{0,1}\,||(p^n(\cdot,b))^{-1}||_\infty^2)\,(||u_n^b - u_m^b||_{H^1}\\ &\quad + \sqrt{t_0}\,(||p^n - p^m||_{0,1}\,||u^{[m]}||_{0,2} + ||f^n - f^m||_{0,1}))^2)\\ &\quad \cdot \exp\{(C\,||p^n||_{0,2} + 1)\,t_0\}. \end{aligned} \tag{3.126}$$

The quantities $||p^n||_{0,2}$, $||f^n||_{0,2}$, $||p^n||_{1,1}$ and $||f^n||_{1,1}$ are bounded according to (3.71)–(3.72). Hence $||u^{[m]}||_{0,2}$ is bounded according to (3.97). It follows that

$$w^{n,m} \longrightarrow 0 \text{ in } C([0,t_0];H^1(a,b)) \text{ as } n,\ m \to \infty. \tag{3.127}$$

Hence $\left(u^{[n]}\right)$ is Cauchy in $C([0,t_0];H^1(a,b)) \cap C^1([0,t_0];L^2(a,b))$ with limit u. u is necessarily a continuous solution of (3.1)–(3.3). On the other hand, $\left(u^{[n]}\right)$ is bounded in $W^{1,\infty}([0,t_0];H^1(a,b)) \cap L^\infty([0,t_0];H^2(a,b))$ by estimates (3.97)–(3.98). Both $W^{1,\infty}([0,t_0];H^1(a,b))$ and $L^\infty([0,t_0];H^2(a,b))$ are the conjugates of separable Banach spaces, hence both spaces are weak* sequentially compact by Alaoglu's theorem. Therefore the sequence $\left(u^{[n]}\right)$ contains a subsequence, say (v^n), that is weak* convergent both in $W^{1,\infty}([0,t_0];H^1(a,b))$ and in $L^\infty([0,t_0];H^2(a,b))$. By boundedness according to (3.100), (3.102), we may also assume that the sequences $(v_x^n(\cdot,a))$, $(v_x^n(\cdot,b))$ are weakly convergent in $H^1(0,t_0)$. Otherwise we extract a subsequence of (v^n). Weak* convergence in $L^\infty([0,t_0];H^2(a,b))$ and in

$W^{1,\infty}([0,t_0];H^1(a,b))$ implies strong convergence in $L^2([0,t_0]\times[a,b])$ by Rellich's theorem. Hence the sequence (v^n) converges strongly in $L^2([0,t_0]\times[a,b])$ to its unique weak* limit $v\in W^{1,\infty}([0,t_0];H^1(a,b))\cap L^\infty([0,t_0];H^2(a,b))$. Since the sequence $\left(u^{[n]}\right)$ also converges strongly in $L^2([0,t_0]\times[a,b])$, we have $u\equiv v$. Thus we have shown (3.117). Suppose v_a is the weak limit of $(v_x^n(\cdot,a))$ in $H^1(0,t_0)$. By Rellich's theorem, $(v_x^n(\cdot,a))$ converges strongly in $L^2(0,t_0)$ to its weak limit v_a in $H^1(0,t_0)$. However, (v_x^n) is weak* convergent to u_x in $L^\infty([0,t_0];H^1(a,b))$, hence weakly convergent to u_x in $L^2([0,t_0];H^1(a,b))$. Since the linear map $h\mapsto h(a)$ from $L^2([0,t_0];H^1(a,b))$ to $L^2(0,t_0)$ is bounded, the sequence $(v_x^n(\cdot,a))$ converges weakly to $u_x(\cdot,a)$ in $L^2(0,t_0)$. Weak and strong limits must coincide, hence necessarily $u_x(\cdot,a)\equiv v_a$. Since the same argumentation works for $u_x(\cdot,b)$, we have shown that u is boundary-regular. Finally, uniqueness was already proven in (3.21). □

Corollary 3.17. *Given $\tilde u^b\in H^2(0,t^*)$ for some $t^*>0$. Let $t_0\in(0,t^*]$, and define $u^b\overset{\text{def}}{=}\tilde u^b|_{[0,t_0]}$. Suppose that the functions f, p, u^0 and u^b satisfy* (3.13)–(3.18). *Let $M\ge ||p||_{1,1}+||p||_{0,2}+||f||_{1,1}+||f||_{0,2}$, and let $p^a\overset{\text{def}}{=}p(\cdot,a)$, $p^b\overset{\text{def}}{=}p(\cdot,b)$. Then there exists a constant $C>0$ and continuous functions F on $\mathbb{R}^5$ and G on $\mathbb{R}^6$ such that, for $x_i\in\mathbb{R}$, $2\le i\le 6$,*

$$F(0,0,x_3,x_4,x_5)=x_3=G(0,0,x_3,x_4,x_5,x_6), \tag{3.128}$$

and, for $0\le t\le t_0$, the solution u of the boundary-initial value problem (3.1)–(3.3) *obeys the estimates:*

$$||u||^2_{0,2,[t]} \le F(t,||\tilde u^b||_{H^2[t]},||u^0||^2_{H^2},||(p^b)^{-1}||_\infty,M)\overset{\text{def}}{=}\Phi(t), \tag{3.129}$$

$$\begin{aligned}||u||^2_{1,1,[t]} \le\ & G(t,||\tilde u^b||_{H^2[t]},||u_0||^2_{H^1}+||p(0)\,u_x^0+f(0)||^2_{H^1},\\ & ||(p^b)^{-1}||_\infty,\Phi(t^*),M),\end{aligned} \tag{3.130}$$

and

$$\begin{aligned}||u_x(\cdot,a)||^2_{H^1} \le\ & ||(p^a)^{-1}||_\infty\, G(t_0,||\tilde u^b||_{H^2[t_0]},||u_0||^2_{H^1}+||p(0)\,u_x^0+f(0)||^2_{H^1},\\ & ||(p^b)^{-1}||_\infty,\Phi(t^*),M)\end{aligned} \tag{3.131}$$

$$\begin{aligned}||u_x(\cdot,b)||_{H^1} \le\ & C\,||(p^b)^{-1}||_\infty\,\big(||\tilde u^b||_{H^2[t_0]}\\ & +\left(1+(M+2)\,||(p^b)^{-1}||_\infty\right)(M+1)\sqrt{t_0}\big).\end{aligned} \tag{3.132}$$

The constant C and the functions F and G can be chosen to depend on $||\tilde u^b||_{H^2}$ and t^* only.

Proof. Fix t_0 and choose sequences $\left(p^k\right)$, $\left(f^k\right)$, $\left(u_k^b\right)$ and $\left(u_k^0\right)$ as in the proof of Theorem 3.16 . Construct the sequence $\left(u^{[k]}\right)$ of solutions for (3.123). We may assume that $\left(u^{[k]}\right)$ is weak* convergent in $W^{1,\infty}([0,t_0];H^1(a,b))\cap L^\infty([0,t_0];H^2(a,b))$ with weak* limit u. By Lemma 3.14, the quantities $||f^k||_{0,2}$, $||p^k||_{0,2}$ obey the uniform bound in (3.72), whereas the quantities $||f^k||_{H^{1,1}[t]}$, $||p^k||_{H^{1,1}[t]}$ obey the estimate in (3.73) if k is large enough. Lemma 3.15 yields bounds for $||u^{[k]}||_{0,2,[t]}$,

$||u^{[k]}||_{1,1,[t]}$ in terms of t, $||u_k^b||_{H^2[t]}$ and the quantities $||u_k^0||_{H^1}$, $||(p^k(\cdot,b))^{-1}||_\infty$, M. Similar bounds follow from (3.100) and (3.102) for the boundary quantities $||u^{[k]}(\cdot,a)||_{H^1}$ and $||u^{[k]}(\cdot,b)||_{H^1}$. Taking the lim inf as $k \to \infty$, we obtain estimates of the form (3.129)–(3.132). The dependencies of F, G and C are readily seen in Lemma 3.15. □

3.7. Proof of Theorem 3.5 – Regularity of Solutions

Lemma 3.18. *Suppose that the conditions of Theorem 3.16 hold. Then the solution u of the boundary-initial value problem* (3.1)–(3.3) *is a strongly right-continuous map into $H^2(a,b)$, i.e. u satisfies*

$$\lim_{t\downarrow t'} ||u(t) - u(t')||_{H^2} = 0 \quad \textit{for every } t' \in [0,t_0). \tag{3.133}$$

Proof. By Corollary 3.17,

$$\limsup_{t\downarrow 0} ||u(t)||_{H^2} \le ||u^0||_{H^2}. \tag{3.134}$$

Hence the claim is true for $t' = 0$ by the weak continuity result of Lemma 3.3 (cf. [3, p. 253]). To prove the claim for any other $t' \in [0,t_0)$, we consider $w(t) \stackrel{\text{def}}{=} u(t+t')$ on $[0, t_0 - t']$. Then Corollary 3.17 holds for w. Therefore the claim follows from estimate (3.134), applied to w. □

Theorem 3.19. *Suppose that the conditions of Theorem 3.16 hold. Then the solution u of the boundary-initial value problem* (3.1)–(3.3) *has the regularity property* (3.19).

Proof. We shall show that u is a strongly left-continuous map on $(0,t_0]$ into $H^2(a,b)$. Then we can deduce from Lemma 3.18 and Eq. (3.1) that (3.19) holds. To this end, define the functions v, $\tilde{p}$ and $\tilde{f}$ on $[0,t_0]\times[a,b]$ by

$$v(t,x) \stackrel{\text{def}}{=} u(t_0 - t, a + b - x), \tag{3.135}$$

$$\tilde{p}(t,x) \stackrel{\text{def}}{=} p(t_0 - t, a + b - x), \tag{3.136}$$

$$\tilde{f}(t,x) \stackrel{\text{def}}{=} -f(t_0 - t, a + b - x), \tag{3.137}$$

and define the functions v^0 and v^b by

$$v^0(x) \stackrel{\text{def}}{=} u(t_0, a + b - x), \quad x \in [a,b], \tag{3.138}$$

$$v^b(t) \stackrel{\text{def}}{=} u(t_0 - t, a), \quad t \in [0,t_0]. \tag{3.139}$$

Then v is the unique solution of the boundary-initial value problem

$$v_t(t,x) = \tilde{p}(t,x)\, v_x(t,x) + \tilde{f}(t,x), \quad t \in [0,t_0], x \in [a,b], \tag{3.140}$$

$$v(0,x) = v^0(x), \quad x \in [a,b], \tag{3.141}$$

$$v(t,b) = v^b(t), \quad t \in [0,t_0]. \tag{3.142}$$

$\tilde{f}$ and $\tilde{p}$ are boundary-regular, and $\tilde{p}$ satisfies (3.14). Also, (3.15) holds for v^0. Lemma 3.3 implies that

$$u_t(t,a) = p(t,a)\, u_x(t,a) + f(t,a) \quad \text{pointwise in } t. \tag{3.143}$$

By Theorem 3.16, u is boundary-regular. Hence Eq. (3.143) proves that v^b satisfies (3.16). Finally, the compatibility conditions (3.17)–(3.18) are valid for $\tilde{f}$, $\tilde{p}$, v^0 and v^b by virtue of Eq. (3.143) and continuity of u. Therefore, by Lemma 3.18, v is strongly right-continuous from $[0, t_0)$ into $H^2(a, b)$. □

4. Solvability of the equations of fiber spinning

In this section we shall prove existence, uniqueness and regularity of solutions for the equations of fiber spinning, Eqs. (2.9)–(2.17). The main result of this section is Theorem 4.2. Throughout we shall use the notation of Section 3. In addition we shall assume that the exit and solidification temperatures T_E and T_S ($T_E > T_S > 0$) are fixed, so that any dependence on these quantities may be omitted.

4.1. Statement of the Main Result

Definition 4.1. *We shall call a vector field* (A, z, v)*, defined on* $[0, t_0] \times [T_S, T_E]$*, a solution of* (2.9)–(2.17) *if and only if*

$$A, z, v \in W^{1,\infty}([0, t_0]; H^1(T_S, T_E)) \cap L^\infty([0, t_0]; H^2(T_S, T_E)), \tag{4.1}$$

$$A,\ z,\ v \text{ satisfy Eqs. } (2.9)\text{–}(2.11), \tag{4.2}$$

$$A \text{ satisfies Eqs. } (2.12),\ (2.16) \text{ pointwise}, \tag{4.3}$$

$$z \text{ satisfies Eqs. } (2.13),\ (2.17) \text{ pointwise}, \tag{4.4}$$

$$v \text{ satisfies Eqs. } (2.14),\ (2.15) \text{ pointwise}. \tag{4.5}$$

Theorem 4.2. *Given initial values* A^0*,* z^0 *on* $[T_S, T_E]$ *and boundary values* A_E*,* v_E*,* v_S *on* $[0, t^*]$ *for some* $t^* > 0$ *such that*

$$A^0, z^0 \in H^2(T_S, T_E), \tag{4.6}$$

$$A_E \in H^2(0, t^*), \tag{4.7}$$

$$v_E, v_S \in W^{1,\infty}(0, t^*), \tag{4.8}$$

$$A^0 > 0 \text{ on } [T_S, T_E], \tag{4.9}$$

$$z^0_T < 0 \text{ on } [T_S, T_E]. \tag{4.10}$$

Also suppose that the compatibility conditions

$$A^0(T_E) = A_E(0), \quad z^0(T_E) = 0, \tag{4.11}$$

$$\dot{A}_E(0) = \frac{\alpha T_E}{\sqrt{A^0(T_E)}} A^0_T(T_E) + (v_S(0) - v_E(0)) \left(\int_{T_S}^{T_E} \frac{z^0_T(T)}{A^0(T)}\, dT \right)^{-1}, \tag{4.12}$$

$$\frac{\alpha T_E}{\sqrt{A^0(T_E)}} z^0_T(T_E) + v_E(0) = 0 \tag{4.13}$$

are satisfied. Then there exists a $t_0 \in (0, t^*]$ *such that the boundary-initial value problem* (2.9)–(2.17) *has a unique solution* (A, z, v) *on* $[0, t_0] \times [T_S, T_E]$. *This solution* (A, z, v) *has the properties:*

$$A, z \in C^1([0, t_0]; H^1(T_S, T_E)) \cap C([0, t_0]; H^2(T_S, T_E)), \tag{4.14}$$

$$A, z \in W^{2,\infty}([0, t_0]; L^2(T_S, T_E)), \tag{4.15}$$

$$v \in W^{1,\infty}([0, t_0]; H^1(T_S, T_E)) \cap C([0, t_0]; H^2(T_S, T_E)), \tag{4.16}$$

$$A,\ z,\ v \text{ are boundary-regular.} \tag{4.17}$$

Moreover, if

$$v_E, v_S \in C^1([0, t_0]), \tag{4.18}$$

then the solution (A, z, v) *has the additional properties:*

$$A, z \in \bigcap_{k=0}^{2} C^k([0, t_0]; H^{2-k}(T_S, T_E)), \tag{4.19}$$

$$v \in C^1([0, t_0]; H^1(T_S, T_E)) \cap C([0, t_0]; H^2(T_S, T_E)). \tag{4.20}$$

Remark 4.3

(a) Conditions (4.9)–(4.10) are in agreement with the expected physical behavior of the flow.

(b) Theorem 4.2 asserts the existence of classical solutions to the equations of fiber spinning.

(c) Solving the momentum balance (2.10) for v yields

$$v(t, T) = v_S(t) + (v_E(t) - v_S(t)) \int_{T_S}^{T} \frac{z_T(t, \tau)}{A(t, \tau)}\, d\tau \left(\int_{T_S}^{T_E} \frac{z_T(t, T)}{A(t, T)}\, dT \right)^{-1}. \tag{4.21}$$

Hence if we differentiate Eq. (4.21) w.r.t. T, we find

$$\frac{v_T(t, T)}{z_T(t, T)} A(t, T) = (v_E(t) - v_S(t)) \left(\int_{T_S}^{T_E} \frac{z_T(t, T)}{A(t, T)}\, dT \right)^{-1}. \tag{4.22}$$

Therefore the compatibility conditions (4.11)–(4.13) are necessary for (4.14) to be true.

(d) We will prove (4.14), (4.15) and (4.17) for A and z, while eliminating Eq. (2.10) by (4.21)–(4.22) in Eqs. (2.9), (2.11).

(e) Theorem 4.2 and its proof can be modified in an obvious fashion to hold for the boundary condition (2.18). It can also be extended to the case of a nonconstant tensile viscosity η^* (see Eq. (2.19)).

From now on we shall assume that t^*, A^0, z^0, A_E, v_E and v_S are given once and for all such that hypotheses (4.6)–(4.13) hold.

4.2. The Solution Operator

Definition 4.4. *Let $h = h(t,T)$ be a boundary-regular function with domain $[0,t'] \times [T_S, T_E]$. Then define the energy functional $\mathcal{E}$ by*

$$\mathcal{E}(h) \stackrel{\text{def}}{=} \left(||h||_{0,2}^2 + ||h||_{1,1}^2 + +||h_T(\cdot, T_S)||_{H^1}^2 + ||h_T(\cdot, T_E)||_{H^1}^2 \right)^{\frac{1}{2}}. \tag{4.23}$$

Definition 4.5. *For $L > 0$ and $t' \in (0, t^*]$, let $S(t', L)$ be the set of vector fields $(B, \xi)^T$ on $[0,t'] \times [T_S, T_E]$ such that*

$$B \text{ and } \xi \text{ are boundary-regular on } [0,t'] \times [T_S, T_E], \tag{4.24}$$

$$\mathcal{E}(B)^2 + \mathcal{E}(\xi)^2 \le L^2, \tag{4.25}$$

$$B(0,T) = A^0(T) \text{ and } B(t, T_E) = A_E(t), \tag{4.26}$$

$$\xi(0,T) = z^0(T) \text{ and } \xi(t, T_E) = 0. \tag{4.27}$$

Remark 4.6 For large L, $S(t^*, L)$ is clearly not empty. Hence in this case $S(t', L) \neq \emptyset$ for every $t' \in (0, t^*]$.

Definition 4.7. *We shall say that $t' \in (0, t^*]$ is admissible if and only if the set $S(t', L)$ is nonempty and all pairs $(B, \xi)^T \in S(t', L)$ have the properties*

$$B(t,T) > 0 \quad \text{and} \quad \int_{T_S}^{T_E} \frac{\xi_T(t,\tau)}{B(t,\tau)}\, d\tau \neq 0 \quad \text{for } t \in [0,t'],\ T \in [T_S, T_E]. \tag{4.28}$$

Lemma 4.8. *Suppose $S(t^*, L)$ is nonempty. Then there exists $t_0 \in (0, t^*]$ such that every $t \in (0, t_0]$ is admissible.*

Proof. Given $(B, \xi)^T \in S(t', L)$ for some $t' \in (0, t^*]$. Denote the embedding constant of the embedding $H^1(T_S, T_E) \hookrightarrow L^\infty(T_S, T_E)$ by C_1. Let $A_{\min}$ and $A_{\max}$ be the minimum and maximum values, resp., of A^0 over $[T_S, T_E]$. Both values are positive by (4.9). Fix $\mu \in (0, A_{\min})$. Since

$$B(t,T) - A^0(T) = \int_0^t B_t(s,T)\, ds, \tag{4.29}$$

we have

$$|B(t,T) - A^0(T)| \le C_1\, L\, t'. \tag{4.30}$$

Hence

$$|B(t,T) - A^0(T)| \le \mu \quad \text{for } t \in [0,t'],\ T \in [T_S, T_E], \tag{4.31}$$

and also

$$\frac{\displaystyle\int_{T_S}^{T_E} |z_T^0|\, dT}{A_{\max} + \mu} > \frac{t' \sqrt{T_E - T_S}\, L}{A_{\min} - \mu}, \tag{4.32}$$

if only t' is small enough. Since

$$\int_{T_S}^{T_E} |\xi_T(t,T) - z_T^0(T)|\, dT \le \int_0^t \int_{T_S}^{T_E} |\xi_{Tt}(s,T)|\, dT\, ds \le t' \sqrt{T_E - T_S}\, L, \tag{4.33}$$

we derive that

$$\left|\int_{T_S}^{T_E} \frac{\xi_T}{B}\, dT\right| \geq \frac{\displaystyle\int_{T_S}^{T_E} |z_T^0|\, dT}{A_{\max} + \mu} - \frac{t' \sqrt{T_E - T_S}\, L}{A_{\min} - \mu} > 0. \tag{4.34}$$

Hence (4.28) is proven. The claim follows. □

Definition 4.9. *Let $t' \in (0, t^*]$ be admissible. Define the solution operator $\Sigma_{t',L}$ on $S(t', L)$ by*

$$\Sigma_{t',L} : \begin{pmatrix} B \\ \xi \end{pmatrix} \longmapsto \begin{pmatrix} Y \\ \zeta \end{pmatrix}, \tag{4.35}$$

where $Y = Y(t, T)$ and $\zeta = \zeta(t, T)$ are the solutions (in the sense of Definition 3.1) of the boundary-initial value problems, stated on $[0, t'] \times [T_S, T_E]$,

$$Y_t = \frac{\alpha\, T}{\sqrt{B}} Y_T + (v_S - v_E) \left(\int_{T_S}^{T_E} \frac{\xi_T}{B}\, dT\right)^{-1}, \tag{4.36}$$

$$Y(0, T) = A^0(T), \tag{4.37}$$

$$Y(t, T_E) = A_E(t), \tag{4.38}$$

and

$$\zeta_t = \frac{\alpha\, T}{\sqrt{B}} \zeta_T + v_S + (v_E - v_S) \int_{T_S}^{T} \frac{\xi_T}{B}\, d\tau \left(\int_{T_S}^{T_E} \frac{\xi_T}{B}\, dT\right)^{-1}, \tag{4.39}$$

$$\zeta(0, T) = z^0(T), \tag{4.40}$$

$$\zeta(t, T_E) = 0. \tag{4.41}$$

Remark 4.10 The coefficients and inhomogeneous terms in (4.36) and (4.39) satisfy conditions (3.13)–(3.14), resp. By Theorem 3.5, $\Sigma_{t',L}$ is a well-defined map from $S(t', L)$ into $\left(C^1([0, t']; H^1(T_S, T_E)) \cap C([0, t']; H^2(T_S, T_E))\right)^2$ if t' is admissible.

4.3. Proof of Theorem 4.2 – Existence, Uniqueness and Regularity of Solutions

Lemma 4.11. *Suppose $S(t^*, L)$ is nonempty. Then there exists an admissible $t_0 \in (0, t^*]$ such that, for all $t \in (0, t_0]$, the range of the solution operator $\Sigma_{t,L}$ is contained in $S(t, L)$.*

Proof. Throughout we shall use the quantities C_1, $A_{\min}$, $A_{\max}$ and μ of Lemma 4.8. Let $t_0 < t_1$ be two distinct values in $(0, t^*]$. Suppose t_1 is chosen such that $C_1\, L\, t_1 \leq \mu$ and such that estimate (4.32) holds with $t' = t_1$. If $(B, \xi)^T \in S(t_\#, L)$ for some $t_\# \in (0, t_0]$, it follows that, for $t \in [0, t_\#]$,

$$|B(t, T) - A^0(T)| \leq C_1\, L\, t_1 \leq \mu, \tag{4.42}$$

and

$$\left| \int_{T_S}^{T_E} \frac{\xi_T}{B} \, dT \right| \geq \frac{\displaystyle\int_{T_S}^{T_E} |z_T^0| \, dT}{A_{\max} + \mu} - \frac{t_1 \sqrt{T_E - T_S} \, L}{A_{\min} - \mu} > 0. \tag{4.43}$$

Now define

$$\mathcal{P}(t,T) \stackrel{\text{def}}{=} \frac{\alpha \, T}{\sqrt{B(t,T)}}, \tag{4.44}$$

$$\mathcal{F}_1(t,T) \stackrel{\text{def}}{=} (v_S(t) - v_E(t)) \left(\int_{T_S}^{T_E} \frac{\xi_T(t,T)}{B(t,T)} \, dT \right)^{-1}, \tag{4.45}$$

$$\mathcal{F}_2(t,T) \stackrel{\text{def}}{=} v_S(t) + \frac{v_E(t) - v_S(t)}{\displaystyle\int_{T_S}^{T_E} \frac{\xi_T(t,T)}{B(t,T)} \, dT} \int_{T_S}^{T} \frac{\xi_T(t,\tau)}{B(t,\tau)} \, d\tau. \tag{4.46}$$

Then it is readily seen that $\mathcal{P}$, $\mathcal{F}_1$ and $\mathcal{F}_2$ are boundary-regular (this observation has already been made in Remark 4.10). Moreover, by assumption, both the functions $\mathcal{P}$, $\mathcal{F}_1$, A^0, A_E and $\mathcal{P}$, $\mathcal{F}_2$, z^0, $z_E \equiv 0$ satisfy compatibility conditions on $[0, t_\#] \times [T_S, T_E]$, analogous to (3.17)–(3.18). (4.42) yields

$$|(\mathcal{P}(t,T))^{-1}| \leq (\alpha \, T_S)^{-1} \sqrt{A_{\max} + \mu}. \tag{4.47}$$

We also observe that, by virtue of estimates (4.42)–(4.43), the quantities $\mathcal{E}(\mathcal{P})$, $\mathcal{E}(\mathcal{F}_1)$, $\mathcal{E}(\mathcal{F}_2)$, $||\mathcal{P}(0) \, B_T(0) + \mathcal{F}_1(0)||_{H^1}$ and $||\mathcal{P}(0) \, \xi_T(0) + \mathcal{F}_2(0)||_{H^1}$ are bounded by expressions involving only t_1, L and the boundary/initial values. Since these estimates hold for $(B, \xi) \in S(t_\#, L)$ if $t_\# \in (0, t_0]$, Corollary 3.17 implies the claim if only t_0 is chosen small enough relative to L. □

Lemma 4.12. *Suppose $S(t, L)$ is nonempty. Then the metric $d(\cdot,\cdot)$, defined for $(B,\xi)^T$, $(\hat{B},\hat{\xi})^T \in S(t,L)$ by*

$$d\left((B,\xi)^T, (\hat{B},\hat{\xi})^T\right) \stackrel{\text{def}}{=} \left(\left\| B - \hat{B} \right\|_{0,1}^2 + \left\| \xi - \hat{\xi} \right\|_{0,1}^2 \right)^{\frac{1}{2}}, \tag{4.48}$$

renders $S(t,L)$ a complete metric space.

Proof. d is clearly a well-defined metric on $S(t,L)$. Suppose the sequence $(v^n) = (B^n, \xi^n)$ is Cauchy in this metric. Then v^n converges strongly to a point v_0 in $\left(L^2([0,t] \times [T_S, T_E])\right)^2$ as $n \to \infty$. On the other hand, since $\mathcal{E}(B^n)^2 + \mathcal{E}(\xi^n)^2 \leq L^2$ for all n, a subsequence of (v^n), say (w^n), is weak* convergent in

$$\left(W^{1,\infty}([0,t]; H^1(T_S,T_E)) \cap L^\infty([0,t]; H^2(T_S,T_E))\right)^2$$

with weak* limit $w_0 \in S(t,L)$. The convergence is strong in $\left(L^2([0,t] \times [T_S, T_E])\right)^2$ by Rellich's theorem. Hence we deduce that $v_0 \equiv w_0$. □

Theorem 4.13. *Suppose $S(t^*, L)$ is nonempty. Then there exists an admissible $t_0 \in (0, t^*]$ such that the solution operator $\Sigma_{t_0,L}$ is a contraction map on $S(t_0, L)$ w.r.t. the metric d given in (4.48).*

Proof. By Lemma 4.11, for large L, the range of $\Sigma_{t_0,L}$ is contained in $S(t_0, L)$ for all $t_0 \in (0, t^*]$ that are sufficiently small relative to L. Throughout $\kappa = \kappa(t_0, L)$ will be a continuous, positive generic function that stays bounded as $t_0 \to 0$. The arguments of κ will be omitted. Given $(B, \xi)^T$, $(\hat{B}, \hat{\xi})^T \in S(t_0, L)$. Set

$$(Y, \zeta)^T \overset{\text{def}}{=} \Sigma_{t_0,L}\left((B, \xi)^T\right), \quad (\hat{Y}, \hat{\zeta})^T \overset{\text{def}}{=} \Sigma_{t_0,L}\left((\hat{B}, \hat{\xi})^T\right), \tag{4.49}$$

and define

$$\gamma \overset{\text{def}}{=} \left(\int_{T_S}^{T_E} \frac{\xi_T}{B}\, dT\right)^{-1}, \quad \hat{\gamma} \overset{\text{def}}{=} \left(\int_{T_S}^{T_E} \frac{\hat{\xi}_T}{\hat{B}}\, dT\right)^{-1}. \tag{4.50}$$

Estimates (4.31)–(4.34) yield that

$$|\gamma(t) - \hat{\gamma}(t)| \leq \kappa \left(||\xi_T(t) - \hat{\xi}_T(t)||_2 + ||B(t) - \hat{B}(t)||_2\right). \tag{4.51}$$

Therefore

$$\begin{aligned}
&\left|\int_{T_S}^{T_E} \int_{T_S}^{T} \left(\frac{\xi_T}{B}\gamma - \frac{\hat{\xi}_T}{\hat{B}}\hat{\gamma}\right) d\tau \left(\zeta - \hat{\zeta}\right) dT\right| \\
\leq\ & \kappa\, ||\zeta(t) - \hat{\zeta}(t)||_2 \int_{T_S}^{T_E} \left|\frac{(\xi_T - \hat{\xi}_T)\,\gamma\,\hat{B} + (\hat{B} - B)\,\hat{\xi}_T\,\hat{\gamma} + (\gamma - \hat{\gamma})\,\hat{B}\,\hat{\xi}_T}{B\,\hat{B}}\right| dT \\
\leq\ & \kappa\, ||\zeta(t) - \hat{\zeta}(t)||_2 \left(||\xi_T(t) - \hat{\xi}_T(t)||_2 + ||B(t) - \hat{B}(t)||_2\right).
\end{aligned} \tag{4.52}$$

Next we obtain

$$\begin{aligned}
&\left|\int_{T_S}^{T_E} T \left(\frac{\zeta_T}{\sqrt{B}} - \frac{\hat{\zeta}_T}{\sqrt{\hat{B}}}\right) \left(\zeta - \hat{\zeta}\right) dT\right| \\
\leq\ & \kappa\, ||\zeta(t) - \hat{\zeta}(t)||_2 \left(||B(t) - \hat{B}(t)||_2 + ||\zeta_T(t) - \hat{\zeta}_T(t)||_2\right).
\end{aligned} \tag{4.53}$$

Combining estimates (4.52)–(4.53), we derive

$$\begin{aligned}
\frac{d}{dt} \int_{T_S}^{T_E} \left(\zeta - \hat{\zeta}\right)^2 dT \ \leq\ & \kappa \left(||\zeta(t) - \hat{\zeta}(t)||_{H^1}^2\right. \\
& \left. + ||\xi(t) - \hat{\xi}(t)||_{H^1}^2 + ||B(t) - \hat{B}(t)||_2^2\right).
\end{aligned} \tag{4.54}$$

We observe that

$$\begin{aligned}
&\int_{T_S}^{T_E} T\left(\frac{\zeta_{TT}}{\sqrt{B}}-\frac{\hat{\zeta}_{TT}}{\sqrt{\hat{B}}}\right)\left(\zeta_T-\hat{\zeta}_T\right)dT \\
\le\ &\int_{T_S}^{T_E} -\frac{\partial}{\partial T}\left(\frac{T}{2\sqrt{B}}\right)\left(\zeta_T-\hat{\zeta}_T\right)^2+\frac{\hat{\zeta}_{TT}}{\sqrt{B\,\hat{B}}}\left(\sqrt{\hat{B}}-\sqrt{B}\right)\left(\zeta_T-\hat{\zeta}_T\right)dT \\
\le\ &\kappa\left(||\zeta_T(t)-\hat{\zeta}_T(t)||_2^2+||B(t)-\hat{B}(t)||_{H^1}\,||\zeta_T(t)-\hat{\zeta}_T(t)||_2\right). \qquad (4.55)
\end{aligned}$$

For this result we have used that $\zeta_T(\cdot,T_E)\equiv\hat{\zeta}_T(\cdot,T_E)$. Hence (after a few calculations similar to (4.52)–(4.55)) we obtain an estimate of the form

$$\begin{aligned}
\frac{d}{dt}||\zeta(t)-\hat{\zeta}(t)||_{H^1}^2 \ \le\ &\kappa\left(||\zeta(t)-\hat{\zeta}(t)||_{H^1}^2\right.\\
&\left.+||\xi(t)-\hat{\xi}(t)||_{H^1}^2+||B(t)-\hat{B}(t)||_{H^1}^2\right). \qquad (4.56)
\end{aligned}$$

We integrate (4.56) over $[0,t_0]$ and apply Gronwall's inequality to deduce the estimate

$$\left\|\zeta-\hat{\zeta}\right\|_{0,1}^2\le\kappa\,t_0\left(d\left((B,\xi)^T,(\hat{B},\hat{\xi})^T\right)\right)^2. \qquad (4.57)$$

In a completely analogous fashion we can derive estimate (4.57) with Y, $\hat{Y}$ taking the place of ζ, $\hat{\zeta}$. Hence, if t_0 is small enough, the solution operator $\Sigma_{t_0,L}$ is a contraction in the metric d. □

Proof of Theorem 4.2. By Theorem 4.13, the solution operator $\Sigma_{t_0,L}$ has a unique fixed point (A,z) in $S(t_0,L)$ if L is large and t_0 is small relative to L. It is readily seen from Eqs. (4.36), (4.39) that (A,z,v) (with v given by (4.21)) is the unique solution of the boundary-initial value problem (2.9)–(2.17). Moreover, A and z are boundary-regular and satisfy the regularity condition (4.14) (see Theorem 3.5). The remaining results for v, (4.16)–(4.17), follow now immediately from its explicit representation (4.21). Also, if hypothesis (4.18) holds, the additional regularity results (4.19)–(4.20) are obvious consequences of Eq. (4.21) and the model equations (2.9), (2.11). □

References

[1] Adams R. A., *Sobolev Spaces*, Academic Press, New York (1975).

[2] Kase S. & T. Matsuo, "Studies on melt spinning. I. Fundamental equations on the dynamics of melt spinning", J. Polym. Sci. A **3** (1965), 2541.

[3] Kato T., *Perturbation Theory for Linear Operators*, 2nd corr. ed., Springer, Berlin (1995).

[4] Lions J. L. & E. Magenes, *Non-Homogeneous Boundary Value Problems and Applications I*, Springer, New York (1972).

[5] Majda A., *Compressible Fluid Flow and Systems of Conservation Laws in Several Space Variables*, Springer, New York (1984).

[6] Pazy A., *Semigroups of Linear Operators and Applications to Partial Differential Equations*, Springer, New York (1983).

[7] Pearson J. R. A., *Mechanics of Polymer Processing*, Elsevier, London (1985).

[8] Petrie C. J. S., *Elongational Flows*, Pitman, London (1979).

[9] Renardy M., W. J. Hrusa & J. A. Nohel, *Mathematical Problems in Viscoelasticity*, Longman & Wiley, New York (1987).

[10] Renardy M. & R. Rogers, *An Introduction to Partial Differential Equations*, Springer, New York (1992).

[11] Ziabicki A., "Physical fundamentals of the fiber-spinning processes", in: *Man-Made Fibers*, ed. H. F. Mark, S. M. Atlas, & E. Cernia, Interscience, New York (1967).

[12] Ziabicki A., *Fundamentals of Fibre Formation*, Wiley, London (1976).

Thomas Hagen and Michael Renardy
Department of Mathematics
Virginia Polytechnic Institute and State University
Interdisciplinary Center for Applied Mathematics
Research Group on **M**odeling and **A**nalysis of
Fluids, **I**nterfaces and **A**morphous Polymers
Blacksburg, Virginia 24061-0123, USA
thagen@math.vt.edu
renardym@math.vt.edu

Progress in Nonlinear Differential Equations
and Their Applications, Vol. 35
© 1999 Birkhäuser Verlag Basel/Switzerland

On Radial Solutions of Quasilinear Boundary Value Problems

D. D. Hai and K. Schmitt

Abstract. In this paper we establish the existence of positive radial solutions for a class of quasilinear elliptic boundary problems. Both the positone and nonpositone cases are considered.

1. Introduction

We consider the existence of positive solutions for the class of boundary value problems

$$\begin{cases} \left(r^{N-1}\phi(u')\right)' + \lambda r^{N-1} f(u) = 0, \; 0 < r < 1 \\ u'(0) = u(1) = 0, \end{cases} \tag{1}$$

where ϕ is an odd, increasing homeomorphism of $\mathbb{R}$, λ is a positive parameter and f is a given nonlinear term which eventually assumes positive values and is either such that $f(0) < 0$ or $f(0) > 0$.

Problem (1) arises in the study of positive radial solutions for quasilinear elliptic boundary value problems of the form

$$\begin{cases} \operatorname{div}(A(\nabla u)) + \lambda f(u) = 0 \;\text{ in }\; \Omega \\ u = 0 \;\text{ on }\; \partial\Omega, \end{cases} \tag{2}$$

where Ω is the unit ball in $\mathbb{R}^N$.

Problem (2), in the semilinear case, i.e. $A(x) = x$, for general domains, has a long history, and many interesting results are available for various types of perturbation terms (see e.g. [3], [4], [10], [19]). The more general case has received much attention during the past decade particularly for the case of the so-called p-Laplacian, i.e. the case

$$A(\nabla u) = |\nabla u|^{p-2}\nabla u, \; p > 1,$$

(see e.g. [7], [13], [16], [17], [18]). In case the function A satisfies no longer a power growth law, most of the results available concern problem (1), i.e. problem (2) with the domain Ω a ball and solutions which are radial (see the above mentioned references). It appears that the most natural setting for such boundary value problems is the setting of Orlicz-Sobolev spaces and some results for such problems have become available (see e.g. [14] and the there cited references).

Existence results for (1) when $f(0) = 0$ and f is ϕ superlinear at ∞ have recently been considered in [7], [9], [11], [12], [13], [15]. Here we are interested in the cases when $f(0) < 0$ or $f(0) > 0$, and f is ϕ superlinear at ∞ (in a sense to be made precise below). In the former case, we shall prove the existence of a positive, decreasing solution to (1) for λ small. The asymptotic behavior of the solution obtained will also be established. This extends some of the results in [6] (see also [18], [20]) where $\phi(x) = x$, and complements those in [15] and [7] where similar problems were considered on an annulus. In the latter case, we shall derive the existence of a positive number λ^* such that (1) has at least two positive solutions for $\lambda < \lambda^*$, at least one for $\lambda = \lambda^*$, and none for $\lambda > \lambda^*$. Thus, our result complements corresponding ones in [2], [7], [9]. We refer to [1], [10], [17], [18], [20], [21] for literature on problem (2) with $A(x) = x$, as well as more general problems on bounded domains. Our main tools used in this study are the use of degree theory and sub- and supersolutions methods.

Concerning notation, we shall denote by $|\cdot|_0$ the usual supremum norm in $C[0,1]$, and since we are interested in positive solutions only we assume that $f(u) = f(0)$ for $u \leq 0$.

2. The case when $f(0)<0$

We make the following assumptions:

$(A.1)$ ϕ is an odd, increasing homeomorphism on $\mathbb{R}$, and for each $\sigma > 0$ there exists a positive number $M_\sigma > 0$ such that

$$\phi(\sigma x) \leq M_\sigma \phi(x) \tag{3}$$

for every $x \in \mathbb{R}$.

$(A.2)$ $f : \mathbb{R}^+ \to \mathbb{R}$ is continuous and

$$\lim_{x\to\infty} \frac{f(x)}{\phi(x)} = \infty. \tag{4}$$

$(A.3)$ There exists $\theta \in (0,1)$ such that

$$N \liminf_{x\to\infty} \frac{F(\theta x)}{x f(x)} > \max\left(N \limsup_{x\to\infty} \frac{\Phi(x)}{x\phi(x)} - 1, 0\right). \tag{5}$$

Here $F(x) = \int_0^x f(\tau)d\tau$, $\Phi(x) = \int_0^x \phi(\tau)d\tau$.

Our main result in this section is:

Theorem 2.1. *Let (A.1)–(A.3) hold and suppose that ϕ^{-1} is concave on $\mathbb{R}^+$. Then there exists a positive number $\bar\lambda$ such that for $0 < \lambda < \bar\lambda$, (1.1) has a positive solution u_λ which is decreasing with respect to r and satisfies*

$$\lim_{\lambda\to 0} u_\lambda(r) = \infty \tag{6}$$

uniformly for r in compact subsets of $[0,1)$.

Remark 2.1. *1. Condition (A.3) was employed in [13] to establish the existence of positive solutions to (1) when $f(0) = 0$.*

2. In the case when $\phi(x) = |x|^{p-2}x$, (A.3) becomes

$$\liminf_{x\to\infty} \frac{F(\theta x)}{xf(x)} > \max\left(\frac{N-p}{Np}, 0\right). \tag{7}$$

In particular, if $f(x) = x^q$, (5) requires that $q < \frac{Np}{\max(N-p,0)} - 1$.

3. In the case when $\phi(x) = x$ and $N \geq 2$, the existence part of Theorem 2.1 was established in [6] under the assumptions that f is nondecreasing on $\mathbb{R}$ and

$$\lim_{x\to\infty}\left(NF(\theta x) - \frac{N-2}{2}xf(x)\right)\left(\frac{x}{f(x)}\right)^{\frac{N}{2}} = \infty \tag{8}$$

for some $\theta \in (0,1)$. We note that (8) can be derived from (7) when $p = 2$ and when f is nondecreasing.

In order to prove this result, we shall need some preliminary considerations.

Lemma 2.2. *Let u be a solution of*

$$\begin{cases} \left(r^{N-1}\phi(u')\right)' = -\lambda r^{N-1}h(r) \\ u'(0) = 0 = u(1), \end{cases} \tag{9}$$

where h is a continuous function with $h \geq -m$ and m is a nonnegative number. Then:

1. *$u(r) \geq u(s) - 2\phi^{-1}\left(\frac{\lambda ms}{N}\right)$ for $0 \leq r \leq s \leq 1$.*
2. *$u'(r) \leq \phi^{-1}\left(\frac{\lambda rm}{N}\right)$ for $r \in [0,1]$.*

Further, there exists a constant $K = K(\lambda, m)$ such that: If u is any solution of (9) with $u(r) < 0$, for some $r \in [0,1)$, then

$$|u|_0 \leq K.$$

Proof. By integrating the equation (9), we obtain

$$\begin{aligned} u(r) &= \int_r^1 \phi^{-1}\left(\frac{\lambda}{s^{N-1}}\int_0^s \tau^{N-1}h(\tau)d\tau\right)ds \\ &= \int_r^1 \phi^{-1}\left(\frac{\lambda}{s^{N-1}}\int_0^s \tau^{N-1}\tilde{h}(\tau)d\tau - \frac{\lambda ms}{N}\right)ds, \end{aligned}$$

where $\tilde{h}(\tau) = h(\tau) + m \geq 0$. Since ϕ^{-1} is concave on $\mathbb{R}^+$ we have that,

$$\phi^{-1}(x-y) \geq \phi^{-1}(x) - 2\phi^{-1}(y), \quad \text{for } x, y \geq 0,$$

and so

$$u(r) \geq \tilde{u}(r) - 2\phi^{-1}\left(\frac{\lambda mr}{N}\right),$$

where

$$\tilde{u}(r) = \int_r^1 \phi^{-1}\left(\frac{\lambda}{s^{N-1}}\int_0^s \tau^{N-1}\tilde{h}(\tau)d\tau\right)ds.$$

Since $(r^{N-1}\phi(\tilde{u}'))' \leq 0$ and $\tilde{u}(1) = 0$, it follows that $\tilde{u}' \leq 0$. Hence

$$u(r) \geq \tilde{u}(s) - 2\phi^{-1}\left(\frac{\lambda ms}{N}\right) \geq u(s) - 2\phi^{-1}\left(\frac{\lambda ms}{N}\right) \text{ for } 0 \leq r \leq s \leq 1.$$

The other conclusions follow by integrating equation (9) once and using the imposed and obtained bounds.

Lemma 2.3. *Let*

$$\delta = \liminf_{x\to\infty} \frac{F(x)}{xf(x)}, \ \Gamma = \limsup_{x\to\infty} \frac{\Phi(x)}{x\phi(x)} \ \text{and } \Phi_*(x) = \int_0^x \phi^{-1}(s)ds.$$

Then there exist $\tilde{\delta} \in (0, \delta)$, *and* $\tilde{\Gamma} \in [\Gamma, 1]$ *with*

$$N\tilde{\delta} > N\tilde{\Gamma} - 1 > 0$$

such that:

$$\Phi_*^{-1}(x) \geq K_1 x^{1-\tilde{\Gamma}}, \tag{10}$$

$$F(x) \leq K_2 x^{\frac{1}{\tilde{\delta}}}, \tag{11}$$

$$f(x) \leq K_3 F(x)^{1-\tilde{\delta}}, \tag{12}$$

for large x, *where* K_i, $i = 1, 2, 3$ *are positive constants.*

Proof. Let $\tilde{\delta} \in (0, \delta)$ and $\tilde{\Gamma} \in [\Gamma, 1]$ be such that $N\tilde{\delta} > N\tilde{\Gamma} - 1 > 0$ and

$$F(x) \geq \tilde{\delta} x f(x), \ \Phi(x) \leq \tilde{\Gamma} x\phi(x)$$

for large x (see condition $(A.3)$). Rewriting these inequalities as differential inequalities, solving them and using some manipulations, we obtain (10) and (11). This implies

$$f(x) \leq K x^{\frac{1}{\tilde{\delta}}-1},$$

where $K = \frac{K_2}{\tilde{\delta}}$ and therefore

$$f(x) \leq K^{\tilde{\delta}}(f(x)x)^{1-\tilde{\delta}} \leq K_3 F(x)^{1-\tilde{\delta}}$$

for large x, and (12) follows.

For each $\mu \geq 0$, define $A_\mu v = u$, where u is the solution of

$$\begin{cases} (r^{N-1}\phi(u'))' = -\lambda r^{N-1}(f(v) + \mu), \ 0 < r < 1 \\ u'(0) = u(1) = 0. \end{cases}$$

Then

$$A_\mu : C[0, 1] \to C[0, 1]$$

is completely continuous and u is a solution of (1) if and only if u is a fixed point of $A \equiv A_0$.

Lemma 2.4. *For each $\lambda > 0$, small enough, there exists a positive number R_λ, with $\lim_{\lambda\to 0} R_\lambda = \infty$, such that all solutions of*

$$u = \zeta A u,\ 0 \le \zeta \le 1,$$

satisfy $|u|_0 \neq R_\lambda$.

Proof. Let u satisfy $u = \zeta A u$ for some $\zeta \in [0,1]$. Then

$$u(r) = \zeta \int_r^1 \phi^{-1}\left(\frac{\lambda}{s^{N-1}} \int_0^s \tau^{N-1} f(u) d\tau\right) ds,$$

then an easy calculation, using the fact that ϕ is odd and increasing, shows that

$$|u(r)| \le \int_r^1 \phi^{-1}\left(\frac{\lambda}{s^{N-1}} \int_0^s \tau^{N-1} |f(u)| d\tau\right) ds,$$

which, in turn, implies that

$$|u|_0 \le \phi^{-1}\left(\frac{\lambda f_s(|u|_0)}{N}\right), \tag{13}$$

where

$$f_s(x) = \max_{0\le y\le x} |f(y)|.$$

Hence

$$\frac{\phi(|u|_0)}{f_s(|u|_0)} \le \frac{\lambda}{N}. \tag{14}$$

Using assumption $(A.2)$ we conclude that

$$\frac{f_s(x)}{\phi(x)} \to \infty,\ \text{as } x \to \infty,$$

hence, let $\bar\lambda > 0$ be defined by

$$\frac{2\bar\lambda}{N} = \frac{\phi(K_0)}{f_s(K_0)},$$

where $K_0 > 0$. Then for $\lambda < \bar\lambda$, there exists $R_\lambda > K_0$ such that

$$\frac{f_s(R_\lambda)}{\phi(R_\lambda)} = \frac{N}{2\lambda} \tag{15}$$

(see condition $(A.2)$). Hence we immediately obtain that $|u|_0 \neq R_\lambda$ and, since (15) holds, we deduce that $\lim_{\lambda\to 0} R_\lambda = \infty$.

The next lemma shows that A_μ has no fixed points for large μ.

Lemma 2.5. *There exists a positive number μ_0 (depending on λ) such that if A_μ has a fixed point, then $\mu \le \mu_0$.*

Proof. Let u satisfy $u = A_\mu u$ and suppose that μ is large, say $\mu \geq 2m > 0$, where $-m$ is a lower bound for f. Then, upon integrating for, say, $\frac{1}{2} \leq r \leq 1$,

$$\begin{aligned} u(r) &= \int_r^1 \phi^{-1}\left(\frac{\lambda}{s^{N-1}}\int_0^s \tau^{N-1}(f(u)+\mu)d\tau\right)ds \\ &\geq \int_r^1 \phi^{-1}(\frac{\lambda\mu s}{2N})ds \\ &\geq \phi^{-1}\left(\frac{\lambda\mu r}{2N}\right)(1-r) \end{aligned} \tag{16}$$

Hence we may assume that for μ large, u is a solution which is nonincreasing on the interval $\left[0, \frac{1}{2}\right]$. Therefore, using the first equality in (16), we may conclude

$$\begin{aligned} u\left(\tfrac{1}{2}\right) &\geq \int_{\frac{1}{2}}^1 \phi^{-1}\left(\frac{\lambda}{s^{N-1}}\int_0^{\frac{1}{2}} \tau^{N-1}f(u)d\tau\right)ds \\ &\geq \int_{\frac{1}{2}}^1 \phi^{-1}\left(\frac{\lambda}{s^{N-1}2^N N}f_i\left(u\left(\frac{1}{2}\right)\right)\right)ds \\ &\geq \frac{1}{2}\phi^{-1}\left(\frac{\lambda}{2N}f_i\left(u\left(\frac{1}{2}\right)\right)\right), \end{aligned} \tag{17}$$

where $f_i(x) = \inf_{y\geq x} f(y)$. We hence obtain from (17)

$$2u\left(\frac{1}{2}\right) \geq \phi^{-1}\left(\frac{\lambda}{2N}f_i\left(u\left(\frac{1}{2}\right)\right)\right). \tag{18}$$

And hence

$$\frac{M_2\phi\left(u\left(\frac{1}{2}\right)\right)}{f_i\left(u\left(\frac{1}{2}\right)\right)} \geq \frac{\lambda}{2N}. \tag{19}$$

The hypotheses on f imply that

$$\frac{f_i(x)}{\phi(x)} \to \infty, \text{ as } x \to \infty$$

and thus (16) and (19) together yield a contradiction to μ not being bounded.

Lemma 2.6. *There exists a function $H : \mathbb{R}^+ \to \mathbb{R}$, with $\lim_{d\to\infty} H(d) = \infty$, such that any fixed point u of A_μ satisfies*

$$u(r) \geq H(|u|_0)(1-r),\ r \in [0,1].$$

Proof. The proof below follows closely the proof of Lemma 3.2 in [13]. Nevertheless, some differences exist; hence, in spite of the similarities, we provide the details. Let us denote by $\Phi_*(x) = \int_0^x \phi^{-1}(s)ds$ (see Lemma 2.3).

If u be a fixed point of A_μ. Then u solves

$$\left(r^{N-1}(\phi(u'))\right)' = -\lambda r^{N-1}(f(u)+\mu). \tag{20}$$

(In what follows we shall denote by C_k, $k = 1, 2, \dots$ various constants independent of u and μ.) Multiplying (20) by ru' gives

$$\begin{aligned} \left[r^N \Phi_*(\phi(u')) + \lambda r^N F(u)\right]' &= r^{N-1}\phi(u')(u') - N r^{N-1}\Phi(u') \\ &\quad + \lambda N r^{N-1} F(u) - \lambda \mu r^N u'. \end{aligned} \tag{21}$$

We choose $a > 0$ such that

$$N \liminf_{x\to\infty} \frac{F(\theta x)}{x f(x)} > a > N \limsup_{x\to\infty} \frac{\Phi(x)}{x\phi(x)} - 1,$$

then, multipling (20) by au, we obtain

$$(ar^{N-1}\phi(u')u)' = ar^{N-1}\phi(u')u' - \lambda a r^{N-1} f(u)u - \lambda a \mu r^{N-1} u. \tag{22}$$

Hence, adding equations (21) and (22), yields

$$\begin{aligned} \psi'(r) &= r^{N-1}[(a+1)\phi(u')u' - N\Phi(u') - \lambda\mu r u'] \\ &\quad + \lambda r^{N-1}[NF(u) - a(f(u)+\mu)u], \end{aligned}$$

where

$$\psi(r) = r^N \Phi_*(\phi(u')) + \lambda r^N F(u) + a r^{N-1}\phi(u')u.$$

The choice of a and Lemma 2.3 imply

$$NF(\theta x) - a(f(x) + \lambda\mu)x \geq C_1 F(\theta x) - C_2 \geq C_3 F(x) - C_2, \tag{23}$$

for $x \geq -2\phi^{-1}(\lambda m)$, and

$$(a+1)\phi(x)x - N\Phi(x) - \lambda\mu x \geq -C_4, \tag{24}$$

for $x \geq 0$. Here we assume that θ is sufficiently close to 1 so that the last inequality in (23) holds. Using Lemma 2.2 and (23)–(24), we deduce that

$$\psi'(r) \geq r^{N-1}(C_3 \lambda F(u) - C_5),$$

or, after integrating,

$$\psi(r) \geq C_3 \lambda \int_0^r s^{N-1} F(u) ds - C_5.$$

Let $0 < \lambda < 1$, $|u|_0 = d \gg 1$ and $r_0 \in (0,1)$ be such that $u(r_0) = \bar{\theta} d$, where $\theta < \bar{\theta} < 1$. For $r \geq r_0$,

$$\psi(r) \geq C_3 \lambda \frac{r_0^N}{N} F(\theta d) - C_6 \geq C_7 \lambda \frac{r_0^N}{N} F(d) - C_6. \tag{25}$$

We next establish an estimate on r_0. By Lemma 2.3 we have the inequality

$$-u'(r) = \phi^{-1}\left(\frac{\lambda}{r^{N-1}} \int_0^r \tau^{N-1} f(u) d\tau\right) \leq \phi^{-1}\left(\frac{\lambda K_3 r}{N} F(d)^{1-\tilde{\delta}}\right), \ r \leq r_0,$$

from which it follows that

$$\begin{aligned} (1-\bar{\theta})d &\leq \int_0^{r_0} \phi^{-1}\left(\frac{\lambda K_3 r}{N} F(d)^{1-\tilde{\delta}}\right) dr \\ &= \frac{N}{\lambda K_3 F(d)^{1-\tilde{\delta}}} \Phi_*\left(\frac{\lambda K_3 r_0}{N} F(d)^{1-\tilde{\delta}}\right). \end{aligned}$$

Hence, by Lemma 2.3,

$$\begin{aligned} r_0 &\geq \frac{N}{\lambda K_3 F(d)^{1-\tilde{\delta}}} \Phi_*^{-1}\left(\frac{\lambda K_3(1-\bar{\theta})dF(d)^{1-\tilde{\delta}}}{N}\right) \\ &\geq C_8 \lambda^{-\tilde{\Gamma}} F(d)^{\tilde{\delta}-\tilde{\Gamma}}, \end{aligned}$$

which, together with (25), implies that

$$\begin{aligned} \psi(r) &= r^N \Phi_* \left(\phi(u')\right) + \lambda r^N F(u) + a r^{N-1} \phi(u') u \\ &\geq C_9 \lambda^{1-N\tilde{\Gamma}} F(d)^{N(\tilde{\delta}-\tilde{\Gamma})+1} \geq C_9 F(d)^{\gamma}, \ r \geq r_0, \end{aligned}$$

where $\gamma = N(\tilde{\delta} - \tilde{\Gamma}) + 1 > 0$.

Since $\lim_{x\to\infty} \frac{F(x)}{x\phi(x)} = \infty$, (see (3) and (5)) we have

$$\psi(r) \leq C_{10} \tilde{F}(|u(r)| + |u'(r)|) + C_{11},$$

where $\tilde{F}(x) = F(x) + mx$, and therefore

$$|u(r)| + |u'(r)| \geq H_1(d),$$

where

$$H_1(d) = \tilde{F}^{-1}\left(\frac{C_9 F(d)^{\gamma} - C_{11}}{C_{10}}\right).$$

Since

$$|u| \leq u + 4\phi^{-1}(\lambda m), \ |u'| \leq -u' + 2\phi^{-1}(\lambda m),$$

by Lemma 2.2 it follows that

$$-u' + u \geq H_1(d) - 6\phi^{-1}(\lambda m) \equiv H_2(d).$$

Solving this differential inequality, we obtain

$$u(r) \geq H_2(d)(1 - e^{-(1-r)}), \ r \geq r_0,$$

and since $H_2(d) \to \infty$ as $d \to \infty$, we get, for large enough d,

$$u(r) \geq \frac{H_2(d)}{e}(1 - r), \ r \geq r_0,$$

and since $u(r_0) = \bar{\theta} d$, the lemma follows.

Lemma 2.7. *For each $\nu > 0$, there exists $\bar{R}_\nu > 0$ such that any solution of*

$$u = A_\mu u \ ,$$

with $\lambda \geq \nu$, satisfies $|u|_0 < \bar{R}_\nu$.

Proof. Let u satisfy $u = A_\mu u$ with $\lambda \geq \nu$. If $|u|_0 \gg 1$, then $u\left(\frac{1}{2}\right) \gg 1$ by Lemma 2.6. On the other hand,

$$\begin{aligned} u\left(\tfrac{1}{2}\right) &\geq \int_{\frac{1}{2}}^{1} \phi^{-1}\left(\frac{\lambda}{s^{N-1}} \int_0^{\frac{1}{2}} \tau^{N-1} f(u) d\tau\right) ds \\ &\geq \phi^{-1}\left(\tfrac{\nu}{N 2^{N+1}} f_i\left(u\left(\tfrac{1}{2}\right)\right)\right), \end{aligned}$$

or

$$\frac{f_i\left(u\left(\frac{1}{2}\right)\right)}{\phi\left(u\left(\frac{1}{2}\right)\right)} \leq \frac{N2^N}{\nu},$$

which is impossible, for $u\left(\frac{1}{2}\right)$ large enough (see condition $(A.2)$).

We are now ready to prove Theorem 2.1.

Proof of Theorem 2.1. It follows from Lemmas 2.5 and 2.7 that

$$\deg(I - A, B(0, \bar{R}_\lambda), 0) = \deg(I - A_\mu, B(0, \bar{R}_\lambda), 0) = 0,$$

where $\overline{R}_\lambda$ is given by Lemma 2.7. On the other hand, Lemma 2.4 implies that,

$$\deg(I - A, B(0, R_\lambda), 0) = 1.$$

Hence it follows from the excision property of the Leray-Schauder degree that there exists a solution u of (1) with

$$R_\lambda < |u|_0 < \overline{R}_\lambda.$$

Since $\lim_{\lambda\to\infty} R_\lambda = \infty$, it follows from Lemma 2.6 that u is positive if λ is sufficiently small. To verify that u is decreasing on $[0, 1]$ follows easily.

3. The case when $f(0)>0$

Our main result in this section is the following theorem:

Theorem 3.1. *Let (A.1)–(A.3) hold and suppose that $f(x) > 0$ for $x \geq 0$. Then there exists a positive number λ^* such that (1) has at least two solutions for $0 < \lambda < \lambda^*$, at least one for $\lambda = \lambda^*$, and none for $\lambda > \lambda^*$.*

We again need some preliminary results.

Lemma 3.2. *There exists a positive number $\tilde{\lambda}$ such that (1) has no solution for $\lambda > \tilde{\lambda}$.*

Proof. Let u be a solution of (1). Since $f : \mathbb{R} \to \mathbb{R}^+$ solutions will be nonnegative and nonincreasing. Hence using the integral equation we obtain

$$\begin{aligned} u\left(\tfrac{1}{2}\right) &\geq \frac{1}{2}\phi^{-1}\left(\lambda \int_0^{\frac{1}{2}} \tau^{N-1} f(u) d\tau\right) \\ &\geq \frac{1}{2}\phi^{-1}\left(\frac{\lambda}{N2^N} f_i\left(u\left(\frac{1}{2}\right)\right)\right), \end{aligned} \tag{26}$$

or

$$\frac{\phi(2u(\frac{1}{2}))}{f_i(u(\frac{1}{2}))} \geq \frac{\lambda}{N2^N}. \tag{27}$$

On the other hand, since there exists a positive number ε such that $f_i(x) \geq \varepsilon$ for every x, it follows from (26) that

$$u\left(\frac{1}{2}\right) \geq \frac{1}{2}\phi^{-1}\left(\frac{\lambda\varepsilon}{N2^N}\right). \tag{28}$$

Combining (26) and (28), we reach a contradiction to (27), if λ is sufficiently large, since $\lim_{x\to\infty} \frac{\phi(x)}{f_i(x)} = 0$.

Using Lemma 2.7 with $\mu = 0$, we obtain:

Lemma 3.3. *For each $\nu > 0$, there exists a positive constant C_ν such that all solutions u of (1) with $\lambda \geq \nu$ satisfy $|u|_0 < C_\nu$.*

Let $\Lambda = \{\lambda > 0 : (1) \text{ has a solution}\}$ and let $\lambda^* = \sup \Lambda$. We then have:

Lemma 3.4. $0 < \lambda^* < \infty$ *and* $\lambda^* \in \Lambda$.

Proof. Using the Leray-Schauder continuation theorem, it follows that (1) has a solution for λ small, and so $\lambda^* > 0$; further Lemma 3.2 implies that $\lambda^* < \infty$. We next verify that $\lambda^* \in \Lambda$. Let $\{\lambda_n\} \subset \Lambda$ be such that $\lambda_n \to \lambda^*$, and let $\{u_n\}$ be the corresponding solutions of (1). Using Lemma 3.3 and equation (3) we deduce that $\{u_n\}$ is bounded in $C^1[0,1]$. Hence $\{u_n\}$ has a subsequence which converges to a function $u \in C[0,1]$. By standard limiting process, it follows that u is a solution of (1).

The following result is similar to Lemma 9 of [15].

Lemma 3.5. *Let $0 < \lambda < \lambda^*$ and let u_{λ^*} be a solution of (1) for λ equals λ_*. Then there exists a positive number ε such that $u_{\lambda^*} + \varepsilon$ is an upper solution of (1).*

For $v \in C[0,1]$, let $A_\lambda v = u$, where u satisfies

$$\begin{cases} \left(r^{N-1}(\phi(u'))\right)' = -\lambda r^{N-1} f(v), \ 0 < r < 1 \\ u'(0) = u(1) = 0. \end{cases} \tag{29}$$

Define

$$\tilde{f}(v) = \begin{cases} f(u_{\lambda^*} + \varepsilon) & \text{if } \ v \geq u_{\lambda^*} + \varepsilon \\ f(v) & \text{if } \ 0 \leq v \leq u_{\lambda^*} + \varepsilon \\ f(0) & \text{if } \ v \leq 0. \end{cases}$$

and let $\tilde{A}_\lambda$ be the solution operator of (28) defined by $\tilde{f}$.

Lemma 3.6. *Let u be a solution of $u = \tilde{A}_\lambda u$. Then $0 \leq u \leq u_{\lambda^*} + \varepsilon$.*

Proof. Let u satisfy $u = \tilde{A}_\lambda u$. Then

$$\begin{cases} \left(r^{N-1}(\phi(u'))\right)' = -\lambda r^{N-1} \tilde{f}(u), \ 0 < r < 1 \\ u'(0) = u(1) = 0. \end{cases}$$

It follows easily that $u \geq 0$. We claim that $u \leq u_{\lambda^*} + \varepsilon$. If not, we may suppose that there exists $\bar{r} \in (0,1)$ such that $u(\bar{r}) > u_{\lambda^*}(\bar{r}) + \varepsilon$. Then there exist numbers

r_0 and r_1 with $0 \leq r_0 < r_1 < 1$ such that $u'(r_0) = u'_{\lambda^*}(r_0)$, $u(r_1) = u_{\lambda^*}(r_1) + \varepsilon$, and $u > u_{\lambda^*} + \varepsilon$ on (r_0, r_1).

Hence

$$\left(r^{N-1}(\phi(u'))\right)' = -\lambda r^{N-1} f(u_{\lambda^*} + \varepsilon),\ r \in (r_0, r_1].$$

Since

$$\left(r^{N-1}(\phi(u'_{\lambda^*}))\right)' \leq -\lambda r^{N-1} f(u_{\lambda^*} + \varepsilon),$$

we deduce that

$$\left(r^{N-1}(\phi(u') - \phi(u'_{\lambda^*}))\right)' \geq 0,\ r \in (r_0, r_1]$$

and so $u - (u_{\lambda^*} + \varepsilon)$ is nondecreasing on $(r_0, r_1]$.

Consequently,

$$u(r) < u_{\lambda^*}(r) + \varepsilon,\ r \in (r_0, r_1],$$

which is a contradiction to the assumption made.

Proof of Theorem 3.1. Let $0 < \lambda < \lambda^*$. Since 0 is a lower solution (subsolution) and u_{λ^*} is an upper solution (supersolution) for (1), there exists a solution u_λ of (1) with $0 \leq u_\lambda \leq u_{\lambda^*}$ (see [9]). We next establish the existence of a second solution. Define

$$B = \{u \in C[0,1] : -\varepsilon < u(x) < u_{\lambda^*}(x) + \varepsilon,\ x \in [0,1]\}.$$

Then B is open and $u_\lambda \in B$. Since $\tilde{A}_\lambda$ is bounded for λ in compact intervals,

$$\deg(I - \tilde{A}_\lambda, B(u_\lambda, R), 0) = 1, \tag{30}$$

if R is sufficiently large.

If there exists $u \in \partial B$ such that $u = \tilde{A}_\lambda u$, then u is a second solution of (1). Suppose that $u \neq \tilde{A}_\lambda u$, for $u \in \partial B$. Then $\deg(I - \tilde{A}_\lambda, B, 0)$ is well defined, and, since $\tilde{A}_\lambda$ has no fixed point in $B(u_\lambda, R)\backslash B$, we conclude from Lemma 3.6 that

$$\deg(I - A_\lambda, B, 0) = \deg(I - \tilde{A}_\lambda, B, 0) = 1.$$

On the other hand, it follows from Lemmas 3.2 and 3.3 that

$$\deg(I - A_\lambda, B(0, \tilde{R}), 0) = 0,$$

for some large number $\tilde{R}$.

Hence

$$\deg(I - A_\lambda, B(0, \tilde{R})\backslash B, 0) = -1,$$

and the existence of a second solution to (1) follows.

4. Remarks

1. We note here that in proofing Theorems 2.1 and 3.1 it was only in the former that the hypothesis that ϕ be a convex function was needed and that several of the preliminary results needed in proving that theorem did not require that hypothesis either. Thus the open question remains whether Theorem 2.1 remains valid without this hypothesis.
2. In a certain sense the results in this paper together with those in [11], [12], and [13] give a fairly complete picture about positive solutions of equation (1). In summary, the results in this paper together with those in the above quoted papers, yield the following table of results for boundary value problems of the form (1) whose nonlinear terms ϕ and f satisfy conditions $(A.1)$, $(A.2)$, and $(A.3)$ (the first column provides the additional conditions imposed on f and ϕ, the second the conclusion about the intervals of the parameter λ for which positive solutions exist, and the third the reference where the result is obtained (solution shall always mean positive solution)):

$f:[0,\infty)\to(0,\infty)$	$\exists\lambda^*>0$, such that $\forall$ $\lambda\in(0,\lambda^*)$, $\exists$ two solutions for $\lambda=\lambda^*$, $\exists$ a solution	Theorem 3.1
$f:[0,\infty)\to[-a,\infty)$ $f(0)<0,\ -a\le f(0)$ ϕ convex	$\exists\lambda^*>0$, such that for $0<\lambda<\lambda^*$, $\exists$ a solution	Theorem 2.1
$f:(0,\infty)\to(0,\infty)$ $\lim_{s\to0}\frac{f(s)}{\phi(s)}=0$	$\exists$ a solution $\forall\lambda>0$	[13]
$f:(0,\infty)\to(0,\infty)$ $\lim_{s\to0}\frac{f(s)}{\phi(s)}=\mu>0$	$\exists\lambda^*>0$, such that $\exists$ a solution $\forall\lambda\in(0,\lambda^*)$	[11], [12]

References

[1] W. Allegretto, P. Nistri, and P. Zecca, "Positive solutions of elliptic nonpositone problems", Diff. Integral Equations **5** (1992), 95–101.

[2] H. Amann, "On the number of solutions of asymptotically superlinear two point boundary value problems", Arch. Rational Mech. Anal. **55** (1974), 207–213.

[3] H. Amann, "Fixed point equations and nonlinear eigenvalue problems in ordered Banach spaces", SIAM Rev. **18** (1976), 620–709.

[4] H. Amann, A. Ambrosetti, and G. Mancini, "Elliptic equations with noninvertible linear part and bounded nonlinearities", Math. Z. **158** (1978), 179–194.

[5] A. Ambrosetti, D. Arcoya and B. Buffoni, "Positive solutions for some semipositone problems via bifurcation theory", Diff. Integral Equations **7** (1994), 655–663.

[6] A. Castro and R. Shivaji, "Nonnegative solutions for a class of radially symmetric nonpositone problems", Proc. A. M. S. **106** (1989), 735–740.

[7] H. Dang, R. Manásevich and K. Schmitt, "Positive radial solutions of some nonlinear partial differential equations", Math. Nachr. **186** (1997), 101–113.

[8] H. Dang and K. Schmitt, "Existence of positive solutions of semilinear elliptic equations in annular domains", Diff. Integral Equations **7** (1994), 747–758.

[9] H. Dang, K. Schmitt and R. Shivaji, "On the number of solutions for boundary value problems involving the p-Laplacian", Electronic J. Diff. Equations **1** (1996), 1–9.

[10] D. DeFigueiredo, P. Lions, and R. Nussbaum, "A priori estimates and existence of positive solutions of semilinear elliptic equations", J. Math. Pures et Appl. **61** (1982), 41–63.

[11] M. Garcia-Huidobro, R. Manásevich, and K. Schmitt, "On principal eigenvalues of p-Laplacian like operators", J. Diff. Equations **130** (1996), 235–246.

[12] M. Garcia-Huidobro, R. Manásevich, and K. Schmitt, "Some bifurcation results for a class of p-Laplacian like operators", Diff. Integral Equations **10** (1997), 51–66.

[13] M. Garcia-Huidobro, R. Manásevich, and K. Schmitt, "Positive radial solutions of quasilinear elliptic partial differential equations on a ball", Nonlinear Anal., to appear.

[14] M. Garcia-Huidobro, R. Manásevich, V. Le, and K. Schmitt, "On principal eigenvalues for quasilinear elliptic differential operators: An Orlicz-Sobolev space setting", Nonlinear Analysis and Differential Equations, to appear.

[15] D. Hai, K. Schmitt, and R. Shivaji, "Positive solutions of quasilinear boundary value problems", J. Math. Anal. Appl. **217** (1998), 672–686.

[16] V. Le and K. Schmitt, *Global Bifurcation in Variational Inequalities: Applications to Obstacle and Unilateral Problems*, Springer-Verlag, New York, 1997.

[17] V. Le and K. Schmitt, "On boundary value problems for degenerate quasilinear elliptic equations and inequalities", J. Diff. Equations **144** (1998), 170–218.

[18] R. Manásevich and K. Schmitt, "Boundary value problems for quasilinear second order differential equations?", pp. 79–119 in *Non Linear Analysis and Boundary Value Problems for Ordinary Differential Equations*, F. Zanolin editor, Springer, Wien, 1976.

[19] R. Schaaf and K. Schmitt, "Asymptotic behavior of positive solution branches of elliptic problems with linear part at resonance", Z. Angew. Math. Phys. **43** (1992), 645–675.

[20] K. Schmitt, "Positive solutions of semilinear elliptic equations", pp. 447–500 in *Topological Methods in Differential Equations and Inclusions*, Granas/Frigon editors, Kluwer, Boston, 1995.

[21] J. Smoller and A. Wasserman, "Existence of positive solutions for semilinear elliptic equations in general domains", Arch. Rational Mech. Anal. **98** (1987), 229–249.

D. D. Hai
Department of Mathematics
Mississippi State University
Mississippi State, MS 39762

K. Schmitt
Department of Mathematics
University of Utah
Salt Lake City, UT 84112

Progress in Nonlinear Differential Equations
and Their Applications, Vol. 35
© 1999 Birkhäuser Verlag Basel/Switzerland

Operator Valued Fourier Multipliers

Matthias Hieber

Dedicated to Prof. Herbert Amann on the Occasion of his 60th Birthday

1. Introduction

Results on Fourier multipliers are important tools in the study of partial differential equations. They represent a major step, for example, when establishing a priori estimates for solutions of parabolic evolution equations of Agmon-Douglis-Nirenberg type [1]. Among the results known, the classical theorem of Mikhlin, which guarantees the boundedness of translations-invariant operators on $L^p(\mathbb{R}^n)$, is of special importance.

Besides scalar-valued symbols, in many applications there occur symbols which take values in $\mathcal{L}(X,Y)$, where $\mathcal{L}(X,Y)$ denotes the space of all bounded, linear operators from some Banach space X to another Banach space Y. This is the case, for example, in questions concerning the problem of maximal regularity for parabolic problems (see e.g. [3], [13], or [15]), in elliptic and parabolic problems in non-smooth domains (allowing corners and edges) (see e.g. [10] or [17]) or in questions related to square function estimates (see e.g. [20]).

Starting from this situation, one would like to extend the classical results on scalar-valued Fourier multipliers due to Marcinkiewicz, Mikhlin, Stekhin and Lizorkin to the case of operator-valued symbols. A (unpublished) result due to Pisier states that this is impossible, in general. In fact, he proved that if Mikhlin's theorem holds true on $L^p(\mathbb{R};X)$ for $\mathcal{L}(X)$-valued symbols, then X has to be necessarily isomorphic to a Hilbert space. On the other hand, spaces of Hölder continuous functions or L^p spaces are of particular interest in differential equations. Hence, in order to establish Fourier multiplier results on spaces of the above kind we may follow the subsequent strategies: change the L^p-scale to a different scale of function spaces or find a substitute for Mikhlin's smoothness condition or take into account special structures of the operators under consideration.

In this paper, we present results in two possible directions indicated above: in fact, it was Amann [3] who recently proved a result for $\mathcal{L}(X)$-valued Fourier multipliers in the scale of vector-valued Besov spaces for arbitrary Banach spaces X. On the other hand, we state a generalization of Stekhin's theorem to $\mathcal{L}(X)$-valued symbols, where X is a UMD space.

2. Notations

We start this section by recalling basic facts on vector-valued distributions and Fourier transforms. It is well known that the proofs of the assertions listed below are straightforward modifications of the ones for the scalar case.

Let $X := (X, |\cdot|)$ be a Banach space. We denote by $\mathcal{S}(\mathbb{R}^n; X)$ the Schwartz space of smooth rapidly decreasing X-valued functions on $\mathbb{R}^n$. Then

$$\mathcal{D}(\mathbb{R}^n; X) \overset{d}{\hookrightarrow} \mathcal{S}(\mathbb{R}^n; X) \overset{d}{\hookrightarrow} \mathcal{E}(\mathbb{R}^n; X),$$

where $\mathcal{D}(\mathbb{R}^n; X)$ is the space of all X-valued C^∞-functions on $\mathbb{R}^n$ with compact support, as usual equipped with the inductive limit topology, and $\mathcal{E}(\mathbb{R}^n; X) := C^\infty(\mathbb{R}^n; X)$ equipped with the bounded convergence topology.

The space $\mathcal{S}'(\mathbb{R}^n; X)$ of X-valued temperate distributions is defined by $\mathcal{S}'(X) := \mathcal{L}(\mathcal{S}(\mathbb{R}^n); X)$. Then

$$\mathcal{E}'(\mathbb{R}^n; X) \hookrightarrow \mathcal{S}'(\mathbb{R}^n; X) \hookrightarrow \mathcal{D}'(\mathbb{R}^n; X).$$

Finally the space $\mathcal{O}_M(\mathbb{R}^n; X)$ of all X-valued slowly increasing smooth functions on $\mathbb{R}^n$ consists of all $u \in \mathcal{E}(\mathbb{R}^n; X)$ such that, given $\alpha \in \mathbb{N}^n$, there exists $m_\alpha \in \mathbb{N}$ and $C_\alpha > 0$ such that

$$|\partial^\alpha u(x)| \leq C_\alpha (1 + |x|^2)^{m_\alpha}, x \in \mathbb{R}^n.$$

Observe that $\mathcal{S}(\mathbb{R}^n;X) \hookrightarrow \mathcal{O}_M(\mathbb{R}^n;X) \hookrightarrow \mathcal{S}'(\mathbb{R}^n;X)$. We also set $\mathcal{S}(\mathbb{R}^n) := \mathcal{S}(\mathbb{R}^n;\mathbb{K})$, $\mathcal{S}'(\mathbb{R}^n) := \mathcal{S}'(\mathbb{R}^n;\mathbb{K})$ and $\mathcal{O}_M(\mathbb{R}^n) := \mathcal{O}_M(\mathbb{R}^n;\mathbb{K})$. Given $u \in L^1(\mathbb{R}^n; X)$ we define the *Fourier transform* of u by

$$\mathcal{F}u(\xi) := \hat{u}(\xi) := \int_{\mathbb{R}^n} e^{-i<x,\xi>} u(x)dx, \quad \xi \in \mathbb{R}^n,$$

where $< x, \xi > := \sum_{j=1}^n x_j \xi_j$. The *Fourier inversion theorem* guarantees that the Fourier transform $\mathcal{F}$ is an isomorphism on $\mathcal{S}(\mathbb{R}^n; X)$ and that

$$\mathcal{F}^{-1}u = (2\pi)^{-n}\check{\hat{u}} = (2\pi)^{-n}\hat{\check{u}}, \quad u \in \mathcal{S}(\mathbb{R}^n; X) \tag{2.1}$$

where $\check{u}(x) := u(-x)$ for $x \in \mathbb{R}^n$ denotes the reflection of u. The Fourier transform $\hat{u} := \mathcal{F}u$ of $u \in \mathcal{S}'(\mathbb{R}^n; X)$ is defined by

$$\hat{u}(\varphi) := u(\hat{\varphi}), \quad \varphi \in \mathcal{S}(\mathbb{R}^n).$$

Defining $\check{u}(\varphi) := u(\check{\varphi})$ for $u \in \mathcal{D}'(\mathbb{R}^n; X)$ and $\varphi \in \mathcal{D}(\mathbb{R}^n)$ it follows that $\mathcal{F}$ is an isomorphism on $\mathcal{S}'(\mathbb{R}^n; X)$ and that (2.1) holds for $u \in \mathcal{S}'(\mathbb{R}^n; X)$.

Let $\varphi \in \mathcal{S}(\mathbb{R}^n)$ and $u \in \mathcal{S}'(\mathbb{R}^n; X)$. Then the convolution $u * \varphi$ of u and φ is defined by

$$(u * \varphi)(x) := u(\tau_x \check{\varphi}), \quad x \in \mathbb{R}^n,$$

where $\tau_a \varphi(x) := \varphi(x - a)$ for $x \in \mathbb{R}^n$ and $a \in \mathbb{R}^n$. Moreover, the convolution theorem states that

$$(u * \varphi)\hat{} = \hat{u}\hat{\varphi}.$$

We define dilation σ_t for $t > 0$ by

$$\sigma_t\varphi(x) := \varphi(tx), \quad (\sigma_t u)(\varphi) := t^{-n}u(\sigma_{1/t}\varphi)$$

for $\varphi \in \mathcal{S}(\mathbb{R}^n)$ and $u \in \mathcal{S}(\mathbb{R}^n; X)$. Then

$$\mathcal{F} \circ \sigma_t = t^{-n}\sigma_{1/t} \circ \mathcal{F}, \quad \partial^\alpha \circ \sigma_t = t^{|\alpha|}\sigma_t \circ \partial^\alpha, \quad \alpha \in \mathbb{N}^n, t > 0. \tag{2.2}$$

Finally let H be a Hilbert space. Then $L^2(\mathbb{R}^n; H)$ is a Hilbert space with respect to the inner product

$$(u|v)_2 := (u|v)_{L^2(\mathbb{R}^n;H)} := \int_{\mathbb{R}^n} u \cdot v dx.$$

Plancherel's Theorem states that

$$(\hat{u}|\hat{v})_2 = (2\pi)^n (u|v)_2$$

for $u, v \in \mathcal{S}(\mathbb{R}^n; H)$. It follows that $(2\pi)^{-n/2}\mathcal{F}$ is a unitary operator on $L^2(\mathbb{R}^n; H)$.

3. Some Classical Techniques in the Vector-Valued Setting

In this section we collect several results on interpolation of operators acting on vector-valued L^p spaces. We start with the Calderón-Zygmund decomposition of an L^1-function. Throughout this section, let X be a Banach space and assume that μ is a positive Borel measure on $\mathbb{R}^n$ such that $\mu(\mathbb{R}^n) > 0$.

3.1. Proposition. *Let X be a Banach space, let $f \in L^1(\mathbb{R}^n; X)$ and let $\alpha > 0$ be given. Then there exist $f, b_i \in L^1(\mathbb{R}^n; X)$, $i \in \mathbb{N}$, a sequence of balls (B_i) and a constant $c > 0$ such that*

i) $f = g + \sum_i b_i$
ii) $|g(x)| \leq c\alpha$, *for a.a.* x
iii) $\operatorname{supp} b_i \subset B_i$ *and* $x \in \mathbb{R}^n$ *is contained in at most N balls B_i*
iv) $\|b_i\|_1 \leq c\alpha|B_i|$ *for all* i
v) $\int_{\mathbb{R}^n} b_i(x)d\mu(x) = 0$ *for all* i
vi) $\sum_i |B_i| \leq \frac{c}{\alpha}\|f\|_1$.

We observe that conditions v) and vi) imply that

$$\|\sum_i b_i\|_1 \leq c \cdot \|f\|_1 \text{ and } \|g\|_1 \leq (1 + c)\|f\|_1 \tag{3.1}$$

Proof. The scalar proof given in [20] literally applies to the X-valued case. □

Interpolation is a basic technique when dealing with singular integral operators acting in L^p-spaces. The following result of Marcinkiewicz is of fundamental importance.

3.2. Theorem. *(Marcinkiewicz). Let X, Y be Banach spaces and let $p_0, p_1 \in [1, \infty]$. Let T be a mapping from $L^{p_0}(\mathbb{R}^n; X) + L^{p_1}(\mathbb{R}^n; X)$ into the space of measurable functions on Y. Suppose that*

i) *T is sublinear, i.e.*

$$|T(f_1+f_2)(x)|_Y \le |Tf_1(x)|_Y + |Tf_2(x)|_Y \quad (f \in L^{p_0}+L^{p_1}, x \text{ a.e. })$$

ii) *T is of weak type (p_0,p_0) and (p_1,p_1), i.e. for $\alpha>0$ there exist constants $C_1, C_2 > 0$ such that*

$$|\chi_{\{x\in\mathbb{R}^n;\|Tf(x)\|_Y>\alpha\}}|_{p_i} \le \frac{C_i}{\alpha}\|f\|_{L^{p_i}(\mathbb{R}^n,X)} \quad (i=0,1)$$

Then, for $p \in (p_0,p_1)$, there exists a constant C_p (depending only on C_0, C_1 and p) such that

$$\|Tf\|_{L^p(\mathbb{R}^n;Y)} \le C_p\|f\|_{L^p(\mathbb{R}^n;X)} \quad (f \in L^p(\mathbb{R}^n;X)).$$

We remark that estimates of weak type (p_i,p_i) for T are often formulated equivalently in terms of weak-L^{p_i}-spaces. In fact, for $p \in [1,\infty)$ and a measurable function $f:\mathbb{R}^n \to X$ we set

$$[f]_p := \left(\sup_{\alpha>0} \alpha^p \mu(\{x\in\mathbb{R}^n; |f(x)|>\alpha\})\right)^{\frac{1}{p}}$$

and define

$$L^p_w(\mathbb{R}^n;X) := \{f:\mathbb{R}^n \to X \text{ measurable }; [f]_p < \infty\}$$

Let $p_0,p_1 \in [1,\infty]$ and let T be a sublinear map from $L^{p_0}(\mathbb{R}^n;X)+L^{p_1}(\mathbb{R}^n;X)$ into the space of measurable functions on Y. Assume that there exist constants $c_0, C_1 > 0$ such that

$$[Tf]_{p_i} \le |f|_{p_i} \qquad (p_i = 0,1).$$

Then, the Marcinkiewicz interpolation theorem says that there exists a constant C_p (depending only on C_0, C_1 and p) such that

$$\|Tf\|_{L^p(\mathbb{R}^n;Y)} \le C_p\|f\|_{L^p(\mathbb{R}^n;X)} \qquad (f \in L^p(\mathbb{R}^n;X)).$$

It is well known that, besides the space $L^1_w(\mathbb{R}^n)$, as we just saw, the function space BMO plays an important rôle in the theory of interpolation. We say that an $L^1_{\text{loc}}(\mathbb{R}^n;X)$ function f belongs to $BMO(\mathbb{R}^n;X)$ if there exists a constant $C>0$ such that

$$\frac{1}{|B|}\int_B |f(x)-f_B|_X dx \le C \tag{3.2}$$

holds for all balls B. Here $f_B := \frac{1}{|B|}\int_B f dx$ denotes the mean value of f over B. The smallest constant C for which (3.2) is satisfied is denoted by $\|f\|_{BMO}$.

Observe that an equivalent definition of BMO arises if we replace the mean value f_B in the above definition by an arbitrary constant C^I_B. Starting from the above definition of BMO it is natural to consider the sharp maximal function $f^\#$ associating to $f \in L^1_{\text{loc}}(\mathbb{R}^n;X)$ the function

$$f^\#(x) := \sup_{x\in B}\frac{1}{|B|}\int_B |f(x)-f_B|_X dx,$$

where the supremum is taken over all balls B containing x. Then $f \in BMO(\mathbb{R}^n; X)$ if and only $f^\# \in L^\infty(\mathbb{R}^n; X)$.

Consider now a function $f \in L^p(\mathbb{R}^n; X)$ for $1 < p < \infty$. Then it is easy to see that

$$|f^\#|_p \leq C_p |f|_p$$

for some constant C_p. The converse inequality, however, is more complicated to prove.

3.3. Proposition. *Let $1 < p < \infty$ and $f \in L^p(\mathbb{R}^n)$. Then*

$$|f|_p \leq C_p |f^\#|_p$$

for a suitable constant C_p.

For a proof see [20, Thm.IV.2.2].

Combining Proposition 3.3 with the vector-valued Marcinkiewicz interpolation theorem we obtain the following result.

3.4. Proposition. *Let X and Y be Banach spaces and let $T \in \mathcal{L}(L^r(\mathbb{R}^n; X), L^r(\mathbb{R}^n; Y))$ for some $r \in [1, \infty)$. Suppose that there exists a constant C_∞ such that*

$$\|Tf\|_{BMO(\mathbb{R}^n;Y)} \leq C_\infty \|f\|_{L^\infty(\mathbb{R}^n;X)}.$$

Then there exists constants C_p such that

$$\|Tf\|_{L^p(\mathbb{R}^n;Y)} \leq C_p \|f\|_{L^p(\mathbb{R}^n;X)} \quad (r \leq p \leq \infty).$$

Proof. We define the sublinear operator $S : L^r(\mathbb{R}^n; X) \to L^r(\mathbb{R}^n; \mathbb{R})$ by

$$(Sf)(x) := h^\#(x),$$

where $h(x) := |Tf(x)|_Y$. By assumption there exist constants M_r, M_∞ such that

$$\begin{aligned} |Sf|_r &\leq M_r \|f\|_{L^r(\mathbb{R}^n;X)}, \quad (f \in L^r) \\ |Sf|_\infty &\leq M_\infty \|f\|_{L^\infty(\mathbb{R}^n;X)}, \quad (f \in L^r \cap L^\infty). \end{aligned}$$

The Marcinkiewicz interpolation theorem implies that

$$|Sf|_p \leq M_p \|f\|_{L^p(\mathbb{R}^n;X)} \quad (f \in L^r \cap L^p, r \leq p \leq \infty)$$

for suitable constants M_p. By Proposition 3.3

$$\|Tf\|_{L^p(\mathbb{R}^n;Y)} = |h|_p \leq C_p |h^\#|_p \leq C_p M_p \|f\|_{L^p(\mathbb{R}^n;X)} \quad (r \leq p < \infty). \qquad \square$$

The above proposition says, roughly speaking, that a bounded operator on L^r which maps L^∞ continuously into BMO acts also as a bounded operator on L^p for $p \in [r, \infty]$. It may be viewed as the counterpart to a consequence of the Marcinkiewicz interpolation theorem which states that a bounded operator on L^r which maps L^1 into L^1_w acts as a bounded operator on L^p for $p \in (1, r)$. Both results will be useful in the following sections.

4. Singular Integrals

In this section we examine the L^p-boundedness of singular integral operators of the form

$$(Tf)(x) = \int_{\mathbb{R}^n} K(x,y)f(y)d\mu(y), \tag{4.1}$$

where the kernel K is singular near $x = y$, and (4.1) is meaningful only in some limiting sense. More precisely, we assume that for $f \in L^2(\mathbb{R}^n; X)$ with compact support, the integral in (4.1) converges absolutely for a. a. $x \in (\operatorname{supp} f)^{\complement}$ and that (4.1) holds for these x, where the kernel K is assumed to belong to $L^1_{\text{loc}}(\mathbb{R}^n \times \mathbb{R}^n, \mathcal{L}(X,Y))$.

4.1. Theorem. *Let X and Y be Banach spaces and let T be defined as in (4.1). Assume that there exists a constant $C > 0$ such that*

i) $\|Tf\|_{L^2(\mathbb{R}^n;Y)} \le C\|f\|_{L^2(\mathbb{R}^n;X)}$ *for all* $f \in L^2(\mathbb{R}^n; X)$

ii) *There exists a constant $r > 1$ such that*

$$\int_{B(y_1,r\delta)^{\complement}} |K(x,y_1) - K(x,y_2)|_{\mathcal{L}(X,Y)} d\mu(x) \le C \text{ whenever } y_2 \in B(y_1,\delta)$$

for all $y_1 \in \mathbb{R}^n, \delta > 0$.

Then there exists a constant C_p (depending only on C and p) such that

$$\|Tf\|_{L^p(\mathbb{R}^n;Y)} \le C_p\|f\|_{L^p(\mathbb{R}^n;X)} \text{ for } f \in L^p \cap L^2 \text{ with } 1 < p < 2.$$

The proof of the scalar-valued version of Theorem 4.1 given in [20] or [8] extends without any difficulties to the situation considered above. However, since Theorem 4.1 is a key step for the following results, we nevertheless present a proof of it.

Proof. We show that the mapping $f \mapsto Tf$ is of weak type $(1,1)$, that is

$$\mu\{x \in \mathbb{R}^n; |Tf(x)| > \alpha\} \le \frac{C'}{\alpha}\int |f| d\mu \quad \text{for} \quad f \in L^1 \cap L^2 \text{ and } \alpha > 0, \tag{4.2}$$

where the constant C' depends only on C. The Marcinkiewicz interpolation theorem 3.2 then yields the assertion.

Fix $\alpha > 0$ and consider the Calderón-Zygmund decomposition of f as $f = g + b$ given in Proposition 3.1. Then (4.2) holds provided we show

$$\mu\{x; |Tg(x)| > \frac{C'}{2}\alpha\} + \mu\{x; |Tb(x)| > \frac{C'\alpha}{2}\} \le \frac{C'}{\alpha}\int |f| d\mu. \tag{4.3}$$

Observe first that $g \in L^2$. Indeed, $\int |g(x)|^2 dx = \int |g(x)||g(x)| dx \le C\alpha \int |g(x)| dx$ $\le C(1+c)\alpha\|f\|_{L^1(\mathbb{R}^n;X)}$ by Proposition 3.1.ii) and (3.1). Chebycheff's inequality and assumption i) implies that

$$\mu\{x; |Tg(x)| > \frac{C'}{2\alpha}\} \le \left(\frac{C'\alpha}{2}\right)^{-2} \|Tg\|_2^2 \le \frac{4C^2}{C'^2} \cdot \frac{1}{\alpha^2}\|g\|_2^2 \le \frac{C'}{\alpha}\|f\|_1.$$

Therefore we proved (4.3) for the part which involves Tg. Consider now the term Tb. For $r > 1$ denote by B_i^r the ball with the same center as B_i, but whose radius is expanded by r. Then

$$\mu(\bigcup_i B_i^r) \leq \sum_i \mu(B_i^r) \leq r \sum_i \mu(B_i) \leq \frac{Cr}{\alpha}\|f\|_1.$$

It therefore remains to show that

$$\mu(\{x; |Tb(x)| > \frac{C'}{2}\alpha\} \cap (\bigcup B_i^r)^{\complement}) \leq \frac{C'}{\alpha}\|f\|_1. \tag{4.4}$$

To this end note that by hypothesis ii)

$$\int_{(B_i^r)^{\complement}} |K(x,y_1) - K(x,y_i)|_{\mathcal{L}(X,Y)} d\mu(x) \leq C \quad (y_1 \in B_i), \tag{4.5}$$

where y_i denotes the center of B_i and B_i^r. For $x \notin B_i^r$ the representation (4.1) is valid for $(Tb_i)(x)$ since by Proposition 3.1.iii) the function b_i is supported in B_i. Hence

$$\begin{aligned}
\int_{(\bigcup B_i^r)^c} |Tb(x)| d\mu(x) &\leq \sum_i \int_{(B_i^r)^{\complement}} |Tb_i(x)| d\mu(x) \\
&= \sum_i \int_{(B_i^r)^{\complement}} \Big| \int_{B_i} [K(x,y) - K(x,y_i)] b_i(y) d\mu(y) \Big| d\mu(x) \\
&\leq \sum_i \int_{B_i} \int_{(B_i^r)^{\complement}} |K(x,y) - K(x,y_i)|_{\mathcal{L}(X,Y)} d\mu(x) |b_i(y)| d\mu(y) \\
&\leq C \sum_i \int_{B_i} b_i(y)| d\mu(y) \\
&\leq C \cdot \sum_i c\alpha |B_i| \leq Cc^2 \|f\|_1,
\end{aligned}$$

by Proposition 3.1.v), (4.5), Proposition 3.1.iv) and (3.1). We therefore obtain (4.4). This shows that the mapping $f \mapsto Tf$ is of weak type $(1,1)$. The Marcinkiewicz interpolation theorem completes the proof. □

4.2. Remark. a) Since $L^p \cap L^2$ is dense in $L^p(1 < p < \infty)$, the operator T in Theorem (4.1) has a unique bounded extension to L^p for $1 < p < 2$.

b) It is worthwile to point out the following generalization of Theorem 4.1. Assume that T defined as in (4.1) satisfies the condition

$$\|Tf\|_{L^r(\mathbb{R}^n;Y)} \leq C\|f\|_{L^r(\mathbb{R}^n;X)}, \qquad f \in L^r(\mathbb{R}^n;X)$$

for some $r \in (1,\infty)$ and condition ii) of Theorem 4.1. Then T admits a bounded extension to L^p for $1 < p < r$. The proof follows the lines of the proof given above and is left to the reader.

Consider now the special case of convolution operators. More specifically, let X and Y be Banach spaces and suppose that $K \in L^1_{\mathrm{loc}}(\mathbb{R}^n \setminus \{0\}, \mathcal{L}(X,Y))$. Then the integral

$$Tf(x) := \int_{\mathbb{R}^n} K(x-y)f(y)dy \tag{4.6}$$

is well defined for $f \in L^\infty(\mathbb{R}^n; X)$ with compact support and $x \in (\operatorname{supp} f)^{\complement}$. For operators of the form (4.6) the condition ii) of Theorem 4.1 is equivalent to the condition

$$\int_{|x|>c|y|} |K(x-y) - K(x)|dx \le C \quad (y \neq 0, c > 1). \tag{4.7}$$

The following result due to Benedek, Calderón and Panzone says that operators of the form (4.6) are bounded operators on L^p, $1 < p < \infty$, provided this holds for some $r \in [1, \infty]$ and condition (4.7) is satisfied.

4.3. Theorem. *(Benedek, Calderón, Panzone). Suppose that* $T \in \mathcal{L}(L^r(\mathbb{R}^n; X), L^r(\mathbb{R}^n; Y))$ *for some* $r \in [1, \infty)$. *Assume that* T *may be represented by (4.6) for* $f \in L^\infty(\mathbb{R}^n; X)$ *with compact support and* $x \in (\operatorname{supp} f)^{\complement}$ *and that (4.7) is satisfied for some* $c > 1$. *Then* T *admits a bounded extension to* L^p *for* $p \in (1, \infty)$; *i. e. there exists a constant* C_p *such that*

$$\|Tf\|_{L^p(\mathbb{R}^n;Y)} \le C_p \|f\|_{L^p(\mathbb{R}^n;X)} \quad (1 < p < \infty).$$

Proof. We note that for $p \in (1, r)$ the assertion follows from Remark 4.2.b) since for translation invariant operators of the form (4.6), the condition ii) of Theorem 4.1 is equivalent to (4.7).

Hence it remains to prove the case where $p \in (r, \infty)$. Our plan is to show that T maps L^∞ boundedly into BMO and to apply then Proposition 3.4.

For a ball $B \subset \mathbb{R}^n$ denote by B_2 the ball with the same center as B but double the radius. For $f \in L^\infty(\mathbb{R}^n; X)$ with compact support write $f = f_1 + f_2$ with $f_1 = f\chi_{B_2}$, $f_2 = \chi_{B_2^c}$. Then $Tf = Tf_1 + Tf_2$. Choosing r' so that $\frac{1}{r'} + \frac{1}{r} = 1$ and since $f_1 \in L^r$ we obtain for Tf_1

$$\begin{aligned}\int_B |Tf_1(x)|_Y dx \le |B|^{\frac{1}{r'}} \|Tf_1\|_{L^r(\mathbb{R}^n;X)} &\le C_r |B|^{\frac{1}{r'}} \|f_1\|_{L^r(\mathbb{R}^n;X)} \\ &\le C_r 2^{\frac{n}{r}} |B| \|f\|_{L^\infty(\mathbb{R}^n;X)}\end{aligned}$$

In order to estimate Tf_2 we use (4.7). Define $C_B = \int_{B_2^c} K(x_0 - y)f(y)dy$, where x_0 denotes the center of B. For $x \in B$ we have

$$Tf_2(x) - C_B = \int_{B_2^c} [K(x-y) - K(x_0 - y)]f(y)dy$$

and therefore

$$|Tf_2(x) - C_B|_Y \le C\|f\|_{L^\infty(\mathbb{R}^n;X)} \tag{4.8}$$

Combining (4.8) with the estimate for Tf_1 it follows that

$$\frac{1}{|B|}\int_B |Tf(x) - C_B|_Y dx \le (2^{\frac{n}{r}} + 1)\max(C, C_r)\|f\|_{L^\infty(\mathbb{R}^n;X)},$$

which gives

$$\|Tf\|_{BMO(\mathbb{R}^n;Y)} \le C'\|f\|_{L^\infty(\mathbb{R}^n;X)},$$

first for functions of compact support and by approximation for all $f \in L^r \cap L^\infty$. Proposition 3.4 yields the assertion for $p \in [r, \infty)$. □

An interesting application of Theorem 4.3 concerns the problem of maximal L^p regularity for parabolic evolution equations. In fact, let $1 < p < \infty$, $J = [0, T]$ for some $T > 0$, and let $-A$ be the generator of a bounded analytic semigroup T on some Banach space X. Then $A \in MR(p; X)$ and we say that there maximal L^p-regularity for the problem $u'(t) + Au(t) = f(t), u(0) = u_0$ if for each $f \in L^p(J; X)$ there exists a unique $u \in W^{1,p}(J; X) \cap L^p(J; D(A))$ satisfying the above initial value problem. In other words, $A \in MR(p; X)$ if and only if $R \in \mathcal{L}(L^p(J; X))$, where R is defined as $Rf := A\int_0^t T(t-s)f(s)ds$. Rewriting R as a convolution operator on $\mathbb{R}$ with the operator-valued kernel k defined as

$$k(t) = \begin{cases} AT(t) & t > 0 \\ 0, & elsewhere \end{cases}$$

it is easy to verify that k fulfills (4.7). Hence, Plancherel's theorem and Theorem 4.3 imply the following result.

4.4. Corollary. *Let $-A$ be the generator of a bounded analytic semigroup on some Banach space X. If $A \in MR(p; X)$ for some $p \in (1, \infty)$, then $A \in MR(q; X)$ for all $q \in (1, \infty)$.*

For further information on singular integrals with Banach space valued kernels we refer to [16].

5. Scalar-Valued Fourier Multipliers

Let X be a Banach space and $m \in \mathcal{O}_M(\mathbb{R}^n)$. The convolution theorem implies that we have

$$u \mapsto \mathcal{F}^{-1}m\mathcal{F}u := \mathcal{F}^{-1}(m\hat{u}) = \mathcal{F}^{-1}m * u \in \mathcal{L}(\mathcal{S}(\mathbb{R}^n; X)). \tag{5.1}$$

We then say that

$$T_m u := \mathcal{F}^{-1}m\mathcal{F}u, \quad u \in \mathcal{S}(\mathbb{R}^n; X)$$

is a *translations-invariant operator* with *symbol m*.

It is a classical problem to ask for conditions which have to be imposed on m so that (5.1) holds true if $\mathcal{S}(\mathbb{R}^n; X)$ is replaced by an X-valued function space $F(\mathbb{R}^n; X)$ satisfying

$$\mathcal{S}(\mathbb{R}^n; X) \hookrightarrow F(\mathbb{R}^n; X) \hookrightarrow \mathcal{S}'(\mathbb{R}^n; X). \tag{5.2}$$

Choosing $X = \mathbb{K}$, the celebrated theorems of Marcinkiewicz, Mikhlin, Mikhlin-Hörmander, Stekhin and Lizorkin give satisfactory answers for the L^p-case. Multiplier results for functions which take values in Banach spaces are for example important in the Dore-Venni approach to regularity properties of solutions of the inhomogeneous Cauchy problem (see [9]).

Let $F(\mathbb{R}^n; X)$ be a Banach space satisfying (5.2). Following the definition for the scalar-case, we call a function $m \in L^\infty(\mathbb{R}^n)$ a *Fourier multiplier for* $F(\mathbb{R}^n; X)$ if $m\hat{u} \in \mathcal{S}'(\mathbb{R}^n; X)$ for $u \in F(\mathbb{R}^n; X)$, $\mathcal{F}^{-1}(m\hat{u}) \in F(\mathbb{R}^n; X)$ and there exists a constant $C > 0$ such that

$$\|\mathcal{F}^{-1}(m\hat{u})\|_{F(\mathbb{R}^n;X)} \leq C\|u\|_{F(\mathbb{R}^n;X)}, \quad (u \in F(\mathbb{R}^n; X)). \tag{5.3}$$

We denote by $\mathcal{M}_p(\mathbb{R}^n; X) := \mathcal{M}_{L^p}(\mathbb{R}^n; X)$, $1 \leq p \leq \infty$, the space of all $L^p(\mathbb{R}^n; X)$-Fourier multipliers. Equipped with the norm

$$\|m\|_{\mathcal{M}_p(\mathbb{R}^n;X)} := \|\mathcal{F}^{-1} m \mathcal{F}\|_{\mathcal{L}(L^p(\mathbb{R}^n;X))}$$

the space $\mathcal{M}_p(\mathbb{R}^n; X)$ becomes a Banach space.

For the time being let $F(\mathbb{R}^n; X) = L^p(\mathbb{R}^n, \mathbb{K})$. If $m : \mathbb{R}^n \to \mathbb{K}$ is sufficiently smooth, then, using Plancherel's theorem, one can show that decay assumptions on the derivatives of m of the form

$$\max_{|\alpha|\leq j} \sup_{\xi\in\mathbb{R}^n} |\xi|^\alpha |D^\alpha m(\xi)|,$$

where $j = \min\{k \in \mathbb{N}, k > \frac{n}{2}\}$, imply (4.7). Hence, by Theorem 4.3 the operator T initially defined on $L^2(\mathbb{R}^n)$ by

$$\widehat{(Tf)}(\xi) = m(\xi)\mathcal{F}f(\xi)$$

extends to a bounded operator on $L^p(\mathbb{R}^n)$, $1 < p < \infty$. More precisely, let $j = \min\{k \in \mathbb{N}, k > \frac{n}{2}\}$ and consider the Banach space

$$\begin{aligned} \mathcal{M}_M &:= \{m : \mathbb{R}^n \to \mathbb{C}; m \in C^j(\mathbb{R}^n \setminus \{0\}), |m|_M < \infty\}, \\ \mathcal{M}_L &:= \{m : \mathbb{R}^n \to \mathbb{C}; m \in C^j(\mathbb{R}^n \setminus \{0\}), |m|_L < \infty\}, \end{aligned}$$

where the norms $|\cdot|_M$ and $|\cdot|_L$ are defined by

$$\begin{aligned} |m|_M &:= \max_{|\alpha|\leq j} \sup_{\xi\in\mathbb{R}^n\setminus\{0\}} |\xi|^\alpha |D^\alpha m(\xi)| \\ |m|_L &:= \max_{|\alpha|\in\{0,1\}} \sup_{\xi\in\mathbb{R}^n\setminus\{0\}} |(\xi_1 \frac{\partial}{\partial\xi_1})^{\alpha_1} \cdots (\xi_n \frac{\partial}{\partial\xi_n})^{\alpha_n} m(\xi)|. \end{aligned}$$

Then the following holds.

5.1. Theorem. *(Mikhlin). Let* $1 < p < \infty$. *Then* $\mathcal{M}_M \hookrightarrow \mathcal{M}_p(\mathbb{R}^n; \mathbb{K})$.

5.2. Theorem. *(Lizorkin). Let* $1 < p < \infty$. *Then* $\mathcal{M}_L \hookrightarrow \mathcal{M}_p(\mathbb{R}^n; \mathbb{K})$.

5.3. Theorem. *(Stekhin). Let $1 < p < \infty$ and let $m : \mathbb{R} \to \mathbb{R}$ be a function of bounded variation. Then $m \in \mathcal{M}_p$ and there exists a constant C_p depending only on p, such that*

$$\|m\|_{\mathcal{M}_p(\mathbb{R};\mathbb{K})} \leq C_p \max(|m(0)|, Var\, m). \tag{5.4}$$

For proofs and generalizations of the results stated above, and for more information on this topic we refer to [19], [20], [21] [12] and [5]. We only mention here that the proof of Propostion 5.3 is based on the fact that the Hilbert transform acts boundedly on $L^p(\mathbb{R})$. The boundedness of the Hilbert transform also plays an important role in the extension of Mikhlin's theorem to Banach space valued functions. Therefore, we next examine the Hilbert transform in $L^p(\mathbb{R}^n; X)$.

For $\varepsilon > 0$, we define $(1/t)_\varepsilon \in L^1_{\text{loc}}(\mathbb{R})$ by

$$(1/t)_\varepsilon(\tau) := \tau^{-1}\chi_{[|\tau|\geq\varepsilon]}(\tau), \quad \tau \in \mathbb{R}$$

so that

$$< (1/t)_\varepsilon, \varphi > \lim_{\varepsilon\to 0} \int\limits_{|\tau|\geq\varepsilon} \frac{\varphi(\tau)}{\tau} d\tau \quad \text{for } \varphi \in \mathcal{S}(\mathbb{R}). \tag{5.5}$$

Then there exists a unique temperate distribution $pv(1/t)$, called the *principal value* of $1/t$, such that

$$< pv(1/t), \varphi >:= \lim_{\varepsilon\to 0} \int\limits_{|\tau|\geq\varepsilon} \frac{\varphi(\tau)}{\tau} d\tau, \qquad \varphi \in \mathcal{S}.$$

For $u \in \mathcal{S}(\mathbb{R}^n; X)$ we define the *Hilbert transform* Hu of u by

$$Hu := \frac{1}{\pi} pv(1/t) * u, \quad u \in \mathcal{S}(\mathbb{R}^n; X).$$

The symbol m of the translation invariant operator H is given by

$$(pv(1/t))\hat{} = -i\pi sgn.$$

For the time being, assume that for some $p \in [1, \infty)$ we have $\|Hu\|_{L^p(\mathbb{R};X)} \leq C\|u\|_{L_p(\mathbb{R};X)}$ for $u \in \mathcal{S}(\mathbb{R}; X)$. Then there exists a unique extension of H to a bounded operator on $L^p(\mathbb{R}; X)$. This extension, also denoted by H, is called the *Hilbert transform* on $L^p(\mathbb{R}; X)$. Since $\widehat{Hu} = -isgn(\cdot)\hat{u}$, it follows from Plancherel's theorem that

$$(Hu|Hv)_2 = \frac{1}{2\pi}(\widehat{Hu}|\widehat{Hv})_2 = \frac{1}{2\pi}(\hat{u}|\hat{v})_2 = (u|v)_2$$

for $u, v \in \mathcal{S}(\mathbb{R}; X)$. By density, H is an isomorphism on $L^2(\mathbb{R}; X)$ satisfying $H^{-1}v = -Hv$ for $v \in L^2(\mathbb{R}; X)$. We summarize these observations in part a) of the following Proposition.

5.4. Proposition. *Let X be a Banach space.*

a) *The Hilbert transform is a unitrary operator on $L^2(\mathbb{R}; X)$ provided X is a Hilbert space.*
b) *Suppose that the Hilbert transform is bounded on $L^p(\mathbb{R}; X)$ for some $p \in (1, \infty)$. Then it is bounded on $L^q(\mathbb{R}; X)$ for all $q \in (1, \infty)$.*

Proof. The assertion b) is an immediate consequence of Theorem 4.3 since

$$\int\limits_{|x|>r|y|} |K(x-y) - K(x)|dx = \int\limits_{|x|>r|y|} |\frac{1}{x-y} - \frac{1}{x}|dx \le C. \qquad \square$$

We remark that the Hilbert transform is not bounded on $L^p(\mathbb{R}; X)$, in general. This fact together with the above Proposition 5.4 motivates the following definition: we call X a UMD space if the Hilbert transform is bounded on $L^p(\mathbb{R}; X)$ for some (and then for all) $p \in (1, \infty)$. The following result is of particular importance in the Dore-Venni approach to the problem of maximal L^p-regularity sketched in the remarks before Corollary 4.4.

5.5. Theorem. *(Bourgain, Burkholder, McConnell, Zimmermann). Let X be a UMD space and let $1 < p < \infty$. Let $m \in C^{n+1}(\mathbb{R}^n \setminus \{0\}; \mathbb{C})$ and assume that there is a constant C such that*

$$|\partial^\alpha m(\xi)| \le C(1 + |\xi|)^{-|\alpha|}, \quad \alpha \in \mathbb{N}^n, \quad |\alpha| \le n+1, \xi \in \mathbb{R}^n \setminus \{0\}.$$

Then $m \in \mathcal{M}_p(\mathbb{R}^n; X)$.

6. Operator-Valued Fourier Multipliers

The above described version of the Mikhlin multiplier theorem is, however, not sufficient for certain purposes. In many applications (see e.g. [3, Section 8]) there occur naturally operators T defined through symbols which take values in $\mathcal{L}(X, Y)$, where X, Y are Banach spaces. In order to generalize the concept of Fourier multipliers described in the previous section to operator-valued symbols, we first discuss convolution of vector-valued distributions.

Let X, Y, Z be Banach spaces and let $X \times Y \to Z$, $(x, y) \mapsto x \bullet y$ be a multiplication (i.e. a continuous bilinear map of norm at most 1). Suppose that $m \in \mathcal{E}(\mathbb{R}^n; X)$. Then pointwise multiplication (with respect to the multiplication above) with a regular Y-valued distribution u is defined by

$$m \bullet u(x) := m(x) \bullet u(x), \text{ a.a. } \quad x \in \mathbb{R}^n.$$

For $u_1 \in \mathcal{D}(\mathbb{R}^n; X), u_2 \in \mathcal{D}(\mathbb{R}^n; Y)$ we define convolution (induced by the multiplication above) by

$$u_1 * \bullet u_2(x) := \int\limits_{\mathbb{R}^n} u_1(x-y) \bullet u_2(y)dy \quad (x \in \mathbb{R}^n).$$

Given $m \in \mathcal{O}_M(\mathbb{R}^n; X)$ we set

$$T_m u := \mathcal{F}^{-1} m_{\bullet} \mathcal{F} u := \mathcal{F}^{-1}(m \bullet \hat{u}), \quad u \in \mathcal{S}'(\mathbb{R}^n; Y).$$

It then follows from [3, Thm. 2.1] that we have

$$T_m \in \mathcal{L}(\mathcal{S}(\mathbb{R}^n; Y), \mathcal{S}(\mathbb{R}^n; Z)) \cap \mathcal{L}(\mathcal{S}'(\mathbb{R}^n; Y), \mathcal{S}'(\mathbb{R}^n; Z)).$$

The convolution theorem says that $T_m u = \mathcal{F}^{-1}(m) * \bullet u$ for $u \in \mathcal{S}(\mathbb{R}^n; Y)$. Hence we set

$$T_m u := \mathcal{F}^{-1}(m) * \bullet u$$

whenever $m \in L^\infty(\mathbb{R}^n; X)$ and $u \in \mathcal{S}'(\mathbb{R}^n; Y)$ are such that the convolution product on the right-hand side above is well defined. We are mainly interested in the case where multiplication is the evaluation map

$$\mathcal{L}(X, Y) \times X \to Y, \qquad (S, f) \mapsto Sf.$$

In the following we always write mu and $m * u$ for $m \bullet u$ and $m * \bullet u$ when no confusion seams likely.

Suppose that $F \in \{BUC, C_0, L^p, B^s_{p,q}; 1 \leq p, q \leq \infty, s \in \mathbb{R}\}$. Then $F(\mathbb{R}^n; X) \hookrightarrow \mathcal{S}'(\mathbb{R}^n; X)$ and we are asking for conditions on $m \in L^\infty(\mathbb{R}^n; \mathcal{L}(X, Y))$ such that

$$T_m f := \mathcal{F}^{-1}(m\hat{f}) := \mathcal{F}^{-1}(m) * f, \quad f \in F(\mathbb{R}^n; X) \tag{6.1}$$

is well defined and belongs to $\mathcal{L}(F(\mathbb{R}^n; X); F(\mathbb{R}^n; Y))$. In this case, m is called an *operator-valued Fourier multiplier* for F.

For the time being let $X = Y$. Then we denote the set of all $\mathcal{L}(X)$-valued Fourier multipliers for F by $\mathcal{M}_F^{\mathcal{L}(X)}(\mathbb{R}^n; X)$. It follows from Plancherel's theorem that

$$\mathcal{M}_{L^2}^{\mathcal{L}(X)}(\mathbb{R}^n; X) = L^\infty(\mathbb{R}^n; \mathcal{L}(X)) \tag{6.2}$$

provided X is a Hilbert space. Again we set $\mathcal{M}_p^{\mathcal{L}(X)}(\mathbb{R}^n; X) := \mathcal{M}_{L^p}^{\mathcal{L}(X)}(\mathbb{R}^n; X)$

The following generalization of Mikhlin's theorem to symbols which take values in $\mathcal{L}(H)$, H being a Hilbert space, is due to J. Schwartz [18].

6.1. Theorem. *(Mikhlin-Schwartz). Let H be a Hilbert space and let $1 < p < \infty$. Suppose that $m \in C^1(\mathbb{R} \setminus \{0\}; \mathcal{L}(H))$ satisfies*

$$\sup_{\alpha \in \{0,1\}} \sup_{\tau \in \mathbb{R} \setminus \{0\}} |\tau|^\alpha |\partial^\alpha m(\tau)|_{\mathcal{L}(H)} =: M < \infty. \tag{6.3}$$

Then $m \in \mathcal{M}_p^{\mathcal{L}(H)}(\mathbb{R}; H)$ and there exists a constant C (depending only on p and H) such that $\|m\|_{\mathcal{M}_p^{\mathcal{L}(H)}(\mathbb{R}; H)} \leq CM$.

The idea of the proof is most easily described by the following diagram:

$$\begin{array}{ccc} L^p(\mathbb{R}; H) & \xrightarrow{\ T\ } & L^p(\mathbb{R}; H) \\ R \big\downarrow & & \big\uparrow S \\ L^p(\mathbb{R}; l^2(H)) & \xrightarrow{\ T^+\ } & L^p(\mathbb{R}; l^2(H)) \end{array}$$

i.e. we decompose T as

$$T = ST^+R,$$

where the operators R, S and T^+ are defined as

$$\begin{array}{llll} R : L^p(\mathbb{R}; H) & \to L^p(\mathbb{R}; l^2(H)), & (Rf)_n(t) & := f(t)\chi_{[2^n \le t < 2^{n+1}]}(t) \\ S : L^p(\mathbb{R}; l^2(H)) & \to L^p(\mathbb{R}; H), & (SF)(t) & := F_n(t)\chi_{[2^n \le t < 2^{n+1}]}(t) \\ T^+ : L^p(\mathbb{R}; l^2(H)) & \to L^p(\mathbb{R}; l^2(H)), & \widehat{(T^+F)}(\tau) & := m^+(\tau)\widehat{F}(\tau), \end{array}$$

with $m^+(\tau) : l^2(H) \to l^2(H)$ defined by $(x_n) \mapsto (y_n)$ where $y_n := m_n(\tau)x_n$ and $m_n(\tau) := m(\tau)\chi_{[2^n \le |\tau| < 2^{n+1}]}$. The assertion of Theorem 6.1 is proved if we can show that all three operators under consideration act boundedly on the spaces defined above.

The following result is the first major step in this direction. In fact, it implies that $T^+ \in \mathcal{L}(L^p(\mathbb{R}; l^2(H)))$.

6.2. Proposition. *Let H be a Hilbert space and assume that*

$$(m_n)_{n\in\mathbb{Z}} \subset L^\infty(\mathbb{R}; \mathcal{L}(H))$$

satisfies

$$\sup_{n\in\mathbb{Z}} \sup_{\tau\in\mathbb{R}} |m_n(\tau)|_{\mathcal{L}(H)} + \sup_{n\in\mathbb{Z}} Var\, m_n(\tau) < \infty.$$

For $p \in (1, \infty)$ let $T : L^p(\mathbb{R}; l^2(H)) \to L^p(\mathbb{R}; l^2(H))$ be the mapping defined by $(TF)(t) := G(t)$, where

$$\widehat{G_n}(\tau) := m_n(\tau)\widehat{F_n}(\tau), \qquad n \in \mathbb{Z}, \tau \in \mathbb{R}.$$

Then $T \in \mathcal{L}(L^p(\mathbb{R}; l^2(H)))$.

The assertion of Proposition 6.2 is essentially a consequence of the extension of Stekhin's theorem to operator-valued symbols and to functions which take values in $l^2(H)$ (see Theorem 6.6 below).

The boundedness of the operators R and S follows easily from the following Littlewood-Paley inequality for Hilbert space valued functions.

6.3. Proposition. *Let H be a Hilbert space, $1 < p < \infty$ and $f \in L^p(\mathbb{R}; H)$. For $n \in \mathbb{Z}$ let $F_n \in L^p(\mathbb{R}; H)$ be defined by*

$$\widehat{F_n}(\xi) = \widehat{f}(\xi)\chi_{[2^n < |\xi| < 2^{n+1}]}(\xi), \qquad \xi \in \mathbb{R}.$$

For $x \in \mathbb{R}$ and $n \in \mathbb{Z}$ set $F(x) := (F_n(x))_{n\in\mathbb{Z}}$. Then there exist constants $C_1, C_2 > 0$ such that

$$C_1\|F\|_{L^p(\mathbb{R};l^2(H))} \le \|f\|_{L^p(\mathbb{R};H)} \le C_2\|F\|_{L^p(\mathbb{R};l^2(H))}.$$

The proof of the Littlewood-Paley inequalities relies on Proposition 6.2, Theorem 4.1 and Theorem 4.3.

It would be very useful to have a general result for Banach space valued Fourier multipliers for arbitrary p and arbitrary Banach spaces X. However, a result due to Pisier (unpublished) states that if the Mikhlin-Schwartz theorem holds on $L^p(\mathbb{R}^n; X)$ for $\mathcal{L}(X)$-valued symbols, then X is isomorphic to a Hilbert space. Hence, there is no version of the Mikhlin multiplier theorem on the scale of the spaces $L^p(\mathbb{R}^n;X)$ for $1<p<\infty$ and arbitrary X.

On the other hand, there is a simple criterion for an $\mathcal{L}(X)$-valued symbol to belong to $\mathcal{M}_p^{\mathcal{L}(X)}(\mathbb{R}^n;X)$ for $1\le p\le\infty$ and arbitrary Banach spaces X. In fact, we put

$$\mathcal{F}L^1(\mathbb{R}^n;X):=(\{u\in\mathcal{S}'(\mathbb{R}^n;X);\mathcal{F}^{-1}u\in L^1(\mathbb{R}^n;X)\};\|\cdot\|_{\mathcal{F}L^1}),$$

where $\|u\|_{\mathcal{F}L^1}:=\|\mathcal{F}^{-1}u\|_{L^1}$. Then

$$\mathcal{S}(\mathbb{R}^n;X)\overset{d}{\hookrightarrow}\mathcal{F}L^1(\mathbb{R}^n;X)\overset{d}{\hookrightarrow}C_0(\mathbb{R}^n;X)$$

and

$$[m\mapsto T_m]\in\mathcal{L}(\mathcal{F}L^1(\mathbb{R}^n;X);\mathcal{L}(F(\mathbb{R}^n;Z)))$$

provided $F\in\{BUC,C_0,L_p,1\le p\le\infty\}$, thanks to Young's inequality for convolutions in the vector-valued setting. Thus, in order to prove a (simple) multiplier result for $\mathcal{L}(X)$-valued symbols it suffices to establish sufficient conditions for a distribution to belong to $\mathcal{F}L^1(\mathbb{R}^n;X)$.

6.4. Proposition. *(X-valued Mikhlin+ε criterion). Let X be a Banach space and let $m\in W_\infty^{n+1}(\mathbb{R}^n;X)$. Suppose that there exists $\varepsilon>0$ such that*

$$\|m\|_{n+1,\infty}+\max_{|\alpha|\le n+1}\||\xi|^{|\alpha|+\varepsilon}\partial^\alpha m\|_\infty=:M\le\infty. \tag{6.4}$$

Then there exists a constant C (depending only on ε, X and n) such that

$$\|m\|_{\mathcal{F}L^1(\mathbb{R}^n;X)}\le CM.$$

In particular, $m\in\mathcal{M}_F^{\mathcal{L}(X)}(\mathbb{R}^n;X)$ provided $F\in\{L^p,1\le p\le\infty,BUC,C_0\}$

Proof. Fix a radial function $\psi:\mathbb{R}^n\to X$ such that $\psi\in\mathcal{D}(\mathbb{R}^n)$ and $\psi_{|B}\equiv 1$ and $\operatorname{supp}\psi\subset 2B$, where B denotes the unit ball of $\mathbb{R}^n$. We set $\psi_0:=\psi$ and $\psi_j:=\psi(2^{-j}\cdot)-\psi(2^{-j+1}\cdot), j\in\mathbb{N}$. Then $\operatorname{supp}\psi_j\subset[2^{j-1}\le|\xi|\le 2^{j+1}]$ and

$$\sum_{j=0}^\infty\psi_j(\xi)=1\quad\text{for all }\xi\in\mathbb{R}^n.$$

It follows from (2.2) that

$$\|\psi_j a\|_{\mathcal{F}L^1}=\|\tilde{\psi}\sigma_{2^j}a\|_{\mathcal{F}L^1} \tag{6.5}$$

where $\tilde{\psi}:=\psi-\psi(2\cdot)$. Setting $M_j:=\sup_{2^{j-1}\le|\xi|\le 2^{j+1}}\max_{|\alpha|\le n+1}\||\xi|^{|\alpha|}\partial^\alpha a\|_\infty$ for $j\in\mathbb{N}$ it follows by Leibniz's rule that

$$|\partial^\alpha\tilde{\psi}\sigma_{2^j}a(\xi)|\le CM_j\quad\text{a. a.}\quad\xi\in[\frac{1}{2}\le|\xi|\le 2],\ |\alpha|\le n+1$$

for some constant $C>0$. Consequently $\|\partial^\alpha(\tilde{\psi}\sigma_{2^j}a)\|_1\le CM_j$, $|\alpha|\le n+1$. Since, by the Riemann-Lebesgue Lemma

$$x^\alpha\mathcal{F}^{-1}(\tilde{\psi}\sigma_{2^j}a)=\mathcal{F}^{-1}(D^\alpha(\tilde{\psi}\sigma_{2^j}a))\in C_0(\mathbb{R}^n;X),$$

we have $|x^\alpha \mathcal{F}^{-1}(\tilde{\psi}\sigma_{2^j}a)(x)| \leq CM_j$ for $x \in \mathbb{R}^n$ and $|\alpha| \leq n+1$. It follows that

$$\|\tilde{\psi}\sigma_{2^j}a\|_{\mathcal{F}L^1} = \|\mathcal{F}^{-1}(\tilde{\psi}\sigma_{2^j}a)\|_1$$
$$\leq CM_j\Big(\int_{|x|\leq 1} 1dx + \int_{|x|\geq 1} |x|^{-n-1}dx\Big) \leq CM_j \quad (j \in \mathbb{N}).$$

Hence, by (6.5), $\psi_j a \in \mathcal{F}L^1(\mathbb{R}^n; X)$ and $\|\psi_j a\|_{\mathcal{F}L^1} \leq CM_j \quad (j \in \mathbb{N})$. Since $\sum_{j=0}^{\infty} \psi_j a = a$ in $\mathcal{S}'(\mathbb{R}^n; X)$ it follows that

$$\sum_0^{\infty} \|\psi_j a\|_{\mathcal{F}L^1} \leq \|\psi a\|_{\mathcal{F}L^1} + C\sum_{j=1}^{\infty} M_j. \tag{6.6}$$

By the same arguments as above we have

$$\|\psi a\|_{\mathcal{F}L^1} \leq C\|a\|_{n+1,\infty}. \tag{6.7}$$

It follows from the assumption that $M_j \leq M2^{-(j-1)\varepsilon}$ for $j \in \mathbb{N}$, hence by (6.6) and (6.7)

$$\|a\|_{\mathcal{F}L^1} \leq \sum_{j=0}^{\infty} \|\psi_j a\|_{\mathcal{F}L_1} \leq C\|a\|_{n+1,\infty} + CM\sum_{j=1}^{\infty} 2^{-(j-1)\varepsilon} < \infty. \qquad \square$$

We note that in the scalar-valued version of Proposition 6.4 proved in [11, Lemma 4.1], i.e $X = \mathbb{K}$, the order of differentiation can be reduced to $[n/2]+1$, thanks to Plancherel's theorem. The extension to X-valued symbols described above is due to Amann [3, Corollary 4.4].

Many operator-valued symbols arising from applications do *not* fulfill condition 6.4 but they do satisfy the classical Mikhlin condition with the absolute value replaced by the operator norm. Recall that the extension of Mikhlin's theorem to $\mathcal{L}(X)$-valued symbols is restricted to the case of Hilbert spaces. Hence, in order to prove results on operator-valued Fourier multipliers, we either have to change the L^p-scale to another scale of function spaces or we have to find a substitute for Mikhlin's smoothness condition.

In 1997, it was Amann [3] who proved, roughly speaking, that a symbol $a : \mathbb{R}^n \to X$ satisfying Mikhlin's condition with the absolute value replaced by $|\cdot|_X$, induces a bounded translations invariant operator on the scale of vector-valued Besov spaces. More precisely, given $s \in \mathbb{R}$ and $p, q \in [1, \infty]$, let $B^s_{p,q}(\mathbb{R}^n; X)$ be the Besov space of X-valued distributions on $\mathbb{R}^n$. We also define $\tilde{\mathrm{B}}^s_{p,q}$ and $b^s_{p,q}$ to be the closure of $\mathcal{S}$ in $B^s_{p,q}$ and the closure of $B^{s+1}_{p,q}$ in $B^s_{p,q}$, respectively. We say that $m \in C^{n+1}(\mathbb{R}^n \setminus \{0\}; X)$ belongs to $S^0(\mathbb{R}^n; X)$ provided

$$\sup_{\xi \in \mathbb{R}^n \setminus \{0\}} \sup_{|\alpha| \leq n+1} |\partial^\alpha m(\xi)|_X < \infty.$$

Then the following holds.

6.5. Theorem. *(Amann). Let X, Y, Z be Banach spaces and let $\mathcal{B} \in \{B, \tilde{B}, b\}$. Then*

$$[m \mapsto T_m] \in \mathcal{L}(S^0(\mathbb{R}^n; X), \mathcal{L}(\mathcal{B}^s_{p,q}(\mathbb{R}^n; Y), \mathcal{B}^s_{p,q}(\mathbb{R}^n; Z)))$$

for $s \in \mathbb{R}$ and $p, q \in [1, \infty]$. In particular, $m \in \mathcal{M}^{\mathcal{L}(X)}_{\mathcal{B}^s_{p,q}}(\mathbb{R}^n; X)$.

We finally give a operator-valued multiplier result for the L^p scale in form of the following extension of Stekhin's result to operator-valued symbols. It is not surprising that in this context the boundedness of the Hilbert transform in $L^p(\mathbb{R}; X)$ is of fundamental importance. We hence have to restrict ourselves to $\mathcal{L}(X)$-valued symbols where X is a UMD space.

The following result due to J. Schwartz [18], originally proved for $\mathcal{L}(L^q(\Omega))$-valued symbols, $1 < q < \infty$, can be easily extended to the case of UMD spaces.

6.6. Theorem. *(Stekhin-Schwartz). Let X be a UMD space and let $1 < p < \infty$. Suppose that $m \in L^\infty(\mathbb{R}; \mathcal{L}(X))$ is of bounded variation on $\mathbb{R}$. Then*

$$m \in \mathcal{M}^{\mathcal{L}(X)}_{L^p}(\mathbb{R}; X).$$

The above result is not only of interest in its own right, but, as we have seen, it is also of importance in the proof of the Mikhlin-Schwarz theorem.

References

[1] Agmon, S., Douglis, A., Nirenberg, L.: "Estimates near boundary for solutions of elliptic partial differential equations satisfying general boundary conditions", Comm. Pure Appl. Math. **12** (1959), 623–727.

[2] Amann, H.: *Linear and Quasilinear Parabolic Problems.* Birkhäuser, Basel, 1995.

[3] Amann, H.: "Operator-valued Fourier multipliers, vector-valued Besov spaces, and applications", Math. Nachr. **186** (1997), 5–56.

[4] Benedek, A., Calderón, A.P., Panzone, R.: "Convolution operators on Banach space valued functions", Proc. Nat. Acad. Sci. USA **48** (1962), 356–365.

[5] Bergh, J., Löfström, J.: *Interpolation Spaces. An Introduction.* Springer, Berlin, 1976.

[6] Bourgain, J.: "Extension of a result of Benedick, Calderón and Panzone", Ark. Mat. **22** (1983), 91–95.

[7] Burkholder, D.L.: "A geometrical condition that implies the existence of certain singular integrals of Banach-space-valued functions". In: Beckner, W., Calderón, A.P., Fefferman, R. and Jones, P.W.,(eds) Conf. Harm. Anal. in Honour of A. Zygmund, Chicago 1981, 270–286, Belmont, Cal., Wadsworth, 1983.

[8] Christ, M.: *Lectures on Singular Integral Operators.* Conf. Board Math. Sci., Reg. Conf. Series Math., 77, Amer. Math. Soc., Providence, R.I., 1990.

[9] Dore, G. Venni, A.: "On the closedness of the sum of two closed operators", Math. Z. **196** (1987), 189–201.

[10] Grisvard, P.: *Elliptic Problems in Nonsmooth Domains.* Pitman, Boston, 1985.

[11] Hieber, M.: "Integrated semigroups and differential operators on L^p spaces", Math. Ann. **291** (1991), 1–16.

[12] Hörmander, L.: "Estimates for translation invariant operators in L^p spaces", Acta Math. **104** (1960), 93–140.

[13] Lunardi, A.: *Analytic Semigroups and Optimal Regularity in Parabolic Equations.* Birkhäuser, Basel, 1995.

[14] McConnell, T.: "On Fourier multiplier transformations of Banach-space valued functions", Trans. Amer. Math. Soc. **285** (1984), 739–757.

[15] Prüss, J.: *Evolutionary Integral Equations and Applications.* Birkhäuser, Basel, 1993.

[16] Rubio de Francia, J.L., Ruiz, F.J., Torrea, J.L.: "Calderón-Zygmund theory for operator-valued kernels", Adv. Math. **62** (1986), 7–48.

[17] Schulze, B.-W.: *Pseudo-Differential Operators on Manifolds with Singularities.* North-Holland, Amsterdam, 1991.

[18] Schwartz, J.: "A remark on inequalities of Calderon-Zygmund type for vector-valued functions", Comm. Pure Appl. Math. **XIV** (1961), 785–799.

[19] Stein, E.M.: *Topics in Harmonic Analysis Related to Littlewood-Paley Theory.* Princeton University Press, Princeton, 1970.

[20] Stein, E.M.: *Harmonic Analysis: Real-Variables Methods, Orthogonality and Oscillatory Integrals.* Princeton University Press, Princeton, 1993.

[21] Triebel, H.: *Interpolation Theory, Function Spaces, Differential Operators.* North-Holland, Amsterdam, 1978.

[22] Zimmermann, F.: "On vector-valued Fourier multiplier theorems", Studia Math. **93** (1989), 201–222.

Mathematisches Institut I
Englerstr. 2
Universität Karlsruhe
D-76128 Karlsruhe
Germany
matthias.hieber@math.uni-karlsruhe.de

Progress in Nonlinear Differential Equations
and Their Applications, Vol. 35
© 1999 Birkhäuser Verlag Basel/Switzerland

Properties of Pseudoholomorphic Curves in Symplectizations III: Fredholm Theory

H. Hofer[1], K. Wysocki[2] and E. Zehnder[3]

Abstract. We shall study smooth maps $\tilde{u}: \dot{S} \to \mathbb{R} \times M$ of finite energy defined on the punctured Riemann surface $\dot{S} = S \backslash \Gamma$ and satisfying a Cauchy-Riemann type equation $T\tilde{u} \circ j = J(\tilde{u}) \circ T\tilde{u}$ for special almost complex structures J, related to contact forms λ on the compact three manifold M. Neither the domain nor the target space are compact. This difficulty leads to an asymptotic analysis near the punctures. A Fredholm theory determines the dimension of the solution space in terms of the asymptotic data defined by non-degenerate periodic solutions of the Reeb vector field associated with λ on M, the Euler characteristic of S, and the number of punctures. Furthermore, some transversality results are established.

1. Introduction

This paper is devoted to the proofs of some technical results concerning pseudoholomorphic curves in symplectizations. We shall establish a Fredholm theory for embedded finite energy surfaces in the symplectization $\mathbb{R} \times M$, where M is a three-dimensional contact manifold. We shall also prove some transversality results. The results presented here are used in our papers [17, 20, 21, 22]. In order to outline the contents of the paper we first have to recall some basic concepts.

Assume M is a compact oriented three manifold. A contact form λ on M is a 1-form satisfying

$$\lambda \wedge d\lambda = \text{volume form}.$$

Since we assume M to be oriented we can compare its orientation with the one induced by the contact form. The contact form is called positive if both orientations coincide, otherwise it is called negative. Note, that if λ is positive, so is $-\lambda$. So positivity is a property of the plane bundle $\xi = \ker(\lambda)$ called the contact structure associated with λ. In the following we consider positive contact structures only. Given a contact form λ the induced contact structure $\xi \to M$ may be viewed as a symplectic vector bundle equipped with the symplectic form $d\lambda|(\xi \oplus \xi)$. Moreover,

---

[1] Research partially supported by a NSF grant and FIM at Zürich,
[2] Research partially supported by an Australian Research Council grant and FIM at Zürich,
[3] Research partially supported by a TH-project.

λ determines a nowhere vanishing vector field X, called the Reeb vector field. It is defined by

$$i_X d\lambda = 0 \text{ and } i_X\lambda = 1.$$

Hence the contact form λ defines a natural splitting

$$TM = \mathbb{R}X \oplus \xi$$

of the tangent bundle TM into a line bundle with a preferred section X and a symplectic vector bundle ξ. The symplectic vector bundle $(\xi, d\lambda)$ has a preferred class of complex multiplications $J : \xi \to \xi$ compatible with the symplectic structure in the following sense. The bilinear map

$$g_J(m)(h,k) = d\lambda(m)(h, J(m)k)$$

defines a positive definite inner product on each fiber ξ_m of ξ. It is well known that the space of all those J's equipped with the C^∞-topology is contractible. We choose now a compatible complex multiplication J on ξ and define the associated special almost complex structure $\tilde{J}$ on the four manifold $\mathbb{R} \times M$ by

$$\tilde{J}(a,m)(h,k) = \big(-\lambda(m)(k), J(m)\pi k + hX(m)\big),$$

where $(a,m) \in \mathbb{R} \times M$ and $(h,k) \in T_{(a,m)}(\mathbb{R} \times M)$. Here $\pi : TM \to \xi$ is the pointwise projection of TM onto ξ along the Reeb vector field X. Assume now that (S,j) is a compact Riemann surface and $\Gamma \subset S$ a finite subset of S. Referring to Γ as the set of punctures we abbreviate $\dot{S} = S \setminus \Gamma$. We are interested in maps $\tilde{u} : \dot{S} \to \mathbb{R} \times M$ solving the first order elliptic system

$$\tilde{J}(\tilde{u}) \circ T\tilde{u} = T\tilde{u} \circ j. \tag{1}$$

As illustrated in [15, 17, 18, 19, 20, 21] there is a class of solutions of particular interest, namely the so-called finite energy surfaces. To explain this notion let Σ be the set of all smooth functions $\phi : \mathbb{R} \to [0,1]$ satisfying $\phi'(s) \geq 0$. For $\phi \in \Sigma$ we define a 1-form λ_ϕ on $\mathbb{R} \times M$ by

$$\lambda_\phi(a,m)(h,k) = \phi(a)\lambda(m)(k),$$

where $(a,m) \in \mathbb{R} \times M$ and $(h,k) \in T_{(a,m)}(\mathbb{R} \times M)$. Given a solution $\tilde{u} = (a,u)$ of (1), the 2-form $\tilde{u}^* d\lambda_\phi$ is a non-negative integrand. Indeed, taking holomorphic local coordinates $s + it$ on S one computes

$$\tilde{u}^* d\lambda_\phi = \left[\phi'(a)\left(a_s^2 + \lambda(u_s)^2\right) + \phi(a)|\pi u_s|^2\right] ds \wedge dt.$$

We define the E-energy of the solution $\tilde{u}$ to be

$$E(\tilde{u}) = \sup\left\{\int_{\dot{S}} \tilde{u}^* d\lambda_\phi | \phi \in \Sigma\right\}. \tag{2}$$

We recall the

Definition 1.1. *A non-constant solution $\tilde{u} : \dot{S} \to \mathbb{R} \times M$ of* (1) *having finite energy $E(\tilde{u}) < \infty$, is called a (special) finite energy surface.*

Our main interest later on is the description of the neighborhood of an embedded solution $\tilde{u}$, where we allow the complex structure on S and the punctures to vary. We call this the local Fredholm theory of finite energy surfaces. In the special case $\dot{S} = S^2 \setminus \{\infty\} = \mathbb{C}$, a non-constant solution $\tilde{u}$ of (1) having finite energy will be called a (special) finite energy plane.

In the following related problem we are going to study, the almost complex structure on $\mathbb{R} \times M$ will not be $\mathbb{R}$-invariant anymore. Assume λ is a contact form on M and let $f : \mathbb{R} \times M \to (0, \infty)$ be a smooth function satisfying

$$\begin{aligned} \frac{\partial f}{\partial a}(a,u) \geq 0 \quad \text{for all} \quad (a,u) \\ \frac{\partial f}{\partial a}(a,u) \geq \sigma > 0 \quad \text{for all} \quad u \in M, a \in (-a_1, a_1) \end{aligned} \tag{3}$$

and

$$\begin{aligned} f(a,u) = g(u) \quad \text{for all} \quad a \geq \quad a_0 \\ f(a,u) = h(u) \quad \text{for all} \quad a \leqslant -a_0, \end{aligned} \tag{4}$$

where $h, g : M \to (0, \infty)$ are smooth functions such that

$$h < g. \tag{5}$$

The real numbers a_0, a_1 satisfy $0 < a_1 < a_0 < \infty$. In view of hypothesis (3) the 2-form

$$\Omega = d(f\lambda) \tag{6}$$

on $\mathbb{R} \times M$ restricted to $[-a_1, a_1] \times M$ is a symplectic form. Let us introduce the family λ_a of contact forms defined by $\lambda_a = f(a, \cdot)\lambda$. We denote the associated Reeb vector fields by X_a and the associated projections onto the contact bundle ξ by π_a. We observe that the space of compatible complex multiplications for ξ is independent of a (as is the contact structure). As we know from Gromov's work in [14], there is a distinguished class of compatible almost complex structures on a given symplectic manifold (W, τ). Recall that the structure J is called admissible if

$$g_J(w)(h,k) = \tau(h, J(w)k)$$

is a Riemannian metric on W.

In the situation at hand, $\mathbb{R} \times M$, we choose an almost complex structure $\hat{J}$ which on $[-a_1, a_1] \times M$ is compatible with the symplectic structure Ω defined in (6). Let $J : \xi \to \xi$ be a compatible complex multiplication on ξ. Choose an arc $a \to J_a$, $a \in \mathbb{R}$, of compatible complex multiplications on ξ satisfying

$$\begin{aligned} J_a = J_{a_0} \quad \text{for all} \quad a \geq \quad a_0 \\ J_a = J_{-a_0} \quad \text{for all} \quad a \leqslant -a_0. \end{aligned} \tag{7}$$

We assume that the associated almost complex structure $\tilde{J}$ on $\mathbb{R} \times M$ defined by

$$\tilde{J}(a,m)(h,k) = \big(-\lambda_a(m)(k), J_a(m)\pi_a k + hX_a(m)\big) \tag{8}$$

satisfies

$$\hat{J}(a,m) = \tilde{J}(a,m) \quad \text{for all} \quad (a,m) \in \big(\mathbb{R} \setminus [-a_1, a_1]\big) \times M. \tag{9}$$

We shall investigate maps $\tilde{u} : \dot{S} \to \mathbb{R} \times M$ solving the equation

$$\hat{J}(\tilde{u}) \circ T\tilde{u} = T\tilde{u} \circ j \tag{10}$$

and meeting the following energy condition. We introduce the set $\hat{\Sigma}$ consisting of all smooth nondecreasing functions $\phi : \mathbb{R} \to [0,1]$ which have the additional property

$$\phi(s) = \frac{1}{2} \quad \text{for all} \quad s \in [-a_1, a_1]. \tag{11}$$

For $\phi \in \hat{\Sigma}$ we define a 1-form λ_ϕ on $\mathbb{R} \times M$ by

$$\lambda_\phi(a,m)(h,k) = \phi(a)\lambda_a(m)(k). \tag{12}$$

If $\phi'(s) > 0$ for all $s \in \mathbb{R} \setminus [-a_1, a_1]$ then $d\lambda_\phi$ is a symplectic structure on $\mathbb{R} \times M$. One easily verifies that the 2-forms $\tilde{u}^* d\lambda_\phi$ are non-negative integrands and we can define an associated energy $E(\tilde{u})$ as before maximizing the integrals

$$\int_{\dot{S}} \tilde{u}^* d\lambda_\phi$$

by varying $\phi \in \hat{\Sigma}$. A solution of (10) having finite energy will be called a generalized finite energy surface. Again our aim is to describe the generalized finite energy surfaces in the neighborhood of a given one, where we allow the complex structure on S to vary.

Another part of this paper will be devoted to a transversality theory for (special and generalized) finite energy surfaces. To be more precise let us note that unparametrized finite energy surfaces may be viewed as a zero set of a Fredholm map. Then we may expect that for generic data the Fredholm index should be at least 0 or satisfy any other inequality dictated perhaps by some natural symmetries. This will indeed be true. We shall study in detail the question which data have to be made generic in order to establish index inequalities. Such considerations are important in the compactness proofs for finite energy surfaces.

In order to describe our results we need some more preparation. We do not give the most general statements here since they would be overly technical. Instead we describe some consequences which immediately apply to situations studied in [17, 20, 21, 22]. For more general statements we refer the reader to the particular sections.

To understand the local Fredholm theory it is useful to recall some of the properties of finite energy surfaces. The interesting part is the behavior of a finite energy surface near a puncture. A neighborhood of a puncture looks biholomorphically like a neighborhood of infinity in $\mathbb{C}$. Hence in these coordinates the puncture corresponds to ∞. Let us denote by $\mathbb{C}_R$ the complement in $\mathbb{C}$ of the R-ball, $R > 0$. Consider a map

$$\tilde{v} = (b, v) : \mathbb{C}_R \to \mathbb{R} \times M$$

satisfying, in the usual complex coordinates $s + it$,

$$\pi v_s + J(v)\pi v_t = 0$$
$$v^*\lambda \circ i = db$$
$$0 < E(\tilde{v}) < \infty.$$

One can now distinguish the following types of behavior of $\tilde{v}$ at infinity, see [18, 19].

- **removable singularity:** if the $\mathbb{R}$-component of $\tilde{v}$ is bounded near ∞ (the puncture).
- **non-removable singularity:** if the $\mathbb{R}$-component of $\tilde{v}$ is unbounded near ∞.

It follows from Gromov's removable singularity theorem, [14], that in the first case the singularity is removable. We shall assume in the following that all punctures are non-removable. In other words in what follows the notion puncture means a non-removable singularity.

We can distinguish two different types of punctures. A puncture is called

- **negative singularity:** if the $\mathbb{R}$-component is unbounded, but bounded from above.
- **positive singularity:** if the $\mathbb{R}$-component is unbounded, but bounded from below.

For a finite energy surface $\tilde{u} = (a, u)$ we denote in the following by $\Gamma^{\pm}$ the set of positive and negative punctures respectively, so that $\Gamma = \Gamma^- \cup \Gamma^+$. Since $0 < E(\tilde{u}) < \infty$ one can show that

$$T := \lim_{r \to \infty} \int_{S^1} u(re^{2\pi it})^* \lambda$$

exists in $\mathbb{R} \setminus \{0\}$. Let us call T the charge of the puncture. It is positive for a positive and negative for a negative puncture. For a generalized finite energy surface introduced in (3)–(12) we have to replace λ by $g\lambda$ near positive punctures and by $h\lambda$ near negative punctures. It has been proved in [19] that $\Gamma^+ \neq \emptyset$ for finite energy surfaces. This also holds for generalized finite energy surfaces.

Crucial for our purpose is the intimate relation between the asymptotic behavior near the punctures Γ and periodic orbits of the Reeb vector field X.

Definition 1.2. *We call a contact form λ non-degenerate if all periodic solutions of $\dot{x} = X(x)$ are non-degenerate. To be more precise, assume $x : \mathbb{R} \to M$ solves $\dot{x} = X(x)$ and is periodic of period $T \neq 0$. The periodic solution (x, T) is called non-degenerate if for the map $T\eta_T(x(0)) : T_{x(0)}M \to T_{x(0)}M$ the eigenvalue 1 has multiplicity 1. Here η is the flow of the Reeb vector field X. Note that we allow negative periods.*

Clearly, (x, T) is non-degenerate iff $(x, -T)$ is non-degenerate. The following result is proved in [15] and extended in [18].

Theorem 1.3. *Let $\tilde{v} = (a, v) : \mathbb{C}_R \to \mathbb{R} \times M$ be a finite energy surface near a puncture as described above with charge T. Then there exists a sequence $R_k \to \infty$*

and a $|T|$-periodic solution x of $\dot{x} = X(x)$ such that

$$\lim_{k\to\infty} v\big(R_k e^{2\pi i t}\big) = x(Tt)$$

in $C^\infty(\mathbb{R})$. If (x,T) is non-degenerate then the limit exists for $R \to \infty$.

In the non-degenerate case we even have an exponentially fast approach to the periodic orbit x as shown in [18]. This will be used in the Fredholm theory later on. In the case that the asymptotic limit is non-degenerate we have associated with the corresponding periodic orbit a so-called Conley-Zehnder index.

We briefly recall this index, referring to [19] for the details. Consider $\mathbb{C}$ with its coordinates $x+iy$ as a real 2-dimensional vector space. We view it as a symplectic vector space with symplectic form $\omega = dx \wedge dy$. We denote by $Sp(1)$ the group of linear symplectic maps in $\mathbb{R}^2$ and by $Sp(1)^*$ the subset consisting of all symplectic maps having 1 not in the spectrum. Denote by $\Sigma(1)$ the collection of all smooth arcs $\Phi : [0,1] \to Sp(1)$ starting at the identity and ending in $Sp(1)^*$. Moreover we denote by $G(1)$ the space of smooth arcs in $Sp(1)$ starting and ending at the identity. We have two natural maps. The first is the inversion map

$$\Sigma(1) \to \Sigma(1) : \Phi \mapsto \Phi^{-1},$$

where $\Phi^{-1}(t) := \Phi(t)^{-1}$. The second is the product map

$$G(1) \times \Sigma(1) \to \Sigma(1) : (\alpha, \Phi) \mapsto \alpha\Phi,$$

where $(\alpha\Phi)(t) := \alpha(t)\Phi(t)$. Observe that the homotopy classes in $G(1)$ give just the fundamental group of $Sp(1)$. It is well known that the fundamental group is isomorphic to $\mathbb{Z}$. A particular isomorphism, called the Maslov isomorphism, is characterized by $[\sigma] \mapsto 1$, where σ is the loop $t \mapsto e^{2\pi i t}$ and $e^{2\pi i t}$ operates by multiplication. Hence this natural isomorphism, normalized as above defines the map

$$\mu_M : G(1) \to \mathbb{Z}.$$

The Conley-Zehnder index for $\Sigma(1)$ (it can be defined more generally in every even dimension) is now uniquely characterized by a few natural axioms.

Theorem 1.4. *There exists a unique map $\mu : \Sigma(1) \to \mathbb{Z}$ satisfying*

- *μ is homotopy invariant.*
- *$\mu(\alpha\Phi) = \mu(\Phi) + 2\mu_M(\alpha)$ for $\alpha \in G(1)$ and $\Phi \in \Sigma(1)$.*
- *$\mu(\Phi) = -\mu(\Phi^{-1})$ for $\Phi \in \Sigma(1)$.*
- *$\mu(\Phi_0) = 1$, where $\Phi_0(t) = e^{\pi i t}$.*

Let us next define the Conley-Zehnder index of a generalized finite energy surface with non-degenerate asymptotic limits. We follow [19]. Assume $\tilde{u}$ is a generalized finite energy surface with non-degenerate asymptotic limits. Then one can compactify $\dot{S}$ be adding at each puncture a circle S^1 (taking cylindrical holomorphic coordinates near the puncture we compactify $(R,\infty) \times S^1$ by $(R,\infty] \times S^1$.) The map $\tilde{u}$ can be continuously extended over the circles. Denote the M-part of this extended map by $\bar{u}$. Since $\Gamma \neq \emptyset$ the pull-back bundle $\bar{u}^*\xi \to \dot{S}$ as symplectic vector bundle is trivializable. Taking such a symplectic trivialization we obtain induced trivializations over the asymptotic limits associated with the punctures. Let (x,T) be a non-degenerate limit associated with a positive puncture. Then, linearizing the flow associated with X we obtain for $t \in \mathbb{R}$ an induced map $Z_x(t) = T\eta_t(x(0))|\xi_{x(0)} : \xi_{x(0)} \to \xi_{x(t)}$, which is symplectic. The induced distinguished trivialization of $x^*\xi \to \mathbb{R}^2$ is T-periodic. Call it $\Psi(t) : \xi_{x(t)} \to \mathbb{R}^2$. Consider the arc of symplectic matrices $\Phi(t) = \Psi(t)Z(t)\Psi(0)^{-1}, t \in [0,T]$ starting at Id and ending at time T at an element in $Sp(1)^*$, assuming the periodic orbit (x,T) to be non-degenerate. Reparametrizing we may assume that $T = 1$. Then the arc $\Phi(t)$ has a well-defined Conley-Zehnder index (which depends on the trivialization), denoted by $\mu(\Phi) = \mu(x,T)$. We compute the index for every positive puncture and take their sum. Let us call this number μ^+. For a negative puncture with charge T we view the underlying periodic solution as a $|T|$-periodic solution, compute its index and take the sum for all negative punctures to obtain the integer μ^-. Then we define

$$\mu(\tilde{u}) = \mu^+ - \mu^- .$$

This number does not depend on the choice of a trivialization of $\bar{u}^*\xi$ as has been shown in [19]. We call the integer $\mu(\tilde{u})$ the Conley-Zehnder index of $\tilde{u}$.

Our first result concerns special finite energy planes.

Theorem 1.5. *Assume that the special finite energy plane $\tilde{u} = (a,u) : \mathbb{C} \to \mathbb{R} \times M$ is an embedding and has the non-degenerate simply covered asymptotic limit (x,T). Abbreviate the image by $P = x(\mathbb{R}) \subset M$. Then*

(i) *$\mu(\tilde{u}) \geq 2$. If $\mu(\tilde{u}) \leqslant 3$, then $u(\mathbb{C}) \cap P = \emptyset$ and $u : \mathbb{C} \to M \setminus P$ is an embedding.*

(ii) *If $\mu(\tilde{u}) = 3$, then there exist $\epsilon > 0$ and a continuous map*

$$\tilde{\Phi} = (a, \Phi) : (-\varepsilon, \varepsilon) \times \mathbb{C}_{S^1} \to \mathbb{R} \times M$$

having the following properties:

a) *For every $\tau \in (-\varepsilon, \varepsilon)$ the map*

$$\tilde{\Phi}(\tau, \cdot) : \mathbb{C} \to \mathbb{R} \times M$$

is a special finite energy plane asymptotic to (x,T) and

$$\tilde{\Phi}(0, \cdot) = \tilde{u}(\cdot) \ \ \text{for } \tau = 0.$$

b) *The restriction Φ satisfies $\Phi(\tau, \mathbb{C}) \cap P = \emptyset$ and*

$$\Phi : (-\varepsilon, \varepsilon) \times \mathbb{C} \to M \setminus P$$

is an embedding.

c) *The following completeness property holds true. If $\tilde{u}_k = (a_k, u_k)$ is any sequence of special finite energy planes all asymptotic to (x,T) and converging in $C^\infty_{loc}(\mathbb{C})$ to the given plane $\tilde{u}$ and normalized so that $a_k(0) = a(0) = 0$, then for k large enough there exists a sequence $\tau_k \to 0$ such that*

$$\tilde{\Phi}(\tau_k, \mathbb{C}) = \tilde{u}_k(\mathbb{C}).$$

Moreover, if k is fixed the maps $\tilde{\Phi}(\tau_k, \cdot)$ and $\tilde{u}_k$ differ only by a biholomorphic reparametrization of $\mathbb{C}$ of the form $z \mapsto bz + c$ with constants $b, c \in \mathbb{C}$.

In Theorem 1.5 we consider special finite energy planes. In view of the $\mathbb{R}$-invariance of the special almost complex structure $\tilde{J}$ introduced above, we conclude that given a solution $\tilde{u} = (a, u)$ of $\tilde{J}(\tilde{u}) \circ T\tilde{u} = T\tilde{u} \circ j$, the map $\tilde{u}_c = (a + c, u)$ is also a solution for every $c \in \mathbb{R}$. Consequently, we obtain from Theorem 1.5 an embedding

$$\mathbb{R} \times (-\varepsilon, \varepsilon) \times \mathbb{C} \to \mathbb{R} \times M$$

defined by

$$(c, \tau, z) \mapsto \big(a(\tau, z) + c, \Phi(\tau, z)\big).$$

So, up to biholomorphic reparametrizations of the plane we have a two-dimensional family, where one dimension accounts for the $\mathbb{R}$-action.

The second result is analogous to Theorem 1.5 but concerns now generalized finite energy surfaces.

Theorem 1.6. *Assume the data $f : \mathbb{R} \times M \to (0, \infty)$, $\hat{J}$, and g, h are as described in (5)–(9). Let $\tilde{u} = (a, u) : \mathbb{C} \to \mathbb{R} \times M$ be an embedded generalized finite energy plane with a non-degenerate, simply covered, asymptotic limit (x, T) having Conley-Zehnder index $\mu = \mu(u) = 3$. Then there exists a continuous map*

$$\tilde{\Phi} = (a, \Phi) : B^2_\varepsilon \times \mathbb{C}_{S^1} \to \mathbb{R} \times M$$

with $\Phi(b, +\infty e^{2\pi i t}) = x(Tt)$ for $b \in B^2_\varepsilon$, such that the restriction to $B^2_\varepsilon \times \mathbb{C}$ of $\tilde{\Phi}$ is a smooth embedding and for every parameter $\tilde{\Phi}(b, \cdot)$ is a generalized finite energy plane asymptotic to (x, T). Further, the following completeness property holds. If $\tilde{u}_k$ is a sequence of generalized finite energy planes asymptotic to (x, T) converging in C^∞_{loc} to $\tilde{u}$, then for k large enough there exists a sequence $b_k \to 0$ in B^2_ε such that

$$\tilde{\Phi}(b_k, \mathbb{C}) = \tilde{u}_k(\mathbb{C}).$$

Moreover $\tilde{u}_k$ and $\tilde{\Phi}(b_k, \cdot)$ (for fixed k) only differ by a biholomorphic parametrization of the form $z \mapsto bz + c$.

We should mention that in case that $\mu(\tilde{u}) = 2$ we have a one parameter family of solutions whereas if $\mu(\tilde{u}) = 1$ the distinguished solution $\tilde{u}$ is isolated. In order to proceed further, let us introduce the notion of an unparametrized special or generalized embedded finite energy surface.

Definition 1.7. *A submanifold C of $\mathbb{R} \times M$ is called an embedded unparametrized special finite energy surface if there exists a Riemann surface (S, j), punctures $\Gamma \subset S$ and an embedding $\tilde{u} : \dot{S} \to \mathbb{R} \times M$ whose image is $\tilde{u}(\dot{S}) = C$ and which satisfies the equation (1), namely $T\tilde{u} \circ j = \tilde{J}(\tilde{u}) \circ T\tilde{u}$, and the estimate $0 < E(\tilde{u}) < \infty$. We shall call C an embedded unparametrized generalized finite energy surface if the above map $\tilde{u}$ satisfies the corresponding equation (10).*

In the following we would like to describe the neighborhood of an embedded generalized finite energy surface. The discussion is also valid for the case of special finite energy surfaces. In fact, there is basically no difference between the special and generalized case, except that in the special case solutions always appear in one-dimensional families due to the $\mathbb{R}$-action (with rare exceptions like the cylinder over a periodic orbit). We point out, however, that in the special case the $\mathbb{R}$-invariance of the almost complex structure allows to decide whether the projection of a finite energy surface into M is an embedding by algebraic means, see [19]. We shall concentrate in the following on the study of generalized embedded finite energy surfaces.

We consider a fixed surface C_0. If another surface C is close enough to C_0 (in a sense to be made precise), we describe C as the graph of a section Φ of the normal bundle of C_0 in $\mathbb{R} \times M$ satisfying a differential constraint. We shall formulate this constraint as a first order elliptic system for Φ. Its Fredholm index turns out to depend on the number of punctures, the Euler characteristic of S and the normal Conley-Zehnder index, which we introduce below. Since we have to change later on the holomorphic structure on a given surface in a specific way we recall briefly the definitions. For us a Riemann surface $S_{\mathcal{A}}$ is a compact Hausdorff space S together with a maximal atlas $\mathcal{A}$ consisting of charts $\phi_\alpha : U_\alpha \to \mathbb{C}$ such that the transition maps are biholomorphic maps in $\mathbb{C}$. This holomorphic structure induces a natural smooth structure $S_{\mathcal{B}}$ on S defined by the maximal atlas $\mathcal{B} \supset \mathcal{A}$ consisting of charts whose transition maps are smooth. Using the holomorphic charts one defines in a natural way the smooth $1-1$ tensor j on S by setting $j[U_\alpha, \phi_\alpha, x, v] = [U_\alpha, \phi_\alpha, x, iv]$ where $[\,]$ denotes the tangent vector at the point $x \in S$ in S. Lang's notation. Summarizing, starting with a holomorphic atlas $\mathcal{A}$ we obtain in a natural way a smooth complex structure $j_{\mathcal{A}}$ satisfying $j_{\mathcal{A}}^2 = -\operatorname{Id}$. Conversly of course, with a smooth structure $\mathcal{B}$ on S and a smooth $1-1$ tensor $j_{\mathcal{B}}$ satisfying $j_{\mathcal{B}}^2 = -\operatorname{Id}$ we can construct a natural holomorphic atlas $\mathcal{A} \subset \mathcal{B}$ such that the associated differentiable structure and the $j_{\mathcal{A}}$ are the given ones. By abuse of notation we shall always simply write (S, j) or $(S, j, \mathcal{B})$ for both interpretations.

There are different ways of approaching our problem. The aim is to study the moduli space of finite energy surfaces

$$\begin{gathered}\tilde{u} : S \setminus \Gamma \to \mathbb{R} \times M \\ T\tilde{u} \circ j = \tilde{J}(\tilde{u}) \circ T\tilde{u} \\ 0 < E(\tilde{u}) < \infty,\end{gathered}$$

where (S, j) is a Riemann surface and $\Gamma \subset S$ a non-empty finite set of points. We shall allow j and Γ to vary and are in principle only interested in the image $C = \tilde{u}(S \setminus \Gamma) \subset \mathbb{R} \times M$. Therefore the object of the study are equivalence classes $[(\tilde{u}, (S, j), \Gamma)]$, where $(\tilde{u}, (S, j), \Gamma)$ and $(\tilde{v}, (S', j'), \Gamma')$ are equivalent if there exists a biholomorphic map

$$\phi : (S, j) \to (S', j')$$

satisfying $\phi(\Gamma) = \Gamma'$ and $\tilde{u} = \tilde{v} \circ \phi$. The collection of all $[(\tilde{u}, (S, j), \Gamma)]$ is shown to have a natural manifold structure if $\tilde{u}$ is assumed to be an immersion. It is possible to extend our study to the case where $\tilde{u}$ is merely somewhere injective and we refer to the forthcoming paper [23] in which we shall use a quite different approach.

The dimension of the local moduli space around $[(\tilde{u}, (S, j), \Gamma)]$ turns out to depend on the number of punctures, $\sharp\Gamma$, the Euler characteristics of the closed surface S and the normal Conley-Zehnder indices associated with the asymptotics near the punctures which will be defined below. In order to describe the local moduli space near $[(\tilde{u}_0, (S_0, j_0), \Gamma_0)]$ where $\tilde{u}_0$ is an embedding or at least immersion we identify $[(\tilde{u}_0, (S_0, j_0), \Gamma_0)]$ with $C_0 = \tilde{u}_0(S \setminus \Gamma_0)$. If C is another surface in the moduli space near C_0 we shall describe it as a graph of a section Φ of the normal bundle of C_0 in $\mathbb{R} \times M$ satisfying a differential constraint. Our method produces a finite energy surface

$$\tilde{u} : S \setminus \Gamma \to \mathbb{R} \times M$$

and a smooth complex structure j on $S \setminus \Gamma$ only such that

$$\begin{aligned} T\tilde{u} \circ j &= \tilde{J}(\tilde{u}) \circ T\tilde{u} \\ 0 < E(\tilde{u}) &< \infty \\ \tilde{u}(S \setminus \Gamma) &= C. \end{aligned}$$

Now observe that S has the complex structure given by j_0 and therefore an induced smooth structure on S and $S \setminus \Gamma$. It turns out that j on $S \setminus \Gamma$ is smooth for the structure on $S \setminus \Gamma$ but only has a Hölder continuous extension over Γ. However by standard results as shown in the Appendix, there exists a complex structure j' on the underlying topological space S such that $j' = j$ on $S \setminus \Gamma$ and such that the induced differentiable structure on S is different from the original one but coincides on $S \setminus \Gamma$ with the original one. This way we obtain the quadruple $(\tilde{u}, (S, j), \Gamma)$ describing the neighboring curve C.

Note that given C we can always take a parametrization in which the punctures are at preassigned positions. Indeed, if $(\tilde{u}, (S, j), \Gamma)$ parametrizes C, so does the quadruple

$$\tilde{u} \circ \phi, (S, T\phi^{-1} \circ j \circ T\phi), \phi^{-1}(\Gamma),$$

where $\phi : S \to S$ is any diffeomorphism.

The embeddedness assumption allows us to express the usual Conley-Zehnder index introduced above in terms of the normal Conley-Zehnder index, the number of punctures and the Euler characteristic of the underlying Riemann surface S. We will present this relationship first.

Since the solutions near a puncture converge asymptotically to a cylinder over a periodic orbit (perhaps multiply covered) the complex normal bundle can be extended over the circles at infinity by $x^*\xi$. This extended complex normal bundle has distinguished symplectic structures compatible with the complex structures at the circles at infinity. This symplectic structure has a unique (up to homotopy) extension to a compatible symplectic structure over the whole bundle. Trivializing this normal bundle now symplectically we can again define a Conley-Zehnder index, called the normal Conley-Zehnder index and denoted by $\mu_N(\tilde{u})$ or $\mu_N(C)$ depending on the situation.

Denote for a compact Riemann surface (S, i) by χ_S its Euler characteristic and by $\sharp\Gamma_S$ the number of punctures. We have the following relationship between the integers μ, μ_N, χ_S and $\sharp\Gamma_S$.

Theorem 1.8. *Let $\tilde{u} : \dot{S} \to \mathbb{R} \times M$ be an embedded generalized finite energy surface with non-degenerate asymptotic limits. Then the following formula holds:*

$$\mu(\tilde{u}) = \mu_N(\tilde{u}) - 2(\sharp\Gamma_S - \chi_S).$$

For later reference we formulate the following consequence for an embedded finite energy plane. It is defined on $\mathbb{C} = \dot{S} = S^2 \setminus \{\infty\}$ and $\chi_S = 2$.

Corollary 1.9. *Let $\tilde{u} : \mathbb{C} \to \mathbb{R} \times M$ be an embedded generalized finite energy plane with non-degenerate asymptotic limit. Then*

$$\mu(\tilde{u}) = \mu_N(\tilde{u}) + 2.$$

Before we proceed let us give a proof of Theorem 1.8.

Proof. Put $C = \tilde{u}(\dot{S})$. The pull-back bundle

$$\tilde{u}^*(T(\mathbb{R} \times M)) \tag{13}$$

admits the following decomposition into complex line bundles

$$\tilde{u}^*(T(\mathbb{R} \times M), \hat{J}) = TC \oplus N_C.$$

We note that TC and N_C are complex line bundles, with the complex structure induced by $\hat{J}$. Take the arc $a \to J_a$, $a \in \mathbb{R}$, of admissible complex multiplications for $\xi \to M$, so that the associated almost complex structure $\tilde{J}$ coincides with $\hat{J}$ on $\big(\mathbb{R} \setminus [-a_1, a_1]\big) \times M$. Observe that $\tilde{J}$, defined by (8) is admissible for Ω on $[-a_1, a_1] \times M$. The space of admissible almost complex structures for a given symplectic form is contractible, see [14, 24]. Consider the complex vector bundle $\tilde{u}^*(T(\mathbb{R} \times M, \tilde{J})$. Outside of a compact set this vector bundle is identical to the one defined in (13), while on the compact set it is a deformation of the first one, keeping the structure fixed on the boundary. It admits the following decomposition into complex line bundles

$$\tilde{u}^*(T(\mathbb{R} \times M), \tilde{J}) = \tilde{u}^*\tilde{\eta} \oplus \tilde{u}^*\tilde{\xi}.$$

Here $\tilde{\xi}$ is the complex line bundle with fiber ξ_u over (a, u) and $\tilde{\eta}$ is the complex line bundle with fiber $\mathbb{R} \times \mathbb{R}X_a(u)$ over (a, u), where X_a is the Reeb vector field associated with λ_a. Disk like neighborhoods of a puncture look conformally like a half-infinite cylinder. We may view such a cylinder near a positive puncture as $[0, \infty) \times S^1$ and near a negative puncture as $(-\infty, 0] \times S^1$. The images of circles $\{s\} \times S^1$ under $\tilde{u}$ for $s \to \pm\infty$ converge to a periodic orbit of the Reeb vector field $X_\pm$. Asymptotically the images of these cylinders look like the cylinders over the corresponding periodic orbit. Note also that the periodic solution might be multiply covered. We compactify $\dot{S}$ by adding to each cylindrical end a circle to obtain a compact surface S' with boundary. As a consequence of the asymptotic studies in [18] the bundles TC and N_C can be extended over the circles at infinity in the obvious way. We denote the extended bundles by the same letter. The same is true for the bundles in the second decomposition. Note that the extensions of TC and $\tilde{u}^*\tilde{\eta}$ over the circles at infinity coincide in view of the asymptotic behavior of $\tilde{u}$. The same holds for N_C and $\tilde{u}^*\tilde{\xi}$. Denote by $\bar{S}'$ the surface S' with the opposite orientation, where S' carries the orientation induced by (S, i). We build the oriented double $W = S' \cup \bar{S}'$. Over W the complex line bundles TC (viewed as bundle over S') and $\tilde{u}^*\tilde{\eta}$ (viewed as bundle over $\bar{S}'$) constitute a complex line bundle denoted by ℓ_1. Clearly ℓ_1 is obtained by gluing the line bundles along the circles at infinity via the identity map. Similarly we obtain a line bundle ℓ_2. The bundle $\ell_1 \oplus \ell_2$ over W is isomorphic to the pull-back of $(T(\mathbb{R} \times M), \tilde{J})$ as well as to the pull-back of $(T(\mathbb{R} \times M), \hat{J})$, by the doubling w of the map $\tilde{u}$. Hence

$$\begin{aligned} c_1(\ell_1 \oplus \ell_2) &= c_1(w^*(T(\mathbb{R} \times M), \tilde{J})) \\ &= c_1(w^*\tilde{\eta} \oplus w^*\tilde{\xi}) \\ &= c_1(w^*\tilde{\eta}) + c_1(w^*\tilde{\xi}). \end{aligned}$$

Clearly, the bundle $\tilde{\eta}$ admits the global non-vanishing section $(a, u) \to X_a(u)$ and is therefore trivial. So is its pull-back, hence $c_1(w^*\tilde{\eta}) = 0$. The complex line bundle $w^*\tilde{\xi} \to W$ also admits a nowhere vanishing section. Indeed, take any section s over S', which is transversal to the zero section and does not vanish over the asymptotic circles. Take the same section over $\bar{S}'$. This defines a global section of $w^*\tilde{\xi}$ over W with non-degenerate zeroes. The zeroes occur in pairs. Indeed, if z is a zero in S', it will not lie on an asymptotic circle by assumption. Hence z viewed as an element in $\bar{S}'$ is a second zero. However its local index is just the opposite, since S' and $\bar{S}'$ have the opposite orientation. Consequently the total sum of the local indices is zero showing that the first Chern class $c_1(w^*\tilde{\xi})$ vanishes. This proves that $c_1(\ell_1 \oplus \ell_2) = 0$. Hence

$$c_1(\ell_1) = -c_1(\ell_2).$$

Next we compute $c_1(\ell_1)([W])$ for the fundamental class $[W]$ of W. Consider the nowhere vanishing section $z \to (-1, 0)$ over $\bar{S}'$. Identifying $\dot{S}$ and C this section induces a vector field on the asymptotic circles of S' which is inward looking at negative punctures and outward looking at positive punctures. This vector field can be extended over S' to a vector field Y, transversal to the zero section and having algebraically precisely $\chi_S - \sharp\Gamma_S$ many zeroes. Lifting Y to a section of TC we have constructed a section of ℓ_1 also having algebraically precisely $\chi_S - \sharp\Gamma_S$ many zeroes. Hence

$$c_1(\ell_2)([W]) = -c_1(\ell_1)([W]) = \sharp\Gamma_S - \chi_S.$$

Recall that ℓ_2 consists of N_C over S' and $\tilde{u}^*\tilde{\xi}$ over $\bar{S}'$. It follows immediately from [19] (Theorem 3.10 and the discussion starting with Lemma 5.3), that

$$\mu_N(\tilde{u}) - \mu(\tilde{u}) = 2c_1(\ell_2)([W]) = 2(\sharp\Gamma_S - \chi_S).$$

The proof of Theorem 1.8 is complete. □

Next we describe a result from our Fredholm theory. Assume as before that $f : \mathbb{R} \times M \to (0, \infty)$ is a smooth function satisfying

$$\begin{aligned} \frac{\partial f}{\partial a}(a,u) &\geq 0 \ \text{ for all } \ (a,u) \in \mathbb{R} \times M \\ \frac{\partial f}{\partial a}(a,u) &\geq \sigma > 0 \ \text{ for all } \ (a,u) \in [-a_1, a_1] \times M \\ f(a,u) &= g(u) \ \text{ for all } \ a \geq a_0 \\ f(a,u) &= h(u) \ \text{ for all } \ a \leqslant -a_0. \end{aligned}$$

For a contact form λ on M we define an arc of contact forms $a \mapsto \lambda_a = f(a, \cdot)\lambda$. Let $\hat{J}^0$ be an almost complex structure on $\mathbb{R} \times M$ which is compatible with the symplectic structure $\Omega = d(f\lambda)$ on $[-a_1, a_1] \times M$ and which otherwise coincides with the almost complex structure $\tilde{J}^0$ associated with the arc $a \mapsto J^0_a$, $a \in \mathbb{R}$, of admissible complex multiplications for $\xi \to M$. Again we assume this arc to be independent of a if $|a| \geq a_0 > a_1$.

Following A. Floer in [9], we introduce a class of perturbations of $\hat{J}^0$, which arises from perturbing $\hat{J}^0$ on the piece $(-a_1, a_1) \times M$. In order to do so, we consider the Fréchet space $\mathcal{C}^\infty$ consisting of all smooth maps

$$(a,u) \mapsto Y_a(u),$$

where $(a,u) \in \mathbb{R} \times M$, and $Y_a(u) \in \operatorname{Hom}(T_{(a,u)}(\mathbb{R} \times M))$ satisfies

$$\begin{aligned} Y_a(u) &= 0 \ \text{ for all } \ |a| \geq a_1 \\ Y_a(u) \circ \hat{J}^0(a,u) + \hat{J}^0(a,u) \circ Y_a(u) &= 0 \\ \Omega(Y_a(u)h, k) + \Omega(h, Y_a(u)k) &= 0 \end{aligned} \tag{14}$$

for all $h, k \in T_{(a,u)}(\mathbb{R} \times M)$. For a given sequence $\varepsilon = (\varepsilon_k)$ satisfying $\varepsilon_k > 0$, $\varepsilon_k \to 0$ as $k \to \infty$, we denote by $\mathcal{C}_\varepsilon$ the linear subspace of $\mathcal{C}^\infty$ consisting of all Y which, in addition to (14), meet the estimate

$$\|Y\|_\varepsilon := \sum_{k=0}^{\infty} \varepsilon_k \|Y\|_{C^k} < \infty.$$

The space $\mathcal{C}_\varepsilon$ is dense in $\mathcal{C}^\infty$ provided the sequence ε converges sufficiently fast to 0, see [9]. Choose now an admissible almost complex structure $\hat{J}$ as described above which is close to $\hat{J}^0$. Then we find a unique Y satisfying (14) with $\|Y\|_\epsilon$ small, such that

$$\hat{J}(a,u) = \hat{J}^0(a,u)\exp(-\hat{J}^0(a,u)Y_a(u)). \tag{15}$$

For $\delta > 0$ sufficiently small we denote by $\mathcal{U}_\delta$ the set of all $\hat{J}$ obtained via formula (15) with $Y \in \mathcal{C}_\varepsilon$ satisfying $\|Y\|_\varepsilon < \delta$. The inverse map

$$B_\delta^{\mathcal{C}_\varepsilon} \to \mathcal{U}_\delta : Y \mapsto \hat{J}$$

defines a global chart for $\mathcal{U}_\delta$, turning it into a Banach manifold diffeomorphic to an open subset in a separable Banach space.

Given the non-degenerate periodic orbits (x_j, T_j) for $j = 1, 2, \dots, m$, of the Reeb vector field associated with the contact form $g\lambda$ and given the non-degenerate periodic orbits $(y_k, T_k^{'})$, $k = 1, 2, \dots, m'$, of the Reeb vector field belonging to the contact form $h\lambda$ we consider the set

$$\mathcal{M}(x_1, \dots, x_m; y_1, \dots, y_{m'})$$

consisting of all pairs $(C, \hat{J})$, where C is an unparametrized immersed generalized energy surface for the almost complex structure $\hat{J} \in \mathcal{U}_\delta$ having m positive punctures asymptotic to the (x_j, T_j) and m' negative punctures asymptotic to the periodic orbits (y_k, T_k'). Then the following statement holds true.

Theorem 1.10. *The set*

$$\mathcal{M} := \mathcal{M}(x_1, \dots, x_m; y_1, \dots, y_{m'})$$

carries in a natural way the structure of a separable Banach manifold. The projection map

$$\eta : \mathcal{M} \to \mathcal{U}_\delta : (C, \hat{J}) \mapsto \hat{J}$$

is a Fredholm map whose Fredholm index near C is

$$\operatorname{Ind}(C) = \mu(C) + \sharp\Gamma - \chi_S,$$

where χ_S is the Euler characteristic of the Riemann surface S and C is parametrized by $\dot{S} = S \setminus \Gamma$.

We observe that the regular values $\alpha \in \mathcal{U}_\delta$ for the projection are precisely the data for which the fiber $\eta^{-1}(\alpha)$ is a smooth finite-dimensional manifold whose dimension agrees with the Fredholm index. Since the set of all possible pairs of asymptotic data in the non-degenerate case is countable and for each case the set of regular values for the projection is residual we can take a countable intersection, which again will be residual.

Next we give an application of Theorem 1.10 which is useful in our construction of a global surface of section for the Hamiltonian flow on a strictly convex three-dimensional energy surface in $\mathbb{R}^4$, see [21]. The set up is the following. Consider the standard symplectic space $\mathbb{R}^4 = \mathbb{R}^2 \times \mathbb{R}^2$, whose coordinates we denote by $x = (q_1, p_1, q_2, p_2)$, equipped with the symplectic form

$$\omega = \Sigma_{j=1}^2 dq_j \wedge dp_j.$$

Thus $\omega(h, k) = \langle Jh, k \rangle$, with the matrix

$$J = \begin{pmatrix} J_0 & 0 \\ 0 & J_0 \end{pmatrix}, \quad \text{where } J_0 = \begin{pmatrix} 0 & -1 \\ 1 & 0 \end{pmatrix}.$$

Observe that $\omega = d\lambda_0$, the one-form λ_0 being given by

$$\lambda_0 = \frac{1}{2}\sum_{j=1}^{2}(q_j dp_j - p_j dq_j).$$

Now consider the sphere $S^3 \subset \mathbb{C}^2$ equipped with the standard contact form λ_{st}. The standard contact form λ_{st} on S^3 is the restriction of λ_0 to S^3,

$$\lambda_{\mathrm{st}} = \lambda_{0\,|S^3}.$$

Let $g, h : S^3 \to (0, \infty)$ be smooth functions such that

$$h < g \text{ on } S^3,$$

and let $f : \mathbb{R} \times S^3 \to (0, \infty)$ be a smooth function satisfying

$$\begin{aligned} f(a,u) &= h(u) \ \text{ for } \ a \leqslant -1 \\ f(a,u) &= g(u) \ \text{ for } \ a \geq \ \ 1 \\ \frac{\partial f}{\partial a}&(a,u) \geq 0 \\ \frac{\partial f}{\partial a}(a,u) \geq \sigma > 0 \ &\text{ for all } \ (a,u) \in \left[-\frac{1}{2}, \frac{1}{2}\right] \times M. \end{aligned}$$

In addition, we assume that h and g have the following properties. The Reeb vector field $X_{g\lambda_{\mathrm{st}}}$ associated with the form $g\lambda_{\mathrm{st}}$ possesses a non-degenerate periodic orbit (x_0, T_0) having Conley-Zehnder index equal to 3. We assume that all periodic orbits (x, T) of the Reeb vector field $X_{h\lambda_{\mathrm{st}}}$ with periods $0 < T \leqslant T_0$ are non-degenerate and have index $\mu(P) \geq 3$.
Fix an admissible complex multiplication $J : \xi \to \xi$, where ξ is a contact structure associated with λ_{st} and choose a smooth arc $a \mapsto J_a$ of admissible complex structures of ξ such that $J_a = J$ for $|a| > 1$. Set $\lambda_a = f(a, \cdot)\lambda_{\mathrm{st}}$ and denote the corresponding Reeb vector field by X_a. With this notation we define the associated almost complex structures $\tilde{J}$ on $\mathbb{R} \times S^3$ by

$$\tilde{J}(a,m)(h,k) = \big(-\lambda_a(m)k, J_a(m)\pi_a k + hX_a(m)\big),$$

for $h, k \in T_{a,m}(\mathbb{R} \times S^3)$. From Theorem 1.10 we deduce the following corollary.

Corollary 1.11. *Suppose that the functions f, h, g are as described above. Then there exists an almost complex structure $\hat{J}$ compatible with $\Omega = d(f\lambda_{st})$ on $\left[-\frac{1}{2}, \frac{1}{2}\right] \times S^3$ and otherwise coinciding with a structure $\tilde{J}$ such that the following holds. Every embedded generalized finite energy surface $\tilde{u} : \dot{S} \to \mathbb{R} \times S^3$ having precisely one positive puncture which is asymptotic to (x_0, T_0) possesses at most one negative puncture. In addition, if there exists a negative puncture, it is necessarily asymptotic to a non-degenerate periodic orbit (x, T) of the Reeb vector field associated with $h\lambda_{st}$ having index $\mu(x, T) = 3$.*

In order to sketch the proof of the corollary we note that in the situation at hand the condition $\text{Ind}(C) \geq 0$ becomes, in view of the formula in Theorem 1.10,

$$\mu(P_0) - \mu^-(\tilde{u}) - \chi_{S^2} + \sharp\Gamma^+ + \sharp\Gamma^- \geq 0.$$

By assumption, $\mu(P_0) = 3$ and $\mu(P) \geq 3$ for all negative punctures. Since $\chi_{S^2} = 2$ and $\sharp\Gamma^+ = 1$ we deduce

$$3(1 - \sharp\Gamma^-) \geq 3 - \mu^-(\tilde{u}) \geq (1 - \sharp\Gamma^-).$$

Consequently, $\sharp\Gamma^- \leqslant 1$ as claimed. If $\sharp\Gamma^- = 1$, then $\mu^-(\tilde{u}) = \mu(P)$ for the single negative puncture asymptotic to P and, inserting $\sharp\Gamma^- = 1$ into the inequality above, we conclude $\mu(P) = 3$ as claimed.

Studying transversality questions, injectivity properties of the finite energy planes are crucial. We begin with a definition.

Definition 1.12. *A finite energy surface $\tilde{u} = (a, u) : \dot{S} \to \mathbb{R} \times M$ is called somewhere injective, if there exists a point $z_0 \in \dot{S}$ for which*

$$T\tilde{u}(z_0) \neq 0 \quad \text{and} \quad \tilde{u}(z_0) \notin \tilde{u}(\dot{S} \setminus \{z_0\}).$$

Similarly we call the M-part u somewhere injective if (1.12) *holds for u replacing $\tilde{u}$.*

We shall prove later on the following result.

Theorem 1.13. *Let $\tilde{u} : \dot{S} \to \mathbb{R} \times M$ be a somewhere injective finite energy surface with non-degenerate asymptotic limits and assume there exists a point $z_0 \in \dot{S}$ with*

$$\pi \circ Tu(z_0) \neq 0. \tag{16}$$

Then the M-part u is somewhere injective. In fact, the subset

$$\left\{ z \in \dot{S} \mid \pi \circ Tu(z) \neq 0 \quad \text{and} \quad u(z) \notin u(\dot{S} \setminus \{z\}) \right\}$$

is open and dense in $\dot{S}$.

We note that the hypothesis (16) is necessary as the simple example of a cylinder over a periodic orbit shows. The next result is the boundary version of the above theorem.

Theorem 1.14. *Assume S is a compact Riemann surface with boundary ∂S and a finite set of punctures $\Gamma \subset S \setminus \partial S$. Let $\dot{S} = S \setminus \Gamma$. Assume $\tilde{u} : \dot{S} \to \mathbb{R} \times M$ is a somewhere injective finite energy surface having non-degenerate asymptotic limits. Assume further that $\tilde{u}$ is an embedding near the boundary in the sense that $\tilde{u}|\partial S$ is an embedding and that there exists an open neighborhood U of ∂S in $\dot{S}$ satisfying*

$$\tilde{u}^{-1}(\tilde{u}(U)) = U.$$

If $\pi \circ Tu(z_0) \neq 0$ for some $z_0 \in \dot{S}$, then the set

$$\{z \in \dot{S} \mid \pi \circ Tu(z) \neq 0 \text{ and } u(z) \notin u(\dot{S} \setminus \{z\})\}$$

is open and dense in $\dot{S}$.

Finally, we shall describe an application of the Fredholm theory which is useful in connection with the so-called disk filling method. This method gives rise to a mixed boundary value problem. As before let M be a compact three manifold equipped with a contact form λ. Assume F is an embedded oriented disk with boundary positively transversal to ξ. We assume that F has precisely one singular point q, which has to belong to the interior, is non-degenerate and positively elliptic, see [1, 6, 12, 19] for facts concerning the characteristic foliation. Further we suppose that the characteristic foliation on F has no limit cycle. This implies that all characteristics emerge from $\{q\}$ and leave the disk transversally through the boundary. Let $\tilde{J}$ be the almost complex structure associated with a smooth admissible complex multiplication $J : \xi \to \xi$ via the formula

$$\tilde{J}(a,u)(h,k) = \big(-\lambda(u)(k), J(u)\pi k + hX(u)\big).$$

We study the following mixed boundary value problem for a map $\tilde{u} = (a,u) : D \setminus \Gamma \to \mathbb{R} \times M$

$$\begin{aligned} \tilde{u}_s + \tilde{J}(\tilde{u})\tilde{u}_t = 0 \text{ on } D \setminus \Gamma, \\ a|\partial D = 0 \text{ and} \\ u : \partial D \to F \setminus \{q\} \text{ has winding number 1.} \end{aligned} \tag{17}$$

Here $D \subset \mathbb{C}$ is the closed unit disk and Γ is a finite, possibly empty subset of $D \setminus \partial D$, which may depend on $\tilde{u}$. In addition, we impose the energy requirement

$$E(\tilde{u}) < \infty, \tag{18}$$

where the energy is obtained by maximizing the integrals

$$\int_{D\setminus\Gamma} \tilde{u}^* d(\phi\lambda)$$

over the set of all smooth maps $\phi : \mathbb{R} \to [0,1]$, satisfying $\phi'(s) \geq 0$. We assume that the set Γ consists of negative punctures and that the associated asymptotic limits are non-degenerate and contractible periodic orbits of X. Given such a periodic orbit (x,T) and a disk map $v_x : D \to M$ satisfying

$$v_x\big(e^{2\pi i t/T}\big) = x(t)$$

we can compute its Conley-Zehnder index $\mu(x, v_x)$ with respect to the homotopy class of the given disk. Given a solution $\tilde{u}$ of the above problem and for every asymptotic limit x_j a disk map v_j, we can glue the v_j at the punctures to u to obtain a disk map

$$\bar{u} = \tilde{u}\#\{v_1, \ldots, v_N\} : D \to \mathbb{R} \times M,$$

$(N = \sharp\Gamma)$ which coincides near the boundary with u. We call the map $\bar{u}$ homotopic to a small disk around the puncture $\{q\}$, if it can be homotoped through a family of disk mappings which map the boundary ∂D to $F \setminus \{q\}$, to a disk mapping whose image lies in a neighborhood of q with small diameter. The result is the following

Theorem 1.15. *Consider a compact 3-manifold (M,λ) equipped with a contact form λ. Then there exists a dense subset $\mathcal{J}_{\text{gen}}$ of the set $\mathcal{J}$ of compatible almost complex structures on ξ such that for every $J \in \mathcal{J}_{\text{gen}}$ the following statement holds true. There exists no embedded solution*

$$\tilde{u} \colon D \setminus \Gamma \to \mathbb{R} \times M$$

of the problem (17) *satisfying* (18) *and having the following additional properties:*

(i) $\Gamma \neq \emptyset$ *and* $\pi \circ Tu(z_0) \neq 0$ *for some* $z_0 \in D \setminus \Gamma$.
(ii) *The asymptotic limits* (x_j, T_j) *belonging to the punctures in* Γ *are non-degenerate.*
(iii) *There exist disc maps* $v_j \colon D \to M$ *satisfying* $v_j(e^{2\pi it}) = x_j(T_j t)$ *and* $\mu(x_j, v_j) \geq 3$. *Moreover,* $\tilde{u}\#\{v_1, \ldots, v_N\} \colon D \to \mathbb{R} \times M$ *is homotopic to a small disc around* e.

The statement will follow from the Fredholm theory for solutions $\tilde{u}$ of the mixed boundary value problem above. The corresponding Fredholm index leads, in the generic situation, to the inequality

$$\sharp\Gamma - \sum_{j=1}^{\sharp\Gamma} \mu(x_j, v_{x_j}) + 1 \geq 0.$$

If $\mu(x_j, v_{x_j}) \geq 3$ for all j, we deduce $1 \geq 2\sharp\Gamma$ so that indeed $\sharp\Gamma = 0$. Theorem 1.15 will be useful in the compactness arguments of [20] and [22].

2. Linear Fredholm theory

In view of the structure of the non linear operators encountered later on we shall set up the Fredholm theory in Hölder spaces instead of in Sobolev spaces. In order to describe the linear model we start with a compact Riemann surface S and a finite subset $\Gamma \subset S$ called punctures, and abbreviate the punctured surface by $\dot{S} = S \setminus \Gamma$. The set Γ is the union of the negative punctures Γ^- and the positive punctures Γ^+,

$$\Gamma = \Gamma^- \cup \Gamma^+.$$

Positive and negative punctures are distinguished by special cylindrical coordinate systems defined as follows. Let $z_0 \in \Gamma^+$ be a positive puncture. Take any holomorphic chart $h : D \subset \mathbb{C} \to U \subset S$ around z_0 satisfying $h(0) = z_0$ and introduce holomorphic polar coordinates in a punctured neighborhood of z_0:

$$\sigma : [0,\infty) \times S^1 \to U \setminus \{z_0\}$$

by $(s,t) \mapsto h\big(e^{-2\pi(s+it)}\big)$. Then $\sigma(s,t) \to z_0$ as $s \to \infty$ and $[0,\infty) \times S^1$ is equipped with the standard complex structure. Similarly, near a negative puncture z_0 we shall introduce a biholomorphic coordinate system

$$\sigma : (-\infty, 0] \times S^1 \to U \setminus \{z_0\}$$

by $(s,t) \mapsto h(e^{2\pi(s+it)})$ such that this time $\sigma(s,t) \to z_0$ as $s \to -\infty$. In the following we shall refer to these distinguished coordinates as to the special coordinates. Coordinate changes have the following crucial property, proved by using Riemann's removable singularity theorem.

Lemma 2.1. *Assume σ and τ are special coordinates around a positive puncture. Then there exists $R > 0$ such that*

$$h = \tau^{-1} \circ \sigma : [R, \infty) \times S^1 \to [0, \infty) \times S^1$$

is a holomorphic embedding, and lifting h to $[R, \infty) \times \mathbb{R} \to [0, \infty) \times \mathbb{R}$ there are unique constants $c \in \mathbb{R}$ and $0 \leqslant d < 1$ such that asymptotically as $s \to \infty$

$$D^\beta \left[h(s,t) - (s + c, t + d)\right] \to (0,0)$$

uniformly in t, for every multi-index β.

A similar statement holds true for negative punctures. The lemma guarantees that the definitions later on in punctured neighborhoods of punctures are independent of the choice of the special coordinates used.

Let V be a finite-dimensional normed space, $\alpha \in (0,1)$ and $R \geq 0$. For a function $u : [0, \infty) \times S^1 \to V$ we denote by $\|u\|^*_{\alpha,R}$ the following semi norm of u,

$$\|u\|^*_{\alpha,R} = \sup \left\{ \frac{|u(s,t) - u(s',t')|}{|(s-s', t-t')|^\alpha} \right\}.$$

The sup is taken over all

$$(s,t) \neq (s',t') \text{ and } s, s' \geq R, \ |(s-s', t-t')| \leqslant 1.$$

Then we define the norm

$$\|u\|_{\alpha,R} = \sup\{|u(s,t)| \mid s \geq R, t \in \mathbb{R}\} + \|u\|^*_{\alpha,R}.$$

A function $u : [R_0, \infty) \times S^1 \to V$ will, in the following discussion, often be viewed as periodically extended to $\mathbb{R}$ in the second argument, so that $u : [R_0, \infty) \times \mathbb{R} \to V$. We call u of class C^α_0 at infinity if for some $R \geq R_0$ we have $\|u\|_{\alpha,R} < \infty$ and

$$\|u\|_{\alpha,R} \to 0 \ \text{ as } \ R \to \infty.$$

Next consider that a function u defined on $\dot{S}$ near a positive puncture z_0. Then u is of class C^α_0 near z_0 if for some special coordinates σ the composition $u \circ \sigma$ is of class C^α_0 near ∞. As a consequence of the above lemma the definition does not depend on the choice of holomorphic coordinates near the puncture. Similarly one treats negative punctures. Let $u : \dot{S} \to V$ be a continuous map. We say that u belongs to $C^\alpha_0(\dot{S}, V)$ if for every point $z \in \dot{S}$ there exists a smooth chart $\phi : V(z) \to W(0) \subset \mathbb{C}$ such that $u \circ \phi^{-1}$ is of class C^α_{loc} and for every positive puncture $u \circ \sigma$ of class C^α_0 at ∞, and for every negative puncture the composition $u \circ \sigma$ is of class C^α_0 at $-\infty$. Clearly, $C^\alpha_0(\dot{S}, V)$ is a Banachable space. Similarly we define $C^{1,\alpha}_0(\dot{S}, V)$ to consist of all maps $u : \dot{S} \to V$ of class C^1 such that $u \in C_0^{\alpha}(\dot{S}, V)$ and u belongs to $C_{loc}^{1,\alpha}$ on $\dot{S}$ and moreover near a positive puncture

$$\frac{\partial(u \circ \sigma)}{\partial s} \text{ and } \frac{\partial(u \circ \sigma)}{\partial t}$$

are of class C_0^{α} at infinity, where (s,t) are the standard holomorphic coordinates on the cylinder. Similarly we treat negative punctures. The space $C_0^{1,\alpha}(\dot{S}, V)$ is again a Banachable space. The spaces $C_0^{k,\alpha}(\dot{S}, V)$ are defined analogously.

Consider a punctured Riemann surface $\dot{S}$ with complex structure $j = j(z)$ and assume that $z \mapsto i(z)$ is a smooth map associating with $z \in \dot{S}$ a complex multiplication on $\mathbb{R}^2$. If the set of punctures Γ is non-empty, the tangent bundle $T\dot{S} \to \dot{S}$ is trivializable. We are interested in particular trivializations. We call a trivialization $\Psi : T\dot{S} \to \dot{S} \times \mathbb{R}^2$ admissible if near a puncture (let's assume a positive one) we have asymptotically as $s \to \infty$,

$$D^{\beta}\Big[pr_2 \circ \Psi(\sigma(s,t)) \circ T\sigma(s,t) - A(t)\Big] \to 0$$

uniformly in t for all multi-indices β. Here $t \mapsto A(t)$ is a smooth loop of complex linear isomorphisms. Such a trivialization always exists. We assume, in addition, that near a puncture z_0 we have

$$D^{\beta}\left[i(\sigma(s,t)) - \tilde{i}(t)\right] \to 0$$

as $s \to \infty$, uniformly in t for all multi-indices β. Here $t \mapsto \tilde{i}(t)$ is a smooth loop of complex multiplications on $\mathbb{R}^2$.

Let $\Lambda^2 \to \dot{S}$ be real vector bundle consisting of 2-vectors in $T\dot{S}$. Since $\dim T_z\dot{S} = 2$ at each point $z \in \dot{S}$, the space Λ_z^2 is one-dimensional and spanned by $v \wedge w$, where $\{v, w\}$ is a basis of $T_z\dot{S}$. We define a special vector bundle $\mathcal{A}_0 \to \dot{S}$ as follows. The fiber over $z \in \dot{S}$ consists of all complex anti-linear maps $\varphi : (T_zS^2, j(z)) \to (\mathbb{R}^2, i(z))$,

$$\varphi \circ j(z) = -i(z) \circ \varphi.$$

Fixing an admissible trivialization Ψ of $T\dot{S}$ we let $C \in \mathrm{Hom}_{\mathbb{R}}(\dot{S} \times \mathbb{R}^2, \mathcal{A}_0)$ be a smooth map. We call C admissible if for the special coordinates near a puncture z_0 (say positive) asymptotically as $s \to \infty$,

$$D^{\beta}\Big[(C(\sigma(s,t))h)\frac{\partial\sigma}{\partial s}(s,t) - \tilde{C}(t)h\Big] \to 0$$

in C^{∞} (for every multi-index β). Here $\tilde{C}(t)$ is a linear map in $\mathcal{L}(\mathbb{R}^2)$, depending smoothly on a parameter $t \in S^1$. Moreover, $\tilde{C}(t)$ is symmetric with respect to the metric in $\mathbb{R}^2$ given by

$$g_t = \omega \circ (\mathrm{Id} \times \tilde{i}(t)).$$

Note that in $\mathbb{R}^2$ every complex multiplication is compatible with the symplectic form $\omega = dx \wedge dy$. In particular, $t \mapsto \tilde{i}(t)\tilde{C}(t)$ is a time-dependent linear Hamiltonian vector-field on $\mathbb{R}^2$. We point out that $\tilde{C}(t)$ is independent of the choice of σ up to a phase-shift in view of Lemma 2.1.

Definition 2.2. *Let C be as described above. We call C non-degenerate at the punctures, if for every puncture the linear Hamiltonian system*

$$\dot{x}(t) = \tilde{i}(t)\tilde{C}(t)x(t) \quad \textit{and} \quad x(0) = x(1)$$

only admits the trivial solution.

Let C be an admissible element in $\mathrm{Hom}_{\mathbb{R}}(\dot{S}\times\mathbb{R}^2, \mathcal{A}_0)$, which is non-degenerate at the punctures. We define the bounded linear operator

$$L_C : C_0^{1,\alpha}(\dot{S}, \mathbb{R}^2) \to C_0^{\alpha}(\mathcal{A}_0)$$

by

$$(L_C u)(z) = Tu(z) + i(z)(Tu)(z)j(z) + (Cu)(z),$$

where $(Cu)(z) = C(z)u(z)$.

Theorem 2.3. *If C is admissible and non-degenerate at the punctures, then L_C is a Fredholm operator.*

The result is well known in the L^p-setting, and we refer to M. Schwarz [29] for full details. The $W^{k,p}$-spaces are defined as follows. Let V be a finite-dimensional vector space. We say that u belongs to $W^{k,p}(\dot{S}, V)$ if u belongs to $W^{k,p}_{loc}$ and if for every puncture the map $u \circ \sigma$ belongs to (for example in the case of a positive puncture) to $W^{k,p}([0,\infty) \times S^1, V)$. In order to define $W^{k,p}(\mathcal{A}_0)$ we call a section $u : \dot{S} \to \mathcal{A}_0$ of class $W^{k,p}$ if it belongs locally to $W^{k,p}$ and if near every puncture z_0 the map

$$(s,t) \to u(\sigma(s,t))\left(\frac{\partial\sigma}{\partial s}(s,t)\right)$$

is of class $W^{k,p}([0,\infty) \times S^1, V)$ (in the case of a positive puncture). For the proof of the following result we refer to M. Schwarz, [29].

Theorem 2.4. *Given $k \in \mathbb{N}$ and $1 < p < \infty$, the operator*

$$L_C : W^{k+1,p}(\dot{S}, \mathbb{R}^2) \to W^{k,p}(\mathcal{A}_0)$$

is Fredholm, provided C is admissible and non-degenerate at the punctures.

From Theorem 2.4 we shall derive the Fredholm result in a C^α-setting. In what follows we denote by L and $\tilde{L}$ the operator L_C defined on $C_0^{1,\alpha}(\dot{S}, \mathbb{R}^2)$ and $W^{1,p}(\dot{S}, \mathbb{R}^2)$ respectively.

Assume first that $u \in C_0^{1,\alpha}(\dot{S}, \mathbb{R}^2)$ solves $Lu = 0$. Then the standard elliptic regularity theory shows that u is smooth. Near a puncture z_0 we put $v(s,t) =$

$u(\sigma(s,t))$. Then $Lu = 0$ becomes

$$\begin{aligned}
&\frac{\partial v}{\partial s} + i(\sigma(s,t))\frac{\partial v}{\partial t} + (C(\sigma(s,t))v)\left(\frac{\partial \sigma}{\partial s}(s,t)\right) \\
&= (Tu)\sigma_s + i(\sigma)(Tu)\sigma_t + (C(\sigma)u(\sigma))\sigma_s \\
&= [Tu + i(Tu)j + Cu]\,(\sigma)(\sigma_s) \\
&= \big[L_C u\big](\sigma)(\sigma_s) \\
&= 0.
\end{aligned}$$

By the properties of C and i, as $s \to \infty$

$$[C(\sigma(s,t))h]\,\sigma_s(s,t) \to \tilde{C}(t)h$$
$$i(\sigma(s,t)) \to \tilde{i}(t),$$

with convergence for all the derivatives uniformly in t. Since $\tilde{C}$ is non-degenerate and v is bounded we conclude from the equation above for v using a well-known argument (see e.g. in [17]) that v converges with all its derivatives exponentially fast to 0 as $s \to \infty$ considering a positive puncture resp. as $s \to -\infty$ in case of a negative puncture. Hence $v \in W^{k,p}$, so that in view of our definition, $u \in W^{k,p}(\dot{S}, \mathbb{R}^2)$. Consequently u belongs to the kernel of $\tilde{L}$ and so $\ker L \subset \ker \tilde{L}$. In view of Theorem 2.4, the operator $\tilde{L}$ is Fredholm from which we conclude that the kernel of L is finite-dimensional. Similar arguments show that $\ker \tilde{L} \subset \ker L$ and we have proved that

$$\ker L = \ker \tilde{L}.$$

Lemma 2.5. *If Y is a topological complement of* $\ker L$ *in $C_0^{1,\alpha}(\dot{S}, \mathbb{R}^2)$, then there exists a constant $c > 0$ such that*

$$\|Lu\|_{0,\alpha} \geq c\|u\|_{1,\alpha} \qquad \textit{for all } u \in Y.$$

Proof. Arguing by contradiction we find a sequence $(u_k) \subset Y$ satisfying

$$\|u_k\|_{1,\alpha} = 1, \qquad \|Lu\|_{0,\alpha} \to 0.$$

By the local Hölder theory for elliptic operators we deduce for a suitable subsequence that $u_k \to u$ in $C^{1,\alpha}_{\mathrm{loc}}(\dot{S}, \mathbb{R}^2)$ and $Lu = 0$. We claim that $u_k \to u$ in Y. This will then contradict the fact that $u \notin Y$ in view of $Lu = 0$. Recalling the definition of $C_0^{1,\alpha}$, the claim follows from the $C^{1,\alpha}_{\mathrm{loc}}$ convergence together with the following

Lemma 2.6. *Assume that u_k is a bounded sequence in $C^{1,\alpha}(\dot{S}, \mathbb{R}^2)$ satisfying $Lu_k \to 0$ in $C_0^{\alpha}(\dot{S}, \mathbb{R}^2)$. Then, given $\varepsilon > 0$, a puncture z_0 and a special coordinate system σ near z_0, there exist k_0 and R_0 such that for $k \geq k_0$ and $R \geq R_0$,*

$$\|u_k \circ \sigma\|_{1,\alpha,R} \leqslant \epsilon.$$

Proof. [Proof of Lemma 2.6] Arguing indirectly we find a puncture z_0, special coordinates σ near z_0 and $\varepsilon > 0$ such that for a suitable subsequence still denoted by (u_k) we have the estimates $\|u_k \circ \sigma\|_{1,\alpha,k} \geq \varepsilon$. For every k we find a number $s_k \geq k$ such that on $Q_k = [s_k - 1, s_k + 1] \times [-1, 1]$ we have the estimates

$$\|u_k \circ \sigma\|_{1,\alpha,Q_k} \geq \frac{\epsilon}{2}.$$

Now, define the sequence v_k of translated maps by

$$v_k(s,t) = u_k \circ \sigma(s_k + s, t).$$

Abbreviating

$$f_k(s,t) = \frac{\partial v_k}{\partial s} + i\,(\sigma(s_k + s, t))\,\frac{\partial v_k}{\partial t} + [C(\sigma(s_k + s, t))v_k]\,\frac{\partial \sigma}{\partial s}(s_k + s, t)$$

we conclude from $Lu_k \to 0$ in C_0^α that $f_k \to 0$ in C_0^α. Since the sequence is $C^{1,\alpha}$-bounded we find by the local Hölder estimates for elliptic operators a subsequence v_k and a bounded map v such that

$$\begin{gathered} v_k \to v \;\text{ in } C^{1,\alpha}_{\text{loc}}(\mathbb{R} \times S^1, \mathbb{R}^2) \\ v \neq 0. \end{gathered} \tag{19}$$

Moreover,

$$v_s + \tilde{i}(t)v_t + \tilde{C}(t)v = 0. \tag{20}$$

We shall show that (20) implies that $v = 0$ in contradiction to (19). Due to the assumption that $\tilde{C}$ is non-degenerate and since v is bounded, v converges exponentially fast to 0 uniformly in t as $s \to \pm\infty$. This implies that $v(\cdot, t) \in L^2(\mathbb{R}, \mathbb{R}^2)$. Taking a Fourier transform with respect to s in (20) we obtain

$$\tilde{i}(t)\hat{v}_t(s,t) + \tilde{C}(t)\hat{v}(s,t) - is\hat{v}(s,t) = 0.$$

Let A be the unbounded operator in L^2 defined by

$$(Ax)(t) = -\tilde{i}(t)\dot{x}(t) - \tilde{C}(t)x(t).$$

The operator is self-adjoint with respect to the L^2-inner product

$$(h,k) = \int_{S^1} \omega(h(t), \tilde{i}(t)k(t))dt,$$

$\omega = dx \wedge dy$. Denote by $|\cdot|$ the norm associated with $(\cdot,\cdot)$. Since A is self-adjoint it follows

$$0 = |A\hat{v}(s,\cdot) - is\hat{v}(s,\cdot)| \geq c|s||\hat{v}(s,\cdot)|$$

for some positive constant c. Hence $\hat{v}(s,\cdot) = 0$ for all $s \in \mathbb{R}$, from which one concludes that $v(s,t) = 0$ for all $(s,t) \in \mathbb{R} \times S^1$. This contradiction concludes the proof of the lemma. □

With the proof of Lemma 2.6, the proof of Lemma 2.5 is complete. □

So far we have proved that the kernel of L is finite-dimensional and that the restriction of L onto a topological complement is a linear isomorphism of Y onto the image $R(L)$, in view of Lemma 2.5. In particular, $R(L)$ is closed. We shall prove next that the codimensions of the images of L and $\tilde{L}$ are the same. This then proves that L is a Fredholm operator, as claimed in Theorem 2.3. Moreover, $\operatorname{Ind}(L) = \operatorname{Ind}(\tilde{L})$, since we already know that $\ker L = \ker \tilde{L}$.

We first consider the case where $\tilde{L}$ is surjective and deduce that L is surjective. The smooth compactly supported sections f are in the image of $\tilde{L}$. If $\tilde{L}u = f$, then $\tilde{L}u = 0$ near the punctures and we conclude that u is smooth and decays exponentially fast as $s \to \infty$ or $s \to -\infty$, in our special coordinates σ. Consequently, u belongs to $C_0^{1,\alpha}(\dot{S}, \mathbb{R}^2) = \operatorname{dom}(L)$. Since $R(L)$ is closed it is sufficient to show that L has a dense image. Therefore it is sufficient to prove that $f \in C_0^\alpha(\dot{S}, \mathbb{R}^2)$ with compact support belongs to the image of L. To prove this we take a sequence of smooth and compactly supported sections satisfying $f_n \to f$ in C_0 and $\|f_n\|_\alpha \leqslant C\|f\|_\alpha$, with C independent of n. In view of the remark above we find a sequence u_n of smooth solutions of $Lu_n = f_n$, with $u_n \in C_0^{1,\alpha}(\dot{S}, \mathbb{R}^2)$. From Lemma 2.5 we conclude that $\|u_n\|_{1,\alpha} \leqslant C'\|f_n\|_\alpha \leqslant C''\|f\|_\alpha$. Consequently, by the Arzelà-Ascoli theorem we find a subsequence such that $u_n \to u$ in C^1_{loc} and, moreover, $u \in C^{1,\alpha}$ with $\|u\|_{1,\alpha} < \infty$. In addition, u solves the equation $Lu = f$. Since $f \in C_0^\alpha$ has compact support we conclude $u \in C_0^{1,\alpha}(\dot{S}, \mathbb{R}^2)$. We have proved that L is surjective, provided that $\tilde{L}$ is surjective.

If $\tilde{L}$ is not surjective we denote by d the codimension of its image and choose smooth compactly supported maps $y_1, \ldots, y_d$ forming a basis of the topological complement X of the range $R(\tilde{L})$ so that $L^p(\mathcal{A}_0) = R(\tilde{L}) \oplus X$ and $\dim X = d < \infty$. Define the operator

$$\hat{L} : W^{1,p}(\dot{S}, \mathbb{R}^2) \times X \to L^p(\mathcal{A}_0)$$

by $\hat{L}(u, y) = \tilde{L}u + y$ and the operator

$$\bar{L} : C_0^{1,\alpha}(\dot{S}, \mathbb{R}^2) \times X \to C_0^\alpha(\mathcal{A}_0)$$

by $\bar{L}(u, y) = Lu + y$. By construction, the operator $\hat{L}$ is surjective and has the kernel $\ker \hat{L} = \ker \tilde{L} \times \{0\} = \ker L \times \{0\} \subset \ker \bar{L}$. We claim

$$\ker \hat{L} = \ker \bar{L} \qquad \text{and} \qquad R(L) \cap X = \{0\}.$$

Indeed an element $(u, y) \in \ker \bar{L}$ solves the equation $Lu = -y$ and hence is, in view of the elliptic regularity theory, smooth. Since y is compactly supported, u has an exponential decay near every puncture so that $u \in W^{1,p}(\dot{S}, \mathbb{R}^2)$. Hence $(u, y) \in \ker \hat{L}$ and therefore $u \in \ker \tilde{L}$ and $y = 0$ as claimed. It follows that $\operatorname{codim} R(L) \geq \dim X$. Since $R(L)$ is closed, $R(L) \cap X = \{0\}$ and $\dim X < \infty$, the range $R(\bar{L})$ is closed. From the surjectivity of $\hat{L}$ one deduces arguing as before the surjectivity of $\bar{L}$. Hence $\operatorname{codim} R(L) \leqslant d$ and so $\operatorname{codim} R(L) = \operatorname{codim}(\tilde{L})$. Summing up we have proved

Proposition 2.7. *Given an admissible C which is non-degenerate near the punctures the operators*

$$L_C : C_0^{1,\alpha}(\dot{S}, \mathbb{R}^2) \to C_0^{\alpha}(\mathcal{A}_0)$$

and

$$\tilde{L}_C : W^{1,p}(\dot{S}, \mathbb{R}^2) \to L^p(\mathcal{A}_0)$$

are both Fredholm operators with the same index and identical kernel.

Next we present the Fredholm index in terms of algebraic invariants. Assume C is an admissible vector bundle homomorphism

$$C : \dot{S} \times \mathbb{R}^2 \to \mathcal{A}_0,$$

which is non-degenerate at the punctures. We denote by $\Gamma^+ = \{z_j^+ \mid j \in \{1,\dots,m\}\}$ and $\Gamma^- = \{z_i^- \mid i \in \{1,\dots,m'\}\}$ the sets of positive and of negative punctures of $\dot{S}$. By $\tilde{C}_k^{\pm}$ we denote the asymptotic limits of C at the punctures. Recall the definition of the Conley-Zehnder index μ given in the introduction. We shall denote it in the following by μ_N, since the Conley-Zehnder index, which will appear later for the linearizations of the nonlinear problems, will be the normal Conley-Zehnder index.

We define the total Conley-Zehnder index of L_C and C respectively by

$$\mu_N(C) = \mu_N(L_C) = \sum_{j=1}^{m} \mu_N(\tilde{C}_j^+) - \sum_{i=1}^{m'} \mu_N(\tilde{C}_i^-).$$

Let us denote by Γ_S the set of punctures of $\dot{S}$ and by $\sharp\Gamma_S$ its cardinality. Further denote by χ_S the Euler characteristic of S. In view of Proposition 2.7, the index of L_C agrees with the index of $\tilde{L}_C$. Therefore, the following index formula is a special case of Theorem 3.3.11 due to M. Schwarz in [29].

Theorem 2.8. *Assume the vector bundle homomorphism $C \subset Hom_{\mathbb{R}}(\dot{S} \times \mathbb{R}^2, \mathcal{A}_0)$ is admissible and non-degenerate at the punctures. Then the integers* $\mathrm{Ind}(L_C)$, *$\mu_N(L_C)$, $\sharp\Gamma_S$ and χ_S are related by the following formula*

$$\mathrm{Ind}(L_C) = \mu_N(L_C) + \chi_S - \sharp\Gamma_S.$$

Next we shall study local properties of nontrivial functions $u \in \ker(L_C)$. The function $u \in C_0^{1,\alpha}(\dot{S}, \mathbb{R}^2) \cap C^{\infty}(\dot{S}, \mathbb{R}^2)$ solves the equation $L_C u = 0$. Consider a point $z \in \dot{S}$ and choose a holomorphic chart φ around z satisfying $\varphi(z) = 0$. In these coordinates the function $v := u \circ \varphi^{-1}$ satisfies an equation of the form

$$v_s + i(s,t)v_t + C(s,t)v = 0,$$

where $C(s,t)$ and $i(s,t)$ belong to $\mathcal{L}(\mathbb{R}^2)$ and $i(s,t)^2 = -\mathrm{Id}$. Assume now that z is a zero of u so that $u(z) = 0$. Then $v(z) = 0$ and we may assume without loss of generality that $i(z) = i$, identifying $\mathbb{R}^2$ with $\mathbb{C}$ via $(s,t) \mapsto s+it$. As a consequence of the similarity principle, see [1, 11, 24], there exists a holomorphic map η defined on an open neighborhood of 0 in $\mathbb{C} = \mathbb{R}^2$, satisfying $\eta(0) = 0$ and a map $z \mapsto \delta(z)$ of class $W^{1,p}_{loc}$ for all $1 < p < \infty$, such that

$$\begin{aligned} v(z) &= \delta(z)\eta(z) \\ \delta(z) \circ i &= i(z) \circ \delta(z) \\ \delta(z) &\in \mathcal{GL}_{\mathbb{R}}(\mathbb{C}). \end{aligned}$$

This implies that u admits only isolated zeroes unless it vanishes identically. We can associate with a zero z' its local index $o(z') \in \mathbb{Z}$. The local index is always a positive integer by the similarity principle, see [1, 11, 19]. We shall show that the total sum of the indices is finite, and moreover is related to some asymptotic invariants, which in turn are related to $\mu_N(L)$.

The map $u : \dot{S} \to \mathbb{C}$ has, as we will see, a special behavior at its punctures. Assume z_0 is a positive puncture and σ a special coordinate system around z_0. Then $v = u \circ \sigma$ satisfies a differential equation of the kind

$$v_s + i(s,t)v_t + C(s,t)v = 0 \quad \text{on} \quad [R, \infty) \times S^1, \tag{21}$$

where $i(s,t) \to \tilde{i}(t)$ and $C(s,t) \to \tilde{C}(t)$ in the sense already described above. Recall that $\tilde{C}(t)$ is symmetric with respect to the inner product $\omega \circ (\mathrm{Id} \times \tilde{i}(t))$ of $\mathbb{R}^2$. Moreover, by the non-degeneracy condition, the only solution of the linear Hamiltonian system

$$\dot{x}(t) = \tilde{i}(t)\tilde{C}(t)x(t) \quad \text{and} \quad x(0) = x(1)$$

is the trivial solution. Arguing as in [18], one verifies the following asymptotic behavior at a positive puncture:

$$v(s,t) = e^{\int_R^s \mu(\rho) d\rho}\left[e(t) + r(s,t)\right], \tag{22}$$

where $\mu : [R, \infty) \to \mathbb{R}$ is a smooth function converging to a negative number μ as $s \to \infty$. The number μ is an eigenvalue of the operator $A : H^{1,2}(S^1, \mathbb{R}^2) \subset L^2(S^1, \mathbb{R}^2) \to L^2(S^1, \mathbb{R}^2)$ defined by

$$Ah = -\tilde{i}(t)\frac{d}{dt}h - \tilde{C}(t)h. \tag{23}$$

The operator A is self-adjoint in $L^2(S^1, \mathbb{R}^2)$ equipped with the inner product

$$(u, v) = \int_{S^1} \omega(u(t), \tilde{i}(t)v(t))\, dt.$$

The function e in (22) is a non-vanishing eigenvector of A corresponding to the eigenvalue μ, and the remainder term r is a smooth function which converges to 0 including all its derivatives as $s \to \infty$. We can study the behavior of u at negative punctures and find similarly that the asymptotic behavior is given by an eigenvector associated now with a positive eigenvalue. From the asymptotic formula it follows immediately that the zeros of u stay away from the punctures unless, of course, u vanishes identically. Since the eigenfunction e in the asymptotic formula

(22) nowhere vanishes, the map $t \mapsto e(t)$ from S^1 into $\mathbb{R}^2 \setminus \{0\}$ defines a winding number, denoted by wind$(e) \in \mathbb{Z}$. Denote by wind(u) the difference between the sum of all winding numbers wind(e) associated with positive punctures and the sum over all winding numbers associated with negative punctures. A degree argument gives now the following Riemann-Roch type result.

Proposition 2.9. *Assume $u \in \ker(L_C)$ and $u \neq 0$. Denote by $\mathcal{O}$ the set of zeros of u. Then $\mathcal{O}$ is a finite set and*

$$\Sigma_{z \in \mathcal{O}} o(z) = \text{wind}(u). \tag{24}$$

Proof. In view of the asymptotic formula (22) we can extend the map $u : \dot{S} \to \mathbb{R}^2$ over a cap near a puncture $z \in \Gamma$ so that $\bar{u}(z) = 0$ and this isolated zero has local degree $\pm$wind(e). Equation (24) is a reformulation of the fact that the extended map $\bar{u}$ from S into $\mathbb{R}^2$ has degree zero. □

In view of the similarity principle and the fact that the local degree at the zero of a holomorphic map is non-negative we conclude from Proposition 2.9 that for $u \neq 0$ we necessarily have

$$\text{wind}(u) \geq 0.$$

At this point it is useful to recall the relationship between $\mu_N(L_C)$ and wind(u) established in [19]. To explain the necessary background let z_0 be a puncture, say a positive one. Set $v = u \circ \sigma$ for some special holomorphic coordinates near z_0. Then v is a solution of the equation (21). Let A be the self-adjoint operator defined by (23). With any eigenvector of A we can associate its winding number. It turns out that this winding number only depends on the eigenvalue. Moreover, to every integer k there exist precisely two eigenvalues (counting multiplicities) having winding number k. Following [19] we define $\alpha(A)$ to be the largest winding number occurring for a negative eigenvalue of A. We set $p(A) = 1$ if there exist precisely two negative eigenvalues with winding number equal to $\alpha(A)$. Otherwise we put $p(A) = 0$. The following proposition, proved in [19], relates the integers $\mu_N(A), \alpha(A)$ and $p(A)$.

Proposition 2.10.

$$\mu_N(A) = 2\alpha(A) + p(A).$$

Now we are in the position to compare $\mu_N(L)$ with wind(L). For $z \in \Gamma$ we denote by $\mu_N(z)$ the Conley-Zehnder index of the asymptotic problem and by $w(z)$ the winding number wind(e) for the corresponding eigenvector in formula (22) at the puncture z. Then, in view of Proposition 2.10, we have for $z \in \Gamma^+$

$$2w(z) \leqslant 2\alpha(z) = \mu_N(z) - p(z), \tag{25}$$

and for $z \in \Gamma^-$,

$$2w(z) \geq 2(\alpha(z) + p(z)) = \mu_N(z) + p(z). \tag{26}$$

Combining (25) with (26) we obtain

$$\begin{aligned} 2\mathrm{wind}(u) &= 2\sum_{z\in\Gamma^+} w(z) - 2\sum_{z\in\Gamma^-} w(z) \\ &\leqslant \sum_{z\in\Gamma^+}\mu_N(z) - \sum_{z\in\Gamma^-}\mu_N(z) - \sum_{z\in\Gamma} p(z) \end{aligned}$$

and so, using the definition of $\mu_N(L_C)$, we arrive at the inequality

$$2\,\mathrm{wind}(u) \leqslant \mu_N(L_C) - \sharp\Gamma_1. \tag{27}$$

Here Γ_1 consists of all punctures having $p(A) = 1$. From (24) and (27) we obtain

$$2\sum_{\{z|u(z)=0\}} o(z) \leqslant \mu_N(L_C) - \sharp\Gamma_1.$$

We apply this formula to the example $S = S^2$ and $\dot{S} = S\setminus\{\infty\} = \mathbb{C}$. We assume that $\mu_N(L_C) = 1$. Since $p(A) \in \{0,1\}$ we conclude from Proposition 2.10 that $\sharp\Gamma_1 = 1$ and it follows from (2) that a non trivial element u of $\ker(L_C)$ does not have any zero. By Theorem 2.8, the Fredholm index is equal to

$$\mathrm{Ind}(L_C) = \mu_N(L_C) + 1 = 2$$

so that $\ker(L_C)$ must have at least dimension 2. We claim that its dimension cannot be larger. Indeed, assume u_1, u_2 and u_3 are three linearly independent elements in the kernel of L_C. After possibly relabeling we find constants a_1, a_2 such that the function $u := u_3 + a_1u_1 + a_2u_2$ vanishes at some point $z_0 \in \dot{S}$. But the function u belongs to $\ker(L_C)$ so that it is either identically zero or does not vanish at any point. Since $u(z_0) = 0$ we conclude that $u(z) \equiv 0$, contradicting the linear independence of u_1, u_2, u_3. Summarizing we have proved

Theorem 2.11. *Consider $\mathbb{C} = S^2\setminus\{\infty\}$ and let L_C be the Fredholm operator of Theorem 2.8. If $\mu_N(L_C) = 1$, then the operator L_C is surjective and its kernel has dimension equal to 2. Every non zero element u in the kernel satisfies*

$$u(z) \neq 0 \qquad \textit{for all } z \in \mathbb{C}.$$

We point out that no genericity assumptions are required in Theorem 2.11. Our setting applies to another class of linear problems we are interested in. Assume $\dot{D} = D\setminus\Gamma$ is a punctured disk. Here Γ is a finite subset of $D\setminus\partial D$. We assume that all the punctures in Γ are negative. We can define the space $C_0^{1,\alpha}(\dot{D},\mathbb{R}^2)$ to consist of all maps $u : \dot{D} \to \mathbb{R}^2$ which are in $C^{1,\alpha}_{loc}$ and at the punctures of class $C_0^{1,\alpha}$. We have to impose a boundary condition on ∂D. Let $z \mapsto \eta(z)$ be a smooth loop, which associates with $z \in \partial D$ a one-dimensional linear subspace of $\mathbb{R}^2$. Consider the subspace $C^{1,\alpha}_{0,\eta}(\dot{D},\mathbb{R}^2)$ of $C_0^{1,\alpha}(\dot{D},\mathbb{R}^2)$ consisting of all u such that for all $z \in \partial D$

$$u(z) \in \eta(z).$$

Let C be a smooth section of $\mathrm{Hom}(\dot{D}\times\mathbb{R}^2)$ such that near a puncture z_0 (assumed to be negative)

$$D^{\beta}\left[(C(\sigma(s,t))h)\frac{\partial\sigma}{\partial s}-\tilde{C}(t)h)\right]\to 0$$

as $s\to-\infty$ for all $h\in\mathbb{R}^2$, where $\tilde{C}(t)$ is a linear map in $\mathcal{L}(\mathbb{R}^2)$, depending smoothly on t. Moreover $\tilde{C}(t)$ is symmetric with respect to the metric in $\mathbb{R}^2$

$$g_t=\omega\circ(\mathrm{Id}\times\tilde{i}(t)),$$

where $\tilde{i}$ is as described before. We also define $C_0^{\alpha}(\dot{D},\mathbb{R}^2)$ as above. We may view $z\mapsto\eta(z)$ as a loop of Lagrangian subspaces, where $\eta(z)\subset(\mathbb{R}^2,\omega)$. We denote by $\mu_M(\eta)$ its Maslov index, see [1]. Define the operator

$$L_C: C^{1,\alpha}_{0,\eta}(\dot{D},\mathbb{R}^2)\to C_0^{\alpha}(\dot{D},\mathbb{R}^2)$$

by

$$(L_C u)(z)=u_s(z)+i(z)u_t(z)+C(z)u(z).$$

The following result can be proved along the same lines as in M. Schwarz [29]. Important in the computations of the index is the addition formula under the gluing operation, Theorem 3.2.12 in [29] and the examples in Section 3.3.1 of [29]. The Maslov boundary conditions at the positive end require only a small modification of the arguments in there.

Theorem 2.12. *The operator L_C is Fredholm provided C is non-degenerate at the punctures. Moreover its Fredholm index is given by the formula*

$$\mathrm{Ind}(L_C)=\mu_M(\eta)-\sum_{i=1}^{m}\mu_N(\tilde{C}_i)-m+1,$$

where $m=\sharp\Gamma$ is the number of negative punctures and $\mu_N(\tilde{C}_i)$ is the normal Conley-Zehnder index at the ith puncture.

An analogous statement holds true for a punctured Riemann surface with boundary.

3. The Cauchy-Riemann operator for unparametrized curves

Given a pseudoholomorphic embedding the aim is to describe the curves nearby as zeros of a nonlinear section of a normal bundle. This paragraph is devoted to the conceptual framework of the analytical setting used later on. Already in his paper [14] Gromov describes the neighboring curves near an embedded closed curve as a section of the normal bundle, which is a solution of his normal Cauchy-Riemann operator. His operator is in principle equivalent to our operator which however is defined on a curve which is not closed. In his paper [31], R. Ye formalized Gromov's approach.

We consider a compact Riemann surface (S,j) equipped with the complex structure j, hence $j^2=-\mathrm{Id}$. If $\Gamma\subset S$ is a non-empty finite subset we abbreviate by $\dot{S} = S \setminus \Gamma$ the punctured surface. By $\Lambda \to \dot{S}$ we denote the 1-dimensional vector bundle of 2-vectors on $\dot{S}$. The fiber Λ_z over $z \in \dot{S}$ is spanned by $h \wedge (jh)$, where $h \neq 0$ belongs to $T_z\dot{S}$. Let (W, J) be a smooth 4-manifold equipped with the almost complex structure J, hence $J^2 = -\mathrm{Id}$. By $\Omega \to W$ we shall denote the smooth vector bundle over W whose fiber Ω_m over the point $m \in W$ consists of the span of all 2-vectors $v \wedge w$ with tangent vectors $v, w \in T_mW$. Recall that $v \wedge w = -w \wedge v$. Since $\dim T_mW = 4$, the dimension of the fiber Ω_m is equal to 6. The almost complex structure J defines a fiberwise linear map $\check{J} : \Omega \to \Omega$ defined at every point $m \in W$ by

$$\check{J} : v \wedge w \mapsto (Jv) \wedge (Jw)$$

and extended linearly. One verifies readily that $\check{J}$ is a well-defined linear map $\Omega_m \to \Omega_m$ and is an involution on Ω satisfying $\check{J}^2 = \mathrm{Id}$. Consequently, the eigenvalues of $\check{J} : \Omega_m \to \Omega_m$ are ± 1 and so the fiber Ω_m splits,

$$\Omega_m = \Omega_m^- \oplus \Omega_m^+,$$

into the direct sum of the eigenspace Ω_m^- belonging to -1 and the eigenspace Ω_m^+ belonging to 1. One verifies readily (e.g. by considering a canonical basis e, Je, f, Jf on T_mW) that $\dim \Omega_m^- = 2$ and we obtain the smooth 2-dimensional bundle

$$\Omega^- \to W$$

having the fiber Ω_m^- over $m \in W$. Next we introduce the 2-dimensional vector bundle

$$\Xi \to \dot{S} \times W$$

whose fiber over the point $(z, m) \in \dot{S} \times W$ is given by $\Xi_{(z,m)} = \mathrm{Hom}_{\mathbb{R}}(\Lambda_z, \Omega_m^-)$. With a given map $u : \dot{S} \to W$ we associate its graph defined by $\mathrm{graph}(u) : \dot{S} \to \dot{S} \times W$, explicitly $z \mapsto (z, u(z))$, and introduce the pull-back bundle

$$\mathrm{graph}\,(u)^*\Xi \to \dot{S}$$

whose fiber over $z \in \dot{S}$ is defined by $\mathrm{graph}(u)^*\Xi_z := \mathrm{Hom}_{\mathbb{R}}(\Lambda_z, \Omega_{u(z)}^-)$. Now define for $z \in \dot{S}$ the linear map

$$\bar{\eta}(u)(z) : \Lambda_z \to \Omega_{u(z)}^-$$

by

$$\bar{\eta}(u)(z)(h \wedge k) := (Tu \cdot h) \wedge (Tu \cdot k) - (JTu \cdot h \wedge JTu \cdot k).$$

Here we used the abbreviations $Tu \equiv Tu(z)$ and $J \equiv J(u(z))$. Recalling $J^2 = -\mathrm{Id}$ one verifies that $\check{J}\,[\bar{\eta}(u)(z)(h \wedge k)] = -\bar{\eta}(u)(z)(h \wedge k)$ so that indeed $\bar{\eta}(u)(z) \in \mathrm{Hom}(\Lambda_z, \Omega_{u(z)}^-)$. Consequently, $\bar{\eta}(u)$ is a section of the bundle $\mathrm{graph}(u)^*\Xi \to \dot{S}$. We abbreviate this bundle in the following by

$$u^*\Xi = \mathrm{graph}(u)^*\Xi \to \dot{S}. \tag{28}$$

In order to interpret the condition $\bar{\eta}(u) = 0$ geometrically we observe that the condition $a \wedge b = (Ja) \wedge (Jb)$ is equivalent to the condition $Ja, Jb \in \text{span}\{a, b\}$. It follows immediately

Proposition 3.1. *Consider an immersion $u : \dot{S} \to W$. Then*

$$\bar{\eta}(u)(z) = 0 \ \text{ if and only if } \ J(u(z))Tu(z)(T_z\dot{S}) = Tu(z)(T_z\dot{S}).$$

Hence $\bar{\eta}(u) = 0$ characterizes the complex embeddings into (W, J). For a pseudoholomorphic curve $u : \dot{S} \to W$ which, by definition, satisfies $Tu \circ j = J(u) \circ Tu$ we conclude from the proposition that $\bar{\eta}(u) = 0$. Conversely, if for an embedding $\bar{\eta}(u) = 0$, then $Tu \circ \bar{j} = J(u) \circ Tu$ for the almost complex structure $\bar{j}$ induced on the punctured surface $\dot{S}$ by

$$\bar{j} = (Tu)^{-1} \circ J(u) \circ (Tu).$$

We shall use now the operator $u \to \bar{\eta}(u)$ in order to describe the complex curves near a given embedded pseudoholomorphic curve. We consider the special case

$$W = \dot{S} \times \mathbb{R}^2$$

to which the analysis later on will be reduced. Corresponding to the splitting of the tangent space $T_{(z,x,y)} = T_z\dot{S} \times \mathbb{R}^2$ we represent the almost complex structure $J = J(z, x, y)$ in matrix form

$$J = \begin{pmatrix} j & \tilde{\Delta} \\ \Delta & i \end{pmatrix},$$

where we have abbreviated $j = j(z, x, y) : T_z\dot{S} \to: T_z\dot{S}$ and $\tilde{\Delta} = \tilde{\Delta}(z, x, y) : T_z\dot{S} \to \mathbb{R}^2$ and so on. At the points $(z, 0, 0) \in \dot{S} \times \{0\}$ we require that

$$J(z, 0, 0) = \begin{pmatrix} j(z) & 0 \\ 0 & i(z) \end{pmatrix}$$

with the complex structure $j(z)$ on $\dot{S}$ from above and a complex multiplication $i(z)$ on $\mathbb{R}^2$. In particular, $\tilde{\Delta}(z, 0, 0) = 0$ and $\Delta(z, 0, 0) = 0$ for all $z \in \dot{S}$.

As a reference map we take the pseudoholomorphic curve

$$u_0 : \dot{S} \to \dot{S} \times \mathbb{R}^2, \quad z \mapsto (z, 0),$$

The nearby curves will be defined as graphs over u_0. Hence we consider maps of the form

$$u : \dot{S} \to \dot{S} \times \mathbb{R}^2, \quad z \mapsto (z, v(z))$$

with $v : \dot{S} \to \mathbb{R}^2$, so that $u = \text{graph}(v)$. In particular, $u_0 = \text{graph}(0)$. Recall now the notation (28). If $v : \dot{S} \to \mathbb{R}^2$, we define the section

$$H(v) = \bar{\eta}(\text{graph}(v))$$

of the bundle $\text{graph}(v)^*\Xi \to \dot{S}$. Since the reference section $u_0 = \text{graph}(0)$ is a pseudoholomorphic curve, we conclude from Proposition 3.1 that

$$H(0) = 0.$$

If $v : \dot{S} \to \mathbb{R}^2$ is small and solves

$$H(v) = 0$$

then $\text{graph}(v) = u : \dot{S} \to \dot{S} \times \mathbb{R}^2$ is a complex submanifold near $\dot{S} \times \{0\}$, in view of Proposition 3.1. In order to apply the implicit function theorem later on we study the linearization of H at the point $v = 0$. Given $v : \dot{S} \to \mathbb{R}^2$ we shall differentiate the curve $\tau \to H(\tau v)$ in τ at the point $\tau = 0$.

Differentiating $J(z, x, y)^2 = -\text{Id}$ in the (x, y) variable and denoting this differentiation by a prime, we obtain, in view of $\tilde{\Delta}(z, 0, 0) = 0$, $\Delta(z, 0, 0) = 0$ at the point $(z, 0, 0)$ the identities

$$\begin{aligned} (j'(z)h)j(z) + j(z)(j'(z)h) &= 0 \\ (\Delta'(z)h)j(z) + i(z)(\Delta'(z)h) &= 0, \end{aligned} \tag{29}$$

for all $h \in \mathbb{R}^2$ and $z \in \dot{S}$. In order to recall the Cauchy-Riemann operator from the previous paragraph we take the vector bundle $\mathcal{A}_0 \to \dot{S}$ whose fiber over $z \in \dot{S}$ consists of the complex antilinear maps $\varphi : (T_z\dot{S}, j(z)) \to (\mathbb{R}^2, i(z))$. If $v : \dot{S} \to \mathbb{R}^2$ is given, the linear map $(Lv)(z) : T_z\dot{S} \to \mathbb{R}^2$ is defined by

$$(Lv)(z) = Tv(z) + i(z)Tv(z)j(z) + [\Delta'(z)v(z)]\, j(z).$$

In view of (29) the map $(Lv)(z)$ is complex antilinear so that Lv is a section of $\mathcal{A}_0$.

Proposition 3.2. *If $v : \dot{S} \to \mathbb{R}^2$ and $z \in \dot{S}$ we define*

$$(H'(0)v)(z) = \frac{d}{d\tau} H(\tau v)(z)_{|\tau=0}.$$

Then

$$(H'(0)v)(z)(h \wedge k) = [h, 0] \wedge [0, (Lv)(z) \cdot k] - [k, 0] \wedge [0, (Lv)(z) \cdot h]$$

for all $h, k \in T_z\dot{S}$, where $[a, b] \in T_z\dot{S} \times \mathbb{R}^2$ are the tangent vectors.

Proof. Since the set of punctures is not empty there exists a nowhere vanishing vector field V on $\dot{S}$. We consider the arc $\tau \mapsto H(\tau v)$, $\tau \in \mathbb{R}$ and recall that

$$\begin{aligned} H(\tau v)(V \wedge jV) = {} & [V, \tau Tv \cdot V] \wedge [jV, \tau Tv \cdot jV] \\ & - J(\text{gr}(\tau v))[V, \tau Tv \cdot V)] \wedge J(\text{gr}(\tau v))[jV, \tau Tv \cdot jV], \end{aligned}$$

abbreviating $\text{gr}(\tau v) \equiv \text{graph}(\tau v)$. We now differentiate $\tau \mapsto H(\tau v)(z)(V(z) \wedge j(z)V(z))$ at the point $\tau = 0$ and abbreviate $J(z) = J(z, 0, 0)$, $j(z) = j(z, 0, 0)$ and so on. In addition, we denote by $A(z, x, y)$ the first column of J,

$$A(z, x, y) = \begin{pmatrix} j(z, x, y) \\ \Delta(z, x, y) \end{pmatrix}.$$

Denoting by $'$ the derivative with respect to the variables (x, y) we obtain

$$\begin{aligned}(H'(0)v)(z)(V(z) \wedge j(z)V(z))& \\ &= [0, Tv(z) \cdot V(z)) \wedge [j(z)V(z), 0] \\ &\quad + [V(z), 0] \wedge [0, Tv(z) \cdot j(z)V(z)] \\ &\quad - J(z)[0, Tv(z) \cdot V(z)] \wedge J(z)[j(z)V(z), 0] \\ &\quad - J(z)[V(z), 0] \wedge J(z)[0, Tv(z) \cdot j(z)V(z)] \\ &\quad + ((A'(z)v(z))V(z)) \wedge [V(z), 0] \\ &\quad - [j(z)V(z), 0] \wedge (A'(z)v(z))j(z)V(z).\end{aligned} \tag{30}$$

In order to simplify the formula we claim that

$$[h, 0] \wedge [k, 0] = [jh, 0] \wedge [jk, 0] \tag{31}$$

for $h, k \in T_z\dot{S}$. Indeed, representing $k = ah + bjh$ for $a, b \in \mathbb{R}$ we obtain for the left-hand side $b[h, 0] \wedge [jh, 0]$. We also obtain the same result for the right-hand side. From (29) and (31) we deduce

$$\begin{aligned}[j(z)V(z), 0] \wedge [(j'(z)h)j(z)V(z), 0]& \\ &= -[j(z)V(z), 0] \wedge [j(z)(j'(z)h)V(z), 0] \\ &= -[V(z), 0] \wedge [(j'(z)h)V(z), 0]\end{aligned} \tag{32}$$

for every $h \in T_z\dot{S}$. Using (29) and (32) we can rewrite the formula (30) as follows

$$\begin{aligned}\big((H'(0)v\big)(z)(V(z) \wedge (j(z)V(z))& \\ &= [V(z), 0] \wedge [0, -i(z)(Lu)(z) \cdot V(z)] \\ &\quad - [j(z)V(z), 0] \wedge [0, (Lu)(z) \cdot V(z)].\end{aligned}$$

A reformulation gives the desired formula and finishes the proof of Proposition 3.2. □

In order to derive a more convenient representation of $H'(0)$ we introduce the vector bundle $\Theta \to \dot{S}$ whose fiber Θ_z over $z \in \dot{S}$ is the subspace of 2-vectors in the tangent space $T_z\dot{S} \times \mathbb{R}^2$ on which $\check{J}$ acts as multiplication by -1. By $\Xi_0 \to \dot{S}$ we denote the vector bundle having fiber $\mathrm{Hom}_{\mathbb{R}}(\Lambda_z, \Theta_z)$ over $z \in \dot{S}$. We define a smooth vector bundle map $\alpha : \mathcal{A}_0 \to \Xi_0$, $a \mapsto \alpha(a)$ by

$$\alpha(a)(z)(h \wedge k) = [h, 0] \wedge [0, a(z)k] - [k, 0] \wedge [0, a(z)h].$$

Then α is a natural isomorphism of vector bundles, and for the derivative $H'(0)$ we find

$$H'(0)v = \alpha(Lv),$$

which allows us to identify $H'(0)$ with L. Thus the results of the previous section concerning the Fredholm theory will be applicable to $H'(0)$ once a suitable functional analytic set up is specified in which $H'(0)$ shows the proper behavior near the punctures.

4. The asymptotic model problem and special coordinates

Given a compact surface (S, j) and the punctures $\Gamma \subset S$, we consider a finite energy surface $\tilde{u} : \dot{S} = S \backslash \Gamma \to \mathbb{R} \times M$ which is an embedding and which is non-degenerate at the punctures. The aim of this section is the construction of a trivialization $\Phi : \dot{S} \times B_\varepsilon(0) \to \mathbb{R} \times M$ of the normal bundle of $\tilde{u}$ in $\mathbb{R} \times M$ satisfying $\Phi(z, 0) = \tilde{u}(z)$ for $z \in \dot{S}$. Here $B_\varepsilon(0) \subset \mathbb{R}^2$ is a small open ball centered at the origin. In the analysis later on we shall need to control the map Φ near the punctures $z \in \Gamma$. For this purpose we shall use the precise exponential asymptotic behavior of $\tilde{u}$ in a neighborhood of a non-degenerate puncture described in convenient coordinates in [18].

Before we get involved in computations in local coordinates we recall the notation and introduce the natural metrics . If (M, λ) is a three manifold equipped with the contact form λ, then the tangent space at $m \in M$ splits,

$$T_m M = \mathbb{R} X_m \oplus \xi_m.$$

The Reeb vector field X on M is defined by $i_X \lambda = 1$ and $i_X d\lambda = 0$, and $\xi \to M$ is the contact structure defined by $\xi = \ker \lambda$. It is a symplectic plane bundle with respect to the 2-form $d\lambda$. As usual the map

$$\pi : T_m M \to \xi_m$$

is the projection onto the contact plane along the Reeb vector field. We choose an almost complex structure J on ξ which is compatible with $d\lambda$ in the sense that

$$g_m(a, b) := d\lambda(a, Jb), \quad a, b \in \xi_m$$

defines an inner product on ξ_m. On the 4-manifold $\mathbb{R} \times M$ we shall use the splitting of the tangent space

$$T_{(a,m)}(\mathbb{R} \times M) = T_a\mathbb{R} \times T_m M = \mathbb{R} \times T_m M$$

identifying $T_a\mathbb{R}$ with $\mathbb{R}$. Associated with J we have the special almost complex structure $\hat{J}$ on the 4-manifold $\mathbb{R} \times M$ acting on the vectors $(\alpha, k) \in T_{(a,m)}(\mathbb{R} \times M)$ as

$$\hat{J}(a, m)(\alpha, k) = \big(-\lambda_m(k), J_m \pi k + \alpha X_m\big).$$

The inner product on $T_{(a,m)}(\mathbb{R} \times M)$ is defined in a natural way by

$$\langle (\alpha, k), (\beta, h) \rangle = \alpha \cdot \beta + \lambda(k)\lambda(h) + d\lambda(\pi k, J\pi h),$$

inducing the norm $|\cdot|$,

$$|(\alpha, k)|^2 := \langle (\alpha, k), (\alpha, k) \rangle.$$

Consider, for simplicity, the embedded pseudoholomorphic curve $\tilde{u} = (a, u) : \mathbb{C} \to \mathbb{R} \times M$. It satisfies $T\tilde{u} \circ i = \hat{J}(\tilde{u}) \circ T\tilde{u}$ or, with $z = s + it$,

$$\tilde{u}_s + \hat{J}(\tilde{u})\tilde{u}_t = 0.$$

Explicitly, the function a and the map u are related as follows:

$$a_s = \lambda(u_t),\ a_t = -\lambda(u_s),\ \pi u_s + J(u)\pi u_t = 0.$$

At points where $\pi Tu(z) \neq 0$ we define the unit vectors $n(z)$ and $m(z) \in T_{\tilde{u}(z)}(\mathbb{R} \times M)$ which are orthogonal to $T_{\tilde{u}(z)}\tilde{u}(\mathbb{C})$ by

$$\begin{aligned}
n(z) &= \frac{1}{|\tilde{u}_s|}\left(|\pi u_s|, -\lambda(u_t)\frac{\pi u_s}{|\pi u_s|} + \lambda(u_s)\frac{\pi u_t}{|\pi u_t|}\right),\\
m(z) &= \hat{J}(\tilde{u}(z)) \cdot n(z)\\
&= \frac{1}{|\tilde{u}_t|}\left(0, |\pi u_s| \cdot X - \lambda(u_s)\frac{\pi u_s}{|\pi u_s|} - \lambda(u_t)\frac{\pi u_t}{|\pi u_t|}\right).
\end{aligned}$$

Lemma 4.1. *Assume $\pi Tu(z) \neq 0$, then*

(i) $|n(z)| = |m(z)| = 1$
(ii) $\langle n(z), m(z)\rangle = 0$
(iii) $n(z), m(z) \in \left(T_{\tilde{u}(z)} \textit{image}\ (\tilde{u})\right)^{\perp} \subset T_{\tilde{u}(z)}(\mathbb{R} \times M)$.

Proof. Observing that $|\hat{J}v| = |v|$ and $\langle \hat{J}v, v\rangle = 0$ for all $v \in T(\mathbb{R} \times M)$ and using the fact that $T_{\tilde{u}(z)}\tilde{u} = \text{span}\{\tilde{u}_s, \tilde{u}_t\}$ with

$$\begin{aligned}
\tilde{u}_s &= \left(-\lambda(u_t), \lambda(u_s)X(u) + \pi u_s\right)\\
\tilde{u}_t &= \left(-\lambda(u_s), \lambda(u_t)X(u) + \pi u_t\right),
\end{aligned}$$

the lemma follows readily from the definitions and the equations for $\tilde{u}$. □

In order to introduce canonical coordinates in a tubular neighborhood of a periodic orbit $P \subset M$ of the Reeb vector field X_λ we denote by λ_0 the standard contact form on $S^1 \times \mathbb{R}^2$, where

$$\lambda_0 = d\vartheta + xdy$$

and (ϑ, x, y) are the coordinates on $S^1 \times \mathbb{R}^2$. Let us recall the following lemma from [18].

Lemma 4.2. *Let (M, λ) be a 3-dimensional contact manifold, and assume (x, T) is a T periodic solution of the corresponding Reeb vector field $\dot{x} = X_\lambda(x)$ on M. Let τ be the minimal period such that $T = k\tau$ for some positive integer k. Then there exist a tubular neighborhood $U \subset S^1 \times \mathbb{R}^2$ of $S^1 \times \{0\}$ and an open neighborhood $V \subset M$ of $P = \{x(t) | t \in \mathbb{R}\}$, and a diffeomorphism $\phi : U \to V$ mapping $S^1 \times \{0\}$ onto P such that*

$$\phi^*\lambda = f\lambda_0$$

with a positive smooth function $f : U \to \mathbb{R}$ satisfying

$$f(\vartheta, 0, 0) = \tau \ \ \textit{and} \ \ df(\vartheta, 0, 0) = 0$$

for all $\vartheta \in S^1$.

These coordinates will be very useful in describing the behavior of a finite energy surface near a puncture p. As an aside we would like to point out that there are infinitely many different isotopy classes of such coordinates. Indeed, we observe that given coordinates ϕ_1 and ϕ_2 as above we find a rotation $r: S^1\times\mathbb{R}^2\to S^1\times\mathbb{R}^2$ which in the universal covering has the form $(\vartheta,x,y)\mapsto(\vartheta+c,x,y)$ for some $c\in\mathbb{R}$ and which satisfies for all ϑ

$$\phi_1\circ r(\vartheta,0,0)=\phi_2(\vartheta,0,0).$$

Define $\alpha=\phi_2^{-1}\circ\phi_1\circ r$ and denote by $\alpha'(\vartheta)$ the derivative with respect to the $\mathbb{R}^2$-variable at $(\vartheta,0,0)$. Then, for every ϑ, the map

$$\alpha'(\vartheta):\mathbb{R}^2\to\mathbb{R}^2$$

is a linear isomorphism preserving the orientation. Hence we obtain a map

$$S^1\to\mathcal{GL}^+(\mathbb{R}^2):\vartheta\mapsto\alpha'(\vartheta)$$

which is homotopic to a map $S^1\to U(1)=S^1$, as usual identifying $\mathbb{R}^2$ with $\mathbb{C}$. The homotopy classes of the latter maps are classified by their degrees or winding numbers. Hence we can define a relative invariant $d(\phi_2,\phi_1)\in\mathbb{Z}$. It is the winding number of a loop $S^1\to U(1)$ homotopic to α' associated with $\phi_2^{-1}\circ\phi_1\circ r$. The following lemma will not be used later on.

Lemma 4.3. *Let (M,λ) and ϕ be as in Lemma 4.2. For a given integer k there exist coordinates ψ having the properties as described in Lemma 4.2 and satisfying $d(\psi,\phi)=k$.*

Proof. Identify $\mathbb{R}^2$ with $\mathbb{C}$ and $S^1=\mathbb{R}/\mathbb{Z}$ with the unit circle in $\mathbb{C}$. We define the vector bundle isomorphism $\eta: S^1\times\mathbb{C}\to S^1\times\mathbb{C}$ by

$$\eta(z,h)=(z,z^{-k}h),\quad |z|=1.$$

Consider $\eta^*\lambda_0$. Clearly $\eta^*\lambda_0(\vartheta,0,0)=\lambda_0(\vartheta,0,0)$. Using now the usual Moser-type proof of Gray's theorem, see [1], [13] and [25], we find a diffeomorphism χ defined near $S^1\times\{0\}$ which is isotopic to the inclusion and has the properties

$$\chi^*\eta^*\lambda_0=g\lambda_0 \text{ and } \chi(\vartheta,0,0)=(\vartheta,0,0).$$

Here g is a smooth function defined near $S^1\times\{0\}$ satisfying $g(\vartheta,0,0)=1$ and $dg(\vartheta,0,0)=0$. Take the coordinates ϕ from Lemma 4.2 and define the map

$$\psi=\phi\circ\eta\circ\chi.$$

Then ψ has the same properties as those listed for ϕ in Lemma 4.2 and $d(\psi,\phi)=k$. This concludes the proof of the lemma. □

Let us recall the properties of the special coordinates in more detail. For simplicity we abbreviate $\lambda=f\lambda_0$, where f is given as in Lemma 4.2. Without loss of generality we work in $\mathbb{R}\times\mathbb{R}^2$ with coordinates (ϑ,x,y) instead of $S^1\times\mathbb{R}^2$. All functions are 1-periodic in the variable ϑ. The Reeb vector field X associated with λ is represented in $\mathbb{R}^3$ by

$$X(\vartheta, x, y) = \begin{pmatrix} X_1 \\ X_2 \\ X_3 \end{pmatrix} = \frac{1}{f^2} \begin{pmatrix} f + xf_x \\ f_y - xf_\vartheta \\ -f_x \end{pmatrix}. \tag{33}$$

The contact plane ξ_m at $m = (\vartheta, x, y)$ is spanned by the vectors $g_1 = (0, 1, 0)$ and $g_2 = (-x, 0, 1)$. The diffeomorphism $\phi : U \to V$ of Lemma 4.2 induces a complex multiplication $J : \xi \to \xi$ compatible with $d\lambda$. In the basis $\{g_1, g_2\}$ of ξ we shall represent J as follows:

$$\begin{aligned} J(\vartheta, x, y)(g_1) &= \alpha g_1 + \beta g_2 \\ J(\vartheta, x, y)(g_2) &= \gamma g_1 + \delta g_2. \end{aligned}$$

The above coefficients $\alpha, \beta, \gamma, \delta$ are smooth and 1-periodic in the variable ϑ. With the complex multiplication $J : \xi \to \xi$ we have associated the usual almost complex structure $\hat{J}$,

$$\hat{J}(a, m)(h, v) = \big(-\lambda(m)(v), J(m)\pi v + hX(m)\big),$$

where $(a, m) \in \mathbb{R} \times S^1 \times \mathbb{R}^2$, $m = (\vartheta, x, y)$, and $(h, v) \in \mathbb{R} \times \mathbb{R}^3$. The map $\pi : \mathbb{R}^3 \to \xi$ is the projection along X onto ξ and so is given by

$$\pi(v) = v - \lambda(v)X.$$

The almost complex structure $\hat{J}$ does not depend on the variable a. It has the following matrix representation with respect to the standard basis of $\mathbb{R}^4$,

$$\hat{J}(\vartheta, x, y) = \begin{pmatrix} 0 & -f & 0 & -xf \\ X_1 & xf(\beta X_2 + \delta X_3) & -x\beta & x^2 f(\beta X_2 + \delta X_3) - x\delta \\ X_2 & -f(\alpha X_2 + \gamma X_3) & \alpha & -xf(\alpha X_2 + \gamma X_3) + \gamma \\ X_3 & -f(\beta X_2 + \delta X_3) & \beta & -xf(\beta X_2 + \delta X_3) + \delta \end{pmatrix} \tag{34}$$

where the coefficients of the matrix are evaluated at the point (ϑ, x, y).

Assume now that we are given a special or generalized embedded finite energy surface $\tilde{u} : \dot{S} \to \mathbb{R} \times M$. We denote by $\tilde{J}$ the given almost complex structure on $\mathbb{R} \times M$. Our aim is to construct a trivialization

$$\Phi : \dot{S} \times B_\varepsilon \to \mathbb{R} \times M$$

of the normal bundle of $\tilde{u}$ in $\mathbb{R} \times M$, where $B_\epsilon = \{z = (x, y) \in \mathbb{R}^2 \mid |z| < \varepsilon\}$ for some ϵ. The pseudoholomorphic curve $\tilde{u}$ is represented by

$$\Phi(z, 0) = \tilde{u}(z)$$

for all $z \in \dot{S}$. We cannot expect the map Φ to be an embedding. This is due to the fact that the asymptotic limits might be multiply covered. However, Φ will be an immersion inducing an embedding of a neighborhood of $\dot{S} \times \{0\}$. At the punctures, the neighborhood, if projected to $\mathbb{R}^2$, shrinks exponentially fast to 0.

We start with the construction of Φ near a puncture $z_0 \in \Gamma \subset S$ which we assume for notational convenience to be positive. Near z_0 we describe $\dot{S}$ in our special cylindrical holomorphic coordinates

$$\sigma : [0, \infty) \times S^1 \to D \setminus \{z_0\} \subset \dot{S}$$

mapping onto the small punctured disc $D \setminus \{z_0\}$ around $\{z_0\}$. Recall that $z_0 = \lim \sigma(s,t)$ as $s \to \infty$. We denote by P_0 the non-degenerate periodic orbit which asymptotically is associated with the puncture according to Theorem 1.3 by $\lim u(\sigma(s,t)) = P_0(Tt)$ as $s \to \infty$. The period T is not necessarily minimal and we set $T = k\tau$, where τ is the minimal period of P_0 and k a positive integer. Denote by

$$\psi : S^1 \times V \subset S^1 \times \mathbb{R}^2 \to U \subset M$$

the special coordinates introduced in Lemma 4.2 describing a tubular neighborhood U of P_0 in M such that $\psi(S^1 \times \{0\}) = P_0$. We introduce the coordinate map

$$\Psi = \mathrm{Id}_{\mathbb{R}} \times \psi^{-1} : \mathbb{R} \times U \to \mathbb{R} \times S^1 \times \mathbb{R}^2.$$

In these local coordinates, the given pseudoholomorphic curve $\tilde{u}$ is represented by

$$\tilde{v}(s,t) = \Psi \circ \tilde{u} \circ \sigma(s,t)$$

for $s \geq s_0$. Working on the covering $\mathbb{R}$ of S^1, the map $\tilde{v}$ becomes a function $\tilde{v} = (a, v) : [s_0, \infty) \times \mathbb{R} \to \mathbb{R}^4$ and we denote the coefficients in the following way:

$$\begin{aligned}\tilde{v}(s,t) &= (a(s,t), v(s,t)) \\ &= (a(s,t), \vartheta(s,t), x(s,t), y(s,t)).\end{aligned}$$

The functions $a, x, y : \mathbb{R}^2 \to \mathbb{R}$ are 1-periodic in t while $\vartheta : \mathbb{R}^2 \to \mathbb{R}$ satisfies $\vartheta(s, t+1) = \vartheta(s,t) + k$. Along the image of the periodic orbit $P_0 \subset M$ under ψ^{-1}, namely $\mathbb{R} \times \{0\} \subset \mathbb{R}^3$, the Reeb vector field is given by $X(\vartheta, 0, 0) = \frac{1}{\tau}(1, 0, 0)$ so that the 1-periodic solution $P_0(Tt)$, for $0 \leqslant t \leqslant 1$, is mapped onto the solution $\hat{P}_0(kt) = (kt, 0, 0)$, $0 \leqslant t \leqslant 1$, of the Reeb vector field in the local coordinates. Reformulating the condition of non-degeneracy of P_0 in our local coordinates we linearize $T \cdot X$ along $\hat{P}_0(kt)$ and project it onto the contact planes along $\hat{P}_0(kt)$, which are the (x,y)-planes $\mathbb{R}^2$, and find the 1-periodic matrix function

$$F_\infty(t) := \frac{1}{\tau^2} \begin{pmatrix} f_{xy} & f_{yy} \\ -f_{xx} & -f_{xy} \end{pmatrix} \in \mathcal{L}(\mathbb{R}^2). \tag{35}$$

The coefficients of the matrix $F_\infty(t)$ are evaluated at $(\vartheta, x, y) = (kt, 0, 0) = \hat{P}_0(kt)$. Since by assumption P_0 is non-degenerate, the boundary value problem for $z(t) \in \mathbb{R}^2$,

$$\dot{z}(t) = F_\infty(t) z(t) \quad \text{and} \quad z(0) = z(1)$$

admits only the trivial solution $z(t) \equiv 0$. If $J(m)$ is the complex structure in the contact plane ξ_m we define the 1-periodic linear maps along the periodic solution $\hat{P}_0(kt)$ by

$$J(t) := J(kt, 0, 0)$$

and introduce

$$S_\infty(t) = J(t)F_\infty(t).$$

With the 1-periodic matrix function $S_\infty(t)$ we associate the asymptotic operator $A_\infty : H^{1,2}(S^1, \mathbb{R}^2) \subset L^2(S^1, \mathbb{R}^2) \to L^2(S^1, \mathbb{R}^2)$ defined by

$$A_\infty := -J(t)\frac{d}{dt} + S_\infty(t).$$

The operator A_∞ is self-adjoint with respect to the inner product of $L^2(S^1, \mathbb{R}^2)$,

$$(h_1, h_2) := \int_0^1 \omega(h_1(t), J(t)h_2(t))\, dt,$$

where $\omega = dx \wedge dy$ on $\mathbb{R}^2$. Recall that, in view of the admissibility assumption, $\omega(a, J(t)b)$ is an inner product on $\mathbb{R}^2$ for every t. We recall from [18] the asymptotic behavior of $\tilde{v} : [s_0, \infty) \times S^1 \to \mathbb{R}^4$ near the puncture $z_0 \in \Gamma$.

Theorem 4.4. *Consider a non-degenerate finite energy surface $\tilde{u} : \dot{S} \to \mathbb{R} \times M$ and assume $z_0 \in \Gamma$ is a positive puncture. Near $z_0 = \lim_{s\to\infty} \sigma(s,t)$, the map $\tilde{v} = \Psi \circ \tilde{u} \circ \sigma : \mathbb{R} \times S^1 \to \mathbb{R}^4$ has, for large s, the following properties. Either there exists a constant $c \in \mathbb{R}$ such that*

$$\tilde{v}(s,t) = (a(s,t), \vartheta(s,t), x(s,t), y(s,t)) = (Ts + c, kt, 0, 0),$$

$(s,t) \in [0, \infty) \times \mathbb{R}$ or, combining $z(s,t) = (x(s,t), y(s,t)) \in \mathbb{R}^2$,

$$a(s,t) = Ts + c + \hat{a}(s,t)$$
$$\vartheta(s,t) = kt + d + \hat{\vartheta}(s,t)$$
$$z(s,t) = e^{\int_{s_0}^{s} \mu(\rho)d\rho}[e(t) + \hat{z}(s,t)].$$

The functions $\hat{a}(s,\cdot)$, $\hat{\vartheta}(s,\cdot)$ and $\hat{z}(s,\cdot)$ converge to 0 as $s \to \infty$, while $\mu(\rho) \to \mu$ and $\mu < 0$ as $\rho \to \infty$. The precise convergence is as follows:

$$|D^\alpha \hat{a}(s,t)| \leqslant M_\alpha e^{-\lambda s}$$
$$\left|D^\alpha \hat{\vartheta}(s,t)\right| \leqslant M_\alpha e^{-\lambda s}$$

for all derivatives $\alpha = (\alpha_1, \alpha_2)$, with constants M_α and a constant $\lambda > 0$. Moreover,

$$D^\alpha \hat{z}(s,t) \to 0$$
$$D^j \left[\mu(s) - \mu\right] \to 0$$

as $s \to \infty$ uniformly in t, for all derivatives. The number μ is a negative eigenvalue of the operator A_∞ and $e(t) \in \mathbb{R}^2$ is a corresponding eigenvector, hence satisfies $e(t) = e(t+1) \neq 0$ for all t.

In order to construct Φ near the positive puncture z_0 we consider the second alternative in Theorem 4.4 in which $\pi Tu(z) \neq 0$ for z near z_0. We take the unit vector fields $n(s,t)$ and $m(s,t) = \hat{J}(\tilde{v}(s,t))n(s,t)$ which are normal to the tangent space $T_{\tilde{v}(s,t)}\tilde{v}$ in $T_{\tilde{v}(s,t)}\mathbb{R}^4 = \mathbb{R}^4$ as described above. Recall that the inner product is given by

$$\langle(\alpha,k),(\beta,h)\rangle = \alpha\beta + \lambda(k)\lambda(h) + d\lambda(\pi k, J\pi h)$$

for $(\alpha,k),(\beta,h) \in T_{\tilde{v}(s,t)}\mathbb{R}^4$, where λ is of the special form $f\lambda_0$. We define $\Phi : \dot{S} \times B_\varepsilon(0) \to \mathbb{R} \times M$ for a small ball $B_\varepsilon(0) \subset \mathbb{R}^2$ centered at the origin by

$$\Phi(\sigma(s,t),x,y) = \Psi^{-1}[\tilde{v}(s,t) + xn(s,t) + ym(s,t)].$$

Then $\Phi(\sigma(s,t),0,0) = \Psi^{-1}(\tilde{v}(s,t)) = \tilde{u} \circ \sigma(s,t)$.

On $\mathbb{R} \times M$ we are given the distinguished almost complex structure $\tilde{J}$. Near the ends it has the form $\tilde{J}(\alpha,h) = \big(-\lambda(h), J\pi h + \alpha X\big)$ for $(\alpha,h) \in T(\mathbb{R} \times M)$. This structure $\tilde{J}$ induces the special almost complex structure $\bar{J}$ on $\dot{S} \times \mathbb{R}^2$ with $\bar{J}(z,x,y) \in \mathcal{L}(T_z\dot{S} \times \mathbb{R}^2)$ defined by

$$\bar{J} = (T\Phi)^{-1} \circ \tilde{J}(\Phi) \circ T\Phi.$$

We shall represent $\bar{J}$ in local coordinates. We describe a punctured neighborhood of the (positive) puncture z_0 using the special cylindrical coordinates $\sigma : [0,\infty)\times\mathbb{R} \to D\backslash\{z_0\} \subset \dot{S}$ satisfying $\sigma(s,t) \to z_0$ as $s \to \infty$. In $\mathbb{R}\times M$ we choose the above special coordinates $\Psi = \mathrm{Id}_{\mathbb{R}} \times \psi^{-1}$ into $\mathbb{R}^4$. The map ψ^{-1} is defined near the asymptotic limit $P_0 \subset M$, recalling that $u(\sigma(s,t)) \to P_0(Tt)$. The desired coordinate map is the composition

$$\Gamma = \Psi \circ \Phi \circ (\sigma \times \mathrm{Id}_{\mathbb{R}^2}),$$

so that $\Gamma(s,t,x,y) = \Psi \circ \Phi(\sigma(s,t),x,y) \in \mathbb{R}^4$. In these coordinates the induced almost complex structure $\bar{J}$ becomes the matrix

$$\bar{J}_0(p) = (T_p\Gamma)^{-1} \cdot \hat{J}(\Gamma(p)) \cdot T_p\Gamma \in \mathcal{L}(\mathbb{R}^4),$$

where we have abbreviated $p = (s,t,x,y)$. The matrix $\hat{J}$ represents the almost complex structure $\tilde{J}$ in the Ψ coordinates

$$\hat{J} = (T\Psi) \circ \tilde{J}(\Psi^{-1}) \circ (T\Psi)^{-1}$$

and was computed already above. Explicitly, the map Γ is given by

$$\begin{aligned}\Gamma(s,t,x,y) &= \Psi \circ \Phi(\sigma(s,t),x,y)\\ &= \tilde{v}(s,t) + xn(s,t) + ym(s,t).\end{aligned}$$

Therefore, the tangent map $T_p\Gamma$ is the matrix having the following columns,

$$T_p\Gamma = \big(\tilde{v}_s + xn_s + ym_s \mid \tilde{v}_t + xn_t + ym_t \mid n \mid m\big).$$

The entries are evaluated at the points (s,t). Recall that $\tilde{v}$ is a pseudoholomorphic curve. Hence it solves $\tilde{v}_s + \hat{J}(\tilde{v})\tilde{v}_t = 0$ so that $T\tilde{v} \circ J_0 = \hat{J}(\tilde{v}) \circ T\tilde{v}$. Using also $\hat{J}(\tilde{v})n = m$ and $\hat{J}(\tilde{v})m = -n$ one verifies readily that at the points $p = (s,t,0,0)$ the almost complex structure $\bar{J}_0$ splits:

$$\bar{J}_0(s,t,0,0) = \begin{pmatrix} J_0 & 0 \\ 0 & J_0 \end{pmatrix}.$$

The matrix J_0 is the standard almost complex structure in $\mathbb{R}^2$ introduced above. Since σ is holomorphic and hence a solution of $T\sigma \circ J_0 = j(\sigma) \circ T\sigma$ we have proved

Lemma 4.5. *The induced almost complex structure $\bar{J}$ on $\dot{S} \times B_\varepsilon(0)$ splits along $\dot{S} \times \{0\}$ near the punctures,*

$$\bar{J}(z,0) = j(z) \oplus i \ \in \mathcal{L}(T_z\dot{S} \times \mathbb{R}^2).$$

In order to find the asymptotic behavior of $\bar{J}_0(s,t,x,y)$ as $s \to \infty$ we first study the behavior of $n(s,t)$ and $m(s,t)$ as $s \to \infty$. Recall the eigenfunction $e(t) = (e_1(t), e_2(t)) \in \mathbb{R}^2$ of the asymptotic operator A_∞ in Theorem 4.4. We introduce $f(t) = J(t)e(t) = (f_1(t), f_2(t)) \in \mathbb{R}^2$ so that $|f(t)| = |e(t)|$. Introducing the (always with respect to our special inner product) normalized vectors

$$\hat{e} = \frac{e(t)}{|e(t)|}, \qquad \hat{f} = \frac{f(t)}{|f(t)|},$$

we define the asymptotic normals

$$n_\infty(t) = (0,0,\hat{e}(t)) \in \mathbb{R}^4$$
$$m_\infty(t) = (0,0,\hat{f}(t)) \in \mathbb{R}^4.$$

Lemma 4.6. *As $s \to \infty$,*

$$D^\alpha[n(s,t) - n_\infty(t)] \to 0$$
$$D^\alpha[m(s,t) - m_\infty(t)] \to 0$$

uniformly in t for all derivatives. Note that $\operatorname{span}\{n_\infty(t), m_\infty(t)\}$ *agrees with the contact plane ξ along the periodic orbit $\hat{P}_0(kt)$ of the Reeb vector field.*

Proof. Recall that $n(s,t)$ is given by

$$\begin{aligned} n(s,t) &= \frac{1}{|\tilde{u}_s|}\left(|\pi u_s|, -\lambda(u_t)\frac{\pi u_s}{|\pi u_s|} + \lambda(u_s)\frac{\pi u_t}{|\pi u_t|}\right) \\ &= \frac{1}{|\tilde{u}_s|}\left(|\pi u_s|, -a_s\frac{\pi u_s}{|\pi u_s|} - a_t\frac{\pi u_t}{|\pi u_t|}\right). \end{aligned}$$

Abbreviating $z=(x,y)\in\mathbb{R}^2$ and the Reeb vector field $X=(X_1,Y)=(X_1,X_2,X_3)$, the projections of u_s and u_t onto the contact planes in TM are explicitly given by

$$\begin{aligned} \pi u_s &= u_s - \lambda(u_s)X(u) = (\vartheta_s, x_s, y_s) + a_t(X_1, X_2, X_3) \\ &= (\vartheta_s + a_tX_1, z_s + a_tY) \end{aligned}$$

and

$$\begin{aligned}\pi u_t &= u_t - \lambda(u_t)X(u) = (\vartheta_t, x_t, y_t) - a_s(X_1, X_2, X_3)\\ &= (\vartheta_t - a_s X_1, z_t - a_s Y).\end{aligned}$$

In what follows we denote by $B(s,t)$ various functions having all partial derivatives uniformly bounded and by $R(s,t)$ various functions which satisfy $D^\alpha R(s,t) \to 0$ uniformly in t for all derivatives as $s \to \infty$. It follows from Theorem 4.4 on the asymptotic behavior of the components of $u = (\vartheta, x, y)$ that

$$\begin{aligned}X_1(\vartheta, x, y) &= X_1(\vartheta, 0, 0) + \int_0^1 \frac{d}{d\rho} X_1(\vartheta, \rho z) d\rho\\ &= \frac{1}{\tau} + \left[\int_0^1 DX_1(\vartheta, \rho z) d\rho\right] \cdot z\\ &= \frac{1}{\tau} + e^{\int_{s_0}^s \mu(\rho) d\rho} \left[\int_0^1 DX_1(\vartheta, \rho z) d\rho\right] \cdot [e(t) + \hat{z}(s,t)]\\ &= \frac{1}{\tau} + e^{\int_{s_0}^s \mu(\rho) d\rho} B(s,t).\end{aligned} \tag{36}$$

Here we have used the fact that all partial derivatives of the function f used in the definition of X, (33), are uniformly bounded. In order to derive a similiar asymptotic formula for the component Y of the Reeb vector field X we introduce matrices

$$\begin{aligned}F(s,t) &= \int_0^1 DY(\vartheta(s,t), \rho z(s,t)) d\rho,\\ F(t) &:= \frac{1}{\tau^2}\begin{pmatrix} f_{xy} & f_{yy}\\ -f_{xx} & -f_{xy}\end{pmatrix}.\end{aligned}$$

In view of the definition of Y,

$$D^\alpha \left[F(s,t) - F(t)\right] \to 0, \qquad \text{as } s \to \infty,$$

uniformly in t for all muli-indices α. Now we can derive the asymptotic formula for the Y component of the Reeb vector field X,

$$\begin{aligned}Y(\vartheta, x, y) = Y(\vartheta, z) &= Y(\vartheta, 0) + \left[\int_0^1 DY(\vartheta, \rho z) d\rho\right] \cdot z\\ &= e^{\int_{s_0}^s \mu(\rho) d\rho} F(s,t)\left[e(t) + \hat{z}(s,t)\right]\\ &= e^{\int_{s_0}^s \mu(\rho) d\rho}\left[F(t)e(t) + R(s,t)\right].\end{aligned} \tag{37}$$

Recall that e is the eigenfunction of A_∞ corresponding to the eigenvalue μ,

$$\begin{aligned}\mu e &= -J(t)\dot{e} + S_\infty(t) e\\ &= -J(t)\left[\dot{e} + J(t) S_\infty(t) e\right]\\ &= -J(t)\left[\dot{e} - F_\infty(t) e\right]\\ &= -J(t)\left[\dot{e} - TF(t) e\right].\end{aligned}$$

Using (36) and (37) we shall determine the asymptotic formulae for the components of πu_s and πu_t. We start with the component $\vartheta_s + a_t X_1$ of πu_s,

$$\begin{aligned}\vartheta_s + a_t X_1 &= \vartheta_s + \frac{a_t}{\tau} + a_t e^{\int_{s_0}^s \mu(\rho)d\rho} B(s,t) \\ &= \vartheta_s + \frac{a_t}{\tau} + e^{\int_{s_0}^s \mu(\rho)d\rho} R(s,t)\end{aligned} \tag{38}$$

since $D^\alpha a \to 0$ uniformly in t as $s \to \infty$ for every multi-index α such that $|\alpha| \geq 1$. The function a and the map u are related by the equation $a_t = -\lambda(u_s)$ from which we conclude

$$\begin{aligned}a_t = -\lambda(u_s) &= -f(\vartheta_s + xy_s) \\ &= -\tau\vartheta_s - \tau x y_s - \left[\int_0^1 Df(\vartheta, \rho z)d\rho\right] \cdot z(\vartheta_s + xy_s) \\ &= -\tau\vartheta_s + e^{\int_{s_0}^s \mu(\rho)d\rho} R(s,t).\end{aligned} \tag{39}$$

Combining (38) with (39) find

$$\vartheta_s + a_t X_1 = e^{\int_{s_0}^s \mu(\rho)d\rho} R(s,t). \tag{40}$$

For the second component of πu_s we find, using (37) and the fact that all partial derivatives of a_t converge to 0, the following asymptotic formula

$$\begin{aligned}z_s + a_t Y &= e^{\int_{s_0}^s \mu(\rho)d\rho} \left[\mu(s)e(t) + \hat{z}(s,t) + \hat{z}_s(s,t)\right] \\ &\quad + a_t e^{\int_{s_0}^s \mu(\rho)d\rho} \left[F(t)e(t) + R(s,t)\right] \\ &= e^{\int_{s_0}^s \mu(\rho)d\rho} \left[\mu(s)e(t) + R(s,t)\right].\end{aligned}$$

Since all derivatives of $\hat{\mu}(s) = \mu(s) - \mu$ converge to 0,

$$z_s + a_t Y = e^{\int_{s_0}^s \mu(\rho)d\rho} \left[\mu e(t) + R(s,t)\right]. \tag{41}$$

Similar calculations for the components of πu_t lead to

$$\begin{aligned}\vartheta_t - a_s X_1 &= e^{\int_{s_0}^s \mu(\rho)d\rho} B(s,t) \\ z_t - a_s Y &= e^{\int_{s_0}^s \mu(\rho)d\rho} B(s,t).\end{aligned} \tag{42}$$

Next we derive the asymptotic formula for $|\pi u_s|$. Recall that $|\cdot|_J^2 = fd\lambda_0(\cdot, J\cdot)$ and that $d\lambda_0 = dx \wedge dy = \langle J_0 \cdot, \cdot\rangle$, where $\langle\cdot,\cdot\rangle$ is the standard inner product in $\mathbb{R}^2$ and J_0 the standard symplectic matrix in $\mathbb{R}^2$. Using Theorem 4.4 and (37) we compute

$$\begin{aligned}|\pi u_s|_J^2 &= fd\lambda_0(\pi u_s, J\pi u_s) = fd\lambda_0(\pi u_s, \pi u_t) = f\langle J_0(z_s + a_tY), z_t - a_sY\rangle\\
&= f\left[\langle J_0 z_s, z_t\rangle - a_s\langle J_0 z_s, Y\rangle - a_t\langle J_0 z_t, Y\rangle\right]\\
&= fe^{2\int_{s_0}^s \mu(\rho)d\rho}\left[\mu(s)\langle J_0 e(t), \dot e(t)\rangle - T\mu(s)\langle J_0 e(t), F(t)e(t)\rangle + R(s,t)\right]\\
&= fe^{2\int_{s_0}^s \mu(\rho)d\rho}[\mu(s)\langle J_0 e(t), \dot e(t) - TF(t)e(t)\rangle + R(s,t)]\\
&= fe^{2\int_{s_0}^s \mu(\rho)d\rho}\left[\mu\langle J_0 e, \dot e(t) - TF(t)e(t)\rangle + R(s,t)\right]\\
&= e^{2\int_{s_0}^s \mu(\rho)d\rho}\left[\mu d(f\lambda_0)(e(t), \dot e(t) - TF(t)e(t)) + R(s,t)\right]\\
&= e^{2\int_{s_0}^s \mu(\rho)d\rho}\left[\langle \mu e, -J(t)(\dot e - TF(t)e)\rangle_J + R(s,t)\right]\\
&= e^{2\int_{s_0}^s \mu(\rho)d\rho}\left[|\mu|^2|e|_J^2 + R(s,t)\right]\\
&= e^{2\int_{s_0}^s \mu(\rho)d\rho}|\mu|^2|e|_J^2\left[1 + R(s,t)\right].\end{aligned}$$

Thus

$$|\pi u_s|_J = e^{\int_{s_0}^s \mu(\rho)d\rho}|\mu||e|_J\left[1 + R(s,t)\right], \tag{43}$$

and so the norm $|\tilde u_s|$ can be written as

$$\begin{aligned}|\tilde u_s|^2 &= a_s^2 + \lambda(u_s)^2 + |\pi u_s|_J^2\\
&= a_s^2 + a_t^2 + |\pi u_s|_J^2\\
&= T^2 + R(s,t) = T^2(1 + R(s,t)).\end{aligned}$$

Hence,

$$|\tilde u_s| = T\left[1 + R(s,t)\right]. \tag{44}$$

We are ready to show that

$$D^\alpha[n(s,t) - n_\infty(t)] \to 0$$

as $s \to \infty$ for all multi-indices α. From (43) and (44) we deduce

$$\begin{aligned}\frac{|\pi u_s|}{|\tilde u_s|} &= e^{\int_{s_0}^s \mu(\rho)d\rho}\frac{|\mu||e|_J(1 + R(s,t))}{T(1 + R(s,t))}\\
&= e^{\int_{s_0}^s \mu(\rho)d\rho}B(s,t).\end{aligned}$$

Hence $D^\alpha(|\pi u_s|/|\tilde u_s|) \to 0$ for all α. Next we will show that $-\pi u_s/|\pi u_s|$ converges with all derivatives to the normalized eigenfunction $\hat e$. Indeed, using equations (40) and (43) we obtain

$$\frac{\vartheta_s + a_t X_1}{|\pi u_s|_J} = \frac{R(s,t)}{|\mu|\,|e|_J(1 + R(s,t))} = R(s,t),$$

with R having all derivatives converging to 0 as $s \to \infty$. From equations (41) and (43) we get

$$\begin{aligned}\frac{z_s + a_t Y}{|\pi u_s|} &= \frac{\mu e + R(s,t)}{|\mu||e|_J(1 + r(s,t))} \\ &= -\hat{e} + R(s,t)\end{aligned}$$

which, in view of $a_s \to T$, implies

$$D^\alpha \left[-a_s \frac{\pi u_s}{|\tilde{u}_s||\pi u_s|_J} - \hat{e} \right] \to 0,$$

as claimed. Finally, it follows from equations (42) and the asymptotic behavior of a that also $D^\alpha[a_t \pi u_t / |\pi u_t|]$ converges to 0 as $s \to \infty$, uniformly in t for all α. Summing up we have proved that $D^\alpha [n(s,t) - n_\infty(s,t)] \to 0$ for all multi-indices α. Similar calculations show that the second statement of the lemma is also true. This finishes the proof of Lemma 4.6. □

Abbreviating the tangent map $A(s,t,x,y) = T_p \Gamma$ where $p = (s,t,x,y)$ we conclude from Lemma 4.6 and Theorem 4.4 that

$$A(s,t,x,y) \to A(t,x,y)$$

as $s \to \infty$ uniformly on bounded sets of (t,x,y) including all derivatives. The limit is the matrix

$$A(t,x,y) = \begin{pmatrix} T & 0 & 0 & 0 \\ 0 & k & 0 & 0 \\ 0 & x\hat{e}_1'(t) + y\hat{f}_1'(t) & \hat{e}_1(t) & \hat{f}_1(t) \\ 0 & x\hat{e}_2'(t) + y\hat{f}_2'(t) & \hat{e}_2(t) & \hat{f}_2(t) \end{pmatrix}.$$

The map Γ itself has (always in the same sense of convergence as above) the following asymptotic behavior as $s \to \infty$,

$$\Gamma(s,t,x,y) \to (\infty, kt, x\hat{e}(t) + y\hat{f}(t)).$$

Since the almost complex structure $\hat{J}$ does not depend on the first coordinate a we deduce in view of (34) the existence of a smooth matrix function $\hat{J}_\infty(t,x,y)$ such that

$$\hat{J}(\Gamma(s,t,x,y)) \to \hat{J}_\infty(t,x,y)$$

as $s \to \infty$. If $(x,y) = (0,0)$ we find, in particular,

$$\hat{J}_\infty(t,0,0) = \hat{J}(kt,0,0) = \begin{pmatrix} \begin{matrix} 0 & -\tau \\ \frac{1}{\tau} & 0 \end{matrix} & 0 \\ 0 & J(t) \end{pmatrix},$$

where τ is the minimal period of the asymptotic periodic orbit P_0. Introducing the matrix

$$N(t) = \left(\hat{e}(t) \mid \hat{f}(t)\right) \in \mathcal{L}(\mathbb{R}^2)$$

we conclude from $\hat{f}(t) = J(t)\hat{e}(t)$ that $J(t)N(t) = N(t)J_0$. Consequently, as $s \to \infty$ and $(x, y) \to (0, 0)$

$$\begin{aligned}\bar{J}_0(s,t,x,y) &\to A^{-1}(t,0,0) \cdot \hat{J}(kt,0,0) \cdot A(t,0,0) \\ &= \begin{pmatrix} J_0 & 0 \\ 0 & N^{-1}(t) \cdot J(t) \cdot N(t) \end{pmatrix} = \begin{pmatrix} J_0 & 0 \\ 0 & J_0 \end{pmatrix}.\end{aligned}$$

We can construct, of course, local trivializations Φ as above near every puncture $z_0 \in \Gamma$. But not every choice of local trivializations near the punctures allows an extension to a trivialization of the normal bundle of $\tilde{u}$ over all $z \in \dot{S}$. However, we can always find local trivializations along the asymptotic cylinders in the appropriate homotopy class such that the desired extension can be carried out. Indeed, fix an integer k, and let Φ be as above. Then define the new trivialization $\tilde{\Phi}$ by

$$\begin{aligned}\tilde{\Phi}(\sigma(s,t),x,y) &= \Phi(\sigma(s,t),\xi,\eta), \\ (\xi,\eta) &= e^{-ktJ_0}(x,y).\end{aligned}$$

In our construction above this amounts to replace the normal vectors $n(s,t)$ and $m(s,t) \in \mathbb{R}^4$ by

$$\tilde{n}(s,t) = e^{kt\bar{J}_0} n(s,t)$$

and

$$\tilde{m}(s,t) = e^{kt\bar{J}_0} m(s,t),$$

where $\bar{J}_0 = \operatorname{diag}(J_0, J_0)$. The normal vectors $\tilde{n}$ and $\tilde{m}$ have the same properties as n and m listed in Lemma 4.1. In view of Lemma 4.6, as $s \to \infty$,

$$\tilde{n}(s,t) \to \tilde{n}_\infty = e^{kt\bar{J}_0} n_\infty(t) = (0, 0, e^{ktJ_0}\hat{e}_1(t)),$$

and analogously for $\tilde{m}(s,t)$. The convergence is uniformly in all derivatives $\alpha = (\alpha_1, \alpha_2)$. The function $e^{ktJ_0}\hat{e}_1(t)$ is an eigenvector corresponding to the above eigenvalue $\mu < 0$ for the operator $\tilde{A}_\infty = U(t)A_\infty U(t)^{-1}$, where $U(t) = e^{ktJ_0}$.

In view of the map $\tilde{\Phi}$ we can choose a local trivialization having a prescribed rotation k relative to a given one. Consequently, if we choose the local trivialization near one puncture in the appropriate homotopy class, the desired extension can be carried out and we obtain a trivialization $\Phi : \dot{S} \times B_\varepsilon(0) \to \mathbb{R} \times M$ of the normal bundle of $\tilde{u}$ satisfying $\Phi(z, 0) = \tilde{u}(z)$ for all $z \in \dot{S}$.

We summarize the discussion of this section. In the coordinates Φ found, all the relevant functions have near the punctures along the cylinders limits as $s \to \infty$, uniformly in t with all derivatives. In particular we have proved the following theorem in which, for simplicity, only the case $k = 0$ is explicitely formulated.

Theorem 4.7. *Consider the embedded finite energy surface*

$$\tilde{u} : (\dot{S}, j) \to (\mathbb{R} \times M, \tilde{J})$$

with the almost complex structure $\tilde{J}$ *(special or generalized) as in the introduction. Assume the punctures are non-degenerate. Then there exists an immersion*

$$\Phi : \dot{S} \times B_\varepsilon(0) \to \mathbb{R} \times M,$$

where $B_\varepsilon(0) \subset \mathbb{R}^2$ *is a small neighborhood of the origin, satisfying* $\Phi(z, 0) = \tilde{u}(z)$ *for all* $z \in \dot{S}$. *Moreover,* Φ *restricted to a sufficiently small neighborhood of* $\dot{S} \times \{0\}$ *is an embedding.*

The almost complex structure $\bar{J} = (T\Phi)^{-1} \circ \tilde{J}(\Phi) \circ (T\Phi)$ *induced on* $\dot{S} \times B_\varepsilon(0)$ *splits along* $\dot{S} \times \{0\}$ *near the punctures,*

$$\bar{J}(z, 0) = j(z) \oplus i \in \mathcal{L}(T_z\dot{S} \times \mathbb{R}^2).$$

Introducing in a punctured neighborhood of the (positive) puncture $z_0 \in \Gamma$ *the cylindrical holomorphic coordinates* $z = \sigma(s,t) \to z_0$ *as* $s \to \infty$, *and in a tubular neighborhood of the associated asymptotic limit* P_0 *in* M *the coordinates* Ψ *of* $\mathbb{R} \times M$, *the almost complex structure converges as follows. If* $\bar{J}_0(s,t,x,y) = (T\sigma \oplus \mathrm{Id}_{\mathbb{R}^2})^{-1} \circ \bar{J}(\sigma(s,t), x, y) \circ (T\sigma \oplus \mathrm{Id}_{\mathbb{R}^2})$, *then*

$$D^\alpha \left[\bar{J}_0(s,t,x,y) - A(t,x,y)^{-1} \hat{J}_\infty(t,x,y) A(t,x,y)\right] \to 0$$

as $s \to \infty$ *uniformly on bounded sets of* (t,x,y), *for all derivatives* α. *The smooth matrix functions* A *and* $\hat{J}_\infty \in \mathcal{L}(\mathbb{R}^4)$ *are described above. Moreover,*

$$\bar{J}_0(s,t,0,0) = \begin{pmatrix} J_0 & 0 \\ 0 & J_0 \end{pmatrix}$$

with the standard symplectic structure J_0 *on* $\mathbb{R}^2$.

5. The implicit function theorem for the local model

If $\tilde{u}_0 : (\dot{S}, j) \to (\mathbb{R} \times M, \tilde{J})$ is an embedded finite energy surface non-degenerate at the punctures $\Gamma \subset S$, we constructed in Section 4 a smooth trivialization of the normal bundle over $\tilde{u}_0(\dot{S})$:

$$\Phi : \dot{S} \times \mathbb{R}^2 \to \mathbb{R} \times M, \quad \Phi(z, 0) = \tilde{u}_0(z), \quad \text{for all } z \in \dot{S}.$$

The induced almost complex structure $\bar{J}$ on $T(\dot{S} \times \mathbb{R}^2)$ is defined by

$$\bar{J} = (T\Phi)^{-1} \circ \tilde{J}(\Phi) \circ T\Phi.$$

Corresponding to the splitting $T(\dot{S} \times \mathbb{R}^2) = T\dot{S} \times \mathbb{R}^2$ we represent $\bar{J}$ in matrix form,

$$\bar{J}(z,x,y) = \begin{pmatrix} j(z,x,y) & \tilde{\Delta}(z,x,y) \\ \Delta(z,x,y) & i(z,x,y) \end{pmatrix},$$

where $j(z,x,y) \in \mathcal{L}(T_z\dot{S},T_z\dot{S})$ and $\Delta(z,x,y) \in \mathcal{L}(T_z\dot{S},\mathbb{R}^2)$ and so on. Along $\dot{S}\times\{0\} \subset \dot{S}\times\mathbb{R}^2$ we know that near the punctures $\Gamma \subset S$ the structure $\bar{J}$ splits:

$$\bar{J}(z,0) = \begin{pmatrix} j(z) & 0 \\ 0 & i \end{pmatrix}, \quad z \in \dot{S}.$$

Here $i = J_0 \in \mathcal{L}(\mathbb{R}^2,\mathbb{R}^2)$. We would like to study finite energy surfaces near $\tilde{u}_0$ in the local model $(\dot{S}\times\mathbb{R}^2,\bar{J})$ in which the reference surface becomes $z \mapsto (z,0)$, $z \in \dot{S}$. We look for finite energy surfaces represented by graphs, $u = \text{graph}(v) : z \mapsto (z,v(z))$, where $z \in \dot{S}$ and $v : \dot{S} \to \mathbb{R}^2$. However, we do not keep the almost complex structure $j(z)$ on $\dot{S}$ fixed. Instead we first look for an embedded complex submanifold of the form $(\text{graph}(v)) \subset \dot{S}\times\mathbb{R}^2$ requiring that the tangent spaces are invariant under $\bar{J}$:

$$\bar{J}(u(z))\left[Tu(z)(T_z\dot{S})\right] = Tu(z)(T_z\dot{S})$$

for all $z \in \dot{S}$. In order to prove the existence of such complex submanifolds graph(v) for v in a neighborhood of 0, we shall design an implicit function theorem based on our Fredholm theory in Section 2. We first recall the Cauchy-Riemann operator L from Section 3. It maps functions $v : \dot{S} \to \mathbb{R}^2$ linearly into the space of sections of the bundle $\mathcal{A}_0$, and is defined by

$$(Lv)(z) = Tv(z) + i \circ Tv(z) \circ j(z) + (\Delta'(z)v(z)) \circ j(z) \tag{45}$$

for every $z \in \dot{S}$. Here $\Delta'(z) = \Delta'(z,0) \in \mathcal{L}(\mathbb{R}^2,\mathcal{L}(T_z\dot{S},\mathbb{R}^2))$ is the derivative of $\Delta(z,x,y)$ with respect to the variable $(x,y) \in \mathbb{R}^2$ at $(x,y) = (0,0)$. In view of $\bar{J}^2 = -\text{Id}$, the map $(\Delta'(z)h) \circ j(z) \in \mathcal{L}(T_z\dot{S},\mathbb{R}^2)$, for $h \in \mathbb{R}^2$, is complex antilinear $(T_z\dot{S},j(z)) \to (\mathbb{R}^2,i)$. Hence Lv is indeed a section of $\mathcal{A}_0$. In order to apply Fredholm theory we shall show, recalling the concepts of Section 2, that

$$C(z)h := (\Delta'(z)h) \circ j(z)$$

is admissible and non-degenerate at the punctures and shall prove

Theorem 5.1. *Assume $\tilde{u}_0 : (\dot{S},j) \to (\mathbb{R}\times M,\tilde{J})$ is an embedded non-degenerate finite energy surface. Then the operator*

$$L : C_0^{1,\alpha}(\dot{S},\mathbb{R}^2) \to C_0^{\alpha}(\mathcal{A}_0)$$

defined by (45) *is a Fredholm operator and*

$$\text{Ind}(L) = \mu(\tilde{u}_0) - \chi_S + \sharp\Gamma_S.$$

The integer $\mu(\tilde{u}_0)$ is the Conley-Zehnder index of $\tilde{u}_0$ and χ_S is the Euler characteristic of S, and $\Gamma_S = \Gamma \subset S$ are the punctures such that $\dot{S} = S \setminus \Gamma$.

In order to prove Theorem 5.1 we begin by verifying the admissibility and the non-degeneracy conditions. We pick a (positive) puncture $z_0 \in \Gamma_S$ and the usual special cylindrical coordinates $z = \sigma(s,t)$ in a punctured neighborhood of

$z_0 \in S$, where $z_0 = \lim \sigma(s,t)$ as $s \to \infty$ and where $\sigma(s,t)$ depends 1-periodically in $t \in \mathbb{R}$. Define

$$W(s,t) \in \mathcal{L}(\mathbb{R}^2, \mathbb{R}^2)$$

by

$$W(s,t)h := (\Delta'(\sigma(s,t))h) \circ j(\sigma(s,t)) \cdot T\sigma \frac{\partial}{\partial s}.$$

Lemma 5.2. *As $s \to \infty$,*

$$D^\alpha \left[W(s,t)h - W(t)h\right] \to 0$$

uniformly in t for every $h \in \mathbb{R}^2$, for all derivatives $\alpha = (\alpha_1, \alpha_2)$.

Proof. We shall compute $\Delta'(\sigma(s,t))$ in the local coordinates introduced in the previous section. We recall that near the puncture $z_0 \in S$ and near its asymptotic limit P_0 the map $\tilde{u}_0$ is represented as a map $\tilde{v} = \tilde{v}(s,t) : \mathbb{R}^2 \to \mathbb{R}^4$. The map Γ is defined by $\Gamma(s,t,x,y) = \tilde{v}(s,t) + xn(s,t) + ym(s,t)$. Abbreviating the Jacobian matrix of Γ at the point $(s,t,x,y) \in \mathbb{R}^4$ by $A(s,t,x,y)$, the almost complex structure $\bar{J}$ is, in the local coordinates represented by the matrix $\bar{J}_0$, defined by

$$A(s,t,x,y) \cdot \bar{J}_0(s,t,x,y) = \hat{J}(\Gamma(s,t,x,y)) \cdot A(s,t,x,y). \tag{46}$$

With respect to $(s,t,x,y) \in \mathbb{R}^4 = \mathbb{R}^2 \times \mathbb{R}^2$, the matrix $\bar{J}_0$ splits:

$$\bar{J}_0(s,t,x,y) = \begin{pmatrix} j_0(s,t,x,y) & \tilde{\Delta}_0(s,t,x,y) \\ \Delta_0(s,t,x,y) & i_0(s,t,x,y) \end{pmatrix}.$$

We recall

$$\bar{J}_0(s,t,0) = \begin{pmatrix} J_0 & 0 \\ 0 & J_0 \end{pmatrix}.$$

Denote by $\Delta_0'(s,t)$ the derivative of $\Delta_0(s,t,x,y)$ with respect to $(x,y) \in \mathbb{R}^2$ at $(x,y) = (0,0)$. Since σ are holomorphic coordinates, $j(\sigma)T\sigma = T\sigma \cdot J_0$. Consequently,

$$\begin{aligned}(\Delta'(\sigma(s,t))h) \circ j(\sigma) \cdot T\sigma &= (\Delta'(\sigma(s,t))h) \circ T\sigma \circ J_0 \\ &= (\Delta_0'(s,t)h)J_0.\end{aligned}$$

Therefore, defining $W_0(s,t) \in \mathcal{L}(\mathbb{R}^2, \mathbb{R}^2)$ by

$$W_0(s,t)h := (\Delta_0'(s,t)h) \cdot J_0 \begin{pmatrix} 1 \\ 0 \end{pmatrix},$$

we have to show that $W_0(s,t) \to W_0(t)$ as $s \to \infty$. In order to simplify the notation we drop the subscript 0 which refers to the local coordinates and simply write $\Delta(s,t) \equiv \Delta_0(s,t)$ and $j(s,t) \equiv j_0(s,t,0,0)$ and $W(s,t) \equiv W_0(s,t)$. We represent the 2×2 matrices Δ and j as

$$\Delta = \begin{pmatrix} \Delta_{11} & \Delta_{12} \\ \Delta_{21} & \Delta_{22} \end{pmatrix} \quad \text{and} \quad j = \begin{pmatrix} j_{11} & j_{12} \\ j_{21} & j_{22} \end{pmatrix}.$$

Moreover, we introduce the matrix elements $A = (A_{ns})$ and $\hat{J} = (\hat{J}_{ns}(\Gamma))$, where $n, s = 1, 2, 3, 4$. In order to compute Δ we apply both sides of (46) to the vector $(1, 0, 0, 0) \in \mathbb{R}^4$. Looking at the third and fourth coordinate we obtain the equations

$$\sum_{s=1}^{2} A_{3s} j_{s1} + n_3 \Delta_{11} + m_3 \Delta_{21} = \sum_{s=1}^{4} \hat{J}(\Gamma)_{3s} A_{s1}$$
$$\sum_{s=1}^{2} A_{4s} j_{s1} + n_4 \Delta_{11} + m_4 \Delta_{21} = \sum_{s=1}^{4} \hat{J}(\Gamma)_{4s} A_{s1}.$$

We have used the components of the vectors $n(s,t) = (n_1(s,t), \ldots, n_4(s,t))$, and similarly $m(s,t) \in \mathbb{R}^4$. The solution is the following:

$$\begin{pmatrix} \Delta_{11} \\ \Delta_{21} \end{pmatrix} = N^{-1} \begin{pmatrix} M_{11} \\ M_{21} \end{pmatrix}.$$

Here we used the abbreviations:

$$M_{11} = -\sum_{s=1}^{2} A_{3s} j_{s1} + \sum_{s=1}^{4} \hat{J}_{3s}(\Gamma) A_{s1}$$
$$M_{21} = -\sum_{s=1}^{2} A_{4s} j_{s1} + \sum_{s=1}^{4} \hat{J}_{4s}(\Gamma) A_{s1}$$

and

$$N(s,t) = \begin{pmatrix} n_3(s,t) & m_3(s,t) \\ n_4(s,t) & m_4(s,t) \end{pmatrix}.$$

Similarly, applying both sides of (46) to the vector $(0, 1, 0, 0)$ we find

$$\begin{pmatrix} \Delta_{21} \\ \Delta_{22} \end{pmatrix} = N^{-1} \begin{pmatrix} M_{21} \\ M_{22} \end{pmatrix},$$

where now

$$M_{21} = -\sum_{s=1}^{2} A_{3s} j_{s2} + \sum_{s=1}^{4} \hat{J}_{3s}(\Gamma) A_{s2}$$
$$M_{22} = -\sum_{s=1}^{2} A_{4s} j_{s2} + \sum_{s=1}^{4} \hat{J}_{4s}(\Gamma) A_{s2}.$$

Consequently,

$$\Delta = \begin{pmatrix} \Delta_{11} & \Delta_{12} \\ \Delta_{21} & \Delta_{22} \end{pmatrix} = N^{-1} \begin{pmatrix} M_{11} & M_{12} \\ M_{21} & M_{22} \end{pmatrix} = N^{-1} M.$$

Differentiating the last equality with respect to (x, y) at the point $(s, t, 0, 0)$ we find for $h = (h_1, h_2) \in \mathbb{R}^2$

$$\Delta'(s,t,0,0)h = h_1 N^{-1} \partial_x \begin{pmatrix} M_{11} & M_{12} \\ M_{21} & M_{22} \end{pmatrix} + h_2 N^{-1} \partial_y \begin{pmatrix} M_{11} & M_{12} \\ M_{21} & M_{22} \end{pmatrix}.$$

Applying $(\Delta'(s,t,0,0)h)J_0$ to the vector $(1,0)$ we obtain for $W(s,t) \in \mathcal{L}(\mathbb{R}^2,\mathbb{R}^2)$ the following formula

$$W(s,t)h = N^{-1}\begin{pmatrix} \partial_x M_{12} & \partial_y M_{12} \\ \partial_x M_{22} & \partial_y M_{22} \end{pmatrix} h.$$

We now take the limit of $W(s,t)$ as $s \to \infty$. In view of Lemma 4.6,

$$D^\alpha \left[N(s,t) - N(t)\right] \to 0$$

as $s \to \infty$ uniformly in t for all derivatives α, where $N(t) = (\hat{e}(t) \mid \hat{f}(t))$. Using also the asymptotic behavior of $A(s,t,x,y)$ and $\hat{J}(\tilde{v}(s,t))$ as $s \to \infty$ determined above, one can verify that indeed $D^\alpha \left[W(s,t)h - W(t)h\right] \to 0$ as $s \to \infty$ uniformly in t for every $h \in \mathbb{R}^2$ and for all derivatives α. The limit map $W(t) \in \mathcal{L}(\mathbb{R}^2)$ is the following matrix

$$W(t) = -N^{-1}(t)J(t)F_\infty(t)N(t) + N^{-1}(t)J(t)N_t(t).$$

We recall from Section 4 that $J(t) = J(kt,0,0) \in \mathcal{L}(\mathbb{R}^2,\mathbb{R}^2)$ is the almost complex structure J in the contact plane $\mathbb{R}^2$ along the periodic solution $\hat{P}_0(kt)$ and $F_\infty(t) = T_\pi \circ dX(kt,0,0) \circ \pi \in \mathcal{L}(\mathbb{R}^2,\mathbb{R}^2)$ is the projection of the linearized Reeb vector field onto the contact plane along the periodic solution. The proof of Lemma 5.2 is complete. □

Lemma 5.3. *$W(t) \in \mathcal{L}(\mathbb{R}^2,\mathbb{R}^2)$ is symmetric with respect to the Euclidean inner product in $\mathbb{R}^2$.*

Proof. Along the periodic solution $\hat{P}_0(kt)$, we have, in our coordinates, $d\lambda = \omega$. Since $J(t)$ is compatible with $d\lambda$ we conclude $J^*(t)J_0J(t) = J_0$. Recalling that $J(t)\hat{e}(t) = \hat{f}(t)$, we therefore find

$$N(t)J_0 = J(t)N(t). \tag{47}$$

We claim that

$$N^*(t)J_0N(t) = \det N(t) \cdot J_0 = \frac{1}{\tau}J_0. \tag{48}$$

Indeed, the first equality follows from $J(t)\hat{e}(t) = \hat{f}(t)$. Recalling that $f(kt,0,0) = \tau$, the second equality follows from the normalization

$$1 = |\hat{e}|^2 = f(t,0,0)(J_0\hat{e},\ J(t)\hat{e}(t)) = \tau(J_0\hat{e},\ \hat{f}(t)) = \tau\det(N(t)),$$

hence proving the claim. Differentiation of (48) with respect to t gives the identity

$$N_t^*J_0N = -N^*J_0N_t$$

and, using (47) and (48), we compute

$$\begin{aligned} (N^{-1}JN_t)^* &= (J_0N^{-1}N_t)^* = -N_t^*(N^{-1})^*J_0 \\ &= -\tau N_t^*J_0N = \tau N^*J_0N_t = J_0N^{-1}N_t \\ &= N^{-1}JN_t. \end{aligned}$$

Therefore, the second term in $W(t)$ is symmetric. Using the fact that $(J_0 F_\infty(t))^* = JF_\infty(t)$ one similarly shows by again using (47) and (48) that the term $N^{-1}JF_\infty N$ in $W(t)$ is also symmetric. This finishes the proof of Lemma 5.3. □

Lemma 5.4. *The asymptotic operator $A_\infty = -J_0\frac{d}{dt} - W(t)$ has a trivial kernel.*

Proof. We have to show that the problem $x'(t) = J_0 W(t)x(t)$ and $x(0) = x(1)$ admits only the trivial solution. Arguing by contradiction we assume $x(t)$ to be a non-trivial solution and define the 1-periodic function $y(t)$ by

$$y(t) = N(t)x(t), \quad 0 \leqslant t \leqslant 1.$$

Differentiation leads to

$$\begin{aligned} y' &= N_t x + N x' = N_t x + N J_0 W x \\ &= N_t x + N J_0 \left[-N^{-1} J F_\infty N + N^{-1} J N_t \right] x \\ &= N_t x + \left[-N J_0^2 N^{-1} F_\infty N + N J_0^2 N^{-1} N_t \right] x \\ &= N_t x + \left[F_\infty N - N_t \right] x \\ &= F_\infty y. \end{aligned}$$

We conclude that the periodic solution P_0 associated with the puncture z_0 is degenerate, in contradiction to the non-degeneracy assumption on $\tilde{u}_0$. This finishes the proof of Lemma 5.4. □

With the lemmata above we have verified that the operator L is admissible and non-degenerate in the sense of Section 2. Moreover, at the punctures the asymptotic data are those of $\tilde{u}_0$. We therefore deduce from Theorem 2.8 that L is Fredholm, and

$$\operatorname{Ind}(L) = \mu_N(\tilde{u}_0) + \chi_S - \sharp\Gamma_S.$$

By Theorem 1.8,

$$\mu_N(\tilde{u}_0) = \mu(\tilde{u}_0) + 2(\sharp\Gamma_S - \chi_S).$$

The proof of Theorem 5.1 is complete. □

In order to describe the analytical setting in the model space $(\dot{S} \times \mathbb{R}^2, \bar{J})$ we use the intrinsic conceptual framework described in Section 3.

We introduced there a two-dimensional vector bundle $\Theta \to \dot{S} \times \mathbb{R}^2$, whose fiber over $\Theta_{(z,x,y)}$ is the two-dimensional real vector space consisting of the two-vectors in $T_z\dot{S} \times \mathbb{R}^2$ on which $\check{\bar{J}}(z,x,y)$ acts via multiplication by -1. Using Θ we introduced the bundle $\Xi \to \dot{S} \times \mathbb{R}^2$ whose fiber over (z,x,y) consists of all real linear maps $T_z\dot{S} \wedge T_z\dot{S} \to \Theta_{(z,x,y)}$. Recall now that $\dot{S} = S \setminus \Gamma$ and $\Gamma \neq \emptyset$. Then there exists a complex trivialization Ψ of the tangent bundle of $\dot{S} \times \mathbb{R}^2$,

$$\Psi : (T(\dot{S} \times \mathbb{R}^2), \bar{J}) \to (\dot{S} \times \mathbb{R}^2) \times \mathbb{R}^4,$$

where $\mathbb{R}^4 = \mathbb{R}^2 \times \mathbb{R}^2$ is equipped with the almost complex structure $J_0 \oplus J_0$. In order to distinguish admissible trivializations we consider a puncture $z_0 \in \Gamma$

which we assume to be positive, and introduce cylindrical holomorphic coordinates $\sigma : [0,\infty) \times S^1 \to \dot{S}$ for which $z_0 = \lim \sigma(s,t)$ as $s \to \infty$. We call Ψ admissible, if near the puncture z_0 the linear map

$$(T\sigma(s,t) \oplus \mathrm{Id}_{\mathbb{R}^2})^{-1} \cdot \Psi(\sigma(s,t), x, y) \cdot (T\sigma(s,t) \oplus \mathrm{Id}_{\mathbb{R}^2})$$

converges uniformly on bounded sets of (t,x,y) as $s \to \infty$ including all its derivatives to a smooth map

$$(t,x,y) \mapsto \Psi_\infty(t,x,y) \ \in \mathcal{GL}(\mathbb{R}^4).$$

The admissible trivialization Ψ induces a trivialization $\hat{\Psi}$ of the bundle $\Theta \to \dot{S}\times\mathbb{R}^2$,

$$\hat{\Psi} : \Theta \to (\dot{S} \times \mathbb{R}^2) \times \Theta_0,$$

where Θ_0 is the two-dimensional real vector space of two-vectors in $\mathbb{C}^2$ (viewed as a linear real vector space) on which $(J_0 \oplus J_0)^\vee$ acts as multiplication by -1. Given two admissible trivializations

$$A, B :\ \Theta \to (\dot{S} \times \mathbb{R}^2) \times \Theta_0$$

we consider the composition of the bundle maps

$$C := A \circ B^{-1} : (\dot{S} \times \mathbb{R}^2) \times \Theta_0 \to (\dot{S} \times \mathbb{R}^2) \times \Theta_0.$$

Abbreviating with $C(z,x,y) \in \mathcal{L}(\Theta_0)$ the projection onto the second component we note that near a puncture z_0

$$D^\beta \left[C(\sigma(s,t),x,y) - \tilde{C}(t,x,y) \right] \to 0$$

as $s \to \infty$ uniformly in (t,x,y) on bounded sets for all derivatives β. The map $(t,x,y) \mapsto \tilde{C}(t,x,y) \in \mathcal{L}(\Theta_0)$ is smooth.

If $v \in C_0^{1,\alpha}(\dot{S},\mathbb{R}^2)$ and $f \in C_0^\alpha(\dot{S},\Theta_0)$, then the composition

$$z \mapsto C(z, v(z)) \cdot f(z) \ \in \Theta_0$$

belongs to $C_0^\alpha(\dot{S},\Theta_0)$. Moreover, the map $v \mapsto Cv$ defines a smooth map

$$C_0^{1,\alpha}(\dot{S},\mathbb{R}^2) \to \mathcal{L}(C_0^\alpha(\dot{S},\Theta_0)).$$

Consider now the set of pairs $\{v, f\}$. Here $v \in C_0^{1,\alpha}(\dot{S},\mathbb{R}^2)$ and $z \mapsto f(z) \in \Theta_{(z,v(z))}$ is a continuous map for which in an admissible trivialization $\hat{\Psi}$

$$z \mapsto \mathrm{pr}_2 \circ \hat{\Psi}(z, v(z)) \cdot f(z)$$

belongs to $C_0^\alpha(\dot{S},\Theta_0)$. Thus we obtain a Banach space bundle structure on

$$\bigcup_{v \in C_0^{1,\alpha}(\dot{S},\mathbb{R}^2)} \{v\} \times C_0^\alpha(\mathrm{graph}(v)^*\Theta).$$

The fibers are the sections of $\text{graph}(v)^*\Theta \to \dot{S}$ of class C_0^α, where $\text{graph}(v)(z) = (z, v(z)) \in \dot{S} \times \mathbb{R}^2$. Similarly we define the Banach space bundle structure on

$$\bigcup_{v \in C_0^{1,\alpha}(\dot{S},\mathbb{R}^2)} \{v\} \times C_0^\alpha(\text{graph}(v)^*\Xi),$$

where the section $f : \dot{S} \to \text{graph}(v)^*\Xi$ belongs to $C_0^\alpha(\text{graph}(v)^*\Xi)$ if f is locally of class C^α, satisfies $f(z) \in \Xi_{(z,v(z))}$ and, at the (positive) puncture, the map

$$(s,t) \mapsto pr_2 \circ \hat{\Psi}(\sigma(s,t), v(s,t)) \circ f(\sigma(s,t))\left(\frac{\partial \sigma}{\partial s} \wedge \frac{\partial \sigma}{\partial t}\right)$$

belongs to the class $C_0^\alpha([0,\infty) \times S^1, \Theta_0)$. In order to simplify the notation we abbreviate

$$\mathcal{B} = C_0^{1,\alpha}(\dot{S}, \mathbb{R}^2) \quad \text{and} \quad \mathcal{E} = \bigcup_{v \in \mathcal{B}} \{v\} \times C_0^\alpha(\text{graph}(v)^*\Xi).$$

Recall the map H from Section 3. Let $v \in \mathcal{B}$ and abbreviate $u = \text{graph}(v) : \dot{S} \to (\dot{S}, \mathbb{R}^2)$. Define

$$H(v)(h \wedge k) = Tu \cdot h \wedge Tu \cdot k - \bar{J}(u)Tu \cdot h \wedge \bar{J}(u)Tu \cdot k,$$

for $h, k \in T_z\dot{S}$. Since the wedge product is bilinear and since the product of two α-Hölder maps is again an α-Hölder map, we conclude that $H(v) \in C_0^\alpha(\text{graph}(v)^*\Xi)$. Hence

$$H : \mathcal{B} \to \mathcal{E}$$

is a section of the bundle $\mathcal{E} \to \mathcal{B}$. Moreover, it is smooth. Recalling the smooth vector bundle map $\alpha : \mathcal{A}_0 \to \Xi_0$ from Section 3, we conclude from Proposition 3.2 and Theorem 5.1,

Proposition 5.5. *The linearization of H at the point $v = 0$ is a Fredholm operator represented by*

$$H'(0)v = \alpha(Lv), \quad v \in C_0^{1,\alpha}(\dot{S}, \mathbb{R}^2),$$

where the operator L is defined by (45). The Fredholm index is equal to $\mu(\tilde{u}_0) - \chi_S + \sharp\Gamma_S$.

Having established the Fredholm property of the section H we can establish by means of the implicit function theorem the existence of families of embedded complex curves near the image of the finite energy surface $\tilde{u}_0$ in the local model.

Theorem 5.6. *Assume $\tilde{u}_0 : (\dot{S}, j) \to (\mathbb{R} \times M, \tilde{J})$ is a non-degenerate finite energy surface. Consider the above trivialization Φ of the normal bundle to $\tilde{u}_0$:*

$$\Phi : (\dot{S} \times \mathbb{R}^2, \bar{J}) \to (\mathbb{R} \times M, \tilde{J}), \quad \Phi(z, 0) = \tilde{u}_0(z)$$

where $\bar{J} = (T\Phi)^{-1} \cdot \tilde{J}(\Phi) \cdot (T\Phi)$. Assume that

$$d := \mu(\tilde{u}_0) - \chi_S + \sharp\Gamma_S > 0.$$

Then there exists a smooth injective map

$$E : B^d_\varepsilon(0) \subset \mathbb{R}^d \to C^{1,\alpha}_0(\dot{S}, \mathbb{R}^2)$$

satisfying $E(0) = 0$ *and* $H(E(\tau)) = 0$. *For every* $\tau \in B^d_\varepsilon(0)$ *the map* $v_\tau := E(\tau) : \dot{S} \to \mathbb{R}^2$ *is smooth and* $u_\tau :=$ *graph*(v_τ) *is a complex submanifold of* $(\dot{S} \times \mathbb{R}^2, \bar{J})$ *satisfying*

$$Tu_\tau(z) \cdot \bar{j}_\tau(z) = \bar{J}(u_\tau(z)) \cdot Tu_\tau(z), \quad z \in \dot{S}.$$

Here $\bar{j}_\tau$ *is a smooth almost complex structure on* $\dot{S}$ *which has a continuous extension to an* α*-Hölder continuous almost complex structure on* S *satisfying at the punctures*

$$\bar{j}_\tau(z_0) = j(z_0), \quad \text{for all } z_0 \in \Gamma,$$

and for all $\tau \in B^d_\varepsilon(0)$.

Proof. By assumption, the Fredholm operator $H'(0)$ is surjective and has a d-dimensional kernel $X := \ker H'(0)$. Choosing a topological complement Y we have the splitting $C^{1,\alpha}_0(\dot{S}, \mathbb{R}^2) = X \oplus Y$. Introducing a basis we can identify $X \equiv \mathbb{R}^d$ and find, recalling that $H(0 \oplus 0) = 0$, by means of the implicit function theorem a unique smooth map $B_\varepsilon(0) \subset \mathbb{R}^d \to Y$, $\tau \mapsto w(\tau)$ such that $E(\tau) := \tau \oplus w(\tau)$ satisfies $H(E(\tau)) = 0$ and $E(0) = 0$. We fix now $\tau \in \mathbb{R}^d$ and abbreviate $v = E(\tau) \in C^{1,\alpha}_0(\dot{S}, \mathbb{R}^2)$ and define $u(z) = (z, v(z)) \in \dot{S} \times \mathbb{R}^2$. By Proposition 3.1,

$$Tu(z)\bar{j}(z) = \bar{J}(u(z))Tu(z), \quad z \in \dot{S} \tag{49}$$

with an α-Hölder continuous almost complex structure $\bar{j}$ on $\dot{S}$. In order to prove that $z \mapsto v(z)$ is smooth we use the fact, see Appendix, that locally around $z \in \dot{S}$ there exists a $C^{1,\alpha}$ diffeomorphism $\varphi : \mathbb{C} \to U(z) \subset \dot{S}$ satisfying

$$\bar{j}(\varphi(\xi)) \cdot T\varphi(\xi) = T\varphi(\xi) \circ i.$$

Consequently,

$$T(u \circ \varphi)(\xi) \circ i = \bar{J}(u \circ \varphi)(\xi) \circ T(u \circ \varphi)(\xi).$$

Hence, by the elliptic regularity theory, the composition $(u \circ \varphi)(\xi) = (\varphi(\xi), v \circ \varphi(\xi))$ is smooth. In particular φ is a smooth local diffeomorphism. It follows that $v : \dot{S} \to \mathbb{R}^2$ is a smooth map, hence $\bar{j}$ is a smooth almost complex structure on $\dot{S}$. Representing $\bar{J}$ according to the splitting $T(\dot{S} \times \mathbb{R}^2) = T\dot{S} \times \mathbb{R}^2$ in matrix form,

$$\bar{J}(z, x, y) = \begin{pmatrix} j(z, x, y) & \tilde{\Delta}(z, x, y) \\ \Delta(z, x, y) & i(z, x, y) \end{pmatrix},$$

the equation (49) becomes, in view of $Tu(z) = (1, Tv(z))$,

$$\begin{aligned} \bar{j}(z) &= j(u(z)) + \tilde{\Delta}(u(z))Tv(z) \\ Tv(z)\bar{j}(z) &= \Delta(u(z)) + i(u(z))Tv(z). \end{aligned} \tag{50}$$

Inserting the expression for $\bar{j}(z)$, the smooth map $v : \dot{S} \to \mathbb{R}^2$ solves the nonlinear system

$$Tv(z)j(u(z)) - i(u(z))Tv(z) = \Delta(u(z)) - Tv(z)\tilde{\Delta}(u(z))Tv(z).$$

By construction, $u(z)$ is close to $(z,0) \in \dot{S} \times \{0\}$ in $C_0^{1,\alpha}(\dot{S}, \mathbb{R}^2)$. Since $u(z) = (z, v(z))$ and $v \in C_0^{1,\alpha}(\dot{S}, \mathbb{R}^2)$ and since near the punctures $j(z,0) = j(z)$, $\Delta(z,0) = 0$ and $\tilde{\Delta}(z,0) = 0$ we conclude from (50) that $\bar{j}$ is smooth on $\dot{S}$ and has a Hölder continuous extension to the punctures $z_0 \in \Gamma$ and $\bar{j}(z_0) = j(z_0)$ independent of τ. This finishes the proof of Theorem 5.6. □

In the special case of a non-degenerate finite energy plane, where $S = S^2$ and $\dot{S} = S^2 \setminus \{\infty\} = \mathbb{C}$, the following stronger statement holds true.

Theorem 5.7. *Consider a non-degenerate finite energy plane $\tilde{u}_0 : (\mathbb{C}, i) \to (\mathbb{R} \times M, \tilde{J})$ and assume $\mu(\tilde{u}_0) = 3$. Let*

$$\Phi : (\mathbb{C} \times \mathbb{R}^2, \bar{J}) \to (\mathbb{R} \times M, \tilde{J}), \quad \Phi(z,0) = \tilde{u}_0(z)$$

be the trivialization along $\tilde{u}_0$ as constructed above. Then there exists a smooth embedding

$$u : B_\varepsilon^2(0) \times \mathbb{C} \to \mathbb{C} \times \mathbb{R}^2$$

of the form $u(\tau, z) = (z, v(\tau, z))$ satisfying $v(0,z) = 0$ for $\tau = 0$. The maps $v_\tau : \mathbb{C} \to \mathbb{R}^2$ and $u_\tau : \mathbb{C} \to \mathbb{C} \times \mathbb{R}^2$, by defined for a fixed τ by $v_\tau(z) = v(\tau, z)$ and $u_\tau(z) = (z, v_\tau(z))$, solve the equation

$$H(v_\tau) = 0, \quad \tau \in B_\varepsilon^2(0)$$

and, in particular,

$$Tu_\tau(z) \cdot \bar{j}_\tau(z) = \bar{J}(u_\tau(z)) \cdot Tu_\tau(z), \quad z \in \mathbb{C}$$

for a smooth almost complex structure $\bar{j}_\tau$ on $\mathbb{C}$ satisfying $\bar{j}_\tau(z) \to i$ as $|z| \to \infty$.

Proof. By Corollary 1.9 we have $\mu_N(\tilde{u}_0) = \mu(\tilde{u}_0) - 2 = 1$. Therefore, in view of Theorem 2.11, the Fredholm operator $H'(0)$ is surjective and has a two-dimensional kernel. Picking a basis A_1 and A_2 of the kernel we introduce the map $\delta : \mathbb{R}^2 \to \ker H'(0)$ defining

$$\delta(\tau) = \tau_1 A_1 + \tau_2 A_2, \quad \tau = (\tau_1, \tau_2) \in \mathbb{R}^2.$$

By Theorem 2.11,

$$\delta(\tau)(z) \neq 0 \quad \text{for all } z \in \mathbb{C}, \tau \neq 0.$$

Choose now a topological complement Y so that $C_0^{1,\alpha}(\mathbb{C}, \mathbb{R}^2) = \ker H'(0) \oplus Y$. Applying the implicit function theorem we find a unique smooth map $\tau \mapsto \gamma(\tau) \in Y$ defined near $\tau = 0$, and satisfying $\gamma(0) = 0$ and

$$H(\delta(\tau) \oplus \gamma(\tau)) = 0.$$

Introduce the map $v : B^2_\varepsilon(0) \times \mathbb{C} \to \mathbb{R}^2$,

$$v(\tau, z) = \delta(\tau)(z) + \gamma(\tau)(z)$$

and the map $u : B^2_\varepsilon(0) \times \mathbb{C} \to \mathbb{C} \times \mathbb{R}^2$,

$$u(\tau, z) = (z, v(\tau, z)).$$

We claim that u is smooth. Indeed, for fixed τ the map $v(\tau)$ defined by $v(\tau)(z) = v(\tau, z)$ is smooth by elliptic regularity theory. Moreover, the map $\tau \mapsto v(\tau)$ from B into the fixed Hölder space $C^{1,\alpha}_0$ is smooth. Carrying out the Fredholm theory and the implicit function theory in different Hölder spaces shows that the map from B into $C^{k,\alpha}_0$ is smooth for every k. Consequently, u is smooth. We shall prove that u is an embedding if $\varepsilon > 0$ is sufficiently small. Differentiating the zero section $\tau \mapsto H(v(\tau)) = 0$ in τ we obtain

$$H'(v(\tau)) \cdot (Dv(\tau) \cdot \Delta\tau) = 0$$

for all $\Delta\tau \in \mathbb{R}^2$. The linear operator $H'(v(\tau))$ is of the same type as $H'(0)$. It has, in particular, the same asymptotic behavior governed by the non-degenerate periodic orbit P_0 and hence the same Fredholm index as $H'(0)$. Therefore, we can apply Theorem 2.11 again in order to conclude that the element $Dv(\tau) \cdot \Delta\tau$ in the kernel of $H'(v(\tau))$ either vanishes identically or does not have any zeros. Fix a point $\hat{z} \in \mathbb{C}$. Then clearly $(Dv(\tau) \cdot \Delta\tau)(\hat{z}) \neq 0$ if τ is sufficiently small and $\Delta\tau \neq 0$. Consequently,

$$(Dv(\tau) \cdot \Delta\tau)(z) \neq 0 \quad \text{for all } z \in \mathbb{C},\ \tau \text{ small and } \Delta\tau \neq 0.$$

Differentiating the map $(\tau, z) \mapsto (z, v(\tau, z)) = u(\tau, z)$ we obtain

$$Du(\tau, z)(\Delta\tau, \Delta z) = \big[\Delta z, Tv(\tau)(z) \cdot \Delta z\big] + (Dv(\tau) \cdot \Delta\tau)(z)$$

for $\Delta\tau \in \mathbb{R}^2$ and $\Delta z \in \mathbb{C}$. Thus $Du(\tau, z) : \mathbb{R}^2 \times \mathbb{C} \to \mathbb{C} \times \mathbb{R}^2$ is a linear isomorphism at every point $(\tau, z) \in B^2_\varepsilon(0) \times \mathbb{C}$ if $\varepsilon > 0$ is small. It remains to show that u is injective. Introduce the complex submanifolds of $(\mathbb{C} \times \mathbb{R}^2, \bar{J})$,

$$C_\tau = u(\tau, \mathbb{C}), \quad \text{then } C_0 = u(0, \mathbb{C}) = \mathbb{C} \times \{0\}.$$

Fix $\tau \neq 0$. We shall first prove that $C_\tau \cap C_0 = \emptyset$. We argue by contradiction and assume that $C_\tau \cap C_0 \neq \emptyset$. Then there exists a point $\hat{z} \in \mathbb{C}$ such that $v(\tau, \hat{z}) = 0$. In order to determine the asymptotic behavior of L_C we use the holomorphic polar coordinates $z = \sigma(s, t) = e^{2\pi(s+it)}$ and abbreviate the composition by $w(s, t) = v(\sigma(s, t))$. By construction, applying the implicit function theorem to the spaces $C^{k,\alpha}_0$ for all k, we have $D^\alpha w(s, t) \to 0$ as $s \to \infty$ uniformly in t, for all derivatives α. In view of Theorem 4.7 and Lemma 5.2, the equation $L_C(v)(z) = 0$ becomes

$$w_s(s, t) + iw_t(s, t) + C_\tau(s, t)w(s, t) = 0,$$

where $C_\tau(s, t) \in (\mathbb{R}^2, \mathbb{R}^2)$ satisfies $D^\alpha[C_\tau(s, t)h - \tilde{C}(t)h] \to 0$ as $s \to \infty$ uniformly in t, for every $h \in \mathbb{R}^2$ and all derivatives α. The map $\tilde{C}(t)$ is independent of τ. Moreover, as before, the problem $\dot{x}(t) = i\tilde{C}(t)x(t)$, $x(0) = x(1)$ admits the trivial solution only. Consequently v is in the kernel of a Fredholm operator L_C which is of the same type as $H'(0)$. In particular, the asymptotic behavior is the same and hence also the index. Since $v(\tau, \hat{z}) = 0$ we therefore conclude from Theorem 2.11 that $v(\tau, z) \equiv 0$ for all $z \in \mathbb{C}$ so that $C_\tau = C_0$. This, however, contradicts the fact that $\tau \mapsto v(\tau)$ is injective and we conclude that $C_\tau \cap C_0 = \emptyset$ if $\tau \neq 0$. By the same argument

$$C_\tau \cap C_{\tau'} = \emptyset \quad \text{if } \tau \neq \tau'.$$

The proof of Theorem 5.7 is complete. □

6. Fredholm theory with exponential weights

This section is devoted to the extension of the Fredholm theory to situations in which the asymptotic limits associated with the punctures are not necessarily non-degenerate. Proceeding as in Section 2 we first introduce the appropriate function spaces. The spaces $C_0^\alpha(\dot{S}, \mathbb{R}^2)$ and $C_0^{1,\alpha}(\dot{S}, \mathbb{R}^2)$ of functions converging to 0 at the punctures have been introduced in Section 2, and we shall define spaces of functions which converge to 0 at the exponential rate approaching the punctures.

Let $\varepsilon > 0$. A continuous map $u : [0, \infty) \times S^1 \to \mathbb{R}^2$ is called of class $C_0^{\alpha,\varepsilon}$ near $s = \infty$ if $e^{\varepsilon s} u$ is of class C_0^α near ∞. Consequently, the map $u : \dot{S} \to \mathbb{R}^2$ is of class $C_0^{\alpha,\varepsilon}$ near a positive puncture $z_0 \in S$, if for some holomorphic coordinates $\sigma : [0, \infty) \times S^1 \to D_{z_0}$ onto the punctured neighborhood $D_{z_0} \subset \dot{S}$, the composition $u \circ \sigma$ belongs to the class $C_0^{\alpha,\varepsilon}$ near ∞. We treat the negative punctures similarly, calling $u : (-\infty] \times S^1 \to \mathbb{R}^2$ of class $C_0^{\alpha,\varepsilon}$ if $e^{-\varepsilon s} u$ belongs to C_0^α near $(-\infty)$. If z_0 is a negative puncture, then the map $u : \dot{S} \to \mathbb{R}^2$ belongs to $C_0^{\alpha,\varepsilon}$ near z_0, if for some cylindrical holomorphic coordinates $\sigma : (-\infty, 0] \times S^1 \to D_{z_0}$, the composition $u \circ \sigma$ belongs to $C_0^{\alpha,\varepsilon}$ near $(-\infty)$.

Finally, a continuous map $u : \dot{S} \to \mathbb{R}^2$ belongs to $C_0^{\alpha,\varepsilon}(\dot{S}, \mathbb{R}^2)$ if it meets the following requirements. At every point $z \in \dot{S}$ there exists a chart $\varphi : V(z) \to W(0) \subset \mathbb{C}$ such that the composition $u \circ \varphi^{-1}$ is of class C_{loc}^α and if near the punctures the map u is of class $C_0^{\alpha,\varepsilon}$ as introduced above. Clearly, $C_0^{\alpha,\varepsilon}(\dot{S}, \mathbb{R}^2)$ is a Banachable space.

Next, the space $C_0^{1,\alpha,\varepsilon}(\dot{S}, \mathbb{R}^2)$ will be defined analogously. It consists of C^1-maps $u : \dot{S} \to \mathbb{R}^2$ belonging to $C_{\text{loc}}^{1,\alpha}$ and such that the first derivatives near a positive puncture

$$\frac{\partial(u \circ \sigma)}{\partial s}, \ \frac{\partial(u \circ \sigma)}{\partial t} : [0, \infty) \times S^1 \to \mathbb{R}^2$$

belong to $C_0^{\alpha,\varepsilon}$ near ∞, while near a negative puncture, the derivatives

$$\frac{\partial(u \circ \sigma)}{\partial s}, \ \frac{\partial(u \circ \sigma)}{\partial t} : (-\infty, 0] \times S^1 \to \mathbb{R}^2$$

belong to $C_0^{\alpha,\varepsilon}$near $(-\infty)$. Again, $C_0^{1,\alpha,\varepsilon}(\dot{S}, \mathbb{R}^2)$ is Banachable space.

As before we denote by $j = j(z)$ a complex structure on the Riemann surface S. Let $z \mapsto i(z)$ be a smooth map associating with $z \in \dot{S}$ a complex multiplication on $\mathbb{R}^2$. We require that near a positive puncture, the complex multiplication $i(z)$ has the following asymptotic behavior in our holomorphic cylindrical coordinates σ:

$$D^\beta \left[i(\sigma(s,t)) - \tilde{i}(t)\right] \to 0$$

as $s \to \infty$, uniformly in $t \in S^1$, for all derivatives β. The map $t \mapsto \tilde{i}(t)$ is a smooth loop of complex multiplications of $\mathbb{R}^2$. The corresponding asymptotic behavior is required near a negative puncture.

Recall the bundle $\mathcal{A}_0 \to \dot{S}$. The fiber over the point $z \in \dot{S}$ consists of all complex antilinear maps $(T_z\dot{S}, j(z)) \to (\mathbb{R}^2, i(z))$. Consider a smooth map $C \in \operatorname{Hom}_{\mathbb{R}}(\dot{S} \times \mathbb{R}^2, \mathcal{A}_0)$. We shall require that the map C has the following asymptotic behavior near the punctures. For every positive puncture z_0 there exists a smooth loop $\tilde{C}(t) \in \mathcal{L}(\mathbb{R}^2)$ such that, in the holomorphic cylindrical coordinates $\sigma : [0,\infty) \times S^1 \to D_{z_0} \subset \dot{S}$ around the puncture $z_0 = \lim \sigma(s,t)$, we have

$$D^\beta \Big\{ [C(\sigma(s,t))h] \cdot \frac{\partial \sigma}{\partial s}(s,t) - \tilde{C}(t)h \Big\} \to 0$$

as $s \to \infty$, uniformly in t for all derivatives β and for every $h \in \mathbb{R}^2$. In addition, we require $\tilde{C}(t)$ to be symmetric with respect to the inner product on $\mathbb{R}^2$ defined by

$$g_t = \omega \circ (\operatorname{Id} \times \tilde{i}(t)),$$

$\omega = dx \wedge dy$. An analogous and corresponding asymptotic behavior for C is required near every negative puncture. The smooth map C induces the bounded linear operator, denoted by the same letter,

$$C : C_0^{1,\alpha,\varepsilon}(\dot{S}, \mathbb{R}^2) \to C_0^{\alpha,\varepsilon}(\mathcal{A}_0),$$

and defined by

$$(Cu)(z) = C(z)u(z), \quad z \in \dot{S}.$$

With the smooth bundle map $C \in \operatorname{Hom}_{\mathbb{R}}(\dot{S} \times \mathbb{R}^2, \mathcal{A}_0)$ having the asymptotic properties as required above, we associate the familiar linear operator

$$L_C : C_0^{1,\alpha,\varepsilon}(\dot{S}, \mathbb{R}^2) \to C_0^{\alpha,\varepsilon}(\mathcal{A}_0),$$
$$(L_C u)(z) = Tu(z) + i(z) \circ Tu(z) \circ j(z) + (Cu)(z).$$

We shall investigate the Fredholm property of this operator. Let $\varepsilon > 0$ and $u \in C_0^{1,\alpha,\varepsilon}(\dot{S}, \mathbb{R}^2)$. We consider a positive puncture z_0, where we introduce the distinguished holomorphic coordinates $\sigma : [0,\infty) \times S^1 \to D_{z_0} \subset \dot{S}$, and define $v \in C_0^{1,\alpha}(D_{z_0}, \mathbb{R}^2)$ by

$$v(\sigma(s,t)) = e^{\lambda(s)} u(\sigma(s,t))$$

for all $(s,t)\in[0,\infty)\times S^1$, with a smooth function $\lambda:[0,\infty)\to\mathbb{R}$ satisfying $\lambda\geq 0$ and

$$\begin{aligned}\lambda(s)&=0 \quad \text{if } 0\leqslant s\leqslant 1\\ \lambda(s)&=\varepsilon s \quad \text{if } s\geq 2.\end{aligned}\tag{51}$$

Recall that $T\sigma\circ i=j(\sigma)\circ T\sigma$. Abbreviating $w(s,t)=v(\sigma(s,t))$ we compute

$$\begin{aligned}&\frac{\partial w}{\partial s}+i(\sigma)\frac{\partial w}{\partial t}+\big[C(\sigma)w\big]\cdot\frac{\partial\sigma}{\partial s}-\frac{d}{ds}\lambda(s)w(s,t)\\ &\quad=Tv(\sigma)\frac{\partial\sigma}{\partial s}+i(\sigma)Tv(\sigma)\frac{\partial\sigma}{\partial t}+\big[C(\sigma)v(\sigma)\big]\cdot\frac{\partial\sigma}{\partial s}-\frac{d}{ds}\lambda(s)v(\sigma)\\ &\quad=\big[Tv+i\circ Tv\circ j+Cv\big](\sigma)\cdot\frac{\partial\sigma}{\partial s}-\frac{d}{ds}\lambda(s)v(\sigma)\\ &\quad=e^{\lambda(s)}\big[Tu+i\circ Tu\circ j+Cu\big](\sigma)\cdot\frac{\partial\sigma}{\partial s}.\end{aligned}\tag{52}$$

We have proved,

$$e^{\lambda(s)}\big[L_Cu\big](\sigma)\cdot\frac{\partial\sigma}{\partial s}=\big[L_Cv\big](\sigma)\cdot\frac{\partial\sigma}{\partial s}-\frac{d}{ds}\lambda(s)v(\sigma).$$

Changing the bundle map C inside D_{z_0} we introduce C_ε by

$$\big[C_\varepsilon(\sigma(s,t))h\big]\cdot\frac{\partial\sigma}{\partial s}=\big[C(\sigma(s,t))h\big]\cdot\frac{\partial\sigma}{\partial s}-\frac{d}{ds}\lambda(s)h\tag{53}$$

for all $(s,t)\in[0,\infty)$ and $h\in\mathbb{R}^2$. In particular $C_\varepsilon(\sigma(s,t))=C(\sigma(s,t))$ if $0\leqslant s\leqslant 1/2$. Since $\frac{d}{ds}\lambda(s)=\varepsilon$ for $s\geq 2$, the asymptotic behavior of C_ε near the negative puncture z_0 becomes, in view of (6),

$$D^\beta\Big\{[C_\varepsilon(\sigma(s,t))h]\cdot\frac{\partial\sigma}{\partial s}-(\tilde{C}(t)-\varepsilon)h\Big\}\to 0$$

as $s\to\infty$, uniformly in t for all derivatives β and $h\in\mathbb{R}^2$. We carry out this modification of C inside the punctured neighborhoods D_{z_0} of every positive puncture $z_0\in\Gamma$. Choosing analogously an appropriate cut-off function we also modify C inside the neighborhoods D_{z_0} of all negative punctures, such that

$$D^\beta\Big\{[C_\varepsilon(\sigma(s,t))h]\cdot\frac{\partial\sigma}{\partial s}-(\tilde{C}(t)+\varepsilon)h\Big\}\to 0$$

as $s\to-\infty$. Away from the punctured neighborhoods D_{z_0} of $z_0\in\Gamma$ we set $C_\varepsilon(z)=C(z)$. We shall abbreviate the matrices

$$\tilde{C}_\varepsilon^\pm(t):=\tilde{C}(t)\mp\varepsilon\,\mathrm{Id}$$

corresponding to the positive resp. negative punctures. Now, we assume that there exists $\varepsilon>0$ such that all the loops $\tilde{C}_\varepsilon^\pm(t)$ are non-degenerate in the usual sense that the equations for $x(t)\in\mathbb{R}^2$, $\dot{x}(t)=\tilde{i}(t)\tilde{C}_\varepsilon^\pm(t)x(t)$, $x(0)=x(1)$ admit only the trivial solution $x\equiv 0$. Define

$$L_{C_\varepsilon}:C_0^{1,\alpha}(\dot{S},\mathbb{R}^2)\to C_0^\alpha(\mathcal{A}_0)$$

by

$$(L_{C_\varepsilon} v)(z) = Tv(z) + i(z) \circ Tv(z) \circ j(z) + (C_\varepsilon v)(z).$$

In view of the results of Section 2, the operator L_{C_ε} is a Fredholm operator. Its index equals

$$\begin{aligned}\operatorname{Ind}(L_{C_\varepsilon}) &= \sum_{j=1}^{m} \mu_N(\tilde{C}^+_{j,\varepsilon}) - \sum_{j=1}^{m'} \mu_N(\tilde{C}^-_{j,\varepsilon}) + \chi_S - \sharp\Gamma_S \\ &= \mu_N(L_{C_\varepsilon}) + \chi_S - \sharp\Gamma_S.\end{aligned}$$

The integers $\mu_N(\tilde{C}^+_{j,\varepsilon})$ and $\mu_N(\tilde{C}^-_{j,\varepsilon})$ are the normal Conley-Zehnder indices of the asymptotic limits corresponding to the m positive and the m' negative punctures. We recall from Proposition 2.10, that

$$\mu_N(\tilde{C}^+_{j,\varepsilon}) = 2\alpha(\tilde{C}^+_{j,\varepsilon}) + p(\tilde{C}^+_{j,\varepsilon}),$$

where $\alpha(\tilde{C}^+_{j,\varepsilon})$ is the largest winding number occurring for the negative eigenvalues of the asymptotic linear operator $A_\infty = -\tilde{i}\frac{d}{dt} - \tilde{C}^+_{j,\varepsilon}(t)$. The integer $p(\tilde{C}^+_{j,\varepsilon})$ is equal to 1 if there are precisely two negative eigenvalues having winding numbers equal to $\alpha(\tilde{C}^+_{j,\varepsilon})$. Otherwise $p(\tilde{C}^+_{j,\varepsilon}) = 0$.

Theorem 6.1. *Assume that C_ε is non-degenerate at all punctures. Then the operator*

$$L_C : C_0^{1,\alpha,\varepsilon}(\dot{S}, \mathbb{R}^2) \to C_0^{\alpha,\varepsilon}(\mathcal{A}_0),$$

is Fredholm, and

$$\operatorname{Ind}(L_C) = \operatorname{Ind}(L_{C_\varepsilon}).$$

Proof. It suffices to show that the dimensions of the kernels resp. cokernels of the two operators L_C and L_{C_ε} agree. Assume that $u \in \ker L_C$. Then $u \in C_0^{1,\alpha,\varepsilon}(\dot{S}, \mathbb{R}^2)$ and $L_C u = 0$. Define $v \in C_0^{1,\alpha}(\dot{S}, \mathbb{R}^2)$ by the modifications above in the punctured neighborhoods D_{z_0} of the punctures. For example, if z_0 is a positive puncture, we set $v(\sigma(s,t)) = e^{\lambda(s)} u(\sigma(s,t))$. Then it follows from the computation (52) and the definition (53) that $L_{C_\varepsilon} v = 0$. Therefore, $\dim(\ker L_C) \leqslant \dim(\ker L_{C_\varepsilon})$. Similarly one proves that $\dim(\ker L_{C_\varepsilon}) \leqslant \dim(\ker L_C)$, so that the dimensions of the kernels do agree. Using the same modifications of the functions one verifies that L_C is surjective if L_{C_ε} is surjective. In the general case one uses the arguments of Proposition 2.7 in order to verify that the dimensions of the cokernels of L_{C_ε} and L_C agree. This finishes the proof of Theorem 6.1. □

In order to give an application of Theorem 6.1 we consider complex curves in $(\dot{S} \times \mathbb{R}^2, \bar{J})$. We recall that the almost complex structure $\bar{J}$ has, with respect to the splitting $T(\dot{S} \times \mathbb{R}^2) = T\dot{S} \times \mathbb{R}^2$, the matrix form

$$\bar{J}(z,x,y) = \begin{pmatrix} j(z,x,y) & \tilde{\Delta}(z,x,y) \\ \Delta(z,x,y) & i(z,x,y) \end{pmatrix},$$

and,

$$\bar{J}(z,0) = \begin{pmatrix} j(z) & 0 \\ 0 & i(z) \end{pmatrix}, \quad z \in \dot{S},$$

where $j(z)$ is an almost complex structure on $\dot{S}$ and $i(z)$ a complex multiplication in $\mathbb{R}^2$. We are interested in complex curves in $\dot{S} \times \mathbb{R}^2$ presented by graphs $u(z) = (z, v(z))$ for functions $v : \dot{S} \to \mathbb{R}^2$. The condition determining a complex curve is the following:

$$\bar{J}(u(z))\big[Tu(z)(T_z\dot{S})\big] = Tu(z)(T_z\dot{S}). \tag{54}$$

With $\bar{J}$ we associate the Cauchy-Riemann type operator for functions $v : \dot{S} \to \mathbb{R}^2$

$$(L_C v)(z) = Tv(z) + i(z) \circ Tv(z) \circ j(z) + (Cv)(z) \tag{55}$$

where

$$(Cv)(z) = \big[\Delta'(z)v(z)\big] \circ j(z).$$

We recall that $\Delta(z,x,y) \in \mathcal{L}(T_z\dot{S}, \mathbb{R}^2)$ and $\Delta'(z)$ is the derivative of $\Delta(z,x,t)$ in the variables (x,y) at $(x,y)=0$. We shall describe an implicit function theorem for situations in which C is possibly degenerate at the punctures.

For the analytical setting we proceed as in Section 5. We denote by $\Theta \to \dot{S} \times \mathbb{R}^2$ the 2-dimensional real vector bundle whose fiber over (z,x,y) is the space of all two-vectors in $T_z(\dot{S} \times \mathbb{R}^2) = T_z\dot{S} \times \mathbb{R}^2$ on which $\check{\bar{J}}(z,x,y)$ acts via multiplication by -1. Using Θ we introduce the bundle $\Xi \to \dot{S} \times \mathbb{R}^2$ whose fiber over (z,x,y) consists of all real linear maps from Λ_z into $\Theta_{(z,x,y)}$, where Λ_z is the 1-dimensional space of two-vectors on $T_z\dot{S}$. Let

$$\Psi : (T(\dot{S} \times \mathbb{R}^2), \bar{J}) \to ((\dot{S} \times \mathbb{R}^2) \times \mathbb{C}^2, i \oplus i)$$

be a smooth complex trivialization admissible in the sense that near a puncture (here a positive one)

$$(T\sigma(s,t) \oplus \mathrm{Id}_{\mathbb{R}^2})^{-1} \cdot \Psi(\sigma(s,t),x,y) \cdot (T\sigma(s,t) \oplus \mathrm{Id}_{\mathbb{R}^2})$$

converges as $s \to \infty$ uniformly on bounded (t,x,y)-sets, including all its derivatives, to a smooth map

$$(t,x,y) \mapsto \Psi_\infty(t,x,y) \in \mathcal{GL}(\mathbb{R}^4).$$

The trivialization Ψ induces the trivialization

$$\hat{\Psi} : \Theta \to (\dot{S} \times \mathbb{R}^2) \times \Theta_0,$$

where Θ_0 is the 2-dimensional real vector space of two-vectors in $\mathbb{C}^2$ (viewed as a real vector space) on which $\check{i}$ operates as multiplication by -1.

Choose $v \in C_0^{1,\alpha,\varepsilon}(\dot{S}, \mathbb{R}^2)$ and denote by $\mathrm{gr}(v)(z) = (z, v(z))$ the graph of v. Then a section $f : \dot{S} \to \Xi$ belongs to $C_0^{\alpha,\varepsilon}(\mathrm{gr}(v)^*\Xi)$ if f belongs to C^α_{loc}, satisfies

$f(z) \in \Xi_{(z,v(z))}$, and if in addition, near every positive puncture, the map

$$(s,t) \mapsto pr_2 \circ \hat{\Psi}(\sigma(s,t), v(s,t)) \circ f(\sigma(s,t))\left(\frac{\partial \sigma}{\partial s} \wedge \frac{\partial \sigma}{\partial t}\right)$$

belongs to $C_0^{\alpha,\varepsilon}([0,\infty) \times S^1, \Theta_0)$ at ∞. As above, $\sigma : [0,\infty) \times S^1 \to D_{z_0}$ are the holomorphic cylindrical coordinates near the positive puncture z_0. The analogous requirement holds true near the negative punctures. Now define the bundle $\mathcal{E} \to \mathcal{B}$ by

$$\mathcal{B} = C_0^{1,\alpha,\varepsilon}(\dot{S}, \mathbb{R}^2), \qquad \mathcal{E} = \bigcup_{v \in \mathcal{B}} \{v\} \times C_0^{\alpha,\varepsilon}(\mathrm{gr}(v)^*\Xi).$$

If $v \in \mathcal{B}$ and $z \in \dot{S}$, we define the map $\bar{\eta}(v)(z)$ as in Section 3 as the linear map

$$h \wedge k \mapsto Tv(z)h \wedge Tv(z)k - \bar{J}(\mathrm{gr}(v)(z))Tv(z)h \wedge \bar{J}(\mathrm{gr}(v)(z))Tv(z)k.$$

Now, the section $H : \mathcal{B} \to \mathcal{E}$ is defined by

$$H(v) = \bar{\eta}(v), \quad v \in \mathcal{B}.$$

Since the wedge product is bilinear and the product of two functions of class $C_0^{\alpha,\varepsilon}$ belongs to the class $C_0^{\alpha,\varepsilon}$, it indeed follows that $H(v) \in C_0^{\alpha,\varepsilon}(\mathrm{gr}(v)^*\Xi)$.

Lemma 6.2. *The section H of the bundle $\mathcal{E} \to \mathcal{B}$ is smooth.*

We omit the technical proof. The condition $H(v) = 0$ is equivalent to the condition (54) for $u = \mathrm{graph}(v)$. Since $H(0) = 0$ we are tempted to apply the implicit function theorem. For the derivative of H at $v = 0$ we have

$$H'(0)v = \alpha(L_C v),$$

with the linear operator L_C defined by (55) and the vector bundle map $\alpha : \mathcal{A}_0 \to \Xi_0$ defined by

$$\alpha(a) \cdot (h \wedge k) = [h, 0] \wedge [0, a(z) \cdot k] - [k, 0] \wedge [0, a(z) \cdot h]$$

for every $h, k \in T_z\dot{S}$. We therefore conclude that $H'(0)$ is Fredholm whenever there exists $\varepsilon > 0$ such that the operator L_C in Theorem 6.1 is Fredholm. Applying the implicit function theorem we therefore deduce

Theorem 6.3. *Assume there exists an $\epsilon > 0$ such that L_C is a surjective Fredholm operator. Choose a topological complement Y of the kernel such that*

$$C_0^{1,\alpha,\varepsilon}(\dot{S}, \mathbb{R}^2) = \ker(L_C) \oplus Y.$$

Then there exists a unique smooth map

$$w : \ker(L_C) \to Y, \quad \tau \mapsto w(\tau)$$

defined in an open neighborhood $U \subset \ker(L_C)$ of $\tau = 0$, satisfying $w(0) = 0$ such that $v(\tau) = \tau + w(\tau)$ solves

$$H(v(\tau)) = 0, \quad \text{for all } \tau \in U.$$

7. Compactness and completeness

We consider the embedded generalized finite energy surface $\tilde{u}_0 : (\dot{S}, j) \to (\mathbb{R} \times M, \tilde{J})$ as described in the introduction. Recall that M is equipped with the contact form λ. Moreover, recall that $f : \mathbb{R} \times M \to (0, \infty)$ is a smooth function satisfying

$$f(a,m) = \begin{cases} g(m) & a \geq \ \ a_0 \\ h(m) & a \leqslant -a_0 \end{cases}$$

and hence is independent of a at the ends. The function f induces the one-form $f\lambda$ on $\mathbb{R} \times M$ and the family $\lambda_a = f(a, \cdot)\lambda$ of contact forms on M. By $X_{f\lambda}$ we denote the associated family of Reeb vector fields on M.

In the previous sections we introduced the special coordinates $\Phi : \dot{S} \times B_\varepsilon \to \mathbb{R} \times M$ near $\tilde{u}_0(\dot{S})$ satisfying $\Phi(z, 0) = \tilde{u}_0(z)$ for all $z \in \dot{S}$, so that $\tilde{u}_0(\dot{S})$ is represented by

$$\tilde{u}_0(\dot{S}) = \{\Phi(z,0) \mid z \in \dot{S}\}.$$

We look for conditions on nearby surfaces $\tilde{u} : \dot{S} \to \mathbb{R} \times M$ whose images can be represented by

$$\tilde{u}(\dot{S}) = \{\Phi(z, v(z)) \mid z \in \dot{S}\},$$

for smooth functions $v : \dot{S} \to \mathbb{R}^2$.

Theorem 7.1. *Assume (S, j_0) is a compact Riemann surface. Let $\Gamma = \{z_1, \dots, z_N\} \subset S$ be a non-empty finite set and denote by $\dot{S} = S \setminus \Gamma$ the punctured surface. Consider an embedded generalized finite energy surface*

$$\tilde{u}_0 : (\dot{S}, j_0) \to (\mathbb{R} \times M, \tilde{J})$$

for which the punctures z_j correspond to non-degenerate periodic orbits (x_j, T_j) of the Reeb vector fields, $1 \leqslant j \leqslant N$. Denote by

$$\Phi : (\dot{S} \times B_\varepsilon, \bar{J}) \to (\mathbb{R} \times M, \tilde{J})$$

the special coordinates introduced in Section 4 satisfying

$$\Phi(z, 0) = \tilde{u}_0(z) \quad \text{for all } z \in \dot{S}.$$

Assume that

$$\tilde{u}_k : (\dot{S}, j_k) \to (\mathbb{R} \times M, \tilde{J}),$$

where j_k are complex structures on $\dot{S}$ which have C^0-extension over Γ, is a sequence of embedded generalized finite energy surfaces satisfying

$$\tilde{u}_k \to \tilde{u}_0 \quad \text{in } C^{1,\alpha}_{loc}(\dot{S}, \mathbb{R} \times M)$$

for some $0 < \alpha < 1$. We assume, in addition, that the nature of the punctures Γ (positive and negative respectively) for every map $\tilde{u}_k$ is the same as for $\tilde{u}_0$. Moreover, at the punctures the asymptotic limits of $\tilde{u}_k$ and $\tilde{u}_0$ are the same periodic *orbits. Then there exists an integer k_0 and for every $k \geq k_0$ there exists a smooth map $v_k : \dot{S} \to \mathbb{R}^2$ satisfying $v_k(\dot{S}) \subset B_\varepsilon$ and representing the image of $\tilde{u}_k$ by*

$$\tilde{u}_k(\dot{S}) = \{\Phi(z, v_k(z)) \mid z \in \dot{S}\}.$$

In addition,

$$v_k \to 0 \qquad \text{in } C_0^{l,\alpha}(\dot{S}, \mathbb{R}^2)$$

for every integer l, and

$$j_k \to j_0 \qquad \text{in } \ C^\infty_{loc}(\dot{S}) \cap C^0(S).$$

Proof. In the following we shall abbreviate

$$C_0 = \tilde{u}_0(\dot{S}) \quad \text{ and } \quad C_k = \tilde{u}_k(\dot{S}).$$

Since $\tilde{u}_k \to \tilde{u}_0(\dot{S})$ in $C_0^{1,\alpha}$, we find for every compact set $K \subset \dot{S}$ a compact $K_k \subset \dot{S}$ such that

$$\tilde{u}_k(K) = \{\Phi(z, v_k(z)) \mid z \in K_k\}$$

provided k is large enough. The maps $v_k : \dot{S} \to \mathbb{R}^2$ solve the first order elliptic system $H(v_k) = 0$, and

$$v_k \to 0 \qquad \text{in } \ C^{1,\alpha}_{\text{loc}}(\dot{S}, \mathbb{R}^2).$$

In view of the elliptic regularity theory,

$$v_k \to 0 \qquad \text{in } \ C^\infty_{\text{loc}}(\dot{S}, \mathbb{R}^2).$$

Therefore, $j_k \to j_0$ in $C^\infty_{\text{loc}}(\dot{S})$. The difficult part of the proof concerns the behavior near the punctures. Define γ_0 as the infimum of the periods of the contractible periodic orbits of the Reeb vector fields $X_{h\lambda}$ and $X_{g\lambda}$. Choose $\epsilon_0 > 0$ such that $2\epsilon_0 < \gamma_0$. Since $E(\tilde{u}_0) < \infty$, we can take punctured disks D_{z_j} around the punctures $z_j \in \Gamma$ which are mutually disjoint and satisfy

$$\int_{D_{z_j}} \tilde{u}_0^* \, d(f\lambda) \leqslant \epsilon_0. \tag{56}$$

Recalling that the asymptotic limits of $\tilde{u}_0$ and $\tilde{u}_k$ at the punctures agree we conclude from (56) and Stokes' theorem that

$$\int_{D_{z_j}} \tilde{u}_k^* \, d(f\lambda) \leqslant 2\epsilon_0, \quad \text{if } \ k \geq k_0 \tag{57}$$

for some large k_0. We focus on the puncture z_j which we assume to be a positive one. Introducing cylindrical coordinates we take a sequence of biholomorphic maps

$$\psi_k^j : ([0, \infty) \times S^1, i) \to (D_{z_j}, j_k)$$

such that the composition

$$\varphi_k^j = \tilde{u}_k \circ \psi_k^j : ([0, \infty) \times S^1, i) \to (\mathbb{R} \times M, \tilde{J})$$

is a pseudoholomorphic half cylinder satisfying

$$\varphi_k^j([0,\infty)\times S^1)\subset \tilde{u}_k(D_{z_j})\subset C_k.$$

In view of (57), and the assumptions of the theorem, the pseudoholomorphic maps φ_k^j have the properties

$$\begin{aligned}\int_{[0,\infty)\times S^1}(\varphi_k^j)^*\,d(f\lambda)&\leqslant 2\epsilon_0\\ E(\varphi_k^j)&\leqslant C\\ \varphi_k^j(s,t)\to x_j(T_jt)&\quad \text{as } s\to\infty.\end{aligned}\tag{58}$$

The periodic orbit (x_j,T_j) is the asymptotic limit associated with the puncture z_j. We claim that as $k\to\infty$,

$$\varphi_k^j\to\varphi_j\qquad \text{in } C^\infty_{\text{loc}}([0,\infty)\times S^1,\mathbb{R}\times M)\tag{59}$$

for a pseudoholomorphic map $\varphi_j:[0,\infty)\times S^1\to\mathbb{R}\times M$ satisfying

$$\text{image}\,(\varphi_j)\subset\tilde{u}_0(\dot{S})=C_0.$$

In order to prove (59) we first show that the gradients of the maps φ_k^j are bounded. If the gradients are unbounded on $[\delta,\infty)\times S^1$ for some $\delta>0$, then the standard bubbling off procedure for which we refer to [15] produces a special or generalized finite energy plane $\tilde{v}:\mathbb{C}\to\mathbb{R}\times M$ having an energy

$$\int_{\mathbb{C}}\tilde{v}^*d(f\lambda)\geq\gamma_0.$$

However, by (58) the energy is necessary bounded from above by $2\varepsilon_0$ which contradicts our choice $2\varepsilon_0<\gamma_0$. So, if the gradients are not bounded they have to blow up at the boundary of $[0,\infty)\times S^1$. In this case a more subtle bubbling off analysis leads, however, again to a contradiction, see [1]. We conclude that the gradients are bounded. By the elliptic regularity theory we obtain C^∞_{loc} bounds and the claim (59) follows by the Arzelà-Ascoli theorem. From (59) we deduce, in particular, that

$$\int_{\{0\}\times S^1}(\varphi_k^j)^*(f\lambda)\to\int_{\{0\}\times S^1}(\varphi_j)^*(f\lambda)\tag{60}$$

as $k\to\infty$. Since

$$\lim_{s\to\infty}\int_{\{s\}\times S^1}(\varphi_k^j)^*(f\lambda)=T_j,$$

we conclude using Stokes' theorem and (58) that the left-hand side of (60) is at least $T_j-2\varepsilon_0\geq\gamma_0-2\varepsilon_0>0$. Consequently, φ_j is a non-constant finite energy half cylinder, and

$$\lim_{s\to\infty}\int_{\{s\}\times S^1}(\varphi_j)^*(f\lambda)=T_j.\tag{61}$$

Lemma 7.2. *Consider the above pseudoholomorphic maps*

$$\varphi_k^j = (a_k^j, w_k^j) : [0, \infty) \times S^1 \to \mathbb{R} \times M.$$

Then given any S^1-invariant neighborhood W of the smooth loop $[t \to x_j(T_j t)]$ in the loop space $C^\infty(S^1, M)$, there exists an integer k_0 and $R > 0$ such that the loops $t \to w_k^j(s,t)$ are contained in W,

$$[w_k^j(s, \cdot)] \in W,$$

for all $k \geq k_0$ and $s \geq R$.

Proof. Following the argument in [17] we first establish a uniform energy decay estimate. By construction,

$$\int_{[0,\infty)\times S^1} (\varphi_j)^* \, d(f\lambda) \leqslant 2\epsilon_0.$$

Hence, if $\delta \leqslant \varepsilon_0$ there exists $R > 0$ such that

$$\int_{[R,\infty)\times S^1} (\varphi_j)^* \, d(f\lambda) \leqslant \frac{\delta}{2}.$$

We claim that given $\delta > 0$ there exist k_0 and R_0 such that

$$\int_{[R,\infty)\times S^1} (\varphi_k^j)^* \, d(f\lambda) \leqslant \delta \quad \text{for all} \quad k \geq k_0 \tag{62}$$

and $R \geq R_0$. In order to prove the claim we compute, for fixed k,

$$\begin{aligned}
\int_{[R,\infty)\times S^1} (\varphi_k^j)^* \, d(f\lambda) &= \lim_{s_0\to\infty} \int_{[R,s_0)\times S^1} (\varphi_k^j)^* \, d(f\lambda) \\
&= \lim_{s_0\to\infty} \left[\int_{\{s_0\}\times S^1} (\varphi_k^j)^* \, (f\lambda) - \int_{\{R\}\times S^1} (\varphi_k^j)^* \, (f\lambda) \right] \\
&= T_j - \int_{\{R\}\times S^1} (\varphi_k^j)^* \, (f\lambda).
\end{aligned}$$

As above, T_j is the charge of the puncture z_j. Now, using the fact that $\varphi_k^j \to \varphi_j$ in C^∞_{loc} as $k \to \infty$, and using (61) we find

$$\begin{aligned}
\lim_{k\to\infty} \int_{[R,\infty)\times S^1} (\varphi_k^j)^* \, d(f\lambda) &= T_j - \int_{\{R\}\times S^1} (\varphi_j)^* \, (f\lambda) \\
&= \int_{[R,\infty)\times S^1} (\varphi_j)^* \, d(f\lambda) \\
&\leqslant \frac{\delta}{2},
\end{aligned}$$

and the claim (62) follows.

Let $\mathcal{T} \subset C^\infty(S^1, M)$ be the collection of loops corresponding to the contractible periodic orbits of the Reeb vector field $X_{f\lambda}$, and denote by $x_{T_j} \in \mathcal{T}$ the distinguished loop $[t \mapsto x_j(T_j t)]$. By assumption, (x_j, T_j) is non-degenerate. We therefore find two disjoint, S^1-invariant open sets V_1 and $V_2 \subset C^\infty(S^1, M)$ having the properties that $\mathcal{T} \subset (V_1 \cup V_2)$ and $V_1 \cap \mathcal{T} = S^1 * x_{T_j}$. We claim that there exists k_0 and R such that

$$[w_k^j(s, \cdot)] \subset V_1 \tag{63}$$

if $k \geq k_0$ and $s \geq R$. Recall that for k fixed there exists s_k such that $w_k^j(s, \cdot) \subset V_1$ if $s \geq s_k$. In view of (62) for every $\varepsilon > 0$ there exists $k_0(\varepsilon)$ and $R_0(\varepsilon)$ such that

$$\int_{[R,\infty)\times S^1} (w_k^j)^* \, d(f\lambda) \leqslant \varepsilon$$

if $k \geq k_0$ and $R \geq R_0$. Hence arguing indirectly and assuming (63) to be false we find sequences $R_k \to \infty$ and $\varepsilon_k \to 0$ such that $[w_k^j(R_k, \cdot)] \notin V_1 \cup V_2$ and

$$\int_{[R_k,\infty)\times S^1} (\varphi_k^j)^* k \, d(f\lambda) \leqslant \varepsilon_k.$$

Defining the rescaled sequence $\tilde{v}_k$ of pseudoholomorphic maps by

$$\tilde{v}_k(s,t) = \varphi_k^j(R_k + s, t)$$

we conclude for $\tilde{v}_k = (a_k, v_k)$ that

$$[v_k(0, \cdot)] \notin V_1 \cup V_2 \tag{64}$$

and $\tilde{v}_k \to \tilde{v}$ in $C^\infty_{\text{loc}}(\mathbb{R}\times S^1, \mathbb{R}\times M)$, for a non vanishing pseudoholomorphic cylinder having vanishing $d(f\lambda)$-energy. According to the classification in [19] the map $\tilde{v}$ is a cylinder over a periodic orbit y of the Reeb vector field so that $[y] \in \mathcal{T}$. It follows from (64) that $[y] \notin V_1 \cup V_2$ contradicting $\mathcal{T} \subset V_1 \cup V_2$ and hence proving the claim (63). The same argument applies to every S^1-invariant open neighborhood $W \subset V_1$ of x_{T_j} and the lemma is proved. □

In view of Lemma 7.2 we can describe the asymptotic behavior of $\varphi_k^j(s,t)$ for $k \geq k_0$ and $s \geq s_0$ in the local coordinates near the limit (x_j, T_j) as described in Theorem 4.4 in a uniform way. From the proof of Theorem 4.9 in [18] it follows that the maps φ_k^j approach $x_j(\mathbb{R})$ in the normal direction in an exponential way, uniformly in k. Using this one now readily verifies that for k large there are maps $v_k : \dot{S} \to \mathbb{R}^2$ representing $\tilde{u}_k(\dot{S}) = \{\Phi(z, v_k(z)) \mid z \in \dot{S}\}$ and satisfying $v_k \to 0$ in $C_0^{l,\alpha}(\dot{S}, \mathbb{R}^2)$ as $k \to \infty$, for every given integer l. Hence, in view of formula (50), $j_k \to j_0$ in $C^\infty_{\text{loc}}(\dot{S})$. The proof of Theorem 7.1 is complete. □

8. Proofs of Theorems 1.5 and 1.6

The results in the previous section allow us to describe the finite energy surfaces in the neighborhood of a given one in the associated local model. The case of finite energy planes is of special interest. We point out that in this case we do not have to impose any genericity requirements for the almost complex structure since the linearization of H at a solution is automatically surjective as soon as the index is right. This is not necessarily true for the general energy surfaces constructed later on.

Proof of Theorem 1.5.

The almost complex structure $\hat{J}$ on $\mathbb{R}\times M$ considered in Theorem 1.5 is $\mathbb{R}$-invariant and the corresponding finite energy planes are the so-called special planes from the introduction. We recall from Theorem 4.3 in [19] that the index of every non-degenerate finite energy plane $\tilde{u}:\mathbb{C}\to\mathbb{R}\times M$, embedded or not, always satisfies $\mu(\tilde{u})\geq 2$. We shall assume now that $\tilde{u}_0:\mathbb{C}\to\mathbb{R}\times M$ is a non-degenerate embedded (special) finite energy plane with index $\mu(\tilde{u}_0)=3$. Its asymptotic limit is the periodic solution (x,T) of the Reeb vector field X_λ. The image will be abbreviated by $P=x(\mathbb{R})\subset M$. We assume the limit to be simply covered. Under this assumption, the map of the local coordinates near $\tilde{u}_0$, constructed in Section 4,

$$\Phi:\mathbb{C}\times B\to\mathbb{R}\times M,\quad \Phi(z,0)=\tilde{u}_0(z)\quad\text{for all}\quad z\in\mathbb{C},\tag{65}$$

is an embedding. The set $B\subset\mathbb{R}^2$ is an open ball centered at the origin. In the local coordinates (65) we have, in view of Theorem 5.6, the embedding

$$u:B\times\mathbb{C}\to\mathbb{C}\times\mathbb{R}^2$$

of the form

$$u(\tau,z)=(z,v(\tau,z))$$

with $v(\tau,\cdot):\mathbb{C}\to\mathbb{R}^2$. For every τ, the map $u_\tau(\cdot):=u(\tau,\cdot):\mathbb{C}\to\mathbb{C}\times\mathbb{R}^2$ is an embedded complex curve near $\mathbb{C}\times\{0\}$ satisfying

$$Tu_\tau(z)\bar{j}_\tau(z)=\bar{J}(u_\tau(z))\cdot Tu_\tau(z),\qquad z\in\mathbb{C}.$$

The almost complex structure $\bar{J}$ on $\mathbb{C}\times\mathbb{R}^2$ is the induced structure

$$\bar{J}=(T\Phi)^{-1}\circ\hat{J}(\Phi)\circ T\Phi.$$

By construction, the pseudoholomorphic curves $u_\tau:\mathbb{C}\to\mathbb{C}\times B$ are embedded and have disjoint images,

$$u_\tau(\mathbb{C})\cap u_{\tau'}(\mathbb{C})=\emptyset\quad\text{if }\tau\neq\tau'.\tag{66}$$

Reparametrizing the plane $\mathbb{C}$ we take an embedding $\psi:B\times\mathbb{C}\to B\times\mathbb{C}$ of the form $(\tau,z)\mapsto(\tau,\psi_\tau(z))$, where the smooth diffeomorphisms $\psi_\tau:\mathbb{C}\to\mathbb{C}$ satisfy

$$\bar{j}_\tau(\psi_\tau(z))\cdot T\psi_\tau(z)=T\psi_\tau(z)\circ i\qquad\text{for all}\quad z\in\mathbb{C}.$$

We postpone the construction of ψ to the Appendix. Using the map Φ into $\mathbb{R}\times M$ we define the embedding

$$\tilde{\Phi}:B\times\mathbb{C}\to\mathbb{R}\times M$$

by

$$\tilde{\Phi}(\tau,z)=\Phi\circ u\circ\psi(\tau,z)\quad\text{for}\quad(\tau,z)\in B\times\mathbb{C}.$$

By construction, $\tilde{\Phi}(\tau) := \tilde{\Phi}(\tau, \cdot) : \mathbb{C} \to \mathbb{R} \times M$ is an embedded finite energy plane near $\tilde{u}_0$ satisfying

$$T\tilde{\Phi}(\tau) \circ i = \hat{J}(\tilde{\Phi}(\tau)) \circ T\tilde{\Phi}(\tau), \qquad \tau \in B,$$

and having the asymptotic limit (x, T). Hence, $\mu(\tilde{\Phi}(\tau)) = \mu(\tilde{u}_0) = 3$ and $E(\tilde{\Phi}(\tau)) = E(\tilde{u}_0) = T$ for all $\tau \in B$. Moreover, in view of (66),

$$\tilde{\Phi}(\tau)(\mathbb{C}) \cap \tilde{\Phi}(\tau')(\mathbb{C}) = \emptyset \qquad \text{if } \ \tau \neq \tau'. \tag{67}$$

We are interested in the projections of $\tilde{\Phi}(\tau)$ into the manifold M. Using the $\mathbb{R}$-invariance of the structure $\hat{J}$ on $\mathbb{R} \times M$ we shall prove

Lemma 8.1. *Fix $\tau \in B$ and denote the associated finite energy plane $\tilde{\Phi}(\tau) = \tilde{u} = (a, u) : \mathbb{C} \to \mathbb{R} \times M$. Then $u(\mathbb{C}) \cap P = \emptyset$ and*

$$u : \mathbb{C} \to M \setminus P$$

is an embedding.

Proof. Define for $c \in \mathbb{R}$ the translated map

$$\tilde{u}_c(z) = \big(a(z) + c, u(z)\big) \in \mathbb{R} \times M.$$

Since $\hat{J}$ is $\mathbb{R}$-invariant, with $\tilde{u}$ also $\tilde{u}_c$ is a finite energy plane having the same asymptotic limit. We claim that

$$\tilde{u}_c(\mathbb{C}) \cap \tilde{u}(\mathbb{C}) = \emptyset \qquad \text{for every } \ c \neq 0. \tag{68}$$

Indeed, assume $c > 0$ and small. Then also $\tilde{u}_c$ belongs to a family $\tilde{\Phi}$ in view of the completeness statement Theorem 7.1. Therefore, $\tilde{u}_c(\mathbb{C}) \cap \tilde{u}(\mathbb{C}) = \emptyset$ by (67). In view of the asymptotic behavior of non-degenerate finite energy planes described in Theorem 4.4, the intersection points for $c \geq \varepsilon > 0$ must be contained in a compact subset of $\mathbb{C}$. Hence, using the intersection theory of pseudoholomorphic curves, see [26, 27, 30], the intersection number $\operatorname{int}(\tilde{u}_\tau(\mathbb{C}), \tilde{u}(\mathbb{C})) \geq 0$ is well defined. The intersection points are isolated and the local index is positive. By the homotopy invariance of the intersection number and the fact that for small $c > 0$ there are no intersection points we conclude that the intersection index vanishes for all $c > 0$. Hence, in view of the positivity of the index, the claim (68) is proved for $c > 0$. The same argument applies for $c < 0$. These arguments also show that

$$\tilde{u}_c(\mathbb{C}) \cap \tilde{u}_b(\mathbb{C}) = \emptyset \qquad \text{if } \ c \neq b,$$

by fixing for example b and then applying the whole construction to $\tilde{u}_b$ replacing our original plane $\tilde{u}_0$.

From (68) we shall conclude that the map $u : \mathbb{C} \to M$ is injective. Assume for two points $z_1, z_2 \in \mathbb{C}$ that $u(z_1) = u(z_2)$. Then $\tilde{u}(z_1) = \tilde{u}_c(z_2)$ for $c = a(z_1) - a(z_2)$. Hence $a(z_1) = a(z_2)$ and since $\tilde{u}$ is an embedding, $z_1 = z_2$ as claimed. Recall now from page 293 in [19] the inequality

$$0 \leqslant \operatorname{wind}_\pi(u) \leqslant \frac{1}{2}\mu(\tilde{u}) - 1$$

for the winding number $\text{wind}_\pi(u)$, which is a nonnegative integer. Since $\mu(\tilde{u}) = \mu(\tilde{u}_0) = 3$, we conclude $\text{wind}_\pi(u) = 0$. Hence, by Corollary 4.2 in [19],

$$\pi \circ Tu(z) \neq 0 \qquad \text{for every } z \in \mathbb{C}, \tag{69}$$

so that $u : \mathbb{C} \to M$ is an injective immersion. Assume $u(\mathbb{C}) \cap P \neq \emptyset$. Then at every intersection point $m = u(z)$ the intersection is transversal since in view of (69) we have $T_m M = \mathbb{R}X(m) \oplus \pi \circ T_m u(\mathbb{C})$. Because u is asymptotic to P it necessarily has self intersections contradicting the injectivity of u and we have proved that $u(\mathbb{C}) \cap P = \emptyset$. In view of the asymptotic behavior near the non-degenerate limit orbit, the injective immersion $u : \mathbb{C} \to M \setminus P$ must be an embedding and the proof of Lemma 8.1 is complete. □

For our map $\tilde{\Phi}$ we shall use in the following the notation $\tilde{\Phi}(\tau, z) = (a(\tau, z), u(\tau, z))$ so that

$$\tilde{\Phi}(\tau) = (a(\tau), u(\tau)) : \mathbb{C} \to \mathbb{R} \times M$$

for every $\tau \in B$. Since all the energy planes $\tilde{\Phi}(\tau)$ have the same asymptotic limit and the same covering numbers, we know from Theorem 4.11 in [19] that

$$\text{either} \quad u(\tau_1)(\mathbb{C}) = u(\tau_2)(\mathbb{C}) \quad \text{or} \quad u(\tau_1)(\mathbb{C}) \cap u(\tau_2)(\mathbb{C}) = \emptyset.$$

Continuing the proof of Theorem 1.5 we shall next single out a 1-dimensional submanifold in the parameter space B given by $\tau \mapsto \tau(\varepsilon) \in \mathbb{R}^2$ in order to define the desired embedding

$$w : I_{\varepsilon_0} \times \mathbb{C} \to M \setminus P \ \text{ by } \ w(\varepsilon, z) = u(\tau(\varepsilon), z),$$

where $I_{\varepsilon_0} = \{|\varepsilon| < \varepsilon_0\}$ is an interval. We first normalize the original plane $\tilde{u}_0 = (a_0(z), u_0(z))$. By a translation in $\mathbb{R}$ and a biholomorphic map of $\mathbb{C}$ we arrange that a_0 takes on its minimum at $z = 0$ and

$$a_0(0) = \min_{z \in \mathbb{C}} a_0(z) = 0. \tag{70}$$

Since the map $\tilde{\Phi} = (a, u) : B \times \mathbb{C} \to \mathbb{R} \times M$ is an immersion, the linearized map satisfies at every point $(\tau, z) \in B \times \mathbb{C}$,

$$\begin{aligned} d_\tau a(\tau, z)(\mathbb{R}^2) + d_z a(\tau, z)(\mathbb{C}) &= \mathbb{R} \\ d_\tau u(\tau, z)(\mathbb{R}^2) + T_z u(\tau, z)(\mathbb{C}) &= T_{u(\tau, z)} M. \end{aligned} \tag{71}$$

In view of the normalization (70) we conclude from (71) that at $(\tau, z) = (0, 0)$, $d_\tau a(0, 0)(\mathbb{R}^2) = \mathbb{R}$. Therefore, there exists a unique 1-dimensional submanifold of $\mathbb{R}^2$ given on an interval by $\varepsilon \mapsto \tau(\varepsilon) \in \mathbb{R}^2$ and defined, at $z = 0$, by

$$a(\tau(\varepsilon), 0) = 0, \quad |\varepsilon| < \varepsilon_0$$

and $\tau(0) = 0$. We shall show that the map $w(\varepsilon, z) = u(\tau(\varepsilon), z)$ from $I_{\varepsilon_0} \times \mathbb{C}$ into $M \setminus P$ is an embedding. In order to prove that it is an immersion we claim that at every point (ε, z), abbreviating $\tau(\varepsilon) = \tau$, we have

$$d_\tau u(\tau, z) \cdot \tau'(\varepsilon)\mathbb{R} + T_z u(\tau, z)(\mathbb{C}) = T_{u(\tau, z)} M, \tag{72}$$

so that $\frac{\partial}{\partial\varepsilon}w(\varepsilon,z)$ is transversal to the surface $w(\varepsilon,\mathbb{C})$. Indeed, fix $\varepsilon\in I_{\varepsilon_0}$, set $\tau=\tau(\varepsilon)$, and consider the translated family $\tilde{u}(\tau)_c=(a(\tau)+c,u(\tau))$, for $c\in\mathbb{R}$ and small. By the completeness statement, Theorem 7.1, the family $\tilde{u}(\tau)_c$ has as member of $\tilde{\Phi}$ the representation $\tilde{u}(\tau)_c(z)=(a(\tau(c),z),u(\tau(c),z))$, with a smooth function $c\mapsto\tau(c)\in\mathbb{R}^2$ satisfying $\tau(0)=\tau$. The functions have the properties that $d_\tau a(\tau(c),z)\cdot\dot{\tau}(c)=1$ and $d_\tau u(\tau(c),z)\cdot\dot{\tau}(c)=0$. Therefore the vectors $\tau'(\varepsilon)$ and $\dot{\tau}(0)$ at $c=0$ are linearly independent so that $\mathbb{R}^2=\mathbb{R}\tau'(\varepsilon)+\mathbb{R}\dot{\tau}(0)$ and the formula (72) follows from (71). In order to verify that the immersion $w:I_{\varepsilon_0}\times\mathbb{C}\to M\setminus P$ is injective we argue as in Lemma 8.1. Assuming that

$$u(\tau_1,z_1)=u(\tau_2,z_2),\quad \text{where } \tau_1=\tau(\varepsilon_1) \text{ and } \tau_2=\tau(\varepsilon_2),$$

we have

$$\tilde{u}(\tau_1)(z_1)=\tilde{u}(\tau_2)_c(z_2),$$

with $c=a(\tau_1,z_1)-a(\tau_2,z_2)$. Recalling that $\tilde{u}(\tau)(\mathbb{C})\cap\tilde{u}(\tau')(\mathbb{C})=\emptyset$ if $\tau\neq\tau'$ and $\tilde{u}(\tau)_c(\mathbb{C})\cap\tilde{u}(\tau')(\mathbb{C})=\emptyset$ if $c\neq 0$ and $\tau\in B$ we conclude using the homotopy property and the positivity of the intersection index that $\tau_1=\tau_2$ and $z_1=z_2$, hence proving the injectivity of w. In view of the asymptotic behavior of the finite energy planes near a non-degenerate limit, the injective immersion $w:I_{\varepsilon_0}\times\mathbb{C}\to M\setminus P$ must be an embedding.

Finally, the completeness assertion of Theorem 1.5 for a sequence $\tilde{u}_k\to\tilde{u}_0$ in $C^\infty_{\text{loc}}(\mathbb{C},\mathbb{R}\times M)$, if normalized so that $a_k(0)=0$, follows from Theorem 7.1. The proof of Theorem 1.5 is complete. □

Proof of Theorem 1.6.

The almost complex structure $\tilde{J}$ in Theorem 1.6 is not translation invariant anymore. The statement of Theorem 1.6 is therefore weaker and follows from the implicit function theorem in the local model, Theorem 5.6, together with the completeness result of Theorem 7.1. □

9. Proof of Theorem 1.10

Recall from the Introduction that U_δ denotes the distinguished space of admissible almost complex structures $\hat{J}$ on $\mathbb{R}\times M$. With (x_j,T_j),$j=1,2,\dots,m^+$, we denote non-degenerate periodic orbits of the Reeb vector field $X_{h\lambda}$ on M while with (y_j,T_j'), $j=1,2,\dots,m^-$, we denote non-degenerate periodic orbits of the Reeb vector field $X_{g\lambda}$. We shall consider the collection of pairs $(\hat{J},C)$, denoted by

$$\mathcal{M}=\mathcal{M}_g(x_1,\dots,x_{m^+};y_1,\dots,y_{m^-}),$$

where $\hat{J}\in\mathcal{U}_\delta$ and $C\subset\mathbb{R}\times M$ is the image of an immersed generalized finite energy surface for $\hat{J}$ with m^+ positive and m^- negative punctures at which it is asymptotic to $x_1,\dots,x_{m^+}$ and $y_1,\dots,y_{m^-}$ respectively. Moreover, the underlying non punctured model for C is assumed to have genus g. To be more precise, for every $(\hat{J}, C) \in \mathcal{M}$ there exists a closed Riemann surface (S, j) of genus g, a finite subset $\Gamma \subset S$, where $\Gamma = \{z_1, z_2, \dots, z_{m^+}, z_1', z_2', \dots, z_{m^-}'\}$ and an immersion

$$\tilde{u} : S \setminus \Gamma \to \mathbb{R} \times M$$

satisfying

$$\begin{aligned} T\tilde{u} \circ j &= \hat{J}(\tilde{u}) \circ T\tilde{u} \\ 0 < E(&\tilde{u}) < \infty \\ \tilde{u}(S &\setminus \Gamma) = C, \end{aligned}$$

moreover $\tilde{u}$ has z_j, $j = 1, \dots, z_{m^+}$, as positive punctures at which it is asymptotic to (x_j, T_j) and z_k', $k = 1, \dots, m^-$, as negative punctures at which it is asymptotic to (y_k, T_k'). We note that $C \subset \mathbb{R} \times M$ is an immersed $\hat{J}$-complex curve,

$$\hat{J}(T_pC) = T_pC, \quad \text{for every} \quad p \in C.$$

Next we introduce in a natural way a Banach manifold structure on $\mathcal{M}$ which is separable. For this purpose we make use of the local model in Section 4 which can be constructed for immersions the same way as for embeddings.

We choose a reference pair $(\hat{J}_0, C_0) \in \mathcal{M}$. To describe a neighborhood in $\mathcal{M}$ we choose a parametrization $\tilde{u}_0 : (\dot{S}, j_0) \to (\mathbb{R} \times M, \hat{J}_0)$ and the associated local coordinates $\Phi_{\hat{J}_0}$

$$\Phi_{\hat{J}_0} : \dot{S} \times B \to \mathbb{R} \times M, \quad \Phi_{\hat{J}_0}(z, 0) = \tilde{u}_0(z) \quad \text{for} \quad z \in \dot{S},$$

where $B = B_\varepsilon^2(0) \subset \mathbb{R}^2$ for some $\varepsilon > 0$. We shall abbreviate $\Phi \equiv \Phi_{\hat{J}_0}$. The almost complex structure $\hat{J} \in \mathcal{U}_\delta$ induces the almost complex structure $\bar{J}$ on $\dot{S} \times B$ by

$$\bar{J} = \left(T\Phi\right)^{-1} \circ \hat{J} \circ T\Phi.$$

The pull-back of the reference structure $\hat{J}_0$ is denoted by $\bar{J}_0$. Define the manifold $\hat{\mathcal{B}}$ by

$$\hat{\mathcal{B}} = \mathcal{U}_\delta \times \mathcal{B}, \quad \text{with} \ \mathcal{B} = C_0^{1,\alpha}(\dot{S}, B),$$

and the Banach space bundle $\hat{\mathcal{E}}$ over $\hat{\mathcal{B}}$ by

$$\hat{\mathcal{E}} = \bigcup_{(\hat{J}, v) \in \hat{\mathcal{B}}} \{(\hat{J}, v)\} \times C_0^\alpha\left(u^* \Xi_{\bar{J}}\right).$$

where, as above, $\bar{J} = \left(T\Phi\right)^{-1} \circ \hat{J} \circ T\Phi$. Abbreviating the graph of v by $u(z) = (z, v(z)) \in \dot{S} \times B$ we define the (now parametrized) smooth section H of $\hat{\mathcal{E}} \to \hat{\mathcal{B}}$ by

$$\begin{aligned} H(\hat{J}, v)(z)(h \wedge k) = &(h, Tv(z)h) \wedge (k, Tv(z)k) \\ &- \bar{J}(u(z)) \cdot (h, Tv(z)h) \wedge \bar{J}(u(z)) \cdot (k, Tv(z)k). \end{aligned}$$

By construction of Φ, the pair $(\hat{J}_0, 0)$ is a solution of $H(\hat{J}, v) = 0$. We shall linearize the smooth map H at the point $(\hat{J}_0, 0)$ into the direction (Y, g) and find

$$\begin{aligned} H'(\hat{J}_0, 0)(Y, g)(z)(h \wedge k) &= (h, 0) \wedge (0, Lg(z)k) - (k, 0) \wedge (0, Lg(z)h) \\ &\quad - (T\Phi(z,0)^{-1} Y(\tilde{u}_0(z)) T\Phi(z,0)(h,0)) \wedge (j_0 z, 0)(k, 0) \\ &\quad - (j_0(z)h, 0) \wedge (T\Phi(z,0)^{-1} Y(\tilde{u}_0(z)) T\Phi(z,0)(k,0)). \end{aligned} \tag{73}$$

Recall that the linear operator L above, obtained by linearizing only in the v-direction, is the Fredholm operator from Section 5, defined by

$$(Lg)(z) = Tg(z) + i_0 \circ Tg(z) \circ j_0(z) + \left[\Delta'(z)g(z)\right] \circ j_0(z)$$

for every $z \in \dot{S}$. To simplify the notation we shall abbreviate in the following discussion

$$A(Y)(z) = \left(T\Phi(z,0)\right)^{-1} \circ Y(\tilde{u}_0(z)) \circ T\Phi(z,0), \; z \in \dot{S}.$$

Recalling now from the introduction that

$$Y(\tilde{u}_0) \circ \hat{J}_0(\tilde{u}_0) + \hat{J}_0(\tilde{u}_0) \circ Y(\tilde{u}_0) = 0,$$

we conclude from the definition of $\bar{J}_0$,

$$A(Y)(z) \circ \bar{J}_0(z, 0) = -\bar{J}_0(z, 0) \circ A(Y)(z), \; z \in \dot{S}.$$

So, $A(Y)$ is $\bar{J}_0$-complex anti-linear. Corresponding to the splitting $T\dot{S} \times \mathbb{R}^2$ of the tangent space of $\dot{S} \times B$ we represent $A(Y)$ in the matrix form

$$A(Y) = \begin{pmatrix} a & c \\ b & d \end{pmatrix}.$$

Since, by construction of Φ, we have $\bar{J}_0(z, 0) = j_0(z) \oplus i_0(z)$ and since $A(Y)$ is $\bar{J}_0$-complex anti-linear, the linear maps $a(z), b(z)$ have the properties

$$\begin{aligned} a(z) \cdot j_0(z) &= -j_0(z) \cdot a(z) \\ b(z) \cdot j_0(z) &= -i_0(z) \cdot b(z), \end{aligned}$$

$z \in \dot{S}$. Recall the bundle map $\alpha : \mathcal{A}_0 \to \Xi_0$ from (3),

$$\alpha(A_0)(z)(h \wedge k) = (h, 0) \wedge (0, A_0(z)k) - (k, 0) \wedge (0, A_0(z)h).$$

Introduce

$$\Theta(Y)(z) := b(z) \circ j(z) \in \mathcal{L}(T_z\dot{S}, \mathbb{R}^2).$$

To simplify the notation, we drop the index 0 in what follows and simply write $j \equiv j_0$ and $i \equiv i_0$.

Proposition 9.1. *The derivative of H at $(\hat{J}, 0)$ is given by the formula*

$$H'(\hat{J}, 0)(z)(Y, g) = \alpha\left[(Lg)(z) + \Theta(Y)(z)\right]. \tag{74}$$

We start the proof of Proposition 9.1 with

Lemma 9.2. *With a and b as above we have*

(i) $(jk,0)\wedge(ah,0)=(jh,0)\wedge(ak,0)$
(ii) $(jk,0)\wedge(0,bh)-(jh,0)\wedge(0,bk)=(h,0)\wedge(0,bjk)-(k,0)\wedge(0,bjh)$

for all $h,k\in T_z\dot{S}$.

Proof. [Proof of Lemma 9.2] After substituting $k=\gamma h+\delta jh$ into both sides of the equation (i) we see that it is enough to show that

$$(h,0)\wedge(ah,0)=(jh,0)\wedge(jah,0).$$

Hence we write $ah=\gamma h+\delta jh$. Then

$$(h,0)\wedge(ah,0)=\delta(h,0)\wedge(jh,0)$$

and the right side is equal to

$$(jh,0)\wedge(ajh,0)=-\delta(jh,0)\wedge(h,0)=\delta(h,0)\wedge(jh,0)$$

proving the first equality. The second equality is verified in a similar way, and the proof of Lemma 9.2 is complete. □

Using the above lemma we find

$$\begin{aligned}
&A(Y)(h,0)\wedge(jk,0))+(jh,0)\wedge(A(Y)(k,0))\\
&\qquad=-(jk,0)\wedge(A(Y)(h,0))+(jh,0)\wedge(A(Y)(k,0))\\
&\qquad=-(jk,0)\wedge(ah,0))-(jk,0)\wedge(0,bh)\\
&\qquad\quad+(jh,0)\wedge(ak,0)+(jh,0)\wedge(0,bk)\\
&\qquad=(jh,0)\wedge(0,bk)-(jk,0)\wedge(0,bh)\\
&\qquad=(k,0)\wedge(0,bjh)-(h,0)\wedge(0,bjk).
\end{aligned}$$

Inserting this equality into (73) we arrive at the formula

$$\begin{aligned}
&H'(\hat{J}_0,0)(Y,g)(z)(h\wedge k)\\
&\quad=\alpha((Lg(z))(h\wedge k))\\
&\qquad-(A(Y)(h,0))\wedge(j(z)k,0)-(j(z)h,0)\wedge(A(Y)(k,0))\\
&\quad=\alpha((Lg(z))(h\wedge k)\\
&\qquad+(h,0)\wedge(0,b(z)j(z)k)-(k,0))\wedge(0,b(z)j(z)h)\\
&\quad=\alpha((Lg(z)+b(z)\circ j(z))(h\wedge k).
\end{aligned}$$

Recalling $\Theta(Y)(z)=b(z)\circ j(z)$, the formula (74) in Proposition 9.1 is proved. □

The next statement explains our choice of the parameter space $\mathcal{U}_\delta$ of admissible almost complex structures on $\mathbb{R}\times M$.

Lemma 9.3. *The linear operator*

$$H'(\hat{J}_0,0):T_{\hat{J}_0}\mathcal{U}_\delta\times C_0^{1,\alpha}(\dot{S},\mathbb{R}^2)\to C_0^{\alpha}(graph(0)^*\Xi_{\bar{J}_0})$$

is surjective and the kernel splits.

Proof. Since $L : C_0^{1,\alpha}(\dot{S},\mathbb{R}^2) \to C_0^\alpha(\mathcal{A}_0)$ is Fredholm, the operator $H'(\hat{J}_0,0)$ has, in view of Lemma 9.2, a closed image and the kernel splits. In order to prove that the operator is surjective it suffices to show that

$$(Y,g) \mapsto (Lg) + \Theta(Y),$$

viewed as an operator

$$T_{\hat{J}_0}\mathcal{U}_\delta \times W^{1,2}(\dot{S},\mathbb{R}^2) \to L^2(\mathcal{A}_0), \tag{75}$$

is surjective. Assume that p is orthogonal to the image. Then p is, in particular, orthogonal to the image of L and therefore a weak solution of its adjoint equation which is an elliptic system having smooth coefficients. Hence p is smooth by the elliptic regularity theory. Moreover, in view of the similarity principle for the Cauchy-Riemann operator, it has the unique continuation property. This means that if p vanishes on some open subset of $\dot{S}$, then $p=0$. By assumption, p is also orthogonal to $\Theta(Y)\circ j$ so that

$$0 = \langle \Theta(Y), p\rangle_{L^2} \qquad \text{for all } Y \in T_{\hat{J}_0}\mathcal{U}_\delta. \tag{76}$$

Assume $p \neq 0$, then we find an open set $Q \subset \dot{S}$ such that for the parametrization $\tilde{u}_0 = (a_0,u_0) : \dot{S} \to \mathbb{R}\times M$ we have

$$a_0(Q) \subset \left[-\frac{1}{2}a_1, \frac{1}{2}a_1\right] \quad \text{and} \quad p(z)\neq 0 \ \text{ for } \ z \in Q.$$

Recall that the real number $a_1 > 0$ enters the definition of $\mathcal{U}_\delta$. Take a smooth map h compactly supported in Q and satisfying

$$\langle p,h\rangle_{L^2} > 0.$$

Then, by the definition of $\Theta(Y)$, we find a smooth map Y^* satisfying $\Theta(Y^*) = h$. By the definition of $\mathcal{U}_\delta$ we find a $Y \in T_{\hat{J}_0}\mathcal{U}_\delta$ which is C^∞ close to Y^* and compactly supported so that

$$\langle \Theta(Y), p\rangle_{L^2} > 0.$$

This contradicts (76) and hence proves that $p=0$. We have proved that the image of (75) is dense and since the operator has closed image, it must be onto. This finishes the proof of Lemma 9.3 □

From Lemma 9.3 we deduce by means of the implicit function theorem that the subset of solutions $H(\hat{J},v) = 0$ near $(\hat{J}_0,0)$ is a submanifold of $\hat{\mathcal{E}}$. Given the solution $(\hat{J},v)$, the map $z \mapsto \Phi_{\hat{J}_0}(z,v(z))$ from $\dot{S}$ into $\mathbb{R}\times M$ is a parametrization of C for an element $(\hat{J},C)\in\mathcal{M}$ near $(\hat{J}_0,C_0)$. The map

$$\left\{(\hat{J},v) \mid H(\hat{J},v)=0\right\} \to \left\{(\hat{J},C)\in\mathcal{M} \mid C = \Phi_{\hat{J}_0}(\mathrm{graph}(v)(\dot{S}))\right\} \tag{77}$$

defines a neighborhood basis of $(\hat{J}_0,C_0)$ in $\mathcal{M}$ by varying $\hat{J}$ in a neighborhood of $\hat{J}_0$ in $\mathcal{U}_\delta$ and v in a neighborhood basis of $v=0$ in $C_0^{1,\alpha}(\dot{S},\mathbb{R}^2)$. Carrying out the construction at every point of $\mathcal{M}$, the associated maps (77) define a manifold structure on $\mathcal{M}$ compatible with this topology.

Lemma 9.4. *The manifold $\mathcal{M}$ is separable.*

Proof. Consider the space $E \subset C^2_{\mathrm{loc}}(\dot{S}, \mathbb{R}^N)$ consisting of the maps $\tilde{u} : \dot{S} \to \mathbb{R}^N$ in this class. It carries in a natural way the structure of a separable Fréchet space. Embedding $M \subset \mathbb{R}^{N-1}$ for some large N we view $\mathbb{R} \times M$ as a subset of $\mathbb{R}^N$ and identify $F = C^2_{\mathrm{loc}}(\dot{S}, \mathbb{R} \times M)$ with the subspace of E consisting of those maps $\tilde{u}$ whose images are contained in $\mathbb{R} \times M$. Hence F is a separable metric space, and so is $\mathcal{U}_\delta \times F$. Denote by $\mathcal{A} \subset \mathcal{U}_\delta \times F$ the subset of all points $(\hat{J}, \tilde{u})$ where $\tilde{u}$ parametrizes an immersed surface $C = \tilde{u}(\dot{S})$ with $(\hat{J}, C) \in \mathcal{M}$. Then $\mathcal{A}$ is separable. Take now a dense sequence $(\hat{J}_k, \tilde{u}_k)$ in $\mathcal{A}$. Then, by the elliptic regularity theory the maps $\tilde{u}_k$ are smooth. Define $C_k = \tilde{u}_k(\dot{S})$. We claim that the sequence $((\hat{J}_k, C_k))$ is dense in $\mathcal{M}$. We choose any $(\hat{J}_0, C_0) \in \mathcal{M}$. Hence there is a parametrization $\tilde{u}_0 \in F$ satisfying $C_0 = \tilde{u}_0(\dot{S})$ and $(\hat{J}_0, \tilde{u}_0) \in \mathcal{A}$. Therefore, there exists a sequence $(k_j) \subset \mathbb{N}$ satisfying

$$\begin{aligned} \hat{J}_{k_j} &\to \hat{J}_0 \qquad \text{in } \mathcal{U}_\delta \\ \tilde{u}_{k_j} &\to \tilde{u}_0 \qquad \text{in } F. \end{aligned}$$

Abbreviating $C_{k_j} = \tilde{u}_{k_j}(\dot{S})$ we now conclude from the completeness statement, Theorem 7.1, that

$$(\hat{J}_{k_j}, C_{k_j}) \to (\hat{J}_0, C_0) \quad \text{in } \mathcal{M}.$$

This completes the proof of Lemma 9.4 □

We are ready to finish the proof of Theorem 1.10. We consider the projection map

$$\eta : \mathcal{M} \to \mathcal{U}_\delta, \quad (\hat{J}, \dot{C}) \mapsto \hat{J},$$

which is a smooth Fredholm map. Its index agrees with the index of the operator L_C. It has been computed in Theorem 5.1. Therefore,

$$\mathrm{Ind}(T_{(\hat{J},C)}\eta) = \mathrm{Ind}(L_C) = \mu(C) - \chi_S + m^+ + m^-. \tag{78}$$

This completes the proof of Theorem 1.10. □

Theorem 1.10 has many useful corollaries based on the Sard-Smale theorem for Fredholm maps. From this theorem we conclude that the set $\mathcal{R} \subset \mathcal{U}_\delta$ consisting of regular values of the projection $\eta : \mathcal{M} \to \mathcal{U}_\delta$ is a residual set in the sense of Baire and hence $\mathcal{R}$ is, in particular, dense in $\mathcal{U}_\delta$. If $\hat{J} \in \mathcal{R}$, then the derivative of η is surjective. So,

$$\mathrm{Ind}(L_C) \geq 0, \tag{79}$$

and $\eta^{-1}(\hat{J}) \subset \mathcal{M}$ is a manifold of dimension $\mathrm{Ind}(L_C)$.

We shall illustrate the use of (79) with the example considered in Corollary 1.11. We assume $M = S^3$ and $S = S^2$, the Riemann sphere. As for the punctures we assume $\Gamma^+ = \{\infty\}$ so that $m^+ = 1$. Denoting the associated asymptotic limit by x and the limits of negative punctures Γ^- by y_k, for $k = 1, \ldots, m^-$, we have, by definition, $\mu(C) = \mu(x) - \sum_{k=1}^{m^-} \mu(y_k)$, so that in view of (78) we obtain the inequality

$$\mu(x) - \sum_{k=1}^{m^-} \mu(y_k) - 1 + m^- \geq 0, \tag{80}$$

provided that $\hat{J}$ is generic. Assuming $\mu(x) = 3$ and $\mu(y_k) \geq 3$, for $k = 1, \dots, m^-$, we deduce from (80) that $m^- \leqslant 1$, so that there can be at most one negative puncture. If there exists a negative puncture we call its asymptotic limit y, and inserting $m^- = 1$ in (80),

$$3 = \mu(x) \geq \mu(y) \geq 3.$$

So, $\mu(y) = \mu(x) = 3$ and $\operatorname{Ind}(L_C) = 0$. In this special example, a component of the set $\mathcal{M}$ consists of a single element $(\hat{J}, C)$ where $C \subset \mathbb{R} \times S^3$ is an immersed pseudoholomorphic cylinder connecting the periodic orbit x of $X_{h\lambda}$ with the periodic orbit y of $X_{g\lambda}$.

In view of the above example it is now easy to finish the proof of Corollary 1.11. We observe that the set of all asymptotic data for all possible $\mathcal{M}$ is countable. Since a countable intersection of residual sets in $\mathcal{U}_\delta$ is again a residual set and hence a dense set in $\mathcal{U}_\delta$, the proof of Corollary 1.11 is complete. □

10. Proof of Theorem 1.13

We consider the somewhere injective finite energy surface $\tilde{u} = (a, u) : \dot{S} = S \setminus \Gamma \to \mathbb{R} \times M$ having non-degenerate asymptotic limits and assume that there exists a point $\hat{z} \in \dot{S}$ satisfying

$$\pi \circ Tu(\hat{z}) \neq 0. \tag{81}$$

We conclude from (81) by means of the similarity principle that the set of points z solving $\pi \circ Tu(z) = 0$ is discrete, see [18]. Moreover, from the asymptotic behavior of $\tilde{u}$ described in [18] we deduce that $\pi \circ Tu(z) \neq 0$ for z in a neighborhood of the punctures $\Gamma \subset S$. Therefore, the subset

$$A_0 := \{z \in \dot{S} \mid \pi \circ Tu(z) = 0\}$$

is a finite set. Consequently, the image

$$M_0 = u(A_0) \subset M$$

is a finite set. Denote by $x_1, \dots, x_n$ the asymptotic limits of $\tilde{u}$ corresponding to Γ. Then, again by the results in [18], the set of points $z \in \dot{S}$ satisfying

$$u(z) \in P := \bigcup_{j=1}^{n} x_j(\mathbb{R})$$

is a finite set. Now define the subset $S_0 \subset \dot{S}$ by

$$S_0 = \{z \in \dot{S} \mid \pi \circ Tu(z) \neq 0 \text{ and } u(z) \notin u(\dot{S} \setminus \{z\})\}.$$

Then S_0 is open in $\dot{S}$ and we shall prove that S_0 is dense in $\dot{S}$. We shall argue by contradiction and hence can assume in view of the above discussion that there is a point $z_0 \in \dot{S}$ and $\varepsilon > 0$ such that every $z \in B_\varepsilon(z_0)$ has the following properties:

$$\begin{gathered} \pi \circ Tu(z) \neq 0 \\ u(z) = u(\zeta) \ \text{ for some } \ \zeta \in \dot{S} \ \text{ and } \zeta \neq z \\ u(z) \notin M_0 \\ u(z) \notin P. \end{gathered} \tag{82}$$

Our aim is to prove that $\tilde{u}$ is not somewhere injective in contradiction to our assumption. Take $z \in \overline{B_\varepsilon(z_0)}$ and define the set of points

$$\Sigma(z) = \{\zeta \in \dot{S} \mid u(\zeta) = u(z) \ \text{ and } \zeta \neq z\}.$$

Then $\Sigma(z) \neq \emptyset$, and $\pi \circ Tu(\zeta) \neq 0$ for all $\zeta \in \Sigma(z)$ in view of (82). This implies that $\Sigma(z)$ consists of isolated points which in view of (82) stay away from the punctures. Consequently, $\Sigma(z)$ is a finite set. For $\varepsilon > 0$ small enough, the map

$$u : \overline{B_\varepsilon(z_0)} \to M$$

is a smooth embedding. If $\Sigma(z_0) = \{\zeta_0, \ldots, \zeta_m\}$ there exists $\delta > 0$ such that $u_{|\overline{B_\delta(\zeta_j)}}$ is an embedding for $j = 0, \ldots, m$. By a compactness argument we find an $\varepsilon > 0$ so small that

$$u\big(\overline{B_\varepsilon(z_0)}\big) \subset \bigcup_{j=0}^{m} u(B_\delta(\zeta_j)).$$

Hence there exists j_0 satisfying

$$u(z_0) \in \operatorname{int}\big(\overline{u(B_\delta(\zeta_{j_0}))} \cap \overline{u(B_\varepsilon(z_0))}\big).$$

Without loss of generality we may assume that $j_0 = 0$ so that $\zeta_{j_0} = \zeta_0$. Hence, perhaps taking a smaller $\varepsilon > 0$ we conclude

$$\begin{gathered} \overline{u(B_\varepsilon(z_0))} \subset u(B_\delta(\zeta_0)) \\ u(z_0) = u(\zeta_0) \ \text{ and } \ z_0 \neq \zeta_0. \end{gathered}$$

By the inverse function theorem we find a smooth map φ satisfying

$$\varphi(z_0) = \zeta_0 \quad \text{ and } \quad u = u \circ \varphi \ \text{ on } \ \overline{B_\varepsilon(z_0)}.$$

Using the differential equations for u and $\tilde{u}$ one verifies that φ is holomorphic as a map $B_\varepsilon(z_0) \to \dot{S}$ which, moreover, satisfies

$$d(a - \varphi^* a) = 0 \quad \text{ on } \ B_\varepsilon(z_0).$$

Consequently there exists a constant $c \in \mathbb{R}$ such that

$$a(z) + c = a \circ \varphi(z), \ z \in B_\varepsilon(z_0).$$

Defining $\tilde{u}_c(z) = (a(z) + c, u(z))$ we infer that

$$\tilde{u}_c(z) = \tilde{u} \circ \varphi(z), \ z \in B_\varepsilon(z_0). \tag{83}$$

We claim that $c = 0$. We set $z_1 = \zeta_0$ and choose a smooth path $\gamma(t)$ connecting $z_0 = \gamma(0)$ with $z_1 = \gamma(1)$ and satisfying $u(\gamma(t)) \notin M_0$ for all $t \in [0,1]$. By the similarity principle there exists a unique continuation φ along γ satisfying $\tilde{u}_c(z) = \tilde{u}(\varphi(z))$. In particular, $\tilde{u}_c(\gamma(t)) = \tilde{u} \circ \varphi(\gamma(t))$. The end point $z_2 = \varphi(\gamma(1))$ has the property

$$u(z_2) = u(z_1) = u(z_0)$$

so that $a(z_0) = a(z_1) + c = a(z_0) + 2c$. If z_2 agrees with z_0 or z_1 the claim $c = 0$ follows. If not, we repeat the procedure. There exists a biholomorphic map φ_1 defined near z_1 and satisfying $\varphi_1(z_1) = z_2$. By continuation,

$$\tilde{u}_c(z) = \tilde{u} \circ \varphi_1(z) \text{ for } z \text{ near } z_1.$$

We now continue φ_1 along the path $\gamma_1(t) = \varphi_1(\gamma(t))$ connecting z_1 with z_2 and find $z_3 = \varphi_1(\gamma_1(1))$ with $u(z_3) = u(z_0)$. Since there are only a finite number of points having image $u(z_0)$ we arrive after a finite number of steps at a point z_m obtained already before so that $z_m = z_j$ with $m > j$. By construction,

$$a(z_j) + (m - j)c = a(z_m) = a(z_j).$$

So, $c = 0$ as claimed. Summarizing we have established the existence of $z_0 \neq \zeta_0$ on $\dot{S}$ satisfying

$$\varphi(z_0) = \zeta_0 \quad \text{and} \quad \tilde{u}(z) = \tilde{u} \circ \varphi(z)$$

for $z \in B_\varepsilon(z_0)$. Introduce now the subset

$$S_1 = \{z \in \dot{S} \mid T\tilde{u}(z) \neq 0 \text{ and } z \text{ satisfies condition } (*)\}.$$

By definition, z satisfies condition $(*)$ if there exists a point $z' \neq z$ and if there exist two sequences (z_k), $z_k \neq z$ and (z'_k), $z'_k \neq z'$, satisfying:

$$z_k \to z$$
$$z'_k \to z'$$
$$\tilde{u}(z_k) = \tilde{u}(z'_k).$$

Clearly, the set S_1 is closed in the set $S_2 = \{z \in \dot{S} \mid T\tilde{u}(z) \neq 0\}$ and we claim that it is also open in S_2. Indeed, choosing a Darboux chart locally near $u(z) = u(z')$ on M one shows, by means of the similarity principle and proceeding as in the proof of Theorem 4 in [11] that there are open neighborhoods of z and z' having the same image in $\mathbb{R} \times M$ under the map $\tilde{u}$. Consequently, the set S_1 is either empty or agrees with S_2. From (83) we conclude $S_1 \neq \emptyset$ so that $S_1 = S_2$. This, however, contradicts our assumption that $\tilde{u}$ is somewhere injective which requires the existence of a point $z_0 \in \dot{S}$ satisfying $T\tilde{u}(z_0) \neq 0$ and $\tilde{u}(z_0) \in \tilde{u}(S \setminus \{z_0\})$. The proof of Theorem 1.13 is complete. □

The proof of Theorem 1.14 for the punctured disk $\dot{D} = D \setminus \Gamma$ with boundary ∂D follows along the same lines and we omit the details.

11. Proof of Theorem 1.15

We recall the mixed boundary value problem from the introduction. The compact oriented three manifold M is equipped with the contact form λ defining the contact structure ξ and the Reeb vector field X on M. The almost complex structures J on ξ considered are compatible with $d\lambda$ in the sense that $d\lambda(h, J(m)k)$ for $h, k \in \xi_m$ is an inner product at every point $m \in M$. Given to us is an oriented compact embedded disk $F \subset M$ having the following properties. The boundary ∂F is positively transversal to ξ. The singular foliation of F induced by ξ possesses precisely one singular point $e \in F$ in the interior. It is non-degenerate and positively elliptic (repeller). All characteristics of F emerge from e and leave the disk transversally through the boundary ∂F. We recall that associated with J on ξ we have the usual $\mathbb{R}$-invariant almost complex structure $\tilde{J} : T(\mathbb{R} \times M) \to T(\mathbb{R} \times M)$ defined by

$$\tilde{J}(a, m)(\alpha, k) = \big(-\lambda(m)k, J(m)\pi k + \alpha X(m)\big) \tag{84}$$

for $(a, m) \in \mathbb{R} \times M$ and $(\alpha, k) \in T(\mathbb{R} \times M)$. The map $\pi : TM \to \mathbb{R}X \oplus \xi \to \xi$ denotes the projection map. In the following we shall identify M with

$$\{0\} \times M \subset \mathbb{R} \times M.$$

Let $D \subset \mathbb{C}$ be the closed unit disk $D = \{z \in \mathbb{C} \mid |z| \leqslant 1\}$. We denote by (x_j, T_j), $j = 1, \dots, m$, non-degenerate and contractible periodic orbits of the Reeb vector field X on M.

With these data given, we shall study the following special finite energy surfaces. We consider smooth maps $\tilde{u} = (a, u) : \dot{D} \to \mathbb{R} \times M$ which are immersions and embeddings near the boundary and satisfy the following properties:

$$\begin{gathered} T\tilde{u} \circ j = \tilde{J}(\tilde{u}) \circ T\tilde{u} \quad \text{on } \dot{D} \\ E(\tilde{u}) < \infty \\ \pi \circ Tu(z_0) \neq 0 \quad \text{for some } z_0 \in \dot{D}. \end{gathered}$$

Here $\dot{D} = D \setminus \Gamma$ for some finite subset $\Gamma = \{z_1, \dots, z_m\} \subset D \setminus \partial D$ (depending on $\tilde{u}$). The boundary conditions require

$$\begin{gathered} a_{|\partial D} = 0 \\ u : \partial D \to F \setminus (\{e\} \cup \partial F). \end{gathered}$$

In addition, $u : \partial D \to u(\partial D) \subset F$ has winding number equal to 1 with respect to e. Finally, the punctures z_j are all negative and the associated asymptotic limits are the periodic orbits (x_j, T_j) of X for $j = 1, \dots, m$ introduced above.

Proceeding as in the previous paragraph we shall describe the solutions near a given one keeping the data F, Γ and (x_j, T_j) fixed. Recalling the Introduction we observe that neighboring surfaces C' are described by finite energy maps $\tilde{w} : D \setminus \Gamma \to \mathbb{R} \times M$ where $D \setminus \Gamma$ is equipped with an induced complex structure j. The structure has a C^α extension over Γ. There exists a new holomorphic structure $\mathcal{A}$ on D which restricts to j over $D \setminus \Gamma$, see Appendix. By the uniformization theorem there exists a holomorphic map $\Phi : (D, i) \to (D, \mathcal{A})$. The underlying differentiable structure of $\mathcal{A}$ coincides with the standard structure on $D \setminus \Gamma$. Now defining $\Gamma' = \Phi^{-1}(\Gamma)$ and $\tilde{v} = \tilde{w} \circ \Phi$ we see that C' is parametrized by a map $\tilde{v} : D \setminus \Gamma' \to \mathbb{R} \times M$, where now D is equipped with i.

In order to apply the Sard-Smale theorem later on we vary $\tilde{J}$ in the parameter space $\mathcal{U}_\delta$. In contrast to the situation of Theorem 1.10 the structures $\tilde{J}$ constructed here are all $\mathbb{R}$-invariant and given by the formula (84), where the $d\lambda$-compatible structure J varies on ξ in a neighborhood of a fixed J_0. We consider the Fréchet space of $\mathcal{C}^\infty$ consisting of all smooth maps

$$m \mapsto Y(m) \in \mathrm{Hom}_{\mathbb{R}}(\xi_m),\ m \in M$$

satisfying on ξ_m

$$Y(m) \circ J_0(m) + J_0(m) \circ Y(m) = 0. \tag{85}$$

It follows by using the assumptions that $d\lambda(\cdot, J_0(m)\cdot)$ is an inner product on ξ_m,

$$d\lambda(Y(m)h, k) + d\lambda(h, Y(m)k) = 0$$

for all $h, k \in \xi_m$. Let $\varepsilon = (\varepsilon_k)$ be a sequence of positive numbers converging to 0. We denote by $\mathcal{C}_\varepsilon$ the linear subspace of $\mathcal{C}^\infty$ consisting of all Y satisfying (85) and

$$\|Y\|_\varepsilon := \sum_{k=1}^{\infty} \varepsilon_k \|Y\|_{C^k} < \infty.$$

If the sequence $\varepsilon = (\varepsilon_k)$ converges sufficiently fast to 0 one can verify that $\mathcal{C}_\varepsilon$ is dense in $\mathcal{C}^\infty$, see [9].

For $\delta > 0$ we now denote by $\mathcal{U}_\delta$ the set of all almost complex structures $\tilde{J} : T(\mathbb{R} \times M) \to T(\mathbb{R} \times M)$ defined by formula (84) in which we insert

$$J(m) = J_0(m) \exp\bigl(-J_0(m)Y(m)\bigr),$$

with $Y \in \mathcal{C}_\varepsilon$ satisfying $\|Y\|_\varepsilon < \delta$. The map

$$Y \mapsto \tilde{J} \in \mathcal{U}_\delta$$

constitutes a global chart for $\mathcal{U}_\delta$ defining a separable Banach manifold structure.

Proceeding as in the previous paragraph we introduce the set

$$\mathcal{M} = \bigl\{(\tilde{J}, C)\bigr\},$$

where $\tilde{J} \in \mathcal{U}_\delta$ and $C = \tilde{u}(\dot{D})$ with a $\tilde{J}$-holomorphic finite energy surface having the properties as described above. Again we shall show that $\mathcal{M}$ carries a natural Banach manifold structure with an underlying separable topology. We start with the local model describing $(\tilde{J}, C) \in \mathcal{M}$ in a neighborhood of a fixed $(\tilde{J}_0, C_0) \in \mathcal{M}$, where $C_0 = \tilde{u}_0(\dot{D})$.

Using the constructions in [1] and [15] for the coordinates near the boundary $\tilde{u}_0(\partial D)$ of C_0, and extending these coordinates to a tubular neighborhood of $C_0 \subset \mathbb{R} \times M$ proceeding as in Section 4 we obtain the map Φ of local coordinates around C_0:

$$\Phi : \dot{D} \times B \to \mathbb{R} \times M, \quad \Phi(z, 0) = \tilde{u}_0(z)$$

for $z \in \dot{D}$. The set B is an open disc $B_\varepsilon(0) \subset \mathbb{R}^2$. In $\dot{D} \times B$ the piece of surface F near $\tilde{u}_0(\partial D)$ lies on $\partial \dot{D} \times (\mathbb{R} \times \{0\})$, where $\mathbb{R} \times \{0\} \subset \mathbb{R}^2$. We introduce the set $\mathcal{B}$ consisting of all functions $v : \dot{D} \to B \subset \mathbb{R}^2$ of class $C^{1,\alpha}_{\mathrm{loc}}$ such that v is of class $C^{1,\alpha}_0$ near the punctures Γ and satisfies the boundary conditions

$$v(\partial D) \subset \mathbb{R} \times \{0\} \subset \mathbb{R}^2. \tag{86}$$

Then we introduce the Banach manifold

$$\hat{\mathcal{B}} = \mathcal{U}_\delta \times \mathcal{B}$$

and the Banach space bundle $\hat{\mathcal{E}} \to \hat{\mathcal{B}}$ by

$$\hat{\mathcal{E}} = \bigcup_{(\tilde{J},v) \in \hat{\mathcal{B}}} \{(\tilde{J}, v)\} \times C^\alpha_0(\mathrm{graph}(v)^* \Xi_{\bar{J}}),$$

where the almost complex structure $\bar{J}$ on $\dot{D} \times B$ is induced by $\tilde{J}$, i.e.

$$\tilde{J}(\Phi(z,b)) \circ T\Phi(z,b) = T\Phi(z,b) \circ \bar{J}(z,b),$$

for all $(z,b) \in \dot{D} \times B$. As before, we define the smooth section H of $\hat{\mathcal{E}} \to \hat{\mathcal{B}}$ by the formula

$$\begin{aligned} H(\tilde{J}, v)(z)(h,k) &= (h, Tv(z)h) \wedge (k, Tv(z)k) \\ &\quad - \bar{J}(u(z))(h, Tv(z)h) \wedge \bar{J}(u(z))(k, Tv(z)k). \end{aligned}$$

We abbreviated $u(z) = (z, v(z))$, where $z \in \dot{D}$. The function $v : \dot{D} \to \mathbb{R}^2$ belongs to $\mathcal{B}$. Proceeding as in the proof os Theorem 1.10 we have to show that the derivative $H'(\tilde{J}_0, 0)$ at the reference solution $(\tilde{J}_0, 0)$ in the direction (Y, g) is surjective, where Y is now in a smaller space. As before

$$H'(\tilde{J}_0, 0)(Y, g) = \alpha(Lg + \Theta(\tilde{Y})), \tag{87}$$

where $\tilde{Y} : T(\mathbb{R} \times M) \to T(\mathbb{R} \times M)$ is related to Y as follows. With respect to the splitting $T_{(a,m)}(\mathbb{R} \times M) = (\mathbb{R}, \mathbb{R}X(m) + \xi_m)$ we set $\tilde{Y}(a,m)(\alpha, \beta, \gamma) = (0, 0, Y(m)\gamma)$. In view of (85) the linear map $\tilde{Y}$ satisfies

$$\tilde{J}_0 \circ \tilde{Y} = -\tilde{Y} \circ \tilde{J}_0. \tag{88}$$

The linear operator L above, obtained by linearizing H only in the v-direction, is the Fredholm operator from Section 5,

$$(Lg)(z) = Tg(z) + i_0 \circ Tg(z) \circ j_0(z) + \left[\Delta'(z) g(z)\right] \circ j_0(z),$$

for every $z \in \dot{D}$. Abbreviating

$$A(Y)(z) = \left(T\Phi(z,0)\right)^{-1} \circ \tilde{Y}(\tilde{u}_0(z)) \circ T\Phi(z,0)$$

we conclude from the definition of $\bar{J}_0$ together with (88)

$$A(Y)(z) \circ \bar{J}_0(z,0) = -\bar{J}_0(z,0) \circ A(Y)(z), \tag{89}$$

so that $A(Y)$ is $\bar{J}_0$-complex anti-linear. Corresponding to the splitting $T\dot{D}\times\mathbb{R}^2$ of the tangent space of $\dot{D}\times B$ we represent $A(Y)$ in the matrix form

$$A(Y) = \begin{pmatrix} a & c \\ b & d \end{pmatrix}. \tag{90}$$

Since, by construction of Φ, we have $\bar{J}_0(z,0) = j_0(z)\oplus i_0(z)$ and since $A(Y)$ is $\bar{J}_0$-complex anti-linear, the linear maps $a(z), b(z)$ have the properties

$$\begin{aligned} a(z)\cdot j_0(z) &= -j_0(z)\cdot a(z) \\ b(z)\cdot j_0(z) &= -i_0(z)\cdot b(z), \end{aligned}$$

$z\in\dot{D}$. The relation between $b(z)\in\mathcal{L}(T_z\dot{D},\mathbb{R}^2)$ and $\Theta(\tilde{Y})$ is given by $\Theta(\tilde{Y})(z) = b(z)\circ j_0(z)$, where $j_0(z)$ is the left upper corner of $\bar{J}_0(z,0)$ in our local model $\dot{D}\times\mathbb{R}^2$. Proceeding as in the proof of Theorem 1.10 we have to verify that the linear operator

$$(Y,g)\mapsto Lg+\Theta(\tilde{Y}) \tag{91}$$

viewed as an operator

$$T_{\tilde{J}_0}\mathcal{U}_\delta\times W^{1,2}_{\text{bdc}}(\dot{D},\mathbb{R}^2)\to L^2(\mathcal{A}_0)$$

is surjective. The space $W^{1,2}_{\text{bdc}}(\dot{D},\mathbb{R}^2)$ is the Sobolev space with the Lagrange boundary condition on $\partial\dot{D}$ induced by the requirement (86).

Assume that p is orthogonal to the image of the operator (91). Then p solves the adjoint equation (to L) and hence is smooth. As a consequence of the similarity principle it vanishes identically, if it vanishes on some open subset. By assumption,

$$0 = \langle\Theta(Y),p\rangle_{L^2} \tag{92}$$

for every $Y\in T_{\tilde{J}_0}\mathcal{U}_\delta$. By the assumptions of Theorem 1.15, the section $\pi\circ Tu_0$ does not vanish identically and invoking Theorem 1.14, the set

$$\dot{D}_0 = \big\{z\in\dot{D}\mid \pi\circ Tu_0(z)\neq 0 \ \text{ and } \ u_0(z)\notin u_0(\dot{D}\setminus\{z\})\big\}$$

is therefore open and dense in $\dot{D}$. Take a point $z_0\in\dot{D}_0$ close to a puncture in Γ and recall that close to a puncture, the normal bundle of $C_0 = \text{image}(\tilde{u}_0)$ and ξ almost coincide in view of our construction in Section 4. Arguing by contradiction we assume now $p\neq 0$. Then we can also assume that p does not vanish in a neighborhood of z_0. Using local coordinates near the puncture from Section 4 we shall construct $\tilde{Y}$ supported near z_0 and satisfying

$$\langle\Theta(\tilde{Y}),p\rangle_{L^2} > 0.$$

This then will contradict (92) and will prove that $p=0$, so that $H'(\tilde{J}_0,0)$ is indeed surjective.

In order to construct $\tilde{Y}$ in the smaller class defined by Y having the above property we make use of the local model from Section 4. Recall the map Γ :

$[0, \infty) \times S^1 \times \mathbb{R}^2 \to \mathbb{R}^4$ introduced there,

$$\Gamma(s,t,x,y) = \tilde{v}(s,t) + xn(s,t) + ym(s,t),$$

where $\tilde{v} = (a, v)$ represents our solution $\tilde{u}_0$ near a puncture in local coordinates and n and m are normal vectors. In this coordinates the solution $\tilde{u}_0$ is represented near a puncture as $(s,t,0,0)$ Now take an arbitrary complex anti-linear map $Y : \xi_{v(s,t)} \to \xi_{v(s,t)}$. The corresponding map $\tilde{Y}(\tilde{v}(s,t)) : T_{\tilde{v}(s,t)}\mathbb{R}^4 \to T_{\tilde{v}(s,t)}\mathbb{R}^4$ satisfies $\tilde{Y}(1) = 0$, $\tilde{Y}(X) = 0$ and $\tilde{Y}(k) = Yk$ for all $k \in \xi_{v(s,t)}$. In our local coordinates the map $A(Y)$ in (90) becomes $A_0(Y)(s,t) : \mathbb{R}^4 \to \mathbb{R}^4$, where $A_0(Y)(s,t) = (T\Gamma)^{-1} \circ \tilde{Y}(\tilde{v}(s,t)) \circ T\Gamma$. We want to compute the map $b_0(s,t)$ corresponding to b. Since $A_0(Y)(s,t)$ is $\bar{J}_0$-complex anti-linear, it suffices to calculate $A_0Y(s,t)e_1$, where $e_1 = (1,0,0,0)$. For s large $\pi v_s(s,t)$ and $\pi v_t(s,t) \neq 0$. So they form a basis for the contact plane $\xi_{v(s,t)}$. With respect to this basis the complex anti-linear map Y may be represented as a 2×2 matrix

$$Y(v(s,t)) = \begin{pmatrix} Y_1(s,t) & Y_2(s,t) \\ Y_2(s,t) & -Y_1(s,t) \end{pmatrix}.$$

Hence

$$\begin{aligned} \tilde{Y}(\tilde{v}(s,t)) \circ (T\Gamma)(s,t,0)e_1 &= \tilde{Y}(\tilde{v}(s,t))\tilde{v}_s \\ &= Y(v(s,t))\pi v_s = Y_1(s,t)\pi v_s + Y(s,t)\pi v_t. \end{aligned}$$

The tangent space $T_{\tilde{v}(s,t)}\mathbb{R}^4 = \mathbb{R}^4$ is spanned by the tangent vectors $\tilde{v}_s$, $\tilde{v}_t$ and the normal vectors n and m. Consequently,

$$Y_1(s,t)\pi v_s + Y_2(s,t)\pi v_t = \alpha\tilde{v}_s + \beta\tilde{v}_t + \gamma n + \delta m$$

with

$$(\alpha,\beta,\gamma,\delta) = \frac{|\pi v_s|}{|\tilde{v}_s|}\left(\frac{|\pi v_s|}{|\tilde{v}_s|}Y_1,\ \frac{|\pi v_s|}{|\tilde{v}_s|}Y_2,\ [-Y_1 a_s - Y_2 a_t],\ [Y_1 a_t - Y_2 a_s]\right).$$

This implies that $A_0(Y)(s,t)e_1 = (T\Gamma)^{-1} \circ \tilde{Y}(\tilde{v}(s,t)) \circ (T\Gamma)e_1 = (\alpha,\beta,\gamma,\delta)$ and since $b_0(s,t)$ is J_0 complex anti-linear we obtain

$$b_0(s,t) = \begin{pmatrix} \gamma & \delta \\ \delta & -\gamma \end{pmatrix} = \frac{|\pi v_s|}{|\tilde{v}_s|}\begin{pmatrix} -a_s & -a_t \\ a_t & -a_s \end{pmatrix}\begin{pmatrix} Y_1 & Y_2 \\ Y_2 & -Y_1 \end{pmatrix}.$$

So, the map $\Theta_0(Y)(s,t) = b_0(s,t) \circ J_0$ is given by

$$\Theta_0(Y)(s,t) = \frac{|\pi v_s|}{|\tilde{v}_s|}\begin{pmatrix} a_t & -a_s \\ a_s & a_t \end{pmatrix}\begin{pmatrix} Y_1 & Y_2 \\ Y_2 & -Y_1 \end{pmatrix} \tag{93}$$

We denote the local representative of the given smooth section p near the puncture by $p_0 : [0,\infty)\times S^1 \to \mathcal{L}(\mathbb{R}^2,\mathbb{R}^2)$. It is complex anti-linear, $J_0 \cdot p_0(s,t) = -p_0(s,t) \cdot J_0$. We assume that at the point (s_0,t_0) we have $p_0(s_0,t_0) \neq 0$. It is now sufficient to find a complex anti-linear map $Y_0 : \xi_{v(s_0,t_0)} \to \xi_{v(s_0,t_0)}$ satisfying

$$\left\langle \Theta_0(Y_0)(s_0,t_0)\begin{pmatrix}1\\0\end{pmatrix}, p_0(s_0,t_0)\begin{pmatrix}1\\0\end{pmatrix}\right\rangle > 0.$$

But this follows immediately from the explicit representation (93) if we choose s_0 sufficiently large, in view of the asymptotics of the function a in Theorem 4.4. This finishes the construction of the map Y and $\tilde{Y}$. We have proved that $H'(\tilde{J}_0, 0)$ is surjective.

Arguing as before, $\mathcal{M}$ is a separable Banach manifold. By the Sard-Smale theorem, the set $\mathcal{R} \subset \mathcal{U}_\delta$ consisting of regular values of the projection $\eta : \mathcal{M} \to \mathcal{U}_\delta$ is a residual set in $\mathcal{U}_\delta$. Constructing such a set $\mathcal{R} = \mathcal{R}^m$ for every number m of punctures we take the countable intersection $\widehat{\mathcal{R}} = \bigcap \mathcal{R}^m$ which is a Baire subset of $\mathcal{U}_\delta$. We choose $\tilde{J} \in \widehat{\mathcal{R}}$ and consider the special finite energy surface

$$\tilde{u} : \dot{D} \to (\mathbb{R} \times M, \tilde{J})$$

postulated in Theorem 1.15. Since $\tilde{J} \in \widehat{\mathcal{R}}$, the associated linearized operator L in the local model is surjective. Therefore, its Fredholm index is non-negative. In view of Theorem 2.11, the index of L is given by the formula

$$\mathrm{Ind}(L) = \mu_M(\eta) - \sum_{j=1}^{m} \mu_N(\tilde{C}_j) - m + 1. \tag{94}$$

In view of the special boundary condition (86), the loop η of lines in $\mathbb{R}^2$ in our case is simply $\eta(z) = \mathbb{R} \times \{0\}$ for all $|z| = 1$, so that the Maslov index vanishes,

$$\mu_M(\eta) = 0. \tag{95}$$

The index $\mathrm{Ind}(L)$ above is expressed in terms of the normal Conley-Zehnder indices with respect to a special trivialization of the normal bundle over C_0 which over the punctures has a continuation to the contact planes ξ along the limits (x_j, T_j) of the punctures. Using the assumptions of Theorem 1.15 we shall compute $\mathrm{Ind}(L)$ in terms of the Conley-Zehnder indices of the limits, defined with respect to symplectic trivializations of the contact bundle ξ over the given disks.

We recall that if (x_j, T_j) is the asymptotic limit associated with the puncture $z_j \in \Gamma$, we have a disk map $v_j : D \to M$ satisfying

$$v_j\left(e^{2\pi i t}\right) = x_j(T_j t),$$

and we denoted the Conley-Zehnder index computed with respect to a complex trivialization of ξ over disk maps by

$$\mu_j(x_j, v_j) \equiv \mu_j(\xi).$$

The integer $\mu_j(\xi)$ depends only on the homotopy class of the given disk map. Given the solution $\tilde{u} = (a, u) : \dot{D} \to (\mathbb{R} \times M, \tilde{J})$ of the theorem, we can glue the disk maps v_j at the punctures $z_j \in \Gamma$ and obtain a disk map

$$w := \tilde{u} \# \{v_1, \dots, v_m\} : D \to \mathbb{R} \times M.$$

This map coincides near the boundary ∂D with the map $u : D \to M$. Theorem 1.15 requires that w can be homotoped through mappings which map the boundary ∂D into the punctured disk $F \setminus \{e\} \subset M$, to a disk mapping whose image lies in a small punctured neighborhood of the singularity e. This assumption allows us to prove

Lemma 11.1. *Consider the solution* $\tilde{u} = (a, u) : D\backslash\Gamma \to \mathbb{R}\times M$ *as in Theorem 1.15 and recall that* $m = \sharp\Gamma$. *Then*

$$\mathrm{Ind}(L) = m - \sum_{j=1}^{m} \mu_j(\xi) + 1.$$

Proof. Recall that in view of the definition of the almost complex structure $\tilde{J}$ the tangent space

$$T_{(a,m)}(\mathbb{R}\times M) = (\mathbb{R}\times\mathbb{R}X)\oplus\xi$$

splits into the complex subspaces $\mathbb{C}X$ and ξ. All our trivializations in what follows are assumed to be complex. We choose a trivialization of the bundle $v_j^*\xi \to D$ and trivialize $v_j^*(\mathbb{C}X) \to D$ by means of the mapping

$$\begin{aligned} v_j^*(\mathbb{C}X) &\to D\times\mathbb{C} \\ \big(z, \alpha X(v_j(z))\big) &\mapsto (z, \alpha). \end{aligned}$$

This gives us a trivialization of the bundle $v_j^*T(\mathbb{R}\times M) \to D$. Since the normal bundle of $\tilde{u}(\dot{D})$ has a continuation to the limit orbits, we can extend the trivializations of $v_j^*\xi$ for $j = 1, \dots, m$, to a trivialization of $\tilde{u}(\dot{D})$ in $\mathbb{R}\times M$. Similarly, the trivializations (11) can be extended continuously to a trivialization of the tangent bundle of $\tilde{u}(\dot{D})$. This way we obtain a trivialization of the bundle

$$(\tilde{u}\#\{v_1, \dots, v_m\})^*T(\mathbb{R}\times M) \to D \tag{96}$$

having the property that along the boundary ∂D it splits into trivializations of the tangent bundle of $\tilde{u}(\dot{D})$ and the normal bundle of $\tilde{u}(\dot{D})$. The image of ∂D is given by $y(t) = \tilde{u}(e^{2\pi it})$, $0 \leqslant t \leqslant 1$. Hence, $\dot{y}(t)$ also defines a trivialization of the tangent bundle of $\tilde{u}(\dot{D})$ over ∂D. However, it does not agree with the trivialization for (96) we have just constructed. Indeed, the trivialization just constructed is homotopic to the trivialization defined by

$$e^{2\pi i(m-1)t}\cdot\dot{y}(t),$$

since the winding number of $y(t)$ is equal to 1 and since we have m punctures. The dot in the formula stands for the complex multiplication in $T(\tilde{u}(\dot{D}))$ induced by $\tilde{J}$.

Consider now the loop $T_{\tilde{u}(e^{2\pi it})}F$ of tangent planes of F along $\tilde{u}(e^{2\pi it})$. These planes are totally real. The image under our trivialization of (96) gives in view of (11) a loop of real planes in $(\mathbb{C}\oplus\mathbb{C}, i\oplus i)$ homotopic to

$$\hat{L}(t) = e^{2\pi i(1-m)t}\mathbb{R}\oplus\hat{N}(t). \tag{97}$$

Here $\hat{N}(t)$ is the loop of lines in $\mathbb{C}$ corresponding to the normal part. For the Maslov index of $\hat{L}$ we obtain

$$\mu_2(\hat{L}) = \mu_1(e^{2\pi i(1-m)t}\mathbb{R}) + \mu_1(\hat{N}). \tag{98}$$

Using the homotopy requirement for the map $w : D \to \mathbb{R}\times M$ we can compute the Maslow index of the loop of totally real planes $T_{\tilde{u}(e^{2\pi it})}F$. It is equal to 2 according to the computations in [1] and [15]. Therefore, $\mu_2(\hat{L}) = 2$ and hence we obtain from (98) the formula $2 = 2(1-m) + \mu_1(\hat{N})$. So,

$$\mu_1(\hat{N}) = 2m. \tag{99}$$

The trivializations of $v_j^*\xi$, extended to a trivialization of the normal bundle of $\tilde{u}(\dot{D})$, show that the integer

$$\mu_1(\hat{N}) - \sum_{j=1}^{m} \mu_j(\xi) \tag{100}$$

agrees with the following integer μ. Given any trivialization of the normal bundle of $\tilde{u}(\dot{D}) \subset \mathbb{R}\times M$ which extends to ξ at the asymptotic limits as in Section 4, then we can define the associated integer $\mu =$ normal Maslov index (of TF) minus the normal Conley-Zehnder index. Consequently, we can compute using (94), (95) and (99),

$$\begin{aligned}
\mathrm{Ind}(L) &= 0 - \sum_{j=1}^{m} \mu_N(\tilde{C}_j) - m + 1\\
&= \mu_1(\hat{N}) - \sum_{j=1}^{m} \mu_j(\xi) - m + 1 = 2m - \sum_{j=1}^{m} \mu_j(\xi) - m + 1\\
&= m - \sum_{j=1}^{m} \mu_j(\xi) + 1.
\end{aligned}$$

The proof of the lemma is complete. □

We finish the proof of Theorem 1.15. In view of Lemma 11.1 we know, if $\tilde{J} \in \widehat{\mathcal{R}} \subset \mathcal{U}_\delta$, then

$$m - \sum_{j=1}^{m} \mu_j(\xi) + 1 \geq 0.$$

If $\mu_j(\xi) \geq 3$ for every puncture z_j, we deduce $m + 1 \geq 3m$ and hence $m = 0$. This finishes the proof of Theorem 1.15. □

We would like to point out that under the assumption $\mu_j(\xi) \geq 2$ for all j we obtain for the number m of punctures in Theorem 1.15 the estimate $1 \geq m$. This leaves the possibility of the existence of a single puncture whose asymptotic limit (x, T) has index $\mu(x, v) = 2$.

12. Appendix

In this appendix we shall collect some results about almost complex structures refered to in the previous paragraphs. In addition, we carry out the details of the proof of Theorem 1.5.

$C^{1,\alpha}$-charts and consequences

We begin with a classical result.

Theorem 12.1. *Given an almost complex structure j of class C^α, $\alpha \in (0,1)$, defined on a neighborhood of $0 \in \mathbb{C}$ and a vector $v \in \mathbb{C}$, there exists a map $f : U(0) \to \mathbb{C}$ defined on a neighborhood of 0 and of class $C^{1,\alpha}$ satisfying*

$$Tf \circ i = j(f) \circ Tf \text{ and } f(0) = 0,\ Tf(0) \cdot 1 = v.$$

For the proof we refer to S.S. Chern [4] or L. Ahlfors and L. Bers [2], see also A. Nijenhuis and W. Woolf [28]. We shall deduce two corollaries.

Corollary 12.2. *Let S be a smooth orientable closed surface and j an almost complex structure on S of class C^α. Then there exist an atlas (ϕ_τ), such that $\phi_\tau : U_\tau \to \mathbb{C}$ satisfies*

$$i \circ T\phi_\tau = T\phi_\tau \circ j.$$

The charts are of class $C^{1,\alpha}$ for the given smooth structure on S. Moreover the transition maps $\phi_{\tau_1} \circ \phi_{\tau_2}^{-1}$ are holomorphic. Therefore the new atlas defines a holomorphic structure on S with an underlying smooth structure so that the identity map of S between the two structures is a $C^{1,\alpha}$-diffeomorphism.

Proof. We just have to note that the transition maps are $C^{1,\alpha}$-solutions of the standard Cauchy Riemann equation $\bar{\partial} u = 0$. By the interior elliptic regularity theory the transition maps are holomorphic. Having this, all other statements are now tautological consequences of the respective definitions. □

Corollary 12.3. *Assume that S is a smooth closed oriented surface and Γ a finite subset. Let j be a smooth almost complex structure on $S \setminus \Gamma$, which has a Hölder continuous extension over Γ which is compatible with the orientation. Then there exists a smooth differentiable structure on the underlying C^0-surface S, inducing the old differentiable structure on $S \setminus \Gamma$, so that j has a smooth extension over Γ for the new structure.*

Proof. Denote the original smooth structure, i.e. the maximal C^∞-atlas by $\mathcal{D}_1$ and the smooth structure induced by the holomorphic structure associated to j by $\mathcal{D}_2$. Then note that

$$\mathrm{Id} : (S, \mathcal{D}_1) \to (S, \mathcal{D}_2)$$

is a a $C^{1,\alpha}$-diffeomorphism. However at points, where j is smooth on some open neighborhood for the old structure the map Id is smooth. In our case we note that only Γ can contain non-smooth points for the identity map. □

From Corollary 12.3 we can conclude the following

Theorem 12.4. *Let λ be a contact form on the closed three-manifold M and $J:\xi\to\xi$ a compatible multiplication for the associated contact structure. Denote by $\tilde{J}$ the associated $\mathbb{R}$-invariant almost complex structure on $\mathbb{R}\times M$. Let $\tilde{u}:S\setminus\Gamma\to\mathbb{R}\times M$ be an embedding, so that $T\tilde{u}(z)(T_zS)$ is a complex line for every $z\in S\setminus\Gamma$, where S is a compact oriented surface and $\Gamma\subset S$ a finite subset. Moreover, assume that the energy*

$$E(\tilde{u})=sup_{\varphi\in\Sigma}\int_{S\setminus\Gamma}\tilde{u}^*d(\varphi\lambda)$$

is finite and the almost complex structure $\bar{j}$ defined on $S\setminus\Gamma$ by

$$\tilde{J}\circ T\tilde{u}=T\tilde{u}\circ\bar{j}$$

has a C^α-extension over Γ. Then S admits a complex structure j inducing $\bar{j}$ on $S\setminus\Gamma$. In particular, $\tilde{u}:S\setminus\Gamma\to\mathbb{R}\times M$ is a finite energy surface.

Construction of the map ψ in the proof of Thorem 1.5

Recall that we have an embedding

$$u:B\times\mathbb{C}\to\mathbb{C}\times\mathbb{R}^2$$

of the form

$$u(\tau,z)=(z,v(\tau,z)).$$

Here $B\subset\mathbb{R}^2$ is an open ball centered at the origin. The map $\tau\mapsto v(\tau):=v(\tau,\cdot)$, $B\to C_0^{k,\alpha}(\mathbb{C},\mathbb{R}^2)$, is smooth and satisfies the equation $H(v(\tau))=0$ and $v(0)\equiv 0$. For every $\tau\in B$, the map $u_\tau(\cdot):=u(\tau,\cdot):\mathbb{C}\to\mathbb{C}\times\mathbb{R}^2$ is an embedded complex curve near $\mathbb{C}\times\{0\}$ satisfying

$$Tu_\tau(z)\bar{j}_\tau(z)=\bar{J}(u_\tau(z))\cdot Tu_\tau(z),\ \ z\in\mathbb{C}.$$

We want to construct a smooth map $\psi:B\times\mathbb{C}\to B\times\mathbb{C}$ of the form $(\tau,z)\mapsto(\tau,\psi_\tau(z))$ such that $\psi_\tau:=\psi(\tau,\cdot):\mathbb{C}\to\mathbb{C}$ satisfies

$$\bar{j}_\tau(\psi_\tau(z))\cdot T\psi_\tau(z)=T\psi_\tau(z)\circ i\quad\text{for all }\ z\in\mathbb{C}.\tag{101}$$

By construction of our special coordinates we have $\bar{j}_0=i$ for $\tau=0$. We know from the asymptotic behavior that every j_τ extended by i over ∞ is of class C^α near ∞ and of course smooth in the complement of ∞. By the first part of the appendix we find for every τ a natural holomorphic structure on S^2 associated to $\bar{j}_\tau$. By the uniformisation theorem there exists for every τ a uniquely determined biholomorphic map $\psi_\tau:(S^2,i)\to(S^2,\bar{j}_\tau)$ such that ψ_τ fixes the points 0, 1 and ∞. Of course the underlying smooth structures on (S^2,i) and $(S^2,\bar{j}_\tau)$ are different. By elliptic regularity theory, using that the almost complex structures $\bar{j}_\tau$ are smooth for the standard differentiable structure in the complement of ∞ we see that ψ_τ is smooth away from ∞, if the domain and the target space are equipped with the standard structure. Define

$$\psi:B\times\mathbb{C}\to B\times\mathbb{C}$$
$$(\tau,z)\mapsto\psi(\tau,z)=(\tau,\psi_\tau(z)).$$

The map ψ shows already some of the desired properties. In particular, it solves equation (101) for every τ fixed, and the induced self-map ψ_τ of $\mathbb{C}$ is smooth. It remains to show that the dependence on the parameter is smooth. This will follow by means of an implicit function theorem.

Lemma 12.5. *For small positive δ, the map $v : B \to C_0^{k,\alpha,\delta}(\mathbb{C}, \mathbb{R}^2)$, $\tau \mapsto v(\tau)$ is smooth for all k.*

Proof. Recall the notations and the local coordinates from Section 4. With respect to the splitting $\mathbb{C} \times \mathbb{R}^2$, the almost complex structures $\bar{J}$ has the matrix form

$$\bar{J}(z,x,y) = \begin{pmatrix} j(z,x,y) & \tilde{\Delta}(z,x,y) \\ \Delta(z,x,y) & i(z,x,y) \end{pmatrix}$$

and satisfies, for every $z \in \mathbb{C}$,

$$\bar{J}(z,0) = \begin{pmatrix} i & 0 \\ 0 & i \end{pmatrix}.$$

Hence equation (12) becomes

$$\bar{j}_\tau(z) = j(u_\tau(z)) + \tilde{\Delta}(u_\tau(z))Tv_\tau(z) \tag{102}$$

$$Tv_\tau(z)\bar{j}_\tau(z) = \Delta(u_\tau(z)) + i(u_\tau(z))Tv_\tau(z). \tag{103}$$

We shall reformulate these equations in the holomorphic polar coordinates $z = \sigma(s,t) = e^{2\pi(s+it)}$. Substituting (102) into (103) the map v_τ solves the equation

$$\begin{aligned} Tv_\tau(\sigma)j(u_\tau(\sigma)) \circ T\sigma - i(u_\tau(\sigma))Tv_\tau(\sigma) \circ T\sigma \\ = \Delta(u_\tau(\sigma)) \circ T\sigma - Tv_\tau(\sigma)\tilde{\Delta}(u_\tau(\sigma))Tv_\tau(\sigma) \circ T\sigma. \end{aligned}$$

Recall from Section 4 that the almost complex structure $\bar{J}$ on $\mathbb{C} \times \mathbb{R}^2$ induces the almost structures $\bar{J}_0$ on $\mathbb{R} \times S^1 \times \mathbb{R}^2$ via

$$\bar{J}_0(s,t,x,y) = T(\sigma \times \mathrm{Id}_{\mathbb{R}^2})^{-1} \circ \bar{J}(\sigma(s,t),x,y) \circ T(\sigma \times \mathrm{Id}_{\mathbb{R}^2}). \tag{104}$$

With respect to the splitting $\mathbb{R} \times S^1 \times \mathbb{R}^2$ it has the form

$$\bar{J}_0(s,t,x,y) = \begin{pmatrix} j_0(s,t,x,y) & \tilde{\Delta}_0(s,t,x,y) \\ \Delta_0(s,t,x,y) & i_0(s,t,x,y) \end{pmatrix}$$

and satisfies $j_0(s,t,0.0) = \imath$ and $i_0(s,t,0,0) = i$ for every $(s,t) \in \mathbb{R} \times S^1$. It follows from (104) that the above block matrices transform according to the equations

$$\begin{aligned} \jmath_0(s,t,x,y) &= (T\sigma)^{-1} \circ j(\sigma,x,y) \circ T\sigma \\ \tilde{\Delta}_0(s,t,x,y) &= (T\sigma)^{-1} \circ \tilde{\Delta}(\sigma,x,y) \\ \Delta_0(s,t,x,y) &= \Delta(\sigma,x,y) \circ T\sigma \\ i_0(s,t,x,y) &= i(\sigma,x,y). \end{aligned}$$

Abbreviating the composition by $w(s,t) = v_\tau(\sigma(s,t))$, setting $\hat{w}(s,t) = (s,t,w(s,t))$ and using the above identities one derives the following equation for w

$$Tw \circ j_0(\hat{w}) - i_0(\hat{w}) \circ Tw = \Delta_0(\hat{w}) - Tw \circ \tilde{\Delta}_0(\hat{w})Tw.$$

Now,

$$\begin{aligned}
j_0(s,t,w) &= i + \left[\int_0^1 j_0'(s,t,\rho w)d\rho\right] w \\
i_0(s,t,w) &= i + \left[\int_0^1 i_0'(s,t,\rho w)d\rho\right] w \\
\Delta_0(s,t,w) &= \Delta_0'(s,t,0,0) + \left[\int_0^1 \rho\left[\int_0^\rho \Delta_0''(s,t,rw)dr\right] w\,d\rho\right] w.
\end{aligned}$$

A similar identity holds also for $\tilde{\Delta}_0$. Substituting these identities into (12) we see that w solves the following the equation

$$Tw + i \circ Tw \circ i + (\Delta_0'(s,t,0,0)w) \circ i + R_\tau(s,t)w = 0.$$

By construction, $D^\alpha w(s,t) \to 0$ as $s \to \infty$ uniformly in t, for every derivative α. Hence, in view of Theorem 4.7 and Lemma 5.2 the perturbation $R_\tau(s,t) \in \mathcal{L}(\mathbb{R}^2,\mathbb{R}^2)$ satisfies $D^\alpha[R_\tau(s,t)h] \to 0$ as $s \to \infty$ uniformly in t for every $h \in \mathbb{R}^2$ and every derivative α. Also, $D^\alpha[(\Delta_0'(s,t,0,0)h)e_2 - \tilde{C}(t)h] \to 0$ as $s \to \infty$ uniformly in t for every $h \in \mathbb{R}^2$ and every α, where $e_2 = ie_1$ and $e_1 = (1,0)$ is the vector of the standard basis of $\mathbb{R}^2$. The linear map $\tilde{C}(t)$ does not depend on τ and has the property that the problem $\dot{x}(t) = i\tilde{C}(t)x(t)$, $x(0) = x(1)$ admits only the trivial solution. Consequently, proceeding as in [18] the solution $w = w_\tau$ has, for every fixed τ the following asymptotic representation for large s, provided it does not vanish identically,

$$w_\tau(s,t) = e^{\int_{s_0}^s \mu_\tau(\rho)d\rho}[e_\tau(t) + \hat{w}_\tau(s,t)],$$

where $D^j[\mu_\tau(s) - \mu_\tau(\infty)] \to 0$ as $s \to \infty$ for every derivative j with some $\mu_\tau(\infty) < 0$, and where $D^\alpha \hat{w}_\tau(s,t) \to 0$ as $s \to \infty$ uniformly in t, for every derivative α. The number $\mu_\tau(\infty)$ is a negative eigenvalue of the operator $A_\infty = -i\frac{d}{dt} - \tilde{C}(t)$, and $e_\tau(t)$ an associated eigenvector. Since the spectrum consists of isolated points independent of τ we conclude that for all multi-indices α

$$D^\alpha w_\tau(s,t) = O(e^{-\lambda s}) \quad \text{as } s \to \infty$$

for some $\lambda > 0$ independent of τ. Consequently, there exists a $\delta > 0$ such that for τ small $v_\tau \in C_0^{k,\alpha,\delta}(\mathbb{C},\mathbb{R}^2)$ for every k. In order to prove that $\tau \mapsto v(\tau) = v_\tau$ is smooth as a map from B into $C_0^{k+1,\alpha,\delta}$ we consider the operator H from Section 6 but now in different spaces. We consider H as a section of the bundle $\mathcal{E} \to \mathcal{B}$, where $\mathcal{B} = C_0^{k+1,\alpha,\delta}(\mathbb{C},\mathbb{R}^2)$ and $\mathcal{E} = \bigcup_{v\in\mathcal{B}}\{v\} \times C_0^{k,\alpha,\delta}(\mathrm{gr}(v)^*\Sigma)$. By definition, $v(\tau)$ solves $H(v(\tau)) = 0$ and $v(0) = 0$. For $\delta > 0$ small the linearization of H is a surjective Fredholm operator whose index is equal to 2 in view of Theorem 2.11 and Theorem 6.1. Applying Theorem 6.3 there exists a unique smooth $\tau \mapsto w(\tau) \in C_0^{k+1,\alpha,\delta}$ satisfying $H(w(\tau)) = 0$ and $w(0) = 0$. By uniqueness, $v(\tau) = w(\tau)$ and the proof of the lemma complete. $\square$

We define the map

$$F : B \times \mathbb{C} \times \mathbb{C} \times C_0^{k+1,\alpha,\delta}(\mathbb{C},\mathbb{C}) \to C_0^{k,\alpha,\delta}(\mathbb{C},\mathbb{C}) \times \mathbb{C} \times \mathbb{C}$$

as follows

$$F(\tau, a, b, \phi) = \left(\frac{\partial}{\partial s} f + \bar{j}_\tau(f) \frac{\partial}{\partial t} f, b + \phi(0), a + b + \phi(1) - 1 \right),$$

where

$$f(z) = az + b + \phi(z).$$

Here, as usual, $z = s + it \in \mathbb{C}$. We shall use the short notation $F(\tau, a, b, \phi) = F(\tau, f)$. The map F is smooth on account of Lemma 12.5. If $\tau = 0$, then $\bar{j}_\tau(z) = i$ and hence $F(0, \mathrm{id}) = 0$. By means of the implicit function theorem we shall construct a unique smooth map $\tau \mapsto f_\tau$ for small τ satisfying $F(\tau, f_\tau) = 0$ and $f_0 = \mathrm{id}$ at $\tau = 0$. Applying the implicit function theorem to every k we can conclude that the map $(\tau, z) \mapsto f_\tau(z) := f(\tau, z)$ is smooth and satisfies

$$\frac{\partial}{\partial s} f_\tau + \bar{j}_\tau(f_\tau) \frac{\partial}{\partial t} f_\tau = 0, \quad f_\tau(0) = 0,\ f_\tau(1) = 1.$$

Since for small τ the map f_τ is close to the identity map, we have $a_\tau \neq 0$ so that $f_\tau(\infty) = \infty$. By Riemann's removable singularity theorem, the map $f_\tau : \mathbb{C} \subset (S^2, i) \to \mathbb{C} \subset (S^2, \bar{j}_\tau)$ has a removable singularity, so that f_τ can be extended to a holomorphic map satisfying $f_\tau(0) = 0$, $f_\tau(1) = 1$ and $f_\tau(\infty) = \infty$. In view of the uniqueness in the uniformisation theorem we conclude that $\psi_\tau = f_\tau$ and so, the map ψ has all the desired properties.

In order to apply the implicit function theorem we have to verify that the partial derivative D_* of F in the variable (a, b, ϕ) at the point $(0, 1, 0, 0)$ is an isomorphism. Explicitely, $D_*F(0, \mathrm{id}) : \mathbb{C} \times \mathbb{C} \times C_0^{k+1,\alpha,\delta} \to C_0^{k,\alpha,\delta} \times \mathbb{C} \times \mathbb{C}$ is given by

$$D_*F(0, \mathrm{id})(\alpha, \beta, \psi) = (\psi_s + i\psi_t, \alpha + \psi(0), \alpha + \beta + \psi(1)),$$

where we have used $\bar{j}_\tau = i$ if $\tau = 0$. Clearly $D_*F(0, \mathrm{id})$ is an isomorphism, if the map $L : C_0^{k+1,\alpha,\delta} \to C_0^{k,\alpha,\delta}$ defined by $L(\psi) = \psi_s + i\phi_t$ is an isomorphism. The verification of this fact will be based on our studies in Section 6. The asymptotic operator in our case is

$$A_\infty = -\frac{d}{dt} + \delta\,\mathrm{Id}\,.$$

If $\delta > 0$ is small, 0 is not in the spectrum so that L is a Fredholm operator. In view of Theorem 6.1 its index is computed using the formula

$$\mathrm{Ind}(L) = \mu_N + \chi_{S^2} - \sharp\Gamma_{S^2} = \mu(A_\infty) + 1,$$

where $\mu(A_\infty)$ is the Conley-Zehnder index of the constant loop $\delta\,\mathrm{Id}$. The eigenvalues of A_∞ are $\lambda = 2\pi k + \delta$, have multiplicity 2 and eigenvectors $x(t) = e^{2\pi kit}x(0)$. The largest negative eigenvalue is $-2\pi + \delta$ and the winding number α of its eigenvector is equal -1. Therefore, by Proposition 2.10, $\mu(A_\infty) = 2\alpha + p = -2 + 1 = -1$.

Hence $\mathrm{Ind}(L) = 0$. Since the kernel consists of bounded holomorphic functions in $C_0^{k,\alpha,\delta}$ we conclude by Liouville's theorem that $\dim(\ker L) = 0$. Consequently, L is an isomorphism and hence also $D_*F(0, \mathrm{id})$. This completes the construction of the desired map ψ.

Acknowledgements. We would like to thank Katrin Wehrheim for suggestions and improvements and Matthias Schwarz for helpful discussions.

References

[1] C. Abbas and H. Hofer. *Holomorphic curves and global questions in contact geometry.* To appear in Birkhäuser.

[2] L. Ahlfors and L. Bers. "Riemann's mapping theorem for variable metrics." Ann. of Math. **72** (1960), 385–404.

[3] D. Bennequin. "Entrelacements et équations de Pfaff." Astérisque, 107–108:83–161, 1983.

[4] S.S. Chern. "An elementary proof of the existence of isothermal parameters on a surface." Proc. Amer. Math. Soc. **6** (1955), 771–782.

[5] Y. Eliashberg. "Classification of overtwisted contact structures on three manifolds." Inv. Math., pages 623–637, 1989.

[6] Y. Eliashberg. "Filling by holomorphic discs and its applications." London Math. Society Lecture Notes, pages 45–67, 1991. Series 151.

[7] Y. Eliashberg. "Contact 3–manifolds, twenty year since J. Martinet's work." Ann. Inst. Fourier **42** (1992), 165–192.

[8] Y. Eliashberg. "Legendrian and transversal knots in tight contact manifolds". In: Topological methods in modern mathematics. Publish or Perish, 1993.

[9] A. Floer. "The unregularised gradient flow of the symplectic action." Comm. Pure Appl. Math. **41** (1988), 775–813.

[10] A. Floer. "Symplectic fixed points and holomorphic spheres." Comm. Math. Phys. **120** (1989), 575–611.

[11] A. Floer, H. Hofer, and D. Salamon. "Transversality results in the elliptic Morse theory for the action functional." Duke Math. Journal **30** (1995), No. 1, 251–292.

[12] E. Giroux. "Convexité en topologie de contact." Comm. Math. Helvetici **66** (1991), 637–677.

[13] J.W. Gray. "Some global properties of contact structures." Ann. of Math. **2(69)** (1959), 421–450.

[14] M. Gromov. "Pseudoholomorphic curves in symplectic manifolds." Invent. Math. **82** (1985), 307–347.

[15] H. Hofer. "Pseudoholomorphic curves in symplectizations with application to the Weinstein conjecture in dimension three." Invent. Math. **114** (1993), 515–563.

[16] H. Hofer, V. Lizan, and J.-C. Sikorav. *On generecity for holomorphic curves in 4-dimensional almost complex manifolds.* Preprint.

[17] H. Hofer, K. Wysocki, and E. Zehnder. "A characterisation of the tight three-sphere." Duke Math. Journal **81** (1995), 159–226.

[18] H. Hofer, K. Wysocki, and E. Zehnder. "Properties of pseudoholomorphic curves in symplectizations I: Asymptotics." Ann. I. H. P. Analyse Non Linéaire **13** (1996), 337–379.

[19] H. Hofer, K. Wysocki, and E. Zehnder. "Properties of pseudoholomorphic curves in symplectizations II: Embedding controls and algebraic invariants." Geometrical and Functional Analysis **5**, No. 2 (1995), 270–328.

[20] H. Hofer, K. Wysocki, and E. Zehnder. "Unknotted periodic orbits for Reeb flows on the 3-sphere." Topological Methods in Nonlinear Analysis, Journal of the Juliusz Schauder Center **7** (1996), 219–244.

[21] H. Hofer, K. Wysocki, and E. Zehnder. "The dynamics on a strictly convex energy surface in $\mathbf{R}^4$". To appear in Ann. of Math.

[22] H. Hofer, K. Wysocki, and E. Zehnder. "A characterization of the tight three-sphere II". Preprint.

[23] H. Hofer, K. Wysocki, and E. Zehnder. "Properties of pseudoholomorphic curves in symplectizations V: Fredholm theory with singularities". In preparation.

[24] H. Hofer and E. Zehnder. *Hamiltonian Dynamics and Symplectic Invariants.* Birkhäuser, 1994.

[25] J. Martinet. "Formes de contact sur les variétés de dimension 3." Springer LNM **209** (1971), 142–163.

[26] D. McDuff. "The local behavior of J-holomorphic curves in almost complex 4-manifolds." J. Diff. Geom. **34** (1991), 143–164.

[27] M. Micallef and B. White. "The structure of branch points in minimal surfaces and in pseudoholomorphic curves." Ann. of Math. **141** (1994), 35–85.

[28] A. Nijenhuis and W. Woolf. "Some integration problems in almost-complex and complex manifolds." Annals of Mathematics **77**, No. 3 (1963), 424-489.

[29] M. Schwarz. *Cohomology Operations from S^1-Cobordisms in Floer Homology.* Ph.D thesis. ETH-Zürich, 1996.

[30] J.-C. Sikorav. "Singularities of J-holomorphic curves." Math. Zeit. **226** (1997), 359–373.

[31] R. Ye. "Filling by holomorphic disks in symplectic 4-manifolds." Trans. Amer. Math. Soc. **350**, 1 (1998), 213–250.

H. Hofer
Courant Institute
251 Mercer Street
New York
NY 10012, USA

K. Wysocki
Department of Mathematics
University of Melbourne
Parkville
Victoria 3052, Australia

E. Zehnder
Department of Mathematics
ETH Zürich
CH-8092 Zürich
Switzerland

Progress in Nonlinear Differential Equations
and Their Applications, Vol. 35
© 1999 Birkhäuser Verlag Basel/Switzerland

Variational Principles for Minimal Surfaces

Cordula Imbusch and Michael Struwe

Abstract. Extending the variational approach to Plateau's problem developed in [6], we show that similar results can be obtained for surfaces normalized by a 3-point condition. Moreover, we supply further details for the regularity results in [6].

1. Background

In [6] a general theory of critical points and a Morse theory was developed for the classical Plateau problem for parametric minimal surfaces X of the type of the 2-disc

$$B = \{w = (u, v) \in \mathbb{R}^2; u^2 + v^2 < 1\},$$

spanning a curve Γ in $\mathbb{R}^n$.

By definition, a surface $X\colon B \to \mathbb{R}^n$ is a minimal surface spanning Γ if $X \in C^2(B; \mathbb{R}^n) \cap C^0(\overline{B}; \mathbb{R}^n)$ is harmonic, $X|_{\partial B}\colon \partial B \to \Gamma$ is a homeomorphism, and if X is conformal; that is, if

$$|X_u|^2 = |X_v|^2, \langle X_u, X_v\rangle = 0 \text{ in } B, \tag{1}$$

where $X_u = \frac{\partial X}{\partial u}$, etc., and where $\langle \cdot, \cdot \rangle$ denotes the scalar product in $\mathbb{R}^n$.

If Γ is smooth ($\Gamma \in C^r$ for some $r \geq 3$ is enough), as shown in [6], minimal surfaces X spanning Γ equivalently can be characterized as follows. Let $\gamma\colon S^1 \cong \mathbb{R}/2\pi \to \Gamma$ be a smooth (C^r-) parametrization of Γ with $\gamma' \neq 0$. Identifying $S^1 = \mathbb{R}/2\pi \cong \partial B$ and letting h denote harmonic extension from ∂B to B, admissible comparison surfaces can be represented via γ as

$$X = h(\gamma \circ x) = \Xi(x), \tag{2}$$

where $x\colon \mathbb{R} \to \mathbb{R}$ is continuous, non-decreasing, and satisfies $x(\phi+2\pi) = x(\phi)+2\pi$. That is, in polar coordinates $w = re^{i\phi}$ on B, we have

$$X(e^{i\phi}) = \gamma(x(\phi)) \text{ on } \partial B,$$

and X is harmonic. Minimal surfaces of finite area spanning Γ then are precisely those maps $X = \Xi(x)$ with finite Dirichlet integral

$$D(X) = \frac{1}{2}\int_B |\nabla X|^2 \, dw$$

which are conformal. By Hildebrandt's regularity result [3], any such minimal surface X is smooth on $\overline{B}$; therefore, finally, (1) is equivalent to orthogonality

$$\langle X_r, X_\phi \rangle = 0 \text{ on } \partial B. \tag{3}$$

In [6] now, minimal surfaces are shown to uniquely correspond to critical points (in the sense defined by (4) below) of the functional

$$E(x) = D(\Xi(x))$$

on the closed convex set

$$M = \{x = id + \xi \colon \mathbb{R} \to \mathbb{R}\ ;\ x \text{ monotone}, \xi \in H^{1/2,2} \cap C^0(\mathbb{R}/2\pi)\},$$

endowed with the metric inherited from $H^{1/2,2} \cap C^0(\mathbb{R}/2\pi)$, where

$$|\xi|^2_{H^{1/2}} = \int_0^{4\pi} \int_0^{4\pi} \frac{|\xi(\phi) - \xi(\phi')|^2}{|\phi - \phi'|^2}\, d\phi d\phi'.$$

However, since $D(X \circ g) = D(X)$ for any X and any conformal diffeomorphism $g \colon B \to B$, in order to ensure the compactness necessary for a satisfactory theory of critical points, admissible comparison maps need to be normalized with respect to the action of the Möbius group

$$\mathcal{G} = \{g \colon w \mapsto e^{i\phi_0} \frac{a + w}{1 + \overline{a}w}; a \in \mathbb{C}, |a| < 1, \phi_\circ \in \mathbb{R}\}$$

of the disc.

In contrast to the classical 3-point normalization [1], leading to the class

$$M^* = \{x \in M;\ x(\frac{2\pi k}{3}) = \frac{2\pi k}{3}, \forall k \in \mathbb{Z}\},$$

in [6] the normalized class

$$M^\dagger = \{x \in M; \int_0^{2\pi} \xi(\phi) d\phi = \int_0^{2\pi} \xi(\phi) \sin\phi\, d\phi = \int_0^{2\pi} \xi(\phi) \cos\phi\, d\phi = 0\}$$

was proposed and it was shown that generalized Lusternik-Schnirelman and Morse theories hold for E in the class $M^\dagger$, and that the critical points of E on $M^\dagger$ uniquely correspond to the minimal surfaces spanning Γ, normalized with respect to $\mathcal{G}$.

In this note we show that the latter bi-unique correspondence with minimal surfaces spanning Γ also holds for the critical points of E in M^*; see Theorem 2.1 below. As is clear from [6], Proposition 2.9, this assertion is equivalent to a regularity result for critical points of E in M^*; see Proposition 2.2. However, although this result removes one obstruction for developing a critical point theory for E in M^*, it cannot resolve the fundamental difficulties for setting up a Morse theory in M^*, pointed out in [6], p. 62.

Our second aim here is to provide further details for the proof of Proposition 2.2 in the case $M^\dagger$, as in [4], in particular, concerning the normalization of testing functions with respect to $\mathcal{G}$, which was only sketched in [6].

2. Results

As shown in [6], for any $x \in M$ the functional E extends to a C^1-functional on the space $x + H^{1/2,2} \cap L^\infty(\mathbb{R}/2\pi)$ and

$$\langle dE(x), \xi \rangle = \int_B \langle \nabla X, \nabla (h(\frac{d}{d\phi}\gamma(x) \cdot \xi)) \rangle \, dw = \int_{\partial B} \langle \partial_n X, \frac{d}{d\phi}\gamma(x) \cdot \xi \rangle do$$

for all $\xi \in H^{1/2,2} \cap L^\infty(\mathbb{R}/2\pi)$, where $X = \Xi(x)$ and where n is the unit outward normal on ∂B. Note that for x and ξ as above there holds $\frac{d}{d\phi}\gamma(x) \cdot \xi \in H^{1/2,2}(\partial B)$. Since $X \in H^{1,2}(B)$ is harmonic, on the other hand, by results of Lions-Magenes [5], Theorem II. 7.4, we have $\partial_n X \in H^{-1/2,2}(\partial B)$, the dual space of $H^{1/2,2}(\partial B)$, and the representation of $\langle dE(x), \xi \rangle$ as a boundary integral is well defined in the sense of distributions.

By definition, moreover, $x \in M$ is critical for E if

$$\langle dE(x), x - y \rangle \leq 0 \tag{4}$$

for all $y \in M$ with $||x - y||_T < 1$. Observe that the latter restriction may be dropped. Indeed, for any $y \in M$ by convexity of M also $y_t = (1-t)x + ty \in M$ for any $t \in [0,1]$, with $x - y_t = t(x-y)$ and $\langle dE(x), x - y \rangle = t^{-1} \langle dE(x), x - y_t \rangle$. Choosing $t = (1 + ||x-y||_T)^{-1}$, by (4) the latter is non-positive. Hence, x is critical for E in M if and only if (4) holds for all $y \in M$.

Similarly, $x \in M^*$ (respectively $M^\dagger$) is critical for E in M^* (respectively $M^\dagger$) if and only if (4) holds for all $y \in M^*$ (respectively $M^\dagger$). Observe that critical points of E in M that belong to M^* (respectively $M^\dagger$) are also critical for E in M^* (respectively $M^\dagger$).

Theorem 2.1. *Suppose x is critical for E in either M, M^*, or $M^\dagger$. Then $X = \Xi(x)$ is a minimal surface spanning Γ.*

It is clear that, conversely, any (suitably normalized) minimal surface X corresponds to a critical point x of E in M (or M^*, or $M^\dagger$) such that $\Xi(x) = X$.

As explained in [6], Proposition 2.9, Theorem 2.1 is a consequence of the following regularity result.

Proposition 2.2. *Suppose x is critical for E in either M, M^*, or $M^\dagger$. Then $X = \Xi(x) \in H^{2,2}(B, \mathbb{R}^n)$.*

For the cases M and $M^\dagger$ the above assertions were established in [6], Propositions 2.9 and 2.10. Their proof is achieved by inserting

$$y = x - \tau^2(x - x(\phi_0)), \tag{5}$$

respectively

$$y = x + \varepsilon \partial_{-h} \partial_h x \tag{6}$$

as testing functions in (4), where $\tau \in C_0^\infty(B_{2r}(e^{i\phi_0}))$ is a cut-off function such that $0 \leq \tau \leq 1, \tau = 1$ on $B_r(e^{i\phi_0}), \phi_0 \in \mathbb{R}$, and where $\partial_h x(\phi) = h^{-1}(x(\phi + h) - x(\phi))$. In the case of M^* and $M^\dagger$ these comparison functions have to be normalized by composition with some $g \in \mathcal{G}$ to achieve $y \circ g \in M^*, M^\dagger$, respectively.

3. Normalization

The Möbius group $\mathcal{G}$ acts on M, as follows. For $g \in \mathcal{G}$ there exists a smooth map $\hat{g}\colon \mathbb{R} \to \mathbb{R}$ such that

$$g(e^{i\phi}) = e^{i\hat{g}(\phi)}, \phi \in \mathbb{R}. \tag{7}$$

In fact, for $g_{\phi_0,a} \in \mathcal{G}$, given by

$$g_{\phi_0,a}(w) = e^{i\phi_0}\frac{a+w}{1+\bar{a}w}, w \in B,$$

where $\phi_0 \in \mathbb{R}$ and $a = \alpha + i\beta \in B \subset \mathbb{R}^2 \cong \mathbb{C}, |a| < 1$, a corresponding map $\hat{g}_{\phi_0,a}$ is given by

$$\hat{g}_{\phi_0,a}(\phi) = \phi + \phi_0 + 2\arctan\left(\frac{\beta\cos\phi - \alpha\sin\phi}{1+\alpha\cos\phi+\beta\sin\phi}\right), \tag{8}$$

as follows immediately from the identity

$$g_{\phi_0,a}(e^{i\phi}) = e^{i(\phi+\phi_0)}\frac{1+ae^{-i\phi}}{1+\bar{a}e^{i\phi}} = e^{i(\phi+\phi_0)}\frac{\zeta}{\bar{\zeta}},$$

where $\zeta = 1 + ae^{-i\phi} = 1 + \alpha\cos\phi + \beta\sin\phi + i(\beta\cos\phi - \alpha\sin\phi)$.

Observe that $\hat{g}$ is uniquely determined by g up to an integer multiple of 2π; moreover, $\hat{g}$ is non-decreasing and satisfies $\hat{g}(\phi + 2\pi) = \hat{g}(\phi) + 2\pi$ for all $\phi \in \mathbb{R}$.

Finally, let $x \in M, g \in \mathcal{G}$ with corresponding maps $X = \Xi(x), \hat{g}\colon \mathbb{R} \to \mathbb{R}$. Also let $\gamma\colon \partial B = S^1 \to \Gamma$ be represented by $\hat{\gamma}\colon \mathbb{R} \to \mathbb{R}^n$, satisfying $\gamma(e^{i\phi}) = \hat{\gamma}(\phi)$; that is, we briefly distinguish the representations of γ in standard Euclidean and polar coordinates. Similarly, let $x\colon \partial B \to \partial B$ be given by $x(e^{i\phi}) = e^{i\hat{x}(\phi)}$. Then we have $X(e^{i\phi}) = \gamma(x(e^{i\phi})) = \hat{\gamma}(\hat{x}(\phi))$ and $(X \circ g)(e^{i\phi}) = X(e^{i\hat{g}(\phi)}) = \hat{\gamma}(\hat{x}(\hat{g}(\phi)))$ for all ϕ. This being understood, we usually indiscriminately write γ, x, g for $\hat{\gamma}, \hat{x}, \hat{g}$ and we summarize the preceding in the simple indentity $\Xi(x) \circ g = \Xi(x \circ g)$ for all $x \in M, g \in \mathcal{G}$.

In this section we establish the following two results.

Proposition 3.1. *For any $x \in M$ there exist conformal transformations $g^*, g^\dagger \in M$ such that $x \circ g^* \in M^*, x \circ g^\dagger \in M^\dagger$, respectively.*

Moreover, let $x \in M^*$ or $M^\dagger$, respectively, and let y be either of the two comparison functions defined in (5) or (6) above.

Proposition 3.2. *i) There is $r_0 = r_0(x) > 0$ such that for $0 < r < r_0$ and any $w_0 = e^{i\phi_0} \in \partial B$ the function y defined by (5) may be normalized by composition with a suitable element $g \in \mathcal{G}$ so that $y \circ g \in M^*$ or $M^\dagger$, respectively, and there exists C such that $|g - id| \le Cr$ as $r \to 0$.*
ii) Similarly, if $x \in M^*$ there is $h_0 = h_0(x) > 0$ such that for $0 < h < h_0, 0 < 2\varepsilon < h^2$ and any $g_h \in \mathcal{G}$ such that $|g_h - id| \le h$, the function

$$y = (1 - 2\varepsilon h^{-2})x + \varepsilon h^{-2}(x \circ g_h + x \circ g_h^{-1}) \tag{9}$$

may be normalized by composition with some $g \in \mathcal{G}$ so that $y \circ g \in M^*$ and $|g - id| \le Ch$ as $h \to 0$.

Observe that for g_h given by $\hat{g}_h(\phi) = \phi + h$ the function y defined by (9) agrees with the function defined by (6). For $x \in M^\dagger$, in fact, it suffices to consider the latter as comparison functions. Although we conjecture that this is also true for $x \in M^*$, our final argument in this case so far requires being able to choose our testing function in the larger class (9); see the discussion following our estimate (23). Note that for $x \in M^\dagger$ and $0 < 2\varepsilon \le h^2$ the function y defined by (6) automatically belongs to $M^\dagger$; see [6], p. 67.

Proof. [Proof of Proposition 3.1] i) The case M^* is classical. Indeed, let $\phi_1 < \phi_2 < \phi_3 < \phi_1 + 2\pi$ be points such that $x(\phi_k) = \frac{2\pi k}{3}, k = 1, 2, 3$. The unique element $g^* \in \mathcal{G}$ with

$$g^*(e^{i\frac{2\pi k}{3}}) = e^{i\phi_k}, k = 1, 2, 3$$

then achieves the desired normalization.

ii) To handle the case $M^\dagger$ denote as $\overline{F} = (F^1, F^2, F^3)^t \colon M \times \mathcal{G} \to \mathbb{R}^3$ the map

$$\overline{F}(x, g) = \int_0^{2\pi} (x \circ g - id) \begin{pmatrix} 1 \\ \sin\phi \\ \cos\phi \end{pmatrix} d\phi,$$

and for $\phi_0 \in \mathbb{R}, a = \alpha + i\beta \in B \subset \mathbb{C}$ denote as $g_{\phi_0,a} \in \mathcal{G}$ the map given by

$$g_{\phi_0,a}(w) = e^{i\phi_0}\frac{a + w}{1 + \bar{a}w}, w \in B.$$

Observe that, given $y \in M$ we may simply choose

$$\phi_y = -\frac{1}{2\pi}\int_0^{2\pi} (y - id)\, d\phi$$

to achieve the first normalization condition

$$F^1(y, g_{\phi_y,0}) = \int_0^{2\pi} (y(\phi + \phi_y) - \phi)\, d\phi = \int_0^{2\pi} (y - id)\, d\phi + 2\pi\phi_y = 0.$$

Moreover, for any $\phi_0 \in \mathbb{R}$ there holds

$$\begin{pmatrix} F^2(y, g_{\phi_0,0}) \\ F^3(y, g_{\phi_0,0}) \end{pmatrix} = \int_0^{2\pi} (y(\phi + \phi_0) - \phi) \begin{pmatrix} \sin\phi \\ \cos\phi \end{pmatrix} d\phi$$
$$= \int_0^{2\phi} (y - id) \begin{pmatrix} \sin(\phi - \phi_0) \\ \cos(\phi - \phi_0) \end{pmatrix} d\phi = \begin{pmatrix} \cos\phi_0 & -\sin\phi_0 \\ \sin\phi_0 & \cos\phi_0 \end{pmatrix} \begin{pmatrix} F^2(y, id) \\ F^3(y, id) \end{pmatrix}.$$

Hence it suffices to find $a \in B$ solving the equations

$$F^2(x, g_{0,a}) = F^3(x, g_{0,a}) = 0;$$

the desired normalization will then be achieved by $g^\dagger = g_{0,a} \circ g_{\phi_y,0}$, where $y = x \circ g_{0,a} \in M$. To simplify the notation, let $F \colon B \to \mathbb{R}^2 \cong \mathbb{C}$ be the map given by

$$F(a) = \begin{pmatrix} F^2(x, g_{0,a}) \\ F^3(x, g_{0,a}) \end{pmatrix} = \int_0^{2\pi} (x \circ g_{0,a} - id) \begin{pmatrix} \sin\phi \\ \cos\phi \end{pmatrix} d\phi.$$

In Lemma 3.3 below we establish that F continuously extends to $\overline{B}$ with $F(a) = 2\pi a$ for $a \in \partial B$. Thus, the rescaled map $\frac{1}{2\pi} F_{|\partial B}$ induces a homeomorphism of ∂B of topological degree 1. The existence of an element $a \in B$ solving the equation $F(a) = 0$ then follows from degree theory; see for instance [2], Theorem 3.1 (d1) and (d6). □

Lemma 3.3. *The above map F continuously extends to $\overline{B}$ with $F(a)=2\pi a, a\in\partial B$.*

Proof. By conformality, for any $g \in \mathcal{G}$ the Dirichlet integral $D(g)$ equals the area $A(g)$ of the surface covered by g; that is,

$$D(g) = \frac{1}{2} \int_B |\nabla g|^2 \, dw = A(g) = \pi.$$

Hence $\mathcal{G}$ is bounded in $H^{1,2}(B;\mathbb{R}^2)$. By compactness of the trace operator $H^{1,2}(B;\mathbb{R}^2) \ni g \mapsto g_{|\partial B} \in L^2(\partial B;\mathbb{R}^2)$ and continuity of x, therefore the set of maps $(x \circ g_{0,a})_{a\in B}$ is relatively compact in $L^2(\partial B;\mathbb{R}^2)$. Moreover, as $a \to a_0 = e^{i\phi_0} \in \partial B$ we have

$$g_{0,a}(w) = \frac{a+w}{1+\overline{a}w} \to a_0$$

locally uniformly on $\overline{B} \setminus \{-a_0\}$, and therefore a representant $\hat{g}_{0,a}(\phi) \to \phi_0$, locally uniformly in $\phi \in]-\phi_0, 2\pi - \phi_0[$.

We can thus continuously extend F to $\overline{B}$ and

$$\begin{aligned} F(e^{i\phi_0}) &= \int_0^{2\pi-\phi_0} (x(\phi_0) - \phi) \begin{pmatrix} \sin\phi \\ \cos\phi \end{pmatrix} d\phi + \int_{2\pi-\phi_0}^{2\pi} (x(2\pi+\phi_0) - \phi) \begin{pmatrix} \sin\phi \\ \cos\phi \end{pmatrix} d\phi \\ &= \int_0^{2\pi} (x(\phi_0) - \phi) \begin{pmatrix} \sin\phi \\ \cos\phi \end{pmatrix} d\phi + 2\pi \int_{2\pi-\phi_0}^{2\pi} \begin{pmatrix} \sin\phi \\ \cos\phi \end{pmatrix} d\phi \\ &= 2\pi \begin{pmatrix} \cos\phi_0 \\ \sin\phi_0 \end{pmatrix} = 2\pi e^{i\phi_0}, \end{aligned}$$

as claimed. □

For the proof of Proposition 3.2 we use the representation (8) of an element $g_{\phi_0,a} \in \mathcal{G}$, where $\phi_0 \in \mathbb{R}, a = \alpha + i\beta \in B \subset \mathbb{C}$.

Introducing ϕ_0, α, and β as coordinates for $\mathcal{G}$, and identifying $\overline{F}$ with the map $\tilde{F}\colon \{M - id\} \times \mathbb{R} \times B \to \mathbb{R}^3$, given by

$$\tilde{F}(\xi, \phi_0, a) = \overline{F}(x, g) = \int_0^{2\pi} \xi \circ \hat{g} \begin{pmatrix} 1 \\ \sin\phi \\ \cos\phi \end{pmatrix} d\phi + \int_0^{2\pi} (\hat{g} - id) \begin{pmatrix} 1 \\ \sin\phi \\ \cos\phi \end{pmatrix} d\phi$$

for $\xi = x - id, x \in M, \hat{g} = \hat{g}_{\phi_0,a}, \phi_0 \in \mathbb{R}, a \in B$, we now assert

Lemma 3.4. *$\tilde{F}$ extends to a map $\tilde{F} \in C^1(H^{1/2,2}(\mathbb{R}/2\pi) \times \mathbb{R} \times B;\mathbb{R}^3)$ and at any $\xi = x - id$, where $x \in M$, any $\phi_0 \in \mathbb{R}$, any $a = \alpha + i\beta \in B$, the differential $L = (\partial_{\phi_0}\tilde{F}, \partial_\alpha\tilde{F}, \partial_\beta\tilde{F})(\xi, \phi_0, a)\colon \mathbb{R}^3 \to \mathbb{R}^3$ is regular.*

Proof. Replacing x by $\tilde{x} = x \circ g_{\phi_0,a}$, if necessary, it suffices to verify the assertion for $\phi_0 = \alpha = \beta = 0$. After a change of coordinates, for any $\phi_0 \in \mathbb{R}, a \in B$ we have

$$\tilde{F}(\xi, \phi_0, a) = \int_0^{2\pi} \xi \begin{pmatrix} 1 \\ \sin(\hat{g}^{-1}(\phi)) \\ \cos(\hat{g}^{-1}(\phi)) \end{pmatrix} \frac{d\phi}{\hat{g}'(\phi)} + \int_0^{2\pi} (\hat{g} - id) \begin{pmatrix} 1 \\ \sin\phi \\ \cos\phi \end{pmatrix} d\phi,$$

where $\hat{g} = \hat{g}_{\phi_0,a}$.

Thus, $\tilde{F}$ is an affine function of ξ and therefore extends smoothly to $H^{1/2,2}(\mathbb{R}/2\pi) \times \mathbb{R} \times B$, as claimed. Moreover, denoting $d = \left(\frac{\partial}{\partial\phi_0}, \frac{\partial}{\partial\alpha}, \frac{\partial}{\partial\beta}\right)$ we have

$$L = d\tilde{F}(\xi, 0) = \int_0^{2\pi} \left\{ \left[\begin{pmatrix} 1 \\ \sin\phi \\ \cos\phi \end{pmatrix} - \xi \begin{pmatrix} 0 \\ \cos\phi \\ -\sin\phi \end{pmatrix} \right] d\hat{g} - \begin{pmatrix} 1 \\ \sin\phi \\ \cos\phi \end{pmatrix} d\hat{g}'\xi \right\} d\phi.$$

Using that

$$d\hat{g} = (1, -2\sin\phi, 2\cos\phi), d\hat{g}' = (0, -2\cos\phi, -2\sin\phi)$$

on account of (8), we thus obtain

$$L = 2 \begin{pmatrix} \pi & 0 & 0 \\ 0 & a_1 - \pi & -a_2 \\ 0 & a_2 & a_1 + \pi \end{pmatrix},$$

where

$$a_1 = \int_0^{2\pi} \xi \sin 2\phi \, d\phi, \quad a_2 = \int_0^{2\phi} \xi \cos 2\phi \, d\phi. \tag{10}$$

Lemma 3.4 thus is an immediate consequence of the following result. □

Lemma 3.5. *For any $x = \xi + id \in M$ and with a_1, a_2 given by* (10), *there holds $a_1^2 + a_2^2 < \pi^2$.*

Proof. Observe that for any $\phi_0 \in \mathbb{R}$ there holds

$$\begin{pmatrix} a_1' \\ a_2' \end{pmatrix} = \int_0^{2\pi} \xi(\phi + \phi_0) \begin{pmatrix} \sin 2\phi \\ \cos 2\phi \end{pmatrix} d\phi = \int_0^{2\pi} \xi(\phi) \begin{pmatrix} \sin 2(\phi - \phi_0) \\ \cos 2(\phi - \phi_0) \end{pmatrix} d\phi$$
$$= \begin{pmatrix} \cos 2\phi_0 & -\sin 2\phi_0 \\ \sin 2\phi_0 & \cos 2\phi_0 \end{pmatrix} \begin{pmatrix} a_1 \\ a_2 \end{pmatrix}.$$

By suitable choice of ϕ_0 and replacing x by $x(\cdot + \phi_0)$, if necessary, we may therefore always assume that $a_2 = 0$ and $a_1 \geq 0$. Also noting that

$$a_1 = \int_0^{2\pi} (x(\phi) - \phi) \sin 2\phi \, d\phi = \int_0^{2\pi} x(\phi) \sin 2\phi \, d\phi + \pi,$$

it therefore suffices to show that

$$\int_0^{2\pi} x(\phi) \sin 2\phi \, d\phi < 0$$

for any $x \in M$.

Using monotonicity of x and the symmetries of the function $\phi \mapsto \sin 2\phi$, the latter integral can be estimated

$$\int_0^{2\pi} x(\phi)\sin 2\phi\, d\phi = \int_0^{\pi/2} (x(\phi) - x(\phi + \frac{\pi}{2}))\sin 2\phi\, d\phi$$
$$+ \int_\pi^{3\pi/2} (x(\phi) - x(\phi + \frac{\pi}{2}))\sin 2\phi\, d\phi \le 0,$$

and equality holds if and only if

$$x(\phi) = x(\phi + \frac{\pi}{2}) \text{ for } \phi \in]0, \frac{\pi}{2}[\cup]\pi, \frac{3\pi}{2}[. \tag{11}$$

By continuity of x in the latter case we also have (11) for $\phi \in [0, \frac{\pi}{2}] \cup [\pi, \frac{3\pi}{2}]$, thus arriving at the contradiction

$$x(0) = x(\frac{\pi}{2}) = x(\pi) = x(\frac{3\pi}{2}) = x(2\pi) = x(0) + 2\pi.$$

The proof is complete. □

Proof. [Proof of Proposition 3.2] i) In the case $M^\dagger$ the asserted result is an immediate consequence of Lemma 3.4 and the implicit function theorem. In fact, it is not hard to see that there even holds $|g - id| = O(r/\sqrt{|ln\, r|})$ as $r \to 0$; see [4], Proposition 6.6. However, for our purposes the estimate $|g - id| \le Cr$ for small $r > 0$ suffices.

In the case $x \in M^*$ we argue as follows. Let y be given by (5), where $w_0 = e^{i\phi_0} \in \partial B$. If $B_{2r}(w_0)$ does not contain any of the points $e^{i\frac{2\pi k}{3}}$, $k = 1, 2, 3$, then $y \in M^*$ and we are done. Otherwise, suppose $1 \in B_{2r}(w_0)$ and let $w_{1,2} = e^{i\phi_{1,2}}$, $\phi_1 < 0 < \phi_2$, be the endpoints of the arc $\partial B \cap B_{2r}(w_0)$. Note that the assumption $1 \in B_{2r}(w_0)$ implies $|\phi_0| < 3r$ and hence $|\phi_{1,2}| < |\phi_0| + 3r < 6r$. Moreover, from the relation

$$y(\phi_1) = x(\phi_1) < 0 < x(\phi_2) = y(\phi_2) \tag{12}$$

and continuity of y we conclude that there exists $\overline{\phi} \in]\phi_1, \phi_2[$ such that $y(\overline{\phi}) = 0$. Using (8) and the fact that the three vectors

$$d\hat{g}_{0,0}(0) = (1, 0, 2),\ d\hat{g}_{0,0}(2\pi/3) = (1, -\sqrt{3}, -1),\ d\hat{g}_{0,0}(4\pi/3) = (1, \sqrt{3}, -1)$$

are linearly independent, for sufficiently small $r < r_0$ we then can determine a map $g = g_{\phi_y,a} \in \mathcal{G}$ such that $\hat{g}(0) = \overline{\phi}, \hat{g}\left(\frac{2\pi k}{3}\right) = \frac{2\pi k}{3}$, $k = 1, 2$, with $|\phi_y| + |a| \le C|\overline{\phi}| \le Cr$, as desired.

ii) For $x \in M^*$ and y given by (9) with $|g_h - id| \le h, 0 < 2\varepsilon < h^2$, similar to (12) we have

$$\begin{aligned}
&y\left(\tfrac{2\pi k}{3} - h\right)\\
&= \left(1 - \tfrac{2\varepsilon}{h^2}\right) x\left(\tfrac{2\pi k}{3} - h\right) + \tfrac{\varepsilon}{h^2}\left(x\left(g_h\left(\tfrac{2\pi k}{3} - h\right)\right)\right) + x\left(g_h^{-1}\left(\tfrac{2\pi k}{3} - h\right)\right)\\
&\le x\left(\tfrac{2\pi k}{3}\right) = \tfrac{2\pi k}{3}\\
&\le \left(1 - \tfrac{2\varepsilon}{h^2}\right) x\left(\tfrac{2\pi k}{3} + h\right) + \tfrac{\varepsilon}{h^2}\left(x\left(g_h\left(\tfrac{2\pi k}{3} + h\right)\right)\right) + x\left(g_h^{-1}\left(\tfrac{2\pi k}{3} + h\right)\right)\\
&= y\left(\tfrac{2\pi k}{3} + h\right), k = 1, 2, 3.
\end{aligned}$$

Thus, for sufficiently small $h < h_0$ there exists $g \in \mathcal{G}$ with $|g - id| < Ch$ such that $y \circ g \in M^*$, as claimed. □

4. Estimates

Finally, we establish the regularity estimates required for the proof of Proposition 2.2. Let x be a critical point of E in M^* or $M^\dagger$, and let y be a comparison function, g an element of $\mathcal{G}$ such that $y \circ g \in M^*$ or $M^\dagger$, respectively, as determined in Proposition 3.2. Also let $X = \Xi(x), Y = \Xi(y)$. Recall that $X \circ g = \Xi(x \circ g)$, and similarly for Y.

Finally, note that by harmonicity of X there holds $\int_B \langle \nabla X, \nabla \varphi \rangle \, dw = 0$ for all $\varphi \in C_0^\infty(B; \mathbb{R}^n)$. Hence for any $Z \in H^{1,2}(B; \mathbb{R}^n)$ the integral

$$\int_B \langle \nabla X, \nabla Z \rangle \, dw = \int_{\partial B} \langle \partial_n X, Z \rangle \, do$$

depends only on the trace $Z|_{\partial B} \in H^{1/2,2}(\partial B; \mathbb{R}^n)$. We will use this observation to replace a comparison surface Y by a more convenient extension of its boundary values, whenever necessary.

Our first aim is to derive the following Dirichlet growth estimate for X; see [6], Lemma 5.4.

Lemma 4.1. *There exist constants $\alpha > 0, C$ such that for all $r > 0, w_0 \in \partial B$ there holds*

$$\int_{B \cap B_r(w_0)} |\nabla X|^2 \, dw \le Cr^{2\alpha}.$$

Proof. As in [6], the desired estimate will follow by inserting the function y given by (5) as comparison function.

On ∂B, for any $z \in M$, letting $Z = \Xi(z)$ we have

$$X - Z = \gamma \circ x - \gamma \circ z = \frac{d}{d\phi}\gamma(x)(x - z) - \int_x^z \int_x^{x'} \frac{d^2}{d\phi^2}\gamma(x'') \, dx'' \, dx'; \qquad (13)$$

compare [6], p. 50. In particular, for $z = y \circ g$ we obtain

$$\frac{d}{d\phi}\gamma(x)(x - y \circ g) = X - Y \circ g + \psi,$$

where

$$\psi = \int_x^{y\circ g} \int_x^{x'} \frac{d^2}{d\phi^2}\gamma(x'')\,dx''\,dx'$$

Thus, from (4) we have

$$0 \geq \langle dE(x), x - y\circ g\rangle = I + II,$$

where

$$\begin{aligned} I &= \int_{\partial B} \langle \partial_n X, X - Y\circ g\rangle\,do \\ &= \int_B \langle \nabla X, \nabla(X-Y)\rangle\,dw + \int_B \langle \nabla X, \nabla(Y - Y\circ g)\rangle\,dw =: I_1 + I_2 \end{aligned}$$

and

$$II = \int_{\partial B} \langle \partial_n X, \psi\rangle do.$$

We now estimate each term separately.

Estimate for I_2: By conformal invariance,

$$I_2 = \int_B \langle \nabla X, \nabla(Y - Y\circ g)\rangle\,dw = \int_B \langle \nabla(X - X\circ g^{-1}), \nabla Y\rangle\,dw.$$

Adding

$$\frac{X - X\circ g^{-1}}{2} + \frac{X + X\circ g^{-1}}{2} - X = 0$$

to Y, the latter may be estimated

$$\begin{aligned} I_2 = \int_B \langle \nabla(X - X\circ g^{-1}), \nabla Y\rangle\,dw = \frac{1}{2}\int_B |\nabla(X - X\circ g^{-1}|^2\,dw \\ + \frac{1}{2}\int_B \left(|\nabla X|^2 - |\nabla(X\circ g^{-1})|^2\right)\,dw + \int_B \langle \nabla(X - X\circ g^{-1}), \nabla(Y - X)\rangle\,dw \\ \geq \frac{1-\delta}{2}\int_B |\nabla(X - X\circ g^{-1})|^2\,dw - \frac{1}{2\delta}\int_B |\nabla(X-Y)|^2\,dw. \end{aligned}$$

Here we also used the fact that by conformal invariance

$$\int_B |\nabla X|^2\,dw = \int_B |\nabla(X\circ g^{-1})|^2\,dw;$$

moreover, we employed Young's inequality $2ab \leq \delta a^2 + \delta^{-1}b^2$ for $0 < \delta, a, b \in \mathbb{R}$.

Estimate for I: In case y is given by (5), on ∂B we have

$$\begin{aligned} X - Y &= \gamma\circ x - \gamma\circ(x - \tau^2(x - x_0)) \\ &= \frac{d}{d\phi}\gamma(x)\tau^2(x - x_0) - \int_x^{x-\tau^2(x-x_0)} \int_x^{x'} \frac{d^2}{d\phi^2}\gamma(z'')dz''dz'. \end{aligned} \tag{14}$$

Also let $X_0 = \gamma(x_0)$, satisfying

$$X - X_0 = \frac{d}{d\phi}\gamma(x)(x - x_0) - \int_x^{x_0}\int_x^{x'} \frac{d^2}{d\phi^2}\gamma(x'')\,dx''\,dx' \tag{15}$$

on ∂B.

Observe that, by continuity of X and x, for sufficiently small $r < r_0$ we may use (15) to define a unique extension $x \in H^{1,2}(B \cap B_{3r}(w_0))$ satisfying the estimates

$$C^{-1}|x - x_0| \leq |X - X_0| \leq C|x - x_0|, \tag{16}$$

$$C^{-1}|\nabla x| \leq |\nabla X| \leq C|\nabla x| \tag{17}$$

with a uniform constant C. Indeed, multiplying (15) by $\frac{d}{d\phi}\gamma(x)/|\frac{d}{d\phi}\gamma(x)|^2$ and observing that

$$\left|\int_x^{x_0}\int_x^{x'} \frac{d^2}{d\phi^2}\gamma(x'')\,dx''\,dx'\right| \leq C|x - x_0|^2 \leq o(1)|x - x_0|,$$

where $o(1) \to 0$ as $r \to 0$, the first estimate immediately follows. The second estimate is obtained in the same way from the differentiated equation (15).

Using (14) and (15) we now modify Y in B by letting

$$\begin{aligned}\tilde{Y} = X - \tau^2(X - X_0) - \tau^2 &\int_x^{x_0}\int_x^{x'} \frac{d^2}{d\phi^2}\gamma(x'')\,dx''\,dx' \\ + &\int_x^{x-\tau^2(x-x_0)}\int_x^{x'} \frac{d^2}{d\phi^2}\gamma(x'')\,dx''\,dx'.\end{aligned}$$

Observe that $Y = \tilde{Y}$ on ∂B; thus, by our above remark there holds

$$\int_B \langle \nabla X, \nabla Y\rangle\,dw = \int_B \langle \nabla X, \nabla \tilde{Y}\rangle\,dw.$$

Moreover, we have the estimates

$$|X - \tau^2(X - X_0) - \tilde{Y}| \leq C\tau^2|x - x_0|^2 \leq C\tau^2|X - X_0|^2$$

and

$$|\nabla(X - \tau^2(X - X_0)) - \nabla\tilde{Y}| \leq C|\nabla\tau|\tau|X - X_0| + C\tau^2|\nabla X||X - X_0|,$$

as is easily verified.

Thus, for sufficiently small $r > 0$ we can estimate

$$\begin{aligned}I_1 &= \int_B \langle \nabla X, \nabla(X - Y)\rangle\,dw = \int_B \langle \nabla X, \nabla(X - \tilde{Y})\rangle\,dw \\ &\geq \int_B (1 - C\tau^2|X - X_0|)|\nabla X|^2\tau^2\,dw - C\int_B |\nabla X||X - X_0||\nabla\tau|\tau\,dw \\ &\geq \frac{7}{8}\int_B |\nabla X|^2\tau^2\,dw - C\int_B |X - X_0|^2|\nabla\tau|^2\,dw.\end{aligned}$$

Moreover, observing that harmonicity implies that $D(X-Y) \le D(X-\tilde{Y})$, and choosing $\delta = \frac{3}{4}$, we can bound the last term in our estimate for I_2 by

$$\begin{aligned}\frac{1}{2\delta}\int_B |\nabla(X-Y)|^2\,dw &\le \frac{1}{2\delta}\int_B |\nabla(X-\tilde{Y})|^2\,dw\\ &\le \frac{2}{3}\int_B ((1+C\tau^2|X-X_0|)|\nabla X|\tau^2 + C\tau|\nabla\tau||X-X_0|)^2\,dw\\ &\le \frac{3}{4}\int_B |\nabla X|^2\tau^2\,dw + C\int_B |X-X_0|^2|\nabla\tau|^2\,dw,\end{aligned}$$

whence

$$I \ge \frac{1}{8}\int_B |\nabla X|^2\tau^2\,dw + \frac{1}{8}\int_B |\nabla(X - X\circ g^{-1}|^2\,dw - C\int_B |X-X_0|^2|\nabla\tau|^2\,dw.$$

Estimate for II: Split

$$\psi = \int_{x\circ g}^{y\circ g}\int_x^{x'} \frac{d^2}{d\phi^2}\gamma(x'')\,dx''\,dx' + \int_x^{x\circ g}\int_x^{x'} \frac{d^2}{d\phi^2}\gamma(x'')\,dx''\,dx' =: \psi_1 + \psi_2$$

and denote

$$II = \int_{\partial B}\langle \partial_n X, \psi\rangle\,do = \int_{\partial B}\langle \partial_n X, \psi_1\rangle\,do + \int_{\partial B}\langle \partial_n X, \psi_2\rangle\,do =: II_1 + II_2.$$

Estimate for II_1: Extending x to balls $B_{r_0}(w_0)\cap B$ as above and using a partition of unity, we extend x to a neighborhood of ∂B in B in such a way that (17) and the analogous estimates

$$C^{-1}|x\circ g - x| \le |X\circ g - X| \le C|x\circ g - x| \tag{18}$$

$$C^{-1}|\nabla(x\circ g - x)| \le |\nabla(X\circ g - X)| \le C|\nabla(x\circ g - x)| \tag{19}$$

hold. Also let $y = x - \tau^2(x-x_0)$. Thus, $\mathrm{supp}(\psi_1) \subset g^{-1}(B\cap B_{2r}(w_0))$, and we may use the above estimates to bound

$$|\nabla\psi_1| \le C|\nabla(y\circ g - x\circ g)||x\circ g - x| + C(|\nabla(y\circ g)| + |\nabla x|)|y\circ g - x\circ g|$$

in B. It follows that

$$\begin{aligned}II_1 &= \int_{\partial B}\langle\partial_n X, \psi_1\rangle\,do = \int_B \langle \nabla X, \nabla\psi_1\rangle\,dw\\ &\le C\int_B |\nabla X||\nabla(y\circ g) - \nabla(x\circ g)||x\circ g - x|\,dw\\ &\quad + C\int_B |\nabla X|(|\nabla x| + |\nabla(y\circ g)|)|y\circ g - x\circ g|\,dw = II_{11} + II_{12}.\end{aligned}$$

By conformal invariance then, and since $g\to id$ uniformly as $r\to 0$ on account of Proposition 3.2 and since therefore, by continuity of x, also $x\circ g^{-1}\to x$, we can estimate

$$\begin{aligned}
II_{11} &= C\int_B |\nabla(X\circ g^{-1})||\nabla(y-x)||x-x\circ g^{-1}|\,dw \\
&\le C||x-x\circ g^{-1}||_{L^\infty(\partial B)}\Big(\int_B |\nabla X||\nabla(y-x)|\,dw \\
&\quad + \int_B |\nabla(X-X\circ g^{-1}|^2\,dw + \int_B |\nabla(y-x)|^2\,dw\Big) \\
&\le o(1)\Big(\int_B (|\nabla X|^2+|\nabla x|^2\Big)\tau^2\,dw + \int_B |\nabla(X-X\circ g^{-1})|^2\,dw \\
&\quad + \int_B |x-x_0|^2|\nabla\tau|^2\,dw\Big),
\end{aligned}$$

where $o(1)\to 0$ as $r\to 0$.

Similarly,

$$\begin{aligned}
II_{12} &\le C\int_B |\nabla(X\circ g^{-1})|(|\nabla(x\circ g^{-1})|+|\nabla y|)|y-x|\,dw \\
&\le C||x-x_0||_{L^\infty(B_{3r}(x_0))}\Big(\int_B |\nabla(X-X\circ g^{-1})|^2\,dw \\
&\quad + \int_{B\cap B_{3r}(w_0)} (|\nabla X|^2+|\nabla x|^2+|x-x_0|^2|\nabla\tau|^2)\,dw\Big),
\end{aligned}$$

where $||x-x_0||_{L^\infty(B_{3r}(w_0))}\to 0$ as $r\to 0$.

Estimate for II_2: Fix $\eta\in C_0^\infty(B)$ such that $0\le\eta\le 1$ and $\eta(w)=1$ for $|w|<1-r_0$ and extend ψ_2 to B as

$$\psi_2 = (1-\eta)\int_x^{x\circ g}\int_x^{x'} \frac{d^2}{d\phi^2}\gamma(x'')\,dx''\,dx',$$

where x on $B\setminus B_{1-r_0}(0)$ is defined above.

We have

$$|\psi_2|\le C|x\circ g-x|^2\le C|X\circ g-X|^2;$$

moreover, near ∂B (omitting η) we compute

$$\begin{aligned}
|\nabla\psi_2| &= |\nabla(x\circ g)\int_x^{x\circ g}\frac{d^2}{d\phi^2}\gamma(x')\,dx' - \nabla x(x\circ g-x)\frac{d^2}{d\phi^2}\gamma(x)| \\
&= |\nabla(x\circ g-x)\int_x^{x\circ g}\frac{d^2}{d\phi^2}\gamma(x')\,dx' + \nabla x\int_x^{x\circ g}(\frac{d^2}{d\phi^2}\gamma(x')-\frac{d^2}{d\phi^2}\gamma(x))\,dx'| \\
&\le C|\nabla(x\circ g-x)||x\circ g-x| + C|\nabla x||x\circ g-x|^2 \\
&\le C|\nabla(X\circ g-X)||X\circ g-X| + C|\nabla X||X\circ g-X|^2.
\end{aligned} \tag{20}$$

Hence

$$\begin{aligned} II_2 &= \int_{\partial B} \langle \partial_n X, \psi_2 \rangle \, do = \int_{\partial B} \langle \nabla X, \nabla \psi_2 \rangle \, dw \\ &\leq C \int_B |\nabla X| |\nabla X \circ g - X| |X \circ g - X| \, dw \\ &\quad + C \int_B (|\nabla X|^2 |X \circ g - X|^2 + |\nabla X| |\nabla \eta| |X \circ g - X|^2) \, dw = II_{21} + II_{22}. \end{aligned}$$

By Young's inequality, we have

$$\begin{aligned} II_{21} &\leq C ||X \circ g - X||^2_{L^\infty} \int_B |\nabla X|^2 \, dw + \frac{1}{16} \int_B |\nabla (X - X \circ g^{-1})|^2 \, dw, \\ II_{22} &\leq C ||X \circ g - X||^2_{L^\infty} \int_B (1 + |\nabla X|^2) \, dw. \end{aligned}$$

Conclusion. Together with our bounds for I and II_1, the above estimates finally yield that for sufficiently small $r < r_0$ there holds

$$0 \geq I + II \geq \frac{1}{16} \int_B |\nabla X|^2 \tau^2 \, dw - C \int_B |X - X_0|^2 |\nabla \tau|^2 \, dw - C ||X \circ g - X||^2_{L^\infty}.$$

Now by Proposition 3.2 we have $|g - id| \leq Cr$ as $r \to 0$. Therefore, by the Courant Lebesgue Lemma [1], p. 102, we can estimate

$$\begin{aligned} ||X \circ g - X||^2_{L^\infty} &\leq C \sup_{|\phi - \phi'| \leq r} |X(e^{i\phi}) - X(e^{i\phi'})|^2 \\ &\leq C \sup_{w_1 \in \partial B} \int_{B_r(w_1)} |\nabla X|^2 \, dw / |ln \, r|. \end{aligned}$$

As in [6], p. 71 f., thus for sufficiently small $r > 0$ and all $w_0 \in \partial B$ we find

$$\begin{aligned} &\int_{B \cap B_r(w_0)} |\nabla X|^2 \, dw \\ &\quad \leq C r^{-2} \int_{B \cap B_{3r} \setminus B_r(w_0)} |X - X_0|^2 \, dw + \frac{1}{2} \sup_{w_1 \in \partial B} \int_{B \cap B_r(w_1)} |\nabla X|^2 \, dw \\ &\quad \leq C \int_{B \cap B_{3r} \setminus B_r(w_0)} |\nabla X|^2 \, dw + \frac{1}{2} \sup_{w_1 \in \partial B} \int_{B \cap B_r(w_1)} |\nabla X|^2 \, dw. \end{aligned}$$

In particular, if we choose $w_0 \in \partial B$ such that

$$\int_{B \cap B_r(w_0)} |\nabla X|^2 \, dw = \sup_{w_1 \in \partial B} \int_{B \cap B_r(w_1)} |\nabla X|^2 \, dw =: \phi(r),$$

from the preceding estimate we obtain that

$$\int_{B \cap B_r(w_0)} |\nabla X|^2 \, dw \leq C_1 \int_{B \cap B_{3r} \setminus B_r(w_0)} |\nabla X|^2 \, dw. \tag{21}$$

Upon adding C_1 times the expression on the left of (21) to both sides, thus "filling the hole" on the right following an idea of Widman [7], and dividing by $1 + C_1$, we obtain the inequality

$$\phi(r) \le \vartheta \int_{B\cap B_{3r}(w_0)} |\nabla X|^2 dw \le \vartheta\phi(3r)$$

for any $r \le r_0$, where $\vartheta = \frac{C_1}{1+C_1} < 1$. By iteration we then obtain the Dirichlet growth condition

$$\phi(r) \le Cr^{2\alpha}, 0 < r < r_0, \tag{22}$$

where $\alpha = -\frac{1}{2}\frac{ln\vartheta}{ln2} > 0$. Replacing C by $2D(X)r_0^{-2\alpha}$, if necessary, inequality (22) also holds for $r \ge r_0$. □

Proof. [Proof of Proposition 2.2] i) In the case of the normalization $x \in M^\dagger$ the proof of Proposition 2.2 can now be completed as in [6], pp. 67–69 and p. 73 f.

ii) In the case of the normalization M^*, the comparison function y given by (9) in general does not belong to M^*. In this case we argue as follows.

By Proposition 3.1. ii) for sufficiently small $h < h_0$ and any $\varepsilon < \frac{1}{2}h^2$ there exists $g \in \mathcal{G}$ with $|g - id| \le Ch$ such that $y \circ g \in M^*$. Hence, if $x \in M^*$ is critical, we have

$$0 \ge \langle dE(x), x - y \circ g\rangle = \langle dE(x), x - y\rangle + \langle dE(x), y - y \circ g\rangle =: I + II.$$

We proceed to estimate these terms. In order not to overburden the exposition with (even more) technicalities, we restrict ourselves to the case where y is given by (6), that is, where $\hat{g}_h(\phi) = \phi + h$. Adapting the estimates to the general case is tedious but presents no further conceptual difficulties.

For y as in (6), as shown in [6], p. 67 ff., the contribution from the first term can be estimated

$$I = -\varepsilon \int_{\partial B} \left\langle \partial_n X, \frac{d}{d\phi}\gamma(x)\partial_{-h}\partial_h x \right\rangle do \ge \frac{\varepsilon}{2}\int_B |\nabla\partial_h X|^2 dw - C\varepsilon \int_B |\nabla X|^2 |\partial_h X|^2 dw.$$

(The ε here has a different meaning than in [6], p. 68.)

In order to bound the second term we need to derive a convenient representation for $\frac{d}{d\phi}\gamma(x)(y - y \circ g)$. In fact, by (13), applied to $z = y$ and $z = y \circ g$, on ∂B we have

$$\frac{d}{d\phi}\gamma(x)(y - y \circ g) = Y - Y \circ g + \psi_1,$$

where $Y = \Xi(y)$ and

$$\begin{aligned}\psi_1 &= \int_x^{y\circ g}\int_x^{x'} \frac{d^2}{d\phi^2}\gamma(x'')\,dx''\,dx' - \int_x^{y}\int_x^{x'} \frac{d^2}{d\phi^2}\gamma(x'')\,dx''\,dx' \\ &= \left(\int_{x\circ g}^{y\circ g}\int_x^{x'} \frac{d^2}{d\phi^2}\gamma(x'')\,dx''\,dx' - \int_x^{y}\int_x^{x'} \frac{d^2}{d\phi^2}\gamma(x'')\,dx''\,dx'\right) \\ &\quad + \int_x^{x\circ g}\int_x^{x'} \frac{d^2}{d\phi^2}\gamma(x'')\,dx''\,dx' = \psi_{11} + \psi_{12}.\end{aligned}$$

We split $II = II_1 + II_2$, where

$$II_1 = \int_{\partial B} \langle \partial_n X, \psi_1\rangle\, do, \quad II_2 = \int_{\partial B} \langle \partial_n X, Y - Y\circ g\rangle\, do.$$

Estimate for II_2: Recall that by harmonicity of X in this expression we may replace Y by any function $\tilde{Y} \in H^{1,2}(B;\mathbb{R}^n)$ such that $\tilde{Y} = Y$ on ∂B. In order to construct a suitable such extension, observe that, letting $X_h(e^{i\phi}) = X(e^{i(\phi+h)})$ on ∂B, we have

$$\varepsilon\partial_{-h}\partial_h X = \frac{\varepsilon}{h^2}(X_h - X) + \frac{\varepsilon}{h^2}(X_{-h} - X).$$

By (13), with $z = x_h = x(\cdot + h)$, we may express

$$X_h - X = \frac{d}{d\phi}\gamma(x)(x_h - x) + \int_x^{x_h}\int_x^{x'} \frac{d^2}{d\phi^2}\gamma(x'')\,dx''\,dx',$$

and similarly for X_{-h}.

On the other hand, for $z = y$ equation (13) yields

$$Y = X + \varepsilon\frac{d}{d\phi}\gamma(x)\partial_{-h}\partial_h x + \int_x^{y}\int_x^{x'} \frac{d^2}{d\phi^2}\gamma(x'')\,dx''\,dx'.$$

Hence we may let

$$\begin{aligned}\tilde{Y} &= X + \varepsilon\partial_{-h}\partial_h X + (1-\eta)\Big(\int_x^{y}\int_x^{x'} \frac{d^2}{d\phi^2}\gamma(x'')\,dx''\,dx' \\ &\quad - \frac{\varepsilon}{h^2}\int_x^{x_h}\int_x^{x'} \frac{d^2}{d\phi^2}\gamma(x'')\,dx''\,dx' - \frac{\varepsilon}{h^2}\int_x^{x_{-h}}\int_x^{x'} \frac{d^2}{d\phi^2}\gamma(x'')\,dx''\,dx'\Big) \\ &=: X + \varepsilon\partial_{-h}\partial_h X + \psi_2,\end{aligned}$$

where $\eta \in C_0^\infty(B)$ satisfies $0 \le \eta \le 1, \eta \equiv 1$ on $B_{1-r_0}(0)$, as in the proof of Lemma 4.1.

Similar to our estimate for I_2 in the proof of Lemma 4.1 then we have

$$\begin{aligned}
II_2 &= \int_{\partial B} \langle \partial_n X, \tilde{Y} - \tilde{Y} \circ g\rangle \, do = \int_B \langle \nabla X, \nabla(\tilde{Y} - \tilde{Y} \circ g)\rangle \, dw \\
&= \int_B \langle \nabla(X - X \circ g^{-1}), \nabla \tilde{Y}\rangle \, dw \\
&= \frac{1}{2} \int_B |\nabla(X - X \circ g^{-1}|^2 \, dw + \int_B \langle \nabla(X - X \circ g^{-1}), \nabla(\tilde{Y} - X)\rangle \, dw \\
&= \frac{1}{2} \int_B |\nabla(X - X \circ g^{-1})|^2 \, dw + \int_B \langle \nabla(X - X \circ g^{-1}), \nabla(\varepsilon \partial_{-h} \partial_h X)\rangle \, dw \\
&\quad + \int_B \langle \nabla(X - X \circ g^{-1}), \nabla \psi_2\rangle \, dw \\
&\geq \frac{1}{4} \int_B |\nabla(X - X \circ g^{-1})|^2 \, dw - 2\varepsilon^2 \int_B |\nabla \partial_{-h} \partial_h X|^2 \, dw - 2 \int_B |\nabla \psi_2|^2 \, dw \\
&= II_{21} + II_{22} + II_{23}.
\end{aligned}$$

But

$$\int_B |\nabla \partial_{-h} \partial_h X|^2 \, dw \leq 2h^{-2} \int_B |\nabla \partial_h X|^2 \, dw$$

and, choosing $\varepsilon \leq \frac{1}{8} h^2$, we can estimate

$$|II_{22}| \leq \frac{\varepsilon}{4} \int_B |\nabla \partial_h X|^2 \, dw.$$

Moreover, we easily verify that

$$|\psi_2| \leq C|\varepsilon \partial_{-h} \partial_h x|^2 + C\varepsilon \left(|\partial_h x|^2 + |\partial_{-h} x|^2\right) \leq C\varepsilon \left(|\partial_h x|^2 + |\partial_{-h} x|^2\right).$$

Near ∂B, where $\eta = 0$, we also have

$$\begin{aligned}
|\nabla \psi_2| &\leq C(|\nabla y| + |\nabla x|)|y - x| \\
&\quad + C\frac{\varepsilon}{h^2}(|\nabla x_h| + |\nabla x_{-h}| + |\nabla x|)(|x - x_h| + |x - x_{-h}|) \\
&\leq \frac{C\varepsilon}{h}(|\nabla x_h| + |\nabla x_{-h}| + |\nabla x|)(|\partial_h x| + |\partial_{-h} x|),
\end{aligned}$$

and hence, observing that $\varepsilon \leq \frac{1}{8} h^2$,

$$\begin{aligned}
|II_{23}| &\leq 2 \int_B |\nabla \psi_2|^2 \, dw \leq C\varepsilon^2 \int_B (|\nabla \eta|^2)(|\partial_h x|^2 + |\partial_{-h} x|^2) \, dw \\
&\quad + C\varepsilon \int_B (1 - \eta)^2 (|\nabla x_h|^2 + |\nabla x_{-h}|^2 + |\nabla x|^2)(|\partial_h x|^2 + |\partial_{-h} x|^2) \, dw.
\end{aligned}$$

Estimate for II_1: Similar to our bounds for II in the proof of Lemma 4.1 we now estimate

$$II_1 = \int_{\partial B} \langle \partial_n X, \psi_1\rangle \, do = \int_{\partial B} \langle \partial_n X, \psi_{11}\rangle \, do + \int_{\partial B} \langle \partial_n X, \psi_{12}\rangle \, do = II_{11} + II_{12}.$$

Again we use the extension of x to an r_0-neighborhood of ∂B and truncate with $1-\eta$ to define extensions of ψ_{11}, ψ_{12} to B.

Then we have

$$\begin{aligned}
|\psi_{11}| &= \left| \int_{x\circ g}^{y\circ g}\int_x^{x'} \frac{d^2}{d\phi^2}\gamma(x'')\,dx''\,dx' - \int_x^y\int_x^{x'} \frac{d^2}{d\phi^2}\gamma(x'')\,dx''\,dx' \right| \\
&\le C(|y\circ g - x\circ g||x\circ g - x| + |y\circ g - x\circ g|^2 + |y-x|^2) \\
&\le C\varepsilon|\partial_{-h}\partial_h x\circ g||x\circ g - x| + C\varepsilon^2(|\partial_{-h}\partial_h x\circ g|^2 + |\partial_{-h}\partial_h x|^2), \\
&\le o(1)\varepsilon(|\partial_{-h}\partial_h x\circ g| + |\partial_{-h}\partial_h x|),
\end{aligned}$$

where we observe that

$$||x\circ g - x||_{L^\infty} + ||\varepsilon\partial_{-h}\partial_h x||_{L^\infty} = o(1) \to 0$$

as $h\to 0, \varepsilon \le \frac{1}{8}h^2$.

Moreover,

$$\begin{aligned}
|\nabla\psi_{11}| = |(\nabla y\circ g - \nabla x\circ g)&\int_x^{x\circ g}\frac{d^2}{d\phi^2}\gamma(x')\,dx' \\
&+ \nabla y\circ g\int_{x\circ g}^{y\circ g}\frac{d^2}{d\phi^2}\gamma(x')\,dx' - \nabla y\int_x^y \frac{d^2}{d\phi^2}\gamma(x')\,dx' \\
&\qquad - \nabla x[(y\circ g - x\circ g) - (y-x)]\frac{d^2}{d\phi^2}\gamma(x)| \\
\le C\varepsilon|\nabla\partial_{-h}\partial_h x\circ g||x\circ g - x| &+ C|\nabla y\circ g - \nabla y||y\circ g - x\circ g| \\
+ C\varepsilon|\nabla\partial_{-h}\partial_h x|&(|y\circ g - x\circ g| + |y-x|) \\
+ C|\nabla x|&(|y\circ g - x\circ g|^2 + |y-x|^2 + |x\circ g - x|^2),
\end{aligned}$$

where we have expanded

$$-\nabla y\int_x^y \frac{d^2}{d\phi^2}\gamma(x')\,dx' = -\nabla y\int_{x\circ g}^{y\circ g}\ldots + \nabla y\left(\int_{x\circ g}^{y\circ g}\ldots - \int_x^y\ldots\right)$$

and then combined the second term with the last term in the expansion of $\nabla\psi_{11}$ to obtain

$$\begin{aligned}
&\Bigg|\nabla y\left(\int_{x\circ g}^{y\circ g}\frac{d^2}{d\phi^2}\gamma(x')dx' - \int_x^y\frac{d^2}{d\phi^2}\gamma(x')dx'\right) \\
&\quad - \nabla x\Big((y\circ g - x\circ g) - (y-x)\Big)\frac{d^2}{d\phi^2}\gamma(x)\Bigg| \\
&\le C|\nabla y - \nabla x|(|y\circ g - x\circ g| + |y-x|) \\
&\quad + \left|\nabla x\int_{x\circ g}^{y\circ g}\left(\frac{d^2}{d\phi^2}\gamma(x') - \frac{d^2}{d\phi^2}(x)\right)dx'\right| + \left|\nabla x\int_x^y\left(\frac{d^2}{d\phi^2}\gamma(x') - \frac{d^2}{d\phi^2}\gamma(x)\right)dx'\right| \\
&\le C\varepsilon|\nabla\partial_{-h}\partial_h x|(|y\circ g - x\circ g| + |y-x|) \\
&\quad + C|\nabla x||y\circ g - x\circ g|(|y\circ g - x\circ g| + |x\circ g - x|) + C|\nabla x||y-x|^2.
\end{aligned}$$

Observing that $|\nabla\partial_{-h}\partial_h x| \le h^{-1}(|\nabla\partial_h x| + |\nabla\partial_{-h}x|), h^{-1}|y-x| \le \varepsilon h^{-2}(|\partial_h x| + |\partial_{-h}x|) \le |\partial_h x| + |\partial_{-h}x|$, and recalling that $\varepsilon \le \frac{1}{8}h^2$, thus we find

$$\begin{aligned}
|\nabla\psi_{11}| &\le C\varepsilon(|\nabla\partial_h x\circ g| + |\nabla\partial_{-h}x\circ g|)|h^{-1}(x\circ g - x)| \\
&\quad + C\varepsilon(|\partial_h x\circ g| + |\partial_{-h}x\circ g|)|h^{-1}(\nabla y\circ g - \nabla y)| \\
&\quad + C\varepsilon(|\nabla\partial_h x| + |\nabla\partial_{-h}x|)(|\partial_h x\circ g| + |\partial_{-h}x\circ g| + |\partial_h x| + |\partial_{-h}x|) \\
&\quad + C\varepsilon|\nabla x|\left(|\partial_h x\circ g|^2 + |\partial_{-h}x\circ g|^2 + |\partial_h x|^2 + |\partial_{-h}x|^2 + |h^{-1}(x\circ g - x)|^2\right).
\end{aligned}$$

In conclusion from this, for any $\delta > 0$ we can bound

$$\begin{aligned}
II_{11} &\le \int_B |\nabla X|(|\nabla\eta||\psi_{11}| + (1-\eta)|\nabla\psi_{11}|)\,dw \\
&\le \frac{\varepsilon}{16}\int_{B\setminus B_{1-r_0}(0)} |\nabla\partial_h x|^2\,dw + \delta\varepsilon\int_{B\setminus B_{1-r_0}(0)} |h^{-1}(\nabla y\circ g - \nabla y)|^2\,dw \\
&\quad + C\varepsilon\int_B |\nabla X|^2\,dw + C(\delta)\varepsilon\int_{B\setminus B_{1-r_0}(0)} (|\nabla X|^2 + |\nabla x|^2) \\
&\qquad \cdot (|h^{-1}(x\circ g - x)|^2 + |\partial_h x\circ g|^2 + |\partial_{-h}x\circ g|^2 + |\partial_h x|^2 + |\partial_{-h}x|^2)\,dw
\end{aligned}$$

Similarly, we estimate

$$|\psi_{12}| = \left|\int_x^{x\circ g}\int_x^{x'} \frac{d^2}{d\phi^2}\gamma(x'')\,dx''\,dx'\right| \le C|x\circ g - x|^2$$

and

$$\begin{aligned}
|\nabla\psi_{12}| &= \left|\nabla\left(\int_x^{x\circ g}\int_x^{x'} \frac{d^2}{d\phi^2}\gamma(x'')\,dx''\,dx'\right)\right| \\
&= \left|(\nabla x\circ g - \nabla x)\int_x^{x\circ g} \frac{d^2}{d\phi^2}\gamma(x')\,dx' + \nabla x\int_x^{x\circ g}\left(\frac{d^2}{d\phi^2}\gamma(x') - \frac{d^2}{d\phi^2}\gamma(x)\right)dx'\right| \\
&\le C|\nabla(x\circ g - x)||x\circ g - x| + C|\nabla x||x\circ g - x|^2,
\end{aligned}$$

as in (20). By Young's inequality we find

$$\begin{aligned}
|II_{12}| &\le C\int_{B\setminus B_{1-r_0}(0)} |\nabla X|(|\nabla(x\circ g - x)||x\circ g - x| + |\nabla\eta||x\circ g - x|^2)\,dw \\
&\quad + C\int_{B\setminus B_{1-r_0}(0)} |\nabla X||\nabla x||x\circ g - x|^2\,dw \\
&\le \delta\int_{B\setminus B_{1-r_0}(0)} |\nabla(x\circ g - x)|^2\,dw \\
&\quad + C(\delta)\int_{B\setminus B_{1-r_0}(0)} \left(1 + |\nabla X|^2 + |\nabla x|^2\right)|x\circ g - x|^2\,dw
\end{aligned}$$

$$\leq \delta\varepsilon \int_{B\setminus B_{1-r_0}(0)} \left|\frac{\nabla(x\circ g - x)}{h}\right|^2 dw$$
$$+ C(\delta)\varepsilon \int_{B\setminus B_{1-r_0}(0)} (1+|\nabla X|^2+|\nabla x|^2)|h^{-1}(x\circ g - x)|^2\, dw.$$

Observe that we may replace x by X on account of (17)–(18). Thus, we can bound

$$\begin{aligned}|II_1| &\leq |II_{11}| + |II_{12}| \\ &\leq \frac{\varepsilon}{8}\int_B |\nabla\partial_h X|^2\, dw + \delta\varepsilon \int_B \left|\frac{\nabla(X\circ g - X)}{h}\right|^2 dw \\ &\quad + C(\delta)\varepsilon \int_B (1+|\nabla X|^2)(1+|h^{-1}(X\circ g - X)|^2 + |\partial_h X|^2)\, dw,\end{aligned}$$

where we simply denote $|\partial_h X|, |\partial_h X\circ g|$, etc., as $|\partial_h X|$.

Conclusion. Together with our bounds for II_2 it now follows that

$$\begin{aligned}0 &\geq I + II \geq I - |II_1| - |II_{22}| - |II_{23}| \\ &\geq \frac{\varepsilon}{8}\int_B |\nabla\partial_h X|^2\, dw - \delta\varepsilon \int_B \left|\frac{\nabla(X\circ g - X)}{h}\right|^2 dw \\ &\quad - C(\delta)\varepsilon \int_B (1+|\nabla X|^2)(1+|\partial_h X|^2 + |h^{-1}(X\circ g - X)|^2)\, dw.\end{aligned}$$

Replacing δ by $\delta/8$ and dividing by $\varepsilon/8$, thus we obtain the estimate

$$\begin{aligned}\int_B |\nabla\partial_h X|^2\, dw \leq &\delta \int_B \left|\frac{\nabla(X\circ g - X)}{h}\right|^2 dw \\ &+ C(\delta)\int_B (1+|\nabla X|^2)(1+|\partial_h X|^2 + |h^{-1}(X\circ g - X)|^2)\, dw.\end{aligned}$$

Invoking Lemma 4.1 and a result of Morrey, as in [6], p. 73 f., we can bound the last of these terms

$$\begin{aligned}C(\delta)&\int_B (1+|\nabla X|^2)(1+|\partial_h X|^2 + |h^{-1}(X\circ g - X)|^2)\, dw \\ &\leq \frac{1}{2}\int_B |\nabla\partial_h X|^2\, dw + \delta\int_B \left|\frac{\nabla(X\circ g - X)}{h}\right|^2 dw + C(\delta)\int_B (1+|\nabla X|^2)\, dw,\end{aligned}$$

whence we obtain the bound

$$\int_B |\nabla\partial_h X|^2\, dw \leq \delta \int_B \left|\frac{\nabla(X\circ g - X)}{h}\right|^2 dw + C(\delta)\int_B (1+|\nabla X|^2)\, dw \tag{23}$$

for any $\delta > 0$, uniformly in h, as $h \to 0$. Remark that in order to apply Morrey's lemma [6], Lemma 5.3, in the present context we may simply extend X to a function $\tilde{X} \in H_0^{1,2}(B_2(0); \mathbb{R}^n)$ by reflection in ∂B and truncation; that is, we let $\tilde{X} = X$ in B and

$$\tilde{X}(w) = \varphi(w) X\left(\frac{w}{|w|^2}\right), w \notin B,$$

where $\varphi \in C_0^\infty(B_2(0))$ is a cut-off function satisfying $0 \leq \varphi \leq 1, \varphi \equiv 1$ on B. In the final estimate (23) for the extended function $\tilde{X}$ on $B_2(0)$ all terms are equivalent (up to a multiplicative constant) to the corresponding integrals on B. This yields the desired estimate (23) for X.

We conjecture that an estimate like (23) for a harmonic function X on the disc already implies that $X \in H^{2,2}$, with the corresponding bounds. In the absence of a rigorous result in this regard, here, however, we proceed as follows.

Introduce difference quotients with respect to $\mathcal{G}$ by letting

$$\int_B |\nabla \partial_h^{\mathcal{G}} X|^2 \, dw = \sup_{|g-id| \leq h} \int_B \left| \frac{\nabla(X \circ g - X)}{h} \right|^2 dw.$$

Repeating the above analysis with a general comparison function y of type (9), where $g_h \in \mathcal{G}$ satisfies $|g_h - id| \leq h$ and $0 \leq \varepsilon \leq \frac{1}{2}h^2$, and using the group property of $\mathcal{G}$, we arrive at the inequality analogous to (23),

$$\int_B \left| \frac{\nabla(X \circ g_h - X)}{h} \right|^2 dw \leq \delta \int_B \left| \frac{\nabla(X \circ g - X)}{h} \right|^2 dw + C(\delta) \int_B (1 + |\nabla X|^2) \, dw$$

for some $g \in \mathcal{G}$ such that $|g - id| \leq C_1 h$ and any $\delta > 0$, with constants $C_1, C(\delta)$ which are uniform in h as $h \to 0$. (We may assume $C_1 \in \mathbb{N}$.)

Taking the supremum with respect to g_h on the left and decomposing the first term on the right as a telescopic sum by writing $g = \tilde{g}_1 \circ \ldots \circ \tilde{g}_{C_1}$ as a C_1-fold product of transformations $\tilde{g}_k \in \mathcal{G}$ satisfying $|\tilde{g}_k - id| \leq h, 1 \leq k \leq C_1$, we then obtain the bound

$$\int_B |\nabla \partial_h^{\mathcal{G}} X|^2 \, dw \leq C\delta \int_B |\nabla \partial_h^{\mathcal{G}} X|^2 \, dw + C(\delta) \int_B (1 + |\nabla X|^2) \, dw$$

for any $\delta > 0$, uniformly in $h > 0$. (Here we also used conformal invariance of Dirichlet's integral again to simplify the representation of the telescopic sum in the second term.) Choosing $\delta > 0$ sufficiently small, there results the estimate

$$\int_B |\nabla \partial_h X|^2 \, dw \leq \int_B |\nabla \partial_h^{\mathcal{G}} X|^2 \, dw \leq C \int_B (1 + |\nabla X|^2) \, dw,$$

uniformly in $h \to 0$. Finally, letting $h \to 0$, from [6], Lemma 5.2 we obtain the desired conclusion. □

References

[1] R. Courant: *Dirichlet's principle, conformal mapping and minimal surfaces*, New York, Interscience, 1950.

[2] K. Deimling: *Nonlinear functional analysis*, Springer, Berlin, etc., 1985.

[3] S. Hildebrandt: "Boundary behavior of minimal surfaces", Arch. Rat. Mech. Anal. **35** (1969), 47–82.

[4] C. Imbusch: *Eine Anwendung des Mountain-Pass-Lemmas auf den Fragenkreis des Plateauschen Problems und eine Alternative zur Drei-Punkte-Bedingung, Diplomarbeit, Bonn, Februar 1997.*

[5] J.-L. Lions, E. Magenes: *Non-homogeneous boundary value problems and applications I*, Springer Grundlehren 181, Berlin-Heidelberg-New York, 1972.

[6] M. Struwe: "Plateau's problem and the calculus of variations", Mathematical Notes **35**, Princeton, New Jersey, 1988.

[7] K.O. Widman: "Hölder continuity of solutions of elliptic systems", Manusc. Math. **5** (1971), 299–308.

Cordula Imbusch and Michael Struwe
Mathematik
ETH-Zentrum
CH-8092 Zürich
struwe@@math.ethz.ch

Progress in Nonlinear Differential Equations
and Their Applications, Vol. 35
© 1999 Birkhäuser Verlag Basel/Switzerland

Hysteresis Operators in Phase-Field Models

Pavel Krejčí and Jürgen Sprekels

Dedicated to Prof. Dr. Herbert Amann on the occasion of his 60th birthday

Abstract. Phase-field systems as mathematical models to forecast the evolution of processes involving phase transitions have drawn a considerable interest in recent years. However, while being capable of capturing many of the experimentally observed phenomena, they are only of restricted value in modelling hysteresis effects occurring during phase transition processes. To overcome this shortcoming, a new approach to phase-field models has been proposed recently by the authors, which is based on the mathematical theory of hysteresis operators developed in the past fifteen years. The approach leads to nonlinearly coupled systems of partial differential equations containing hysteretic nonlinearities at different places. For such systems, well-posedness and thermodynamic consistency results have been proved which are reported in this paper. We note that, due to the lack of smoothness (hysteresis operators are, as a rule, non-differentiable) in the system, the method of proof has to be different from the techniques usually employed for classical phase-field systems.

1. Introduction

The theory of hysteresis operators developed in the past fifteen years (let us at least refer to the monographs [13], [20], [26], [4], [14] devoted to this subject) has proved to be a powerful tool for solving mathematical problems in various branches of applications such as solid mechanics, material fatigue, ferromagnetism, phase transitions, and many others. In this paper, we report on an approach using hysteresis operators to classical *phase-field models* for phase transitions and their generalizations which has recently been proposed by the authors in [15], [16].

The motivation for such an approach is quite obvious: in nature, many phase transitions are accompanied by hysteresis effects (rather, they are the driving mechanisms behind their occurrence). On the other hand, the nonconvex free energy functionals (typically, double-well potentials) usually considered in phase-field models may induce hysteresis effects by themselves (cf., for instance, Chapter 4 in [4]); however, they are by far too simplistic to give a correct account of the complicated loopings due to the storage and deletion of internal memory that are observed in thermoplastic materials or ferromagnets. Another motivation comes from the fact that hysteresis operators also arise quite naturally already in simple classical phase-field models. We demonstrate this for the well-known model for melting and solidification which is usually referred to as the *relaxed Stefan problem with undercooling and overheating* (see [9], [24], [25], for instance).

To fix things, suppose that the phase transition takes place in some open and bounded container $\Omega \subset \mathbb{R}^N$ during the time period $[0,T]$, where $T > 0$ is some final time. Then the mathematical problem consists in finding real-valued functions $\theta = \theta(x,t)$ (absolute temperature) and $\chi = \chi(x,t)$ (phase fraction, the *order parameter* of the phase transition) in $\Omega \times]0,T[$. The function χ is allowed to take values in the interval $[0,1]$ only, where $\chi = 1$ corresponds to the liquid phase, $\chi = 0$ to the solid phase and $\chi \in]0,1[$ to the mushy region. The evolution of the system is governed by the balance of internal energy

$$U_t = -\operatorname{div} q + \psi\,, \tag{1.1}$$

where $U = U(\theta,\chi)$ is the internal energy, q is the heat flux which we assume here to obey Fourier's law

$$q = -\kappa \nabla\theta\,, \tag{1.2}$$

with a constant heat conduction coefficient $\kappa > 0$, and ψ is the heat source density, and by the melting/solidification law

$$\mu\,\chi_t \in -\partial_\chi F(\theta,\chi)\,, \tag{1.3}$$

where $F = F(\theta,\chi)$ is the free energy, ∂_χ is the partial subdifferential with respect to χ and $\mu > 0$ is a fixed relaxation coefficient. In order to ensure the thermodynamical consistency of the model, we have to require that

$$\theta(x,t) > 0 \quad \text{a.e. in } \Omega \times]0,T[\,, \tag{1.4}$$

and that the Clausius-Duhem inequality $S_t \ge -\operatorname{div}\left(\frac{q}{\theta}\right) + \frac{\psi}{\theta}$ is satisfied, which in view of (1.1), (1.2) and (1.4) is certainly the case if only

$$U_t \le \theta\, S_t \quad \text{a.e.}\,, \tag{1.5}$$

where $S := \frac{1}{\theta}(U - F)$ denotes the entropy.

A standard choice [9] for F and U is given by

$$F := F_0(\theta) + \lambda(\chi) + \theta\, I(\chi) - \frac{L}{\theta_c}(\theta - \theta_c)\,\chi\,, \tag{1.6}$$

$$U := c_V\,\theta + \lambda(\chi) + L\,\chi\,, \tag{1.7}$$

where

$$F_0(\theta) := c_V\,\theta(1 - \log\theta)\,, \tag{1.8}$$

$$\lambda(\chi) := \alpha\,\chi(1-\chi)\,. \tag{1.9}$$

Here, I is the indicator function of the interval $[0,1]$, and L (latent heat), θ_c (melting temperature), c_V (specific heat) and $\alpha < L$ (limit of undercooling/overheating) are positive constants (see Fig. 1). Note that the graph $\Gamma_\theta(\chi) :=$

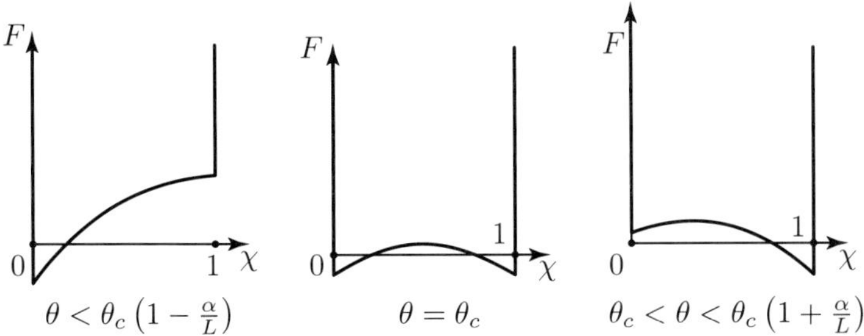

FIGURE 1. Free energy F at different temperatures θ.

$\theta\,\big(I(\chi) - L\,\chi/\theta_c\big) + L\,\chi + \alpha\,\chi(1 - \chi)$ is just the *double-obstacle potential* considered in a number of recent papers. We refer the reader to [2], [3], [8], [12].

The differential inclusion (1.3) then reads

$$\mu\,\chi_t + \lambda'(\chi) - \frac{L}{\theta_c}(\theta - \theta_c) \in -\partial_\chi I(\chi)\,, \tag{1.10}$$

or, equivalently (see Fig. 2),

$$\chi \in [0,1]\,, \quad \Big(\mu\,\chi_t + \lambda'(\chi) - \frac{L}{\theta_c}(\theta - \theta_c)\Big)(z - \chi) \geq 0 \quad \forall z \in [0,1]\,. \tag{1.11}$$

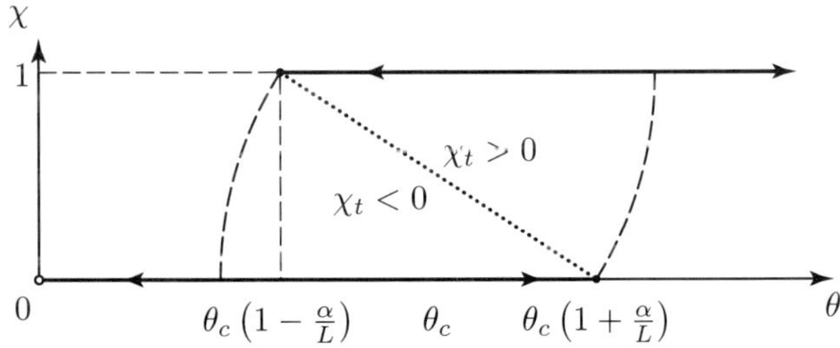

FIGURE 2. A $(\theta - \chi)$-diagram corresponding to (1.11).

It is easy to see that every solution (θ, χ) of (1.1), (1.2), (1.6)–(1.9), (1.11) for which (1.4) holds, formally satisfies the Clausius-Duhem inequality. Indeed, we have for $\chi \in [0,1]$, $I(\chi) \equiv 0$ and $S = c_V \log\theta + \frac{L}{\theta_c}\chi$, hence

$$U_t - \theta S_t = \Big(\lambda'(\chi) - \frac{L}{\theta_c}(\theta - \theta_c)\Big)\chi_t \le 0, \tag{1.12}$$

according to (1.11).

We now introduce the auxiliary variable

$$w(x,t) := \frac{1}{\mu}\int_0^t \left(\frac{L}{\theta_c}(\theta - \theta_c) - \lambda'(\chi)\right)(x,\tau)\,d\tau. \tag{1.13}$$

Then inequality (1.11) may be rewritten as

$$\chi \in [0,1], \quad (\chi_t - w_t)(z - \chi) \ge 0 \quad \forall z \in [0,1]. \tag{1.14}$$

At this point, the notion of hysteresis operators comes into play. Variational inequality (1.14) is known to have a unique solution $\chi \in W^{1,1}(0,T)$ for every $w \in W^{1,1}(0,T)$ and initial condition $\chi(0) = \chi^0 \in [0,1]$. According to [13], [26], [4], [14], the solution operator s_Z of (1.14), called *stop operator* (where the subscript Z stands for the convex constraint $Z = [0,1]$), that is,

$$\chi = s_Z[\chi^0, w], \tag{1.15}$$

is a special and fundamental *hysteresis operator*.

The hysteretic input-output behaviour of the stop operator is illustrated in Fig. 3. Along the upper (lower) threshold line $\chi = 1$, ($\chi = 0$), the process is irre-

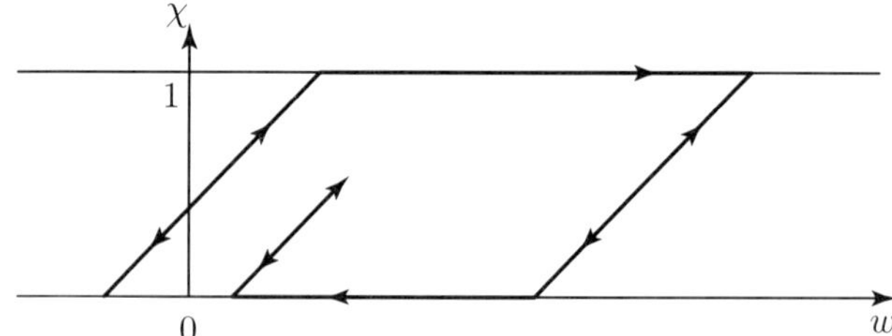

FIGURE 3. A diagram of the stop operator (1.15).

versible and can only move to the right (to the left, respectively), while in between, motions in both directions are admissible. This is similar to Prandtl's model of perfect elastoplasticity, where the horizontal parts of the diagram correspond to plastic yielding and the intermediate lines can be interpreted as linearly elastic trajectories.

Identity (1.15) enables us to eliminate χ from (1.13) and to rewrite the system (1.1)–(1.3) in the form

$$\mu w_t = \frac{L}{\theta_c}(\theta - \theta_c) - \lambda'\big(s_Z[\chi^0, w]\big), \tag{1.16}$$

$$\left(c_V\,\theta + \lambda\,(s_Z[\chi^0, w]) + L\,s_Z[\chi^0, w]\right)_t - \kappa\,\Delta\theta = \psi\,. \tag{1.17}$$

We thus obtain in a natural way a system of equations for an order parameter w and the absolute temperature θ involving hysteresis operators. In the next section we state precisely the problem for more general classes of such systems.

2. Statement of the problem

We consider the system of equations in $\Omega \times]0, T[$

$$\mu\, w_t + f_1[w] + f_2[w]\,\theta = 0\,, \tag{2.1}$$

$$(\theta + F_1[w])_t - \Delta\theta = \psi(x, t, \theta)\,, \tag{2.2}$$

coupled with the initial conditions

$$w(x, 0) = w^0(x)\,, \quad \theta(x, 0) = \theta^0(x)\,, \quad \text{for } x \in \Omega\,, \tag{2.3}$$

and with the Neumann boundary condition

$$\frac{\partial\theta}{\partial n}(x, t) = 0 \quad \text{for } (x, t) \in \partial\Omega \times]0, T[\,, \tag{2.4}$$

where $\partial\theta/\partial n$ denotes the outward normal derivative of θ. We assume that $T > 0$, $\mu > 0$ are given numbers and that $\Omega \subset \mathbb{R}^N$ is a given bounded domain with Lipschitz boundary $\partial\Omega$.

At the first glance, the system (2.1)–(2.4) does not seem to be very difficult from the mathematical point of view. In fact, if f_1, f_2, F_1 were real-valued functions having suitable properties (smoothness, monotonicity, and the like), then this would be true. However, in our case f_1, f_2, F_1 will be hysteresis operators and thus, in particular, non-smooth. Also, when dealing with these operators, we will always have to account for the full history of the inputs which makes the theory less obvious.

We now formulate precisely the assumptions on the mappings f_1, f_2, F_1, ψ.

Hypothesis 2.1. The operators $f_1, f_2 : C[0, T] \to C[0, T]$ are causal and Lipschitz continuous, and f_2 is bounded, i. e., there exists some $K_1 > 0$ such that

$$\left|f_i[w_1](t) - f_i[w_2](t)\right| \le K_1 \max_{0 \le s \le t} |w_1(s) - w_2(s)|\,, \quad i = 1, 2\,, \tag{2.5}$$

$$\left|f_2[w](t)\right| \le K_1\,, \tag{2.6}$$

whenever $w_1, w_2, w \in C[0, T]$ and $t \in [0, T]$.

Hypothesis 2.2. The mapping $F_1 : W^{1,2}(0,T) \to W^{1,2}(0,T)$ is causal, and there exist a constant $K_2 > 0$ and a function $\varphi : \mathbb{R}^+ \to \mathbb{R}^+$ such that

$$\left|\frac{d}{dt}F_1[w](t)\right| \leq K_2\,|\dot{w}(t)| \quad \text{a.e. in }]0,T[\,, \quad \forall\, w \in W^{1,2}(0,T)\,, \tag{2.7}$$

$$\begin{aligned} &\big|F_1[w_1](t) - F_1[w_2](t)\big| \leq \varphi(R)\,\|w_1 - w_2\|_{W^{1,2}(0,t)}\,, \\ &\forall R > 0\,, \quad \forall\, w_1, w_2 \in W^{1,2}(0,T) \quad \text{with} \\ &\max\big\{\|w_i\|_{W^{1,2}(0,T)}\;;\; i = 1,2\big\} \leq R\,, \end{aligned} \tag{2.8}$$

where for $1 \leq p < \infty$ we denote

$$\|w\|_{W^{1,p}(0,t)} := |w(0)| + \Big(\int_0^t |\dot{w}(s)|^p\,ds\Big)^{1/p} \quad \forall\, t \in]0,T]\,. \tag{2.9}$$

Besides, we assume that ψ satisfies

$$\psi_0 := \psi(\,\cdot\,,\,\cdot\,,0) \in L^q(\Omega\times]0,T[)\,, \quad |\psi_\theta(x,t,\theta)| \leq K_2 \text{ a.e.}, \tag{2.10}$$

for some $q > \dfrac{r_N^2}{r_N - 1}$, where $r_N := \max\Big\{2, 1 + \frac{N}{2}\Big\}$.

Hypothesis 2.3. It holds

$$\psi_0(x,t) \geq 0 \quad \text{a.e. in } \Omega\,\times]0,T[\,, \tag{2.11}$$

$$F_1[w](t) \geq 0 \quad \forall\; w \in W^{1,2}(0,T)\,, \quad \forall t \in [0,T]\,, \tag{2.12}$$

and there exist operators $F_2, g : W^{1,2}(0,T) \to W^{1,2}(0,T)$ and a constant $K_3 > 0$ such that

$$0 \leq g[w]_t\,w_t \leq K_3\,w_t^2\,, \tag{2.13}$$

$$F_i[w]_t - f_i[w]\,g[w]_t \leq 0\,, \tag{2.14}$$

for any $w \in W^{1,2}(0,T)$ and a.e. $t \in]0,T[\,$, $i = 1,2$.

Let us mention that property (2.13) is called *piecewise* ([26]) or *local* ([14]) *monotonicity* .

Remark 2.4. The domains of definition of the operators f_i, F_i, g can be extended in a natural way to functions depending on both x and t, as they appear in (2.1), (2.2): it suffices to keep the same symbols and to put

$$f_i[w](x,t) := f_i[w(x,\,\cdot\,)](t) \quad \text{for } x \in \Omega\,,\ t \in]0,T[\,, \tag{2.15}$$

(similarly for F_i and g) for every function w such that $w(x\,,\,\cdot\,)$ belongs to the original domain of definition for a.e. $x \in \Omega$.

Remark 2.5. System (1.16), (1.17) is a special case of (2.1), (2.2) (up to the constants c_V, κ): we simply have to put $g[w] := s_Z[\chi^0, w]$, $f_1[w] := \lambda'(g[w]) + L$, $F_1[w] := \lambda(g[w]) + Lg[w]$, $f_2[w] := -L/\theta_c$, $F_2[w] := -Lg[w]/\theta_c$. Then Hypotheses 2.1–2.3 hold.

Remark 2.6. Equations (2.1), (2.2) may be regarded as a phase-field system for the free energy functional $F = F[w,\theta] := \theta(1-\log\,\theta)+F_1[w]+F_2[w]\,\theta$. In the classical case, the relaxation law (1.3) with χ replaced by w is combined with identities of the form $f_i[w] = \delta_w\, F_i[w]$, $i = 1, 2$, where δ_w denotes the variation with respect to w, in order to make the model comply with the Second Principle of Thermodynamics. However, since hysteresis operators are, as a rule, *non-differentiable*, we cannot hope to have these identities, as the variation $\delta_w\, F_i[w]$ of F_i with respect to w does not exist. In this regard, the situation is entirely different from classical phase-field models. On the other hand, inequality (2.14) is a typical condition which guarantees the thermodynamical consistency of hysteresis operators also in other areas of applications. It is fulfilled, in particular, for operators of the form

$$f_i[w] \;:=\; \mathcal{P}_i\big[g[w]\big]\,, \quad F_i[w] \;:=\; \mathcal{U}_i\big[g[w]\big]\,, \tag{2.16}$$

where $\mathcal{P}_i$ is a hysteresis operator with a *clockwise admissible hysteresis potential* $\mathcal{U}_i$ in the sense of Section 2.5 in [4]. Note that in this case the "dissipation" over a closed cycle (i.e. $u(t_1) = u(t_2)$, $\mathcal{P}_i[u](t_1) = \mathcal{P}_i[u](t_2)$, $\mathcal{U}_i[u](t_1) = \mathcal{U}_i[u](t_2)$) is positive and equal to the integral

$$\int_{t_1}^{t_2} \mathcal{P}_i[u](t)\frac{du(t)}{dt}\,dt\,,$$

which, in geometrical terms, represents the area of the corresponding hysteresis loop, see Fig. 4. A classical example is the *Prandtl-Ishlinskii operator*

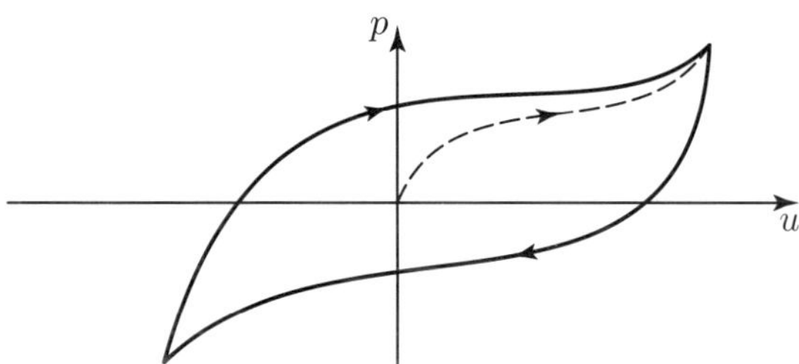

FIGURE 4. Clockwise admissibility for $p = \mathcal{P}_i[u]$

$$\mathcal{P}_i[u] \;:=\; \int_0^\infty h_i(r)\, s_{Z_r}[u]\, dr, \quad \mathcal{U}_i[u] \;:=\; \frac{1}{2}\int_0^\infty h_i(r)\, s_{Z_r}^2[u]\, dr, \tag{2.17}$$

where s_{Z_r} is the stop operator with characteristic set $Z_r = [-r,r]$, and where h_i are given nonnegative density functions.

Also here (1.5) follows from (2.14) if θ is positive. Indeed, if we define the internal energy $U = U[w,\theta] := \theta + F_1[w]$ and the entropy $S = S[w,\theta] := \log\theta - F_2[w]$, then we formally obtain

$$U_t \;-\; \theta\, S_t \;=\; F_1[w]_t \;+\; \theta\, F_2[w]_t \;\leq\; -\,\mu\, w_t\, g[w]_t \;\leq\; 0\,, \tag{2.18}$$

so that (1.5) is satisfied. We shall see later (cf. Theorem 2.10) that Hypothesis 2.3 ensures the positivity of θ, as well. In conclusion, inequality (2.14), which reflects the fundamental energy dissipation properties of the hysteresis operators f_i, takes over the role of the identity $f_i[w] = \delta_w F_i[w]$ which is meaningless here. We should also notice that for constant temperature (2.18) just means that F decreases in time.

Remark 2.7. Thinking in terms of classical models, the system (2.1), (2.2) can be regarded as a phase-field model of *Caginalp type*, see [5], [4] and the references cited there. One can also consider a hysteresis counterpart of the *Penrose-Fife model* of phase transitions (cf. [21], [4], [6], [7], [11], [10], [18], [19], [23]), in which (2.1) has to be replaced by

$$\mu w_t + f_1[w]/\theta + f_2[w] = 0. \tag{2.1$'$}$$

The Penrose-Fife-type model has been studied in detail in [16]. We refer the reader to that paper.

The main results for the system (2.1)–(2.4) are the following.

Theorem 2.8 (Existence). *Let $\Omega \subset \mathbb{R}^N$ be a bounded domain with Lipschitz boundary, and let Hypotheses 2.1, 2.2 hold. Then for every $\theta^0 \in W^{1,2}(\Omega)\cap L^\infty(\Omega)$ and $w^0 \in L^\infty(\Omega)$ problem (2.1)–(2.4) has a solution $(w,\theta) \in (L^\infty(\Omega\times]0,T[))^2$ such that $\theta_t, \Delta\theta \in L^2(\Omega\times]0,T[)$, $w_t \in L^\infty(\Omega\times]0,T[)$, and such that (2.1), (2.2) are satisfied almost everywhere.*

Theorem 2.9 (Uniqueness and Continuous Dependence). *Let the hypotheses of Theorem 2.8 hold. Let $w_i^0 \in L^\infty(\Omega)$, $\theta_i^0 \in W^{1,2}(\Omega)\cap L^\infty(\Omega)$ and $\psi_i : \Omega\times]0,T[\times\mathbb{R} \to \mathbb{R}$, $i = 1,2$, be given functions. Let each of the functions $\psi = \psi_1$, $\psi = \psi_2$ satisfy (2.10), and let there exist a function $d_\psi \in L^2(\Omega\times]0,T[)$ such that for a.e. $(x,t,\vartheta^i) \in \Omega\times]0,T[\times\mathbb{R}$, $i = 1,2$, we have*

$$\left|\psi_1(x,t,\vartheta^1) - \psi_2(x,t,\vartheta^2)\right| \le d_\psi(x,t) + K_2|\vartheta^1 - \vartheta^2|. \tag{2.19}$$

Let (w_1,θ_1), (w_2,θ_2) be solutions to (2.1)–(2.4) corresponding to the data w_1^0, θ_1^0, ψ_1 and w_2^0, θ_2^0, ψ_2, respectively. Then there exists a constant $C > 0$ such that, for all $t \in [0,T]$,

$$\begin{aligned}\int_0^t \int_\Omega |\theta_1 - \theta_2|^2(x,\tau)\,dx\,d\tau \le C\Big[t\Big(\|w_1^0 - w_2^0\|^2_{L^2(\Omega)} \\ + \|\theta_1^0 - \theta_2^0\|^2_{L^2(\Omega)}\Big) + \int_0^t \int_\Omega d_\psi^2(x,\tau)\,dx\,d\tau\Big],\end{aligned} \tag{2.20}$$

$$\begin{aligned}\int_\Omega \|w_1 - w_2\|^2_{W^{1,2}(0,T)}(x)\,dx \le C\Big[\|w_1^0 - w_2^0\|^2_{L^2(\Omega)} \\ + \|\theta_1^0 - \theta_2^0\|^2_{L^2(\Omega)} + \int_0^T \int_\Omega d_\psi^2(x,t)\,dx\,dt\Big].\end{aligned} \tag{2.21}$$

Theorem 2.10 (Thermodynamic Consistency). *Let Hypothesis* 2.3 *and the assumptions of Theorem* 2.8 *be fulfilled. Assume that* $\theta^0(x) \geq \delta$ *a.e. in* Ω *for some constant* $\delta > 0$. *Then there exists some* $\tilde{K} > 0$ *such that the solution* (w, θ) *to (2.1)–(2.4) satisfies* $\theta(x,t) \geq \delta e^{-\tilde{K}t}$ *a.e., that is, (1.4) and (2.18) hold almost everywhere.*

3. Solution operator of the order parameter equation

In this section, we study equation (2.1) with given θ and unknown w separately. Neglecting the space dependence for the moment being, we write it in the form

$$\mu \dot{w} + f_1[w] + f_2[w]\,\theta = 0\,, \quad w(0) = w^0\,. \tag{3.1}$$

We have the following result.

Lemma 3.1 (Existence). *Let Hypothesis* 2.1 *hold, and let* $\theta \in L^1(0,T)$ *and* $w^0 \in \mathbb{R}$ *be given. Then there exists a solution* $w \in W^{1,1}(0,T)$ *of (3.1) such that (3.1) holds a.e., together with the estimate*

$$|\dot{w}(t)| \leq C_1\Big(1 + |w^0| + \|\theta\|_{L^1(0,t)} + |\theta(t)|\Big)\,, \tag{3.2}$$

where $C_1 > 0$ *is a constant independent of* w^0 *and* θ.

Sketch of the proof (for details see [15]*).* For each $t \in]0,T]$ put $C_0[0,t] := \{w \in C[0,t];\, w(0) = w^0\}$, and

$$H_0[w](t) := w^0 - \frac{1}{\mu}\int_0^t \big(f_1[w] + f_2[w]\,\theta\big)(s)\,ds\,. \tag{3.3}$$

The causality and Lipschitz continuity of f_1 and f_2 in Hypothesis 2.1 ensure that the sequence of *successive approximations*

$$w_0(t) := w^0\,, \quad w_k(t) := H_0[w_{k-1}](t) \quad \text{for } k \geq 1\,, t \in [0,T] \tag{3.4}$$

converges uniformly to a fixed point $w \in C_0[0,T]$ of H_0 satisfying (3.1) almost everywhere.

It remains to verify (3.2). To this end, let

$$f^0(t) := f_1[0](t) \tag{3.5}$$

be the image of the constant function $w \equiv 0$ under f_1. From the identity

$$w(t) - w^0 = -\frac{1}{\mu}\int_0^t \big(f_1[w] - f^0 + f^0 + f_2[w]\,\theta\big)(s)\,ds\,, \tag{3.6}$$

and from Hypothesis 2.1, which entails, in particular, that

$$\big|f_1[w] - f^0\big|(t) \leq K_1\Big(|w^0| + \max_{0 \leq s \leq t}\{|w(s) - w^0|\}\Big)\,, \tag{3.7}$$

we infer that

$$|w(t) - w^0| \le \frac{K_1}{\mu} \int_0^t \Big(\max_{0 \le s \le \tau} \{|w(s) - w^0|\} + |w^0| + \frac{1}{K_1}\big|f^0(\tau)\big| + \big|\theta(\tau)\big|\Big)\, d\tau\,. \tag{3.8}$$

Thus,

$$|w(t) - w^0| \le \frac{K_1}{\mu} \int_0^t e^{\frac{K_1}{\mu}(t-\tau)} \Big(|w^0| + \frac{1}{K_1}\big|f^0(\tau)\big| + |\theta(\tau)|\Big)\, d\tau\,, \tag{3.9}$$

and (3.2) follows from (3.1), (3.7) and (3.9). □

Lemma 3.2 (Uniqueness and Continuous Dependence). *Let Hypothesis* 2.1 *hold. Then to every* $M > 0$ *there exists a constant* $C_M > 0$ *such that for every* $\theta_1, \theta_2 \in L^1(0,T)$, $\|\theta_i\|_{L^1(0,T)} \le M$, $i = 1,2$, *the corresponding solutions* w_1, w_2 *of (3.1) with initial conditions* w_1^0, w_2^0, *respectively, satisfy for a. e.* $t \in]0,T[$ *the estimates*

$$|w_1(t) - w_2(t)| \le C_M\Big(|w_1^0 - w_2^0| + \int_0^t |\theta_1 - \theta_2|(s)\, ds\Big)\,, \tag{3.10}$$

$$|\dot w_1(t) - \dot w_2(t)| \le C_M\Big(|w_1^0 - w_2^0| + \int_0^t |\theta_1 - \theta_2|(s)\, ds\Big) \times \big(1 + |\theta_1(t)|\big) + \frac{K_1}{\mu}\big|\theta_1(t) - \theta_2(t)\big|\,. \tag{3.11}$$

Proof. We refer to the proof of Lemma 3.2 in [15]. □

Lemmas 3.1 and 3.2 enable us to introduce the solution operator $\mathcal{P}_p : \mathbb{R} \times L^p(0,T) \to W^{1,p}(0,T)$ of equation (3.1) for every $1 \le p \le \infty$ through the formula

$$w = \mathcal{P}_p[w^0, \theta]\,. \tag{3.12}$$

$\mathcal{P}_p$ is obviously causal, and it satisfies according to Lemmas 3.1, 3.2 for every $t \in [0,T]$ the following estimates.

Proposition 3.3. *Let Hypothesis* 2.1 *hold. Then there exist a constant* $C_2 > 0$ *and a function* $\gamma : \mathbb{R}^+ \to \mathbb{R}^+$ *such that for every* $R > 0$ *and every* (w^0, θ), (w_1^0, θ_1), $(w_2^0, \theta_2) \in \mathbb{R} \times L^p(0,T)$ *and* $t \in [0,T]$ *satisfying* $\max\big\{\|\theta_i\|_{L^p(0,t)};\, i = 1,2\big\} \le R$, *we have*

$$\big\|\mathcal{P}_p[w^0, \theta]\big\|_{W^{1,p}(0,t)} \le C_2\big(1 + |w^0| + \|\theta\|_{L^p(0,t)}\big)\,, \tag{3.13}$$

$$\|\mathcal{P}_p[w_1^0, \theta_1] - \mathcal{P}_p[w_2^0, \theta_2]\|_{W^{1,p}(0,t)} \le \gamma(R)\big(|w_1^0 - w_2^0| + \|\theta_1 - \theta_2\|_{L^p(0,t)}\big)\,. \tag{3.14}$$

4. Existence and thermodynamic consistency

This section is devoted to the proofs of Theorems 2.8 and 2.10 (for the proof of Theorem 2.9, we refer the reader to [15]). Using the operator $\mathcal{P}_p$ introduced in (3.12), we can formally rewrite the problem (2.1)–(2.4) as a single equation,

$$\big(\theta + \mathcal{V}_p[w^0\,,\,\theta]\big)_t - \Delta\theta = \psi(x,t,\theta)\,, \tag{4.1}$$

coupled with initial and boundary conditions (2.3), (2.4), where we have put

$$\mathcal{V}_p[w^0\,,\,\theta](x,t) := F_1\Big[\mathcal{P}_p[w^0(x),\theta(x,\,\cdot\,)]\Big](t)\,, \tag{4.2}$$

for $x\in\Omega$, $t\in[0,T]$ and $p\in[1,\infty]$. The natural domains of definition of $\mathcal{V}_p$ are the spaces $\mathcal{D}_p^t := L^p(\Omega)\times L^p(\Omega\times\,]0,t[)$ for $p\in[1,\infty]$ and $t\in\,]0,T]$. From Proposition 3.3 and Hypothesis 2.2 we see that $\mathcal{V}_p$ maps $\mathcal{D}_p^t$ into $L^p(\Omega;W^{1,p}(0,t))$. Moreover, since for every $p>r$ it holds $\mathcal{V}_r\big|_{\mathcal{D}_p^t} = \mathcal{V}_p$, we may simply write $\mathcal{V}$ in place of $\mathcal{V}_p$, with an implicitly given domain of definition. The operator $\mathcal{V}$ has the following properties.

Theorem 4.1. *Let Hypotheses* 2.1, 2.2 *hold. Then there exist a constant* $C_3>0$ *and a function* $\tilde\psi:\mathbb{R}^+\to\mathbb{R}^+$ *such that for every* $R>0$, $p\in[1,\infty]$, $(w^0,\theta)\in\mathcal{D}_p^T$, $(w_i^0,\theta_i)\in\mathcal{D}_\infty^T$, $i=1,2$ *satisfying* $\max\big\{\|w_i^0\|_{L^\infty(\Omega)}\,,\,\|\theta_i\|_{L^\infty(\Omega\times]0,T[)} : i=1,2\big\}\le R$, *and every* $t\in\,]0,T]$, *it holds*

$$\|\mathcal{V}[w^0\,,\,\theta]_t\|_{L^p(\Omega\times]0,t[)} \le C_3\Big(1+\|w^0\|_{L^p(\Omega)}+\|\theta\|_{L^p(\Omega\times]0,t[)}\Big)\,, \tag{4.3}$$

$$\begin{aligned}\Big\|\mathcal{V}[w_1^0\,,\,\theta_1] &- \mathcal{V}[w_2^0\,,\,\theta_2]\Big\|_{L^2(\Omega;\,L^\infty(0,t))} \\ &\le \tilde\psi(R)\Big(\|w_1^0-w_2^0\|_{L^2(\Omega)}+\|\theta_1-\theta_2\|_{L^2(\Omega\times]0,t[)}\Big)\,.\end{aligned} \tag{4.4}$$

Proof. It suffices to use Lemma 3.1, Proposition 3.3, Hypothesis 2.2 and to integrate over Ω. □

According to the above considerations, we reformulate Theorem 2.8 in the following way.

Theorem 4.2. *Let the hypotheses of Theorem 2.8 hold. Then, for every* $w^0\in L^\infty(\Omega)$ *and* $\theta^0\in L^\infty(\Omega)\cap W^{1,2}(\Omega)$, *there exists* $\theta\in L^\infty(\Omega\times]0,T[)$ *such that* $\theta_t,\Delta\theta\in L^2(\Omega\times]0,T[)$ *and such that the equation*

$$\big(\theta + \mathcal{V}[w^0\,,\,\theta]\big)_t - \Delta\theta = \psi(x,t,\theta) \tag{4.5}$$

is satisfied almost everywhere, together with the initial and boundary conditions (2.3), (2.4).

Note that equation (4.5) does not have the general form considered by Visintin [26], since the operator $\mathcal{V}$ is not piecewise monotone. We present here a simple and direct proof of Theorem 4.2 which is based on well-known properties of linear parabolic equations of the following type

$$u_t - \Delta u + u = g, \quad (x,t) \in \Omega \times]0,T[, \tag{4.6}$$

$$u(x,0) = u^0(x), \quad x \in \Omega, \tag{4.7}$$

$$\frac{\partial u}{\partial n}(x,t) = 0, \quad (x,t) \in \partial\Omega \times]0,T[, \tag{4.8}$$

where $\Omega \subset \mathbb{R}^N$ is a bounded domain with a Lipschitz boundary and g, u^0 are given functions.

The following well-known result can be found in §7 of Chapter III in [17], for the much more general case of variable discontinuous coefficients and anisotropic norms.

Lemma 4.3.

(i) *Let* $p \in [1,\infty[$, $g \in L^p(\Omega\times]0,T[)$ *and* $u^0 \in L^p(\Omega)$. *Then the solution* u *of* (4.6)–(4.8) *satisfies, for every* $t \in [0,T]$,

$$\left|u(\,\cdot\,,t)\right|_p^p \le \left|u^0\right|_p^p + \int_0^t \left|g(\,\cdot\,,\tau)\right|_p^p d\tau, \tag{4.9}$$

where $|\cdot|_p$ *denotes the norm in* $L^p(\Omega)$.

(ii) *Let* r_N *and* q *be as in Hypothesis* 2.2. *Then there exists some* $K_\infty > 0$ *such that for every* $u^0 \in L^\infty(\Omega)$ *and* $g \in L^q(\Omega\times]0,T[)$ *the solution* u *of* (4.6)–(4.8) *satisfies*

$$\|u\|_\infty \le K_\infty \max\left\{|u^0|_\infty, \|g\|_q\right\}, \tag{4.10}$$

where $\|\cdot\|_q$ *denotes the norm in* $L^q(\Omega\times]0,T[)$.

Proof of Theorem 4.2. We construct the solution by successive approximation. We define the sequence $\{\theta^k\}_{k=1}^\infty$ recursively as the solutions to the linear problems

$$\theta_t^k - \Delta\theta^k + \theta^k = \Psi_k(x,t), \quad (x,t) \in \Omega \times]0,T[, \tag{4.11}$$

$$\frac{\partial \theta^k}{\partial n}(x,t) = 0, \quad (x,t) \in \partial\Omega \times]0,T[, \tag{4.12}$$

$$\theta^k(x,0) = \theta^0(x), \quad x \in \Omega, \tag{4.13}$$

where, with $\theta^0(x,t) \equiv \theta^0(x)$,

$$\begin{aligned} \Psi_k(x,t) \; := \; & \theta^{k-1}(x,t) + \psi(x,t,\theta^{k-1}(x,t)) \\ & - \mathcal{V}[w^0, \theta^{k-1}]_t(x,t), \quad k \in \mathbb{N}. \end{aligned} \tag{4.14}$$

It follows via induction from Hypothesis 2.2, Lemma 4.3 **(i)** and Proposition 4.1 that $\theta^k \in L^\infty(0,T;\, L^q(\Omega))$. Besides, there exists some constant $C_4 > 0$, independent of k, such that

$$\Big|\theta^k(\,\cdot\,,t)\Big|_q^q \le C_4\Big(1 + \int_0^t |\theta^{k-1}(\,\cdot\,,\tau)|_q^q\,d\tau\Big), \quad \forall\, k \in \mathbb{N}, \forall\, t \in [0,T]. \tag{4.15}$$

By induction, we therefore have

$$\Big|\theta^k(\,\cdot\,,t)\Big|_q^q \le C_4\, e^{C_4\, t}. \tag{4.16}$$

Hence, $\{\theta^k\}_{k=1}^\infty$ is bounded in $L^q(\Omega\times\,]0,T[)$. From Lemma 4.3 **(ii)**, we infer that $\{\theta^k\}_{k=1}^\infty$ is bounded in $L^\infty(\Omega\times\,]0,T[)$, say

$$\|\theta^k\|_\infty \le C_5. \tag{4.17}$$

Consequently,

$$\|\theta_t^k\|_2^2 + \|\Delta\theta^k\|_2^2 \le C_6, \tag{4.18}$$

for some constant $C_6 > 0$ which is independent of k.

To prove the convergence of $\{\theta^k\}$ as $k \to \infty$, we integrate (4.11) with respect to t and subtract the resulting identities corresponding to $k+1$ and k. This yields

$$\begin{aligned} &(\theta^{k+1} - \theta^k)(x,t) - \Delta\int_0^t (\theta^{k+1} - \theta^k)(x,\tau)\,d\tau \\ &+ \int_0^t (\theta^{k+1} - \theta^k)(x,\tau)\,d\tau \\ = \;& -\Big(\mathcal{V}[w^0,\theta^k] - \mathcal{V}[w^0,\theta^{k-1}]\Big)(x,t) + \int_0^t \Big(\psi\big(x,\tau,\theta^k(x,\tau)\big) \\ & - \psi\big(x,\tau,\theta^{k-1}(x,\tau)\big)\Big)\,d\tau + \int_0^t (\theta^k - \theta^{k-1})(x,\tau)\,d\tau. \end{aligned} \tag{4.19}$$

Multiplying (4.19) by $(\theta^{k+1} - \theta^k)(x,t)$, integrating over Ω, and using (2.10), (4.4), and (4.17), we find that there is some $C_7 > 0$ such that for all $k \in \mathbb{N}$ and $t \in [0,T]$ it holds

$$\begin{aligned} &\int_\Omega |\theta^{k+1} - \theta^k|^2(x,t)\,dx + \frac{d}{dt}\int_\Omega \Big(\Big|\nabla\int_0^t (\theta^{k+1} - \theta^k)(x,\tau)\,d\tau\Big|^2 \\ &+ \Big|\int_0^t (\theta^{k+1} - \theta^k)(x,\tau)\,d\tau\Big|^2\Big)\,dx \\ \le \;& C_7\int_0^t\int_\Omega |\theta^k - \theta^{k-1}|^2(x,\tau)\,dx\,d\tau. \end{aligned} \tag{4.20}$$

Integrating (4.20) from 0 to $\bar{t}$, we obtain that

$$\|\theta^{k+1} - \theta^k\|^2_{L^2(\Omega\times\,]0,\bar{t}[)} \le C_7\int_0^{\bar{t}} \|\theta^k - \theta^{k-1}\|^2_{L^2(\Omega\times\,]0,t[)}\,dt, \tag{4.21}$$

for any $\bar{t} \in [0,T]$. Hence, for all $t \in [0,T]$,

$$\|\theta^{k+1} - \theta^k\|^2_{L^2(\Omega\times]0,t[)} \leq \|\theta^1 - \theta^0\|^2_2 \, \frac{(C_7\, t)^k}{k!} \quad \forall\ k \in \mathbb{N}. \tag{4.22}$$

The series $\sum_{k=0}^{\infty} \Big(\frac{(C_7\, T)^k}{k!}\Big)^{1/2}$ is convergent. This means that $\{\theta^k\}_{k=1}^{\infty}$ is a Cauchy sequence in $L^2(\Omega\times]0,T[)$ which, by (4.17), is bounded in $L^\infty(\Omega\times]0,T[)$. Hence there is some $\theta \in L^\infty(\Omega\times]0,T[)$ such that $\theta^k \to \theta$ strongly in $L^2(\Omega\times]0,T[)$ and weakly-star in $L^\infty(\Omega\times]0,T[)$. From (4.18) it follows that θ_t, $\Delta\theta \in L^2(\Omega\times]0,T[)$, and also, that $\theta^k_t \to \theta_t$, $\Delta\theta^k \to \Delta\theta$, both weakly in $L^2(\Omega\times]0,T[)$. Moreover, by (4.3), (4.4), $\mathcal{V}[w^0, \theta^k]_t \to \mathcal{V}[w^0, \theta]_t$, also weakly in $L^2(\Omega\times]0,T[)$. Passing to the limit in (4.11)–(4.14) we see that θ is a solution of (4.5), (2.3), (2.4). This completes the proof of Theorem 4.2 (and, consequently, of Theorem 2.8). □

Proof of Theorem 2.10. By Hypotheses 2.2, 2.3, we have, for a.e. $(x,t) \in \Omega\times]0,T[$,

$$\begin{aligned} \theta_t - \Delta\theta &= \psi(x,t,\theta) - F_1[w]_t \\ &\geq \psi_0(x,t) - K_2|\theta| - f_1[w]\, g[w]_t \\ &\geq -K_2|\theta| + \frac{g[w]_t}{\mu\, w_t} \cdot f_1[w]\big(f_1[w] + f_2[w]\,\theta\big) \\ &\geq \theta \cdot \Big(- K \operatorname{sign}(\theta) + \frac{g[w]_t}{\mu\, w_t}\, f_1[w]\, f_2[w]\Big). \end{aligned} \tag{4.23}$$

Hypotheses 2.1 and 2.3 imply that

$$0 \leq \frac{g[w]_t}{\mu\, w_t} \leq \frac{K_3}{\mu} \quad \text{a.e.}, \quad \big|f_2[w]\big| \leq K_1, \tag{4.24}$$

and from (3.7), (3.9), it follows that

$$\big|f_1[w](x,t)\big| \leq C_8\Big(1 + \int_0^t \big|\theta(x,\tau)\big|\, d\tau\Big) \quad \text{a.e.}, \tag{4.25}$$

with some constant $C_8 > 0$ which is independent of x.

Since $\theta \in L^\infty(\Omega\times]0,T[)$, (4.23) can be written as

$$\theta_t - \Delta\theta + a(x,t)\,\theta \geq 0 \quad \text{in } \Omega\times]0,T[, \tag{4.26}$$

with some $a \in L^\infty(\Omega\times]0,T[)$. For the sake of definiteness, put

$$\tilde{K} := \|a\|_\infty. \tag{4.27}$$

Let us test (4.26) with $\vartheta(x,t) := \big(\delta\, e^{-\tilde{K}t} - \theta(x,t)\big)^+$. This yields

$$\int_0^t \int_\Omega \Big(\big(\delta\, e^{-\tilde{K}\tau} - \vartheta\big)_t \vartheta - \big|\nabla\vartheta\big|^2 + a\,\big(\delta\, e^{-\tilde{K}\tau} - \vartheta\big)\,\vartheta\Big)\, dx\, d\tau \geq 0, \tag{4.28}$$

whence

$$\int_\Omega \left|\vartheta(x,t)\right|^2 dx + \int_0^t \int_\Omega \Big(|\nabla\vartheta|^2 + \delta(\tilde{K} - a(x,\tau))\, e^{-\tilde{K}\tau}\,\vartheta\Big)\, dx\, d\tau$$
$$\leq \int_\Omega \left|\vartheta(x,0)\right|^2 dx + \tilde{K}\int_0^t \int_\Omega \left|\vartheta(x,\tau)\right|^2 dx\, d\tau\,. \tag{4.29}$$

By assumption, $\vartheta(x,0) \equiv 0$. Gronwall's lemma then implies $\vartheta \equiv 0$, and the proof is complete. □

5. Concluding remarks

Let us recall that the system (2.1)–(2.2) was motivated by the time-relaxed Stefan problem (1.1)–(1.3) for the free energy F defined in (1.6). Apparently, this form for F does not account for interfacial energies and leads to a model with sharp interfaces. In many classical phase field models, interfacial energies are explicitly incorporated by introducing a quadratic gradient term of the order parameter χ, that is, one considers a free energy functional of the form

$$F[\chi,\theta] := F_0(\theta) + \lambda(\chi) + \theta\, I(\chi) - \frac{L}{\theta_c}(\theta - \theta_c)\,\chi + \frac{\gamma}{2}\left|\nabla\chi\right|^2, \tag{5.1}$$

with some constant $\gamma > 0$. Then the variational inequality (1.11) must be replaced by

$$\Big(\mu\,\chi_t - \gamma\,\Delta\chi + \lambda'(\chi) - \frac{L}{\theta_c}(\theta - \theta_c)\Big)(z-\chi) \geq 0\,, \quad \forall\, z \in [0,1]\,, \tag{5.2}$$

together with $\chi \in [0,1]$. Introducing the new variable

$$w(x,t) := \frac{1}{\mu}\int_0^t \left(\frac{L}{\theta_c}(\theta - \theta_c) - \lambda'(\chi) + \gamma\,\Delta\chi\right)(x,\tau)\, d\tau \tag{5.3}$$

in place of (1.13), we still obtain the variational inequality (1.14) for w. However, this time the relaxation law (1.16) has to be replaced by

$$\mu\, w_t - \Delta(s_Z[\chi^0, w]) = \frac{L}{\theta_c}(\theta - \theta_c) - \lambda'(s_Z[\chi^0, w])\,. \tag{5.4}$$

We then have to study the system (1.17), (5.4), together with appropriate initial and boundary conditions. This observation motivates an investigation of phase-field models of the general form

$$\mu\, w_t - \Delta(g[w]) + f_1[w] + f_2[w]\,\theta = 0\,, \tag{5.5}$$

$$(\theta + F_1[w])_t - \Delta\theta = \psi(x,t,\theta)\,, \tag{5.6}$$

coupled with suitable initial and boundary conditions, where the Hypotheses 2.1 to 2.3 hold. At the present time, no existence results seem to be known for systems of this form.

References

[1] Besov, O. V., Il'in, V. P., Nikol'skii, S. M.: *Integral Representation of Functions and Embedding Theorems* (Russian). Moscow, Nauka, 1975.

[2] Blowey, J. F., Elliott, C. M.: Curvature dependent phase boundary motion and double obstacle problems. In: *Degenerate Diffusion*. W.M. Ni, L.A. Peletier, and J.L. Vázquez eds., pp. 19–60, IMA Vol. Math. Appl., **47**, Springer, New York, 1993.

[3] Blowey, J. F., Elliott, C. M.: A phase-field model with double obstacle potential. In: *Motion by Mean Curvature and Related Topics*. G. Buttazzo and A. Visintin eds., pp. 1–22, De Gruyter, Berlin, 1994.

[4] Brokate, M., Sprekels, J.: *Hysteresis and Phase Transitions*. Appl. Math. Sci. **121**, Springer-Verlag, New York, 1996.

[5] Caginalp, G.: "An analysis of a phase field model of a free boundary." Arch. Rational Mech. Anal. **92** (1986), 205–245.

[6] Colli, P., Sprekels, J.: "On a Penrose-Fife model with zero interfacial energy leading to a phase-field system of relaxed Stefan type." Ann. Mat. Pura Appl. (4) **169** (1995), 269–289.

[7] Colli, P., Sprekels, J.: "Stefan problems and the Penrose-Fife phase-field model." Adv. Math. Sci. Appl. **7** (1997), 911–934.

[8] Colli, P., Sprekels, J.: "Global solutions to the Penrose-Fife phase-field model with zero interfacial energy and Fourier law." Adv. Math. Sci. Appl. (to appear).

[9] Frémond, M., Visintin, A.: "Dissipation dans le changement de phase. Surfusion. Changement de phase irréversible." C. R. Acad. Sci. Paris Sér. II Méc. Phys. Chim. Sci. Univers Sci. Terre **301** (1985), 1265–1268.

[10] Kenmochi, N., Niezgódka, M.: "Systems of nonlinear parabolic equations for phase change problems." Adv. Math. Sci. Appl. **3** (1993/94), 89–117.

[11] Klein, O.: "A semidiscrete scheme for a Penrose-Fife system and some Stefan problems in $\mathbb{R}^3$." Adv. Math. Sci. Appl. **7** (1997), 491–523.

[12] Klein, O.: *Existence and Approximation Results for Phase-Field Systems of Penrose-Fife Type and Some Stefan Problems*. Ph.D.-thesis, Humboldt University, Berlin, 1997.

[13] Krasnosel'skii, M. A., Pokrovskii, A. V.: *Systems with Hysteresis*. Springer-Verlag, Heidelberg, 1989 (Russian edition: Nauka, Moscow, 1983).

[14] Krejčí, P.: "Hysteresis, convexity and dissipation in hyperbolic equations." Gakuto Int. Series Math. Sci. & Appl. **8**, Gakkōtosho, Tokyo, 1996.

[15] Krejčí, P., Sprekels, J.: "A hysteresis approach to phase-field models." Nonlinear Analysis TMA (to appear).

[16] Krejčí, P., Sprekels, J.: "Hysteresis operators in phase-field models of Penrose-Fife type." Appl. Math. (to appear).

[17] Ladyzhenskaya, O. A., Solonnikov, V. A., Ural'tseva, N. N.: *Linear and Quasilinear Equations of Parabolic Type*. American Mathematical Society, 1968. (Russian edition: Nauka, Moscow, 1967).

[18] Laurençot, Ph.: "Solutions to a Penrose-Fife model of phase-field type." J. Math. Anal. Appl. **185** (1994), 262–274.

[19] Laurençot, Ph.: "Weak solutions to a Penrose-Fife model for phase transitions." Adv. Math. Sci. Appl. **5** (1995), 117–138.

[20] Mayergoyz, I. D.: *Mathematical Models for Hysteresis.* Springer Verlag, New York, 1991.

[21] Penrose, O., Fife, P.C.: "Thermodynamically consistent models of phase field type for the kinetics of phase transitions." Physica D, **43** (1990), 44–62.

[22] Protter, M. H., Weinberger, H. F.: *Maximum Principle in Differential Equations.* Prentice Hall, Englewood Cliffs, 1967.

[23] Sprekels, J., Zheng, S.: "Global smooth solutions to a thermodynamically consistent model of phase-field type in higher space dimensions." J. Math. Anal. Appl. **176** (1993), 200–223.

[24] Visintin, A.: "Stefan problem with phase relaxation." IMA J. Appl. Math. **34** (1985), 225–245.

[25] Visintin, A.: "Supercooling and superheating effects in phase transitions." IMA J. Appl. Math. **35** (1985), 233–256.

[26] Visintin, A.: *Differential Models of Hysteresis.* Springer-Verlag, New York, 1994.

Pavel Krejčí and Jürgen Sprekels
Weierstrass Institute for Applied
Analysis and Stochastics
Mohrenstrasse 39
D-10117 Berlin, Germany

Progress in Nonlinear Differential Equations
and Their Applications, Vol. 35
© 1999 Birkhäuser Verlag Basel/Switzerland

Regularity for a Class of Sums of Noncommuting Operators

Alessandra Lunardi

Dedicated to Herbert Amann on the occasion of his 60th birthday

Abstract. We give some optimal regularity results in interpolation spaces for operators of the type $A^2 + B^2$, where A and B are noncommuting generators of semigroups in a Banach space. Our main examples are the Heisenberg Laplacian and the Grushin operator, for which we prove regularity results in suitable Hölder and fractional Sobolev spaces.

1. Introduction

It is well known that interpolation theory is a fundamental tool in the study of regularity in PDE's. The monograph [2] is an example of a book on PDE's where interpolation plays a central role.

Also the regularity results of this paper are obtained through the characterization of interpolation spaces.

We start with two linear operators A and B, generators of strongly continuous semigroups e^{tA}, e^{tB} in a Banach space X. Characterizations of the real interpolation spaces $(X, D(A) \cap D(B))_{\theta,p}$ are well known in the case where A and B commute. See e.g. [14, §1.13, 1.14]. Nothing seems to be known in the noncommutative case, except in the trivial case where the commutator is dominated in some sense by A or B.

In this paper we consider the simplest (from our point of view) nontrivial case of noncommuting operators; we assume that the commutator $[A, B]$ has a closed extension Z which in its turn generates a semigroup and commutes with both A and B. We show that

$$(X, D(A) \cap D(B))_{\theta,p} = (X, D(A))_{\theta,p} \cap (X, D(B))_{\theta,p}, \quad 0 < \theta < 1, \ 1 \leq p \leq \infty$$

with equivalent norms, as in the commutative case. We show also that this space contains $(X, D(Z))_{\theta/2,p}$, with continuous embedding. Moreover we prove a similar characterization of $(X, K^2)_{\theta,p}$, where $K^2 = D(A^2) \cap D(AB) \cap D(BA) \cap D(B^2)$.

Examples of such couples of operators are easily found among the realizations A, B of vector fields $\mathcal{A}$, $\mathcal{B}$ in $L^p(\mathbb{R}^n)$, $1 \leq p < \infty$. For instance we may take as $\mathcal{A}$

and $\mathcal{B}$ the derivatives with respect to x and y in the Heisenberg group $\mathbb{R}^3$,

$$\mathcal{A}f(x,y,z) = f_x + 2yf_z, \ \ \mathcal{B}f(x,y,z) = f_y - 2yf_z,$$

where the commutator is $f \mapsto -4f_z$, or else the vector fields in $\mathbb{R}^2$

$$\mathcal{A}f(x,y) = f_x, \ \ \mathcal{B}f(x,y) = xf_y,$$

and in this case the commutator is $f \mapsto f_y$.

Of main interest are the sums of squares $\mathcal{L} = \mathcal{A}^2 + \mathcal{B}^2$. In the first example $\mathcal{L}$ is the Heisenberg Laplacian Δ_H,

$$\Delta_H f(x,y,z) = f_{xx} + f_{yy} + 4yf_{xz} - 4xf_{yz} + 4(x^2+y^2)f_{zz}, \tag{1.1}$$

a well-studied hypoelliptic operator, of great importance in several fields. See e.g. [9]. In the second example $\mathcal{L}$ is the Grushin operator,

$$\mathcal{L}f(x,y) = f_{xx} + x^2 f_{yy},$$

and it is one of the simplest examples of hypoelliptic operators without an underlying Lie group structure.

We take advantage of the homogeneity properties of the vector fields $\mathcal{A}$ and $\mathcal{B}$ to prove that the semigroup $T(t)$ generated by the realization of either the Heisenberg Laplacian or the Grushin operator satisfies

$$\|T(t)\|_{L(L^p, W^{m,p})} \leq \frac{C}{t^{m/2}}, \ \ 0 < t \leq 1, \ \ m \in \mathbb{N}. \tag{1.2}$$

where L^p is the usual Lebesgue space in $\mathbb{R}^3$ or in $\mathbb{R}^2$, and $W^{m,p}$ is the nonisotropic Sobolev space canonically associated to $\mathcal{A}$ and $\mathcal{B}$.

Once (1.2) is established, we use again a general interpolation theorem to characterize $D_L(\theta,p) = (X, D(L))_{\theta,p}$ and $D_L(\theta+1,p) = \{f \in D(L) : Lf \in (X, D(L))_{\theta,p}\}$ where L is the realization of Δ_H in $L^p(\mathbb{R}^3)$ or the realization of the Grushin operator in $L^p(\mathbb{R}^2)$. Such a theorem reads as follows: if $T(t)$ is a semigroup in a Banach space X with generator A, and $E \subset X$ is a Banach space such that $T(t) \in L(X,E)$ for $t > 0$, and $\|T(t)\|_{L(X,E)} \leq Ct^{-m\beta}$ for $0 < t \leq 1$, with $m \in \mathbb{N}$, $\beta \in (0,1)$, then $(X, D(A^m))_{\theta,p} \subset (X,E)_{\theta/\beta,p}$ for $0 < \theta < \beta$. It was proved for $m = 1$ in [1], and its generalization to any m is not difficult.

With the aid of this theorem, $D_L(\theta,p)$ and $D_L(\theta+1,p)$ are characterized as suitable nonisotropic fractional Sobolev spaces. As usual, these characterizations yield optimal regularity results for the solution of

$$\lambda u - Lu = f, \tag{1.3}$$

with $\lambda > 0$, in such fractional Sobolev spaces. In the case of the Heisenberg Laplacian we take also $p = \infty$, working in $BUC_H(\mathbb{R}^3)$ (the space of the uniformly continuous functions with respect to the Heisenberg distance) rather than in $L^\infty(\mathbb{R}^3)$. The corresponding interpolation spaces $D_L(\theta,\infty)$ and $D_L(\theta+1,\infty)$ are characterized as nonisotropic Hölder spaces, and this characterization yields optimal Schauder type estimates for the solution of (1.3). Note that, as in the case of the usual Laplacian, the domain of L in $BUC_H(\mathbb{R}^3)$ is not a nonisotropic BUC^2 space, but it is larger, so that to characterize $D_L(\theta,\infty)$ it is not sufficient to characterize the interpolation spaces between $BUC_H(\mathbb{R}^3)$ and the nonisotropic BUC^2 space; we need the help of the abstract theorem mentioned above.

2. Two interpolation theorems

2.1. Interpolation between domains of noncommuting operators

Let X be a Banach space. We shall be concerned with two densely defined linear operators $A : D(A) \subset X \mapsto X$, $B : D(B) \subset X \mapsto X$, which generate contraction semigroups in X. Therefore, if L denotes either A or B,

$$\rho(L) \supset (0, +\infty), \quad \|R(\lambda, L)\|_{L(X)} \leq 1, \quad \lambda > 0. \tag{2.1}$$

We assume that the commutator $[A, B]$ has a closed extension $Z : D(Z) \subset X \mapsto X$ which, in its turn, generates a contraction semigroup, and that Z commutes with both A and B, in the sense that

$$R(\lambda, Z)R(\mu, A) = R(\mu, A)R(\lambda, Z); \quad R(\lambda, Z)R(\mu, B) = R(\mu, B)R(\lambda, Z), \quad \lambda, \mu > 0. \tag{2.2}$$

this implies that $D(Z)$ is invariant under $R(\mu,A)$ and $R(\mu,B)$, and that $ZR(\mu,A) = R(\mu,A)Z$, $ZR(\mu, B) = R(\mu, B)Z$ on $D(Z)$.

Let $\alpha = (\alpha_1, \dots, \alpha_k)$ be any multi-index. As usual we set $|\alpha| = \sum_{i=1}^k \alpha_i$. Following [14], we define the spaces K^m by

$$K^m = \bigcap_{|\alpha|=m} D(A^{\alpha_1} B^{\alpha_2} \dots A^{\alpha_{k-1}} B^{\alpha_k}).$$

K^m is endowed with the intersection norm

$$\|x\|_{K^m} = \|x\| + \sum_{|\alpha|=m} \|A^{\alpha_1} B^{\alpha_2} \dots A^{\alpha_{k-1}} B^{\alpha_k} x\|.$$

In particular,

$$K^1 = D(A) \cap D(B), \quad K^2 = D(A^2) \cap D(AB) \cap D(BA) \cap D(B^2).$$

Note that if m is even then $K^m \subset D(Z^{m/2})$.

Finally we assume that there is a dense subspace D such that

$$R(\mu, B)R(\lambda, A)(D) \subset K^2, \quad R(\mu, A)R(\lambda, B)(D) \subset K^2, \quad \lambda, \mu > 0. \tag{2.3}$$

We shall use the interpolation functor $(\cdot, \cdot)_{\theta,p}$. If X, Y are Banach spaces such that $Y \subset X$ with continuous embedding, and $0 < \theta < 1$, $1 \leq p \leq \infty$, the interpolation spaces $(X, Y)_{\theta,p}$ are defined by

$$\begin{aligned} &(X, Y)_{\theta,p} = \{x \in X : t \mapsto t^{-\theta} K(t, x, X, Y) \in L^p_*(0, 1)\}, \\ &\|x\|_{(X,Y)_{\theta,p}} = \|t \mapsto t^{-\theta} K(t, x, X, Y)\|_{L^p_*(0,1)} \end{aligned}$$

where

$$K(t, x, X, Y) = \inf_{x=a+b,\, a\in X,\, b\in Y} \|a\|_X + t\|b\|_Y,$$

and if I is an interval contained in $(0,+\infty)$, $L^p_*(I)$ is the Lebesgue space with respect to the measure dt/t.

We recall that if L satisfies (2.1) and $0<\sigma<1$ the space $D_L(\sigma,p)$ is the set of all $x\in X$ such that $\lambda\mapsto\varphi(\lambda)=\lambda^\sigma\|LR(\lambda,L)x\|\in L^p_*(0,+\infty)$. It is endowed with the norm $\|x\|+\|\varphi\|_{L^p_*(0,+\infty)}$, and it coincides with the interpolation space $(X,D(L))_{\sigma,p}$. See [14, §1.14].

Throughout the section we shall use two commutation formulas stated in the next lemma.

Lemma 2.1. *For every $x\in X$ and λ, $\lambda_0>0$, $R(\lambda_0,B)R^2(\lambda,A)R(\lambda_0,B)x\in D(Z)$, and*

$$ZR(\lambda_0,B)R^2(\lambda,A)R(\lambda_0,B)x=[R(\lambda,A),R(\lambda_0,B)]x. \tag{2.4}$$

Moreover $R(\lambda,B)$ maps $D(A)\cap D(Z)$ into $D(A)$, $R(\lambda,A)$ maps $D(B)\cap D(Z)$ into $D(B)$, and for every $\lambda>0$

$$\begin{aligned}&\text{(i)}\quad [A,R(\lambda,B)]=R^2(\lambda,B)Z,\quad \text{on } D(A)\cap D(Z),\\ &\text{(ii)}\quad [B,R(\lambda,A)]=-R^2(\lambda,A)Z,\quad \text{on } D(A)\cap D(Z).\end{aligned} \tag{2.5}$$

Proof. Let $x_n\in D$, $x_n\to x$ as $n\to\infty$, and set $u_n=R(\lambda,A)R(\lambda_0,B)x_n$. Then $u_n\in K^2$, and $(\lambda_0-B)(\lambda-A)u_n=x_n$, so that

$$(\lambda-A)(\lambda_0-B)u_n=Zu_n+x_n,$$

which means that

$$u_n=R(\lambda_0,B)R(\lambda,A)(Zu_n+x_n)=ZR(\lambda_0,B)R(\lambda,A)u_n+R(\lambda_0,B)R(\lambda,A)x_n.$$

Therefore, both sequences

$$y_n=R(\lambda_0,B)R(\lambda,A)u_n\quad\text{and}\quad Zy_n=u_n-R(\lambda_0,B)R(\lambda,A)x_n$$

converge. This implies that $\lim_{n\to\infty}R(\lambda_0,B)R(\lambda,A)u_n=R(\lambda_0,B)R(\lambda,A)u\in D(Z)$, and $ZR(\lambda_0,B)R(\lambda,A)u=u-R(\lambda_0,B)R(\lambda,A)x$. The first statement follows.

Let us prove that the second statement holds. Thanks to (2.4), for every $x\in D(A)\cap D(Z)$, $\lambda>0$, we have

$$\begin{aligned}&\lambda^2R(\lambda_0,B)R^2(\lambda,A)R(\lambda_0,B)Zx\\ &=\lambda^2R(\lambda,A),R(\lambda_0,B)x-\lambda^2R(\lambda_0,B)R(\lambda,A)x\\ &=\lambda AR(\lambda,A),R(\lambda_0,B)x-\lambda R(\lambda_0,B)AR(\lambda,A)x\end{aligned}$$

so that, letting $\lambda\to\infty$,

$$A(\lambda R(\lambda,A)R(\lambda_0,B)x)\quad\text{converges to}\quad R(\lambda_0,B)^2Zx+R(\lambda_0,B)Ax.$$

It follows that $R(\lambda_0,B)x\in D(A)$ and (2.5)(i) is true. (2.5)(ii) may be proved similarly. □

Proposition 2.2. *Let $1 \le p \le \infty$, $0 < \sigma < 1$. Then, under the above assumptions,*

$$(i)\quad D(A) \cap D(B) \subset D_Z(1/2, \infty),$$

$$(ii)\quad D_A(\sigma, p) \cap D_B(\sigma, p) \subset D_Z(\sigma/2, p).$$

Proof. For $x \in D(A) \cap D(B)$ set

$$\varphi(\lambda) = \lambda^4 R(\lambda, A) R^2(\lambda, B) R(\lambda, A)x, \quad \lambda > 0.$$

Then

$$\begin{aligned} \|\varphi(\lambda) - x\| &\le \lambda^3 \|R(\lambda, A)R^2(\lambda, B)(\lambda R(\lambda, A)x - x)\| \\ &\quad + \lambda^2 \|R(\lambda, A)R(\lambda, B)(\lambda R(\lambda, B)x - x)\| \\ &\quad + \lambda \|R(\lambda, A)(\lambda R(\lambda, B)x - x)\| + \|(\lambda R(\lambda, A)x - x)\| \\ &\le 2\lambda(\|Ax\| + \|Bx\|) \end{aligned}$$

and by (2.5)

$$\begin{aligned} \|Z\varphi(\lambda)\| &= \lambda^4 \|R(\lambda, A)R(\lambda, B)x - R(\lambda, B)R(\lambda, A)x\| \\ &= \|\lambda^3 R(\lambda, A)(\lambda R(\lambda, B)x - x) + \lambda^2(\lambda R(\lambda, A)x - x) + \lambda^2 x \\ &\quad -\lambda^3 R(\lambda, B)(\lambda R(\lambda, A)x - x) - \lambda^2(\lambda R(\lambda, B)x - x) - \lambda^2 x\| \\ &\le 2\lambda(\|Ax\| + \|Bx\|). \end{aligned}$$

Taking $\lambda = t^{-1/2}$ we get

$$\begin{aligned} t^{-1/2} K(t, x, X, D(Z)) &\le t^{-1/2}(\|x - \varphi(t^{-1/2})\| + t\|\varphi(t^{-1/2})\| + t\|Z\varphi(t^{-1/2})\|) \\ &\le C(\|Ax\| + \|Bx\| + t\|x\|) \end{aligned}$$

so that $x \in (X, D(Z))_{1/2,\infty}$, with norm less than $C\|x\|_{D(A)\cap D(B)}$.

The same procedure gives (ii). □

Theorem 2.3. *For $0 < \theta < 1$, $\theta \ne 1/2$, $1 \le p \le \infty$, we have*

$$(X, K^1)_{\theta,p} = D_A(\theta, p) \cap D_B(\theta, p),$$

$$\begin{cases} (X, K^2)_{\theta,p} = D_A(2\theta, p) \cap D_B(2\theta, p), \ \theta < 1/2, \\ (X, K^2)_{\theta,p} = \{x \in K^1 \cap D_Z(\theta, p) : Ax, Bx \in D_A(2\theta - 1, p) \cap D_B(2\theta - 1, p)\}, \end{cases}$$

with equivalence of the respective norms.

Proof. The proof is in three steps. We prove that

(i) for $0 < \sigma < 1$, $(X, K^1)_{\sigma,p} = D_A(\sigma, p) \cap D_B(\sigma, p)$,
(ii) $(X, K^2)_{1/2,1} \subset K^1 \subset (X, K^2)_{1/2,\infty}$,
(iii) for $0 < \sigma < 1$, $(K^1, K^2)_{\sigma,p} = \{x \in K^1 \cap D_Z((\sigma + 1)/2, p) : Ax, Bx \in D_A(\sigma, p) \cap D_B(\sigma, p)\}$.

Then the statement will be a consequence of the Reiteration Theorem ([14, §1.10.2]).

Step (i) The embedding $\subset$ is obvious, since K^1 is continuously embedded both in $D(A)$ and in $D(B)$. To prove the converse, for every $x \in D_A(\sigma,p) \cap D_A(\sigma,p)$ we split x as $x = u(\lambda) + x - u(\lambda)$, where

$$u(\lambda) = \lambda^4 R(\lambda^2, Z)R(\lambda, A)R(\lambda, B)x, \quad \lambda > 0. \tag{2.6}$$

Then

$$\begin{aligned} \|u(\lambda) - x\| &\leq \|\lambda^3 R(\lambda^2, Z)R(\lambda, A)(\lambda R(\lambda, B)x - x)\| \\ &\quad + \|\lambda^2 R(\lambda^2, Z)(\lambda R(\lambda, A)x - x)\| + \|\lambda^2 R(\lambda^2, Z)x - x\| \\ &\leq \|BR(\lambda, B)x\| + \|AR(\lambda, A)x\| + \|ZR(\lambda^2, Z)x\|. \end{aligned} \tag{2.7}$$

Let us estimate $\|u(\lambda)\|_{K^1}$. We have

$$\|u(\lambda)\| \leq \|x\|,$$

$$Au(\lambda) = \lambda^4 AR(\lambda, A)R(\lambda, B)R(\lambda^2, Z)x$$

so that

$$\begin{aligned} \|Au(\lambda)\| &\leq \|\lambda^2 AR(\lambda, A)R(\lambda, B)(\lambda^2 R(\lambda^2, Z)x - x)\| \\ &\quad + \|\lambda AR(\lambda, A)(\lambda R(\lambda, B)x - x)\| + \|\lambda AR(\lambda, A)x\| \\ &\leq 2\lambda\|ZR(\lambda^2, Z)x\| + 2\lambda\|BR(\lambda, B)x\| + \lambda\|AR(\lambda, A)x\|. \end{aligned}$$

Similarly, using (2.5)(ii) we get

$$Bu(\lambda) = \lambda^4 R(\lambda^2, Z)R(\lambda, A)BR(\lambda, B)x - \lambda^4 R(\lambda, A)^2 R(\lambda, B)ZR(\lambda^2, Z)x \tag{2.8}$$

so that

$$\|Bu(\lambda)\| \leq \lambda(\|BR(\lambda, B)x\| + \|ZR(\lambda^2, Z)x\|).$$

Therefore,

$$\begin{aligned} &\lambda^\sigma K(\lambda^{-1}, x, X, K^1) \leq \lambda^\sigma(\|x - u(\lambda)\| + \lambda^{-1}\|u(\lambda)\|_{K^1}) \\ &\quad \leq C\lambda^\sigma(\lambda^{-1}\|x\| + \|AR(\lambda, A)x\| + \|BR(\lambda, B)x\| + \|ZR(\lambda^2, Z)x\|). \end{aligned}$$

It follows that $\lambda \mapsto \lambda^\sigma K(\lambda^{-1}, x, X, K^1) \in L^p_*(1, \infty)$ and its norm is less than $C(\|x\| + \|x\|_{D_Z(\sigma/2,p)} + \|x\|_{D_A(\sigma,p)} + \|x\|_{D_B(\sigma,p)}) \leq C'(\|x\|_{D_A(\sigma,p)} + \|x\|_{D_B(\sigma,p)})$. Consequently, $\lambda \mapsto \lambda^{-\sigma} K(\lambda, x, X, K^1) \in L^p_*(0,1)$, with norm less than $C'(\|x\|_{D_A(\sigma,p)} + \|x\|_{D_B(\sigma,p)})$, so that $D_A(\sigma,p) \cap D_B(\sigma,p) \subset (X, K^1)_{\sigma,p}$.

Step (ii) We recall that the embedding $(X, K^2)_{1/2,1} \subset K^1$ is equivalent to the existence of $C > 0$ such that $\|\varphi\|_{K^1} \leq C\|\varphi\|^{1/2}\|\varphi\|^{1/2}_{K^2}$ for all $\varphi \in K^2$. See e.g. [14, §1.10.1]. Such a constant C obviously exists thanks to the inequalities $\|\varphi\|_{D(A)} \leq C\|\varphi\|^{1/2}_{D(A^2)}\|\varphi\|^{1/2}$, $\|\varphi\|_{D(B)} \leq C\|\varphi\|^{1/2}_{D(B^2)}\|\varphi\|^{1/2}$.

To prove the other embedding we modify slightly the procedure of step 1. For every $f \in K^1$ we define $v(\lambda)$ by

$$v(\lambda) = \lambda^6 R^2(\lambda^2, Z)R(\lambda, A)R(\lambda, B)x = \lambda^2 R(\lambda^2, Z)u(\lambda), \tag{2.9}$$

where $u(\lambda)$ is defined by (2.6). Arguing as in estimate (2.7) we get easily

$$\|v(\lambda) - x\| \leq C\lambda^{-1}(\|x\|_{D_Z(1/2,\infty)} + \|Ax\| + \|Bx\|) \leq C'\lambda^{-1}\|x\|_{K^1}, \quad \lambda > 0.$$

Let us estimate $\|v(\lambda)\|_{K^2}$. By (2.5)(i) we have

$$A^2v(\lambda) = \lambda^6 AR(\lambda, A)R(\lambda, B)R^2(\lambda^2, Z)Ax + \lambda^6 AR(\lambda, A)R^2(\lambda, B)ZR^2(\lambda^2, Z)x \tag{2.10}$$

so that

$$\|A^2v(\lambda)\| \le 2\lambda(\|Ax\| + \|x\|_{D_Z(1/2,\infty)}).$$

Moreover, by (2.5)(ii),

$$ABv(\lambda) = \lambda^6 R^2(\lambda^2, Z)AR(\lambda, A)R(\lambda, B)Bx - \lambda^6 AR(\lambda, A)R^2(\lambda, B)ZR^2(\lambda^2, Z)x \tag{2.11}$$

so that, again

$$\|ABv(\lambda)\| \le 2\lambda(\|Bx\| + \|x\|_{D_Z(1/2,\infty)}).$$

Applying twice (2.5)(ii) we get

$$\begin{aligned}B^2v(\lambda) = {} & \lambda^6 R(\lambda,A)BR(\lambda,B)R^2(\lambda^2,Z)Bx - 2\lambda^6 R^2(\lambda,A)R(\lambda,B)ZR^2(\lambda^2,Z)Bx \\ & -2\lambda^6 R^3(\lambda, A)R(\lambda, B)Z^2R^2(\lambda^2, Z)x\end{aligned} \tag{2.12}$$

so that

$$\|B^2v(\lambda)\| \le 2\lambda\|Bx\| + 4\lambda\|Bx\| + 4\lambda\|x\|_{D_Z(1/2,\infty)}.$$

Finally,

$$\|Zv(\lambda)\| \le \lambda\|x\|_{D_Z(1/2,\infty)}.$$

Summing up we get

$$\|v(\lambda)\|_{K^2} \le C(\|x\| + \lambda(\|x\|_{K^1} + \|x\|_{D_Z(1/2,\infty)})) \le C'(\|x\| + \lambda\|x\|_{K^1}).$$

Therefore,

$$K(\lambda, x, X, K^2) \le \|x - v(\lambda^{-1/2})\| + \lambda\|v(\lambda^{-1/2})\|_{K^2} \le C\lambda^{1/2}\|x\|_{K^1}, \quad 0 < \lambda \le 1.$$

This implies that $x \in (X, K^2)_{1/2,\infty}$, with norm less than $C\|x\|_{K^1}$.

Step (iii) Let us denote by $\mathcal{B}$ the right-hand space, endowed with its obvious norm. To prove the embedding $(K^1, K^2)_{\sigma,p} \subset \mathcal{B}$ we remark that A and B are bounded operators both from K^1 to X and from K^2 to K^1. By interpolation, for every $\sigma \in (0,1)$ A and B are bounded from $(K^1, K^2)_{\sigma,p}$ to $(X, K^1)_{\sigma,p} = D_A(\sigma, p) \cap D_B(\sigma, p)$. Therefore for every $x \in (K^1, K^2)_{\sigma,p}$, Ax and Bx are in $D_A(\sigma, p) \cap D_B(\sigma, p)$. Moreover $K^1 \subset D_Z(1/2, \infty)$, $K^2 \subset D(Z)$ so that $(K^1, K^2)_{\sigma,p} \subset (D_Z(1/2, \infty), D(Z))_{\sigma,p} = D_Z((1 + \sigma)/2, \infty)$. The embedding $\subset$ follows.

The embedding $\supset$ is proved by the method of steps (i) and (ii). Indeed, for every x in the right-hand side space we define v by (2.9) and using (2.5)(i) we get

$$\begin{aligned}& Av(\lambda) - Ax \\ & = \lambda^6 R(\lambda, A)R(\lambda, B)R^2(\lambda^2, Z)Ax - Ax + \lambda^6 R(\lambda, A)R^2(\lambda, B)ZR^2(\lambda^2, Z)x \\ & = \lambda^4 R(\lambda, A)R(\lambda, B)R(\lambda^2, Z)(\lambda^2 R(\lambda^2, Z)Ax - Ax) \\ & \quad + \lambda^2 R(\lambda, A)R(\lambda, B)(\lambda^2 R(\lambda^2, Z)Ax - Ax) + \lambda R(\lambda, A)(\lambda R(\lambda, B)Ax - Ax) \\ & \quad + \lambda R(\lambda, A)Ax - Ax + \lambda^6 R(\lambda, A)R^2(\lambda, B)ZR^2(\lambda^2, Z)x\end{aligned}$$

so that

$$\|Av(\lambda) - Ax\| \\ \le C(\|AR(\lambda, A)Ax\| + \|BR(\lambda, B)Ax\| + \|ZR(\lambda^2, Z)Ax\| + \lambda\|ZR(\lambda^2, Z)x\|).$$

Similarly, by (2.5)(ii)

$$\begin{aligned} &Bv(\lambda) - Bx \\ &= \lambda^6 R(\lambda, A)R(\lambda, B)R^2(\lambda^2, Z)Bx - Bx - \lambda^6 R^2(\lambda, A)R(\lambda, B)ZR^2(\lambda^2, Z)x \\ &= \lambda^4 R(\lambda, A)R(\lambda, B)R(\lambda^2, Z)(\lambda^2 R(\lambda^2, Z)Bx - Bx) \\ &\quad + \lambda^2 R(\lambda, A)R(\lambda, B)(\lambda^2 R(\lambda^2, Z)Bx - Bx) + \lambda R(\lambda, A)(\lambda R(\lambda, B)Bx - Bx) \\ &\quad + \lambda R(\lambda, A)Bx - Bx - \lambda^6 R(\lambda, A)^2 R(\lambda, B)ZR^2(\lambda^2, Z)x, \end{aligned}$$

so that, again

$$\|Bv(\lambda) - Bx\| \\ \le C(\|AR(\lambda, A)Bx\| + \|BR(\lambda, B)Bx\| + \|ZR(\lambda^2, Z)Bx\| + \lambda\|ZR(\lambda^2, Z)x\|).$$

Therefore,

$$\|v(\lambda) - x\|_{K^1} \le C(\|x\|_{K^1} + \|AR(\lambda, A)Ax\| + \|AR(\lambda, A)Bx\| + \|BR(\lambda, B)Ax\| \\ + \|BR(\lambda, B)Bx\| + \lambda\|ZR(\lambda^2, Z)x\|\|ZR(\lambda^2, Z)Ax\| + \|ZR(\lambda^2, Z)Bx\|).$$

Moreover, using (2.10), (2.11), (2.12) and arguing as in step (ii), we get

$$\|A^2 v(\lambda)\| + \|ABv(\lambda)\| + \|B^2 v(\lambda)\| + \|Zv(\lambda)\| \\ \le C\lambda(\|AR(\lambda, A)Ax\| + \|AR(\lambda, A)Bx\| + \|BR(\lambda, B)Ax\| + \|BR(\lambda, B)Bx\| \\ + \|ZR(\lambda^2, Z)Ax\| + \|ZR(\lambda^2, Z)Bx\| + \lambda\|ZR(\lambda^2, Z)x\|),$$

so that $\lambda^\sigma K(\lambda^{-1}, x, K^1, K^2) \le \lambda^\sigma \|x - v(\lambda\|_{K^1} + \lambda^{\sigma-1}\|v(\lambda)\|_{K^2} \in L^p_*(1, \infty)$. With the change of variable $\lambda \mapsto \lambda^{-1}$ we get $x \in (K^1, K^2)_{\sigma,p}$, and this finishes the proof. □

Remark 2.4. The technique of Theorem 2.3 may be easily extended to prove embedding results for the spaces (X, K^m). Take for instance $m = 4$ (this will be used later). Then

$$\begin{aligned} &(a) \quad (X, K^4)_{\sigma,p} \subset D_A(4\sigma, p) \cap D_B(4\sigma, p), \quad \sigma \in (0, 1/4), \\ &(b) \quad (X, K^4)_{\sigma,p} \subset \{x \in K^1 \cap D_Z(2\sigma, p) : \\ &\qquad Ax,\ Bx \in D_A(4\sigma - 1, p) \cap D_B(4\sigma - 1, p)\},\ \sigma \in (1/4, 1/2), \\ &(c) \quad (X, K^4)_{\sigma,p} \subset \{x \in K^2 :\ A^2x,\ ABx,\ BAx,\ B^2x \in (X, K^2)_{2\sigma-1,p}\}, \\ &\qquad \sigma \in (1/2, 3/4) \cup (3/4, 1). \end{aligned} \tag{2.13}$$

In fact, it is easy to see that there is $C > 0$ such that $\|f\|_{K^2} \le C\|f\|^{1/2}\|f\|^{1/2}_{K^4}$, so that by the Reiteration Theorem $(X, K^4)_{\sigma,p} \subset (X, K^2)_{2\sigma,p}$ if $\sigma < 1/2$, $(X, K^4)_{\sigma,p} \subset (K^2, K^4)_{2\sigma-1,p}$ if $\sigma > 1/2$. In the first case Theorem 2.3 yields the embeddings (a) and (b). In the latter case we observe that, denoting by L any of the operators A^2, AB, BA, B^2, L is bounded both from K^2 to X and from K^4 to K^2. By interpolation, it is bounded from $(K^2, K^4)_{2\sigma-1,p}$ to $(X, K^2)_{2\sigma-1,p}$, so that (c) holds.

Theorem 2.3 and Remark 2.4 may be extended to the case of n operators $A_1, \dots, A_n$, under the following assumptions: for every i, j the commutator $[A_i, A_j]$ has a closed extension Z_{ij} which commutes with each A_k, and there is a dense subset D such that $R(\mu, A_i)R(\lambda, A_j)(D) \subset K^2$ for every $i, j = 1, \dots, n$, λ, $\mu > 0$.

2.2. An embedding result

The following theorem gives a useful condition to characterize $(X, D(L^m))_{\theta,p}$, where L is the generator of a semigroup and m is any integer. The idea of the proof is taken from [1], where a similar procedure was used for a special L in the case $m = 1$.

We recall that if $D \subset E \subset X$ are Banach spaces and $\beta \in (0,1)$, E is said to belong to $J_\beta(X, D)$ if there is $C > 0$ such that $\|x\|_E \leq C\|x\|_X^{1-\beta}\|x\|_D^\beta$, for all $x \in D$. In this case, the Reiteration Theorem implies that $(X, E)_{\alpha,p} \supset (X, D)_{\alpha\beta,p}$, for every $\alpha \in (0,1)$ and $p \in [1, \infty]$.

Theorem 2.5. *Let $T(t)$ be a semigroup in X with generator $L : D(L) \mapsto X$. Assume that there exists a Banach space $E \subset X$ and $m \in \mathbb{N}$, $0 < \beta < 1$, $\omega \in \mathbb{R}$, $C > 0$ such that*

$$\|T(t)\|_{L(X,E)} \leq \frac{Ce^{\omega t}}{t^{m\beta}}, \quad t > 0, \tag{2.14}$$

and for every $x \in X$, $t \mapsto T(t)x$ is measurable with values in E. Then $E \in J_\beta(X, D(L^m))$, so that $(X, D(L^m))_{\theta,p} \subset (X, E)_{\theta/\beta,p}$, for every $\theta \in (0, \beta)$, $p \in [1, \infty]$.

Proof. It is sufficient to prove the statement in the case $\omega = 0$. Indeed, if $\omega \neq 0$ the semigroup $T(t)e^{-\omega t}$ generated by $L - \omega I$ satisfies (2.14) with $\omega = 0$, and $D((L - \omega I)^m = D(L^m)$ with equivalent norms, so that $(X, D((L - \omega I)^m))_{\theta,p} = (X, D(L^m))_{\theta,p}$ with equivalent norms.

So, let $\omega = 0$. Let $x \in D(L^m)$, $\lambda > 0$ and set $(\lambda I - L)^m x = y$. Then $x = (R(\lambda, L))^m y$ so that

$$x = \frac{(-1)^{m-1}}{(m-1)!}\frac{d^{m-1}}{d\lambda^{m-1}}R(\lambda, L)y = \frac{1}{(m-1)!}\int_0^\infty e^{-\lambda s}s^{m-1}T(s)y\, ds,$$

the integral converges with values in E thanks to (2.14), and

$$\begin{aligned}\|x\|_E &\leq \frac{C}{(m-1)!}\int_0^\infty e^{-\lambda s}s^{m(1-\beta)-1}ds\|y\| = \frac{C\Gamma(m(1-\beta))}{(m-1)!}\lambda^{m\beta-m}\|y\| \\ &= \frac{C\Gamma(m(1-\beta))}{(m-1)!}\lambda^{m\beta-m}\left\|\sum_{r=0}^m \binom{m}{r}\lambda^{m-r}(-1)^r L^r x\right\| \leq C'\sum_{r=0}^m \lambda^{m\beta-r}\|L^r u\|, \ \lambda > 0.\end{aligned}$$

Let us recall that $D(L^r)$ belongs to $J_{m/r}(X, D(L^m))$ (see e.g [14, §1.14.3]) so that there is C such that $\|x\|_{D(L^r)} \leq C\|x\|_{D(L^m)}^{r/m}\|x\|_X^{1-r/m}$. Using such inequalities and then $ab \le (a^2+b^2)/2$ we get

$$\|x\|_E \le C\lambda^{m\beta}(\lambda^{-m}\|u\|_{D(L^m)} + \|u\|), \quad \lambda > 0,$$

so that taking the minimum for $\lambda > 0$

$$\|x\|_E \le C\|u\|^{1-\beta}\|u\|^{\beta}_{D(L^m)}$$

and the statement holds. □

3. Examples

3.1. The Heisenberg Laplacian

The Heisenberg Laplacian in $\mathbb{R}^3$ is the simplest example of a sublaplacian in a stratified Lie group. Such operators have been thoroughly studied in [6] and in [12].

The sum

$$(x, y, z) \oplus (x', y', z') = (x + x', y + y', z + z' + 2(yx' - xy')),$$

gives $\mathbb{R}^3$ the structure of noncommutative Lie group, whose Haar measure is the usual Lebesgue measure. The distance in $\mathbb{R}^3$ associated to the Heisenberg group is $d(p, \widetilde{p}) = |(-\widetilde{p}) \oplus p|$, where $|\cdot|$ is the pseudonorm defined by $|(x, y, z)| = ((x^2 + y^2)^2 + z^2)^{1/4}$.

We shall consider the usual spaces $L^p(\mathbb{R}^3)$, $1 \le p \le \infty$, as well as functional spaces adapted to the distance d and the sum $\oplus$. They will be denoted by the subindex H.

For instance, the space of the "uniformly continuous" functions is defined by

$$BUC_H(\mathbb{R}^3) = \{f \in L^\infty(\mathbb{R}^3) : \lim_{p\to 0} \|f(\cdot \oplus p) - f\|_\infty = 0\}.$$

The "derivatives" D_j, $j = 1, 2, 3$ are defined by

$$D_j f(p) = \lim_{h\to 0} \frac{f(p \oplus he_j) - f(p)}{h}, \tag{3.1}$$

for all functions $f : \mathbb{R}^3 \mapsto \mathbb{R}$ such that such limits exist.

Let $k \in \mathbb{N}$, let $\alpha = (\alpha_1, \dots, \alpha_k) \in \mathbb{N}^k$ be a multi-index with $\alpha_i \in \{1, 2\}$, and set $|\alpha| = k$, $D_\alpha = D_{\alpha_1} D_{\alpha_2} \cdots D_{\alpha_k}$. D_α is said to be a derivative of order k.

It is convenient to define the BUC^k_H spaces by

$$\begin{cases} BUC^k_H(\mathbb{R}^3) = \{f \in BUC_H(\mathbb{R}^3) : \exists D_\alpha f \in BUC_H(\mathbb{R}^3),\ |\alpha| \le k\}, \\ \|f\|_{BUC^k_H} = \|f\|_\infty + [f]_{BUC^k_H} = \|f\|_\infty + \sum\limits_{|\alpha|=k} \|D_\alpha f\|_\infty. \end{cases} \tag{3.2}$$

Due to the equalities $[D_1, D_2] = -4D_3$, if $f \in BUC^{2k}_H(\mathbb{R}^3)$ with $k \in \mathbb{N}$ then f is k times continuously differentiable with respect to z, with bounded derivatives.

Similarly, for $k \in \mathbb{N}$, $p \in [1, \infty)$ the Sobolev spaces $W_H^{k,p}(\mathbb{R}^3)$ are defined by

$$\begin{cases} W_H^{k,p}(\mathbb{R}^3) = \{f \in L^p(\mathbb{R}^3) : \exists D_\alpha f \in L^p(\mathbb{R}^3),\ |\alpha| \le k\}, \\ \|f\|_{W_H^{k,p}} = \|f\|_{L^p} + \sum_{|\alpha|=k} \|D_\alpha f\|_{L^p}. \end{cases} \tag{3.3}$$

Other functional spaces associated to the Heisenberg group are the following. For detailed descriptions of such spaces we refer to [6], [10], [12].

For $0 < \theta < 1$ the Hölder space with respect to the distance d and the sum $\oplus$ is

$$\begin{cases} C_H^\theta(\mathbb{R}^3) = \left\{ f \in L^\infty(\mathbb{R}^3) : [f]_{C_H^\theta} = \sup_{p \in \mathbb{R}^3,\, h \neq 0} \dfrac{|f(p \oplus h) - f(p)|}{|h|^\theta} < +\infty \right\}, \\ \|f\|_{C_H^\theta} = \|f\|_\infty + [f]_{C_H^\theta}. \end{cases} \tag{3.4}$$

For $0 < \theta < 1$, $k \in \mathbb{N}$ we set

$$\begin{cases} C_H^{\theta+k}(\mathbb{R}^3) = \{f \in BUC_H^k(\mathbb{R}^3) : D_\alpha f \in C_H^\theta(\mathbb{R}^3),\ |\alpha| \le k\}, \\ \|f\|_{C_H^{\theta+k}} = \|f\|_{BUC_H^k} + [f]_{C_H^{\theta+k}} = \|f\|_{BUC_H^k} + \displaystyle\sum_{|\alpha|=k} [D_\alpha f]_{C_H^\theta}. \end{cases} \tag{3.5}$$

Now we define the Sobolev space $W_H^{\theta+k,p}(\mathbb{R}^3)$ by

$$W_H^{\theta+k,p}(\mathbb{R}^3) = \left\{ f \in W_H^{k,p}(\mathbb{R}^3) : \int_{\mathbb{R}^3} \int_{\mathbb{R}^3} \frac{|D_\alpha f(\xi \oplus h) - D_\alpha f(\xi)|^p}{|h|^{4+\theta p}} d\xi\, dh < \infty \text{ for } |\alpha| = k \right\},$$

It is endowed with the norm

$$\|f\|_{W_H^{\theta+k,p}} = \|f\|_{L^p} + \sum_{|\alpha|=k} \|D_\alpha f\|_{L^p} + \sum_{|\alpha|=k} \left(\int_{\mathbb{R}^3} \int_{\mathbb{R}^3} \frac{|D_\alpha f(\xi \oplus h) - D_\alpha f(\xi)|^p}{|h|^{4+\theta p}} d\xi\, dh \right)^{1/p}.$$

Let L be the realization of any derivative $D_j, j = 1, 2, 3$, in $X = L^p(\mathbb{R}^3)$, $1 \le p < \infty$, or in $X = BUC_H(\mathbb{R}^3)$. It is not difficult to see that L generates the contraction semigroup

$$(T_j(t)f)(p) = f(p \oplus te_j), \quad f \in X,\ p \in \mathbb{R}^3. \tag{3.6}$$

Let us denote respectively by A, B the realizations of D_1, D_2, in X. Then A and B satisfy the assumptions of §2.1. Indeed for every smooth f the commutator $[D_1, D_2]f$ is nothing but $-4D_3 f$, whereas $D_j D_3 f = D_3 D_j f$ for $j = 1, 2$. Coming back to the realizations A, B it is not hard to see that Z is the realization of $-4D_3$ in X, and that $R(\lambda, Z)$ commutes both with $R(\mu, A)$ and with $R(\mu, B)$. If $X = L^p(\mathbb{R}^3)$, we take $C_0^\infty(\mathbb{R}^3)$ (the space of the smooth functions with compact support) as D. If $X = BUC_H(\mathbb{R}^3)$ we take as D the intersection $\cap_{k \in \mathbb{N}} BUC_H^k(\mathbb{R}^3)$.

So, the results of §2.1 may be applied. As a consequence of Theorem 2.3 and Remark 2.4 we get

Corollary 3.1. *For $0<\theta<1$, $\theta\neq 1/2$, $1\leq p<\infty$, we have*

$$(L^p(\mathbb{R}^3), W^{2,p}_H(\mathbb{R}^3))_{\theta,p} = W^{2\theta,p}_H(\mathbb{R}^3),$$

$$(BUC_H(\mathbb{R}^3), BUC^2_H(\mathbb{R}^3))_{\theta,\infty} = C^{2\theta}_H(\mathbb{R}^3),$$

with equivalence of the respective norms.

Proof. Let $\theta<1/2$, $1\leq p<\infty$. By 2.3, $(L^p(\mathbb{R}^3), W^{2,p}_H(\mathbb{R}^3))_{\theta,p} = D_A(2\theta,p)\cap D_B(2\theta,p)$, and by Proposition 2.2(ii) it is embedded in $D_Z(\theta,p)$. Recall now that if L generates a bounded semigroup e^{tL}, then $D_L(\theta,p)$ consists of the elements $x\in X$ such that $t\mapsto\varphi(t)=t^{-\theta}\|e^{tL}x-x\|\in L^p_*(0,\infty)$, and the norm $x\mapsto\|x\|+\|\varphi\|_{L^p_*(0,\infty)}$ is equivalent to the $D_L(\theta,p)$-norm. Then the equality $(L^p(\mathbb{R}^3), W^{2,p}_H(\mathbb{R}^3))_{\theta,p} = W^{2\theta,p}_H(\mathbb{R}^3)$ follows arguing as in the isotropic case. See e.g. [14, §2.5.1]. The rest of the statement follows. □

Corollary 3.2. *For $0<\theta<1$, $\theta\neq 1/4,\ 1/2,\ 3/4$, $1\leq p<\infty$, we have*

$$(L^p(\mathbb{R}^3), W^{4,p}_H(\mathbb{R}^3))_{\theta,p} \subset W^{4\theta,p}_H(\mathbb{R}^3),$$

$$(BUC_H(\mathbb{R}^3), BUC^4_H(\mathbb{R}^3))_{\theta,\infty} \subset C^{4\theta}_H(\mathbb{R}^3),$$

with continuous embeddings.

The realization L of the Heisenberg Laplacian in $X=L^p(\mathbb{R}^3)$, $1\leq p<\infty$, or in $X=BUC_H(\mathbb{R}^3)$ generates a contraction analytic semigroup $T(t)$ thanks to [6, Thm. 3.1]. An explicit representation formula of $T(t)$ is in [7], but we will not use it. In the case $1<p<\infty$ it is possible to show (see e.g. [6, Cor. 4.13]) that the domain of its generator A coincides with $W^{2,p}_H(\mathbb{R}^3)$, but we will not need this characterization in the sequel. What we will use are the inclusions $W^{2,p}_H(\mathbb{R}^3)\subset D(A)$, $W^{4,p}_H(\mathbb{R}^3)\subset D(A^2)$, if $X=L^p(\mathbb{R}^3)$, $1\leq p<\infty$, and $BUC^2_H(\mathbb{R}^3)\subset D(A)$, $BUC^4_H(\mathbb{R}^3)\subset D(A^2)$, if $X=BUC_H(\mathbb{R}^3)$.

Theorem 3.3. *Let $0<\theta<1$, $\theta\neq 1/2$. If $X=L^p(\mathbb{R}^3)$, $1\leq p<\infty$, then*

$$D_L(\theta,p)=W^{2\theta,p}_H(\mathbb{R}^3),\quad D_L(\theta+1,p)=W^{2\theta+2,p}_H(\mathbb{R}^3),$$

with equivalence of the respective norms. If $X=BUC_H(\mathbb{R}^3)$, then

$$D_L(\theta,\infty)=C^{2\theta}_H(\mathbb{R}^3),\quad D_L(\theta+1,\infty)=C^{2\theta+2}_H(\mathbb{R}^3),$$

with equivalence of the respective norms.

Proof. If $1 < p < \infty$ and $X = L^p(\mathbb{R}^3)$ the first part of the statement is an immediate consequence of Theorem 2.3, since we know from [6, Cor. 4.13] that $D(L) = W_H^{2,p}(\mathbb{R}^3)$ and $D(L^2) = W_H^{4,p}(\mathbb{R}^3)$. But for $p = 1$ only the inclusion $D(L) \supset W_H^{2,1}(\mathbb{R}^3)$ is known to be true (see [6, Thm. 3.8]). So we follow a unified approach which lets us consider both $p = 1$ and $p > 1$ because we do not need any characterization of $D(L)$.

In any case, the embeddings $W_H^{2\theta,p}(\mathbb{R}^3) \subset D_L(\theta,p)$ and $W_H^{2\theta+2,p}(\mathbb{R}^3) \subset D_L(\theta+1,p)$ are consequences of Theorem 2.3, since $W_H^{2,p}(\mathbb{R}^3) \subset D(L)$ and $W_H^{4,p}(\mathbb{R}^3) \subset D(L^2)$.

Let us prove the embeddings $\supset$.

We use a homogeneity argument such as in [6, §3]. For $r > 0$ denote by δ_r the dilation $(x,y,z) \mapsto (rx, ry, r^2 z)$. Note that each D_α is homogeneous of degree $|\alpha|$ with respect to the family of dilations δ_r, that is

$$D_\alpha(f \circ \delta_r) = r^{|\alpha|}(D_\alpha f) \circ \delta_r, \;\; f \in W_H^{|\alpha|,p}(\mathbb{R}^3), \; r > 0.$$

From

$$L(f \circ \delta_r) = r^2 (Lf) \circ \delta_r, \;\; r > 0, \; f \in D(L)$$

it follows easily that for $1 \le p \le \infty$

$$T(t)f = (T(r^2 t)(f \circ \delta_{1/r})) \circ \delta_r, \;\; f \in L^p(\mathbb{R}^3), \; r > 0, \; t > 0.$$

Thanks to the homogeneity of D_α, this implies that for every multi-index α and for $0 < \sigma < 1$

$$\|D_\alpha T(t) f\|_{L^p} = r^{|\alpha|} \|D_\alpha (T(r^2 t)(f \circ \delta_{1/r})) \circ \delta_r\|_{L^p}$$

so that choosing $r = 1/\sqrt{t}$ we get

$$\|D_\alpha T(t) f\|_{L^p} = t^{-|\alpha|/2} \|D_\alpha T(1)\|_{L(L^p)} \|f\|_{L^p}, \;\; t > 0. \tag{3.7}$$

The fact that $D_\alpha T(1) \in L(L^p)$ is again a consequence of homogeneity, see [6, Cor. 3.5, 3.6].

It follows that for every $\omega > 0$, $k \in \mathbb{N}$ there is C such that

$$\|T(t)\|_{L(L^p, W_H^{k,p})} \le \frac{Ce^{\omega t}}{t^{k/2}}, \;\; t > 0,$$

$$\|T(t)\|_{L(BUC_H, BUC_H^k)} \le \frac{Ce^{\omega t}}{t^{k/2}}, \;\; t > 0.$$

We take $k = 4$ and we apply Theorem 2.5 with $X = L^p(\mathbb{R}^3)$, $Y = W_H^{4,p}(\mathbb{R}^3)$, obtaining

$$D_L(\theta,p) = (L^p, D(L^2))_{\theta/2,p} \subset (L^p, W_H^{4,p})_{\theta/2,p} \subset W_H^{2\theta,p},$$

and

$$D_L(\theta+1,p) = (L^p, D(L^2))_{(\theta+1)/2,p} \subset (L^p, W_H^{4,p})_{(\theta+1)/2,p} \subset W_H^{2\theta+2,p},$$

the last inclusions following from Remark 2.4.

Similarly, applying Theorem 2.5 with $X = BUC_H(\mathbb{R}^3)$, $Y = BUC_H^4(\mathbb{R}^3)$, we get the second part of the statement. □

The space $D_L(\theta,p)$ may be characterized also using arguments from harmonic analysis on stratified Lie groups. See [12, Thms. 10, 12].

The characterizations of $D_L(\theta,p)$ and $D_L(\theta+1,p)$ yield optimal regularity results for the hypoelliptic equation (1.3). Indeed, they may be read as follows:

Corollary 3.4. *Let $\lambda>0$, $f\in W^{\alpha,p}_H(\mathbb{R}^3)$ with $0<\alpha<1$ or $1<\alpha<2$, $1\le p<\infty$. Then (1.3) has a unique solution $u\in W^{\alpha+2,p}_H(\mathbb{R}^3)$, and there is $C>0$, independent of f, such that*

$$\|u\|_{W^{\alpha+2,p}_H(\mathbb{R}^3)}\le C\|f\|_{W^{\alpha,p}_H(\mathbb{R}^3)}.$$

If $f\in C^{\alpha}_H(\mathbb{R}^3)$ then (1.3) has a unique solution $u\in C^{\alpha+2}_H(\mathbb{R}^3)$, and there is $C>0$, independent of f, such that

$$\|u\|_{C^{\alpha+2}_H(\mathbb{R}^3)}\le C\|f\|_{C^{\alpha}_H(\mathbb{R}^3)}.$$

Moreover, the characterizations of $D_L(\theta,p)$ and $D_L(\theta+1,p)$ give also several optimal regularity results for the evolution problem

$$\begin{cases} u_t(t,p)-\Delta_H u(t,p)=g(t,p), & 0<t\le T,\ p\in\mathbb{R}^3,\\ u(0,p)=u_0(p), & p\in\mathbb{R}^3, \end{cases}\tag{3.8}$$

through the known regularity theorems for abstract parabolic equations, see e.g. [4], [5], [13], [11, Ch. 4].

4. The Grushin operator

Let now $X=L^p(\mathbb{R}^2)$, $1\le p\le\infty$. The Grushin operator is defined by

$$\mathcal{L}u(x,y)=u_{xx}(x,y)+x^2u_{yy}(x,y).$$

As all the (possibly degenerate) second order symmetric elliptic operators with regular coefficients in $\mathbb{R}^n$, the realization L of $\mathcal{L}$ in $L^p(\mathbb{R}^2)$, $1\le p\le\infty$, generates a contraction semigroup, which is strongly continuous if $p<\infty$, analytic if $1<p<\infty$. See [3, Ex. 3.2.11, Thm. 1.4.1, Thm. 1.4.2].

$\mathcal{L}$ may be written as $\mathcal{A}^2+\mathcal{B}^2$, where

$$\mathcal{A}u(x,y)=u_x(x,y),\ \ \mathcal{B}u(x,y)=xu_y(x,y).$$

So we define the realizations of $\mathcal{A}$ and $\mathcal{B}$ in $X=L^p(\mathbb{R}^2)$, $1\le p<\infty$ by

$$D(A)=\{f\in L^p(\mathbb{R}^2):\ f_x\in L^p(\mathbb{R}^2)\},\ \ Af=f_x,$$
$$D(B)=\{f\in L^p(\mathbb{R}^2):\ xf_y\in L^p(\mathbb{R}^2)\},\ \ Bf=xf_y.$$

They are easily seen to generate the strongly continuous contraction groups

$$e^{tA}f(x,y)=f(x+t,y),\ \ t\in\mathbb{R},$$
$$e^{tB}f(x,y)=f(x,tx+y),\ \ t\in\mathbb{R}.$$

The closed extension of their commutator is the realization Z of $\partial/\partial y$ in X, which generates the translation group

$$e^{tZ}f(x,y)=f(x,y+t),\ \ t\in\mathbb{R}.$$

So, A and B satisfy the assumptions of §2.1 (as D we may take again $C_0^\infty(\mathbb{R}^2)$).

The spaces K^m are nonisotropic weighted Sobolev spaces. Also in this case we define, for $m \in \mathbb{N}$

$$W_G^{m,p}(\mathbb{R}^2) = K^m.$$

In particular,

$$K^2 = \{f \in L^p(\mathbb{R}^2) : \ f_y,\, f_{xx},\, xf_{xy},\, x^2 f_{yy} \in L^p(\mathbb{R}^2)\}.$$

Using the semigroup characterization ([14, §1.13.2]) we see immediately that the interpolation spaces $(X, D(A))_{\theta,p}$, $(X, D(B))_{\theta,p}$, $(X, D(Z))_{\theta,p}$ are given respectively by

$$(X,D(A))_{\theta,p} = \left\{ f \in L^p(\mathbb{R}^2) : [f]_{\theta,p}^p = \int_{\mathbb{R}^3} \frac{|f(x+h,y) - f(x,y)|^p}{|h|^{1+\theta p}} dx\,dy\,dh < \infty \right\},$$

with norm equivalent to

$$f \mapsto \|f\|_{L^p} + [f]_{\theta,p},$$

$$(X, D(B))_{\theta,p} = \left\{ f \in L^p(\mathbb{R}^2) : \ [[f]]_{\theta,p}^p = \int_{\mathbb{R}^3} \frac{(|x|^\theta |f(x,y+h) - f(x,y)|)^p}{|h|^{1+\theta p}} dx\,dy\,dh < \infty \right\},$$

with norm equivalent to

$$f \mapsto \|f\|_{L^p} + [[f]]_{\theta,p},$$

$$(X,D(A))_{\theta,p} = \left\{ f \in L^p(\mathbb{R}^2) : [[[f]]]_{\theta,p}^p = \int_{\mathbb{R}^3} \frac{|f(x,y+h) - f(x,y)|^p}{|h|^{1+\theta p}} dx\,dy\,dh < \infty \right\},$$

with norm equivalent to

$$f \mapsto \|f\|_{L^p} + [[[f]]]_{\theta,p}.$$

So it is convenient to define

$$W_G^{\theta,p}(\mathbb{R}^2) = D_A(\theta,p) \cap D_B(\theta,p),$$

$$W_G^{\theta+1,p}(\mathbb{R}^2) = \{f \in W_G^{1,p}(\mathbb{R}^2) : f_x, xf_y \in D_A(\theta,p) \cap D_B(\theta,p), f_y \in D_Z((\theta+1)/2,p)\},$$

$$W_G^{\theta+2,p}(\mathbb{R}^2) = \{f \in W_G^{2,p}(\mathbb{R}^2) : \ f_y,\, f_{xx},\, xf_{xy},\, x^2 f_{yy} \in D_A(\theta,p) \cap D_B(\theta,p)\},$$

$$W_G^{\theta+3,p}(\mathbb{R}^2) = \{f \in W_G^{3,p}(\mathbb{R}^2) : \ f_y \in D_Z((\theta+1)/2, p),$$

$$f_{xxx},\, f_{xy},\, f_{xxy},\, xf_{yy},\, xf_{yxx},\, x^2 f_{xyy},\, x^3 f_{yyy} \in D_A(\theta,p) \cap D_B(\theta,p)\},$$

endowing them with the intersection norm. Theorem 2.3 and Remark 2.4 imply

$$(L^p(\mathbb{R}^2), W_G^{2,p}(\mathbb{R}^2))_{\theta,p} = W_G^{2\theta,p}(\mathbb{R}^2), \ \ \theta \neq 1/2,$$

$$(L^p(\mathbb{R}^2), W_G^{4,p}(\mathbb{R}^2))_{\theta,p} \subset W_G^{4\theta,p}(\mathbb{R}^2), \ \ \theta \neq 1/4, 1/2, 3/4.$$

Theorem 4.1. *Let $0 < \theta < 1$. Then*

$$D_L(\theta,p) = W_G^{2\theta,p}(\mathbb{R}^2), \ \ D_L(\theta+1,p) = W_G^{2\theta+2,p}(\mathbb{R}^2),$$

with equivalence of the respective norms.

Proof. We follow the proof of Theorem 3.3. The embeddings $W_G^{2\theta,p}(\mathbb{R}^2) \subset D_L(\theta,p)$ and $W_G^{2\theta+2,p}(\mathbb{R}^2) \subset D_L(\theta+1,p)$ are consequences of Theorem 2.3, since $W_G^{2,p}(\mathbb{R}^{2n+1}) \subset D(L)$ and $W_G^{4,p}(\mathbb{R}^{2n+1}) \subset D(L^2)$.

To prove the embeddings $\supset$ we introduce the usual family of dilations, $\delta_r(x,y) = (rx, ry)$. As easily seen, both A and B are homogeneous with degree 1 with respect to the dilations δ_r. The same arguments of the proof of Theorem 3.3 yield that for every $k \in \mathbb{N}$ and $\omega > 0$ there is C such that

$$\|T(t)\|_{L(L^p, W_G^{k,p})} \leq \frac{Ce^{\omega t}}{t^{k/2}}, \quad t > 0.$$

So, as in the proof of Theorem 3.3, we may apply Theorem 2.5 with $X = L^p(\mathbb{R}^2)$, $Y = W_G^{4,p}(\mathbb{R}^2)$, $0 < \theta < 1$, obtaining

$$D_L(\theta,p) \subset (L^p, W_G^{4,p})_{\theta/2,p} \subset W_G^{2\theta,p}(\mathbb{R}^2),$$

$$D_L(\theta+1,p) \subset (L^p, W_G^{4,p})_{(\theta+1)/2,p} \subset W_G^{2\theta+2,p}(\mathbb{R}^2),$$

and the statement follows. □

As in the case of the Heisenberg Laplacian, the characterizations of $D_L(\theta,p)$ and $D_L(\theta+1,p)$ give optimal regularity results for the hypoelliptic equation (1.3).

Corollary 4.2. *Let $\lambda > 0$, $f \in W_G^{\alpha,p}(\mathbb{R}^2)$ with $0 < \alpha < 1$ or $1 < \alpha < 2$, $1 \leq p < \infty$. Then (1.3) has a unique solution $u \in W_G^{\alpha+2,p}(\mathbb{R}^2)$, and there is $C > 0$, independent of f, such that*

$$\|u\|_{W_G^{\alpha+2,p}(\mathbb{R}^2)} \leq C\|f\|_{W_G^{\alpha,p}(\mathbb{R}^{23})}.$$

References

[1] J.O.-O. Adeyeye: "Characterization of real interpolation spaces between the domain of the Laplace operator and $L_p(\Omega)$; Ω polygonal and applications", J. Maths. Pures Appl. **67** (1988), 263–290.

[2] H. Amann: *Linear and Quasilinear Parabolic Problems*, Vol. I, Birkhäuser Verlag (1995).

[3] E.B. Davies: *Heat kernels ad spectral theory*, Cambridge Univ. Press (1990).

[4] G. Da Prato, P. Grisvard: "Sommes d'opérateurs linéaires et équations différentielles opérationelles", J. Maths. Pures Appliquées **54** (1975), 305–387.

[5] G. Di Blasio: "Linear parabolic evolution equations in L^p spaces", Ann. Mat. Pura Appl. **138** (1984), 55–104.

[6] G.B. Folland: "Subelliptic estimates and function spaces on nilpotent Lie groups", Ark. Mat. **13** (1975), 161–207.

[7] B. Gaveau: "Principe de moindre action, propagation de la chaleur et estimées sous elliptiques sur certains groupes nilpotents", Acta Math. **139** (1977), 95–153.

[8] L. Hörmander: "Hypoelliptic differential equations of second order", Acta Math. **119** (1967), 147–171.

[9] R. Howe: "On the role of the Heisenberg group in harmonic analysis", Bull. Amer. Math. Soc. **32** (1980), 821–843.

[10] S.G. Krantz: "Lipschitz spaces on stratified Lie groups", Trans. Amer. Math. Soc. **269** (1982), 29–66.

[11] A. Lunardi: *Analytic Semigroups and Optimal Regularity in Parabolic Problems*, Birkhäuser Verlag, Basel (1995).

[12] K. Saka: "Besov spaces and Sobolev spaces on a nilpotent Lie group", Tohoku Math. J. **31** (1979), 383–437.

[13] E. Sinestrari: "On the abstract Cauchy problem in spaces of continuous functions", J. Math. Anal. Appl. **107** (1985), 16–66.

[14] H. Triebel: *Interpolation Theory, Function Spaces, Differential Operators*, North-Holland, Amsterdam (1978).

Alessandra Lunardi
Dipartimento di Matematica
Università di Parma
Via D'Azeglio 85/A
I-43100 Parma, Italy
`lunardi@prmat.math.unipr.it`

Progress in Nonlinear Differential Equations
and Their Applications, Vol. 35
© 1999 Birkhäuser Verlag Basel/Switzerland

On Stationary Incompressible Navier-Stokes Flows with Fast Decay and the Vanishing Flux Condition

Tetsuro Miyakawa

Dedicated to Professor Herbert Amann on his 60th birthday

Abstract. Stationary incompressible Navier-Stokes flows $\{w, p\}$ are considered on the entire space $\mathbb{R}^n$ ($n \geq 3$) and on a smooth exterior domain Ω in $\mathbb{R}^n$ ($n \geq 3$), where w is a fluid velocity and p the associated pressure. In the case of the entire space the existence is proved for a flow $\{w, p\}$ such that

$$\begin{aligned} |\nabla^j w| &\leq c(1+|x|)^{1-n-j} \qquad (j = 0, 1, 2), \\ |\nabla^k p| &\leq c(1+|x|)^{-n-k} \qquad (k = 0, 1), \end{aligned} \tag{$*$}$$

provided that the external force potential $F = (F_{jk})$ is sufficiently small and decays appropriately as $|x| \to \infty$. For the exterior problem it is shown that if F decays appropriately, then the vanishing flux condition

$$\int_{\partial\Omega} \nu \cdot (T[w,p] - w \otimes w + F) dS = 0,$$

with $T[w,p] = (T_{jk}[w,p])$, $T_{jk}[w,p] = -\delta_{jk}p + \partial_j w_k + \partial_k w_j$, and ν the unit outward normal to the boundary $\partial\Omega$, gives rise to the decay estimates $(*)$ and vice versa.

1. Statement of the results

This paper studies pointwise asymptotic behavior of stationary incompressible Navier-Stokes flows on the entire space $\mathbb{R}^n$ and on a smooth exterior domain in $\mathbb{R}^n$, $n \geq 3$. We first deal with the stationary problem in $\mathbb{R}^n$:

$$-\Delta w + w \cdot \nabla w = f - \nabla p, \qquad \nabla \cdot w = 0, \qquad \lim_{|x| \to \infty} w = 0, \tag{NS}$$

and establish the existence of solutions which decay as $|x| \to \infty$ more rapidly than the fundamental solution of the linear Stokes system, assuming that f is written as

$$f = \nabla \cdot F = (f_1, \ldots, f_n), \qquad f_j = \sum_{k=1}^{n} \partial_k F_{kj}, \qquad \partial_k = \partial/\partial x_k,$$

in terms of some smooth functions $F = \{F_{jk}\}_{j,k=1}^n$. Employing the Stokes fundamental solution $E = \{E_{jk}\}_{j,k=1}^n$ and $Q = \{Q_j\}_{j=1}^n$ ([5, 13]):

$$E_{jk}(x) = \frac{1}{2\omega_{n-1}}\left[\frac{\delta_{jk}}{(n-2)|x|^{n-2}} + \frac{x_j x_k}{|x|^n}\right], \qquad Q_j(x) = \frac{x_j}{\omega_{n-1}|x|^n},$$

with ω_{n-1} the surface area of the unit sphere in $\mathbb{R}^n$, we consider problem (NS) in the form of the convolution integral

$$w = (\nabla E)\cdot(F - w\otimes w). \tag{IE}$$

The associated pressure p, uniquely determined up to addition of constants, is given by

$$p = Q\cdot(\nabla\cdot F - w\cdot\nabla w) = (\nabla Q)\cdot(F - w\otimes w). \tag{P}$$

We know (see [1, 10]) that if $|F| \le c(1+|x|)^{1-n}$ and $|\nabla F| \le c(1+|x|)^{-n}$ with a small constant $c > 0$, then (IE) possesses a unique solution $\{w, p\}$ satisfying

$$|w| \le C(1+|x|)^{2-n}, \qquad |\nabla w| \le C(1+|x|)^{1-n}, \qquad |p| \le C(1+|x|)^{1-n}. \tag{1.1}$$

Notice that in (1.1) the decay rate of w is the same as that of the Stokes fundamental solution E, and so (1.1) looks a natural result. However, we shall show in this paper that (1.1) is not in general the optimal result. In fact, we prove the following, which improves a result given in [7, 8].

Theorem 1.1. (i) *Let $n \ge 3$ and let $F = \{F_{jk}\}_{j,k=1}^n$ satisfy*

$$|F| \le C_0(1+|x|)^{2-2n} \qquad \text{and} \qquad |\nabla F| \le C_1(1+|x|)^{1-2n}. \tag{1.2}$$

If C_0 is sufficiently small, then (IE) *admits a unique solution $\{w, p\}$ such that*

$$\begin{aligned} &|w| \le C(1+|x|)^{1-n}, \quad |\nabla w| \le C(1+|x|)^{-n},\\ &|p| \le C(1+|x|)^{-n}, \qquad |\nabla p| \le C(1+|x|)^{-n-1}. \end{aligned} \tag{1.3}$$

(ii) *Let $n \ge 4$ and suppose that F satisfies condition* (1.2). *Then for an arbitrary solution $\{w, p\}$ of* (IE), *property* (1.1) *implies property* (1.3).

(iii) *In addition to* (1.2), *suppose further that*

$$|\nabla^2 F| \le C_2(1+|x|)^{-2n}. \tag{1.4}$$

Then the function w treated in (i) *and* (ii) *satisfies*

$$|\nabla^2 w| \le C(1+|x|)^{-n-1}. \tag{1.5}$$

When $n = 3$, the asymptotic behavior is studied in [2] for solutions with finite Dirichlet integral, which can be obtained, even for large data, via the Galerkin approximation as given in [5, 6, 13]. It is shown in [2] that there exist solutions which behave like $|x|^{-1+\delta}$, $0 < \delta < 1$. However, the optimality of this decay rate is an open problem. Stability properties of solutions obtained in Theorem 1.1 are discussed in [8, 9] with the aid of the theory of Hardy spaces ([7, 12]).

We next consider the exterior stationary problem:

$$-\Delta w + w\cdot\nabla w = \nabla\cdot F - \nabla p \ \text{ in } \ \Omega, \quad \nabla\cdot w = 0 \ \text{ in } \ \Omega, \\ w|_{\partial\Omega} = w^*, \qquad \lim_{|x|\to\infty} w = 0. \tag{1.6}$$

Here, Ω is an exterior domain in $\mathbb{R}^n$, $n \geq 3$, with smooth boundary $\partial\Omega$; $w^* \in C^2(\partial\Omega)$; and the function $F = \{F_{jk}\}_{j,k=1}^n$ satisfies $|F_{jk}| \leq c(1+|x|)^{1-n}$, $|\nabla F_{jk}| \leq c(1+|x|)^{-n}$. The existence of a stationary flow satisfying (1.1) is shown in [1, 10] when c is sufficiently small. In this paper we shall show that, in contrast to the case of the entire space, solutions to (1.6) with the same decay rate as in Theorem 1.1 can exist under a very restricted condition, *i.e.*, only in the case where the solution $\{w,p\}$ satisfies the *vanishing flux condition*

$$\int_{\partial\Omega} \nu\cdot(T[w,p] - w\otimes w + F)dS = 0, \tag{1.7}$$

where ν is the unit outward normal to $\partial\Omega$ and

$$T[w,p] = (T_{jk}[w,p])_{j,k=1}^n, \qquad T_{jk}[w,p] = -\delta_{jk}p + \partial_j w_k + \partial_k w_j,$$

is the stress tensor associated with the flow $\{w,p\}$. To be more precise, the result is:

Theorem 1.2. (i) *Suppose that*

$$|F| \leq C(1+|x|)^{2-2n} \qquad \text{and} \qquad |\nabla F| \leq C(1+|x|)^{1-2n}. \tag{1.8}$$

If w^*, w and F are small in the sense as will be specified in the proof, then (1.1) *and* (1.7) *together imply the estimates*

$$|w(x)| \leq C(1+|x|)^{1-n}, \qquad |\nabla w(x)| \leq C(1+|x|)^{-n}. \tag{1.9}$$

When $n \geq 4$, (1.1) *and* (1.7) *together imply* (1.9) *irrespective of the size of w^*, w and F.*

(ii) *If F satisfies the additional condition:*

$$|\nabla^2 F| \leq C(1+|x|)^{-2n}, \tag{1.10}$$

then the function w treated in (i) *and the associated pressure p satisfy*

$$|\nabla^2 w(x)| \leq C(1+|x|)^{-n-1}, \\ |\nabla p| \leq C(1+|x|)^{-n-1}, \quad |p| \leq C(1+|x|)^{-n} \tag{1.11}$$

provided that (1.1) *and* (1.7) *hold.*

(iii) *Conversely, if w satisfies* (1.9), *then $\{w,p\}$ satisfies* (1.7).

A result similar to Theorem 1.2 (ii) is given in [4] in the L^p-framework. Since (NS) is written as $\nabla\cdot(T[w,p] - w\otimes w + F) = 0$, application of the divergence theorem shows that condition (1.7) holds in the case of the entire spaces with $\partial\Omega$ replaced by an arbitrary sphere. This fact will play a crucial role in proving Theorem 1.1 (ii). Contrary to the case of the exterior problem, Theorem 1.1 asserts *the existence* of a solution with fast decay in the case of the entire spaces. Stability properties of solutions given in Theorem 1.2 are discussed in [1].

2. Proof of Theorem 1.1

We first prove (i) and (ii) for the velocity field w and the associated pressure p. Estimate (1.3) for the pressure-gradient ∇p will be deduced later.

(i) We invoke the iteration scheme

$$w_{k+1} = (\nabla E)\cdot(F - w_k \otimes w_k) \quad (k \geq 0), \qquad w_0 = (\nabla E)\cdot F, \tag{2.1}$$

to find a uniform (in k) estimate for

$$\|w_k\| = \sup(1+|y|)^{n-1}|w_k(y)|$$

in terms of

$$\|F\| = \sup(1+|y|)^{2n-2}|F(y)|.$$

Since $|\nabla E(x-y)| \leq M|x-y|^{1-n}$ and $n \geq 3$, we see from (1.2) and (2.1) that

$$\begin{aligned} |w_{k+1}| &\leq M(\|F\| + \|w_k\|^2)\int |x-y|^{1-n}(1+|y|)^{2-2n}dy \\ &\leq M_1(\|F\| + \|w_k\|^2)(1+|x|)^{1-n}. \end{aligned}$$

Hence

$$\|w_{k+1}\| \leq M_1(\|F\| + \|w_k\|^2) \tag{2.2}$$

with a constant $M_1 > 0$ independent of k. Here, we assume that

$$\|F\| < \frac{1}{4M_1^2}, \tag{2.3}$$

to get from (2.2)

$$\|w_k\| \leq K \equiv \frac{1-\sqrt{1-4M_1^2\|F\|}}{2M_1} < \frac{1}{2M_1}. \tag{2.4}$$

Similarly, from

$$\begin{aligned} v_k \equiv w_{k+1} - w_k &= -(\nabla E)\cdot(w_k \otimes w_k - w_{k-1}\otimes w_{k-1}) \\ &= -(\nabla E)\cdot(v_{k-1}\otimes w_k + w_{k-1}\otimes v_{k-1}) \qquad (k \geq 1), \end{aligned}$$

we obtain

$$\|v_k\| \leq M_1(\|w_k\| + \|w_{k-1}\|)\|v_{k-1}\| \leq 2M_1K\|v_{k-1}\|,$$

so that

$$\|v_k\| \leq (2M_1K)^{k-1}\|v_1\|.$$

Since $2M_1K < 1$ by (2.4), it follows that a function w exists satisfying

$$\lim_{k\to\infty}\|w_k - w\| = 0, \qquad \|w\| \leq K,$$

and

$$w = (\nabla E)\cdot(F - w\otimes w).$$

This proves the existence of a unique solution w of (IE) satisfying the first estimate of (1.3) under the assumption (2.3).

We next deduce the second estimate of (1.3) for the solution w obtained above. To do so, observe first that the regularity theory for (NS) as given for instance in [1, 5, 13] shows that our solution w is a classical solution of (NS). Thus, we can apply $\partial_\ell = \partial/\partial x_\ell$, $\ell = 1, \dots, n$, to (NS) and obtain

$$-\Delta w_\ell + \nabla p_\ell = \nabla \cdot (F_\ell - w_\ell \otimes w - w \otimes w_\ell),$$
$$\nabla \cdot w_\ell = 0,$$

where $w_\ell = \partial_\ell w$ and $F_\ell = \partial_\ell F$. Hence we can formally write

$$w_\ell = E \cdot \{\nabla \cdot (F_\ell - w_\ell \otimes w - w \otimes w_\ell)\} = (\nabla E) \cdot (F_\ell - w_\ell \otimes w - w \otimes w_\ell). \quad (2.5)$$

Recall that

$$|w| \le K(1 + |x|)^{1-n}, \qquad |w_\ell| \le C(1 + |x|)^{1-n} \quad (2.6)$$

by (1.1) and the foregoing argument. The argument below and (2.6) will justify the representation (2.5) as well. We write

$$w_\ell(x) = u_\ell(x) + v_\ell(x), \quad (2.7)$$

where

$$\begin{aligned} u_\ell(x) &= \int_{|x-y|<|x|/2} (\nabla E)(x - y) G_\ell(y) dy, \\ v_\ell(x) &= \int_{|x-y|>|x|/2} (\nabla E)(x - y) G_\ell(y) dy, \\ G_\ell(y) &= (F_\ell - w_\ell \otimes w - w \otimes w_\ell)(y). \end{aligned}$$

Assume that $|x|$ is sufficiently large and introduce the norms

$$\|w_\ell\| = \sup (1 + |y|)^{n-1} |w_\ell(y)|, \qquad \|F_\ell\| = \sup (1 + |y|)^{2n-2} |F_\ell(y)|.$$

Then (2.6) implies

$$\begin{aligned} |u_\ell(x)| &\le M(\|F_\ell\| + \|w\| \cdot \|w_\ell\|) \int_{|x-y|<|x|/2} |x - y|^{1-n} (1 + |y|)^{2-2n} dy \\ &\le M(\|F_\ell\| + K\|w_\ell\|)|x|^{2-2n} \int_{|x-y|<|x|/2} |x - y|^{1-n} dy \\ &= M'(\|F_\ell\| + K\|w_\ell\|)|x|^{3-2n} \le M''(\|F_\ell\| + K\|w_\ell\|)|x|^{-n}. \end{aligned}$$

To estimate v_ℓ, we integrate it by parts to obtain

$$\begin{aligned} v_\ell(x) &= \int_{|x-y|>|x|/2} (\partial_\ell \nabla E)(x - y)(F - w \otimes w)(y) dy \\ &\quad + \int_{|x-y|=|x|/2} \nu_\ell (\nabla E)(x - y) \cdot (F - w \otimes w)(y) dS_y \\ &\equiv I_{1\ell}(x) + I_{2\ell}(x). \end{aligned}$$

Here and in what follows $\nu = (\nu_1, \dots, \nu_n)$ is the unit outward normal. Applying (2.6) gives

$$\begin{aligned} |I_{1\ell}| &\le M(\|F\| + \|w\|^2) \int_{|x-y|>|x|/2} |x-y|^{-n}(1+|y|)^{2-2n} dy \\ &\le M(\|F\| + K^2)|x|^{-n} \int_{|x-y|>|x|/2} (1+|y|)^{2-2n} dy \\ &\le M'(\|F\| + K^2)|x|^{-n}, \end{aligned}$$

and

$$\begin{aligned} |I_{2\ell}| &\le M(\|F\| + \|w\|^2) \int_{|x-y|=|x|/2} |x-y|^{1-n}(1+|y|)^{2-2n} dS_y \\ &= M'(\|F\| + K^2)|x|^{1-n} \int_{|x-y|=|x|/2} (1+|y|)^{2-2n} dS_y \\ &\le M''(\|F\| + K^2)|x|^{1-n}(1+|x|)^{2-2n}|x|^{n-1} \\ &= M''(\|F\| + K^2)(1+|x|)^{2-2n} \le M''(\|F\| + K^2)(1+|x|)^{-n}. \end{aligned}$$

We thus obtain $|v_\ell| \le |I_{1\ell}| + |I_{2\ell}| \le C|x|^{-n}$; and (2.7) implies

$$|w_\ell(x)| = |\partial_\ell w(x)| \le |u_\ell(x)| + |v_\ell(x)| \le C|x|^{-n}$$

for large $|x|$. Estimate for the pressure p is deduced from (P) in exactly the same way as estimating ∇w. This proves (i) except for the estimate of ∇p, which will be deduced later.

(ii) Taking $R > 0$ to be fixed later, we represent w on $D = \{|x| > R\}$ in the form

$$w = (\nabla E) \cdot (\widetilde{F} - \widetilde{w} \otimes \widetilde{w}) + w_0, \tag{2.8}$$

where

$$\widetilde{w}(x) = \begin{cases} w(x) & (|x| > R) \\ 0 & (|x| < R) \end{cases}$$

and

$$\begin{aligned} w_0 &= \int_{|y|=R} w \cdot T[E, Q] \cdot \nu dS_y + \int_{|y|=R} E \cdot \nu \cdot (T[w,p] - w \otimes w + F) dS_y \\ &\equiv I_1 + I_2. \end{aligned}$$

Applying the divergence theorem to $\nabla \cdot (T[w,p] - w \otimes w + F) = 0$ gives

$$\int_{|y|=R} \nu \cdot (T[w,p] - w \otimes w + F) dS_y = 0.$$

We thus obtain

$$I_2 = \int_{|y|=R} \widetilde{E}(x,y) \cdot \nu_y \cdot (T[w,p] - w \otimes w + F)(y) dS_y$$

with

$$\widetilde{E}(x,y) = E(x-y) - E(x) = \int_0^1 \frac{d}{d\theta} E(x-\theta y)d\theta. \tag{2.9}$$

Since $|\widetilde{E}(x,y)| \leq C|x-y|^{1-n}$ and $|\nabla_x \widetilde{E}(x,y)| \leq C|x-y|^{-n}$, it follows that

$$|w_0| = O(|x|^{1-n}), \qquad |\nabla w_0| = O(|x|^{-n}) \qquad \text{as} \quad |x| \to \infty.$$

Now let

$$[w]_{\delta,D} = \sup_D (1+|y|)^{1+\delta}|w(y)| \quad (0 \leq \delta < 1),$$

and consider this time the linear iteration scheme on D:

$$v_{k+1} = (\nabla E)\cdot(\widetilde{F} - \widetilde{w}\otimes\widetilde{v}_k) + w_0, \qquad v_0 = (\nabla E)\cdot\widetilde{F} + w_0.$$

Since $n \geq 4$, invoking the estimate

$$\int |x-y|^{1-n}(1+|y|)^{-n-\delta}dy \leq C_\delta(1+|x|)^{1-n},$$

which holds provided $\delta > 0$, we see by a suitable choice of $\delta > 0$ that

$$\|v_{k+1}\|_D \leq M([F]_D + [w]_{\delta,D}\|v_k\|_D) + \|w_0\|_D,$$

and

$$\|v_{k+1} - v_k\|_D \leq M[w]_{\delta,D}\|v_k - v_{k-1}\|_D,$$

where

$$\|v\|_D = \sup_D (1+|y|)^{n-1}|v(y)|, \qquad [F]_D = \sup_D (1+|y|)^{2n-2}|F(y)|.$$

Since $n \geq 4$, it follows from (1.1) that a constant $R > 0$ can be chosen so that

$$\max([w]_{\delta,D}, [w]_{0,D}) < \frac{1}{2M}. \tag{2.10}$$

Thus,

$$\|v_{k+1}\|_D \leq \frac{1}{2}\|v_k\|_D + M[F]_D + \|w_0\|_D$$

and

$$\|v_{k+1} - v_k\|_D \leq \frac{1}{2}\|v_k - v_{k-1}\|_D.$$

Hence, a constant $K > 0$ exists so that $\|v_k\|_D \leq K$ for all k, and we can find a function v defined on D such that $\|v\|_D \leq K$, $\lim_{k\to\infty} \|v_k - v\|_D = 0$, and

$$v = (\nabla E)\cdot(\widetilde{F} - \widetilde{w}\otimes\widetilde{v}) + w_0 \qquad \text{in} \quad D.$$

This, together with (2.8), implies $v - w = -(\nabla E)\cdot[\widetilde{w}\otimes(\widetilde{v} - \widetilde{w})]$, so we get from (2.10)

$$[v-w]_{0,D} \leq M[w]_{0,D}[v-w]_{0,D} < \frac{1}{2}[v-w]_{0,D}.$$

Hence, $v = w$ on D, and this proves estimate (1.3) for w. Estimates (1.3) for ∇w and p are deduced in the same way as in the proof of (i); so the details are omitted.

It remains to deduce estimate (1.3) for ∇p. We shall show that if

$$|w| \leq C(1+|x|)^{1-n}, \qquad |\nabla w| \leq C(1+|x|)^{-n}, \tag{2.11}$$

then

$$|\nabla p| \le C(1+|x|)^{-n-1}. \tag{2.12}$$

Taking the divergence of the Navier-Stokes system gives

$$-\Delta p = \partial_j\partial_k(w_j w_k) - \partial_j\partial_k F_{jk} = \partial_j w_k \partial_k w_j - \partial_j\partial_k F_{jk}.$$

Hereafter we will use the summation convention. Let

$$N(x) = \frac{1}{(n-2)\omega_{n-1}|x|^{n-2}}$$

be the fundamental solution of $-\Delta$. Then, for $\ell = 1,\cdots,n$,

$$\begin{aligned}
\partial_\ell p &= (\partial_\ell N)\cdot(\partial_j\partial_k(w_j w_k) - \partial_j\partial_k F_{jk}) = (\partial_k\partial_\ell N)\cdot(w_j\partial_j w_k - \partial_j F_{jk})\\
&= c(w_j\partial_j w_k - \partial_j F_{jk}) + \lim_{\varepsilon\to 0}\int_{|x-y|>\varepsilon}(\partial_k\partial_\ell N)\cdot(w_j\partial_j w_k - \partial_j F_{jk})dy\\
&= c(w_j\partial_j w_k - \partial_j F_{jk}) + \int_{|x-y|>|x|/2}(\partial_k\partial_\ell N)\cdot(w_j\partial_j w_k - \partial_j F_{jk})dy\\
&\quad + \lim_{\varepsilon\to 0}\int_{\varepsilon<|x-y|<|x|/2}(\partial_k\partial_\ell N)\cdot(w_j\partial_j w_k - \partial_j F_{jk})dy\\
&= c(w_j\partial_j w_k - \partial_j F_{jk}) + \int_{|x-y|=|x|/2}(\partial_k\partial_\ell N)\cdot\nu_j(w_j w_k - F_{jk})dS_y\\
&\quad + \int_{|x-y|>|x|/2}(\partial_j\partial_k\partial_\ell N)\cdot(w_j w_k - F_{jk})dy\\
&\quad + \left(\lim_{\varepsilon\to 0}\int_{|x-y|=\varepsilon} + \int_{|x-y|=|x|/2}\right)\nu_k(\partial_\ell N)\cdot(w_j\partial_j w_k - \partial_j F_{jk})dS_y\\
&\quad + \int_{|x-y|<|x|/2}(\partial_\ell N)\cdot(\partial_k w_j\partial_j w_k - \partial_j\partial_k F_{jk})dy\\
&\equiv J_0 + J_1 + J_2 + J_3 + J_4.
\end{aligned}$$

Due to (1.2) and (2.11), J_0, J_1, J_2 and J_4 are estimated as

$$|J_0| \le M(1+|x|)^{1-2n} \le M(1+|x|)^{-n-1},$$

$$
\begin{aligned}
|J_1| &\le M\int_{|x-y|=|x|/2}|x-y|^{-n}(1+|y|)^{2-2n}dS_y\\
&\le M|x|^{-n}(1+|x|)^{2-2n}|x|^{n-1}\le C(1+|x|)^{1-2n}\\
&\le C(1+|x|)^{-n-1},\\
|J_2| &\le M\int_{|x-y|>|x|/2}|x-y|^{-n-1}(1+|y|)^{2-2n}dy\le M|x|^{-n-1},\\
|J_4| &\le M\int_{|x-y|<|x|/2}|x-y|^{1-n}(1+|y|)^{-2n}dy\\
&\le M(1+|x|)^{-2n}|x|\le C(1+|x|)^{1-2n}\le C(1+|x|)^{-n-1}.
\end{aligned}
$$

Finally, we write

$$J_3 = J_{31} + J_{32}$$

where

$$
\begin{aligned}
J_{31} &= c_{k\ell}(w_j\partial_j w_k-\partial_j F_{jk})(x), \qquad c_{k\ell}=-\int_{|\omega|=1}\omega_k(\partial_\ell N)(\omega)dS_\omega,\\
J_{32} &= \int_{|x-y|=|x|/2}\nu_k(\partial_\ell N)(x-y)\cdot(w_j\partial_j w_k-\partial_j F_{jk})(y)dS_y.
\end{aligned}
$$

Direct calculation gives

$$
\begin{aligned}
|J_{31}| &\le C(1+|x|)^{1-2n}\le C(1+|x|)^{-n-1}\\
|J_{32}| &\le M\int_{|x-y|=|x|/2}|x-y|^{1-n}(1+|y|)^{1-2n}dS_y\\
&\le M|x|^{1-n}(1+|x|)^{1-2n}|x|^{n-1}=M(1+|x|)^{1-2n}\\
&\le M(1+|x|)^{-n-1}.
\end{aligned}
$$

Collecting terms gives (2.12), and the proofs of (i) and (ii) are complete.

(iii) We deduce (1.5), assuming (1.2), (1.3) and (1.4). Consider the integral equation

$$
\begin{aligned}
\nabla^2 w &= (\nabla^2 E)\cdot(\nabla\cdot F-w\cdot\nabla w)\\
&\equiv \lim_{\varepsilon\to 0}\int_{|x-y|>\varepsilon}(\nabla^2 E)\cdot(\nabla\cdot F-w\cdot\nabla w)dy \qquad (2.13)\\
&\quad +c(\nabla\cdot F-w\cdot\nabla w),
\end{aligned}
$$

where c is some constant. We first show that $\nabla^2 w\in L^\infty$. Since $\nabla^2 E$ is a Calderón-Zygmund kernel ([11]), it follows from (2.13) that $\nabla^2 w\in L^p$ for all $1<p<\infty$.

Hence, the relation

$$\begin{aligned}\nabla^3 w &= (\nabla^2 E)\cdot(\nabla^2 F - \nabla w\cdot\nabla w - w\cdot\nabla^2 w)\\ &\equiv \lim_{\varepsilon\to 0}\int_{|x-y|>\varepsilon}(\nabla^2 E)\cdot(\nabla^2 F - \nabla w\cdot\nabla w - w\cdot\nabla^2 w)dy\\ &\quad + c(\nabla^2 F - \nabla w\cdot\nabla w - w\cdot\nabla^2 w),\end{aligned}$$

where c is some constant, shows that $\nabla^3 w \in L^p$ for all $1 < p < \infty$. We thus conclude that $\nabla^2 w \in L^\infty$ by the Gagliardo-Nirenberg inequality

$$\|\nabla^2 w\|_\infty \le M\|\nabla^2 w\|_p^{1-n/p}\|\nabla^3 w\|_p^{n/p} \qquad (n<p<\infty),$$

where $\|\cdot\|_p$ is the L^p-norm.

Now we return our attention to (2.13). Observe first that since

$$\nabla^2 w = (\nabla E)\cdot(\nabla^2 F - \nabla w\cdot\nabla w - w\cdot\nabla^2 w),$$

conditions (1.2), (1.3), (1.4) and the boundedness of $\nabla^2 w$ together imply

$$|\nabla^2 w| \le C(1+|x|)^{2-n}.$$

Thus,

$$|w\cdot\nabla^2 w| \le C(1+|x|)^{3-2n}. \tag{2.14}$$

We next write (2.13) as

$$\begin{aligned}(\nabla^2 w)(x) &= c(\nabla\cdot F - w\cdot\nabla w) + \int_{|x-y|>|x|/2}(\nabla^3 E)\cdot(F - w\otimes w)dy\\ &\quad + \lim_{\varepsilon\to 0}\int_{\varepsilon<|x-y|<|x|/2}(\nabla^2 E)\cdot(\nabla\cdot F - w\cdot\nabla w)dy\\ &\quad + \int_{|x-y|=|x|/2}\nu\cdot(\nabla^2 E)\cdot(F - w\otimes w)dS_y\\ &= H_0 + H_1 + H_2 + H_3.\end{aligned}$$

Direct calculation gives

$$|H_0| \le C(1+|x|)^{-n-1}, \qquad |H_1| \le C(1+|x|)^{-n-1}, \qquad |H_3| \le C(1+|x|)^{-n-1}.$$

It thus suffices to deduce the desired estimate for H_2. We write

$$H_2 = H_{21} + H_{22} + H_{23},$$

where

$$\begin{aligned}H_{21} &= c(\nabla\cdot F - w\cdot\nabla w), \qquad c = -\int_{|\omega|=1}\omega(\nabla E)(\omega)dS_\omega,\\ H_{22} &= \int_{|x-y|=|x|/2}\nu(\nabla E)(x-y)(\nabla\cdot F - w\cdot\nabla w)(y)dS_y,\\ H_{23} &= \int_{|x-y|<|x|/2}(\nabla E)(x-y)(\nabla^2 F - w\cdot\nabla^2 w - \nabla w\cdot\nabla w)(y)dy.\end{aligned}$$

We easily see by (1.2) and (1.4) that

$$|H_{21}| \leq C(1+|x|)^{1-2n} \leq C(1+|x|)^{-n-1},$$

and

$$|H_{22}| \leq C|x|^{1-n}(1+|x|)^{1-2n}|x|^{n-1} = C(1+|x|)^{1-2n} \leq C(1+|x|)^{-n-1}.$$

Furthermore, it follows from (1.2), (1.4) and the boundedness of $\nabla^2 w$ that

$$|H_{23}| \leq C(1+|x|)^{1-2n} + C(1+|x|)^{2-n} \leq C(1+|x|)^{2-n}.$$

Hence, $|\nabla^2 w| \leq C(1+|x|)^{2-n}$, so $|w \cdot \nabla^2 w| \leq C(1+|x|)^{3-2n}$, and

$$|H_{23}| \leq C(1+|x|)^{1-2n} + C(1+|x|)^{4-2n} \leq C(1+|x|)^{4-2n}.$$

We thus have

$$|\nabla^2 w(x)| \leq C_1(1+|x|)^{-n-1} + C_2(1+|x|)^{4-2n}. \tag{2.15}$$

This proves (1.5) for $n \geq 5$. When $n = 3, 4$, we have $|w \cdot \nabla^2 w| \leq C(1+|x|)^{5-3n}$ by (2.15). This implies $|H_{23}| \leq C(1+|x|)^{6-3n}$, and therefore

$$|\nabla^2 w| \leq C(1+|x|)^{-n-1} + C(1+|x|)^{6-3n}. \tag{2.16}$$

This proves (1.5) for $n = 4$. When $n = 3$, we have $|w \cdot \nabla^2 w| \leq C(1+|x|)^{7-4n}$ by (2.16). This implies $|H_{23}| \leq C(1+|x|)^{8-4n} = C(1+|x|)^{-4}$, and therefore

$$|\nabla^2 w| \leq C(1+|x|)^{-4}.$$

This proves (1.5) for $n = 3$, and the proof of (iii) is complete. □

3. Proof of Theorem 1.2

Without loss of generality we may assume in this section that

$$\overline{\Omega} \subset \{x : |x| > 1\}. \tag{3.1}$$

(i) We again invoke the Stokes fundamental solution $\{E, Q\}$ in order to represent the solution w in the form

$$\begin{aligned} w &= E \cdot (\nabla \cdot (F - w \otimes w)) + \int_{\partial\Omega} E \cdot \nu \cdot T[w,p] dS \\ &\quad + \int_{\partial\Omega} w^* \cdot T[E,Q] \cdot \nu dS \\ &= (\nabla \cdot E) \cdot (F - w \otimes w) \\ &\quad + \int_{\partial\Omega} E \cdot \nu \cdot (T[w,p] - w^* \otimes w^* + F) dS \\ &\quad + \int_{\partial\Omega} w^* \cdot T[E,Q] \cdot \nu dS \\ &\equiv I_1 + I_2 + I_3. \end{aligned} \tag{3.2}$$

We easily see that $|I_3| = O(|x|^{1-n})$ and $|\nabla I_3| = O(|x|^{-n})$ as $|x| \to \infty$, and so $\nabla I_3 \in L^r$ for all $1 < r \le \infty$, since I_3 is smooth up to the boundary (see [14]). By using (1.7), the single layer potential I_2 is written as

$$I_2 = \int_{\partial\Omega} \widetilde{E}(x,y) \cdot \nu_y \cdot (T[w,p] - w^* \otimes w^* + F) dS_y$$

in terms of $\widetilde{E}$ defined by (2.9). Since we are assuming (1.7), it follows as in the proof of Theorem 1.1 that

$$|I_2 + I_3| = O(|x|^{1-n}) \qquad \text{and} \qquad |\nabla(I_2 + I_3)| = O(|x|^{-n}). \tag{3.3}$$

Introducing the norms

$$\|w\| = \sup_{\Omega} |y|^{n-1} |w(y)|$$

and

$$[F] = \sup_{\Omega} |y|^{2n-2} |F(y)|, \quad [\nabla F] = \sup_{\Omega} |y|^{2n-1} |\nabla F(y)|,$$

we estimate the functions $\{w_k\}$ defined by the iteration scheme:

$$w_{k+1} = (\nabla E) \cdot (F - w_k \otimes w_k) + w_0, \qquad (w_0 = I_2 + I_3).$$

Due to (3.3), the argument as employed in the proof of Theorem 1.1 (i) applies with no substantial modification to show that if (1.7) holds and if $[F]$ and $\|w_0\|$ are sufficiently small, there is a function v such that $\lim_{k\to\infty} \|w_k - v\| = 0$ and

$$v = (\nabla E) \cdot (F - v \otimes v) + w_0, \qquad |v| \le C(1 + |x|)^{1-n}.$$

That $v = w$ in Ω is proved in the same way as in Theorem 1.1 if $\sup_{\Omega} |y|^{n-2} |w(y)|$ is sufficiently small. We thus conclude that w satisfies the first estimate of (1.9). To deduce the second estimate of (1.9) for our solution w, it suffices to estimate, for large $|x|$, the singular integral:

$$\begin{aligned} V &\equiv \lim_{\varepsilon\to 0} \int_{\Omega\cap\{|x-y|>\varepsilon\}} (\nabla^2 E)(x-y)(F - w \otimes w)(y) dy \\ &= \int_{\Omega\cap\{|x-y|>|x|/2\}} + \lim_{\varepsilon\to 0} \int_{\Omega\cap\{\varepsilon<|x-y|<|x|/2\}} \equiv u_1 + u_2. \end{aligned}$$

Since $\|w\|$ is finite, we see that

$$\begin{aligned} |u_1(x)| &\le M([F] + \|w\|^2) \int_{|x-y|>|x|/2} |x-y|^{-n} (1 + |y|)^{2-2n} dy \\ &\le M([F] + \|w\|^2) |x|^{-n} \int (1 + |y|)^{2-2n} dy = C|x|^{-n}. \end{aligned}$$

To estimate u_2, we integrate it by parts to get

$$u_2(x) = J_0 + J_1 + J_2, \tag{3.4}$$

where

$$\begin{aligned}
J_0(x) &= \int_{\Omega\cap\{|x-y|<|x|/2\}} (\nabla E)(x-y)(\nabla\cdot F - w\cdot\nabla w)(y)dy \\
J_1(x) &= \int_{|x-y|=|x|/2} \nu\cdot(\nabla E)(x-y)(F-w\otimes w)(y)dS_y \\
J_2(x) &= \lim_{\varepsilon\to 0}\int_{|x-y|=\varepsilon} \nu\cdot(\nabla E)(x-y)(F-w\otimes w)(y)dS_y.
\end{aligned}$$

Note that any integral on $\partial\Omega$ does not appear in (3.4), since $|x|$ is assumed sufficiently large. We easily see that

$$\begin{aligned}
|J_1(x)| &\le M([F]+\|w\|^2)|x|^{1-n}\int_{|x-y|=|x|/2}(1+|y|)^{2-2n}dS_y \\
&\le M([F]+\|w\|^2)|x|^{1-n}(1+|x|)^{2-2n}|x|^{n-1} \\
&= M([F]+\|w\|^2)(1+|x|)^{2-2n} \le M([F]+\|w\|^2)|x|^{-n}.
\end{aligned}$$

On the other hand, introducing the polar coordinates $y = x+\varepsilon\omega$ with $\omega\in S^{n-1}$, and noting that ∇E is a homogeneous odd function of degree $1-n$, we obtain

$$J_2(x) = c\cdot(F-w\otimes w)(x), \qquad c = -\int_{S^{n-1}}\omega\cdot(\nabla E)(\omega)d\omega.$$

Hence,

$$|J_2(x)| \le C([F]+\|w\|^2)(1+|x|)^{2-2n} \le C(1+|x|)^{-n},$$

because $n\ge 3$. Finally, since we already know that $|\nabla w| \le C(1+|x|)^{1-n}$ and $|w|\le C(1+|x|)^{1-n}$, the function J_0 is estimated as

$$\begin{aligned}
|J_0(x)| &\le M\int_{|x-y|<|x|/2}|x-y|^{1-n}(1+|y|)^{2-2n}dy \\
&\le M(1+|x|)^{2-2n}\int_{|x-y|<|x|/2}|x-y|^{1-n}dy \\
&\le M(1+|x|)^{3-2n} \le M(1+|x|)^{-n}.
\end{aligned}$$

It thus follows from (3.4) that $|u_2(x)|\le M(1+|x|)^{-n}$, and therefore

$$|V(x)| \le |u_1(x)|+|u_2(x)| \le M(1+|x|)^{-n}.$$

We have thus deduced the second estimate of (1.9), and this proves the first assertion of (i).

We next prove the second assertion of (i). Since $n\ge 4$, it follows from (1.1) that

$$\limsup_{|x|\to\infty}|x|^{1+\varepsilon}|w(x)| = 0 \quad \text{for some } \varepsilon>0. \tag{3.5}$$

Bearing this in mind, we consider on $D=\{|y|>R\}$ the formula:

$$w = (\nabla E)\cdot(\widetilde{F}-\widetilde{w}\otimes\widetilde{w}) + w_0 \equiv w_1 + w_0, \tag{3.6}$$

where

$$w_1 = (\nabla E) \cdot (\widetilde{F} - \widetilde{w} \otimes \widetilde{w}) = \int_{|y|>R} (\nabla E)(x-y)(F - w \otimes w)(y)dy,$$

$$w_0 = \int_{|y|=R} E \cdot \nu \cdot (T[w,p] - w \otimes w + F)dS_y + \int_{|y|=R} w \cdot T[E,Q] \cdot \nu dS_y,$$

and $R > 0$ is taken sufficiently large so that $\Omega \supset \overline{D}$. Note that (see [14]) w_0 is smooth on compact subsets of $\overline{D}$. Since (1.7) implies

$$\int_{|x|=R} \nu \cdot (T[w,p] - w \otimes w + F)dS = 0,$$

it follows that $|w_0| \le C(1+|x|)^{1-n}$ and $|\nabla w_0| \le C(1+|x|)^{-n}$. Consider the *linear* iteration scheme on D:

$$v_{k+1} = (\nabla E) \cdot (\widetilde{F} - \widetilde{w} \otimes \widetilde{v}_k) + w_0 \qquad (k = 0, 1, 2, \dots).$$

Since $n \ge 4$, writing

$$\|w\|_D = \sup_{|y|>R} |y|^{1-n}|w(y)|,$$

we get the estimates

$$\|v_{k+1}\|_D \le M([F]_D + [w]_{\varepsilon,D}\|v_k\|_D) + \|w_0\|_D \tag{3.7}$$

where

$$[F]_D = \sup_D |y|^{n+\varepsilon}|F(y)|, \qquad [w]_{\varepsilon,D} = \sup_D |y|^{1+\varepsilon}|w(y)|.$$

Here we fix $R > 0$ so that

$$M[w]_{\varepsilon,D} < \frac{1}{2},$$

which is possible by (3.5). Then, by (3.7) the sequence $\{v_k\}$ converges in the norm $\|\cdot\|_D$ to a function v such that

$$v = (\nabla E) \cdot (\widetilde{F} - \widetilde{w} \otimes \widetilde{v}) + w_0.$$

Since

$$v - w = -(\nabla E) \cdot [\widetilde{w} \otimes (\widetilde{v} - \widetilde{w})],$$

we obtain

$$[v - w]_{0,D} \le M'[w]_{0,D}[v - w]_{0,D}.$$

Since we may assume that

$$M'[w]_{0,D} < \frac{1}{2},$$

we conclude that $v = w$ on D, and this proves the first estimate of (1.9). The second estimate of (1.9) is deduced as above, and so the proof of (i) is complete.

(ii) First we show the boundedness of $\nabla^2 w$. By the regularity theory as given in [5, 13], we may assume that w is smooth on $\overline{\Omega}$. Direct calculation gives

$$\begin{aligned}\nabla^2 w &= (\nabla^2 E)\cdot(\nabla\cdot F - w\cdot\nabla w) + \int_{\partial\Omega}(\nabla^2 E)\cdot\nu\cdot T[w,p]dS \\ &\quad + \int_{\partial\Omega} w^*\cdot(\nabla^2 T[E,Q])\cdot\nu dS.\end{aligned}$$

The boundary integrals on the right-hand side above are obviously bounded; so we need only show the boundedness of the singular integral

$$\begin{aligned}v &= (\nabla^2 E)\cdot(\nabla\cdot F - w\cdot\nabla w) \\ &= \lim_{\varepsilon\to 0}\int_{\Omega\cap\{|x-y|>\varepsilon\}}(\nabla^2 E)\cdot(\nabla\cdot F - w\cdot\nabla w)dy + c(\nabla\cdot F - w\cdot\nabla w) \\ &\equiv v' + c(\nabla\cdot F - w\cdot\nabla w),\end{aligned}$$

where c is an appropriate constant. Since $\nabla^2 E$ is a Calderón-Zygmund kernel ([11]), our assumption on F and w implies $v\in L^p(\Omega)$; so $\nabla^2 w\in L^p(\Omega)$ for all $1<p<\infty$.

We next apply ∇ to v' and integrate by parts, to get

$$\begin{aligned}\nabla v' &= -\int_{\partial\Omega}\nu\cdot(\nabla^2 E)\cdot(\nabla\cdot F - w\cdot\nabla w)dS \\ &\quad + \lim_{\varepsilon\to 0}\int_{\Omega\cap\{|x-y|>\varepsilon\}}(\nabla^2 E)\cdot(\nabla^2 F - \nabla w\cdot\nabla w - w\cdot\nabla^2 w)dy.\end{aligned}$$

The boundary integral is in $L^p(\Omega)$ for all $1<p<\infty$. Since $\nabla^2 w\in L^p(\Omega)$ for all $1<p<\infty$, this shows that $\nabla v'\in L^p(\Omega)$, and so $\nabla v\in L^p(\Omega)$ for all $1<p<\infty$. Hence, v is bounded on Ω by the Sobolev embedding theorem; and so $\nabla^2 w$ is bounded on Ω.

Now, assume $|x|$ is sufficiently large. Then,

$$\begin{aligned}\nabla^2 w = \lim_{\varepsilon\to 0}&\int_{\Omega\cap\{|x-y|>\varepsilon\}}(\nabla^2 E)\cdot(\nabla\cdot F - w\cdot\nabla w)dy \\ &+ \int_{\partial\Omega}(\nabla^2 E)\cdot\nu\cdot T[w,p]dS \\ &+ \int_{\partial\Omega} w^*\cdot(\nabla^2 T[E,Q])\cdot\nu dS + c(\nabla\cdot F - w\cdot\nabla w)\end{aligned}$$

$$
\begin{aligned}
&= \int_{\Omega\cap\{|x-y|>|x|/2\}} (\nabla^3 E)\cdot(F - w\otimes w)dy \\
&\quad + \int_{\partial\Omega} (\nabla^2 E)\cdot\nu\cdot(T[w,p] - w^*\otimes w^* + F)dS \\
&\quad + \int_{\partial\Omega} w^*\cdot(\nabla^2 T[E,Q])\cdot\nu dS \\
&\quad + \int_{|x-y|=|x|/2} (\nabla^2 E)\cdot\nu\cdot(F - w\otimes w)dS_y \\
&\quad + \lim_{\varepsilon\to 0}\int_{\varepsilon<|x-y|<|x|/2} (\nabla^2 E)\cdot(\nabla\cdot(F - w\otimes w))dy \\
&\quad + c(\nabla\cdot F - w\cdot\nabla w) \\
&\equiv W_1 + W_2 + W_3 + W_4 + W_5 + W_6.
\end{aligned}
$$

We easily see that

$$
\begin{aligned}
|W_1| &\le M\int_{|x-y|>|x|/2} |x-y|^{-n-1}(1+|y|)^{2-2n}dy \le M(1+|x|)^{-n-1}, \\
|W_3| &\le M(1+|x|)^{-n-1}, \qquad |W_6| \le C(1+|x|)^{-n-1},
\end{aligned}
$$

and

$$
\begin{aligned}
|W_4| &\le M\int_{|x-y|=|x|/2} |x-y|^{-n}(1+|y|)^{2-2n}dy \\
&\le M|x|^{-n}(1+|x|)^{2-2n}|x|^{n-1} = M(1+|x|)^{1-2n} \le M(1+|x|)^{-n-1}.
\end{aligned}
$$

On the other hand, by condition (1.7) we can write

$$
W_2 = \int_{\partial\Omega} (\nabla^2\widetilde{E})\cdot\nu\cdot(T[w,p] - w^*\otimes w^* + F)dS.
$$

Since $|\nabla^2\widetilde{E}(x,y)| \le M|x-y|^{-n-1}$, it follows that

$$
|W_2| \le M(1+|x|)^{-n-1}.
$$

It thus remains to estimate W_5. An integration by parts gives

$$
W_5 = W_{51} + W_{52} + W_{53},
$$

where

$$
\begin{aligned}
W_{51} &= c(\nabla\cdot F - w\cdot\nabla w)), \qquad c = -\int_{|\omega|=1} \omega(\nabla E)(\omega)dS_\omega \\
W_{52} &= \int_{|x-y|=|x|/2} \nu(\nabla E)\cdot(\nabla\cdot F - w\cdot\nabla w))dS_y \\
W_{53} &= \int_{|x-y|<|x|/2} (\nabla E)\cdot(\nabla^2 F - w\cdot\nabla^2 w - \nabla w\cdot\nabla w)dy.
\end{aligned}
$$

It is easy to see that

$$|W_{51}| \le M(1+|x|)^{1-2n} \le M(1+|x|)^{-n-1},$$

and

$$|W_{52}| \le M(1+|x|)^{1-2n} \le M(1+|x|)^{-n-1}.$$

Furthermore, $\nabla^2 w$ is bounded; thus,

$$|W_{53}| \le M \int_{|x-y|<|x|/2} |x-y|^{1-n}(1+|y|)^{1-n} dy \le M(1+|x|)^{2-n}.$$

Collecting terms gives $|\nabla^2 w| \le M(1+|x|)^{2-n}$, so $|w\cdot\nabla^2 w| \le M(1+|x|)^{3-2n}$. Hence,

$$|W_{53}| \le M(1+|x|)^{4-2n}, \quad \text{and therefore} \quad |\nabla^2 w| \le M(1+|x|)^{4-2n}.$$

This proves (1.11) for $\nabla^2 w$ when $n \ge 5$. When $n = 3, 4$, we get $|w \cdot \nabla^2 w| \le M(1+|x|)^{5-3n}$, so

$$|W_{53}| \le M(1+|x|)^{6-3n}, \quad \text{and therefore} \quad |\nabla^2 w| \le M(1+|x|)^{6-3n}.$$

This proves (1.11) for $\nabla^2 w$ when $n = 4$. When $n = 3$, we get $|w\cdot\nabla^2 w| \le M(1+|x|)^{7-4n}$, so

$$|W_{53}| \le M(1+|x|)^{8-4n}, \quad \text{and therefore} \quad |\nabla^2 w| \le M(1+|x|)^{8-4n}.$$

This proves (1.11) for $\nabla^2 w$ when $n = 3$. Now, the desired estimate for ∇p is immediately obtained from

$$\nabla p = \Delta w - w\cdot\nabla w + \nabla\cdot F.$$

The estimate for p is then deduced by integrating the estimate for ∇p. Finally, (iii) is proved in [1, 3]. This completes the proof of Theorem 1.2. □

Acknowledgment

The present work was initiated while the author was visiting SFB 256 at University of Bonn, Germany, in 1996. The author thanks Professor Stefan Hildebrandt for his kind invitation to Bonn. The work was also partially supported by Grant-in-Aid for Scientific Research, No. 08454031 and No. 09874032, from the Japan Ministry of Education.

References

[1] W. Borchers and T. Miyakawa, "On stability of exterior stationary Navier-Stokes flows". Acta Math. **174** (1995), 1–72.

[2] J. Frehse and M. Ruzicka, "Weighted estimates for the stationary Navier-Stokes equations". In *Mathematical Theory in Fluid Mechanics*, ed. by G. P. Galdi *et al.*, Research Notes in Math. # 354, Longman, Harlow, Essex, 1996, pp. 1–29.

[3] H. Kozono and H. Sohr, "On stationary Navier-Stokes equations in unbounded domains". Ricerche di Mat. **152** (1993), 69–86.

[4] H. Kozono, H. Sohr and M. Yamazaki, "Representation formula, net force and energy relation to the stationary Navier-Stokes equations in 3-dimensional exterior domains". Kyushu J. Math. **51** (1997), 239–260.

[5] O. A. Ladyzhenskaya, *The Mathematical Theory of Viscous Incompressible Flow.* Gordon & Breach, New York, 1969.

[6] J. Leray, "Etude de diverses équations intégrales nonlinéaires et de quelques problèmes que pose l'hydrodynamique". J. Math. Pures Appl. **12** (1933), 1–82.

[7] T. Miyakawa, "Hardy spaces of solenoidal vector fields, with applications to the Navier-Stokes equations". Kyushu J. Math. **50** (1996), 1–64.

[8] T. Miyakawa, "On L^1-stability of stationary Navier-Stokes flows in $\mathbb{R}^n$". J. Math. Sci., Univ. of Tokyo **4** (1997), 67–119.

[9] T. Miyakawa, "Application of Hardy space techniques to the time-decay problem for incompressible Navier-Stokes flows in $\mathbb{R}^n$". Preprint, Kobe University, 1997.

[10] A. Novotny and M. Padula, "Note on decay of stationary solutions of 3D Navier-Stokes equations in an exterior domain". Diff. Int. Eq. **8** (1995), 1833–1842.

[11] E. M. Stein, *Singular Integrals and Differentiability Properties of Functions.* Princeton University Press, Princeton, 1970.

[12] E. M. Stein, *Harmonic Analysis.* Princeton University Press, Princeton, 1993.

[13] R. Témam, *Navier-Stokes Equations.* North-Holland, Amsterdam, 1977.

[14] M. Wiegner, "Schauder estimates for boundary layer potentials". Math. Meth. Appl. Sci. **16** (1993), 877–894.

Tetsuro Miyakawa
Department of Mathematics
Faculty of Science
Kobe University
Rokko, Kobe 657, Japan

Progress in Nonlinear Differential Equations
and Their Applications, Vol. 35
© 1999 Birkhäuser Verlag Basel/Switzerland

Non-stationary Viscous Flows with a Cylindrical Free Surface

V.V. Pukhnachov

To Professor Herbert Amann on the occasion of his 60th birthday

Abstract. The motion of a viscous incompressible capillary liquid with the free boundary in the absence of external forces is considered. Suppose that initially the liquid is bounded by a cylindrical free surface and that the longitudinal component of the initial velocity field depends linearly on a longitudinal coordinate, while the other components and the pressure are independent of this coordinate. Then the Navier-Stokes equations have a solution where the velocity field keeps the same structure and the free surface remains a cylindrical one. The solution gives a pithy example of the partially invariant solution for the Navier-Stokes equations describing a free boundary flow. As a corollary, the primary 3-D problem is reduced to a 2-D one.

The analogy between the problem of the motion of a "flat drop with distributed mass sources" is drawn. The question of a local solvability in time to this problem is discussed. A sufficient condition for blowing-up of the solution is formulated.

A further reduction of the problem occurs for rotationally symmetric and planar flows. In both cases, sufficient conditions for a global solvability in time are obtained and the asymptotic behaviour of solutions for a large time is analyzed. Besides, the family of exact solutions in the 1-D problem is found.

1. Non-stationary free boundary problem for the Navier-Stokes equations

Let us consider the problem of non-stationary motion of an incompressible viscous homogeneous liquid entirely bounded by a free surface. The mathematical formulation of the problem consists of determining of the domain $\Omega_t \subset \mathbb{R}^3$, $t \in (0,T)$, and the solution $\vec{v}(x,t) = (v_1, v_2, v_3)$ (velocity vector), $p(x,t)$ (pressure) of the Navier-Stokes equations in this domain,

$$\vec{v}_t + \vec{v} \cdot \nabla \vec{v} = -\nabla p/\rho + \nu \Delta \vec{v}, \quad \nabla \cdot \vec{v} = 0, \tag{1.1}$$

satisfying the initial conditions

$$\vec{v}(x,0) = \vec{v}_0(x), \quad x \in \Omega_0, \tag{1.2}$$

where Ω_0 is given, and boundary conditions

$$\vec{v}\cdot\vec{n} = V_n, \quad x \in \partial\Omega_t \equiv \Gamma_t, \quad t \in (0,T), \tag{1.3}$$

$$p\vec{n} - 2\rho\nu D\cdot\vec{n} = 2\sigma H\vec{n}, \quad x \in \Gamma_t, \quad t \in (0,T). \tag{1.4}$$

Here ν is the kinematic coefficient of viscosity, ρ is the density (both are positive constants); $\vec{n}$ is the unit outer normal vector to Γ_t, V_n is the velocity of the displacement of Γ_t in the direction of $\vec{n}$, $D = [\nabla\vec{v} + (\nabla\vec{v})^*]/2$ is the strain tensor, $\sigma = \text{const} \geq 0$ is the surface tension coefficient, H is the mean curvature of the surface Γ_t (it is assumed that $H > 0$ if this surface is convex outside the liquid). Also, the compatibility conditions, $\nabla\cdot\vec{v}_0 = 0$ if $x \in \Omega_0$, and

$$D_0\cdot\vec{n}_0 - (\vec{n}_0\cdot D_0\cdot\vec{n}_0)\vec{n}_0 = 0, \quad x \in \Gamma_0 \tag{1.5}$$

are assumed to be fulfilled. Here $D_0 = [\nabla\vec{v_0} + (\nabla\vec{v_0})^*]/2, \vec{n_0}$ is the unit outer normal vector to Γ_0.

Problem (1.1)–(1.4), with Ω_0 being a bounded domain, was investigated by V.A. Solonnikov [1–3]. Its solvability on a finite interval of time was established in the work [1] in the case $\sigma = 0$ and in the work [2] in the case $\sigma > 0$. The global existence theorem in time was proved in [3] under some additional assumptions ($\sigma > 0$, an appropriate norm of the vector-function $\vec{v_0}$ is small, the domain Ω_0 is close to a ball).

Below problem (1.1)–(1.4) is considered in the case of an jet-like domain Ω_0 with the cylindrical boundary $\Gamma_0 = \{x : (x_1,x_2) \in \gamma_0, x_3 \in \mathbb{R}\}$ where γ_0 is a bounded planar curve. Let us suppose that $v_{0,1}$ and $v_{0,2}$ do not depend on x_3 while $v_{0,3} = x_3 f_0(x_1,x_2)$. Then problem (1.1)–(1.4) has a solution where Γ_t is the cylindrical surface for all $t \in (0,T)$. It turns out that this solution is the partially invariant one of the system (1.1) in the sense of L.V. Ovsiannikov [4].

2. Invariance of the free boundary conditions

A regular method was proposed in [5] for obtaining invariant solutions of system (1.1) adjusted before to the conditions at the free boundary. It is well known that system (1.1) admits the 10-parameter local Lie group G_{10} generated by the operators $X_0 = \partial/\partial_t$, $X_k = \partial/\partial x_k$, $Y_k = t\,\partial/\partial x_k + \partial/\partial v_k$, $X_{kl} = x_l\partial/\partial x_k - x_k\partial/\partial x_l + v_l\partial/\partial v_k - v_k\partial/\partial v_l$, where $k = 1,2,3$; $l = 2,3$; $k < l$. (We use the terminology of group analysis of differential equations following the monograph [6].)

Theorem 1 [5,7]. *If the free boundary Γ_t is a non-degenerate invariant manifold of some subgroup G of the group G_{10}, then conditions (1.3), (1.4) fulfilled at this surface are also invariant with respect to G.*

It should be noted that system (1.1) is invariant under an infinite Lie pseudogroup containing G_{10} as a subgroup [7]. Nevertheless, the replacement of G_{10} group by a larger subgroup of this pseudogroup within the conditions of Theorem 1 is impossible. However, in the case $\sigma = 0$ in condition (1.4), the analog of

Theorem 1 is valid if G_{10} is changed to the group $G_{11} = G_{10} \otimes G_1$, where G_1 is one-parameter group of dilatation with operator

$$Z = 2t\,\partial/\partial t + \sum_{i=1}^{3}(x_i\partial/\partial x_i - v_i\partial/\partial v_i) - 2p\,\partial/\partial p.$$

Theorem 1 and its analog are the base for the construction of exact (invariant and partially invariant) solutions of free boundary problems for system (1.1). Various examples of such solutions are presented in the papers [5, 8] and in the monograph [7].

3. A new example of a partially invariant solution

Let us consider the two-parameter group $G_2 \subset G_{10}$ generated by the operators X_3 and Y_3. The full system of its invariants is $I_1 = x_1$, $I_2 = x_2$, $I_3 = t$, $I_4 = v_1$, $I_5 = v_2$, $I_6 = p$. The rank of the matrix $(\partial I_j/\partial v_k)$ is equal to 3 which is less than the number of unknown functions in system (1.1) (here $j = 1, \dots, 6$; $k = 1, \dots, 4$; $v_4 = p$). This means (see [6]) that the system has no invariant solutions with respect to the group G_2. But we are able to obtain its partially invariant solution by setting invariants I_4, I_5, I_6 to be functions of invariants I_1, I_2, I_3, i.e.

$$v_1 = v_1(x_1, x_2, t), \quad v_2 = v_2(x_1, x_2, t), \quad p = p(x_1, x_2, t). \tag{3.1}$$

Substitution of (3.1) into system (1.1) leads to an overdetermined system. The latter one is compatible if

$$v_3 = x_3 f(x_1, x_2, t), \tag{3.2}$$

where f is a new unknown function. The resulting system of equations which connects only the invariants of the group G_2 has the following form:

$$\vec{v}_t + \vec{v}\cdot\nabla\vec{v} = -\nabla p/\rho + \nu\Delta\vec{v}, \quad \nabla\cdot\vec{v} + f = 0,$$
$$f_t + \vec{v}\cdot\nabla f = \nu\Delta f - f^2. \tag{3.3}$$

Here and further $\vec{v} = (v_1, v_2)$; ∇ and Δ stand for the gradient and the Laplacian in $(x_1, x_2) = x$, respectively.

Now we are going to formulate a free boundary problem for system (3.3). To this end, we note that any invariant manifold of the group G_2 in the space of independent variables in system (1.1) may be written in the form $F(x_1, x_2, t) = 0$. If we take it as a free boundary Γ_t in the flow associated with the solution (3.1), (3.2), we are sure, due to Theorem 1, that the free boundary conditions (1.3), (1.4) can be written in terms of the invariants of group G_2.

Let Ω_t denote the domain in $\mathbb{R}^3$ bounded by the cylindrical surface Γ_t and let ω_t be the section of Ω_t with the plane $x_3 = 0$. The set $\omega_t \subset \mathbb{R}^2$ (which is a priori unknown) is the domain of definition for the solution of system (3.3). In consequence of (3.1), (3.2), conditions (1.3), (1.4) produce the following conditions at the free boundary $\gamma_t = \partial\omega_t$:

$$\vec{v}\cdot\vec{n} = V_n, \quad \partial f/\partial n = 0,$$

$$p\vec{n} - 2\rho\nu D \cdot \vec{n} = \sigma K \vec{n}, \quad x \in \gamma_t, \quad t \in (0,T). \tag{3.4}$$

Now $\vec{n}$ is the unit outer normal vector to γ_t, V_n is the velocity of the displacement of curve γ_t in the direction of $\vec{n}$, D is the strain tensor corresponding to the vector $\vec{v}$, K is the curvature of γ_t.

Relations (3.3), (3.4) have to be supplemented by initial conditions: ω_0 is given,

$$\vec{v}(x,0) = \vec{v}_0(x), \quad f(x,0) = f_0(x), \quad x \in \omega_0. \tag{3.5}$$

By virtue of the second equation of system (3.3), functions $\vec{v}_0$ and f_0 are connected by the equality

$$\nabla \cdot \vec{v}_0 + f_0 = 0, \quad x \in \omega_0. \tag{3.6}$$

Besides, relation (1.5) induces a compatibility condition of the initial and boundary values of the function $\vec{v}$,

$$\vec{s}_0 \cdot D_0 \cdot \vec{n}_0 = 0, \quad x \in \gamma_0, \tag{3.7}$$

where $\vec{s}_0$ and $\vec{n}_0$ are unit tangential and normal vectors to the curve γ_0, respectively, and $D_0(x) = D(x,0)$.

Thus, the solution $\vec{v}$, p, f of the two-dimensional free boundary problem (3.3)–(3.5) describes, according to (3.1), (3.2), a three dimensional non-stationary motion of a viscous fluid with a cylindrical free surface.

4. On the solvability of problem (3.3)–(3.7)

In the following we assume that the domain ω_0 is bounded. Let us consider the problem of finding a function f as a solution of the second initial boundary value problem for the parabolic equation at known $\vec{v}$ and ω_t. If $f_0 = 0$ in the condition (3.5) then it follows immediately from (3.3), (3.4) that $f = 0$ if $x \in \omega_t$, $t \in (0,T)$. In this case problem (3.3)–(3.5) turns into the plane analog of the problem of a finite liquid mass motion. The solvability of the latter problem was studied by V.A. Solonnikov in the papers [1–3]. Methods developed in these works allow us to investigate the problem (3.3)–(3.5). However, the consideration of the common situation demands very cumbersome constructions. We restrict ourselves to the analysis of a partial case when $f_0 = C = \text{const}$ in the condition (3.5). In this case the unique solution of the initial boundary value problem for the determination of the function f is

$$f = C/(1 + Ct). \tag{4.1}$$

Problem (3.3)–(3.5), where the function f is known and given by formula (4.1), is equivalent to the problem of the motion of a "flat drop" in the presence of mass sources distributed uniformly in the domain ω_t with density $-f$. We formulate one of the results concerning its solvability.

Theorem 2. *Let $\omega_0 \in \mathbb{R}^2$ be a bounded domain with the boundary $\gamma_0 \in C^{2+\alpha}$, $0 < \alpha < 1$. Let $v_0(x) \in C^{2+\alpha}(\bar{\omega}_0)$, and $f_0 = C = \text{const}$, satisfy the compatibility conditions (3.6), (3.7). Then there exists $T > 0$ so that the problem (3.3)–(3.5) with $\sigma = 0$ has a unique solution $\vec{v} \in C^{2+\alpha,1+\alpha/2}(\bar{Q}_T)$, $\nabla p \in C^{\alpha,\alpha/2}(\bar{Q}_T)$, $f = C(1+Ct)^{-1}$ where $Q_T = \{x,t : x \in \omega_t,\ 0 < t < T\}$ and the lateral boundary of the domain Q_T is a surface of Hölder class $C^{2+\alpha,1+\alpha/2}$ in $\mathbb{R}^2 \times \mathbb{R}$.*

The proof of Theorem 2 is almost literally the same as the proof of Theorem 5 in [1]. Firstly, problem (3.3)–(3.5) is transformed into a fixed domain problem by introducing Lagrangian coordinates. The only difference between our problem and the 2-D analog of the problem studied in [1] is the nonhomogeneity in the continuity equation: its right side is equal to $-C(1+Ct)^{-1}$. In a next step the method of succesive approximations is applied. Its convergence for sufficiently small $T > 0$ is based on precise estimates for the solution of the linearized problem corresponding to (3.3)–(3.5) which are obtained in [9].

5. Blowing-up solutions

A remarkable property of problem (3.3)–(3.5) is the presence of blowing-up solutions. These solutions describe flows with an unbounded expansion of the cross section ω_t of the liquid cylinder Ω_t in finite time t_*. It is interesting to note that the viscous terms are not able to prevent the blow-up though they provide the space regularity of the solution. (The simplest example of this situation for the 1-D analog of problem (3.3)–(3.5) was given in [8].)

Let us introduce the following notations:

$$S(t) = \int_{\omega_t} dx, \quad M(t) = \int_{\omega_t} f dx, \quad \bar{f} = M/S, \quad g = f - \bar{f} \tag{5.1}$$

so that $S(t)$ is the area of the domain ω_t, $\bar{f}(t)$ is the mean value of $f(x,t)$ over ω_t and

$$\int_{\omega_t} g dx = 0. \tag{5.2}$$

Theorem 3. *Let us suppose that $\bar{f}(0) = -k < 0$. Then the lifespan t_* of the solution of problem (3.3)–(3.5) is less or equal to $1/k$.*

The assertion of Theorem 3 follows immediately from the identity

$$\frac{d\bar{f}}{dt} = -\bar{f}^2 - \frac{2}{S}\int_{\omega_t} g^2 dx. \tag{5.3}$$

To prove (5.3), we determine the derivatives dS/dt and dM/dt. From the second equation of (3.3) and the first condition of (3.4) we find:

$$\frac{dS}{dt} = \int_{\gamma_t} V_n d\gamma_t = \int_{\gamma_t} \vec{v} \cdot \vec{n}\, d\gamma_t = \int_{\omega_t} \nabla \cdot \vec{v}\, dx = -\int_{\omega_t} f dx = -M. \tag{5.4}$$

Further, using the second and the third equations of system (3.3), the first and the second boundary conditions (3.4), definition (5.1), and equality (5.2) we have

$$\frac{dM}{dt} = \int_{\omega_t} f_t dx + \int_{\gamma_t} V_n f d\gamma_t = \int_{\omega_t} (\nu\Delta f - f^2 - \vec{v}\cdot\nabla f)dx + \int_{\gamma_t} V_n f\, d\gamma_t =$$

$$= \nu\int_{\gamma_t} \frac{\partial f}{\partial n} d\gamma_t - \int_{\omega_t} f^2 dx + \int_{\omega_t} f\nabla\cdot\vec{v}\, dx - \int_{\omega_t} \nabla\cdot(f\vec{v})dx + \int_{\gamma_t} f\vec{v}\cdot\vec{n}\, d\gamma_t =$$

$$= -2\int_{\omega_t} f^2 dx = -2S\bar{f}^2 - 2\int_{\omega_t} g^2 dx.$$

This equality together with (5.4) and relation

$$\frac{d\bar{f}}{dt} = \frac{1}{S}\frac{dM}{dt} - \frac{M}{S^2}\frac{dS}{dt} = \frac{1}{S}\frac{dM}{dt} + \bar{f}^2$$

implies (5.3). Since $\bar{f}(0) = -k < 0$ and, due to (5.3), $d\bar{f}/dt \leq -\bar{f}^2$, we obtain the estimate

$$\bar{f} \leq -k(1-kt)^{-1} \quad \text{for} \quad t \in [0, t_*) \tag{5.5}$$

which proves Theorem 3.

Rewriting (5.4) in the form $dS/dt = -S\bar{f}$ and adapting inequality (5.5) we derive the estimate from below for the area of the domain ω_t,

$$S(t) \geq \frac{S(0)}{1-kt}, \quad 0 \leq t < t_*.$$

We remark that the statement concerning the blow-up of solutions of problem (3.3)–(3.5) takes place even for $\bar{f}(0) = 0$ but $g(x,0) \neq 0$ though the lower estimate of $S(t)$ becomes more complicated in this case.

Let now the function $f_0(x)$ in (3.5) satisfy

$$0 < l \leq f_0 \leq m < \infty \quad \text{as} \quad x \in \bar{\omega}_0,$$

where l and m are constants. Then a priori estimates of function $f(x,t)$,

$$\frac{l}{1+lt} \leq f(x,t) \leq \frac{m}{1+mt} \quad \text{as} \quad x \in \bar{\omega}_t, \quad t \in [0, t_*)$$

hold according to the maximum principle. As a consequence we have two-sided bounds for $S(t)$,

$$\frac{S(0)}{1+mt} \leq S(t) \leq \frac{S(0)}{1+lt}, \quad 0 \leq t < t_*.$$

6. Axialsymmetric jet

System (3.3) describes the partially invariant solution (3.1), (3.2) of the Navier-Stokes equations (1.1) that inherits certain group properties of the original system. That gives us the possibility to construct invariant solutions. One of the most interesting solutions can be interpreted as the motion of an expanding (or contracting) and rotating axialsymmetric jet.

Let us denote by r, θ the polar coordinates in the plane x_1, x_2. If ω_0 is the disk $r < a$ and $u_0 = u_0(r)$, $v_0 = v_0(r)$ where u and v stand for the projections of the vector $\vec{v}$ on the axis r and θ, respectively, then problem (3.3)–(3.5) has a rotationally symmetric solution $u = u(r,t)$, $v = v(r,t)$, $p = p(r,t)$, $f = f(r,t)$ corresponding to the free boundary $r = s(t)$.

The remarkable fact is that the functions u, f and s might be found independently of v and p. The problem of finding these functions has the form

$$\begin{gathered} f_t + uf_r + f^2 = \nu r^{-1}(rf_r)_r, \\ u_r + r^{-1}u + f = 0 \quad \text{if} \quad 0 < r < s(t), \quad t \in (0,T), \\ f_r[s(t),t] = 0, \quad \frac{ds}{dt} = u[s(t),t], \quad t \in (0,T), \\ s(0) = a > 0, \quad f(r,0) = f_0(r), \quad r \in [0,a]. \end{gathered} \tag{6.1}$$

If problem (6.1) is solved, then the function v can be determined as the solution of the following linear problem:

$$\begin{gathered} v_t + uv_r + r^{-1}uv = \nu[r^{-1}(rv_r)_r - r^{-2}v], \\ \text{if} \quad 0 < r < s(t), \quad t \in (0,T), \\ (r^{-1}v)_r = 0 \quad \text{if} \quad r = s(t), \quad v(r,0) = v_0(r) \quad \text{if} \quad r \in [0,a]. \end{gathered}$$

Finally, if the functions u, v and s are known, the function p is defined by a quadrature from the relations

$$\begin{gathered} u_t + uu_r - r^{-1}v^2 = -p_r/\rho + \nu[r^{-1}(ru_r)_r - r^{-2}u], \\ \text{if} \quad 0 < r < s(t), \quad t \in (0,T), \\ p - 2\rho\nu u_r = \sigma/s(t) \quad \text{if} \quad r = s(t), \quad t \in [0,T]. \end{gathered}$$

The effective investigation of the free boundary problem (6.1) is based on the passage to the new spatial (Lagrangian) coordinate η so that

$$\frac{dr}{dt} = u(r,t), \quad r = \eta \quad \text{if} \quad t = 0.$$

Then the function $w(\eta,t) = f[r(\eta,t),t]$ is the solution of the initial boundary value problem for the integro-differential equation in the rectangular $\Pi_t = \{\eta, t : 0 < \eta < a,\ 0 < t < T\}$:

$$\begin{gathered} w_t + w^2 = \nu\eta^{-1} \exp\Big[\int_0^t w(\eta,\tau)d\tau\Big]\Big\langle\eta^{-1}\Big\{\int_0^\eta 2\zeta\Big[\exp\int_0^t w(\zeta,\tau)d\tau\Big]d\zeta\Big\}\times \\ \times \exp\Big[\int_0^t w(\eta,\tau)d\tau\Big]w_\eta\Big\rangle_\eta \quad \text{if} \quad (\eta,t) \in \Pi_T, \\ w_\eta(a,t) = 0 \quad \text{if} \quad t \in (0,T), \quad w(\eta,0) = f_0(\eta) \quad \text{if} \quad \eta \in [0,a]. \end{gathered} \tag{6.2}$$

Theorem 4. *Let $f_0(\eta)$ satisfy the conditions $f_0 \in C^{2+\alpha}[0,a]$ with $\alpha \in (0,1)$, $f_0'(0) = f_0'(a) = 0$. Then problem (6.2) has a unique solution $w \in C_e^{2+\alpha,1+\alpha/2}(\bar{\Pi}_T)$ where T is an arbitrary positive number if $\min\limits_{0 \le \eta \le a} f_0(\eta) \equiv C \ge 0$ and $T < -C^{-1}$ in the case $C < 0$.*

Here $C_e^{2+\alpha,1+\alpha/2}(\bar{\Pi}_T)$ is the class of functions $w(\eta,t) \in C^{2+\alpha,1+\alpha/2}(\bar{\Pi}_T)$ admitting the even continuation with respect to η into the domain $\mathrm{P}_T = \{\eta, t : -a < \eta < 0,\ \ 0 < t < T\}$ so that the continued function $w_e(\eta,t)$ belongs to the Hölder class $C_e^{2+\alpha,1+\alpha/2}(\bar{\Pi}_T \cup \bar{\mathrm{P}}_T)$.

We do not perform the proof of this theorem for the following reason: it has no essential distinctions from the proof of Theorem 1, Chapter 7 of the monograph [7], where the similar problem has been considered for the equation

$$w_t + w^2 = \nu \exp[\int_0^t w(\eta,\tau)d\tau]\{\exp[\int_0^t w(\eta,\tau)d\tau]w_\eta\}_\eta.$$

The last equation describes the expansion or contraction of a plane jet with rectlinear parallel free boundaries (see Section 7).

In conclusion, we note that problem (6.2) has a family of exact solutions,

$$w = 2\phi(t) \equiv C(1+Ct)^{-1}, \tag{6.3}$$

where $C = \text{const.}$ (Corresponding to (6.3), the function f appeared already in Section 4.) For this solution, according to (6.1),

$$u = -r\phi(t), \quad s = a(1+Ct)^{-1/2}. \tag{6.4}$$

If we set $v = 0$, we obtain from (3.1), (3.2) the velocity field

$$v_1 = -x_1\phi(t), \quad v_2 = -x_2\phi(t), \quad v_3 = -x_3\phi(t). \tag{6.5}$$

The following distribution of pressure conforms to (6.5) in the solution of problem (3.3)–(3.5):

$$p = \frac{3\rho}{2}[s^2(t) - (x_1^2 + x_2^2)]\phi^2(t) - 2\rho\nu\phi(t) + \frac{\sigma}{s(t)}, \tag{6.6}$$

where $\phi(t)$ and $s(t)$ are given by formulas (6.3), (6.4).

It should be noted that the velocity field (6.5) is a potential one. Besides, it does not depend on the viscosity coefficient. Hence, the solution (6.4), (6.5) is a rare example of a potential motion of a viscous fluid which is consistent with natural free boundary conditions. When $\nu = 0$ in (6.6), this solution turns into the known L.V. Ovsiannikov solution [10].

7. Symmetric deformation of a strip

Let us denote $x_1 = x$, $x_3 = y$ and consider a solution of system (3.3)

$$v_1 = -\int_0^x f(z,t)dz, \quad v_2 = 0, \quad f = f(x,t), \quad p = p(x,t). \tag{7.1}$$

It is easy to check that solution (7.1) is compatible with conditions (3.4) on the free boundary of the strip $S_t = \{x, y : \ |x| < s(t), \quad y \in \mathbb{R}\}$ where function $s(t)$ should be determined together with $f(x,t)$. Further we suppose that f is an even function of x. This fact allows to interpret the solution (7.1) as a symmetric deformation of a viscous strip.

The mathematical statement of the problem is to find a function $s(t)$, $0 < t < T$, and a solution $f(x,t)$ of the equation

$$f_t + f^2 - f_x \int_0^x f(z,t)dz = \nu f_{xx} \tag{7.2}$$

in the domain $Q_T = \{x, t: \ 0 < x < s(t), \quad 0 < t < T\}$ satisfying the boundary conditions

$$f_x(0,t) = 0, \quad f_x[s(t),t] = 0, \tag{7.3}$$
$$\frac{ds}{dt} = -\int_0^s f(z,t)dz, \quad t \in (0,T)$$

and initial conditions

$$f(x,0) = f_0(x), \quad x \in [0,a] \quad s(0) = a > 0. \tag{7.4}$$

Here $f_0(x)$ is a given function. If the function f is determined, the function p can be reconstructed by an integration.

An effective method for investing problem (7.2)–(7.4) is the transition to Lagrangian coordinates defined by means of the transformation $x = X(\xi,t)$ where X solves the Cauchy problem

$$\frac{dX}{dt} = -\int_0^X f(z,t)dz, \quad t \in (0,T), \quad X(\xi,0) = \xi.$$

As a result, the domain Q_T transforms into the rectangular $\Pi_T = \{\xi, t: \ 0 < \xi < a, \quad 0 < t < T\}$. Problem (7.2)–(7.4) can then be rewritten as follows:

$$w_t + w^2 = \nu \exp[\int_0^t w(\xi,\tau)d\tau]\{\exp[\int_0^t w(\xi,\tau)d\tau]w_\xi\}_\xi \quad \text{in} \quad \Pi_T, \tag{7.5}$$

$$w_\xi(0,t) = w_\xi(a,t) = 0 \quad \text{for} \quad 0 < t < T, \tag{7.6}$$

$$w(\xi,0) = f_0(\xi) \quad \text{for} \quad 0 < \xi < a. \tag{7.7}$$

After solving problem (7.5)–(7.7), the function f is defined uniquely by relations

$$w(\xi,t) = f[X(\xi,t),t], \quad X_\xi = \exp[-\int_0^t w(\xi,\tau)d\tau], \quad X(0,t) = 0.$$

Theorem 5 [7]. *Let $f_0(\xi)$ satisfy the conditions $f_0 \in C^{2+\alpha}[0,a]$ with $0 < \alpha < 1$, $f_0'(0) = f_0'(a) = 0$*
Then problem (7.5)–(7.7) has a unique solution $\omega \in C^{2+\alpha,1+\alpha/2}(\bar{\Pi}_T)$ where T is an arbitrary positive number if $\min_{0 \le \xi \le a} f_0(\xi) \equiv C \ge 0$ and $T < -C^{-1}$ in the case $C < 0$.

The proof of Theorem 5 follows the scheme developed in the monograph [11]. The presence of non-local terms in (7.5) does not bring in principal difficulties. The first estimate in a series of a priori estimates of the solution (the two-sided estimate of values of the function ω in $\bar{\Pi}_T$) is obtained by a comparison from above and below with "flat solutions" $w_\chi(t) = (t + \chi)^{-1}$ where $\chi = \text{const}$ [8]. Meanwhile, the same comparison provides the universal asymptotics of the solution in a "favorable" case $f_0(\xi) > 0, \quad 0 \le \xi \le a$:

$$w(\xi, t) = t^{-1}[1 + O(t^{-1})] \quad \text{as} \quad t \to \infty \tag{7.8}$$

uniformly in ξ on $[0, \mathrm{a}]$.

Theorem 5 is a direct analog of Theorem 4 for flows with rotational symmetry. The latter problem has also a family of flat solutions (6.3). Analogously to (7.8), globally existing solutions of problem (6.2) have the same asymptotic behaviour as (7.8).

Let us return to problem (7.5)–(7.7). In view of Theorem 3, if the mean value $\bar{f}_0$ of function f_0 on $[0, a]$ is negative then the solution $w(\xi, t)$ of the problem blows up in a finite time t_*. The simplest example of this situation demonstrates the flat solution $w_\chi(t)$ with $\chi < 0$. However, as shown in [12], the flat blow-up profile is not stable in the class of blowing up solutions. Moreover, if the initial function $f_0(\xi)$ is monotone decreasing and satisfies certain additional assumptions (cf. [12]) then corresponding to $w(\xi, t)$ the function $f(x, t)$ has the asymptotics

$$f(x, t) \sim -\frac{\cos^2[x\sqrt{\beta(t_* - t)}]}{t_* - t} \quad \text{for} \quad t \simeq t_*, \tag{7.9}$$

where $\beta = \beta(f_0) > 0$ is a constant. This formula describes a stable blow-up pattern in problem (7.5)–(7.7). Relevant to (7.9) the coordinate of the free boundary is

$$s(t) \sim \frac{\pi}{2\sqrt{\beta(t_* - t)}}$$

in contrast to the flat solution where $s = at_*(t_* - t)^{-1}$.

8. Exact solutions of the problem of a strip deformation

Let us suppose that the function $f_0(x)$ in the initial condition (7.4) has the form

$$f_0 = A_0 + B_0 \cos(\pi n x/a), \tag{8.1}$$

where A_0, B_0 are arbitrary constants and n is an arbitrary integer. Then problem (7.2)–(7.4) has the exact solution

$$f = A(t) + B(t)\cos[\frac{\pi n x}{s(t)}], \tag{8.2}$$

where functions A, B and s are determined as the solution of the Cauchy problem

$$\frac{dA}{dt} = -A^2 - B^2, \quad \frac{dB}{dt} = -[2A + \nu(\frac{\pi n}{s})^2]B, \quad \frac{ds}{dt} = -As, \tag{8.3}$$

$$A(0) = A_0, \quad B(0) = B_0, \quad s(0) = a. \tag{8.4}$$

The properties of the solution of the problem (8.3), (8.4) are determined by the sign of A_0. First, we assume that $A_0 \le 0$. If $A_0 < 0$ or $A_0 = 0$ but $B_0 \ne 0$ then in accordance with the results of Sections 5, 7 there exists such t_* $(0 < t_* < \infty)$ that $A \to -\infty, s \to \infty$ as $t \to t_*^-$. Moreover, we obtain from (8.3) that $A \le -c(t_* - t)^{-1}$ for $t \in [0, t_*]$ where $c = \text{const} > 0$ and $|B| \to \infty$ as $t \to t_*^-$ if $B_0 \neq 0$.

In order to study the behaviour of unbounded solutions, we observe that system (8.3) is invariant under the scaling transformation

$$(t, A, B, s) \to (k^2 t, k^{-2} A, k^{-2} B, ks).$$

This property allows us to introduce new variables

$$\lambda = s^2 A, \quad \mu = s^2 B$$

and to reduce system (8.3) to the first order equation

$$\frac{d\mu}{d\lambda} = \frac{\mu(4\lambda + \zeta)}{3\lambda^2 + \mu^2}, \tag{8.5}$$

where $\zeta = \nu(\pi n)^2$, and two quadratures,

$$s\frac{d\lambda}{ds} = \frac{3\lambda^2 + \mu^2}{\lambda}, \quad s\frac{ds}{dt} = -\lambda. \tag{8.6}$$

The justification of this reduction is guaranteed by the monotonicity of the functions $A(t)$, $s(t)$.

Since now we are interested in the case $\lambda \to -\infty$, $|\mu| \to \infty$ as $t \to t_*^-$, it makes sense to change variables $\lambda = -1/p$, $\mu = 1/q$ in (8.5). We obtain

$$\frac{dq}{dp} = \frac{pq(4 - \zeta p)}{p^2 + 3q^2}. \tag{8.7}$$

An elementary analysis shows that all non-trivial small solutions of the equation (8.7) have one of two forms:

$$q = \pm p + O(p^2) \quad \text{as} \quad p \to 0,$$

where the positive sign corresponds to $B_0 > 0$ and the negative sign corresponds to $B_0 < 0$.

Using this relation, the definition of function p and q, and equations (8.6) we obtain the representation of functions A, B, and s near the singular point:

$$A = -\frac{1}{2(t_* - t)} + O(1), \quad B = \frac{\operatorname{sign} B_0}{2(t_* - t)} + O(1),$$

$$s = \frac{\pi}{2\sqrt{\beta(t_* - t)}} + O(\sqrt{t_* - t}) \quad \text{as} \quad t \to t_*^-$$

if $B_0 \neq 0$; here $\beta = \beta(A_0, B_0, a, n) = \text{const} > 0$. Substitution of the last relations into (8.2) leads to the asymptotics

$$f = -\frac{1}{2(t_* - t)}\{1 - sgnB_0 \cos[2nx\sqrt{\beta(t_* - t)}]\} + O(1) \quad \text{as} \quad t \to t_*^-$$

of the exact blowing-up solutions to problem (7.2)–(7.4). If $n = 1$ and $B_0 < 0$ this asymptotics is similar to (7.9). The obtained blow-up pattern is stable within the class of solutions (8.2). In the case $B_0 = 0$, we have the flat solution $A = A_0(1 + A_0 t)^{-1}$, $B = 0$, $s = a(1 + A_0 t)^{-1}$ which is evidently unstable.

Let us assume now that $A_0 > 0$. In this case, due to the inequality $dA/dt \leq -A^2$ and the condition $A(0) = A_0$, we have the lower bound for A, $A \geq A_0(1 + A_0 t)^{-1}$ for $t > 0$. Together with (8.4), it is sufficient to obtain the following estimates of the functions s and B:

$$s \leq \frac{a}{1 + A_0 t}, \quad |B| \leq \frac{|B_0|}{(1 + A_0 t)^2} \exp\Big\{ - \frac{\zeta[(1 + A_0 t)^3 - 1]}{3a^2 A_0} \Big\} \quad \text{for} \quad t > 0. \quad (8.8)$$

Returning now to the first equation (8.3) and treating it as the Riccati equation, on the base of (8.8) we conclude that in the case $A_0 > 0$ the Cauchy problem (8.3), (8.4) has a global solution in time and the function A has the asymptotic behaviour

$$A = \frac{1}{t}[1 + O(\frac{1}{t})] \quad \text{as} \quad t \to \infty.$$

Thus we see that the condition $A_0 = \bar{f}(0) > 0$ is not only necessary but also sufficient for a global existence of the special solutions (8.2) of problem (7.2)–(7.4). We do not know if the condition $\bar{f}(0) > 0$ guarantees global solvability of problem (7.2)–(7.4) in a general situation. A partial answer to this question is given in the next section.

9. Global existence in time of non-positive solutions of problem (7.2)–(7.4)

First of all we note that the scaling transformation

$$(x, t, f, s) \to (ax, a^2\nu^{-1}t, a^{-2}\nu f, as)$$

gives a possibility to restrict our consideration of problem (7.2)–(7.4) to the case $a = 1$, $\nu = 1$. In the following we suppose that this transformation is performed.

As we already know, the condition

$$\bar{f}(0) = \int_0^1 f_0(x)dx > 0 \quad (9.1)$$

is necessary for the global solvability of problem (7.2)–(7.4). Below we assume that the given function $f_0(x)$ satisfies (9.1). Saying nothing about sign-definiteness of the function $f_0(x)$ we are not able to obtain an estimate of $|f(x, t)|$ in the domain $\bar{Q}_T$ independently of T. However, under some additional assumptions, we can estimate the $L_2(0, s(t))$ – norm of $f(\cdot, t)$ uniformly in $t > 0$.

Let us introduce the new space variable

$$z = x \exp[\int_0^t \varphi(\tau)d\tau], \quad (9.2)$$

where

$$\varphi(t) = \frac{1}{s(t)} \int_0^s f(x, t)dx \quad (9.3)$$

and write

$$u(z, t) = f(x, t) - \varphi(t). \quad (9.4)$$

Due to (9.2), (9.3), and the last relation (7.3) we have

$$s(t) = \exp[-\int_0^t \varphi(\tau)d\tau]. \tag{9.5}$$

Hence the a priori unknown domain Q_T turns into the rectangular $\Pi_T = \{z, t : 0 < z < 1,\ 0 < t < T\}$ in the plane $z,\ t$. Problem (7.2)–(7.4) is now reformulated as follows: find the solution $u(z,t)$, $\varphi(t)$ of the system

$$u_t - \exp(2\int_0^t \varphi d\tau)u_{zz} - u_z \int_0^z u(\xi,t)d\xi +$$

$$+u^2 + 2\varphi u - 2\int_0^1 u^2(\xi,t)d\xi = 0 \quad \text{in} \quad \Pi_T, \tag{9.6}$$

$$\frac{d\varphi}{dt} = -\varphi^2 - 2\int_0^1 u^2(\xi,t)d\xi \quad \text{in} \quad (0,T) \tag{9.7}$$

under the boundary and initial conditions

$$u_z(0,t) = u_z(1,t) = 0 \quad \text{for} \quad 0 < t < T, \tag{9.8}$$

$$u(z,0) = u_0(z) \quad \text{for} \quad 0 \le z \le 1, \quad \varphi(0) = \varphi_0. \tag{9.9}$$

Here $\varphi_0 = \bar{f}(0)$ (see (9.1)) and $u_0(z) = f_0(z) - \varphi_0$.

Further $\|v\|_p$ denotes the norm of the function $v(z)$ in $L_p(0,1)$ and $\|w(\cdot,t)\|_p$ denotes the $L_p(0,1)$-norm of the function $w(z,t)$ at fixed t. We put $q(t) = \|u(\cdot,t)\|_2$ and $q_0 = \|u_0\|_2$ so that q_0 is a given non-negative constant.

Lemma. *Assume that*

$$\varphi_0 > 0, \quad q_0 < \min\left(\frac{2}{3\pi}, \frac{\varphi_0}{\sqrt{2}}\right). \tag{9.10}$$

Then the solution of problem (9.6)–(9.9) satisfies the inequalities

$$q(t) \le q_0(1+\varphi_0 t)^{-2\delta}, \tag{9.11}$$

$$\frac{\varphi_0\delta}{1+\varphi_0 t} \le \varphi \le \frac{\varphi_0}{1+\varphi_0 t} \tag{9.12}$$

for any $t \in [0,T]$, where $\delta = (\varphi_0^2 - 2q_0^2)^{1/2}/\varphi_0$.

Proof. Since $\varphi_0 = \varphi(0)$ is positive, there exists $\tau \in (0,T]$ so that

$$\varphi(t) > 0 \quad \text{for} \quad 0 \le t \le \tau. \tag{9.13}$$

Under the assumptions (9.10) we shall establish that inequality (9.13) holds true for $\tau = T$. We will prove simultaneously that the function $q(t) = \|u(\cdot,t)\|_2$ is monotone decreasing on $[0,\tau]$. To this end, we use the energy identity

$$\frac{d}{dt}\int_0^1 u^2 dz + 4\varphi\int_0^1 u^2 dz + 2\exp(2\int_0^t \varphi d\tau)\int_0^1 u_z^2 dz + 3\int_0^1 u^3 dz = 0 \tag{9.14}$$

for the solution of problems (9.6)–(9.9) and the elementary inequalities

$$\|v\|_3^3 \leq \|v_z\|_2\,\|v\|_2^2, \tag{9.15}$$

$$\|v\|_2 \leq \pi\|v_z\|_2 \tag{9.16}$$

which hold for function $v(z) \in W_2^1(0,1)$ having mean value zero on $(0,1)$. On the account of (9.2)–(9.4) this latter property takes place for the function $u(z,t)$ at arbitrary $t \in [0,T]$.

Using inequalities (9.15), (9.16) for the function $u_0(z)$ and the hypothesis $q_0 < 2/3\pi$, we have

$$2\|u_{0,z}\|_2^2 - 3\|u_0\|_3^3 \geq (2-3\pi q_0)\|u_{0,z}\|_2^2 \geq 0.$$

Then the identity (9.14) considered at the moment $t=0$ shows that

$$\left.\frac{dq^2}{dt}\right|_{t=0} + 4\varphi_0 q_0^2 \leq 0$$

and $dq/dt \leq 0$ for $t=0$ as $\varphi_0 > 0$. Let us only consider the interesting case $u_0 \neq 0$. Then the last inequality turns into the strict one, $dq/dt < 0$ for $t=0$. Therefore, there exists ε, $0 < \varepsilon \leq \tau$ so that the function $q(t)$ is strictly decreasing for $0 \leq t \leq \varepsilon$. Hence, taking into account inequalities (9.13), (9.15), (9.16) we obtain

$$2\exp(2\int_0^t \varphi d\tau)\int_0^1 u_z^2 dz + 3\int_0^1 u^3 dz \geq 2\|u_z\|_2^2 - 3\|u\|_3^3 \geq$$
$$[2-3\pi q(t)]\|u_z\|_2^2 \geq (2-3\pi q_0)\|u_z\|_2^2 \geq \pi^{-2}(2-3\pi q_0)q^2(t)$$

as $t \in [0,\varepsilon]$. Combining the last inequality and the identity (9.14) we conclude that

$$\frac{dq^2}{dt} + 4\varphi q^2 \leq 0, \quad 0 \leq t \leq \varepsilon. \tag{9.17}$$

This means that, firstly,

$$q^2(t) \leq q_0^2\exp(-4\int_0^t \varphi d\tau), \quad 0 \leq t \leq \varepsilon \tag{9.18}$$

and, secondly, that either $dq/dt < 0$ for $t=\varepsilon$ or $q(\varepsilon)=0$ (in the latter case, the equality $dq/dt = 0$ at $t=\varepsilon$ is possible).

In the former case the function $q(t)$ continues to decrease strictly if t is greater than ε (but sufficiently close to ε), hence, inequality (9.18) is fulfilled also for these t. Assume now that $q=0$ for $t=\varepsilon$. As follows from (9.6)–(9.8), the function $u_z = V$ solves the problem

$$V_t - \exp(2\int_0^t \varphi d\tau)V_{zz} - V_z\int_0^z u(\xi,t)d\xi + (u+2\varphi)V = 0$$
$$\text{in}\quad R_{\varepsilon,T} = \{z,t:\ 0<z<1, \quad \varepsilon<t<T\},$$
$$V(0,t) = V(1,t) = 0, \quad \varepsilon<t<T, \quad V(z,\varepsilon)=0, \quad 0 \leq z \leq 1.$$

It is clear that $V=0$ in $\bar{R}_{\varepsilon,T}$. Therefore, $u=U(t)$ which is compatible with the mean zero condition for u only if $U=0$. As a consequence, we get $q=0$ for

$t \in [\varepsilon, T]$ and both relations (9.17),(9.18) are satisfied as equalities for t greater than ε. Thus, the set of values $t \in [0, \tau]$ for which inequality (9.17) holds is open in $[0, \tau]$. The completeness of this set is evident. Consequently, it coincides with the whole interval $[0, \tau]$.

In the case $\varphi_0 > 0$, $q_0 < 2/3\pi$, the estimate (9.18) can be continued on the whole interval $[0, \tau]$ of the supposed existence of a solution of problem (9.6)–(9.9), provided (9.13) holds true. We will show that it is fulfilled if additionally the condition $q_0 < \varphi/\sqrt{2}$ is true.

Thanks to estimate (9.18) (which together with (9.17) is valid now for any $t \in [0, \tau]$) and to relation (9.7) we arrive at the inequality

$$\frac{d\varphi}{dt} \geq -\varphi^2 - 2q_0^2 \exp(-4\int_0^t \varphi d\tau).$$

It transforms into the second order differential inequality for the function

$$\psi(t) = \exp(\int_0^t \varphi d\tau),$$

namely,

$$\ddot{\psi} \geq -\frac{2q_0^2}{\psi^3} \tag{9.19}$$

(here and below a dot denotes the derivative of a function depending only on t). The initial data to (9.19) is obtained from (9.9) and the definition of the function ψ; they have the form

$$\psi(0) = 1, \quad \dot{\psi}(0) = \varphi_0 \tag{9.20}$$

Since $\dot{\psi}(0) = \varphi\psi > 0$ for $0 \leq t \leq \tau$, inequality (9.19) is equivalent to

$$\dot{\psi}\ddot{\psi} \geq -2q_0^2 \frac{\dot{\psi}}{\psi^3}.$$

Integration of this inequality with the initial conditions (9.20) gives

$$\dot{\psi}^2 \geq \varphi_0^2 - 2q_0^2 + \frac{2q_0^2}{\psi^2}.$$

From here, due to the conditions $q_0 < \varphi_0/\sqrt{2}$ and (9.20), we obtain the lower bound for the function ψ:

$$\psi \geq [(\varphi_0^2 - 2q_0^2)t^2 + 2\varphi_0 t + 1]^{1/2}, \quad 0 \leq t \leq \tau. \tag{9.21}$$

The lower bound of $\dot{\psi}$ is evident,

$$\dot{\psi}^2 \geq (\varphi_0^2 - 2q_0^2)^{1/2}. \tag{9.22}$$

Let us derive upper estimates for ψ and $\dot{\psi}$. According to (9.7),

$$\dot{\varphi} \leq -\varphi^2 \tag{9.23}$$

which is equivalent to

$$\ddot{\psi} \leq 0$$

by the definition of ψ. Integrating this inequality twice and using the initial conditions (9.20) we conclude that

$$\dot{\psi} \leq \varphi_0, \quad \psi \leq 1 + \varphi_0 t \quad \text{for} \quad t \in [0, \tau]. \tag{9.24}$$

Recall now that $\varphi = \dot{\psi}/\psi$. Hence we obtain from (9.22), (9.24) the lower estimate

$$\varphi \geq \frac{\varphi_0 \delta}{1 + \varphi_0 t}, \quad 0 \leq t \leq \tau, \tag{9.25}$$

of φ. Since the right-hand side of (9.25) does not depend on τ, it can be continued on the whole interval $[0, T]$. Using (9.23) and (9.25) we get the upper bound for φ:

$$\varphi \leq \frac{\varphi_0}{1 + \varphi_0 t}, \quad 0 \leq t \leq \tau, \tag{9.26}$$

which is also independent of τ; meaning that in fact (9.26) is true for $t \in [0, T]$. Thus, inequalities (9.12) are proved. Finally, inequalities (9.25) and (9.18) (in which one can consider $0 \leq t \leq T$) provide the required inequality (9.11), and the lemma is proved. □

For further investigation of problem (9.6)–(9.9) by well-known methods of the theory of quasilinear parabolic equations [11], it is important to note that the estimates (9.11), (9.12) hold true uniformly in T. (In connection with problems of this type an appropriate technique was developed in Chapter 7 of the monograph [7].) Here we formulate only the final result.

Theorem 6. *Assume that the function $u_0(z)$ and the constant φ_0 satisfy the following conditions: (i) $\varphi_0 > 0$; (ii) $u_0 \in C^{2+\alpha}[0,1]$ with $0 < \alpha < 1$, $u_0'(0) = u_0'(1) = 0$; (iii) the mean value of u_0 on $(0,1)$ equals to zero; (iv) $\|u_0\|_2 < \min(2/3\pi, \varphi_0/\sqrt{2})$. Then problem (9.6)–(9.9) has a unique solution $u \in C^{2+\alpha,\, 1+\alpha/2}(\bar{\Pi}_T)$, $\varphi \in C^1[0,T]$ for arbitrary $T > 0$.*

Acknowledgement

The author is grateful to Professors V.A. Galaktionov and J.L. Vazquez for stimulating discussions and for the chance to read their paper [12] as a manuscript.

References

[1] Solonnikov, V.A.," Solvability of the problem on motion of a viscous incompressible liquid bounded by a free surface", Izv. Akad. Nauk SSSR Ser. Mat. **41** No. 6 (1977), 1388–1424 (in Russian).

[2] Solonnikov, V.A., "Solvability of the problem on evolution of isolated volume of a viscous capillary liquid", Zap. Nauchn. Sem. Leningrad. Otdel. Mat. Inst. Steklov **140** (1984), 179–186 (in Russian).

[3] Solonnikov, V.A., "On non-stationary motion of a finite liquid mass bounded by free surface", Zap. Nauchn. Sem. Leningrad. Otdel. Mat. Inst. Steklov **152** (1986), 137–157 (in Russian).

[4] Ovsiannikov, L.V., "Partial invariance", Soviet Math. Dokl. **186** No. 1 (1969), 22–25 (in Russian).

[5] Pukhnachov, V.V., "Invariant solutions of the Navier-Stokes equations describing free boundary motions", Soviet Math. Dokl. **202** No. 2 (1972), 303–305 (in Russian).

[6] Ovsiannikov, L.V., *Group Analysis of Differential Equations.* Academic Press, 1982.

[7] Andreev, V.K., Kaptsov, O.V., Pukhnachov, V.V. and Rodionov, A.A., *Application of Group-Theoretic Methods in Hydrodynamics.* Kluwer, Dordrecht, 1998.

[8] Pukhnachov, V.V., "Non-stationary motion of a viscous fluid with a free boundary described by partially invariant solutions of the Navier-Stokes equations", Dinamica Sploshnoi Sredy, Akad. Nauk SSSR Sibirsk. Otdel., Inst. Gidrodinamiki, Novosibirsk, **10** (1972), 125–137 (in Russian).

[9] Solonnikov, V.A., "Estimates of solutions of one initial boundary value problem for linear non-stationary system of the Navier-Stokes equations", Zap. Nauchn. Sem. Leningrad. Otdel. Mat. Inst. Steklov **59** (1976), 178–254 (in Russian).

[10] Ovsiannikov, L.V., "General equations and examples". In: *Problem on non-stationary motion of a liquid with free boundary.* Nauka, Novosibirsk (1967), 5–75 (in Russian).

[11] Ladyzhenskaya, O.A., Solonnikov, V.A. and Uraltseva N.N., *Linear and Quasilinear Equations of Parabolic Type.* Amer. Math. Soc., Providence, RI, 1968.

[12] Galaktionov, V.A. and Vazquez, J.L., "Blow-up of a class of solutions with free boundary for the Navier-Stokes equations", (to appear in Advances in Differential Equations).

V.V. Pukhnachov
Lavrentyev Institute of Hydrodynamics
Novosibirsk
630090 Russia

Progress in Nonlinear Differential Equations
and Their Applications, Vol. 35
© 1999 Birkhäuser Verlag Basel/Switzerland

A Note on Heteroclinics for a Hamiltonian System of Double Pendulum Type

Paul H. Rabinowitz*

Dedicated to Herbert Amann for his 60th birthday

1. Introduction

Consider the Hamiltonian system

$$\ddot{q} + V'(q) = 0 \tag{HS}$$

where V satisfies

(V_1) $V \in C^2(\mathbb{R}^2, \mathbb{R})$ and is 1-periodic in q_1, q_2, the components of q,

and

(V_2) $V(0) = 0 > V(x)$, $x \in \mathbb{R}^2 \backslash \mathbb{Z}^2$.

By (V_2), (HS) can be considered on $\mathbb{R}^2/\mathbb{Z}^2 = T^2$ or on the covering space $\mathbb{R}^2$. The latter point of view will mainly be adopted here. Under (V_1)–(V_2), (HS) can be viewed as a model system of double pendulum type. Indeed the ideas employed here apply equally well to the Lagrangian system associated with

$$\sum_{i,j=1}^{2} a_{ij}(q)\dot{q}_i\dot{q}_j - V(q)$$

where $A(q) = (a_{ij}(q))$ is a positive definite matrix for all $q \in \mathbb{R}^2$ and satisfies the periodicity and smoothness assumptions of (V_1). Moreover such Lagrangian systems include the actual double pendulum. However to minimize technicalities, we prefer to work with the simpler system (HS).

For each $k \in \mathbb{Z}^2 \backslash \{0\}$, let

$$\begin{aligned} F_k = \{ q \in W^{1,2}_{\text{loc}}(\mathbb{R}^2, \mathbb{R}) \mid & \text{ there is a } T = T(q) > 0 \\ & \text{such that } q(t+T) = q(t) + k \text{ for all } t \in \mathbb{R} \}. \end{aligned} \tag{1.1}$$

Viewed on T^2, F_k is a set of closed curves of homotopy type k. For $q \in F_k$, set

$$I_k(q) = \int_0^{T(q)} L(q)dt \tag{1.2}$$

---

* This research was sponsored by the NSF under grant #MCS 8110556.

where

$$L(q) = \frac{1}{2}|\dot{q}|^2 - V(q),$$

the Lagrangian associated with (HS). Taking $k = (k_1, k_2)$ with k_1, k_2 relatively prime and setting

$$c_k = \inf_{q \in F_k} I_k(q), \tag{1.3}$$

it is not difficult to show that either (i) there is a $q_k \in F_k$ such that $I_k(q_k) = c_k$ and q_k is a solution of (HS) or (ii) if

$$G_k = \{q \in W^{1,2}_{\text{loc}}(\mathbb{R}^2, \mathbb{R}) \mid q(-\infty) = 0,\ q(\infty) = k\}, \tag{1.4}$$

$$I(q) = \int_{-\infty}^{\infty} L(q)dt,$$

and

$$\overline{c_k} = \inf_{G_k} I(q), \tag{1.5}$$

there is a heteroclinic solution Q_k of (HS) with $Q_k \in G_k$ and $I(Q_k) = \overline{c_k} = c_k$. See e.g. Bolotin-Negrini [1] or Rabinowitz [6] for details. As was shown in [1] and [6], a sufficient condition for (i) to occur is

$$c_k < \overline{c_k}. \tag{1.6}$$

Heuristically (1.6) can be interpreted as saying that there is a mountain range surrounding 0 making $I(q) > I(p) + \alpha$ for each $q \in G_k$ and some $p \in F_k$ and $\alpha > 0$. A family of examples where (1.6) is satisfied can be found e.g. in [1] or [2].

Suppose the geometrical condition (1.6) holds. Let

$$\mathcal{P}_k = \{q \in F_k \mid I_k(q) = c_k\}.$$

If $p \in \mathcal{P}_k$, so is

$$\tau_\theta p(t) = p(t - \theta) \tag{1.7}$$

for all $\theta \in \mathbb{R}$ as is $p + j$ for all $j \in \mathbb{Z}^2$. It is not difficult to show $\mathcal{P}_k \cap \mathbb{Z}^2 = \emptyset$. See e.g. [6]. Consider the component of $\mathbb{R}^2 \setminus \bigcup_{p \in \mathcal{P}_k} p(\mathbb{R})$ which contains 0. Its boundary consists of a pair of elements of $\mathcal{P}_k$: v^+ and v^- (modulo the translations given by (1.7)). Normalizing v^+, v^- appropriately, it was shown in [6] (and also implicitly in [1]) that when (V_1)–(V_2) and (1.6) hold, there are four heteroclinic solutions of (HS) emanating from 0 and asymptotic to v^+ and v^-. More precisely, there is a solution Q^+ of (HS) with $Q^+(t) \to 0$ as $t \to -\infty$ and $Q^+(t) - v^+(t) \to 0$ as $t \to \infty$ and a solution $P^+(t)$ with $P^+(t) \to 0$ as $t \to -\infty$ and $P^+(t) - v^-(t) \to 0$ as $t \to \infty$. Likewise there exist a second pair of solutions of (HS), $Q^-(t)$, $P^-(t)$ heteroclinic to 0 and to $v^+(-t)$ and $v^-(-t)$ respectively. These results are proved in [1] using a Jacobi metric and approximation arguments in the spirit of earlier work on heteroclinic geodesics on T^2 by Morse [5] and Hedlund [4]. A more direct variational approach was taken in [6]. The existence of chaotic solutions to (HS) under the basic framework of $(V_1)-(V_2)$ and (1.6) plus some additional hypothesis has been studied in [2]. A higher dimensional analogue of [2] which requires an additional symmetry has been treated in [3].

The main goal of this paper is to show that if (1.6) is strengthened, (HS) possesses an additional heteroclinic solution. More precisely there is a solution $Q^*(t)$ with either $Q^*(t) - v^-(t) \to 0$ as $t \to -\infty$ and $Q^*(t) - v^+(t) \to 0$ as $t \to \infty$ or $Q^*(t) - v^+(t) \to 0$ as $t \to -\infty$ and $Q^*(t) - v^-(t) \to 0$ as $t \to \infty$. Such solutions are like the heteroclinic geodesics determined by Morse and Hedlund. This result is based on the variational approach of [6]. Therefore the variational construction of [6] will be recalled in §2 and in fact will be simplified a bit. A crucial lower bound for the 'renormalized functional' will also be obtained. Then in §3, the new heteroclinic, Q^*, will be constructed.

2. A variational framework

The variational method used in [6] to construct the heteroclinics $Q^\pm$, $P^\pm$ will be sketched since it will be required both for some estimates on $Q^\pm$, $P^\pm$ and to obtain Q^*.

Let $\mathcal{R}$ denote the closed region in $\mathcal{R}^2$ bounded by v^+ and v^-. As was shown in [6], $\mathbb{Z}^2 \cap \mathcal{R} = \mathbb{Z}k$ and there is a $u_1 \in G_k$ heteroclinic to 0 and k with $u_1(\mathbb{R}) \subset \mathcal{R}$. Indeed $I(u_1) = \overline{c_k}$. For $i \in \mathbb{N}$, set $u_i = u_1 + (i-1)k$ and for $i \in -\mathbb{N}$, set $u_i = u_1 + ik$. Let $\mathcal{R}^\pm$ denote the region between $v^\pm$ and $\bigcup_{i\neq 0} u_i(\mathbb{R})$.

Each of the regions $\mathcal{R}^\pm$ will be further subdivided. As was shown in [6], there are a pair of solutions $z^\pm$ of (HS) with the properties that (i) $q^\pm(0) \in v^\pm(\mathbb{R})$, (ii) $z^\pm(\infty) = 0$, and (iii) $z^\pm$ minimizes

$$\int_0^\infty L(q)dt$$

over the class of $W^{1,2}_{\text{loc}}$ curves satisfying (i) and (ii). The curves $z^\pm$ may not be unique in which case the representative with the largest t for which $q(0) = v^\pm(t)$ is chosen. Set $z_0^\pm = z^\pm$, and $z_i^\pm = z^\pm + ik$, $i \in \mathbb{Z}$.

The time parametrization on $v^\pm$ can now be normalized by taking $v^\pm(0) = z^\pm(0) = z_0^\pm(0)$. Let $\mathcal{R}_1^\pm$ be the region bounded by $z_0^\pm(\mathbb{R}^+)$, $u_1(\mathbb{R})$, $z_1^\pm(\mathbb{R}^+)$, and $v^\pm([0, T^\pm])$ where $T^\pm = T(v^\pm)$ in (1.1). Finally set $\mathcal{R}_i^\pm = \mathcal{R}_1^\pm + (i-1)k$ for $i \in \mathbb{N}$ and $\mathcal{R}_i^\pm = \mathcal{R}_1^\pm + ik$ if $i \in -\mathbb{N}$.

The heteroclinics Q^+, P^+ lie in $\mathcal{R}^+$, $\mathcal{R}^-$ respectively and are obtained from a minimization problem that will be described next. Let $q \in W^{1,2}_{\text{loc}}(\mathbb{R}, \mathbb{R}^2)$ such that q satisfies

$(\Gamma_1^\pm)$ $\quad q(-\infty) = 0$,

$(\Gamma_2^\pm)$ $\quad q(\mathbb{R}) \subset \mathcal{R}^\pm$,

$(\Gamma_3^\pm)$ $\quad q$ intersects $z_i^\pm(\mathbb{R}^+)$ for all $i \in \mathbb{N}$.

For $i \in \mathbb{N}$, let $t_i = t_i(q)$ denote the smallest value of t such that $q(t) \in z_i^\pm(\mathbb{R}^+)$. Define $s_i = s_i(q) \in \mathbb{R}^+$ via $z_i^\pm(s_i) = q(t_i)$. It is further required that

$(\Gamma_4^\pm)$ $\quad s_{i+1}(q) \le s_i(q)$ $i \in \mathbb{N}$.

Define

$$\Gamma^{\pm} = \{q \in W^{1,2}_{\rm loc} \mid q \text{ satisfies } (\Gamma^{\pm}_1)\text{–}(\Gamma^{\pm}_4)\}.$$

Remark 2.1. In [6], it was further required that if $q \in \Gamma^{\pm}$, $q|_{-\infty}^{t_1(q)}$ lies in $\mathcal{R}^{\pm}_1$ and $q|_{t_i(q)}^{t_{i+1}(q)}$ lies in $\mathcal{R}^{\pm}_{i+1}$, $i \geq 1$. As will be seen from the existence proof for $\overline{Q}$ in §3, this assumption can be dropped.

Now a functional J_+ will be defined in $\Gamma^{\pm}$. For $q \in \Gamma^{\pm}$, set

$$a_1(q) = \int_{-\infty}^{t_1(q)} L(q)dt - c_k \tag{2.2}$$

and for $i \geq 2$, let

$$a_i(q) = \int_{t_{i-1}(q)}^{t_i(q)} L(q)dt - c_k. \tag{2.3}$$

For $q \in \Gamma^{\pm}$, define

$$J_+(q) = \sum_1^{\infty} a_i(q) \tag{2.4}$$

and set

$$c^{\pm} = \sup_{q \in \Gamma^{\pm}} J_{\pm}(q). \tag{2.5}$$

In [6], it was proved that:

Theorem 2.6. *Suppose V satisfies (V_1)–(V_2) and (1.6) holds. Then there is a $Q^+ \in \Gamma^+$ (resp. $P^+ \in \Gamma^-$) such that $J_+(Q^+) = c^+$ (resp. $J_+(P^+) = c^-$). Moreover Q^+ (resp. P^+) is a solution of (HS) with $Q^+(t)$ (resp. $P^+(t)$) $\to 0$ as $t \to -\infty$ and $Q^+(t) - v^+(t)$ (resp. $P^+(t) - v^-(t)$) $\to 0$ as $t \to \infty$.*

An analogous result gives the heteroclinics Q^-, P^- mentioned in the introduction as minimizers of $J_-(q)$ over $\Lambda^{\pm}$ where

$$\Lambda^{\pm} = \{q \in W^{1,2}_{\rm loc}(\mathbb{R}, \mathbb{R}^2) \mid q \text{ satisfies } (\Lambda^{\pm}_1)\text{–}(\Lambda^{\pm}_4)\},$$

$(\Lambda^{\pm}_i) = (\Gamma^{\pm}_i)$, $i = 1, 2$,
$(\Lambda^{\pm}_3)$ q intersects $z^{\pm}_i(\mathbb{R}^+)$, $i \in -\mathbb{N}$
and
$(\Lambda^{\pm}_4)$ $s_{i-1}(q) \leq s_i(q)$, $i \in -\mathbb{N}$ and $z_i(s_i(q)) = q(t_i)$, $i \in -\mathbb{N}$.
Now for $i \in -\mathbb{N}$, $t_i(q)$ denotes the smallest value of t such that $q(t_i) \in z_i(\mathbb{R}^+)$. The function $J_-(q)$ is defined as

$$J_-(q) = \sum_{-\infty}^{-1} a_i(q) \quad \text{where} \quad a_{-1}(q) = \int_{-\infty}^{t_i(q)} L(q)dt - c_k$$

and for $i \leq -2$

$$a_i(q) = \int_{t_{i+1}(q)}^{t_i(q)} L(q)dt - c_k.$$

Remark 2.7. Since the numbers $a_i(q)$, $i \in \mathbb{Z}$ may be positive or negative, it is not evident that any of the four minimization values, $J_+(Q^+)$, $J_+(P^+)$, $J_-(Q^-)$, $J_-(P^-)$ are positive. The next result shows that at least a pair of these numbers are positive.

Proposition 2.8.
$$J_+(Q^+) + J_-(Q^-) \geq \bar{c}_k - c_k;$$
$$J_+(P^+) + J_-(P^-) \geq \bar{c}_k - c_k.$$

Proof. Let $\epsilon > 0$. By the definition of $J_\pm$,

$$J_+(Q^+) + J_-(Q^-) = \sum_{i=1}^{\infty} a_i(Q^+) + \sum_{i=-\infty}^{-1} a_i(Q^-). \tag{2.9}$$

Since each of the series converge absolutely – see [6] – for n sufficiently large,

$$\sum_{i \geq -n} |a_i(Q^+)| + \sum_{i \leq -n} |a_i(Q^-)| < \epsilon. \tag{2.10}$$

Gluing the curves $Q^+\big|_{-\infty}^{t_n(Q^+)}$, $z_n^+\big|_{s_n(Q^+)}^{s_{-n}(Q^-)}$, and $(Q^-(-t) + 2nk)\big|_{-t_{-n}(Q^-)}^{\infty}$ produces a function φ joining 0 to $2nk$, i.e. $\varphi \in G_{2nk}$. For n large, $s_n(Q^+)$, $s_{-n}(Q^-)$ are each near 0. Therefore by (2.10),

$$\Big| \int_{\mathbb{R}} L(\varphi)dt - 2nc_k - J_+(Q^+) - J_-(Q^-) \Big| < 2\epsilon. \tag{2.11}$$

Consider $\varphi(t) + k$. For t near $-\infty$, $\varphi(t) + k$ lies in the region bounded by $\varphi(\mathbb{R})$ and $\bigcup_1^{2n} u_i(\mathbb{R})$ while for t near ∞, $\varphi(t) + k$ lies outside this region. Therefore for some $\bar{t} < \bar{s}$, $\varphi(\bar{t}) + k = \varphi(\bar{s})$, i.e. there are a pair of points on the curve φ which are a distance k apart. By the definition of c_k,

$$\int_{\bar{t}}^{\bar{s}} L(\varphi)dt - c_k \geq 0. \tag{2.12}$$

Excising $\varphi\big|_{\bar{t}}^{\bar{s}}$ from $\varphi(\mathbb{R})$ and gluing $\varphi\big|_{-\infty}^{\bar{t}}$ to $\varphi\big|_{\bar{s}}^{\infty} - k$ and using (2.11) produces a new curve $\varphi_1 \in G_{(2n-1)k}$ with

$$\int_{-\infty}^{\infty} L(\varphi_1)dt - (2n-1)c_k \leq J_+(Q^+) + J_-(Q^-) + 2\epsilon. \tag{2.13}$$

Repeating the argument that generates φ_1 from φ another $(2n-2)$ times gives $\bar{\varphi} \in G_k$ such that

$$\int_{-\infty}^{\infty} L(\bar{\varphi})dt - c_k \leq J_+(Q^+) + J_-(Q^-) + 2\epsilon. \tag{2.14}$$

Since $\bar{\varphi} \in G_k$,

$$\int_{-\infty}^{\infty} L(\bar{\varphi})dt \geq \bar{c}_k, \tag{2.15}$$

and therefore

$$\bar{c}_k - c_k \leq J_+(Q^+) + J_-(Q^-) + 2\epsilon. \tag{2.16}$$

Now letting $\epsilon \to 0$ gives the first inequality in the statement of Proposition 2.8. The second follows similarly. □

Remark 2.17. A solution q of (HS) will be said to be a *minimal solution* of (HS) if for all $\sigma < s$ and for all $w \in W^{1,2}_{\mathrm{loc}}$ for which $w(\alpha) = q(\sigma)$, $w(\beta) = q(s)$ for some α, β, then

$$\int_\sigma^s L(q)dt < \int_\alpha^\beta L(w)dt, \tag{2.18}$$

unless $\beta - \alpha = s - \sigma$ and w coincides with q. Thus q minimizes the Lagrangian integrated over any pair of points on $q(\mathbb{R})$.

Lemma 2.19. *Q^+ is a minimal solution of (HS).*

Proof. The variational characterization of Q^+ implies (2.18) with $\leq$ holds for all $\sigma < s$. Moreover if there were equality, excising $Q^+\big|_\sigma^s$ from Q^+ and replacing it by $w\big|_\alpha^\beta$ shows the resulting function Q belongs to Γ^+ and $J_+(Q) = J_+(Q^+) = c^+$. Therefore Q is a solution of (HS) by the argument of the proof of Theorem 2.6. But Q coincides with Q^+ for $t \notin [\sigma, s]$. Therefore by the basic uniqueness theorem for ordinary differential equations, $Q^+(t) \equiv Q(t)$. □

Remark 2.20. Likewise Q^-, $P^\pm$ are minimal solutions of (HS).

Remark 2.21. An argument similar to that of Lemma 2.19 shows that if $Q \in \Gamma^+$ is another solution of (HS), then $Q \cap Q^+ = \{0\}$.

3. A new heteroclinic solution

This section establishes the existence of a solution of (HS) that is heteroclinic to v^- and v^+ or to v^+ and v^- provided that (1.6) is strengthened. Two observations are needed to formulate the result. First observe that either

$$J_+(Q^+) + J_-(P^-) \geq \bar{c}_k - c_k \tag{3.1}$$

or

$$J_+(P^+) + J_-(Q^-) \geq \bar{c}_k - c_k \tag{3.2}$$

because if both bail, Proposition 2.8 is violated. In what follows it will be assumed that (3.1) holds. Then there is a solution of (HS) heteroclinic to v^- (or P^-) and V^+ (or Q^+).

Next consider the class of $W^{1,2}_{\mathrm{loc}}$ curves ψ joining a point $x^- \in v^-(\mathbb{R})$ to a point $x^+ \in v^+(\mathbb{R})$. Further normalize matters by requiring that x^- lies between $v^-(0)$ and $v^-(T^-)$. Consider the variational problem of minimizing

$$\int L(\psi)dt \tag{3.3}$$

over the class of such curves. Straightforward arguments such as in [6] show that there is a minimizer φ to this problem with φ either being the chain consisting of z_0^- and z_0^+ or there is an $S \in (0,\infty)$ with $\varphi \in C^2([0,S],\mathbb{R}^2)$. Note that the curve φ lies between v^-, v^+, $z_0^- \cup z_0^-$, and $z_1^- \cup z_1^-$ for otherwise by replacing a segment of φ by a corresponding segment of one of these curves, (3.3) would be violated. Set

$$b \equiv \int L(\varphi)dt. \tag{3.4}$$

Note that

$$b \leq \int_0^\infty (L(z^-) + L(z^+))dt. \tag{3.5}$$

The strengthened version of (1.6) we require is

$$\bar{c}_k - c_k > b + c_k. \tag{3.6}$$

As with (1.6), (3.6) can be interpreted heuristically as a statement that the mountain range surrounding 0 is 'high' enough.

The curve φ need not be a minimizer of the above variational problem. All that is needed is a φ joining v^- and v^+ so that (3.6) holds. Indeed as will be seen in Lemma 3.15, if $\varphi(0) = v^-(\alpha)$ and $\varphi(S) = v^+(\beta)$, it suffices that

$$\int_0^\alpha L(v^-)dt + \int_0^S L(\varphi)dt + \int_\beta^{T^+(v^+)} L(v^+)dt < \bar{c}_k - c_k.$$

A precise existence theorem can now be stated.

Theorem 3.7. *If (V_1)–(V_2), (3.1), and (3.6) are satisfied, there is a solution, Q^*, of (HS) such that $Q^*(t) - v^-(t) \to 0$ as $t \to -\infty$ and $Q^*(t) - v^+(t) \to 0$ as $t \to \infty$.*

The proof of Theorem 3.7 parallels that of Theorem 1.12 of [6]. First a class of admissible functions, Γ^*, will be introduced. Then a functional, J^* is defined on Γ^* and Q^* is obtained as the infimum of J^* on Γ^*. Since several of the details here are the same as their counterparts in [6], when appropriate we refer to [6].

Let $\mathcal{R}^\pm$, u_i, $z_i^\pm$ be as in §2. Set

$$\Gamma^* = \{q \in W^{1,2}_{\text{loc}}(\mathbb{R},\mathbb{R}^2) \mid q \text{ satisfies } (\Gamma_1^*)\text{–}(\Gamma_4^*)\}$$

where

(Γ_1^*) $q(t)$ lies in $\mathcal{R}^-$ for $t \leq 0$, and in $\mathcal{R}^+$ for $t \geq 0$.

(Γ_2^*) $q(0) \in u_1(\mathbb{R})$.

(Γ_3^*) q crosses $z_i^-(\mathbb{R}^+)$ for all $i \leq 0$ and $z_i^+(\mathbb{R}^+)$ for all $i \geq 1$.

For $i > 0$ define $t_i = t_i(q)$ to be the smallest value of $t \geq 0$ such that $q(t) \in z_i^+(\mathbb{R}^+)$ and for $i \leq 0$, let t_i be the largest value of $t < 0$ such that $q(t_i) \in z_i^-(\mathbb{R}^+)$. Define the corresponding numbers s_i by $q(t_i) = y(s_i)$ with $y = z_i^+$ if $i \geq 1$ and $y = z_i^-$ if $i \leq 0$. Finally we require

$$(\Gamma_4^*) \quad \begin{array}{ll} s_{i+1}(q) \leq s_i(q) & \text{for } i \geq 1 \\ s_{i-1}(q) \leq s_i(q) & \text{for } i \leq 0. \end{array}$$

Next for $i \in \mathbb{N}$, define

$$a_i(q) = \int_{t_{i-1}(q)}^{t_i(q)} L(q)dt - c_k$$

and for $i \in -\mathbb{N}$,

$$a_i(q) = \int_{t_i(q)}^{t_{i+1}(q)} L(q)dt - c_k.$$

Set

$$J^*(q) = \sum_{|i| \in \mathbb{N}} a_i(q) \tag{3.8}$$

and

$$c^* = \inf_{q \in \Gamma^*} J^*(q). \tag{3.9}$$

As in [6, Proposition 3.6], if $q \in \Gamma^*$,

$$J^*(q) \geq -\int_0^\infty (L(z^+) + L(z^-))dt \equiv -K \tag{3.10}$$

and if $J^*(q) < \infty$,

$$\sum_{|i| \in \mathbb{N}} |a_i(q)| \leq J^*(q) + 2K. \tag{3.11}$$

Indeed to verify (3.10), note that e.g. for $i > 0$, since $s_{i+1} \leq s_i$, gluing $q\Big|_{t_i}^{t_{i+1}}$ to $z_{i+1}^+\Big|_{s_{i+1}}^{s_i}$ yields a function φ whose endpoints differ by k. Hence extending φ as in (1.1), $\varphi \in F_k$ so

$$\int_{t_i}^{t_{i+1}} L(q)dt + \int_{s_{i+1}}^{s_i} L(z_{i+1}^+)dt \geq c_k$$

or

$$a_{i+1}(q) \geq -\int_{s_{i+1}}^{s_i} L(z_{i+1}^+)dt. \tag{3.12}$$

A similar estimate holds for each $i \in \mathbb{Z}$ and adding them produces the basic lower bound (3.10).

As in Proposition 3.12 of [6], if $J^*(q) < \infty$,

$$t_{i+1}(q) - t_i(q) \to T^+ \tag{3.13}$$

and

$$\|q - v^+\|_{L^\infty[t_i, t_{i+1}]} \to 0 \quad \text{as } i \to \infty. \tag{3.14}$$

A similar statement with T^-, v^- obtains for $i \to -\infty$. Thus any $q \in \Gamma^*$ for which $J^*(q) < \infty$ is finite possesses the asymptotic behavior required for Theorem 3.7.

The next result shows in part the role of (3.6).

Lemma 3.15. $c^* \leq b + c_k$.

Proof. If φ is the minimizer of (3.3), suppose $\varphi(0) = v^-(\alpha)$ and $\varphi(S) = v^+(\beta)$. Define q as follows:

$$\begin{aligned} q(t) &= v(t), \quad t \le \alpha \\ &= \varphi(t-\alpha), \quad \alpha \le t \le \alpha + S \\ &= v + (t + \beta - \alpha - S), \quad t \ge \alpha + S. \end{aligned} \tag{3.16}$$

Set $p(t) = \tau_\theta q(t)$. Choosing θ so that $p(0) \in u_1(\mathbb{R})$, p belongs to Γ^* and

$$\begin{aligned} J^*(q) = a_1(q) = \int_0^\alpha L(v^-)dt + + \int_0^S L(\varphi)dt + \int_\beta^{T^+} L(v^*)dt - c_k \\ \le b + c_k. \end{aligned} \tag{3.17}$$

Hence the Lemma follows from (3.9) □

As a useful consequence of Lemma 3.15, note that any minimizing sequence (q_m) for (3.9) can be assumed to satisfy $J^*(q_m) \le b + c_k$.

Let $\rho_k = \bar{c}_k - 2c_k - b$ so $\rho_k > 0$ by (3.6).

Proposition 3.18. *Suppose $q \in \Gamma^*$. If $q(t)$ is near 0 or k, $J^*(q) \ge c_k + b + \rho_k/2$.*

Proof. By 'near' is meant small compared to ρ_k. Thus suppose that $q(t)$ is near 0 for some t, e.g. $t = \sigma$. We distinguish between three possibilities:

(i) $\sigma \le t_0(q)$,

(ii) $t_0(q) < \sigma \le 0$,

(iii) $0 < \sigma$.

For each of these cases, it will be shown that modifying q slightly produces a new curve $p = p^+ \cup p^-$ with $p^+ \in \Gamma^+$, $p^- \in \Lambda^-$, and

$$J^*(q) - J_+(p^+) - J_-(p^-) \ge -\epsilon \tag{3.19}$$

where e.g. $0 < \epsilon < \frac{1}{2}\rho_k$. By the variational problems that determine Q^+, P^- and by Proposition 2.8

$$J_+(p^+) + J_-(p^-) \ge J_+(Q^+) + J_-(P^-) \ge \bar{c}_k - c_k. \tag{3.20}$$

Combining (3.19)–(3.20) shows

$$J^*(q) \ge \bar{c}_k - c_k - \epsilon = \rho_k + c_k + b - \epsilon \ge c_k + b + \frac{\rho_k}{2}. \tag{3.21}$$

Now suppose that (i) occurs. Let $s \in \mathbb{R}$ be such that $q(0) = u_1(s)$. Replace $q\big|_\sigma^0$ by e.g. $w\big|_0^\infty$ glued to $u_1\big|_{-\infty}^s$ where w minimizes

$$\int_0^\infty L(\varphi)dt$$

over the class of $W^{1,2}_{\text{loc}}$ curves with $w(0) = q(\sigma)$ and $w(\infty) = 0$. Since u_1 is a minimal solution of (HS),

$$\int_\sigma^0 L(q)dt \ge \int_0^\infty L(w)dt + \int_0^s L(u_1)dt - \epsilon. \tag{3.22}$$

Now define p^- to be the curve obtained by gluing $q\big|_{-\infty}^{\sigma}$ to $w\big|_0^{\infty}$ so $q^- \in \Lambda^-$. Take p^+ to be the curve obtained by gluing $u_1\big|_{-\infty}^{s}$ to $q\big|_0^{\infty}$ so $p^+ \in \Gamma^+$. Finally note that (3.19) follows from (3.22).

Next suppose that (ii) occurs. Let p^- be obtained by gluing $q\big|_{-\infty}^{t_0(q)}$ to $z_0^-\big|_{s_0(q)}^{\infty}$ where s_0 is defined via $z_0^-(s_0(q)) = q(t_0(q))$. Therefore $p^- \in \Lambda^-$. Let p^+ be obtained by gluing $u_1\big|_{-\infty}^{s}$ to $q\big|_0^{\infty}$ where $q(0) = u_1(s)$ as in case (i). Therefore $p^+ \in \Gamma^+$. Due to the minimality of z_0^- and u_1,

$$\int_{t_0(q)}^{0} L(q)dt \geq \int_{s_0(q)}^{\infty} L(z_0^-)dt + \int_{-\infty}^{s} L(u_1)dt - \epsilon \tag{3.23}$$

and (3.23) implies (3.19) for $p^- \cup p^+$.

Finally suppose that (iii) obtains. Let p^- be as in case (ii) and let p^+ denote $w\big|_{-\infty}^{0}$ glued to $q\big|_{\sigma}^{\infty}$ where w minimizes

$$\int_{-\infty}^{0} L(p)dt$$

over curves p joining 0 to $q(\sigma)$. As above

$$\int_{t_0(q)}^{\sigma} L(q)dt \geq \int_{s_0(q)}^{\infty} L(z_0^-)dt - \epsilon \tag{3.24}$$

from which (3.19) follows. □

Similar arguments hold if $q(\sigma)$ is near k.

Now we turn to the proof of Theorem 3.7. It and more are consequences of

Theorem 3.25. *If (V_1)–(V_2), (3.1), and (3.6) are satisfied, there is a $Q^* \in \Gamma^*$ such that $J^*(Q^*) = c^*$. Moreover Q^* is a minimal solution of (HS).*

Proof. Let (q_m) be a minimizing sequence for (3.9). By Lemma 3.15, it can be assumed that

$$J^*(q_m) \leq b + c_k. \tag{3.26}$$

The minimizing sequence can be assumed to satisfy some further conditions:

(Γ_5^*) (i) $q_m(t) \in \mathcal{R}_1^+$ for $t \in [0, t_1(q_m)]$,

(ii) $q_m(t) \in \mathcal{R}_1^-$ for $t \in [t_0(q_m), 0]$,

(iii) $q_m(t) \in \mathcal{R}_{i+1}^+$ for $t \in [t_i(q_m), t_{i+1}(q_m)]$ and $i \geq 1$

(iv) $q_m(t) \in \mathcal{R}_i^-$ for $t \in [t_i(q_m), t_{i+1}(q_m)]$ and $i \leq -1$

(v) $q_m(t)$ lies between v^+ and $Q^+ + k$ for $t \geq t_1(q_m)$

(vi) $q_m(t)$ lies between v^- and P^- for $t \leq t_{-1}(q_m)$.

Conditions (iii) and (v), (iv) and (vi) can be combined but will be verified in the order given. In fact only (i), (iii), and (v) will be proved; the remaining cases follow similarly.

If (i) is violated, there is an interval $[\underline{\sigma},\bar{\sigma}] \subset (0,t_1(q_m))$ with $q_m(\underline{\sigma}), q_m(\bar{\sigma}) \in \overline{z_0^+(\mathbb{R}^+)}$. But then using the minimality of z_0^+, $q_m\big|_{\underline{\sigma}}^{\bar{\sigma}}$ can be replaced by the corresponding portion of z_0^+ and the new function $\bar{q}_m$ has $J^*(\bar{q}_m) \le J^*(q_m)$.

To verify (iii), if it fails, as in (i) there is an interval $[\underline{\sigma},\bar{\sigma}] \subset [t_i(q_m), t_{i+1}(q_m)]$ such that $q_m(\underline{\sigma}), q_m(\bar{\sigma}) \in \overline{z_i^+(\mathbb{R}^+)}$. But then the argument of (i) applies.

Remark. The arguments of (i) and (iii) can be used in [6] to simplify the proof of Theorem 3.5 there.

Lastly to verify (v), if it is violated, there is an interval $[\underline{\sigma},\bar{\sigma}] \subset (t_1(q),\infty]$ such that $q_m(\underline{\sigma}) \in Q^+ + k$, $q_m\big|_{\underline{\sigma}}^{\bar{\sigma}}$ lies between $Q^+ + k$ and $\bigcup_{i\ge 2} u_i(\mathbb{R})$, and $q_m(\bar{\sigma}) \in Q^+ + k$ (if $\bar{\sigma} < \infty$). Define $\underline{t},\bar{t}$ by $Q^+(\underline{t}) + k = q_m(\underline{\sigma})$, $Q^+(\bar{t}) + k = q_m(\bar{\sigma})$. The minimality property of Q^+ implies:

$$\int_{\underline{\sigma}}^{\bar{\sigma}} L(q_m)dt > \int_{\underline{t}}^{\bar{t}} L(Q^+ + k)dt = \int_{\underline{t}}^{\bar{t}} L(Q^+)dt \tag{3.27}$$

(with an appropriate modification if $\bar{\sigma} = \infty$). Thus replacing $q_m\big|_{\underline{\sigma}}^{\bar{\sigma}}$ by $(Q^+ + k)\big|_{\underline{t}}^{\bar{t}}$ decreases J^*.

Now consider the minimizing sequence (q_m) satisfying (Γ_1^*)–(Γ_5^*). As in Lemma 4.2 of [6], there is a $\beta > 0$ such that

$$t_i(q) - t_{i-1}(q) \ge \beta \tag{3.28}$$

for all $i \in \mathbb{Z}$ and $q \in \Gamma^*$. The bound (3.22) combined with (Γ_2^*) implies (q_m) is bounded in $W^{1,2}_{\text{loc}}$ (see Lemma 4.5 of [6]). Hence it can be assumed that q_m converges weakly in $W^{1,2}_{\text{loc}}$ and strongly in L^∞_{loc} to a function Q^*. It remains to show that:

(A) $Q^* \in \Gamma$

(B) $J^*(Q^*) = c^*$

(C) Q^* is a minimal solution of (HS).

Proof of (A). The L^∞_{loc} convergence of q_m implies (Γ_1^*). Likewise (Γ_2^*) follows since by Proposition 3.18 and (3.25), q_m cannot get close to 0 or k. To check (Γ_3^*), note that for each $i \in \mathbb{Z}$, there is an $A_i > 0$ such that

$$|t_i(q_m)| \le A_i. \tag{3.29}$$

Indeed as in [6], the only way in which $t_i(q_m)$ can be infinite, for fixed $i \in \mathbb{Z}$, is if $q_m(\sigma_m) \to ik$ or $(i-1)k$ along some sequence $\sigma_m \in [t_{i-1}(q_m), t_i(q_m)]$. But Proposition 3.18 and (3.25) show q_m avoids a neighborhood of 0 and k and (Γ_5^*) (ii)–(vi) does the same for the remaining values of i.

Lastly (Γ_4^*) will be verified. This is the most difficult case. Suppose that (Γ_4^*) fails for Q^*. In particular assume it fails for a smallest $i \ge 1$. (The argument for $i < 0$ is the same.) Thus

$$s_{i+1}(Q^*) > s_i(Q^*). \tag{3.30}$$

As in Proposition 2.8, consider $Q^*(t)+k$. For $t>0$ and small, $Q^*(t)+k$ lies inside the region bounded by $\bigcup_{i\ge 1}\overline{u_i(\mathbb{R})}$ and $Q^+(\mathbb{R}^+)$ while by (3.30), $Q^*(t_i(Q^*))+k$ lies outside this region. Indeed for this argument, t small can be replaced by $t_{i-1}(Q^*)$ if $i>1$. Suppose $i>1$. The case of $i=1$ is handled similarly. There are points $\xi\in(t_{i-1}(Q^*),t_i(Q^*))$ and $\eta\in(t_i(Q^*),t_{i+1}(Q^*))$ such that $Q^*(\xi)+k=Q^*(\eta)$. The L^∞_{loc} convergence of q_m to Q^* implies $q_m(t_{i-1}(Q^*))+k$ is inside the region bounded by $\bigcup_{i\ge 1}\overline{u_i(\mathbb{R})}$ and $q_m(\mathbb{R}^+)$ and $q_m(t_i(Q^*))+k$ is outside this region for all large m. Hence there are points $\xi_m\in(t_{i-1}(Q^*),t_i(Q^*))$ and $\eta_m\in(t_i(Q^*),t_{i+1}(Q^*))$ such that $q_m(\xi_m)+k=q_m(\eta_m)$. Consider $q_m\big|_{\xi_m}^{\eta_m}$. By the definition of c_k,

$$\int_{\xi_m}^{\eta_m} L(q_m)dt\ge c_k.$$

We claim there is a $\delta>0$ such that for all large m,

$$\int_{\xi_m}^{\eta_m} L(q_m)dt\ge c_k+\delta. \tag{3.31}$$

Assuming (3.31) for the moment, excising $q_m\big|_{\xi_m}^{\eta_m}$ from q_m and shifting $q_m\big|_{\eta_m}^{\infty}$ by $-k$ produces $\widehat{q}_m\in\Gamma^*$ and by (3.31),

$$J^*(\widehat{q}_m)\le J^*(q_m)-\delta \tag{3.32}$$

contrary to (q_m) being a minimizing sequence. Hence Q^* satisfies (Γ_4^*).

To verify (3.31), observe that if it fails, without loss of generality,

$$\int_{\xi_m}^{\eta_m} L(q_m)dt\to c_k.$$

This implies $\xi_m\to\bar\xi$, $\eta_m-\xi_m\to T^+$, and

$$q_m\big|_{\xi_m}^{\eta_m}\to v^+\big|_{\bar\xi}^{\bar\xi+T^+} \tag{3.33}$$

in L^∞. But

$$q_m\big|_{\xi_m}^{\eta_m}\to Q^+\big|_{\bar\xi}^{\bar\xi+T^+} \tag{3.34}$$

while by (3.30),

$$q_m(t_{i+1}(Q^*))\to Q^*(t_{i+1}(Q^*))=z_{i+1}^+(s_{i+1}(Q^*)). \tag{3.35}$$

Now (3.30) shows q_m can be modified for e.g. $t\ge\eta_m$ so that the resulting function, $\bar q_m$ lies in Γ^*, $a_{i+1}(\bar q_m)$ is small, and $a_j(\bar q_m)=0$ for $j>i+1$. Since $a_j(\bar q_m)=a_j(q_m)$ for $j\le i$, while (3.35) shows there is a $\gamma>0$ such that

$$a_{i+1}(q_m)\ge\gamma, \tag{3.36}$$

it follows that

$$J^*(\bar q_m)<J^*(q_m)-\frac{\gamma}{2} \tag{3.37}$$

contrary to (q_m) being a minimizing sequence. Thus (3.31) must hold. □

Proof of (B). First it will be shown that $J^*(Q^*) < \infty$. As in [6, Lemma 4.39],

$$\int_{t_{-n}(Q^*)}^{t_n(Q^*)} L(Q^*)dt \le \varliminf_{m\to\infty} \int_{t_{-n}(q_m)}^{t_n(q_m)} L(q_m)dt \tag{3.38}$$

so by (3.11),

$$\sum_{-n}^{n} a_i(Q^*) \le \varliminf_{m\to\infty} \sum_{-n}^{n} a_i(q_m) \le \varliminf_{m\to\infty} \sum_{-n}^{n} |a_i(q_m)| \le b + c_k + K. \tag{3.39}$$

Now

$$\sum_{-n}^{n} a_i(Q^*) \equiv \Sigma_n^+ + \Sigma_n^- \tag{3.40}$$

where Σ_n^- denotes the sum over those i for which $a_i(Q^*) < 0$ and Σ_n^+ denotes the sum over those i for which $a_i(Q^*) \ge 0$. By the argument which led to (3.10), $\Sigma_n^- \ge -K$ or

$$-\Sigma_n^- \le K. \tag{3.41}$$

Therefore by (3.39)–(3.41)

$$\Sigma_n^+ = \sum_{-n}^{n} a_i(Q^*) - \Sigma_n^- \le b + c_k + 2K. \tag{3.42}$$

It follows that

$$\sum_{-n}^{n} |a_i(Q^*)| \le b + c_k + 3K \tag{3.43}$$

independently of n so $J^*(Q^*) < \infty$.

Now to show that $J^*(Q^*) = c^*$, let $\epsilon > 0$. Since $Q^* \in \Gamma^*$ and $J^*(Q^*) < \infty$, $s_n(Q^*) \to 0$ as $|n| \to \infty$. In particular, $s_n(Q^*) \le \epsilon$ if $|n| \ge \ell$. It can also be assumed that

$$J^*(Q^*) \le \sum_{-n}^{n} a_i(Q^*) + \epsilon \tag{3.44}$$

for $n \ge \ell$. By (3.39), there is a $\varphi = \varphi(\epsilon)$ such that if $m \ge \varphi(\epsilon)$,

$$\sum_{-\ell}^{\ell} a_i(Q^*) \le \sum_{-\ell}^{\ell} a_i(q_m) + \epsilon = J^*(q_m) - \sum_{|i|>\ell} a_i(q_m) + \epsilon. \tag{3.45}$$

For m possibly still larger,

$$J^*(q_m) \le c^* + \epsilon. \tag{3.46}$$

Combining (3.44)–(3.46) yields

$$J^*(Q^*) \le c + 3\epsilon - \sum_{|i|>\ell} a_i(q_m). \tag{3.47}$$

By (3.12),

$$-\sum_{|i|>\ell} a_i(q_m) \leq \int_0^{s_\ell(q_m)} L(z_0^+)dt + \int_0^{s_{-\ell}(q_m)} L(z_0^-)dt \tag{3.48}$$

and by (Γ_5^+)(v)–(vi),

$$s_\ell(q_m) \leq s_\ell(Q^+); \;\; s_{-\ell}(q_m) \leq s_{-\ell}(P^-). \tag{3.49}$$

For ℓ large enough,

$$s_\ell(Q^t) + s_{-\ell}(P^-) < \epsilon. \tag{3.50}$$

Hence

$$J^*(Q^*) \leq c^* + 3\epsilon + \int_0^\epsilon (L(z_0^+) + L(z_0^-))dt. \tag{3.51}$$

Now ϵ arbitrary implies $J^*(Q^*) = c^*$. □

Proof of (C). That Q^* is a solution of (HS) follows as in [6] and likewise the minimality of Q^* as in Remark 2.17. □

Remark 3.52. The minimality of Q^* implies that Q^* is a simple curve in $\mathbb{R}^2$. Likewise as in [6], there are no points $-\infty < \sigma < s < \infty$ such that $Q^*(\sigma) + k = Q^*(s)$. Indeed this latter fact implies that Q^* viewed on the torus is a simple curve.

Remark 3.53. Since Q^* asymptotes to $v^\pm$ as $t \to \pm\infty$ and $v^\pm$ are each zero energy solutions of (HS) – see [6] – Q^* is also a zero energy solution of (HS).

References

[1] Bolotin, S. V. and P. Negrini, "Variational criteria for nonintegrability", to appear Russian J. of Math. Phys.

[2] Bolotin, S. V. and P. H. Rabinowitz, "A variational construction of chaotic trajectories for a Hamiltonian system on a torus", to appear Boll. Un. Mat. Ital.

[3] Bolotin, S. V. and P. H. Rabinowitz, "A variational construction of chaotic trajectories for a reversible Hamiltonian system", to appear J. Diff. Eq.

[4] Hedlund, G. A., "Geodesics on a two dimensional Riemannian manifold with periodic coefficients", Ann. of Math. **33** (1932), 719–739.

[5] Morse, M., "A fundamental class of geodesics on any closed surface of genus greater than one", Trans. Amer. Math. Soc. **26** (1924), 25–60.

[6] Rabinowitz, P. H., "Heteroclinics for a Hamiltonian system of double pendulum type", Top. Meth. in Nonlin. Analysis **9** (1997), 41–76.

Paul H. Rabinowitz
University of Wisconsin
Madison

Progress in Nonlinear Differential Equations
and Their Applications, Vol. 35
© 1999 Birkhäuser Verlag Basel/Switzerland

Stability Properties Related to the Magnetic Bénard Problem

B. Scarpellini

Introduction

In two papers ([Sc.4], [Sc-v.W]), the stability of periodic equilibria of the Boussinesq equations was investigated. The specific point of view taken in these papers was to look at perturbations which are pulslike, or more generally, square integrable over the finite plate $\Omega = \mathbb{R} \times (-\frac{1}{2}, \frac{1}{2})$ and thus not periodic. A number of results were obtained, among others results which relate instability under square integrable perturbations with instability under periodic perturbations. The purpose of this note is to investigate the magnetic Bénard problem from this point of view and to extend some of the indicated results to the present case. We thereby adopt a split attitude. Part of our presentation is expository, partly for reasons of place, partly of reasons for readability. We thereby rely heavily on [Sc.1], [Sc.2], [Sc.3] in which the basic instability theory under square integrable perturbations is presented. Proofs which are mere variations of proofs presented in [Sc.1], [Sc.2], [Sc.3] will in general be omitted. If however a proof of a statement requires a new technical argument then we present it in detail; this arises especially in connection with the magnetic Bénard problem. The reason for studying the magnetic Bénard problem are twofold. On the one hand it is of considerable physical interest as can be seen from the books of Chandrasekhar [Ch] and Roberts [Ro] where it finds considerable attention. On the other hand it is tempting to apply the methods developped in [Sc.2], [Sc.3], [Sc-v.W] to the stability questions arising in connection with the magnetic Bénard problem in order to see what can be attained. As far as the technical side is concerned we note that the material contained in [Sc.1], [Sc.2], [Sc.3] will be presented in a monograph under preparation [Sc.5] in which all proofs are carried out in great detail. For reasons of place we cannot give an exhaustive list of papers on the magnetic Bénard problem. We refer temporarily to the book of B. Stranghan [St] for a list of texts related to energy stability; further papers will be referred to in the course of our digression.

Notation. $\mathbb{R}$, $\mathbb{C}$ denote real and complex numbers with ζ^* the complex conjugate of $\zeta \in \mathbb{C}$. For X, Y Banach spaces, $\| \ \|_X$, $\| \ \|_Y$ denote their respective norms; for X fixed in a context we write $\| \ \|$ instead of $\| \ \|_X$. For a multiindex $\alpha = (\alpha_1, \dots, \alpha_p)$ we set $D^\alpha = \partial_1^{\alpha_1} \cdots \partial_p^{\alpha_p}$ with $\partial_j = \partial_{x_j}$ the derivative with respect to x_j; we set $|\alpha| = \sum |\alpha_j|$. For $\Omega \subseteq \mathbb{R}^n$ a domain, $C^p(\Omega)$ is the set of functions f on Ω having continuous derivatives up to order p, and $C_0^p(\Omega)$ is the subset of $f \in C^p(\Omega)$ having compact support on Ω; likewise with $C^p(\overline{\Omega})$, $C_0^p(\overline{\Omega})$. For $n = 2, 3$ and Ω having the segment property (Adams [Ad], pg. 54) an $f \in \mathcal{L}^2(\Omega)$ is in $H^p(\Omega)$ iff $\lim_n \|f - f_n\|_{\mathcal{L}^2} = 0$ for some sequence $f_n \in C_0^p(\overline{\Omega})$ which is a Cauchy sequence in $H^p(\Omega)$. The scalar product on $H^p(\Omega)$ is given by

$$(u, v)_p = \sum (D^\alpha u, D^\alpha v)_0, \quad |\alpha| \leq p \tag{0.1}$$

where $(u, v)_0 = \int_\Omega uv^* dx^n$. We have $\mathcal{L}^2(\Omega) = H^0(\Omega)$ and set $\| \ \|_{H^p} = \| \ \|_{H^p(\Omega)}$ if no confusion arises. For any vector $u = (u_1, \dots, u_n)$ in $\mathcal{L}^2(\Omega)^n$ we set

$$\|u\|_{\mathcal{L}^2}^2 = \sum \|u_j\|_{\mathcal{L}^2}^2 \tag{0.2}$$

without danger of confusion; likewise with Sobolev norms. The scalar product on $H^p(\Omega)^n$ is given by

$$\langle u, v \rangle_p = \sum_1^n (u_j, v_j)_p, \tag{0.3}$$

for vectors $u = (u_1, \dots, u_n)$, $v = (v_1, \dots, v_n)$ in $H^p(\Omega)^n$. Evidently, $\langle \ , \ \rangle_p$ depends on the number n of components and on Ω; since these are kept fixed within a context we do not display them by a special label.

1. Navier-Stokes on an infinite plate

1.1. Stress-free boundary conditions

While we aim at the magnetic Bénard problem (MBP for short), we reconsider the Navier-Stokes (briefly NS-) equations on an infinite plate $\Omega = \mathbb{R}^2 \times (-\frac{1}{2}, \frac{1}{2})$. The stability aspect has been treated at length in [Sc.1]–[Sc.3] and is the main theme of [Sc.5]. There, the velocity field is subject to Dirichlet boundary conditions. This choice was motivated by Chandrasekhar [Ch], where Dirichlet conditions are listed as the physically most realistic ones. Mathematically, Dirichlet conditions pose serious problems since they are not easily accessible to Fourier analysis. One may ask as to what extent the methods in [Sc.1]–[Sc.3] extend to the case of stress-free boundary conditions. While physically less relevant for fluids, this case will be important for the MBP where the magnetic field will be subject to stress free conditions. We will therefore briefly consider the NS-equations on a plate, subject to stress free boundary conditions. It turns out that due to a technical circumstance, the theory simplifies enormously if compared with [Sc.1]–[Sc.3]. To start with, we refer henceforth to $\Omega = \mathbb{R}^2 \times (-\frac{1}{2}, \frac{1}{2})$ as the infinite plate. We assume from the outset that we are given an equilibrium solution $u_0 = (u_1^0, u_2^0, u_3^0)$, p_0 of the NS-equations which is L_1-periodic in x, L_2-periodic in y, sufficiently smooth and satisfies $\operatorname{div} u_0 = 0$, but not necessarily any specific boundary conditions. If we insert the substitution $u = u_0 + v$, $\nabla p = \nabla p_0 + \nabla \pi$ into the NS-equations and recall that u_0, p_0 is by assumption an equilibrium solution then we obtain after a rearrangement of terms:

$$v_t = \nu \triangle v - (u_0 \nabla)v - (v \nabla)u_0 - \nabla \pi - (v \nabla)v, \quad \operatorname{div} v = 0. \tag{1.1}$$

Here and below we set

$$(u \nabla)v = \Big(\sum u_j \, \partial_j v_1, \sum u_j \, \partial_j v_2, \sum u_j \, \partial_j v_3 \Big)$$

where $u = (u_1, u_2, u_3)$, $v = (v_1, v_2, v_3)$. In (1.1), ν is the coefficient of viscosity. Matemathically we are free to impose on v in (1.1) any boundary condition. Here we stick to the stress free case and require

$$\partial_z v_1 = \partial_z v_2 = v_3 = 0 \text{ on } \partial_\Omega = \mathbb{R}^2 \times \Big\{ -\frac{1}{2}, \frac{1}{2} \Big\}. \tag{1.2}$$

As it stands, equation (1.1) has a rather formal character which admits various interpretations. Here we consider two of them. In the first, v ranges loosely speaking over a subspace of $\mathcal{L}^2(\Omega)^3$, while in the second it ranges over fields which are L_1-periodic in x, L_2-periodic in y. In this section we concentrate on the first interpretation and fix to this effect a functional setting. We set $L^2(\Omega) = \mathcal{L}^2(\Omega)^3$ and define

$$\begin{aligned} &E \subseteq L^2(\Omega) \text{ is the } \mathcal{L}^2\text{-closure of the set of } f \in H^1(\Omega)^2 \times H_0^1(\Omega) \\ &\text{such that } \operatorname{div} f = 0, \text{ and } P \text{ is the orthogonal projection} \\ &\text{from } L^2(\Omega) \text{ onto } E; \end{aligned} \tag{1.3}$$

for a discussion see [Sc.2], Section 1.5. A “Dirichlet”-Laplacian $\triangle_d$ is then introduced according to:

$$\operatorname{dom}(\triangle_d) = H^2(\Omega) \cap H_0^1(\Omega) \text{ and } \triangle_d f = \triangle f, \ f \in \operatorname{dom}(\triangle_d). \tag{1.4}$$

The following is well known

Proposition 1.1. *$\triangle_d$ is selfadjoint on $L^2(\Omega)$ and $\leq -\varepsilon$ for some $\varepsilon > 0$.*

Remark. The ε stems from Poincaré's inequality.

Preparatory to the next step we stipulate:

$$f \in \widetilde{H}^2(\Omega) \text{ iff } f \in H^2(\Omega) \text{ and } \partial_z f = 0 \text{ on } \partial\Omega. \tag{1.5}$$

Here $\partial_z f$ is up to a sign the derivative $\partial_n f$ with respect to the outer unit normal at $\partial\Omega$ and understood in the sense of traces; see Section 1.2, 1.5 in [Sc.2] for details. We now define

$$\operatorname{dom}(\widetilde{\triangle}) = \widetilde{H}^2(\Omega) \text{ and } \widetilde{\triangle} f = \triangle f \text{ for } f \in \operatorname{dom}(\widetilde{\triangle}). \tag{1.6}$$

The “Neumann”-Laplacian $\widetilde{\triangle}$ has well-known properties, among others:

Proposition 1.2. *$\widetilde{\triangle}$ is selfadjoint on $L^2(\Omega)$ and $\widetilde{\triangle} \leq 0$.*

The operators $\triangle_d$, $\widetilde{\triangle}$ give rise to an operator A' defined by:

$$\begin{aligned}&\operatorname{dom}(A') = \operatorname{dom}(\widetilde{\triangle})^2 \times \operatorname{dom}(\triangle_d) \text{ and for } f = (u,v,w) \\ &\text{in } \operatorname{dom}(A') \text{ we set } A'f = (\triangle u, \triangle v, \triangle w).\end{aligned} \tag{1.7}$$

We express (1.7) more compactly by writing: $A' = \operatorname{diag}(\widetilde{\triangle}, \widetilde{\triangle}, \triangle_d)$. By Propositions 1.1, 1.2 we have

Proposition 1.3. *A' is selfadjoint on $L^2(\Omega)$ and ≤ 0.*

The basic statement relating A' with P is

Lemma 1. *$PA' \subseteq A'P$.*

Remark. The full proof is somewhat lenghty; we content us to outline the main steps, and rely strongly on [Sc.2].

Proposition 1.4. *(a) If $u \in H^1(\Omega)$ then $u \in H_0^1(\Omega)$ iff $\int_\Omega(\varphi\partial_z u + u\partial_z\varphi)dx^n = 0$ for all $\varphi \in H^1(\Omega)$, (b) if $u, \varphi \in H^1(\Omega)$ then $\int_\Omega(u\partial_s\varphi + \varphi\partial_s u)dx^n = 0$ for $s \in \{x,y\}$, (c) $u \in \widetilde{H}^2(\Omega)$ iff $u \in H^2(\Omega)$ and $\partial_z u \in H_0^1(\Omega)$.*

Remarks. Clauses (a), (c) are asserted by Proposition 1.1 in [Sc.2]; clause (b), not explicitely stated there follows via Gauss' formula (1.9) in [Sc.2] by the same arguments used in the proof of (a).

Proposition 1.5. *$f \in E$ iff $\langle f, \nabla p\rangle = 0$ for all $p \in H^1(\Omega)$.*

Remark. This is asserted at the end of Section 1.5 and again by Proposition 3.6 in [Sc.2].

Proposition 1.6. *If $f \in (H^1(\Omega)^2 \times H_0^1(\Omega)) \cap E$ then* $\operatorname{div} f = 0$.

Proof. Let $f = (u,v,w)$ satisfy the assumption. By Proposition 1.5 we then have that

$$(u, \partial_x p)_0 + (v, \partial_y p)_0 + (w, \partial_z p)_0 = 0, \quad \forall p \in H^1(\Omega).$$

By (a), (b) of Proposition 1.4 we can transport the derivatives to the left so as to get

$$-(\partial_x u, p)_0 - (\partial_y v, p)_0 - (\partial_z w, p)_0 = 0, \quad \forall p \in H^1(\Omega).$$

From this the claim follows. □

Corollary. *$f \in \operatorname{dom}(A') \cap E$ iff $f \in \operatorname{dom}(A')$ and* $\operatorname{div} f = 0$.

Proof. Immediate from (1.7) and Proposition 1.6. □

For use below we introduce two complete orthonormal systems φ_k, $k \geq 0$ and ψ_p, $p \geq 1$ on $\mathcal{L}^2(I)$, $I = [-\frac{1}{2}, \frac{1}{2}]$. The $\varphi_k' s$ are given by

$$\begin{aligned} \varphi_{2k} &= (-1)^k \sqrt{2} \cos 2\pi k z \ \text{ for } \ k \geq 1, \ \varphi_0 \equiv 1, \\ \varphi_{2k+1} &= (-1)^k \sqrt{2} \sin(2k+1)\pi z, \ k \geq 1. \end{aligned} \tag{1.8}$$

We set $\Lambda_p = p^2\pi^2$, $p \geq 0$ and define the $\psi_p' s$ according to

$$\Lambda_p^{\frac{1}{2}} \psi_p = \varphi_p' \ \text{ whence } \ \psi_p' = -\Lambda_p^{\frac{1}{2}} \varphi_p, \ p \geq 1. \tag{1.9}$$

We note

$$\psi_p = 0 \ \text{ and } \ \partial_z \varphi_k = 0 \ \text{ on } \ \partial I = \{ -\frac{1}{2}, \frac{1}{2} \}. \tag{1.10}$$

Proposition 1.7. *If $f \in \operatorname{dom}(A') \cap E$ then $A'f \in E$.*

Proof. Let $f = (u, v, w)$ satisfy the assumption; by the above corollary we have $\operatorname{div} f = 0$. From Proposition 1.2 in [Sc.2] it follows that the set of functions given by

$$L_N = \sum_0^N A_k(x,y)\varphi_k(z), \quad A_k \in C_0^\infty(\mathbb{R}^2), \tag{1.11}$$

is dense in $H^1(\Omega)$, $\| \ \|_{H^1}$. By Proposition 1.5 it thus suffices to show:

$$\langle f, \nabla L_N \rangle = 0 \ \text{ for all } L_N \text{ given by (1.11)}. \tag{1.12}$$

By definition we have

$$\langle A'f, \nabla L_N \rangle = (\widetilde{\triangle} u, \partial_x L_N)_0 + (\widetilde{\triangle} v, \partial_y L_N)_0 + (\triangle_d w, \partial_z L_N)_0.$$

According to (1.9), (1.10) we have that $\partial_x L_N, \partial_y L_N \in \operatorname{dom}(\widetilde{\triangle})$ and $\partial_z L_N \in \operatorname{dom}(\triangle_d)$ whence

$$\langle A'f, \nabla L_N \rangle = (u, \partial_x \triangle L_N)_0 + (v, \partial_y \triangle L_N)_0 + (w, \partial_z \triangle L_N)_0.$$

By Proposition 1.4 and since $w \in H_0^1(\Omega)$ we conclude from this equation

$$\langle A'f, \nabla L_N \rangle = -(\operatorname{div} f, \triangle L_N)_0 = 0$$

since $\operatorname{div} f = 0$ as noted. The claim now follows from (1.12). □

Slightly more elaborate is the proof of

Proposition 1.8. *If $f \in \operatorname{dom}(A')$ then $Pf \in \operatorname{dom}(A')$.*

Proof. First we note that if $u \in H^p(\Omega)$ then $\widehat{u} = \int_{-1/2}^{1/2} u\,dz$ is in $H^p(\mathbb{R}^2)$. This entails that any $f = (u, v, w)$ in $\mathrm{dom}(A')$ admits a decomposition

$$f = g + h, \quad g = (\widehat{u}, \widehat{v}, 0), \quad g, h \in \mathrm{dom}(A'). \tag{1.13}$$

It thus suffices to show that Pg, Ph are in $\mathrm{dom}(A')$. As to Ph we invoke Proposition 1.8 in [Sc.2] according to which there is a (unique) solution $p \in \widetilde{H}^2(\Omega)$ of $\widetilde{\triangle} p = \mathrm{div}\, h$ with $\widehat{p} = 0$ and such that

$$(1 - P)h = Qh = \nabla p, \quad \nabla p \in H^1(\Omega)^2 \times H_0^1(\Omega). \tag{1.14}$$

Since $\mathrm{div}\, h \in H^1(\Omega)$ we have by standard regularity theorems that p is even in $H^3(\Omega)$ whence $\partial_z p \in H^2(\Omega) \cap H_0^1(\Omega)$ according to (c) of Proposition 1.4. By (iii) of Proposition 1.1 in [Sc.2] we have that $\partial_x p$ is in $\widetilde{H}^2(\Omega)$ iff

$$\int_\Omega \big(\varphi\, \partial_z^2(\partial_x p) + (\partial_z \varphi)\,(\partial_z \partial_x p)\big)\, dx^3 = 0 \tag{1.15}$$

for all $\varphi \in H^1(\Omega)$. By Adams [Ad, pg. 54], it suffices to verify (1.15) for all $\varphi \in C_0^\infty(\mathbb{R}^3)$. Since $p \in H^3(\Omega)$ and according to Proposition 1.4 the integral in (1.15) is equal to the integral

$$-\int_\Omega \big((\partial_x \varphi)\,(\partial_z^2 p) + (\partial_z \partial_x \varphi)\,(\partial_z p)\big)\, dx^3 \tag{1.16}$$

which vanishes by Proposition 1.1, (iii) in [Sc.2] and since $p \in \widetilde{H}^2(\Omega)$. With $\varphi \in C_0^\infty(\mathbb{R}^2)$ arbitrary we thus have that $\partial_x p \in \widetilde{H}^2(\Omega)$. Likewise we find $\partial_y p \in \widetilde{H}^2(\Omega)$ whence $Qh \in \mathrm{dom}(A')$ and thus $Ph \in \mathrm{dom}(A')$ by (1.14) and the subsequent remarks. In order to prove $Qg \in \mathrm{dom}(A')$ we note that $g = (A, B, 0)$ with $A, B \in H^2(\mathbb{R}^2)$. We set $\nu = \nu(\alpha, \beta) = \alpha^2 + \beta^2$ and $\widehat{A}(\alpha, \beta)$, $\widehat{B}(\alpha, \beta)$ the Fourier transforms of A, B. It follows from Section 1.5 in [Sc.2] that $Qg = (C, D, 0)$ with the Fourier transforms $\widehat{C}$, $\widehat{D}$ given by

$$\widehat{C} = \alpha\,(\alpha \widehat{A} + \beta \widehat{B})\,\nu^{-1}, \quad \widehat{D} = \beta\,(\alpha \widehat{A} + \beta \widehat{B})\,\nu^{-1}. \tag{1.17}$$

Since $A, B \in H^2(\mathbb{R}^2)$, both $(1 + \nu)\widehat{A}$ and $(1 + \nu)\widehat{B}$ are in $\mathcal{L}^2(\mathbb{R}^2)$. From this and (1.17) one easily infers that $(1 + \nu)\widehat{C}$, $(1 + \nu)\widehat{D}$ are in $\mathcal{L}^2(\mathbb{R}^2)$ what entails $C, D \in H^2(\mathbb{R}^2)$ by standard Fourier theory; thus $Qg \in \mathrm{dom}(A')$ as claimed, concluding the proof. □

Proof of Lemma 1. Let $f, g \in \mathrm{dom}(A')$. By Proposition 1.8 we have that $Pf, Pg \in \mathrm{dom}(A')$ and thus by Proposition 1.7 the relation $A'Pf, A'Pg \in E$. From these remarks and since A' is selfadjoint we obtain two chains of equations

$$\begin{aligned} \langle A'Pf, g\rangle &= \langle PA'Pf, g\rangle = \langle A'Pf, Pg\rangle = \langle f, PA'Pg\rangle, \\ \langle PA'f, g\rangle &= \langle A'f, Pg\rangle = \langle f, A'Pg\rangle = \langle f, PA'Pg\rangle. \end{aligned}$$

Thus $\langle A'Pf, g\rangle = \langle PA'f, g\rangle$ for all $g \in \mathrm{dom}(A')$. Since $\mathrm{dom}(A')$ is dense in $L^2(\Omega) = \mathcal{L}^2(\Omega)^3$, the claim follows. □

Remark. By Lemma 1 the space E reduces to A'. We denote the restriction of A' to E by $A' \restriction E$; similar use of the symbol $\restriction$ will be made in other cases. We also set henceforth

$$A'_S = A' \restriction E \tag{1.18}$$

and note that A'_S is Stokes operator in the context of stress free boundary conditions. Next we consider the linear perturbation in (1.1) due to the equilibrium solution u_0. That is we define an operator T_0 acting on $v \in H^1(\Omega)$ according to

$$T_0 v = -(u_0 \nabla)v - (v\nabla)u_0. \tag{1.19}$$

Lemma 2. *T_0 is bounded relative to A' with relative bound zero.*

Sketch of Proof. Due to the definition of A' via (1.7) it is easily seen that the proof reduces to the assertion: (a) to $\varepsilon > 0$ there is K_ε such that

$$\|\nabla u\|^2_{\mathcal{L}^2} \le \varepsilon \|\widetilde{u}\|^2_{\mathcal{L}^2} + K_\varepsilon \|u\|^2_{\mathcal{L}^2}, \quad \forall u \in \widetilde{H}^2(\Omega),$$

(b) likewise with $\triangle_d$ in place of $\widetilde{\triangle}$. Since the proofs of (a), (b) coincide, we sketch that of (a); see also the corollary to Theorem 1.1 in [Sc.2] concerning (b). From Gauss's formula (1.9) in [Sc.2] one obtains the familiar relation

$$\int_\Omega \big((\nabla v)(\nabla u) + v \,\triangle u\big)\, dx^3 = 0, \quad v \in H^1(\Omega), \quad u \in \widetilde{H}^2(\Omega) \tag{1.20}$$

which implies

$$\|\nabla u\|^2_{\mathcal{L}^2} = \big|\langle u, \widetilde{\triangle} u\rangle_0\big| \le \|u\|_{\mathcal{L}^2} \|\widetilde{\triangle} u\|_{\mathcal{L}^2} \le \varepsilon \|\widetilde{\triangle} u\|_{\mathcal{L}^2} + \varepsilon^{-1} \|u\|^2_{\mathcal{L}^2}$$

for any $u \in \widetilde{H}^2(\Omega)$, proving (b). The claim of the lemma now follows by a straightforward amalgamation of (a), (b) and from the assumption that u_0 is L_1, L_2-periodic and smooth. □

Corollary. *$PT_0 \restriction E$ is bounded relative to A'_S with relative bound zero.*

Proof. To $\varepsilon > 0$ seek K_ε via Lemma 2. For $f \in \mathrm{dom}(A'_S) = \mathrm{dom}(A') \cap E$ we then have:

$$\|PT_0 f\|_{\mathcal{L}^2} \le \varepsilon \|A' f\|^2_{\mathcal{L}^2} + K_\varepsilon \|f\|^2_{\mathcal{L}^2}.$$

Since $A' = A'_S$ on E, the claim follows. □

From this corollary and standard semigroup theory one infers

Proposition 1.9. *$A'_S + PT_0$ generates a holomorphic semigroup on E.*

With Proposition 1.9 at disposal, one can handle the nonlinearity $(v\nabla)v$ in (1.1) in all well-known way. For $\lambda > 0$ sufficiently large, one can define fractional powers B^α_λ of $B_\lambda = \lambda - (A'_S + PT_0)$ giving rise to fractional power spaces $X^\alpha = \mathrm{dom}(B^\alpha_\lambda)$ with norm $\|x\|_\alpha = \|B^\alpha_\lambda x\|_{\mathcal{L}^2}$; since these norms are equivalent for different λ's ($\gg 0$) one may fix a $\lambda \gg 0$ once and for all. It can be shown by interpolation theory (see also Theorem 1.6.1, pg. 39 in [He]) that if $\alpha \in (\frac{3}{4}, 1)$

then X^α is continuously embedded in $C^0(\overline{\Omega})^3$; based on this fact a simple analysis shows (see [Pa], Lemma 3.4 for the argument):

Proposition 1.10. *If $\alpha \in (\frac{3}{4}, 1)$ then $N(u,v) = (u\nabla)v$ is a bounded bilinear mapping from $(X^\alpha)^2$ into $C^0(\overline{\Omega})^3$.*

Based on Proposition 1.5, 1.9, 1.10 we may put equation (1.1) into an abstract form, i.e.:

$$v_t = (A'_S + PT_0)v + PN(u,v), \quad v \in E. \tag{1.21}$$

In this form it is subject to standard existence and uniqueness theorems; see [Pa], Theorem 3.1, pg. 196. The main question to be asked in the present context is that of the stability resp. instability of the trivial solution $v_0 \equiv 0$ of (1.21). More precisely one defines the given periodic equilibrium u_0 (Section 1.1) as Ljapounov-stable resp. instable if $v_0 \equiv 0$ as an equilibrium solution of (1.21) is Ljapounov-stable resp. instable. The basic result in this connection is the theorem below which for lack of space has to be stated without proof:

Theorem 1.1. *(a) If $\sigma(A'_S + PT_0)$ lies strictly in the left half plane then $v_0 \equiv 0$ is a Ljapounov-stable equilibrium of (1.21), (b) if $\lambda \in \sigma(A'_S + PT_0)$ with $\mathrm{re}(\lambda) > 0$ exists then $v_0 \equiv 0$ is Ljapounov-instable.*

Remarks. Part (a) is well known and a consequence of semigroup theory (D. Henry [He], Chapter 5). Part (b) however is quite difficult to prove. In case of Dirichlet boundary conditions a proof has been given in [Sc.3]; a very detailed account of it is to be found in [Sc.5]. In case of stress free boundary conditions a careful examination shows that apart from a minor deviation, the proof of Theorem 1.1 (b) is essentially the same as in the Dirichlet case. Theorem 1.1 causes us to look more closely at the spectrum $\sigma(A'_S + PT_0)$ and to seek a description in terms of so-called Θ-periodic spectra analoguous to that obtained in [Sc.2].

1.2. Θ-periodic spectra

In order to express $\sigma(A'_S + PT_0)$ in terms of Θ-periodic spectra we recall some notations from [Sc.2]. With L_1, L_2 the periods of the equilibrium solution u_0 (Section 1.1) we set $Q_L = (0, L_1) \times (0, L_2)$ and $Q = Q_L \times (-\frac{1}{2}, \frac{1}{2})$; we put $M = [0, 2\pi]^2$ and let $\dot{M}$ be M without the corners $(0,0)$, $(2\pi, 0)$, $(0, 2\pi)$, $(2\pi, 2\pi)$. Given $\Theta = (\Theta_1, \Theta_2)$ in M we let $C^0_\Theta(Q)$ be the subset of $f \in C^0(\overline{\Omega})$ such that:

$$f(x + jL_1, y + kL_2, z) = \mathrm{e}^{i(j\Theta_1 + k\Theta_2)} f(x, y, z) \tag{1.22}$$

for all $(x,y,z) \in \overline{\Omega}$, $j, k \in \mathbb{Z}$. With $\Theta \in M$ one associates Sobolev spaces $H^p_\Theta(Q)$, $H^1_{\Theta,0}(Q)$, $\widetilde{H}^2_\Theta(Q)$ in a standard way ([Sc.2], Section 2). E.g., $f \in \widetilde{H}^2_\Theta(Q)$ iff $f \in \mathcal{L}^2(Q)$ and if there is a sequence $f_n \in C^2_\Theta(Q)$ with: (a) $\lim_n \|f - f_n\|_{\mathcal{L}^2} = 0$, (b) f_n, $n \geq 1$ is Cauchy in $H^2(Q)$, $\|\ \|_{H^2}$, (c) $\partial_z f_n = 0$ on $\partial\Omega$. We note that

$\mathcal{L}^2(Q) = H^0_\Theta(Q)$. Setting $L^2(Q) = \mathcal{L}^2(Q)^3$ we define:

$$E_\Theta \subseteq L^2(Q) \text{ is the } \mathcal{L}^2\text{-closure of the set of } f \in H^1_\Theta(Q)^2 \times H^1_{\Theta,0}(Q) \tag{1.23}$$
with $\operatorname{div} f = 0$; P_Θ is the orthogonal projection from $L^2(Q)$ onto E_Θ.

The Θ-periodic Dirichlet Laplacian $\triangle_\Theta$ is given by

$$\operatorname{dom}(\triangle_\Theta) = H^2_\Theta(Q) \cap H^1_{\Theta,0}(Q) \text{ and } \triangle_\Theta f = \triangle f \text{ for } f \in \operatorname{dom}(\triangle_\Theta). \tag{1.24}$$

Proposition 1.11. *$\triangle_\Theta$ is selfadjoint and $\leq -\varepsilon$ for an ε independent of $\Theta \in M$.*

A proof can easily be given in terms of Fourier series; for an exposition see [Sc.1, Sc.2]. The Θ-periodic Neumann-Laplacian $\widetilde{\triangle}_\Theta$ is defined according to

$$\operatorname{dom}(\widetilde{\triangle}_\Theta) = \widetilde{H}^2_\Theta(Q), \quad \widetilde{\triangle}_\Theta f = \triangle f, \quad f \in \operatorname{dom}(\widetilde{\triangle}_\Theta). \tag{1.25}$$

Proposition 1.12. *$\widetilde{\triangle}_\Theta$ is selfadjoint and ≤ 0, ($\Theta \in M$).*

The remarks subsequent to Propositions 1.11 apply also to Proposition 1.12. In terms of $\triangle_\Theta$, $\widetilde{\triangle}_\Theta$ we can define the counterpart A'_Θ to A' in (1.7) according to

$$\operatorname{dom}(A'_\Theta) = \operatorname{dom}(\widetilde{\triangle}_\Theta)^2 \times \operatorname{dom}(\triangle_\Theta) \text{ and } A'_\Theta f = (\widetilde{\triangle}_\Theta u, \widetilde{\triangle}_\Theta v, \widetilde{\triangle}_\Theta w) \tag{1.26}$$
for $f = (u, v, w)$ in $\operatorname{dom}(A'_\Theta)$.

From Proposition 1.11, 1.12 we infer

Proposition 1.13. *A'_Θ is selfadjoint on $L^2(Q)$ and ≤ 0.*

The main result about A'_Θ is

Lemma 3. *$P_\Theta A'_\Theta \subseteq A'_\Theta P_\Theta$, $\Theta \in M$.*

Remarks. The proof is very similar to that of Lemma 1; we therefore content us to stress the main parts. There is a minor difference between the cases $\Theta \in \dot{M}$ and $\Theta \in M - \dot{M}$, the latter representing the periodic case which may be considered as settled by the existing literature.

Proposition 1.14. *(a) an $u \in H^1_\Theta(Q)$ is in $H^1_{\Theta,0}(Q)$ iff*

$$\int_Q (\varphi \partial_z u + u \partial_z \varphi)\, dx^3 = 0, \quad \forall \varphi \in H^1(Q),$$

(b) if $u, \varphi \in H^1_\Theta(Q)$ then

$$\int_Q (u \partial_s \varphi^* + \varphi^* \partial_s u)\, dx^3 = 0, \quad s \in \{x, y\}, \quad \Theta \in M.$$

Hint of proof. Both statements follow from Gauss's formula (2.2) in [Sc.2] by arguments analoguous to those in the proof of Proposition 1.1 in [Sc.2]. □

Remark. Formula (2.2) in [Sc.2] contains a misprint: "$\triangle u$" has to be replaced by "$\operatorname{div} u$".

Proposition 1.15. *$f \in E_\Theta$ iff $\langle f, \nabla p\rangle = 0$, $\forall p \in H^1_\Theta(Q)$.*

Proof. If $\Theta \in \dot{M}$ this follows from Proposition 2.6 [Sc.2] and the subsequent remarks, if $\Theta \in M - \dot{M}$, i.e. in the periodic case the assertion is known to hold. □

Proposition 1.16. *If $f \in (H^1_\Theta(Q)^2 \times H^1_{\Theta,0}(Q)) \cap E$ then $\operatorname{div} f = 0$.*

Proof. The proof, based on Proposition 1.14, is the same as that of Proposition 1.6. □

Corollary. *$f \in \operatorname{dom}(A'_\Theta) \cap E_\Theta$ iff $f \in \operatorname{dom}(A'_\Theta)$ and $\operatorname{div} f = 0$.*

In order to prove counterparts to Proposition 1.7, 1.8 we need some notation. With L_1, L_2 the periods of u_0 and $\Theta = (\Theta_1, \Theta_2)$ in M we set

$$\widehat{\alpha} = (2\pi\alpha + \Theta_1)\, L_1^{-1}, \quad \widehat{\beta} = (2\pi\beta + \Theta_2)\, L_2^{-1}, \quad \alpha, \beta \in \mathbb{Z}. \tag{1.27}$$

A slight ambiguity in (1.27) is avoided in that α always appears as first, β as second index of Fourier coefficients. We also set

$$e_{\alpha\beta} = \mathrm{e}^{i(\widehat{\alpha}x + \widehat{\beta}y)}(L_1 L_2)^{-\frac{1}{2}}, \quad (\alpha, \beta \in \mathbb{Z}). \tag{1.28}$$

Recalling (1.8), (1.9) we have that both $e_{\alpha\beta}\varphi_j$ and $e_{\alpha\beta}\psi_k$, $\alpha, \beta \in \mathbb{Z}$, $j \geq 0$, $k \geq 1$ are complete orthonormal systems on $\mathcal{L}^2(Q)$ giving rise to Fourier series with different properties ([Sc.2], Section 2.1, 2.2).

Proposition 1.17. *If $f \in \operatorname{dom}(A'_\Theta) \cap E_\Theta$ then $A'_\Theta f \in E_\Theta$.*

Proof. By Proposition 2.2 in [Sc.2] we have that the set of "polynomials"

$$L_N = \sum_N a_{\alpha\beta j}\, e_{\alpha\beta}\, \varphi_j, \quad |\alpha|, |\beta| \leq N, \quad 0 \leq j \leq N \tag{1.29}$$

is dense in $H^1_\Theta(Q)$, $\| \ \|_{H^1}$. Now let $f = (u, v, w)$ satisfy the assumption. By this remark and Proposition 1.15 it suffices to show that $\langle A'_\Theta f, \nabla L_N\rangle = 0$ for all L_N of the form (1.29). This is achieved via Proposition 1.16 by repeating the arguments in the proof of Proposition 1.7 as they stand. □

Proposition 1.18. *If $f \in \operatorname{dom}(A'_\Theta)$ then $P_\Theta f \in \operatorname{dom}(A'_\Theta)$.*

Proof. For $\alpha, \beta \in \mathbb{Z}$ and $j \geq 0$ we set $\widehat{\Lambda}(\alpha, \beta) = \widehat{\alpha}^2 + \widehat{\beta}^2 + \Lambda_j$ where $\Lambda_j = (j\pi)^2$; we assume first $\Theta = (\Theta_1, \Theta_2) \in \dot{M}$ whence $\widehat{\alpha}^2 + \widehat{\beta}^2 > 0$. Recalling Proposition 1.11, 1.12 the following is easily established: (a) $e_{\alpha\beta}\varphi_j$ is an eigenvector of $\widetilde{\triangle}_\Theta$ to the eigenvalue $-\widehat{\Lambda}(\alpha, \beta)$ for $\alpha, \beta \in \mathbb{Z}$, $j \geq 0$, (b) $e_{\alpha\beta}\psi_k$ is an eigenvector of $\triangle_\Theta$ to the eigenvalue $-\widehat{\Lambda}(\alpha, \beta)$ for $\alpha, \beta \in \mathbb{Z}$, $k \geq 1$. Now let $f = (u, v, w)$ be in $\mathrm{dom}(A'_\Theta)$, thus $u, v \in \mathrm{dom}(\widetilde{\triangle}_\Theta)$, $w \in \mathrm{dom}(\triangle_\Theta)$. From (a), (b) and standard spectral theory it then follows that u, v, w admit Fourier expansions

$$u = \sum a_{\alpha\beta j}\, e_{\alpha\beta}\, \varphi_j, \quad v = \sum b_{\alpha\beta j}\, e_{\alpha\beta}\, \varphi_j, \quad w = \sum c_{\alpha\beta k}\, e_{\alpha\beta}\, \psi_k$$

$(\alpha, \beta \in \mathbb{Z},\ j \geq 0,\ k \geq 1)$ subject to the condition

$$\sum \widehat{\Lambda}_j(\alpha, \beta)^2 \left(|a_{\alpha\beta j}|^2 + |b_{\alpha\beta j}|^2 + |c_{\alpha\beta j}|^2\right) < \infty \tag{1.30}$$

where for simplicity we have set $c_{\alpha\beta 0} = 0$, $\alpha, \beta \in \mathbb{Z}$. Next we introduce

$$\gamma_{\alpha\beta j} = i\,\widehat{\alpha}\, a_{\alpha\beta_j} + i\,\widehat{\beta}\, b_{\alpha\beta j} - \Lambda_j^{\frac{1}{2}}\, c_{\alpha\beta j}.$$

and let $(1 - P)f = (U, V, W)$. By formulas (2.24) in [Sc.2], Section 2.4 it follows that U, V, W have Fourier expansions

$$U = \sum A_{\alpha\beta j}\, e_{\alpha\beta}\, \varphi_j, \quad V = \sum B_{\alpha\beta j}\, e_{\alpha\beta}\, \varphi_j, \quad W = \sum C_{\alpha\beta k}\, e_{\alpha\beta}\, \psi_k \tag{1.31}$$

$(\alpha, \beta \in \mathbb{Z},\ j \geq 0,\ k \geq 1)$ where

$$A_{\alpha\beta j} = -i\,\widehat{\alpha}\,\widehat{\Lambda}_j(\alpha, \beta)^{-1}\,\gamma_{\alpha\beta j}, \quad B_{\alpha\beta j} = -i\,\widehat{\beta}\,\widehat{\Lambda}_j(\alpha, \beta)^{-1}\,\gamma_{\alpha\beta j}, \tag{1.32}$$

$$\text{and } C_{\alpha\beta k} = -\Lambda_k^{\frac{1}{2}}\,\widehat{\Lambda}_k(\alpha, \beta)^{-1}\,\gamma_{\alpha\beta j}.$$

Based on (1.30), (1.32) it is now straightforward to show:

$$\sum \widehat{\Lambda}_j(\alpha, \beta)^2 \left(|A_{\alpha\beta j}|^2 + |B_{\alpha\beta j}|^2 + |C_{\alpha\beta j}|^2\right) < \infty, \tag{1.33}$$

$(C_{\alpha\beta 0} = 0,\ \alpha, \beta \in \mathbb{Z})$. From (1.33) and standard spectral theory it follows that $U, V \in \mathrm{dom}(\widetilde{\triangle}_\Theta)$, $W \in \mathrm{dom}(\triangle_\Theta)$ whence $(1 - P)f \in \mathrm{dom}(A'_\Theta)$, proving the claim for $\Theta \in \dot{M}$. If $\Theta = 0$, i.e. in the periodic case, all arguments remain the same except that A_{000}, B_{000} do not appear in (1.31), (1.32). □

Proof of Lemma 3. Based on the above propositions, one proceeds by the same arguments as in the proof of Lemma 1. □

Remarks. By Lemma 3, E_Θ is an invariant subspace of the selfadjoint operator A'_Θ; this gives rise to the definition:

$$A'_S(\Theta) = A'_\Theta \restriction E_\Theta, \quad \Theta \in M. \tag{1.34}$$

We call $A'_S(\Theta)$ the Θ-periodic Stokes operator under stress free boundary conditions. We now recall the operator T_0 formally given by (1.19), but provide it now with the domain $H^1(Q)$; which domain for T_0 is at stake will always be understood from the context.

Lemma 4. *Given ε there is K_ε such that*

$$\|T_0 v\|^2_{\mathcal{L}^2} \leq \varepsilon \, \|A'_\Theta v\|^2_{\mathcal{L}^2} + K_\varepsilon \, \|v\|^2_{\mathcal{L}^2}, \quad v \in \mathrm{dom}(A'_\Theta), \quad \Theta \in M.$$

Proof. Due to Definition (1.26) and the smoothness of the equilibrium solution u_0 it is easily seen that it suffices to prove: (a) given ε there is K_ε such that

$$\|\nabla u\|^2_{\mathcal{L}^2} \leq \varepsilon \, \|\widetilde{\triangle}_\Theta u\|^2_{\mathcal{L}^2} + K_\varepsilon \, \|u\|^2_{\mathcal{L}^2}, \quad u \in \mathrm{dom}(\widetilde{\triangle}_\Theta), \quad \Theta \in M,$$

(b) likewise with $\triangle_\Theta$. Now based on Gauss's formula (2.2) in [Sc.2] (see remark following Proposition 1.14) one straightforwardly infers

$$\|\nabla u\|^2_{\mathcal{L}^2} \leq \left|(\widetilde{\triangle} u, u)_0\right| \leq \varepsilon \, \|\widetilde{\triangle}_\Theta u\|^2_{\mathcal{L}^2} + \varepsilon^{-1} \, \|u\|^2_{\mathcal{L}^2}$$

for any $u \in \mathrm{dom}(\widetilde{\triangle}_\Theta)$, $\Theta \in M$, what proves (a). The proof of (b) is analogous. □

Corollary. *Given $\varepsilon > 0$ there is K_ε such that*

$$\|P_\Theta \, T_0 \, v\|_{\mathcal{L}^2} \leq \varepsilon \, \|A'_S(\Theta) \, v\|^2_{\mathcal{L}^2} + K_\varepsilon \, \|v\|^2_{\mathcal{L}^2}$$

for $v \in \mathrm{dom}(A'_S(\Theta))$, $\Theta \in M$.

Proof. Immediate from Lemma 4. □

Proposition 1.19. *For $\Theta \in M$, $A'_S(\Theta) + P_\Theta T_0$ has compact resolvents and generates a holomorphic semigroup on E_Θ.*

Proof. This follows from the above corollary by standard semigroup theory and the fact that A'_Θ has compact resolvents. □

Based on Proposition 1.19 we are in a position to state one of two formulas which express $\sigma(A'_S + PT)$ in terms of $\sigma(A'_S(\Theta) + P_\Theta T_0)$:

Theorem 1.2.

$$\sigma(A'_S + PT_0) = \text{ closure of } \bigcup_{\Theta \in \dot{M}} \sigma\big(A'_S(\Theta) + P_\Theta T_0\big). \tag{$*$}$$

Remark. Formula $(*)$ is proved for Dirichlet conditions in [Sc.2]. There, $(*)$ is not explicitely stated but its proof occurs as part of the proof of another, more sophisticated formula, provided by Theorems 4.1–4.3 in [Sc.2]. We postpone a discussion of Theorem 1.2 to the end of the next paragraph.

1.3. The influence of the corners

In order to state the second spectral formula we have to digress on the periodic case which is associated with any of the four corners of $M = [0, 2\pi]^2$. We use the label "per" to characterize the periodic case which means "L_1-periodic in x, L_2-periodic in y" whenever L_1, L_2 are fixed. We write $H^p_{\text{per}}(Q)$, $\widetilde{H}^2_{\text{per}}(Q)$, $H^1_{\text{per},0}(Q)$ instead of $H^p_{(0,0)}(Q)$, $\widetilde{H}^2_{(0,0)}(Q)$, $H^1_{(0,0),0}(Q)$ and also E_{per}, P_{per}, A'_{per}, $A'_S(\text{per})$ in place of $E_{(0,0)}$, $P_{(0,0)}$, $A'_{(0,0)}$, $A'_S((0,0))$. We now fix $\delta \in [0, 2\pi]$ and stipulate:

$$E_{0\delta} \subseteq L^2(Q) \text{ is the } \mathcal{L}^2\text{-closure of the } f = (u, v, w) \text{ in } H^1_{\text{per}}(Q)^2 \times H^1_{\text{per},0}(Q) \tag{1.35}$$

with $\operatorname{div} f = 0$ and such that

$$(\delta)\ L_1^{-1} \cos\delta \int_Q u\, dx^3 + L_2^{-1} \sin\delta \int_Q v\, dx^3 = 0.$$

We let $P_{0\delta}$ be the orthogonal projection from $L^2(Q)$ onto $E_{0\delta}$ and define $A'_S(0,\delta)$ according to

$$f \in \operatorname{dom}(A'_S(0,\delta)) \text{ iff } f \in \operatorname{dom}(A'_S(\text{per})) \text{ and } f \text{ satisfies } (\delta) \text{ in } (1.35), \tag{1.36}$$

and $A'_S(0,\delta)f = A'_S(\text{per})f$ for such f.

It turns out that $A'_S(0,\delta)$ is selfadjoint on $E_{0\delta}$, is ≤ 0 and has compact resolvents. Moreover one has that

Proposition 1.20. *Given ε there is K_ε such that*

$$\|P_{0\delta}\, T_0\, v\|^2_{\mathcal{L}^2} \le \varepsilon\, \|A'_S(0,\delta)\, v\|^2_{\mathcal{L}^2} + K_\varepsilon\, \|v\|^2_{\mathcal{L}^2}$$

for $v \in \operatorname{dom}(A'_S(0,\delta))$, $\delta \in [0, 2\pi]$.

Together with the foregoing remark and by standard semigroup theory this implies

Proposition 1.21. *$A'_S(0,\delta) + P_{0\delta}T_0$ has compact resolvents and generates a holomorphic semigroup on $E_{0\delta}$.*

The announced spectral formula is provided by

Theorem 1.3.

$$\sigma\big(A'_S + P\,T_0\big) = \bigcup_{\Theta \in \dot{M}} \sigma\big(A'_S(\Theta) + P_\Theta\, T_0\big) \cup \bigcup_{\delta \in [0,2\pi]} \sigma\big(A'_S(0,\delta) + P_{0\delta}\, T_0\big). \tag{$*$}$$

Remarks. Theorems 1.1–1.3 are the main results which can be obtained in the present context. We have yet to say something about their proofs. Statements corresponding to Theorems 1.2–1.3 but for Dirichlet conditions, i.e. with $v = 0$ on $\partial\Omega$, have been proved in [Sc.2], (Theorems 4.1–4.3) while the result corresponding to Theorem 1.1 has been proved in [Sc.3], (Theorem 1). The question arises if the proofs for stress free boundaries are more difficult, remain the same or get simpler if compared with the Dirichlet case. In fact, it is the latter which occurs;

it turns out that stress free boundaries greatly simplify the arguments. The reason for this is Lemma 3 which implies that the Stokes operator $A'_S(\Theta)$ is just the restriction of the selfadjoint operator A'_Θ to its invariant subspace E_Θ ((1.34)), and that A'_Θ is built up from the Laplacians $\widetilde{\triangle}_\Theta$, $\triangle_\Theta$ which admit a complete analysis in terms of Fourier series. This circumstance then extends to A'_Θ. E.g. one finds that $(A'_\Theta - \lambda)^{-1}$, $\Theta \in M$ $(\lambda > 0)$ admits a holomorphic extension into a complex neighbourhood $\mathcal{M}$ of $M = [0, 2\pi]^2$. Now the difficulties associated with the corners $M - \dot{M}$, a source of considerable troubles in [Sc.2, Sc.3], still remain, but now they enter only via the projection operators P_Θ, $\Theta \in M$, which exhibit a simgular behaviour near the corners. In fact P_Θ is only holomorphic in a complex neighbourhood $\widetilde{\mathcal{M}} \subseteq \mathcal{M}$ of $\dot{M}$ which includes a real punctuated disc $S_\varepsilon = \{\Theta \in \mathbb{R}^2,\ 0 < |\Theta| < \varepsilon\}$. In order to recall a basic property of P_Θ, established in [Sc.2], Section 4.4 we set $\Theta(r, \delta) = (r \cos \delta, r \sin \delta)$, $r \in (0, \varepsilon)$, $\delta \in [0, 2\pi]$. The relevant statement then is

Proposition 1.22. $\lim_{r \to 0} P_{\Theta(r,\delta)} = P_{0\delta}$ *in the uniform sense for* $\delta \in [0, 2\pi]$.

The principal consequence of this proposition is

Lemma 5. $P_{0\delta} A'_{per} \subseteq A'_{per} P_{0\delta}$ $(\delta \in [0, 2\pi])$.

Outline of proof. From Lemma 3 and by holomorphic extension one obtains

$$P_\Theta (A'_\Theta - 1)^{-1} = (A'_\Theta - 1)^{-1} P_\Theta, \quad \Theta \in \widetilde{\mathcal{M}},$$

whence with $\Theta(r, \delta)$ as above,

$$P_{\Theta(r,\delta)} \left(A'_{\Theta(r,\delta)} - 1\right)^{-1} = \left(A'_{\Theta(r,\delta)} - 1\right)^{-1} P_{\Theta(r,\delta)}.$$

Letting $r \to 0$, one infers from Proposition 1.22 and the remarks preceding it, that

$$P_{0\delta} (A'_{\mathrm{per}} - 1)^{-1} = (A'_{\mathrm{per}} - 1)^{-1} P_{0\delta},$$

from which the assertion immediately follows. □

Remarks. By (1.35) and Lemma 5, $E_{0\delta}$ is an invariant subspace of A'_{per}; thus $A'_S(0, \delta)$ defined via (1.36) has the representation $A'_S(0, \delta) = A'_{\mathrm{per}} \restriction E_{0\delta}$. From this and Lemma 4 one easily deduces Proposition 1.20. By Lemmas 3–5 and Proposition 1.20 it is now an elaborate but completely straightforward task to check that all steps in [Sc.2, Sc.3] involved in the proofs of Theorems 4.1–4.3 and Theorem 1 respectively carry over without change to the present context so as to yield proofs of Theorems 1.1–1.3. The only point to be observed is that the operator family $R_\Theta = A_\Theta^{-1} P_\Theta$ in [Sc.2], Section 4 (where $A_\Theta = \mathrm{diag}(\triangle_\Theta, \triangle_\Theta, \triangle_\Theta)$) has now to be replaced by $(A'_\Theta - 1)^{-1} P_\Theta$ since $\|(A'_\Theta)^{-1}\|_\infty$ is not bounded in a neighbourhood of $\Theta = 0$. With this stipulation, the above remarks as to the proofs of Theorems 1.1–1.3 apply. We conclude with a question, amply discussed in [Sc.2], Section 4.5, [Sc-v.W], Section 5 and [Sc.5], Section 2, which in the present context reads as follows:

$$\sigma\left(A'_{\mathrm{per}} + P_{\mathrm{per}} T_0\right) \subseteq \sigma\left(A'_S + P T_0\right). \quad (?). \tag{1.37}$$

The importance of this question is supported by the papers of A. Mielke [Mi] and Kagei-v. Wahl [Ka-v.W] from which it issues that there is a strong connection between (1.37) and the notion of Eckhaus instability; see [Sc.4], Section 2 for a discussion. For Dirichlet boundary conditions, (1.37) can only be proved for certain classes of equilibrium solutions such as rolls; whether (1.37) holds in general is still open. In case of stress free boundary conditions we are in a better situation. We recall the equilibrium solution $u_0 = (u_1^0, u_2^0, u_3^0)$ in Section 1.1 which defines the operator T_0 formally via (1.19) and which is provided in the present context with $H^1_{\text{per}}(Q)$ as domain. It was assumed that u_0 is L_1, L_2-periodic in x, y resp., sufficiently smooth and such that $\operatorname{div} u_0 = 0$; for convenience we set $u_j^0 = u_j$, $j \leq 3$.

Lemma 6. *Let the equilibrium solution u_0 also satisfy stress free boundary conditions. Then (1.37) holds.*

Proof. Let $\sigma \in \sigma(A'_S(\text{per}) + P_{\text{per}} T_0)$. We distinguish two cases: (a) $\sigma = 0$, (b) $\sigma \neq 0$. We dismiss the case (a) since the arguments, albeit in the stress free context, remain the same as in the Dirichlet case settled by Lemma 4.6 in [Sc.2]. We thus assume (b). This means that there is a $\varphi = (\varphi_1, \varphi_2, \varphi_3)$ in $\operatorname{dom}(A'_S(\text{per}))$ such that

$$P_{\text{per}}\,(A'_{\text{per}} + T_0)\,\varphi = \sigma\,\varphi \tag{1.38}$$

where use of (1.34) for $\Theta = 0$ has been made. This implies that there is $p \in H^1_{\text{per}}(Q)$ with

$$(A'_{\text{per}} + T_0)\,\varphi = \sigma\,\varphi + \nabla p. \tag{1.39}$$

The first component in (1.39) written out more explicitely yields the equation

$$\triangle\varphi_1 - (u_0\,\nabla)\varphi_1 - (\varphi\,\nabla)u_1 = \sigma\,\varphi_1 + \partial_x p. \tag{1.40}$$

Since $\operatorname{div}\varphi = \operatorname{div} u_0 = 0$ one obtains by a simple computation:

$$(u_0\,\nabla)\varphi_1 = \operatorname{div}(\varphi\,u_0), \quad (\varphi\,\nabla)u_1 = \operatorname{div}(u_1\,\varphi). \tag{1.41}$$

By taking care of (1.40), (1.41) and the fact that both φ and u_0 satisfy stress free boundary conditions it follows that an application of $\int_Q dx^3$ to the left-hand side of (1.40) yields zero whence $\int_Q \varphi_1\,dx^3 = 0$ since $\sigma \neq 0$. It follows that φ satisfies (δ) in (1.35) for any $\delta \in [0, 2\pi]$ whence $P_{0\delta}\varphi = \varphi$. Since $P_{0\delta} P_{\text{per}} = P_{0\delta}$, an application of $P_{0\delta}$ to both sides of (1.39) then yields

$$P_{0\delta}\,(A'_{\text{per}} + T_0)\,\varphi = \sigma\,\varphi. \tag{1.42}$$

Since $A'_S(0,\delta) = A'_{\text{per}} \restriction E_{0\delta}$ by Lemma 5 we can rewrite (1.42) as follows:

$$\big(A'_S(0,\delta) + P_{0\delta}\,T_0\big)\,\varphi = \sigma\,\varphi.$$

We thus have that $\sigma \in \sigma(A'_S(0,\delta) + P_{0\delta} T_0)$; the assertion now follows from Theorem 1.3. □

Remark. With Lemma 5 at disposal one can proceed in the same way as in [Sc.4], Section 2 in order to establish for any equilibrium solution u_0, subject to the above conditions, a relationship between instability via Theorem 1.1 and Eckhaus instability. We have to refer to [Sc.4], Section 2 for a discussion.

2. The magnetic Bénard problem

2.1. The magnetic Boussinesq problem

In case of the magnetic Boussinesq equations one is in a situation which differs from the standard Boussinesq equations by the presence of an additional complication: the fluid is now conducting and besides the temperature there is also a magnetic field which interacts with the fluid. We first proceed in a formal way and let $u = (u_1, u_2, u_3)$ be the velocity field of the fluid, p the pressure, T the temperatur and $H = (H_1, H_2, H_3)$ the magnetic field. As to boundary conditions we only assume for the moment that $T(x, y, -\frac{1}{2}) = T_0$, $T(x, y, \frac{1}{2}) = T_1$ with $T_0 > T_1$; as before, $\Omega = \mathbb{R}^2 \times (-\frac{1}{2}, \frac{1}{2})$ is the infinite plate. In order to save place we assume from the outset that we are given a sufficiently smooth equilibrium solution of the basic equations namely $u_0 = (u_1, u_2, u_3)$, p_0, τ_0 and $H_0 = (H_1, H_2, H_3)$ which are supposed to be L_1-periodic in x, L_2-periodic in y. In order to investigate its stability one is led to a standard step which amounts to perform the substitution:

$$u = u_0 + v, \quad p = p_0 + \pi, \quad T = \tau_0 + \vartheta, \quad H = H_0 + h \tag{2.1}$$

where $v = (v_1, v_2, v_3)$, $h = (h_1, h_2, h_3)$.

By inserting (2.1) into the basic equations and by taking care of the fact that u_0, p_0, τ_0, H_0 are equilibrium solutions, we end up with the following set of equations:

$$\begin{aligned} \partial_t v &= \nu \triangle v + \nabla P + S\,(\nabla\, h) H_0 + S\,(H_0\, \nabla) v + g\, \alpha\, \vartheta\, k \\ &\quad - (u_0\, \nabla) v - (v\, \nabla) u_0 - (v\, \nabla) v + S(h\, \nabla) h, \\ \partial_t \vartheta &= \kappa \triangle \vartheta - (u_0\, \nabla) \vartheta - (v\, \nabla) \tau_0 - (v\, \nabla) \vartheta, \\ \partial_t h &= r \triangle h + (H_0\, \nabla) v - (v\, \nabla) H_0 + (h\, \nabla) u_0 - (u_0\, \nabla) h \\ &\quad + (h\, \nabla) v - (v\, \nabla) h, \\ \operatorname{div} v &= 0, \quad \operatorname{div} h = 0, \quad P = \pi + S\, H_0\, h + \frac{S}{2}\, h^2. \end{aligned} \tag{2.2}$$

Here, $k = (0, 0, 1)^{\mathrm{t}}$, while ν, κ, α, g, S, r denote a series of physical constants > 0 intimately related to the physical interpretation of (2.2); for simplicity we set $\nu = \kappa = r = 1$ in what follows. As to the physical background of (2.2) we refer to the books of Chandrasekhar [Ch] and P. Roberts [Ro]. Among the equilibrium solutions u_0, p_0, τ_0, H_0 there is a favourite one, given by

$$\begin{aligned} &u_0 = 0, \quad p_0(z) = a\, z + b\, z^2, \quad \tau_0(z) = \frac{1}{2}(T_0 + T_1) - z\,(T_0 + T_1), \\ &H_0 = (0, 0, B) \end{aligned} \tag{2.3}$$

Here, a, b are constants and H_0 is a constant magnetic field pointing in the z-direction; we refer to (2.3) as the "trivial" equilibrium. The stability of (2.3) will be investigated in a subsequent paper.

2.2. Remarks on boundary conditions

While the steps leading from the basic equations to (2.2) are purely formal, some care is needed when it comes to interpret (2.2) as an evolution equation. It is here where the boundary conditions enter. On v and ϑ we henceforth impose Dirichlet boundary conditions:

$$v = (v_1, v_2, v_3) = 0 \text{ and } \vartheta = 0 \text{ on } \partial\Omega = \Big\{ -\frac{1}{2}, \frac{1}{2} \Big\} \times \mathbb{R}^2. \tag{2.4}$$

As to $h = (h_1, h_2, h_3)$ the situation is less clear. If we set $h = 0$ on $\partial\Omega$ then the system (2.2) is overdetermined: we have nine equations for eight unknowns. This has been observed by Lassner [La] who suggested to add a gradient to the right-hand side of the equations for $\partial_t h$; for an interpretation of q as a magnetic pressure see [La]. In Galdi [Ga] the energy (in-) stability of (2.3) is investigated, with h subject to Dirichlet conditions; the ∇q suggested by Lassner does not appear. However the results in [Ga] remain valid despite this omission: since only scalar products between divergence free fields are taken, ∇q would in any case cancel and leave the results unaffected. It turns out that if we impose stress free boundary conditions on $h = (h_1, h_2, h_3)$ then the above difficulties disappear. We thus require

$$\partial_z h_1 = \partial_z h_2 = h_3 = 0 \text{ on } \partial\Omega. \tag{2.5}$$

Condition (2.5) is associated with the case of a completely conducting wall (see [Ch], [Ro]). It turns out that under conditions (2.4), (2.5) the condition $\operatorname{div} h = 0$ can be dispensed with. The reason is that in this setting the right-hand side of (2.2) leaves the subspace of divergence free fields invariant. This will be made precise below. Although our main interest is in the boundary conditions (2.4), (2.5) we will also keep an eye on the case where h is subject to Dirichlet conditions. We use "MBE" as short for "magnetic Boussinesq (or Bénard) equations".

2.3. MBE on an infinite plate

Our first task is to look at (2.2) as a system on the infinite plate. That is we impose on (2.2) a functional setting which is in accordance with the boundary conditions (2.4), (2.5) and transforms (2.2) into an evolution equation. To this end we recall some facts about the NS-equations subject to Dirichlet boundary conditions; we refer to [Sc.2] for details. These equations are controlled by the Stokes operator A_S defined as follows. With $\triangle_d$ given by (1.4) we set $A = \operatorname{diag}(\triangle_d, \triangle_d, \triangle_d)$ (see the remark subsequent to (1.7)) and recall E, P defined via (1.3). We then stipulate

$$\operatorname{dom}(A_S) = \operatorname{dom}(A) \cap E \text{ and } A_S f = P A f \text{ for } f \in \operatorname{dom}(A_S). \tag{2.6}$$

The following is well known (Romanov [Rom]):

Proposition 2.1. *A_S is selfadjoint on E and $\leq -d$ for some $d > 0$.*

Next we define operators T_j, $j = 1, 2, 3$ as follows:

$$\begin{aligned} T_1(v, \vartheta, h) &= S\,(h\,\nabla)H_0 + S\,(H_0\,\nabla)h + g\,\alpha\,\vartheta\,k \\ &\quad - (u_0\,\nabla)v - (v\,\nabla)u_0, \\ T_2(v, \vartheta, h) &= -(u_0\,\nabla)\vartheta - (\vartheta\,\nabla)\tau_0, \\ T_3(v, \vartheta, h) &= (H_0\,\nabla)v - (v\,\nabla)H_0 + (h\,\nabla)u_0 - (u_0\,\nabla)h. \end{aligned} \tag{2.7}$$

These definitions are supplied by the stipulations: $\operatorname{dom}(T_j) = H^1(\Omega)^7$, $j = 1, 2, 3$. Finally we introduce nonlinearities

$$\begin{aligned} &N_1(v, h) = S(h\,\nabla)h - (v\,\nabla v), \quad N_2(v, \vartheta) = -(v\,\nabla)\vartheta, \\ &N_3(v, h) = (h\,\nabla)v - (v\,\nabla)h; \end{aligned} \tag{2.8}$$

we postpone a digression on the domains of definition of the N_j. Based on (1.3), (2.7), (2.8) one can rewrite the system (2.2) more concisely as follows:

$$\begin{aligned} \partial_t v &= A_S\,v + P\,T_1(v, \vartheta, h) + P\,N_1(v), \\ \partial_t \vartheta &= \kappa\,\triangle_d \vartheta + T_2(v, \vartheta) + N_2(v, \vartheta), \\ \partial_t h &= A'_S\,h + T_3(v, h) + N_3(v, h), \end{aligned} \tag{2.9}$$

where A'_S is given by (1.18). An even more compact representation is obtained by defining T, $\mathcal{A}_S$, $\mathcal{P}$ via

$$\begin{aligned} &T(v, \vartheta, h) = (T_1(v, \vartheta, h), T_2(v, \vartheta), T_3(v, h)) \\ &\mathcal{A}_S = \operatorname{diag}(A_S, \kappa\,\triangle_d, r\,A'_S), \quad \mathcal{P} = \operatorname{diag}(P, 1, 1). \\ &N(v, \vartheta, h) = (N_1(v, h), N_2(v, \vartheta), N_3(v, h)) \end{aligned} \tag{2.10}$$

Setting $w = (v_1, v_2, v_3, \vartheta, h_1, h_2, h_3)$ or more shortly $w = (v, \vartheta, h)$, (2.9) now assumes the form

$$\partial_t w = \mathcal{A}_S\,w + \mathcal{P}\,T\,w + \mathcal{P}\,N(w); \tag{2.11}$$

by definition $\operatorname{dom}(\mathcal{A}_S) = \operatorname{dom}(A_S) \times \operatorname{dom}(\triangle_d) \times \operatorname{dom}(A'_S)$ and $\operatorname{dom}(T) = H^1(\Omega)^7$. In order to recognize (2.11) as an evolution equation we assume henceforth:

$$u_0 = (u_1, u_2, u_3) = 0 \ \text{ on } \ \partial\Omega. \tag{2.12}$$

Proposition 2.2. *If $v \in \operatorname{dom}(A_S)$, $h \in \operatorname{dom}(A'_S)$ then $T_3(v, \vartheta, h)$ is in $H^1(\Omega)^2 \times H^1_0(\Omega) \cap E$.*

Proof. By assumption H_0, u_0 are smooth, L_1, L_2-periodic in x, y resp. and, subject to the basic equations, satisfy $\operatorname{div} H_0 = \operatorname{div} u_0 = 0$; moreover u_0 satisfies (2.12). Since $v \in \operatorname{dom}(A_S)$, $h \in \operatorname{dom}(A'_S)$ we have that $T_3(v, \vartheta, h)$ is in $H^1(\Omega)^3$. With $\partial_j = \partial_{x_j}$ and $x_1 = x$, $x_2 = y$, $x_3 = z$ the third components of $(h\nabla)u_0$, $(u_0\nabla)h$ are given by

$$\sum h_j\,\partial_j u_3, \quad \sum u_j\,\partial_j h_3. \tag{2.13}$$

Since $\operatorname{div} u_0 = 0$ and by virtue of (2.12) we have that $\partial_z u_3 = 0$ on $\partial\Omega$. From this fact and by (2.12) it follows that both expressions in (2.13) are in $H^1_0(\Omega)$. By the same remarks, with v and H_0 in place of u_0 and h resp. we infer that the third components of $(v\nabla)H_0$ and $(H_0\nabla)v$ are in $H_0^1(\Omega)$. Next we recall the identity

$$\operatorname{rot}(F\times G) = (G\,\nabla)F - (F\,\nabla)G + F\operatorname{div}G - G\operatorname{div}F. \tag{2.14}$$

By proper specialization of (2.14) and since $\operatorname{div}H_0$, $\operatorname{div}u_0$, $\operatorname{div}v$, $\operatorname{div}h$ are all $=0$ we infer $\operatorname{div}T_3(v,\vartheta,h)=0$, proving the claim. □

Concerning the nonlinearity N_3 we have

Proposition 2.3. *If $v\in\operatorname{dom}(A_S)$, $h\in\operatorname{dom}(A'_S)$ then $N_3(v,h)\in H^1(\Omega)^2\times H_0^1(\Omega)\cap E$.*

Proof. We first note two facts which are easily proved by standard approximation arguments. Given $f\in H^2(\Omega)\cap H_0^1(\Omega)$, $g\in H^2(\Omega)$ then: (a) $f\partial_j g\in H_0^1(\Omega)$, for $j\le 3$, (b) $g\partial_x f, g\partial_y f\in H_0^1(\Omega)$. Likewise it is routine to show that since $v\in\operatorname{dom}(A_S)$, $h\in\operatorname{dom}(A'_S)$ we have $N_3(v,h)\in H^1(\Omega)^3$. Now the third components of $(h\nabla)v$ and $(v\nabla)h$ are given by

$$\sum h_j\,\partial_j v_3, \quad \sum v_j\,\partial_j h_3. \tag{2.15}$$

From (a), (b) above we infer that the two terms in (2.15) are in $H_0^1(\Omega)$ whence $N_3(v,h)\in H^1(\Omega)^2\times H_0^1(\Omega)$. From (2.14) and since $\operatorname{div}v=\operatorname{div}h=0$ we infer $\operatorname{div}N_3(v,h)=0$, proving the claim. □

Remarks. Propositions 2.2, 2.3 evidently justify the omission of P in the third equation of (2.9).

Proposition 2.4. *Given ε there is K_ε such that: if $w\in\operatorname{dom}(\mathcal{A}_S)$ then*

$$\|\mathcal{P}\,T\,w\|^2_{\mathcal{L}^2}\le\varepsilon\,\|\mathcal{A}_S\,w\|^2_{\mathcal{L}^2}+K_\varepsilon\,\|w\|^2_{\mathcal{L}^2}.$$

Hint of proof. Since $\mathcal{A}_S=\operatorname{diag}(A_S,\kappa\triangle_d,A'_S)$ the claim easily follows from the corresponding statements regarding A_S, $\kappa\triangle_d$, A'_S respectively. As to A_S we refer to Corollary 2.2 in [Sc.2], while A'_S is handled by the Corollary to Lemma 2; as to $\kappa\triangle_d$ we may refer e.g. to the proof of Lemma 2. □

Proposition 2.5. *$\mathcal{A}_S+\mathcal{P}T$ is a holomorphic semigroup generator on $E\times\mathcal{L}^2(\Omega)\times E$.*

With Proposition 2.5 at disposal one can interprete the nonlinearities N_j, $j\le 3$, or likewise $\mathcal{N}$, properly. In fact, for $\lambda>0$ sufficiently large one introduces fractional powers $(\lambda-\mathcal{A}_S-PT)^\alpha$ for $\alpha\ge 0$ in the usual way. These give rise to fractional power spaces X_α; one has $X_0=E\times\mathcal{L}^2(\Omega)\times E$ and $X_1=\operatorname{dom}(\mathcal{A}_S)$. By established arguments one can view $\mathcal{N}(w)$ as the diagonal $\mathcal{N}'(w,w)$ of a bilinear map $\mathcal{N}'(\ ,\ )$ which maps $(X_\alpha)^2$ boundedly into $E\times\mathcal{L}^2(\Omega)\times E$ provided that $\alpha\in(\frac{3}{4},1)$. The invariance expressed by Proposition 2.3 remains unaffected due to straightforward approximation arguments. Based on these remarks and Proposition 2.5 we can interprete (2.11) as an evolution equation in a well-known way; see Pazy [Pa], pg. 196 to this effect. In particular, the notion of Ljapounov stability and instability of the trivial solution $w_0=0$ of (2.11) can be introduced. There is a basic result in this connection, analoguous to Theorem 1.1, which has to be stated without proof, namely:

Theorem 2.1. *(a) If $\sigma(\mathcal{A}_S + \mathcal{P}T)$ lies strictly in the left half plane then $w_0 = 0$ is Ljapounov stable, (b) if $\lambda \in \sigma(\mathcal{A}_S + \mathcal{P}T)$ with* $\mathrm{re}(\lambda) > 0$ *exists then $w_0 = 0$ is Ljapounov instable.*

Remarks. The remarks subsequent to Theorem 1.1 apply here too. The proof of (b), while involved, follows closely the lines of [Sc.3]. It is based among others on concepts encountered below. As to the properties of the properly defined operator "rotrot" we refer to R. Picard [Pi].

2.4. Θ-periodic spectra

As in Chapter 1 we are led to analyze the structure of $\sigma(\mathcal{A}_S + \mathcal{P}T)$ in terms of Θ-periodic spectra. The material in Section 1.2, 1.3 can thereby be taken over as it stands. We yet have to supply the definition of Θ-periodic Dirichlet Stokes operator $A_S(\Theta)$, amply discussed in [Sc.1], [Sc.2]. To this effect we recall (1.24) and define A_Θ according to

$$\begin{aligned}&\mathrm{dom}(A_\Theta) = \mathrm{dom}(\triangle_\Theta)^3 \ \text{ and } \ A_\Theta f = (\triangle f_1, \triangle f_2, \triangle f_3) \\ &\text{for } \ f = (f_1, f_2, f_3) \ \text{ in } \ \mathrm{dom}(A_\Theta),\end{aligned} \tag{2.16}$$

i.e. $A_\Theta = \mathrm{diag}(\triangle_\Theta, \triangle_\Theta, \triangle_\Theta)$. With P_Θ, E_Θ as in (1.23) and with Proposition 1.16 at disposal we stipulate:

$$\begin{aligned}&\mathrm{dom}(A_S(\Theta)) = \mathrm{dom}(A_\Theta) \cap E \ \text{ and } \ A_S(\Theta) f = P_\Theta A_\Theta f \\ &\text{for } \ f \in \mathrm{dom}(A_S(\Theta)),\end{aligned} \tag{2.17}$$

where $\Theta \in M$, $M = [0, 2\pi]^2$. Since Lemma 3 is not available for A_Θ, $A_S(\Theta)$ is considerably more difficult to handle than $A'_S(\Theta)$ in Section 1.2. In [Sc.1] it is shown:

Proposition 2.6. *If $\Theta \in \dot{M}$ then $A_S(\Theta)$ is selfadjoint on E_Θ, $\leq -d$ for some Θ-independent $d > 0$ and $A_S(\Theta)^{-1}$ is compact.*

Remark. We recall that $\dot{M}$ is $M = [0, 2\pi]^2$ minus the corners; if $\Theta \in M - \dot{M}$ then we are in the periodic case for which Proposition 2.6 is known to hold (Kirchgässner-Kielhöfer [K-K]).

The operators $A_S(\Theta)$, $\triangle_\Theta$, $A'_S(\Theta)$ give rise to a Θ-periodic "magnetic" Stokes operator $\mathcal{A}_S(\Theta)$ given by

$$\begin{aligned}&\mathrm{dom}(\mathcal{A}_S(\Theta)) = \mathrm{dom}(A_S(\Theta)) \times \mathrm{dom}(\triangle_\Theta) \times \mathrm{dom}(A'_S(\Theta)), \\ &\mathcal{A}_S(\Theta)\, w = (A_S(\Theta)\, v, \triangle_\Theta\, \vartheta, A'_S(\Theta)\, h)\end{aligned} \tag{2.18}$$

where $w = (v, \vartheta, h)$ is in $\mathrm{dom}(\mathcal{A}_S(\Theta))$. By Propositions 1.11, 2.6, Lemma 3 and with Q given prior to (1.22) we then have

Proposition 2.7. *$\mathcal{A}_S(\Theta)$ is selfadjoint on $E_\Theta \times \mathcal{L}^2(Q) \times E_\Theta$ and ≤ 0, $(\Theta \in M)$ and has compact resolvents.*

Next we recall the operators T_j, $j \le 3$ given by (2.7). In the present context we set $\operatorname{dom}(T_j) = H^1(Q)^7$, $j \le 3$. We do not characterize the so defined T_j's by an extra label but read off their interpretation from the context. In analogy to Proposition 2.2 we have

Proposition 2.8. *If $h \in \operatorname{dom}(A'_S(\Theta))$, $v \in \operatorname{dom}(A_S(\Theta))$ (some $\Theta \in M$) then $T_3(v, \vartheta, h)$ is in $H^1_\Theta(Q)^2 \times H^1_{\Theta,0}(Q)$ and satisfies* $\operatorname{div} T_3(v, \vartheta, h) = 0$.

Briefly, $T_3(v, \vartheta, h) \in E_\Theta$ under the stated assumptions. The proof is omitted since it is analogous to that of Proposition 2.2.

We now cast the T_j, $j \le 3$ into a single operator T with $\operatorname{dom}(T) = H^1(Q)^7$ given by

$$\begin{aligned} &T\,w = (T_1(v, \vartheta, h),\, T_2(v, \vartheta, h),\, T_3(v, \vartheta, h)) \\ &\text{for } \ w = (v, \vartheta, h) \ \text{ in } \ \operatorname{dom}(T). \end{aligned} \tag{2.19}$$

Recalling P_Θ in (1.23) we then set

$$\mathcal{P}_\Theta = \operatorname{diag}(P_\Theta, 1, 1). \tag{2.20}$$

From Proposition 2.8 and (2.20) we infer

Proposition 2.9. *If $w \in \operatorname{dom}(\mathcal{A}_S(\Theta))$, (some $\Theta \in M$) then $\mathcal{P}_\Theta T w$ is in $E_\Theta \times \mathcal{L}^2(Q) \times E_\Theta$.*

Lemma 7. *Given ε there is K_ε such that if $\Theta \in \dot{M}$ and $w \in \operatorname{dom}(\mathcal{A}_S(\Theta))$ then*

$$\|\mathcal{P}_\Theta\, T\, w\|^2_{\mathcal{L}^2} \le \varepsilon\, \|\mathcal{A}_S(\Theta)\, w\|^2_{\mathcal{L}^2} + K_\varepsilon\, \|w\|^2_{\mathcal{L}^2}. \tag{2.21}$$

Remarks. The Lemma is a straightforward consequence of the definition of $\mathcal{A}_S(\Theta)$ via (2.18) and the fact that for each constituent in (2.18) a corresponding statement is available: Corollary 2.2 in [Sc.2] for $A_S(\Theta)$ and Lemma 4 for $A'_S(\Theta)$, while $\triangle_\Theta$ is a special case of $A'_S(\Theta)$. The restriction to $\dot{M}$ is for convenience only; if $\Theta \in M - \dot{M}$ then we are in the periodic case for which the Lemma holds on standard grounds. From these remarks, Propositions 2.7, 2.9 and Lemma 7 we infer by standard semigroup theory

Lemma 8. *$\mathcal{A}_S(\Theta) + \mathcal{P}_\Theta T$ generates a holomorphic semigroup on $E_\Theta \times \mathcal{L}^2(Q) \times E_\Theta$ ($\Theta \in M$) and has compact resolvents.*

These preparations enable use to state the first basic result in the present context, namely

Theorem 2.2.

$$\sigma\big(\mathcal{A}_S + \mathcal{P}\,T\big) = \text{closure}\Big(\bigcup_{\Theta \in \dot{M}} \sigma\big(\mathcal{A}_S(\Theta) + \mathcal{P}_\Theta\, T\big)\Big). \tag{2.22}$$

We relegate a brief discussion on the proof of Theoreom 2.2 to the end of the paragraph. Theorem 2.2 is the counterpart in the MHE-context to Theorem 2.1. There is also a counterpart to Theorem 1.3 whose formulation calls for some preliminaries. First we recall the convention in Section 1.3 according to which the label "per" denotes the periodic case associated with any $\Theta \in M - \dot{M}$; we also recall $E_{0\delta}$, $P_{0\delta}$ in (1.35). With A_{per} thus given by (2.16) we follow (4.38) in [Sc.2] and define for $\delta \in [0, 2\pi]$:

$$f \text{ is in } \operatorname{dom}(A_S(0,\delta)) \text{ iff } f \text{ is in } \operatorname{dom}(A_{\text{per}}) \cap E_{\text{per}} \text{ and satisfies } (\delta) \text{ in (1.35); for such } f \text{ we set } A_S(0,\delta)f = P_{0\delta}A_{\text{per}}f. \tag{2.23}$$

In [Sc.2], Section 4.4 it is shown

Proposition 2.10. *$A_S(0,\delta)$ is selfadjoint on $E_{0\delta}$, $\leq -d$ for some δ-independent $d > 0$ and $A_S(0,\delta)^{-1}$ is compact.*

As before we amalgamate the operators $A_S(0,\delta)$, $\triangle_{\text{per}}$, $A'_S(0,\delta)$ (given by (1.36)) into a single one $\mathcal{A}_S(0,\delta)$ given by

$$\begin{aligned} &\operatorname{dom}(\mathcal{A}_S(0,\delta)) = \operatorname{dom}(A_S(0,\delta)) \times \operatorname{dom}(\triangle_{\text{per}}) \times (A'_S(0,\delta)) \text{ and} \\ &\mathcal{A}_S(0,\delta)\, w = (A_S(0,\delta)\, v, \triangle_{\text{per}}\, \vartheta,\ A'_S(0,\delta)\, h) \end{aligned} \tag{2.24}$$

where $w = (v, \vartheta, h)$ is in $\operatorname{dom}(\mathcal{A}_S(0,\delta))$. Due to its diagonal structure, $\mathcal{A}_S(0,\delta)$ inherits the properties of its constituents i.e. we have

Proposition 2.11. *$\mathcal{A}_S(0,\delta)$ is selfadjoint on $E_{0\delta} \times \mathcal{L}^2(Q) \times E_{0\delta}$, is ≤ 0 and has compact resolvents.*

Next we have to focus on the T_j in (2.7). First we note:

Proposition 2.12. *Let $v \in \operatorname{dom}(A_S(per))$, $h \in \operatorname{dom}(A'_S(per))$; then $T_3(v,\vartheta,h) \in E_{0\delta}$ for any $\delta \in [0, 2\pi]$.*

Proof. By setting $\Theta = 0$ in Proposition 2.8 we infer that $T_3(v,\vartheta,h)$ is in $H^1_{\text{per}}(Q)^2 \times H^1_{\text{per},0}(Q)$ and satisfies $\operatorname{div} T_3(v,\vartheta,h) =)$ whence $T_3(v,\vartheta,h) \in E_{\text{per}}$. The first two components of $(H_0\nabla)v$ and $(v\nabla)H_0$ are given by

$$a_j = \sum_k H_k\, \partial_k v_j, \quad b_j = \sum_k v_k\, \partial_k H_j, \quad j = 1, 2$$

where $\partial_1 = \partial_x$, $\partial_2 = \partial_y$, $\partial_3 = \partial_z$. Since $\operatorname{div} H_0 = \operatorname{div} v = 0$ we have that

$$a_j = \sum_k \partial_k (H_k v_j), \quad b_j = \sum_k \partial_k (v_k H_j), \quad j = 1, 2,$$

whence

$$\int_Q a_j\, dx^3 = \sum_k \int_Q \partial_k (H_k v_j)\, dx^3, \quad \int_Q b_j\, dx^3 = \sum_k \int_Q \partial_k (v_k H_j)\, dx^3. \tag{2.25}$$

The terms in $\sum_k$ with $k = 1, 2$ vanish since $H_k v_j$, $v_k H_j$ are L_1, L_2-periodic in x, y resp.; the terms with $k = 3$ vainsh since v satisfies Dirichlet boundary conditions. Thus the left-hand sides of "=" in (2.25) vanish. Next we consider the first two components of $(u_0\nabla)h$ and $(h\nabla)u_0$ resp. which are given by

$$c_j = \sum_k h_k\, \partial_k u_j, \quad d_j = \sum_k u_k \partial_k h_j, \quad j = 1, 2.$$

By the same arguments as before we find that $\int_Q c_j dx^3$ and $\int_Q d_j dx^3$ vanish. These facts entail that $T_3(v,\vartheta,h)$ satisfies (δ) in (1.35) for any $\delta \in [0,2\pi]$; thus $T_3(v,\vartheta,h)$ is in $E_{0\delta}$, $\delta \in [0,2\pi]$ by the introductory remarks. □

Recalling $P_{0\delta}$ in (1.35) we set for $\delta \in [0,2\pi]$:

$$\mathcal{P}_{0\delta} w = (P_{0\delta} v, \vartheta, h) \ \text{ for } \ w = (v,\vartheta,h) \in \mathcal{L}^2(Q)^7. \tag{2.26}$$

From Definitions (2.19), (2.24) and Proposition 2.12 we conclude

Proposition 2.13. *If $w \in \operatorname{dom}(\mathcal{A}_S(0,\delta))$ for some $\delta \in [0,2\pi]$ then $\mathcal{P}_{0\delta} T w$ is in $E_{0\delta} \times \mathcal{L}^2(Q) \times E_{0\delta}$.*

We still need a counterpart to Lemma 7; this is provided by

Lemma 9. *Given ε there is K_ε such that if $\delta \in [0,2\pi]$, $w \in \operatorname{dom}(\mathcal{A}_S(0,\delta))$ then*

$$\|\mathcal{P}_{0\delta} T\, w\|^2_{\mathcal{L}^2} \le \varepsilon \, \|\mathcal{A}_S(0,\delta) w\|^2_{\mathcal{L}^2} + K_\varepsilon \, \|w\|^2_{\mathcal{L}^2}. \tag{2.27}$$

Remarks. The remarks subsequent to Lemma 7 apply here too. Indeed Lemma 9 is a consequence of (2.24) and the fact that a corresponding statement holds for each constituent in (2.24): for $A_S(0,\delta)$ the statement is established by (4.46) in [Sc.2], for $A'_S(0,\delta)$ we have Proposition 1.20 at disposal while $\triangle_{\text{per}}$ is handled by standard arguments. Lemma 9 then follows straightforwardly by an amalgamation of the three cases. From Proposition 2.11, 2.13 and Lemma 9 we infer by standard arguments from semigroup theory:

Lemma 10. *For any $\delta \in [0,2\pi]$, $\mathcal{A}_S(0,\delta) + \mathcal{P}_{0\delta} T$ generates a holomorphic semigroup on $E_{0\delta} \times \mathcal{L}^2(Q) \times E_{0\delta}$ and has compact resolvents.*

After these preliminaries we are in a position to state the counterpart to Theorem 1.3, i.e.:

Theorem 2.3.

$$\sigma\big(\mathcal{A}_S + \mathcal{P} T\big) = \bigcup_{\Theta \in \dot{M}} \sigma\big(\mathcal{A}_S(\Theta) + \mathcal{P}_\Theta\, T\big) \cup \bigcup_{\delta \in [0,2\pi]} \sigma\big(\mathcal{A}_S(0,\delta) + \mathcal{P}_{0\delta}\, T\big).$$

Remarks. As in case of Theorems 1.2, 1.3 we have to add a remark on the proofs of Theorems 2.2, 2.3. These proofs have been checked in detail; they are structured after the proofs given in [Sc.2], and much more explicitely in [Sc.5], for the Dirichlet case. These proofs, routine as they are, cannot be reproduced here, even approximately, due to their length. A few indications have to suffice. Starting point is (2.18) which defines $\mathcal{A}_S(\Theta)$ in terms of $A_S(\Theta)$, $\triangle_\Theta)$, $A'_S(\Theta)$. About these constituents we know everything, be it on the basis of [Sc.1], [Sc.2] in case of $A_S(\Theta)$, be it via Section 1.2 in case of $A'_S(\Theta)$. Due to its diagonal structure, $\mathcal{A}_S(\Theta)$ inherits all properties of its constituents, in particular those concerning its extension into the complex. We define $\mathcal{P}'_\Theta$, I' via

$$\mathcal{P}'_\Theta\, w = (P_\Theta v, \vartheta, P_\Theta h), \quad I'\, w = (0,0,h) \ \text{ for } \ w = (v,\vartheta,h).$$

It turns out that $\mathcal{R}'_\Theta = (\mathcal{A}_S(\Theta) - I')^{-1}\mathcal{P}'_\Theta$ is the correct substitute for R_Θ in Lemma 4.1 of [Sc.2]. We introduce T' differing from T in (2.19) in that T_3 is replaced by T'_3 given by

$$T'_3(v,\vartheta,h) = T_3(v,\vartheta,h) + h.$$

A statement which expresses the relative boundedness of T' with respect to $\mathcal{A}_S(\Theta) - I'$ is then easily deduced from Lemma 7. With $\mathcal{R}'_\Theta$, T', $\mathcal{A}_S(\Theta) - I'$ as above we then proceed in very much the same way as with R_Θ, T_0, $A_S(\Theta)$ in [Sc.2] in order to obtain ultimatively Theorems 2.2, 2.3. One of the major questions to be asked in connection with Theorem 2.3 is that about the relationship between $\sigma(\mathcal{A}_S + \mathcal{P}T)$ and $\sigma(\mathcal{A}_S(\text{per}) + \mathcal{P}_{\text{per}}T)$. A partial answer in this direction is provided by

Theorem 2.4. *Let $\lambda \neq 0$ be in $\sigma(\mathcal{A}_S(per)+\mathcal{P}_{per}T)$ and $w = (v,\vartheta,h)$ an eigenvector for λ such that: (a) v satisfies (δ) in (1.35) for some $\delta \in [0,2\pi]$. Then $\lambda \in \sigma(\mathcal{A}_S + \mathcal{P}T)$.*

Proof. By assumption $w \in \text{dom}(\mathcal{A}_S(\text{per}))$ and

$$\big(\mathcal{A}_S(\text{per}) + \mathcal{P}_{\text{per}}\, T\big)\, w = \lambda\, w. \tag{2.28}$$

By definition of $\mathcal{A}_S(\text{per})$ via (2.18) for $\Theta = 0$ and since $A_S(\text{per})$, $A'_S(\text{per})$ are given by (2.17), (1.34) respectively for $\Theta = 0$ (the periodic case), (2.28) splits into three equations:

$$\begin{aligned} \mathcal{P}_{\text{per}}\, A_{\text{per}}\, v + P_{\text{per}}\, T_1 &= \lambda\, v \\ \triangle\vartheta + T_2 &= \lambda\,\vartheta \\ P_{\text{per}}\, A'_{\text{per}}\, h + T_3 &= \lambda\, h, \quad T_j = T_j(v,\vartheta,h). \end{aligned} \tag{2.29}$$

Now let $T_3 = (a_1,a_2,a_3)$ and $h = (h_1,h_2,h_3)$. We recall Proposition 2.8 which holds for $\Theta = 0$ and the proof of Proposition 2.2; together they yield

$$\int_Q a_j\, dx^3 = 0,\ j = 1,2 \ \text{ and } \ \operatorname{div} T_3 = 0. \tag{2.30}$$

By virtue of Lemma 3 and since $\mathcal{P}_{\text{per}}h = h$ the third equation in (2.29) reduces to

$$\triangle h_j + a_j = \lambda\, h_j, \quad j = 1,2,3. \tag{2.31}$$

Due to the boundary conditions imposed on h we have that

$$\int_Q \triangle h_j\, dx^3 = 0 \ \text{ for } \ j = 1,2. \tag{2.32}$$

From (2.30), (2.31), (2.32) and since $\lambda \neq 0$ we obtain

$$\int_Q h_j\, dx^3 = 0 \ \text{ for } \ j = 1,2. \tag{2.33}$$

By (2.33) we thus have $P_{0\delta}h = h$ for any $\delta \in [0, 2\pi]$ and thus for the δ in (a) of the theorem. Now $P_{0\delta}T_3 = T_3$ by Proposition 2.12 and $P_{0\delta}P_{\text{per}} = P_{0\delta}$ since $E_{0\delta} \subseteq E_{\text{per}}$; this permits us to rewrite the third equation in (2.29) as follows:

$$P_{0\delta}\, A'_{\text{per}}\, h + T_3 = \lambda\, h. \tag{2.34}$$

Now $P_{0\delta}v = v$ by assumption. An application of $P_{0\delta}$ to the first equation of (2.29) then yields:

$$P_{0\delta}\, A_{\text{per}}\, v + P_{0\delta}\, T_1 = \lambda\, v. \tag{2.35}$$

By combining (2.34), (2.35) with the second equation in (2.29) we obtain

$$P_{0\delta}\, A_{\text{per}}\, v + P_{0\delta}\, T_1\, v = \lambda\, v, \quad \triangle v + T_2 = \lambda\, \vartheta, \quad P_{0\delta}\, A'_{\text{per}}\, h + T_3 = \lambda\, h. \tag{2.36}$$

These equations can be expressed more concisely as follows:

$$\big(\mathcal{A}_S(0,\delta) + \mathcal{P}_{0\delta}\, T\big)\, w = \lambda\, w. \tag{2.37}$$

The claim now follows from (2.37) and Theorem 2.3. □

Remarks. Theorem 2.3 has many applications which will be exploited elsewhere. E.g. if $\lambda \neq 0$ is real it is easily seen that assumption (a) is satisfied. Likewise, by a straightforward application to the arguments in [Sc-v.W], Section 5 it can be shown that (a) holds if the equilibrium solution is a roll, i.e. depending on z only. Moreover one can prove with the aid of Theorem 2.3 the equivalence of Ljapounov instability as emphasized by Theorem 2.1 with Eckhaus instability; we refer to [Sc.4] for a discussion in connection with the Boussinesq equation. As indicated in Section 2.2 we will say something on the case where the magnetic field is subject to Dirichlet conditions, i.e. $h = 0$ on $\partial\Omega$. The question then arises as to what extent the theory now differs from the case of stress free boundary conditions for h. In fact the situation now becomes simpler since the digression on the stress free NS-equations (in Chapter 1) becomes superfluous. The steps leading from the NS-equations to the MBE are in the Dirichlet case essentially the same as those leading from the NS-equations to the Boussinesq equations; see [Sc.3] for an outline and [Sc.5] for a more detailed account. As a result one ends up with Theorems 2.2, 2.3 which thus remain valid in the Dirichlet case. There is however a difficulty concerning the counterpart to Theorem 2.4. To explain this we thus assume Dirichlet conditions both for the velocity field and the magnetic field. We assume that $\lambda \neq 0$ is a real eigenvalue of $\mathcal{A}_S(\text{per}) + \mathcal{P}_{\text{per}}T$, interpreted now in terms of Dirichlet conditions. We let $w = (v, \vartheta, h)$ be an eigenvector of λ which without loss of generality may be taken to be real. It is then easy to find $\delta, \delta' \in [0, 2\pi]$ such that v, h satisfy (δ) resp. (δ') in (1.35). However there is no reason to expect $\delta = \delta'$. But it is precisely this equality which is needed in order to be able to infer that w is in $E_{0\delta} \times \mathcal{L}^2(Q) \times E_{0\delta}$ and thus $\lambda \in \sigma(\mathcal{A}_S(\text{per}) + \mathcal{P}_{\text{per}}T)$. The same difficulty arises in similar contexts. Briefly, we cannot prove Theorem 2.4 in the pure Dirichlet case. This shows that stress free boundary conditions for the magnetic field have both mathematical and physical advantages.

References

[Ad] R. Adams: *Sobolev Spaces.* Academic Press, New York (1975).

[Ch] S. Chandrasekhar: *Hydrodynamic and Hydromagnetic Stability.* Dover Publications (1961).

[Ga] G. Galdi: "Nonlinear stability of the magnetic Bénard problem via a generalized energy method". Arch. Rat. Mech. Anal. **110** (1985), 167–186.

[He] D. Henry: "Geometric theory of semilinear parabolic equations". Lecture Notes in Math. **840**, Springer, Berlin (1981).

[Ka-v.W] Y. Kagei–W. von Wahl: "The Eckhaus criterion for convection solutions of the Boussinesq equations". Int. J. Non-linear Mechanics **32** (1997), 563–620.

[K-K] K. Kirchgässner–H. Kielhöfer: "Stability and bifurcation in fluid dynamics". Rocky Mountain J. Math. **3** (1973), 275–317.

[La] G. Lassner: "Über ein Rand-Anfangswertproblem der Magnetohydrodynamik". Arch. Rat. Mech. Anal. **25** (1967), 388–405.

[Mi] A. Mielke: "Mathematical analysis of sideband instabilities with applications to Rayleigh-Bénard convections". J. Nonlinear Sci. **7** (1997), 57–99.

[Pa] A. Pazy: *Semigroups of Linear Operators and Partial Differential Equations.* Springer, New York (1993).

[Pi] R. Picard: "On a selfadjoint extension of curl in exterior domains". To appear.

[Ro] P. Roberts: *An introduction to Magnetohydrodynamics.* Longmans (1974).

[Rom] V. A. Romanov: "Stability of plane parallel Couette flows". Functional analysis and its applications **7** (1973), 137–146.

[Sc.1] B. Scarpellini: "Regularity properties of some Stokes operators on an infinite plate". Pitman Res. Notes Math. **314** (1994), 221–231.

[Sc.2] B. Scarpellini: "$\mathcal{L}^2$-perturbations of space periodic equilibria of Navier-Stokes". Zeitschr. f. Analysis und Anwendungen **14**, no. 4 (1995), 779–828.

[Sc.3] B. Scarpellini: "The principle of linearized stability for space periodic equilibria of Navier-Stokes on an infinite plate". Analysis **15** (1995), 359–394.

[Sc.4] B. Scarpellini: "Stability properties of the Boussinesq equations". To appear in the NSE-6 Conference report, Amann et. al (Eds.) (1998).

[Sc.5] B. Scarpellini: *Stability, Instability and Direct Integrals.* To appear in Pitman Res. Notes in Math.

[Sc-v.W] B. Scarpellini–W. von Wahl: "Stability properties of the Boussinesq equations". To appear ZAMP, March 1998.

[St] B. Stranghan: *The Energy Method, Stability and Nonlinear Convection.* Appl. Math. Sciences, Springer, New York (1992)

Progress in Nonlinear Differential Equations
and Their Applications, Vol. 35
© 1999 Birkhäuser Verlag Basel/Switzerland

Remarks on the Asymptotic Bounds for the Energy Dissipation in the Couette Problem

B.J. Schmitt and M. Wiegner

Dedicated to Professor Herbert Amann on the occasion of his 60th birthday

1. Introduction

The establishment of rigorous bounds for turbulent fluid flow has been a continuous challenge in the analysis of the Navier-Stokes equations. A few years ago, the interest in this topic has been reenhanced by Doering and Constantin's so-called background flow method [DC1]. By picking up an idea of Hopf [Ho] and decomposing the flow field into a stationary "background flow" part which carries the boundary values and a fluctuating part they could estimate the rate of energy dissipation in the Couette system for the whole range of Reynolds numbers. In [DC2] they stated more generally a variational principle which produces upper bounds for the average dissipation rate if only the background flow satisfies a certain spectral constraint. Their upper bound computed explicitly however is based on a one-dimensional background flow, as does the result of [GGHL] where an improved profile function is used. A systematic decrease of this bound by about 16 percent was further obtained in [NGH1] due to an additional balance parameter. In [NGH2, NGH3] the same authors could establish an even smaller bound, their proof, however, is based on numerical analysis of the spectral constraint. It turned out that for $\mathrm{Re} \to \infty$ their bound exactly recovers the one derived in the Howard-Busse optimum theory [Hw, B1, B2]. A closer investigation of the connection between the results obtained from the background flow method on one hand and the Howard-Busse approach on the other hand has been done by Kerswell [K].

In the present paper we are interested in the improvement of the non-numerical, mathematically rigorous bound given in [NGH1]. We shall again make use of a one-dimensional background flow, but with a profile function which is different from the quadratic one used in [GGHL, NGH1]. The main progress is gained by computing exactly the best constant in a one-sided Poincaré inequality with weights, and by exploiting the solenoidality of the flow field more efficiently than before. The details will be given in Section 4. In Section 2 and 3 we have to recall some details of the background flow method and its application to the Couette system in order to make clear where our technical improvements apply.

In the final section we give some additional comments on how our result compares with the ones mentioned above.

Because of the well-known problem concerning global existence and regularity of solutions to the Navier-Stokes equations a remark on the class of solutions considered here is appropriate. Throughout this paper we assume that the solutions are sufficiently regular to perform all of the desired manipulations of the Navier-Stokes system, and we also assume that the solutions exist globally in time. The underlying function space will be $L^2(\Omega)$, and in the case of the Couette system the periodicity in the plane variables x and y has to be respected in the usual way, see Section 3 below. In fact our result continues to hold for weak solutions (of Leray-Hopf type) satisfying the energy inequality at time $t = 0$, since this can be used as a starting point in Section 2 instead of the energy balance identity (2.2) below.

2. Background flow and balance parameter

In this section and in the following one we briefly recall the derivation of bounds for the energy dissipation along the lines of the background flow method [DC1, DC2] taking into account the idea of the balance parameter [NGH1], and its application to the Couette system.

We consider three-dimensional fluid flow in a region Ω which is described in terms of the velocity field $u = u(x, y, z, t)$ by the Navier-Stokes equation

$$\left.\begin{array}{rcl} u_t - \nu\Delta u + u\cdot\nabla u + \nabla p &=& 0, \\ \operatorname{div} u &=& 0 \end{array}\right\} \tag{2.1}$$

in $\Omega \times (0, \infty)$. Here p denotes the pressure, and $\nu > 0$ is the kinematic viscosity. The fluid is driven from the boundary Σ of the fluid region:

$$u|\Sigma = U^{\Sigma}.$$

According to the background flow method we decompose $u = U + v$ into a fixed background flow part U which carries the boundary values, and a fluctuating part v:

$$\operatorname{div} U = \operatorname{div} v = 0, \qquad U|\Sigma = U^{\Sigma}, \qquad v|\Sigma = 0.$$

Apart from that U is arbitrary, in particular it need not solve the Navier-Stokes equation. By testing the resulting equation for v with v one obtains the energy balance

$$\frac{1}{2}\|v\|_t^2 = -\nu\|\nabla v\|^2 - (v, (\nabla U)_{\mathrm{sym}} v) - (U_t - \nu\Delta U + U\cdot\nabla U, v). \tag{2.2}$$

Here $(\cdot\,,\,\cdot)$ denotes the scalar product in $L^2(\Omega)$ with associated norm $\|\cdot\|$. $A_{\mathrm{sym}} = \frac{1}{2}(A + A^{\mathrm{T}})$ is the symmetric part of a quadratic matrix A. By multiplying (2.2) with a balance parameter $a > 1$ and adding

$$\nu\|\nabla u\|^2 = \nu\|\nabla U\|^2 + \nu\|\nabla v\|^2 - 2(\Delta U, v) \tag{2.3}$$

we arrive at

$$\nu\|\nabla u\|^2+\frac{a}{2}\|v\|_t^2 = \nu\|\nabla U\|^2-a\Big\{\nu\frac{a-1}{a}\|\nabla v\|^2+ +(v,(\nabla U)_{\mathrm{sym}}v)+(U_t-\nu\frac{a-2}{a}\Delta U+U\cdot\nabla U,v)\Big\}. \tag{2.4}$$

The parameter a will be fixed later. In order to estimate the dissipation we must control the functional

$$H_{U,a}(w):=\nu\frac{a-1}{a}\|\nabla w\|^2+(w,(\nabla U)_{\mathrm{sym}}w)+(U_t-\nu\frac{a-2}{a}\Delta U+U\cdot\nabla U,w)$$

which is defined on the space $\mathcal{V}=\{w\in H_0^{1,2}(\Omega)\mid \operatorname{div} w=0\}$. We say that U fulfills the *spectral constraint* iff the quadratic part of $H_{U,a}$
i) is strictly positive on $\mathcal{V}\setminus\{0\}$ or
ii) is nonnegative on $\mathcal{V}$ and $U_t-\nu\frac{a-2}{a}\Delta U+U\cdot\nabla U$ is a gradient.
In either case there exists a minimizer $V\in\mathcal{V}$ for $H_{U,a}$ which can be found by solving the corresponding Euler-Lagrange problem; in case *ii)* one can choose $V=0$. Furthermore we have

$$\min_{w\in\mathcal{V}}H_{U,a}(w)=\frac{1}{2}(U_t-\nu\frac{a-2}{a}\Delta U+U\cdot\nabla U,V).$$

By averaging (2.4) over $(0,T)$ and letting $T\to\infty$ we then obtain

$$\limsup_{T\to\infty}\langle\nu\|\nabla u\|^2\rangle_T\le\limsup_{T\to\infty}\langle\nu\|\nabla U\|^2-\frac{a}{2}(U_t-\nu\frac{a-2}{a}\Delta U+U\cdot\nabla U,V)\rangle_T,$$

here the abbreviation $\langle f\rangle_T=T^{-1}\int_0^T f(t)\,dt$ is used. If U is independent of time and satisfies the spectral constraint we therefore have found a finite bound for the long time limit ε of the averaged energy dissipation rate:

$$\varepsilon:=\limsup_{T\to\infty}\langle\frac{\nu}{|\Omega|}\|\nabla u\|^2\rangle_T\le\frac{\nu}{|\Omega|}\|\nabla U\|^2-\frac{a}{2|\Omega|}(-\nu\frac{a-2}{a}\Delta U+U\cdot\nabla U,V). \tag{2.5}$$

3. Application to the Couette problem

In the Couette problem the fluid volume is a layer $\mathbb{R}^2\times(0,h)$ with thickness $h>0$. All quantities are assumed to be periodic in the plane variables x and y with periodicity $0<L_x<\infty$ and $0<L_y<\infty$, respectively. To this end the underlying function spaces have to be modified accordingly, i.e. the functions are regarded in the box $\Omega=(0,L_x)\times(0,L_y)\times(0,h)$ with periodic behaviour in x and y. For a more detailed description of the exact choice of function spaces see e.g. [SW, Sc]. The boundary of the fluid volume corresponds to $\Sigma=(0,L_x)\times(0,L_y)\times\{0,h\}$.

The boundary condition for the Couette problem then reads

$$U^{\Sigma}(x,y,z)=\begin{cases}U_0e_x & \text{for } z=h,\\ 0 & \text{for } z=0.\end{cases} \tag{3.1}$$

Here $e_x=(1,0,0)^{\mathrm{T}}$ is the direction of the shearing while its intensity is given by $U_0\ge0$. The Reynolds number $\mathrm{Re}=\frac{U_0h}{\nu}$ serves as control parameter.

The background flow is assumed to be one-dimensional:

$$U(x,y,z) = \varphi(z)e_x = U_0\Phi(\frac{z}{h})e_x,$$

with a profile function Φ satisfying $\Phi(0) = 0$, $\Phi(1) = 1$ and, for reasons of symmetry, $\Phi(1-\xi) = 1 - \Phi(\xi)$, $\xi \in [0,1]$.

For this type of background flow the solutions of the Euler-Lagrange equation derived from the functional $H_{U,a}$ are known explicitly [NGH1]:

$$V(x,y,z) = \frac{a-2}{2(a-1)}U_0(\xi - \Phi(\xi))e_x|_{\xi=z/h}.$$

From (2.5) one therefore obtains the bound ($\|\cdot\|$ here denotes the norm in $L^2(0,1)$)

$$\varepsilon \le \frac{\nu U_0^2}{h^2}\Big(\frac{a^2}{4(a-1)}\|\Phi'\|^2 - \frac{(a-2)^2}{4(a-1)}\Big) \tag{3.2}$$

provided that the spectral constraint is fulfilled:

$$\forall 0 \ne w \in \mathcal{V}: \qquad \nu\frac{a-1}{a}\|\nabla w\|^2 + \int_\Omega \varphi'(z)w_1 w_3\, d(x,y,z) > 0. \tag{3.3}$$

A sufficient condition for this to hold is

$$U_0 h\hat{K}_0(\Phi')m < \nu\frac{a-1}{a},$$

where

$$\hat{K}_0(\Phi') := \sup\{\int_0^1 |\Phi'(\xi)||f(\xi)|^2\, d\xi \mid \|f'\| = 1,\ f(0) = 0,\ f(1) = 0\}$$

and m is chosen such that

$$h^2 m\hat{K}_0(\Phi')\|\nabla w\|^2 \ge \int_\Omega |\Phi'(\frac{z}{h})||w_1||w_3|\, d(x,y,z) \tag{3.4}$$

for every $w \in \mathcal{V}$, see Lemma 4.1 below. For the non-dimensionalized dissipation rate $c_\varepsilon = \frac{\varepsilon h}{U_0^3}$ the result therefore reads as follows:

$$c_\varepsilon \le \mathrm{Re}^{-1}\Big(1 + (\|\Phi'\|^2 - 1)\frac{a^2}{4(a-1)}\Big) \tag{3.5}$$

if

$$\mathrm{Re} < \frac{a-1}{a}(m\hat{K}_0(\Phi'))^{-1}. \tag{3.6}$$

In order to deal with high Reynolds numbers one therefore must choose the profile function Φ such that $\hat{K}_0(\Phi')$ becomes so small that (3.6) is satisfied. This is done by concentrating Φ' to boundary layers with thickness $\frac{\delta}{2} \in (0, \frac{1}{2}]$:

$$\Phi(\xi) = \begin{cases} \frac{1}{2}\theta(\frac{2\xi}{\delta}) & \text{for } 0 \le 2\xi \le \delta, \\ \frac{1}{2} & \text{for } \delta \le 2\xi \le 1 \end{cases}$$

where $\theta(0) = 0$, $\theta(1) = 1$ and $\delta \in (0, 1]$. Again Φ is assumed to be antisymmetric. A scaling argument shows that

$$\hat{K}_0(\Phi') = \frac{\delta}{4} K_0(\theta')$$

where

$$K_0(\theta') := \sup\{\int_0^1 |\theta'(\eta)||f(\eta)|^2 \, d\eta \;\mid\; \|f'\| = 1,\; f(0) = 0\}. \tag{3.7}$$

Since $\|\Phi'\|^2 = \frac{1}{\delta}\|\theta'\|^2$ we obtain from (3.5), (3.6):

$$c_\varepsilon \mathrm{Re} \leq 1 + (\frac{1}{\delta}\|\theta'\|^2 - 1)\frac{a^2}{4(a-1)}$$

if

$$\delta \mathrm{Re} < \frac{a-1}{a}\frac{4}{mK_0(\theta')}.$$

If Re is large and θ considered as fixed then

$$\delta \to \frac{a-1}{a}\frac{4}{mK_0(\theta')\mathrm{Re}}$$

gives the best bound, namely

$$c_\varepsilon \mathrm{Re} \leq -\frac{(a-2)^2}{4(a-1)} + \frac{a^3}{16(a-1)^2} mK_0(\theta')\|\theta'\|^2 \mathrm{Re}.$$

The right-hand side attains its minimum for

$$a_{\min} = \frac{3}{2}\Big(1 + \sqrt{1 + \frac{32}{9(mK_0(\theta')\|\theta'\|^2\mathrm{Re} - 4)}}\Big) \tag{3.8}$$

which yields the bound

$$\begin{aligned} c_\varepsilon \mathrm{Re} &\leq \frac{9}{16} mK_0(\theta')\|\theta'\|^2 \frac{(1+D)(1+5D)}{(1+3D)^2}\mathrm{Re} \\ &\leq \frac{27}{64} mK_0(\theta')\|\theta'\|^2 \mathrm{Re}, \\ D &= \sqrt{1 + \frac{32}{9(mK_0(\theta')\|\theta'\|^2\mathrm{Re} - 4)}} \end{aligned} \tag{3.9}$$

provided that

$$\mathrm{Re} > \frac{4}{mK_0(\theta')}.$$

4. The asymptotic bound

By letting $\mathrm{Re} \to \infty$ in (3.8), (3.9) we obtain $a_{\min}(\mathrm{Re}) \downarrow 3$ and the following asymptotic bound for the non-dimensionalized dissipation rate:

$$\limsup_{\mathrm{Re}\to\infty} c_\varepsilon(\mathrm{Re}) \le \frac{27}{64} m K_0(\theta')\|\theta'\|^2. \tag{4.1}$$

In this section we shall show that $m = \frac{3\sqrt{3}}{16}$ is admissible in (3.4) and that for $\theta(\eta) = 1 - (1-\eta)^{3.4}$ we have $K_0(\theta')\|\theta'\|^2 \le 0.33521$. The total asymptotic bound therefore reads

$$\limsup_{\mathrm{Re}\to\infty} c_\varepsilon(\mathrm{Re}) \le 0.0460.$$

This improves the bound $\frac{3}{32\sqrt{2}}$ given in [NGH1] by more than 30 percent. There it was made use of the quadratic profile $\theta(\eta) = 1-(1-\eta)^2$, the estimate $K_0(\theta')\|\theta'\|^2 \le \frac{4}{9}$ (following [GGHL]) and $m = \frac{1}{2\sqrt{2}}$ (from [DC1]). Of course our technical improvement applies to the case (3.9) of finite Reynolds numbers, too.

Lemma 4.1. *For every $w \in \mathcal{V}$ (i.e. $\operatorname{div} w = 0$, $w = 0$ for $z = 0$ and for $z = h$, w periodic in x and y) we have*

$$\int_\Omega |\Phi'(\tfrac{z}{h})||w_1 w_3|\, d(x,y,z) \le \frac{3\sqrt{3}}{16} h^2 \hat{K}_0(\Phi')\|\nabla w\|^2.$$

Proof. By application of Young's inequality and the definition of $\hat{K}_0(\Phi')$ it follows that

$$\begin{aligned} \int_\Omega |\Phi'(\tfrac{z}{h})||w_1 w_3|\, d(x,y,z) \le \qquad\qquad\qquad\qquad\qquad\qquad\qquad (4.2)\\ \le\ & h\int_0^{L_x}\int_0^{L_y}\int_0^1 |\Phi'(\xi)|\Big(\frac{\gamma}{2}|w_1(x,y,h\xi)|^2 + \frac{1}{2\gamma}|w_3(x,y,h\xi)|^2\Big)\, d\xi\, dy\, dx \\ \le\ & h\int_0^{L_x}\int_0^{L_y} \hat{K}_0(\Phi')\int_0^1 \Big(\frac{\gamma}{2}|\partial_z w_1(x,y,h\xi)|^2 \\ & \qquad + \frac{1}{2\gamma}|\partial_z w_3(x,y,h\xi)|^2\Big) h^2\, d\xi\, dy\, dx \\ =\ & h^2\hat{K}_0(\Phi')\Big(\frac{1}{2\sqrt{3}}\|\partial_z w_1\|^2 + \frac{\sqrt{3}}{2}\|\partial_z w_3\|^2\Big), \end{aligned}$$

by choosing $\gamma = \frac{1}{\sqrt{3}}$. Because of $\operatorname{div} w = 0$ and $w|\Sigma = 0$ we can further write

$$\begin{aligned} \|\partial_z w_3\|^2 &= \frac{3}{8}\|\partial_z w_3\|^2 + \frac{1}{4}(-\partial_x w_1 - \partial_y w_2, \partial_z w_3) + \frac{3}{8}\|\partial_x w_1 + \partial_y w_2\|^2 \\ &= \frac{3}{8}\|\partial_z w_3\|^2 + \frac{1}{4}(-\partial_z w_1, \partial_x w_3) + \frac{1}{4}(-\partial_z w_2, \partial_y w_3) \\ &\qquad + \frac{3}{8}\|\partial_x w_1\|^2 + \frac{3}{4}(\partial_y w_1, \partial_x w_2) + \frac{3}{8}\|\partial_y w_2\|^2 \end{aligned}$$

$$\leq \frac{3}{8}(\|\partial_z w_3\|^2 + \|\partial_x w_1\|^2 + \|\partial_y w_1\|^2 + \|\partial_x w_2\|^2 + \|\partial_y w_2\|^2)$$
$$+ \frac{1}{24}\|\partial_z w_1\|^2 + \frac{3}{8}\|\partial_x w_3\|^2 + \frac{1}{8}\|\partial_z w_2\|^2 + \frac{1}{8}\|\partial_y w_3\|^2,$$

again by means of Young's inequality. The right-hand side of (4.2) can therefore be estimated by

$$h^2 \hat{K}_0(\Phi')\Big(\frac{3\sqrt{3}}{16}(\|\partial_z w_3\|^2 + \|\partial_x w_1\|^2 + \|\partial_y w_1\|^2 + \|\partial_x w_2\|^2 + \|\partial_y w_2\|^2$$
$$+ \|\partial_z w_1\|^2 + \|\partial_x w_3\|^2) + \frac{\sqrt{3}}{16}(\|\partial_z w_2\|^2 + \|\partial_y w_3\|^2)\Big)$$
$$\leq \frac{3\sqrt{3}}{16} h^2 \hat{K}_0(\Phi')\|\nabla w\|^2.$$

□

The improvement of the constant m is therefore due to the exploitation of all nine entries of ∇w, instead of only five in [DC1]. Next we compute $K_0(\theta')$ explicitly for the family of profile functions $\theta_\alpha(\eta) = 1 - (1-\eta)^\alpha$, $\alpha > 0$.

Lemma 4.2. *For $\theta_\alpha(\eta) = 1 - (1-\eta)^\alpha$, $\alpha > 0$ we have*

$$K_0(\theta'_\alpha) = \frac{4\alpha}{(\alpha+1)^2\lambda^2_{\alpha+1}}$$

where λ_j denotes the first positive zero of the Bessel function $J_{-1/j}$ of order $-\frac{1}{j}$, $j > 1$.

Proof. Recalling the definition (3.7) of $K_0(\theta')$ and using the fact that $\theta'_\alpha > 0$ in $(0,1)$ we see by means of variational methods that $K_0(\theta'_\alpha) = \mu^{-1}$, $\mu > 0$ being the smallest eigenvalue of the Euler-Lagrange problem

$$\begin{aligned} -f''(\eta) &= \mu\theta'_\alpha(\eta)f(\eta), \qquad \text{in } \eta \in (0,1), \\ f(0) = 0,\ & f'(1) = 0. \end{aligned} \tag{4.3}$$

The latter boundary condition is nothing else but the transversality condition for the free boundary point $\eta = 1$. Further we can assume that $f(1) > 0$. By substituting $g(\tau) = \tau^{-1/(\alpha+1)} f(1 - \gamma\tau^{2/(\alpha+1)})$ and choosing

$$\gamma = \Big(\frac{(\alpha+1)^2}{4\alpha\mu}\Big)^{\frac{1}{\alpha+1}}$$

we can rewrite (4.3) as

$$\tau^2 g''(\tau) + \tau g'(\tau) = (\frac{1}{(\alpha+1)^2} - \tau^2)g(\tau)$$

whence

$$g(\tau) = c_1 J_{-1/(\alpha+1)}(\tau) + c_2 J_{1/(\alpha+1)}(\tau),$$

Here J_n denotes as usual the Bessel function of order n. The boundary conditions $f(1) > 0$, $f'(1) = 0$ imply that

$$c_1 = \Gamma(1 - \frac{1}{\alpha+1})2^{\frac{1}{\alpha+1}} f(1), \quad c_2 = 0.$$

From the positivity of the first eigenfunction f of (4.3) in $(0, 1)$ and the condition $f(0) = 0$ we conclude that $\gamma^{-(\alpha+1)/2}$ must coincide with the first zero $\lambda_{\alpha+1}$ of $J_{-1/(\alpha+1)}$. In other words, $\mu = (\alpha+1)^2\lambda^2_{\alpha+1}/(4\alpha)$. □

Consequently

$$K_0(\theta'_\alpha)\|\theta'_\alpha\|^2 = \frac{4\alpha^3}{(2\alpha-1)(\alpha+1)^2\lambda^2_{\alpha+1}}$$

for those profile functions θ_α. The expression on the left-hand side can never be smaller than $\frac{1}{3}$ because we may choose $f(\eta) = \frac{\theta(\eta)}{\|\theta'\|}$ in (3.7):

$$K_0(\theta') \geq \frac{1}{\|\theta'\|^2}\int_0^1 |\theta'(\eta)||\theta(\eta)|^2\, d\eta \geq \frac{1}{\|\theta'\|^2}\int_0^1 \frac{1}{3}(\theta(\eta)^3)'\, d\eta = \frac{1}{3}\frac{1}{\|\theta'\|^2}.$$

For $\alpha = 3.4$ however the above expression evaluates to ≤ 0.33521 which is therefore almost optimal.

Our result now follows by making use of $\theta = \theta_{3.4}$ in (3.9) and applying Lemma 4.1 and 4.2:

Theorem 4.3. *Let u be x, y-periodic in the layer $\mathbb{R}^2 \times (0, h)$ with periodicity L_x and L_y, respectively. If u is a solution of (2.1) with Couette boundary data (3.1) then the non-dimensional average dissipation rate*

$$c_\varepsilon = \limsup_{T\to\infty} \frac{\nu}{L_x L_y U_0^3}\langle\|\nabla u\|^2\rangle_T$$

is bounded in terms of the Reynolds number $\mathrm{Re} = \frac{U_0 h}{\nu}$ *by*

$$c_\varepsilon \leq 0.046\Big(1 - \frac{(D-1)(7D+1)}{3(3D+1)^2}\Big) \quad \textit{with} \quad D = \sqrt{1 + \frac{32}{0.98\mathrm{Re} - 36}} \quad \textit{if} \quad \mathrm{Re} > 73.24.$$

In particular we have found the rigorous (non-numerical) asymptotic bound mentioned above:

$$\limsup_{\mathrm{Re}\to\infty} c_\varepsilon(\mathrm{Re}) \leq 0.0460.$$

5. Remarks

For sufficiently large Reynolds numbers $\mathrm{Re} > 73.24$ we have proved in Theorem 4.3 a non-numerical rigorous upper bound on the energy dissipation rate which lowers the one given in [NGH1] by about 30 percent. The occurrence of decimal fractions in Theorem 4.3 is just for convenience of comparison; it might as well have been written in terms of $\lambda_{4.4}$, the first zero of the Bessel function $J_{-5/22}$. The range $\mathrm{Re} < 73.24$ could have been treated, too, in Section 3 by choosing $\delta = 1$ and adapting the balance parameter a and the profile function θ_α in a way similar to [NGH1]. On the other hand, by considering the Couette flow itself, i.e. the profile $\varphi(z) = U_0 \frac{z}{h}$, as background flow one does not only get the lower bound $c_\varepsilon \geq \mathrm{Re}^{-1}$ for every Re but also the upper bound $c_\varepsilon \leq \mathrm{Re}^{-1}$ as long as $\mathrm{Re} \leq \mathrm{Re}_c \approx 82.65$ which is the energy stability border for the Couette problem [J, DR]. For this reason we have confined ourselves to $\mathrm{Re} > 73.24$ which is good enough for dealing with the interesting range of Reynolds numbers $\mathrm{Re} > \mathrm{Re}_c$, cf. Fig. 1.

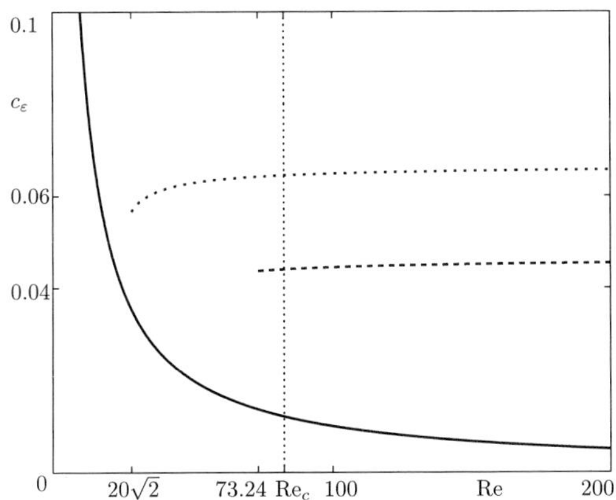

FIGURE 1. Bounds on the non-dimensional energy dissipation rate c_ε. Solid line: lower bound Re^{-1}, for $\mathrm{Re} < \mathrm{Re}_c$ it holds $c_\varepsilon(\mathrm{Re}) = \mathrm{Re}^{-1}$; long dashes: upper bound from Theorem 4.3; short dashes: upper bound obtained in [NGH1].

We have seen in the end of Section 4 that our choice of the profile function almost gives the optimal value for $K_0(\theta')\|\theta'\|^2$. Assuming that $m = \frac{3\sqrt{3}}{16}$ cannot be improved too drastically in Lemma 4.1, either, we stress that a further improvement of the bound should be due to a more refined investigation of the spectral constraint. Instead of simply applying Hölder's inequality to w_1 and w_3 and treating these components separately one should try to solve the constraint as exactly as possible. Indeed, in [NGH2, NGH3] an upper bound for c_ε is presented which is smaller than ours by a factor of about 4; however the admissibility of the chosen background flows is checked numerically, whereas our result is completely non-numerical.

Therefore the derivation of better rigorous upper bounds for c_ε remains an interesting goal, in particular since even the result of [NGH2, NGH3] cannot reflect the bound which is predicted from experiments, see e.g. [DC2] for a discussion of this topic.

Acknowlegdement

The authors would like to thank R. Kaiser (University of Bayreuth) for some helpful discussions on the topic of this paper.

References

[B1] Busse F.H.: "Bounds for turbulent shear flow". J. Fluid Mech. **41** (1970), 219–240.

[B2] Busse F.H.: "The optimum theory of turbulence". Adv. Appl. Mech. **18** (1978), 77–121.

[DC1] Doering C.R., Constantin P.: "Energy dissipation in shear driven turbulence". Phys. Rev. Lett. **69** (1992), 1648–1651.

[DC2] Doering C.R., Constantin P.: "Variational bounds on energy dissipation in incompressible flows: Shear flow". *Phys. Rev. E* **49** (1994), 4087–4099.

[DR] Drazin P.G., Reid W.H.: *Hydrodynamic Stability.* Cambridge University Press: Cambridge (1981).

[GGHL] Gebhardt T., Grossmann S., Holthaus M., Löhden M.: "Rigorous bounds on the plane-shear-flow dissipation rate". *Phys. Rev. E* **51** (1995), 360–365.

[Ho] Hopf E.: "Ein allgemeiner Endlichkeitssatz der Hydrodynamik". Math. Ann. **117** (1941), 764–775.

[Hw] Howard L.N.: "Bounds on flow quantities". Ann. Rev. Fluid Mech. **4** (1972), 473–494.

[J] Joseph D.D.: *Stability of Fluid Motions I, II.* Springer Tracts in Natural Philosophy **27, 28**, Springer: Berlin, Heidelberg, New York (1976).

[K] Kerswell R.R.: "Variational bounds on shear-driven turbulence and turbulent Boussinesq convection". *Physica D* **100** (1997), 355–376.

[NGH1] Nicodemus R., Grossmann S., Holthaus M.: "Improved variational principle for bounds on energy dissipation in turbulent shear flow". Physica D **101** (1997), 178–190.

[NGH2] Nicodemus R., Grossmann S., Holthaus M.: "Variational bound on energy dissipation in turbulent shear flow". Phys. Rev. Lett. **79** (1997), 4170–4173.

[NGH3] Nicodemus R., Grossmann S., Holthaus M.: "Variational bound on energy dissipation in plane Couette flow". Phys. Rev. E **56** (1997), 6774–6786.

[Sc] Scarpellini B.: "$\mathcal{L}^2$-perturbations of space periodic equilibria of Navier-Stokes". Z. Anal. Anwend. **14** (1995), 779–828.

[SW] Schmitt B.J., Wahl W. von: Decomposition of solenoidal fields into poloidal fields, toroidal fields and the mean flow. Applications to the Boussinesq-equations. *The Navier-Stokes Equations II — Theory and Numerical Methods.* Proceedings, Oberwolfach 1991. Lecture Notes in Math. **1530** (eds. J.G. Heywood, K. Masuda, R. Rautmann, S.A. Solonnikov), 291–305. Springer: Berlin, Heidelberg, New York (1992).

B.J. Schmitt and M. Wiegner
RWTH Aachen
Department of Mathematics I
D-52056 Aachen, Germany

Progress in Nonlinear Differential Equations
and Their Applications, Vol. 35
© 1999 Birkhäuser Verlag Basel/Switzerland

A Special Class of Weak Solutions of the Navier-Stokes Equations in Arbitrary Three-dimensional Domains

Hermann Sohr

1. Introduction and notations

Let $\Omega \subseteq \mathbb{R}^3$ be any domain, i.e. any nonempty open connected set in $\mathbb{R}^3$. Here we are mainly interested in unbounded domains. $\partial\Omega$ denotes the boundary of Ω. In Ω we consider the system of the Navier-Stokes equations of the form

$$\begin{aligned} u_t - \Delta u + u\nabla u + \nabla p = f\ , \quad & \operatorname{div} u = 0 \\ u|_{\partial\Omega} = 0\ ,, \quad & u(0) = u_0, \end{aligned} \tag{1.1}$$

describing the unknown velocity field $u = (u_1, u_2, u_3)$ and the unknown pressure p of a flow within $[0,T) \times \Omega$ where $0 < T \leq \infty$; $f = (f_1, f_2, f_3)$ denotes the given force and u_0 the given initial velocity at $t = 0$.

Our aim is to construct weak solutions u of (1.1) which have new properties in particular concerning the asymptotic behavior of u as $t \to \infty$. Our main result, Theorem 2.1, yields an energy inequality which contains a scalar weight function depending on the time t. Further it yields several integrability properties in space and time.

In the equations (1.1), $\nabla = (D_1, D_2, D_3) = (\frac{\partial}{\partial x_1}, \frac{\partial}{\partial x_2}, \frac{\partial}{\partial x_3})$ denotes the gradient concerning the space variables $x = (x_1, x_2, x_3) \in \Omega$, $u_t = \frac{d}{dt}u$ with $t \in [0,T)$ means the time derivative, $\Delta = D_1^2 + D_2^2 + D_3^2$ the Laplacian. The nonlinear term $u\nabla u = u \cdot \nabla u = (u \cdot \nabla)u$ is defined by $u\nabla u = (u_1 D_1 + u_2 D_2 + u_3 D_3)u$ and div $u = D_1 u_1 + D_2 u_2 + D_3 u_3$.

We will use the following notations. $L^q = L^q(\Omega)$, $1 \leq q \leq \infty$, is the usual Lebesgue space with norm $\|u\|_{L^q} = \|u\|_q = (\int_\Omega |u|^q\, dx)^{\frac{1}{q}}$. If $q = 2$, $L^2(\Omega)$ means the Hilbert space with scalar product $\langle u, v\rangle = \langle u, v\rangle_\Omega = \int_\Omega u(x)v(x)\, dx$. All spaces we consider here are real. $W^{k,q} = W^{k,q}(\Omega), k = 0, 1, \ldots$, means the usual Sobolev space with norm $\|u\|_{W^{k,q}}$. Consider $q \geq 1, s \geq 1,\ 0 < T \leq \infty$ and any Banach space X with norm $\|\cdot\|_X$. Then $L^s(0,T;X)$ means the usual space with norm $\|u\|_{L^s(0,T;X)} = \|u\|_{X,s;T} = (\int_0^T \|u\|_X^s\, dt)^{\frac{1}{s}}$. If $X = L^q(\Omega)$ we write $\|u\|_{L^s(0,T;L^q)} = \|u\|_{L^q,s;T} = \|u\|_{q,s;T}$. In case $s = q = 2$ we get the Hilbert space $L^2(0,T;L^2(\Omega)) = L^2((0,T)\times\Omega)$ with scalar product $\langle u, v\rangle = \langle u, v\rangle_{\Omega,T} = \int_0^T \int_\Omega u(t,x)v(t,x)\, dx\, dt$.

Furthermore we need the usual loc-spaces $L^q_{\text{loc}}(\Omega)$, $L^s_{\text{loc}}(0,T;X)$ and $L^s_{\text{loc}}([0,T);X)$; $u \in L^q_{\text{loc}}(\Omega)$ means that $u \in L^q(\Omega')$ for all subdomains $\Omega' \subseteq \Omega$ with $\overline{\Omega}' \subset \Omega, \overline{\Omega}'$ denotes the closure of Ω', and $u \in L^s_{\text{loc}}([0,T);X)$ means that $u \in L^s(0,T';X)$ for all $0 < T' < T$.

The corresponding spaces of vector fields $u = (u_1, \ldots, u_n), n \in \mathbb{N}$, are denoted by $L^q(\Omega)^n$, $W^{k,q}(\Omega)^n, \ldots$. The scalar product in $L^2(\Omega)^n$ is defined by $\langle u, v\rangle = \langle u, v\rangle_\Omega = \int_\Omega u \cdot v \, dx$ where $u \cdot v = u_1 v_1 + \cdots + u_n v_n$.

$C_0^\infty(\Omega)$ means the space of smooth functions with compact support in Ω and $C_0^\infty(\Omega)'$ is the space of distributions in Ω. Correspondingly, we get the spaces $C_0^\infty(\Omega)^n$ and $(C_0^\infty(\Omega)^n)' = (C_0^\infty(\Omega)')^n$ in the vector valued case; $[f,v] = [f,v]_\Omega$ means the value of some $f \in C_0^\infty(\Omega)'$ at $v \in C_0^\infty(\Omega)$. The same notation $[f,v] = [f,v]_\Omega$ is used for $f \in (C_0^\infty(\Omega)^n)'$, $v \in C_0^\infty(\Omega)^n$. Similarly we use the notation $[.,.]_{\Omega,T}$ if Ω is replaced by $(0,T) \times \Omega$.

Let $u \in L^s(0,T;X), 1 \le s < \infty$. Then we write $\frac{d}{dt}u = u_t \in L^s(0,T;X)$ if there are a function $f \in L^s(0,T;X)$ and some $u_0 \in X$ such that $u(t) = u_0 + \int_0^t f(\tau)d\tau$ holds for almost all $(a.a.)$ $t \in [0,T)$. In this case we may redefine u on a subset of $[0,T)$ of measure zero such that u is continuous and $u(0) = u_0$. We say u is continuous almost everywhere (a.e) and set $\frac{d}{dt}u = f$.

Next we introduce some special notations for (1.1): Let $C_{0,\sigma}^\infty(\Omega) = \{v \in C_0^\infty(\Omega)^3; \operatorname{div} v = 0\}$ and let

$$L^2_\sigma(\Omega) = \overline{C_{0,\sigma}^\infty(\Omega)}^{\|\cdot\|_2} \subseteq L^2(\Omega)^3$$

which means the closure in $L^2(\Omega)^3$. With

$$G(\Omega) = \{\nabla p \in L^2(\Omega)^3;\ p \in L^2_{\text{loc}}(\Omega)\}$$

we obtain the orthogonal decomposition $L^2(\Omega)^3 = L^2_\sigma(\Omega) \oplus G(\Omega)$, each $f \in L^2(\Omega)^3$ has the unique representation $f = f_0 + \nabla p$ with $f_0 \in L^2_\sigma(\Omega), \nabla p \in G(\Omega)$, and the Helmholtz projection $P : L^2(\Omega)^3 \to L^2_\sigma(\Omega)$ is defined by $Pf = f_0$.

The subspace $W^{1,2}_{0,\sigma}(\Omega) \subseteq W^{1,2}(\Omega)^3$ is defined by

$$W^{1,2}_{0,\sigma}(\Omega) = \overline{C_{0,\sigma}^\infty(\Omega)}^{\|\cdot\|_{W^{1,2}}}$$

where $\|u\|_{W^{1,2}} = \|u\|_2 + \|\nabla u\|_2$ with

$$\nabla u = (D_j u_k)^3_{j,k=1},\ \|\nabla u\|_2 = (\sum_{j,k=1}^3 \|D_j u_k\|_2^2)^{\frac{1}{2}} = \langle \nabla u, \nabla u\rangle^{\frac{1}{2}}.$$

The Stokes operator $A : D(A) \to L^2_\sigma(\Omega), D(A) \subseteq L^2_\sigma(\Omega)$ is defined by

$$D(A) = \{u \in W^{1,2}_{0,\sigma}(\Omega); v \mapsto \langle \nabla u, \nabla v\rangle \text{ continuous in } \|v\|_2\}$$

and by $\langle Au, v\rangle = \langle \nabla u, \nabla v\rangle$, $u \in D(A), v \in W^{1,2}_{0,\sigma}(\Omega)$.

A is positive selfadjoint (see [BM92], [KO94], [GS89], [Ma84]) and all properties of A which we need can be derived from the spectral representation $A = \int_0^\infty \lambda \, dE_\lambda$ where E_λ $(\lambda \ge 0)$ are projections in $L^2_\sigma(\Omega)$. In particular we need the resolvent $(\mu+A)^{-1} = (\mu I+A)^{-1} = \int_0^\infty (\mu+\lambda)^{-1} dE_\lambda$, for $\mu > 0$, I means the identity, the fractional powers $A^\alpha = \int_0^\infty \lambda^\alpha \, dE_\lambda$ with $D(A^\alpha) = \{u \in L^2_\sigma(\Omega);\ \int_0^\infty \lambda^{2\alpha} \, d\|E_\lambda u\|_2^2 < \infty\}$, $-1 \le \alpha \le 1$, and the operators $e^{-tA} = \int_0^\infty e^{-t\lambda} \, dE_\lambda$ with $t \ge 0$.

We need the following properties of the operator norms $\|\cdot\|$:

$$\|A^\alpha(\mu+A)^{-1}\| \le \frac{1}{\mu^{1-\alpha}}, \quad 0 \le \alpha \le 1,\ \mu > 0, \tag{1.2}$$

$$\|A^\alpha e^{-tA}\| \le \frac{1}{t^\alpha}, \quad 0 \le \alpha \le 1,\ t \ge 0. \tag{1.3}$$

The operators $J_k = (I + k^{-1}A^{\frac{1}{2}})^{-1}$ with $k \in \mathbb{N}$ (Yosida approximation) are used for certain "smoothing" procedures, see [So83], [So84], [SWW86]. It holds that $\|J_k\| \le 1$ and $v = \lim_{k\to\infty} J_k v$ in $L^2_\sigma(\Omega)$ for each $v \in L^2_\sigma(\Omega)$.

If $\alpha = \frac{1}{2}$ we get $D(A^{\frac{1}{2}}) = W^{1,2}_{0,\sigma}(\Omega)$ and $\langle \nabla u, \nabla v\rangle = \langle A^{\frac{1}{2}}u, A^{\frac{1}{2}}v\rangle$ for all $u, v \in D(A^{\frac{1}{2}})$. For each $v \in L^2_\sigma(\Omega)$, $t \mapsto e^{-tA}v$ is strongly continuous and it holds

$$\lim_{t\to\infty} \|e^{-tA}v\|_2 = 0; \tag{1.4}$$

the function $t \mapsto e^{-tA}v$ is denoted by $e^{-\bullet A}v$. Finally we note the interpolation property, see [Fr69]; 1, (9.9),

$$\|u\|_q \le C\,(\|u\|_{q_0} + \|u\|_{q_1}) \tag{1.5}$$

where $1 \le q_0 < q < q_1 < \infty$, $\frac{1}{q} = \alpha\frac{1}{q_0} + (1-\alpha)\frac{1}{q_1}$, $0 < \alpha < 1$.

2. Weak solutions and main theorem

Since we do not know up to now whether the equations (1.1) have a solution u in the classical sense at least if f, u_0 and Ω are sufficiently smooth, our aim is to improve the properties of weak solutions which exist globally in time but which are maybe not unique and not smooth. We need the space

$$C_0^\infty([0,T); C^\infty_{0,\sigma}(\Omega)) = \{u|_{[0,T)\times\Omega};\, u \in C_0^\infty((-1,T)\times\Omega)^3,\ \operatorname{div} u = 0\};$$

$u|_{[0,T)\times\Omega}$ means the restriction to $[0,T)\times\Omega$. Setting $u(t) = u(t,\cdot)$ we get $u(t) = 0$ near T but $u(0) \ne 0$ in general. f in (1.1) will have always the form $f = \operatorname{div} F$ where $F = (F_{jk})_{j,k=1}^3$ means a matrix function and where $\operatorname{div} F$ is defined by $\operatorname{div} F = (D_1F_{j1} + D_2F_{j2} + D_3F_{j3})_{j=1}^3$; thus div applies to the lines of F.

Let $f = \operatorname{div} F$ with $F \in L^2_{\text{loc}}([0,T); L^2(\Omega)^9)$ and $u_0 \in L^2_\sigma(\Omega)$. Then a measurable function u in $[0,T)\times\Omega$ is called a **weak solution** of the Navier-Stokes equations (1.1) if the following holds:

$$\begin{aligned} &a) \quad u \in L^\infty_{\text{loc}}([0,T);\ L^2_\sigma(\Omega)) \cap L^2_{\text{loc}}([0,T);\ W^{1,2}_{0,\sigma}(\Omega)),\\ &b) \quad -\langle u, v_t\rangle_{\Omega,T} + \langle \nabla u, \nabla v\rangle_{\Omega,T} - \langle uu, \nabla v\rangle_{\Omega,T} \\ &\qquad = \langle u_0, v(0)\rangle_\Omega - \langle F, \nabla v\rangle_{\Omega,T} \end{aligned} \tag{2.1}$$

for all $v \in C_0^\infty([0,T); C^\infty_{0,\sigma}(\Omega))$.

Here we use the notation $uu = (u_j u_k)_{j,k=1}^3$ and since $\operatorname{div} u = 0$ we get $\operatorname{div}(uu) = u\nabla u$ and $[u\nabla u, v]_{\Omega,T} = [\operatorname{div}(uu), v]_{\Omega,T} = -\langle uu, \nabla v\rangle_{\Omega,T}$; further, $[f, v]_{\Omega,T} = [\operatorname{div} F, v]_{\Omega,T} = -\langle F, \nabla v\rangle_{\Omega,T}$. A distribution p in $(0,T)\times\Omega$ such that (1.1) holds in the sense of distributions together with a weak solution u is called an associated pressure of u.

If a weak solution u of (1.1) additionally satisfies the condition

$$uu \in L^2_{\mathrm{loc}}([0,T);\ L^2(\Omega)^9), \tag{2.2}$$

then for $0 < T' < T$ we get (see [SvW85])

$$\begin{aligned}\langle u\nabla u, u\rangle_{\Omega,T'} &= \langle \operatorname{div}(uu), u\rangle_{\Omega,T'} \\ &= \langle u, \frac{1}{2}\nabla|u|^2\rangle_{\Omega,T'} = -\langle \operatorname{div} u, \frac{1}{2}|u|^2\rangle_{\Omega,T'} = 0\end{aligned} \tag{2.3}$$

and in (1.1) we may take the scalar product with $u(t)$ and integrate from 0 to t. This yields the energy inequality (even equality in this case)

$$\frac{1}{2}\|u(t)\|_2^2 + \int_0^t \|\nabla u\|_2^2\, d\tau \ \le\ \frac{1}{2}\|u_0\|_2^2 - \int_0^t \langle F, \nabla u\rangle d\tau \tag{2.4}$$

for a.a. $t \in [0,T)$. For general weak solutions u, (2.4) is an additional property. It follows the estimate

$$\|u\|^2_{2,\infty;T'} + \|\nabla u\|^2_{2,2;T'} \le \|u_0\|_2^2 + \|F\|^2_{2,2;T'} \tag{2.5}$$

for $0 < T' < T$.

We know, see e.g. [Ma84], that a weak solution $u : [0,T) \to L^2_\sigma(\Omega)$ is weakly continuous (after redefinition on a set of measure zero). In this case (2.4) even holds for all $t \in [0,T)$ if it holds for a.a. $t \in [0,T)$.

For our purpose it is convenient to give the operators P and $A^{-\frac{1}{2}}P$ extended meanings similar as in the theory of distributions. This will enable us to find a (nonlinear) evolution equation for u which has some advantages. For any distribution $f \in (C_0^\infty(\Omega)^3)'$, Pf simply means the restriction $Pf = f|_{C^\infty_{0,\sigma}(\Omega)}$ to $C^\infty_{0,\sigma}(\Omega)$. The extended operator

$$A^{-\frac{1}{2}}P : \hat{D}(A^{-\frac{1}{2}}P) \longrightarrow L^2_\sigma(\Omega)$$

is defined as follows. $\hat{D}(A^{-\frac{1}{2}}P)$ contains all those $f \in (C_0^\infty(\Omega)^3)'$ such that the expression

$$[f, v] = [Pf, v] = [Pf, A^{-\frac{1}{2}}A^{\frac{1}{2}}v] = [Pf, A^{-\frac{1}{2}}w]$$

with $v \in C^\infty_{0,\sigma}(\Omega)$ and $w = A^{\frac{1}{2}}v$ yields a continuous function in $\|w\|_2$. Then $A^{-\frac{1}{2}}Pf \in L^2_\sigma(\Omega)$ is defined for $f \in \hat{D}(A^{-\frac{1}{2}}P)$ by the relation

$$[Pf, A^{-\frac{1}{2}}w] = \langle A^{-\frac{1}{2}}Pf, w\rangle$$

using the Riesz representation theorem. If in particular $f \in L^2(\Omega)^3$ and $Pf \in D(A^{-\frac{1}{2}})$, then $A^{-\frac{1}{2}}Pf$ has the original meaning. Thus we get $D(A^{-\frac{1}{2}}P) \subseteq \hat{D}(A^{-\frac{1}{2}}P) \subseteq (C_0^\infty(\Omega)^3)'$.

There are several constructions of weak solutions in the literature, see e.g. [Le33], [Ho51], [La69], [Te77], [Ma84], [CKN82], [BM92], [He80]. Since we are mainly interested in the asymptotic behaviour of weak solutions we need a special construction which is based on the Yosida approximation, see [So83], [So84], [SWW86] for this approach.

In our main theorem below we only treat the case $T = \infty$. We need a special assumption on the data f, u_0 which ensures a certain decay of u in space and time. In particular we are looking for exponents $q > 1$ and $\rho > 1$ as small as possible such that

$$\|u\|_{q,\rho;\infty} = \left(\int_0^\infty \|u(t)\|_q^\rho \, dt \right)^{\frac{1}{\rho}} < \infty.$$

The smaller q and ρ, the "stronger" the decay in space and time direction and therefore the better the result; this means under d) in the next theorem that $\frac{3}{q} + \frac{2}{\rho}$ is close to $\frac{5}{2}$.

Theorem 2.1. *Let $\Omega \subseteq \mathbb{R}^3$ be any domain, suppose $u_0 \in D(A^{-\frac{1}{2}})$ and $f = \operatorname{div} F$ with $F \in L^2(0,\infty; L^2(\Omega)^9) \cap L^1(0,\infty; L^2(\Omega)^9)$. Then there exists a weak solution u of the Navier-Stokes equations (1.1) such that $u : [0,\infty) \to L^2_\sigma(\Omega)$ is weakly continuous, satisfies $u(0) = u_0$ and the following properties:*

a) $$\frac{1}{2}\|\phi(t)u(t)\|_2^2 + \int_0^t \|\phi \nabla u\|_2^2 d\tau \le \frac{1}{2}\|\phi(0)u_0\|_2^2 - \int_0^t \langle \phi^2 F, \nabla u \rangle d\tau + C(\phi, u_0, F) \tag{2.6}$$

for $0 \le t < \infty$ and all continuous functions $\phi : [0,\infty) \to \mathbb{R}$ for which there is some $\gamma \in (1,\infty)$ with $\frac{d}{dt}\phi^2 \in L^\gamma(0,\infty;\mathbb{R})$; $C(\phi, u_0, F) > 0$ is a constant; if $\|\phi F\|_{2,2;\infty} < \infty$ it follows

$$\|\phi u\|_{2,\infty;\infty}^2 + \|\phi \nabla u\|_{2,2;\infty}^2 \le \|\phi(0)u_0\|_2^2 + \|\phi F\|_{2,2;\infty}^2 + 2C(\phi, u_0, F).$$

b) $(A^{-\frac{1}{2}}u)_t$, $A^{\frac{1}{2}}u$, $A^{-\frac{1}{2}}Pu\nabla u$, $A^{-\frac{1}{2}}P \operatorname{div} F \in L^s(0,\infty; L^2_\sigma(\Omega))$,

$$(A^{-\frac{1}{2}}u)_t + A^{\frac{1}{2}}u + A^{-\frac{1}{2}}Pu\nabla u = A^{-\frac{1}{2}}Pf \quad \textit{for a.a. } t \in [0,\infty)$$

and

$$\|(A^{-\frac{1}{2}}u)_t\|_{2,s;\infty} + \|A^{\frac{1}{2}}u\|_{2,s;\infty} \le C(s, u_0, F) \tag{2.7}$$

for $1 < s \le \frac{4}{3}$ with some constant $C(s, u_0, F) > 0$.

c) $A^{-\frac{1}{2}}u : [0,\infty) \to L^2_\sigma(\Omega)$ *is strongly continuous, it holds* $A^{-\frac{1}{2}}Pu\nabla u \in L^1(0,\infty; L^2_\sigma(\Omega))$ *and*

$$\lim_{t\to\infty} \|A^{-\frac{1}{2}}u(t)\|_2 = 0. \tag{2.8}$$

d) $u \in L^\rho(0,\infty; L^q(\Omega)^3)$ *for* $2 \le q \le 6,\ 1 < \rho < \infty$ *with* $\frac{3}{2} \le \frac{3}{q} + \frac{2}{\rho} < \frac{5}{2}$ *and it holds*

$$\|u\|_{q,\rho;\infty} \le C(q,\rho,u_0,F) \tag{2.9}$$

with some constant $C(q,\rho,u_0,F) > 0$.

Remarks. This theorem holds for general domains Ω, bounded or not, smooth or not, it is however only interesting for unbounded domains; the result is new to our knowledge for general domains. For special domains, in particular for exterior domains with smooth boundary there exists a rather complete theory on the asymptotic behaviour of weak solutions, see [BM91], [BM92], [GS91], [GM86], [KO93], [KO94], [SWW86], [SvW85], [W87]. Energy inequalties with weights depending on the space variables have been treated in [FS95] and [F96].

The proof of this theorem will be given in Section 5 using the approximation procedure in Section 4. In the next section we will prepare some preliminary facts mainly concerning the linearized equations (1.1).

Theorem 2.1 is applicable to $\phi(t) = (1+t)^\alpha,\ t \ge 0$ with $0 \le \alpha < \frac{1}{2}$. This yields the following

Corollary 2.1. *Let* $\Omega \subseteq \mathbb{R}^3$ *be any domain, suppose* $u_0 \in D(A^{-\frac{1}{2}})$, $f = \operatorname{div} F$ *with* $F \in L^2(0,\infty; L^2(\Omega)^9) \cap L^1(0,\infty; L^2(\Omega)^9)$ *and* $\int_0^\infty \|(1+t)^\alpha F(t)\|_2^2 dt < \infty$ *with* $0 \le \alpha < \frac{1}{2}$. *Then there exists a weak solution* u *of the Navier-Stokes equations (1.1) with the properties*
a) $\int_0^\infty \|(1+t)^\alpha \nabla u\|_2^2 dt < \infty$,
b) $\|u(t)\|_2 \le \frac{C}{(1+t)^\alpha}\ ,\ \ t \ge 0$,
where $C = C(\alpha, u_0, F) > 0$ *is a constant.*

Remark. The property b) has been first proved for a special weak solution by Borchers and Miyakawa [BM92] using a completely different approach. An easy suffcient condition for $u_0 \in D(A^{-\frac{1}{2}})$ is given in (3.24).

In the last section we will apply our method to prove a theorem on local in time strong solutions for arbitrary domains under very weak assumptions on the data u_0, F.

3. The Stokes equations

The time dependent Stokes equations

$$\begin{aligned} u_t - \Delta u + \nabla p = f, \quad \operatorname{div} u = 0 \\ u|_{\partial\Omega} = 0, \quad u(0) = u_0 \end{aligned} \tag{3.1}$$

are again considered in $[0,T) \times \Omega$ with $0 < T \le \infty$ and any domain $\Omega \subseteq \mathbb{R}^3$. In this linear case it is convenient to use the following definition of weak solutions u.

Let $u_0 \in L^2_\sigma(\Omega), 1 \le s < \infty$ and $f = \operatorname{div} F$ with $F \in L^s_{\text{loc}}([0,T); L^2(\Omega)^9)$. Then a measurable function u in $[0,T) \times \Omega$ is called a **weak solution** of (3.1) if the following holds:

a) $u \in L^\infty_{\text{loc}}([0,T); L^2_\sigma(\Omega)) \cap L^2_{\text{loc}}([0,T); W^{1,2}_{0,\sigma}(\Omega))$

b) $-\langle u, v_t\rangle_{\Omega,T} + \langle \nabla u, \nabla v\rangle_{\Omega,T} \;=\; \langle u_0, v(0)\rangle_\Omega - \langle F, \nabla v\rangle_{\Omega,T}$ (3.2)

for all $v \in C^\infty_0([0,T); C^\infty_{0,\sigma}(\Omega))$.

Our first lemma yields energy equalities for (3.1); it essentially rests on Yosida's smoothing procedure.

Lemma 3.1. *Let $\Omega \subseteq \mathbb{R}^3$ be any domain, let $0 < T \le \infty$, $u_0 \in L^2_\sigma(\Omega)$, $f = \operatorname{div} F$ with $F \in L^2_{\text{loc}}([0,T); L^2(\Omega)^9)$ and let u be a weak solution of (3.1). Then u is uniquely determined and it holds*

$$\begin{aligned}\frac{1}{2}\|\phi(t)\, u(t)\|_2^2 + \int_0^t \|\phi \nabla u\|_2^2\, d\tau &= \frac{1}{2}\|\phi(0)\, u_0\|_2^2 \\ &\quad - \int_0^t \langle \phi^2 F, \nabla u\rangle\, d\tau + \frac{1}{2}\int_0^t \phi_\tau^2\, \|u\|_2^2\, d\tau\end{aligned} \tag{3.3}$$

for a.a. $t \in [0,T)$ and all continuous functions $\phi : [0,T) \to \mathbb{R}$ such that $\phi_t^2 \in L^1_{\text{loc}}([0,T);\mathbb{R})$. In particular it holds

$$\frac{1}{2}\|u(t)\|_2^2 + \int_0^t \|\nabla u\|_2^2\, d\tau = \frac{1}{2}\|u_0\|_2^2 - \int_0^t \langle F, \nabla u\rangle\, d\tau \tag{3.4}$$

for a.a. $t \in [0,T)$ and it follows

$$\|u\|^2_{2,\infty;T'} + \|\nabla u\|^2_{2,2;T'} \le \|u_0\|_2^2 + \|F\|^2_{2,2;T'} \tag{3.5}$$

for $0 < T' < T$.

Proof. Using $\|\nabla u\|_2 = \|A^{\frac{1}{2}} v\|_2$ for $v \in D(A^{\frac{1}{2}})$ we get $\|\nabla A^{-\frac{1}{2}} v\|_2 = \|v\|_2$ for $v \in D(A^{-\frac{1}{2}})$; therefore $\nabla A^{-\frac{1}{2}}$ is bounded and $\|\nabla A^{-\frac{1}{2}}\| \le 1$ with the operator norm $\|\cdot\|$. From $\langle A^{-\frac{1}{2}} P \operatorname{div} F, w\rangle_\Omega = [\operatorname{div} F, A^{-\frac{1}{2}} A^{\frac{1}{2}} v]_\Omega = -\langle F, \nabla v\rangle_\Omega$ with $w = A^{\frac{1}{2}} v$, $v \in C^\infty_{0,\sigma}(\Omega)$ we see that $|\langle A^{-\frac{1}{2}} P \operatorname{div} F, w\rangle_\Omega| \le \|F\|_2\, \|\nabla v\|_2 = \|F\|_2\, \|w\|_2$. This shows that $A^{-\frac{1}{2}} P\mathrm{div}$ is a bounded operator with norm

$$\|A^{-\frac{1}{2}} P\mathrm{div}\,\| \le 1. \tag{3.6}$$

We use the Yosida approximation and write $u_k = J_k u$, $u_{0k} = J_k u_0$ with $J_k = (I + k^{-1} A^{\frac{1}{2}})^{-1}$, $k \in \mathbb{N}$. Writing $w = w_k = (I + k^{-1} A^{\frac{1}{2}}) v$ and inserting $v = J_k w$ in (3.2) yields

$$\begin{aligned}-\langle u_k, w_t\rangle_{\Omega,T} + \langle A u_k, w\rangle_{\Omega,T} \;&=\; \langle u_{0k}, w(0)\rangle_\Omega + \\ &\quad\; \langle A^{\frac{1}{2}} J_k A^{-\frac{1}{2}} P \operatorname{div} F,\; w\rangle_{\Omega,T}.\end{aligned}$$

We obtain $u_{k,t}$, Au_k, $A^{\frac{1}{2}} J_k A^{-\frac{1}{2}} P \operatorname{div} F \in L^2(0,T'; L^2_\sigma(\Omega))$ for $0 < T' < T$,

$$\frac{d}{dt} u_k + A u_k = A^{\frac{1}{2}} J_k A^{-\frac{1}{2}} P \operatorname{div} F \;\text{ for a.a. } t \in [0,T) \tag{3.7}$$

and $u_k(0) = u_{k0}$. Now we take the scalar product with u_k and integrate from 0 to t. Then we use that $|\langle A^{-\frac{1}{2}}P \operatorname{div} F, A^{\frac{1}{2}}u_k\rangle| \le \frac{1}{2}\|F\|_2^2 + \frac{1}{2}\|\nabla u_k\|_2^2$ and get the inequality

$$\|u_k(t)\|_2^2 + \int_0^t \|\nabla u_k\|_2^2\, d\tau \le \|u_0\|_2^2 + \int_0^t \|F\|_2^2\, d\tau.$$

Multiplying (3.7) with ϕ, taking the skalar produkt with ϕu_k, integrating from 0 to t, letting $k \to \infty$ and using Lebesgue's theorem leads to (3.3) and (3.4). To prove the uniqueness assertion we consider the case $F = 0$ and $u_0 = 0$ and conclude that $u = 0$. This proves the lemma. □

Our next aim is to prove an explicit representation of weak solutions u. For this purpose we need the following lemma which is a special case of a result by Cannarsa and Vespri [CV86], (1.13).

Lemma 3.2. *Let $\Omega \subseteq \mathbb{R}^3$ be any domain, let $0 < T \le \infty, 1 < s < \infty$, $f \in L^s_{\mathrm{loc}}([0,T); L^2_\sigma(\Omega))$ and let $u : [0,T) \to L^2_\sigma(\Omega)$ be defined by*

$$u(t) = \int_0^t e^{-(t-\tau)A} f(\tau)d\tau, \quad 0 \le t < T. \tag{3.8}$$

Then $u_t, Au \in L^s_{\mathrm{loc}}([0,T); L^2_\sigma(\Omega)), u : [0,T) \to L^2_\sigma(\Omega)$ is strongly continuous, $u(0) = 0$,

$$u_t + Au = f \quad a.e.\ in\ [0,T) \tag{3.9}$$

and

$$\|u_t\|_{2,s;T'} + \|Au\|_{2,s;T'} \le C\|f\|_{2,s;T'} \tag{3.10}$$

for $0 < T' < T$ with some constant $C = C(s) > 0$ only depending on s. If $\|f\|_{2,s;T} < \infty$ it follows

$$\|u_t\|_{2,s;T} + \|Au\|_{2,s;T} \le C\|f\|_{2,s;T} \tag{3.11}$$

Proof. The proof in [CV86] rests on the estimates $\|(\mu + A)^{-1}\| \le \mu^{-1}, \mu > 0$, $\|e^{-tA}\| \le 1$ and $\|tAe^{-tA}\| \le 1$, $t \ge 0$. A completely different proof rests on the Dore-Venni theory [DV87] in the extended version given in [GS91] and [PS90]. [GS91] also contains applications to the Stokes operator A. □

We have to treat also nonzero initial values u_0. This case can be reduced to $u(0) = 0$ if we subtract from u above the function $e^{-\bullet A}u_0 : [0,T) \to L^s_\sigma(\Omega)$. This procedure requires to consider norms of the forms $\|A^\beta e^{-\bullet A}u_0\|_{2,s;T}$ and $\|e^{-\bullet A}u_0\|_{q,s;T}$. For this purpose we need the following lemma.

Lemma 3.3. *Let $\Omega \subseteq \mathbb{R}^3$ be any domain with Stokes operator A.*
a) Suppose $0 \le \alpha \le \frac{1}{2}$, $2 \le q \le 6$ with $2\alpha + \frac{3}{q} = \frac{3}{2}$, $u_0 \in D(A^\alpha)$. Then $u_0 \in L^q(\Omega)^3$ and

$$\|u_0\|_q \le C\|A^\alpha u_0\|_2 \tag{3.12}$$

where $C = C(q,\alpha) > 0$ is a constant.

b) Suppose $0 \leq \alpha \leq \frac{1}{2}$, $\frac{6}{5} \leq q \leq 2$, $2\alpha + \frac{3}{2} = \frac{3}{q}$, $u_0 \in L^2_\sigma(\Omega) \cap L^q(\Omega)^3$. *Then* $u_0 \in D(A^{-\alpha})$ *and*

$$\|A^{-\alpha} u_0\|_2 \leq C \|u_0\|_q \tag{3.13}$$

where $C = C(q, \alpha) > 0$ *is a constant.*

Proof. a) Consider the expression $\|u_0\|_{(\frac{1}{2}-\frac{2}{3}\alpha)^{-1}}$. In the case $\alpha = \frac{1}{2}$ we get from Sobolev's embedding lemma [Fr69], Part 1, (9.7) that $\|u_0\|_6 \leq C\|\nabla u\|_2 = C\|A^{\frac{1}{2}} u\|_2$ with some $C > 0$; for $\alpha = 0$ we get $\|u_0\|_2 = \|A^0 u_0\|_2$. An interpolation argument now leads to (3.12); see [GS89], [BM91] for details.
b) (3.13) follows from (3.12) by a duality argument. We get $|\langle u_0, A^{-\alpha} v\rangle| \leq \|u_0\|_q \|A^{-\alpha} v\|_{q'}$ with $v \in D(A^{-\alpha})$, $\frac{1}{q} + \frac{1}{q'} = 1$, use (3.12) with $2\alpha + \frac{3}{q'} = \frac{3}{2}$ and get $\|A^{-\alpha} v\|_{q'} \leq C\|v\|_2$, $|\langle u_0, A^{-\alpha} v\rangle| \leq C\|u_0\|_q \|v\|_2$. This proves (3.13) since $A^{-\alpha}$ is selfadjoint. The proof of the lemma is complete. □

Let $0 \leq \alpha \leq \frac{1}{2}$. Then the interpolation property

$$\|A^\alpha v\|_2 \leq \|A^{\frac{1}{2}} v\|_2^{2\alpha} \|v\|_2^{1-2\alpha}, \quad v \in D(A^{\frac{1}{2}}) \tag{3.14}$$

follows from the spectral representation by using Hölder's inequality. Combining with (3.12) yields the inequality

$$\|u_0\|_q \leq C\|\nabla u_0\|_2^{2\alpha} \|u_0\|_2^{1-2\alpha} \leq C(\|u_0\|_2 + \|\nabla u_0\|_2) \tag{3.15}$$

for $u_0 \in D(A^{\frac{1}{2}})$, $0 \leq \alpha \leq \frac{1}{2}$, $2 \leq q \leq 6$ with $2\alpha(\frac{1}{2} - \frac{1}{3}) + (1 - 2\alpha)\frac{1}{2} = \frac{1}{q}$ which means $2\alpha + \frac{3}{q} = \frac{3}{2}$; (3.15) is a special case of Sobolev's embedding lemma [Fr69], Part 1, (9.7).

The next lemma rests on simple calculations.

Lemma 3.4. *Let* $\Omega \subseteq \mathbb{R}^3$ *be any domain, let* $u_0 \in D(A^{-\frac{1}{2}})$, $1 \leq T \leq \infty, 1 < s \leq 2$, $2 \leq q \leq 6$ *and* $1 < \rho < \infty$ *such that*

$$\frac{3}{2} \leq \frac{3}{q} + \frac{2}{\rho} < \frac{5}{2}.$$

Then

$$\|A^{\frac{1}{2}} e^{-\bullet A} u_0\|_{2,s;T} \leq C(\|u_0\|_2 + \|A^{-\frac{1}{2}} u_0\|_2) \tag{3.16}$$

and

$$\|e^{-\bullet A} u_0\|_{q,\rho;T} \leq C(\|u_0\|_2 + \|A^{-\frac{1}{2}} u_0\|_2) \tag{3.17}$$

where $C = C(s, q, \rho) > 0$ *is a constant.*

Proof. Taking $\alpha \in [0, \frac{1}{2}]$ with $2\alpha + \frac{3}{q} = \frac{3}{2}$ and using (3.12) yields $\|e^{-\bullet A}u_0\|_{q,\rho;T} \leq C\|A^\alpha e^{-\bullet A}u_0\|_{2,\rho;T}$. Let $\rho^* = \frac{1}{\alpha}$. Then from the assumptions we get $\rho^* \geq \rho$, $\rho^* \geq 2$, $(\alpha + \frac{1}{2})\rho > 1$. Now we use (3.14), (1.3) and obtain

$$\|A^\alpha e^{-\bullet A}u_0\|_{2,\rho;T} \leq \left(\int_0^1 \|A^\alpha e^{-tA}u_0\|_2^\rho \, dt\right)^{\frac{1}{\rho}} + \left(\int_1^\infty \|A^\alpha e^{-tA}u_0\|_2^\rho \, dt\right)^{\frac{1}{\rho}}$$

$$\leq \left(\int_0^1 \|A^\alpha e^{-tA}u_0\|_2^{\rho^*} \, dt\right)^{\frac{1}{\rho^*}} + \left(\int_1^\infty \|A^{\alpha+\frac{1}{2}} e^{-tA}A^{-\frac{1}{2}}u_0\|_2^\rho \, dt\right)^{\frac{1}{\rho}}$$

$$\leq \left(\int_0^\infty \|A^{\frac{1}{2}} e^{-tA}u_0\|_2^2 \, \|e^{-tA}u_0\|_2^{\rho^*-2} \, dt\right)^{\frac{1}{\rho^*}} + \left(\int_1^\infty t^{-(\alpha+\frac{1}{2})\rho} \, dt\right)^{\frac{1}{\rho}} \|A^{-\frac{1}{2}}u_0\|_2$$

$$\leq \|u_0\|_2^{1-2\alpha} \left(\int_0^\infty \langle Ae^{-2tA}u_0, u_0\rangle \, dt\right)^{\frac{1}{\rho^*}} + C(\alpha,\rho)\|A^{-\frac{1}{2}}u_0\|_2$$

$$= \|u_0\|_2^{1-2\alpha} \left(\lim_{T'\to\infty} \frac{1}{2}\langle (I - e^{-2T'A})u_0, u_0\rangle\right)^{\frac{1}{\rho^*}} + C(\alpha,\rho)\|A^{-\frac{1}{2}}u_0\|_2$$

$$\leq \|u_0\|_2 + C(\alpha,\rho)\|A^{-\frac{1}{2}}u_0\|_2,$$

where $C(\alpha, \rho) > 0$ is a constant. This proves (3.16).

To prove (3.17) we argue similarly and get

$$\|A^{\frac{1}{2}} e^{-\bullet A}u_0\|_{2,s;T} \leq \left(\int_0^1 \|A^{\frac{1}{2}} e^{-tA}u_0\|_2^s \, dt\right)^{\frac{1}{s}} + \left(\int_1^\infty \|Ae^{-tA}A^{-\frac{1}{2}}u_0\|_2^s \, dt\right)^{\frac{1}{s}}$$

$$\leq \left(\int_0^\infty \|A^{\frac{1}{2}} e^{-tA}u_0\|_2^2 \, dt\right)^{\frac{1}{2}} + \left(\int_1^\infty t^{-s} \, dt\right)^{\frac{1}{s}} \|A^{-\frac{1}{2}}u_0\|_2$$

$$\leq \|u_0\|_2 + C(s)\|A^{-\frac{1}{2}}u_0\|_2$$

where $C(s) > 0$ is a constant. This proves (3.17). □

The theorem below describes completely the weak solutions of the linear equations (3.1) in the case $s \in (1, 2]$ of "small" exponents in time direction; this case is interesting since it ensures a "strong" decay as $t \to \infty$.

Theorem 3.5. *Let $\Omega \subseteq \mathbb{R}^3$ be any domain, let $1 \leq T \leq \infty$, $u_0 \in D(A^{-\frac{1}{2}})$ and $f = \operatorname{div} F$ with $F \in L^2(0, T; L^2(\Omega)^9) \cap L^1(0, T; L^2(\Omega)^9)$. Then the function $u : [0, T) \to L^2_\sigma(\Omega)$ defined by*

$$u(t) = e^{-tA}u_0 + A^{\frac{1}{2}} \int_0^t e^{-(t-\tau)A}A^{-\frac{1}{2}}Pf(\tau) \, d\tau, \; 0 \leq t < T, \tag{3.18}$$

is the only weak solution of (3.1) for each $s \in (1, 2]$; u has the following properties:

a) $(A^{-\frac{1}{2}}u)_t$, $A^{\frac{1}{2}}u \in L^s(0,T;L^2_\sigma(\Omega))$ *for all* $s \in (1,2]$, $A^{-\frac{1}{2}}u$ *is strongly continuous from* $[0,T)$ *to* $L^2_\sigma(\Omega)$, *it holds* $(A^{-\frac{1}{2}}u)(0) = A^{-\frac{1}{2}}u_0$,

$$(A^{-\frac{1}{2}}u)_t + A^{\frac{1}{2}}u = A^{-\frac{1}{2}}Pf \quad \text{for a.a, } t \in [0,T) \tag{3.19}$$

and

$$\|(A^{-\frac{1}{2}}u)_t\|_{2,s;T} + \|A^{\frac{1}{2}}u\|_{2,s;T} \leq C(\|u_0\|_2 + \|A^{-\frac{1}{2}}u_0\|_2 + \|F\|_{2,s;T}) \tag{3.20}$$

where $C = C(s) > 0$ *is a constant.*

b) $u \in L^\rho(0,T;L^q(\Omega)^3)$ *for all* $s \in (1,2]$, $\rho \geq s$, $2 \leq q \leq 6$ *such that* $\frac{3}{q} + \frac{2}{\rho} = \frac{1}{2} + \frac{2}{s}$ *and it holds*

$$\|u\|_{q,\rho;T} \leq C(\|u_0\|_2 + \|A^{-\frac{1}{2}}u_0\|_2 + \|F\|_{2,s;T}) \tag{3.21}$$

where $C = C(q,\rho,s) > 0$ *is a constant. It follows*

$$\|u\|_{q,\rho;T} \leq C\left(\|u_0\|_2 + \|A^{-\frac{1}{2}}u_0\|_2 + \|F\|_{2,2;T} + \|F\|_{2,1;T}\right) \tag{3.21$'$}$$

for all $\rho > 1$, $2 \leq q \leq 6$ *such that* $\frac{3}{2} \leq \frac{3}{q} + \frac{2}{\rho} < \frac{5}{2}$; $C = C(q,\rho) > 0$ *is a constant.*

c) $\|A^{-\frac{1}{2}}u(t)\|_2 \leq \|A^{-\frac{1}{2}}u_0\|_2 + \|F\|_{2,1;t}$ *for all* $0 \leq t < T$ *and if* $T = \infty$ *it holds*

$$\lim_{t\to\infty} \|A^{-\frac{1}{2}}u(t)\|_2 = 0 \tag{3.22}$$

Remarks. a) By interpolation, similar as in (1.5), we conclude for $s \in (1,2]$ that

$$\|F\|_{2,s;T} \leq C(\|F\|_{2,2;T} + \|F\|_{2,1;T}) < \infty. \tag{3.23}$$

b) From (3.13) we get that $u_0 \in D(A^{-\frac{1}{2}})$ and $\|A^{-\frac{1}{2}}u_0\|_2 \leq C\|u_0\|_{\frac{6}{5}}$ if $u_0 \in L^2_\sigma(\Omega) \cap L^{\frac{6}{5}}(\Omega)^3$. Therefore,

$$u_0 \in L^2_\sigma(\Omega) \cap L^{\frac{6}{5}}(\Omega) \text{ implies } u_0 \in D(A^{-\frac{1}{2}}) \tag{3.24}$$

and we obtain a sufficient criterion for the assumption on u_0 in the theorem above.

Proof. Using (3.6) and (3.23) we obtain $\|A^{-\frac{1}{2}}Pf\|_2 = \|A^{-\frac{1}{2}}P \operatorname{div} F\|_2 \leq \|F\|_2$ and $A^{-\frac{1}{2}}Pf \in L^s(0,T;L^2_\sigma(\Omega))$ for all $s \in (1,2]$. Setting $\hat{u}(t) = u(t) - e^{-tA}u_0$, $t \in [0,T)$, we get from (3.18) that $A^{-\frac{1}{2}}\hat{u}(t) = \int_0^t e^{-(t-\tau)A}A^{-\frac{1}{2}}Pf\,d\tau$ and Lemma 3.2 now yields the properties (3.9), (3.10) and (3.11) with u, f replaced by $A^{-\frac{1}{2}}\hat{u}$, $A^{-\frac{1}{2}}Pf$. This proves the inequality

$$\begin{aligned}&\|(A^{-\frac{1}{2}}u)_t\|_{2,s;T} - \|(e^{-\bullet A}A^{-\frac{1}{2}}u_0)_t\|_{2,s;T} + \|A^{\frac{1}{2}}u\|_{2,s;T} - \|A^{\frac{1}{2}}e^{-\bullet A}u_0\|_{2,s;T}\\ &\leq \|(A^{-\frac{1}{2}}\hat{u})_t\|_{2,s;T} + \|A^{\frac{1}{2}}\hat{u}\|_{2,s;T} \leq C\|A^{-\frac{1}{2}}Pf\|_{2,s;T}\end{aligned}$$

and using (3.17) we obtain (3.20). (3.19) follows from (3.9) when we use that $(e^{-\bullet A}A^{-\frac{1}{2}}u_0)_t + A(e^{-\bullet A}A^{-\frac{1}{2}}u_0) = 0$; u is a weak solution.

To prove b) we choose $\alpha \in [0,\frac{1}{2}]$ with $2\alpha + \frac{3}{q} = \frac{3}{2}$ and obtain the relation

$$1 - \left(\alpha + \frac{1}{2}\right) + \frac{1}{\rho} = \frac{1}{s}. \tag{3.25}$$

Using (3.12) and (1.3) we get for $t \in [0,T)$ that

$$\begin{aligned} \|u(t) - e^{-tA}u_0\|_q &\leq C\|A^\alpha(u(t) - e^{-tA}u_0)\|_2 \\ &= C\|\int_0^t A^{\alpha+\frac{1}{2}}e^{-(t-\tau)A}A^{-\frac{1}{2}}Pf\,d\tau\|_2 \\ &\leq C\int_0^t (t-\tau)^{-(\alpha+\frac{1}{2})}\,\|A^{-\frac{1}{2}}Pf\|_2\,d\tau \end{aligned}$$

with C from (3.12). The integral kernel of the last expression is of power type and can be estimated by the Hardy-Littlewood theorem, see [Tr78], 1.18.9, Theorem 3; there we need condition (3.25). This yields

$$\begin{aligned} &\|u - e^{-\bullet A}u_0\|_{q,\rho;T} \leq C\|A^\alpha(u - e^{-\bullet A}u_0)\|_{2,\rho;T} \\ &\leq C'\|A^{-\frac{1}{2}}Pf\|_{2,s;T} \leq C'\|F\|_{2,s;T} \end{aligned}$$

with $C' = C'(q,\rho,s) > 0$. Combining this inequality with (3.17), (3.23) yields the desired estimate (3.21); (3.21)$'$ is a consequence.

To prove c) we use the estimate

$$\begin{aligned} \|A^{-\frac{1}{2}}(u(t) - e^{-tA}u_0)\|_2 &= \|\int_0^t e^{-(t-\tau)A}A^{-\frac{1}{2}}Pf\,d\tau\|_2 \leq \int_0^t \|A^{-\frac{1}{2}}Pf\|_2\,d\tau \\ &\leq \|F\|_{2,1;t} \end{aligned}$$

and if $T = \infty$, we use for $0 < t_0 < t$ the representation

$$A^{-\frac{1}{2}}u(t) = e^{-(t-t_0)A}A^{-\frac{1}{2}}u(t_0) + \int_{t_0}^t e^{-(t-\tau)A}A^{-\frac{1}{2}}Pf\;d\tau.$$

Let $\varepsilon > 0$ be given. Then we choose $t_0 > 0$ so that $\int_{t_0}^\infty \|A^{-\frac{1}{2}}Pf\|_2\,d\tau \leq \frac{\varepsilon}{2}$ and $t' > t_0$ so that $\|e^{-(t-t_0)A}A^{-\frac{1}{2}}u(t_0)\|_2 \leq \frac{\varepsilon}{2}$ for $t \geq t'$; here we use (1.4). This yields $\|A^{-\frac{1}{2}}u(t)\|_2 \leq \varepsilon$ if $t \geq t'$. This proves (3.22). The uniqueness of u defined by (3.18) follows from Lemma 3.1 if we apply it to the case $F = 0,\ u_0 = 0$. The proof is complete. □

4. Approximation of the Navier-Stokes equations

We modify the Navier-Stokes equations (1.1) and replace the nonlinear term $u\nabla u$ by the "smoothing" term $(J_k u)\nabla u$ where $J_k u = (I + k^{-1}A^{\frac{1}{2}})^{-1}u,\ k \in \mathbb{N}$, means the Yosida approximation of u. Thus we obtain instead of (1.1) the "smoothed" Navier-Stokes equations (see [So83], [So84]):

$$\begin{aligned} u_t - \Delta u + (J_k u)\nabla u + \nabla p = f, \quad \text{div } u = 0 \\ u|_{\partial\Omega} = 0, \qquad u(0) = u_0. \end{aligned} \tag{4.1}$$

A weak solution u of these equations is defined exactly in the same way as for the original equations (1.1) in Section 2 if we replace the nonlinear term $\langle uu, \nabla v\rangle_{\Omega,T}$ in (2.1) by the expression $\langle (J_k u)u, \nabla v\rangle_{\Omega,T}$.

The equations (4.1) have better properties than the original equations (1.1). The next lemma shows that there exists a unique weak solution $u = u_k$ of (4.1). Letting $k \to \infty$ we will obtain a weak solution u of (1.1) with the desired properties in Theorem 2.1.

It can be shown that each $u = u_k$ is indeed a smooth and classical solution of (4.1) together with some pressure p, at least for $t > 0$, if the data f, u_0 and the domain Ω are sufficiently smooth. This becomes clear since $J_k u$ will satisfy Serrin's uniqueness and regularity condition, see Section 6. Therefore we may apply the complete regularity theory as developed by Galdi and Maremonti [GM88].

The nonlinear term $(J_k u)\nabla u$ satisfies the relations $(J_k u)\nabla u = \operatorname{div}(J_k u)u$ with $(J_k u)u = ((J_k u)_l u_k)_{l,k=1}^3$ and

$$\begin{aligned} \langle (J_k u)\nabla u, u\rangle_{\Omega,T} &= \langle \operatorname{div}\, [(J_k u)u],\, u\rangle_{\Omega,T} = -\langle (J_k u)u, \nabla u\rangle_{\Omega,T} \\ &= -\langle J_k u, \frac{1}{2}\nabla |u|^2\rangle_{\Omega,T} = \langle \operatorname{div}\, (J_k u), \frac{1}{2}|u|^2\rangle_{\Omega,T} = 0. \end{aligned} \tag{4.2}$$

In the literature there are several procedures to construct weak solutions of (1.1), mostly used is the Galerkin method, see [Le33], [Ho51], [La69], [He80], [Te77]. The Yosida approximation has been introduced in [So83], [So84], [SWW86] for this purpose; see [BM92], [CKN82], [Ma84] for further possibilities.

Lemma 4.1. *Let $\Omega \subseteq \mathbb{R}^3$ be any domain, let $0 < T \le \infty$, $u_0 \in L^2_\sigma(\Omega)$ and $f =$ div F with $F \in L^2_{\text{loc}}([0,T); L^2(\Omega)^9)$. Then for each $k \in \mathbb{N}$ there exists a unique weak solution*

$$u = u_k \in L^\infty_{\text{loc}}([0,T); L^2_\sigma(\Omega)) \cap L^2_{\text{loc}}([0,T);\ W^{1,2}_{0,\sigma}(\Omega))$$

of the equations (4.1), i.e. it holds (2.1) for all $v \in C_0^\infty([0,T); C^\infty_{0,\sigma}(\Omega))$ with uu replaced by $(J_k u)u$.

Proof. For simplicity we may assume that $0 < T < \infty$ and $F \in L^2(0,T; L^2(\Omega)^9)$; since $T > 0$ is arbitrary the assertion follows in the general case. Suppose first we already have such a weak solution u for fixed $k \in \mathbb{N}$ with $u \in L^\infty(0,T; L^2_\sigma(\Omega)) \cap L^2(0,T; W^{1,2}_{0,\sigma}(\Omega))$. We shortly write $\|u\|_T = \|u\|_{2,\infty;T} + \|\nabla u\|_{2,2;T}$. We estimate the nonlinear expression $(J_k u)u$, use Hölder's inequality and (3.12), (3.15), (1.2); with constants C, C', C''_k, C''' this yields the following inequalities:

$$\begin{aligned} &\|(J_k u)u\|_{2,2;T} \le C\|J_k u\|_{4,8;T}\ \|u\|_{4,\frac{8}{3};T}, \\ &\|J_k u\|_{4,8;T} \le C'\|A^{\frac{3}{8}} J_k u\|_{2,8;T} \le C''_k\|u\|_{2,8;T} \le C''_k T^{\frac{1}{8}}\|u\|_T, \\ &\|u\|_{4,\frac{8}{3};T} \le C'''\|A^{\frac{3}{8}}u\|_{2,\frac{8}{3};T} \le C'''\left(\int_0^T \|A^{\frac{1}{2}}u\|_2^2\ \|u\|_2^{\frac{2}{3}}\,dt\right)^{\frac{3}{8}} \\ &\le C'''\|A^{\frac{1}{2}}u\|_{2,2,T}^{\frac{3}{4}}\ \|u\|_{2,\infty;T}^{\frac{1}{4}} \le C'''\|u\|_T. \end{aligned}$$

The given weak solution u of (4.1) is also a weak solution of the linear equations (3.1) if we replace $f =$ div F by the expression div $(F - (J_k u)u)$. Now we can apply Theorem 3.5 for the linear case and obtain the representation (3.18) with f replaced by div $(F-(J_k u)u)$; this has the form $u = Gu$ with

$$(Gu)(t) = e^{-tA}u_0 + A^{\frac{1}{2}} \int_0^t e^{-(t-\tau)A} A^{-\frac{1}{2}} P\text{div}\,(F-(J_k u)u)\, d\tau, \tag{4.3}$$

$0 \le t < T$.

In the next step we show that we can solve the equation $u = Gu$ in the space with norm $\|u\|_T$ by Banach's fixed point theorem if T is sufficiently small. Similar integral equations have been treated in the literature, see e.g. [Sol77], p. 522. To solve (4.3), we need some preparations.

Consider now any $u \in L^\infty(0,T; L^2_\sigma(\Omega)) \cap L^2(0,T; W^{1,2}_{0,\sigma}(\Omega))$. Then we conclude from Theorem 3.5 that $\tilde{u} = Gu$ is a weak solution of (3.1) with f replaced by div $(F-(J_k u)u)$ and we may apply Lemma 3.1. If moreover $Gu = u$, u will be the desired weak solution of (4.1). Applying (3.5) and using the estimates of $(J_k u)u$ we obtain

$$\begin{aligned} \|Gu\|_T &\le 2(\|u_0\|_2 + \|F - (J_k u)u\|_{2,2;T}) \\ &\le 2(\|u_0\|_2 + \|F\|_{2,2;T} + C_k T^{\frac{1}{8}} \|u\|_T^2) \end{aligned} \tag{4.4}$$

with some constant $C_k > 0$. Now we choose some $\overline{T}$ with $0 < \overline{T} \le T$, put $z = \|u\|_{\overline{T}}$, $\hat{G}(z) = az^2 + b$ with $a = 2C_k \overline{T}^{\frac{1}{8}}$, $b = 2(\|u_0\|_2 + \|F\|_{2,2;\overline{T}})$ and assume that

$$4ab < 1. \tag{4.5}$$

Sufficient for (4.5) is the condition

$$\overline{T} < [16 C_k (\|u_0\|_2 + \|F\|_{2,2;T}]^{-8}. \tag{4.6}$$

Using (4.5) we see that the minimal root z_1 of the equation $\hat{G}(z) - z = az^2 - z + b = 0$ has the form $z_1 = 2b[1 + (1-4ab)^{\frac{1}{2}}]^{-1} > 0$ and a calculation shows,

$$z \in [0, z_1] \text{ implies } \hat{G}(z) \in [0, z_1]. \tag{4.7}$$

Consider the closed subset $D_{\overline{T}} \subseteq L^\infty(0,\overline{T}; L^2_\sigma(\Omega)) \cap L^2(0,T; W^{1,2}_{0,\sigma}(\Omega))$ defined by $\|u\|_{\overline{T}} \le z_1$. From (4.4) we get $\|Gu\|_{\overline{T}} \le \hat{G}(\|u\|_{\overline{T}})$ and from (4.7) we conclude, $u \in D_{\overline{T}}$ implies $Gu \in D_{\overline{T}}$.

Next we estimate $\|Gu - Gw\|_{\overline{T}}$ with $u, w \in D_{\overline{T}}$. For this purpose we replace $u_0, F - (J_k u)u$ in (4.4) by $0, -[(J_k u)u - (J_k w)w] = (J_k w - J_k u)u + (J_k w)(w-u)$. This yields in the same way as above the inequality

$$\begin{aligned} \|Gu - Gw\|_{\overline{T}} &\le 2C_k \overline{T}^{\frac{1}{8}} [\|u\|_{\overline{T}} \|u-w\|_{\overline{T}} + \|u-w\|_{\overline{T}} \|w\|_{\overline{T}}] \\ &\le 2az_1 \|u-w\|_{\overline{T}} \end{aligned}$$

and it holds $2az_1 < 1$ because of (4.5). Therefore, Banach's fixed point theorem is applicable and yields a unique $u \in D_{\overline{T}}$ with $u = Gu$, u is the desired solution for Lemma 4.1 first only on the interval $[0, \overline{T})$. □

We can extend u to the whole interval $[0, T)$ if $\overline{T} < T$. For this purpose we apply Lemma 3.1 with F replaced by $F - (J_k u)u$ and use (3.4), (4.2). This yields the inequality (3.5) without the term $(J_k u)u$, we get

$$\|u\|_{2,\infty;\overline{T}}^2 + \|\nabla u\|_{2,2;\overline{T}}^2 \leq \|u_0\|_2^2 + \|F\|_{2,2;T}^2 \tag{4.8}$$

and therefore it holds $\|u(t)\|_2 \leq C,\ t \in [0, \overline{T}]$ with C not depending on $\overline{T}$. This enables us to repeat the procedure above for some interval $[\overline{T}, \overline{T'}]$ instead of $[0, \overline{T}]$. We may extend u to some interval $[0, \overline{T}']$ with $\overline{T} < \overline{T}' \leq T$ and so on. This yields the desired solution u of Lemma 4.1 and the proof is complete.

We need some further properties of the solutions $u = u_k$ of Lemma 4.1 in particular for the case $T = \infty$. Let now $F \in L^2(0, \infty; L^2(\Omega)^9)$. As mentioned above, $u = u_k$ becomes a weak solution of (3.1) with F replaced by $F - (J_k u)u$, we may apply Lemma 3.1 (since $\|(J_k u)u\|_{2,2;T} < \infty$) and Theorem 3.5.

In particular it holds equality (3.3) with F replaced by $F - (J_k u)u$; because of (4.2) the term with $(J_k u)u$ disappears. This yields

$$\begin{aligned} \frac{1}{2}\|\phi(t)u(t)\|_2^2 + \int_0^t \|\phi\nabla u\|_2^2 d\tau &= \frac{1}{2}\|\phi(0)u_0\|_2^2 \\ &\quad - \int_0^t \langle \phi^2 F, \nabla u\rangle d\tau + \frac{1}{2}\int_0^t \phi_\tau^2\ \|u\|_2^2 d\tau \end{aligned} \tag{4.9}$$

for $0 \leq t < \infty$, ϕ as in Lemma 3.1 and for all $u = u_k$. Now we use (4.8) with $\overline{T} = \infty$ written in the form

$$\|u\|_{2,\infty;\infty} + \|\nabla u\|_{2,2;\infty} \leq 2(\|u_0\|_2 + \|F\|_{2,2;\infty}). \tag{4.10}$$

In order to apply Theorem 3.5 we first estimate the term $(J_k u)u$. Similar as in the proof above we use (3.12), (3.15), (1.2) and obtain with exponent $\frac{4}{3}$ in time direction the estimate

$$\begin{aligned} \|(J_k u)u\|_{2,\frac{4}{3};\infty} &\leq C\|J_k u\|_{4,\frac{8}{3};\infty}\ \|u\|_{4,\frac{8}{3};\infty} \\ &\leq C'(\|u\|_{2,\infty;\infty} + \|\nabla u\|_{2,2;\infty})^2 \\ &\leq C''(\|u_0\|_2 + \|F\|_{2,2;\infty})^2 \end{aligned} \tag{4.11}$$

where the constants C, C', C'' now do not depend on k, u_0, F, Ω.

Let $2 \leq q \leq 6$, $2 \leq \rho < \infty$, $\frac{3}{q} + \frac{2}{\rho} = \frac{3}{2}$. Then we choose $\alpha = \frac{1}{2}(\frac{3}{2} - \frac{3}{q})$ so that $\alpha\rho = 1$, use (3.15), obtain

$$\begin{aligned} \|u\|_{q,\rho;\infty} &= \left(\int_0^\infty \|u\|_q^\rho dt\right)^{\frac{1}{\rho}} \leq C\left(\int_0^\infty \|\nabla u\|_2^{2\alpha\rho}\ \|u\|_2^{(1-2\alpha)\rho} dt\right)^{\frac{1}{\rho}} \\ &\leq C\|\nabla u\|_{2,2;\infty}^{\frac{2}{\rho}}\ \|u\|_{2,\infty;\infty}^{1-\frac{2}{\rho}} \leq C\left(\|u\|_{2,\infty;\infty} + \|\nabla u\|_{2,2;\infty}\right) \end{aligned}$$

and therefore

$$\|u\|_{q,\rho;\infty} \leq C\left(\|u_0\|_2 + \|F\|_{2,2;\infty}\right) \tag{4.12}$$

which also holds if $q = 2$, $\rho = \infty$; $C > 0$ does not depend on k.

Next we assume that $u_0 \in D(A^{-\frac{1}{2}})$ and $F \in L^2(0,\infty;L^2(\Omega)^9) \cap L^1(0,\infty;L^2(\Omega)^9)$ as in Theorem 2.1.

Let $s = \frac{4}{3}$, $\rho \geq s$, $2 \leq q \leq 6$ with $\frac{3}{q} + \frac{2}{\rho} = \frac{1}{2} + \frac{2}{s} = 2$. Then we apply Theorem 3.5 with F replaced by $F - (J_k u)u$ and use (4.11). This yields the estimate

$$\|u\|_{q,\rho;\infty} \leq C[\|u_0\|_2 + \|A^{-\frac{1}{2}}u_0\|_2 + \|F\|_{2,s;\infty} + (\|u_0\|_2 + \|F\|_{2,2;\infty})^2] \tag{4.13}$$

with some constant C not depending on k, u_0, F, Ω. The exponent 2 on the right is caused by the nonlinear structure of $(J_k u)u$.

Similar as in (4.11) we get the inequality

$$\|(J_k u)u\|_{2,1;\infty} \leq C\|J_k u\|_{4,\frac{8}{3};\infty}\|u\|_{4,\frac{8}{5};\infty}$$

and since $\frac{3}{4} + \frac{2}{\frac{8}{3}} = \frac{3}{2}$, $\frac{3}{4} + \frac{2}{\frac{8}{5}} = 2$, we may use (4.12), (4.13) and obtain

$$\|(J_k u)u\|_{2,1;\infty} \leq C(u_0, F) \tag{4.14}$$

with $C(u_0,F) > 0$ only depending on u_0, F. Using (4.11), (4.14) and the interpolation argument as in (1.5) we get

$$\|(J_k u)u\|_{2,s;\infty} \leq C(u_0, F) \quad \text{for all } s \in [1, \frac{4}{3}] \tag{4.15}$$

with $C(u_0,F) > 0$ only depending on u_0, F. Using again Theorem 3.5 with F replaced by $F - (J_k u)u$ and (4.15) we get from (3.21) that

$$\|u\|_{q,\rho;\infty} \leq C(q,\rho,u_0,F) \tag{4.16}$$

for all $1 < \rho \leq 2$, $2 \leq q \leq 6$ such that $2 \leq \frac{3}{q} + \frac{2}{\rho} < \frac{5}{2}$; the constant $C(q,\rho,u_0,F)$ is independent of k.

Combining (4.12), where $\frac{3}{q} + \frac{2}{\rho} = \frac{3}{2}$, $2 \leq q \leq 6$, $2 \leq \rho \leq \infty$, with estimate (4.16), where $2 \leq \frac{3}{q} + \frac{2}{\rho} < \frac{5}{2}$, $2 \leq q \leq 6$, $1 < \rho \leq 2$, and using again the interpolation as in (1.5) we now obtain the inequality

$$\|u\|_{q,\rho;\infty} \leq C(q,\rho,u_0,F) \tag{4.17}$$

even for $2 \leq q \leq 6$, $1 < \rho < \infty$ such that $\frac{3}{2} \leq \frac{3}{q} + \frac{2}{\rho} < \frac{5}{2}$; $C(q,\rho,u_0,F) > 0$ does not depend on k. Note that $\rho = \infty$ is included if $q = 2$.

In particular let $q = 2$, then (4.17) holds for each exponent ρ with $2 < \rho \leq \infty$. This yields

$$\|u\|_{2,\rho;\infty} \leq C(\rho,u_0,F) \tag{4.18}$$

for $2 < \rho \leq \infty$ with $C(\rho,u_0,F) > 0$ not depending on k.

Consider again estimate (4.9) with some ϕ so that $\phi_t^2 \in L^\gamma(0,\infty;\mathbb{R})$ with $1 < \gamma < \infty$. Then from (4.18) we get with $\frac{1}{\gamma} + \frac{1}{\gamma'} = 1$

$$\begin{aligned} |\int_0^t \phi_\tau^2 \|u\|_2^2 \, d\tau| &\leq (\int_0^\infty |\phi_t^2|^\gamma \, dt)^{\frac{1}{\gamma}} (\int_0^\infty \|u\|_2^{2\gamma'} \, dt)^{\frac{1}{\gamma'}} \\ &\leq C(\phi,u_0,F) \end{aligned} \tag{4.19}$$

with some constant $C(\phi,u_0,F) > 0$ not depending on k.

In the next step we use (4.15) and apply Theorem 3.5, (3.20), with f replaced by $F-(J_k u)u$. Similar as above this yields for $1 < s \le \frac{4}{3}$ and $u = u_k$ the estimate

$$\|(A^{-\frac{1}{2}}u)_t\|_{2,s;\infty} + \|A^{\frac{1}{2}}u\|_{2,s;\infty} \le C(s,u_0,F) \tag{4.20}$$

where $C(s,u_0,F) > 0$ does not depend on k.

Finally we consider for later use a bounded subdomain $\Omega_0 \subseteq \Omega$ with $\overline{\Omega}_0 \subseteq \Omega$. Let A_0, P_0 be the operators corresponding to A, P with Ω replaced by Ω_0. Then we get the embedding $W^{1,2}_{0,\sigma}(\Omega_0) \subseteq W^{1,2}_{0,\sigma}(\Omega)$ in a natural way and for $v \in D(A^{\frac{1}{2}})$, $w \in L^2_\sigma(\Omega_0)$ we get $A_0^{-\frac{1}{2}} w \in D(A^{\frac{1}{2}})$ and

$$\begin{aligned} |\langle A_0^{-\frac{1}{2}} P_0 A^{\frac{1}{2}} v, w\rangle| &= |\langle v, A^{\frac{1}{2}} A_0^{-\frac{1}{2}} w\rangle| \le \|v\|_2 \|\nabla A_0^{-\frac{1}{2}} w\|_2 \\ &\le \|v\|_2 \|w\|_2 . \end{aligned}$$

This shows $A_0^{-\frac{1}{2}} P_0 A^{\frac{1}{2}}$ is a bounded operator with norm $\|A_0^{-\frac{1}{2}} P_0 A^{\frac{1}{2}}\| \le 1$. We obtain the representation

$$\begin{aligned} A_0^{-\frac{1}{2}} P_0 u(t) = A_0^{-\frac{1}{2}} P_0 A^{\frac{1}{2}} A^{-\frac{1}{2}} u(t) &= A_0^{-\frac{1}{2}} P_0 A^{\frac{1}{2}} \Big(A^{-\frac{1}{2}} u_0 + \int_0^t (A^{-\frac{1}{2}} u)_\tau \, d\tau\Big) \\ &= (A_0^{-\frac{1}{2}} P_0 A^{\frac{1}{2}}) A^{-\frac{1}{2}} u_0 + \int_0^t (A_0^{-\frac{1}{2}} P_0 A^{\frac{1}{2}})(A^{-\frac{1}{2}} u)_\tau \, d\tau . \end{aligned}$$

Using (4.20) yields

$$\begin{aligned} \|(A_0^{-\frac{1}{2}} P_0 u)_t\|_{2,s;\infty} &= \|(A_0^{-\frac{1}{2}} P_0 A^{\frac{1}{2}})(A^{-\frac{1}{2}} u)_t\|_{2,s;\infty} \\ &\le \|(A^{-\frac{1}{2}} u)_t\|_{2,s;\infty} \le C(s,u_0,F) \end{aligned} \tag{4.21}$$

for $1 < s \le \frac{4}{3}$; $C(s,u_0,F)$ is chosen as in (4.20).

5. Proof of Theorem 2.1

Consider the weak solutions $u = u_k$ of Lemma 4.1. In the preceding section we proved the uniform boundedness of u_k in several norms. This enables us to treat the limit as $k \to \infty$ which leads to the desired weak solution of Theorem 2.1. We carry out this procedure in several steps.

From (4.10), (4.17), (4.20) we conclude that there is a constant $C > 0$ not depending on $k \in \mathbb{N}$ with

$$\begin{aligned} &\|\nabla u_k\|_{2,2;\infty} \le C, \quad \|u_k\|_{q,\rho;\infty} \le C \ \text{ with } q, \rho \text{ as in (4.17)}, \\ &\|(A^{-\frac{1}{2}} u_k)_t\|_{2,s;\infty} + \|A^{\frac{1}{2}} u_k\|_{2,s;\infty} \le C \ \text{ with } s \text{ as in (4.20)}. \end{aligned}$$

The Banach spaces under these norms are reflexive and therefore we find a subsequence of (u_k) which converges weakly in these spaces to some u contained in all of them. For simplicity we may assume that (u_k) itself has this property.

Next we use a well-known compactness argument, see [Te77],Ch.III,2.2. Let $\Omega_0 \subseteq \Omega$ be a bounded Lipschitz subdomain (see [Te77] for the definition) with

$\overline{\Omega_0} \subseteq \Omega$, and consider the Banach spaces $H_0 = W^{1,2}(\Omega_0)^3$, $H_1 = L^2(\Omega_0)^3$; let H_2 be the completion of H_1 under the norm $\|A_0^{-\frac{1}{2}} P_0 v\|_2$ with A_0, P_0 as in (4.21). Since $A_0^{-\frac{1}{2}} P_0$ is bounded (because of the boundedness of Ω_0), we get the continuous embedding $H_1 \subseteq H_2$. Since Ω_0 is a bounded Lipschitz domain, the embedding $H_0 \subseteq H_1$ is compact, see [Te77], [Fr69].

Consider $0 < T < \infty$ and $1 < s \leq \frac{4}{3}$ as in (4.20). Then the embedding

$$\{v \in L^2(0,T;H_0)\ ;\ v_t \in L^s(0,T;H_2)\} \subseteq L^2(0,T;H_1)$$

is compact; the norm of the space on the left-hand side is defined by

$$\|v\|_{L^2(0,T;H_0)} + \|v_t\|_{L^s(0,T;H_2)}$$

see [Te77], Ch. III, Theorem 2.1.

From (4.10) and (4.21) we obtain that

$$\|u_k\|_{L^2(0,T;H_0)} + \|(u_k)_t\|_{L^2(0,T;H_2)} \leq C$$

with some $C > 0$ not depending on k. Due to the compactness property above we conclude that (u_k) converges strongly in each $L^2(0,T;L^2(\Omega_0)^3)$ for all $0 < T < \infty$ and all $\Omega_0 \subseteq \Omega$ defined above.

We conclude (taking a sequence $\Omega_0^1 \subseteq \Omega_0^2 \subset \cdots \subseteq \Omega_0^j \subseteq \cdots$ of such subdomains with $\Omega = \cup_{j=1}^{\infty} \Omega_0^j$ and a sequence $0 < T_1 < T_2 < \cdots < T_j < \cdots$ with $T_j \to \infty$) that the following holds:

(u_k) has a subsequence such that $(u_k(t))$ converges to $u(t)$ strongly in each $L^2(\Omega_0)^3$ and weakly in $L^2_\sigma(\Omega)$ for a.a. $t \in [0,\infty)$. For simplicity we may assume that (u_k) itself has this property.

Moreover it holds $u \in L^\infty(0,\infty;L^2_\sigma(\Omega))$ because of (4.10). Since $(A^{-\frac{1}{2}}u)_t \in L^s(0,\infty;L^2_\sigma(\Omega))$ with $1 < s \leq \frac{4}{3}$ as shown above, we may assume (after redefinition of u on a subset of $[0,\infty)$ of measure zero) that $A^{-\frac{1}{2}}u$ is strongly continuous in $L^2_\sigma(\Omega)$. Since $u \in L^\infty(0,\infty;L^2_\sigma(\Omega))$ we conclude that $u : [0,\infty) \to L^2_\sigma(\Omega)$ is weakly continuous.

Using these properties we may let $k \to \infty$ in (4.9) first for a.a. $t \in [0,\infty)$ and we may use (4.19). This yields the inequality (2.6) for a.a. $t \in [0,\infty)$. Since u is a weakly continuous, (2.6) even holds for all $t \in [0,\infty)$.

Consider now any test function $v \in C_0^\infty([0,\infty);C_{0,\sigma}^\infty(\Omega))$ in the definition of weak solutions in (2.1) with $T = \infty$. Then the support of v is contained in some $[0,T) \times \Omega_0$ with Ω_0 as above and $0 < T < \infty$. Since (u_k) converges strongly in $L^2(0,T;\Omega_0)$, $\langle u_k u_k, \nabla v\rangle_{\Omega,\infty}$ converges to $\langle uu, \nabla v\rangle_{\Omega,\infty}$ and we may let $k \to \infty$ in each term of (2.1). This shows that u is a weak solution of (1.1).

It remains to prove the property (2.8). For this purpose we use again Theorem 3.5 with F replaced by $F - (J_k u)u$ and we apply (4.14). This shows that $\|uu\|_{2,1;\infty} < \infty$ and we may use (3.23) with f replaced by $\operatorname{div}(F - uu)$. This shows (2.8). The proof of Theorem 2.1 is complete.

6. Strong solutions

Another application of the theory in Sections 3 and 4 is the construction of strong solutions under very weak assumptions on u_0 and F. Up to now we only know the existence of such solutions in a certain initial interval $[0,T)$ where $T>0$ depends on u_0 and F.

Let $\Omega \subseteq \mathbb{R}^3$ be any domain, let $0<T\leq\infty$, $u_0\in L^2_\sigma(\Omega)$ and $f=$ div F with $F\in L^2_{\text{loc}}([0,T);L^2(\Omega)^9)$. A weak solution u of (1.1) is called a **strong solution** if there are exponents $q>3, \rho>2$ such that

$$u\in L^\rho(0,T;L^q(\Omega)^3) \text{ with } \frac{3}{q}+\frac{2}{\rho}=1. \tag{6.1}$$

(6.1) is called the Serrin condition [Se63] and we know that a weak solution satisfying this condition is uniquely determined and smooth on the open interval $(0,T)$ if u_0, F and Ω are sufficiently smooth; see [Se63], [Ma84], [SvW85]; the most general regularity result is contained in [GM88].

Our next theorem concerns strong solutions, it extends a result by Fujita and Kato [FK64] in particular to general domains. Other types of strong solutions mainly for more special domains are treated in [Te77], [La69], [He80], [KO93], [KO94], [Sol77].

Theorem 6.6. *Let $\Omega\subseteq\mathbb{R}^3$ be any domain, let $0<T\leq\infty, u_0\in D(A^{\frac14})$ and $f=$ div F with $F\in L^4(0,T;L^2(\Omega)^9)$. Then there is a constant $C>0$ not depending on u_0,F,T and Ω with the following property: If $0<T<\infty$ and*

$$\|F\|_{2,4;T}+\|A^{\frac14}u_0-e^{-2TA}A^{\frac14}u_0\|_2^{\frac18}\|A^{\frac14}u_0\|_2^{\frac78}\leq C, \tag{6.2}$$

or if $T=\infty$ and

$$\|F\|_{2,4;\infty}+\|A^{\frac14}u_0\|_2\leq C \tag{6.3}$$

then there exists a strong solution u of the Navier-Stokes equations (1.1) in $[0,T)\times\Omega$ with the additional properties

$$\nabla u\in L^4(0,T;L^2(\Omega)^9) \text{ and } u\in L^8(0,T;L^4(\Omega)^3).$$

Remarks. Since $\lim_{T\to0}e^{-2TA}A^{\frac14}u_0=A^{\frac14}u_0$ in the strong sense we get $\lim_{T\to0}\|A^{\frac14}u_0-e^{-2TA}A^{\frac14}u_0\|_2=0$ for $u_0\in D(A^{\frac14})$; therefore the condition (6.2) can always be satisfied with some $T>0$ without any smallness assumption on u_0. It is interesting that the constant C in (6.2) does not depend on the domain Ω. Therefore, the interval $[0,T)$ of existence only depends on the norms in (6.2). If these norms are uniformly small for all domains $\Omega\subseteq\mathbb{R}^3$, then $[0,T)$ can be chosen independently of Ω.

In order to get a strong solution u on the whole interval $[0,\infty)$ we need the smallness condition (6.3).

Proof. The proof rests on Banach's fixed point theorem similar as in Section 4 and on the estimates in Theorem 3.5. Now we solve the integral equation (4.3) without approximation, i.e. $(J_k u)u$ in (4.3) is replaced by uu; furthermore we set $\hat{u}(t) = u(t) - e^{-tA}u_0$, $t \geq 0$, and replace u by $\hat{u} + e^{-\bullet A}u_0$. Thus we solve instead of (4.3) the integral equation $\hat{u} = G\hat{u}$ with

$$(G\hat{u})(t) = A^{\frac{1}{2}} \int_0^t e^{-(t-\tau)A} A^{-\frac{1}{2}} P \operatorname{div}(F - (\hat{u} + e^{-\tau A}u_0)(\hat{u} + e^{-\tau A}u_0))d\tau, \quad (6.4)$$

$0 \leq t < T$. Let $\mathcal{A}(u_0, T) = \|A^{\frac{1}{4}}u_0 - e^{-2TA}A^{\frac{1}{4}}u_0\|_2^{\frac{1}{8}} \|A^{\frac{1}{4}}u_0\|_2^{\frac{7}{8}}$.

A similar calculation as in the proof of Lemma 3.4 yields the estimate

$$\|e^{-\bullet A}u_0\|_{4,8;T} \leq C\mathcal{A}(u_0, T) \quad (6.5)$$

for $0<T<\infty$; here $C>0$ is a constant not depending on u_0,T,Ω. If $T=\infty$ we get

$$\|e^{-\bullet A}u_0\|_{4,8;\infty} \leq C\|A^{\frac{1}{4}}u_0\|_2 \quad (6.6)$$

with C as before. Similarly we obtain

$$\|A^{\frac{1}{2}}e^{-\bullet A}u_0\|_{2,4;T} \leq \|A^{\frac{1}{4}}u_0\|_2 \quad (6.7)$$

for $0 < T \leq \infty$.

To solve $\hat{u} = G\hat{u}$, we now use the norm $\|\hat{u}\|_T = \|(A^{-\frac{1}{2}}\hat{u})_t\|_{2,4;T} + \|A^{\frac{1}{2}}\hat{u}\|_{2,4;T}$. Instead of (4.4) we obtain, using Theorem 3.5 and (6.6), the estimate

$$\|G\hat{u}\|_T \leq C_1\|F\|_{2,4;T} + C_1(\|\hat{u}\|_T + \mathcal{A}(u_0, T))^2 \quad (6.8)$$

with some $C_1 > 0$ not depending on F, u_0, T, Ω. We set $z = \|\hat{u}\|_T + \mathcal{A}(u_0, T)$, $\hat{G}(z) = az^2 + b$ with $a = C_1$, $b = C_1\|F\|_{2,4;T} + \mathcal{A}(u_0, T)$ and obtain

$$\|G\hat{u}\|_T + \mathcal{A}(u_0, T) \leq \hat{G}(\|\hat{u}\|_T + \mathcal{A}(u_0, T)). \quad (6.9)$$

As in (4.5) we need the condition $4ab < 1$; the condition (6.2) with $C = \frac{1}{4}(C_1^2 + C_1)^{-1}$ is sufficient for $4ab < 1$. If $T = \infty$ we have to put $e^{-2TA} = 0$ and (6.3) is sufficient for $4ab < 1$.

Let $z_1 > 0$ be the minimal root of $\hat{G}(z) - z = az^2 - z + b = 0$ and let D_T be defined by $\|\hat{u}\|_T + \mathcal{A}(u_0, T) \leq z_1$. Then the same argument as in the proof of Lemma 4.1 shows that Banach's fixed point theorem is applicable to G and D_T; this yields the solution $u \in D_T$ which is the desired strong solution of (1.1). From $\|\hat{u}_T\| < \infty$ and (6.7) we get

$$\|A^{\frac{1}{2}}u\|_{2,4;T} \leq \|A^{\frac{1}{2}}e^{-\bullet A}u_0\|_{2,4;T} + \|\hat{u}\|_T \leq \|A^{\frac{1}{4}}u_0\|_2 + \|\hat{u}\|_T < \infty$$

Further we get $\|(A^{-\frac{1}{2}}\hat{u})_t\|_{2,4;T} < \infty$ and therefore $\|(A^{-\frac{1}{2}}u)_t\|_{2,4;T} < \infty$ if $u_0 \in D(A^{-\frac{1}{2}})$. From (3.21) we obtain (with $\hat{u}(0) = 0$ and with f replaced by $\hat{u}_t + A\hat{u}$) the estimate

$$\|\hat{u}\|_{4,8;T} \leq C\|(A^{-\frac{1}{2}}\hat{u})_t + A^{\frac{1}{2}}\hat{u}\|_{2,4;T} \leq C\|\hat{u}\|_T < \infty.$$

This proves the theorem. □

The proof additionally shows that $\|(A^{-\frac{1}{2}}\hat{u})_t\|_{2,4;T} < \infty$ and if $u_0 \in D(A^{-\frac{1}{2}})$ this yields that

$$(A^{-\frac{1}{2}}u)_t \in L^4(0,T;L^2_\sigma(\Omega)).$$

References

[BM91] W. Borchers, T. Miyakawa. "Algebraic L^2 decay for Navier-Stokes flows in exterior domains." Hiroshima Math J. **21** (1991), 621–640.

[BM92] W. Borchers, T. Miyakawa. "L^2-decay for Navier-Stokes flows in unbounded domains, with applications to exterior stationary flows." Arch. Rat. Mech. Anal. **118** (1992), 273–295.

[CKN82] L. Caffarelli, R. Kohn, L. Nirenberg. "Partial regularity of suitable weak solutions of the Navier-Stokes equations." Comm Pure Appl. Math. **35** (1982), 771–831.

[CV86] P. Cannarsa, V. Vespri. "On maximal L^p regularity for the abstract Cauchy problem." Bolletino U. M. I., 5-B (1986), 165–175.

[DV87] G. Dore, A. Venni. "On the closedness of the sum of two operators." Math. Z. **196** (1987), 189–201.

[F96] R. Farwig. "Partial regularity and weighted energy estimates of global weak solutions of the Navier-Stokes equations in unbounded domains." Progress in partial differential equations: Pitman Research Notes in Mathematics Series **345** (1196), 205–215.

[FS95] R. Farwig, H. Sohr. "Weighted energy inequalities for the Navier-Stokes equations in exterior domains." Applicable Analysis **58** (1995), 157–173.

[Fr69] A. Friedman. *Partial Differential Equations.* Holt, Rinehart and Winston, 1969.

[FK64] H. Fujita, T. Kato. "On the Navier-Stokes initial value problem." Arch. Rat. Mech. Anal. **16** (1964), 269–315.

[GM86] G. Galdi, P. Maremonti. "Monotonic decreasing and asymptotic behaviour of the kinematic energy for weak soltions of the Navier-Stokes equations in exterior domains." Arch. Rat. Mech. Anal. **94** (1986), 253–266.

[GM88] G. Galdi, P. Maremonti. "Sulla regolarità delle soluzioni deboli al sistema di Navier-Stokes in domini arbitrari." Annali dell' Università di Ferrara **34** (1988), 59–73.

[GS89] Y. Giga, H. Sohr. "On the Stokes operator in exterior domains." J. Fac. Sci. Univ. Tokyo, I.A. Math. **36** (1989), 103–130.

[GS91] Y. Giga, H: Sohr. "Abstract L^p estimates for the Cauchy problem with applications to the Navier-Stokes equations in exterior domains." J. Funct. Anal. **102** (1991), 72–94.

[He80] J. Heywood. "The Navier-Stokes equations: On the existence, regularity and decay of solutions." Indiana Univ. Math. J. **29** (1980), 639–681.

[Ho51] E. Hopf. "Über die Anfangswertaufgabe für die hydrodynamischen Grundgleichungen." Math. Nachr. **4** (1951), 213–231.

[KO93] H. Kozono, T. Ogawa. "Global strong solutions and its decay properties for the Navier-Stokes equations in three dimensional domains with non-compact boundaries." Math. Z. **216** (1994), 1–30.

[KO94] H. Kozono, T. Ogawa. "On stability of Navier-Stokes flows in exterior domains." Arch. Rat. Mech. Anal. **128** (1994), 1–31.

[La69] O. A. Ladyzhenskaya. *The Mathematical Theory of Viscous Incompressible Flow.* Gordon and Breach, New York, 1969.

[Le33] J. Leray. "Étude de diverses équations intégrales non linéaires et de quelques problèmes que pose l' hydrodynamique." J. Math. Pures Appl. **12** (1933), 1–82.

[Ma84] K. Masuda. "Weak solutions of Navier-Stokes equations." Tokoku Math. J. **36** (1984), 623–646.

[PS90] J. Prüss, H. Sohr. "On operators with bounded imaginary powers in Banach spaces." Math. Z. **203** (1990), 429–452.

[Se63] J. Serrin. *The initial value problem for the Navier-Stokes equations.* Nonlinear problems, University of Wisconsin press, Madison, 1963.

[So83] H. Sohr. "Zur Regularitätstheorie der instationären Gleichungen von Navier-Stokes." Math. Z. **184** (1983), 359–375.

[So84] H. Sohr. "Optimale lokale Existenzsätze für die Gleichungen von Navier-Stokes." Math. Ann. **267** (1984), 107–123.

[SvW85] H. Sohr, W. von Wahl. "A new proof of Leray's structure theorem and the smoothness of weak solutions of Navier-Stokes equations for large $|x|$." Bayreuther Math. Schr. **20** (1985), 153–204.

[SWW86] H. Sohr, W. von Wahl, M. Wiegner. *Zur Asymptotik der Gleichungen von Navier-Stokes.* Nachr. Akad. Wiss. Göttingen, 1986, 45–59.

[Sol77] V.A. Solonnikov. "Estimates for solutions of non stationary Navier-Stokes equations." J. Soviet Math. **8** (1977), 467–529.

[Te77] R. Teman. *Navier-Stokes Equations.* North Holland, Amsterdam-New York-Oxford, 1977.

[Tr78] H. Triebel. *Interpolation Theory, Function Spaces, Differential Operators.* North Holland, Amsterdam-New York-Oxford, 1978.

[W87] M. Wiegner. "Decay results for weak solutions of the Navier-Stokes equations on $\mathbb{R}^n$." J. London Math. Soc. **35** (1987), 303–313.

Hermann Sohr
Fachbereich Mathematik/Informatik
Universität Paderborn
D-33095 Paderborn, Germany
hsohr@uni-paderborn.de

Progress in Nonlinear Differential Equations
and Their Applications, Vol. 35
© 1999 Birkhäuser Verlag Basel/Switzerland

On Quasistationary Approximation in the Problem of Motion of a Capillary Drop

V.A. Solonnikov

Dedicated to Professor H. Amann on the occasion of his 60th birthday

1. Introduction

The present paper is concerned with the problem of the motion on an isolated liquid mass subjected to capillary forces on the free boundary. In an exact formulation, it consists in the determination of the domain $\Omega_t \subset \mathbb{R}^3$ filled with a viscous incompressible liquid at the moment $t > 0$, of the velocity vector field $\vec{v}(x,t) = (v_1, v_2, v_3)$ and of a scalar pressure $p(x,t)$ satisfying in Ω_t the Navier-Stokes equations

$$\vec{v}_t + (\vec{v} \cdot \nabla)\vec{v} - \nu\nabla^2\vec{v} + \nabla p = 0, \quad \nabla \cdot \vec{v} = 0 \quad (x \in \Omega_t) \tag{1.1}$$

and the initial and boundary conditions

$$\vec{v}\Big|_{t=0} = \vec{v}_0(x), \quad x \in \Omega_0 \equiv \Omega, \tag{1.2}$$

$$\vec{n} \cdot T(\vec{v},p)\vec{n} - \sigma H\vec{n}\Big|_{\Gamma_t} = 0, \tag{1.3}$$

$$\vec{v} \cdot \vec{n}\Big|_{\Gamma} = V_n \ . \tag{1.4}$$

Here $\vec{n}$ is the unit outer normal to $\Gamma_t \equiv \partial\Omega_t$, $T(\vec{v},p) = -pI + \nu S(\vec{v})$ is the stress tensor, $S(\vec{v}) = \left(\frac{\partial v_i}{\partial x_j} + \frac{\partial v_j}{\partial x_i}\right)_{i,j-1,2,3}$ is the double deformation tensor, H is the double mean curvature of Γ_t, which is negative for convex surfaces, and V_n is the velocity of the motion of Γ_t in the direction of the exterior normal $\vec{n}$. The domain Ω_0, as well as the initial velocity $\vec{v}_0(x)$, $x \in \Omega_0$, are given.

This problem has been studied in the papers [1–5] where it has been shown that it is solvable on a certain finite time interval, but if Ω_0 is close to a ball and $\vec{v}_0(x)$ is small, then the solution is defined for all positive values of time, and the limiting regime as $t \to +\infty$ is the rotation of the liquid as a rigid body about a certain axis which is moving with a constant speed. This speed, the direction of the axis and a total angular momentum of the liquid are completely determined by initial the data. It has been also shown that Ω_t tends to a certain equilibrium figure, as $t \to \infty$. The analysis of the behaviour of the liquid for large positive t

in two- and three-dimensional case has been carried out in the papers [4] and [5], respectively. In the latter paper also selfgravitation forces have been taken into account but they can be easily excluded simply by setting a gravitational constant κ equal to zero. The case $\sigma = 0$ has been studied in [6].

A crucial role in the proof of the solvability of problem (1.1)–(1.4) on a finite time interval was played by a passage to Lagrangean coordinates which made it possible to write (1.1)–(1.4) as a nonlinear problem in a fixed domain Ω_0.

In a recent paper of M. Günther and G. Prokert [7], the same problem is studied in a quasistationary approximation, when the full nonstationary Navier-Stokes system is replaced by stationary Stokes equations and the initial condition for $\vec{v}(x,t)$ is removed. The problem then reduces to find the domain Ω_t (which is given for $t = 0$), the velocity $\vec{v}(x,t)$ and the pressure $p(x,t)$ satisfying the equations

$$\begin{gathered} -\nu\nabla^2\vec{v}(x,t) + \nabla p(x,t) = 0, \quad \nabla\cdot\vec{v}(x,t) = 0, \quad x \in \Omega_t, \\ T\vec{n} - \sigma H\vec{n}\Big|_{\Gamma_t} = 0, \quad \vec{v}\cdot\vec{n}\Big|_{\Gamma_t} = V_n. \end{gathered} \tag{1.5}$$

Under the assumption that the domain Ω_0 is star-shaped it is shown in [7] that problem (1.5) is solvable on a finite time interval, and if Ω_0 is close to a ball, then the solution is defined for all $t > 0$ and it tends to the equilibrium rest state as $t \to \infty$.

In this paper we show that the method of Lagrangean coordinates is applicable also to problem (1.5) in an arbitrary domain whose boundary possesses a certain regularity.

The relation between Eulerian coordinates $x \in \Omega_t$ and Lagrangean coordinates $\xi \in \Omega_0 \equiv \Omega$ is given by the formula

$$\vec{x} = \vec{\xi} + \int_0^t \vec{u}(\xi,\tau)dt = \vec{X}(\xi,t) \tag{1.6}$$

where $\vec{u}(\xi,t) = (u_1,u_2,u_3)$ is the velocity vector field written as a function of ξ, t and $\vec{x}$ and $\vec{\xi}$ are radii-vectors corresponding to the points $x \in \Omega_t$ and $\xi \in \Omega$, respectively. This formula defines a mapping $x = X(\xi,t)$ of Ω onto Ω_t. From the equation $\nabla\cdot\vec{v} = 0$ it follows that this mapping is invertible and the determinant of its Jacobi matrix equals 1. Under transformation (1.6), equations (1.5) take the form

$$-\nu\nabla_u^2\vec{u} + \nabla_u q = 0, \quad \nabla_u\cdot\vec{u} = 0,$$

$$\vec{n}\cdot T_u(\vec{u},q)\vec{n} - \sigma\Delta(t)\vec{X}(\xi,t)\Big|_{\Gamma_0} = 0$$

where $q(\xi,t) = p(X(\xi,t),t)$, $\nabla_u = \mathcal{A}\nabla = \left(\sum_{j=1}^n A_{ij}\frac{\partial}{\partial\xi_j}\right)_{i,j=1,2,3}$, A_{ij} is the cofactor of the element

$$a_{ij} = \delta_{ij} + \int_0^t \frac{\partial u_i(\xi,\tau)}{\partial\xi_j}d\tau$$

of the Jacobi matrix $J = (\frac{\partial X}{\partial \xi})$ (since $\det J = 1$, the matrix $\mathcal{A} = (A_{ij})_{i,j=1,2,3}$ coincides with $(J^{-1})^{\top}$). Further, $\vec{n}$ is an exterior normal to the surface $\Gamma_t = X(\Gamma_0)$, $\Delta(t)$ is the Laplace-Beltrami operator on Γ_t, and $T_u(\vec{u}, q)$ is a transformed stress tensor:

$$T_u(\vec{u}, q) = -pI + \nu S_u(\vec{u}), \quad (S_u(\vec{u}))_{ij} = \sum_{k=1}^{3} \left(A_{ik} \frac{\partial u_j}{\partial \xi_k} + A_{jk} \frac{\partial u_i}{\partial \xi_k} \right).$$

We have made use of the well-known formula $H\vec{n} = \Delta(t)\vec{x} = \Delta(t)\vec{X}(\xi, t)$.

It is convenient to write problem (1.5) in a slightly different equivalent form. The boundary conditions can be written as two equations

$$\nu \Pi_0 \Pi S_u(\vec{u})\vec{n} = 0,$$

$$\vec{n}_0 \cdot T_u(\vec{u}, q)\vec{n} - \sigma \vec{n}_0 \cdot \Delta(t)\vec{X}(\xi, t) = 0, \quad \xi \in \Gamma_0 ,$$

where $\vec{n}_0(\xi)$ is a normal to Γ_0 at the point ξ, and Π_0 and Π are the projections onto the tangential plane to Γ_0 at ξ and the tangential plane to Γ_t at the point $X(\xi, t)$, respectively, i.e.,

$$\Pi_0 \vec{f}(\xi) = \vec{f}(\xi) - \vec{n}_0(\vec{n}_0 \cdot \vec{f}(\xi)),$$

$$\Pi \vec{f}(\xi) = \vec{f}(\vec{\xi}) - \vec{n}(\vec{n} \cdot \vec{f}(\xi)).$$

Clearly, if $\vec{n}_0 \cdot \vec{n} > 0$ (which is the case for small t), then these two conditions are equivalent to $T_u\vec{n} - \sigma \Delta(t)\vec{X} = 0$.

Further, since the operator corresponding to a stationary elliptic problem

$$-\nu \nabla^2 \vec{v} + \nabla p = \vec{f}, \quad \nabla \cdot \vec{v} = g, \quad T\vec{n}\Big|_{\Gamma_t} = \vec{h} \tag{1.7}$$

has a non-trivial six-dimensional kernel V_0 consisting of the vectors of rigid displacement

$$\vec{v}_0 = \vec{a} + \vec{b} \times \vec{x}\,, \quad p_0 = 0$$

where $\vec{a}$ and $\vec{b}$ are arbitrary constant vectors, it is convenient to introduce into the Stokes equations a linear term

$$\ell(\vec{v}) = \sum_{k=1}^{6} (\vec{v}, \vec{\varphi}_k)\vec{\varphi}_k(x)$$

where $\vec{\varphi}_k$, $k = 1, \ldots, 6$ are linearly independent vector fields from V_0 and $(\vec{v}, \vec{\varphi}_k) = \int_{\Omega_t} \vec{v}(x)\vec{\varphi}_k(x)\,dx$. According to the Schmidt lemma (see e.g. [8]), the problem

$$-\nu \nabla^2 \vec{v} + \ell(\vec{v}) + \nabla p = \vec{f}, \quad \nabla \cdot \vec{v} = g, \quad T\vec{n}\Big|_{\Gamma_t} = \vec{h}$$

is uniquely solvable for arbitrary data $(\vec{f}, g, \vec{h})$, but if the data satisfy necessary orthogonality conditions, then $\ell(\vec{v}) = 0$, and $(\vec{v}, p)$ is a solution of (1.7). In the problem (1.5) the orthogonality conditions reduce to

$$\int_{\Gamma_t} H\vec{n} \cdot \vec{\varphi}\,dS = \int_{\Gamma_t} \Delta(t)\vec{x} \cdot \vec{\varphi}\,dS = 0, \quad \forall \vec{\varphi} \in V_0,$$

and they are fulfilled. So instead of (1.5) we consider the problem

$$
\begin{gathered}
-\nu\nabla_u^2\vec{u}+\sum_{k=1}^{6}\lambda_k(\vec{u})\vec{\varphi}_k(X(\xi,t))+\nabla_u q=0,\\
\nabla_u\cdot\vec{u}=0\quad(\xi\in\Omega,\ t>0),\\
\Pi_0\Pi S_u(\vec{u})\vec{n}\Big|_{\Gamma_0}=0,\\
\vec{n}_0\cdot T_u(\vec{u},q)\vec{n}-\sigma\vec{n}_0\cdot\Delta(t)\vec{X}(\xi,t)\Big|_{\Gamma_0}=0
\end{gathered}
\tag{1.8}
$$

where

$$\lambda_k(\vec{u})=\int_\Omega\vec{u}(\eta,t)\vec{\varphi}_k(X(\eta,t))\,d\eta-M_k(t).$$

The solution of this problem determines the solution $\vec{v}(x,t)=\vec{u}\circ X^{-1}(x,t)$ of problem (1.5) satisfying the conditions

$$\int_{\Omega_t}\vec{v}\cdot\vec{\varphi}_k(x)\,dx=M_k(t)$$

which are equivalent to the prescription of total linear and angular momenta of the liquid.

Problem (1.8) is solved in the Hölder spaces $C^\ell(\Omega)$ (ℓ is a positive non-integer) with the norm defined by

$$|u|_{C^\ell(\Omega)}=\sum_{|j|<\ell}\sup_\Omega|D^ju(x)|+[u]_\Omega^{(\ell)},$$

$$[u]_\Omega^{(\ell)}=\sum_{|j|=[\ell]}\sup_{x,y\in\Omega}|x-y|^{-\ell+[\ell]}|D^ju(x)-D^ju(y)|.$$

The spaces $C^\ell(\Gamma_0)$ are defined in a similar way by means of local maps.

The main result of the paper is the following

Theorem 1. *Let $\Gamma_0\in C^{\ell+3}$, $\ell>0$. Then problem (1.8) has a unique solution $\vec{u}\in C(0,T,C^{\ell+2}(\Omega))$, $q\in C(0,T;C^{\ell+1}(\Omega))$ on a certain finite time interval $(0,T)$, and it satisfies the conditions*

$$\int_\Omega\vec{u}(\xi)\vec{\varphi}_k(X(\xi,t))\,d\xi=M_k(t),\quad k=1,\dots,6,\quad\forall t\in(0,T)\tag{1.9}$$

and the inequality

$$\sup_{t<T}|\vec{u}(\cdot,t)|_{C^{\ell+2}(\Omega)}+\sup_{t<T}|q(\cdot,t)|_{C^{\ell+1}(\Omega)}\le C\Big(|H_0|_{C^{\ell+2}(\Gamma_0)}+\sum_{k=1}^{6}\sup_{t<T}|M_k(t)|\Big)\tag{1.10}$$

where $H_0=\vec{n}_0\cdot\Delta(0)\vec{\xi}$ is the doubled mean curvature of Γ_0. The free boundary Γ_t belongs to $C^1(0,T;C^{\ell+2})\cap C(0,T;C^{\ell+3})$ (i.e., it can be locally represented by the equation $\eta_3=\phi(\eta',t)$ where $\eta'=(\eta_1,\eta_2)\in\omega\subset\mathbb{R}^2$ and $\phi\in C^1(0,T;C^{\ell+2}(\omega))\cap C(0,T;C^{\ell+3}(\omega))$).

The paper is organized as follows. In Section 2 we consider a model linear problem in the half-space. In Section 3 we study the linear problem

$$\begin{aligned} -\nu\nabla^2\vec{v} + \ell(\vec{v}) + \nabla p = \vec{f}(\xi,t), \quad \nabla\cdot\vec{v} = g(\xi,t) \quad (\xi\in\Omega,\ t>0), \\ \nu\Pi_0 S(\vec{v})\vec{n}_0\Big|_{\Gamma_0} = \vec{b}(\xi,t), \quad \vec{b}\cdot\vec{n} = 0, \\ \vec{n}_0\cdot T(\vec{v},p)\vec{n}_0 - \sigma\vec{n}_0\cdot\int_0^t \Delta\vec{v}(\xi,\tau)\,d\tau\Big|_{\Gamma_0} = b(\xi,t) + \int_0^t B(\xi,\tau)d\tau \end{aligned} \tag{1.11}$$

where $\Delta = \Delta(0)$ is the Laplace-Beltrami operator on Γ_0. We obtain Schauder estimates for this problem and prove the existence of a unique solution in the Hölder spaces. Finally, in Section 4 problem (1.8) is analysed. It is written in the form

$$\begin{aligned} -\nu\nabla^2\vec{u} + \nabla q + \ell(\vec{u}) = \vec{\ell}_1(\vec{u},q) + \sum_{k=1}^{6} M_k(t)\,\vec{\varphi}_k(X(\xi,t)), \\ \nabla\cdot\vec{u} = \ell_2(\vec{u}), \\ \nu\Pi_0 S(\vec{u})\vec{n}_0\Big|_{\Gamma_0} = \vec{\ell}_3(\vec{u}), \\ \vec{n}_0\cdot T(\vec{u},q)\vec{n}_0 - \sigma\vec{n}_0\cdot\Delta\int_0^t \vec{u}\,d\tau\Big|_{\Gamma_0} = \ell_4(\vec{u}) + \int_0^t \ell_5(\vec{u})d\tau + \vec{n}_0\cdot\Delta(t)\xi\Big|_{\Gamma_0} \end{aligned} \tag{1.12}$$

where ℓ_i are differential expressions containing the integrals $\int_0^t \frac{\partial u_i}{\partial \xi_i}d\tau$ and $\int_0^t u_i d\tau$ as multipliers, and, as a result, their norms are estimated by the norm of $\vec{u}$ multiplied by a small constant, if t is small. These estimates make it possible to prove the solvability of problem (1.8) by the method of successive approximations.

It is a pleasure for me to express my thanks to Professor V.V. Pukhnachov for his encouragement and many fruitful discussions.

2. Model problems in the half-space

In this section we consider the boundary value problem

$$-\nu\nabla^2\vec{v} + \nabla p = \vec{f}(x,t), \quad \nabla\cdot\vec{v} = g(x,t), \quad x\in\mathbb{R}^3_+ = \{x_3>0\},\ t>0,$$

$$\nu\left(\frac{\partial v_j}{\partial x_3} + \frac{\partial v_3}{\partial x_j}\right)\Big|_{x_3=0} = b_j(x',t), \quad j=1,2, \tag{2.1}$$

$$-p + 2\nu\frac{\partial v_3}{\partial x_3} + \sigma\Delta'\int_0^t v_3(x,\tau)d\tau\Big|_{x_3=0} = b_3(x',t) + \int_0^t B(x',\tau)d\tau, \tag{2.2}$$

and a "penalized" problem with condition (2.2) replaced by

$$\begin{aligned} -p+2\nu\frac{\partial v_3}{\partial x_3} &+\sigma\Delta'\int_0^t v_3 d\tau-\varepsilon\Delta'\int_0^t(-p+2\nu\frac{\partial v_3}{\partial x_3})d\tau\Big|_{x_3=0} \\ &=b_3(x,t)+\int_0^t B(x',\tau)dx'+\varepsilon\sum_{j=1}^2\frac{\partial}{\partial x_j}\int_0^t B_j d\tau. \end{aligned} \tag{2.3}$$

Here ε is a small positive parameter, $x'=(x_1,x_2)\in\mathbb{R}^2$, $\Delta'=\frac{\partial^2}{\partial x_1^2}+\frac{\partial^2}{\partial x_2^2}$; the functions $\vec{f},g,b_k,B,B_j$ are smooth and have compact supports.

We obtain solutions of both problems in an explicit form. First of all, we extend $\vec{f}$ and g from $\mathbb{R}^3_+$ into the whole space $\mathbb{R}^3$ with the preservation of a certain finite smoothness and in such a way that the extended functions also have compact supports (we may, for instance, use the Hesteness-Whiteney procedure), and we consider the volume potentials

$$\vec{w}(x,t)=\int_{\mathbb{R}^3}U(x-y)\vec{f}(y,t)dy+\int_{\mathbb{R}^3}\nabla E(x-y)g(y,t)dy, \tag{2.4}$$

$$s(x,t)=\int_{\mathbb{R}^3_+}\nabla E(x-y)\cdot\vec{f}(y,t)dy-\nu g(x) \tag{2.5}$$

where $E(x)=-\frac{1}{4\pi|x|}$ is a fundamental solution of the Laplace equation and $U(x)$ is a fundamental matrix of solutions of the Stokes system with the elements

$$U_{km}(x)=\frac{1}{8\pi\nu}\left(\frac{\delta_{km}}{|x|}+\frac{x_k x_m}{|x|^3}\right),\quad k,m=1,2,3.$$

These potentials decay at infinity like $|x|^{-1}$ and $|x|^{-2}$, respectively, and they satisfy the equations

$$-\nu\nabla^2\vec{w}+\nabla s=\vec{f}(x,t),\quad \nabla\cdot\vec{w}=g(x,t),\quad x\in\mathbb{R}^3.$$

For $\vec{u}=\vec{v}-\vec{w}$, $q=p-s$ we obtain the problem

$$-\nu\nabla^2\vec{u}+\nabla q=0,\quad \nabla\cdot u=0,\quad x\in\mathbb{R}^3_+, \tag{2.6}$$

$$\nu\left(\frac{\partial u_j}{\partial x_3}+\frac{\partial u_3}{\partial x_j}\right)\Big|_{x_3=0}=d_j(x',t),\quad j=1,2, \tag{2.7}$$

$$-q+2\nu\frac{\partial u_3}{\partial x_3}+\sigma\Delta'\int_0^t u_3 d\tau\Big|_{x_3=0}=d_3(x',t)+\int_0^t D(x',\tau)d\tau, \tag{2.8}$$

or (2.6), (2.7) and

$$\begin{aligned} -q+2\nu\frac{\partial u_3}{\partial x_3} &+\sigma\Delta'\int_0^t u_3 d\tau+\varepsilon\Delta'\int_0^t(q-2\nu\frac{\partial u_3}{\partial x_3})d\tau\Big|_{x_3=0} \\ &= d_3+\int_0^t D(x',\tau)d\tau+\varepsilon\sum_{j=1}^2\frac{\partial}{\partial x_j}\int_0^t D_j(x',\tau)d\tau\ , \end{aligned} \tag{2.9}$$

instead of (2.8). Here

$$d_j = b_j - \nu \left(\frac{\partial w_j}{\partial x_3} + \frac{\partial w_3}{\partial x_j} \right) \Big|_{x_3=0},$$

$$d_3 = b_3 + \left(s - 2\nu \frac{\partial w_3}{\partial x_3} \right) \Big|_{x_3=0},$$

$$D = B - \sigma \Delta' w_3 \Big|_{x_3=0}$$

or

$$D = B - \sigma \Delta w_3 \Big|_{x_3=0}, \quad D_j = B_j + \frac{\partial}{\partial x_j} \left(-s + 2\nu \frac{\partial w_3}{\partial x_3} \right) \Big|_{x_3=0},$$

respectively.

After the Fourier transform

$$\widetilde{u}(\xi) = \int_{\mathbb{R}^2} e^{-ix' \cdot \xi} u(x') dx', \quad \xi = (\xi_1, \xi_2) \in \mathbb{R}^2,$$

we reduce (2.6)–(2.8) and (2.6), (2.7), (2.9) to boundary value problems for the system of ordinary differential equations in the interval $x_3 > 0$:

$$\begin{gathered} -\nu \frac{d^2 \widetilde{u}_j}{dx_3^2} + \nu \xi^2 \widetilde{u}_j + i\xi_j \widetilde{q} = 0, \quad j = 1, 2, \\ -\nu \frac{d^2 \widetilde{u}_3}{dx_3^2} + \nu \xi^2 \widetilde{u}_3 + \frac{\partial \widetilde{q}}{\partial x_3} = 0, \\ \frac{d\widetilde{u}_3}{dx_3} + i\xi_1 \widetilde{u}_1 + i\xi_2 \widetilde{u}_2 = 0, \quad x_3 > 0, \\ \widetilde{u}_1, \widetilde{u}_2, \widetilde{u}_3, \widetilde{q} \to 0 \quad (x_3 \to +\infty), \end{gathered} \tag{2.10}$$

$$\nu \left(\frac{d\widetilde{u}_j}{dx_3} + i\xi_j \widetilde{u}_3 \right) \Big|_{x_3=0} = \widetilde{d}_j, \tag{2.11}$$

$$-\widetilde{q} + 2\nu \frac{\partial \widetilde{u}_3}{\partial x_3} - \sigma \xi^2 \int_0^t \widetilde{u}_3(\xi, \tau) d\tau \Big|_{x_3=0} = \widetilde{d}_3 + \int_0^t \widetilde{D}(\xi, \tau) d\tau \tag{2.12}$$

or

$$\begin{aligned} -\widetilde{q} + 2\nu \frac{d\widetilde{u}_3}{dx_3} \quad &- \quad \sigma \xi^2 \int_0^2 \widetilde{u}_3 d\tau + \varepsilon \xi^2 \int_0^t \left(-\widetilde{q} + 2\nu \frac{d\widetilde{u}_3}{dx_3} \right) d\tau \Big|_{x_3=0} \\ &= \quad \widetilde{d}_3 + \int_0^t \widetilde{D}(\xi, \tau) d\tau + \varepsilon \sum_{j=1}^2 i\xi_j \int_0^t \widetilde{D}_j(\xi, \tau) d\tau. \end{aligned} \tag{2.13}$$

It is easy to verify that the solution of system (2.10) satisfying the boundary conditions $\vec{\widetilde{u}}\Big|_{x_3=0} = \vec{\widetilde{a}} = (\widetilde{a}_1, \widetilde{a}_2, \widetilde{a}_3)$ has the form

$$\begin{aligned} \vec{\widetilde{u}} &= \vec{\widetilde{a}} e^{-|\xi| x_3} + \left(\frac{i\xi_1}{|x|}, \frac{i\xi_2}{|\xi|}, -1 \right) (i\xi_1 \widetilde{a}_1 + i\xi_2 \widetilde{a}_2 - |\xi| \widetilde{a}_3) x_3 e^{-|\xi| x_3}, \\ \widetilde{q} &= -2\nu (i\xi_1 \widetilde{a}_1 + i\xi_2 \widetilde{a}_2 - |\xi| \widetilde{a}_3) e^{|\xi| x_3}. \end{aligned}$$

We find $\widetilde{a}_k$ from the boundary conditions (2.11)–(2.13). Equations (2.11) imply

$$\nu\left(-|\xi|\widetilde{a}_j+\frac{i\xi_j}{|\xi|}(i\xi_1\widetilde{a}_1+i\xi_2\widetilde{a}_2)\right)=d_j,\quad j=1,2$$

and, as a consequence,

$$\widetilde{a}_j=-\frac{1}{\nu|\xi|}\widetilde{d}_j+\frac{1}{2\nu|\xi|^2}\frac{i\xi_j}{|\xi|}(i\xi_1\widetilde{d}_1+i\xi_2\widetilde{d}_2),\quad j=1,2. \tag{2.14}$$

Furthermore, since $-q+2\nu\frac{d\widetilde{u}_3}{dx_3}\Big|_{x_3=0}=-2\nu|\xi|\widetilde{a}_3$, we obtain from (2.12)

$$-2\nu|\xi|\widetilde{a}_3(\xi,t)-\sigma\xi^2\int_0^t\widetilde{a}_3(\xi,\tau)d\tau=\widetilde{d}_3+\int_0^t\widetilde{D}d\tau,$$

i.e.

$$\frac{da_3}{dt}+\frac{\sigma|\xi|}{2\nu}a_3=-\frac{1}{2\nu|\xi|}\left[\left(\frac{d}{dt}\right)\widetilde{d}_3+\widetilde{D}\right],\quad t>0,\quad a_3\Big|_{t=0}=-\frac{1}{2\nu|\xi|}\widetilde{d}_3(0).$$

Hence,

$$\begin{aligned}\widetilde{a}_3(\xi,t)&=-\frac{\widetilde{d}_3(\xi,t)}{2\nu|\xi|}+\frac{\sigma}{4\nu^2}\int_0^t\widetilde{d}_3(\xi,\tau)e^{-\frac{\sigma}{2\nu}|\xi|(t-\tau)}d\tau\\&\quad-\frac{1}{2\nu|\xi|}\int_0^t\widetilde{D}(\xi,\tau)e^{-\frac{\sigma}{2\nu}|\xi|(t-\tau)}d\tau.\end{aligned} \tag{2.15}$$

Similarly, condition (2.13) is equivalent to

$$\begin{aligned}&-2\nu|\xi|\widetilde{a}_3-\sigma|\xi|^2\int_0^t\widetilde{a}_3(\xi,\tau)\,d\tau-2\nu\varepsilon|\xi|^3\int_0^t\widetilde{a}_3(\xi,\tau)\,d\tau\\&\qquad=\widetilde{d}_3+\int_0^t\widetilde{D}(\xi,\tau)\,d\tau+\varepsilon\sum_{j=1}^2 i\xi_j\int_0^t\widetilde{D}_j(\xi,\tau)\,d\tau,\end{aligned}$$

which yields

$$\begin{aligned}\widetilde{a}_3(\xi,t)&=-\frac{1}{2\nu|\xi|}\widetilde{d}_3(\xi,t)+\left(\frac{\sigma}{4\nu^2}+\frac{\varepsilon|\xi|}{2\nu}\right)\int_0^t\widetilde{d}_3(\xi,\tau)e^{-(\frac{\sigma}{2\nu}|\xi|+\varepsilon\xi^2)(t-\tau)}d\tau\\&\quad-\frac{1}{2\nu|\xi|}\int_0^t(\widetilde{D}(\xi,\tau)+\varepsilon\sum_{j=1}^2 i\xi_j\widetilde{D}_j(\xi,\tau))e^{-(\frac{\sigma}{2\nu}|\xi|+\varepsilon\xi^2)(t-\tau)}d\tau.\end{aligned} \tag{2.16}$$

Formulas (2.14)–(2.16) provide the representation of $a_k(x',t)$ in terms of convolution integrals:

$$a_j(x',t)=-\frac{1}{2\pi\nu}\left(\frac{1}{|x|}*d_j\right)-\frac{1}{4\pi\nu}\sum_{i=1}^2\left(\frac{\delta_{ij}}{|x|}-\frac{x_ix_j}{|x|^3}\right)*d_i,\quad j=1,2, \tag{2.17}$$

$$a_3(x',t) = -\frac{1}{4\pi\nu}\left(\frac{1}{|x|} * d_3\right) - \frac{1}{4\pi\nu}\int_0^t \frac{\partial P(x',t-\tau)}{\partial t} * d_3(x',\tau)d\tau \\ -\frac{1}{4\pi\nu}\int_0^t P(x',t-\tau) * D(x',\tau)d\tau, \tag{2.18}$$

$$\begin{aligned} a_3(x',t) = \\ -\frac{1}{4\pi\nu}\left(\frac{1}{|x|} * d_3\right) - \frac{1}{4\pi\nu}\int_0^t \frac{\partial P(x',t-\tau)}{\partial t} * \Gamma(x,\varepsilon(t-\tau)) * d_3(x',\tau)d\tau \\ -\frac{1}{4\pi\nu}\int_0^t P(x',t-\tau) * \frac{\partial \Gamma(x,\varepsilon(t-\tau))}{\partial t} * d_3(x',\tau)d\tau \\ -\frac{1}{4\pi\nu}\int_0^t P(x',t-\tau) * (\Gamma(x',\varepsilon(t-\tau) * D(x',\tau) \\ +\varepsilon\sum_{i=1}^{2}\Gamma_{x_i}(x',\varepsilon(t-\tau))D_i(x',\tau))d\tau. \end{aligned} \tag{2.19}$$

where "$*$" denotes convolution with respect to x',

$$\Gamma(x,t) = (4\pi t)^{-1}\exp(-\frac{|x'|^2}{4t}),$$
$$P(x',t) = \frac{1}{\left(|x'|^2 + \frac{\sigma^2}{4\nu^2}t^2\right)^{1/2}}.$$

These formulas together with the well-known representation of $\vec{u}, q$ in the form of the double layer potential

$$\vec{u} = \mathcal{V}(x',x_3) * \vec{a}(x',t), \quad q = \vec{R}(x',x_3) * \vec{a}(x',t), \tag{2.20}$$

$$\mathcal{V} = (V_{ij})_{i,j=1,2,3}, \quad V_{ij} = \frac{3}{2\pi}\frac{x_i x_j x_3}{|x|^5}, \quad R_j = \frac{\nu}{\pi}\left(\frac{\delta_{3j}}{|x|^3} - \frac{3x_j x_3}{|x|^5}\right),$$

yield the explicit solution of problems (2.6)–(2.8) and (2.6), (2.7), (2.9). Under our hypotheses, $\vec{a} = O(|x'|^{-1})$ for large $|x'|$, so formulas (2.20) make sense and $\vec{u} = O(|x|^{-1})$, $q = O(|x|^{-2})$ $(|x| \to \infty)$.

Now, we pass to the estimates of solutions of problems (2.1), (2.2) and (2.1), (2.3). We start with the following proposition.

Proposition 1. *Assume that $d_j \in C^{\ell+1}(\mathbb{R}^2), D \in C^{\ell}(\mathbb{R}^2), D_j \in C^{\ell}(\mathbb{R}^2)$. Then functions (2.17)–(2.19) satisfy the inequalities*

$$[a_j]_{\mathbb{R}^2}^{(\ell+2)} \le C_1 \sum_{i=1}^{2}[d_i]_{\mathbb{R}^2}^{(\ell+1)}, \tag{2.21}$$

$$[a_3(\cdot,t)]_{\mathbb{R}^2}^{(\ell+2)} \le C_2\left(\sup_{\tau<t}[d_3(\cdot,\tau)]_{\mathbb{R}^2}^{(\ell+1)} + \sup_{\tau<t}[D(\cdot,\tau)]_{\mathbb{R}^2}^{(\ell)}\right), \tag{2.22}$$

$$
\begin{aligned}
[a_3(\cdot,t)]^{(\ell+2)}_{\mathbb{R}^2} \quad \le \quad & C_3\left(\sup_{\tau<t}[d_3(\cdot,\tau)]^{(\ell+1)}_{\mathbb{R}^2}+\sup_{\tau<t}[D(\cdot,\tau)]^{(\ell)}_{\mathbb{R}^2}\right.\\
& \left. + \sum_{i=1}^{2}\sup_{\tau<t}[D_i(\cdot,\tau)]^{(\ell)}_{\mathbb{R}^2}\right),
\end{aligned}
\tag{2.23}
$$

respectively. The constant C_3 is independent of ε.

Proof. We use the scheme of estimates of convolution integrals proposed by K. K. Golovkin ([9], see also [10], §2). It is based on the theorem on the equivalence of semi-norms $[u]^{(\ell)}_{\mathbb{R}^2}$ and

$$\sum_{i=1}^{2}\sup_{h>0} h^{-\ell}\sup_{\mathbb{R}^2}|\Delta_i^{k_i}(h)u(x')|$$

on the set of functions $u\in C^{\ell}(\mathbb{R}^2)$ (see [11]). Here

$$\Delta_i^{k_i}(h)u(x)=\sum_{m=0}^{k_i}(-1)^{m-k_i}\binom{k_i}{m}u(x+e_ihm)$$

is the k_i-th finite difference of $u(x)$ with respect to x_i, $e_1=(1,0)$, $e_2=(0,1)$, and k_i are arbitrary integers satisfying the condition $k_i>\ell$. This scheme makes the estimate of convolution integrals with weakly singular kernels extremely easy. Consider a finite difference $\Delta_i^{s+m}(h)a_j(x',t)$ of the functions $a_j(x',t)$ (2.17). We have

$$
\begin{aligned}
\Delta_i^{s+m}(h)a_j(x',t) \quad = \quad & -\frac{1}{2\pi\nu}\left(\Delta_i^s(h)\frac{1}{|x'|}*\Delta_i^m(h)d_j(x',t)\right)\\
& -\frac{1}{4\pi\nu}\sum_{r=1}^{2}\left(\Delta_i^s(h)\left(\frac{\delta_{rj}}{|x|}-\frac{x_rx_j}{|x'|^3}\right)*\Delta_i^m(h)d_r(x',t)\right).
\end{aligned}
$$

An arbitrary function $K(x')$ which is homogeneous of order -1 and smooth on the unit circle $|x'|=1$, satisfies the inequality

$$\int_{\mathbb{R}^2}|\Delta_i^s(h)K(y')|dy'\le Ch,\quad s>1,$$

hence, assuming that $s>1, m>l+1$, we obtain

$$|\Delta_i^{s+m}(h)a_j(x',t)|\le Ch\sum_{r=1}^{2}\sup_{\mathbb{R}^2}|\Delta_i^m(h)d_r(x',t)|$$

and

$$
\begin{aligned}
\sup_{h>0}h^{-\ell-2}\quad & \sup_{\mathbb{R}^2}|\Delta_i^{s+m}(h)a_j(x',t)|\\
& \le C\sum_{r=1}^{2}\sup_{h>0}h^{-\ell-1}\sup_{\mathbb{R}^2}|\Delta_i^m(h)d_r(x',t)|.
\end{aligned}
$$

Due to the theorem on the equivalent norms mentioned above, this inequality implies (2.21). The proof of (2.22) is similar; it is based on inequalities

$$\int_0^t\int_{\mathbb{R}^2}|\Delta_i^s(h)P(y',t-\tau)|dy'd\tau\le Ch^2\quad(s>2),$$

$$\int_0^t\int_{\mathbb{R}^2}|\Delta_i^s(h)P_t(y',t-\tau)|dy'd\tau\le Ch\quad(s>1).$$

The proof of (2.23) which is slightly more complicated requires also similar estimates for $\Gamma_{x_k}(x',\varepsilon(t-\tau))$, namely,

$$\varepsilon\int_0^t d\tau\int_{\mathbb{R}^2}|\Delta_i^s(h)\Gamma_{y_k}(y',\varepsilon(t-\tau))|dy'\le Ch\quad(s>1)$$

and the well-known relation $\int_{\mathbb{R}^2}\Gamma(x',\varepsilon(t-\tau))dx'=1$.

We write (2.19) in the form

$$\begin{aligned}a_3(x',t)&=-\frac{1}{4\pi\nu}\left(\frac{1}{|x|}*d_3\right)\\&\quad-\frac{1}{4\pi\nu}\int_0^t\frac{\partial P(x',t-\tau)}{\partial t}*\Gamma(x',\varepsilon(t-\tau))*d_3(x',\tau)d\tau\\&\quad-\frac{1}{4\pi\nu}\int_0^t P(x',t-\tau)*\Gamma(x',\varepsilon(t-\tau))*D(x',\tau)d\tau\\&\quad-\frac{\varepsilon}{4\pi\nu}\sum_{i=1}^2\int_0^t\frac{\partial\Gamma(x',\varepsilon(t-\tau))}{\partial x_i}*P(x',t-\tau)*\\&\quad*\left(D_i(x',\tau)+\frac{\partial d_3(x',\tau)}{\partial x_i}\right)d\tau\\&\equiv\sum_{k=1}^4 a_3^{(k)}(x',t).\end{aligned}\tag{2.24}$$

and evaluate finite differences $\Delta_j^{s_1+s_2+m}(h)a_3^{(k)}(x',t)$.

The first term $a_3^{(1)}(x,t)$ has been already considered above, for the second term we have

$$\begin{aligned}&\Delta_j^{s_1+s_2+m}(h)a_3^{(2)}(x',t)\\&\quad=-\frac{1}{4\pi\nu}\int_0^t\Delta_j^{s_1+s_2}(h)P_t(x',t-\tau)*\Gamma(x',\varepsilon(t-\tau))*\Delta_j^m(h)d_3(x',\tau)d\tau\end{aligned}$$

and, as a consequence,

$$\begin{aligned}&|\Delta_j^{s_1+s_2+m}(h)a_3^{(2)}(x',t)|\\&\quad=\frac{1}{4\pi\nu}\sup_{\tau<t}\sup_{\mathbb{R}^2}|\Delta_j^m(h)d_3(z',\tau)|\cdot\sup_{\tau<t}\int_{\mathbb{R}^2}\Gamma(y',\varepsilon(t-\tau))dy'\\&\cdot\int_0^t d\tau\int_{\mathbb{R}^2}|\Delta_j^{s_1+s_2}(h)\frac{\partial P(x',t-\tau)}{\partial t}|dx'\le Ch\sup_{\tau<t}\sup_{\mathbb{R}^2}|\Delta_j^m(h)d_3(x'\tau)|\end{aligned}$$

In a similar way we show that

$$|\Delta_j^{s_1+s_2+m}(h)a_3^{(3)}(x',t)| \le Ch^2 \sup_{\tau\le t}\sup_{\mathbb{R}^2} |\Delta_j^m(h)D(x',\tau)|;$$

finally, from the formula

$$\begin{aligned}
&\Delta_j^{s_1+s_2+m}(h)a_3^{(4)}(x',t)\\
&\quad = -\frac{\varepsilon}{4\pi\nu}\sum_{i=1}^2 \int_0^t \Delta_j^{s_1}(h)\frac{\partial\Gamma(x,\varepsilon(t-\tau))}{\partial x_i} * \Delta_j^{s_2}(h)P(x',t-\tau)\\
&\qquad * \Delta_j^m(h)H_i(x',\tau)d\tau
\end{aligned}$$

with $H_i = D_i + \frac{\partial d_3}{\partial x_i}$, we find

$$\begin{aligned}
&|\Delta_j^{s_1+s_2+m}(h)a_3^{(4)}(x',t)|\\
&\quad \le \frac{1}{4\pi\nu}\sum_{i=1}^2 \sup_{\tau<t}\sup_{\mathbb{R}^2}|\Delta_j^m(h)H_i(y',\tau)| \sup_{\tau<t}\int_{\mathbb{R}^2}|\Delta_j^{s_2}(h)P(y',t-\tau)|dy'\\
&\qquad \times \varepsilon\int_0^t\int_{\mathbb{R}^2}\left|\Delta_j^{s_1}(h)\frac{\partial\Gamma(x',\varepsilon(t-\tau))}{\partial x_i}\right|dx'd\tau\\
&\quad \le Ch^2\sum_{i=1}^2 \sup_{\tau<t}\sup_{\mathbb{R}^2}|\Delta_j^m(h)H(z',\tau)|.
\end{aligned}$$

All these estimates together imply (2.23). The proposition is proved. □

From (2.21)–(2.23) and from the well-known estimate of the double layer potential (2.20)

$$[\vec{u}]^{(\ell+2)}_{\mathbb{R}^3_+} + [q]^{(\ell+1)}_{\mathbb{R}^3_+} \le C[\vec{a}]^{(\ell+2)}_{\mathbb{R}^2}$$

we immediately obtain the following inequality for the solution of problems (2.6)–(2.8) and (2.6), (2.7), (2.9) which have been just constructed:

$$\begin{aligned}
&\sup_{\tau<t}[\vec{u}(\cdot,\tau)]^{(\ell+2)}_{\mathbb{R}^3_+} + \sup_{\tau<t}[q(\cdot,\tau)]^{(\ell+1)}_{\mathbb{R}^3_+}\\
&\quad \le C\left(\sup_{\tau<t}[\vec{d}(\cdot,\tau)]^{(\ell+1)}_{\mathbb{R}^3_+} + \sup_{\tau<t}[D(\cdot,\tau)]^{(\ell)}_{\mathbb{R}^2_+} + \sum_{i=1}^2\sup_{\tau<t}[D_i(\cdot,\tau)]^{(\ell)}_{\mathbb{R}^2}\right)
\end{aligned} \tag{2.25}$$

(in the case of problem (2.6)–(2.8) the last term should be omitted). Now, we can establish the main result of this section.

Theorem 2. *For arbitrary smooth compactly supported f, g, b_k, B, B_j problems (2.1), (2.2) and (2.1), (2.3) have unique solutions decaying at infinity and the* *following inequality is satisfied:*

$$\sup_{\tau<t}[\vec{u}(\cdot,\tau)]^{(\ell+2)}_{\mathbb{R}^3_+} + \sup_{\tau<t}[p(\cdot,\tau)]^{(\ell+1)}_{\mathbb{R}^3_+} \le C\Big(\sup_{\tau<t}[\vec{f}(\cdot,\tau)]^{(\ell)}_{\mathbb{R}^3_+}$$
$$+\sup_{\tau<t}[g(\cdot,\tau)]^{(\ell+1)}_{\mathbb{R}^3_+} + \sup_{\tau<t}[\vec{b}(\cdot,\tau)]^{(\ell+1)}_{\mathbb{R}^2_+} + \sup_{\tau<t}[B(\cdot,\tau)]^{(\ell)}_{\mathbb{R}^2_+} + \sum_{i=1}^{2}\sup_{\tau<t}[B_i(\cdot,\tau)]^{(\ell)}_{\mathbb{R}^2}\Big). \tag{2.26}$$

In the case of problem (2.1), (2.2) *the last term in the right-hand side should be omitted. The constant C is independent of* ε.

Proof. The solutions have been constructed above. The functions $\vec{f}$ and g can be extended into $\mathbb{R}^3$ in such a way that

$$[\vec{f}(\cdot,\tau)]^{(\ell)}_{\mathbb{R}^3} \le C[\vec{f}(\cdot,\tau)]^{(\ell)}_{\mathbb{R}^3_+}, \qquad [g(\cdot,\tau)]^{(\ell+1)}_{\mathbb{R}^3} \le C[g(\cdot,\tau)]^{(\ell+1)}_{\mathbb{R}^3_+}$$

for all $\tau \le t$, so $\vec{w}$ and s satisfy the inequality

$$[\vec{w}(\cdot,\tau)]^{(\ell+2)}_{\mathbb{R}^3_+} + [s(\cdot,\tau)]^{(\ell+1)}_{\mathbb{R}^3_+} \le C\Big([\vec{f}(\cdot,\tau)]^{(\ell)}_{\mathbb{R}^3_+} + [g(\cdot,\tau)]^{(\ell+1)}_{\mathbb{R}^3_+}\Big)\,. \tag{2.27}$$

Estimate (2.26) follows from (2.25) and (2.27).

The behaviour of $\vec{v}$, p for large $|x|$ can be clarified with the help of formulas (2.4), (2.5), (2.17)–(2.20). The uniqueness of solutions of both problems follows from the fact that homogeneous problems (2.10)–(2.12) and (2.10), (2.11), (2.13) have only trivial solutions. The theorem is proved. $\square$

We observe as a conclusion that our explicit formulas make it possible to estimate the solution of problem (2.1), (2.2) in other norms, for instance, in the norms

$$\left(\sum_{k=1}^{3}\int_0^\infty \frac{dh}{h^{1+2(\ell+2)}}\sup_{\tau<t}\int_{\mathbb{R}^3_+}\left|\Delta_k^m(h)\,\vec{u}(x,\tau)\right|^2 dx\right)^{\frac12}, \qquad m>\ell+2,\ \ \ell>0\,.$$

3. Linear problem

In this section we study a linear problem (1.11). We obtain an a priori estimate of the solution in Hölder norms and then we establish the solvability of the problem. Throughout the whole section, the points of the domain Ω are denoted by x (in contrast to Section 1 and 4).

3.1. A priori estimate

We prove here the following theorem.

Theorem 3. *Assume that $\Gamma_0 \in C^{2+\ell}$. An arbitrary solution of problem* (1.11) $\vec{v} \in C(0,T;C^{2+\ell}(\Omega))$, $p \in C(0,T;C^{1+\ell}(\Omega))$, *satisfies the inequality*

$$\begin{aligned}\sup_{t<T} |\vec{v}(\cdot,t)|_{C^{2+\ell}(\Omega)} &+ \sup_{t<T} |p(\cdot,t)|_{C^{1+\ell}(\Omega)} \\ &\leq C\Big(\sup_{t<T} |\vec{f}(\cdot,t)|_{C^{\ell}(\Omega)} + \sup_{t<T} |g(\cdot,t)|_{C^{\ell+1}(\Omega)} \\ &\quad + \sup_{t<T} |\vec{b}(\cdot,t)|_{C^{\ell+1}(\Gamma_0)} + \sup_{t<T} |b(\cdot,t)|_{C^{\ell+1}(\Gamma_0)} \\ &\quad + \sup_{t<T} |B(\cdot,t)|_{C^{\ell}(\Gamma_0)}\Big) \equiv CN(T)\,.\end{aligned} \tag{3.1}$$

Proof. Using Schauder's method, it is possible to estimate the norms on the left-hand side of (3.1) by the norms of the data and by some weak norm (say, $L_2(\Omega)$-norm) of the solution. Let $x_0 \in \overline{\Omega}$ and $K_r(x_0) = \{x \in \Omega : |x - x_0| < r\}$. We show that for arbitrary $t < T$, $x_0 \in \Omega$ and for arbitrary small $r > 0$ the following estimate holds:

$$\begin{aligned}&\sup_{\tau<t} |\vec{v}(\cdot,\tau)|_{C^{2+\ell}(K_r(x_0))} + \sup_{\tau<t} |p(\cdot,\tau)|_{C^{1+\ell}(K_r(x_0))} \\ &\leq C_1\Big(\sup_{\tau<t} |\vec{f}(\cdot,\tau)|_{C^{\ell}(K_{2r}(x_0))} + \sup_{\tau<t} |g(\cdot,\tau)|_{C^{1+\ell}(K_{2r}(x_0))} \\ &\quad + \sup_{\tau<t} |\vec{b}(\cdot,\tau)|_{C^{1+\ell}(\Gamma_{2r})} + \sup_{\tau<t} |B(\cdot,\tau)|_{C^{\ell}(\Gamma_{2r})} + \sup_{\tau<t} |b(\cdot,\tau)|_{C^{1+\ell}(\Gamma_{2r})}\Big) \\ &\quad + C_2 \sup_{\tau<t} \|\vec{v}(\cdot,\tau)\|_{L_2(\Omega)}, \qquad \forall t < T\end{aligned} \tag{3.2}$$

(there should be no norms of $\vec{b}$, b and B on the right-hand side, if $\Gamma_{2r} \equiv \Gamma_0 \cap K_{2r}(x_0) = \emptyset$).

Clearly, it suffices to prove (3.2) in two cases: $\Gamma_{2r} = \emptyset$ and $x_0 \in \Gamma_0$. In the first case boundary conditions do not play any role, and (3.2) is a well-known interior Schauder estimate for the Stokes system. We consider the second case assuming (without loss of generality) that $x_0 = 0$ and the Ox_3-axis is directed along the interior normal $-\vec{n}_0(x_0)$. As $\Gamma_0 \in C^{2+\ell}$, in the neighbourhood of the origin Γ_0 can be given by the equation

$$x_3 = \phi(x')\,, \qquad x' \in K'_d = \{|x'| < d\}\,,$$

with $\phi \in C^{\ell+2}(K'_d)$, the exterior normal is defined by

$$\vec{n}_0 = \left(\frac{\phi_{x_1}}{\sqrt{1+|\nabla\phi(x')|^2}}, \frac{\phi_{x_1}}{\sqrt{1+|\nabla\phi(x')|^2}}, -\frac{1}{\sqrt{1+|\nabla\phi(x')|^2}}\right)$$

and the Laplace-Beltrami operator Δ is given by the formula

$$\Delta f = \frac{1}{\sqrt{g^{(0)}}} \sum_{\alpha,\beta=1}^{2} \frac{\partial}{\partial x_\alpha} \frac{\widehat{g}^{(0)}_{\alpha\beta}}{\sqrt{g^{(0)}}} \frac{\partial f}{\partial x_\beta} = \sum_{\alpha,\beta=1}^{2} h^{(0)}_{\alpha\beta} \frac{\partial^2 f}{\partial x_\alpha \partial x_\beta} + \sum_{\beta=1}^{2} h^{(0)}_\beta \frac{\partial f}{\partial x_\beta}\,. \tag{3.3}$$

Here $g^{(0)} = \det(g^{(0)}_{\alpha\beta})_{\alpha,\beta=1,2}$,

$$g^{(0)}_{\alpha\beta} = \frac{\partial \vec{r}}{\partial x_\alpha} \cdot \frac{\partial \vec{r}}{\partial x_\beta} = \delta_{\alpha\beta} + \phi_{x_\alpha}\,\phi_{x_\beta}\ ,$$

$\vec{r} = (x_1, x_2, \varphi(x'))$ is a radius-vector of the point $x \in \Gamma_0$, $\widehat{g}^{(0)}_{\alpha\beta}$ are the elements of the associated matrix to $(g^{(0)}_{\alpha\beta})_{\alpha,\beta=1,2}$, i.e., $\widehat{g}^{(0)}_{11} = g^{(0)}_{22}$, $\widehat{g}^{(0)}_{22} = g^{(0)}_{11}$, $\widehat{g}^{(0)}_{12} = \widehat{g}^{(0)}_{21} = -g^{(0)}_{12} = -g^{(0)}_{21}$, and, finally,

$$h^{(0)}_{\alpha\beta} = \frac{\widehat{g}^{(0)}_{\alpha\beta}}{g^{(0)}}\ , \qquad h^{(0)}_{\beta} = \frac{1}{\sqrt{g^{(0)}}} \sum_{\alpha=1}^{2} \frac{\partial}{\partial x_\alpha} \frac{\widehat{g}^{(0)}_{\alpha\beta}}{\sqrt{g^{(0)}}}\ . \tag{3.4}$$

We "rectify" Γ_0 near the origin by the transformation $y = Z(x)$:

$$y_1 = x_1, \qquad y_2 = x_2, \qquad y_3 = x_3 - \phi(x') \quad (|x'| < d)\ . \tag{3.5}$$

In the new coordinates (1.11) takes the form

$$\begin{gathered} -\nu\,\widehat{\nabla}^2\widehat{\vec{v}} + \widehat{\nabla}\widehat{p} = \vec{f}_1\ , \qquad \widehat{\nabla}\cdot\widehat{\vec{v}} = \widehat{g}\ , \\ \nu\,\Pi_0\,\widehat{S}(\widehat{\vec{v}})\,\vec{n}_0\Big|_{y_3=0} = \widehat{\vec{b}}\ , \\ \vec{n}_0\cdot\widehat{T}(\widehat{\vec{v}},\widehat{p})\,\vec{n}_0 - \sigma\,\vec{n}_0\cdot\int_0^t \Delta\,\widehat{\vec{v}}\,d\tau\Big|_{y_3=0} = \widehat{b} + \int_0^t \widehat{B}\,d\tau\ , \end{gathered} \tag{3.6}$$

where

$$\begin{gathered} \widehat{\nabla} = \left(\frac{\partial}{\partial y_1} - \phi_{y_1}\frac{\partial}{\partial y_3},\ \frac{\partial}{\partial y_2} - \phi_2\frac{\partial}{\partial y_3},\ \frac{\partial}{\partial y_3}\right), \qquad \widehat{T}(\vec{v},p) = -p\,I + \nu\,\widehat{S}(\vec{v})\ , \\ \widehat{S}_{ij}(\vec{v}) = \sum_{k=1}^{3}\left(J_{ki}\frac{\partial v_j}{\partial y_k} + J_{kj}\frac{\partial v_i}{\partial y_k}\right), \qquad J_{ki} = \frac{\partial y_k}{\partial x_i} = \delta_{ki} - \delta_{k3}(1-\delta_{3i})\,\phi_{y_i}(y') \end{gathered}$$

and

$$\begin{gathered} \vec{f}_1 = \widehat{\vec{f}} - \sum_{k=1}^{6}\int_\Omega \vec{v}\cdot\vec{\varphi}_k(z)\,dz\,\widehat{\vec{\varphi}}_k(y)\ , \\ \widehat{\vec{v}}(y,t) = \vec{v}(Z^{-1}(y),t)\,, \qquad \widehat{p}(y,t) = p(Z^{-1}(y),t)\,, \qquad \text{etc.} \end{gathered}$$

Let $\zeta_r(y)$ be a smooth cut-off function equal to 1 for $|y| < r/2$ and to zero for $|y| > r$. It follows from (3.6) that $\vec{u} = \zeta_r\widehat{\vec{v}}$ and $q = \zeta_r\widehat{p}$ satisfy the equations

$$\begin{aligned} & -\nu\,\nabla^2\vec{u} + \nabla q = \zeta_r\,\vec{f}_1 + \vec{f}' + \vec{m}_1(\vec{u},q)\ , \\ & \nabla\cdot\vec{u} = \widehat{g}\,\zeta_r + g' + m_2(\vec{u})\ , \\ & \nu\left(\frac{\partial u_3}{\partial y_j} + \frac{\partial u_j}{\partial y_3}\right)\Big|_{y_3=0} = -(\zeta_r\,\widehat{b}_j + b'_j) + m_{3j}(\vec{u})\,, \qquad j = 1,2\ , \\ & T_{33}(\vec{u},q) + \sigma\int_0^t \Delta' u_3(y',\tau)\,d\tau\Big|_{y_3=0} \\ & \qquad = \widehat{b}\,\zeta_r + b' + \int_0^t (\widehat{B}\,\zeta_r + B')\,d\tau + m_4(\vec{u}) + \int_0^t m_5(\vec{u})\,d\tau\ , \end{aligned} \tag{3.7}$$

where

$$\begin{aligned}
&\vec{f}' = -2\,\nu\,\widehat{\nabla}\vec{\widehat{v}}\,\widehat{\nabla}\zeta_r - \nu\,\vec{\widehat{v}}\,\widehat{\nabla}^2\zeta_r + \widehat{p}\,\widehat{\nabla}\zeta_r\ ,\\
&g' = \vec{\widehat{v}}\,\widehat{\nabla}\zeta_r\ ,\\
&b_j' = \nu\Big(\frac{\partial}{\partial y_j} - \phi_{y_j}\frac{\partial}{\partial y_3}\Big)\zeta_r(y')\,(\vec{\widehat{v}}\cdot\vec{n}_0) + \nu(\widehat{\nabla}\zeta_r\cdot\vec{n}_0)\,\widehat{v}_j\\
&\qquad - 2\,\nu\,n_{0j}(\widehat{\nabla}\zeta_r\cdot\vec{n}_0)\,(\vec{\widehat{v}}\cdot\vec{n}_0)\,,\qquad j=1,2\ ,\\
&b' = 2\nu(\widehat{\nabla}\zeta_r\cdot\vec{n}_0)\,(\vec{\widehat{v}}\cdot\vec{n}_0)\ ,\\
&B' = -\sigma\vec{n}_0\cdot\Big(\Delta\,(\zeta_r\,\vec{\widehat{v}}) - \zeta_r\,\Delta\,\vec{\widehat{v}}\Big)
\end{aligned}\tag{3.8}$$

and

$$\begin{aligned}
&\vec{m}_1(\vec{u},q) = -\nu(\nabla^2 - \widehat{\nabla}^2)\,\vec{u} + (\nabla - \widehat{\nabla})\,q\ ,\\
&m_2(\vec{u}) = (\nabla - \widehat{\nabla})\cdot\vec{u}\ ,\\
&m_{3j}(\vec{u}) = \nu\Big(S_{j3}(\vec{u}) + (\widehat{S}(\vec{u})\,\vec{n}_0)_j\Big) - \nu\,n_{0j}(\vec{n}_0\cdot\widehat{S}(\vec{u})\,\vec{n}_0)\Big|_{y_3=0}\ ,\\
&m_4(\vec{u}) = 2\,\nu\Big(S_{33}(\vec{u}) - \vec{n}_0\cdot\widehat{S}(\vec{u})\,\vec{n}_0\Big)\Big|_{y_3=0}\ ,\\
&m_5(\vec{u}) = \sigma\Big(\Delta' u_3 + \vec{n}_0\cdot\Delta\,\vec{u}\Big)\Big|_{y_3=0}\ .
\end{aligned}$$

All the functions in (3.7) can be extended by zero into the domain $|y| > r$, $y_3 > 0$, and (3.7) can be considered as a boundary value problem for $(\vec{u}, q)$ in $\mathbb{R}^3_+$. Since $\phi_{y_j}(y')\Big|_{y'=0} = 0$, the leading coefficients in $\vec{m_1}$, m_2, m_{3j}, m_4, m_5 vanish at the origin, and for these expressions the inequality

$$\begin{aligned}
&[\vec{m}_1]^{(\ell)}_{\mathbb{R}^3_+} + [m_2]^{(\ell+1)}_{\mathbb{R}^3_+} + \sum_{j=1}^{2}[m_{3j}]^{(\ell+1)}_{\mathbb{R}^2} + [m_4]^{(\ell+1)}_{\mathbb{R}^2} + [m_5]^{(\ell)}_{\mathbb{R}^2}\\
&\qquad \le C r\Big([\vec{u}]^{(\ell+2)}_{\mathbb{R}^3_+} + [q]^{(\ell+1)}_{\mathbb{R}^3_+}\Big)\\
&\qquad\quad + C_2(r)\Big(\sum_{|j|<\ell+2}\sup_{\mathbb{R}^3_+}|D^j\vec{u}(y,t)| + \sum_{|j|<\ell+1}\sup_{\mathbb{R}^3_+}|D^j q(y,t)|\Big)
\end{aligned}$$

holds, whereas functions (3.8) satisfy the inequality

$$\begin{aligned}
&[\vec{f}']^{(\ell)}_{\mathbb{R}^3_+} + [g']^{(\ell+1)}_{\mathbb{R}^3_+} + \sum_{j=1}^{2}[b_j']^{(\ell+1)}_{\mathbb{R}^2} + [b']^{(\ell+1)}_{\mathbb{R}^2} + [B']^{(\ell)}_{\mathbb{R}^2}\\
&\qquad\qquad \le C_3(r)\Big(\sum_{|j|<\ell+2}\sup_{\Omega}|D^j\vec{v}| + \sum_{|j|<\ell+1}\sup_{\Omega}|D^j p|\Big)\ .
\end{aligned}$$

Now, we can apply estimate (2.26) to $(\vec{u}, q)$. Choosing r small enough, we obtain

$$\begin{aligned}
&\sup_{\tau<t}[\vec{u}(\cdot,\tau)]^{(\ell+2)}_{\mathbb{R}^3_+} + \sup_{\tau<t}[q(\cdot,\tau)]^{(\ell+1)}_{\mathbb{R}^3_+} \\
&\quad \le C\Bigg(\sup_{\tau<t}[\zeta_r\,\vec{f}(\cdot,\tau)]^{(\ell)}_{\mathbb{R}^3_+} + \sup_{\tau<t}[\zeta_r\,g(\cdot,\tau)]^{(\ell+1)}_{\mathbb{R}^3_+} \\
&\qquad + \sum_{j=1}^{2}\sup_{\tau<t}[\zeta_r\,b_j(\cdot,\tau)]^{(\ell+1)}_{\mathbb{R}^3_+} + \sup_{\tau<t}[\zeta_r\,b(\cdot,\tau)]^{(\ell+1)}_{\mathbb{R}^3_+} + \sup_{\tau<t}[\zeta_r\,B(\cdot,\tau)]^{(\ell)}_{\mathbb{R}^3_+} \\
&\qquad + \sum_{|j|<\ell+2}\sup_{\tau<t}\sup_{\Omega}|D^j\vec{v}(x,\tau)| + \sum_{|j|<\ell+1}\sup_{\tau<t}\sup_{\Omega}|D^j p(x,t)|\Bigg)\,.
\end{aligned}$$

As usual in Schauder's method, the combination of this inequality and interior estimates gives the estimate of $(\vec{v}, p)$ in the whole domain Ω:

$$\begin{aligned}
&\sup_{\tau<t}|\vec{v}(\cdot,\tau)|_{C^{\ell+2}(\Omega)} + \sup_{\tau<t}|p(\cdot,\tau)|_{C^{\ell+1}(\Omega)} \\
&\le C\Bigg(N(t) + \sum_{|j|<\ell+2}\sup_{\tau<t}\sup_{\Omega}|D^j\vec{v}(\cdot,\tau)| + \sum_{|j|<\ell+1}\sup_{\tau<t}\sup_{\Omega}|D^j p(\cdot,\tau)|\Bigg)
\end{aligned} \tag{3.9}$$

(N is the sum of norms on the right-hand side of (3.1)). The last term in (3.9) may be evaluated by expressing $D^j p$ in terms of $\vec{v}$ from the equation and (for $j = 0$) from the boundary conditions. Finally, an application of the interpolation inequalities

$$\sup_{\Omega}|D^j\vec{v}(x,\tau)| \le \varepsilon_1\,|\vec{v}(\cdot,\tau)|_{C^{\ell+2}(\Omega)} + C(\varepsilon_1)\,\|\vec{v}(\cdot,\tau)\|_{L_2(\Omega)}$$

leads to

$$\begin{aligned}
&\sup_{\tau<t}|\vec{v}(\cdot,\tau)|_{C^{\ell+2}(\Omega)} + \sup_{\tau<t}|p(\cdot,\tau)|_{C^{\ell+1}(\Omega)} \\
&\qquad\qquad \le C\Big(N(t) + \sup_{\tau<t}\|\vec{v}(\cdot,\tau)\|_{L_2(\Omega)}\Big)\,.
\end{aligned} \tag{3.10}$$

It remains to evaluate the L_2-norm of $\vec{v}$. Multiplying equation (1.11_1) by $\vec{v}$ and integrating over Ω, we obtain after integration by parts:

$$\begin{aligned}
&\frac{\nu}{2}\,\|S(\vec{v}(\cdot,\tau))\|^2_{L_2(\Omega)} + \sum_{k=1}^{6}|(\vec{v},\vec{\varphi}_k)|^2 = \int_{\Omega} pg\,dx \\
&\quad + \int_{\Omega}(\vec{f} - \nu\,\nabla g)\cdot\vec{v}\,dx + \int_{\Gamma_0}(\vec{b}_0\cdot\Pi_0\,\vec{v} + b\vec{v}\cdot\vec{n}_0)\,dS \\
&\quad + \sigma\int_0^{\tau}\int_{\Gamma_0}(\vec{n}_0\cdot\Delta\vec{v}(x,\lambda) + B(x',\lambda))\,(\vec{v}(x,\tau)\cdot\vec{n}_0(x))\,dS_x\,d\lambda\,.
\end{aligned} \tag{3.11}$$

In virtue of Korn's inequality (in the form presented in [6]), the left-hand side is greater than $C\|\vec{v}(\cdot,\tau)\|^2_{W^1_2(\Omega)}$, so (3.10) and (3.11) imply

$$\sup_{\tau<t} |\vec{v}(\cdot,\tau)|_{C^{\ell+2}(\Omega)} + \sup_{\tau<t} |p(\cdot,\tau)|_{C^{\ell+1}(\Omega)} \leq C\Big(N(t) + \int_0^t |\vec{v}(\cdot,\tau)|_{C^{\ell+2}(\Omega)}\, d\tau\Big) . \tag{3.12}$$

Using the Gronwall lemma, we obtain (3.1). The theorem is proved. □

3.2. Existence theorem

In this section we prove the solvability of problem (1.11) and complete the proof of the following theorem.

Theorem 4. *Assume that* $\Gamma_0 \in C^{\ell+3}$.
For arbitrary $\vec{f} \in C(0,T;C^\ell(\Omega))$, $g \in C(0,T;C^{\ell+1}(\Omega))$, $\vec{b} \in C(0,T;C^{\ell+1}(\Gamma_0))$, $b \in C(0,T;C^{\ell+1}(\Gamma_0))$, $B \in C(0,T;C^\ell(\Gamma_0))$ *problem* (1.11) *has a unique solution* $\vec{v} \in C(0,T;C^{\ell+2}(\Omega))$, $p \in C(0,T;C^{\ell+1}(\Omega))$, *and this solution satisfies inequality* (3.1).

Most probably, this theorem holds under weaker restriction on the boundary: $\Gamma_0 \in C^{\ell+2}$. However $C^{\ell+3}$-smoothness of Γ_0 is required in the proof of the solvability of the nonlinear problem (1.8).

We shall make use of the solvability of the stationary problem

$$\begin{aligned} &-\nu\, \nabla^2 \vec{w} + \nabla s + \sum_{k=1}^{6} (\vec{w},\vec{\varphi}_k)\, \vec{\varphi}_k(x) = \vec{f}(x)\ , \\ &\nabla\cdot\vec{w} = g(x) \qquad (x\in\Omega)\ , \\ &T(\vec{v},s)\,\vec{n}_0\Big|_{\Gamma_0} = \vec{h}\ . \end{aligned} \tag{3.13}$$

The following proposition may be considered as evident.

Proposition 2. *If* $\Gamma_0 \in C^{\ell+2}$, *then for arbitrary* $\vec{f} \in C^\ell(\Omega)$, $g \in C^{\ell+1}(\Omega)$, $\vec{h} \in C^{\ell+1}(\Gamma_0)$ *problem* (3.13) *has a unique solution* $\vec{w} \in C^{\ell+2}(\Omega)$, $s \in C^{\ell+1}(\Omega)$, *and*

$$|\vec{w}|_{C^{\ell+2}(\Omega)} + |s|_{C^{\ell+1}(\Omega)} \leq C\Big(|\vec{f}|_{C^\ell(\Omega)} + |g|_{C^{\ell+1}(\Omega)} + |\vec{h}|_{C^{\ell+1}(\Gamma_0)}\Big) .$$

This proposition makes it possible to restrict oneself with the consideration of problem (1.11) in the case $\vec{f}=0$, $g=0$, $\vec{b}=0$. Moreover, it can be assumed that the function $B(x,t)$ is smooth both in x and in t.

Under these assumptions, we study the "penalized" problem

$$
\begin{aligned}
&-\nu\,\nabla^2\vec{v}_\varepsilon+\nabla p_\varepsilon+\sum_{k=1}^{6}(\vec{v}_\varepsilon,\vec{\varphi}_k)\,\vec{\varphi}_k=0\ ,\\
&\nabla\cdot\vec{v}_\varepsilon=0\ ,\\
&\Pi_0\,S(\vec{v}_\varepsilon)\,\vec{n}_0\Big|_{\Gamma_0}=0\ ,\\
&\vec{n}_0\cdot T(\vec{v}_\varepsilon,p_\varepsilon)\,\vec{n}_0-\varepsilon\int_0^t\Delta(\vec{n}_0\cdot T(\vec{v}_\varepsilon,p_\varepsilon)\,\vec{n}_0)\,d\tau\\
&\qquad\qquad\qquad-\sigma\int_0^t\vec{n}_0\cdot\Delta\vec{v}_\varepsilon\,d\tau\Big|_{\Gamma_0}=\int_0^t B\,d\tau\ ,
\end{aligned}
\tag{3.14}
$$

which can be also regarded as the problem of the determination of the solution $(\vec{v}_\varepsilon,p_\varepsilon)$ of (3.13) with $\vec{h}=h_\varepsilon\,\vec{n}$, where h_ε solves the Cauchy problem on Γ_0

$$
\frac{\partial h_\varepsilon}{\partial t}-\varepsilon\,\Delta h_\varepsilon-\sigma\,A\,h_\varepsilon=B\,,\qquad h_\varepsilon\Big|_{t=0}=0\ ,
\tag{3.15}
$$

and $Ah_\varepsilon=\vec{n}_0\cdot\Delta\vec{v}_\varepsilon$. There holds the following

Proposition 3. *For arbitrary $B\in C^{\ell,\ell/2}(G_T)$, $G_T=\Gamma_0\times(0,T)$, problem (3.15) has a unique solution $h_\varepsilon\in C^{2+\ell,\ell/2+1}(G_T)$, and*

$$
|h_\varepsilon|_{C^{2+\ell,\ell/2+1}(G_T)}\le C(\varepsilon)\,|B|_{C^{\ell,\ell/2}(G_T)}\ .
\tag{3.16}
$$

Proof. In the case $\sigma=0$ the proposition is well known. Consider the operator A. We have

$$
\begin{aligned}
&|Ah|_{C^\ell(\Gamma_0)}\le C|\vec{v}|_{C^{2+\ell}(\Omega)}\le C|h|_{C^{1+\ell}(\Gamma_0)}\ ,\\
&\sup_{\Gamma_0}|Ah|\le C|\vec{v}|_{C^2(\Omega)}\le C|h|_{C^{1+\kappa}(\Gamma_0)}\ ,\qquad \kappa>0\ .
\end{aligned}
\tag{3.17}
$$

Applying the last inequality to derivatives and finite differences of h with respect to the parameter t, we obtain

$$
|Ah|_{C^{\ell/2}(0,\tau;C(\Gamma_0))}\le C|h|_{C^{\ell/2}(0,\tau,C^{1+\kappa}(\Gamma_0))}\ ,\qquad \tau<t\ .
\tag{3.18}
$$

From (3.17) and (3.18) it follows that

$$
|Ah|_{C^{\ell,\ell/2}(G_\tau)}\le C\Big(\sup_{t<\tau}|h(\cdot,t)|_{C^{1+\ell}(\Gamma_0)}+|h|_{C^{\ell/2}(0,\tau;C^{1+\kappa}(\Gamma_0))}\Big)
\tag{3.19}
$$

with the constant independent of $\tau<t$.

In virtue of standard interpolation inequalities for Hölder spaces, the right-hand side of (3.19) does not exceed

$$
\begin{aligned}
&\varepsilon_1\Big(\sup_{t<\tau}|h(\cdot,t)|_{C^{2+\ell}(\Gamma_0)}+|h|_{C^{\ell/2}(0,\tau;C^2(\Gamma_0))}\Big)\\
&\qquad\qquad+C_1(\varepsilon)\Big(\sup_{t<\tau}\sup_{\Gamma_0}|h(x,t)|+\sup_{\Gamma_0}|h|_{C^{\ell/2}(0,\tau)}\Big)\ .
\end{aligned}
$$

If $\ell<1$, then

$$
\sup_{\Gamma_0}|h|_{C^{\ell/2}(0,\tau)}\le\varepsilon_2\sup_{\Gamma_0}\sup_{t<\tau}|h_t'(x,t)|+C_2(\varepsilon_2)\sup_{t<\tau}\sup_{\Gamma_0}|h(x,t)|\ .
$$

Choosing ε_1 and then ε_2 sufficiently small and making use of the estimate

$$\sup_{t<\tau}\sup_{\Gamma_0}|h(x,t)| \le \int_0^\tau \sup_{\Gamma_0}|h'(x,t)|\,dt\ ,$$

we arrive at

$$|Ah|_{C^{\ell,\ell/2}(G_\tau)} \le \varepsilon_3\,|h|_{C^{2+\ell,1+\ell/2}(G_\tau)} + C_3(\varepsilon_3)\int_0^\tau |h|_{C^{2+\ell,1+\ell/2}(G_t)}\,dt\,, \qquad l\in(0,1)\ .$$

This estimate makes it possible to prove the solvability of problem (3.15) by the method of successive approximations: we put $h_0 = 0$ and define h_{m+1}, $m = 0, 1, \dots$, as solutions of the problems

$$\frac{\partial h_{m+1}}{\partial t} - \varepsilon\,\Delta h_{m+1} = -A\,h_m + B\,, \qquad h_{m+1}\Big|_{t=0} = 0\ .$$

Since $h_{m+1} - h_m = H_{m+1}$ is a solution of the problem

$$\frac{\partial H_{m+1}}{\partial t} - \varepsilon\,\Delta H_{m+1} = -A\,H_m\,, \qquad H_{m+1}\Big|_{t=0} = 0\ ,$$

we have

$$\begin{aligned}|H_{m+1}|_{C^{2+\ell,1+\ell/2}(G_\tau)} &\le C|AH_m|_{C^{\ell,\ell/2}(G_\tau)}\\ &\le C\Big(\varepsilon_3\,|H_m|_{C^{2+\ell,1+\ell/2}(G_\tau)} + C_3(\varepsilon_3)\int_0^\tau |H_m|_{C^{2+\ell,1+\ell/2}(Q_t)}\,dt\Big)\ .\end{aligned}$$

Hence, the sum

$$\Sigma_M(\tau) = \sum_{m=0}^{M} |H_m|_{C^{2+\ell,1+\ell/2}(G_\tau)}$$

satisfies the estimate

$$\Sigma_M(\tau) \le |H_1|_{C^{2+\ell,1+\ell/2}(G_\tau)} + C\Big(\varepsilon_3\Sigma_M(\tau) + C_3(\varepsilon_3)\int_0^\tau \Sigma_M(t)\,dt\Big)\ ,$$

and if $C\varepsilon_3 \le \frac{1}{2}$, then

$$\Sigma_M(\tau) \le 2\,|H_1|_{C^{2+\ell,1+\ell/2}(G_\tau)} + 2\,C\,C_3\int_0^\tau \Sigma_M(t)\,dt\ ,$$

which implies

$$\Sigma_M(\tau) \le 2\,e^{2CC_3\tau}\,|H_1|_{C^{2+\ell,1+\ell/2}(G_\tau)}\ .$$

This shows that the sequence $\{h_m\}$ is convergent in the interval $(0,T)$ to the solution h_ε of (3.15), and

$$|h_\varepsilon|_{C^{2+\ell,1+\ell/2}(G_T)} \le 2\,e^{2CC_3T}\,|H_1|_{C^{2+\ell,1+\ell/2}(G_T)} \le C(T)\,|B|_{C^{\ell,\ell/2}(G_T)}\ ,$$

so the proposition is proved for $\ell < 1$.

In the case $\ell > 1$, we can establish the regularity of the solution $h_\varepsilon \in C^{2+\lambda,1+\lambda/2}(G_T)$, $\lambda < 1$, by bootstrap arguments, since, as it is seen from (3.19), Ah_ε satisfies the estimate

$$|Ah_\varepsilon|_{C^{\ell_1,\ell_1/2}(G_T)} \le C\,|h_\varepsilon|_{C^{\ell_1+1+\kappa,(\ell_1+1+\kappa)/2}(G_T)}$$

for arbitrary $\kappa \in (0,1)$. Setting $\ell_1 + 1 + \kappa = 2 + \lambda$ we can conclude that $Ah_\varepsilon \in C^{1+\lambda-\kappa, \frac{1+\lambda-\kappa}{2}}(G_T)$, hence,

$$h_\varepsilon \in C^{2+\ell_2, 1+\ell_2/2}(G_T) \quad \text{with } \ell_2 = \min(\ell, 1+\lambda-\kappa) \quad \text{etc.}$$

Moreover, h_ε satisfies the estimate

$$|h_\varepsilon|_{C^{2+\ell,1+\ell/2}(G_T)} \le C\Big(|B|_{C^{\ell,\ell/2}(G_T)} + |Ah_\varepsilon|_{C^{\ell,\ell/2}(G_T)}\Big)\,.$$

Using (3.19) and interpolation inequalities, we obtain (3.16). The proof is completed. □

It follows from Propositions 2 and 3 that problem (3.14) has a unique solution $\vec{v}_\varepsilon \in C(0,T,C^{3+\ell}(\Omega))$, $p_\varepsilon \in C(0,T,C^{2+\ell}(\Omega))$. The next proposition concerns uniform estimates of this solution (with respect to ε).

Proposition 4. *The solution of problem (3.14) satisfies the inequality*

$$\sup_{t<T} |\vec{v}_\varepsilon(\cdot,t)|_{C^{\ell+2}(\Omega)} + \sup_{t<T} |p_\varepsilon(\cdot,t)|_{C^{1+\ell}(\Omega)} \le C \sup_{t<T} |B(\cdot,t)|_{C^\ell(\Gamma_0)}\,, \tag{3.20}$$

with the constant independent of ε.

Proof. Inequality (3.20) is proved by the procedure presented in 3.1. After introduction of variables (3.5) and of functions $\vec{u} = \zeta_r \vec{v}_\varepsilon$, $q = \zeta_r p_\varepsilon$, we arrive at the problem (3.7) with $\widehat{\vec{f}} = 0$, $\widehat{g} = 0$, $\widehat{b}_j = 0$ and with the last condition replaced by

$$\begin{aligned} T_{33}(\vec{u},q) - \varepsilon \int_0^t \Delta' T_{33}(\vec{u},q)\,d\tau + \sigma \int_0^t \Delta' u_3\,d\tau\Big|_{y_3=0} \\ = b' + \int_0^t (\zeta_r \widehat{B} + B' + \varepsilon B'')\,d\tau + m_4(\vec{u}) \\ + \int_0^t \Big[m_5(\vec{u}) + \varepsilon \sum_{\alpha=1}^2 \frac{\partial}{\partial y_\alpha} m_{6\alpha}(\vec{u},q) + \varepsilon\, m_7(\vec{u},q)\Big]\,d\tau \end{aligned}$$

where

$$\begin{aligned} B'' &= -\Delta\,\vec{n}_0 \cdot \widehat{T}(\zeta_r \vec{v}_\varepsilon, \zeta_r p_\varepsilon)\,\vec{n}_0 + \zeta_r\,\Delta\,\vec{n}_0 \cdot \widehat{T}(\vec{v}_\varepsilon, p_\varepsilon)\,\vec{n}_0\Big|_{y_3=0}\,, \\ m_{6\alpha}(\vec{u},q) &= \sum_{\beta=1}^2 \Big[h_{\alpha\beta}(y')\,\frac{\partial}{\partial y_\beta}\Big(\vec{n}_0(y') \cdot \widehat{T}(\vec{u},q)\,\vec{n}_0(y')\Big) \\ &\quad - h_{\alpha\beta}(0)\,\frac{\partial}{\partial y_\beta}\Big(\vec{n}_0(0) \cdot \widehat{T}(u,q)\,n_0(0)\Big)\Big] \\ &\quad + \frac{\partial}{\partial y_\alpha}\Big[\widehat{S}_{33}(\vec{u}) - S_{33}(\vec{u})\Big]\Big|_{y_3=0}\,, \\ m_7(\vec{u},q) &= -\sum_{\alpha,\beta=1}^2 \frac{\partial h^{(0)}_{\alpha\beta}(y')}{\partial y_\alpha}\,\frac{\partial}{\partial y_\beta}\vec{n}_0 \cdot \widehat{T}(\vec{u},q)\,\vec{n}_0 \\ &\quad + \sum_{\alpha=1}^2 h_\alpha(y')\,\frac{\partial}{\partial y_\alpha}\vec{n}_0 \cdot \widehat{T}(\vec{u},q)\,\vec{n}_0\Big|_{y_3=0}\,. \end{aligned}$$

These expressions satisfy the inequalities

$$[B'']^{(\ell)}_{\mathbb{R}^2} \le C(r)\Big(|\vec v_\varepsilon|_{C^{\ell+2}(\Omega)} + |p_\varepsilon|_{C^{1+\alpha}(\Omega)}\Big)\,,$$
$$[m_{6\alpha}(\vec u, q)]^{(\ell)}_{\mathbb{R}^2} \le C\,r\Big([\vec u]^{(\ell+2)}_{\mathbb{R}^3_+} + [q]^{(\ell+1)}_{\mathbb{R}^3_+}\Big)$$
$$+\,C\Big(\sum_{|j|<\ell+2}\sup_{\mathbb{R}^3_+}|D^j_y\vec u(y,t)| + \sum_{|j|<\ell+1}\sup_{\mathbb{R}^3_+}|D^j_y q(y,t)|\Big)\,,$$
$$[m_7(\vec u, q)]^{(\ell)}_{\mathbb{R}^2} \le C\Big(|\vec u|_{C^{\ell+2}(\mathbb{R}^3_+)} + |q|_{C^{\ell+1}(\mathbb{R}^3_+)}\Big)\,.$$

Hence, using (2.26), we obtain in the case of small r and ε inequality (3.9) with an additional term

$$C(r)\,\varepsilon\,\Big(\sup_{\tau<t}|\vec v_\varepsilon(\cdot,\tau)|_{C^{\ell+2}(\Omega)} + \sup_{\tau<t}|p_\varepsilon(\cdot,\tau)|_{C^{\ell+1}(\Omega)}\Big)$$

on the right-hand side, and if ε is so small that $C(r)\varepsilon < \frac{1}{2}$, we arrive at (3.10).

Finally, in (3.11) we shall have an additional term

$$\varepsilon\int_0^T\!\!\int_{\Gamma_0}\Delta(\vec n\cdot T\vec n)\,(\vec v_\varepsilon\cdot\vec n)\,dS\,d\lambda = -\varepsilon\int_0^T\!\!\int_{\Gamma_0}\Big\langle\nabla_\Gamma(\vec n\cdot T\vec n),\,\nabla_\Gamma(\vec v_\varepsilon\cdot\vec n)\Big\rangle\,dS\,d\lambda\,,$$

where

$$\langle\nabla_\Gamma f,\,\nabla_\Gamma h\rangle = \sum_{\alpha,\beta=1}^{2}\frac{\widehat g_{\alpha\beta}}{g}\,\frac{\partial f}{\partial\eta_\beta}\,\frac{\partial h}{\partial\eta_\alpha}$$

in local coordinates. It is evaluated by

$$\varepsilon\sup_{\lambda<\tau}\sup_{\Gamma_0}|\nabla_\Gamma(\vec v_\varepsilon\cdot\vec n)|\int_0^T\!\!\int_{\Gamma_0}|\nabla_\Gamma(\vec n\cdot T\vec n)|\,dS\,d\lambda\,,$$

so in (3.12) we shall have an additional term $C\varepsilon\sup_{\tau<t}|\vec v(\cdot,\tau)|_{C^{\ell+2}(\Omega)}$ which causes no problems. The proposition is proved. □

It remains to pass to the limit as $\varepsilon\to 0$. For this it is useful to evaluate the derivatives $\vec v_{\varepsilon t}\equiv\dot{\vec v}_\varepsilon$ and $p_{\varepsilon t} = \dot p_\varepsilon$. They are a solution to the problem

$$-\nu\,\nabla^2\dot{\vec v}_\varepsilon + \nabla\dot p_\varepsilon + \sum_{k=1}^{6}(\dot{\vec v}_\varepsilon,\vec\varphi_k)\,\vec\varphi_k = 0\,,$$
$$\nabla\cdot\dot{\vec v}_\varepsilon = 0\,,$$
$$\Pi_0 S(\dot{\vec v}_\varepsilon)\,\vec n_0\Big|_{\Gamma_0} = 0\,,$$
$$\vec n_0\cdot T(\dot{\vec v}_\varepsilon,\dot p_\varepsilon)\,\vec n_0 - \varepsilon\int_0^t\Delta\Big(\vec n_0\cdot T(\dot{\vec v}_\varepsilon,\dot p_\varepsilon)\,\vec n_0\Big)\,d\tau - \sigma\int_0^t\vec n_0\cdot\Delta\dot{\vec v}_\varepsilon\,d\tau\Big|_{\Gamma_0}$$
$$= B(x,0) + \int_0^t B_\tau(x,\tau)\,d\tau\,,$$

similar to (3.14), so they can be evaluated by the norms of B_τ and $B(x,0)$ uniformly with respect to ε by the method presented above.

Hence, there exists a sequence $\{\vec{v}_{\varepsilon_k}, p_{\varepsilon_k}\}$ which is convergent in $C^{2+\ell'}(\Omega) \times C^{1+\ell'}(\Omega)$, $\ell' < \ell$, uniformly with respect to t. If we pass to the limit in (3.14_1)–(3.14_3) and in the identity

$$\int_{\Gamma_0} \Big\{ \vec{n}_0 \cdot T(\vec{v}_{\varepsilon_k}, p_{\varepsilon_k})\, \vec{n}_0 - \sigma \int_0^t \vec{n}_0 \cdot \Delta \vec{v}_{\varepsilon_k}\, d\tau - \int_0^t B\, d\tau \Big\} \eta\, dS$$
$$- \varepsilon \int_{\Gamma_0} \Big\langle \int_0^t \nabla_\Gamma \Big(\vec{n}_0 \cdot T(\vec{v}_{\varepsilon_k}, p_{\varepsilon_k})\, \vec{n}_0 \Big),\, \nabla_\Gamma \eta \Big\rangle\, dS = 0\,,$$

where η is an arbitrary continuously differentiable function on Γ_0, we see that the limiting functions $\vec{v} \in C(0,T; C^{\ell+2}(\Omega))$ and $p \in C(0,T; C^{\ell+1}(\Omega))$ solve (1.11). This completes the proof of Theorem 4.

4. Proof of Theorem 1

To prove the solvability of problem (1.8), we retain in the left-hand side of the equations and boundary conditions only principal linear terms and write (1.8) in the form (1.12) with

$$\begin{aligned}
\vec{\ell}_1(\vec{u}, q) &= \nu(\nabla_u^2 - \nabla^2)\, \vec{u} + (\nabla - \nabla_u)\, q \\
&\quad + \sum_{k=1}^{6} \int_\Omega \vec{u}(\eta, t) \Big[\vec{\varphi}_k(\eta)\, \vec{\varphi}_k(\xi) - \vec{\varphi}_k(X(\eta,t))\, \vec{\varphi}_k(X(\xi,t)) \Big]\, d\eta\,, \\
\ell_2(\vec{u}) &= (\nabla - \nabla_u) \cdot \vec{u}\,, \\
\vec{\ell}_3(\vec{u}) &= \nu\, \Pi_0 \Big(\Pi_0 S(\vec{u})\, \vec{n}_0 - \Pi\, S_u(\vec{u})\, \vec{n} \Big)\,, \\
\ell_4(\vec{u}) &= \vec{n}_0 \cdot \Big(T(\vec{u}, q)\, \vec{n}_0 - T_u(\vec{u}, q)\, \vec{n} \Big) \\
&= -q\, \vec{n}_0 \cdot (\vec{n}_0 - \vec{n}) + 2\nu\, \vec{n}_0 \cdot \Big(S(\vec{u})\, \vec{n}_0 - S_u(\vec{u})\, \vec{n} \Big)\,, \\
\ell_5(\vec{u}) &= \sigma\, \vec{n}_0 \cdot \frac{\partial}{\partial t} \Big[\Big(\Delta(t) - \Delta(0) \Big) \int_0^t \vec{u}\, d\tau \Big] \\
&= \sigma\, \vec{n}_0 \cdot \Big[\Big(\Delta(t) - \Delta(0) \Big)\, \vec{u} + \dot{\Delta}(t) \int_0^t \vec{u}\, d\tau \Big]\,,
\end{aligned} \tag{4.1}$$

where $\dot{\Delta}(t)$ is the operator obtained by differenting the coefficients of $\Delta(t)$ with respect to t. We recall that $\nabla_u = \mathcal{A}\nabla$ where $\mathcal{A} = (A_{ij})_{i,j=1,2,3}$ and A_{ij} is the cofactor of

$$a_{ij}(\xi, t) = \frac{\partial X_i}{\partial \xi_j} = \delta_{ij} + \int_0^t \frac{\partial u_i(\xi, \tau)}{\partial \xi_j}\, d\tau\,.$$

Hence, $\nabla_u - \nabla = \mathcal{B}\nabla$, and the elements of the matrix $\mathcal{B} = A - I$ are polynomials of the second order with respect to $\int_0^t \frac{\partial u_i(\xi,\tau)}{\partial \xi_j}\, d\tau$. The differences $\vec{\varphi}_k(\xi) - \vec{\varphi}_k(X(\xi,t))$ either vanish (if $\vec{\varphi}_k = \text{const}$), or they are linear functions of $\int_0^t u_i\, d\tau$. The vectors

$\vec{n}$ and $\vec{n}_0$ are related to each other by $\vec{n} = \frac{\mathcal{A}\vec{n}_0}{|\mathcal{A}\vec{n}_0|}$, hence, the difference

$$\vec{n} - \vec{n}_0 = \frac{\mathcal{B}\vec{n}_0}{|\mathcal{A}\vec{n}_0|} - \vec{n}_0 \frac{|\mathcal{A}\vec{n}_0| - 1}{|\mathcal{A}\vec{n}_0|} = \frac{\mathcal{B}\vec{n}_0}{|\mathcal{A}\vec{n}_0|} - \vec{n}_0 \cdot \frac{2\,\mathcal{B}\vec{n}_0 \cdot \vec{n}_0 + \mathcal{B}\vec{n}_0 \cdot \mathcal{B}\vec{n}_0}{|\mathcal{A}\vec{n}_0|\,(|\mathcal{A}\vec{n}_0| + 1)}$$

is also expressed in terms of $\mathcal{B}$. Finally, the Laplace-Beltrami operator $\Delta(t)$ is given by the formula

$$\Delta(t)\, f = \frac{1}{\sqrt{g}} \sum_{\alpha,\beta=1}^{2} \frac{\partial}{\partial \eta_\alpha} \frac{\widehat{g}_{\alpha\beta}}{\sqrt{g}} \frac{\partial f}{\partial \eta_\beta} = \sum_{\alpha,\beta=1}^{2} h_{\alpha\beta} \frac{\partial^2 f}{\partial \eta_\alpha\, \partial \eta_\beta} + \sum_{\beta=1}^{2} h_{\alpha\beta} \frac{\partial f}{\partial \eta_\beta}\,, \tag{4.2}$$

where (η_1, η_2) are local coordinates on some part $\Gamma' \subset \Gamma_0$. We choose them to be cartesian coordinates on a tangent plane to Γ_0 at a certain point ξ_0, so that (η_1, η_2, η_3) are cartesian coordinates in $\mathbb{R}^3$ with the origin in ξ_0 and with η_3-axis directed along $-\vec{n}(\xi_0)$. Let

$$\eta_3 = \phi(\eta')\,, \qquad \eta' = (\eta_1, \eta_2), \quad \phi \in C^{\ell+3}(K'_d), \quad K'_d = \{|\eta'| < d\}$$

be the equation of Γ'. Then the radius-vector of the point $\eta \in \Gamma'_t \subset \Gamma_t$, where Γ'_t is the image of Γ' under transformation (1.6), is defined by

$$\vec{r} = \left(\eta_1 + \int_0^t \widehat{u}_1(\eta, \tau)\, d\tau,\ \eta_2 + \int_0^t \widehat{u}_2(\eta, \tau)\, d\tau,\ \phi(\eta') + \int_0^t \widehat{u}_3(\eta, \tau)\, d\tau \right),$$

where $\widehat{u}_i$ are the components of $\vec{u}$ in a local system (η_1, η_2, η_3), i.e., $\widehat{u}_i = \sum_{k=1}^{3} C_{ik} u_k$ and $(C_{ik})_{i,k=1,2,3}$ is a certain orthogonal matrix. The coefficients of the Laplace-Beltrami operator (4.2) are given by the formulas

$$\begin{aligned}
&g = (\det g_{\alpha\beta})_{\alpha,\beta=1,2}\,,\\
&g_{\alpha\beta} = \frac{\partial \vec{r}}{\partial \eta_\alpha} \cdot \frac{\partial \vec{r}}{\partial \eta_\beta} = \delta_{\alpha\beta} + \phi_{\eta_\alpha}\phi_{\eta_\beta} + \phi_{\eta_\alpha} V_{3\beta} + \phi_{\eta_\beta} V_{3\alpha}\\
&\qquad + V_{\alpha\beta} + V_{\beta\alpha} + \sum_{k=1}^{3} V_{k\alpha} V_{k\beta}\,,\\
&\widehat{g}_{11} = g_{22}\,, \qquad \widehat{g}_{22} = g_{11}\,, \qquad \widehat{g}_{12} = \widehat{g}_{21} = -g_{12}\,,\\
&V_{k\alpha} = \int_0^t \frac{\partial \omega_k(\eta', \tau)}{\partial \eta_\alpha}\, d\tau\,, \qquad \omega_k(\eta', \tau) = \widehat{u}_k(\eta, \tau)\Big|_{\eta_3 = \varphi(\eta')}\,.
\end{aligned} \tag{4.3}$$

Hence, $\widehat{g}_{\alpha\beta} - \widehat{g}^{(0)}_{\alpha\beta}$ are also polynomials of the second order with respect to $V_{k\alpha} = \int_0^t \frac{\partial}{\partial \eta_\alpha} \widehat{u}_k(\eta', \phi(\eta'), \tau)\, d\tau$.

The coefficients $h_{\alpha\beta}$, h_α are expressed in terms of $\widehat{g}_{\alpha\beta}$ as in (3.4), i.e.,

$$h_{\alpha\beta} = \frac{\widehat{g}_{\alpha\beta}}{g}\,, \qquad h_\beta = \frac{1}{\sqrt{g}} \sum_{\alpha=1}^{2} \frac{\partial}{\partial \eta_\alpha} \frac{\widehat{g}_{\alpha\beta}}{\sqrt{g}}\,. \tag{4.4}$$

Further we have

$$\dot{\Delta}(t)\, f = \sum_{\alpha,\beta=1}^{2} \dot{h}_{\alpha\beta} \frac{\partial^2 f}{\partial \eta_\alpha\, \partial \eta_\beta} + \sum_{\beta=1}^{2} \dot{h}_\beta \frac{\partial f}{\partial \eta_\beta}\,,$$

where the dot denotes differentiation with respect to t. The coefficients $\dot{h}_{\alpha\beta}$ are expressed in terms of $g_{\alpha\beta}$ and of

$$\dot{g}_{\alpha\beta} = \phi_{\eta_\alpha} \frac{\partial \omega_3}{\partial \eta_\beta} + \phi_{\eta_\beta} \frac{\partial \omega_3}{\partial \eta_\alpha} + \frac{\partial \omega_\alpha}{\partial \eta_\beta} + \frac{\partial \omega_\beta}{\partial \eta_\alpha} + \sum_{k=1}^{3} \left(V_{k\alpha} \frac{\partial \omega_k}{\partial \eta_\beta} + V_{k\beta} \frac{\partial \omega_k}{\partial \eta_\beta} \right)$$

and they belong to $C^{\ell+1}(K_d)$. The $\dot{h}_\beta$ depend also on the second derivatives of φ and ω_k, and belong to $C^\ell(K_d)$. If

$$t \sup_{\tau<t} |\vec{u}(\cdot,\tau)|_{C^{2+\ell}(\Omega)} \leq \delta < 1 \tag{4.5}$$

(so that $|g(\xi,t)| > C > 0$), then

$$|\dot{h}_{\alpha\beta}|_{C^{\ell+1}(K_d)} + |\dot{h}_\beta|_{C^\ell(K_d)} \leq C\, |\vec{\omega}|_{C^{\ell+2}(K_d)} \leq C\, |\vec{u}|_{C^{\ell+2}(\Omega)} \, .$$

All these considerations show that the expressions in (4.1) possess the following properties:

Proposition 5. *Let $\vec{u}(\xi,t)$ and $\vec{u}'(\xi,t)$ satisfy the condition* (4.5). *Then*

$$\begin{aligned}
&|\vec{\ell}_1(\vec{u},q) - \vec{\ell}_1(\vec{u}',q')|_{C^\ell(\Omega)} + |\ell_2(\vec{u}) - \ell_2(\vec{u}')|_{C^{\ell+1}(\Omega)} + |\vec{\ell}_3(u) - \vec{\ell}_3(\vec{u}')|_{C^{\ell+1}(\Gamma_0)} \\
&+ |\ell_4(\vec{u},q) - \ell_4(\vec{u}',q')|_{C^{\ell+1}(\Gamma_0)} + |\ell_5(\vec{u}) - \ell_5(\vec{u}')|_{C^\ell(\Gamma_0)} \\
&\leq C\, t \Big(\sup_{\tau<t} \Big| \vec{u}(\cdot,\tau) - \vec{u}'(\cdot,\tau) \Big|_{C^{\ell+2}(\Omega)} + \sup_{\tau<t} \Big| q(\cdot,\tau) - q'(\cdot,\tau) \Big|_{C^{\ell+1}(\Omega)} \Big) \\
&\quad \times \Big(\sup_{\tau<t} |\vec{u}(\cdot,\tau)|_{C^{\ell+2}(\Omega)} + \sup_{\tau<t} |\vec{u}'(\cdot,\tau)|_{C^{\ell+2}(\Omega)} \Big) \\
&\leq 2\, C\, \delta \Big(\sup_{\tau<t} |\vec{u}(\cdot,\tau) - \vec{u}'(\cdot,\tau)|_{C^{\ell+2}(\Omega)} + \sup_{\tau<t} |q(\cdot,\tau) - q'(\cdot,\tau)|_{C^{\ell+1}(\Omega)} \Big) \, .
\end{aligned} \tag{4.6}$$

This proposition is proved by direct calculations based on the above formulas for $\nabla_u - \nabla$, $\vec{n} - \vec{n}_0$, and for the coefficients of $\Delta(t)$. In the paper [2] these computations are carried out in anisotropic Sobolev spaces which take into account regularity properties of $\vec{u}$ with respect to time. The proof of (4.6) requires only part of these calculations. The condition (4.5) should guarantee the positivity of $g = \det(g_{\alpha\beta})$ and of $|\mathcal{A}\vec{n}_0|$.

Now, let us consider the function

$$\vec{n}_0 \cdot \Delta(t)\, \vec{\xi} = H_0(\xi) + L(\vec{u}) \, ,$$

where $H_0(\xi) = \vec{n}_0 \cdot \Delta(0)\vec{\xi}$ is the doubled mean curvature of Γ_0 and

$$L(\vec{u}) = \vec{n}_0 \cdot \int_0^t \dot{\Delta}(\tau)\, \vec{\xi}\, d\tau \, .$$

In the local coordinates, $\vec{\xi} = (\eta_1, \eta_2, \phi(\eta'))$ and

$$\begin{aligned}\vec{n}_0 \cdot \dot{\Delta}(t)\,\vec{\xi} = & -\frac{1}{\sqrt{1+|\nabla\varphi|^2}}\Big(\dot{\Delta}(t)\,\phi(\eta') - \sum_{\gamma=1}^{2}\phi_{\eta_\gamma}\dot{\Delta}(t)\,\eta_\gamma\Big)\\ = & -\sum_{\alpha,\beta=1}^{2}\frac{\dot{h}_{\alpha\beta}\,\phi_{\eta_\alpha\eta_\beta}}{\sqrt{1+|\nabla\varphi|^2}}\,,\end{aligned}$$

hence, the following proposition holds:

Proposition 6. *If $\vec{u}$, $\vec{u}'$ satisfy* (4.5), *then*

$$|L(\vec{u}) - L(\vec{u}')|_{C^{1+\ell}(\Gamma)} \le C\int_0^t |\vec{u}(\cdot,\tau) - \vec{u}'(\cdot,\tau)|_{C^{\ell+2}(\Omega)}\,d\tau\ .$$

Now, we pass to the proof of the solvability of problem (1.8). We use the method of successive approximations and define $(\vec{u}_{m+1}, q_{m+1})$, $m \in \mathbb{N}$, as the solution of the problem

$$\begin{aligned}&-\nu\,\nabla^2\vec{u}_{m+1} + \sum_{k=1}^{6}(\vec{u}_{m+1}, \vec{\varphi}_k)\,\vec{\varphi}_k + \nabla q_{m+1}\\ &\qquad = \vec{\ell}_1(\vec{u}_m, q_m) + \sum_{k=1}^{6} M_k(t)\,\vec{\varphi}_k(X_m(\xi,t))\ ,\\ &\nabla\cdot\vec{u}_{m+1} = \ell_2(\vec{u}_m)\ ,\\ &\Pi_0\,S(\vec{u}_{m+1})\,\vec{n}_0\Big|_{\Gamma_0} = \ell_3(\vec{u}_m)\ ,\\ &\vec{n}_0\cdot T(\vec{u}_{m+1}, q_{m+1})\,\vec{n}_0 - \sigma\,\vec{n}_0\cdot\Delta\int_0^t \vec{u}_{m+1}(\xi,\tau)\,d\tau\Big|_{\Gamma_0}\\ &\qquad = H_0(\xi) + L(\vec{u}_m) + \ell_4(\vec{u}_m, q_m) + \int_0^t \ell_5(\vec{u}_m)\,d\tau\ ,\end{aligned}\tag{4.7}$$

where

$$\vec{X}_m(\xi,t) = \xi + \int_0^t \vec{u}_m(\xi,\tau)\,d\tau\,.$$

As the initial approximation, we take $\vec{u}_0 = 0$, $q_0 = 0$.

We have to show that all approximations are defined on a certain time interval $(0,T)$ and that the sequences $\{\vec{u}_m\}$ and $\{q_m\}$ are convergent.

First of all,

$$|\vec{u}_1(\cdot,t)|_{C^{2+\ell}(\Omega)} + |q_1(\cdot,t)|_{C^{1+\ell}(\Omega)} \le C\Big(|H_0|_{C^{1+\ell}(\Gamma_0)} + \sum_{k=1}^{6}\sup_{\tau<t}|M_k(t)|\Big)\ .$$

Now, let $\vec{w}_{m+1} = \vec{u}_{m+1} - \vec{u}_m$, $s_{m+1} = q_{m+1} - q_m$. It follows from (4.7) that

$$\begin{aligned}
&-\nu\,\nabla^2 \vec{w}_{m+1} + \sum_{k=1}^{6} (\vec{w}_{m+1}, \vec{\varphi}_k)\,\vec{\varphi}_k + \nabla s_{m+1} = \vec{\ell}_1(\vec{u}_m, q_m) - \vec{\ell}_1(\vec{u}_{m-1}, q_{m-1}) \\
&\qquad\qquad + \sum_{k=1}^{6} M_k(t)\Big[\vec{\varphi}_k(X_m(\xi,t)) - \vec{\varphi}_k(X_{m-1}(\xi,t))\Big]\,, \\
&\nabla\cdot\vec{w}_{m+1} = \ell_2(\vec{u}_m) - \ell_2(\vec{u}_{m-1})\,, \\
&\Pi_0\, S(\vec{w}_{m+1})\,\vec{n}_0\Big|_{\Gamma_0} = \vec{\ell}_3(\vec{u}_m) - \vec{\ell}_3(\vec{u}_{m-1})\,, \\
&\vec{n}_0\cdot T(\vec{w}_{m+1}, s_{m+1})\,\vec{n}_0 - \sigma\,\vec{n}_0\cdot\Delta\int_0^t \vec{w}_{m+1}(\xi,\tau)\,d\tau\Big|_{\Gamma_0} \\
&\qquad = L(\vec{u}_m) - L(\vec{u}_{m-1}) + \ell_4(\vec{u}_m, q_m) - \ell_4(\vec{u}_{m-1}, q_{m-1}) \\
&\qquad\quad + \int_0^t \Big(\ell_5(\vec{u}_m) - \ell_5(\vec{u}_{m-1})\Big)\,d\tau\,.
\end{aligned}$$

Assume that $(\vec{u}_j, q_j)$, $j = 1, \ldots, m$, are defined for $t \in (0,T)$ and that $\vec{u}_j$ satisfy condition (4.5) for $t = T$. Then, by virtue of Theorem 3 and Propositions 5, 6, $\vec{w}_{j+1}$, s_{j+1} satisfy for arbitrary $t < T$ the inequality

$$\begin{aligned}
&\sup_{\tau<t} |\vec{w}_{j+1}(\cdot,\tau)|_{C^{\ell+2}(\Omega)} + \sup_{\tau<t} |s_{j+1}(\cdot,\tau)|_{C^{\ell+1}(\Omega)} \\
&\qquad \le C_1\,\delta\Big(\sup_{\tau<t} |\vec{w}_j(\cdot,\tau)|_{C^{\ell+2}(\Omega)} + \sup_{\tau<t} |s_j(\cdot,\tau)|_{C^{\ell+1}(\Omega)}\Big) \\
&\qquad\quad + C_2 \int_0^t |\vec{w}_j(\cdot,\tau)|_{C^{\ell+2}(\Omega)}\,d\tau\,.
\end{aligned}$$

Hence, for

$$\Sigma_{m+1}(t) = \sum_{j=0}^{m} \Big(\sup_{\tau<t} |\vec{w}_j(\cdot,\tau)|_{C^{\ell+2}(\Omega)} + \sup_{\tau<t} |s_j(\cdot,\tau)|_{C^{\ell+1}(\Omega)}\Big)$$

we have

$$\Sigma_{m+1}(t) \le |\vec{u}_1|_{C^{2+\ell}(\Omega)} + |q_1|_{C^{1+\ell}(\Omega)} + C_1\,\delta\,\Sigma_{m+1}(t) + C_2 \int_0^t \Sigma_{m+1}(\tau)\,d\tau\,.$$

If $C_1\delta < \frac{1}{2}$, then this implies

$$\Sigma_{m+1}(T) \le 2\,C_0\Big(|H_0|_{C^{1+\ell}(\Gamma_0)} + \sum_{k=1}^{m} \sup_{t<T} |M_k(t)|\Big)\,e^{2C_2T} \tag{4.8}$$

and $\vec{u}_{m+1} = \sum_{j=1}^{m+1} \vec{w}_j$ also satisfies condition (4.5), provided that

$$2\,C_0\,T\,e^{2C_2T}\Big(|H_0|_{C^{1+\ell}(\Gamma_0)} + \sum_{k=1}^{n} \sup_{t<T} |M_k(t)|\Big) \le \delta\,. \tag{4.9}$$

It follows that all $\vec{u}_m(\xi,t)$, $q_m(\xi,t)$ are defined for $t \in (0,T)$ and that these functions converge, as $m \to \infty$, to the solution of problem (1.8) which also satisfies inequality (4.8), i.e.,

$$\sup_{t<T} |\vec{u}(\cdot,t)|_{C^{2+\ell}(\Omega)} + \sup_{t<T} |q(\cdot,t)|_{C^{1+\ell}(\Omega)} \leq 2\, C_0\, e^{2C_2 T}\Big(|H_0|_{C^{1+\ell}(\Gamma_0)} + \sum_{k=1}^{6} \sup_{t<T} |M_k(t)|\Big) \, .$$

The uniqueness of the solution also follows from Theorem 3 and Propositions 5 and 6, because the difference of two solutions $\vec{u} - \vec{u}' \equiv \vec{w}$, $q - q' = s$ satisfies the inequality

$$\sup_{\tau<t} |\vec{w}(\cdot,\tau)|_{C^{\ell+2}(\Omega)} + \sup_{\tau<t} |s(\cdot,\tau)|_{C^{\ell+1}(\Omega)} \leq 2\, C_2 \int_0^t \Big(|\vec{w}(\cdot,\tau)|_{C^{2+\ell}(\Omega)} + |s(\cdot,\tau)|_{C^{1+\ell}(\Omega)}\Big)\, d\tau \; ;$$

hence, $\vec{w} = 0$, $s = 0$.

To prove (1.10), we multiply equation (1.8_1) by $\vec{\varphi}_m(X(\xi,t))$ and integrate over Ω. Since

$$\begin{aligned}\int_\Omega (-\nu\, \nabla_u^2 \vec{u} + \nabla_u q) \cdot \vec{\varphi}_m(X(\xi,t))\, d\xi &= \int_{\Omega_t} (-\nu\, \nabla^2 \vec{v} + \nabla p) \cdot \vec{\varphi}_m(x)\, dx \\ &= \frac{\nu}{2} \int_{\Omega_t} S(\vec{v}) : S(\vec{\varphi}_m)\, dx - \int_{\Gamma_t} T(\vec{v},p)\, \vec{n} \cdot \vec{\varphi}_m(x)\, dS_x \\ &= -\int_{\Gamma_t} \Delta(t)\, \vec{x} \cdot \vec{\varphi}_m(x)\, dS_x = 0 \; ,\end{aligned}$$

we obtain

$$\sum_{k=1}^{6} \lambda_k(\vec{u}) \int_\Omega \vec{\varphi}_k(X(\xi,t)) \cdot \vec{\varphi}_m(X(\xi,t))\, d\xi = 0 \; .$$

This implies $\lambda_k(\vec{u}) = 0$, because the determinant of the matrix with the elements

$$\int_\Omega \vec{\varphi}_k(X(\xi,t)) \cdot \vec{\varphi}_m(X(\xi,t))\, d\xi = \int_{\Omega_t} \vec{\varphi}_k(x) \cdot \vec{\varphi}_m(x)\, dx$$

is different from zero.

The free boundary Γ_t is defined as the image of Γ_0 under transformation (1.6), hence, $\Gamma_t \in C^1(0,T;C^{2+\ell})$. From the equation $\sigma\, H = \vec{n} \cdot T(\vec{v},p)\, \vec{n}$ it is easy to conclude that $\Gamma \in C(0,T;C^{3+\ell})$. Theorem 1 is proved. □

References

[1] V.A. Solonnikov, "On an initial-boundary value problem for the Stokes system arising in the study of a problem with a free boundary", Trudy Mat. Inst. Steklov **188** (1990), 150–188;
English transl. in Proc. Steklov Inst. of Math. **3** (1991), 191–239.

[2] V.A. Solonnikov, "Solvability of the problem on evolution of a viscous incompressible liquid bounded by a free surface on a finite time interval", Algebra Anal. **3** (1) (1991), 222–257;
English transl. in St. Petersbg. Math. J. **3** (1) (1992), 189–220.

[3] V.A. Solonnikov, "Unsteady flow of a finite mass of a liquid bounded by a free surface", Zap. Nauchn. Sem. LDMI **152** (1986), 137–157;
English transl. in J. Soviet Math., **40** (5) (1988).

[4] V.A. Solonnikov, "On the evolution of an isolated volume of a viscous capillary liquid for large values of the time", Vestnik Leningr. Univ. Ser **1** (3) (1987), 49–55;
English transl. in Vestn. Leningr. Univ. Math., **20** (3) (1987), 52–58.

[5] V.A. Solonnikov, "On unsteady motion of a finite isolated mass of self-gravitating fluid", Algebra Anal., **1** (1) (1989), 207–249;
English transl. in Leningr. Math. J. **1** (1) (1990), 227–276.

[6] V.A. Solonnikov, "On the transient motion of an isolated volume of viscous incompressible fluid", Izv. Akad. Nauk SSSR, ser. matem. **51** (5) (1987), 1065–1087;
English transl. in Math. USSR Izvestiya **31** (2) (1988), 381–405.

[7] M. Günther and G. Prokert, "Existence results for the quasistationary motion of a free capillary liquid drop", Zeitschrift für Analysis und ihre Anwendungen **16** (2) (1997), 311–348.

[8] M.M. Vainberg and V.A. Trenogin, "Theory of branching of solutions of nonlinear equations", Moscow, "Nauka" (1969), 527p.

[9] K.K. Golovkin, "Certain conditions for the smoothness of a function of several variables and estimates of convolution operators", Doklady Akad. Nauk SSSR **139** (1961), 524–527;
English transl. in Sov. Math. Doklady, **2** (1961), 949–953.

[10] K.K. Golovkin and V.A. Solonnikov, "Estimates of integral operators in translation-invariant norms", Trudy Mat. Inst. Steklov, **70** (1964), 45–58;
English transl. in Amer. Math. Soc. Translations, **61** (2) (1967), 97–112.

[11] K.K. Golovkin, "On equivalent normalization of fractional spaces", Trudy Mat. Inst. Steklov **66** (1962), 364–383.

Progress in Nonlinear Differential Equations
and Their Applications, Vol. 35
© 1999 Birkhäuser Verlag Basel/Switzerland

On the Asymptotic Behaviour of a Landau-Ginzburg Model for the Martensitic Phase Transitions in Shape Memory Alloys

Jürgen Sprekels and Songmu Zheng*

Dedicated to Prof. Dr. Herbert Amann on the occasion of his 60th birthday

Abstract. In this paper we report recent results on the asymptotic behaviour of a system of partial differential equations modelling the dynamics of martensitic phase transitions in shape memory alloys. In this model, the free energy is assumed to be in the Landau-Ginzburg form and nonconvex as function of the shear strain (which is the *order parameter* of the phase transitions), and the material is assumed to be linearly viscous. The system admits a unique global solution; in addition, the orbit is compact in suitable function spaces, and even the existence of a compact maximal attractor can be shown.

1. Introduction

In this paper, we study a system modelling the thermomechanical developments in a one-dimensional heat-conducting viscous solid of constant density ϱ (assumed to be normalized to unity, i.e. $\varrho = 1$) which is subject to heating and loading. More precisely, we consider metallic solids that not only respond to a change of the strain ε by an elastic (possibly nonlinear) stress $\sigma = \sigma(\varepsilon)$, but also to a change of the curvature of their metallic lattices by a couple stress $\mu = \mu(\varepsilon_x)$. We assume that the Helmholtz free energy density F is a potential of Landau-Ginzburg form, that is,

$$F = F(\varepsilon, \varepsilon_x, \theta), \tag{1.1}$$

where θ denotes the absolute temperature. To cover systems modelling first-order stress-induced and temperature-induced solid-solid phase transitions accompanied by hysteresis phenomena, we do not assume that F is a convex function of the order parameter ε.

* Supported by NSF of China, No. 19331040 and the grant No. 96024603 from State Education Commission of China.

A particular class of materials, where both stress-induced and temperature-induced first-order phase transitions occur that lead to a rather spectacular hysteretic behaviour, are the so-called shape memory alloys (for an account of the physical properties of shape memory alloys, we refer the reader to the monograph [4] by Brokate and Sprekels). In these materials the metallic lattice is deformed by shear, and the assumption of a constant density is justified. The shape memory effect itself is due to martensitic phase transitions between different configurations of the crystal lattice, namely *austenite* and *martensitic twins*.

In a series of papers (see [8], [9], for instance) F. Falk has proposed a Landau-Ginzburg theory, using the strain ε as order parameter, to explain the occurrence of the martensitic transitions in shape memory alloys (in this connection, we also refer to the works of I. Müller (cf. [1], [14])). The simplest form for the free energy density F, that accounts quite well for the experimentally observed behaviour and takes couple stresses into account, is given by ([8], [9])

$$F(\varepsilon,\varepsilon_x,\theta) = F_0(\theta) + F_1(\varepsilon)\,\theta + F_2(\varepsilon) + \frac{\delta}{2}\,\varepsilon_x^2\,, \tag{1.2}$$

where

$$F_1(\varepsilon) = \alpha_1\varepsilon^2\,, \quad F_2(\varepsilon) = \alpha_3\varepsilon^6 - \alpha_2\varepsilon^4 - \alpha_1\theta_1\varepsilon^2\,, \tag{1.3}$$

$$F_0(\theta) = -\,C_V\,\theta\,\log\left(\frac{\theta}{\theta_2}\right) + C_V\,\theta + \tilde{C}\,, \tag{1.4}$$

with θ_1, δ, α_i (i=1,2,3), θ_2, C_V, $\tilde{C}$ being positive constants. The positive constant C_V denotes the specific heat. Observe that in the interesting range of temperature, for θ close to θ_1, F is not a convex function of the shear strain ε. In fact, $F(\cdot,\varepsilon_x,\theta)$ may have up to three minima which correspond to the austenitic and the two martensitic phases.

We want to forecast the dynamics of the phase transitions in the one-dimensional situation. To this end, let $\Omega=(0,1)$, and, for $t>0$, $\Omega_t=\Omega\times(0,t)$. Then the balance laws of momentum and internal energy read

$$u_{tt} - \sigma_x + \mu_{xx} = 0, \quad \text{in } \Omega_\infty\,, \tag{1.5}$$

$$U_t + q_x - \sigma\varepsilon_t - \mu\varepsilon_{xt} = 0, \quad \text{in } \Omega_\infty\,, \tag{1.6}$$

and the Second Law of Thermodynamics is expressed by the Clausius-Duhem inequality

$$S_t + \left(\frac{q}{\theta}\right)_x \geq 0, \quad \text{in } \Omega_t\,. \tag{1.7}$$

Here, u, σ, μ, U, q, ε, S and θ denote displacement, stress, couple stress, internal energy density, heat flux, shear strain, entropy density and absolute temperature, in this order. In addition, we have for one-dimensional, homogeneous, thermoviscoelastic materials constitutive relations

$$\varepsilon = u_x\,, \quad \sigma = \frac{\partial F}{\partial\varepsilon} + \gamma\varepsilon_t\,, \quad \mu = \frac{\partial F}{\partial\varepsilon_x}\,, \quad S = -\frac{\partial F}{\partial\theta}\,, \quad U = F + \theta S\,, \tag{1.8}$$

where $\gamma > 0$ denotes the (constant) coefficient of viscosity. For the heat flux q, we assume Fourier's law

$$q = -\,k\,\theta_x\,, \tag{1.9}$$

where $k > 0$ is the heat conductivity (assumed constant). Obviously, this assumption implies the validity of (1.7) so that the second law of thermodynamics is satisfied automatically.

Inserting the constitutive relations into the balance laws (1.5)–(1.6), we obtain the system of partial differential equations

$$u_{tt} - (f_1\theta + f_2)_x - \gamma\varepsilon_{xt} + \delta u_{xxxx} = 0\,, \quad \text{in } \Omega_\infty\,, \tag{1.10}$$

$$C_V\theta_t - k\theta_{xx} - f_1\theta\varepsilon_t - \gamma\varepsilon_t^2 = 0\,, \quad \text{in } \Omega_\infty\,, \tag{1.11}$$

$$\varepsilon = u_x\,, \quad \text{in } \Omega_\infty\,, \tag{1.12}$$

where

$$f_1 = f_1(\varepsilon) = F_1'(\varepsilon)\,, \quad f_2 = f_2(\varepsilon) = F_2'(\varepsilon)\,. \tag{1.13}$$

In addition, we prescribe the initial and boundary conditions

$$u|_{x=0} = \varepsilon_x|_{x=0} = 0\,, \quad \varepsilon|_{x=1} = (\gamma u_{xt} - \delta u_{xxx} + \sigma_1)|_{x=1} = 0\,, \tag{1.14}$$

where

$$\sigma_1 = f_1\theta + f_2\,, \tag{1.15}$$

as well as

$$\theta_x|_{x=0,\,1} = 0\,, \tag{1.16}$$

and

$$u(x,0) = u_0(x)\,, \quad u_t(x,0) = u_1(x)\,, \quad \theta(x,0) = \theta_0(x) > 0\,, \quad x \in \overline{\Omega}\,. \tag{1.17}$$

The mechanical meaning of the boundary conditions is clear. For instance, the second one at $x = 1$ simply means stress-free.

We now employ the idea of Andrews [2] and Pego [17] to simplify the problem by introducing the *velocity potential*

$$p(x,t) = \int_1^x u_t(y,t)\,dy\,. \tag{1.18}$$

Then

$$\varepsilon_t = p_{xx}\,, \quad \text{in } \Omega_\infty\,, \tag{1.19}$$

and (1.10)–(1.11) turn out to be

$$p_t - \gamma p_{xx} + \delta\varepsilon_{xx} - \sigma_1 = 0\,, \quad \text{in } \Omega_\infty\,, \tag{1.20}$$

$$C_V\theta_t - k\theta_{xx} - f_1\theta p_{xx} - \gamma p_{xx}^2 = 0\,, \quad \text{in } \Omega_\infty\,. \tag{1.21}$$

Accordingly, the initial and boundary conditions (1.14), (1.16), (1.17) become

$$p_x|_{x=0} = \varepsilon_x|_{x=0} = 0\,, \tag{1.22}$$

$$p|_{x=1} = p_{xx}|_{x=1} = \varepsilon|_{x=1} = 0\,, \tag{1.23}$$

$$\varepsilon(x,0) = \varepsilon_0(x) := u_{0x}(x)\,, \quad p(x,0) = p_0(x) := \int_1^x u_1(y)dy\,, \quad \theta(x,0) = \theta_0(x)\,. \tag{1.24}$$

It is easy to see that if (u, v, θ) is a smooth solution to (1.10)–(1.17), then (ε, p, θ) is a smooth solution to (1.19)–(1.24) and vice versa. Therefore, it suffices to consider the problem (1.19)–(1.24). In the sequel, we assume without loss of generality that $C_V = 1$.

We now recall some related results in the literature. When $\delta = 0$, Dafermos [6], Dafermos & Hsiao [7], Chen & Hoffmann [5], Jiang [11], proved the global existence of classical solution to the system (1.10)–(1.12) with various boundary conditions for a class of solid-like materials. However, the asymptotic behaviour was not considered in these papers. Recently, on the basis of Dafermos [6] and of Dafermos & Hsiao [7], T. Luo [13] further investigated the asymptotic behaviour of smooth solutions as time tends to infinity for a special class of solid-like materials in which $e = C_V\theta$, $F_2 = 0$ and $\delta = 0$. Racke and Zheng [18], and Shen, Zheng & Zhu [20], respectively, obtained global existence, uniqueness and asymptotic behaviour of weak solutions to (1.10)–(1.12) for $\delta = 0$ in the case when both ends of the rod are thermally insulated and when at least one end is stress-free (or both ends are clamped, respectively).

Concerning the case $\delta > 0$, we refer to Sprekels & Zheng [21] for the case $\delta > 0$, $\gamma = 0$, and to Hoffmann & Zochowski [10] for the case $\delta > 0$, $\gamma > 0$, for global existence and uniqueness results for the above model of shape memory alloys. However, the a priori estimates of the solutions obtained in these papers depended on T. Consequently, no results concerning the asymptotic behaviour as $t \to +\infty$ have been established there. We also refer to Andrews [2], Andrews & Ball [3], and Pego [17], for the (simpler) isothermal and purely viscoelastic case. We now state the main results of this paper.

Theorem 1.1. (cf. Sprekels, Zheng & Zhu [23])
Suppose that ε_0, $p_0 \in H^3$ and $\theta_0 \in H^1$ are given functions satisfying the compatibility conditions $p_{0x}|_{x=0} = \varepsilon_{0x}|_{x=0} = 0$, $p_0|_{x=1} = p_{0xx}|_{x=1} = \varepsilon_0|_{x=1} = 0$, and suppose that $\theta_0 > 0$ in $[0,1]$. Then the following results hold.

(i) *The problem admits a unique global solution (ε, p, θ) satisfying*

$$\begin{aligned}
&\varepsilon \in C(\mathbb{R}^+; H^3)\,, \quad \varepsilon_t \in C(\mathbb{R}^+; H^1) \cap L^2(\mathbb{R}^+; H^2)\,,\\
&p \in C(\mathbb{R}^+; H^3) \cap L^2(\mathbb{R}^+; H^4)\,, \quad p_t \in C(\mathbb{R}^+; H^1) \cap L^2(\mathbb{R}^+; H^2)\,,\\
&\theta \in C(\mathbb{R}^+; H^1)\,, \quad \theta_x \in L^2(\mathbb{R}^+; H^1)\,, \quad \theta_t \in L^2(\mathbb{R}^+; L^2)\,,\\
&\theta(x,t) > 0 \quad \forall\,(x,t) \in [0,1] \times \mathbb{R}^+\,.
\end{aligned} \tag{1.25}$$

(ii) *As $t \to +\infty$, it holds*

$$\|p(\cdot,t)\|_{H^3} \to 0\,, \quad \|p_t(\cdot,t)\|_{H^1} \to 0\,, \tag{1.26}$$

$$\|\delta\varepsilon_{xx} - \sigma_1\|_{H^1} \to 0\,, \quad \|\varepsilon_t(\cdot,t)\|_{H^1} \to 0\,, \quad \|\theta_x(\cdot,t)\| \to 0\,. \tag{1.27}$$

(iii) *In addition, for all $\nu > 0$,*

$$\varepsilon \in C([\nu,+\infty);H^4)\,, \quad p \in C([\nu,+\infty);H^4)\,, \quad \theta \in C([\nu,+\infty);H^3)\,, \tag{1.28}$$

i.e. the orbit is compact in $H^3 \times H^3 \times H^1$.

(iv) $$(\varepsilon(\cdot,t), p(\cdot,t), \theta(\cdot,t)) \to (\overline{\varepsilon}, 0, \overline{\theta})\,, \quad as \quad t \to \infty\,, \quad in \quad H^3 \times H^3 \times H^1\,, \tag{1.29}$$

where $(\overline{\varepsilon},\, \overline{\theta})$ is one of the equilibria for the corresponding stationary problem.

To formulate the existence result for a compact attractor, we first multiply (1.20) by $-p_{xx}$, add the result to (1.21) and integrate with respect to x over Ω to obtain that

$$\frac{d}{dt}\int_0^1 \left(\theta + F_2(\varepsilon) + \frac{1}{2}\,p_x^2 + \frac{\delta}{2}\,\varepsilon_x^2\right)(t)\,dx = 0. \tag{1.30}$$

Thus,

$$E_1(t) := \int_0^1 \left(\theta + F_2(\varepsilon) + \frac{1}{2}\,p_x^2 + \frac{\delta}{2}\,\varepsilon_x^2\right)(t)\,dx = E_1(0)\,, \tag{1.31}$$

where $E_1(0)$ is a constant depending on the initial data. Obviously, this energy conservation indicates that there can be no absorbing ball for initial data varying *in the whole space*. Instead, we should rather consider the dynamics in a restricted set which is invariant for the orbit. In this regard the situation is quite similar to that encountered for the Cahn-Hilliard equation (cf., for instance, [24]).

Now let us consider the space

$$\begin{aligned} H \quad := \quad \Big\{ & (\varepsilon,p,\theta) \in H^3[0,1] \times H^3[0,1] \times H^1[0,1] \;:\; p_x|_{x=0} = \varepsilon_x|_{x=0} = 0\,, \\ & p|_{x=1} = p_{xx}|_{x=1} = \varepsilon|_{x=1} = 0 \Big\}\,, \end{aligned} \tag{1.32}$$

which becomes a Hilbert space when equipped with the usual inner product and norm of $H^3 \times H^3 \times H^1$. Moreover, consider for arbitrary given constants $\beta_1 > 0$, $\beta_2 \in \mathbb{R}$, the set

$$\begin{aligned} H_{\beta_1,\beta_2} \quad := \quad \Big\{ & (\varepsilon,p,\theta) \in H \;:\; \theta > 0\,, \quad \int_0^1 \Big(\log\theta - F_1(\varepsilon)\Big)\,dx > \beta_2\,, \\ & \int_0^1 \left(\theta + F_2(\varepsilon) + \frac{1}{2}\,p_x^2 + \frac{\delta}{2}\,\varepsilon_x^2\right) dx < \beta_1 \Big\}\,. \end{aligned} \tag{1.33}$$

Apparently H_{β_1,β_2} is an open subset of H.

We are now in the position to state our second main theorem.

Theorem 1.2. (cf. Sprekels & Zheng [22])
For every $\beta_1 > 0$, $\beta_2 < 0$, the semigroup $S(t)$ defined by (1.19)–(1.24) maps H_{β_1,β_2} into itself. In addition, it possesses in H_{β_1,β_2} a maximal attractor A_{β_1,β_2} which is compact.

In the next section, we will indicate how the Theorems 1.1 and 1.2 can be proved. To this end, we will use the following notations: L^p, $1 \le p \le \infty$, $W^{m,\infty}$, $m \in \mathbb{N}$, $H^1 \equiv W^{1,2}$, $H_0^1 = W_0^{1,2}$, denote the usual (Sobolev) spaces on (0,1). In addition, $(\cdot,\cdot)$ stands for the inner product in L^2, and $\|\cdot\|_B$ denotes the norm in the space B; we also put $\|\cdot\| := \|\cdot\|_{L_2}$. We denote by $C^k(I,B)$, $k \in \mathbb{N}_0$, the space of k-times continuously differentiable functions from $I \subset \mathbb{R}$ into a Banach space B, and, likewise, by $L^p(I,B)$, $1 \le p \le \infty$, the corresponding Lebesgue-spaces. Finally, ∂_t or $\frac{d}{dt}$ or a subscript t and ∂_x or a subscript x denote the (partial) derivatives with respect to t and x, respectively.

2. Proofs of Theorems 1.1 and 1.2

The general framework to prove the global existence and uniqueness of the solution has been established for instance in Sprekels & Zheng [21] and Hoffmann & Zochowski [10]. We shall focus our attention on the study of the asymptotic behaviour as $t \to +\infty$, that is, we try to derive a priori estimates for ε, p, θ that are uniform with respect to t and, in the case of Theorem 1.2, also with respect to initial data in H_{β_1,β_2}.

To this end, suppose that $(\varepsilon_0, p_0, \theta_0) \in H_{\beta_1,\beta_2}$ for some (given) fixed constants $\beta_1 > 0$, $\beta_2 < 0$. From now on, we will always denote by $C > 0$ a universal positive constant that may depend on β_1, β_2, but neither on the initial data nor on t.

Lemma 2.1. *For any $t > 0$, the following estimates hold.*

$$\|\varepsilon(t)\| + \|\varepsilon(t)\|_{L^6} + \|p_x(t)\| + \|\varepsilon_x(t)\| + \|\theta(t)\|_{L^1} \le C, \tag{2.1}$$

$$\|p(t)\|_{L^\infty} + \|\varepsilon(t)\|_{L^\infty} \le C, \tag{2.2}$$

$$\theta(x,t) > 0 \quad \forall\, (x,t) \in [0,1] \times \mathbb{R}^+, \tag{2.3}$$

$$E_1(t) = E_1(0) < \beta_1 . \tag{2.4}$$

Proof. Applying the maximum principle to (1.21), we obtain (2.3), and (2.4) follows from (1.31). By Young's inequality, $F_2(\varepsilon) \ge C_1\varepsilon^6 - C_2$, from which (2.1) follows. Inequality (2.2) is then a consequence of the boundary conditions and of Poincaré's inequality. □

Lemma 2.2. *For any $t > 0$, the following estimates hold.*

$$\int_0^t \int_0^1 \left(\frac{\theta_x^2}{\theta^2} + \frac{p_{xx}^2}{\theta} \right) dx d\tau \le C \,, \tag{2.5}$$

$$\int_0^t \|p_x(\tau)\|^2 d\tau \le \int_0^t \|p_x(\tau)\|_{L^\infty}^2 d\tau \le C \,, \quad \int_0^t \|p(\tau)\|_{L^\infty}^2 d\tau \le C \,, \tag{2.6}$$

$$E_2(t) := \int_0^1 (\log \theta - F_1(\varepsilon))(t) \, dx \ge E_2(0) > \beta_2 \,, \tag{2.7}$$

$$\int_0^t \|p_x(\tau)\|^{n+2} d\tau \le C \quad \forall \; n \ge 0 \,, \tag{2.8}$$

$$\int_0^1 \theta(t) \, dx \ge C > 0 \,. \tag{2.9}$$

Proof. Multiplication of (1.21) by θ^{-1} and integration with respect to x over Ω yields

$$\frac{d}{dt} \int_0^1 (\log \theta - F_1(\varepsilon))(t) \, dx \; - \; \int_0^1 \left(\frac{k\theta_x^2}{\theta^2} + \frac{\gamma p_{xx}^2}{\theta} \right) (t) \, dx = 0 \,, \tag{2.10}$$

from which (2.7) follows. In addition, since $\log \theta \le \theta - 1$ for all $\theta > 0$, we have

$$\int_0^t \int_0^1 \left(\frac{k\theta_x^2}{\theta^2} + \frac{\gamma p_{xx}^2}{\theta} \right) dx d\tau \le C \,. \tag{2.11}$$

Next, observe that it follows from $p_x|_{x=0} = 0$ that

$$p_x(x,t) = p_x(0,t) + \int_0^x p_{xx}(y,t) \, dy = \int_0^x p_{xx}(y,t) \, dy \,. \tag{2.12}$$

Hence,

$$\begin{aligned} \int_0^t \|p_x(\tau)\|_{L^\infty}^2 d\tau &\le \int_0^t \Big(\int_0^1 |p_{xx}(x,\tau)| \, dx \Big)^2 d\tau \\ \le & \int_0^t \Big(\int_0^1 \sqrt{\theta} \, \frac{|p_{xx}|}{\sqrt{\theta}} \, dx \Big)^2 d\tau \le \int_0^t \Big(\int_0^1 \theta \, dx \Big) \int_0^1 \frac{p_{xx}^2}{\theta} dx \; d\tau \\ \le & \; C \int_0^t \int_0^1 \frac{p_{xx}^2}{\theta} dx d\tau \le C \,. \end{aligned} \tag{2.13}$$

Therefore also

$$\int_0^t \|p_x(\tau)\|^2 d\tau \le \int_0^t \|p_x\|_{L^\infty}^2 d\tau \le C \,. \tag{2.14}$$

Combining (2.6) with (2.1) yields

$$\int_0^t \|p_x(\tau)\|^{n+2} d\tau \le C \quad \forall \; n \ge 0 \,. \tag{2.15}$$

Finally, from (2.10) and Jensen's inequality,

$$- \log\Big(\int_0^1 \theta(t)\,dx\Big) \le - \int_0^1 \log\,\theta(t)\,dx \le C\,, \tag{2.16}$$

which implies (2.9). □

Note that it follows from the Lemmas 2.1 and 2.2 that the semigroup $S(t)$ maps H_{β_1,β_2} into itself. Also, we will see in what follows that (2.8) plays a very important role in deriving uniform a priori estimates.

Lemma 2.3. *For any $t>0$, the following estimates hold.*

$$\frac{d}{dt}\,\|\theta_x(t)\|^2 \,+\, \|\theta_t(t)\|^2 \le C\left(\|\theta(t)\|_{L^\infty}\,\|p_{xx}(t)\|_{L^\infty}^2 \,+\, \|p_{xx}(t)\|_{L^4}^4\right)\,, \tag{2.17}$$

$$\frac{d}{dt}\left(\|p_{xt}(t)\|^2 \,+\, \delta\,\|\varepsilon_{xt}(t)\|^2\right) \,+\, \gamma\,\|\varepsilon_{tt}(t)\|^2 \le C\int_0^1 (\theta^2\varepsilon_t^2+\varepsilon_t^2+\theta_t^2)(t)\,dx\,, \tag{2.18}$$

$$\begin{aligned}&\frac{\gamma}{2}\,\frac{d}{dt}\|\varepsilon_{xt}(t)\|^2 \;-\; \frac{d}{dt}(p_{xt}(t),\varepsilon_{xt}(t)) \;-\; \|p_{xxt}(t)\|^2 \;+\; \frac{\delta}{2}\,\|\varepsilon_{xxt}(t)\|^2\\ \le\;& C\int_0^1 (\theta^2\varepsilon_t^2+\varepsilon_t^2+\theta_t^2)(t)\,dx\,.\end{aligned} \tag{2.19}$$

Proof. For the sake of brevity, we omit the arguments of functions from now on. Multiplying (1.21) by θ_t and integrating with respect to x over Ω, we obtain that

$$\begin{aligned}\frac{k}{2}\,\frac{d}{dt}\,\|\theta_x\|^2 \,+\, \|\theta_t\|^2 \;&=\; \int_0^1 \Big(f_1\,\theta\,\theta_t\,p_{xx} \,+\, \gamma\,\theta_t\,p_{xx}^2\Big)\,dx\\ &\le\; C\left(\|\theta\,p_{xx}\|\,\|\theta_t\| \,+\, \Big(\int_0^1 p_{xx}^4\,dx\Big)^{\frac12}\,\|\theta_t\|\right)\\ &\le\; C\left(\|\theta\|_{L^\infty}^{\frac12}\,\|p_{xx}\|_{L^\infty}\,\Big(\int_0^1 \theta\,dx\Big)^{\frac12}\,\|\theta_t\| \,+\, \|p_{xx}\|_{L^4}^2\,\|\theta_t\|\right)\,,\end{aligned} \tag{2.20}$$

and (2.17) follows from the Cauchy-Schwarz inequality.

Next, we differentiate (1.20) with respect to t, multiply the result by $-\varepsilon_{tt}$, and integrate with respect to x over Ω to obtain

$$\begin{aligned}0 \;&=\; (p_{tt},-p_{xxt}) \,+\, \gamma\,\|\varepsilon_{tt}\|^2 \,+\, (\delta\,\varepsilon_{xt},\varepsilon_{xtt}) \,+\, \int_0^1 \sigma_{1t}\,\varepsilon_{tt}\,dx\\ &=\; (p_{xtt},p_{xt}) \,+\, \gamma\,\|\varepsilon_{tt}\|^2 \,+\, \delta\,(\varepsilon_{xt},\varepsilon_{xtt}) + \int_0^1 \Big(f_1'\,\varepsilon_t\,\theta \,+\, f_2'\,\varepsilon_t\\ &\quad + f_1\,\theta_t\Big)\,\varepsilon_{tt}\,dx.\end{aligned} \tag{2.21}$$

Combining this with (2.2) yields

$$\frac{1}{2}\frac{d}{dt}\left(\|p_{xt}\|^2 + \delta\,\|\varepsilon_{xt}\|^2\right) + \gamma\,\|\varepsilon_{tt}\|^2 \le \frac{\gamma}{2}\,\|\varepsilon_{tt}\|^2 + C\int_0^1 \left(\theta^2\,\varepsilon_t^2 + \varepsilon_t^2 + \theta_t^2\right)dx\,, \tag{2.22}$$

and (2.18) follows.

Finally, we differentiate (1.20) with respect to t, multiply the result by ε_{xxt}, and integrate the result with respect to x over Ω to arrive at

$$\begin{aligned} 0 &= (p_{tt},\varepsilon_{xxt}) - \gamma\,(\varepsilon_{tt},\varepsilon_{xxt}) + \delta\|\varepsilon_{xxt}\|^2 - \int_0^1 \varepsilon_{xxt}\,\sigma_{1t}\,dx \\ &= (p_{xxtt},\varepsilon_t) + \gamma\,(\varepsilon_{xtt},\varepsilon_{xt}) + \delta\|\varepsilon_{xxt}\|^2 - \int_0^1 \varepsilon_{xxt}\,\sigma_{1t}\,dx \\ &= \frac{d}{dt}(p_{xxt},\varepsilon_t) - \|p_{xxt}\|^2 + \frac{\gamma}{2}\frac{d}{dt}\|\varepsilon_{xt}\|^2 + \delta\|\varepsilon_{xxt}\|^2 \\ &\quad - \int_0^1 \varepsilon_{xxt}\,\sigma_{1t}\,dx\,. \end{aligned} \tag{2.23}$$

However, $(p_{xxt},\varepsilon_t) = -(p_{xt},\varepsilon_{xt})$, by integration by parts, and $\sigma_{1t} = f_1'(\varepsilon)\,\varepsilon_t\,\theta + f_1(\varepsilon)\,\theta_t + f_2'(\varepsilon)\,\varepsilon_t$. Hence, (2.19) follows from (2.23) and the Cauchy-Schwarz inequality. □

Now let $\eta \in (0,1)$ be a small positive constant chosen in such a way that

$$E(t) := \|p_{xt}(t)\|^2 + \delta\|\varepsilon_{xt}(t)\|^2 + \frac{\gamma\,\eta}{2}\,\|\varepsilon_{xt}(t)\|^2 - \eta\,(p_{xt}(t),\varepsilon_{xt}(t)) \tag{2.24}$$

is positive definite, i.e., such that there is a constant $C>0$ satisfying

$$E(t) \ge C\left(\|p_{xt}(t)\|^2 + \|\varepsilon_{xt}(t)\|^2\right). \tag{2.25}$$

We then have

Lemma 2.4. *For any $t>0$ it holds that*

$$\begin{aligned} &\frac{d}{dt}E(t) + \frac{\delta\,\eta}{2}\,\|\varepsilon_{xxt}(t)\|^2 + (1-\eta)\,\|\varepsilon_{tt}(t)\|^2 \\ \le\ & C\int_0^1 \left(\theta^2\,\varepsilon_t^2 + \varepsilon_t^2 + \theta_t^2\right)(t)\,dx\,. \end{aligned} \tag{2.26}$$

Proof. Multiplying (2.19) by η, then adding the result up with (2.18), yields (2.26). □

We now introduce the mean value of temperature,

$$\bar\theta(t) := \int_0^1 \theta(x,t)\,dx\,. \tag{2.27}$$

We then obtain the following result.

Lemma 2.5. *For any $t > 0$ it holds*

$$\frac{d}{dt}\,\|(\theta-\bar\theta)(t)\|^2 \,+\, k\,\|\theta_x(t)\|^2 \le C\left(\|p_{xx}(t)\|^2_{L^\infty} \,+\, \|p_{xx}(t)\|^4\right). \tag{2.28}$$

Proof. It follows from the energy equation (1.11) that

$$\bar\theta_t - \int_0^1 f_1\,\theta\,\varepsilon_t\,dx - \gamma\|\varepsilon_t\|^2 = 0\,. \tag{2.29}$$

Thus,

$$(\theta-\bar\theta)_t \,-\, k\,\theta_{xx} - \Big(f_1\,\theta\,\varepsilon_t \,-\, \int_0^1 f_1\,\theta\,\varepsilon_t\,dx\Big) \,-\, \gamma\,(\varepsilon_t^2 - \|\varepsilon_t\|^2) = 0\,. \tag{2.30}$$

Multiplication of (2.30) by $\theta-\bar\theta$ and integration with respect to x yields

$$\begin{aligned} &\frac{1}{2}\,\frac{d}{dt}\,\|\theta-\bar\theta\|^2 \,+\, k\,\|\theta_x\|^2 \\ \le\;& \int_0^1 |\theta-\bar\theta|\,\Big|f_1\,\theta\,\varepsilon_t - \int_0^1 f_1\,\theta\,\varepsilon_t\,dx\Big|\,dx \,+\, \gamma\int_0^1 |\theta-\bar\theta|\,\big|\varepsilon_t^2 - \|\varepsilon_t\|^2\big|\,dx \\ \le\;& \frac{k}{2}\,\|\theta_x\|^2 \,+\, C\left(\|p_{xx}\|^2_{L^\infty} \,+\, \|p_{xx}\|^4\right), \end{aligned} \tag{2.31}$$

from which (2.28) follows. □

Lemma 2.6. *There exists some $\mu > 0$, depending only on β_1 and β_2, such that for any $t > 0$ it holds*

$$\begin{aligned} &\frac{d}{dt}\left(\|(\theta-\bar\theta)(t)\|^2 \,+\, \|\theta_x(t)\|^2 \,+\, \mu\,E(t)\right) \,+\, C\,\big(\|(\theta-\bar\theta)(t)\|^2 \\ &+\|\theta_x(t)\|^2 \,+\, \mu\,E(t)\big) \\ \le\;& C\left(\|p_x(t)\|^2 \,+\, \|p_x(t)\|^6 \,+\, \|p_x(t)\|^8 \,+\, \|p_x(t)\|^{14}\right). \end{aligned} \tag{2.32}$$

Proof. We multiply (2.26) by a positive constant μ (which is yet to be specified), and then add the result up with (2.17) and (2.28) to find that

$$\begin{aligned} &\frac{d}{dt}\left(\|(\theta-\bar\theta)(t)\|^2 + \|\theta_x(t)\|^2 + \mu E(t)\right) + \|\theta_t(t)\|^2 + k\,\|\theta_x(t)\|^2 \\ &\quad + C\mu\left(\|p_{xxxx}(t)\|^2 + \|\varepsilon_{tt}(t)\|^2\right) \\ \le\;& C\left(\|\theta(t)\|_{L^\infty}\,\|p_{xx}(t)\|^2_{L^\infty} + \|p_{xx}(t)\|^4_{L^4}\right) + C\left(\|p_{xx}(t)\|^2_{L^\infty} + \|p_{xx}(t)\|^4\right) \\ &\quad + C\mu\int_0^1 \left(\theta^2 p_{xx}^2 + p_{xx}^2 + \theta_t^2\right)(t)\,dx. \end{aligned} \tag{2.33}$$

Now, we choose μ small enough to cancel the term $\int_0^1 \theta_t^2(t)\,dx$ on the right-hand side of (2.33). Applying Nirenberg's inequality appropriately to the remaining terms on the right-hand side of (2.33), and invoking Young's inequality, we can infer that

$$\begin{aligned}&\frac{d}{dt}\left(\|(\theta-\bar\theta)(t)\|^2+\|\theta_x(t)\|^2+\mu\,E(t)\right)+\|\theta_t(t)\|^2+k\,\|\theta_x(t)\|^2\\&\quad+\frac{C\mu}{2}\left(\|p_{xxxx}(t)\|^2+\|\varepsilon_{tt}(t)\|^2\right)\\&\le\ C\left(\|p_x(t)\|^2+\|p_x(t)\|^6+\|p_x(t)\|^8+\|p_x(t)\|^{14}\right).\end{aligned}\tag{2.34}$$

Next, observe that

$$E(t)\ \le\ C\left(\|p_{xt}(t)\|^2+\|\varepsilon_{xt}(t)\|^2\right)\ \le\ C\left(\|\varepsilon_{tt}(t)\|^2+\|p_{xxxx}(t)\|^2\right),\tag{2.35}$$

and that Poincaré's inequality implies that

$$\|(\theta-\bar\theta)(t)\|\le C\,\|\theta_x(t)\|\,.\tag{2.36}$$

The assertion then follows from (2.34)–(2.36). □

Remark: Observe that (2.32) and (2.8) yield for any $t>0$ that

$$\|(\theta-\bar\theta)(t)\|^2\ +\ \|\theta_x(t)\|^2\ +\ \mu\,E(t)\le C\left(\|\theta_0\|_{H^1}^2+\|p_0\|_{H^3}^2+\|\varepsilon_0\|_{H^3}^2\right)e^{-Ct}\ +\ C\,.\tag{2.37}$$

By virtue of the boundary conditions, and owing to (1.18)–(1.20), it also holds

$$\|\theta(t)\|_{H^1}^2+\|p(t)\|_{H^3}^2+\|\varepsilon(t)\|_{H^3}^2\le C\left(\|(\theta-\bar\theta)(t)\|^2+\|\theta_x(t)\|^2+\mu\,E(t)\right)\ +\ C\,.\tag{2.38}$$

In conclusion, there exists an absorbing ball set in H_{β_1,β_2}. Therefore, the existence of a compact maximal attractor A_{β_1,β_2} in H_{β_1,β_2} (and thus the assertion of Theorem 1.2) would be a consequence of Theorem I.1.1 in R. Temam [24], if the assertion of Theorem 1.1 were true.

According to the above remark, we only need to prove the remaining assertions of Theorem 1.1. In the sequel, we denote by $\hat C>0$ any constant that may depend on the initial data (more precisely, on $\|p_0\|_{H^3}$, $\|\varepsilon_0\|_{H^3}$, and $\|\theta_0\,\|_{H^1}$), but not on t. By (2.37) and (2.38), we have, for all $t\ge 0$,

$$\|(\theta-\bar\theta)(t)\|\ +\ \|\theta(t)\|_{H^1}\ +\ \|p(t)\|_{H^3}\ +\ \|p_t(t)\|_{H^1}\ +\ \|\varepsilon(t)\|_{H^3}\ +\ \|\varepsilon_t(t)\|_{H^1}\le\hat C\,.\tag{2.39}$$

We also have, by (2.5) and (2.39),

$$\int_0^t\|\varepsilon_t(\tau)\|^2\,d\tau\ =\ \int_0^t\|p_{xx}(\tau)\|^2\,d\tau\ \le\ \sup_{0\le\tau\le t}\|\theta(t)\|_{L^\infty}\int_0^t\int_0^1\frac{p_{xx}^2}{\theta}\,dx\,d\tau\ \le\ \hat C\,.\tag{2.40}$$

Similarly, using (2.5) and (2.39) again,

$$\int_0^t\|\theta_x(\tau)\|^2\,d\tau\ \le\ \hat C\,.\tag{2.41}$$

By comparison in (1.21), also

$$\int_0^t \left(\|\theta_t(\tau)\|^2 + \|\theta_{xx}(\tau)\|^2\right) d\tau \le \hat{C}\,, \tag{2.42}$$

so that (2.26) and comparison in (1.20) yield that

$$\int_0^t \left(\|\varepsilon_{xxt}(\tau)\|^2 + \|\varepsilon_{tt}(\tau)\|^2 + \|p_{tt}(\tau)\|^2 + \|p_{xxt}(\tau)\|^2 + \|p_{xxxx}(\tau)\|^2\right) d\tau \le \hat{C}\,. \tag{2.43}$$

With this, (1.25) is completely proved.

Now we proceed to investigate the compactness of the orbit of the solution for $t > 0$ in $H^3 \times H^3 \times H^1$. To this end, we assume that the initial data are so smooth that the solution will have enough smoothness to carry out the following argument (if the initial data belonged just to $H^3 \times H^3 \times H^1$, one could argue via approximation).

Differentiating (1.20) twice with respect to t, we find that

$$p_{ttt} - \gamma\, p_{xxtt} + \delta\varepsilon_{xxtt} - \sigma_{1tt} = 0\,, \tag{2.44}$$

where

$$\sigma_{1tt} = f_1'(\varepsilon)\,\varepsilon_{tt}\,\theta + 2\,f_1'(\varepsilon)\,\varepsilon_t\,\theta_t + f_1(\varepsilon)\,\theta_{tt} + f_2''(\varepsilon)\,\varepsilon_t^2 + f_2'(\varepsilon)\,\varepsilon_{tt}. \tag{2.45}$$

Multiplying (2.44) by p_{tt}, and integrating with respect to x over Ω and by parts, we obtain after a straightforward calculation that

$$0 = \frac{1}{2}\,\frac{d}{dt}\left(\|p_{tt}(t)\|^2 + \delta\|\varepsilon_{tt}(t)\|^2\right) + \gamma\,\|p_{xtt}(t)\|^2 - (\sigma_{1tt}(t), p_{tt}(t))\,. \tag{2.46}$$

Multiplying (2.46) by t^2, and using (2.39) and (2.45), we obtain that

$$\begin{aligned} &\frac{1}{2}\,\frac{d}{dt}\left(t^2\,\|p_{tt}(t)\|^2 + t^2\,\delta\,\|\varepsilon_{tt}\|^2\right) - t\,\left(\|p_{tt}(t)\|^2 + \delta\,\|\varepsilon_{tt}(t)\|^2\right) \\ &+\gamma\,t^2\,\|p_{xtt}(t)\|^2 \\ \le\;& t^2\,\|p_{tt}(t)\|^2 + \hat{C}\,t^2\,\|\sigma_{1tt}(t)\|^2 \\ \le\;& t^2\,\|p_{tt}(t)\|^2 + \hat{C}\,t^2\,\left(\|\varepsilon_{tt}(t)\|^2 + \|\theta_t(t)\|^2 + \|\theta_{tt}(t)\|^2 + \|\varepsilon_t(t)\|^2\right). \end{aligned} \tag{2.47}$$

Hence, it follows from (2.24), (2.40), (2.42), and (2.43), that

$$\begin{aligned} &t^2\,\left(\|p_{tt}(t)\|^2 + \delta\,\|\varepsilon_{tt}(t)\|^2\right) + \int_0^t \tau^2\,\|p_{xtt}(\tau)\|^2\,d\tau \\ \le\;& \hat{C}\left(1 + t^2 + \int_0^t \tau^2\,\|\theta_{tt}(\tau)\|^2\,d\tau\right). \end{aligned} \tag{2.48}$$

On the other hand, differentiating (1.21) with respect to t, we get

$$\theta_{tt} - k\,\theta_{xxt} - (f_1(\varepsilon)\,\theta\,p_{xx} + \gamma\,p_{xx}^2)_t = 0\,. \tag{2.49}$$

Multiplying by θ_{tt} , integrating with respect to x , and invoking (2.39) and Young's inequality, we arrive at

$$\begin{aligned}&\frac{k}{2}\frac{d}{dt}\,\|\theta_{xt}(t)\|^2 \,+\, \|\theta_{tt}(t)\|^2\\&\le \frac{1}{2}\,\|\theta_{tt}(t)\|^2 \,+\, \hat{C}\,\left(\|p_{xx}\|^2 \,+\, \|\theta_t(t)\|^2 + \|p_{xxt}(t)\|^2\right)\,. \qquad (2.50)\end{aligned}$$

Multiplication of (2.50) by t^2 yields

$$\begin{aligned}&\frac{k}{2}\frac{d}{dt}\,\left(t^2\,\|\theta_{xt}(t)\|^2\right) \,-\, k\,t\,\|\theta_{xt}(t)\|^2 \,+\, \frac{t^2}{2}\,\|\theta_{tt}(t)\|^2\\&\le\;\; \hat{C}\,t^2\,\left(\|p_{xx}\|^2 \,+\, \|\theta_t(t)\|^2 \,+\, \|p_{xxt}(t)\|^2\right)\,. \qquad (2.51)\end{aligned}$$

In order to estimate $\int_0^t \tau\,\|\theta_{xt}(\tau)\|^2\,d\tau$, we multiply (2.49) by θ_t , and then integrate with respect to x over Ω , to obtain

$$\begin{aligned}&\frac{1}{2}\frac{d}{dt}\,\|\theta_t(t)\|^2 \,+\, k\,\|\theta_{xt}(t)\|^2\\&\le \frac{1}{2}\,\|\theta_t(t)\|^2 \,+\, \hat{C}\,\left(\|p_{xx}\|^2 \,+\, \|\theta_t(t)\|^2 \,+\, \|p_{xxt}(t)\|^2\right). \qquad (2.52)\end{aligned}$$

Multiplying (2.52) by t , we find that

$$\begin{aligned}&\frac{1}{2}\frac{d}{dt}\,\left(t\,\|\theta_t(t)\|^2\right) \,+\, k\,t\,\|\theta_{xt}(t)\|^2\\&\le\;\; \hat{C}\Big(\|\theta_t(t)\|^2 \,+\, t\,\|\theta_t(t)\|^2 \,+\, t\,\left(\|p_{xx}\|^2 + \|\theta_t(t)\|^2 + \|p_{xxt}(t)\|^2\right)\Big). \qquad (2.53)\end{aligned}$$

Therefore,

$$t\,\|\theta_t(t)\|^2 \,+\, \int_0^t \tau\,\|\theta_{xt}(\tau)\|^2\,d\tau \,\le\, \hat{C}\,(1\,+\,t)\,, \qquad (2.54)$$

so that, combining (2.54) with (2.51),

$$\int_0^t \tau^2\,\|\theta_{tt}(\tau)\|^2\,d\tau \,\le\, \hat{C}\,\left(1\,+\,t^2\right)\,. \qquad (2.55)$$

Hence it follows from (2.48) that

$$\|p_{tt}(t)\|^2 \,+\, \|\varepsilon_{tt}(t)\|^2 \,<\, \hat{C}\,\left(1\,+\,t^{-2}\right)\,. \qquad (2.56)$$

Moreover, by virtue of (2.54) and (2.51),

$$\|\theta_t(t)\|^2 \,\le\, \hat{C}\,\left(1\,+\,t^{-1}\right)\,, \quad \|\theta_{xt}(t)\|^2 \,\le\, \hat{C}\,\left(1\,+\,t^{-2}\right)\,. \qquad (2.57)$$

Thus, it easily follows from the equations (1.19) to (1.21) that for any initial datum in $H^3 \times H^3 \times H^1$ it holds

$$(\varepsilon(\cdot,t), p(\cdot,t), \theta(\cdot,t)) \in H^4 \times H^4 \times H^3 \quad \forall\, t > 0\,. \qquad (2.58)$$

Moreover, we can infer from the previous Lemmas and estimates that for any $\nu > 0$ the triple (ε, p, θ) belongs to $C([\nu, +\infty); H^4 \times H^4 \times H^3)$. From this, the compactness of the orbit in $H^3 \times H^3 \times H^1$ follows, that is, part **(iii)** of Theorem 1.1 is also proved.

It remains to show the asymptotic results asserted in the parts **(ii)** and **(iv)** of Theorem 1.1. In the sequel, a convergence symbol "$\to$" has always to be understood as $t \to +\infty$. We will make use of the following basic lemma from Shen & Zheng [19]:

Lemma 2.7. *Suppose that y and h are nonnegative functions on $(0,\infty)$ such that y' is locally integrable, and such that there exist positive constants A_1, A_2, A_3, A_4 such that*

$$y'(t) \le A_1 y^2(t) + A_2 + h(t) \quad \forall t \ge 0, \tag{2.59}$$

$$\int_0^t y(\tau)\,d\tau \le A_3, \quad \int_0^t h(\tau)\,d\tau \le A_4, \quad \forall t \ge 0. \tag{2.60}$$

Then

$$\lim_{t\to\infty} y(t) = 0. \tag{2.61}$$

We proceed in a series of steps.

Lemma 2.8. *It holds*

$$\|p(t)\|_{H^3} \to 0, \quad \|p_t(t)\|_{H^1} \to 0, \tag{2.62}$$

$$\|\varepsilon_t(t)\|_{H^1} \to 0, \quad \|(\delta\,\varepsilon_{xx} - \sigma_1)(t)\|_{H^1} \to 0, \tag{2.63}$$

$$\|u_t(t)\|_{H^2} \to 0. \tag{2.64}$$

Proof. It follows from (2.18) and (2.39) that

$$\frac{d}{dt}\left(\|p_{xt}(t)\|^2 + \delta\,\|\,\varepsilon_{xt}(t)\|^2\right) + \gamma\,\|\varepsilon_{tt}(t)\|^2 \le \hat{C}\left(\|\varepsilon_t(t)\|^2 + \|\theta_t(t)\|^2\right). \tag{2.65}$$

Combining (2.65) with (2.40), (2.42), (2.43), and Poincaré's inequality, and applying Lemma 2.7, we obtain that

$$\|p_{xt}(t)\|^2 + \|\varepsilon_{xt}(t)\|^2 \to 0. \tag{2.66}$$

Hence, $\|p_{xxx}(t)\|^2 \to 0$, and thus also $\|u_t\|_{H^2} \to 0$.

Next, we differentiate (1.20) with respect to t, then multiply by $\delta\varepsilon_{xx} - \sigma_1$ and integrate with respect to x over Ω. It follows that

$$\begin{aligned}\frac{1}{2}\frac{d}{dt}\|(\delta\varepsilon_{xx} - \sigma_1)(t)\|^2 &= -\,(p_{tt}(t) - \gamma\varepsilon_{tt}(t), \delta\varepsilon_{xx}(t) - \sigma_1(t)) \\ &\le \frac{1}{2}\|(\delta\,\varepsilon_{xx} - \sigma_1)(t)\|^2 + \hat{C}\left(\|p_{tt}(t)\|^2 \right. \\ &\quad \left. + \|\varepsilon_{tt}(t)\|^2\right).\end{aligned} \tag{2.67}$$

Combining (2.67) with (1.20), (2.40), (2.43), and Poincaré's inequality, we can infer from Lemma 2.7 that

$$\|(\delta\,\varepsilon_{xx} - \sigma_1)(t)\|^2 \to 0. \tag{2.68}$$

From (1.20) and (2.66), we also get

$$\|(\delta\varepsilon_{xx} - \sigma_1)_x(t)\|^2 \to 0. \tag{2.69}$$

The assertions of Lemma 2.8 now follow from the above estimates and from Poincaré's inequality. □

Lemma 2.9. *It holds*

$$\|\theta_x(t)\| \to 0\,. \tag{2.70}$$

Proof. We multiply (1.21) by θ_t, integrate with respect to x over Ω, and use (2.39), to get

$$\frac{k}{2}\frac{d}{dt}\|\theta_x(t)\|^2 + \|\theta_t(t)\|^2 \le \frac{1}{2}\|\theta_t(t)\|^2 + \hat{C}\,\|p_{xx}(t)\|^2\,. \tag{2.71}$$

Hence, we can invoke (2.40), (2.41) and Lemma 2.7, to conclude the assertion. □

Concerning the convergence of ε, u, θ, we have the following result.

Lemma 2.10. *It holds*

$$(\varepsilon(\cdot,t),p(\cdot,t),\theta(\cdot,t)) \to (\overline{\varepsilon},0,\overline{\theta}) \quad in\ H^3\times H^3\times H^1\,, \tag{2.72}$$

$$u\to\overline{u} \quad in\ \ H^4\,, \quad with\ \overline{u}(x)=\int_0^x \overline{\varepsilon}(y)\,dy \quad \forall\, x\in[0,1]\,, \tag{2.73}$$

where $(\overline{\varepsilon},\overline{\theta})$ *is one of the equilibria for the corresponding stationary problem,*

$$\delta\,\varepsilon_{xx} - f_1(\varepsilon)\,\theta - f_2(\varepsilon) = 0\,, \tag{2.74}$$

$$\varepsilon_x|_{x=0}=0, \quad \varepsilon|_{x=1}=0, \tag{2.75}$$

$$\theta = \text{const.}\,, \tag{2.76}$$

$$\int_0^1\Big(\theta + F_2(\varepsilon) + \frac{\delta}{2}\,\varepsilon_x^2\Big)\,dx = E_1\,. \tag{2.77}$$

Proof. It follows from (1.30) and (2.10) that, for any $0<\nu<1$,

$$\frac{d}{dt}\int_0^1\Big(\theta-\nu\,\log\theta + F_2(\varepsilon)+\nu F_1(\varepsilon) + \frac{1}{2}p_x^2 + \frac{\delta}{2}\varepsilon_x^2\Big)(t)\,dx \\ +\nu\int_0^1\left(\frac{k\theta_x^2}{\theta^2}+\frac{\gamma p_{xx}^2}{\theta}\right)(t)\,dx = 0. \tag{2.78}$$

Thus, the system (1.19)–(1.21) has a Lyapunov function of the form

$$\int_0^1\Big(\theta-\nu\,\log\theta + F_2(\varepsilon)+\nu F_1(\varepsilon) + \frac{1}{2}p_x^2 + \frac{\delta}{2}\varepsilon_x^2\Big)(t)\,dx\,.$$

Since the orbit is compact, as proved above, it follows from the standard theory of dynamical systems that the ω-limit set is connected, compact and consists of equilibria. Since the corresponding stationary problem admits only a finite number of solutions (see Zhou [26], and also Luckhaus & Zheng [12], Novick-Cohen & Zheng [16], Zheng [25]), (2.72) follows. In view of the boundary condition $u|_{x=0}=0$, we also get (2.73). With this, the proof of Lemma 2.10 and, at the same time, also of Theorem 1.1 and Theorem 1.2, is complete. □

References

[1] M. Achenbach and I. Müller, "Creep and yield in martensitic transformations." Ingenieur-Archiv **53** (1983), 73–83.

[2] G. Andrews, "On the existence of solutions to the equation $u_{tt} = u_{xxt} + \sigma(u_x)_x$." J. Differential Equations **35** (1980), 200–231.

[3] G. Andrews and J. M. Ball, "Asymptotic behaviour and changes of phase in one-dimensional nonlinear viscoelasticity." J. Differential Equations **44** (1982), 306–341.

[4] M. Brokate and J. Sprekels, *Hysteresis and Phase Transitions.* Applied Math. Sci. **121**, Springer-Verlag, New York 1996.

[5] Z. Chen and K. H. Hoffmann, "On a one-dimensional nonlinear thermoviscoelastic model for structural phase transitions in shape memory alloys." J. Differential Equations **112** (1994), 325–350.

[6] C. M. Dafermos, "Global smooth solutions to the initial boundary value problem for the equations of one-dimensional nonlinear thermoviscoelasticity." SIAM J. Math. Anal. **13** (1982), 397–408.

[7] C. M. Dafermos and L. Hsiao, "Global smooth thermomechanical processes in one-dimensional nonlinear thermoviscoelasticity." Nonlinear Analysis, T.M.A. **6** (1982), 435–454.

[8] F. Falk, "Ginzburg-Landau theory of static domain walls in shape-memory alloys." Z. Phys. B – Condensed Matter **51** (1983), 177–185.

[9] F. Falk, "Ginzburg-Landau theory and solitary waves in shape-memory alloys." Z. Phys. B – Condensed Matter **54** (1984), 159–167.

[10] K. H. Hoffmann and A. Zochowski, "Existence of solutions to some non-linear thermoelastic systems with viscosity." Math. Meth. Appl. Sci. **15** (1992), 187–204.

[11] S. Jiang, "Global large solutions to initial boundary value problems in one-dimensional nonlinear thermoviscoelasticity." Quart. Appl. Math. **51** (1993), 731–744.

[12] S. Luckhaus and S. Zheng, "A nonlinear boundary value problem involving nonlocal term." Nonlinear Analysis, T.M.A. **22** (1994), 129–135.

[13] T. Luo, *Qualitative behaviour to nonlinear evolution equations with dissipation.* Ph. D. Thesis, Institute of Mathematics, Academy of Sciences of China, Beijing 1994.

[14] I. Müller and K. Wilmański, "A model for phase transitions in pseudoelastic bodies." Nuovo Cimento B **57** (1980), 283–318.

[15] M. Niezgódka and J. Sprekels, "Existence of solutions for a mathematical model of structural phase transitions in shape memory alloys." Math. Meth. Appl. Sci. **10** (1988), 197–223.

[16] A. Novick-Cohen and S. Zheng, "The Penrose-Fife type equations: Counting the one-dimensional stationary solutions." Proc. Royal Soc. of Edinburgh **126A** (1996), 483–504.

[17] R. L. Pego, "Phase transitions in one-dimensional nonlinear viscoelasticity: admissibility and stability." Arch. Rat. Mech. Anal. **97** (1987), 353–394.

[18] R. Racke and S. Zheng, "Global existence and asymptotic behaviour in nonlinear thermoviscoelasticity." J. Differential Equations **134** (1997), 46–67.

[19] W. Shen and S. Zheng, "On the coupled Cahn-Hilliard equations." Comm. PDE **18** (1993), 701–727.

[20] W. Shen, S. Zheng and P. Zhu, "Global existence and asymptotic behaviour of weak solution to nonlinear thermoviscoelastic system." To appear in *Quart. Appl. Math.*

[21] J. Sprekels and S. Zheng, "Global solutions to the equations of a Ginzburg-Landau theory for structural phase transitions in shape memory alloys." Physica D **39** (1989), 59–76.

[22] J. Sprekels and S. Zheng, "Maximal attractor for the system of a Landau-Ginzburg theory for structural phase transitions in shape memory alloys." Submitted.

[23] J. Sprekels, S. Zheng and P. Zhu, "Asymptotic behaviour of solutions to the system of a Landau-Ginzburg theory for structural phase transitions in shape memory alloys." To appear 1998 in *SIAM J. Math. Anal.*

[24] R. Temam, *Infinite-Dimensional Dynamical Systems in Mechanics and Physics.* Applied Math. Sci. **68**, Springer-Verlag, New York 1988.

[25] S. Zheng, "Nonlinear Parabolic Equations and Hyperbolic-Parabolic Coupled Systems." Pitman Series Monographs and Surveys in Pure and Applied Mathematics **76**, Longman Group Limited, London, 1995.

[26] P. Zhou, "Multiplicity of solutions to a nonlinear boundary value problem." Submitted to J. Math. Anal. Appl.

Jürgen Sprekels
Weierstrass Institute for Applied
Analysis and Stochastics
Mohrenstrasse 39
D-10117 Berlin, Germany

Songmu Zheng
Institute of Mathematics
Fudan University
200433 Shanghai, P. R. China

Progress in Nonlinear Differential Equations
and Their Applications, Vol. 35
© 1999 Birkhäuser Verlag Basel/Switzerland

Decompositions of Function Spaces

Hans Triebel

1. Introduction

Let $\psi(x)$ be a C^∞ function in $\mathbb{R}^n$ with compact support such that

$$1 = \sum_{k \in \mathbb{Z}^n} \psi(x-k) \quad \text{for all} \quad x \in \mathbb{R}^n \tag{1.1}$$

is a resolution of unity in $\mathbb{R}^n$. By [Tri92], 2.4.7, we have the so-called localization principle

$$\|f|F^s_{pq}(\mathbb{R}^n)\| \sim \left(\sum_{k \in \mathbb{Z}^n} \left\|\psi(\cdot - k)f \mid F^s_{pq}(\mathbb{R}^n)\right\|^p \right)^{\frac{1}{p}} \tag{1.2}$$

for all spaces $F^s_{pq}(\mathbb{R}^n)$ where $s \in \mathbb{R}$, $0 < p \le \infty$, $0 < q \le \infty$, with $q = \infty$ if $p = \infty$, (equivalent quasi-norms). In addition nowadays powerful characterizing atomic and subatomic decompositions of $f \in F^s_{pq}(\mathbb{R}^n)$ in elementary building blocks are available. Together with estimates for entropy numbers of related compact embeddings between (weighted) spaces of such type these localizations and decompositions pave the way to a substantial spectral theory for (degenerate) pseudodifferential operators on $\mathbb{R}^n$ and also on bounded domains and even on fractals. Details about this programme may be found in [ET96] and [Tri97].

It is the main aim of this paper to develop a corresponding intrinsic theory of refined localizations, atomic and subatomic decompositions in bounded (smooth) domains Ω in $\mathbb{R}^n$. We hope that this is of interest for its own sake. However our main motivation comes from spectral theory for (degenerate and weighted) differential and pseudodifferential operators on hyperbolic manifolds. As a prototype of hyperbolic manifolds we have in mind may serve the Poincaré n-ball, that means the unit ball $B = \Omega$ in $\mathbb{R}^n$ equipped with the Riemannian metric

$$dS^2 = (1 - |x|^2)^{-2} \sum_{j=1}^n dx_j^2. \tag{1.3}$$

It is quite clear that a theory of function spaces on hyperbolic manifolds, say on the Poincaré n-ball, which is sufficiently powerful for the above indicated purposes, results in a detailed study of the delicate interplay of spaces on $\mathbb{R}^n$ on the one hand,

and weighted and unweighted spaces on Ω on the other hand. A central role on the Euclidean side will be played by spaces of type

$$\widetilde{F}^s_{pq}(\Omega) = \{ f \in F^s_{pq}(\mathbb{R}^n) : \operatorname{supp} f \subset \overline{\Omega} \} \tag{1.4}$$

which can be interpreted either as (closed) subspaces of $F^s_{pq}(\mathbb{R}^n)$ or as subspaces of $D'(\Omega)$. We have to clarify the nature of these spaces, their relations to $F^s_{pq}(\Omega)$, to $\overset{\circ}{F}{}^s_{pq}(\Omega)$, which stands for the completion of $C_0^\infty(\Omega) = D(\Omega)$ in $F^s_{pq}(\Omega)$, and to some weighted spaces in Ω, where powers of $\operatorname{dist}(x, \partial\Omega)$ with $x \in \Omega$, will occur. Furthermore, refined localizations and (sub)atomic decompositions will be derived.

Connections with spaces on hyperbolic manifolds and a related spectral theory will be shifted to later papers.

All definitions, results, comments, and some shorter arguments are collected in Section 2. Proofs of the main assertions are given in Section 3. In the course of these proofs we obtain some new results for the considered spaces which might also be of self-contained interest: equivalent quasi-norms, homogeneity properties and assertions about pointwise multipliers. We collected these results in the separate subsections 3.7 and 3.9 which complement Section 2.

2. Definitions and results

2.1 Basic notation, definitions, properties

2.1.1 Basic notation Let $\mathbb{R}^n$ be Euclidean n-space, put $\mathbb{R} = \mathbb{R}^1$. Let $\mathbb{N}$ be the collection of all natural numbers and $\mathbb{N}_0 = \mathbb{N} \cup \{0\}$. Let $S(\mathbb{R}^n)$ be the Schwartz space of all complex-valued, rapidly decreasing, infinitely differentiable functions on $\mathbb{R}^n$. By $S'(\mathbb{R}^n)$ we denote the dual space of all tempered distributions on $\mathbb{R}^n$. Furthermore, $L_p(\mathbb{R}^n)$ with $0 < p \leq \infty$, is the usual quasi-Banach space with respect to the Lebesgue measure, quasi-normed by

$$\| f \mid L_p(\mathbb{R}^n) \| = \left(\int_{\mathbb{R}^n} |f(x)|^p dx \right)^{\frac{1}{p}}, \tag{2.1}$$

with the usual modification if $p = \infty$. If $\varphi \in S(\mathbb{R}^n)$ then

$$\widehat{\varphi}(\xi) = (F\varphi)(\xi) = (2\pi)^{-\frac{n}{2}} \int_{\mathbb{R}^n} e^{-i\xi x} \varphi(x)\, dx, \quad \xi \in \mathbb{R}^n, \tag{2.2}$$

denotes the Fourier transform of φ. As usual, $\overset{\vee}{\varphi}$ or $F^{-1}\varphi$, stand for the inverse Fourier transform, given by the right-hand side of (2.2) with i in place of $-i$. Whether we use $F\varphi, F^{-1}\varphi$ or $\widehat{\varphi}, \overset{\vee}{\varphi}$ will be a matter of convenience. Of course ξx denotes the scalar product in $\mathbb{R}^n$. Both F and F^{-1} are extended to $S'(\mathbb{R}^n)$ in the standard way. Let $\varphi \in S(\mathbb{R}^n)$ with

$$\varphi(x) = 1 \quad \text{if} \quad |x| \leq 1 \quad \text{and} \quad \varphi(y) = 0 \quad \text{if} \quad |y| \geq \frac{3}{2}. \tag{2.3}$$

We put $\varphi_0 = \varphi, \varphi_1(x) = \varphi(\frac{x}{2}) - \varphi(x)$, and

$$\varphi_k(x) = \varphi_1(2^{-k+1}x), \quad x \in \mathbb{R}^n,\ k \in \mathbb{N}. \tag{2.4}$$

Then, since

$$1 = \sum_{k=0}^{\infty} \varphi_k(x) \quad \text{for all} \quad x \in \mathbb{R}^n, \tag{2.5}$$

the φ_k form a dyadic resolution of unity. Recall that $(\varphi_k \widehat{f})^\vee(x)$ is an entire analytic function on $\mathbb{R}^n$ for any $f \in S'(\mathbb{R}^n)$. In particular, $(\varphi_k \widehat{f})^\vee(x)$ makes sense pointwise.

2.1.2 Definition *Let $s \in \mathbb{R}$. Let either $0 < p < \infty, 0 < q \leq \infty$ or $p = q = \infty$. Then $F^s_{pq}(\mathbb{R}^n)$ is the collection of all $f \in S'(\mathbb{R}^n)$ such that*

$$\|f \mid F^s_{pq}(\mathbb{R}^n)\| = \left\| \left(\sum_{j=0}^{\infty} 2^{jsq} |(\varphi_j \widehat{f})^\vee(\cdot)|^q \right)^{\frac{1}{q}} \mid L_p(\mathbb{R}^n) \right\| \tag{2.6}$$

(with the usual modification if $q = \infty$) is finite.

2.1.3 Remark We assume that the reader is familiar with the theory of these spaces. We refer to [Tri83], [Tri92], and as far as more recent topics are concerned, to [ET96], [RuS96], and [Tri97]. Recall that $F^s_{pq}(\mathbb{R}^n)$ are quasi-Banach spaces which are independent of φ chosen according to 2.1.1 (Banach spaces if $p \geq 1$, $q \geq 1$). We mention that

$$H^s_p(\mathbb{R}^n) = F^s_{p,2}(\mathbb{R}^n), \quad s \in \mathbb{R}, \quad 1 < p < \infty, \tag{2.7}$$

are the usual (fractional) Sobolev spaces with the (classical) Sobolev spaces

$$W^k_p(\mathbb{R}^n) = H^k_p(\mathbb{R}^n), \quad k \in \mathbb{N}_0, \quad 1 < p < \infty,$$

as a subclass. Usually one introduces parallel to $F^s_{pq}(\mathbb{R}^n)$ the Besov spaces $B^s_{pq}(\mathbb{R}^n)$. But for the theory we wish to develop here, the spaces $F^s_{pq}(\mathbb{R}^n)$ have priority. For example, the localization principle (1.2) is valid for B^s_{pq} in place of F^s_{pq} if, and only if, $p = q$, that means in case of $B^s_{pp} = F^s_{pp}$, now with $0 < p \leq \infty$, see [Tri92], 2.4.7. This is also the reason why we incorporated here the (generalized) Hölder-Zygmund spaces

$$\mathcal{C}^s(\mathbb{R}^n) = F^s_{\infty\infty}(\mathbb{R}^n), \quad s \in \mathbb{R}, \tag{2.8}$$

into the F-scale.

2.1.4 Definition *Let Ω be a bounded domain in $\mathbb{R}^n$. Let $s \in \mathbb{R}$. Let $0 < p \leq \infty$, $0 < q \leq \infty$ (with $q = \infty$ if $p = \infty$). Then*

$$F^s_{pq}(\Omega) = \left\{ f \in D'(\Omega) : \text{ there is a } g \in F^s_{pq}(\mathbb{R}^n) \text{ with } g|\Omega = f \right\}, \tag{2.9}$$

$$\|f \mid F^s_{pq}(\Omega)\| = \inf \|g \mid F^s_{pq}(\mathbb{R}^n)\|, \tag{2.10}$$

where the infimum is taken over all $g \in F^s_{pq}(\mathbb{R}^n)$ *such that its restriction* $g|\Omega$ *to* Ω *coincides in* $D'(\Omega)$ *with* f. *Furthermore* $\overset{\circ}{F}{}^s_{pq}(\Omega)$ *is the completion of* $D(\Omega)$ *in* $F^s_{pq}(\Omega)$.

2.1.5 Remark Of course, $D(\Omega) = C_0^\infty(\Omega)$ stands for the collection of all complex C^∞ functions with compact support in Ω, whereas $D'(\Omega)$ is the dual space of complex distributions. Mostly we suppose that the bounded domain Ω has a C^∞ boundary $\partial\Omega$. Sometimes it is sufficient to know that $\partial\Omega$ is Lipschitzian. Then $\partial\Omega$ is a compact $(n-1)$- set, where the notion of d-sets comes from fractal geometry, see [Tri97] for details. Hence the techniques developed in [Tri97] can be applied. However we shall not stress this point in this paper. As said in the Introduction we are mainly interested in the spaces $\widetilde{F}^s_{pq}(\Omega)$ of type (1.4), including their relations to $F^s_{pq}(\Omega)$ and $\overset{\circ}{F}{}^s_{pq}(\Omega)$. Although not crucial for us it sheds some light on what follows if one clarifies first under which conditions $F^s_{pq}(\Omega)$ and $\overset{\circ}{F}{}^s_{pq}(\Omega)$ coincide.

2.1.6 Proposition *Let* Ω *be a bounded* C^∞ *domain in* $\mathbb{R}^n$ *and let* $F^s_{pq}(\Omega)$ *and* $\overset{\circ}{F}{}^s_{pq}(\Omega)$ *be the spaces introduced in Definition 2.1.4. Then*

$$F^s_{pq}(\Omega) = \overset{\circ}{F}{}^s_{pq}(\Omega) \tag{2.11}$$

if, and only if, one of the following two conditions is satisfied:

$$\text{(i)} \quad 0 < p < \infty, \ -\infty < s < \frac{1}{p}, \ 0 < q < \infty; \tag{2.12}$$

$$\text{(ii)} \quad 1 < p < \infty, \ s = \frac{1}{p}, \ 0 < q < \infty. \tag{2.13}$$

2.1.7 Comment Let $C^\infty(\overline{\Omega})$ be the restriction of $S(\mathbb{R}^n)$ to Ω. It is more or less well known that

$$C^\infty(\overline{\Omega}) \quad \text{is not dense in} \, F^s_{p\infty}(\Omega) \quad \text{for} \quad 0 < p \leq \infty \text{ and } s \in \mathbb{R}. \tag{2.14}$$

This is mentioned in some books (including our own ones) but nowhere are detailed arguments. In the course of the proof of the above proposition we prove also (2.14). The rather simple construction in 3.3 imitates the famous Weierstrass function and is near to [Tri97], Section 16, pp. 119–122. It makes clear that just the failure of the spaces $F^s_{p\infty}(\Omega)$ according to (2.14) makes them interesting in fractal geometry. In particular, (2.14) disproves also (2.11) with $q = \infty$ for any $0 < p \leq \infty$ and $s \in \mathbb{R}$.

To prepare what follows we recall the standard abbreviations

$$\sigma_p = n(\frac{1}{p} - 1)_+ \quad \text{and} \quad \sigma_{pq} = n\left(\frac{1}{\min(p,q)} - 1\right)_+, \tag{2.15}$$

where $0 < p \leq \infty$, $0 < q \leq \infty$ and $a_+ = \max(a, 0)$ if $a \in \mathbb{R}$.

2.1.8 Definition *Let Ω be a bounded domain in $\mathbb{R}^n$. Let $0 < p \le \infty$, $0 < q \le \infty$ (with $q = \infty$ if $p = \infty$) and $s > \sigma_p$. Then*

$$\widetilde{F}^s_{pq}(\Omega) = \{f \in F^s_{pq}(\mathbb{R}^n) : \operatorname{supp} f \subset \overline{\Omega}\} \tag{2.16}$$

with

$$\|f \mid \widetilde{F}^s_{pq}(\Omega)\| = \|f \mid F^s_{pq}(\mathbb{R}^n)\|. \tag{2.17}$$

2.1.9 Interpretation and discussion Let $|\partial\Omega| = 0$, where $|\partial\Omega|$ is the Lebesgue measure of $\partial\Omega$. Of course this is always the case if $\partial\Omega$ is C^∞ or Lipschitzian. Since Ω is bounded and $s > \sigma_p$ we have

$$\widetilde{F}^s_{pq}(\Omega) \subset L_1(\Omega). \tag{2.18}$$

As far as the underlying embedding theorems are concerned we refer to [Tri83], 2.7, and [RuS96], 2.2.4. In particular there is no ambiguity to interpret $\widetilde{F}^s_{pq}(\Omega)$ also as a space of distributions in Ω,

$$\widetilde{F}^s_{pq}(\Omega) \subset D'(\Omega). \tag{2.19}$$

Since there is no danger of confusion we shall not distinguish between these two interpretations. On the other hand if s is not necessarily restricted by $s > \sigma_p$ then one should distinguish between

$$\widetilde{F}^s_{pq}(\overline{\Omega}) = \{f \in F^s_{pq}(\mathbb{R}^n) : \ \operatorname{supp} f \subset \overline{\Omega}\} \tag{2.20}$$

and

$$\widetilde{F}^s_{pq}(\Omega) =$$

$$\left\{f \in D'(\Omega) : \text{ there is a } g \in F^s_{pq}(\mathbb{R}^n) \text{ with } \operatorname{supp} g \subset \overline{\Omega} \text{ and } f = g|\Omega\right\}, \tag{2.21}$$

$$\|f \mid \widetilde{F}^s_{pq}(\Omega)\| = \inf \|g \mid F^s_{pq}(\mathbb{R}^n)\|, \tag{2.22}$$

where the infimum is taken over all admitted g in (2.21). These two types of spaces may be different, more precisely,

$$\widetilde{F}^s_{pq}(\Omega) = \widetilde{F}^s_{pq}(\overline{\Omega}) / \{h \in F^s_{pq}(\mathbb{R}^n) : \operatorname{supp} h \subset \partial\Omega\} \tag{2.23}$$

as a factor space. We stick at Definition 2.1.8 and $s > \sigma_p$ in this paper. But we refer to 2.3.1 and to the discussion in 2.5.8 (iv) in comparison with what has been done in literature. On the other hand, spaces of type $\widetilde{F}^s_{pq}(\Omega)$ and their counterparts $\widetilde{B}^s_{pq}(\Omega)$ and $\widetilde{H}^s_p(\Omega)$, played a substantial role in the theory of scales of spaces and their duality. We refer to [Tri78], 2.10.3–2.10.5, 4.3.2, 4.8.1, 4.9.2. We return to this point in 2.3.2 below.

2.1.10 Definition *Let $n \geq 2$, $0 < p \leq \infty$, $0 < q \leq \infty$ (with $q = \infty$ if $p = \infty$) and $s > \sigma_p$. Then $F^s_{pq}(\mathbb{R}^n)$ is said to have the Fubini property if for all $f \in F^s_{pq}(\mathbb{R}^n)$*

$$\|f \,|\, F^s_{pq}(\mathbb{R}^n)\| \sim \sum_{j=1}^{n} \Big\| \, \|f(x_1, \ldots, x_{j-1}, \cdot, x_{j+1}, \ldots, x_n) \,|\, F^s_{pq}(\mathbb{R})\| \Big| \, L_p(\mathbb{R}^{n-1}) \Big\| \tag{2.24}$$

(equivalent quasi-norms).

2.1.11 Explanation Since $s > \sigma_p$ the space $F^s_{pq}(\mathbb{R}^n)$ consists entirely of regular distributions belonging to $L_1(\mathbb{R}^n) + L_\infty(\mathbb{R}^n)$. Hence it makes sense to fix $n-1$ variables $x_1, \ldots, x_{j-1}, x_{j+1}, \ldots, x_n$ and to ask whether

$$f(x_1, \ldots, x_{j-1}, \cdot, x_{j+1}, \ldots, x_n)$$

belongs to $F^s_{pq}(\mathbb{R})$ on the real line $\mathbb{R}$ (a.e. with respect to the $(n-1)$-dimensional Lebesgue measure) and whether the outcome yields an equivalent quasi-norm according to (2.24). It is even sufficient to check (2.24) for smooth functions: If $q < \infty$ then $S(\mathbb{R}^n)$ is dense in $F^s_{pq}(\mathbb{R}^n)$; if $q = \infty$ then one can rely on the Fatou property of $F^s_{pq}(\mathbb{R}^n)$, and in both cases on Fatou's lemma, see [ET96], p.48. As proved in [Tri83], 2.5.13, the spaces $F^s_{pp}(\mathbb{R}^n)$ with $0 < p \leq \infty$, $s > \sigma_p$, have the Fubini property. If $p \neq q$ the situation is more complicated. As for large values of s we refer again to [Tri83], 2.5.13. The case $1 < p < \infty$, $1 < q < \infty$, $s > 0$, may be found in [Kal80], see also [RuS96], p. 70. We need the Fubini property as a technical tool. For this purpose we give in 3.2 a proof of the following extension of the above results, which solves the problem stated in [Tri83], Remark 2.5.13/2 on p. 117. Recall that σ_{pq} has been introduced in (2.15).

2.1.12 Theorem *Let $n \geq 2$. The spaces $F^s_{pq}(\mathbb{R}^n)$ with $0 < p \leq \infty$, $0 < q \leq \infty$ (where $q = \infty$ if $p = \infty$) and $s > \sigma_{pq}$ have the Fubini property.*

2.1.13 Remark The proof relies at least partly on subatomic decompositions without moment conditions. This explains the influence of q. It would be desirable to clarify what happens if $\sigma_p < s \leq \sigma_{pq}$.

2.2 The spaces $\widetilde{F}^s_{pq}(\Omega)$

2.2.1 Resolution of unity Let Ω be a bounded domain in $\mathbb{R}^n$ with Lipschitzian boundary. Although not really needed we may always assume that Ω is connected. This means in case of $n = 1$ that Ω is an open bounded interval. It is not difficult to see that there exist open balls $\{K_{jr} : j \in \mathbb{N}_0; r = 1, \ldots, N_j\}$ and subordinated C^∞ functions $\{\varphi_{jr}(x) : j \in \mathbb{N}_0; r = 1, \ldots, N_j\}$ with the following properties:
(i) There are positive numbers $c_1, \ldots, c_5$ and $N \in \mathbb{N}$ such that $c_1 2^{-j}$ is the radius of K_{jr},

$$c_2 \leq N_j \, 2^{-j(n-1)} \leq c_3, \quad \text{where} \quad j \in \mathbb{N}_0, \tag{2.25}$$

at most N of all these balls have a non-empty intersection,

$$\bigcup_{j\in\mathbb{N}_0}\bigcup_{r=1}^{N_j} K_{jr} = \Omega, \tag{2.26}$$

and for any ball

$$c_4\, 2^{-j} \le \operatorname{dist}(\overline{K}_{jr}, \partial\Omega) \le c_5\, 2^{-j}. \tag{2.27}$$

(ii) There are positive numbers c_γ such that for $j \in \mathbb{N}_0$ and $r = 1, \ldots, N_j$

$$|D^\gamma \varphi_{jr}(x)| \le c_\gamma\, 2^{j|\gamma|} \quad \text{where} \quad \gamma \in \mathbb{N}_0^n \tag{2.28}$$

is a multi-index,

$$\operatorname{supp} \varphi_{jr} \subset K_{jr}, \tag{2.29}$$

and

$$\sum_{j=0}^{\infty}\sum_{r=1}^{N_j} \varphi_{jr}(x) = 1 \quad \text{if} \quad x \in \Omega. \tag{2.30}$$

2.2.2 Theorem *Let Ω be a bounded C^∞ domain in $\mathbb{R}^n$ and let $\{\varphi_{jr}(x) : j \in \mathbb{N}_0; r = 1, \ldots, N_j\}$ be the above resolution of unity. Let $0 < p \le \infty$, $0 < q \le \infty$ (with $q = \infty$ if $p = \infty$) and $s > \sigma_{pq}$ (see (2.15)). Then $f \in L_1(\Omega)$ belongs to $\widetilde{F}^s_{pq}(\Omega)$ if, and only if,*

$$\left(\sum_{j=0}^{\infty}\sum_{r=1}^{N_j} \|\varphi_{jr}\, f \,|\, F^s_{pq}(\mathbb{R}^n)\|^p\right)^{\frac{1}{p}} < \infty \tag{2.31}$$

(usual modification if $p = q = \infty$). Furthermore, (2.31) is an equivalent quasi-norm.

2.2.3 Comment This is the refined localization principle in smooth domains Ω which is the substitute of (1.1) and (1.2) in $\mathbb{R}^n$. As mentioned in 2.1.3 there is no hope to find something like (2.31) with B^s_{pq} in place of F^s_{pq} if $p \ne q$. Maybe one can extend this theorem to bounded domains with Lipschitzian boundary. But the proof given below depends on a characterization of $F^s_{pq}(\Omega)$ in bounded C^∞ domains and $s > \sigma_{pq}$. An extension to $s > \sigma_p$ would be desirable. If s is smaller than σ_p then one has additional difficulties discussed in 2.1.9. But the problem is interesting in any case, see also 2.5.8(iv). The restriction to $s > \sigma_{pq}$ or even to $s > \sigma_p$ makes it possible to use $L_1(\Omega)$ as the largest space in which everything happens, see (2.18). The proof of the above theorem depends on some inequalities of Hardy type for the spaces $\widetilde{F}^s_{pq}(\Omega)$ which we are going to describe next. Let

$$d(x) = \operatorname{dist}(x, \partial\Omega), \quad \text{where} \quad x \in \Omega; \tag{2.32}$$

be the distance of a point $x \in \Omega$ to the boundary $\partial\Omega$.

2.2.4 Definition *Let Ω be a bounded domain in $\mathbb{R}^n$. Let $0 < p \le \infty$ and $s \in \mathbb{R}$. Then $L_p(\Omega, d^{-s})$ is the collection of all complex Lebesgue measurable functions in Ω such that*

$$\|f \mid L_p(\Omega, d^{-s})\| = \left(\int_\Omega d^{-sp}(x) |f(x)|^p \, dx \right)^{\frac{1}{p}} < \infty \tag{2.33}$$

(with the usual modification if $p = \infty$).

2.2.5 Proposition *Let Ω be a bounded C^∞ domain in $\mathbb{R}^n$. Let $0 < p \le \infty$, $0 < q \le \infty$ (with $q = \infty$ if $p = \infty$), and $s > \sigma_p$. There exists a positive number c such that*

$$\|f \mid L_p(\Omega, d^{-s})\| \le c \|f \mid \widetilde{F}^s_{pq}(\Omega)\| \tag{2.34}$$

for all $f \in \widetilde{F}^s_{pq}(\Omega)$.

2.2.6 Remark The reason why $s > \sigma_p$ instead of $s > \sigma_{pq}$ is sufficient comes from the monotonicity of the spaces $\widetilde{F}^s_{pq}(\Omega)$ with respect to the q-index. Hence we must prove (2.34) only for $q = \infty$.

2.2.7 Hardy inequalities Let $1 < p < \infty$ and $s = k \in \mathbb{N}$. By 2.1.3 the estimate (2.34) covers in this case

$$\int_\Omega d^{-kp}(x) |f(x)|^p \, dx \le c \sum_{|\alpha| \le k} \int_\Omega |D^\alpha f(x)|^p \, dx \tag{2.35}$$

for all $f \in \mathring{W}^k_p(\Omega) = \widetilde{W}^k_p(\Omega)$, see also 2.4.2 below. This is an immediate consequence of the well-known classical Hardy inequality. By real and complex interpolation of $\mathring{W}^k_p(\Omega)$ with $1 < p < \infty$ and $k \in \mathbb{N}_0$ one obtains

$$\|f \mid L_p(\Omega, d^{-s})\| \le c \|f \mid \widetilde{H}^s_p(\Omega)\|; \quad 1 < p < \infty, \ s > 0; \tag{2.36}$$

and

$$\|f \mid L_p(\Omega, d^{-s})\| \le c \|f \mid \widetilde{B}^s_{pp}(\Omega)\|; \quad 1 < p < \infty, \ s > 0, \tag{2.37}$$

where $\widetilde{B}^s_{pp} = \widetilde{F}^s_{pp}$. As for the needed interpolation we refer to [Tri78], 4.3.2, Theorem 2 on p. 318 (in the 1995 edition); see also [Tri78], 3.2.6, p. 259, for weighted versions of type (2.37). Let

$$0 < p < \infty, \ 0 < q < \infty, \ \text{and} \ \sigma_{pq} < s < \min(1, \frac{1}{p}), \tag{2.38}$$

see (2.15). By [RuS96], 4.6.3, p. 208, the characteristic function of Ω is a pointwise multiplier in $F^s_{pq}(\mathbb{R}^n)$, and hence we can apply [Tri83], Proposition 1 and Remark 1 on p. 155. This proves (2.34) under the restriction (2.38), where, of course, $\widetilde{F}^s_{pq}(\Omega) = F^s_{pq}(\Omega)$. In other words, Proposition 2.2.5 extends these known special cases in a rather final way. Two additional comments seem to be in order. First, if $0 < p \le \infty$, $0 < q \le \infty$, and $s > \sigma_p$ then there is an obvious counterpart $\widetilde{B}^s_{pq}(\Omega)$ of $\widetilde{F}^s_{pq}(\Omega)$. If $p \ne q$, then assertions of type (2.34) cannot be expected. If

$s > 0$, $1 < p < \infty$, $1 \le q \le \infty$, then the substitute, which is more complicated, may be found in [Tri78], 4.3.2, p. 319. Secondly, it is well known that F and B spaces with $s - \frac{1}{p} \in \mathbb{N}_0$ play a special role. We return to this problem later on for the F-spaces.

2.2.8 Theorem *Let Ω be a bounded C^∞ domain in $\mathbb{R}^n$. Let $0 < p \le \infty$, $0 < q \le \infty$ (with $q = \infty$ if $p = \infty$), and $s > \sigma_{pq}$. Then*

$$\widetilde{F}^s_{pq}(\Omega) = F^s_{pq}(\Omega) \cap L_p(\Omega, d^{-s}) \tag{2.39}$$

(equivalent quasi-norms).

2.2.9 Remark Let $0 < p \le \infty$, $0 < q \le \infty$ (with $q = \infty$ if $p = \infty$), and $s > \sigma_p$. By Definitions 2.1.4 and 2.1.8, and by Proposition 2.2.5 there is a positive number c such that for all $f \in \widetilde{F}^s_{pq}(\Omega)$

$$\|f \,|\, F^s_{pq}(\Omega)\| + \|f \,|\, L_p(\Omega, d^{-s})\| \le c \, \|f \,|\, \widetilde{F}^s_{pq}(\Omega)\|. \tag{2.40}$$

Hence the theorem claims that at least under the additional restriction $s > \sigma_{pq}$ the converse of (2.40) is also true. The question whether Theorems 2.2.8 and 2.2.2 hold for all $s > \sigma_p$ remains open. For example, the following assertion indicates that $s > \sigma_{pq}$ might not be the last word.

2.2.10 Proposition *Let Ω be a bounded C^∞ domain in $\mathbb{R}^n$. Let*

$$0 < p < \infty, \; 0 < q \le \infty, \; \sigma_p < s < \frac{1}{p}. \tag{2.41}$$

Then

$$F^s_{pq}(\Omega) = \widetilde{F}^s_{pq}(\Omega). \tag{2.42}$$

2.2.11 Remark Under the conditions (2.41) for p, q, and s, the characteristic function of Ω is a pointwise multiplier for $F^s_{pq}(\mathbb{R}^n)$. We refer to [Fra86] and [RuS96], 4.6.3, p. 208, and the usual localization technique. Hence, together with Proposition 2.2.5, we have (2.39) also under the condition (2.41). Unfortunately, at least by our arguments below, this is not sufficient to extend also Theorem 2.2.2 to these cases. Finally, by Proposition 2.1.6 we have also

$$\widetilde{F}^s_{pq}(\Omega) = F^s_{pq}(\Omega) = \overset{\circ}{F}{}^s_{pq}(\Omega) \tag{2.43}$$

if

$$0 < p < \infty, \; 0 < q < \infty, \; \sigma_p < s < \frac{1}{p}. \tag{2.44}$$

The question under which conditions $\widetilde{F}^s_{pq}(\Omega)$ and $\overset{\circ}{F}{}^s_{pq}(\Omega)$ coincide will be discussed later on in 2.4.2.

2.2.12 Equivalent quasi-norms By Theorem 2.2.8 any intrinsic equivalent quasi-norm in $F^s_{pq}(\Omega)$, complemented by $\|d^{-s} f \,|\, L_p(\Omega)\|$, gives an intrinsic equivalent quasi-norm in $\widetilde{F}^s_{pq}(\Omega)$. A similar remark applies to Theorem 2.2.2. But one can say a little bit more. This will be formulated in 3.7 as a by-product of the proofs of these two theorems.

2.3 The spaces $F^s_{pq}(\Omega)$

2.3.1 $\widetilde{F}^s_{pq}(\Omega)$ **revisited** The spaces $F^s_{pq}(\Omega)$ have been introduced for all $s \in \mathbb{R}$, $0 < p \leq \infty$, $0 < q \leq \infty$ (with $q = \infty$ if $p = \infty$) whereas we restricted the definition of $\widetilde{F}^s_{pq}(\Omega)$ in 2.1.8 to $s > \sigma_p$. We wanted to avoid the difficulties indicated by (2.20)–(2.23). This is ensured by (2.18), where the second factor on the right-hand side of (2.23) is trivial. However this assertion can be extended to some $s \leq 0$. More precisely:
Let

$$1 < p \leq \infty, \quad 0 < q \leq \infty, \quad (\textit{with } q = \infty \textit{ if } p = \infty), \tag{2.45}$$

and

$$\frac{1}{p} - 1 < s \leq 0. \tag{2.46}$$

Then

$$\left\{ h \in F^s_{pq}(\mathbb{R}^n) : \text{ supp } h \subset \partial\Omega \right\} = \{0\}. \tag{2.47}$$

Since $\partial\Omega$ is a compact set, the spaces in (2.47) are monotone with respect to all three indices s, p, q. Hence to prove the above assertion it is sufficient to assume $1 < p = q < \infty$ and $\frac{1}{p} - 1 < s < 0$. By [Tri83], 2.11.2, p. 178, under these restrictions we have the duality

$$\left(F^s_{pp}(\mathbb{R}^n) \right)' = F^{-s}_{p'\,p'}(\mathbb{R}^n) \tag{2.48}$$

according to the dual pairing $(S(\mathbb{R}^n), S'(\mathbb{R}^n))$. Of course, $\frac{1}{p} + \frac{1}{p'} = 1$, and $B^\sigma_{pp} = F^\sigma_{pp}$. We claim that under the above conditions $1 < p' < \infty$, $0 < -s < \frac{1}{p'}$,

$$\{ \varphi \in S(\mathbb{R}^n) : \varphi(x) = 0 \quad \text{near} \quad \partial\Omega \} \tag{2.49}$$

is dense in $F^{-s}_{p'\,p'}(\mathbb{R}^n)$. This follows by atomic decompositions in the same way as in [Tri97], Step 2 of the proof of Theorem 17.8, where we use that the compact C^∞ manifold $\partial\Omega$ is a $(n-1)$-set. Let $h \in F^s_{pq}(\mathbb{R}^n)$ with supp $h \subset \partial\Omega$. By (2.48) we have

$$\| h \,|\, F^s_{pp}(\mathbb{R}^n) \| = \sup |h(\varphi)| = 0, \tag{2.50}$$

where the supremum is taken over all φ with (2.49) and $\| \varphi \,|\, F^{-s}_{p'\,p'}(\mathbb{R}^n) \| = 1$. This proves (2.47). Returning to (2.23) we can now extend the definition of $\widetilde{F}^s_{pq}(\Omega)$ in 2.1.8 without any ambiguities to

$$0 < p \leq \infty,\ 0 < q \leq \infty, \quad (\text{ with } q = \infty \text{ if } p = \infty\ ), \tag{2.51}$$

and

$$\max\left(\frac{1}{p} - 1, n(\frac{1}{p} - 1) \right) < s < \infty. \tag{2.52}$$

The pointwise multiplier assertion of the characteristic function of the bounded C^∞ *domain* Ω mentioned in Remark 2.2.11 *can also be extended to (2.51) and*

$$\max\left(\frac{1}{p}-1, n(\frac{1}{p}-1)\right) < s < \frac{1}{p}. \tag{2.53}$$

We refer again to [RuS96], 4.6.3, p. 208, where one finds the final version which in turn is based on [Tri83], [Kal85], [Fra86], and the forerunners mentioned there. Now it follows (2.42) with (2.51), (2.53), in place of (2.41). By Proposition 2.1.6 we have also (2.43) now with

$$0 < p < \infty,\ 0 < q < \infty,\ \max\left(\frac{1}{p}-1, n(\frac{1}{p}-1)\right) < s < \frac{1}{p}. \tag{2.54}$$

2.3.2 Scales of spaces Let again Ω be a bounded C^∞ domain in $\mathbb{R}^n$ and let $1 \le p \le \infty$, $1 \le q \le \infty$ (with $q=\infty$ if $p=\infty$) be fixed. We introduce the scales of Banach spaces

$$\left\{\widetilde{F}^s_{pq}(\Omega) \text{ if } s>0\,;\ F^s_{pq}(\Omega) \text{ if } s\le 0\right\}. \tag{2.55}$$

By 2.3.1 this makes sense and there is even a comfortable overlapping of the spaces $\widetilde{F}^s_{pq}(\Omega)$ and $F^s_{pq}(\Omega)$ within the interval $\frac{1}{p}-1 < s < \frac{1}{p}$. These scales are the correct counterparts of the corresponding scales in $\mathbb{R}^n$

$$\left\{F^s_{pq}(\mathbb{R}^n) \quad \text{where} \quad s \in \mathbb{R}\right\}.$$

At least for the special cases $H^s_p = F^s_{p,2}$ and $B^s_{pp} = F^s_{pp}$ with $1<p<\infty$ one has a perfect theory of lifting via fractional powers of regular elliptic operators and of duality. We refer to [Tri78], 4.9.2, p. 335. We need here some duality assertions. By [Tri83], 2.11.2, complemented by [Mar87] and [FrJ90], we have

$$\left(F^s_{pq}(\mathbb{R}^n)\right)' = F^{-s}_{p'\,q'}(\mathbb{R}^n), \quad s\in\mathbb{R}, \tag{2.56}$$

where

$$1 \le p < \infty,\ 1 \le q < \infty \quad (\text{ with } q=1 \text{ if } p=1\ ) \tag{2.57}$$

and

$$\frac{1}{p}+\frac{1}{p'} = \frac{1}{q}+\frac{1}{q'} = 1. \tag{2.58}$$

See also [RuS96], 2.1.5, p. 20, for the final formulation. By the same arguments as in [Tri78], 2.10.5 and 4.8.1 we have

$$\left(\widetilde{F}^s_{pq}(\Omega)\right)' = F^{-s}_{p'\,q'}(\Omega), \quad s>0, \tag{2.59}$$

and p and q as in (2.57).

2.3.3 Theorem *Let* Ω *be a bounded* C^∞ *domain in* $\mathbb{R}^n$. *Let* $\{\varphi_{jr}(x) : j \in \mathbb{N}_0;\ r=1,\ldots,N_j\}$ *be the resolution of unity introduced in 2.2.1. Let*

$$1 < p \le \infty,\ 1 < q \le \infty, \quad (\textit{with } q=\infty \textit{ if } p=\infty), \tag{2.60}$$

and

$$-\infty < s < \frac{1}{p}. \tag{2.61}$$

Then $f \in D'(\Omega)$ belongs to $F^s_{pq}(\Omega)$ if, and only if, it can be represented as

$$f = \sum_{j=0}^{\infty} \sum_{r=1}^{N_j} \varphi_{jr}(x)\, f \quad \textit{(convergence in } D'(\Omega)\textit{)} \tag{2.62}$$

such that

$$\left(\sum_{j=0}^{\infty} \sum_{r=1}^{N_j} \left\| \varphi_{jr}\, f \mid F^s_{pq}(\mathbb{R}^n) \right\|^p \right)^{\frac{1}{p}} < \infty \tag{2.63}$$

(usual modification if $p = q = \infty$). Furthermore, (2.63) is an equivalent norm.

2.3.4 The case $p = \infty$ Restricted to p and q as in (2.60), Theorem 2.2.2 and the above theorem complement each other. By (2.61) and in agreement with Proposition 2.2.10 the admitted values of s are overlapping each other, with exception in case of $p = \infty$. Under the restriction (2.60) there is only one space which is not covered, this is $\mathcal{C}^0 = F^0_{\infty\infty}$ $(= B^0_{\infty\infty})$. We will not dicuss this space in detail. But it is well known that $s = 0$, $p = \infty$ is a very delicate combination, so also in our context. To shed some light on this comment we first remark that $\widetilde{\mathcal{C}}^0(\Omega)$ makes sense, see 2.3.1. We have

$$\mathcal{C}^0(\Omega) \neq \widetilde{\mathcal{C}}^0(\Omega): \tag{2.64}$$

Assume that these two spaces coincide. Then it follows easily that the characteristic function of Ω would be a pointwise multiplier in $\mathcal{C}^0(\mathbb{R}^n)$. But this is not the case, see [RuS96], 4.6.3, Theorem 1, p. 208. One can even construct explicitly functions which belong to $\mathcal{C}^0(\Omega)$, but not to $\widetilde{\mathcal{C}}^0(\Omega)$: Let $n = 1$ and let $\Omega = (0, 1)$ be the unit interval. Then

$$\log x \in \mathcal{C}^0(\Omega) \quad \text{and} \quad \log x \notin \widetilde{\mathcal{C}}^0(\Omega). \tag{2.65}$$

To prove the first assertion we recall that $\log |x|$ belongs to BMO, see [Ste93], pp. 140/1, for a short proof. Hence by embedding, see [Tri83], p. 93, and [RuS96], p. 33, $\log x$ is then also an element of $\mathcal{C}^0(\Omega)$. On the other hand, assume that for any $\varrho(x) \in S(\mathbb{R})$,

$$f(x) = \varrho(x) \log x \text{ if } x > 0 \quad \text{and} \quad f(x) = 0 \text{ if } x \leq 0, \tag{2.66}$$

is an element of $\mathcal{C}^0(\mathbb{R})$, then $x\, f(x)$ would belong to $\mathcal{C}^1(\mathbb{R})$ and hence $x \log x \in \widetilde{\mathcal{C}}^1(\Omega)$. This contradicts (2.39) with $s = 1$ and $p = q = \infty$. Hence we have also the second assertion in (2.65). Otherwise it is left open whether representations of type (2.62), (2.63) with $s = 0, p = q = \infty$, make sense and what is the outcome.

2.3.5 Remark We prove Theorem 2.3.3 using Theorem 2.2.2 and the duality (2.59). This explains the range of p and q in (2.60). One can try to interpolate the spaces in the Theorems 2.2.2 and 2.3.3 and to extend the described characterizations. To stay within the F-spaces one has to use complex methods, especially the complex method $(\cdot,\cdot)_\Theta$ developed in [Tri83], 2.4.4–2.4.7, and in particular 3.3.6, and the references given there, which works also for quasi-Banach spaces of type $F^s_{pq}(\Omega)$. However so far no interpolation results for the complex method $(\cdot,\cdot)_\Theta$ are available which incorporate also spaces of type $\widetilde{F}^s_{pq}(\Omega)$; but we refer to 2.5.8 in this context. For this reason we use Theorem 2.2.2 only in those cases which are covered by Proposition 2.2.10. We avoid most general formulations and complement the above theorem by the following assertion.

2.3.6 Corollary *Let Ω be a bounded C^∞ domain in $\mathbb{R}^n$. Let $\{\varphi_{jr}(x) : j \in \mathbb{N}_0;\ r = 1,\dots,N_j\}$ be the above resolution of unity. Let*

$$\frac{n-1}{n} < p \le 1, \quad p \le q \le \infty, \quad -\infty < s < \frac{1}{p}. \tag{2.67}$$

Then $f \in D'(\Omega)$ belongs to $F^s_{pq}(\Omega)$ if, and only if, it can be represented by (2.62), (2.63) (equivalent quasi-norms).

2.4 The spaces $\overset{\circ}{F}{}^s_{pq}(\Omega)$

2.4.1 Preliminaries The definition of $\overset{\circ}{F}{}^s_{pq}(\Omega)$ has been given in 2.1.4. In Proposition 2.1.6 we clarified under which conditions $\overset{\circ}{F}{}^s_{pq}(\Omega)$ coincides with $F^s_{pq}(\Omega)$. If $s \ge \frac{1}{p}$ then the adequate substitute of (2.11) is the question under which conditions

$$\overset{\circ}{F}{}^s_{pq}(\Omega) = \widetilde{F}^s_{pq}(\Omega) \tag{2.68}$$

hold. Since the sixties, problems of this type attracted a lot of attention. In [Tri78], 4.3.2, p. 320, one finds the necessary references to related early results by J. L. Lions, Magenes, Grisvard and others with respect to classical spaces. In [Tri78], 4.3.2, pp. 317–320, we have given a more or less final discussion of this problem as far as the spaces $H^s_p = F^s_{p,2}$ and B^s_{pq} with $1 < p < \infty$, $1 \le q < \infty$, are concerned. What follows might also be considered as a continuation of this discussion restricted to the F-spaces (as always in this paper). But it is clear that there should be corresponding assertions for the B-spaces. As in Proposition 2.1.6 the spaces $\overset{\circ}{F}{}^s_{p\infty}(\Omega)$ (with the interesting special cases $\overset{\circ}{\mathcal{C}}{}^s(\Omega) = \overset{\circ}{F}{}^s_{\infty\infty}(\Omega)$) have no chance to coincide with $F^s_{p\infty}(\Omega)$ or $\widetilde{F}^s_{p\infty}(\Omega)$. But this has nothing to do with the behaviour at the boundary, but with the fact that the (quasi-)Banach spaces $F^s_{p\infty}(\Omega)$ are not separable. We return to this question in 2.4.4 for the spaces $\mathcal{C}^s(\Omega)$.

2.4.2 Theorem *Let Ω be a bounded C^∞ domain in $\mathbb{R}^n$. Let $0 < p < \infty$ and $0 < q < \infty$. Let*

$$s > \sigma_p \quad \text{and} \quad s - \frac{1}{p} \notin \mathbb{N}_0. \tag{2.69}$$

Then

$$\overset{\circ}{F}{}^{s}_{pq}(\Omega) = \widetilde{F}^{s}_{pq}(\Omega). \tag{2.70}$$

2.4.3 Remark As a warning we must add a remark about the proof. If $q \geq p$ then we combine the one-dimensional case with the Fubini property described in Theorem 2.1.12. If $q < p$ then this argument is not always at our disposal. On the other hand the pointwise multiplier property of the characteristic function of Ω in the $(\frac{1}{p}, s)$- region (2.51), (2.53) can be lifted by $s \mapsto s+m$ with $m \in \mathbb{N}$ from F^s_{pq} to $\overset{\circ}{F}{}^{s+m}_{pq}$. This proves (2.70) in these $(\frac{1}{p}, s)$-regions by the same arguments as in [Tri83], proof of Theorem 2.9.2, pp. 167-8. The outcome is independent of $0 < q < \infty$. This situation suggests to use the complex interpolation method $(\cdot,\cdot)_\Theta$ for suitable spaces $F^{s_0}_{p_0\, p_0}(\Omega)$ and $F^{s_1}_{p_1\, q_1}(\Omega)$ within the strips

$$m < s - \frac{1}{p} < m+1, \quad m \in \mathbb{N}_0, \tag{2.71}$$

in Fig. 1. The difficulties here come from the spaces $\overset{\circ}{F}{}^{s}_{pq}(\Omega)$ and we have to admit that our respective arguments will be somewhat vague and sketchy. A direct proof would be desirable.

2.4.4 Limiting cases We discuss three limiting situations.
(i) Let

$$s - \frac{1}{p} = m \in \mathbb{N}_0 \tag{2.72}$$

as indicated in Fig. 1. As in the proof of (2.11), (2.13) in 3.3, Step 5, and with the same references to [Tri78] it follows that

$$f \in C^\infty(\overline{\Omega}), \quad D^\gamma f | \partial\Omega = 0 \quad \text{if} \quad |\gamma| \leq m-1, \tag{2.73}$$

belongs to $\overset{\circ}{F}{}^{s}_{pq}(\Omega)$ with $1 < p < \infty$. On the other hand by Proposition 2.2.5 there are functions of this type which do not belong to $\widetilde{F}^{s}_{pq}(\Omega)$. Hence

$$\overset{\circ}{F}{}^{s}_{pq}(\Omega) \neq \widetilde{F}^{s}_{pq}(\Omega) \quad \text{if} \quad 1 < p < \infty,\ 0 < q \leq \infty,\ s - \frac{1}{p} \in \mathbb{N}_0. \tag{2.74}$$

(ii) Presumably this assertion holds also for $p \leq 1$. However this is by no means obvious. Take, for example, $0 < p \leq 1$, $s = \frac{1}{p}$, and $0 < q < \infty$. Then by Proposition 2.1.6 the space $\overset{\circ}{F}{}^{s}_{pq}(\Omega)$ is a genuine subspace of $F^s_{pq}(\Omega)$. At first glance this seems to improve the possibility of having (2.70). But this is not the case. Let $n = 1$ and let $\Omega = (0,1)$ be the unit interval. Let $\varrho(x)$ be a C^∞ function on $\mathbb{R}$ with compact support and, say, $\varrho(x) = 1$ near the origin. Let

$$f(x) = \varrho(x)\, |\log|x||^{-\alpha}, \quad x \in \mathbb{R},\ \frac{1}{p} \geq \alpha > \frac{1}{p} - 1. \tag{2.75}$$

We claim that this function belongs to $\overset{\circ}{F}{}^{\frac{1}{p}}_{pq}(\Omega)$ but not to $\widetilde{F}{}^{\frac{1}{p}}_{pq}(\Omega)$ where $0 < p \le 1$ and $0 < q < \infty$. The second assertion follows immediately from Proposition 2.2.5. Since $f(0) = 0$, it remains to prove that $f'(x)$ belongs to $F^{\frac{1}{p}-1}_{pq}(\mathbb{R})$. We have in $S'(\mathbb{R})$

$$f'(x) = \frac{1}{x} \, |\log |x||^{-\alpha-1} \quad \text{near the origin.} \tag{2.76}$$

let φ be a C^∞ function with compact support near 1 such that

$$\sum_{j=0}^{\infty} \varphi(2^j x) = 1, \quad 0 < x < 1, \tag{2.77}$$

is a resolution of unity. Then

$$\begin{aligned} f'(x) &= \sum_{j=0}^{\infty} \left(\varphi(2^j x) + \varphi(-2^j x)\right) \frac{1}{x} \, |\log |x||^{-\alpha-1} \\ &= \sum_{j=0}^{\infty} a_j(x) \, j^{-\alpha-1}, \quad x \, \text{near the origin,} \end{aligned} \tag{2.78}$$

is an atomic decomposition in $F^{\frac{1}{p}-1}_{pq}(\mathbb{R})$ where $a_j(x)$ are the atoms with the correct size and vanishing moment conditions. The needed detail about atomic decompositions may be found in [FrJ90] or also in [Tri97], Ch. 13. Obviously, by (2.75) we have $\{j^{-\alpha-1}\} \in \ell_p$. Hence $f \in F^{\frac{1}{p}}_{pq}(\mathbb{R})$ and also $f \in \overset{\circ}{F}{}^{\frac{1}{p}}_{pq}(\Omega)$. In other words, we have again (2.74) now even with $0 < p < \infty$, $0 < q \le \infty$, and $s = \frac{1}{p}$, at least in case of $\Omega = (0,1)$. Replacing $f(x)$ by $x^m f(x)$ with $m \in \mathbb{N}$ one can extend this assertion to $s - \frac{1}{p} \in \mathbb{N}_0$. Furthermore at least in some cases a generalization from one to n dimensions is possible by using the Fubini property from Theorem 2.1.12.
(iii) Finally we add a remark on the interesting limiting case of the Zygmund spaces $\mathcal{C}^s(\Omega) = F^s_{\infty\infty}(\Omega)$, $s > 0$. By (2.14), $C^\infty(\overline{\Omega})$ is not dense in $\mathcal{C}^s(\Omega)$ (recall that $\mathcal{C}^s(\Omega)$ is isomorphic to ℓ_∞, see [Tri78], Remark 4.9.4/6, p. 343, for references). As another candidate of a suitable substitute of $\overset{\circ}{F}{}^s_{pq}(\Omega)$ in our situation may serve

$$\mathcal{C}^s_0(\Omega) = \{f \in \mathcal{C}^s(\Omega) : \quad D^\gamma f|\partial\Omega = 0 \quad \text{if} \quad |\gamma| < s\} . \tag{2.79}$$

In the critical case $s = m \in \mathbb{N}$ it follows from 2.3.4 (and the Fubini theorem 2.1.12) that the boundary values of the normal derivatives $\frac{\partial^m}{\partial \nu^m} f|\partial\Omega$ do not exist, in general. We can apply Theorem 2.2.8 for all values of s and obtain

$$\mathcal{C}^s_0(\Omega) = \widetilde{\mathcal{C}}^s(\Omega) \quad \text{if, and only if,} \quad 0 < s \notin \mathbb{N}. \tag{2.80}$$

This complements both (2.70) and (2.74).

2.4.5 Proposition *Let Ω be a bounded C^∞ domain in $\mathbb{R}^n$. Let $0 < p < \infty$, $0 < q < \infty$, and*

$$m - 1 + \frac{1}{p} < s < m + \frac{1}{p} \quad \textit{where} \quad m \in \mathbb{N}. \tag{2.81}$$

Then

$$\overset{\circ}{F}{}^{s}_{pq}(\Omega) = \left\{ f \in F^s_{pq}(\Omega) \, : \, D^\gamma f | \partial\Omega = 0 \quad if \quad |\gamma| \le m-1 \right\}. \tag{2.82}$$

2.4.6 Remark This proposition complements both Theorem 2.4.2 and Proposition 2.1.6. The restriction $s > \sigma_p$ is not necessary here.

2.4.7 Corollary *Under the assumptions of Theorem 2.4.2 the characteristic function of Ω is a pointwise multiplier for the subspace*

$$\left\{ f \in F^s_{pq}(\mathbb{R}^n) \, : \, D^\gamma f | \partial\Omega = 0 \quad if \quad |\gamma| < s - \frac{1}{p} \right\} \tag{2.83}$$

of $F^s_{pq}(\mathbb{R}^n)$.

2.4.8 Characteristic functions as pointwise multipliers If $\sigma_p < s < \frac{1}{p}$ then the space in (2.83) coincides with $F^s_{pq}(\mathbb{R}^n)$ and we have the pointwise multiplier assertion mentioned several times, see [Fra86] and [RuS96]. Otherwise the corollary follows from Proposition 2.4.5 and Theorem 2.4.2. Of interest is the fact that the original multiplier $(\frac{1}{p}, s)$-region (2.51), (2.53) is not only lifted by $s \mapsto s+m$ with $m \in \mathbb{N}$ (what is more or less obvious), but enlarged as indicated in Fig. 1. The corollary can be extended to the case $q=\infty$. If $p=q=\infty$ then it follows by (2.80) that the characteristic function of Ω is a pointwise multiplier in the space (2.83) with $p=q=\infty$ if $0 < s \notin \mathbb{N}$. The case $n=1$ and $0<p<\infty$, $q=\infty$, is covered by lifting of (2.51) and (2.53). Finally the case $n>1$ and $0<p<\infty$, $q=\infty$, follows from the one-dimensional case and Theorem 2.1.12.

2.5 Atomic and subatomic decompositions

2.5.1 Aim of this section Atomic decompositions in diverse function spaces have a history of more than twenty years, we refer to [Tri92], 1.9, for a historically-minded report, see also [Tri92], 3.2. In connection with the spaces $B^s_{pq}(\mathbb{R}^n)$ and $F^s_{pq}(\mathbb{R}^n)$ atomic representations go back to [FrJ85], [FrJ90], and [FJW91]. We complemented these results in [Tri97] and [Tri98] by subatomic or quarkonial decompositions. Atomic and subatomic decompositions proved to be a very powerful instrument. Distinguished applications are mapping properties of pseudodifferential operators, as considered in [Tor91] and [Tri92], and estimates of entropy numbers, as done in [Tri97]. As indicated in the Introduction the aim of this paper is to pave the way for an intrinsic theory of weighted function spaces on hyperbolic manifolds including estimates for entropy numbers. As said in the Introduction, the spaces $\widetilde{F}^s_{pq}(\Omega)$ from Definition 2.1.8 will play a central role. Using Theorem 2.2.2, atomic and subatomic decompositions will be shifted from $\mathbb{R}^n$ to Ω. This works for all types of atomic and subatomic decompositions, with and without moment conditions for the related elementary building blocks, and also for the spaces $F^s_{pq}(\Omega)$ as far as they are covered by Theorem 2.3.3 and Corollary 2.3.6. We restrict ourselves here to the simplest case: Subatomic decompositions of the

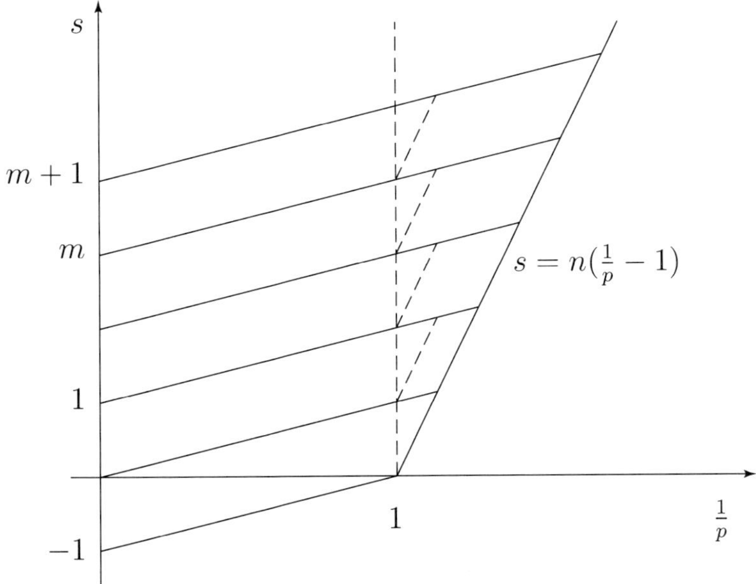

FIGURE 1

spaces $\widetilde{F}^s_{pq}(\Omega)$ for $s > \sigma_{pq}$, without moment conditions. By the techniques developed in [Tri97], Sections 13 and 14, it will be clear how to handle the other possible cases treated there.

2.5.2 Spaces on manifolds By Theorems 2.2.2 and 2.2.8 (and also by Theorem 2.3.3 and Corollary 2.3.6) on the one hand and Riemannian metrics of type (1.3) on the other hand it is quite clear that we have special representations of corresponding F^s_{pq} spaces on related (hyperbolic) manifolds with bounded geometry. The theory of function spaces on manifolds may be found in [Tri92], Ch. 7, with the necessary references. Recently L. Skrzypczak extended this theory systematically, including (invariant) atomic decompositions, [Skr97a], [Skr97b]. Our way (in the following papers) is slightly different: We identify the manifold with the above bounded domain Ω in $\mathbb{R}^n$ equipped with Riemannian metrics of type (1.3) and reduce Riemannian questions to Euclidean problems. We employed just this interplay also in the previous paper [Tri88].

2.5.3 Some notation in $\mathbb{R}^n$ As usual, $\mathbb{Z}^n$ stands for the lattice of all points $x = (x_1, \ldots, x_n)$ in $\mathbb{R}^n$ with integer-valued components x_r. We consider the resolutions of unity in $\mathbb{R}^n$

$$\{\varphi_{jm}(x) \; : \; j \in \mathbb{N}_0, \; m \in \mathbb{Z}^n\} \tag{2.84}$$

for which there are constants $c > 0$ and $c_\gamma > 0$, where $\gamma \in \mathbb{N}_0^n$ are multi-indices, with the following properties:

$$\text{(i)} \qquad \operatorname{supp} \varphi_{jm} \subset B_{jm} = \{y : |y - 2^{-j}m| \le c2^{-j}\}, \tag{2.85}$$

where $j \in \mathbb{N}_0$ and $m \in \mathbb{Z}^n$;

$$\text{(ii)} \qquad |D^\gamma \varphi_{jm}(x)| \le c_\gamma 2^{j|\gamma|} \quad \text{where} \quad j \in \mathbb{N}_0,\ m \in \mathbb{Z}^n,\ \gamma \in \mathbb{N}_0^n; \tag{2.86}$$

$$\text{(iii)} \qquad \sum_{m \in \mathbb{Z}^n} \varphi_{jm}(x) = 1 \quad \text{for all } x \in \mathbb{R}^n \text{ and } j \in \mathbb{N}_0. \tag{2.87}$$

Of course, by (2.87) and (2.85) we have for any $j \in \mathbb{N}_0$ a resolution of unity with a limited number of overlappings. The simplest case is given by (1.1) and its dyadically dilated counterparts

$$\sum_{m \in \mathbb{Z}^n} \psi(2^j x - m) = 1 \quad \text{for all } x \in \mathbb{R}^n \text{ and } j \in \mathbb{N}_0. \tag{2.88}$$

We based subatomic decompositions in [Tri97], Ch. 14, and in [Tri98] on regular resolutions (2.88). This is the first choice, but by no means necessary. Just on the contrary, it is the higher flexibility expressed by (2.85)–(2.87) which is the main advantage of this method. In a slightly different context, the regular case is more or less covered by [FrJ90], Theorem 4.2, Corollary 4.3, and [FJW91], Ch. 7. But we must confess that we missed the point in [Tri97], [Tri98] to make clear this difference which is now of great use in what follows. For this reason we give again a detailed formulation of subatomic decompositions in $\mathbb{R}^n$ but restricted to the cases of interest here, this means to F^s_{pq} with $s > \sigma_{pq}$. But there are obvious counterparts for all the other cases treated in [Tri97]. Let χ_{jm} be the characteristic function of the cube in $\mathbb{R}^n$ centered at $2^{-j}m$ with side-length 2^{-j}. Then we put

$$\chi^{(p)}_{jm}(x) = 2^{\frac{jn}{p}} \chi_{jm}(x) \quad \text{where} \quad x \in \mathbb{R}^n,\ 0 < p \le \infty, \tag{2.89}$$

and $j \in \mathbb{N}_0$, $m \in \mathbb{Z}^n$. As in [Tri97], 13.5, we introduce the space

$$f_{pq} = \left\{ \lambda : \|\lambda \,|\, f_{pq}\| = \left\| \left(\sum_{j=0}^\infty \sum_{m \in \mathbb{Z}^n} |\lambda_{jm} \chi^{(p)}_{jm}(\cdot)|^q \right)^{\frac{1}{q}} \,|\, L_p(\mathbb{R}^n) \right\| < \infty \right\} \tag{2.90}$$

of the sequences

$$\lambda = \{\lambda_{jm} \in \mathbb{C} : j \in \mathbb{N}_0, m \in \mathbb{Z}^n\} \tag{2.91}$$

(usual modification if $q = \infty$). Let $s \in \mathbb{R}$, $0 < p \le \infty$, $\gamma \in \mathbb{N}_0^n$ and $x^{j,m} \in B_{jm}$. Then

$$(\gamma qu)_{jm}(x) = 2^{-j(s-\frac{n}{p})+j|\gamma|}(x - x^{j,m})^\gamma \varphi_{jm}(x) \tag{2.92}$$

are the elementary building blocks we are looking for, related to the balls B_{jm} in (2.85). We called them also $(s,p) - \gamma$-quarks in [Tri97] what might explain the notation. Finally we complement the notation (2.91) by

$$\lambda^\gamma = \left\{\lambda^\gamma_{jm} \in \mathbb{C} : j \in \mathbb{N}_0,\, m \in \mathbb{Z}^n\right\} \quad \text{where} \quad \gamma \in \mathbb{N}^n_0. \tag{2.93}$$

2.5.4 Theorem *Let $0 < p \le \infty$, $0 < q \le \infty$ (with $q = \infty$ if $p = \infty$), $s > \sigma_{pq}$, and let $(\gamma qu)_{jm}(x)$ be the $(s,p)-\gamma$-quarks given by (2.92). There is a number $\varkappa > 0$ with the following property. Let $\mu > \varkappa$. Then $f \in S'(\mathbb{R}^n)$ belongs to $F^s_{pq}(\mathbb{R}^n)$ if, and only if, it can be represented as*

$$f = \sum_{\gamma\in\mathbb{N}^n_0} \sum_{j=0}^{\infty} \sum_{m\in\mathbb{Z}^n} \lambda^\gamma_{jm} (\gamma qu)_{jm}(x) \tag{2.94}$$

with

$$\sup_\gamma 2^{\mu|\gamma|} \|\lambda^\gamma \mid f_{pq}\| < \infty. \tag{2.95}$$

Furthermore, the infimum in (2.95) over all admissible representations (2.94) is an equivalent quasi-norm in $F^s_{pq}(\mathbb{R}^n)$.

2.5.5 Comment The proof of this theorem is completely covered by [Tri97], Ch. 14, [Far97], dealing with anisotropic generalizations, and [Tri98]. The assumptions ensure that the right-hand side of (2.94) converges absolutely in $L_1(\mathbb{R}^n) + L_\infty(\mathbb{R}^n)$. For fixed μ there are optimal sequences λ^γ which depend linearly on f,

$$\lambda^\gamma_{jm} = \left(f, \Psi^{\gamma,\mu}_{jm}\right) \quad \text{(dual pairing)} \tag{2.96}$$

where $\Psi^{\gamma,\mu}_{jm}$ are (dilated and translated) wavelets such that

$$\sup_\gamma 2^{\mu|\gamma|} \|\lambda^\gamma \mid f_{pq}\| \le c(\mu)\, \|f \mid F^s_{pq}(\mathbb{R}^n)\| \tag{2.97}$$

for some $c(\mu) > 0$ which is independent of $f \in F^s_{pq}(\mathbb{R}^n)$. The functions $\Psi^{\gamma,\mu}_{jm}$ can be calculated explicitly. We refer to [Tri97], especially 14.16, [Far97], and [Tri98]. For the if-part one needs only (2.85) and (2.86), but not (2.87).

2.5.6 Some notation in domains Let Ω be a bounded C^∞ domain in $\mathbb{R}^n$. It seems to be quite clear that the combination of Theorem 2.2.2 and Theorem 2.5.4 results in a subatomic decomposition of the covered spaces $\widetilde{F}^s_{pq}(\Omega)$. Since everything converges absolutely in $L_1(\Omega)$ one has also no problem to clip together the decompositions in (2.31) and (2.94), (2.95), and to rearrange them. We introduce the necessary notation. First we combine 2.2.1 with (2.85)–(2.87). There are positive numbers c_l $(l = 1, \ldots, 8)$ and c_γ $(\gamma \in \mathbb{N}^n_0)$, (irregular) lattices

$$\{x^{j,m} : m = 1, \ldots, M_j\} \subset \Omega, \quad \text{where} \quad j \in \mathbb{N}_0, \tag{2.98}$$

and subordinated approximative resolutions of unity $\{\varphi_{jm} : m = 1, \ldots, M_j\}$ with the following properties:

(i)

$$c_1 \le M_j 2^{-jn} \le c_2 \quad \text{where} \quad j \in \mathbb{N}_0; \tag{2.99}$$

$$|x^{j,m_1} - x^{j,m_2}| \ge c_3 2^{-j} \quad \text{where} \quad j \in \mathbb{N}_0 \text{ and } m_1 \ne m_2; \tag{2.100}$$

$$\operatorname{dist}(B_{jm},\partial\Omega) \geq c_4 2^{-j} \quad \text{where} \quad j \in \mathbb{N}_0;\ 1 \leq m \leq M_j \tag{2.101}$$

with

$$B_{jm} = \{y : |y - x^{j,m}| \leq c_5 2^{-j}\} \subset \Omega. \tag{2.102}$$

(ii) $\varphi_{jm}(x)$ are C^∞ functions with

$$\operatorname{supp} \varphi_{jm}(x) \subset B_{jm} \quad \text{where} \quad j \in \mathbb{N}_0,\ 1 \leq m \leq M_j; \tag{2.103}$$

$$|D^\gamma \varphi_{jm}(x)| \leq c_\gamma 2^{j|\gamma|} \quad \text{where} \quad \gamma \in \mathbb{N}_0^n;\ j \in \mathbb{N}_0,\ m = 1,\dots,M_j; \tag{2.104}$$

and

$$c_6 \leq \sum_{m=1}^{M_j} \varphi_{jm}(x) \leq c_7 \quad \text{if} \quad x \in \Omega \text{ with } \operatorname{dist}(x,\partial\Omega) \geq c_8 2^{-j} \tag{2.105}$$

for all $j \in \mathbb{N}_0$. For short, we call $\{\varphi_{jm}\}$ a *family of approximative resolutions of unity in* Ω if one finds lattices $\{x^{j,m}\}$ with all the properties (2.98)–(2.105). The more or less obvious counterpart of f_{pq} in (2.90) is given by

$$f_{pq}^\Omega = \left\{ \lambda : \|\lambda \,|\, f_{pq}^\Omega\| = \left\| \left(\sum_{j=0}^\infty \sum_{m=1}^{M_j} |\lambda_{jm} \chi_{jm}^{(p)}(\cdot)|^q \right)^{\frac{1}{q}} \,|\, L_p(\Omega) \right\| < \infty \right\} \tag{2.106}$$

where

$$\lambda = \{\lambda_{jm} \in \mathbb{C} : j \in \mathbb{N}_0;\ m = 1,\dots,M_j\} \tag{2.107}$$

are the related sequences, $\chi_{jm}^{(p)}(x)$ has the same meaning as in (2.89), where now $\chi_{jm}(x)$ is the characteristic function of B_{jm}. We use also the obvious counterpart λ^γ of (2.93) adapted to (2.107).

2.5.7 Theorem *Let Ω be a bounded C^∞ domain in $\mathbb{R}^n$. Let $0 < p \leq \infty$, $0 < q \leq \infty$ (with $q = \infty$ if $p = \infty$), and $s > \sigma_{pq}$. There are a number $\varkappa > 0$ and a family $\{\varphi_{jm}\}$ of approximative resolutions of unity in Ω with the following property. Let $\mu > \varkappa$ and let $(\gamma qu)_{jm}(x)$ be the $(s,p)-\gamma$-quarks given by (2.92) but now with respect to the above points $x^{j,m}$ and the above functions φ_{jm} in 2.5.6. Then $f \in L_1(\Omega)$ belongs to $\widetilde{F}_{pq}^s(\Omega)$ if, and only if, it can be represented as*

$$f = \sum_{\gamma \in \mathbb{N}_0^n} \sum_{j=0}^\infty \sum_{m=1}^{M_j} \lambda_{jm}^\gamma (\gamma qu)_{jm}(x) \tag{2.108}$$

with

$$\sup_\gamma 2^{\mu|\gamma|} \|\lambda^\gamma \,|\, f_{pq}^\Omega\| < \infty. \tag{2.109}$$

Furthermore, the infimum in (2.109) over all admissible representations (2.108) is an equivalent quasi-norm in $\widetilde{F}_{pq}^s(\Omega)$.

2.5.8 Comments, references, and further perspectives

(i) If one compares (2.108) and (2.94) then the summation in (2.108) is restricted to those newly interpreted $(s,p)-\gamma$-quarks (2.92) with (2.101)–(2.103). This is the best what can be expected and it is of special interest if we have in addition (2.70) or (2.82). In these cases it follows from the $\mathbb{R}^n$-theory that one can replace $(s,p)-\gamma$-quarks in (2.108) by arbitrary (s,p)-atoms with obvious counterparts of (2.92), (2.103), (2.104). Remark that (2.109) ensures that the right-hand side of (2.108) converges always absolutely in $L_1(\Omega)$. For the if-part, (2.105) is not needed.
(ii) In a similar way it should be possible to prove decompositions of type (2.108), (2.109) for the spaces $F^s_{pq}(\Omega)$ as far as they are covered by Theorem 2.3.3 or Corollary 2.3.6. Some moment conditions are coming in. We refer to [Tri97], 14.2, pp. 92–93, for the needed modifications. Some greater care about the convergence of (2.108) is now necessary.
(iii) As mentioned in 2.3.2 the spaces in (2.55) form a scale. At least in some cases this can be used to prove subatomic decompositions for some spaces $F^s_{pq}(\Omega)$ in a different way than indicated in (ii). Let

$$k-1+\frac{1}{p}+(n-1)(\frac{1}{p}-1)_+ < s < k+\frac{1}{p} \quad \text{where} \quad k\in\mathbb{N} \tag{2.110}$$

and $0<p\le\infty$. Then the iterated Laplacian Δ^k maps $\widetilde{F}^s_{pq}(\Omega)$ with $0<q\le\infty$ onto $F^{s-2k}_{pq}(\Omega)$. This follows from 2.4.5 and 2.4.2 and the theory of boundary value problems for elliptic differential operators developed in [FrR95] and in [RuS96], Ch. 3 (there one finds also references to the extensive literature about this subject and also to forerunners). Combining Theorem 2.5.7 with these mapping properties one obtains at least in some spaces $F^{s-2k}_{pq}(\Omega)$ subatomic decompositions of type (2.108), (2.109) with $\Delta^k(\gamma qu)_{jm}(x)$ as elementary building blocks.
(iv) In [Miy90], [CKS92], [CKS93], and [CDS96] the authors study several types of Hardy spaces in domains in $\mathbb{R}^n$ (general, C^∞, Lipschitzian). In our notation these spaces coincide with $F^0_{p,2}(\Omega)$ and $\widetilde{F}^0_{p,2}(\Omega)$, given by (2.21) and with subspaces of $\widetilde{F}^0_{p,2}(\overline{\Omega})$, given by (2.20), where always $0<p\le 1$. There are intrinsic atomic decompositions and, at least in case of the Hardy spaces $F^0_{p,2}(\Omega)$, there are two types of atoms, say, *boundary atoms*, and *interior atoms*, where for the boundary atoms no or only a limited number of moment conditions must be fulfilled. If one follows the path outlined above in (ii) then one arrives precisely at the same type of intrinsic atoms even with the desirable additional information that the distance of the support of the boundary atoms to $\partial\Omega$ is proportional to the diameter of its support (for the interior atoms the situation is even better). Also (iii) can be applied in this way to some Hardy spaces. However these outlined procedures and what has been done in [CKS93] and [CDS96] is quite contrary. We used the results about mapping properties of elliptic operators in [FrR95] to get intrinsic (sub)atomic decompositions, whereas in [CKS93] and [CDS96] the authors constructed first atomic decompositions and applied them afterwards to study the (Dirichlet and Neumann) Laplacian.

(v) We complement what had been said above in (ii) and (iv) by some further speculations. Let $1 \le p < \infty$ and $1 \le q \le \infty$. By 2.3.2 one has the scale (2.55) which combines naturally $\widetilde{F}^s_{pq}(\Omega)$, $s > 0$ with $F^s_{pq}(\Omega)$, $s \le 0$, including liftings. Assume that for any $N \in \mathbb{N}$ and $|s| < N$ one has common lifts with respect to s for all fixed $p's$ and $q's$. Then, in particular, interpolation, real $(\cdot,\cdot)_{\Theta,q}$, complex for Banach spaces $[\cdot,\cdot]_\Theta$, and especially complex for quasi-Banach spaces $(\cdot,\cdot)_\Theta$ can be done first for F^s_{pq} with, say, $s \le 0$, and lifted afterwards within the scales (2.55). Let us furthermore assume that on the basis of [FrR95] and [Joh96] the nowadays available theory of fractional powers of elliptic boundary value operators as suitable pseudodifferential operators, and the techniques developed in [Tri78], 4.9.2, the scale properties in 2.3.2 can be extended to spaces $F^s_{pq}(\Omega)$ and $\widetilde{F}^s_{pq}(\Omega)$, with $0 < p < \infty$ and maybe, with $\min(p,q,1) = \min(p,1)$. Then there is a good chance to extend Theorem 2.3.3 and Corollary 2.3.6, based in turn on Remark 2.3.5, to many other spaces $F^s_{pq}(\Omega)$ covering all values of $0 < p < \infty$. Then one obtains subatomic and atomic decompositions for these spaces of the above type. This may apply especially to $h_p(\Omega) = F^0_{p,2}(\Omega)$ where $0 < p \le 1$.
(vi) Atomic characterizations of all spaces $F^s_{pq}(\Omega)$ and $B^s_{pq}(\Omega)$ in smooth and non-smooth bounded domains Ω in $\mathbb{R}^n$ may be found in [TrW96]. A short description has also been given in [ET96], 2.5. There are again boundary atoms and interior atoms. But in contrast to the boundary atoms in the papers mentioned in (iv) the boundary atoms in [TrW96] have a support in $\overline{\Omega}$, but not necessarily in Ω. But this is quite clear, since the spaces $F^s_{pq}(\Omega)$ with $s > \sigma_p$ have boundary values. In other words, subatomic (and atomic) decompositions obtained and outlined in this paper complement what has been done in [TrW96]. Finally as for invariant atoms for function spaces on manifolds we refer again to [Skr97a,b].

3. Proofs and complements

3.1 Scheme of proofs

The proofs are ordered in the following way:

3.2 Theorem 2.1.12,
3.3 Proposition 2.1.6,
3.4 Proposition 2.2.5,
3.5 Theorem 2.2.8,
3.6 Theorem 2.2.2,
3.8 Theorem 2.3.3,
3.10 Corollary 2.3.6,
3.11 Theorem 2.4.2,
3.12 Proposition 2.4.5,
3.13 Theorem 2.5.7.

Together with the given arguments in Section 2 this covers all assertions. In 3.7 we obtain new equivalent quasi-norms in $\widetilde{F}^s_{pq}(\Omega)$ based on the proofs given in 3.5 and 3.6. In connection with the proof in 3.8 we describe in 3.9 some homogeneity properties of the involved spaces and related pointwise multiplier assertions which might also be of self-contained interest.

3.2 Proof of Theorem 2.1.12 *Step 1* Let $0 < p \leq \infty$, $0 < q \leq \infty$ (with $q = \infty$ if $p = \infty$) and $s > \sigma_{pq}$. We estimate the left-hand side of (2.24) from above by the right-hand side. Let $0 < u < \min(1, p, q)$, $s < N \in \mathbb{N}$, and

$$d^N_{t,u} f(x) = \left(t^{-n} \int_{|h| \leq t} \left| \Delta^N_h f(x) \right|^u dh \right)^{\frac{1}{u}}, \quad t > 0, \tag{3.1}$$

where $\Delta^N_h f(x)$ are the usual (n-dimensional) differences. By [Tri92], 3.5.2, 3.5.3, pp. 193-4,

$$\| f \,|\, L_p(\mathbb{R}^n) \| + \left\| \left(\int_0^1 t^{-sq} d^N_{t,u} f(\cdot)^q \, \frac{dt}{t} \right)^{\frac{1}{q}} \,|\, L_p(\mathbb{R}^n) \right\| \tag{3.2}$$

is an equivalent quasi-norm in $F^s_{pq}(\mathbb{R}^n)$. Let $\Delta^N_{\tau,j} f(x)$ with $j = 1, \ldots, n$ and $x \in \mathbb{R}^n$, $\tau \in \mathbb{R}$, be the one-dimensional differences in $\mathbb{R}^n$ with respect to the jth direction of the coordinates. Replacing N in Δ^N_h by nN one obtains by elementary reasoning

$$\left| \Delta^{nN}_h f(x) \right|^u \leq c \sum_{k=1}^{K} \sum_{j=1}^{n} \left| \Delta^N_{h_j,j} f(x + h^{j,k}) \right|^u \tag{3.3}$$

for suitable numbers $c > 0$ and $K \in \mathbb{N}$. Here the components of $h^{j,k} \in \mathbb{R}^n$ are separately proportional to the components of $h = (h_1, \ldots, h_n) \in \mathbb{R}^n$. A simple way to justify (3.3) is to shift relations of this type to the Fourier side and to use formulas of type (10), (11) in [Tri83], 2.5.13, pp. 115–6, also dealing with the Fubini property. Let $(Mf)(x)$ be the m-dimensional Hardy-Littlewood maximal function in $\mathbb{R}^m$ with $0 \leq m \leq n$. Then by (3.1) with nN in place of N and (3.3) it follows at least a.e.

$$d^{nN}_{t,u} f(x) \leq c \sum_{j=1}^{n} \sum_{l=1}^{L} \left(M \left[\frac{1}{t} \int_{-t}^{t} \left| \Delta^N_{\tau,j} f(\cdots, \cdot + c_l \tau, \cdots) \right|^u d\tau \right] (x) \right)^{\frac{1}{u}} \tag{3.4}$$

for some positive c_l, where the interior integrals are the one-dimensional versions of (3.1) but with shifted arguments, which come from the shifted arguments in (3.3). Furthermore all the remaining integrals, also with respect to dimensions smaller than n, can be estimated from above at least a.e., by some m-dimensional maximal functions. Checking the arguments in Step 1 of the proof of Theorem 3.5.3 in [Tri92], pp. 194–5, it follows that (3.2) (here in the one-dimensional version) remains an equivalent quasi-norm also with the indicated shiftings in (3.4). Furthermore one can (3.2) discretize by $t \to 2^{-l}$ with $l \in \mathbb{N}_0$ and ℓ_q in place of the t-integral. We insert (3.4) in the discretized version of (3.2) and use the vector-valued maximal inequality by Fefferman and Stein with respect to $\frac{q}{u} > 1$ and $\frac{p}{u} > 1$. A formulation and references may be found in [Tri92], 2.2.2, p. 89. Then it follows again by the one-dimensional version of [Tri92], 3.5.3, p. 194, (with the above modification), that (3.2) can be estimated from above by the right-hand side of (2.24).

Step 2 We estimate the right-hand side of (2.24) from above by the left-hand side. Let $f \in F^s_{pq}(\mathbb{R}^n)$ be given by the subatomic decomposition according to Theorem 2.5.4. For $(\gamma qu)_{jm}(x)$ we may assume that they have a product structure with respect to $x_1, \ldots, x_n$. We fix $n-1$ coordinates, say, $x' = (x_2, \ldots, x_n)$, and look at (2.94) as a subatomic decomposition of $f(\cdot, x_2, \ldots, x_n)$. The desired estimate can be done for each $\gamma \in \mathbb{N}^n_0$ separately and we may assume $\gamma = 0$ without restriction of generality. Put for simplicity $\lambda_{jm} = \lambda^\gamma_{jm}$ if $\gamma = 0$. Let χ_{jm} be the characteristic function of B_{jm} in (2.85) or of a somewhat larger cube. Then we have by (2.94) and (2.92)

$$\lambda_{jm}(\gamma qu)_{jm}(x) = \lambda_{jm} 2^{j\frac{n-1}{p}} \varphi_{jm'}(x') \chi_{jm}(x) \varphi_{jm_1}(x_1) 2^{-j(s-\frac{1}{p})}, \qquad (3.5)$$

where $m = (m_1, \ldots, m_n) = (m_1, m') \in \mathbb{Z}^n$. The last factor is just the one-dimensional substitute of (2.92), whereas for fixed $x' = (x_2, \ldots, x_n)$ the coefficients λ_{jm} must be replaced by the corresponding coefficients in (3.5) which can be estimated by

$$|\lambda_{jm}| 2^{j\frac{n-1}{p}} \chi_{jm'}(x_2, \ldots, x_n). \qquad (3.6)$$

Taking the one-dimensional version of f_{pq} in (2.90) with respect to x_1 and afterwards $L_p(\mathbb{R}^{n-1})$ with respect to $x_2, \ldots, x_n$, then one obtains the n-dimensional version of f_{pq} in (2.90). But this is sufficient to get the desired estimate.

3.3 Proof of Proposition 2.1.6 and of (2.14) *Step 1* First wo prove (2.14). Let $0 < p < \infty$ and $q = \infty$. We may assume $s = 0$. By (1.2) it is sufficient to prove that $S(\mathbb{R}^n)$ is not dense in $F^0_{p,\infty}(\mathbb{R}^n)$. Let $\varrho \in S(\mathbb{R}^n)$ be non-trivial such that $\widehat{\varrho}$ has a compact support near the origin. Let

$$f = \sum_{k=0}^\infty e^{i2^k x_1} \varrho(x) \quad \text{where} \quad x \in \mathbb{R}^n. \qquad (3.7)$$

We may assume

$$(\varphi_j \widehat{f})^\vee(x) = e^{i2^j x_1} \varrho(x), \quad \text{with} \quad x \in \mathbb{R}^n, \qquad (3.8)$$

where $\{\varphi_j\}$ are the same functions as in (2.3)–(2.6). For any $\psi \in S(\mathbb{R}^n)$ it follows from (2.6) with $q = \infty$ and (3.8)

$$\|f - \psi \,|\, F^0_{p,\infty}(\mathbb{R}^n)\| \geq \|\varrho | L_p(\mathbb{R}^n)\| > 0. \qquad (3.9)$$

This proves (2.14) if $p < \infty$. In case of $p = q = \infty$, $0 < s < 1$, and $\Omega = (-1, 1)$ the function

$$f(t) = |t|^s, \quad t \in (-1, 1), \qquad (3.10)$$

belongs to $\mathcal{C}^s(\Omega)$, but cannot be approximated in $\mathcal{C}^s(\Omega)$ by smooth functions. There is an n-dimensional counterpart and by lifting, the assertion (2.14) holds for any $s \in \mathbb{R}$ and $p = q = \infty$.

Step 2 To prove Proposition 2.1.6 we may assume $p < \infty$ and $q < \infty$. Since $S(\mathbb{R}^n)$ is dense in $F^s_{pq}(\mathbb{R}^n)$ we have (2.11) if, and only if, any $f \in D(\mathbb{R}^n)$ can be approximated in $F^s_{pq}(\mathbb{R}^n)$ by functions belonging to

$$D_\Omega(\mathbb{R}^n) = \{g \in D(\mathbb{R}^n) : \ g(x) = 0 \quad \text{near} \quad \partial\Omega\} . \tag{3.11}$$

First we prove the following assertion: *Let* $s \in \mathbb{R}$, $0 < p < \infty$, $0 < q_1 < \infty$, *and* $0 < q_2 < \infty$. *Then the desired approximation property holds for* $F^s_{pq_1}(\mathbb{R}^n)$ *if, and only if, it holds for* $F^s_{pq_2}(\mathbb{R}^n)$ *(independence of the third index).* Let $f \in D(\mathbb{R}^n)$ be given and let $g \in D_\Omega(\mathbb{R}^n)$. Then we use an atomic decomposition of $f - g$ in $F^s_{pq_1}(\mathbb{R}^n)$,

$$f - g = \sum_{j=0}^\infty \sum_{m \in \mathbb{Z}^n} \lambda_{jm} a_{jm}(x) = h_1(x) + h_2(x), \tag{3.12}$$

where $a_{jk}(x)$ are related atoms and $\{\lambda_{jk}\} \in f_{pq_1}$. We refer to [Tri97], Ch. 13, for details or to [FrJ90] for the original version. As indicated in (3.12) we split this atomic decomposition in two sums. In $h_2(x)$ we collect all those couples (j, m) for which the balls B_{jm} in (2.85) with a sufficiently large number c have a non-empty intersection with $\partial\Omega$. Let $0 < q_1 < q_2 < \infty$. For given $\varepsilon > 0$ we decompose $h_1 = h_{1,\varepsilon} + h_1^\varepsilon$ with

$$h_{1,\varepsilon} \in D_\Omega(\mathbb{R}^n) \quad \text{and} \quad \|h_1^\varepsilon \,|\, F^s_{pq_1}(\mathbb{R}^n)\| \le \varepsilon.$$

Now it follows

$$\begin{aligned} \|f - g - h_{1,\varepsilon} \,|\, F^s_{pq_1}(\mathbb{R}^n)\| &\le c\,\|h_2 \,|\, F^s_{pq_1}(\mathbb{R}^n)\| + c\,\varepsilon \\ &\sim \|h_2 \,|\, F^s_{pq_2}(\mathbb{R}^n)\| + \varepsilon \\ &\le c'\,\|f - g - h_{1,\varepsilon} \,|\, F^s_{pq_2}(\mathbb{R}^n)\| + c'\,\varepsilon, \end{aligned} \tag{3.13}$$

where, of course, the middle equivalence concerning h_2 is the crucial point in this estimate. This independence of q (equivalence) of $\|h_2 \,|\, F^s_{pq}(\mathbb{R}^n)\|$ for functions of this special structure is one of the major technical discoveries in this context and due to Frazier and Jawerth. We refer to [FrJ90], §11, for a detailed argument. In any case the above claimed independence of the approximation property on q with $q < \infty$ is a consequence of (3.13).

Step 3 Let $0 < p < \infty$, $0 < q < \infty$ and $s < \frac{1}{p}$. We prove (2.11). By Step 2 we have to approximate $f \in D(\mathbb{R}^n)$ by $g \in D_\Omega(\mathbb{R}^n)$. By the technique developed in [Tri97], Corollary 13.9, pp. 81–83, we have for any fixed $j \in \mathbb{N}$ in a 2^{-j}-neighbourhood $(\partial\Omega)_j$ of $\partial\Omega$ the atomic decomposition

$$f(x) = \sum_m \lambda_{jm} a_{jm}(x), \qquad x \in (\partial\Omega)_j, \tag{3.14}$$

where the sum is taken over those balls B_{jm} in (2.85) having a non-empty intersection with, say, $(\partial\Omega)_j$. The number of these balls can be estimated from above by $c2^{j(n-1)}$. Furthermore we may assume

$$|\lambda_{jm}| \le C 2^{j(s - \frac{n}{p})}, \tag{3.15}$$

where $C = C(f)$ is independent of j. This is a consequence of

$$|a_{jm}(x)| \le 2^{-j(s-\frac{n}{p})}.$$

By Step 2 we may also assume $q = p$. By (2.90), now with fixed j we have

$$\begin{aligned} \|\lambda \,|\, f_{pp}\| &= \left(\sum_m |\lambda_{jm}|^p \right)^{\frac{1}{p}} \\ &\le C 2^{j(s-\frac{n}{p})} 2^{j\frac{n-1}{p}} \le C 2^{j(s-\frac{1}{p})}. \end{aligned} \tag{3.16}$$

Since $s < \frac{1}{p}$ we obtain the desired approximation of $f \in D(\mathbb{R}^n)$ in $F^s_{pq}(\mathbb{R}^n)$ by functions belonging to $D_\Omega(\mathbb{R}^n)$. This proves (2.11) under the condition (2.12).

Step 4 Let $0 < p \le 1$, $0 < q < \infty$ and $s = \frac{1}{p}$. Then the trace of $F^{\frac{1}{p}}_{pq}(\mathbb{R}^n)$ on $\partial\Omega$ coincides with $L_p(\partial\Omega)$, [Tri92], 4.4.3, p. 220, (in case of $n = 1$, $f \in F^{\frac{1}{p}}_{pq}(\mathbb{R})$ is continuous on $\overline{\Omega}$). This disproves (2.11) in all cases $0 < p \le 1$, $0 < q < \infty$, $s \ge \frac{1}{p}$. If $1 < p < \infty$, $0 < q < \infty$ and $s > \frac{1}{p}$, then the trace of $F^s_{pq}(\mathbb{R}^n)$ on $\partial\Omega$ coincides with $F^{s-\frac{1}{p}}_{pp}(\partial\Omega)$, see [Tri92], 4.4.2, p. 213, and the references given there. This disproves (2.11) also in these cases.

Step 5 It remains the case $1 < p < \infty$, $s = \frac{1}{p}$, $0 < q < \infty$. By Step 2 we may assume $q = 2$. Then we have the Sobolev spaces $F^s_{pq}(\mathbb{R}^n) = H^s_p(\mathbb{R}^n)$, and (2.11) is known, see [Tri78], Theorem 2.9.3.

3.4 Proof of Proposition 2.2.5 As said in Remark 2.2.6 we may assume $q \ge p$. Then we can use the equivalent quasi-norm (3.1), (3.2). By the technique developed in connection with the proof of Theorem 5.2.2 in [Tri92], p. 245, we may replace the integration in (3.1) over the ball $\{h \in \mathbb{R}^n : |h| \le t\}$ by the integration over the cone

$$K_t^- = \{h = (h', h_n) \in \mathbb{R}^n : -t < h_n < 0,\ |h'| < |h_n|\}. \tag{3.17}$$

Hence we have again the equivalent quasi-norm (3.2) in $F^s_{pq}(\mathbb{R}^n)$ but now based on

$$d^N_{t,u} f(x) = \left(t^{-n} \int_{h \in K_t^-} \left| \Delta^N_h f(x) \right|^u dh \right)^{\frac{1}{u}}, \quad t > 0, \tag{3.18}$$

where we use the same notation $d^N_{t,u}$. By the usual localization technique we may replace Ω in Proposition 2.2.5 by $\mathbb{R}^n_+$ and assume that $f \in \widetilde{F}^s_{pq}(\mathbb{R}^n_+)$ has a compact support, say, near the origin. Let $x \in \mathbb{R}^n_+$ and

$$c_1 x_n < t < c_2 x_n, \tag{3.19}$$

where c_1 and c_2 are suitably chosen positive numbers. Then we have

$$\Delta^N_h f(x) = f(x) \quad \text{if} \quad h \in K_t^- \text{ with } |h_n| \sim t,$$

and

$$|f(x)| \le c\, d^N_{t,u} f(x) \tag{3.20}$$

for these values of t. Inserting this estimate in (3.2), now based on (3.18), we obtain

$$\int_{\mathbb{R}^n_+} x_n^{-sp} |f(x)|^p \, dx \le c \| f \mid F^s_{pq}(\mathbb{R}^n)\|^p \tag{3.21}$$

(usual modification if $p = \infty$). But this is just what we wanted to prove.

3.5 Proof of Theorem 2.2.8 *Step 1* Let $f \in \widetilde{F}^s_{pq}(\Omega)$. Then it follows by Definitions 2.1.4 and 2.1.8, and by Proposition 2.2.5,

$$\| f \mid F^s_{pq}(\Omega)\| + \| f \mid L_p(\Omega, d^{-s})\| \le c \, \| f \mid \widetilde{F}^s_{pq}(\Omega)\|. \tag{3.22}$$

Step 2 We prove the reverse inequality. Again by localization we may assume

$$f \in F^s_{pq}(\mathbb{R}^n_+) \, , \; x_n^{-s} f \in L_p(\mathbb{R}^n_+), \quad \text{and} \quad f(x) = 0 \tag{3.23}$$

if $x \in \mathbb{R}^n_+$ and $|x| > 1$. Let

$$\text{ext}\, f(x) = f(x) \quad \text{if} \quad x \in \mathbb{R}^n_+, \quad \text{ext}\, f(x) = 0 \quad \text{if} \quad x \in \mathbb{R}^n_-. \tag{3.24}$$

We must prove $\text{ext}\, f \in F^s_{pq}(\mathbb{R}^n)$. We use again the quasi-norm (3.2) where we replace now K^-_t in (3.18) by the cone

$$K^+_t = \{ h = (h', h_n) \in \mathbb{R}^n : \; 0 < h_n < t; \; |h'| < h_n \}. \tag{3.25}$$

We have

$$d^N_{t,u} \; \text{ext}\, f(x) = d^N_{t,u} \, f(x) \quad \text{if} \quad x_n > 0 \tag{3.26}$$

and the desired estimate with respect to $L_p(\mathbb{R}^n_+)$. Let $-1 < x_n < 0$. Then

$$d^N_{t,u} \; \text{ext}\, f(x) = 0 \quad \text{if} \quad 0 < t \le c_1 |x_n| \tag{3.27}$$

for some $c_1 > 0$. Hence we may assume $t > c_1|x_n|$. There is a number $c_2 > 0$ such that for $h \in K^+_t$ and $|h| \ge c_2 |x_n|$,

$$|x + h| \sim |h| \sim x_n + h_n \sim h_n, \tag{3.28}$$

that means that all these numbers can be pairwise estimated from above and from below by positive constants which are independent of x and h. The two cases

$$|h| < c_2|x_n| \quad \text{and} \quad |h| \ge c_2 |x_n|$$

are treated separately in the following two steps.

Step 3 Let $|h| \le c_2|x_n|$ according to Step 2. The corresponding part of $d^N_{t,u} f(x)$ in (3.18) with K^+_t in place of K^-_t can be estimated by

$$\left(t^{-n} \int_{h \in K^+_t, \, |h| < c_2 |x_n|} \left| \Delta^N_h \; \text{ext}\, f(x) \right|^u \, dh \right)^{\frac{1}{u}} \le c' \left(M_{c|x_n|} |f|^u \right) (\widetilde{x})^{\frac{1}{u}}, \tag{3.29}$$

$\widetilde{x} = (x', -x_n)$,

with the Hardy-Littlewood maximal function M_a, where the related mean values are restricted to balls of radius at most $a > 0$. Then we have

$$\begin{aligned}\left(\int_{c_1|x_n|}^1 t^{-sq}\left(M_{c|x_n|}|f|^u\right)^{\frac{q}{u}}(\widetilde{x})\,\frac{dt}{t}\right)^{\frac{1}{q}} &\le c_2\,|x_n|^{-s}\left(M_{c|x_n|}\,|f|^u\right)^{\frac{1}{u}}(\widetilde{x})\\ &\le c_3\left(M_{c|x_n|}\left|y_n^{-s}\,f(y)\right|^u\right)^{\frac{1}{u}}(\widetilde{x})\\ &\le c_3\left(M\left|y_n^{-s}\,f(y)\right|^u\right)^{\frac{1}{u}}(\widetilde{x}), \qquad (3.30)\end{aligned}$$

where M stands for the usual maximal function. We choose $u < p$. Then we take the $L_p(\mathbb{R}^n_-)$-quasi-norm which is transformed by $x \mapsto \widetilde{x}$ to $L_p(\mathbb{R}^n_+)$, apply the Hardy-Littlewood maximal inequality with respect to $\frac{p}{u} > 1$ and estimate all these expressions by

$$c\,\|x_n^{-s} f(x)\,|\,L_p(\mathbb{R}^n_+)\|. \tag{3.31}$$

Step 4 Let $|h| \ge c_2\,|x_n|$ and $h \in K_t^+$ according to Step 2. We have now (3.28). Let

$$\chi(t,h) = 1 \text{ if } |h| \le t \qquad \text{and} \qquad \chi(t,h) = 0 \text{ if } |h| > t.$$

Then we have to estimate terms of type (3.29) now with $h \in K_t^+$, $|h| \ge c_2|x_n|$. It is sufficient to deal with integrands of type $|f(x+h)|^u$. Furthermore we replace K_t^+ by $K^+ = K_1^+$ and $|f(x+h)|^u$ by $|f(x+h)|^u\,\chi(t,h)$. We choose $u < q$ and obtain by the triangle inequality for norms

$$\left[\int_{c|x_n|}^1 t^{-sq}\left(t^{-n}\int_{h\in K^+,|h|\ge c'|x_n|}|f(x+h)|^u\,\chi(t,h)\,dh\right)^{\frac{q}{u}}\frac{dt}{t}\right]^{\frac{u}{q}\frac{1}{u}}$$

$$\le\left[\int_{h\in K^+,|h|\ge c'|x_n|}|f(x+h)|^u\left(\int_{c|x_n|}^1 t^{-sq-n\frac{q}{u}}\,\chi(t,h)\,\frac{dt}{t}\right)^{\frac{u}{q}}dh\right]^{\frac{1}{u}}. \tag{3.32}$$

By our assumptions the integration over the inner integral is restricted to $t \ge |h|$ (not to speak about constants). Hence (3.32) can be estimated from above by

$$c\left[\int_{h\in K^+,\,|h|\ge c'|x_n|}|f(x+h)|^u\,|h|^{-su}\,\frac{dh}{|h|^n}\right]^{\frac{1}{u}}$$

$$\le c\,|x_n|^{-s}\left[\int_{h\in K^+_\infty,\,|h|\ge c'}|f(x+|x_n|h)|^u\;|h|^{-su}\,\frac{dh}{|h|^n}\right]^{\frac{1}{u}} \tag{3.33}$$

where we replaced h by $|x_n|h$. Of course, K^+_∞ is the cone (3.25) with $t = \infty$. We apply the L_p-quasi-norm to (3.32) and hence to the right-hand side of (3.33), with respect to $\mathbb{R}^{n-1} \times [-1,0]$. We replace again x by $\widetilde{x} = (x', -x_n)$ and choose $u < p$.

Again by the triangle inequality with respect to $L_{\frac{p}{u}}(\mathbb{R}^{n-1}\times[0,1])$ we estimate the desired term by

$$c\left[\int_{h\in K_\infty^+,\,|h|\geq c''}|h|^{-su}\frac{dh}{|h|^n}\left(\int_{\mathbb{R}^{n-1}\times[0,1]}|f(x+x_nh)|^p\,x_n^{-sp}\,dx\right)^{\frac{u}{p}}\right]^{\frac{1}{u}}. \tag{3.34}$$

In the inner integral we use the transformation $x+x_nh=y$. The respective Jacobian is proportional to $|h|^{-1}$. Using (3.28) we estimate (3.34) by

$$\begin{aligned}&c\left\|y_n^{-s}\,f(y)\,|\,L_p(\mathbb{R}^n_+)\right\|\left[\int_{h\in K_\infty^+,\,|h|\geq c'}|h|^{-su}\,h_n^{su}\,\frac{dh}{|h|^{n+\frac{u}{p}}}\right]^{\frac{1}{u}}\\ &\qquad\leq c''\left\|y_n^{-s}\,f(y)\,|\,L_p(\mathbb{R}^n_+)\right\|.\end{aligned} \tag{3.35}$$

Clipping together Step 2 and the estimates in (3.31) and (3.35) we obtain ext $f\in F^s_{pq}(\mathbb{R}^n)$ and the reverse of (3.22).

3.6 Proof of Theorem 2.2.2 *Step 1* Let $f\in\widetilde{F}^s_{pq}(\Omega)$ with $p<\infty$ (in case of $p=\infty$ and, hence, $q=\infty$, one has to modify the arguments appropriately). By Theorem 2.2.8 we have

$$\|f\,|\,\widetilde{F}^s_{pq}(\Omega)\|^p\leq c\,\|f\,|\,F^s_{pq}(\Omega)\|^p+c\int_\Omega d^{-sp}(x)\left|\sum_{j=0}^{\infty}\sum_{r=1}^{N_j}\varphi_{jr}(x)\,f(x)\right|^p\,dx, \tag{3.36}$$

where $\{\varphi_{jr}\}$ is the resolution of unity according to 2.2.1. We wish to estimate this expression from above by (2.31). By [Tri92], 5.2.2, p. 245, one may characterize $F^s_{pq}(\Omega)$ in terms of $d^N_{t,u}f$ as in (3.1) but with inward cones in place of $|h|\leq t$. Since everything is local we may again replace Ω by $\mathbb{R}^n_+$ and assume $f(x)=0$ if $|x|\geq 1$. Let K_t^+ be given by (3.25). We fix $x\in\mathbb{R}^n_+$ with $x_n\sim 2^{-j}$. Let either $0<t\leq c2^{-j}$ or $t\sim 2^{-k}$ with $k\in\mathbb{N}_0, k<j$. The typical term in (3.1), with K_t^+ in place of $|h|\leq t$, we have to estimate, is given by

$$\begin{aligned}d^N_{t,u}f(x)\leq c\left(t^{-n}\int_{h\in K_t^+}\left|\Delta_h^N\varphi_{jr_j}f(x)\right|^u\,dh\right)^{\frac{1}{u}}&+\\ +c\left(2^{-kn}\sum_{l=k}^{j}\int_{h\in K_t^+}|(\varphi_{lr_l}f)(x+h)|^u\,dh\right)^{\frac{1}{u}}&++\end{aligned} \tag{3.37}$$

where $++$ indicates both replacements of h by mh with $m=1,\ldots,N$ and also the replacement of φ_{lr_l} by neighbouring φ_{lr}. But in any case by construction of the resolution of unity $\{\varphi_{lr}\}$ and the restriction of h to the upward cone K_t^+ it follows that there are at most L terms of this type, where L is independent of x, j and k. Furthermore in case of $0<t\leq 2^{-j}$ one has only the first term on the right-hand

side of (3.37) (and neighbouring terms of the same type). Inserting (3.37) in

$$\left(\int_0^1 t^{-sq} \, d^N_{t,u} f(\cdot)^q \, \frac{dt}{t} \right)^{\frac{1}{q}}, \tag{3.38}$$

the first terms on the right-hand side of (3.37) result in what we wish to have. As for the remaining terms we are much in the same situation as in (3.32)–(3.35). We must replace $|f(x+h)|^u$ on the right-hand side of (3.32) by

$$\sum_{l=k}^{j} |(\varphi_{lr_l} f)(x+h)|^u \sim \left(\sum_{l=k}^{j} |(\varphi_{lr_l} f)(x+h)|^p \right)^{\frac{u}{p}} \tag{3.39}$$

where we used the support properties of φ_{lr_l}. The sum on the right-hand side can be extended to the whole resolution of unity $\{\varphi_{lr}\}$. Then we obtain the obvious counterpart of (3.35). Clipping together all these estimates we get

$$\|f \mid F^s_{pq}(\Omega)\|^p$$
$$\leq c \sum_{j=0}^{\infty} \sum_{r=1}^{N_j} \|\varphi_{jr} f \mid F^s_{pq}(\mathbb{R}^n)\|^p + c \sum_{j=0}^{\infty} \sum_{r=1}^{N_j} \int_\Omega d^{-sp}(x) \, |\varphi_{jr}(x) \, f(x)|^p \, dx, \tag{3.40}$$

where we switched from $\mathbb{R}^n$ to Ω. In Step 3 below we prove the following assertion: *Let p, q, s, u and N as above. There is a number $c > 0$ such that for all $0 < \lambda \leq 1$ and all*

$$f \in F^s_{pq}(\mathbb{R}^n), \quad \operatorname{supp} f \subset B_\lambda = \{y \in \mathbb{R}^n : |y| < \lambda\} \tag{3.41}$$

holds

$$\|f \mid L_p(\mathbb{R}^n)\| \leq c\lambda^s \left\| \left(\int_0^\lambda t^{-sq} d^N_{t,u} f(\cdot)^q \, \frac{dt}{t} \right)^{\frac{1}{q}} \Big| \, L_p(\mathbb{R}^n) \right\|. \tag{3.42}$$

We apply this estimate to $\varphi_{jr} f \in F^s_{pq}(\mathbb{R}^n)$. Then we obtain by (3.36) and (3.40)

$$\|f \mid \widetilde{F}^s_{pq}(\Omega)\|^p \leq c \sum_{j=0}^{\infty} \sum_{r=1}^{N_j} \|\varphi_{jr} \, f \mid F^s_{pq}(\mathbb{R}^n)\|^p. \tag{3.43}$$

Step 2 We prove the reverse inequality to (3.43). By the above local replacement of Ω by $\mathbb{R}^n_+$ we must estimate

$$\left\| \left(\int_0^1 t^{-sq} d^N_{t,u} \varphi_{jr} f(\cdot)^q \, \frac{dt}{t} \right)^{\frac{1}{q}} \Big| \, L_p(\mathbb{R}^n) \right\|^p + \|\varphi_{jr} f \mid L_p(\mathbb{R}^n)\|^p \tag{3.44}$$

where we now choose

$$d^N_{t,u} \varphi_{jr} f(x) = \left(t^{-n} \int_{h \in K^-_t} |\Delta^N_h \varphi_{jr} f(x)|^u \, dh \right)^{\frac{1}{u}}, \tag{3.45}$$

where K_t^- is given by (3.17) and again $u < \min(1, p, q)$. We split the integral over t in (3.44) in $t \in (0, a2^{-j})$ and $t \in (a2^{-j}, 1)$. Let $0 < t < a2^{-j}$, where $a > 0$ is sufficiently small. Let bK_{jr} be a ball concentric with the ball K_{jr} in 2.2.1 and with radius b times the radius of K_{jr}. By mathematical induction we have for some constants c_m

$$(\Delta_h^N f\varphi_{jr})(x) = \sum_{m=0}^{N} c_m (\Delta_h^{N-m} f)(x)(\Delta_h^m \varphi_{jr})(x + (N-m)h). \tag{3.46}$$

We choose $K \in \mathbb{N}$ and $N \in \mathbb{N}$ such that $K > s$ and $N - K > s$. Let $x \in bK_{jr}$, where $b > 0$ is assumed to be small. By (3.46) we have

$$|\Delta_h^N f\varphi_{jr}(x)| \le c \sum_{l=0}^{K} |\Delta_h^{N-l} f(x)| + c \sum_{l=K+1}^{N} \sum_{r=0}^{N} |f(x+rh)| 2^{jl} t^l. \tag{3.47}$$

We insert the first terms on the right-hand side of (3.47) in (3.45) and obtain for the relevant part of the t-integral in (3.44)

$$\sum_{l=0}^{K} \int_0^{a2^{-j}} t^{-sq} d_{t,u}^{N-l} f(x)^q \, \frac{dt}{t}, \qquad x \in bK_{jr}, \tag{3.48}$$

where $d_{t,u}^{N-l} f(x)$ are given by obvious counterparts of (3.45). Recall $N - l \ge N - K > s$. Summing up all these terms with respect to $j \in \mathbb{N}_0$ and $r = 1, \ldots, N_j$ one can estimate the relevant part of the right-hand side of (3.43) from above by $c\|f \mid F^s_{pq}(\Omega)\|^p$ (again switching from $\mathbb{R}^n_+$ to Ω). We have to deal with all the remainig terms. First let again $0 < t < a2^{-j}$, where $a > 0$ is small. Then also for the second terms in (3.47) all arguments $x + lh$ are within dK_{jr}, where $d > 0$ may be assumed to be small. Then it follows for the relevant parts in (3.44)

$$\begin{aligned}
&\left(\int_0^{a2^{-j}} t^{-sq} \ldots \frac{dt}{t} \right)^{\frac{1}{q}} \\
&\quad \le c \left(M|y_n^{-s} f(y)|^u \right)^{\frac{1}{u}} (x)\, 2^{-js} 2^{jK} \left(\int_0^{a2^{-j}} t^{(K-s)q} \frac{dt}{t} \right)^{\frac{1}{q}} \\
&\quad \le c' \left(M|y_n^{-s} f(y)|^u \right)^{\frac{1}{u}} (x), \qquad x \in bK_{jr}.
\end{aligned} \tag{3.49}$$

Of course, M stands again for the maximal function. We used the previous technique and $K > s$. Let now $a2^{-j} < t < 1$ and again $x \in bK_{jr}$. In (3.45) only those $h \in K_t^-$ with $x + lh \in K_{jr}$ are of interest. Furthermore,

$$\left(\int_{a2^{-j}}^{1} t^{-sq} \frac{dt}{t} \right)^{\frac{1}{q}} \le c\, 2^{js}. \tag{3.50}$$

Hence all the terms can be estimated by the right-hand side of (3.49) and we have

$$\left(\int_0^1 t^{-sq}\cdots\frac{dt}{t}\right)\le c\left(M|y_n^{-s}f(y)|^u\right)^{\frac{1}{u}}(x),\quad x\in bK_{jr}. \tag{3.51}$$

Finally we have to estimate (3.45) for $x\in\mathbb{R}^n\backslash bK_{jr}$. We replace again locally Ω by $\mathbb{R}^n_+$. Since the cone K_t^- is downwards it follows that $d^N_{t,u}\varphi_{jr}f(x)$ in (3.45) is zero with possible exception of $x\in cK_{jr}+K_t^+$. This happens for at most L values of r_k in $x\in K_{kr_k}$ where $k=0,\dots,j$. In that case t must be larger than $c2^{-k}$ and we have (3.50) with k in place of j. Since $2^{ks}=2^{js}\,2^{-(j-k)s}$ we obtain by the above arguments

$$\left(\int_0^1 t^{-sq}\cdots\frac{dt}{t}\right)^{\frac{1}{q}}\le c\,2^{-(j-k)s}\left(M|y_n^{-s}f(y)|^u\right)^{\frac{1}{u}}(x),\quad x\in K_{kr_k}. \tag{3.52}$$

Hence for the remaining parts of the right-hand side of (3.43) we have the estimates (3.51) and (3.52). If we fix, say, $x\in K_{kr}$ then by (3.43), (3.44) we must sum the pth powers of (3.52) with respect to $j\ge k$. Since $s>0$ all these terms can be estimated from above by

$$c\left\|\left(M|y_n^{-s}f(y)|^u\right)^{\frac{1}{u}}(\cdot)\,|\,L_p(\mathbb{R}^n)\right\|^p. \tag{3.53}$$

Recall $u<p$. The maximal inequality and what had been said before show that the right-hand side of (3.43) can be estimated by

$$\begin{aligned}\sum_{j=0}^{\infty}\sum_{r=1}^{N_j}\|\varphi_{jr}f\,|\,F^s_{pq}(\mathbb{R}^n)\|^p &\le c\left(\|f\,|\,F^s_{pq}(\Omega)\|^p+\|d^{-s}f\,|\,L_p(\Omega)\|^p\right)\\ &\le c'\|f\,|\,\widetilde{F}^s_{pq}(\Omega)\|^p,\end{aligned} \tag{3.54}$$

where we used Theorem 2.2.8.

Step 3 We prove (3.41), (3.42). There is a number $c>0$ such that

$$\|g\,|\,L_p(\mathbb{R}^n)\|\le c\left\|\left(\int_0^1 t^{-sq}d^N_{t,u}g(\cdot)^q\frac{dt}{t}\right)^{\frac{1}{q}}\,|\,L_p(\mathbb{R}^n)\right\| \tag{3.55}$$

for all

$$g\in F^s_{pq}(\mathbb{R}^n),\quad \operatorname{supp} g\subset B_1=\{y\in\mathbb{R}^n:\ |y|<1\}. \tag{3.56}$$

This follows by standard arguments from the compact embedding of, say, $F^s_{pq}(B_2)$ in all spaces $L_r(B_2)$ with $0<r<p^*$ for some $1<p^*\le\infty$. If f is given by (3.41) then we apply (3.55) to $g(x)=f(\lambda x)$. By (3.1) we have

$$\left[d^N_{t,u}f(\lambda\cdot)\right](x)=\left(t^{-n}\int_{|h|\le t}|\Delta^N_{\lambda h}f(\lambda x)|^u\,dh\right)^{\frac{1}{u}}=\left(d^N_{\lambda t,u}f\right)(\lambda x). \tag{3.57}$$

We insert (3.57) in (3.55) and obtain (3.42).

3.7 Equivalent quasi-norms Let Ω be a bounded C^∞ domain in $\mathbb{R}^n$, and let $0 < p \le \infty$, $0 < q \le \infty$ (with $q = \infty$ if $p = \infty$), and $s > \sigma_{pq}$. Let $d(x)$ be given by (2.32). Furthermore, $d^N_{t,u} f(x)$ has the meaning (3.1) where we always assume $u < \min(1, p, q)$. *Then $f \in L_1(\Omega)$ belongs to $\widetilde{F}^s_{pq}(\Omega)$ if, and only if,*

$$\left\| \left(\int_0^{cd(\cdot)} t^{-sq}\, d^N_{t,u} f(\cdot)^q \frac{dt}{t} \right)^{\frac{1}{q}} \,|\, L_p(\Omega) \right\| + \left\| d^{-s} f \,|\, L_p(\Omega) \right\| < \infty \tag{3.58}$$

(usual modification if $q = \infty$) , where $c > 0$ may be chosen arbitrarily small. Furthermore, (3.58) is an intrinsic quasi-norm for any small $c > 0$. This assertion is covered by the proof in 3.6, especially by Step 2. Hence the differences in (3.1) can be restricted to a ball centered at $x \in \Omega$ with radius $c\, d(x)$. Everything else is covered by $d^{-s} f \in L_p(\Omega)$. This might be of special interest if, in addition, $s - \frac{1}{p} \notin \mathbb{N}_0$. By Theorem 2.4.2 in this case the above assertion applies to $\overset{\circ}{F}{}^s_{pq}(\Omega)$.

3.8 Proof of Theorem 2.3.3 *Step 1* We dualize Theorem 2.2.2 by using (2.59) with (2.57). This will cover the spaces $F^s_{pq}(\Omega)$ with (2.60) and $s < 0$. If $1 < p < \infty$, $1 < q \le \infty$, $0 < s < \frac{1}{p}$, then, by Proposition 2.2.10, the assertion in Theorem 2.3.3 is a special case of Theorem 2.2.2. As for the remainig cases $1 < p < \infty$, $1 < q \le \infty$, $s = 0$, one can use complex interpolation $[\cdot,\cdot]_\Theta$ or $(\cdot,\cdot)_\Theta$ with $\Theta = \frac{1}{2}$ between $F^s_{pq}(\Omega)$ and $F^{-s}_{pq}(\Omega)$ for some $0 < s < \frac{1}{p}$. We do not go into detail, but we mention that the needed pointwise multiplier assertions are covered by the following steps and formalized in 3.9 below. As for the classical interpolation $[\cdot,\cdot]_\Theta$ in the vector-valued case of $\ell_p(A)$ we refer to [Tri78], 1.18.1, p. 121. There is a counterpart for $(\cdot,\cdot)_\Theta$. Concerning the interpolation $(\cdot,\cdot)_\Theta$ of the spaces $F^s_{pq}(\Omega)$ we refer to [Tri83], 2.4.7, p. 69.

Step 2 Let

$$1 \le p < \infty, \quad 1 \le q < \infty \text{ (with } q = 1 \text{ if } p = 1) \quad \text{and} \quad s > 0. \tag{3.59}$$

By Theorem 2.2.2 we may assume that

$$A \,:\, f \in \widetilde{F}^s_{pq}(\Omega) \mapsto \{\varphi_{jr} f\} \in \ell_p\left(\widetilde{F}^s_{pq}(K_{jr})\right) \tag{3.60}$$

is a linear and bounded operator. Of course $\{\varphi_{jr}(x) : j \in \mathbb{N}_0,\, r = 1, \ldots, N_j\}$ is the resolution of unity introduced in 2.2.1. Next we claim that

$$B\{g_{jr}\}(x) = \sum_{j=0}^{\infty} \sum_{r=1}^{N_j} g_{jr}(x) \tag{3.61}$$

is a linear and bounded map,

$$B \,:\, \ell_p\left(\widetilde{F}^s_{pq}(K_{jr})\right) \mapsto \widetilde{F}^s_{pq}(\Omega). \tag{3.62}$$

We apply again Theorem 2.2.2. In 3.9 below we prove that there is a constant $c > 0$ such that for all $j \in \mathbb{N}_0$ and $r = 1, \dots, N_j$,

$$\|\varphi_{jr} g \,|\, \widetilde{F}^s_{pq}(K_{jr})\| \le c \, \|g \,|\, \widetilde{F}^s_{pq}(K_{jr})\| \tag{3.63}$$

for all $g \in \widetilde{F}^s_{pq}(K_{jr})$, see (3.81), (3.82). To justify this somewhat surprising pointwise multiplier assertion one needs (2.28), (2.29), but not (2.30). Taking (3.63) for granted then (3.62) is a consequence of minor technical modifications (of type $j \to j-1, j+1$, etc.) and Theorem 2.2.2. Furthermore,

$$id = B \circ A \qquad (\text{identity in } \widetilde{F}^s_{pq}(\Omega)\ ). \tag{3.64}$$

This is the situation which fits in the scheme of interpolation mentioned in Step 1. Here we dualize (3.64) and obtain

$$id = A' \circ B' \quad (\text{ identity in } F^{-s}_{p'q'}(\Omega)\ ) \tag{3.65}$$

where we used (2.57), (2.59). Similarly we have by (3.60)

$$A' \ : \ \ell_{p'} \left(F^{-s}_{p'q'}(K_{jr}) \right) \mapsto F^{-s}_{p'q'}(\Omega), \tag{3.66}$$

where A' is given by

$$A'\{h_{jr}\}(x) = \sum_{j=0}^{\infty} \sum_{r=1}^{N_j} \varphi_{jr}(x) \, h_{jr}(x) \tag{3.67}$$

since for $f \in \widetilde{F}^s_{pq}(\Omega)$ and $\{h_{jr}\} \in \ell_{p'} \left(F^{-s}_{p'q'}(K_{jr}) \right)$ holds

$$(Af, \{h_{jr}\}) = \sum_j \sum_r (\varphi_{jr} f, h_{jr}) = \left(f, \sum_j \sum_r \varphi_{jr} h_{jr} \right) \tag{3.68}$$

in the understanding of dual pairing. As for B' we have

$$B' \ : \ F^{-s}_{p'q'}(\Omega) \mapsto \ell_{p'} \left(F^{-s}_{p'q'}(K_{jr}) \right) \tag{3.69}$$

where B' is given by

$$B'f = \{f\} \in \ell_{p'} \left(F^{-s}_{p'q'}(K_{jr}) \right) \tag{3.70}$$

since for $f \in F^{-s}_{p'q'}(\Omega)$ and $\{g_{jr}\} \in \ell_p \left(\widetilde{F}^s_{pq}(K_{jr}) \right)$ holds

$$(B\{g_{jr}\}, f) = \sum_{j=0}^{\infty} \sum_{r=1}^{N_j} (g_{jr}, f). \tag{3.71}$$

By (3.65) and the assertions about A' and B' we have

$$\|f \,|\, F^{-s}_{p'q'}(\Omega)\| \le c \left(\sum_{j=0}^{\infty} \sum_{r=1}^{N_j} \|f \,|\, F^{-s}_{p'q'}(K_{jr})\|^{p'} \right)^{\frac{1}{p'}} \le c' \, \|f \,|\, F^{-s}_{p'q'}(\Omega)\|, \quad f \in F^{-s}_{p'q'}(\Omega). \tag{3.72}$$

This is an interesting formula by itself, but not yet the final assertion we wish to have. First we remark that there is a constant $c > 0$ such that for all $j \in \mathbb{N}_0$ and $r = 1, \dots, N_j$,

$$\|\varphi_{jr} f \mid F^{-s}_{p'q'}(K_{jr})\| \le c \, \|f \mid F^{-s}_{p'q'}(K_{jr})\| \tag{3.73}$$

for all $f \in F^{-s}_{p'q'}(K_{jr})$. This follows either by duality from (3.63) or directly from what we have to say in 3.9. By the homogeneity properties in 3.9 for the spaces $F^{-s}_{p'q'}(K_{jr})$ it follows also that K_{jr} on the left-hand side of (3.73) can be replaced by $\mathbb{R}^n$. Then we obtain

$$\left(\sum_{j=0}^{\infty} \sum_{r=1}^{N_j} \|\varphi_{jr} f \mid F^{-s}_{p'q'}(\mathbb{R}^n)\|^{p'} \right)^{\frac{1}{p'}} \le c \, \|f \mid F^{-s}_{p'q'}(\Omega)\|. \tag{3.74}$$

To prove the opposite inequality we remark that in each term in the sum over j and r in (3.72) one may replace f by $f(\varphi_{jr} + +)$, where $++$ indicates terms $\varphi_{j'r'}$ with couples (j', r') near (j, r). The triangle inequality and (3.72) prove the converse inequality to (3.74). The proof is complete.

3.9 Homogeneity and pointwise multipliers Let B_λ as in (3.41) be the ball centered at the origin and of radius λ with $0 < \lambda \le 1$. The spaces

$$\widetilde{F}^s_{pq}(B_\lambda) \; : \; 0 < p \le \infty \, , \; 0 < q \le \infty \; (\; q = \infty \text{ if } p = \infty \;), \quad s > \sigma_{pq}, \tag{3.75}$$

related to Theorem 2.2.2 and the spaces

$$F^s_{pq}(B_\lambda) \; : \; 1 < p \le \infty \, , \; 1 < q \le \infty \; (\; q = \infty \text{ if } p = \infty), \quad s < \frac{1}{p}, \tag{3.76}$$

related to Theorem 2.3.3 have some remarkable homogeneity properties and resulting pointwise multiplier assertions.

(i) *Homogeneity of the spaces in (3.75)* Let $d^N_{t,u} f(x)$ be given by (3.1). We put

$$D^{s,\lambda}_q f(x) = \left(\int_0^\lambda t^{-sq} d^N_{t,u} f(x)^q \frac{dt}{t} \right)^{\frac{1}{q}} . \tag{3.77}$$

We always assume that the numbers $s < N \in \mathbb{N}$ and $0 < u < \min(1, p, q)$ are fixed. By the proof of Theorem 2.2.2 in 3.6 it follows

$$\left\| f \mid \widetilde{F}^s_{pq}(B_\lambda) \right\| \sim \left\| D^{s,\lambda}_q f \mid L_p(\mathbb{R}^n) \right\| , \tag{3.78}$$

where the equivalence constants are *independent* of λ (and, of course, of $f \in \widetilde{F}^s_{pq}(B_\lambda)$). Let $f \in \widetilde{F}^s_{pq}(B_\lambda)$. Then $g(x) = f(\lambda x) \in \widetilde{F}^s_{pq}(B_1)$. By (3.57) we have

$$D^{s,1}_q g(x) = \left(\int_0^1 t^{-sq} d^N_{\lambda t,u} f(\lambda x)^q \frac{dt}{t} \right)^{\frac{1}{q}} = \lambda^s D^{s,\lambda}_q f(\lambda x) \tag{3.79}$$

and by (3.78)

$$\|f(\lambda \cdot) \mid \widetilde{F}^s_{pq}(B_1)\| \sim \lambda^{s - \frac{n}{p}} \|f \mid \widetilde{F}^s_{pq}(B_\lambda)\|, \tag{3.80}$$

where again the equivalence constants are independent of $0 < \lambda \le 1$.

(ii) *Pointwise multipliers for the spaces in (3.75)* As an application we obtain the following multiplier assertion: Let

$$\varphi \in C^\infty(\overline{B_\lambda}), \quad |D^\gamma \varphi(x)| \le c_\gamma \lambda^{-|\gamma|}, \quad \gamma \in \mathbb{N}_0^n. \tag{3.81}$$

Then we have for the spaces in (3.75)

$$\|\varphi f \,|\, \widetilde{F}^s_{pq}(B_\lambda)\| \le c \|f \,|\, \widetilde{F}^s_{pq}(B_\lambda)\|, \tag{3.82}$$

where c depends on finitely many c_γ, but not on λ. This follows from (3.80): The assertion for B_λ is transformed to B_1, where one has a well-known multiplier assertion for $\varphi(\lambda x)$.

(iii) *Homogeneity of the spaces in (3.76)* Let

$$1 \le p < \infty, \quad 1 \le q < \infty, \; (\; q = 1 \text{ if } p = 1), \quad s > 0. \tag{3.83}$$

By the same arguments as in [Tri78], 2.10.5, pp. 234/235, it follows

$$\left(\widetilde{F}^s_{pq}(B_\lambda)\right)' = \left(F^s_{pq}(\mathbb{R}^n)\right)' / \{g \in F^{-s}_{p'q'}(\mathbb{R}^n) : \operatorname{supp} g \subset \mathbb{R}^n \backslash B_\lambda\} = F^{-s}_{p'q'}(B_\lambda). \tag{3.84}$$

We have (2.59) but now with the additional information that $\|f \,|\, F^{-s}_{p'q'}(B_\lambda)\|$ by Definition 2.1.4 is precisely the norm of the dual of $\widetilde{F}^s_{pq}(B_\lambda)$ via the dual pairing $(D(B_\lambda), D'(B_\lambda))$. Then we have (with the usual interpretation)

$$\|f \,|\, F^{-s}_{p'q'}(B_\lambda)\| = \sup \left| \int_{B_\lambda} f(x)\, \varphi(x)\, dx \right| = \sup \lambda^n \left| \int_{B_1} f(\lambda x)\, \varphi(\lambda x)\, dx \right|, \tag{3.85}$$

where the supremum is taken over

$$1 \ge \|\varphi \,|\, \widetilde{F}^s_{pq}(B_\lambda)\| \sim \lambda^{-s+\frac{n}{p}} \|\varphi(\lambda \cdot) \,|\, \widetilde{F}^s_{pq}(B_1)\|, \tag{3.86}$$

where we used (3.80). It follows

$$\|f \,|\, F^{-s}_{p'q'}(B_\lambda)\| \sim \lambda^{n+s-\frac{n}{p}} \|f(\lambda \cdot) \,|\, F^{-s}_{p'q'}(B_1)\|. \tag{3.87}$$

Changing $p', q', -s$ in p, q, s we obtain the cases in (3.76) with $s < 0$. As remarked in Step 1 in 3.8 the cases $0 < s < \frac{1}{p}$ are covered by (i) and $s = 0$ is a matter of interpolation. In other words, in all cases (3.76) we have the homogeneity

$$\|f(\lambda \cdot) \,|\, F^s_{pq}(B_1)\| \sim \lambda^{s-\frac{n}{p}} \|f \,|\, F^s_{pq}(B_\lambda)\|, \tag{3.88}$$

complementing (3.80), where again the equivalence constants are independent of $0 < \lambda \le 1$. Furthermore there is an obvious counterpart of the pointwise multiplier assertion (3.81), (3.82).

3.10 Proof of Corollary 2.3.6

If $\frac{n-1}{n} < p \le 1$, then $\sigma_p < \frac{1}{p}$. Let

$$\frac{n-1}{n} < p \le 1, \quad p \le q \le \infty, \quad \sigma_p < s < \frac{1}{p}. \tag{3.89}$$

Then we have (2.42), and hence Theorem 2.2.2 gives a characterization of $F^s_{pq}(\Omega)$. Recall the complex interpolation formula

$$\left(F^{s_0}_{p_0q_0}(\Omega), F^{s_1}_{p_1q_1}(\Omega)\right)_\Theta = F^s_{pq}(\Omega), \quad 0<\Theta<1, \tag{3.90}$$

with

$$s=(1-\Theta)s_0+\Theta s_1, \quad \frac{1}{p}=\frac{1-\Theta}{p_0}+\frac{\Theta}{p_1}, \quad \frac{1}{q}=\frac{1-\Theta}{q_0}+\frac{\Theta}{q_1}, \tag{3.91}$$

which works for all admitted values of the involved parameters with $p_0<\infty$, $p_1<\infty$. We refer to [Tri92], 1.6.7, p. 45, combined with [Tri92], 1.10.2, p. 70, and [Tri83], 3.3.6, pp. 203–4. By the technique developed in [Tri83] in connection with this interpolation method there is also a vector-valued version for spaces $\ell_p\left(F^s_{pq}(\Omega)\right)$. Then the corollary follows by interpolation of Theorem 2.2.2, restricted to (3.89), and Theorem 2.3.3.

3.11 Proof of Theorem 2.4.2 *Step 1* Let p,q,s be given by (2.51), (2.53). As remarked there, the characteristic function of Ω is a pointwise multiplier for $F^s_{pq}(\Omega)$. Let $m\in\mathbb{N}$. We lift the $(\frac{1}{p},s)$-region (2.51), (2.53) by $s\mapsto s+m$, see also Fig. 1. Then we have (2.70) for these values of p,q,s. As for a more detailed argument of this type we refer to [Tri83], 2.9.2, pp. 167-8. In case of (2.41) we have (2.42), what in this case is of course the same as (2.70). Especially, in the one-dimensional case $n=1$ the proof is complete.

Step 2 Let $n\geq 2$ and $q\geq p$. Then the desired assertion follows from the one-dimensional case and Theorem 2.1.12.

Step 3 Let $n\geq 2$ and $q<p$. Then, in general, we cannot apply Theorem 2.1.12. We wish to use the complex interpolation (3.90), (3.91), where we assume that $(\frac{1}{p_0},s_0)$ and $(\frac{1}{p_1},s_1)$ are in the same strip (2.81) with $m\in\mathbb{N}$ (obvious modification if $m=0$). We suppose in addition that $(\frac{1}{p_0},s_0)$ with $0<q_0\leq\infty$ is covered by Step 1, whereas $(\frac{1}{p_1},s_1)$ with $q_1\geq p_1$ is covered by Step 2. We refer again to Fig. 1. By (3.91) any admitted value of p,q,s in the indicated strip can be reached in this way. First we claim that under the indicated values for the parameters (3.90) holds also with $\widetilde{F}$ in place of F. This follows from (3.90) with $\mathbb{R}^n$ in place of Ω and the fact that the interpolation respects the support property. This can be checked by a technical modification (local means) of formula (13) in [Tri83], 2.4.7, p. 71. Next we claim that (3.90) holds also with $\overset{\circ}{F}$ in place of F. As in (2.82) (what will be proved independently of our arguments here in 3.12) the trace operator

$$tr\ :\ f\mapsto\left\{\frac{\partial^r}{\partial\nu^r}f\,|\,\partial\Omega\ :\ r=0,\ldots,m-1\right\} \tag{3.92}$$

is the same for $F^{s_0}_{p_0q_0}(\Omega)$ and $F^{s_1}_{p_1q_1}(\Omega)$. Here ν is the outer normal with respect to $\partial\Omega$. Recall that by the arguments in Step 2 in 3.3 the trace is independent of q_0

and q_1, respectively. In the notation in [Tri97], 20.2, p. 159, we have

$$tr\, F^{s_j}_{p_j q_j}(\Omega) = \prod_{r=0}^{m-1} \mathbb{B}^{s_j - \frac{1}{p_j} - r}_{p_j p_j}(\partial\Omega); \quad j = 1, 2. \tag{3.93}$$

We suppose that there is also a common linear extension operator ext from the spaces on the right-hand side of (3.93) into $F^{s_j}_{p_j q_j}(\Omega)$. This is not covered by the method in [Tri97] and also not by the arguments in 3.12 below. But at least partly one can use the technique developed in [Jon93a] and [Jon93b] to construct linear extension operators (the result itself cannot be applied immediately, but the method). In case of $n = 2$ one has no problems at this point. We refer to [Tri83], 2.7.2, pp. 131–139, where one finds also how to handle product spaces as on the right-hand side of (3.93). Then $P = id - \text{ext} \circ tr$ is a projection of $F^{\cdot}_{\cdot\cdot}(\Omega)$ onto $\overset{\circ}{F}{}^{\cdot}_{\cdot\cdot}(\Omega)$ and we have a standard situation in interpolation theory (in Banach spaces, extended to quasi-Banach spaces), [Tri78], 1.17.2, p. 119. Taking all these assertions for granted it follows that we have (3.90) also for $\widetilde{F}$ in place of F, and for $\overset{\circ}{F}$ in place of F. Then the remaining cases follow by interpolation of Step 1 and Step 2.

3.12 Proof of Proposition 2.4.5 *Step 1* Recall again that the trace of $f \in F^s_{pq}(\Omega)$ on $\partial\Omega$ is independent of q. This follows also from the technique in Step 2 in 3.3. We have (3.93). In particular any $f \in \overset{\circ}{F}{}^s_{pq}(\Omega)$ belongs to the space on the right-hand side of (2.82).

Step 2 Let f be an element of the right-hand side of (2.82). Since $q < \infty$ this element can be approximated in $F^s_{pq}(\Omega)$ by functions $f_j \in C^\infty(\overline{\Omega})$ with

$$\frac{\partial^r}{\partial\nu^r} f_j|\partial\Omega \longrightarrow 0 \quad \text{in} \quad \mathbb{B}^{s-\frac{1}{p}-r}_{pp}(\partial\Omega) \quad \text{if} \quad j \to \infty, \tag{3.94}$$

according to (3.93). By the atomic technique in [Tri97], 18.6, 18.12, 20.2, it follows that there are functions $g_j \in C^\infty(\overline{\Omega})$ with the same traces

$$\frac{\partial^r}{\partial\nu^r} g_j|\partial\Omega = \frac{\partial^r}{\partial\nu^r} f_j|\partial\Omega,$$

tending to zero in $F^s_{pq}(\Omega)$ if $j \to \infty$. But then we can approximate the above element f by functions $f_j \in C^\infty(\overline{\Omega})$ with $D^\gamma f_j|\partial\Omega = 0$ if $|\gamma| \le m-1$. In other words, it is sufficient to prove that

$$f \in C^\infty(\overline{\Omega}), \quad D^\gamma f|\partial\Omega = 0 \quad \text{if} \quad |\gamma| \le m-1, \tag{3.95}$$

belongs to $\overset{\circ}{F}{}^s_{pq}(\Omega)$. But this follows again by the above-mentioned atomic technique (with and without moment conditions).

3.13 Proof of Theorem 2.5.7 *Step 1* As in (3.41) we denote by B_λ the ball centered at the origin in $\mathbb{R}^n$ and of radius $\lambda > 0$. Let

$$f \in F^s_{pq}(\mathbb{R}^n) \quad \text{with} \quad \text{supp}\, f \subset B_1, \tag{3.96}$$

where p, q, s are restricted as in Theorem 2.5.7. Let φ be a C^∞ function in $\mathbb{R}^n$ with compact support in B_3 and $\varphi(x) = 1$ if $|x| \le 2$. We expand f according to Theorem 2.5.4 and multiply afterwards f with φ. Then we obtain the subatomic decomposition

$$f = \sum_{\gamma \in \mathbb{N}_0^n} \sum_{l=0}^{\infty} \sum_m \lambda_{lm}^\gamma (\gamma qu)_{lm}(x), \tag{3.97}$$

where we may assume that for given $l \in \mathbb{N}_0$ the summation over m is restricted to those $m \in \mathbb{Z}^n$ with $|2^{-l}m| \le c$ for some suitably chosen c adapted to the constants in 2.5.3 and 2.5.4. Furthermore we tacitly incorporated φ in $(\gamma qu)_{lm}(x)$ given by (2.92), now not necessarily based on (2.87) but on 2.5.6 with an approximative resolution of unity in a neighbourhood of B_1 and also slightly distorted $x^{l,m}$ according to our later needs. Let now

$$f \in F_{pq}^s(\mathbb{R}^n) \quad \text{with} \quad \operatorname{supp} f \subset B_\lambda\,; \quad \lambda = 2^{-j}, \tag{3.98}$$

with $j \in \mathbb{N}_0$. We expand $f(\lambda\cdot)$ according to (3.97). We use the homogeneity property described in 3.9(i), in particular (3.80). For our purpose we can replace $\widetilde{F}$ in (3.80) by F, since f with (3.98) belongs also to, say, $\widetilde{F}_{pq}^s(B_{3\lambda})$. This is sufficient for us. Then we obtain an expansion of type (3.97) with $\sum_{l=j}^{\infty}$ in place of $\sum_{l=0}^{\infty}$ where all the coefficients according to (2.92) and (2.90) adapt in the desired way (as it has to be).

Step 2 Now Theorem 2.5.7 is an easy consequence of Step 1, Theorem 2.2.2, and the fact that all the series in question converge absolutely in $L_1(\Omega)$ and, hence, can rearranged as one wishes.

References

[CDS96] Chang, D.-C., Dafni, G., and Stein, E. M. "Hardy spaces, BMO, and boundary value problems for the Laplacian on a smooth domain." Preprint, 1996.

[CKS92] Chang, D.-C., Krantz, S. G., and Stein, E. M. "Hardy spaces and elliptic boundary-value problems." In: Proc. Madison Symp. Complex Analysis. Contemporary Math **137**, AMS, 1992, 119–131.

[CKS93] Chang, D.-C., Krantz, S. G., and Stein, E. M. "H^p theory on a smooth domain in $\mathbb{R}^N$ and elliptic boundary value problems." J. Funct. Anal. **144** (1993), 286–347.

[ET96] Edmunds, D. E. and Triebel, H. *Function Spaces, Entropy Numbers, Differential Operators.* Cambridge Univ. Press, 1996

[Far97] Farkas, W. "Atomic and subatomic decompositions in anisotropic function spaces." Math. Nachr. (to appear).

[FJW91] Frazier, M., Jawerth, B., and Weiss, G. "Littlewood-Paley theory and the study of function spaces." CBMS-AMS Regional Conf. Ser. **79**, 1991.

[Fra86] Franke, J. "On the spaces F_{pq}^s of Triebel-Lizorkin type: Pointwise multipliers and spaces on domains." Math. Nachr. **125** (1986), 29–68.

[FrJ85] Frazier, M. and Jawerth, B. "Decomposition of Besov spaces." Indiana Univ. Math. J. **34** (1985), 777–799.

[FrJ90] Frazier, M. and Jawerth, B. "A discrete transform and decompositions of distribution spaces." J. Funct. Anal. **93** (1990), 34–170.

[FrR95] Franke, J. and Runst, Th. "Regular elliptic boundary value problems in Besov-Triebel-Lizorkin spaces." Math. Nachr. **174** (1995), 113–149.

[Joh96] Johnsen, J. "Elliptic boundary problems and the Boutet de Monvel calculus in Besov and Triebel-Lizorkin spaces." Math. Scand **79** (1996), 25–85.

[Jon93a] Jonsson, A. "Atomic decomposition of Besov spaces on closed sets." In: Function spaces, differential operators and non-linear analysis. Teubner-Texte Math. **113**, Teubner, Leipzig, 1993, 285–289.

[Jon93b] Jonsson, A. "Besov spaces on closed sets by means of atomic decompositions." Research Report **7**, 1993, Univ. Umeå.

[Kal80] Kaljabin, G. A. "The description of functions of classes of Besov-Lizorkin-Triebel type." (Russian) Trudy Mat. Inst. Steklov **156** (1980), 82–109.

[Kal85] Kaljabin, G. A. "Theorems on extensions, multipliers and diffeomorphisms for generalized Sobolev-Liouville classes in domains with Lipschitz boundary." (Russian) Trudy Mat. Inst. Steklov **172** (1985), 173–186.

[Mar87] Marschall, J. "Some remarks on Triebel spaces." Studia Math. **87** (1987), 79–92.

[Miy90] Miyachi, A. "H^p spaces over open subsets of $\mathbb{R}^n$." Studia Math. **95** (1990), 205–228.

[RuS96] Runst, Th. and Sickel, W. *Sobolev Spaces of Fractional Order, Nemytskij Operators, and Nonlinear Partial Differential Equations.* De Gruyter, Berlin, 1996.

[Ste93] Stein, E. M. *Harmonic analysis.* Princeton Univ. Press, 1993.

[Skr97a] Skrzypczak, L. "Atomic decompositions on manifolds with bounded geometry." Preprint, Poznan, 1997.

[Skr97b] Skrzypczak, L. "Besov spaces on symmetric manifolds – the atomic decompositions." Preprint, Poznan, 1997.

[Tor91] Torres, R. H. "Boundedness results for operators with singular kernels on distribution spaces." Memoirs Amer. Math. Soc **90** (1991), 442.

[Tri78] Triebel, H. *Interpolation Theory, Function Spaces, Differential Operators.* North-Holland, Amsterdam, 1978 (sec. edition: Barth, Leipzig, 1995).

[Tri83] Triebel, H. *Theory of Function Spaces.* Birkhäuser, Basel, 1983.

[Tri88] Triebel, H. "On a class of weighted function spaces and related pseudodifferential operators." Studia Math. **90** (1988), 37–68.

[Tri92] Triebel, H. *Theory of Function Spaces, II.* Birkhäuser, Basel, 1992.

[Tri97] Triebel, H. *Fractals and Spectra.* Birkhäuser, Basel, 1997.

[Tri98] Triebel, H. *Gaussian Decompositions in Function Spaces* (to appear).

[TrW96] Triebel, H. and Winkelvoss, H. "Intrinsic atomic characterizations of function spaces on domains." Math. Z. **221** (1996), 647–673.

Hans Triebel
Mathematisches Institut
Friedrich-Schiller-Universität Jena
D-07740 Jena, Germany

Progress in Nonlinear Differential Equations
and Their Applications, Vol. 35
© 1999 Birkhäuser Verlag Basel/Switzerland

Functional Calculus for Unbounded Operators by Integration over the Positive Real Axis. Applications to Elliptic Systems in Half-Cylinders

Wolf von Wahl

Abstract. We survey two methods to introduce functions of unbounded operators $\mathcal{A}$ by integration over the positive real axis. $-\mathcal{A}$ generates an analytic semigroup or a semigroup with singularity at $t = 0$.

1. Position of the Problems. Some Notations

Let $\mathcal{A}$ be an unbounded operator in a Banach space $\mathcal{B}$. In some shadowed sector $\Sigma \subset \mathbb{C}$ as in the figure below $\mathcal{A}$ is supposed to fulfill one of the estimates

$$\|(\lambda + \mathcal{A})^{-1}\| \leq \frac{c_1}{|\lambda|^{1-\alpha/2m} + 1}, \tag{1}$$

$$\|(\lambda + \mathcal{A})^{-1}\| \leq \frac{c_2}{|\lambda| + 1} \tag{2}$$

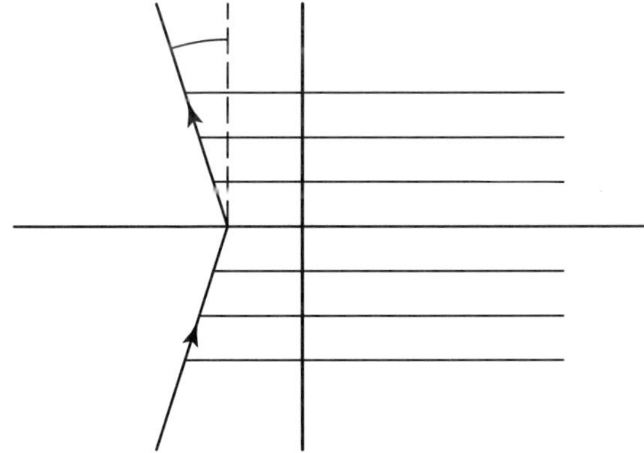

FIGURE 1

where c_1, c_2 are positive constants. In the first case α, m are also fixed positive numbers from $(0,1)$, $\mathbb{N}$, respectively. The curve bounding Σ is denoted by Γ and is to be run through from negative imaginary parts to positive ones. Thus the spectrum of $-\mathcal{A}$ is situated on the left-hand side of Γ, whereas Γ and the domain on the right-hand side of Γ belong to the resolvent set of $-\mathcal{A}$. Our examples consist in elliptic systems

$$\mathcal{A} = a_{\gamma\delta}(x)D^{\gamma+\delta} \tag{3}$$

of order $2m$, $m \in \mathbb{N}$, over a bounded open set Ω of $\mathbb{R}^n$ whose boundary $\partial\Omega$ is smooth. The $N \times N$-matrices $a_{\gamma\delta}$ are Hölder-continuous in $\overline{\Omega}$ to some exponent $\alpha \in (0,1)$ or simply continuous. $N \in \mathbb{N}$ is fixed. The multiindizes γ, δ of $\mathbb{R}^n$ run through all n-tuples $(\beta_1, \dots, \beta_n)$, $\beta_i \in \mathbb{N} \cup \{0\}$, $i = 1, \dots, n$, with $|\beta| = \beta_1 + \dots + \beta_n \leq m$, and in (3) we use the summation convention. As in [8, p. 1] we assume that the matrices $a_{\gamma\delta}$ are hermitian for $|\gamma| = |\delta| = m$ and satisfy the Legendre-Hadamard condition. (1) refers to $\mathcal{B} = C^\alpha(\overline{\Omega})$, $\mathcal{D}(\mathcal{A}) = \{u | u \in (C^{2m+\alpha}(\overline{\Omega}))^N,\ \partial^j u/\partial \nu^j = 0$ on $\partial\Omega$, $j = 0, \dots, m-1\}$, whereas condition (2) is the appropriate one in the case $\mathcal{B} = L^p(\Omega)$, $\mathcal{D}(\mathcal{A}) = (W^{2m,p}(\Omega))^N \cap (\overset{\circ}{W}{}^{m,p}(\Omega))^N$. $\alpha \in (0,1)$ and p, $1 < p < +\infty$, are fixed but arbitrary within the aforesaid limitations. To achieve the situation in Figure 1 we need in general that the term of order zero in (3) is sufficiently positive, for instance of the form $a_{00}u$ with a_{00} hermitian and $a_{00} \geq \lambda I$, $\lambda > 0$ and sufficiently large.

According to [3, I. § 5] one can introduce a functional calculus by setting

$$f(\mathcal{A}) = \frac{1}{2\pi i} \int_\Gamma f(-\lambda)(\lambda + \mathcal{A})^{-1}\, d\lambda$$

if (2) is fulfilled.[1] Under additional requirements on f this is also possible if only (1) is valid ([8, p. 3], [6, pp. 241, 247, 248]). f is supposed to be holomorphic in the right half-plane and to satisfy

(F_1) $|f(-z)| \cdot |z|^{-1+\alpha/2m}$ is integrable over Γ if condition (1) is fulfilled

or

(F_2) $|f(-z)| \cdot |z|^{-1}$ is integrable over Γ if condition (2) is satisfied.

In fact somewhat more is needed to guarantee that the integral in the definition of $f(\mathcal{A})$ does not change its value if Γ is replaced by a curve Γ' of the same shape which lies in the resolvent set of $-\mathcal{A}$. We do not want to go into these details here. Higher regularity of expressions $f(\mathcal{A})u$ for $\mathcal{A}$ elliptic is studied in C^α-spaces in Section 3. Our main interest is in the question under which circumstances this functional calculus can be explained by integration over the positive real axis. To say it more precisely **the first way**: To which functions G, being holomorphic in the

[1] In [1, 3] the estimate (2) is needed only in a smaller sector when compared with Fig. 1. As for the fractional powers to follow they can also be defined if $\mathcal{A}$ has no bounded inverse, cf. [1, 3, I. § 5].

right half-plane and satisfying (F1,2), there is a function $f : (0, +\infty) \to \mathbb{C}$ with

$$G(\mathcal{A}) = \int_0^{+\infty} f(\lambda)(\lambda + \mathcal{A})^{-1}\, d\lambda.$$

As it is well known this question plays a certain rôle when treating the fractional powers $\mathcal{A}^{-\rho}$, $0 < \rho < 1$. For these particular functions the afore-mentioned question was analyzed in [1, 2] if (2) is satisfied but possibly in a sector smaller than Σ introduced before. We study this question under condition (1) as well and from a more general point of view. The answer is that G has to be in the range of Stieltjes's transform, but Stieltjes's transform can be considered on various Banach-spaces. We do so in Section 2. Whereas the previous representation of $G(\mathcal{A})$ rests on Stieltjes's transform and involves the resolvent of $-\mathcal{A}$ only, there is a second possibility to transform the integral over Γ into one over $(0, +\infty)$. It rests on Laplace's transform $\mathcal{L}$ and involves the semigroup $e^{-t\mathcal{A}}$. If $G = \mathcal{L}h$ we have

$$G(\mathcal{A}) = \int_0^{+\infty} e^{-\sigma\mathcal{A}} h(\sigma)\, d\sigma.$$

We compare these two possibilities in Section 2. Laplace's transform may be more widely known than Stieltjes's although they are closely related with each other and each of them with the Fourier-transform (cf. [9, pp. 325, 334], [4]). It should be noted that under condition (1) the semigroup $e^{-t\mathcal{A}}$ behaves singularly at $t = 0$ according to the estimate

$$\|e^{-t\mathcal{A}}\| \leq \frac{ce^{-\frac{1}{c}t}}{t^{\alpha/2m}},\ t > 0$$

which can not be improved (cf. [5]). c is a positive constant. This estimate is an easy consequence of the formula

$$e^{-t\mathcal{A}} = \frac{1}{2\pi i} \int_\Gamma e^{\lambda t}(\lambda + \mathcal{A})^{-1}\, d\lambda,\ t > 0,$$

for the function $f_t(\lambda) = e^{-t\lambda}$. Thus, if we take Laplace's transform under condition (1), we are needing in $G(\mathcal{A}) = \int_0^\infty e^{-\sigma\mathcal{A}} h(\sigma)\, d\sigma$ that h at $\sigma = 0$ behaves less singular than $1/\sigma^{1-\alpha/2m}$. We concentrate our interest on the use of Stieltjes's transform since the resulting formulas only involve the resolvent of $-\mathcal{A}$.

In [2, 3] also the resolvent of $\mathcal{A}^\rho$, ρ as above, was represented as an integral over the positive real axis, this is, in the first way requested. To that end condition (2) was assumed or at least the same behaviour at 0 and $+\infty$ of $(\lambda + \mathcal{A})^{-1}$ as $1/(1 + \lambda)$. In Section 3 we study $(\lambda + \mathcal{A}^\rho)^{-1}$ for $\alpha/2m < \rho < 1$ when only (1) holds. It turns out then that the semigroup $e^{-t\mathcal{A}^\rho}$ at $t = 0$ has a singular behaviour which depends on ρ and increases if ρ decreases. As for elliptic (cf. [3, III. § 2]) systems (3) with m odd one can use $\mathcal{A}^{1/2m}$ to construct particular solutions to the elliptic system (J is the $N \times N$ unit-matrix)

$$J\partial_t^{2m} u - a_{\gamma\delta}(x) D^{\gamma+\delta} u = 0$$

in a half-cylinder with boundary values at the bottom and on the mantle or to Cauchy's problem for the system above.

We introduce some notation. We set $\mathbb{C}_+^* = \mathbb{C} - \{\Re z \leq 0\}$, $\mathbb{C}_-^* = \mathbb{C} - \{\Re z \geq 0\}$. If I is some interval (open, closed, halfopen, bounded, unbounded) then $L^p(I)$, $1 \leq p < +\infty$, is the Banach-space of all almost everywhere defined $f : I \to \mathbb{C}$ with $\int_I (f|^2\, dx < +\infty$. If Γ is a curve in $\mathbb{C}$ and if I is its parameter-interval, then $L^p(\Gamma)$ is the Banach-space of all almost everywhere defined $f : \Gamma \to \mathbb{C}$ with

$$\int_I |f \circ \gamma|^p |\dot{\gamma}|\, dt < +\infty$$

where γ is the parametrization of Γ. If $\mathcal{B}$ is a Banach-space and if $\mathcal{A}$ is a linear operator in $\mathcal{B}$, then $\mathcal{R}(\mathcal{A})$ or $\mathcal{R}(\mathcal{A} : \mathcal{B} \to \mathcal{B})$, if the choice of $\mathcal{B}$ is emphasized, is its range, $\{\lambda | \lambda \in \mathbb{C}, (\mathcal{A} - \lambda)^{-1}$ bounded and everywhere in $\mathcal{B}$ defined$\}$ its resolvent set. c is a positive constant which may change from line to line but does not depend on the functions involved.

2. Transformation of the Integral over Γ into an Integral over $(0, +\infty)$

The first step in analyzing the question raised above is

Theorem 1. *Let the integral operator K be defined by*

$$(Kf)(t) = \int_0^\infty \frac{f(\lambda)}{\lambda + t}\, d\lambda, \ f \in L^2((0, +\infty)).$$

Then $Kf \in L^2((0, +\infty))$ and Kf has a holomorphic extension into $\mathbb{C}_+^*$ which is given by

$$(Kf)(z) = \int_0^\infty \frac{f(\lambda)}{\lambda + z}\, d\lambda.$$

If $G \in \mathcal{R}(K)$ we use G also for its holomorphic extension to $\mathbb{C}_+^*$, and with $G = Kf$ we obtain

$$G(\mathcal{A}) = \int_0^\infty f(\lambda)(\lambda + \mathcal{A})^{-1}\, d\lambda.$$

Proof. The assertion $Kf \in L^2((0,+\infty))$ follows from [9, pp. 368–371], where it is also shown that $\|K\| \le \pi$. The assertion on the holomorphic extension is obvious. Formally we have

$$G(\mathcal{A}) = \frac{1}{2\pi i}\int_{\Gamma} G(-\mu)(\mu+\mathcal{A})^{-1}\,d\mu,$$

$$\int_0^\infty f(\lambda)(\lambda+\mathcal{A})^{-1}\,d\lambda = \frac{1}{2\pi i}\int_0^\infty f(\lambda)\int_{\Gamma}\frac{1}{\lambda-\mu}(\mu+\mathcal{A})^{-1}\,d\mu\,d\lambda,$$

$$= \frac{1}{2\pi i}\int_{\Gamma}\left(\int_0^\infty \frac{f(\lambda)}{\lambda-\mu}\,d\lambda\right)(\mu+\mathcal{A})^{-1}\,d\mu$$

with the inner integral being equal to $G(-\mu)$. The integrations and the exchange of their order are easily justified since $G, f \in L^2(\Gamma), L^2((0,+\infty))$ respectively, and $2(1-\frac{\alpha}{2m}) = 2-\alpha/m > 1$. □

The question is close by for which functions G our Theorem 1 is applicable. The integral equation in Theorem 1 is called Stieltjes's integral equation and the question for the range $\mathcal{R}(K)$ of K was answered in [4, § 6]. To describe the result we are needing Doetsch's transformation N. It acts isometrically from $L^2((0,+\infty))$ onto $L^2((-\infty,+\infty))$ and is given by

$$(Ng)(u) = \frac{1}{\sqrt{2\pi}}\int_0^\infty x^{iu-\frac{1}{2}}g(x)\,dx,\ -\infty < u < +\infty$$

where the integral converges in the quadratic mean l.i.m. $\int_{1/\eta}^{\eta}$ as $\eta \to +\infty$. The inverse N^{-1} of N is found by the formula

$$(N^{-1}h)(x) = \frac{1}{\sqrt{2\pi}}\int_{-\infty}^{+\infty} x^{-iu-\frac{1}{2}}h(u)\,du,\ 0 < x < +\infty.$$

Again the integral converges in the quadratic mean l.i.m. $\int_{-\eta}^{\eta}$ as $\eta \to +\infty$. Now we have

Theorem 2. *G is in $\mathcal{R}(K) = K(L^2((0,+\infty)))$ if and only if*

$$e^{|u|\pi}(NG)(u) \in L^2((-\infty,+\infty)).$$

The essence of Theorem 2 can also be described in a more direct form, cf. [4]. If G is as above then there are functions G_1, G_2 being holomorphic in the upper half-plane and in the lower half-plane respectively such that

$$\int_{-\infty}^{+\infty} |G_1(x+iy)|^2\, dx \le M_1,\ y>0, \qquad \int_{-\infty}^{+\infty} |G_2(x+iy)|^2\, dx \le M_2,\ y<0,$$

with constants M_1, M_2 not depending on y. Moreover G_1 has a.e. boundary values

$$\lim_{y\downarrow 0} G_1(x+iy) = G_1(x) \in L^2((-\infty,+\infty)) \tag{4}$$

and correspondingly

$$\lim_{y\uparrow 0} G_2(x+iy) = G_2(x) \in L^2((-\infty,+\infty)) \tag{5}$$

which, due to the assumptions on G, coincide on $(0,+\infty)$ and have the common values

$$G_1(x) = G_2(x) = G(x).$$

The unique solution f of $G = Kf$ is then given by

$$f(\lambda) = \frac{i}{2\pi}[G_1(-\lambda) - G_2(-\lambda)],\ 0<\lambda<+\infty, \tag{6}$$

with the boundary values $G_1(-\lambda)$, $G_2(-\lambda)$ in the sense of (4, 5) on both sides of the negative real axis.

Important functions of $\mathcal{A}$ are e.g. $\mathcal{A}^{-\rho}$ and $(\mu+\mathcal{A}^\rho)^{-1}$, $\Re\mu>0$, $0<\rho<1$. We have (reflexion-formula for the Γ-function)

$$\int_0^\infty \frac{1}{(z+\lambda)\lambda^\rho}\, d\rho = \frac{\pi}{\sin\pi\rho}\frac{1}{z^\rho},\ z\in\mathbb{C}_+^*. \tag{7}$$

Thus we have found an eigenfunction of K. This eigenfunction however is not a member of any of the spaces $L^p((0,+\infty))$, $1<p<+\infty$. Nevertheless $G = K(1/\lambda^\rho)$ is admissible in Theorem 1 and we obtain the well-known formula

$$\mathcal{A}^{-\rho} = \frac{\sin\pi\rho}{\pi}\int_0^\infty \frac{1}{\lambda^\rho}(\lambda+\mathcal{A})^{-1}\, d\lambda.$$

The situation is similar with the function $G(\lambda) = (\mu+\lambda^\rho)^{-1}$ in the sense that it can be represented as an image of some function under K; the argument however is not in $L^2((0,+\infty))$ if $\rho\le\frac{1}{2}$. It was computed in [1, 3] and we will mention it later. For these reasons it may be useful to consider K on spaces other than $L^2((0,+\infty))$, to characterize its range and to use the information in the sense of Theorem 1, namely for the calculus of functions of $\mathcal{A}$. We introduce the Banach-space $\mathcal{B}_\gamma$, which consists of all continuous $f:(0,+\infty)\to\mathbb{C}$ for which $|x|^\gamma|f(x)|\le D$, $x\in(0,+\infty)$. The exponent γ is taken from $(\alpha/2m,1)$ if the estimate (1) holds, and from $(0,1)$ if (2) holds but otherwise not restricted. $\mathcal{B}_\gamma$ becomes a Banach-space when endowed with the norm

$$\|f\|_{\mathcal{B}_\gamma} = \sup_{x\in(0,+\infty)} |x|^\gamma|f(x)|.$$

As it is evident from (7), K maps $\mathcal{B}_\gamma$ into itself and satisfies $\|K\| = \frac{\pi}{\sin \pi\gamma}$. As for $\mathcal{R}(K)$ and the calculus of functions of $\mathcal{A}$ we have

Theorem 3. *Let g be holomorphic in $\mathbb{C}_+^*$ and satisfy the assumptions (F1,2) for the functional calculus with $\mathcal{A}$. Let $|g(z)| \to 0$ as $|z| \to \infty$ in $\mathbb{C}_+^*$. Let*

$$\int_0^\infty |g(x+iy)| \cdot |x+iy|^{-(1-\alpha/2m)}\, dx \to 0 \text{ as } y \to \pm\infty, \tag{8}$$

if estimate (1) is valid, and let

$$\int_0^\infty |g(x+iy)| \cdot |x+iy|^{-1}\, dx \to 0 \text{ as } y \to \pm\infty \tag{9}$$

if estimate (2) is valid. Let

$$\begin{aligned} h_1(i\tau) &= g(-i\tau),\ \tau > 0 \\ h_2(i\tau) &= g(-i\tau),\ \tau < 0 \end{aligned}$$

and assume h_1 has a continuous continuation into $\Re\, z \ge 0$, $\Im\, z \ge 0$, $z \ne 0$, which is holomorphic in $\Re\, z > 0$, $\Im\, z > 0$. h_2 is supposed to have a continuous continuation into $\Re\, z \ge 0$, $\Im\, z \le 0$, $z \ne 0$, which is holomorphic in $\Re\, z > 0$, $\Im z < 0$. Assume moreover that

$$|z|\,|h_i(z)| \to 0,\ |z| \to 0 \tag{10}$$

$$|z|^\gamma |h_i(z)| \le D, \tag{11}$$

with some constants D, γ and for z in the domain of definition of h_i, $i = 1,2$. γ is from $(\alpha/2m, 1)$ if (1) is valid and from $(0,1)$ if (2) is valid. Then

$$g|(0,+\infty) \in \mathcal{R}(K : \mathcal{B}_\gamma \to \mathcal{B}_\gamma)$$

$$g(-\mu) = \frac{1}{2\pi i}\int_0^\infty \frac{1}{\lambda-\mu}(h_1(\lambda) - h_2(\lambda))\, d\lambda,\ \Re\, \mu < 0 \tag{12}$$

$$g(\mathcal{A}) = \frac{1}{2\pi i}\int_0^\infty (h_1(\lambda) - h_2(\lambda))(\lambda + \mathcal{A})^{-1}\, d\lambda \tag{13}$$

$h_1 - h_2$ is the unique solution of (12) for g given. 0 is in the residual spectrum of $K : \mathcal{B}_\gamma \to \mathcal{B}_\gamma$.

Proof. On using (8, 9) and Cauchy's Theorem we obtain

$$g(\mathcal{A}) = \frac{1}{2\pi i}\int_{-i\infty}^{+i\infty} g(-i\tau)(i\tau + \mathcal{A})^{-1} d(i\tau).$$

Applying Cauchy's Theorem again to $\int\limits_0^{+i\infty} g(-i\tau)\cdot(i\tau+\mathcal{A})^{-1}d(i\tau)$ in $\Re\, z \ge 0$, $\Im\, z \ge 0$ and to $\int\limits_{-i\infty}^{0} g(-i\tau)(i\tau+\mathcal{A})^{-1}d(i\tau)$ in $\Re\, z \ge 0$, $\Im\, z \le 0$ and using (10, 11) we end up with

$$g(\mathcal{A}) = \frac{1}{2\pi i}\int\limits_0^\infty (h_1(\lambda)-h_2(\lambda))(\lambda+\mathcal{A})^{-1}\,d\lambda.$$

The calculations in the proof of Theorem 1 show that

$$g(\mathcal{A}) = \frac{1}{2\pi i}\int\limits_\Gamma \left(\frac{1}{2\pi i}\int\limits_0^\infty \frac{1}{\lambda-\mu}(h_1(\lambda)-h_2(\lambda))\,d\lambda\right) \tag{14}$$

$$= \frac{1}{2\pi i}\int\limits_\Gamma g(-\mu)(\mu+\mathcal{A})^{-1}\,d\mu \tag{15}$$

where the last equation is simply the definition of $g(\mathcal{A})$.

So far we have shown (13). We are going to prove that (14, 15) imply (12). Assume that $\Re\,\lambda < 0$ and that λ is on the left-hand side of Γ, Γ as indicated in Fig. 1. Let moreover $\lambda+\mathcal{A}$ be boundedly invertible. Let $\tilde{g}$ satisfy the assumptions made for g in the present theorem and let

$$\int\limits_\Gamma \tilde{g}(-\mu)(\mu+\mathcal{A})^{-1}\,d\mu = 0.$$

On using the resolvent equation and Cauchy's integral formula we obtain with the aid of our assumptions on $\tilde{g}$ the equation

$$\begin{aligned} -\tilde{g}(-\lambda)(\lambda+\mathcal{A})^{-1} &= \frac{1}{2\pi i}\int\limits_\Gamma \frac{\tilde{g}(-\mu)}{\lambda-\mu}(\mu+\mathcal{A})^{-1}\,d\mu, \\ &= \frac{1}{2\pi i}\int\limits_\Gamma \tilde{g}(-\mu)(\mu+\mathcal{A})^{-1}\,d\mu\cdot(\lambda+\mathcal{A})^{-1}, = 0. \end{aligned}$$

Since we can vary λ within a small open set, we obtain $\tilde{g}\equiv 0$. Thus (12) holds. As for the last assertion we show that $g \in B\gamma$,

$$0 = \int\limits_0^\infty \frac{1}{x+t}g(x)\,dx,\ t>0,$$

implies $g \equiv 0$. More precisely we are going to prove that 0 is in the residual spectrum of K. Taking derivatives with respect to t we obtain

$$\int\limits_0^\infty g(x)\sum_{k=1}^N c_k\frac{1}{(x+t)^k}\,dx = 0,\ N\in\mathbb{N},\ c_1,\dots,c_N\in\mathbb{C},$$

in particular

$$\int_0^\infty g(x)\frac{1}{x+t}\left(\frac{1}{t}-\frac{1}{x+t}\right)\sum_{k=0}^N c_k\left(\frac{1}{x+t}\right)^k dx = 0,\ N\in\mathbb{N},\ c_0,\dots,c_N\in\mathbb{C}.$$

We change variables and introduce the new argument ξ by $x+t=1/\xi$, $x=1/\xi-t$, $dx=-1/\xi^2\,d\xi$. This gives

$$\begin{aligned}0&=\int_0^{1/t} g\left(\frac{1}{\xi}-t\right)\xi\left(\frac{1}{t}-\xi\right)\frac{1}{\xi^2}\cdot\sum_{k=0}^N c_k\xi^k\,d\xi\\ &=\int_0^{1/t}\left(\frac{1}{\xi}-t\right)g\left(\frac{1}{\xi}-t\right)\frac{\xi(\frac{1}{t}-\xi)}{\xi^2(\frac{1}{\xi}-t)}\cdot\sum_{k=0}^N c_k\xi^k\,d\xi=\frac{1}{t}\int_0^{1/t}\left(\frac{1}{\xi}-t\right)g\left(\frac{1}{\xi}-t\right)\sum_{k=0}^N c_k\xi^k\,d\xi.\end{aligned}$$

This implies $g\equiv 0$. In particular 0 is not in the point spectrum of K. Evidently

$$\begin{aligned}\mathcal{R}(K)\subset\mathcal{M}:=\{f|f\in\mathcal{B}_\gamma\cap C^\infty((0,+\infty)),|f^{(n)}(t)|\le\frac{n}{t^{n+\gamma}}\frac{\pi}{\sin\pi\gamma}\|f\|_{\mathcal{B}_\gamma},\\ n\in\mathbb{N},\ t>0\}.\end{aligned}$$

$\mathcal{M}$ is closed. Therefore $\overline{\mathcal{R}(K)}\subset\mathcal{M}$. Now consider the function

$$f(t)=\frac{\sin t^2}{t^\gamma}$$

which is in $\mathcal{B}_\gamma$. Let $\mathcal{N}$ be the subspace of $\mathcal{B}_\gamma$ spanned by f. Since $f'(t)=-\gamma\cdot\frac{\sin t^2}{t^{\gamma+1}}+2t^{1-\gamma}\cos t^2$ we arrive at

$$|f'(\sqrt{(2n+1)\pi})|=2\sqrt{(2n+1)\pi}^{\,1-\gamma},\ n\in\mathbb{N}$$

and $\overline{\mathcal{R}(K)}\cap\mathcal{N}=\{0\}$. We have thus

$$\|h-f\|_{\mathcal{B}_\gamma}\ge c,\ h\in\overline{\mathcal{R}(K)},\ f\text{ as above,}$$

with a positive constant c. The assertion is proved. □

We draw some consequences from Theorem 3. Beside the well-known formula for the resolvent of $\mathcal{A}^\gamma$, but now under the assumption (1) only, Theorem 3 opens the possibility to construct solutions to certain elliptic systems of order $2m$ (with m odd) in a half-cylinder. Again only condition (1) is needed. The elliptic case will be dealt with in Section 3 where we will also provide appropriate references.

Formula (12) corresponds to (6) since

$$\begin{aligned}\frac{1}{2\pi i}(h_1(\lambda)-h_2(\lambda))&=\frac{i}{2\pi}(h_2(\lambda)-h_1(\lambda))\\ h_2(\lambda+i\tau)&=g(-\lambda-i\tau),\ \tau<0,\\ h_1(\lambda+i\tau)&=g(-\lambda-i\tau),\ \tau>0.\end{aligned}$$

As already mentioned, K is unique also if considered in $L^2((0,+\infty))$. Then 0 is an element of the continuous spectrum of K which is in contrast to the case K:

$\mathcal{B}_\gamma \to \mathcal{B}_\gamma$ (cf. Theorem 3). If we take $g(z) = (\eta + z^\gamma)^{-1}$ with γ as before in Theorem 3, (13) furnishes

$$(\eta + \mathcal{A}^\gamma)^{-1} = \frac{\sin \pi\gamma}{\pi} \int_0^\infty \frac{\lambda^\gamma}{\eta^2 + 2\eta\lambda^\gamma \cos \pi\gamma + \lambda^{2\gamma}} (\lambda + \mathcal{A})^{-1}\, d\lambda, \ \eta \geq 0, \tag{16}$$

a formula which was proved already in [2]. Theorem 3 is in some way a detailed version of Theorem 4 to follow.

Theorem 4. *Let* $g \in \mathcal{R}(K : \mathcal{B}_\gamma \to \mathcal{B}_\gamma)$ *with* $\gamma \in (\frac{\alpha}{2m}, 1)$ *if (1) is valid and* $\gamma \in (0,1)$ *if (2) is valid. Its holomorphic continuation into* $\Re\, z > 0$ *is also denoted by* g. *Then* g *satisfies (F1,2). Let* $g = Kf$. *Then*

$$g(\mathcal{A}) = \int_0^\infty f(\lambda)(\lambda + \mathcal{A})^{-1}\, d\lambda. \tag{17}$$

Proof. Observe that f in $g = Kf$ is uniquely determined. This is due to Theorem 3. We have

$$\begin{aligned} |g(z)| &\leq \int_0^\infty \frac{|f(\lambda)|}{|\lambda + z|}\, d\lambda, \leq 2 \int_0^\infty \frac{|f(\lambda)|}{|\lambda| + |z|}\, d\lambda, \\ &\leq 2 \left(\int_0^a \frac{1}{|\lambda|^\gamma(|\lambda| + |z|)}\, d\lambda + \int_a^\infty \frac{1}{|\lambda|^\gamma(|\lambda| + |z|)}\, d\lambda \right) \|f\|_{\mathcal{B}_\gamma}. \end{aligned}$$

a is any fixed positive number. Let (1) be valid. Then we employ $|\lambda| + |z| \geq c|\lambda|^{1-\alpha/2m-\varepsilon}|z|^{\frac{\alpha}{2m}+\varepsilon}$, $\varepsilon > 0$ and sufficiently small. Thus the second integral decays as $|z|^{\frac{\alpha}{2m}+\varepsilon}$ if $|z|$ tends to $+\infty$. If (2) is valid we use $|\lambda| + |z| \geq c|\lambda|^{1-\varepsilon} \cdot |z|^\varepsilon$ for the second integral. As for the first one it is easy to see that it behaves like $|z|^{-\gamma}$. Thus (F1,2) are fulfilled. The proof of (16) is not more than a repetition of the proof of Theorem 1. Using the previous estimates the integrations and the exchange of their order can easily be justified. □

As is well known we can represent $\mathcal{A}^{-\gamma}$ not only by means of Theorem 3 but also in the form

$$\mathcal{A}^{-\gamma} = \frac{1}{\Gamma(\gamma)} \int_0^\infty e^{-\mathcal{A}\sigma} \sigma^{\gamma-1}\, d\sigma$$

with $\gamma > \frac{\alpha}{2m}$ if (1) is assumed and with $\gamma > 0$ in the case of condition (2). This is also a consequence of Theorem 5 to follow. Whereas Theorems 3 and 4 rest on the use of Stieltjes's transform the formula above is a consequence of the closely related Laplace transform. It represents the second possibility to explain the functional calculus of $\mathcal{A}$ by integration over the positive real axis.

Theorem 5. *Let g satisfy the assumptions (F1,2) for the functional calculus with $\mathcal{A}$. Let $\Lambda_0 = \max\{\Re\lambda | \lambda \in \Gamma\}$. Thus $\Lambda_0 < 0$. Let $\mathcal{L}$ operate on the continuous functions $h : (0, +\infty) \to \mathbb{C}$ which are integrable at 0 and satisfy an estimate*

$$|h(\sigma)| \leq K e^{\mu\sigma}$$

for some individual constants $K \geq 0$ and μ with $\mu \in (-\infty, |\Lambda_0|)$. Then

$$\mathcal{L}h(\lambda) = \int_0^\infty e^{-\lambda\sigma} h(\sigma)\, d\sigma$$

is holomorphic in $\Re\lambda > |\Lambda_0|$ and tends to 0 as $|\lambda| \to +\infty$ on every half-ray in $\mathbb{C}$ on which the real parts tend to $+\infty$ if the corresponding points tend to ∞. If for g as above there exists an h as above such that

$$g = \mathcal{L}h$$

and such that $h(\sigma) \cdot \sigma^{-\alpha/2m}$ is integrable at 0, then

$$g(\mathcal{A}) = \int_0^\infty e^{-\mathcal{A}\sigma} h(\sigma)\, d\sigma.$$

h is determined uniquely.

Proof. That for given g the argument h under $\mathcal{L}$ is unique is well known [9, ch. VII]. We have

$$\int_0^\infty e^{-\mathcal{A}\sigma} h(\sigma)\, d\sigma = \frac{1}{2\pi i} \int_\Gamma \Big(\int_0^\infty e^{\lambda\sigma} h(\sigma) d\sigma \Big) (\lambda + \mathcal{A})^{-1}\, d\lambda.$$

We have $g(-\lambda) = \int_0^\infty e^{\lambda\sigma} h(\sigma)\, d\sigma$, $\Re\lambda \leq \Lambda_0$, which proves the assertion. □

When comparing Theorems 4 and 5 the impression is that the range for the admissible g may be wider in Theorem 5 but the formula is more complicated since it involves the semigroup instead of the resolvent only. In fact one can treat more effectively particular functions of $\mathcal{A}$ by using the representation (17) (see (16)). It also opens the possibility to define functions of $\mathcal{A}$ if $-\mathcal{A}$ is not known to generate a semigroup (cf. [1]).

3. Remarks on the Semigroup $e^{-t\mathcal{A}^g}$ and Elliptic Systems. Higher Regularity of $f(\mathcal{A})u$

If $\mathcal{A}$ satisfies (1), $-\mathcal{A}$ generates an analytic semigroup $e^{-t\mathcal{A}}$, and so does $-\mathcal{A}^\gamma$, $0 < \gamma < 1$ (cf. [1, 2, 3]). If $m \in \mathbb{N}$, $\gamma = \frac{1}{2m}$, then $u(t) = e^{-t\mathcal{A}^\gamma} u_0$, $u_0 \in B$, solves

$$\partial_t^{2m} u - \mathcal{A}u = 0,\ t > 0 \tag{18}$$

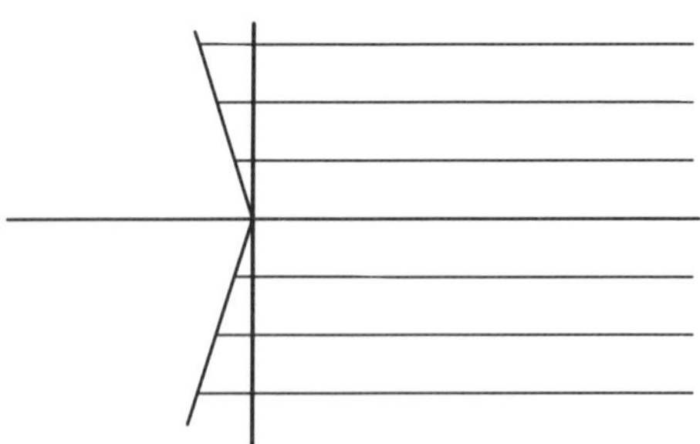

FIGURE 2

as was pointed out in [1, 3]. If $\mathcal{A}$ only satisfies (2), the semigroup $e^{-t\mathcal{A}}$ has a well-known singular behaviour at $t = 0$. This was studied in detail in [5, 6] for elliptic operators of order $2m$ in C^α-spaces. $e^{-t\mathcal{A}^\gamma}u_0$, $u_0 \in \mathcal{B}$, $\frac{\alpha}{2m} < \gamma < 1$, is still a solution of (18) and we want to compute its singular behaviour at 0. The calculations to follow are an application of (16). For $\eta > 0$ we have

$$\begin{aligned}(\eta+\mathcal{A}^\gamma)^{-1} &= \frac{\sin\pi\gamma}{\eta\pi}\int\limits_0^\infty \frac{\lambda^\gamma/\eta}{1+2(\lambda^\gamma/\eta)\cos\pi\gamma+\lambda^{2\gamma}/\eta^2}(\lambda+\mathcal{A})^{-1}\,d\lambda,\\ &= \frac{\eta^{\frac{1}{\gamma}-1}\sin\pi\gamma}{\gamma\pi}\int\limits_0^\infty \frac{\rho^{1/\gamma}}{1+2\rho\cos\pi\gamma+\rho^2}(\rho^{1/\gamma}\eta^{1/\gamma}+\mathcal{A})^{-1}\,d\rho.\end{aligned} \tag{19}$$

Thus $(\eta+\mathcal{A}^\gamma)^{-1}$ has a holomorphic extension to a sector as described in the figure to follow, where in case that (2) is valid an estimate

$$\|(\eta+\mathcal{A}^\gamma)^{-1}\| \le \frac{c}{|\eta|^{1-\alpha/(\gamma 2m)}},\ \eta \neq 0,$$

holds. Since 0 is in the resolvent set of $-\mathcal{A}^\gamma$ the operator-valued function $(\eta+\mathcal{A}^\gamma)^{-1}$ is also holomorphic in a neighbourhood of 0 and we end up with a sector Σ_γ as in Figure 1 for $(\eta+\mathcal{A}^\gamma)^{-1}$. In Σ_γ we have an estimate

$$\|(\eta+\mathcal{A}^\gamma)^{-1}\| \le \frac{c}{|\eta|^{1-\alpha/\gamma 2m}+1}.$$

If the curve bounding Σ_γ is denoted by Γ_γ the angle ϑ_γ it forms with the imaginary axis is different from the corresponding angle ϑ in Figure 1. As it follows from (19) its size is $(1-\gamma)\frac{\pi}{2}+\gamma\vartheta$. The semigroup $e^{-t\mathcal{A}^\gamma}$ satisfies the estimate

$$\|e^{-t\mathcal{A}^\gamma}\| \le \frac{ce^{-\frac{1}{c}t}}{t^{\alpha/\gamma 2m}},\ t>0, \tag{20}$$

whereas

$$\|e^{-t\mathcal{A}}\| \le \frac{ce^{-\frac{1}{c}t}}{t^{\alpha/2m}},\ t>0, \tag{21}$$

these estimates being easy consequences from the functional calculus introduced in Section 1. In particular we see that the semigroup generated by $-\mathcal{A}^\gamma$ shows a more singular behaviour at $t = 0$ when compared with the semigroup generated by $-\mathcal{A}$. This however refers to the case when (2) is valid, i.e. to elliptic systems in C^α-spaces.

If m is odd we can solve elliptic problems in a half-cylinder by using (18). This is seen as follows. Let $\mathcal{A}$ be an elliptic system as described in Section 1. Then there exists a $c_0 > 0$ such that

$$-\eta^* a_{\gamma\delta}(x)\xi^{\gamma+\delta}\eta \geq c_0|\xi|^{2m}|\eta|^2, \ \xi \in \mathbb{R}^n, \ \eta \in \mathbb{C}^N, \ x \in \overline{\Omega},$$

where the summation is taken over all multiindices γ, δ with $|\gamma| = |\delta| = m$. This corresponds to

$$\begin{aligned} I\partial_t^{2m}u - a_{\gamma\delta}(x)D^{\gamma+\delta}u &= 0, \\ -(I\partial_t^{2m}u - a_{\gamma\delta}(x)D^{\gamma+\delta}u) &=: \tilde{\mathcal{A}}u = 0, \end{aligned} \tag{22}$$

where I is the $N \times N$-unit-matrix. $\tilde{\mathcal{A}}$ is elliptic in $[0, +\infty) \times \overline{\Omega} \subset \mathbb{R}^n$: We have

$$(-1)^m\eta^*(-I)t^{2m}\eta + (-1)^m\eta^* a_{\gamma\delta}(x)\xi^{\gamma+\delta}\eta \geq t^{2m}|\eta|^2 + c_0|\xi|^{2m}|\eta|^2, \geq c_0'|t^2 + |\xi|^2)^m|\eta|^2$$

for some constant $c_0' > 0$. If $\mathcal{B} = C^\alpha(\overline{\Omega})$ and if m is odd we have thus found a solution $e^{-t\mathcal{A}^\gamma}u_0$, $\gamma = 1/2m$, of the following Cauchy-problem for the elliptic system $\tilde{\mathcal{A}}$ defined in (22):

$$\begin{cases} u \in \bigcap\limits_{T,\varepsilon,0<\varepsilon<T} C^0([\varepsilon, T], C^{2m+\alpha}(\overline{\Omega})) \cap \\ \qquad \cap \bigcap\limits_{\substack{p,1<p<+\infty \\ T,\varepsilon,0<\varepsilon<T}} C^0([0, T], L^p(\Omega)), \\ u(0, x) = u_0(x) \text{ in the sense above}, \\ \tilde{\mathcal{A}}u = 0 \text{ in all } [\varepsilon, T] \times \overline{\Omega}, \ 0 < \varepsilon < T. \end{cases} \tag{23}$$

The preceeding solution of Cauchy's problem however has a defect in as much as it is not possible to prescribe $\partial_t u, \ldots, \partial_t^{2m-1}u$ arbitrarily at $t = 0$. It should also be mentioned that

$$\|\mathcal{A}u(t)\| \leq \frac{ce^{-\frac{1}{c}t}}{t^{1+\alpha/2m}}, \ t > 0. \tag{24}$$

Thus we have exponential decay as $t \to +\infty$. It is easy to see that for $m = 1$ and u_0 sufficiently regular, the solution u of $\tilde{\mathcal{A}}u = 0$ in (23) is unique provided we assume that it decays as $t \to +\infty$, for instance as in (24).

Since $u(t) \in \mathcal{D}(\mathcal{A})$, $t > 0$, and therefore $\partial^j u/\partial\nu^j = 0$ at $\partial\Omega$, $j = 0, \ldots, m-1$, $t > 0$, it may be more appropiate to denote u as a solution of $\tilde{\mathcal{A}}u = 0$ in the half-cylinder $[0, +\infty) \times \overline{\Omega}$.

The case $m = 1$, this is $2m = 2$, was already treated in [3, III.2] and [1, p. 431] but under the assumption that(2) holds for the non-negative reals and large positive λ respectively. Under these conditions it is also possible to treat inhomogeneous problems, cf. [3, pp. 265–270].

We return to functions $f(\mathcal{A})$ as introduced in Section 1. As for higher regularity of $f(\mathcal{A})u$ in C^α-spaces we have to sharpen the decay-conditions on f at infinity. The result we are needing is Theorem I.1 in [7]: Let $k \in \mathbb{N} \cup \{0\}$, $k < 2m$, $u \in C^{k+\alpha}(\overline{\Omega})$. Then

$$(\lambda + \mathcal{A})^{-1} u \in C^{2m+k+\alpha}(\overline{\Omega}),\ \lambda \in \Sigma,$$
$$\|(\lambda + \mathcal{A})^{-1} u\|_{C^{2m+k+\alpha}(\overline{\Omega})} \leq c|\lambda|^{(k+\alpha)/2m} \cdot \|u\|_{C^{k+\alpha}(\overline{\Omega})}.$$

If we thus take u from $C^{k+\alpha}(\overline{\Omega})$ then

$$f(\mathcal{A})u = \frac{1}{2\pi i} \int\limits_{\Gamma} f(-\lambda)(\lambda + \mathcal{A})^{-1} u \, d\lambda, \quad \in C^{2m+k+\alpha}(\overline{\Omega})$$

provided $|f(-z)||z|^{(k+\alpha)/2m} \in L^1(\Gamma)$. This is one main application of [7, Theorem I.1].

References

[1] Balakrishnan, A. V.: "Fractional powers of closed operators and the semi-groups generated by them." Pacific J. Math. **10**, 419–437(1960).

[2] Kato, T.: "Note on fractional powers of linear operators." Proc. Japan Acad. **36**, 94–96(1960).

[3] Krein, S. G.: *Linear Differential Equations in Banach Space.* American Mathematical Society Translation of Mathematical Monographs, vol. **29**: Providence, R.I. (1971).

[4] Steiner, A.: *Die einseitig unendliche Fouriertransformation und zwei Klassen quasianalytischer Funktionen.* Festband zum 70. Geburtstag von Rolf Nevanlinna, H. P. Künzi und A. Pfluger Hrs.geber, 89–104. Springer: Berlin, Heidelberg, New York (1966).

[5] Wahl, W. von: "Gebrochene Potenzen eines elliptischen Operators und parabolische Differentialgleichungen in Räumen hölderstetiger Funktionen." Nachr. Ak. d. Wiss. Göttingen, II. Math. Phys. Kl. **11**, 231–258(1972).

[6] Wahl, W. von: "Lineare und semilineare parabolische Differentialgleichungen in Räumen hölderstetiger Funktionen." Abh. Math. Sem. Univ. Hamburg **43**, 234–262(1975).

[7] Wahl, W. von: "Neue Resolventenabschätzungen für elliptische Differentialoperatoren und semilineare parabolische Gleichungen." Abh. Math. Sem. Univ. Hamburg **46**, 179–204(1976).

[8] Wahl, W. von: "Ein Einfacher Beweis für die Stetigkeit von $D^{2m}\underline{u}$ bei Parabolischen Systemen $\partial_t \underline{u} + \mathcal{A}\underline{u} = \underline{f}$ der Ordnung $2m$ mit in (t, x) Hölderstetiger Rechter Seite $\underline{f}$." Preprint. University of Bayreuth (1997).

[9] Widder, D. V.: *The Laplace Transform.* Princeton University Press: Princeton (1972).

[10] Yosida, K.: *Functional Analysis.* Grundlehren der mathematischen Wissenschaften 123. Fifth Edition. Springer: Berlin, Heidelberg, New York (1978).

Wolf von Wahl
University of Bayreuth,
Department of Mathematics
D-95440 Bayreuth

Progress in Nonlinear Differential Equations and Their Applications

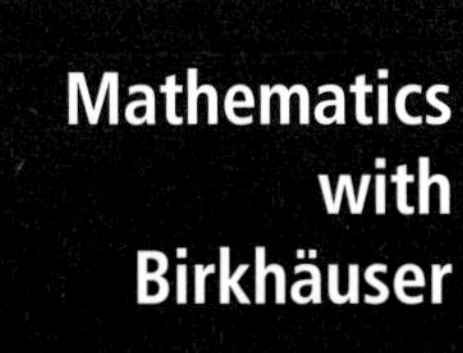

Editor

Haim Brezis, Département de Mathématiques, Université P. et M. Curie 4, Place Jussieu, 75252 Paris Cedex 05, France and Department of Mathematics, Rutgers University New Brunswick, NJ 08903, U.S.A.

"Progress in Nonlinear and Differential Equations and Their Applications" is a book series that lies at the interface of pure and applied mathematics. Many differential equations are motivated by problems arising in such diversified fields as Mechanics, Physics, Differential Geometry, Engineering, Control Theory, Biology, and Economics. This series is open to both the theoretical and applied aspects, hopefully stimulating a fruitful interaction between the two sides. It will publish monographs, polished notes arising from lectures and seminars, graduate level texts, and proceedings of focused and refereed conferences.
We encourage preparation of manuscripts in some form of TeX for delivery in camera-ready copy, which leads to rapid publication, or in electronic form for interfacing with laser printers or typesetters. Proposals should be sent directly to the editor or to: **Birkhäuser Boston, 675 Massachusetts Avenue, Cambridge, MA 02139, USA**

PNLDE 1+2 Partial Differential Equations and the Calculus of Variations. Essays in Honor of Ennio De Giorgi
F. Colombini / A. Marino / L. Modica / S. Spagnolo (Eds)

PNLDE 3 Propagation and Interaction of Singularities in Nonlinear Hyperbolic Problems
M. Beals

PNLDE 4 Variational Methods
H. Berestycki, I. Ekeland, J.-M. Coron (Eds)

PNLDE 5 Composite Media and Homogenization Theory
G. Dal Maso, G. Dell' Antonio (Eds)

PNLDE 6 Infinite Dimensional Morse Theory and Multiple Solution Problems
K.-C. Chang

PNLDE 7 Nonlinear Diffusion Equations and Their Equilibrium States, 3
Proceedings from a Conference held August 20-29, 1989 in Gregynog, Wales
N.G. Lloyd, W.M. Ni, L.A. Peletier, J. Serrin (Eds)

PNLDE 8 An Introduction to G-Convergence
G. Dal Maso

PNLDE 9 Differential Inclusions in Nonsmooth Mechanical Problems. Shocks and Dry Friction
M.D.P. Monteiro Marques

PLNDE 10 Periodic Solutions of Singular Lagrangian Systems
Antonio Ambrossetti and Vittorio Coti Zelati

PNLDE 11 Nonlinear Waves and Weak Turbulence with Applications in Oceanography and Condensed Matter
N. Fitzmaurice, D. Gurarie, F. McCaughan and W.A. Woyczynski (Eds)

PNLDE 12 Seminar on Dynamical Systems. Euler International Mathematical Institute, St. Petersburg, 1991
S. Kuksin

PNLDE 13 Ginzburg-Landau Vortices
F. Bethuel, H. Brezis, F. Hélein

Mathematics with Birkhäuser

NONLINEAR DIFFERENTIAL EQUATIONS AND APPLICATIONS

First published in 1994.

NoDEA is published in 1 volume per year and 4 issues per volume (approx. 500 pages per volume). Back volumes are available. Format: 17 x 24 cm

Prices are available on request.
ISSN 1021-9722 (Printed edition)
ISSN 1420-9004 (Electronic edition)

Aims and Scope:
The purpose of the Journal NoDEA – Nonlinear Differential Equations and Applications – is to encourage the interaction between pure mathematics and applied sciences. In particular, it will provide a forum for research papers on nonlinear differential equations and applications to natural sciences. General theory will include ordinary and partial differential equations (both deterministic and stochastic), variational and topological methods, control theory, qualitative analysis including stability and bifurcation. Applications to natural sciences include the treatment of problems in classical, statistical and quantum mechanics, electromagnetism, population dynamics, chemical kinetics, combustion theory.

Abstracted/ Indexed in:
Current Contents, Mathematical Reviews, Zentralblatt für Mathematik, Mathematics Abstracts

Managing Editor:
L. Salvadori (Chairman), Università di Trento, Italy
Assistant Editor:
E. Pagani, Università di Trento. Italy
Editorial Committee:
A. Ambrosetti/ A. Cellina / R. Conti/ G. Da Prato
Editorial Board:
H. Amann/ J.-P. Aubin/ J.M. Ball/ H. Brezis/ C. Cercignani/ G. Dell'Antonio/ I.J. Duistermaat/ A. Fasano/ A. Friedman/ M. Giaquinta/ A. Halanay/ H. Hofer/ S. Klainerman/ J. Necas/ V.P. Palamodov/ S. Spagnolo/ P. Villaggio/ W. Von Wahl

http://www.birkhauser.ch

Please order through your bookseller or through:

Birkhäuser Verlag AG
P.O Box 133
CH-4010 Basel/Switzerland
Fax: +41/61/205 07 92
e-mail: orders@birkhauser.ch